P9-DUF-496

Fibre Reinforced Cement and Concrete

Fibre Reinforced Cement and Concrete

Proceedings of the Fourth International Symposium held by RILEM (The International Union of Testing and Research Laboratories for Materials and Structures) and organized by the Department of Mechanical and Process Engineering, University of Sheffield, UK.

Sheffield
July 20-23, 1992

EDITED BY
R. N. Swamy

E & FN SPON
An Imprint of Chapman & Hall
London · Glasgow · New York · Tokyo · Melbourne · Madras

Published by E & FN Spon, an imprint of Chapman & Hall, 2-6 Boundary Row, London SE1 8HN

Chapman & Hall, 2-6 Boundary Row, London SE1 8HN, UK

Blackie Academic & Professional, Wester Cleddens Road, Bishopbriggs, Glasgow G64 2NZ, UK

Van Nostrand Reinhold, 115 5th Avenue, New York, NY10003, USA

Chapman & Hall Japan, Thomson Publishing Japan, Hirawacho Nemoto Building, 6F, 1-7-11 Hirakawa-cho, Chiyoda-ku, Tokyo 102, Japan

Chapman & Hall Australia, Thomas Nelson Australia, 102 Dodds Street, South Melbourne, Victoria 3205, Australia

Chapman & Hall India, R. Seshadri, 32 Second Main Road, CIT East, Madras 600 035, India

---

First edition 1992

© 1992 RILEM

Printed in Great Britain by St Edmundsbury Press, Bury St Edmunds, Suffolk

ISBN 0 419 18130 X

Apart from any fair dealing for the purposes of research or private study, or criticism or review, as permitted under the UK Copyright Designs and Patents Act, 1988, this publication may not be reproduced, stored, or transmitted, in any form or by any meas, without the prior permission in writing of the publishers, or in the case of reprographic reproduction only in accordance with the terms of the licences issued by the Copyright Licensing Agency in the UK, or in accordance with the terms of licences issued by the appropriate Reproduction Rights Organization outside the UK. Enquiries concerning reproduction outside the terms stated here should be sent to the publishers at the UK address printed on this page.

The publisher makes no representation, express or implied, with regard to the accuracy of the information contained in this book and cannot accept any legal responsibility or liability for any errors or omissions that may be made.

A catalogue record for this book is available from the British Library.

Library of Congress Cataloging-in-Publication data available.

Publisher's Note
This book has been produced from camera ready copy provided by the individual contributors in order to make the book available for the Symposium.

Other RILEM publications on Fiber Composites

High Performance Fiber Reinforced Cement Composites

Edited by H.W. Reinhardt and A.E. Naaman

Advanced composites and the fundamental understanding of their behaviour is a rapidly expanding area of civil engineering. These materials can be designed to have outstanding combinations of strength (five-to ten-times that of conventional concrete) and energy absorption capacity (up to 1000-times that of plain concrete). Exciting engineering applications are therefore being developed to take advantage of these properties for blast resistant, earthquake resistant and offshore structures, and also for building components such as cladding, pipes, tiles and roofing.

This book provides a compendium of the most recent research advances and reviews presented at an international workshop held under the auspices of RILEM and the American Concrete Institute in Mainz, Germany in June 1991. It includes over 40 contributions from the leading international specialists and researchers from USA, Europe, Japan and elsewhere and is an essential reference for engineers and researchers who need to be at the forefront of developments in cement composites.

Among the materials discussed are composites containing steel, polypropylene, polyacryonitrile, carbon, glass and cellulose fibres. SIFCON is extensively covered. The other main topics considered are: technologies of production; composite optimization; mechanical fracture and interface properties; modelling; structural applications and implications.

RILEM Proceedings 15, Published 1992, 584 pages, ISBN 0 419 39270 4

Vegetable Plants and their Fibres as Building Materials

Edited by H.S. Sobral

In many parts of the world the use of vegetable plants and their fibres as building materials is vital for low-cost housing and other construction, for reinforcing cement-based and concrete products, for soil stabilization and many other applications. However, great technical ingenuity is needed to develop appropriate fabrication techniques and to overcome some of the inherent problems of the materials, such as their poor durability.

This volume brings together a wealth of information and experience on successful applications of plants and fibres. It forms the Proceedings of the Second International RILEM Symposium on Vegetable Plants and their Fibres as Building Materials held in Salvador, Bahia, Brazil in September 1990. It will be of value to those involved in research and use of indigenous fibre materials and to architects, consultants, government departments and development agencies responsible for building programmes throughout the developing world. The objectives of the Symposium were:

* To examine the state-of-the-art in the field of vegetable plants and their fibres as building materials, emphasizing their use, properties, fabrication, new procedures and future developments.
* To make available research results on new techniques for fibre reinforcement and their use in concrete, stabilized clay and other matrices.
* To analyse procedures to make vegetable fibres and wood the usual building materials in developing countries.

RILEM Proceedings 7, Published 1990, 392 pages, ISBN 0 412 39250 X

A full list of RILEM publications available from E & F N Spon is given at the back of the book.

Contents

Preface

Nearly three decades of research and development have not in any way dimmed the engineering potential or the versatility of applications of fibre cement composites in the construction industry. The concept of fibre reinforcement of the cement matrix still remains exciting and innovative, and the ability to transform an essentially brittle concrete to a material able to tolerate damage and develop post-cracking ductility and energy absorption capability continues to enthuse researchers and users equally. In the excitement of the development of this new material, sometimes less than acceptable performance, occasionally, even failures have occurred. However, when designed, fabricated and used intelligently, many practical applications have shown that these composites can withstand the test of time, environment and critical exposure conditions. The scope for the economic and technical exploitation of the material is unlimited provided we design the matrix and the type and geometry of the reinforcing element to suit the performance requirements of the application and the needs of the society where they are used.

RILEM has been right at the forefront of the development of fibre cement composites from the very beginning. Over a period of two decades and through various committee activities, reports, workshops and conferences, RILEM has worked incessantly to achieve the transformation of a laboratory concept to a practical reality. Many individuals from many countries have made significant contributions to this success, and to all of them we owe a great debt of gratitude.

The present Symposium is the fourth in the RILEM series, following those held in London (1975) and in Sheffield (1978 and 1986). The themes of this Symposium include the usual areas of great interest to the researcher and pracitising engineer--fabrication, early age properties, engineering and dynamic behaviour, fracture characteristics, structural behaviour and applications. Ageing and durability of these composites are of particular concern, and therefore form a major theme of these Proceedings.

High performance fibres and reinforcing elements are also increasingly important, for a variety of reasons, and special attention has been given to these new composites. Natural fibres and wood/cellulose fibres are especially attractive to developing countries, and these, together with ferrocement, have also formed popular themes of these Proceedings. The unique feature of this Symposium and these Proceedings is that they show that fibre reinforcement is a global concept that transcends traditional ideas of the matrix and of the fibre, and that it is possible to produce fibre cement composites with optimized performance characteristics for a given set of load, usage and environmental conditions, consistent with cost, service life and durability.

As Chairman of the Organizing and Technical Committees, I would like to record my sincere thanks to all those who have made the Symposium very rewarding. Our thanks are due to all those who reviewed the papers, to the keynote speakers and the authors, and to Mrs Norma Parkes who behind the scenes coordinated the organization of the Symposium.

There is no doubt that fibre cement composites are here to stay. Fibre reinforcement can provide high performance materials to resist critical and stringent load and exposure conditions. On the other hand, it can also provide the medium for housing, schools and infrastructure construction for day to day enhancement of human living. RILEM has shown the way and will continue to do so for many years to come.

R. N. Swamy
Sheffield, May 1992

International Scientific Committee

Dr Narayan Swamy (Chairman), University of Sheffield, UK
Dr Ben Barr, University of Wales College of Cardiff, UK
Professor Arnon Bentur, National Building Research Institute, Israel
Professor Andrej Brandt, Polish Academy of Sciences, Poland
Professor Victor Li, University of Michigan, USA
Professor Sidney Mindess, University of British Columbia, Vancouver, Canada
Dr Pritpal Mangat, University of Aberdeen, UK
Professor Tony Naaman, University of Michigan, USA
Professor V. Ramakrishnan, South Dakota School of Mines and Technology, USA
Dr Peter Robins, University of Loughborough, UK
Dr Pierre Rossi, Laboratoire Central des Ponts et Chaussees, France
Dr Ake Skarendahl, Swedish Cement and Concrete Research Institute, Sweden

Symposium Co-sponsors

American Concrete Institute
American Society for Testing and Materials
The Concrete Society
Institute of Concrete Technology
Japan Concrete Institute

PART ONE
KEYNOTE PAPERS

1 FIBRE REINFORCED CONCRETE – WHERE DO WE GO FROM HERE?

B. I. G. BARR
University of Wales College of Cardiff, UK

Abstract
Two main themes are developed in the paper. The first theme is the need for all working on various aspects of FRC theory and applications to pool resources more effectively and to learn from one another and from workers in parallel fields. The second theme is the need to look more carefully at the future needs of society so that FRC products can compete more effectively and efficiently. The paper reviews briefly the factors affecting the use of FRC, FRC materials and structures and current areas of research and applications. The future aims of researchers and design engineers should be directed towards the better understanding of FRC through modelling and experimental studies together with greter exploitation of FRC resulting from improved performance of existing and new FRC materials. It is recommended that new RILEM Technical Committees be set up with clearly defined goals to achieve these aims.
Keywords: Fibre reinforced concrete, current research, applications, future opportunities.

1 Introduction

Fibre reinforced cements and concretes are firmly established as construction materials. Since the early 1960's extensive research and developments have been carried out with FRC materials leading to a wide range of practical applications. In recent years, a great deal has been learned regarding the limitations of some FRC materials and, in particular, the importance of good design. The Universities and the Construction Industry world-wide are blessed with experts in all facets of FRC theory and applications. Is there anybody bold enough to ask the experts, individually or collectively, where do we go from here?

In considering the question posed by the title, the author is reminded of two apt quotations from the Bible and Shakespeare. In the King James version of the Bible we come across, in Genesis, the following quotation: "And it came to pass, as they journeyed from the east, that they found a plain in the land of Shinar; and they dwelt there. And they said one to another, Go to, let us make brick, and burn them thoroughly. And they had brick for stone, and slime had they for mortar. And they said, Go to, let us build us a city and a

Fibre Reinforced Cement and Concrete. Edited by R. N. Swamy. © 1992 RILEM.
Published by E & FN Spon, 2-6 Boundary Row, London SE1 8HN. ISBN 0 419 18130 X.

tower, whose top may reach unto heaven; and let us make us a name .... And the Lord said, Behold, the people is one, and they begin to do; and now nothing will be restrained from them which they have imagined to do. Go to, let us go down, and there confound their language, that they may not understand one another's speech". The above quotation is a useful reminder of the inability of the experimentalists and the numerical modellers to talk in a common language.

Before attempting to consider the question posed by the title, it is worthwhile to consider some advice from the Bard of Avon. In Macbeth, Shakespeare writes "if you can look into the seeds of time, And say which grain will grow and which will not, Speak then to me ...." If academics possessed the vision they would be industrialists, if industrialists possessed the vision they would be entrepreneurs and if entrepreneurs possessed the vision they would be millionaires. With our limited abilities and facilities one needs to think hard and long before attempting to answer the question given in the title.

The two quotations given above lead directly to the two main themes of the paper. The first theme is the need for all of us working on various aspects of FRC theory and/or practice to pool our resources more effectively and to learn from one another and also from those in different fields of study. The second theme is the need to look more carefully at the future needs of society so that FRC products can be exploited more effectively and efficiently.

2 Factors affecting the use of FRC

It is not possible to consider FRC in isolation from some of the actual and perceived changes which have taken place with cements and concrete. The 1960's saw many changes taking place in cement and concrete technology with rapid developments in the construction industry. During this time there were also major developments taking place in structural analysis with the computer being used increasingly in the analysis of structures.

The main areas of change in cement and concrete technology include the improved methods of cement manufacture and the change in the relative amounts of tricalcium silicate to dicalcium silicate. As a result the 28-day strength has increased steadily over the years and the hydration process is somewhat faster. The second major area of change is that which has occurred in construction. Probably the greatest change in this area is the remarkable growth of the ready-mixed concrete industry together with the more extensive use of precast concrete units.

Unfortunately a number of concrete structures and components have failed the durability test during recent years. The fact that these structures may be only a small percentage of the total number of concrete structures or components which are giving good serviceability does not change the perception of the layman regarding the use of concrete. When durability problems occur the results can be very expensive since demolition or major remedial work may be the only solution. When problems arise with the durability of concrete, it is common practice to blame some of these problems on the changes which have taken place in recent times. The contractor blames the Specification, the designer blames the poor workmanship and both blame the changes in modern cements. The requirements for durability have been summarised in the current British Code (BS 8110) as follows: "to produce a durable structure requires the integration of all aspects of design, materials and construction". This statement is particularly true of FRC materials and structures.

Many changes have also taken place with the relevant Codes of Practice for concrete. When limit state design principles were introduced into Codes, durability was taken into account in terms of minimum compressive strength related to cover and limits on minimum cement content and maximum water-cement ratio related to environmental conditions. Over the years greater emphasis has been placed on designing for durability and this is well illustrated by the increased emphasis on environmental factors in more recent Codes and, in particular, Eurocode 2. Furthermore, the trend towards higher strength (but not ductility) in conjunction with lighter structural elements has resulted in Codes having to adopt a much more careful approach to the serviceability limit states of cracking and deflection. Whereas these potential problems under serviceability conditions may have caused difficulties for the concrete industry, they provided opportunities for the use of FRC materials. Perhaps these opportunities were not fully grasped by the industry.

3 FRC Materials and structures

It is not always appreciated that FRC is simply one example of composite materials. The "Composites" Journal states that it publishes papers on all materials which can be classed as composites and includes the following list - fibre-reinforced metals and plastics, aligned eutectics, fibre-reinforced cement, whisker-reinforced and dispersion strengthened materials, laminates and wood. Composite materials are used by a wide range of professions - from medicine to aeronautical engineering. The matrix and fibres may vary from one application to another but most of the basic engineering principles are the same. Those working with FRC can gain a great deal from researchers working with other composite materials. For example, all researchers working in the field can benefit from the development of testing techniques reported by other colleagues. One of the best example of wasted energy is the number of times that the Isopescu shear test specimen has been discovered "independently" in the last 20 years.

Many materials have been tried and many more will be tried as fibre reinforcement. The fibres used in FRC materials are often divided into two broad categories as follows:

(a) Low modulus, high elongation fibres such as nylon, polypropylene and polyethylene in which the fibres enhance primarily the energy absorption characteristics only.

(b) High strength, high modulus fibres such as steel, glass and asbestos in which the fibres enhance (to some extent) the strength as well as the toughness of the composites.

Although such simplifications may be too crude for the research worker, they may be attractive in explaining the benefits of FRC to the Architect and the Client. There is some merit in having only two grades of reinforcing steel (mild steel and high yield steel in the U.K.) compared with a family of concretes when one is discussing construction methods and materials with a Client.

To illustrate the above point, it is worth considering for a moment steel fibre reinforcement. Steel fibres come in a variety of shapes, e.g. straight, Duoform, crimped, hooked etc. Furthermore, they come in various lengths with various diameters giving a range of aspect ratios. The reasons for the wide range of shapes is obvious and each type will have specific characteristics. However, the adoption of some standard tests early on in the development of FRC technology could have saved a considerable amount of time and energy. The point being made is that a fast and convenient test to allow a comparison to be made could help to identify the market more quickly for any new type of fibre being brought into the market place.

Before leaving this section, some consideration should also be given to FRC structures and structural form in a general sense. In an otherwise excellent book, Gordon (1978) wrote that "leaves are therefore important panel structures, and they seem to make use of most of the known structural devices to increase their stiffness in bending. Nearly all leaves are provided with an elaborate rib structure; the membranes between the ribs are stiffened by being of cellular construction". The above quotation implies that Nature has learned from Science whereas the truth is that Science has been very slow to learn from Nature. Unfortunately, engineers have also been very slow to learn from Nature. A brief consideration of the animal world (bones and feathers) and the plant world (bamboo and reeds) would have led to tubular bridges being developed much earlier. Similarly, the use of fibres to provide toughness should also have been realised from a study of Nature. Structural performance depends upon both the materials being used and the structural shape being adopted. This combination results in the Engineer's Equation:

Right Material + Right Shape = Perfect Structure

4 Where are we now?

This paper is not intended to be a state-of-the-art Report. Such information is available in a number of other publications. The activities during the last six years are reported in detail in the proceedings of the Sheffield Conference (1986), a report prepared by ACI Committee 544 on Fibre Reinforced Concrete (1987), the proceedings of the Cardiff Conference (1989) and the proceedings of the Mainz Conference (1991).

FRC has been used in many areas of application including:

- Airfield Pavements
- Manhole Covers/Slabs etc
- Industrial Floors
- Nuclear Power Industry
- Spillways
- Repair of Concrete Structures
- Refractory Concrete
- Shotcrete
- Impact Loading Situations
- Blast Loading Situations
- Shear Failure Zones in Structures
- Punching Shear
- Deep Beam
- Piles
- Marine Environment

A great deal of research and development is still in progress on FRC materials. The research covers the whole spectrum from theoretical to experimental studies. Space does not allow a full discussion of current activities and only a brief review of some of the current work is given in the following paragraphs.

A great deal of work is going on to develop models for use with FRC materials. Some are difficult and intellectually stimulating and some are more user friendly. Some models include time dependent variables such as ageing, drying, shrinkage and creep whereas others are much simpler in nature. A significant amount of effort is being given to the development of models to predict the fracture behaviour of FRC materials.

At the other end of the spectrum a large amount of experimental studies are also being pursued. Whereas the 1980's saw a significant amount of work on toughness assessment, there has been increasing interest in impact testing in recent years. The effects of curing, temperature and fatigue are also being investigated in some detail. Curing appears to have a significant effect on the strength of FRC materials but not on their toughness. In the case of steel fibres both cryogenic and high temperature have been studied.

Despite the wide range of fibres already available for use in FRC materials, work is still in progress on the development of new fibres including new fibre geometries. Some examples of current developments include Arapree, meshed fabrics, natural fibres and polyacrylonitride fibres. Variations in the other constituents to produce a range of products (high strength by the use of silica fume, lightweight concrete products) have also been studied in recent years. The use of microfillers is very interesting since they can improve workability and also improve the strength.

Generally there is a limit to the amount of fibres which can be added when using normal concrete production techniques. The fibre concentration can be increased dramatically by using other methods of manufacture. An example is SIFCON (Slurry Infiltrated Fibre Concrete) which improves mixing techniques and mechanical properties. Up to 14% by volume of steel fibres can be used using this technique.

A similar technique has been used to produce composites by the impregnation of steel wool with a cementitious slurry.

Durability is another major area where extensive work is being carried out. Durability has already been discussed generally and the durability of FRC is even more important since FRC products are often exposed to more significant environmental exposure conditions.

The relative poor performance of FRC materials in shear is well reported in the literature. This is another area where a significant amount of work is in progress. It is an interesting area of study since the effect of fibres on shear performance is far from clear. An extension of this field of study is the work in progress on combined bending, shear and torsion.

Last, but not least, there is also some very interesting work being carried out on FRC materials as repair materials and as overlays. The maintenance sector of the construction industry should not be overlooked by the FRC community. In 1973, maintenance and repairs accounted for 32% of spending in the U.K. public sector. By 1983 this figure had increased to 49%. Repairs and maintenance will offer the next generation of engineers many opportunities to innovate with materials, techniques and new concepts. The FRC community must make sure that it does not overlook this opportunity.

The above list of current activities is not exhaustive. However, it is sufficient to demonstrate the wide range of skills within the FRC community. Some specialise in experimental studies, some in theoretical studies but few in both. There has been a tendency to be too specialised and perhaps too inward looking. Not only can we learn from other colleagues within our own community but also from researchers in other field. How often do we look at the work of others in ceramics, timber or geotechnical engineering? Geotechnical engineers wish to reinforce the earth. This can be achieved in many ways. One method involves the use of geofabrics and geogrids and a second method involves the use of soil nails i.e. long steel rods (25mm or more in diameter) up to 6m long. Surely there is some scope for cross-fertilization, if not full co-operation, between the FRC community and those involved in ground strengthening.

Our ability to communicate with other members of the building team has not been good in the past. For example, we need to communicate more effectively with Architects. The role of the Architects and Engineers can be illustrated as follows:

ARCHITECT: Aesthetics
Building Regulations
Contracts
Construction
Cracking
Deflections
ENGINEER: Equations

The architect is interested in aesthetics, the engineers in equations and they meet somehow through construction. How many architects know the advantages and disadvantages of GRC? Are architects prejudiced by perception or by the facts?

5 Where do we go from here?

Some suggestions regarding the future direction of FRC may be determined from the recent report by Reinhardt and Naaman (1992) regarding the proceedings of the International Workshop held at Mainz, Germany in 1991 which looked specifically at high performance fibre cement composites. They suggest that the term "high performance" implies an optimum combination of properties for a given application and that the importance of durability should not be underestimated. The production of high performance FRC was considered at Mainz and a proposal was made that a new RILEM TC should be considered to look at special concrete, including SIFCON and compact reinforced composites, in the fresh state.

Three other areas were highlighted by the report of Reinhardt and Naaman - standardization of testing, optimization and modelling. The need for standard test methods is clear and has already been discussed above. The standard tests should be able to characterize existing and new materials and help reduce the large volume of papers which do not allow a ready comparison to be made with other results. Optimization leads directly into that area where there is a conflict between the requirements for strength and those of ductility. The question in simple terms is should composites be designed so that the fibres are allowed to break (aiming for strength) or that the fibres are allowed to pull out (aiming for toughness)?

Reinhardt and Naaman also suggest that there should be a future workshop devoted only to the modelling of fibre-reinforced cementious composites. Modelling allows the engineer to take care of scaling and size effects. It also has the advantage of being able to point the direction of future work towards achieving high strength and tough composites.

Engineering and in particular civil engineering can not take place in a vacuum. Civil engineers provide facilities for society and therefore some thought regarding the future needs of mankind may help point us in the right direction. In the early nineteenth century only 3% of the world population lived in towns and cities. By the 1970's this proportion had increased to over 40% and this trend is continuing. It has been predicted that by the end of this decade some two-thirds of the population of the rich nations will live in towns and that appproximately a quarter of the population of the Third World will also be living in towns. There are many challenges and opportunities for the construction industry as a result of this trend.

The explosion in the world's population and it's re-distribution will take place when there is increasing concern for the environment. Concrete should do well in the future since the raw materials required are available world-wide, the energy requirements for its production are modest and low level skills are required during manufacture. The scope for the return on capital used for innovation is still available in the long term. Whereas we can have some confidence in the volume of opportunities world-wide, the distribution may be different from that in the past. The Pacific basin is likely to assume greater importance in future years.

Predicting the future is not to be recommended. There are numerous examples of eminent people who got it wrong! In 1899 the director of the U.S. Patent Office said that "everything that can be invented has been invented". In 1923 a Nobel prize winner in Physics predicted that "there is no likelihood man can ever tap the power of the atom". In 1905 the U.S.A. President expressed the view that "sensible and responsible women do not want to vote". On a lighter note, the President of Warner Brothers Pictures asked in 1927, "who the hell wants to hear actors talk". In view of the above quotations who will answer the question **"where do we go from here?"**.

6 Conclusions

Our aims and objectives within the FRC community are clear. Our aims include the better understanding of FRC (through modelling and experimental studies), the more widespread use of FRC and the improved performance of existing and new FRC materials. Our objectives in achieving these aims are also relatively easy to determine. A better understanding of FRC can only be achieved by the combined efforts of all members of the FRC community. There is a need for better co-operation and improved communication within the whole range of workers engaged in research and development. This can best be achieved by small groups working together on specific RILEM Technical Committees and that work being discussed at Workshops with the task of producing simple concise recommendations.

Apart from the need to communicate within the FRC community, there is a need to communicate with a range of people outside the community. Many advances have been made by applying the principles developed in one area of research to other topics. This paper has drawn on one simple example, that of the parallel activities taking place in the area of ground strengthening. More importantly, there is a need to talk to other colleagues in the construction industry and to Clients. However, this would be easier to achieve if there was greater consensus in terms of testing and the use of models and some simplification made regarding defining products for applications.

This Symposium could prepare the foundations for the immediate future of FRC. We need to take the opportunity of talking together and deciding how the objectives are to be achieved and in what time scale. Improving on Babel should be readily achieved but we must accept that there will be some disappointment along the way, as stated so eloquently by Shakespeare.

7 References

ACI Committee 544 (1987) Fibre Reinforced Concrete Properties and Applications (eds. S.P. Shah and G.B. Batson) ACI SP-105, American Concrete Institute, Detroit, pp597.

J.E. Gordon (1978) Structures, Pelican Books pp393.

H.W. Reinhardt and A.E. Naaman, (1992) International workshop - High performance fibre reinforced cement composites, Materials and Structures, Vol.25, pp60-62.

H.W. Reinhardt and A.E. Naaman, (1992) (eds) RILEM International workshop High performance fibre reinforced cement composites Mainz, Germany, June 1991. Chapman and Hall.

RILEM Symposium FRC 86 (1986) Developments in Fibre Reinforced Cement and Concrete (eds. R.M. Swamy, R.L. Wagstaffe and D.R. Oakley) RILEM Technical Committee 49-TRF.

R.N. Swamy and B.I.G. Barr (1989) (eds) Fibre Reinforced Cements and Concretes - Recent Developments Elsevier Applied Science, London, pp700.

2 PERFORMANCE DRIVEN DESIGN OF FIBER REINFORCED CEMENTITIOUS COMPOSITES

V. C. LI
Advanced Civil Engineering Materials Research Laboratory,
Department of Civil and Environmental Engineering,
University of Michigan, Ann Arbor, MI, USA

Abstract
This paper describes the performance driven approach in the design of fiber reinforced cementitious composites. This approach is illustrated with structural durability as an example. The identified material property, crack width, is then related quantitatively to material structures -- fiber, matrix and interface properties, by means of micromechanics. It is suggested that tailoring of material structure can lead to controlled crack widths, and hence directly influences the durability of built structures. Success in the performance driven design of fiber reinforced cementitious composites will depend on future research in quantifying links between specific structural performance, material properties, and material structures.
Keywords: Performance, Fiber Concrete, Composites, Design, Micromechanics, Durability, Crack Width.

1 Introduction

The performance of a structure is directly associated with the mechanical and physical properties of the material used to build it. The properties of a material are in turn controlled by its own constituents. Hence details of the material make-up dictate the performance of a built structure. Composites in particular, provide broad latitudes in influencing structural performance because of the possibility of material structure tailoring. (In this paper, when the word *structure* is preceded by *material*, as in *material structure*, we mean the fiber, matrix, and interface of the cementitious composites. Otherwise, the word structure is used in the sense of a built or constructed facility). While this philosophy is well known, its application to fiber reinforced cementitious composites (FRCC) has met with only limited success so far. This paper surveys the advantages of the performance driven design approach, the obstacles in the adoption of such an approach, and some potential solutions offered by recent developments in micromechanics. A practical example -- structural durability, is used to illustrate the concepts described. The discussion is limited to the materials aspects of structural performance, and specifically relate to fiber reinforced cementitious composites. It is

Fibre Reinforced Cement and Concrete. Edited by R. N. Swamy. © 1992 RILEM.
Published by E & FN Spon, 2-6 Boundary Row, London SE1 8HN. ISBN 0 419 18130 X.

hoped that stronger recognition of the performance-property-microstructure relationships will provide more rational and systematic development of FRCC and enhanced usage of this versatile material.

2 The Performance Driven Design Approach

Because a given constructed facility is usually made up of many structural components, with potentially different materials chosen for different components, it is more convenient to discuss here the performance of structural components rather than structural systems. Some structural components that have utilized FRCC include slab on grades, bridge decks, wall panels, facade elements and water-tight structures. In addition, high performance FRCCs may be selectively applied to local parts of a structure. For example, Naaman (1991) suggested their use in beam-column connections in earthquake resistant frames, selected plastic hinge or fuse locations in seismic structures, the lower sections of shear walls or the lower columns in high rise buildings, the disturbed regions near the anchorages at the end of prestressed concrete girders, the high bending and punching shear zones around columns in two-way slab systems, and tie-back anchors. Clearly the diversity of these components and strategic structural locations lead to a diversity of performance requirements. It may also be pointed out that as our understanding of performance-property-microstructure relationships of FRCC increases, and as our confidence in their near and long term performance are enhanced by experience, further applications of FRCC will be found. Increasingly more load-carrying structural members will employ FRCC.

Figure 1 illustrates the performance driven design approach for FRCC. Apart from the performance-property-material structure nodes, a fourth node -- processing, may be included. Processing (as, e.g., pursued by Krenchel and co-workers) is of course crucial in any materials development, but it has been left out here to simplify the discussion and focus on the main theme. The performance of a given structural component may be defined as deflection control, light weight, seismic resistance, dimensional stability, reliability and durability. The properties may include moduli, various strengths (tensile, compressive, flexural, shear, etc.), ductility, toughness, notch sensitivity, density, permeability, coefficient of expansion, and impact, temperature, fatigue and wear resistant properties. The material structure for FRCC generally include the fiber, matrix and interface, although it is clear that each of these have their own microstructures as well. The idea of the performance driven design approach is basically one where the performance and functionalities of a given structure or structural component are specified, and a material must be chosen so that the properties can meet the expected structural demand. Such an approach is of course routinely used. However it has been rare to consider the approach a step further. That is, given the required properties, the fiber, matrix and interface are tailored to optimize the needed properties. In other words, the performance driven design approach ensures a direct link between the material composition and the structural performance. The quantitative link between material properties and the associated material microstructures is often known as micromechanics. Micromechanics takes into account the material structure and local deformation mechanisms in predicting the composite macroscopic behavior.

Because of the increasing availability of a wide range of fibers with generally declining cost, an equally wide range of cement based matrixes with a variety of chemical admixtures, and to a certain extent, controllable interfaces, the properties of an FRCC can significantly vary with different combinations of fibers, matrices, and interfaces. As an example, the flexural strength and fracture toughness of an FRCC can vary over at least one order of magnitude, and strain capacity can vary by two orders of magnitude. It is therefore quite plausible that fiber, matrix and interface properties be tailored in an FRCC with composite properties required for specified structural performance.

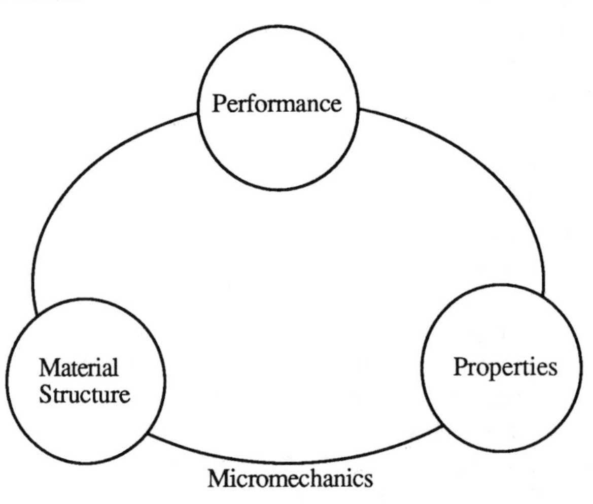

Figure 1: The Performance Driven Design Approach. Micromechanics Provides the Quantitative Link between Material Structure and Properties

While the performance driven design approach is attractive, it has been difficult to implement because most structural engineers do not design materials, whereas materials engineers do not design structures. While there are quantitative linkages between certain structural performance and material properties, some important ones, such as structural durability and related material properties, are often not well established. Apart from a few exceptional cases, most quantitative linkages between material properties and material structures are also weak. As a result, direct linkages between structural performance and material structure are almost non-existent. This phenomenon produces two inhibiting effects: the improper and limited use of FRCC in structures, and the slow development of advanced FRCCs. To overcome these inhibiting effects, it is necessary to launch a fresh approach in FRCC research. Structural performance which can benefit from the special properties of FRCC should be identified, and these properties should be related to the microstructure of the FRCC. Such an approach affords specific guidelines for the engineering of specific FRCC to meet specific performance requirements in a specific structure.

3 Structural Durability

The industrialized world is currently facing an increasingly aggravating infrastructural decay problem. Just in concrete structures alone, it has been estimated that the rehabilitation cost in the U.S. will reach into trillions of dollars over the next twenty years (National Research Council, 1987). It is no wonder, then, whether in considering rehabilitation of an aged structure or in new construction, the issue of structural durability has been a major concern. Interestingly, the study of durability of FRCC has been on the rise in recent years. In general, studies on steel FRCC (Balaguru and Ramakrishnan, 1986; Hoff, 1987; Kosa et al, 1991), polypropylene FRCC (Hannant and Zonsveld, 1980; Swamy and Hussin, 1986), and carbon FRCC (Akihama et al, 1984) indicate that the material durability is either enhanced or unchanged in the presence of fibers. These studies establish the baseline that FRCC can be used as a durable construction material. However, they do not directly address whether structural life will be extended or not by use of FRCC. In the following, an attempt is made to address durability as a structural performance. The related properties are then established. Finally the material structure most suitable for optimizing these properties are discussed in the context of recently developed micromechanical models. This presentation is offered as an illustration of the performance-driven design approach described in the previous section. While the various links are described, it will be clear that the success of this approach awaits further research.

In concrete structures, most durability issues arise because of concrete cover scaling and rebar corrosion. It has been long known that these problems are associated with the permeability of concrete to water and aggressive agents such as chloride ions. Concrete permeability, in turn, is dominated by the presence of cracks in concrete. Improving durability of concrete structures, therefore, requires in part the control of cracks in concrete (see, e.g., Mindess and Young, 1981).

Cracks in concrete can be generated in a number of ways. They exist as processing defects in the cement paste or in the aggregate/cement interface. Cracks can result from shrinkage stresses due to drying or carbonation, in addition to thermal and mechanical loads (see, e.g., Neville and Brooks, 1987). Concrete cover delamination in bridge decks has been associated with cracks generated from the pressure exerted by the expanding corrosion debris of the re-bar on the surrounding concrete. Figure 2, for example, illustrates a bridge deck cutout which exposes large scale delamination at the upper layer of reinforcements. Similarly if the concrete is used as a surface layer, substructure movements can lead to high imposed strains. Because of the low strain capacity of cementitious materials, large cracks may result. Whenever cracks are generated, the migration of aggressive agents into the concrete then depends on the crack width.

Figure 2: Cutout of Bridge Deck Showing Extensive Delamination Deterioration (Photo Courtesy of Dr. K. Maser)

From the above discussion, it seems that a general approach to crack control in FRCC would be to create a composite with pseudo strain-hardening behavior (Aveston et al, 1971; Ali et al, 1975; Baggott, 1983; Krenchel and Jensen, 1980; Laws, 1987; Oakley and Proctor, 1975; Stang and Shah, 1989; Li and Wu, 1991). This results in: 1) high first cracking strength and strain by stabilizing microcracks, and 2) load redistribution capability so that macrocracks can be delayed (Li and Hashida, 1992) and multiple cracking crack widths can be controlled. The following discussion will focus on crack width as the material property governing structural durability, taking advantage of recent studies relating crack width to water penetration in FRCC. This view may be unnecessarily narrow, and as our understanding of links between durability and other composite properties improves, there is no doubt that they should be incorporated into a more comprehensive approach to structural durability design. The following discussion is also not meant to imply that crack width control is the only means of dealing with structural durability, as this complex problem can be attacked from various angles, such as by chemical means. However, since migration of aggressive agents requires a pathway, and since typical concrete has low resistance to cracking, crack width control can be a very effective method of improving structural durability. Our intention, therefore, is to relate material structure to crack width as a controlling property of structural durability (Figure 3).

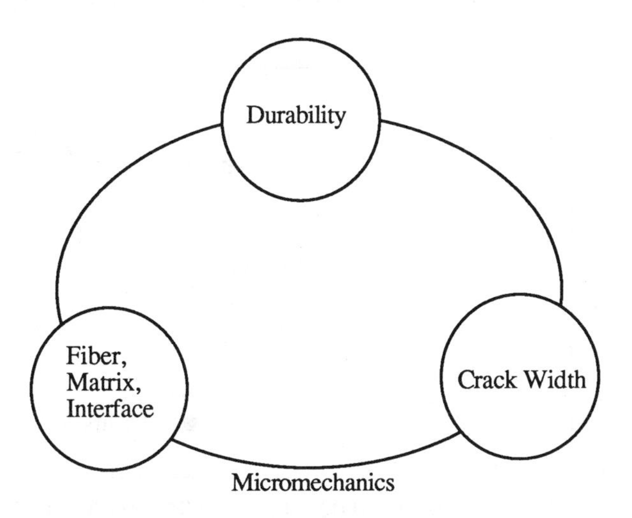

Figure 3: The Performance Driven Design Approach for FRCC Targeted for Structural Durability

4 Relationship Between Water Penetration And Crack Width

Two types of experimental tests which relate water penetration and crack width in FRCC are reviewed here. Although the experimental approaches and investigation emphasizes are different in these investigations, they both suggest that: 1) crack width governs water penetration, and 2) fiber addition has beneficial effects in reducing water penetration.

Keer et al (1989) studied the influence of chemical treatments of cracked and uncracked FRCC cement sheets on water absorption. The cement sheets reinforced with polypropylene networks between 3.0 and 4.7 % undergoes pseudo strain-hardening when loaded beyond its first crack strain. The FRCC sheets were precracked to predetermined strain-levels, resulting in controlled crack widths of .02 - .06 mm in the unloaded state. (However, the number of cracks is likely to be different for the different specimens.) Specimens kept wet on one side by sponges were weighted at fixed time intervals up to 96 hours. Water absorption was calculated as the weight difference (with respect to initial weight) and normalized by the dry weight. For our present purpose, we show in Figure 4 the water absorption as a function of crack width for the untreated specimens. The 1 hour measurement shows a stronger dependence of water absorption on crack width than the 96 hour ones. Hannant and Keer (1983) reported (for the same series of tests) that when crack widths are small, autogenous healing of the concrete could take place under natural weathering conditions. This mechanism will further limit the migration of aggressive agents into the concrete.

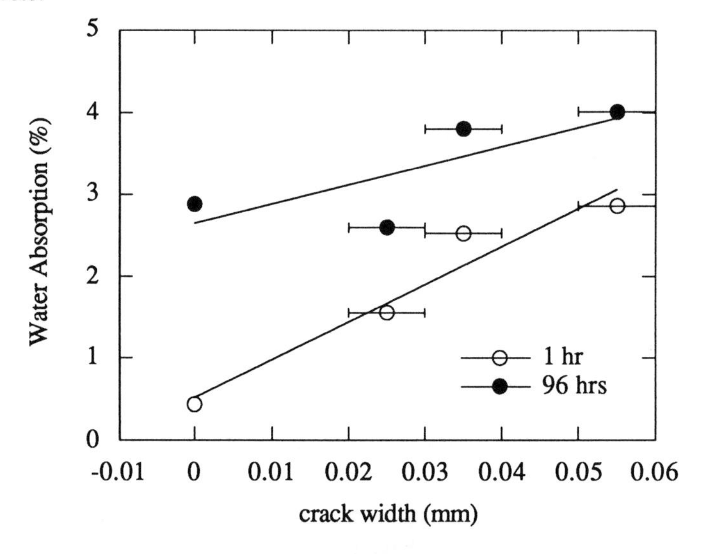

Figure 4: Measured Water Absorption Into Polypropylene FRCC with Controlled Crack Widths (after Keer et al, 1989).

Tsukamoto (1990) studied water tightness of FRCC with less than 2.5 % volume fraction of polymer and steel fibers. The water flow rate on one side of a precracked specimen of controlled crack width was measured while the opposite side was subjected to fluid pressure created by a water column. The results confirm previous findings that the flow rate scales with the third power of crack width (Figure 5) with a threshold value which depends on the fluid pressure. Tsukamoto found that for a given crack width, the presence of fibers appears to reduce the flow rate significantly. This is apparently due to the roughening effect of the fracture surface in the presence of fibers. In addition, because the fibers likely bridge the crack faces after the crack is introduced, the effective crack width felt by the fluid flow may be smaller than that indicated by specimen surface measurements.

The above studies indicate that the water flow rate is highly sensitive to crack width, while the water absorption as measured by Keer et al is less sensitive, at least for long term measurements. In both cases, they found positive influence of fibers on reducing water penetration into cementitious materials. The flow rate test under steady state condition can be considered as measuring the effective permeability of the concrete with a crack (so that Figure 5 may be read as effective permeability as a function of crack width for the various materials, since the pressure head is fixed), whereas the water absorption test is likely measuring a different material property related to moisture migration, presumably via capillary action. These properties may independently or in combination govern structural durability, depending on the given environment.

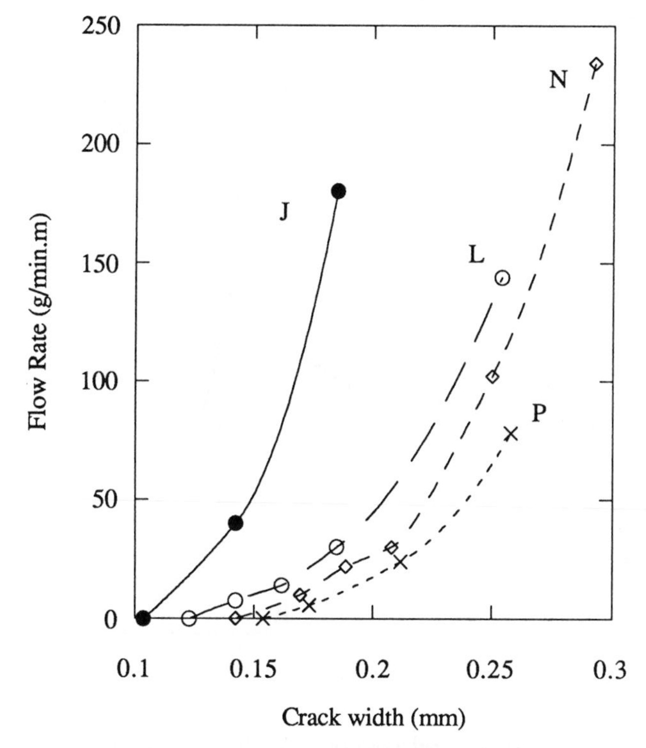

Figure 5: Measured Flow Rate Through FRCCs with Controlled Crack Widths (after Tsukamoto, 1990).
J = Plain Concrete; L = Polyacrylonitril FRC (d_f = 0.1 mm, L_f = 6 mm; V_f = 1.7%); N = Steel FRC (d_f = 0.5 mm, L_f = 30 mm; V_f = 1.0%); P = Polyvinylalcohol FRC (L_f = 24 mm; V_f = 0.8%). Pressure head = 7 mWS/m; Temperature = 20°C.

5 Crack Width Control By Fibers

In conventional concrete structures, crack widths are controlled by the steel re-bars. Crack width in such structures coupled with an ordinary fiber reinforced concrete has been studied by Kanazu et al. (1982), Stang (1991) and Stang and Aarre (1991). Stang showed that crack widths can be controlled to within 0.25 mm, depending on stress in the reinforcing bar. In the following, our attention is placed on FRCCs which exhibit pseudo strain-hardening such that crack widths in the FRCCs are controlled by the reinforcing fibers, to within 0.05 mm and less, without the benefit of the reinforcing steel bar. An idealized tensile stress-strain curve in a pseudo strain-hardening cementitious composite is shown in Figure 6. This curve may be divided into four regions of straining. In region I, the composite is deforming linear elastically and no through-thickness crack occurs. In region II, multiple cracking occurs and the strain increases with additional cracks while the crack width remains constant. In region III, multiple cracking has been completed and further straining results in direct loading of fibers. In region IV, pull-out or failure of fibers occur. In typical design, it seems prudent to limit the strain below ε_{mc}. Our interest in crack width control therefore lies in region II.

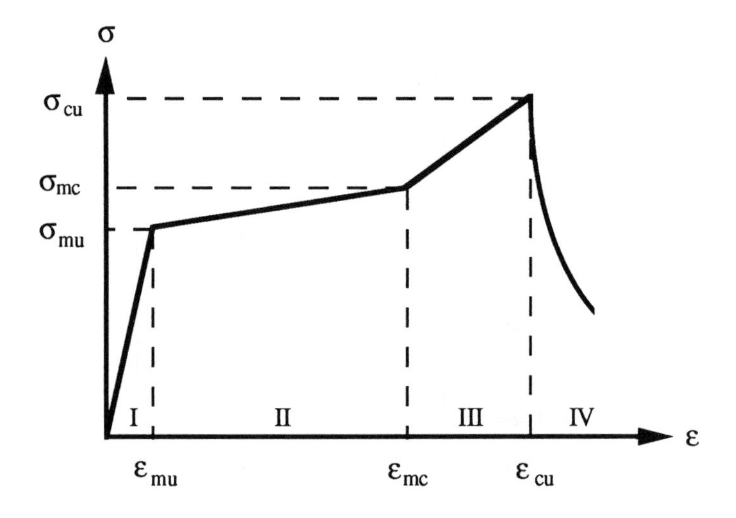

Figure 6: Schematics of Tensile Behavior of a Pseudo Strain-Hardening FRCC

In continuous aligned fiber composites, multiple cracks will be spaced between x' and 2x', where x' is given by (Naaman, 1970; Aveston et al, 1971):

$$x' = \frac{E_m d_f \varepsilon_{mu} V_m}{4\tau V_f} \tag{1}$$

The first crack strain ε_{mu} has also been derived by Aveston et al, and is given as:

$$\varepsilon_{mu} = \left[\frac{24V_f^2 \tau \gamma_m E_f}{E_c E_m^2 d_f V_m}\right]^{\frac{1}{3}} \tag{2}$$

In (1) and (2), E and V are the modulus and volume fractions, respectively. Subscripts m, f and c denote matrix, fiber and composites. d_f and τ are the fiber diameter and interface frictional bond strength. The maximum crack width w can be calculated by integrating the difference in strain between the fiber and the matrix over a matrix block, so that

$$w = \varepsilon_{mu}(1+\alpha)x' \tag{3}$$

where $\alpha = E_m V_m / E_f V_f$.

In practical applications, it is more likely that short random fibers will be used in cementitious composites. However, the above procedure does not apply to this type of composites because of fiber discontinuity. Recently the ascending branch of the crack bridging stress σ_B has been derived as a function of crack opening δ for 3-D random short fiber composites (Li, 1992):

$$\tilde{\sigma}_B(\tilde{\delta}) = \left[2\left(\frac{\tilde{\delta}}{\tilde{\delta}^*}\right)^{1/2} - \frac{\tilde{\delta}}{\tilde{\delta}^*}\right]g \qquad for\ \tilde{\delta} \le \tilde{\delta}^* \tag{4}$$

where $\tilde{\sigma}_B \equiv \frac{\sigma_B}{\sigma_o}$ and $\sigma_o \equiv \frac{1}{2}V_f\tau\left(\frac{L_f}{d_f}\right)$, $\tilde{\delta} \equiv \frac{\delta}{L_f/2}$ and $\tilde{\delta}^* \equiv \left(\frac{2\alpha\tau}{(1+\alpha)E_f}\right)\left(\frac{L_f}{d_f}\right)$. L_f is the fiber length and g is a snubbing factor included to take into account the friction pulley effect for bridging fibers inclined at an angle to the principal tensile stress direction. $\tilde{\delta}^*$ corresponds approximately to the crack opening at the peak bridging stress. The crack opening in region II may be computed if the load at first cracking is known. For a pseudo strain-hardening material, this corresponds to the state when a matrix crack propagates steadily at essentially no increasing remote load, i.e. at the steady state crack strength σ_{ss}, and (Li and Leung, 1992; Li and Wu, 1992),

$$\tilde{\sigma}_{ss} = \left(2\sqrt{\bar{c}_s} - \bar{c}_s\right)g \tag{5}$$

$\bar{c}_s$ is the non-dimensional crack length at initiation of steady state cracking and is defined in terms of a non-dimensionalized material parameter $\bar{K}$:

$$\bar{K} = \frac{2}{\sqrt{\pi}}\bar{c}_s\left(\frac{2}{3}\sqrt{\bar{c}_s} - \frac{1}{2}\bar{c}_s\right) \qquad 0 \le \bar{c}_s \le 1 \tag{6}$$

Li and Leung showed that $\bar{K}$ may be interpreted as a ratio of crack tip fracture energy absorption rate G_{tip} to the energy absorption rate in the fiber bridging zone G_r behind the crack front, hence:

$$\bar{K} = \frac{10}{3\sqrt{\pi}}\left(\frac{G_{tip}}{G_r}\right) \tag{7}$$

These fracture energy terms can be written as a function of material structural parameters:

$$G_r = \frac{5}{24} g\tau V_f d_f \left(\frac{L_f}{d_f}\right)^2 \tilde{\delta}^* \tag{8}$$

and

$$G_{tip} = (E_c / E_m)G_m \tag{9}$$

and G_m is the matrix fracture energy. Once $\bar{K}$ is determined from material structure parameters, it is then possible to calculate the steady state crack width $\tilde{\delta}'$ from (4)-(6). [Indeed, by requiring $\sigma_B = \sigma_{ss}$, it can be seen from (4) and (5) that $\tilde{\delta} / \tilde{\delta}^* = \bar{c}_s \equiv \tilde{\delta}' / \tilde{\delta}^*$.] In (9), the composite modulus E_c has been related to the fiber and matrix moduli and volume fractions in the form:

$$E_c = V_m E_m + \eta V_f E_f \tag{10}$$

where η is a fiber efficiency factor used to reduce the contribution of the fibers to the composite modulus because of fiber randomness to load direction and fiber finite length. A variety of values for η (for a succinct review of efficiency factors, see Bentur and Mindess, 1990) has been derived. Naaman et al (1991) suggested that an arithmetic average of the lower and upper bound for continuous aligned composites, such as derived by Halpin and Tsai (1969) serves to fit experimental data (of steel FRCC) well. More accurate representations of E_c in terms of fiber and matrix parameters can be found in Tandon and Weng (1986), and Wakashima and Tsukamoto (1991). However, given the cumbersome character of these representations, and the relatively forgiving nature of modulus representation at fiber volume fractions (typically less than 10% and usually just a few percent) expected in cementitious composites, the simple form of (10) with η=1 (upper bound) is employed for the calculations presented here.

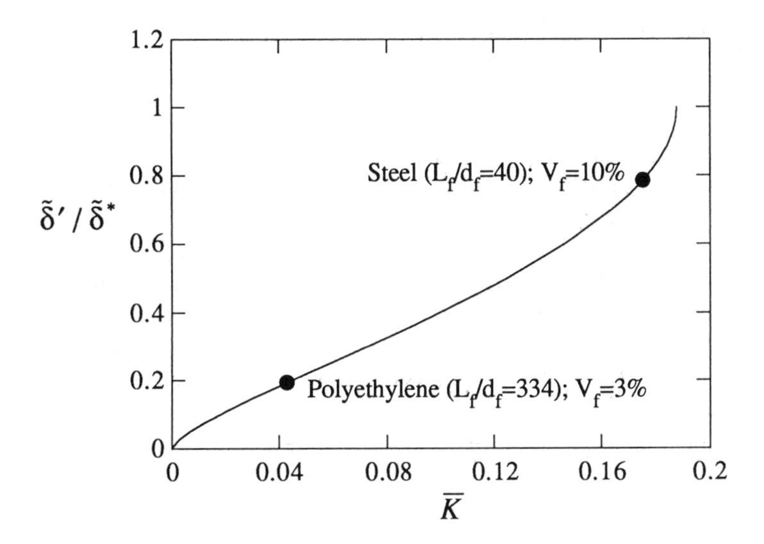

Figure 7: Normalized Crack Width Controlled by $\overline{K}$

Table 1: Fiber, Matrix and Interface Properties used in Composite Crack Width Opening

	Fiber				Matrix			Interface	
Type	E_f (GPa)	L_f (mm)	d_f (mm)	V_f	E_m (GPa)	K_m (MPa$\sqrt{m}$)	ν	g	τ (MPa)
Steel	200	6	0.15	0.1	15	0.2	0.2	2	6.5
Polyethylene	120	12.7	.038	.03	15	0.2	0.2	2	1

Figure 7 shows a plot of $\tilde{\delta}' / \tilde{\delta}^*$ as a function of $\overline{K}$. The non-dimensional form of this plot allows the inclusion of any material (fiber, matrix and interface properties), geometries (fiber length and diameter) and fiber volume fraction to be represented on a single curve. Two specific composites, reinforced with commercially available steel and polymeric fibers, are located on this curve. Properties of these fibers, as well as mortar matrix and interface properties, are tabulated in Table 1. An interesting point about Figure 7 is that when $\overline{K}$ exceeds 0.188, the crack opening is predicted to approach infinity. This $\overline{K}$ value corresponds to $\overline{K}_{crit}$ discussed in Li and Leung (1992) who showed that when $\overline{K} > \overline{K}_{crit}$, multiple cracking cannot be achieved. Thus softening occurs as soon as the first crack appears so that a finite crack width cannot be maintained at the failure load. Details of conditions for multiple cracking can be found in Li and Wu (1992).

While Figure 7 is empowered with dealing with a wide range of material structures, its physical meaning may be obscured by the non-dimensionalization. Figure 8 shows two physical plots for crack width. In Figure 8a, the influences of fiber modulus and interface bond strength on crack width are examined, for a fiber of fixed aspect ratio of 100. Other fiber and matrix parametric values are given in the figure caption. As expected, increasing interface bond reduces the crack width. For each bond strength, choosing a fiber with higher modulus tends to lower the crack width, at least initially. At higher modulus, the crack width tends to increase again. These curves are terminated at a fiber modulus which causes $\overline{K}$ to reach its critical value, at which state no multiple cracking can occur, and the composite once again fail catastrophically with a single large fracture. Figure 8b examines the influence of fiber aspect ratio on crack width, for a fixed fiber modulus, chosen for a steel. For each interface bond strength, the crack width is shown to decrease with aspect ratio. When the aspect ratio is too low, however, conditions on multiple cracking are again not met and these curves indicate crack widths going to infinity. For the range of parametric values covered, Figure 8 seems to suggest that as long as multiple cracking conditions are met, the crack widths are likely limited to less than 0.02 mm. Even smaller crack widths can be achieved with higher fiber volume fractions. These figures are useful for designing composites with controlled crack widths and hence structural durability as we described in the previous section.

For comparison purpose, eqns. (1)-(3) and (4)-(6) have been used to compute the crack widths for the composites with microstructural parameters described in Table 1, for both the continuous aligned case and the short random case. The results are summarized in Table 2.

Table 2: Computed Composite Crack Width in mm x 10^{-3}

Fiber	**Continuous Aligned**	**Discontinuous Random**
Steel	0.72	2.5
Polyethylene	1.13	5.7

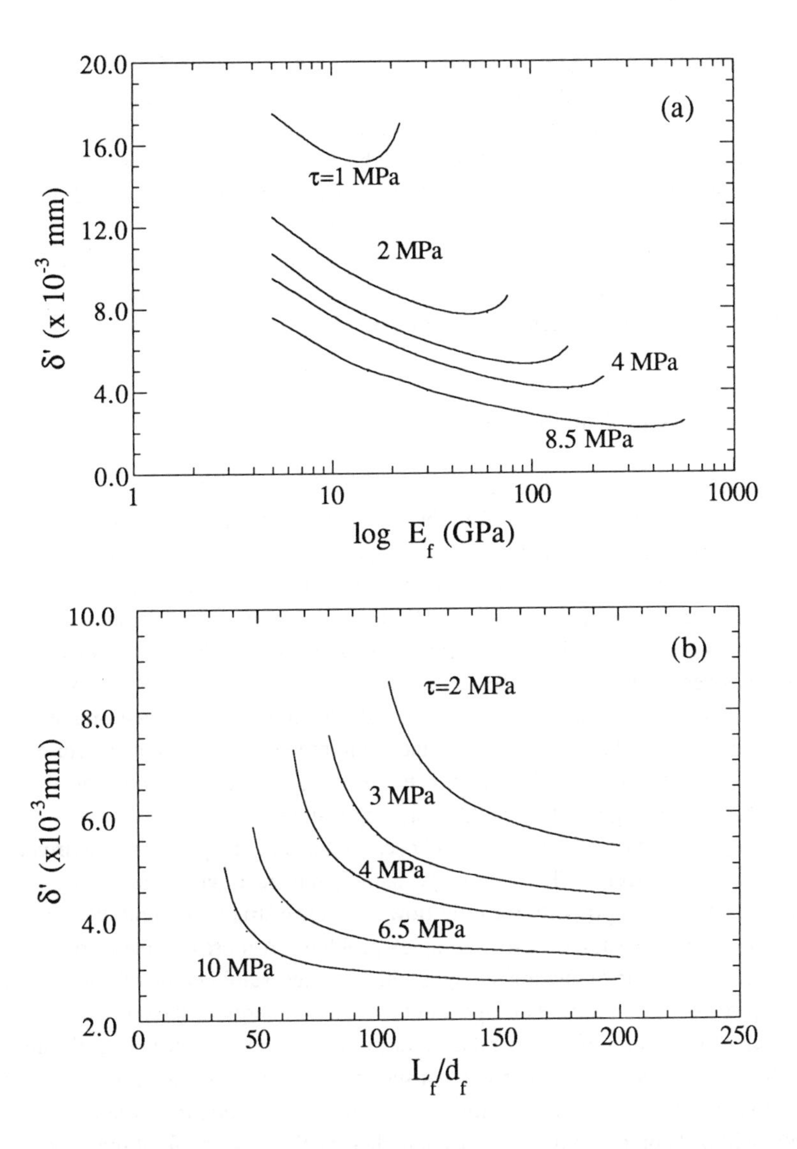

Figure 8: Dependence of Crack Width on (a) Fiber Modulus and Bond Strength, for L_f = 5 mm; d_f = .05 mm; and (b) Fiber Aspect Ratio and Bond Strength, for E_f = 200 GPa; d_f = .15 mm. For both cases E_m = 15 GPa; K_m = 0.2 MPa√m; g = 1; and V_f = 3%.

The calculations confirm that the stiffer steel fiber provides smaller crack widths than the polymer fiber, and that for each fiber type, the continuous aligned case provides smaller crack widths than the short random case. In all cases, these composites provide crack widths of less than 0.01 mm. This is much smaller than the mm scale cracks typically encountered in reinforced concrete structures. The small crack width of such composites more than meets the code specifications, or recommendations of ACI 224R-80, the CEB Model Code and the FIP (0.1 -0.3 mm allowable crack width depending on concrete cover, see, e.g. Mehta, 1986) for reinforced or prestressed concrete structures exposed to aggressive environments. Based on Figure 5, the flow rate and therefore the effective permeability could be expected to be much smaller when this type of material is used in comparison to ordinary concrete.

6 Discussions And Conclusions

In this paper, we present the performance driven design framework for FRCC. This approach links a desired structural performance to specific FRCC properties which in turn are linked to the material structure of the composite. The material structure can then be tailored for specific structural performance in mind. The advantage of the performance driven design approach is that FRCCs with optimized properties can be developed for a given application. In recent years, (see, e.g. Reinhardt and Naaman, 1991) there has been a debate as to the definition of *high performance* FRCC. The present discussion, (in particular Figure 1), suggests that high performance implies different meanings according to the performance requirements of different structures or different structural components. Indeed, it is not feasible, nor economically sensible, to design FRCCs to be high performance for all applications.

The example of structural durability is chosen to illustrate the performance driven design concept. One specific FRCC property, namely the crack width, has been identified as a controlling property for fluid flow and hence structural durability. This addresses only one dimension of a very complex problem. A more complete solution will include at least the first cracking strength which reflects stabilization of microcracks induced by processing and during curing, and the fracture energy of the composite which reflects stabilization of macrocracks induced by loading. Nevertheless, through micromechanical analyses, it is illustrated that specific micromechanical parameters can be chosen to control crack widths to within 0.05 mm and less, such that water penetration will be kept to a minimal, hence lengthening the service life of the structure. Further research, combined with experimental verifications in the field, will be required to validate these ideas. Specifically, additional work will be necessary to identify structural life in a given environment and its relationship to crack widths. Figure 8 (or more generally figure 7) can then be used to locate combinations of micromechanical parameters associated with fiber, matrix and/or interface, and the corresponding composite can be designed for this particular application.

The micromechanical model for crack width described in this paper has assumed that the composite has been properly reinforced so that pseudo strain-hardening behavior results. The detailed conditions for pseudo strain-hardening for continuously aligned and discontinuous random fiber reinforced brittle matrix composites can be found in a review by Li and Wu (1992). Also it should be noted that because a real composite is populated with defects of various sizes and inhomogeneous matrix and interface properties, the various regions (I-IV) indicated in Figure 6 are likely to merge together continuously, rather than sharply divided as shown. The predicted crack widths therefore serve as rough order-of-magnitude estimates, and more detailed experimental verifications are now planned at the University of Michigan.

Although the performance-property-material structure linkages represented in Figure 1 provide a useful framework for FRCC development, it is recognized that many elements of this framework must be strengthened. For example, certain material properties may not be well defined for a given performance. The micromechanisms associated with these properties can be even less understood. Impact resistance and fatigue resistance in FRCC are just two such examples. To successfully utilize performance driven design of FRCC, it will be necessary to clarify the properties associated with the desired structural performance and micromechanical models must be developed to relate such properties to the material microstructures. The significant amount of work on the micromechanics and micromechanisms of important composite properties, such as fracture toughness (see, e.g. Shah, 1990), accomplished over the last decade, sets an excellent platform from which successful development of FRCCs based on the performance driven design approach can be launched.

7 Acknowledgments

Research at the ACE-MRL at the University of Michigan has been sponsored by the Air Force of Scientific Research (Program Manager: Dr. S. Wu) and the National Science Foundation (Program Manager: Dr. K. Chong). Helpful discussions with A. Naaman and H.C. Wu on this work are gratefully acknowledged. The author has also benefited from stimulating discussions with H. Krenchel and H. Stang.

8 References

Akihama, S., Suenaga, T., and Banno, T. (1984) Mechanical properties of carbon fiber reinforced cement composite and the application to large domes. **Kajima Institute of Construction Technology Report** No. 53.

Ali, M.A., Majumdar, A. J., Singh, B. (1975) Properties of glass fiber cement - the effect of fiber length and content, **J. of Mat'ls Sci.**, 10, 1732-1740.

Aveston, J., Cooper, G.A. and Kelly, A. (1971) Single and multiple fracture, in **The Properties of Fiber Composites**, Conf. Proc., NPL, IPC Science and Technology Press Ltd., pp. 15-24.

Aveston, J., Mercer, R.A., and Sillwood, J.M. (1974) Fiber reinforced cements -- scientific foundations for specifications, in **Composites standards testing and design**, Conf. Proc. National Physical Laboratory, IPC Science and Technology Press, Guildford, UK., pp. 93-103.

Baggott, R. (1983) Polypropylene fiber reinforcement of lightweight cementitious matrices. **The International J. of Cem. Comp. & Lightweight Conc.**, 5, 2, 105-114.

Balaguru, P., and Ramakrishnan V. (1986) Mechanical properties of superplasticized fiber reinforced concrete developed for bridge decks and highway pavements, in **Concrete in Transportation**, ACI SP 93, Detroit, Michigan.

Bentur, A. and Mindess, S. (1990) **Fiber Reinforced Cementitious Composites**, Elsevier Applied Science, Essex, England.

Hoff, G.C., (1987) Durability of fiber reinforced concrete in a severe marine environment, in **Proceedings, Katharine and Bryant Mther International Symposium on Concrete Durability**, ACI, Detroit, SP100, V.1, pp. 997-1041.

Hannant, D.J. and Keer, J.G. (1983) Autogenous healing of thin cement based sheets. **Cement and Concrete Research**, V. 13, 357-365.

Hannant, D.J., and Zonsveld, J.J. (1980) Polyolefin fibrous networks in cement matrices for low cost sheeting. **Phil. Trans. of the Royal Soc.**, London, A294, 83-88.

Kanazu, T., Aoyagi, Y. and Nakano, T. (1982) Mechanical behavior of concrete tension members reinforced with deformed bars and steel fibers. **Trans. Japan Concr. Inst.** 4, 395-402.

Keer, J.G., Xu, G. and Filip, R. (1989) Cracking and moisture penetration in fibre cement sheeting, in **Fibre Reinforced Cements and Concretes -- Recent Developments**, eds. R.N. Swamy and B. Barr, Elsevier Applied Science, London, pp. 592-601.

Kosa, K., Naaman, A.E. and Hansen, W. (1991) Durability of fiber reinforced concrete and SIFCON, **ACI Materials J.**, V.88, No. 3, May-June, pp 310-319.

Krenchel, H. (1987) Mix design and testing methods of fibre reinforced concrete. In Proc. of International Symposium on **Fibre Reinforced Concrete**, Ed. V.S. Parameswaran and T.S. Krishnamoorthy, Oxford & IBH Publishing.

Krenchel, H. and Jensen H.W. (1980). Organic reinforcing fibers for cement and concrete, in fibrous concrete, in **The Concrete Society**, Proceeding of the Symposium on Fibrous Concrete, Lancaster, The Construction Press, pp. 87-98.

Krenchel, H. and Hansen, S. (1991) Low porosity cement for high performance concrete and FRC-Materials. In Proc. of International Workshop on **High Performance Fiber Reinforced Cement Composites**, Ed. H. Reinhardt and A. Naaman, Chapman and Hall.

Laws V. (1987) Stress/strain curve of fibrous composites, **J. Mater. Sci. Letters**, 6, 675-678.

Li, V.C. (1992) Post-crack scaling relations for fiber reinforced cementitious composites", ASCE **J. of Materials in Civil Engineering**, V.4, No.1, 41-57.

Li, V.C. and Leung, C.K.Y. (1992) Theory of steady state and multiple cracking of random discontinuous fiber reinforced brittle matrix composites. Accepted for publication in ASCE **J. of Engineering Mechanics**.

Li, V.C. and Hashida, T. (1992) Ductile fracture in cementitious materials? To appear in Proc. 1st Int'l Conf. on **Fracture Mechanics of Concrete Structures**.

Li V.C. and Wu, H.C. (1991) Pseudo strain-hardening design in cementitious composites, in Proc. of International Workshop on **High Performance Fiber Reinforced Cement Composites**, Ed. H. Reinhardt and A. Naaman, Chapman and Hall.

Li, V.C. and Wu, H.C. (1992) Conditions for pseudo strain-hardening in fiber reinforced brittle matrix composites. Submitted for publication in **Applied Mechanics Review**.

Mehta, P.K. (1986) **Concrete-Structure, Properties and Materials**, Prentice-Hall, Inc., New Jersey, U.S.A.

Mindess, S. and Young, J.F. (1981) **Concrete**. Prentice-Hall, Inc., New Jersey, U.S.A.

Naaman, A.E. (1970) **Reinforcing Mechanisms in Ferro-Cement**. MS Thesis, The Massachusetts Institute of Technology, Cambridge, MA.

Naaman, A.E., Otter D. and Najm, H. (1991) Elastic modulus of SIFCON in tension and compression. **ACI Materials J.**, V. 88, No. 6.

Naaman, A.E. (1992) SIFCON: Tailored properties for structural performance, in Proc. of International Workshop on **High Performance Fiber Reinforced Cement Composites**, Ed. H. Reinhardt and A. Naaman, Chapman and Hall.

National Research Council (1987) **Concrete Durability: A Multibillion-dollar Opportunity**. Performed by National Materials Advisory Board of NRC, NMAB-437.

Neville, A.M., and Brooks, J.J. (1987) **Concrete Technology**. Longman Scientific, Essex, England.

Oakley, D.R., and Proctor, B.A. (1975) Tensile stress-strain behavior of glass fiber reinforced cement composites, in **Fiber Reinforced Cement and Concrete**, RILEM Symposium, ed. A. Neville, Construction Press, pp. 347-359.

Reinhardt, H., and Naaman, A. eds, (1991) **High Performance Fiber Reinforced Cement Composites**, Chapman and Hall.

Shah, S.P. (1990) **Toughening Mechanisms in Quasi-brittle Materials**, Kluwer Academic Publishers, Netherlands.

Stang, H. and Shah, S.P. (1989) Damage evolution in FRC materials modelling and experimental observations. In **Fibre Reinforced Cements and Concretes -- Recent Developments**, eds. R.N. Swamy and B. Barr, Elsevier Applied Science, London, pp. 378-387.

Stang, H. (1991) Prediction of crack width in conventionally reinforced FRC. In **Proc. Third International Symposium on Brittle Matrix Composites**. Warsaw, Poland, 17-19 Sep., 1991. (Eds. A.M. Brandt and J.H. Marshall).

Stang, H. and Aarre, T. (1991) Evaluation of crack width in FRC with conventional reinforcement. Submitted for publication in **J. Cem. Concr. Comp.**

Swamy, R.N. and Hussin, M.W. (1986) Effect of curing conditions on the tensile behavior of fibre cement composites. In proceedings, **Developments in Fibre Reinforced Cement and Concrete**, 1986 RILEM FRC Symposium, Vol. 1, 4.9. Eds. Swamy, R.N., Wagstaffe, R.L. and Oakley, D.R., Sheffield, England.

Tandon, G.P. and Weng, G.J. (1986) Average stress in the matrix and effective moduli of randomly oriented composites. **Composites Science and Technology** 27, 111-132.

Wakashima K. and Tsukamoto, H. (1991) Mean-field micromechanics model and its application to the analysis of thermomechanical behavior of composite materials, in press in **Materials Science and Engineering A**.

Tsukamoto, T. (1990) Tightness of fiber concrete, in **Darmstadt Concrete**, V. 5, pp.215-225.

3 "FROM FOREST TO FACTORY TO FABRICATION"

R. S. P. COUTTS

Division of Forest Products, CSIRO, Clayton, Australia

Abstract
This lecture will give an overview of the way fibres, derived from plants, have been used to reinforce cement based products. Time will be spent on explaining the unique structure of plant fibres, how they can be incorporated into existing processes for manufacturing fibre cement products, the properties of natural fibre reinforced cements, current usage and possible trends for the future.
Keywords: Natural fibres, Fibre cement products, Wood pulp fibres, Composite materials.

1 Introduction

By way of an apology for the lack of "theoretical principles" offered in this lecture I would like to state that first and foremost I am a chemist - and by default a pseudo-material scientist.

After emerging from alchemy the pioneering chemists persisted with "smelling, feeling and tasting" their products of reaction. I have carried a little of this philosophy into my transition from chemist to material scientist - only to the extent that I believe that if "you make 'em and break 'em" you can still achieve a successful fibre cement composite with a minimal understanding of the principles governing its performance.

Accepting the above we must still have an appreciation of the stress-strain characteristics and the need for achieving a critical fibre volume. As we shall see, the exact interpretation of the mechanism of strengthening presents considerable theoretical problems for natural fibre reinforced material. Several excellent texts, including those by Hannant (1978) and Piggott (1980), on the theoretical principles of fibre reinforcement are available which deal with such topics. To an audience such as is present today, the need for accurate testing is of fundamental significance for the use and progress of composite materials. Coupled with testing is the need to monitor durability to ensure long service life for the products.

Nature has provided mankind with a multitude of fibrous materials from plant, animal and mineral origins. Most of these materials in their various forms, have been utilised

Fibre Reinforced Cement and Concrete. Edited by R. N. Swamy. © 1992 RILEM.
Published by E & FN Spon, 2-6 Boundary Row, London SE1 8HN. ISBN 0 419 18130 X.

for generating a vaste array of products used to clothe, house and generally serve man in numerous ingenious ways.

Asbestos a naturally occurring mineral fibre "fell into disrepute" in the 1970's. Asbestos at that time was used as the major source of reinforcement in a fibre cement industry valued in billions of dollars. A replacement fibre was needed and, due to the large capital investment in existing manufacturing equipment, it was desirable that any replacement should be compatible with conventional processes. Fibres made from glass, steel, carbon, a range of synthetic organic polymers and plants were under investigation.

It is interesting to reflect on the scientific literature of the mid 1960's through to the early 1980's in which we find extensive studies on fibre cements based on glass, steel and synthetic organic fibres. Such fibres lent themselves to theoretical analysis because they possess homogeneous chemical composition, regular geometrical form and can be obtained with constant dimensions. Advances were made in applying linear elastic fracture mechanics, the rule of mixtures, interfacial phenomena, etc. to such systems. The prime objective of using fibre reinforcement too often was to achieve strength and not fracture toughness. At the present time greater importance is placed on the enhancement of the tensile strain capacity of the matrix, the ability of the fibres to inhibit unstable crack growth.

Natural fibres based on cellulose derived from trees, vegetables and grasses were, shall we say, "a little cumbersome"!! Fibres from even one given species are of varying chemical composition, irregular in their geometrical form and the dimensions of the fibres vary greatly (depending on age, part of plant, growing conditions or how they were extracted).

Although the use of plant fibre as a reinforcement in building products has been known for centuries its main use in developing countries has been to provide cheap, relatively low performance materials. The tendency in developed countries was to neglect the research of natural plant fibres for use in cement composites, that is until the "explosion" of interest, as evidenced by the scientific and patent literature, which occurred in the mid 1980's and is expanding to the present time.

Having stated the above I now hope to convince you of the great potential of natural plant fibre as a source of reinforcement for fibre cement products.

2 "FROM FOREST ...."

I apologise to my fellow researchers (of other natural plant fibres) for the above heading but as we will establish wood pulp provides most of the natural plant fibre for use in commercial fibre cement manufacture (at the present time). In fact in the fibre cement sheeting and roofing industries, of the developed nations, wood pulp fibre is used more than any other fibre (apart from asbestos). Hence, we will spend some time on its unique structure, while at the same time considering the similar nature of other plant fibres which should grow in importance in the near future.

2.1 Plant fibre structure and composition

Plant fibres can be derived from wood, bast, seed, leaf and grass, needless to say there are far to many plants to describe in this lecture. Cook (1980) reported on natural fibres of importance for concrete and cement composites and more recently a comprehensive text "Natural fibre reinforced cement and concrete" (Swamy 1988) has appeared which gives an excellent review of the work done with plant fibres upto the late 1980's.

In general terms it can be said that the fibre cells, which are themselves composite structures, have a cylindrical or ribbon-like shape made up of different layers with a hollow centre (lumen). The fibre can vary in length from less than 1mm to greater than 70mm. The diameter can be from less than 5μm to greater than 50μm. When the fibres are separated from each other they can collapse flat, if the walls of the fibres are thick, or the lumen may remain open. Fibres may develop a spiral twist along their length, which can be of significance during fibre composite failure.

Plant fibres contain cellulose, a natural polymer, as the main material of reinforcement. In a simplified description we can say chains of molecular cellulose are held together by hydrogen bonds to form fibrils, which in turn are held together by amorphous hemicellulose and form microfibrils. The microfibrils are assembled in various layers of differing thickness at different angles of orientation to build up the internal structure of the fibre, the main reinforcing element of interest to our research. Figure 1 shows a schematic structure of wood fibres.

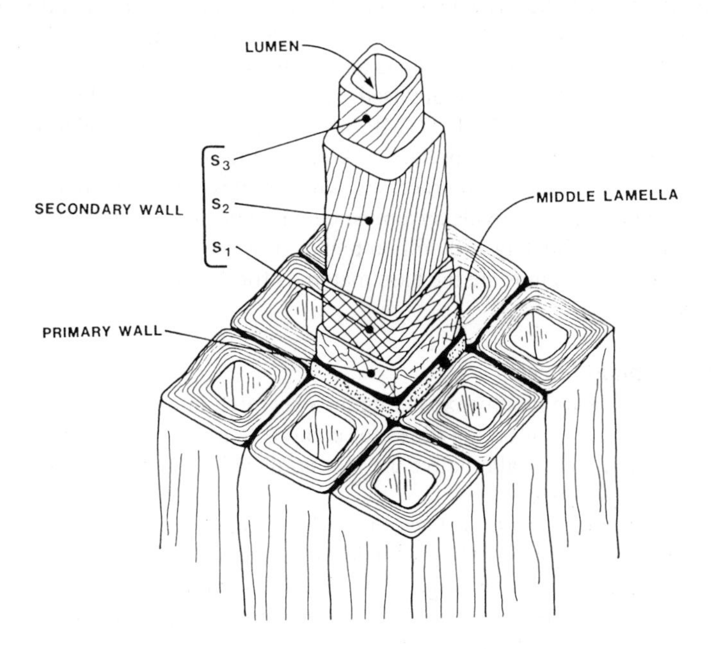

Figure 1. Schematic structure of wood fibre

Fibres are cemented together in the plant by lignin. In much of the work on plant fibres we are really discussing aggregates of fibres, which are often incorrectly called "fibres". As we will see, when we discuss durability, much of the lack of performance of certain "fibres" can be attributed to breakdown of these aggregates, due to the alkalinity of the matrix materials, and not to the fibre cell itself.

2.2 Plant fibre preparation

As stated, much of the work reporting the use of fibres really is referring to aggregates of fibres, thus we find that "fibres" are prepared in Central America, Africa, Asia, India and a number of developing nations by traditional methods, requiring high labour input, rather than by high capital cost technological means.

Retting of bast fibres such as jute, flax, kenaf, etc allows bundles of fibres to be freed from cellular tissue surrounding them by the combined action of bacteria and moisture, then the fibrous material is crushed, washed and dried. Decortication, a process used for leaf plants such as sisal and abaca, involves crushing and scraping the leaf to remove cellular tissue, then washing and drying.

Without doubt wood pulp fibre is the plant fibre of most commercial importance at the present time. A considerable amount of information is available; however, we will only spend a little time examining the basic types of processes used to generate fibres from wood.

Basically bulk wood can be treated mechanically by grinding, which literally tears the fibres apart. The yields can be as high as 97%; however, most of the non-cellulosic matter is still present. Mechanical pulping is an extension of ground wood pulping. Temperature, pressure and chemicals are used, in conjunction with mechanical action, to provide a range of pulps (eg. TMP, CTMP) used for newsprint and toiletries, again yields are high >80%. Chemical pulps such as kraft, neutral sulphite or soda are generated by sophisticated chemical processes, which remove the extractives, hemicelluloses and lignin, to provide an almost pure cellulosic reinforcing fibre in about 45% yield. Approximately 150 million tonnes of wood pulp fibres are produced globally per annum. Such pulps are the base materials for the production of paper and paper products.

2.3 Properties of plant fibres

The developing nations are rapidly expanding their research into indigenous plant fibres due to the need for cheap reinforcing fibres for composite material, particularly in the building industry, and, to find alternative uses for cordage fibres which are losing markets to synthetic polymers.

The more common fibres, such as abaca, sisal, flax, kenaf and bamboo, range in tensile strengths from say 50-500 MPa, with densities of about 1.2-1.5 g/cm^3. The elastic modulus of plant fibres can range between 5-100 GPa. Unfortunately researchers often report properties of the fibre when in fact they are studying aggregates of fibres with very different properties to those of the individual fibre.

Wood fibres vary in properties depending on whether they are softwoods or hardwoods. As there are numerous members one cannot give specific properties to each class. By way of example we will consider *Pinus radiata* (softwood) and *Eucalyptus regnans* (hardwood) as representatives of each class.

The physical dimensions are important in the application of fibres as reinforcement. Hardwood fibres are much shorter (av. 1.0mm) and narrower (av. 20μm) than softwood fibres (av. length 3.5mm). With softwoods there is a difference in fibre diameters between early and late wood (av. diameter 45μm and 13μm resp.). Hardwood fibres generally have a higher relative cell wall thickness than do early wood softwood fibres. This implies hardwood fibres are stiffer and have greater resistance to collapse.

As formulations are usually prepared by mass rather than volume of ingredients the number of fibres must be considered. The number of hardwood fibres per unit mass is always much larger than that of softwoods eg in the same mass of *E.regnans*, as of *P.radiata*, there would be seven or eight times as many fibres.

Theoretical calculations suggest the modulus of elasticity of cellulose could be as high as 150 GPa. Experimental values for single fibres vary with change in fibril angle, drying restraints and defects but range between 10-100 GPa.

Tensile strength of single wood fibres vary for the same reasons as elastic modulus. Page (1970) reported values of 2000 MPa for defect-free black spruce fibres with a zero fibril angle; this could be considered as a maximum value. A more realistic number would be in the range 500-1000 MPa.

The flexibility of a fibre is of great importance during the preparation stage of a composite material and also during the process of composite failure. A point of significance is that mechanical pulps are 20-30 times stiffer than chemical pulps from the same species.

The energy needed to form a fracture surface is called work of fracture, and is related to fracture toughness by consideration of the area of the surface formed during fracture. Gordon and Jeronimidis (1980) showed that, weight for weight, the strength and stiffness of wood along the grain compares well with the best engineering metals, as does its work of fracture across the grain.

Earlier Page *et al* (1971) showed that for single wood fibres a pseudo-plastic deformation took place in tension. Wood fibres are hollow tubes composed of layers of fibrils wound in a steep spiral (see Fig 1), and so behave as a spirally wound fibre reinforced composite tube. Under axial tensile strain such structures can fail by buckling due to the induced shear stresses. Such a failure mechanism can lead to high values of fracture toughness. This process will be seen to be important in wood fibre reinforced cements, in generating fracture toughness in the composite, as fibres fail in tension.

Plant fibres are affected by water. Plant fibres have cellulose as the primary component of the cell wall and the crystalline microfibrils are the elements which give the fibre its tensile strength; however, there are disordered zones which are believed to play a significant role with respect to mechanical properties. Hemicellulose and lignin act as matrix materials in wood fibres and are generally believed to be amorphous. They are hygroscopic thermoplastic substances and so are affected by humidity and temperature, this in turn affects mechanical properties of the fibres.

2.4 Refining wood fibres

When we discuss manufacturing processes the significance of refining (or beating) wood fibres will become apparent. Changes observed in fibre structure, as a result of mechanical action on the fibrous material, can depend on the type of refiner or beater, the refining conditions used, the fibre type (hardwood or softwood) and the pulp type (chemical or mechanical). The main effects which are observed are internal and external fibrillation, fines formation and fibre shorting.

Internal fibrillation can be considered by using a piece of rope as an analogy. Rope is a helical wrap of stands which themselves are helical wraps of fibre aggregates. If we twist the rope in the direction of the helical wrap the rope appears stiffer; likewise, if we twist in the opposite direction the rope unravels (or delaminates) becoming an open structure and "floppy"; such is the case with internal fibrillation of individual fibres.

External fibrillation can be observed by scanning electron microscopy (SEM) and is similar to brushing out the surface microfibrils, making the fibre surface "hairy" and more readily able to bond.

Fines are generated when microfibrils or section of the surface of the fibre are actually removed by the mechanical action of the refiner.

Fibre shortening results from the cutting action of the blades or discs present in the machinery and is an undesirable feature of refining.

Refining plays an important role in producing a large surface area for fibre-fibre and fibre-matrix bonding. More importantly, refining can assist in controlling the drainage rates of processing liquids during the fabrication of products. This is one of wood pulp fibres unique advantages over other fibres in that it emulates asbestos fibres in the ability to control drainage during fibre cement manufacture or conventional equipment.

3 " ...TO FACTORY..."

The utilisation of plant fibres as reinforcement for cement products falls into to two well defined areas - firstly, low cost, low performance materials made by labour intensive techniques, and secondly, high performance materials made by conventional fibre cement technology.

Most research on plant fibre (other than wood pulp) reinforced cements and concrete has been conducted using labour intensive techniques. Extensive studies by scientists from both developing and developed nations have resulted in a considerable number of formulations being prepared with a variety of matrix materials, fibre types, mixing conditions and curing regimes. Most of the research has been involved with fibre concrete and will not be discussed in this presentation, however, some work on sheet material and roofing is under investigation.

Sisal and coir (coconut fibre) are two of the most studied fibres but, bamboo, jute, hemp, various reeds and grasses, have also been considered (Swamy 1984, 1988). The Swedish Cement and Concrete Institute has been involved with sisal products for many years especially with respect to durability (Gram 1983).

Appropriate Technology International (ATI) and Save the Child Federation (SCF) have actively promoted fibre reinforced concrete roofing tiles since the 1970's. Thousands of tiles were made in countries such as Kenya, Sri Lanka and Latin America using indigenous plant fibres as reinforcement. Based on their findings both ATI and

SCF have concluded such fibre reinforced sheets are unsuitable, as currently manufactured, for rural housing (Lola, 1986).

3.1 Wood pulp fibre cements

As this topic is of considerable interest to the present audience we will spend more time with details. Mind you I can appreciate that with such a mixed audience with different *fibre allegiance* the interest may be *for* or *against* !!

The amount of data available on wood fibre reinforced cement (WFRC) products, in the scientific literature, has been limited until recent times due to the fact that manufacturing interests had been responsible for much of the preliminary work and for commercial reasons had retained the knowledge in-house or locked away in patent literature.

Unfortunately, due to the difficulty in handling theoretical treatments involving wood fibres there was less interest from the academic fraternity than in say glass, steel or synthetic organic fibres. As stated at the start of this lecture I have no intention of presenting such theoretical data at this time. It suffices to say the substance will be handled in a pragmatic fashion.

3.2 Fabrication processes

The manufacture of asbestos fibre cement products is a mature industry - Hodgson (1987) suggests that over 1100 sheet machines and 500 pipe machines are in production around the globe, with total capital investment in excess of A$4.6 billion.

The Hatschek process (or wet process) is the most widely used method of production (see Figure 2). The manufacturing techniques are closely related to conventional heavy paper and board making processes. An aqueous slurry of asbestos and cement matrix, about 7-10% solids by weight, is supplied to a holding tank which contains a number of

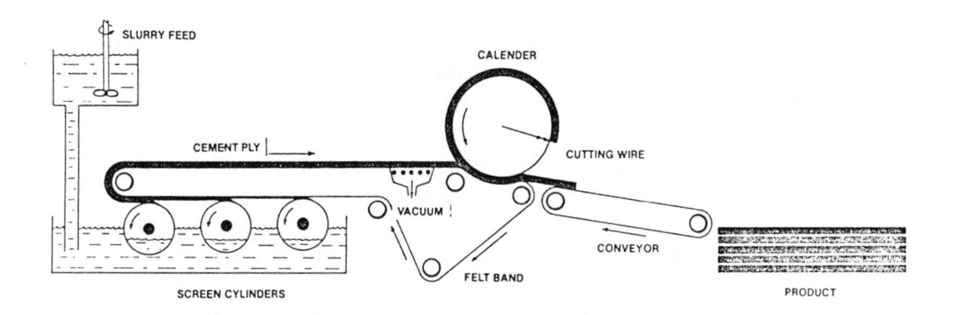

Figure 2. Schematic of Hatschek process

rotating screen cylinders. The cylinders pick up the solid matter removing some of the water in the operation. An endless felt band travels over the top surfaces of the cylinders and picks up a thin layer of formulation from each cylinder. The built-up laminated ply then travels over vacuum de-watering devices which remove most of the water. The formulation is then wound up on a steel calendar, or assimilation roll, until a product of desired thickness is formed. The material is further compressed by pressure rolls which are in contact with the assimilation rolls.

For sheet production the layer built up on the assimilation roll is automatically cut off and drops onto a conveyor to be transferred to a stack for curing. If corrugated roofing is to be made, the flat sheet is taken off to a corrugating station where the sheets are deposited onto oiled steel moulds for shaping.

Pipe machines are similar to the Hatschek process but usually have only one or two vats in series. The pressure imposed on the mandrel by the press rolls is much greater than for sheet production, so as to form a dense product. The machine may be stopped while the mandrel carrying the pipe is set to one side for pipe withdrawl. This process is often referred to as the Mazza process.

The Magnani (or semi-dry) process can be used to prepare pipes and corrugated sheet. This process has the advantage that it can provide a greater thickness of material at the peaks and troughs of the corrugations and so increase the bending strength. The thick slurry (about 50% solids) of this process can flow uniformly and directly onto a felt conveyor which passes over numerous vacuum boxes to dewater the formulation. In the case of corrugated roofing the felt is compressed over a corrugated former by a shaped roller. Pipe formation is similar to the Mazza process.

Injection moulding is now tending to replace the hand moulding of green sheets (from the Hatschek process) for the manufacture of special fittings. A slurry of 40-50% solids is pumped into a permeable mould and then subjected to pressures, in excess of 20 atmospheres, in a hydraulic press via a rubber diaphragm. The mix is dewatered by this process of pressure filtration, and then has sufficient green strength for the product to be demoulded by means of a suction lifting pad, and transferred to a pallet for curing. The operation is very fast.

The formulation of the matrix, and hence the cure of the product, has varied from country to country and between companies within a country. The formulations remain confidential to the company or its licensees and only general details will be discussed here.

The autoclaved curing process has always been favoured in Australia and the USA, and in some European countries for the manufacture of pressure pipes.

In the autoclave process, the matrix is usually a mixture of ordinary Portland cement (OPC) and finely ground sand (silica), or lime and silica. The product, after an initial pre-cure period in air, is cured in an autoclave in a steam environment, say 8 hours at 170-180°C. The cured sheets are virtually at full strength after autoclaving and can be dispatched from the factory in a short time.

By contrast the more traditional air-cured products require 14-28 days of air-curing before they can be dispatched, this involves considerable stock inventory. The process is lower in capital outlay, as no high pressure autoclaves and steam raising plant are required; however, cement is more expensive than silica, and therefore material costs are higher.

3.3 Asbestos alternatives

We have reviewed generally the processes used to manufacture asbestos fibre cement and can note a number of requirements of replacement fibres, if we wish to use existing capital intensive equipment.

For the Hatschek process a replacement fibre must be water dispersable, in a relatively dilute slurry, and able to form a film on the screens. At the same time the fibre must be able to resist chemical attack due to the high alkalinity (~pH 13) of the matrix. If the product is to be autoclaved resistance to temperatures above 170°C is also required. The basic essentials of cost, availability and mechanical performance of the fibre are obvious.

As has been reported glass, steel, carbon, synthetic organic as well as wood pulp fibres have been under examination for use in cement systems. We will look at a comparison of the properties of these fibres as possible asbestos replacements in Table 1.

Table 1. Comparison of properties of fibres for possible asbestos alternatives

Fibre	Alkali resist.	Temp. resist.	Process. ability	Strength	Toughness	Cost
Wood pulp (chemical)	1	1	1	1	1	3
Wood pulp (mechanical)	2	2	2	2	3	3
Polypropylene	1	3	3	3	3	2
PVA	1	3	3	1	1	2
Kevlar	1	1	2	1	1	1
Steel	1	1	3	3	3	2
Glass	3	1	3	3	3	2
Mineral fibre	3	1	3	3	3	3
Carbon	1	1	3	1	1	1

1. High, 2. Medium, 3. Low.

For countries committed to autoclaved products the combination of high alkalinity and high temperature eliminates most fibres apart from wood pulp, steel, carbon and aramid fibres. The cost of the latter two is almost a factor of ten higher than wood pulp and so look unattractive. Steel fibres have processing problems.

If one considers air-curing fibre cement, to eliminate the temperature problems, there are still processing limitations. The inorganic fibres eg steel and glass tend to be too stiff or dense to perform well during film-forming from dilute slurries; while the organic fibres lack a surface suitable for bonding to the matrix and/or introduce drainage problems. Mixtures of organic fibres (mainly PVA) and wood pulp fibres are successfully used to produce air-cured products in Europe.

3.4 Mechanical properties of WFRC

It will be seen that, when suitably pretreated by refining, wood fibres can afford a strong, tough and durable fibre cement, when produced commercially by traditional slurry/dewatered systems followed by autoclaving. Such WFRC formulations can be used for the production of flat sheeting, corrugated roofing, moulded products and low-pressure pipes which traditionally have used asbestos fibre. It will be demonstrated that some times the laboratory experiments are misleading with respect to the manufacturing processes and care must be taken in extrapolating the laboratory results into production.

Since we are limited it time and space those interested in the theoretical presentation of this topic should consider the text edited by Swamy (1988). I will only discuss selected results, relating to products containing wood fibre (and some plant fibres) as the sole reinforcement for cement matrices, in order to give an appreciation of various effects. These studies have included chemical and mechanical pulps of both hardwoods and softwoods.

Although wood fibres are cheap and readily processed they have the disadvantage of being hygroscopic. The composite properties are altered by absorption of water and, for this reason, extensive testing when both wet and dry is required.

WFRC products are generally loaded in bending and so flexural strength has more meaning than tensile strength in the characterisation of these materials.

At CSIRO there was an interest in using high yield pulps (TMP, CTMP), as an alternative to chemical pulps, for reinforcing fibre cements. Such pulps make less demand on the forest resources for a given quantity of pulp (yields twice that of chemical pulps), less problems with effluent treatment, chemical requirements are much lower and processing plants are economical at a smaller scale.

Coutts (1986) reported that mechanical pulps of *P.radiata* in general were unacceptable as cement reinforcement when autoclaved (with MOR's less than the matrix), but, when air-cured had flexural strengths greater than 18 MPa at 8-10% by mass of fibre. This compares poorly when one notes flexural strengths of WFRC's containing chemical pulps of *P. radiata* are in excess of 20 MPa when autoclaved and 30 MPa when air-cured. When autoclaving mechanical pulps the high temperature and alkalinity virtually "chemically pulps" the fibres releasing extractives of polysaccharides and wood acids, which "poison" the matrix near the fibre causing poor interfacial bonds. Air-curing is less drastic, with respect to chemical attack, hence better properties are evident in the final composites.

Although it was documented in the laboratory that kraft wood fibre was effective as a reinforcement in a cement matrix (CSIRO, 1977), the pulp performed poorly on a pilot-plant Hatschek machine because the fibres were unable to form a web capable of retaining cement and silica particles. The open nature of the web permitted rapid drainage, with loss of matrix, resulting in low product strength.

A collaborative project, between CSIRO and James Hardie Industries, starting in

1978 resulted in laboratory data which demonstrated the benefit of refining wood fibres for use in WFRC materials (Coutts and Ridikas, 1982). The breakthrough that made commercial production possible came about by the work of the Hardie's team in adapting the fibre refining step to suit the Hatschek machine. Before launching the product in 1981 over 50,000 sheets had been prepared and tested on the pilot-plant, and about $8 million invested in installing refining equipment in the factories.

The effect of moisture on the strength properties of WFRC composites is of importance, and early on in the research variations in test conditions produced variations in test results. It was evident that standard conditions must be adopted. The flexural strength of laboratory WFRC's can be reduced to as low as 50% of dry strength values. In the case of commercial products the reduction is considerably less but is taken into consideration for product application.

Recently Fördös and Tram (1986) reported WFRC's containing microsilica with excellent strength values ranging between 25-55 MPa. These workers used very high stack compression pressures, of approximately 20 MPa, whereas most results usually report pressures of approximately 2-3 MPa. Recently Coutts and Warden (1990) demonstrated the effect on compaction on the properties of air-cured WFRC and showed that flexural strength increased with casting pressure without resulting in a reduction in fracture toughness.

Few workers have reported the elastic modulus of WFRC materials. Mai *et.al.*(1979) have shown that both tensile and bending moduli are reduced from approximately 13 GPa to 9 GPA as the fibre content increases from 2 to 10% fibre by mass.

Fracture toughness is perhaps the most important property for a building material. Although strength and stiffness are important the ability of a material to absorb impact during handling can decide whether it will find an application in the market place.

Fibre type, pulping method, refining conditions and test conditions all have an effect on the fracture toughness results of a given formulation. As the fibre content increases up to about 8-10% by mass the fracture toughness increases rapidly then starts to taper off. Values of fracture toughness in excess of 60 times matrix values can be obtained with 10% fibre by mass, fracture toughness increases even further when tested wet. This matter will be discussed when we consider bonding and microstructure.

TMP pulps when compared with chemical pulps of the same species tend to produce WFRC materials with fracture toughness values less than half that of the chemical pulp reinforced composite; this has been explained in terms of fibre number and fibre morphology (Coutts, 1986). The variation of fracture toughness values between softwoods and hardwoods can be attributed to fibre length and fibre morphology, which we will see is so important for fibre pull-out which takes place during failure under load.

The physical properties of WFRC's have a considerable influence on their acceptability for use in the construction industry. If a product is strong and tough and has low density it will be preferred by the workers, who handle such products on the building site, compared to similar materials that are dense. At the same time due consideration must be given to water absorption, for as the density is lowered, the void volume increases with an associated potential increase in water absorption. Thus a load on a structure may be considerably increased should the material become wet, with more than 30% increase in weight occurring in some laboratory cases.

High temperature mechanical pulps are very stiff, compared to chemical or low temperature mechanical pulps, and cause poor packing as the fibre content increases. As voidage increases with poor packing so the density decreases and water absorption increases.

Matrix material also affects the density and water absorption. Air-cured WFRC's are more dense than autoclaved materials.

3.4 Bonding and microstructure of WFRC's

As well as the physical properties of fibre and matrix, a major factor which controls the performance of the composite is the type and arrangement of bonds linking the two materials. The interface (the region of intimate contact between fibre and matrix) plays the dual role of transmitting the stresses between the two phases and of increasing the fracture energy of the composite by deflecting cracks and delocalising stress at the crack tip.

The interfacial bond itself can be physical or chemical in nature, or a combination of both. Too strong a bond between fibre and matrix results in a brittle material which is strong, whereas a weak bond results in a tough material lacking strength.

The chemistry and morphology of cement matrices have been well documented and will not be considered further, apart from stating that cement is strongly alkaline and presents metal hydroxy groups at its surface. Similarly cellulosic fibres contain hydroxyl groups at the surface; thus, it is feasible to believe that hydrogen bonding or hydroxide bridges may play a major role in the bonding of WFRC's. Coutts and Kightly (1982, 1984) used SEM and considerable wet/dry testing data to hypothesis that hydrogen bonds are significant in affecting the mechanical behaviour of WFRC composites.

Wet or dry, wood fibre has about the same tensile strength, but its stiffness is considerably lower when wet. Thus a dried WFRC composite has stiff highly contorted fibres locked into a rigid cement matrix which could be bonded at the interface by a large number of hydrogen bonds or hydroxy bridged sites. This system when stressed can transfer the stress from matrix to fibres via the many interfacial bonds, and hence sufficient stress may be passed on to the fibre, after the matrix has cracked, to cause the reinforcing fibre to fracture under tensile load.

On the other hand, in a moist sample, the hydrogen bonds or hydroxy bridges between fibres and between fibre and matrix are destroyed (by insertion of water molecules between the bridges); and, at the same time, the cellulosic fibres become swollen by absorbed water and become less stiff. Under stress this system allows the fibres to move relative to each other and the matrix. However, due to the pressure of swelling and the highly contorted assembledge of fibres, considerable frictional forces are developed. This frictional energy contributes to the observed high increase in fracture toughness. If the forces are built up over sufficient length of a fibre, they may fracture. It is noted that more of the fibre population are pull out from the matrix without failing in tension.

As well as chemical bonding aspects of wood fibres, we must also consider mechanical bonding potential. Most of the theoretical data on fibre reinforcement is based on smooth, cylindrical fibres of uniform shape and dimensions.

Maximum fracture energy is often achieved if frictional energy is dissipated via fibre pull-out. Wood fibres are relatively long compared to their diameter and hence have aspect ratios of 60-200 (depending on whether they are hardwoods or softwoods, early wood or late wood fibres), but, more importantly, the fibres are hollow and can collapse to ribbons and at the same time develop a helical twist along their length (like a cork-screw). When fibres such as these are used to reinforce a brittle matrix, an asymmetrical process will be taking place during pull-out (after interfacial debonding has occurred).

In classical pull-out of straight fibres (glass, steel or synthetic organic) the forces are symmetrically distributed around the fibres; in the case of the contorted wood fibres, the leading edge of the helical fibre can experience considerable compressive forces resulting in a ploughing action, which can damage fibre or matrix, resulting in increased fracture surfaces and hence increased fracture energy.

The effect of refining the fibres results in improved mechanical properties, as well as better processing during manufacture. This phenomenon can again be considered in terms of mechanical bonding, in that the external surface of the fibre is "unwound" and the microfibrils so formed offer extra anchoring points by which the fibres can accept stresses from the matrix.

Bonding and microstructure go "hand in hand", but microstructure is best discussed with visual aids and must wait until the lecture, when a number of scanning electron micrographs will be presented.

3.5 Durability of plant fibre cements

Considerable doubt has been cast on the ability of natural fibres to resist deterioration in cement matrices, yet no evidence has been put forward to support the claims. When poor mechanical performance of the composite has been offered as confirmation of fibre failure, due consideration should be given to the potential of the fibre to "poison" the matrix surrounding the fibre, resulting in weak interfacial bonding. An extensive review by Gram (1983) reflects this uncertainty.

As stated earlier in this lecture too often aggregates of fibres have been used as the reinforcement and the high alkalinity, coupled with cycles of wetting and drying, "pulp" the fibre bundles resulting in "poisoned" cement and weak interfacial bonds, the end result low durability. On the other hand there are many claims which suggest that natural fibre reinforced cement products are durable after 30 years of service.

Sharman and Vautier (1986) have done some excellent work on the durability of autoclaved WFRC products at the Building Research Association of New Zealand. They discuss the possible ageing mechanisms of corrosion, carbonation, moisture stressing and microbiological attack.

Akers and co-workers (1989) have published a series of papers which discuss the ageing behaviour of cellulose fibres both autoclaved and air-cured, in normal environments and accelerated conditions. Testing had taken place which showed exposure of WFRC composites (air-cured or autoclaved) to natural weathering led to an overall increase in flexural strength and elastic modulus after 5 years. The same workers found that air-cured WFRC products when aged, either normally or by accelerated means, showed a marked reduction in fracture toughness. The ageing of autoclaved materials did not result in mineralisation of the fibres and the expected loss in fracture toughness.

The need to use synthetic fibres in air-cured products was apparent, however, with ageing there appears to be an increase in the interfacial bond, between synthetic fibre and cement, which leads to greater fibre failure, rather than pull-out, and thus higher strength but lower fracture toughness.

A general picture is emerging as more studies are conducted that the autoclaved WFRC products are more durable than the air-cured hybrid composites which contain mixtures of cellulose and synthetic organic fibres.

4 "...TO FABRICATION"

4.1 Production rates

The Western world produces about half of the 28 million tonnes of fibre cement products made each year. Only a little over one million tonnes of these materials are asbestos free.

This symposium will appreciate that progress towards asbestos-free cement products still has a long way to go. However, it can be appreciated that some asbestos free systems have been commercialised, the two major fibres to have been adopted are PVA and wood pulp fibre.

Australia adopted the autoclaved process using wood fibres as a viable commercial alternative to asbestos fibres as early as 1980 (Anon 1981). James Hardie Industries have an annual production in Australia, New Zealand and Malaysia of about 350,000 tonnes. They have produced of the order of 3 million tonnes since 1980. Their technology has been introduced into Europe, South Africa and the USA.

Eternit of Denmark , Partek of Finland and Cape Boards of UK would produce another 250,000 tonnes of wood fibre reinforced cement products each year. Other European countries produce a mixed range of products and it is estimated another 150,000 tonnes might be manufactured each year which is reinforced by cellulosic fibres.

PVA fibres require the support of asbestos or wood pulp fibres to aid its dispersion during the manufacturing process, as it tends to "ball-up". Due to the cost of the fibre and to its tendency to cause over drainage properties of the slurries, if present in high loadings, it has been practice to incorporate more support fibre (wood pulp fibre or asbestos) than actual PVA fibre. Thus, of the estimated 380,000 tonnes of PVA fibre reinforced cement products produced each year, mainly in Europe, a considerable amount of wood pulp fibre is involved in the manufacturing process.

It has been estimated that some 8 million tonnes of non-asbestos fibre cement has been produced on conventional machinery by large manufacturing firms. More than 5.5 million tonnes has been reinforced solely by wood pulp fibre. Therefore, in a commercial sense wood pulp fibre, a natural plant fibre, has proved itself the preferred fibre of substitution for asbestos - at the present time.

4.2 Applications

Wood fibre reinforced cement products have a multitude of applications in the building and construction industry which will be highlighted by a number of slides during the lecture. For convenience, and with no personal bias,I will illustrate this section with examples from Australia.

It suffices to say that the autoclaved sheet product has internal and external cladding applications. All products are easy to cut and fix. They are unaffected by sunlight and steam and will not split or rot. The hard wearing surfaces will readily accept PVA and acrylic paints. The fibre cement sheets will not burn and have the following "Early Fire Hazard Indices" tested to Australian Standard AS 1530 Part 3 - 1982: Ignition, 0; Flame spread, 0; Heat evolved, 0; Smoke developed, 0. The product is rodent proof and has been tested by CSIRO and found to be totally unaffected by termites.

The WFRC sheeting is also used as underlay for cork and vinyl tiles. Another application is for bracing sheets for timber stud walls in areas prone to high winds and cyclonic conditions. WFRC roofing shingles are virtually maintenance-free and easily fixed to recreate the appearance and character of a traditional slate or shingle roof.

High-pressure techniques have been used to generate materials with increased density and improved mechanical performance. This product is used in wet areas, such as bathrooms, laundries, external balconies, decking and pool surrounds. The structural properties of this product find use in bridge construction.

The process for manufacturing compressed building products results in a higher ratio of wet to dry strength than occurs for non-compressed products. This suggested to James Hardie Industries that a pipe-making process similar to the compressed sheet process might be used. They have produced low-pressure pipes suitable for drain and irrigation piping. The Mazza process is the basic method of manufacture, along with autoclave curing.

During 1985, the newly developed pressed, corrugated sheet was introduced, when a new $5M press was commissioned. This product was the last of the building products to contain asbestos fibres, and so with this plant operational all fibre cement products in Australia were reinforced with wood pulp - a natural plant fibre.

5 The Future

Two decades of intensive study of glass, steel and synthetic organic polymer fibres in cement systems have resulted in an understanding of the theoretical consequences of using these fibres as replacement for asbestos. This knowledge bank is invaluable and necessary for our progress; however, we have seen that to manufacture a successful asbestos-free product we have used plant fibres - the least recognised of the alternative fibres.

The future of sheet and roofing fibre cement products must be natural plant fibre reinforcement !! This statement is supported by the facts that two of the most important considerations are cost and availability - plant fibres fulfil these conditions - provided adequate performance is also present in the fibre.

We look at those countries which are still producing asbestos fibre cement products such as South America and India and see they possess a wealth of natural fibre which at the present time is often considered waste material. We are now seeing a rapid expansion in the interest of these countries to study and utilise this often untapped source of reinforcement (Swamy 1988).

Too often the plant fibre is used in a partially prepared state, there is more effort needed to examine alternative fibre preparations instead of the traditional methods (eg retting). Novel, cheap methods of pulping plant material, for use in developing countries, must be researched. The high capital cost pulp mills currently in use are beyond the financial capacity of countries attempting to house the millions presently without substantial shelter.

The performance and durability of autoclaved WFRC products would suggest that more manufacturers will choose this route in preference to the air-cured one. In developing countries alternative, low capital cost processes will be needed to manufacture plant fibre reinforced cement products.

6 References

Andonian, R., Mai, Y.M. and Cotterell, B. (1979) Strength and fracture properties of cellulose fibre reinforced cement composites. **Int. J. Cement Comp.**, 1, 151-158.

Akers, S.A.S. Crawford,D. Schultes,K. and Gernka (1989) Micromechanical studies of fresh and weathered fibre cement composites. **Int.J. Cement Comp. Lightweight Concr.**, 11, 117-123. (and refs.therein).

Anon. (1981) New - a wood-fibre cement building board. **CSIRO Industrial Research News**, No 146, 1-2.

Cook, D.J. (1980) Concrete and cement composites reinforced with natural fibres, **Proc. Symp.on Fibrous Concrete** April 1980. Construction Press Ltd., Lancaster, pp. 99-114.

Coutts, R.S.P. and Ridikas, V. (1982) Refined wood fibre cement products. **Appita.**, 35, 395-400.

Coutts, R.S.P. and Kightly, P. (1982) Microstructure of autoclaved refined wood fibre cement mortars. **J.Mat. Sci.**, 17, 1801-1806.

Coutts, R.S.P. and Kightly, P. (1984) Bonding in wood fibre-cement composites. **J. Mat.Sci.**, 19, 3355-3359.

Coutts, R.S.P. (1986) High yield wood pulps as reinforcement for cement products. **Appita.**, 39, 31-35.

Coutts, R.S.P. and Warden, P. (1990) Effects of compaction on the properties of air-cured wood fibre reinforced cement. **Cement & Conc. Composites.**, 12, 151-156.

CSIRO (1977) **Australian Patent** No. 512 457 (filed 15th August).

Fördös, Z. and Tram, B. (1986) Natural fibres as reinforcement in cement based composites. **in RILEM FRC 86** (eds R.N.Swamy, R.L.Wagstaffe and D.R.Oakley), Rilem Technical Committee 49-TFR, paper 2.9.

Gram, H-E. (1983) **Durability of natural fibres in concrete**. Swedish Cement and Concrete Research Institute, Stockholm.

Hannant, D.J. (1978) **Fibre Cements and Fibre Concretes**. John Wiley & Sons, Chitchester.
Hodgson, A.A. (1987) **Alternatives to Asbestos and Asbestos Products**. Anjalena Publications Ltd., Berkshire, England.
Lola, C.R. (1986) Fibre reinforced concrete roofing technology appraisal report. in **RILEM FRC 86**. (eds R.N.Swamy, R.L.Wagstaffe and D.R.Oakley), Rilem Technical Committee 49-TFR, paper 2.12.
Page, D.H. (1970) The chemistry and physics of wood pulp fibres. **Tappi** STAP No. 8, pp 201.
Page, D.H., El-Hosseiny, F. and Winkler, K. (1971) Behaviour of single wood fibres under axial tensile strain. **Nature**. (London), 229, 252-253.
Piggott, M.R. (1980) **Load Bearing Fibre Composites**. Pergman Press, Toronto.
Sharman, W.R. and Vautier, P.B. (1986) Durability studies on wood fibre reinforced cement sheet. in **RILEM FRC 86**. (eds R.N.Swamy, R.L.Wagstaffe and D.R.Oakley), Rilem Technical Committee 49-TFR, paper 7.2.
Swamy, R.N. (Ed) (1984) **New Reinforced Concretes**, Vol 2, Concrete Technology and Design, Surrey University Press.
Swamy, R.N. (Ed) (1988) **Natural Fibre Reinforced Cement and Concrete**, Vol. 5, Concrete Technology and Design, Blackie, London.

PART TWO

NEW FIBRES, FABRICATION, EARLY AGE AND STRENGTH PROPERTIES

4 COMPARISON OF FUNDAMENTAL PROPERTIES OF CONCRETE USING NEW- AND OLD-TYPE STEEL FIBRE

K. KOHNO, J. SUDA, and K. MIYAZAKI
Department of Civil Engineering, University of Tokushima, Japan
N. KAKIMI and M. SUZUKI
Department of Development, Igeta Steel Sheet Co., Ltd, Sakai, Japan

Abstract
In this study, the fundamental properties of steel fibre reinforced concrete (SFRC) using new-type of steel fibre having a flat and convex cross section were investigated and compared with those of SFRC using old-type of steel fibre. These properties evaluated include consistency, flexural strength, tensile strength, compressive strength, impact resistance and durability in sea-water. In addition, the dispersion and orientation factors were discussed using X-ray analysis. The results of these investigations show that flexural strength, tensile strength, impact resistance and sea-water resistance of SFRC using new-type steel fibre are higher than those of SFRC using old-type fibre. The dispersion and orientation factors are better for the new-type of fibre, although the concrete gives from 2 to 3 cm lower slump than SFRC using old-type steel fibre.
Keywords: New-Type Steel Fibre, Flexural Strength, Tensile Strength, Impact Resistance, Sea-Water Resistance, Dispersion.

1 Introduction

Steel fibre reinforced concrete (SFRC) has various excellent properties as a composite material; for instance, flexural strength, tensile strength, shear strength, toughness, impact resistance, crack resistance and resistance to frost damage are improved by the use of steel fibre. SFRC has been used in Japan in tunnel lining, pavement and gunned mortar (1). On the other hand, the workability of SFRC deteriorates with the increase of fibre content in the mixture, and the adoption of a fibre dispenser is necessary to avoid the occurrence of fibre ball during concrete mixing.

Recently, as new-type of steel fibre with a flat and convex cross section has been developed to improve dispersion (2). The fundamental properties of SFRC using new-type steel fibre (NSFRC) have been investigated and compared with those of SFRC using old-type steel fibre (OSFRC).

2 Outline of test procedure

2.1 Materials used

Fibre Reinforced Cement and Concrete. Edited by R. N. Swamy. © 1992 RILEM.
Published by E & FN Spon, 2-6 Boundary Row, London SE1 8HN. ISBN 0 419 18130 X.

(1) Steel fibre :

The shape of new-type steel fibre used in this study was that of a flat and convex cross section to prevent intertwinement among fibres and to facilitate their easier dispersion. Table 1 shows the shape, size and main properties of new-type steel fibre as compared with old-type sheared steel fibre。 The surface area of new-type steel fibre is

Table 1, Shape, size, surface area and tensile strengtg of new-type and old-type steel fibre

Type	Shape	Size (mm)	Surface area (mm²)	Tensile strength (MPa)
New-type steel fibre		Width Depth Length 0.20 x 1.25 x 30	87.5	84.3
Old-type steel fibre		0.50 x 0.50 x 30	60.5	84.3

greater than that of old-one. With the use of new-type steel fibre it is possible to avoid the occurrence of fibre balls as shown in Photo.1。

Photo. 1. Examples of fibre balls grown in SFRC using old-type steel fibre (Fibre content, 1.0 %)

(2) Cement, aggregates and chemical admixture :

Ordinary portland cement (specific gravity = 3.15, specific surface area = 3110 cm²/g and 28 day compressive strength = 42.2 MPa) is used。

Crushed sandstone having a maximum size of 15 mm was used as a coarse aggregate and river sand was used as a fine aggregate。

An air-entraining and water-reducing agent containing ligninsulfonic acid calcium was used as an admixture.

The main properties of cement, aggregates and chemical admixture used are shown in Table 2.

2.2 Mixture proportions of concretes used

The mixture proportions of SFRC used are shown in Table 3. Steel fibre

Table 2. Properties of cement, aggregates and admixture used

Type of material	Properties
Ordinary portland cement	Specific gravity = 3.15, Blaine's value = 3110 cm^2/g, 28-day compressive strength = 42.2 MPa
Fine aggregate *	Specific gravity = 2.60, Absorption = 1.96 %, F.M. = 2.81
Coarse aggregate **	Specific gravity = 2.56, Absorption = 2.28 %, F.M. = 6.54
AEWR agent	Specific gravity = 1.20, Liquid, Rigninsulfonic-acid calcium

Note) * River sand, ** Crushed sandstone

content by volume was 1.0, 1.5 and 2.0 percent. A 12 cm slump of OSFRC was aimed for, and the same mixture proportions were used for NSFRC in the first series. In the second series, mixture proportions involving a slightly higher water content were used to obtain the same slump as that of OSFRC. In the third series, one concrete mixture to be used for pavement, having a low slump of 4 cm, was used for a flexural strength test.

Table 3. Mixture proportions of concretes used

Type of mixture	Fibre content (%vol)	Maximum size (mm)	Slump (cm)	W/C (%)	s/a (%)	Cement content (kg/m^3)
New-type and old-type fibre (Same mixture)	0.0	15	12 *	51.4	50	350
	1.0	15	12	60.0	60	350
	1.5	15	12	62.0	63	350
	2.0	15	12	65.1	66	350
New-type fibre (Mixture with same slump)	1.0	15	12	61.4	60	350
	1.5	15	12	64.0	63	350
	2.0	15	12	67.4	66	350
Paving concrete	1.5	20	4	40.0	65	400

Note) AEWR agent ; 50 cc/cement=1kg, * Slump of OSFRC = 12 cm

2.3 Mixing of concrete and making of specimens

Concrete was mixed in a pan type mixer with 50 litre capacity. First sand and cement were put into the pan of a mixer and the mortar was mixed for 60 seconds, adding water and steel fibre. To add the steel fibre, a fibre dispenser was used to maintain a condition of stability, although with the new-type steel fibre it was possible to avoid the occurrence of fibre balls without a fibre dispenser. The mixing of SFRC was continued for 120 seconds after the addition of crushed sandstone.

Concrete was placed into cylinder molds of ø10 x 20 cm size and beam molds of □10 x 10 x 40 cm size by test items as are shown in Table 4, and was compacted by a vibrating table (100 Hz in frequency and 0.9 mm in amplitude).

Each specimen was cured in a water tank at 20 ± 2°C to the age of 28 days.

2.4 Testing of concrete

Table 4. Test items, shape and size of specimens and curing methods

Test items	Shape and size of specimens	Curing method
Flexural strength	□ 10 x 10 x 40 cm beam	20 ± 2°C in water
Compressive strength	ϕ 10 x 20 cm cylinder	20 ± 2°C in water
Tensile strength	ϕ 10 x 20 cm cylinder	20 ± 2°C in water
Impact resistance	□ 10 x 10 x 40 cm beam	20 ± 2°C in water
Sea-watwe resistance	□ 10 x 10 x 40 cm beam	20 ± 2°C in water, and wet and dry *

Note) * Specimens were cured in sea-watwe (pH = 7.8) for one day, and then cured in a chamber of 60°C for one day as 1 cycle (until 45 cycles).

(1) Slump and air-content of fresh concrete :

The slump test and air-content test of fresh SFRC were performed immediately after the mixing of the concrete with JIS A 1101 and 1128, respectively.

(2) Flexural strength, tensile strength and compressive strength :

A flexural strength test using a beam specimen was performed in accordance with JCI-SF4 " Method of tests for flexural toughness of fibre reinforced concrete " (3).

The tensile strength test and compressive strength test of SFRC were performed using JIS A 1113 and 1108, respectively.

(3) Impact resistance :

The impact resistance was determined by a ball drop test (weight of ball = 11.4 kg and height of drop = 5 cm) (4). The number of blows required to bring about a visible crack (first cracking) and to propagate it to the top surface of the beam (failure cracking) were counted. Three beam specimens of □10 x 10 x 40 cm were used for each mixture.

(3) Sea-water resistance :

The sea-water resistance of SFRC specimens was determined by an accelerated test using a repetition of wet in sea-water and dry, after specimens were cured in a water tank of 20°C up to an age of 28 days. This test involved the following conditions; soaking a specimen in sea-water at 20°C for one day and then drying it in a chamber at 60°C for one day made up one cycle. The changes in the dynamic modulus of elasticity of the specimens were measured under the given cycles.

2.5 Bond strength between fibre and matrix

In order to determine the bond strength between fibre and matrix, tensile mortar specimens were used in accordance with JCI-SF8 " Method of test for bond of fibres " (3).

The cement-sand ratio of the mortar mixture was 1 to 3, and the water cement ratio was 0.50. Mortar was mixed by a mortar mixer and placed into a briquet mold containing four fibers. The bond strength was determined by a small type tensile testing machine using deformation control.

2.6 Investigation of dispersion of fibers in matrix

The state of dispersion of the fibres in the concrete matrix was investigated by X-ray analysis。 Thin test specimens of a 10 x 10 cm size were taken from the concrete beam of □10 x 10 x 40 cm size. They were treated by X-ray portrait analysis under the application of a perspective method, and the dispersion and orientation factors of fibres were calculated.

The image analysis was made on the diagonal stripe part, which comprises the X-ray image divided into mesh and is hardly affected by a mold frame (Fig. 1), according to the following procedures.

The X-ray photograph is input in an image processing system (MIPS) as the image data. The image data is converted into a thin line image by using such approach as black and white inversion, noise elimination, and binary processing and so forth。

Finally numbers of line for each mesh were counted。

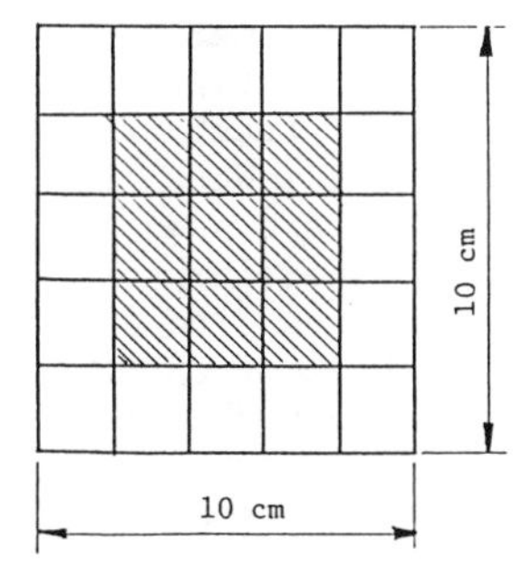

Fig. 1. Mesh division

(1) Dispersion factor calculation :

The dispersion factor of the steel fibre in each mesh is determined by the following equation (1)。

$$\alpha = e^{-\zeta} \qquad (1)$$

$$\zeta = \sqrt{\frac{\Sigma(xi - \mu)^2}{n}} \; / \; \mu$$

α : Dispersion factor
μ : Average number of steel fibres in a specimen
n : Number of specimen (Number of mesh: n = 9)
xi: Number of steel fibres in each mesh specimen

(2) Orientation factor calculation :

An orientation factor is defined as an effective ratio to an axis direction: the ratio of the length of a fibre when it is projected in a certain direction to the original length. The orientation factor was calculated using the following equation (2).

$$\beta = \sqrt{\frac{t^2}{(\frac{\ell}{N})^2 + t^2}} \qquad (2)$$

$$\ell = a/d$$

β : Orientation factor
t : Thickness of a specimen
ℓ : Total length of projected steel fibres
N : Number of steel fibre
a : Total area of projected steel fibres
d : Average thickness of steel fibres using the actual value

3 Test results and discussion

3.1 Effect of fibre type on slump

Kohno et al. (5) reported that the water content of SFRC mixture must increase at a rate of about 8 kg/m^3 with an increment of 0.4 percent fibre content per volume to maintain a constant slump. The mixtures used in this investigation show almost the same results, although the increasing ratio of water tends to be higher in the mixture of 2.0 percent fibre volume.

The effect of fibre type on the slump of fresh SFRC is shown in Fig. 2. When the same mixtures were used for both fibres, the slump of NSFRC became from 2 to 3 cm lower than those of OSFRC, as can be seen in this figure. The reason is considered to be the increase of surface area of new-type steel fibre compared with that of the old-type fibre. Therefore the water content of NSFRC must be increased from 5 to 8 kg/m^3 to maintain the same slump as that of OSFRC. The workability of fresh NSFRC is good during the handling of concrete.

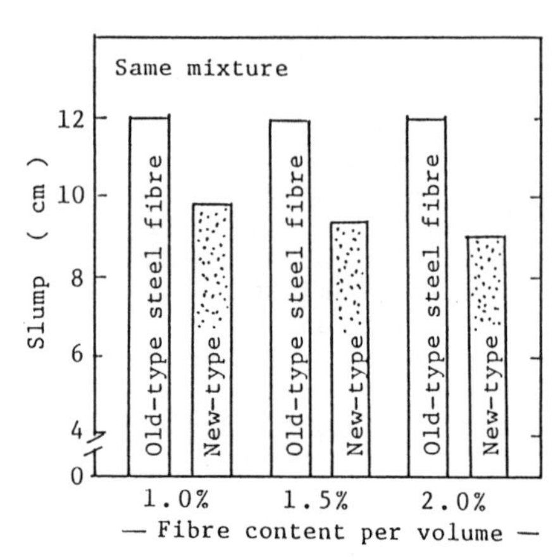

Fig. 2. Effect of fibre type on slump of fresh SFRC

3.2 Dispersion of steel fibre in matrix

The dispersion and orientation factors calculated by the results of X-ray portrait analysis are shown in Table 5. As can be seen in this table, the dispersion factor of NSFRC shows higher values compared with that of OSFRC in each fibre volume. An example of the state of fibre dispersion revealed by the X-ray portrait is shown in Photo. 2.

Table 5. Dispersion factor and orientation factor in SFRC

Fibre content (%vol)	Type of fibre	Dispersion factor	Orientation factor
1.0	New-type	0.761	0.520
	Old-type	0.752	0.696
1.5	New-type	0.763	0.521
	Old-type	0.734	0.649
2.0	New-type	0.800	0.481
	Old-type	0.778	0.651

On the other hand, the orientation factors of NSFRC become smaller than those of OSFRC. These values mean that the dispersion of steel fibre in concrete matrix is improved by the use of new-type steel fibre.

(1) New-type steel fibre

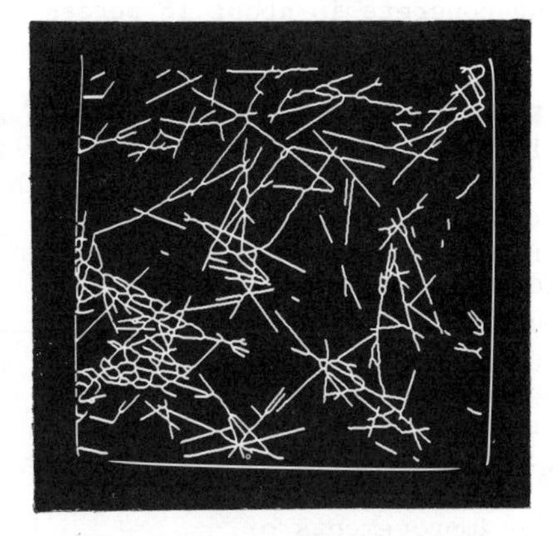

(2) Old-type steel fibre

Photo. 2. Example of fibre dispersion by X-ray portrait
(Fibre content, 1.5 %)

3.3 Effect of fibre type on flexural strength
The effect of fibre type on the flexural strength of SFRC is shown in Fig. 3 in the cases of similar mixtures having a different slump and mixtures requiring a slightly higher water content to maintain same slump, that is to say, mixtures with same slump.

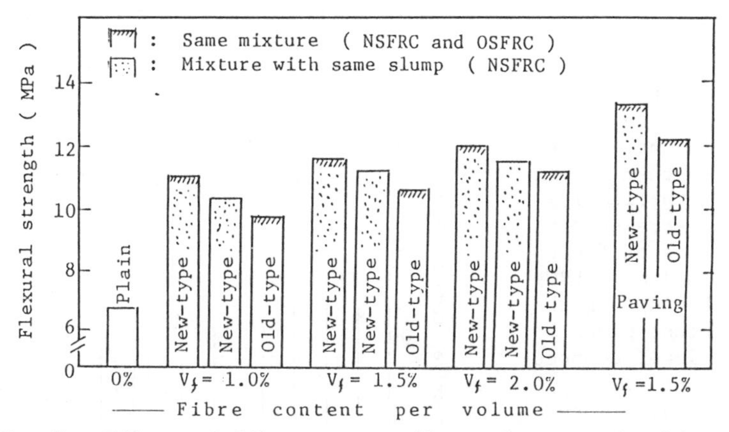

Fig. 3. Effect of fibre type on flexural strength of SFRC

As can be seen in this figure, the flexural strengths of NSFRC clearly increase as compared with those of OSFRC. When a fibre content of 1.0 or 1.5 percent per volume is used, the flexural strength of NSFRC becomes about 10 percent higher in the same mixture and about 5 percent higher in a mixture with the same slump, despite the increased water content. This could be due to bond strength improved and dispersion of the new-type steel fibre.

Moreover, the flexural strength of low-slump NSFRC mixture used for paving concrete is about 15 percent higher than that of OSFRC in the case of the same mixture (see Fig. 3).

3.4 Effect of fibre type on tensile strength

A splitting tensile test using a cylinder specimen of ϕ10 x 20 cm was used in this experiment, although this test is recommended for the first crack tensile strength (6).

Fig. 4 shows the test results for tensile strength of NSFRC and OSFRC in the case of mixtures with the same slump. The tensile strength of NSFRC is several percent higher in comparison with that of OSFRC because of the improvements of the fibre bond and dispersion in the concrete matrix. The tensile strength of SFRC clearly increases with the increment of fibre volume from 1.0 to 2.0 percent.

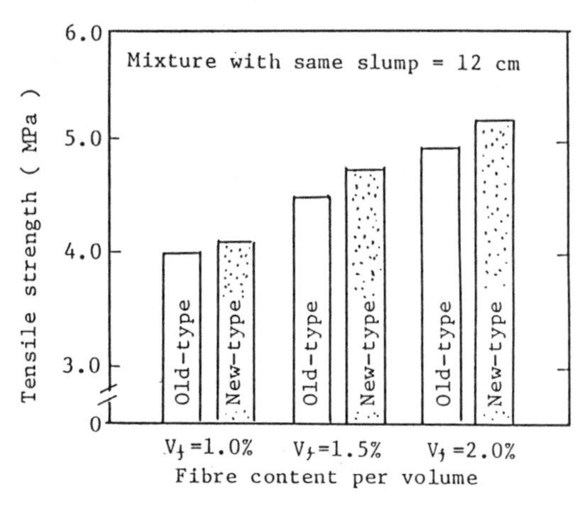

Fig. 4. Effect of fibre type on splitting tensile strength of SFRC

3.5 Effect of fibre type on compressive strength

As can be seen in Fig. 5, which shows the effects of fibre type and fibre content on the compressive strength of SFRC, the strength clearly tends to decrease with the increase of fibre content, and an NSFRC mixture having a higher water content gives a slightly lower compressive strength than OSFRC for each fibre content. However, the effect of fibre type on compressive strength is small in comparison with other strengths (7).

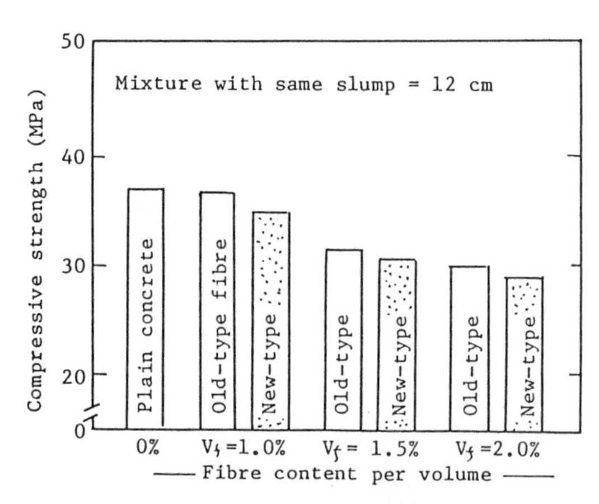

Fig. 5. Effect of fibre type and fibre content on compressive strength of SFRC

3.6 Bond strength between steel fibre and matrix

The bond strength between the steel fibre and mortar matrix is shown in Fig. 6. The bond strength of new-type steel fibre becomes twice the value of OSFRC. The reason is considered to be the increase in surface area and the wavy shape of new-type steel fibre. The surface area is 1.5 times higher than that of old-type fibre.

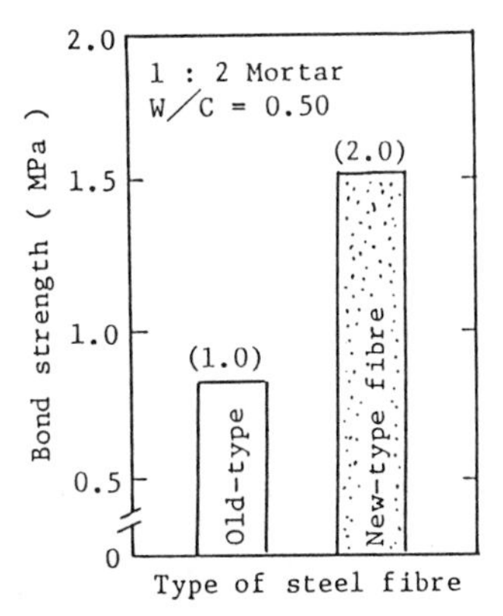

Fig. 6. Bond strength between steel fibre and matrix

3.7 Impact resistance

The results of impact resistance calculated by a ball drop test are shown in Fig. 7. The figures for initial cracking and rupture of SFRC are compared with the values of plain concrete (=1.0).

The impact resistance is clearly improved by the use of steel fibre, especially new-type steel fibre. It is considered that the improvement is caused by higher fibre bond and dispersion properties in the concrete matrix.

The plain concrete beam was ruptured at the same time as cracking, although SFRC resisted crack propergation after the initial cracking of the beam. The use of steel fibre ie effective for the improvement of the impact resistance of concrete

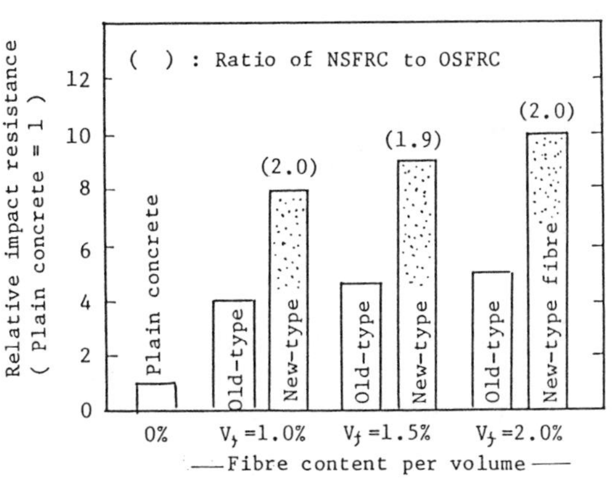

Fig. 7. Impact resistance of SFRC by a ball drop test

3.8 Sea-water resistance

Fig. 8 shows the change in the dynamic modulus of elasticity by the repetition of wet and dry treatment, using SFRC having fibre content of 1.0 percent. As can be seen in this figure, the dynamic modulus of elasticity of plain concrete tends to decrease with repeated cycles, although the values of SFRC show a tendency of increase. These phenomena mean that the sea-water resistance of SFRC is superior to that of plain concrete. The dynamic modulus of elasticity of NSFRC is a little higher than that of OSFRC at the same cycle, as the same mixture is used.

One weak point of SFRC in connection with sea-water is that the spots of steel rust appear in longer cycles in the surface of the concrete.

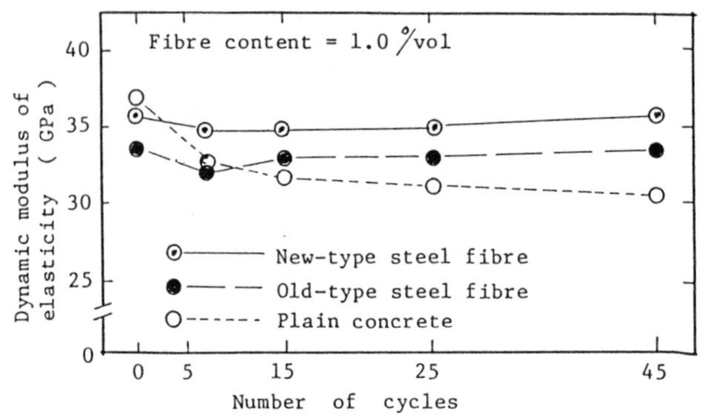

Fig. 8. Change of dynamic modulus of elasticity due to repeated treatment by an alternation of wet (sea-water) and dry conditions

4 Conclusions

The following conclusions can be drawn using the results of this investigation.

(1) When same mixture proportion is used, SFRC using new-type steel fibre gives a 2 to 3 cm lower slump than SFRC using old-type steel fibre.

(2) Flexural strength and tensile strength of SFRC using new-type steel fibre become from 5 to 10 percent higher than those of SFRC using old-type steel fibre

(3) The bond strength between new-type steel fibre and matrix is improved by the increase of surface area and the wavy shape of the fibre.

(4) The impact resistance determined by a ball drop test and the sea-water resistance determined by repetition of wet and dry conditions are improved by the use of new-type steel fibre.

(5) The dispersion and orientation factors of SFRC improve through the use of new-type steel fibre.

References

1. JSCE, Sub-Committee on Study of SFRC, " Recommendation on Design and Construction Works on SFRC " JSCE Concrete Library No.50 (Mar.,1983).
2. K. Kohno, N. Kakimi and M. Suzuki. " New-Type Steel Fibre and Field Test on SFRC " Cement & Concrete No.529, pp. 34-40 (Mar., 1991).
3. Japan Concrete Institute, " JCI Standards for Test Methods of Fibre Reinforced Concrete " JCI (Feb., 1984).
4. K. Kohno and S. Kawasaki, " Vibrating Compaction and Pressure Curing of SFRC " Journal of the Society of Material Science, Japan, Vol.29, No.318, pp. 260-265 (Mar., 1980).
5. K. Kohno, S. Nohda, N. Yamanaka and Y. Inaba, " Investigation on Mix Proportion of SFRC and Tests on Concrete Flags " Proc. of JCI Symposium on SFRC, pp. 72-75 (Nov., 1977).
6. ACI Committee 544, " Measurement of Properties of Fiber Reinforced Concrete ", ACI Journal, pp. 283-289 (July, 1979).
7. K.Kohno, " Review on Fibre Reinforced Concrete " Journal of the Society of Material Science, Japan, Vol.26, No.290, pp. 1061-1071 (Nov., 1977).

5 MICROWAVE PROCESSING OF FIBRE REINFORCED CEMENT COMPOSITES

J. PERA, J. AMBROISE and M. FARHA
Laboratoire des Matériaux Minéraux, INSA Lyon,
Villeurbanne, France

Abstract
The use of microwave energy as a source of heat to accelate curing of fibre reinforced cement composites (F.R.C.C.) was investigated. Tests were performed on composites containing metakaolin in order to prevent the deposition of portlandite at the glass-matrix interface. E Glass and A.R. glass fibres were used at a dosage of 5 % by weight of binder. F.R.C.C. were premixed without vibration or pressing.
Specimens were treated with different levels of microwave power, shortly after casting. The flexural strength was then determined.
When the composites containing 20 or 30 % metakaolin were respectively treated for 120 or 90 minutes on a power level of about 120-200 W, all the calcium hydroxide was consumed. The flexural strengths measured after the microwave treatment were in the range 9-12 MPa.
Keywords : Microwave Energy, Fibres, Composites, Metakaolin, Flexural Strength, Microstructure, Curing.

1 Introduction

The reduction of the curing period in the fibre reinforced cement composites (F.R.C.C.) is an important factor for the productivity and the reduction of workshop area. The hydration of cement is relatively slow at ambient temperatures and for this reason, F.R.C.C. must cure for several days after casting ; for example, they achieve a substantial proportion of their ultimate stength when the cure is carried out for 7 days, in a humidity greater than 95 % relative humidity and at a temperature higher than 15°C.

It is possible to use accelerated curing schedules, either by the use of chemical accelerators or by a higher curing temperature. Because F.R.C.C. are generally of thin section, they are prone to rapid drying and the curing conditions have to be carefully controlled to prevent thermal cracking (Proctor, 1977).

Xuequan and al (1986) used microwave energy to accelera-

Fibre Reinforced Cement and Concrete. Edited by R. N. Swamy. © 1992 RILEM.
Published by E & FN Spon, 2-6 Boundary Row, London SE1 8HN. ISBN 0 419 18130 X.

te the hydration of ordinary portland cement mortars. This new treatment can heat materials both rapidly and uniformly. The amount of energy used must be regulated carefully : too much energy boils the water present in the mortar while too little has no effect (Hutchinson, 1991).

It is therefore interest of use microwave energy to reduce the curing period of F.R.C.C. This paper describes the effects of this heat treatment on the mechanical properties and microstructure of F.R.C.C. containing metakaolin and either E glass or A.R. flass fibres.

2 Experimental

2.1.Materials

White OPC cement was mixed with metakaolin, silica sand and limestone filler to form a mortar with a water to cement ratio varying from 0.50 to 0.55 according to the quantity of metakaolin introduced in the mix. Table 1 gives the different compositions of the mixes.

Table 1. Details of mixes and specimens.
Binder : 60 %.
Silica sand : 30 %.
Limestone filler : 10 %.

Type of binder		W/C	Type of glass fiber
white cement (%)	metakaolin (%)		
80	20	0.50	AR
70	30	0.50	AR
60	40	0.55	E

Metakaolin is produced from kaolin by calcination at 800°C. Its reacts readily with calcium hydroxide and improves the long-term strength properties of F.R.C.C. (Ambroise and al, 1986-88). When the amount of metakaolin in the blended cement is 40 %, E glass fibres can be used.

5 % by weight of 25 mm long E or AR glass fibre strands were added to the mortar. Specimens of F.R.C.C. were then cast in wood moulds :

width : 5 cm ;
length : 16 cm ;
thickness : 3 cm.

The samples were covered with a plastic sheet to prevent the evaporation of water and placed in the oven.

Wood moulds were used instead of foamed polystyrene moulds because they remained rigid, even at high power levels.

2.2. Heat treatment

A 2450 MHZ microwave oven (model CEM-MDS 81) was used. The maximum microwave output was fixed at 800 W. In order to obtain the optimum process conditions, F.R.C.C. specimens were treated with different power levels (120 W-200 W) and time (45-240 minutes). The effect of high power levels (280-640 W) was also investigated on some samples.

2.3. Mechanical and physico-chemical investigations

Flexural tests were performed after heating and at different curing times : 24 h, 3 days, 7 days and 28 days. After demoulding, specimens were placed in polythene bags and stored at 20°C until the desired date.

The internal temperature of the F.R.C.C. was measured during the microwave treatment.

The lime consumption was studied by infra-red spectrometry (KBr discs.), using a Perkin-Elmer FTIR (Fourier transform) spectrometer.

3 Results and discussion

3.1. Flexural strength

Figures 1 to 3 give the effect of microwave curing on the flexural strength of F.R.C.C. Data shown here are the mean values of three duplicated experimental results.

First, the time of treatment was kept constant (45 minutes) and the power level varied from 120 W to 240 W. Figure 2 shows that the best flexural strengths after 24 hours curing were obtained for the following powers :

- 120 W with the binder containing 20 % metakaolin ;
- 160 W with the binder containing 30 % metakaolin ;
- 200 W with the binder containing 40 % metakaolin.

Then, these power levels were kept constant and the time of heating varied from 45 to 240 minutes. Figure 2 gives the flexural strengths obtained just after the treatment. When the time of heating was lower than 60 minutes, the level of strength was very low. The optimal times were :

90 minutes with the binders containing 30 or 40 % metakaolin ;

120 minutes with the binder containing 20 % metakaolin.

These powers and times were kept constant in order to study the influence of curing at 20°C on the flexural strengths. Figure 3 shows a small increase in strength between 1 and 28 days of curing : about 10 %. 90 % of the final strength are obtained 24 hours after the treatment, while only 25 to 30 % of the 28 days flexural strength are reached after 24 hours without microwave curing.

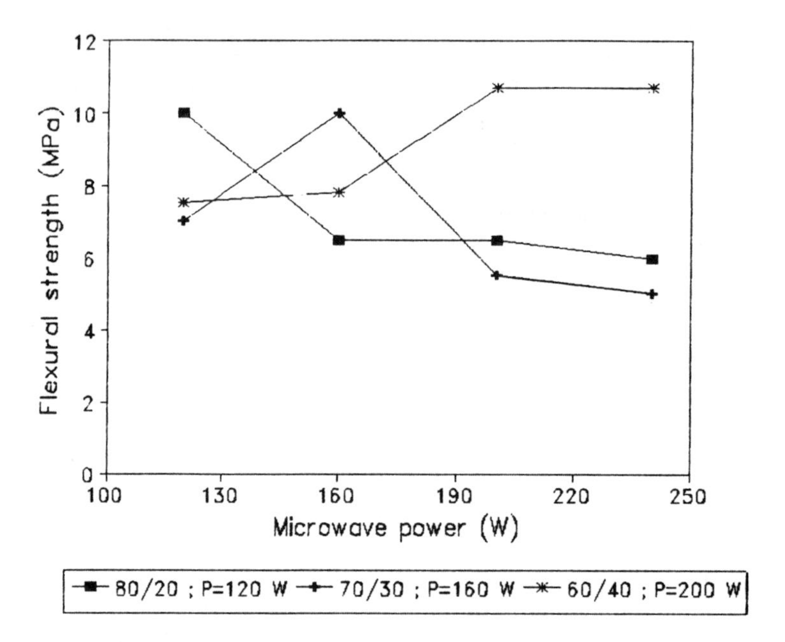

Fig. 1 Flexural strength versus microwave power, 24 hours after treatment (45 minutes).

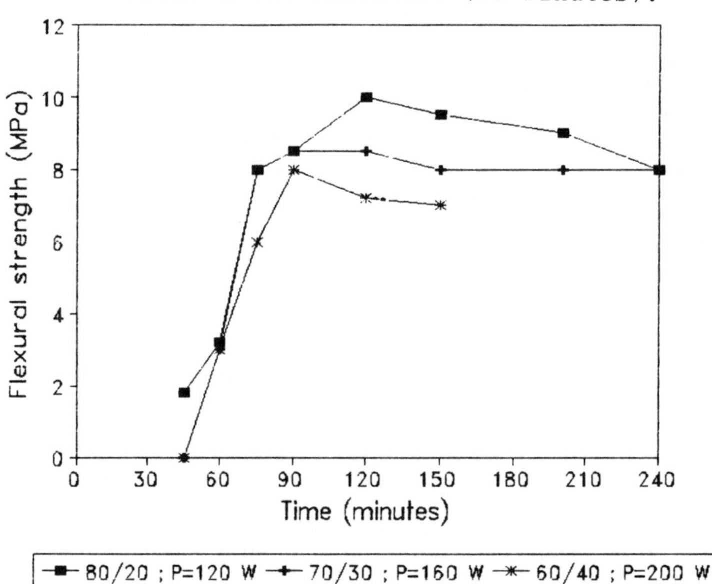

Fig. 2. Influence of treatment time on the flexural strength just after heating.

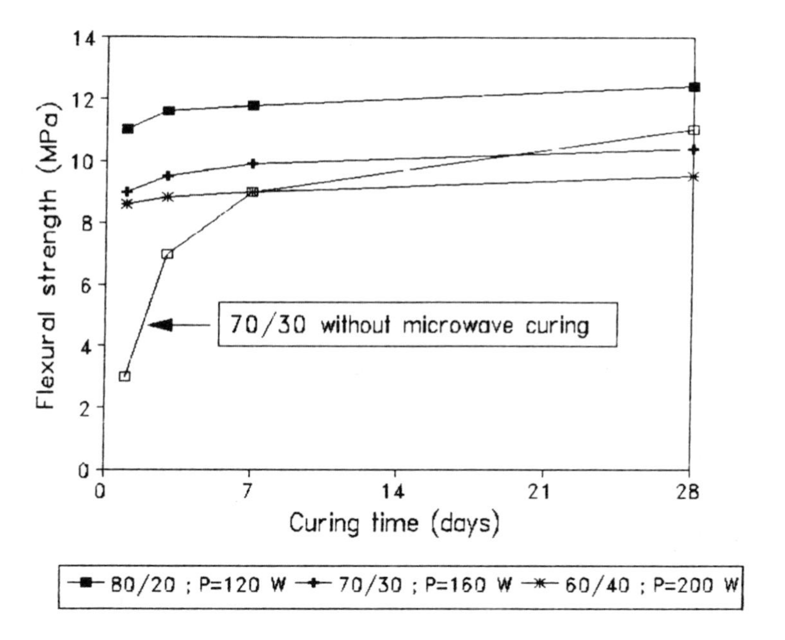

Fig. 3. Influence of curing time on the flexural strength.

When higher power levels were used, the flexural strengths obtained were very low (Table 2). The evaporation of water could not be prevented.

Table 2. Influence of high power levels on the flexural strength

Metakaolin content of the binder (%)	Power level (W)	Time of heating (minutes)	Flexural strength after treatment (MPa)
20	280	60	2.9
	640	20	0.4
30	280	60	2.2.
40	280	60	1.7

When a polymer (FORTON compound) was added at dosage rates of 2 to 15 %, there was a decrease in strength, especially when the polymer content was higher than 5 %

(Figure 4). The incorporation of polymeric materials in F.R.C.C. is not necessary when microwave treatment is used.

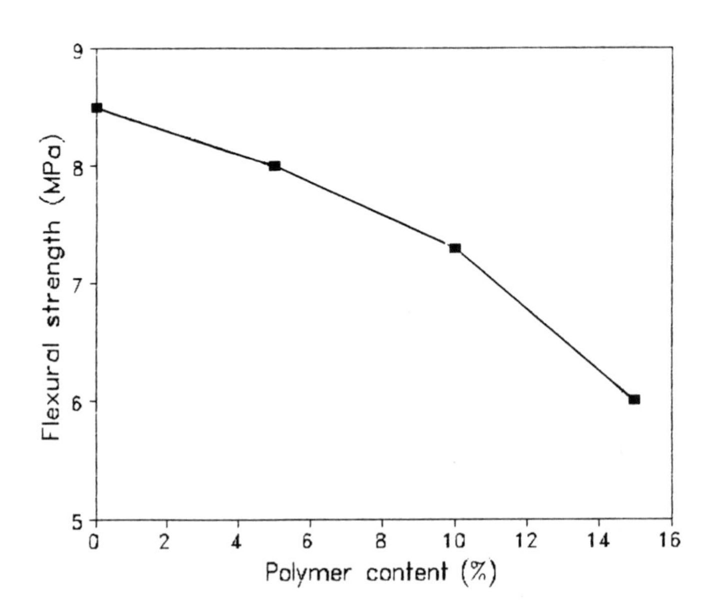

Fig. 4. Influence of a polymeric addition on the flexural strength, just after the treatment. OPC/MK = 80/20. P = 120 W.

3.2. Temperature of F.R.C.C.

The temperature of the specimen containing 30 % metakaolin was measured during the microwave treatment and its evolution is given in figure 5. After 60 minutes of heating, a temperature of 85°C was reached.

3.3. Lime consumption

Infra-red measurements of pure OPC and matrices containing 20 % and 30 % metakaolin are shown in figure 6. It can be seen that all the calcium hydroxide has been consumed in the metakaolin blended cements, just after microwave treatment. The absorbance at 3640 cm^{-1} from the OH ions relevant to $Ca(OH)_2$ has completely disappeared.

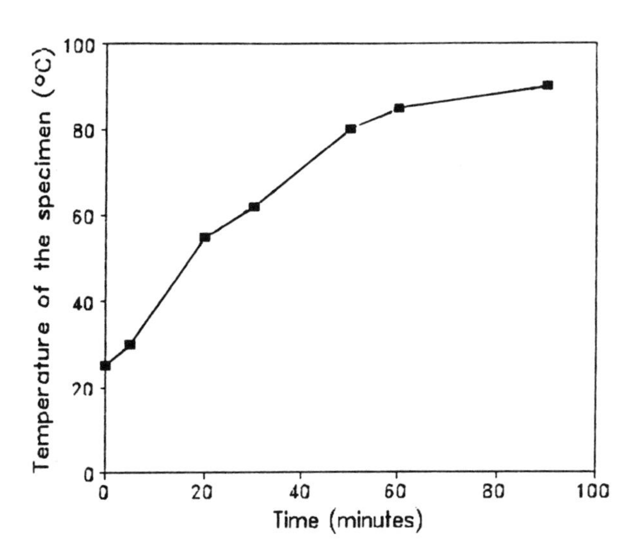

Fig. 5. Evolution of temperature during the microwave treatment. OPC/MK = 70/30. P = 160 W.

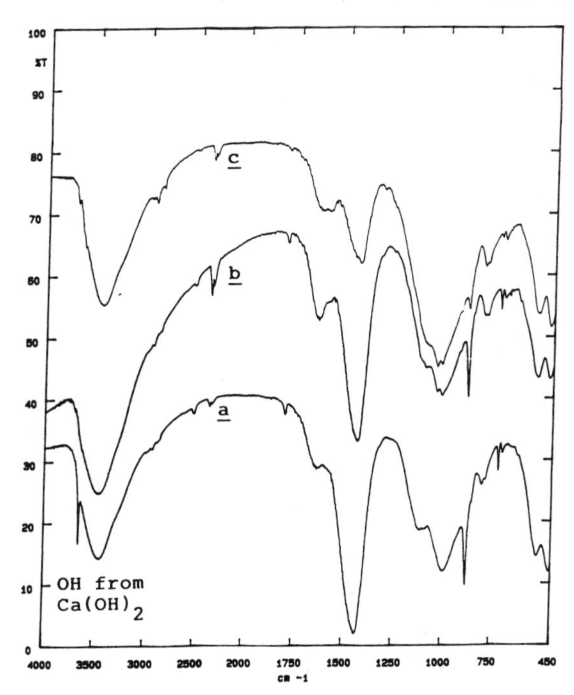

a : 100 % OPC - b: OPC/MK = 80/20 - c : OPC/MK = 70/30.
Fig. 6. I.R. spectra of different matrices.

When data are performed on specimens without microwave curing, the lime consumption is slower as it can be seen in figure 7. All the lime is only consumed after 120 days of curing in a blended cement containing 30 % metakaolin.

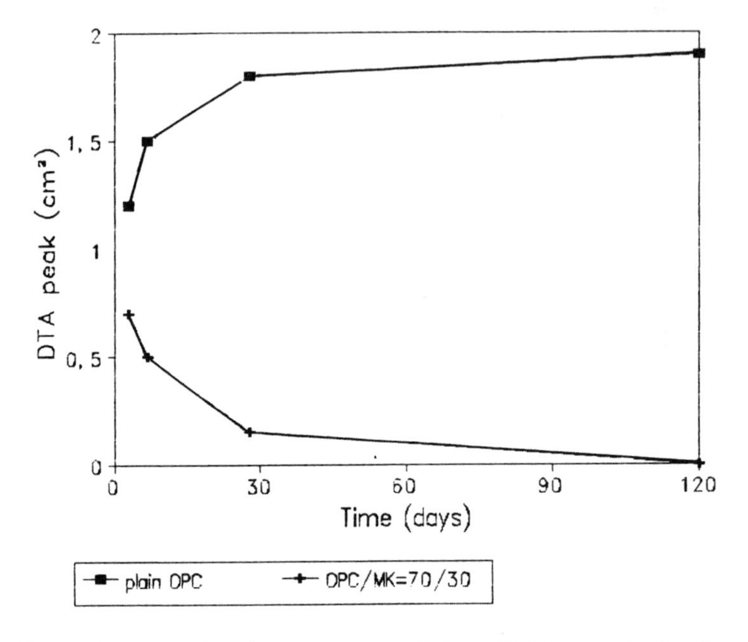

Fig. 7. Unreacted lime measured by differential thermal analysis on specimens without microwave curing.

4 Conclusion

Microwave energy can be used to accelerate the curing of fiber reinforced cement composites. The conditions of treatment depend upon the composition of the matrix : when the amount of metakaolin increases, the power level needed is higher. The evaporation of water must be prevented and very high powers do not lead to good results. The use of a polymeric addition is not necessary. There is no more unreacted lime in composites containing 20 or 30 % metakaolin.

5 References

Ambroise, J., Murat, M., and Péra, J. (1986). Durability of E-Glass Fiber in Portland Cement-Metakaolinite Matrix. Proceedings of the **International Symposium on the Durability of Glass-Fiber Reinforced Concrete**, Chicago, S. Diamond Editor, 1986, pp. 285-292.

Ambroise, J., Dejean, J., Foumbi, J., and Péra, J. (1987). Metakaolin blended cements : an efficient way to improve GRC durability and ductility. Proceedings of the **6th Biennal Congress of the Glass Reinforced Cement Association**, Edinburgh, Glass Reinforced Cement Association, 1987, pp. 19-27.

Ambroise, J., Dejean, J., and Péra J. (1988). Study of fiber-matrix interfaces in metakaolin-blended cement GRC composites. Proceedings of the **Materials Research Society**, Boston, 1988, Vol. 114, pp. 175-180.

Hutchinson, R.G., Chang, J.T., Jennings, H.M., and Brodwin, M.E. (1991). Thermal acceleration of portland cement mortars with microwave energy. **Cement and Concrete Research**, Vol. 21, 1991, pp. 795-799.

Proctor, B.A. (1977). Glass Fibre Reinforced Cement - Principles and Practice. Proceedings of **International Congress on Glass Fibre Reinforced Cement**, Brighton, Glass Reinforced Cement Association, 1977, pp. 51-67.

Xuequan, W., Jianbgo, D., and Mingshu, T. (1987). Microwave curing technique in concrete manufacture. **Cement and Concrete Research**, Vol. 17, 1987, pp. 205-210.

6 WORKABILITY AND DURABILITY OF STEEL FIBER REINFORCED CONCRETE CAST WITH NORMAL PLASTICIZERS

M. UYAN, H. YILDIRIM and A. H. ERYAMAN
Building Materials Department, Istanbul Technical University, Ayazaga Campus, Istanbul, Turkey

Abstract
In this experimental work the profits of using Normalplasticizer is investigated. Fourteen mixtures are made and tested. Acidic attack on specimens is observed. Problems seen during the mixing are noted.

Test results have shown a considerable improvement in the workability and in uniform distribution, a considerable increase in the compressive, flexural strengths, improved resistance against acidic attack.
Keywords: Workability, Fibre Reinforced Concrete, Chemical Attack, Weight-loss, Compressive Strength, Flexural Strength.

1 Introduction

During the last twenty years steel fibre reinforced concrete and concrete admixtures have been questioned and used in many research works [1,2]. In this experimental study, the workability and the durability of steel fiber reinforced concrete (SFRC) cast with water-reducing admixture was investigated. High fiber concentrations cause troubles in workability. Balling and tangling of fibres in the mixer prevent uniform distribution and also cause problems when the concrete is placed [3]. On the contrary, workability problems can be overcome by use of water-reducing agents [4].

The purpose of this project is to determine effects of normal-range water-reducing admixtures on SFRC and investigate durability of SFRC against acidic attack. The practical aim of our research is to investigate suitable SFRC workability for machines related to concrete casting (for example concrete pumps [5,7]).

At the end of the study the effective concentrations of fiber with right quantity of admixtures for suitable workability is determined. The contradicting effects of fibres and plasticizers on fresh and hardened concrete is discussed.

2 Experimental programme

2.1 Materials

Hooked steel fibres glued together in bundles were used. Steel fibres have a nominal length of 50 mm. and a diameter of 0.5 mm. Specific

Fibre Reinforced Cement and Concrete. Edited by R. N. Swamy. © 1992 RILEM.
Published by E & FN Spon, 2-6 Boundary Row, London SE1 8HN. ISBN 0 419 18130 X.

gravity of fiber's steel is 7.009 g/cm^3. The collating of fibres creates an artificial low aspect ratio of approximately 30 when introduced to the mix. That prevents tangling of fibres during the mixing. When the glue is dissolved by the water in the mix, the fibres will be separated as individual fibres with an aspect ratio of 100 [6].

Normal Portland-Pozzolan Cement was used for all mixtures. The specific gravity of cement is 3.04 g/cm^3. The 7th day's flexural and compressive strength of cement are respectively 6.9 N/mm^2 and 33.4 N/mm^2. On the 28th day, flexural strength is 9.5 N/mm^2 and compressive strength is 48.8 N/mm^2.

Granulation of fine and coarse aggregates were close to B16 curve (DIN 1045). Maximum aggregate size was chosen to be 16 mm. due to specimens dimensions (10x10x50cm. Beam) and fiber length (50 mm.). The proportions of sand, fine and coarse aggregates in concrete are respectively 40%, 38%, 22% [1].

A normal-range water-reducing admixture based on ligno-sulphonate was used for this project. Drinking tap water was used for mixing concrete.

Table 1. Basic mix proportions

Item	Quantity (kg/m^3)
Aggregates	1806.75 ~ 1777.3
Cement	350
Water	205
Fibres	0 ~ 77.1
Plasticizer	0 ~ 1.75

2.2 Mixes

Fourteen mixtures in four groups have been made in this project. To compare the properties of the fibrous and/or admixture added concrete with plain concrete, in all mixtures, water-cement ratio was kept 0.586 with a cement content of 350 kg/m^3. See Table 1. The proportions of aggregates were kept unchanged for the same reason. The mixtures were named by initials and numbers where;

PC meant plain concrete,
NP meant concrete cast with a normal-range water-reducing admixture only,
F meant concrete cast with steel fibres.

The subsequent numbers were showing the content of fiber in litre per cubic meter concrete and/or the content of admixture in gramme per kilogramme of cement.

The first group mixture PC was made for reference purpose.
The second group mixtures were NP-3 and NP-5 (normal-range water-reducing admixture was mixed by 0.3% & 0.5% of cement weight).
The third group mixtures were F-5, F-8, F-11 (steel fibres were

mixed by 0.5%, 0.8%, 1.1% of volume of concrete).
The fourth group mixtures were F-5/NP-3, F-8/NP-3, F-11/NP-3, F-5/NP-5, F-8/NP-5, F-11/NP-5. See Figure 1.

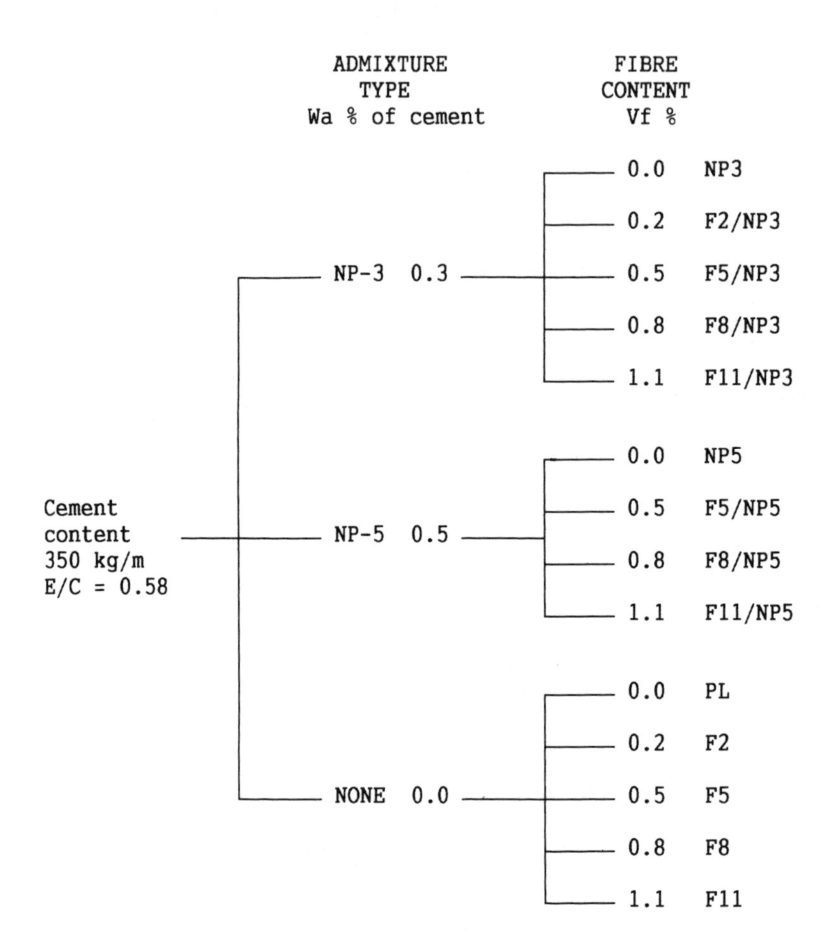

Figure 1 - Investigated Mixtures

At the beginning instead of mixture F-11, F-2 (steel fibres mixed by 0.2% of volume of concrete) had been made but the author decided to change this mixture because of the negligible results in workability and in ultimate strength.

2.3 Batching
All the batches were mixed in a forced-action mixer of 55 litre capacity. Each mixture was made by two batches one after another. Thus, workability tests results were provided from average of two batches. The batching procedure is as follows [8]:

1) Aggregates and cement
2) Dump fibres in without blending
3) 60 percent water
4) Mix 3 minutes
5) Add 40 percent water with admixture during the mix
6) Set 3 minutes
7) Mix 2 minutes

From each mixture, minimum twelve beams 10 x 10 x 50 cm. were made.

3 Testing procedure

The freshly mixed concrete were tested for slump, vebe time, air content, and inverted cone time [9].

The hardened concrete was tested for flexural and compressive strength at 7th, 28th days. The flexural test consists of testing a simply supported beam (10x10x50 cm.) to failure under four point loading on a 42 cm. span. The compressive test was made on the pieces of specimens which are broken into two after the flexural test.

Weight-loss of fully submerged specimens into 1% HCl solution was used as a criterion for durability [10]. The pH of HCl solution was checked every day and necessary quantity of pure HCl was added in such a manner that the acidity of solution was being kept around pH = 1~4.

After a four week period, once every week the spent solutions was replaced with freshly made solutions. Every week, the weak reaction product was removed from the surfaces of concrete specimens with a steel-wire brush, ultrasonic test was made for each specimen and all the specimens were weighted.

4 Test results

Reduced data are given as averages in the graphs given in figures. Each average represents, two readings of two batches for testing fresh concrete (Figures 2, 3, Table 2), three readings of three specimens for testing mechanical properties (Figures 4, 5, Table 3). Readings differed from the corresponding averages with more than 10% for compressive strength and 15% for flexural strength were excluded (Tables 3).

5 Discussion of test results

5.1 Fresh concrete

At the first glance to Figures 2 and 3, an improvement of workability is evident with the increasing admixture content while the increase of fibre content is affecting adversely. But this negative behaviour can be overcome by use of Normalplasticizer. For example, Figures 2.b and 3.b show that the slump of the mixture F-11/NP-5 is 21 % higher than the plain concrete and, in Figures 2.d and 3.d, VeBe time of F-11/NP-5 is almost equivalent to the plain concrete (Table 2).

Table 2. Results of tests on fresh concrete

Mixtures	SLUMP (cm.)	VEBE (sc.)	INVERTED CONE (sc.)	AIR CONTENT (lt/m^3)
PL	3.9	3.2	11.4	12.8
NP-3	9.9	1	*	15.2
NP-5	14.7	0.9	*	20.4
F-2	3.8	3.2	15.5	12.1
F-5	2.1	6.4	21.2	11.9
F-8	1.2	13.9	37.6	6.4
F-11	0.5	14.5	93	4.6
F-2/NP-3	10.5	0.9	7.8	25.5
F-5/NP-3	5.5	2.3	9.1	24.4
F-8/NP-3	3.3	4.1	29.4	15.2
F-11/NP-3	0.8	9.2	55	20
F-5/NP-5	10	1.1	6.2	48.3
F-8/NP-5	9.8	1.2	9.9	68.8
F-11/NP-5	4.7	3.4	21.7	39.6

"*" Inverted Cone Test couldn't been realized because of excessive fluency of concrete

In Figures 2.c, 2.d, an inconvenient change of workability is observed around the fibre content of 0.6% ~ 0.8%. Truly, during the mixing of 0.5% fibre content without plasticizer, fibres were getting stuck on mixer's arms, and during the mixing of 0.8% fibre content, first fibre balling was seen. When the Normalplasticizer was added into the mixture, fibre balling vanished even in the mixtures with highest fibre concentration (F-11). The profit gained in workability with admixture is evident in Figures 3.c, 3.d: VeBe time of F-8 and F-11 is almost equivalent to the plain concrete with the increase of admixture content. Fibres in mixtures F-8 and F-11 which had been tending to pile at the lower end of the inverted cone, caused any problem in presence of Normalplasticizer. An inconvenient observation is that, when the mixing is made in full capacity of mixer with higher fibre concentrations, plasticizer can not prevent fibres to tangle on the surface of fresh concrete. Because, the mixer can not blend easily the mixture in presence of high fibre content.

Another interesting result is that increase of fibre content is decreasing air content in mixtures without plasticizier (Figure 2.a). Adversely, when the Normalplasticizer was added, there was an increase of air content with the increase of fibre content. This can be interpreted as the fact that fibres are preventing the escape of air bubbles captured in the presence of Normalplasticizer (Figures 2.a, 3.a).

As seen in Table 2, fibre content of 0.2% (F-2) has no significant effect on workability according to its results which are close to plain concrete's results.

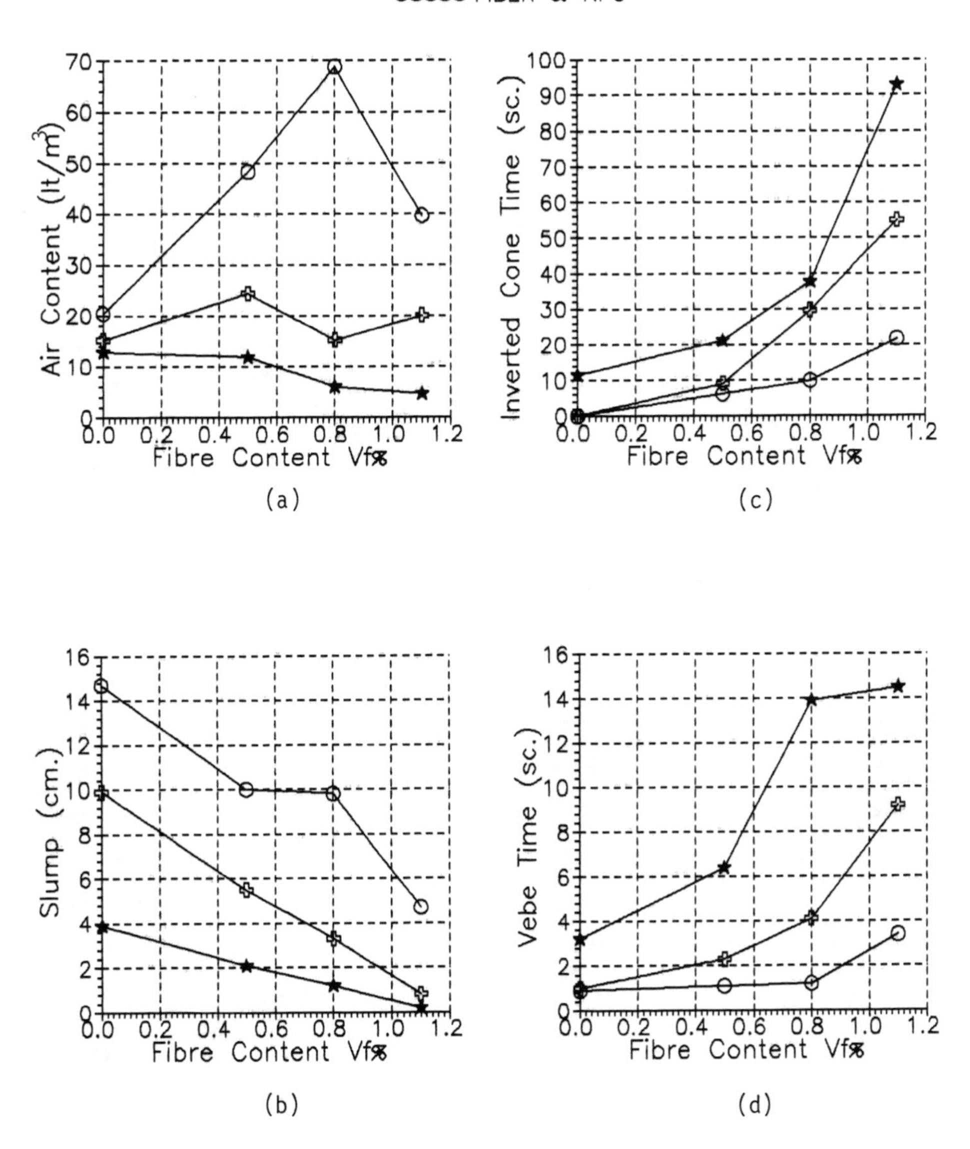
ONLY FIBER
FIBER & NP3
FIBER & NP5
Air Content (lt/m³)
Fibre Content Vf%
(a)
Inverted Cone Time (sc.)
Fibre Content Vf%
(c)
Slump (cm.)
Fibre Content Vf%
(b)
Vebe Time (sc.)
Fibre Content Vf%
(d)

Figure 2

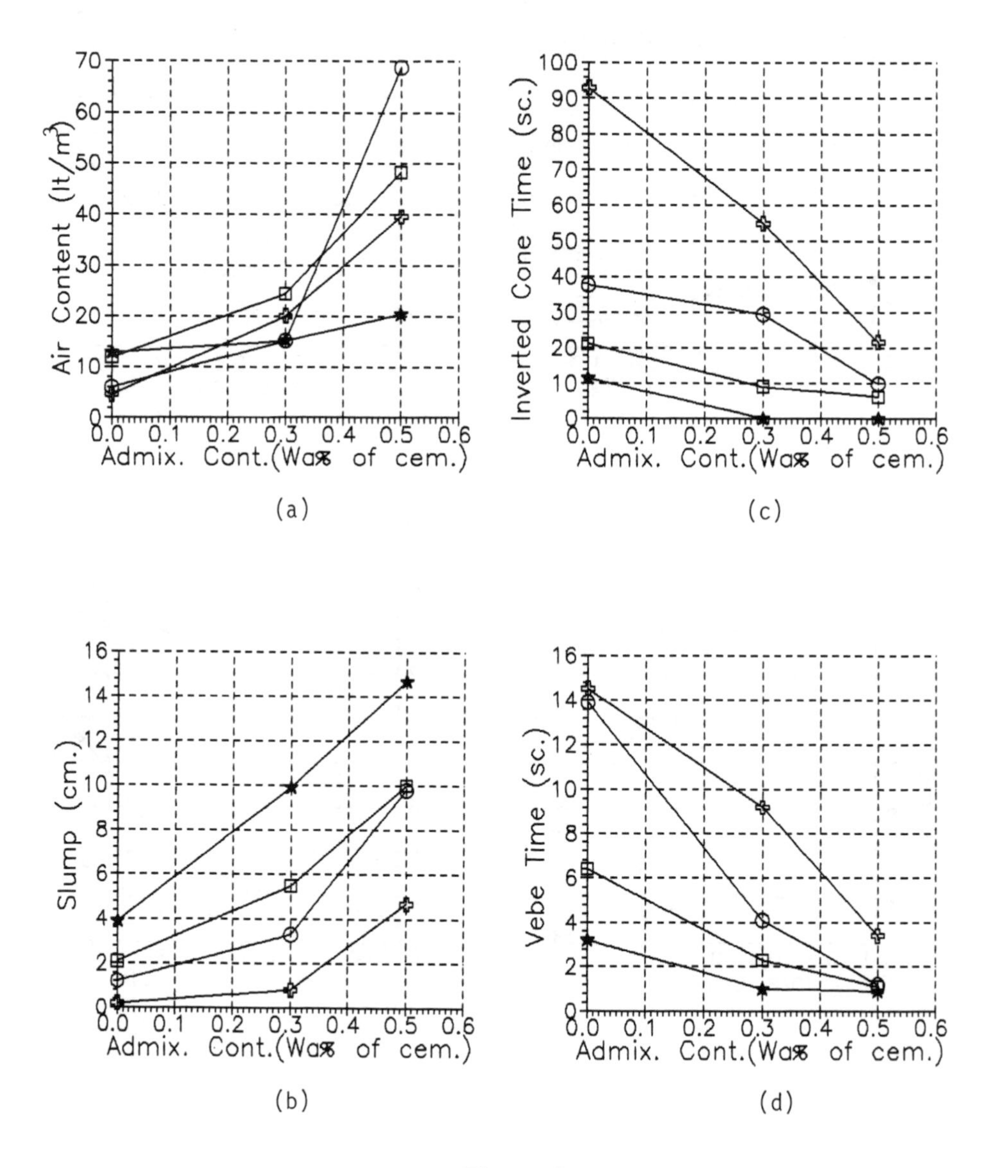

No Fibre
Fibre 0.5%
Fibre 0.8%
Fibre 1.1%
Air Content (lt/m3)
Admix. Cont.(Wa% of cem.)
(a)
Inverted Cone Time (sc.)
Admix. Cont.(Wa% of cem.)
(c)
Slump (cm.)
Admix. Cont.(Wa% of cem.)
(b)
Vebe Time (sc.)
Admix. Cont.(Wa% of cem.)
(d)

Figure 3

Table 3. Results of tests on hardened concrete (N/mm^2)

Mixture	Compressive Strength		Flexural Strength	
	7th day	28th day	7th day	28th day
PL	29.1 **(76)**	38.1 **(100)**	4.3 **(99)**	4.4 **(100)**
NP-3	30.9 **(81)**	43.3 **(113)**	4.6 **(104)**	4.2 **(97)**
NP-5	30.4 **(80)**	42.4 **(111)**	4.1 **(94)**	4.3 **(98)**
F-2	31.0 **(81)**	39.2 **(103)**	4.0 **(92)**	4.0 **(90)**
F-5	35.0 **(92)**	45.3 **(119)**	4.2 **(95)**	4.2 **(95)**
F-8	33.8 **(89)**	45.7 **(120)**	4.8 **(109)**	5.0 **(115)**
F-11	34.8 **(91)**	46.9 **(123)**	5.5 **(125)**	5.4 **(122)**
F-2/NP-3	32.3 **(85)**	40.2 **(105)**	4.3 **(98)**	4.4 **(100)**
F-5/NP-3	35.4 **(93)**	46.6 **(122)**	3.7 **(84)**	4.4 **(101)**
F-8/NP-3	39.2 **(103)**	51.3 **(134)**	4.8 **(111)**	5.5 **(126)**
F-11/NP-3	39.4 **(103)**	50.6 **(133)**	7.1 **(163)**	6.4 **(146)**
F-5/NP-5	34.0 **(89)**	44.2 **(116)**	4.0 **(90)**	4.4 **(100)**
F-8/NP-5	31.0 **(81)**	40.2 **(105)**	4.8 **(109)**	5.9 **(134)**
F-11/NP-5	40.2 **(106)**	51.9 **(136)**	5.1 **(117)**	4.7 **(107)**

() **Percentages of improvement in strength regarding the values of compression and flexure at 28th day.**

5.2 Hardened concrete

In Figures 4 and 5, it is obvious that both, fibres and admixture separately are improving mechanical properties. Normalplasticizer is increasing compressive strength about 12% (always regarding 28th day compressive or flexural strength of plain concrete) meanwhile, with the augmentation of fibre content, the increase of compressive strength is 23% (For example mixture F-11) totalling 35%, which is equal to the growth of compressive strength for mixtures F-11/NP-3 or F-11/NP-5. See Table 3. In graphics 5.a and 5.b, augmentation of each curve is approximately 12%. So the use of Normalplasticizer with fibres affects the compressive strength increase. Figures 5.a, 5.b show that this increase is a result of individual effect of Normalplasticizer which is providing a good workability and an easy placement as determined by the tests on fresh concrete.

Table 3 shows that there is no significant change between the 7th and 28th day's flexural strength. Only with the fibre content augmentation, a little improvement is noticed. This is because, on 28th day, hardened concrete does not allow fibres to slip trough the matrix easily. Figures 4.c, 4.d and 5.c, 5.d shows how does fiber concentration affect on flexural strength. Again, fibre content of 0.6% ~ 0.8% is the area of effective improvements in flexure.

In Figure 5.d a remarkable event is taking place: Normalplasticizer

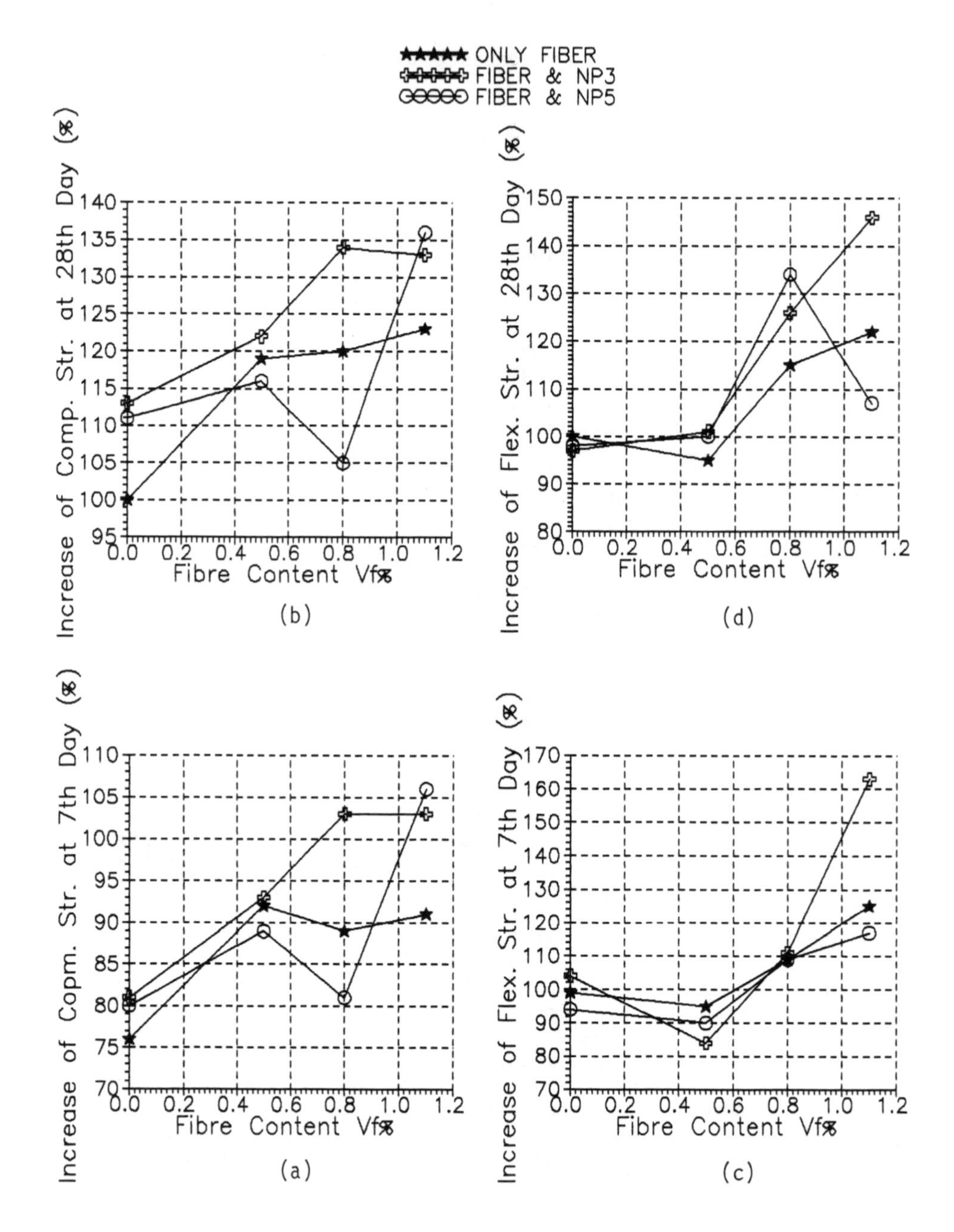
ONLY FIBER
FIBER & NP3
FIBER & NP5
Increase of Comp. Str. at 28th Day (%)
Fibre Content Vf%
(b)
Increase of Flex. Str. at 28th Day (%)
Fibre Content Vf%
(d)
Increase of Copm. Str. at 7th Day (%)
Fibre Content Vf%
(a)
Increase of Flex. Str. at 7th Day (%)
Fibre Content Vf%
(c)

Figure 4

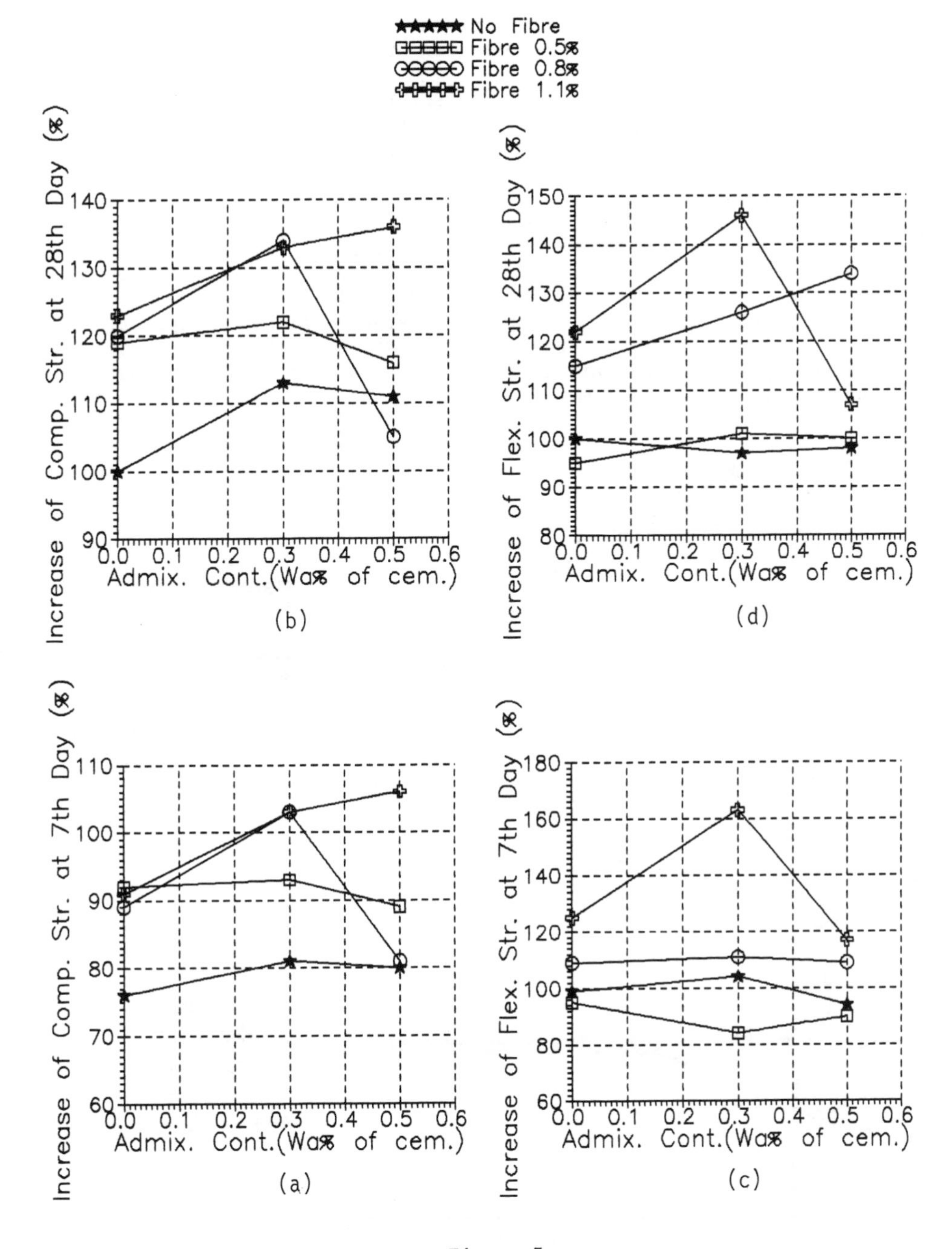
No Fibre
Fibre 0.5%
Fibre 0.8%
Fibre 1.1%
Increase of Comp. Str. at 28th Day (%)
Admix. Cont.(Wa% of cem.)
(b)
Increase of Flex. Str. at 28th Day (%)
Admix. Cont.(Wa% of cem.)
(d)
Increase of Comp. Str. at 7th Day (%)
Admix. Cont.(Wa% of cem.)
(a)
Increase of Flex. Str. at 7th Day (%)
Admix. Cont.(Wa% of cem.)
(c)

Figure 5

which had shown no effect on the flexural strength of plain concrete (NP-3, NP-5), increased the augmentation of ultimate strength of fibrous concrete.

As seen in Table 3, 0.2% fibre content (F-2) has no considerable aid to the mechanical properties. During the work, when this is noticed, mixture F-2 is replaced by mixture F-11.

5.3 Durability

As seen in Figure 6.a, with fibre content augmentation, weight loss curves become more horizontal. Because the easy split of particles of aggregate from the matrix is prevented by fibres. Ultrasonic test shows that corrosive effect of Hydrochloric acid is taking place only on the surface of specimens. There is no sign of fissure propagation directed into the specimen. See Fig. 6.b. Normalplasticizer improves endurance against acidic attack by providing a good placement. But endurance of fibrous specimens in weight-loss is higher then specimens cast with plasticizer.

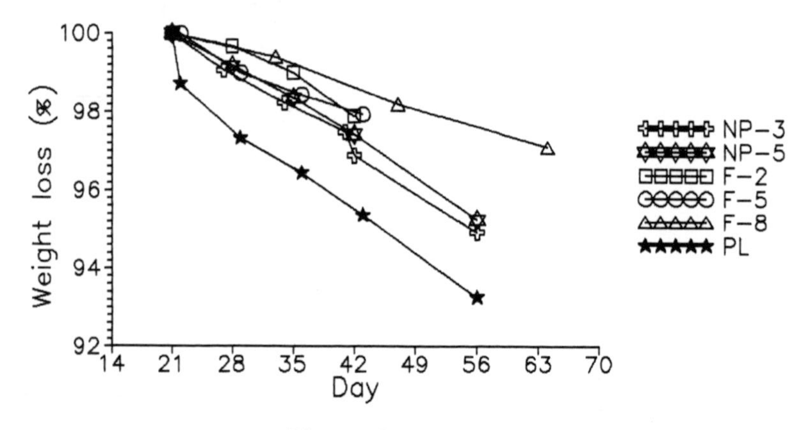

Figure 6.a

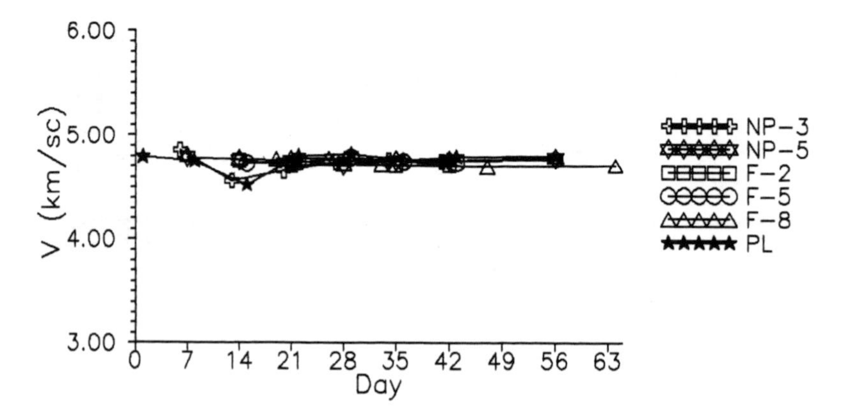

Figure 6.b

6 Conclusions

The following conclusions are based on test results contained in this report:

1) Increase in workability is observed by the use of Normalplasticizer. Absence of fibre balling and Inverted Cone Test results are the best convenient proof of this improvement.
2) In compression a remarkable increase equal to the total of individual influences of Fibres and Normalplasticizer is obtained.
3) By the use of Normalplasticizer, especially in flexural strength a notable improvement is determined because of uniform distribution of fibres throughout the matrix and easy placement of the mix.
4) With the use of Normalplasticizer, effective fibre concentration is estimated between 0.6% ~ 1.1%.
5) Resistance of fibrous concrete against acidic attack is higher than concrete cast with plasticizer according weight-loss test.

Briefly, Normalplasticizers are improving fibrous concrete's characteristics in many ways.

7 References

ACI Committee 544, (Nov. 1973) State-On-The-Art Report on Fibre Reinforced Concrete. ACI Journal, Proceedings V.70, No 11, pp. 729-744.

ACI Committee 544, (Jul. 1978) **Measurement of Properties of Fibre Reinforced Concrete.** ACI Journal, Proceedings V-75, No 7, pp. 283-289.

ACI Publication SP-44, (1975) **Fibre Reinforced Concrete.** ACI Publication, Detroit, pp. 544.

ACI 304-2R Pumping Steel Fibre Reinforced Concrete. Refer. 10.

ASTM Specification C 192.

Concrete Society (1980). Fibrous Concrete. CI80. pp. 1-49

El-Refai, F. E. and E. H. Morsy (Jul. 1986) Some Properties of Fibre Reinforced Concrete with Superplasticizer in **Developments in Fibre Reinforced Cement and Concrete.** 4. Rilem Symposium. Volume 1, pp. 189-197.

Kaden, R. A. (1975) Pumping Fibrous Concrete for Spillway Test. ACI Publication SP-44, pp. 497-510.

Mehta, P.K. (1985) Chemical Admixtures of Low Water Cement Ratio Concretes Containing Latex or Silica Fume as Admixtures in Technology of Concrete When Pozzolans, Slags and Chemical Admixtures Are Used, Symposium '85. Mexico. pp. 325-341.

Ramakrishnan, V., T. Brandshang, W. V. Coyle, E. K. Schrader (May 1980) **Comparative Evaluation of Concrete Reinforced with Straight Steel Fibres and Fibres with Deformed Ends Glued Together into Bundles.** ACI Journal, Proceedings V-77, No 3, pp. 135-144.

7 PLASTIC SHRINKAGE CHARACTERISTICS OF FIBRE REINFORCED CEMENT COMPOSITES

A. KHAJURIA and P. BALAGURU
Rutgers, The State University, Piscataway, NJ, USA

Abstract
This paper deals with the contribution of fibres to the reduction of shrinkage cracks during the initial setting stage of concrete. The fibres were either polymeric fibres or steel fibres. In the case of polymeric fibres, the fibres were made of nylon6, polypropylene and polyester. The fibre length was 19 mm in all cases except in one case when a mix of different fibre lengths was used. Nylon6 and polyester fibres were single filament type whereas polypropylene was a fibrillated type. The fibre content varied from 0.02 to 0.1 volume percent. In the case of steel fibres, low-carbon hooked steel fibres were used. Three fibre lengths of 30 mm, 50 mm, and 60 mm, and two fibre contents of 45 kg/m^3 and 60 kg/m^3 were evaluated. The tests were conducted by using 600 x 900-mm rectangular slabs that were 19 mm thick. The slabs were subjected to rapid drying right after casting.

The contribution of fibres was evaluated using the crack area. In all cases, the crack area of unreinforced matrix is compared with the crack area of reinforced matrix. The test results show that: fibres with higher modulus of elasticity are less effective for 19 mm-long fibres, fibre count (number of fibres/kg) plays a role in the reduction of cracking, and fibres are more effective when the cracking in plain matrix is extensive.
Keywords: Polymeric Fibre, Steel Fibre, Shrinkage, Cement-Mortar, Cracking, Crack Reduction.

1 Introduction

Fibres are used in concrete primarily to improve the mechanical properties of the matrix, in the area of ductility and energy absorption. In the early stages of FRC development, only steel fibres were used. Polymeric fibres were introduced in the late 1970s. Since then, the use of these fibres is steadily increasing. The primary contribution of polymeric fibres is in the area of shrinkage crack reduction of fresh concrete. Since the volume fraction of fibres in use is normally low (approximately one tenth of one percent), the mechanical properties of concrete are seldom affected by the fibres, as reported by ACI Committee 544 (1988).

This paper deals with the evaluation of fibre reinforced concrete

Fibre Reinforced Cement and Concrete. Edited by R. N. Swamy. © 1992 RILEM.
Published by E & FN Spon, 2-6 Boundary Row, London SE1 8HN. ISBN 0 419 18130 X.

having steel and polymeric fibres in reducing the shrinkage cracks of concrete during the initial setting period. The details of the constituent materials, mixture proportions, test procedures, experimental results, analysis of test results and the conclusions are presented. The tests were conducted using the procedure recommended by Kraai (1985). The experimental results and their analysis indicate that the fibres can provide considerable improvement in reducing shrinkage cracks.

2 Experimental program

The experimental program was designed to evaluate the performance of various fibre types and their volume fraction. In case of polymeric fibres, three types of fibres were used. They were designated as N6, PP, and PY. The length of these fibres was the same in all cases (19 mm) except in one case in which a mixture of different fibre lengths was used. The fibre content was varied from 0.45 kg/m^3 to 0.90 kg/m^3. In the case of steel fibres, low carbon hooked-end fibres were used at 45 and 60 kg/m^3. The lengths of the fibre used were 30 mm, 50 mm, and 60 mm. The matrix consisted of a cement-sand mortar mix or a lightweight concrete mix. The overall experimental program is shown in Table 1.

3 Materials, mixture proportions and specimen preparation

3.1 Materials

The constituent materials used consisted of:

ASTM Type I cement
Natural sand and lightweight aggregate
Tap water
Polymeric or steel fibres

The sand used had a fineness modulus of 2.60 and satisfied ASTM C33 (1984) requirements. The lightweight aggregate (expanded shale) had a maximum size of 9 mm.

Three types of polymeric fibres, evaluated in this investigation, were designated as N6 (nylon6), PP (polypropylene), and PY (polyester). The physical properties of these fibres, taken from manufacturers' pamphlets, are presented in Table 2. The fibre lengths were 19 mm for all except one mix. In one case, a mix of equal proportions of 3 fibre lengths of 19 mm, 25 mm, and 38 mm was used.

The steel fibres were made of low carbon steel and were hooked at the ends. These had the following geometrical dimensions:

length (ℓ) = 30.0 mm; diameter (ϕ) = 0.5 mm; Aspect ratio (ℓ/ϕ)= 60
length (ℓ) = 50.0 mm; diameter (ϕ) = 0.5 mm; Aspect ratio (ℓ/ϕ)= 100
length (ℓ) = 60.0 mm; diameter (ϕ) = 0.8 mm; Aspect ratio (ℓ/ϕ)= 75

Table 1. Experimental program

Mix designation	Cement-sand mortar mix	Lightweight concrete
CON (zero fibre content)	X	X
N6-45 (19 mm long, 0.45 kg/m^3)	X	X
N6-60 (19 mm long, 0.60 kg/m^3)	X	X
N6-90 (19 mm long, 0.90 kg/m^3)	X	
N6M-60 (equal proportion of 19 mm, 25 mm, and 38 mm long fibres, 0.60 kg/m^3)	X	
PP-60 (19 mm long, 0.60 kg/m^3)	X	
PP-90 (19 mm long, 0.90 kg/m^3)	X	
PY-60 (19 mm long, 0.60 kg/m^3)	X	
HS3-45 (30 mm long, 45 kg/m^3)	X	
HS3-60 (30 mm long, 60 kg/m^3)	X	
HS5-45 (50 mm long, 45 kg/m^3)	X	
HS5-60 (50 mm long, 60 kg/m^3)	X	
HS6-45 (60 mm long, 45 kg/m^3)	X	
HS6-60 (60 mm long, 60 kg/m^3)	X	

N6 - Nylon6, PP - Polypropylene
PY - Polyester, HS - Hooked-end steel

Table 2. Physical properties of polymeric fibres

Property	Fibre type		
	Nylon6	Polyester	Polypropylene
Tensile strength (MPa)	897	897 - 1,104	553 - 759
Young's modulus (MPa)	5,175	17,250	3,450
Water absorption (%)	4.5	-	Nil
Specific gravity	1.16	1.34	0.9
Melting point (^oC)	242	257	160 - 170
Ultimate elongation (%)	20	-	15

Note: Information provided by manufacturers

3.2 Mixture proportions

The cement mortar matrix consisted of 1 part cement and 1.5 part sand. The water-cement ratio was 0.5. The rich cement matrix was used in order to induce cracks even for specimens containing 0.9 kg/m^3 of polymeric fibres and 60 kg/m^3 of steel fibres.

For lightweight concrete, the mixture had a proportion of one part (cement): 0.5 part (sand): 0.5 part (lightweight aggregate). The water-cement ratio was 0.5.

3.3 Specimen preparation

The test specimen used for shrinkage test was 600 x 900-mm rectangular slab that was 19 mm thick. These specimens were cast on wooden molds lined with plexiglas rims (Fig 1). The base consisted of 13 mm thick plywood with a tile board glued to the top. A thin polyethylene sheet was placed on top of the tile board to eliminate friction (or adhesion) between the mortar and the tile board. A strip of 13 x 25-mm hardware cloth was placed along the perimeter of the specimen to provide edge restraints. The edge restraints minimize the movement of the slab from the edges, thus creating potential for cracks in the slab. The hardware cloth was nailed to the base using roofing nails.

The mortar for shrinkage slab was mixed using a 0.085 m^3 capacity laboratory mixer. The mixing procedure used is as follows:

Place sand, cement and water in the mixer
Mix for 3 minutes
Rest for 3 minutes
Add fibres
Mix for an additional 10 minutes
Discharge the mixture to make test specimens

A longer duration of 10 minutes of mixing was used in order to insure uniform distribution of fibres in the mix.

In the case of lightweight concrete, coarse aggregate and sand were mixed with 2/3 of water for 1 minute before cement and the rest of the water were added to the mixer. This step replaces step 1 of the mixing procedure for cement mortar. In all cases, the mix had a flowing consistency.

All the specimens were cast using a table vibrator. The concrete or mortar mixture with or without the fibres was poured on the mold which was placed on the vibrating table. The specimens were screeded along the longer direction using aluminum screed. Screeding provided the final test surface.

4 Test method

After casting, the slabs were placed on a flat surface and subjected to wind velocity of 19 to 23 km/h using high velocity fans. The wind accelerated the drying process. The laboratory temperature was between 21 to 24°C and the relative humidity was about 50%. The crack widths

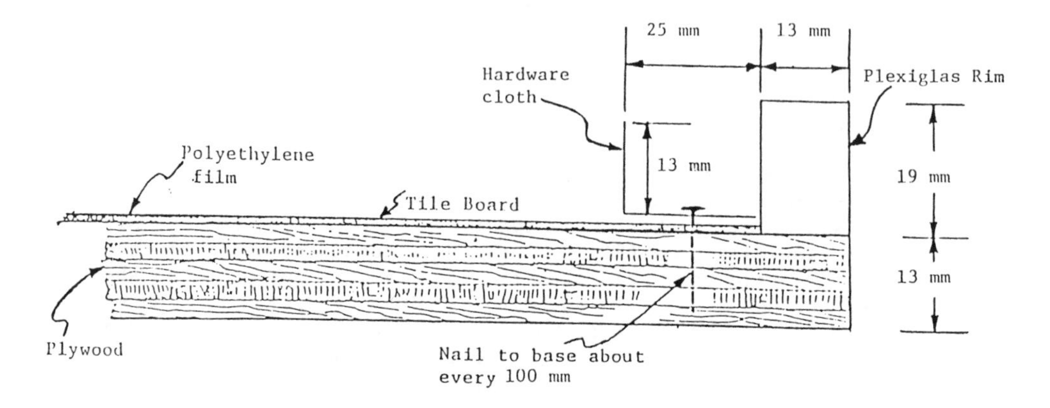

Fig. 1 Cross-section of Rectangular Slab Used for Evaluating Plastic Shrinkage

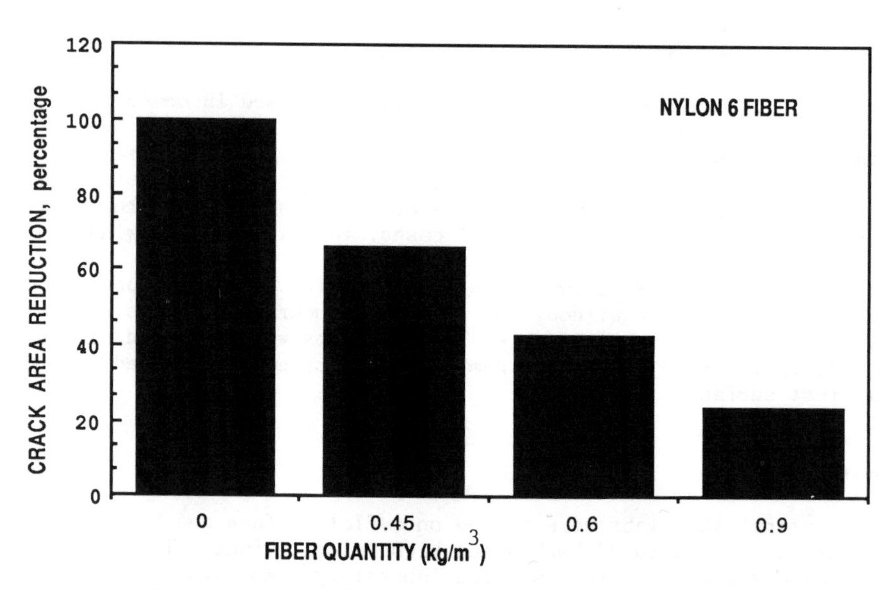

Fig 2 Comparison of Crack Reduction Potential of Nylon 6 Fiber

and lenghts were measured after 24 hours in order to index the crack area. The cracks started developing after 3 to 3 1/2 hours.

The weighted crack values were obtained by dividing the cracks into 4 categories. Cracks that were wider than 3 mm were designated as large cracks and they were multiplied by a 'weight' of 3. The cracks that were about 2 mm wide were multiplied by 2. These cracks were designated as medium cracks. The cracks that were 1 mm and less than 0.5 mm wide were designated as small and hairline cracks respectively. The weightage factors for small and hairline cracks were 1 and 0.5 respectively.

The weighted crack values for the 4 different size cracks were added up to a single number which was used for comparison. The control slab (no fibres) crack value was taken as 100%. The weighted crack value of the other panels was expressed as a percentage of the control.

5 Test results and discussion

For the fibre quantities used, there was no problem in mixing. Fibres mixed well with the matrix leaving no clumps of fibres. This was true for all types of fibres. Hence, the uniformity of fiber distribution is not a variable.

5.1 Polymeric fibres

The results of polymeric fibres are presented in Tables 3 and 4 and Figs. 2 to 5. Table 3 presents the results for cement mortar whereas Table 4 deals with the lightweight concrete. Fig. 2 presents the influence of fiber volume fraction on crack reduction. Fig. 3 provides a comparative evaluation of the three fiber types at 0.6 kg/m^3 fibre loading. This figure also shows the behavior of mortar containing the mix of three fibre lengths of nylon6 fibres. Fig. 4 provides a comparison of nylon6 and polypropylene fibres at 0.9 kg/m^3. Fig. 5 shows the variation of crack reduction with respect to fiber volume fraction for lightweight concrete.

A careful study of Tables 3 and 4 and Figs. 2 to 5 lead to the following observations:

The addition of fibres even at 0.45 kg/m^3 results in some crack reduction.
Increase in fibre content of N6 fibres consistently improves the performance of cement mortar, Fig. 2.
Addition of N6 fibres not only results in lower crack area but also results in less wide cracks, Table 3 .
N6 fibres performed slightly better than the other two fibres at fibre contents of 0.6 and 0.9 kg/m^3, Figs. 3 and 4. Note that N6 fibres have a higher fibre count for a given weight.
Increase in fibre content provides consistent decrease in the shrinkage crack area for lightweight concrete, Fig. 5.

Table 3. Cracking characteristics: polymeric and normal weight mortar

Mix designation	Crack length (cm)				Weighted average	Percentage of control
	Large	Medium	Small	Hairline		
CON	0	87.5	75.0	112.5	306.25	100
N6-45	0	12.5	135.0	87.75	203.75	67
N6-60	0	10.0	100.0	25.0	132.50	43
N6-90	0	0	12.5	125.0	75.00	24
N6M-60	0	0	67.5	162.5	148.75	49
PP-60	0	25.0	87.5	137.5	206.25	67
PP-90	0	35.0	100.0	35.0	187.50	61
PY-60	0	50.0	97.5	100.5	258.75	84

Table 4. Cracking characteristics: nylon6 fibres and lightweight concrete

Mix designation	Crack length (cm)				Weighted average	Percentage of control
	Large	Medium	Small	Hairline		
CON	25.0	15.0	10.0	112.5	171.25	100
N6-45	0	12.5	12.5	75.0	75.0	44
N6-60	0	12.5	12.5	25.0	50.0	29

Table 5. Cracking characteristics: steel fibres and normal weight mortar

Mix designation	Crack Length (cm)				Weighted average	Percentage of control
	Large	Medium	Small	Hairline		
CON	30.0	70.0	50.0	30.0	295.0	100
HS3-45	0	55.0	40.0	20.0	160.0	54
HS5-45	0	0	30.0	40.0	50.0	17
HS6-45	0	50.0	42.0	60.0	172.0	58
HS3-60	0	10.0	30.0	30.0	65.0	22
HS5-60	0	0	0	10.0	5.0	2
HS6-60	0	0	80.0	20.0	90.0	31

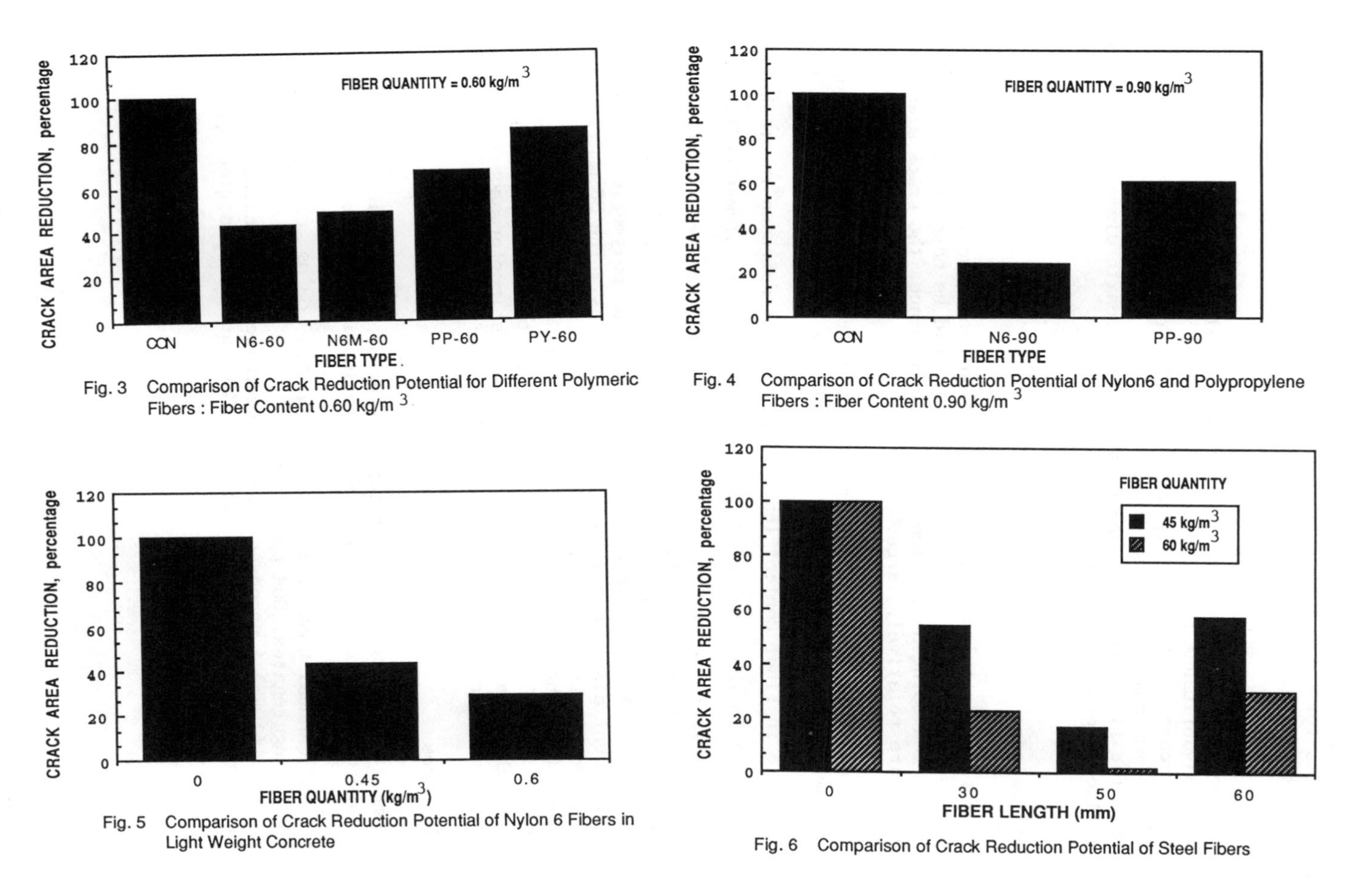

Fig. 3 Comparison of Crack Reduction Potential for Different Polymeric Fibers : Fiber Content 0.60 kg/m ^{3}

Fig. 4 Comparison of Crack Reduction Potential of Nylon6 and Polypropylene Fibers : Fiber Content 0.90 kg/m ^{3}

Fig. 5 Comparison of Crack Reduction Potential of Nylon 6 Fibers in Light Weight Concrete

Fig. 6 Comparison of Crack Reduction Potential of Steel Fibers

5.2 Steel fibres

The results for steel fibres are presented in Table 5 and Fig. 6. The results show that:

- 50 mm long fibres perform better than the other two lengths.
- An increase in the quantity of fibre, for all lengths, resulted in reduced shrinkage of the crack area.
- An increase in the quantity of fibres also results in the reduction of the maximum size of the crack.
- Both the fibre count and the aspect ratio (length/diameter) seem to influence crack reduction.

6 Conclusions

The experiment results of the investigation presented in the paper lead to the following general conclusions. It sould be noted that the study was conducted using a rich cement mortar and hence, the crack areas reported are relatively large.

- Addition of fibres lead to a reduction in cracking. This is true for all three polymeric fibre types and the steel fibres tested in this program.
- Increase in fiber content leads to consistent decrease in crack reduction in the range of 0.45 to 0.90 kg/m^3 for polymeric fibres.
- Increase in fibre content from 45 to 60 kg/m^3 of steel fibres also leads to further reduction of cracking.
- Increase in fibre count for a given weight seems to result in better performance in the area of crack reduction for polymeric fibres.
- Fibre count also plays a role for steel fibres. The 30 mm fibres provided better results than 60 mm fibres.
- Aspect (length/diameter) ratio plays a role in crack reduction for steel fibres. For the three fibre lengths tested, the fibres with the highest aspect ratio of 100, had the best performance. The other two fibre lengths had aspect ratios of 60 and 75.

7 References

ACI Committee 544 (1988) Measurements of properties of fibre reinforced concrete. **American Concrete Institute Materials Journal**, 85, pp 583-593.

ASTM (1984) **Annual Book of ASTM Standards**. Vol. 04.02, Concrete and Mineral Aggregates, 904 pp.

Kraai, P.P. (1985) A proposed test to determine the cracking potential due to drying shrinkage of concrete. **Concrete Construction**, pp 775-778.

8 FREE AND RESTRAINED SHRINKAGE OF FIBRE REINFORCED CONCRETE WITH LOW POLYPROPYLENE FIBRE CONTENT AT EARLY AGE

K. KOVLER, J. SIKULER and A. BENTUR
Faculty of Civil Engineering, National Building Research Institute, Technion, Haifa, Israel

Abstract
The performance of concretes reinforced with low contents of polypropylene fibres is expected to be improved by reducing sensitivity to cracking. The concepts of an improved method for measuring the cracking sensitivity using a modified ring test for restrained shrinkage evaluation is described. The test was used to study the influence of polypropylene fibre content and type on the cracking sensitivity of poorly cured concrete, either fresh concrete without any curing or concretes cured for 2 days prior to being exposed to drying conditions. The fibres were more effective in the case of the fresh concrete. Fibres of improved structure were required to provide reduced cracking sensitivity for the 2 days cured concrete.
Keywords: Free and Restrained Shrinkage, Weight Loss, Deformation, Thermal Expansion, Polypropylene Fibres, Early Age.

1 Introduction

It is known that very low polypropylene (PP) fibre content reinforcement can be effective in improving the cracking resistance of the fresh concrete. The performance of the fibres is usually evaluated by both, free and restrained shrinkage tests.

The results published on free shrinkage of concretes containing polypropylene fibres is not always consistent. Reduction in plastic shrinkage by as much as 25% and 45% has been reported by Zollo (1984), Zollo and Ilter (1986). A more modest reduction of about 10% has been noted by Swamy and Stavrides (1976) and Kubota and Sakane (1967). This is in agreement with free shrinkage studies of concretes with other types of fibres, where only a small reduction in shrinkage strains, if at all, was registered - Reinsdorf (1985), Hannant (1978).

Measurement of the free shrinkage deformation only is not sufficient for estimation of material performance, since it is the cracking sensitivity which is of practical interest. While some tests may show a reduction in free shrinkage due to the presence of the fibres, this is not necessarily an indication for an overall reduction in the cracking tendency, which is a function of both, the free shrinkage and the reinforcing effect of the fibres - Bentur and

Fibre Reinforced Cement and Concrete. Edited by R. N. Swamy. © 1992 RILEM.
Published by E & FN Spon, 2-6 Boundary Row, London SE1 8HN. ISBN 0 419 18130 X.

Mindess (1990). The cracking tendency can only be judged on the basis of restrained shrinkage tests.

Many researchers have conducted restrained shrinkage tests using ring specimens. In such tests ring-shaped test pieces of concrete are cast between two rigid (usually steel) rings, with the inner ring (core) providing the restraint when the specimen is kept in a drying environment. Ring specimens can be exposed to various drying conditions, such as low relative humidity at room temperature, mild drying in a heated oven, or a wind tunnel. The extent of cracking depends on the restraint condition and drying environment. The restraint produces tensile stresses in the concrete in the ring, which reach a maximum value at the interface with the inner surface of the specimen, i.e. the restraining core. If this stress is sufficiently high, cracking may occur.

A drawback of restrained shrinkage tests of fresh and hardened concrete, by means of conventional ring specimens with steel core, is in their relatively low crack sensitivity. In order to increase maximum tensile stress in the concrete ring, Dahl (1986) suggested a modified ring test. In this method steel ribs are welded to the outer steel ring to provide additional restraint. However, other ways may be suggested to increase maximum tensile stresses and crack sensitivity of restrained shrinkage tests. Concepts of cracking and restrained shrinkage were also discussed in other publications of Leshchinsky (1989) and Grzybowski and Shah (1989).

The present paper describes the concepts of a modified ring test method of improved cracking sensitivity, developed in order to achieve cracks within a relatively short period of time. This method was applied to the study of the influence of polypropylene fibres of different kind and contents on the cracking sensitivity of fresh concrete (exposed immediately to drying), and a poorly cured-hardened concrete (exposed to drying after 2 days of curing). The effectiveness of the polypropylene fibres for these two types of deficient curing was evaluated.

2 Materials and Methods

The materials used in this study were ordinary portland cement having standard compressive strength of 30 MPa, specific gravity of 3.1 g/cm^3; Coarse aggregate - crushed dolomite gravel with a maximum particle size of 6 mm, specific gravity of 2.75 g/cm^3; Fine aggregate - quartz sand from natural source having fineness modulus of 1.76, specific gravity of 2.63 g/cm^3; Water/cement ratio - w/c=0.57; Polypropylene fibres were of Israeli production, fibrillated type, with a length of 25.4 mm and thickness of 40 μm (labelled here type A); Fibre content V_f was varied from 0 up to 0.3% and in some cases mixes with up to 0.6% volume content were made; Concrete composition was: 1:2:2 (cement:sand:gravel).

Most of the specimens were cast at a temperature of T = 20°C and air relative humidity of RH = 65%. They were immediately exposed to intensive drying in aerodynamic tunnel with wind velocity of 10 km/h at T = 30°C and RH = 40%. Other specimens were cured under cover for 3, 6, 20, 48 and 72 hours and then exposed to drying. Part of the specimens were cast at the temperature of drying.

Free shrinkage tests were carried out in special beam molds (40x40x500 mm) which did not require demoulding. Displacement of the free end of the beam was registered by a gage with accuracy of 0.001 mm. Simultaneously, weight loss of the specimen was registered, with accuracy of 0.1 g.

For restrain shrinkage tests ring-shaped specimens with outer diameter b = 236 mm and thickness h = 43 mm were used. Inner diameter was varied from a = 125 mm up to 187 mm. The restraint was provided by steel or perspex core. Crack width was measured by means of optical microscope with 33^x magnification.

3 Cracking Sensitivity of the Restrained Shrinkage Ring Test

Usually, restrained shrinkage tests by means of conventional concrete ring specimens are carried out with circular steel cores at inner/outer radii ratio of the ring in the range of 0.47-0.80 - Leschinsky (1980). Such tests, under normal laboratory drying conditions, take a period of several weeks or even months. For example, the experiments by Grzybowski and Shah (1989) showed that, in hardened concrete specimens cured for 4 days at 20°C and 100% RH and then dried at 20°C and 50% RH, cracks appeared after a period of 6 weeks drying. It is possible to somewhat shorten the period of drying, e.g. by a temperature increase of 10-20°C, but it is still difficult to obtain conditions for cracking within days or hours. In other words, conventional ring specimens with steel core do not allow to evaluate crack resistance of concrete within a reasonably short time, as may be required in a test of practical significance, and not intended only for research purposes. Preliminary tests in this work indicated that more than a week is required to obtain cracks with "conventional" ring specimens.

In order to increase the cracking sensitivity, i.e. reduce the time at which cracks appear in the ring tests, several approaches were evaluated, based on analytical treatment to determine the tangential tensile stresses developed in the test. Three main means were analyzed to predict the stress enhancement that they may bring about, to develop a ring test with maximum cracking sensitivity:

(a) Changing the ratio between the outer and inner diameter of the ring to develop an "optimal" geometry which allows maximum stress to develop, but provides, at the same time, sufficiently large dimensions for casting a specimen with reasonbly large aggregate size.

(b) Providing additional stress raising effect by temperature-induced stresses. For that purpose inner cores of materials with a high coefficient of thermal expansion were analyzed, and a temperature gradient of 5 to 15°C was induced by simultaneously applying drying conditions in an environmental chamber in which the temperature was higher than the one at which the specimen was placed prior to drying. Inner core materials applicable for this purpose were plastic materials, which also have a lower modulus of elasticity. Therefore, the analysis in this case took into account simultaneously the influence of lowering the rigidity of the core (which lowers the stresses induced in the ring) and the increase in the coefficient of thermal expansion (which increases the stresses induced in the ring).

(c) Introduction of a wedge in the ring test which acts as a stress raiser. The effect of the wedge was considered on top of the stress raising influence described in (a) and (b).

A detailed analysis of these three mechanisms was carried out and the analytical results were verified experimentally. A detailed account of this work is provided in a separate publication by Kovler, Sikuler and Bentur (1992).

However, some of the trends are shown in Fig. 1, emphasizing the influence of geometry and termperature differences when using a steel and a perspex core. The results of the analysis, shown in Fig. 1, are for hardened concrete (represented by a high modulus of elasticity of the concrete) and for a simulation of fresh, or early age concrete, which was obtained by assuming a very low modulus of elasticity of the concrete, E_0 in comparison with modulus elasticity of core material E_1.

The following cases were analyzed: hardened concrete and perspex core (E_0/E_1) = 10), hardened concrete and steel core ($E_0/E_1 = 0.1$), fresh concrete and perspex core ($E_0/E_1 = 1$) and at last fresh concrete and steel core ($E_0/E_1 = 0.01$). The coefficients of thermal expansion for concrete and steel were equalled to $\alpha_0 = \alpha_1 = 1x10^{-5}$/grad; for perspex: $\alpha_1 = 7.5x10^{-5}$/grad.

On the basis of the trends shown in Fig. 1, "optimal" test was suggested, using a ring with an inner and outer diameter of 186 mm and 236 mm, respectively, and an inner core of perspex, to provide temperature stress raising influence by subjecting the specimen to a temperature difference of 10°C immediately with the exposure to drying conditions. Fig. 1 shows the efficiency of raising stresses by this mean, and in practice this method served to shorten times for crack appearance from several weeks in the conventional ring test, to several days or hours in the perspex core-temperature difference type of ring test.

4 Influence of Low Volume Polypropylene Fibre Reinforcement on Cracking Sensitivity

The optimal ring test method, described in the previous section, was used to evaluate the influence of polypropylene fibres in enhancing the crack sensitivity of the concrete. Three variables were studied.

(a) The influence of the polypropylene fibres in fresh concrete, i.e. concrete exposed to drying immediately after casting. Casting was carried out at 20°C and drying at 30°C/40% RH + wind (see section 2).

(b) The influence of polypropylene on hardened concretes, which were water-cured for up to 3 days at 20°C before being exposed to drying conditions identical to those described in (a).

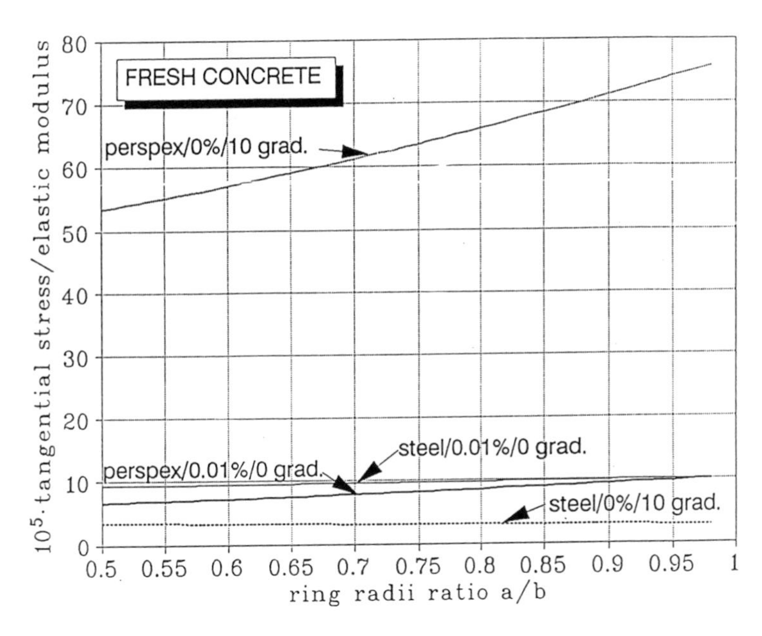

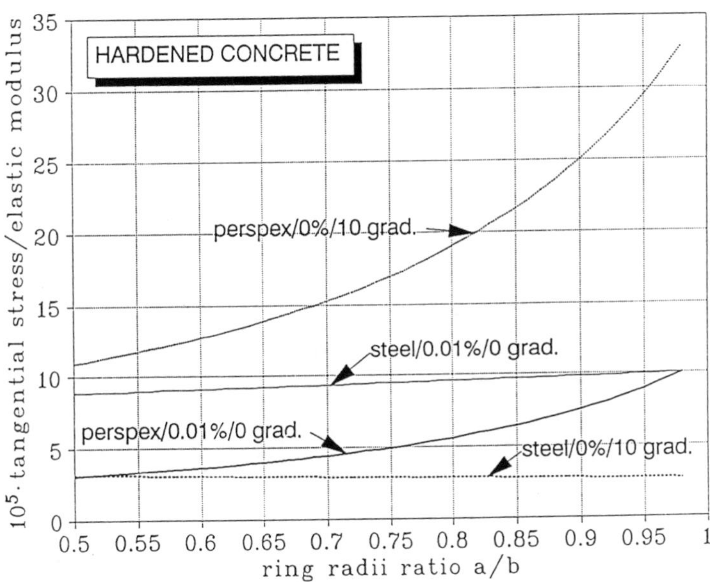

Fig. 1. The influence of geometry (a/b - ratio of inner and outer diameter), elastic and thermal properties of inner core (steel vs. perspex) on the tangential stresses (ratio of tangential stress to concrete modulus of elasticity) developed during restrained shrinkage test.

(c) The influence of the type of polypropylene fibre on the cracking sensitivity of fresh concrete exposed to drying as in (a). Three types of fibrillated fibres were evaluated: Fibres with smooth surface with a film thickness of 40 μm (type A) and 20 μm (type B) and a fibre identical to type A, but with a surface that was roughened by mechanical means (type C).

The performance of each of the systems was studied by determining weight loss and shrinkage strains during drying in unrestrained conditions, time of appearance of cracks in restrained shrinkage tests, and the mode of crack growth in restrained shrinkage tests containing a wedge.

4.1 Effect of Fibre Content

The presence of polypropylene fibres had hardly any influence on the behavior in free shrinkage conditions. Weight loss and free shrinkage curves were practically independent of fibre content in the range of 0 to 0.3% (Figs. 2 and 3).

In contrast to the lack of the influence of fibres on the behaviour under free shrinkage conditions, it was found to have a marked effect on the cracking sensitivity of fresh concrete, i.e. concrete exposed to drying immediately after casting. This is clearly demonstrated in Table 1, which presents the time of appearance of a crack in a restrained shrinkage test. If the fibre content is 0.3% or more, no cracks could be observed. It should be noted that, in the range of 0 to 0.2% fibres, there does not seem to be a positive influence of the fibres, as the cracking time seems to be in the range of 80 to 120 minutes, practically independent of fibre content in the range of 0 to 0.2%. However, when crack width is analyzed in terms of crack width vs. time curve observed in restrained shrinkage with a wedge (Fig. 4), it can be clearly seen that crack width is reduced considerably as the fibre content increases in the range of 0 to 0.2%.

Table 1

Effect of polypropylene fibre content on the time of crack appearance (minutes) in fresh concretes exposed to restrained shrinkage.

Fibre content (% vol.)	0	0.05	0.1	0.2	0.3	0.4	0.5
Time of cracking (mins.)	80	107	90	120	-	-	-

The positive influence of the polypropylene fibres is reduced drastically in the case of concretes which have been exposed to restrained shrinkage after being water cured for 2 days, as clearly seen in table 2 which presents the time after which cracks occurred in the restrained shrinkage test. The time values range between 13 to 21 minutes, and are practically independent of fibre contents, to levels of up to 0.5% fibre content by volume. This implies that, for these conditions, the polypropylene fibres are largely ineffective.

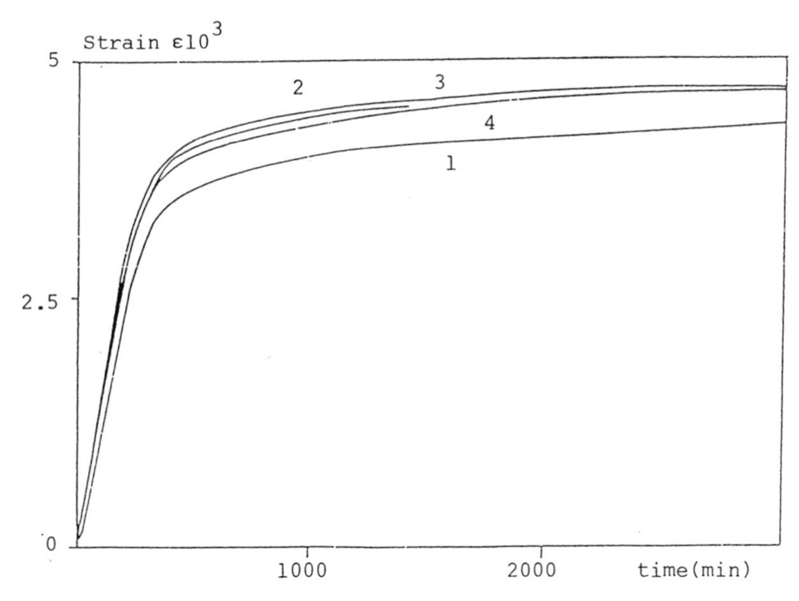

Fig. 2. The influence of polypropylene fibre content on the weight loss-time curves in free drying conditions, applied immediately after casting (20°C/65% RH) -
1 - V_f = 0%; 2 - V_f = 0.05%; 3 - V_f 0.1%; 4 - V_f = 0.3%.

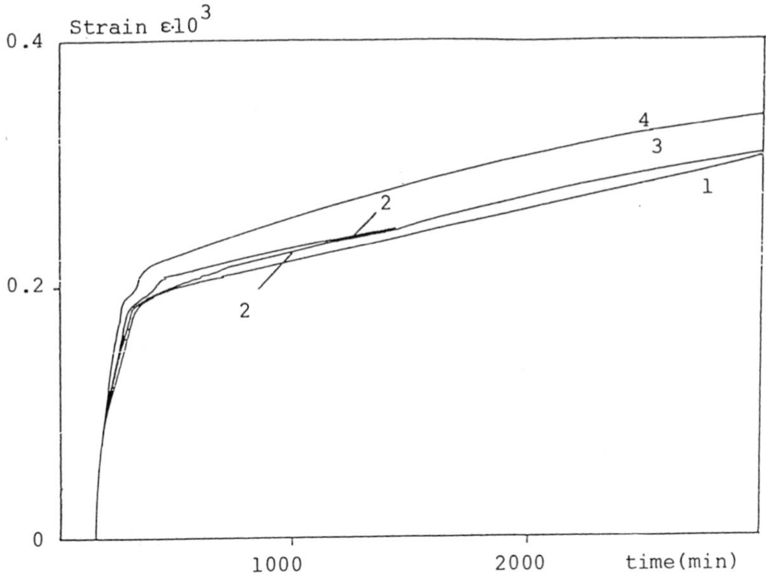

Fig. 3. The influence of polypropylene fibre content on the free shrinkage strain-time curves in free drying conditions applied immediately after casting (20°C/65% RH) -
1 - V_f = 0%; 2 - V_f = 0.05%; 3 - V_f = 0.2%; 4 - V_f = 0.3%.

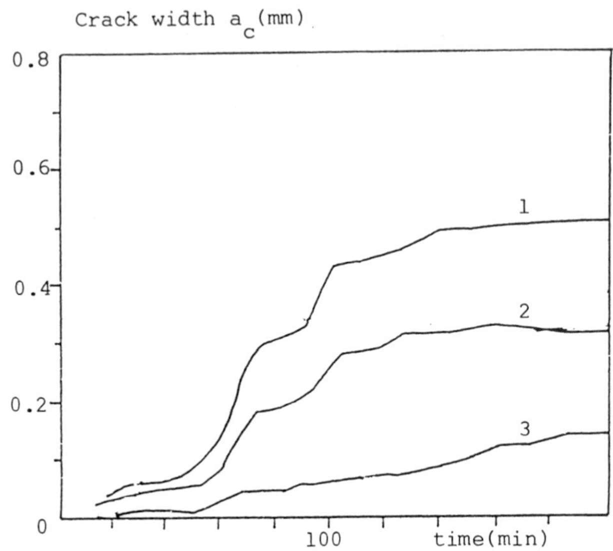

Fig. 4. Crack development (crack width) in restrained shrinkage of fresh concrete with different contents of polypropylene fibres - 1 - V_f = 0%; 2 - V_f = 0.1%; 3 - V_f = 0.2%.

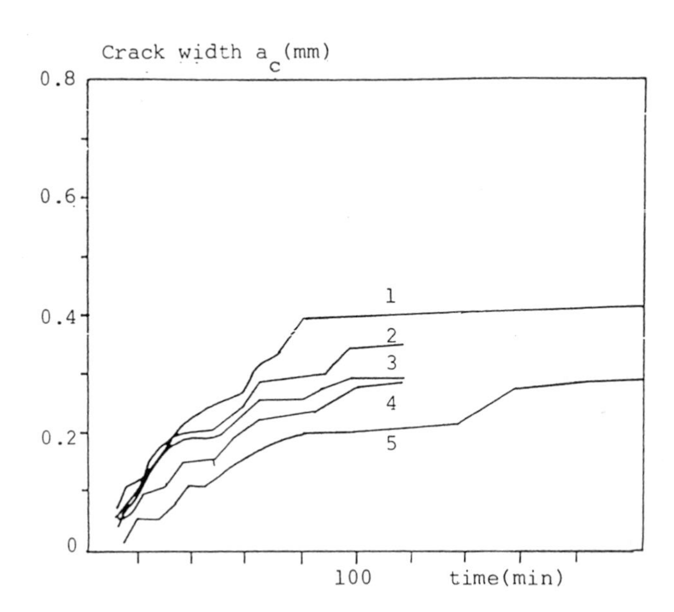

Fig. 5. Crack development (crack width) in restrained shrinkage of concretes cured under water for 2 days prior to exposure to drying - 1 - V_f = 0%; 2 - V_f = 0.05%; 3 - V_f = 0.1%; 4 - V_f = 0.2%; 5 - V_f = 0.3%.

Yet, when considering the crack width developed with time in the restrained shrinage test (Fig. 5), it can be seen that the increase in fibre content, although not effective in suppressing cracks, leads to a considerable reduction in their width, by about 50%, as the fibre content increases from 0 to 0.3%. Although this is significant, it is not as dramatic as the decrease in crack width in the fresh concrete exposed to restrained shrinkage (Fig. 4).

Table 2

Effect of polypropylene fibre content on the time of crack appearance (minutes) in concrete cured under water for 2 days prior to exposure to restrained shrinkage.

Fibre content (% vol.)	0	0.05	0.1	0.2	0.3	0.4	0.5
Time of cracking (mins.)	20	15	13	20	14	10	21

4.2 Effect of Fibre Type

The effectiveness of fibre type was evaluated for fresh and 2-days cured concrete that were exposed to restrained shrinkage. The fibre content was 0.05% for the fresh concrete and 0.1% for the 2-days cured concrete.

The results for the time of cracking (Table 3) indicate that the fibre of smaller thickness (type B) and the fibre of roughened surface (type C) were effective in eliminating cracking in the fresh concrete. It should be noted that this data indicates that 0.05% of fibres of the type B or C led to crack suppression in the fresh concrete, whereas with type A, 0.3% of fibres was needed to achieve this performance

Table 3

Effect of polypropylene fibre type on the cracking time (mins.) of fresh concrete and 2-days cured concrete exposed to restrained shrinkage.

Material	Fresh Concrete $V_f = 0.05\%$			2-days cured concrete $V_f = 0.1\%$		
Fibre type	A	B	C	A	B	C
Time of cracking (mins.)	107	-	-	13	13	13

Table 3 does not indicate any superiority of one of the fibre types with respect to the cracking performance of the concrete cured for 2 days and exposed to restrained shrinkage. However, evaluation of the width and length of the cracks developed after 5 hours of restrained shrinkage (Table 4) clearly shows that, although cracks appeared at the same time with all three fibres, the cracks observed with type C fibres were smaller (crack and length) by an order of magnitude. This implies that the improved bond that can be potentially achieved by roughening of the surface of the fibre is crucial with respect to its performance regarding the improvement of cracking resistance of hardened concrete.

Table 4

Crack width and length observed after 5 hours of restrained shrinkage of concretes cured for 2 days and reinforced with 0.1% polypropylene fibres of three different types.

Crack parameter	Fibre Type		
	A	B	C
Crack width, mm	0.35	0.35	0.03
Crack length, mm	15*	13	2**

* through macrocrack across the ring.
** microcrack on the surface of the concrete.

5 Conclusions

1. A restrained shrinkage test method was developed which is of improved sensitivity and can provide cracking data within a relatively short period of time, of several hours.

2. Low volume of polypropylene fibres was found to be extremely effective in suppressing restrained shrinkage cracks in fresh concrete (exposed immediately after casting to drying) but it could not eliminate restrained shrinkage cracks in 2-days cured concrete, but only reduce their width.

3. The effectiveness of the polypropylene fibres in hardened concrete (cured for 2 days) could be significantly improved by roughening its surface. It did not eliminate cracking but reduced their width and length by an order of magnitude.

6 References

Bentur, A. and Mindess, S. (1990), Fibre reinforced cementitious composites, Elsevier Science Publishers, London & N.Y., 449 pp.

Dahl, P.A. (1986), Influence of fiber reinforcement on plastic shrinkage and cracking. In Brittle Matrix Composites - 1, ed. A.M. Brandt & I.H. Marshall. Proc. European Mechanical Colloquium 204. Elsevier Applied Science, pp. 435-41.

Grzybowski, M. and Shah, S.P. (1989), Model to predict cracking in fibre reinforced concrete due to restrained shrinkage. Mag. of Concr. Res., V.41, No.148, 125-35.

Hannant, D.J. (1978), Fibre Cements and Fibre Concretes, John Wiley and Sons, Ltd., Chichester, 219 pp.

Kovler, K., Sikuler, J. and Bentur, A. (1992), Restrained shrinkage tests of fiber reinforced concrete ring specimens: effect of core thermal expansion, in preparation.

Kubota, H. and Sakane, K. (1967), A study on the improvement of cement mortars by admixing polymer emulsion and synthetic fiber. In Synthetic Resins in Building Construction. Proc. RILEM Symp., Paris, pp. 115-26.

Leshchinsky, M.Yu. (1980), Ispytanye betona (concrete testing). Stroyizdat Publ., Moscow, 360 pp (in Russian).

Reinsdorf, S. (1985), Glasfaserbewehrter Mortel und Beton. Bauforschung Baupraxis, N.154, 1-49.

Swamy, R.N. and Stavrides, H. (1976), Influence of fiber reinforcement on restrained shrinkage. J. Amer. Concr. Inst., V.76, 443-60.

Zollo, R.F. (1984), Collated fibrillated polypropylene fibers in FRC. In Fibre Reinforced Concrete, ed. G.C. Hoff. ACI SP-81, American Concrete Inst., Detroit, pp. 397-409.

Zollo, R.F. and Ilter, J.A. (1986), Plastic and drying shrinkage in concrete containing fibrillated polypropylene fibres. Developments in Fibre Reinforced Cement and Concrete, ed. R.N. Swamy, R.L. Wagstaffe & D.R. Oakley. Proc. RILEM Symp., Sheffield, Paper 4.5.

9 FIBRE EFFECT ON CRACKING OF CONCRETE DUE TO SHRINKAGE

H. M. S. ABDUL-WAHAB and H. K. AHMAD
University of Technology, Baghdad, Iraq

Abstract
The effect of using fibres to control cracking due to restrained shrinkage in large concrete members is investigated experimentally. A total of 44 specimens were tested representing sections of square columns 400x400mm with thick steel cores providing the internal restraint. Four types of fibres were used, namely: straight, hooked and flat steel fibres, and nylon fibres. The fibre content, core diameter and specimen height were varied. Observations were continued for nearly 18 months. The inclusion of fibres resulted in significant reductions in shrinkage and cracking, particularly with the steel types.

Keywords: Cracking , Crack width, Drying shrinkage, Fibre reinforced concrete, Nylon fibres, Restrained shrinkage, Steel fibres, Tests.

1 Introduction

Cracking due to restrained drying shrinkage is a major problem in concrete structures. In large sized concrete members, as in columns, differential shrinkage between the surface and the interior concrete causes tensile stresses to develop at the surface. The larger shrinkage at the surface causes cracks to develop that may, with time penetrate deeper into the concrete[2]. In hot climates, as in Iraq, the tendency to cracking in precast and cast-in-situ reinforced concrete columns has been recognized as a serious hazard[1]. Extensive longitudinal cracks are initiated at the points of maximum tensile stress due to shrinkage, usually at midpoints of the faces of rectangular columns. With further drying, the cracks spread along the length and increase in width.

One possible method to control or reduce the adverse effects of cracking due to restrained shrinkage in concrete members is the addition of fibres[2,3,8]. However, the shrinkage of fibre reinforced concrete is influenced by a combination of factors such as specimen size, fibre type, fibre content and age of concrete when drying begins[3,4,8].

Fibre Reinforced Cement and Concrete. Edited by R. N. Swamy. © 1992 RILEM.
Published by E & FN Spon, 2-6 Boundary Row, London SE1 8HN. ISBN 0 419 18130 X.

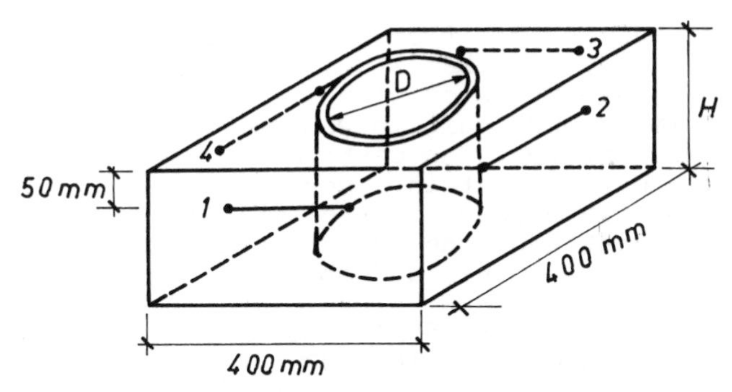

Fig.1. Test specimen.

Results obtained from studies on the effects of fibres on cracking due to restrained shrinkage using the ring-type specimens have shown that the presence of fibres considerably reduce shrinkage and the average crack width and delays the formation of the first crack[3-8]. The amount of fibre reinforcement was found to be the most significant factor in controlling and reducing cracking due to shrinkage[5,6].

The aim of this study was to experimentally investigate the effect of using different types and amounts of fibres on the durability and cracking tendency in FRC rectangular column sections with stiff cores due to shrinkage. The main parameters considered were the fibre type, fibre volume fraction, core diameter and height of column section. Observations were continued for nearly 18 months under the prevailing local climatic conditions in Baghdad where the annual temperature variation is between -8.5°C to 58°C and the relative humidity varying between 5% -100% .

2 Experimental programme

A total of 44 specimens of 400mm square column sections as shown in Fig.1 were tested. Thick steel pipes with different diameters were cast at the centre of the specimens to act as stiff cores so that the cracking tendency in the thin shell at the surface of the column can be observed. The thick steel pipes had three outer diameters, namely 220, 270 and 320mm. The height of specimens was 100mm or 200mm. The wall thickness of the steel pipes was between 7.3mm - 9.3mm of ductile steel.

Ordinary Portland cement was used with natural sand within zone 2 and coarse aggregate (natural gravel) with maximum size of 19mm complying with BS 882-1973. The mix proportion was 1:2:4 of cement : sand : coarse aggregate with a water : cement ratio of 0.53 giving a medium workability of 60mm to 80mm slump. Four types of fibres were used, namely : straight steel fibres with length, l = 20mm and diameter, d = 0.4mm, flat steel fibres with l = 30mm and 0.5mm thickness, hooked steel fibres with l = 30mm and d = 0.5mm, and nylon fibres with l = 20mm and d = 0.05mm.

The specimens were cast in wooden moulds and compacted using a poker vibrator. The casting and testing of the specimens was carried out in a covered area in the shade. The specimens were exposed to the ambient conditions, but not to direct heat from the sun or to rain. The duration of the period of measurements and observations extended to nearly 18 months.

Control specimens of 150x150x150mm cubes, 150mm diameter x 300mm high cylinders and 100x100x400mm prisms were cast with each batch of casting. They were used to determine the cube compressive strength, tensile splitting strength, modulus of rupture, modulus of elasticity, and free shrinkage.

After 24 hours from casting, the wooden moulds were removed and demec points for a 200mm mechanical extensometer were fixed at selected positions on the sides of the specimens as shown in Fig.1. The first (initial) readings for shrinkage and strains were taken at the age of 2 days for all the specimens. Continuous inspection and observation followed to determine the age at which the first crack appeared and to record the formation and development of first and subsequent cracks. Periodic measurements of strains and free shrinkage (linear) were made for all the specimens, as well as recording the temperature and relative humidity for the duration of tests.

The control specimens were tested at 28 days and the results of the properties of concrete are given in Table 1.

Table 1 : Properties of Concrete.

Group / Property	Plain concrete	FRC 1% str.st. fibre	FRC 1% nylon fibre
Cube comp. strength,MPa	26.3	24.2	21.56
Cylinder compressive strength, MPa.	20.5	20.75	21.5
Tensile split.strength,MPa	3.02	7.85	8.44
Modulus of rupture,MPa.	2.5	2.15	2.97
Modulus of elasticity,GPa	16.5	25.89	17.82
Shrinkage at 526 days,x10^{-5}	55	43	46

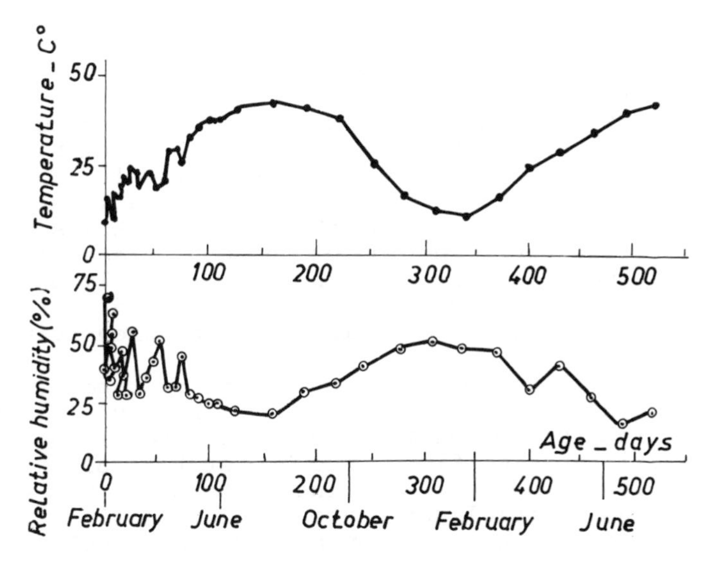

Fig.2. Variation in temperature and R.H.during test period

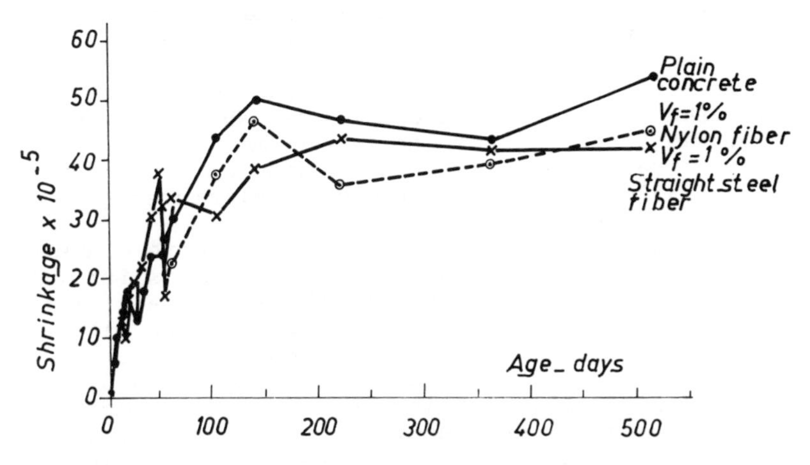

Fig.3. Effect of fibre on free drying shrinkage.

3 Results and discussion

Fig.2 shows the temperature and relative humidity during the test period. The temperature ranged between 9°C to 45°C and the relative humidity varied between 17% - 71% . Fig.3 shows the free shrinkage results obtained from prisms of plain concrete and FRC with volume fraction, V_f = 1% of straight steel fibres or nylon fibres. After 526 days of drying, the final free shrinkage was not significantly affected by the presence of fibres and ranged between 43 - 55 x 10^{-5} . This conforms with the conclusion made by other research workers that the addition of fibres does not appreciably change the free shrinkage[3,5,8].

Table 2 gives details of all the specimens tested, and the age at which the first and second crack appeared. It is clear that cracking occurs mostly in the first 10 - 50 days of age. If no cracks appear then, they are less likely to appear later on. The position of the first crack was invariably at the thinnest section of the concrete shell. Other cracks usually appeared in the diametrically opposite location. Also, it is evident that the inclusion of fibres have a direct influence in delaying, or even preventing the appearance of cracks.

The effect of the fibre content on the restrained shrinkage strains for the four types of fibres used is shown in Fig.4 , which is typical for the three core diameters tested. For all types of fibre, the higher the volume fraction, V_f the lower the recorded final strains at the sides of the specimens. Initial shrinkage during the first three months exhibited a wide variation as a direct result of the variation in temperature and relative humidity. However, the final values of shrinkage for the long duration of tests clearly illustrate the influence of fibre content. Also, it is evident that steel fibres have a much greater effect in reducing shrinkage than nylon fibres. Fig.5 further shows the effect of 1% volume fraction of the four types of fibres, with different core diameters. When no cracks appeared, as for D = 220mm and D = 270mm, the variations in shrinkage strains were limited. As soon as crcks developed, as for D = 320mm, the influence of steel fibres became more pronounced than that for nylon fibres.

Fig.6 shows the effect of varying the core diameter on shrinkage using all four types of fibre. It is for fibre content of 1% volume fraction and it is typical for the other values. The final shrinkage appear to be little influenced by the core diameter provided that no cracking takes place. But with a larger core diameter, and therefore a much thinner shell, cracking is more likely, and positive strains (expansion) are recorded. Fig.7 further illustrate the effect of the core diameter with variable volume fraction of hooked steel fibres but with specimen height of 100mm. For the largest core diameter of D = 320mm, the higher the fibre content the lower the final strains. With the medium core diameter of D = 270mm, the inclusion of steel fibres actually prevented cracking. For the smallest core diameter no cracking appeared, but shrinkage was appreciably reduced by the steel fibres.

Considering the results shown in Figures 5 and 7 , it appears that the effect of reducing the specimens height from 200mm to 100mm was to increase the cracking tendency. This is mainly due to the reduction in the size of the concrete shell. All the other factors exhibited similar effects.

Table 2. Details of specimens and location of cracks

Specimen	Height H (mm)	Core dia.D (mm)	Type of fibre	Fibre vol. (%)	Age (days) 1st crack	Age (days) 2nd crack	Location point No
A	200	220	None	-	-	-	-
AS1	200	220	Str.st.	0.5	-	-	-
AS2	200	220	=	1.0	-	-	-
AS3	200	220	=	1.5	-	-	-
AF1	200	220	Flat st	0.5	-	-	-
AF2	200	220	=	1.0	-	-	-
AH1	200	220	Hooked	0.5	-	-	-
AH2	200	220	=	1.0	-	-	-
AN1	200	220	Nylon	0.5	-	-	-
AN2	200	220	=	1.0	-	-	-
AN3	200	220	=	1.5	-	-	-
AH3	100	220	None	-	-	-	-
AH4	100	220	Hooked	0.5	-	-	-
AH5	100	220	=	1.0	-	-	-
B	200	270	None	-	-	-	-
BS1	200	270	Str.st	0.5	-	-	-
BS2	200	270	=	1.0	-	-	-
BS3	200	270	=	1.5	-	-	-
BF1	200	270	Flat st	0.5	-	-	-
BF2	200	270	=	1.0	-	-	-
BF3	200	270	=	1.5	-	-	-
BH1	200	270	Hooked	0.5	-	-	-
BH2	200	270	=	1.0	-	-	-
BN1	200	270	Nylon	0.5	-	-	-
BN2	200	270	=	1.0	-	-	-
BN3	200	270	=	1.5	-	-	-
BH3	100	270	None	-	42	45	2,4
BH4	100	270	Hooked	0.5	-	-	-
BH5	100	270	=	1.0	-	-	-
C	200	320	None	-	11	20	2,4
CS1	200	320	Str.st.	0.5	21	28	2,4
CS2	200	320	=	1.0	26	35	2,4
CS3	200	320	=	1.5	42	124	2,4
CF1	200	320	Flat st	0.5	16	22	2,4
CF2	200	320	=	1.0	19	25	2,4
CF3	200	320	=	1.5	25	42	2,4
CH1	200	320	Hooked	0.5	42	45	2,4
CH2	200	320	=	1.0	48	52	2,4
CN1	200	320	Nylon	0.5	14	19	2,4
CN2	200	320	=	1.0	16	19	2,4
CN3	200	320	=	1.5	19	42	2,4
CH3	100	320	None	-	36	38	2,4
CH4	100	320	Hooked	0.5	42	44	2,4
CH5	100	320	=	1.0	44	48	2,4

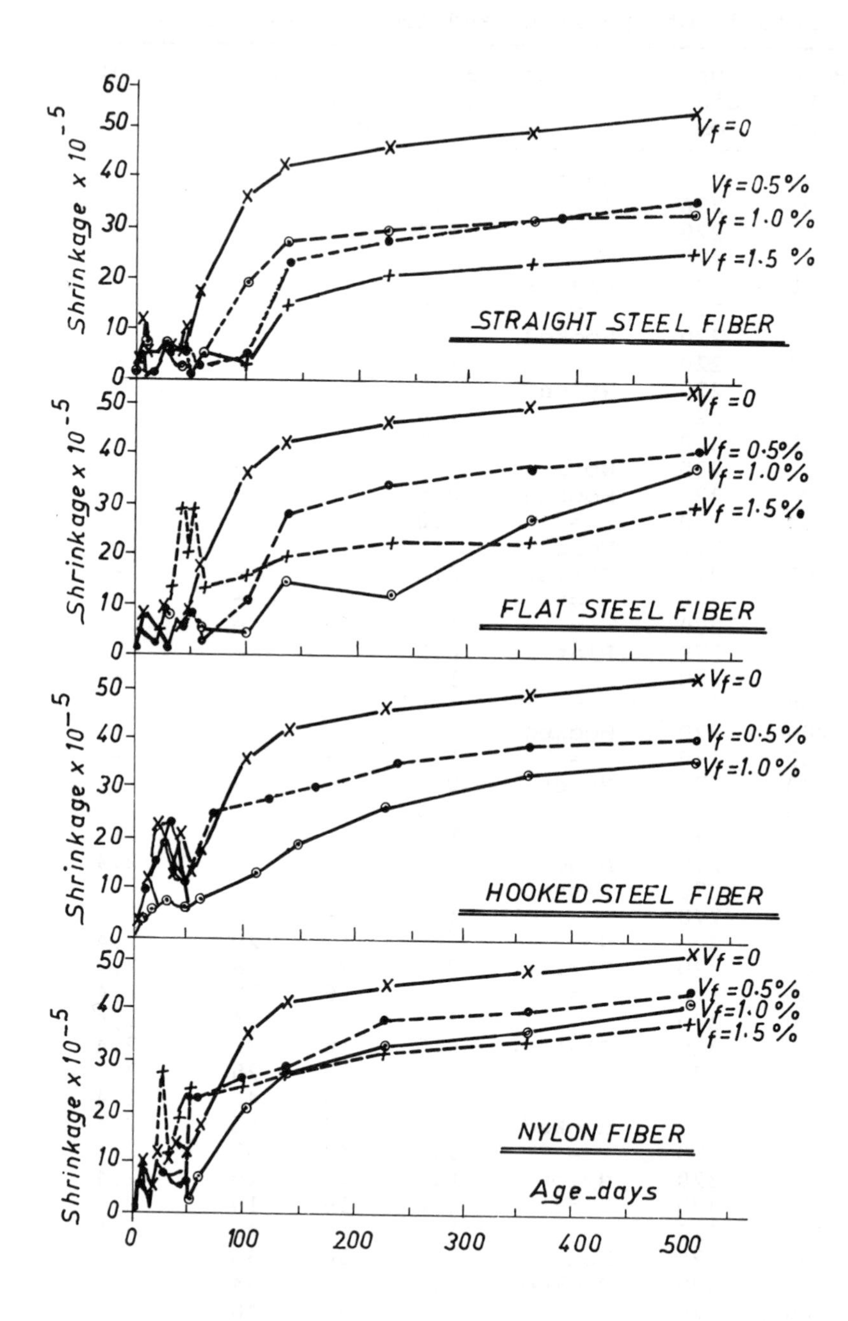

Fig.4. Effect of fibre content on restrained shrinkage.
(D= 270mm; H= 200mm)

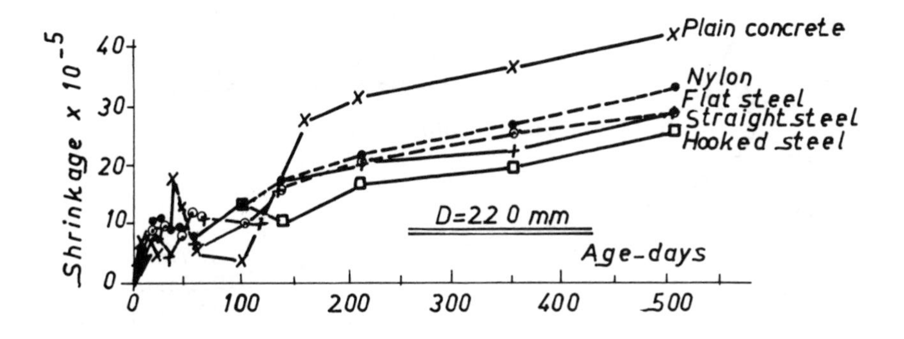

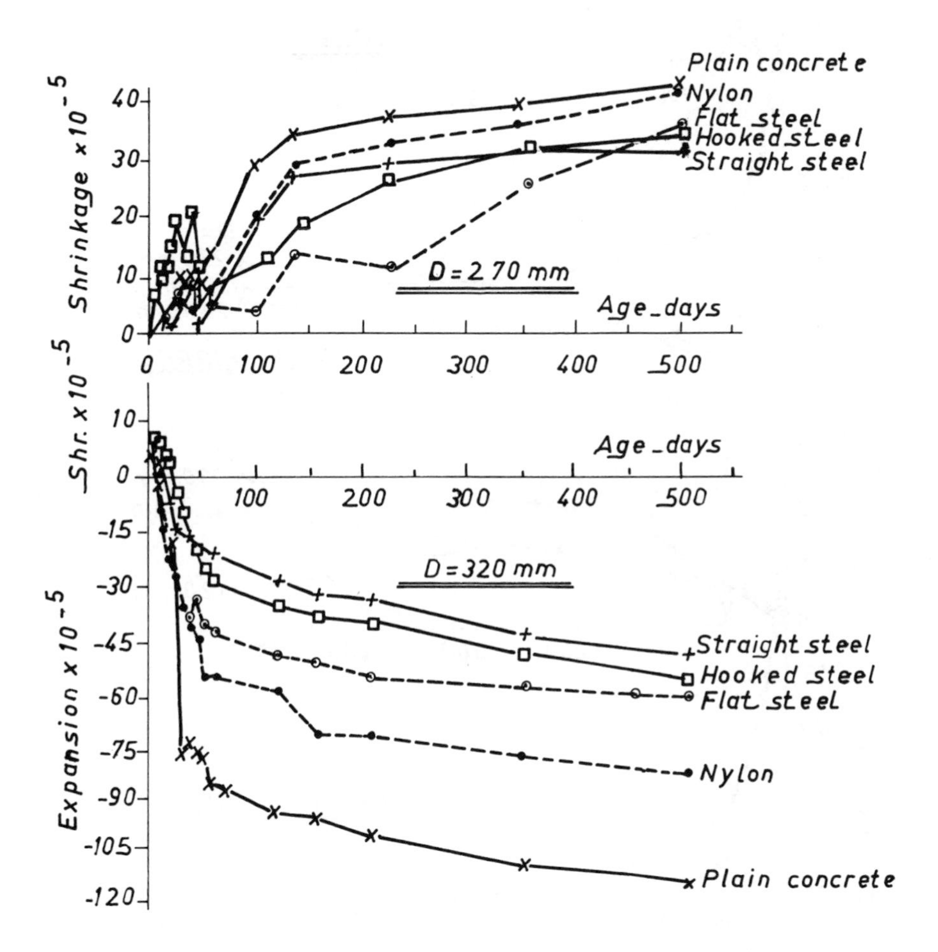

Fig.5. Effect of type of fibre on shrinkage variation for different core diameters.(V_f= 1%; H= 200mm)

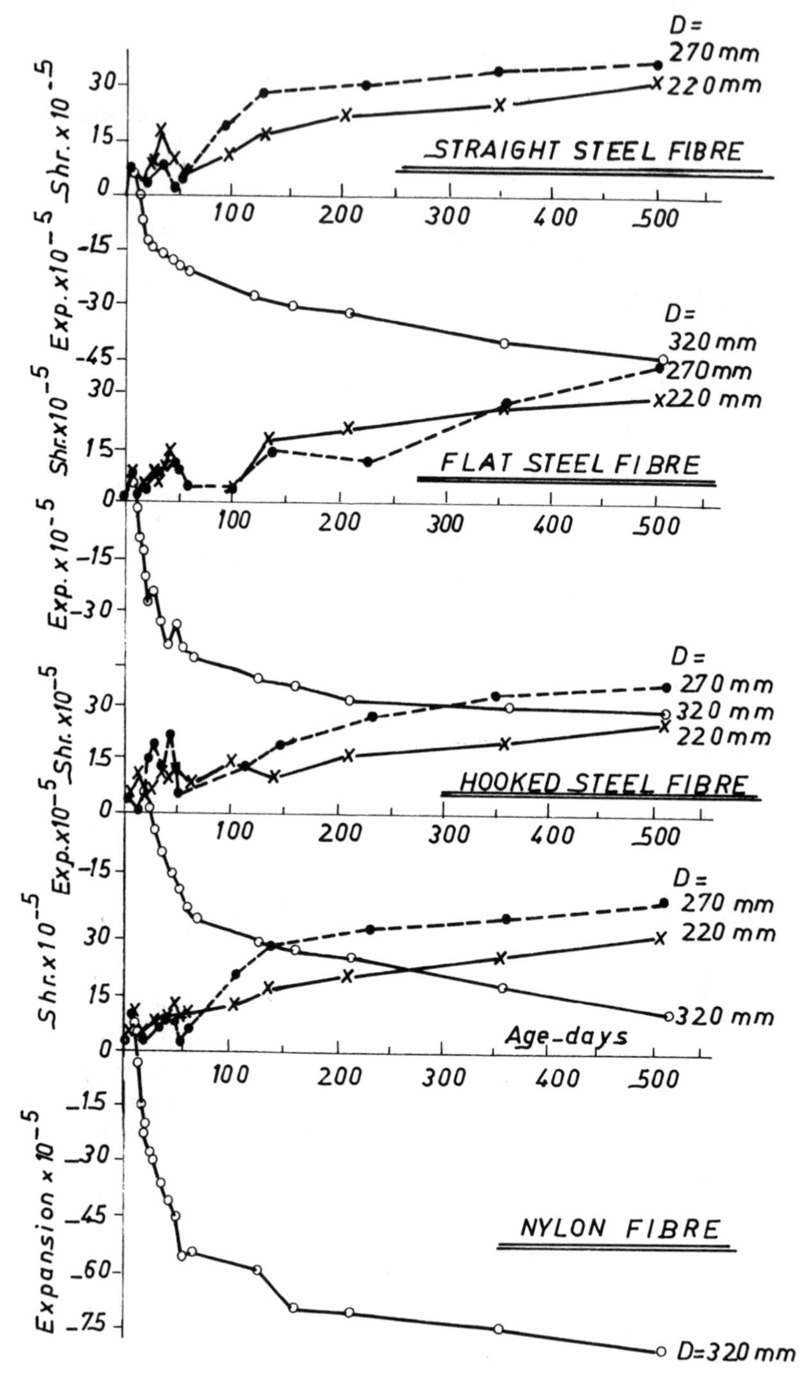

Fig.6. Effect of core diameter on shrinkage variation.
(V_f= 1%; H= 200mm)

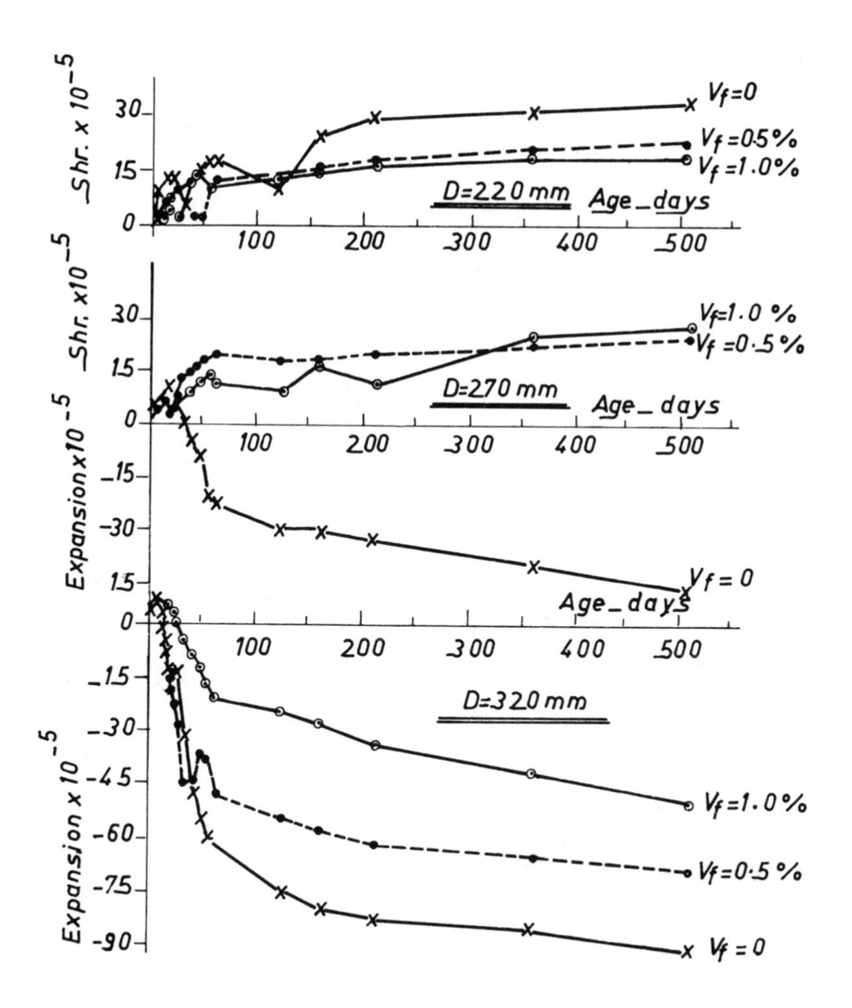

Fig.7. Effect of core diameter on strain variation.
(H= 100mm; hooked steel fibre)

For the specimens with D = 320mm that underwent extensive cracking, the crack width was greatly influenced by the fibre type and fibre content as shown in Fig.8. The crack width was significantly reduced with all four types of fibre used. Generally, the higher the fibre content, the smaller the crack width. However, steel fibres appear to have a greater influence in reducing the crack width than nylon, with straight steel fibres exhibiting the highest efficiency.

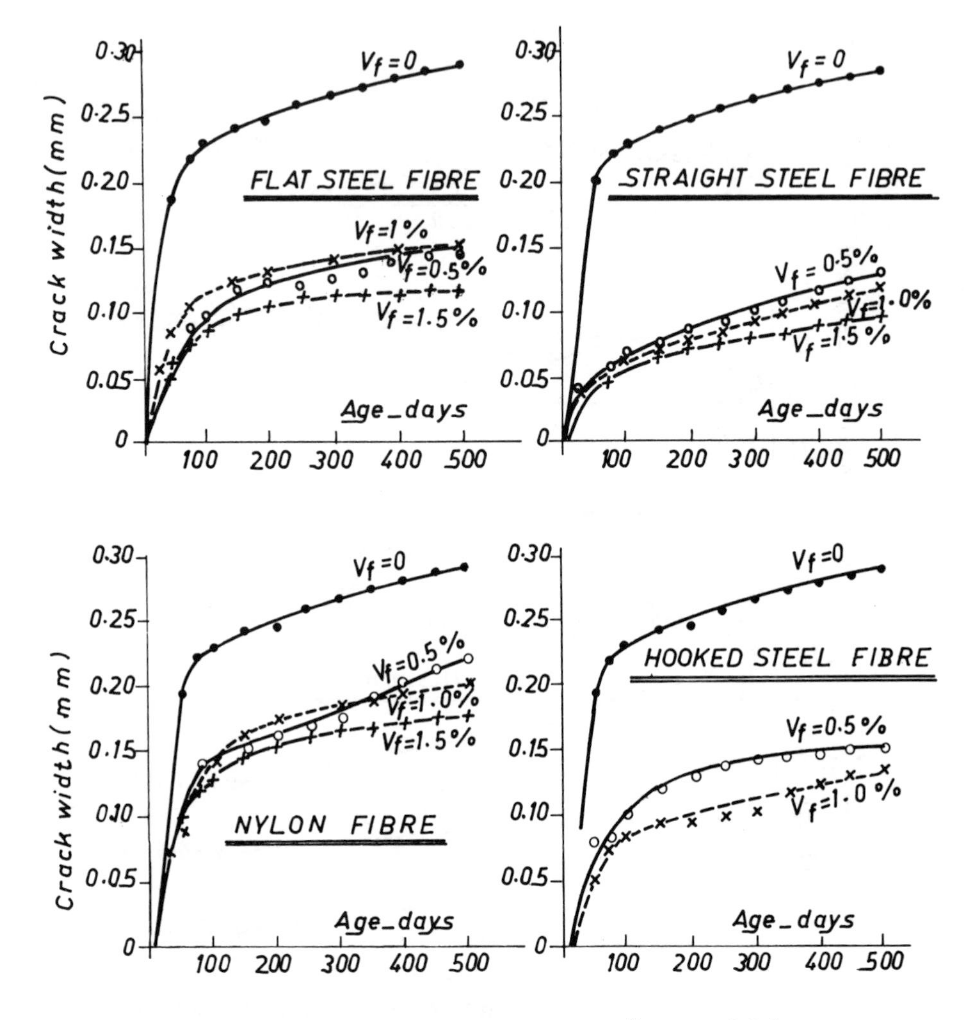

Fig.8. Effect of fibre content on crack width.
(D= 320mm; H= 200mm)

4 Conclusions

From the results of this investigation, the following conclusions can be drawn.

1. The type of fibre have a direct influence on restrained drying shrinkage in concrete columns, as well as on the control of cracking. Steel fibres have substantially greater effect than nylon fibres, but straight or hooked steel fibres may be more efficient than the flat steel type.

2. Increasing the fibre content resulted in a general reduction in drying shrinkage. But a more significant influence of the fibre content was observed when the specimens were cracked.
3. The stiffness of the core of the specimen and the core diame ter had a direct effect on the cracking tendency. The inclusion of fibres influenced this tendency significantly by preventing cracking in some cases or controlling crack development in others.
4. The crack widths were substantially reduced by the inclusion of all four types of fibre used. Nevertheless, steel fibres were evidently more effective than nylon fibres.

5 Acknowledgments

The experimental work reported in this paper was carried out by the authors at the structures laboratory of the Building and Construction Engineering Department, University of Technology, Baghdad. The authors wish to acknowledge the assistance of the technical staff and the facilities made available to them.

6 References

1. Abdul-Wahab, H.M.S. and Ahmad, H.K.(1990) Cracking due to shrinkage in reinforced concrete columns, Al-Muhandis (The Engineer), Iraqi Engineers Society,103, 3-16.
2. ACI Committee 207 (1990) Effect of restraint, volume change and reinforcement on cracking of mass concrete, (ACI 207.2R). ACI Materials J.,87 (3), 271-295.
3. Chern,J.G., and Young, C.H.(1990) Study of factors influencing drying shrinkage of steel fibre reinforced concrete, ACI Materials J., 87 (2), 123-129.Materials Journal, Vol. 87, No.2, March-April 1990, pp.123-129.
4. Clastres, P. and Debicki, G.(1989) Shrinkage cracking test of a cement paste reinforced by synthetic fibres, International Conference on Recent Developments in Fibre Reinforced Cements and Concretes, Cardiff (U.K.).
5. Grzybowski, M. and Shah, S.P.(1989) Model to predict cracking in fibre reinforced concrete due to restrained shrinkage, Mag. of Concrete Research, 41 (148), 125-135.
6. Grzybowski, M. and Shah, S.P.(1990) Shrinkage cracking of fibre reinforced concrete, ACI Materials J., 87 (2), 138-148.
7. Krenchel,H. and Shah,S.(1987) Restrained shrinkage tests with PP-fibre reinforced concrete, Fibre Reinforced Concrete Properties and Applications, SP-105,ACI,141-158.
8. Swamy,R.N. and Stavrides,H.(1979) Influence of fibre reinforcement on restrained shrinkage and cracking, ACI Journal, 76 (3), 443-460.

10 PROPERTIES OF GFRC MORTARS WITH DIFFERENT POZZOLANIC ADDITIVES

J. MADEJ
Beton Vuis Ltd, Bratislava, Czech and Slovak Federal Republic

Abstract
Three types of glass fibres of different alkali resistance were used in plain and blended OPC mortars as to investigate the properties of fresh and hardened mortars, including durability investigation of E-glass fibres in the matrix. The differences in the properties of fresh mixes, depending both on the type of the fibres, as well as on the particular type of pozzolana, were observed. Individual types of glass fibres differed from each other: in relation to an air-entrainment a positive effect of E-glass fibres compared to other types of glass fibres studied (REZAL 5, Cem-FIL 1) was observed. An increase in tensile strength compared to plain OPC mortar is presented when the pozzolans are used, mainly in the case of metakaolinite, condensed silica fume and blast-furnace slag.
Keywords: Durability, Glass Fibres, Microstructure, Mortars, Pozzolans, Strength Increase.

1 Introduction

An interest of building industry in Czecho-Slovakia for application of fibre reinforced composites (FRC) has increased during the last years. Besides the steel and polypropylene fibres, several types of glass fibres have been developed and investigated, as stated by Komloš et al. (1988, 1991).

Pozzolans such as fly ash, silica fume, slags or natural pozzolans have been successfully used as the additives for cement mortars and concrete; the general knowledge in this field is summarized by Uchikawa (1986), Malhotra (1989) and many others. The physico-mechanical properties of concrete with pozzolanic additives have been studied, mainly in combination with other admixtures (plasticizers, superplasticizers, polymers etc.). The improved durability of E-glass fibres was confirmed when combined with some mineral powders such as fly ash, metakaolinite or condensed silica fume - Leonard and Bentur (1984) Ambroise et al. (1986), Murat and Al Cheikh (1989).

The objective of this research programme was to compare several natural and artificial pozzolans in relation to their influence on the properties of fresh and hardened mortars, being combined with different types of glass fibres.

Fibre Reinforced Cement and Concrete. Edited by R. N. Swamy. © 1992 RILEM.
Published by E & FN Spon, 2-6 Boundary Row, London SE1 8HN. ISBN 0 419 18130 X.

2 Experimental

2.1 Materials

Cement

Ordinary Portland cement PC 400 (according to the Czecho-Slovak Standard ČSN 72 2121), corresponding to ASTM type I, was used.

Sand

Natural silica sand (sized 0.09 to 1.5 mm) was used for preparation of fresh mortars and the specimens.

Glass fibres

Three different types of glass fibres were used:

- E-glass fibres, non-alkali resistant glass fibres produced in Czecho-Slovakia
- REZAL 5, low-zirconium dioxide content glass fibres produced in Czecho-Slovakia
- Cem-FIL 1, alkali resistant glass fibres, a product of Pilkington Brothers Ltd, Great Britain.

The length of glass fibres was 6 mm.

Pozzolans

7 different additives were used:

- low-calcium fly ash
- zeolite, with the main component clinoptilolite
- condensed silica fume, a by-product from FeSi production
- metakaolinite, thermal activated product of kaolinite
- ground granulated blast-furnace slag
- diatomaceous earth
- thermal activated clay.

Chemical analysis and fundamental physico-chemical properties of cement and the pozzolans are given in Table 1.

Superplasticizer

An aqueous naftalene formaldehyde-based superplasticizer (44% of dry solid) of a Czecho-Slovak origin was used.

2.2 Mixture proportions

The cement mortars with a cement:(sand+pozzolana) ratio of 1:1 were prepared. The water to cement ratio was kept constant, w/c=0.40. The superplasticizer was used in the amount of 1.0 wt% of dry solid from cement weight. The amounts of pozzolanic additives in the case of normal and glass fibres reinforced cement (GFRC) mortars were 20 wt% of the mass of sand, except for activated clay, where a dosage ought to be decreased to 10 wt%. The reason of the reduced dosage in case of activated clay was a strong negative effect on the workability of fresh mixes.

2.3 Preparation and casting of specimens

The mortars were prepared using a laboratory mortar mixer with a total volume of 3 liters (the apparatus corresponds to the Czecho-Slovak Standard ČSN 72 2117). The mixing procedure was as follows : sand and glass fibres were homogenized together and mixed with the half volume of mixing water and superplasticizer ("premix procedure") for a time of 60 sec. After than cement, pozzolanic additive and the rest of mixing water were added, the procedure was in accordance with ČSN 72 2117.

Fresh mixtures
The properties of fresh mixtures-workability according to the Czecho-Slovak Standard ČSN 72 2441 (See Figure 1), density, and air-content, were measured.

Hardened mortars
The mortar specimens- prisms 40 x 40 x 160 mm - were prepared. After 24 hours' storage in an air with a relative humidity of 95% and a temperature 20° C, the specimens were demoulded and stored in water for the next 27 days.

2.4 Testing of specimens
The mass, the measurements of resonate frequency (from which dynamic moduli of elasticity were calculated), the tensile strength (central point loading test) and compressive strength of specimens were tested.

Table 1. Chemical composition and physical properties of cement and pozzolans

		Pozzolans						
Component	Cement	Fly Ash	Zeo-lite	Silica Fume	Kaoli-nite	Slag	Diato-mite	Activ. clay
SiO_2	19.58	53.83	66.60	95.57	47.65	39.52	69.80	53.94
Al_2O_3	7.58	28.00	12.15	0.11	38.05	6.79	18.35	21.87
Fe_2O_3	3.61	6.99	1.45	0.20	0.20	0.98	3.52	12.21
CaO	61.96	3.26	6.27	0.56	0.20	38.24	0.75	15.02
MgO	1.85	2.75	1.40	0.91	0.52	9.35	0.31	2.29
SO_3	2.13	0.80	0.04	0.32	-	0.26	0.43	0.16
Na_2O	0.28	0.76	0.64	0.12	-	0.38	-	2.38
K_2O	0.91	1.97	2.84	0.78	-	0.78	-	0.74
Ign. loss	1.01	1.58	8.11	1.36	13.89	1.61	5.31	11.22
Density, $kg.m^{-3}$	3148	2256	2242	2493	2560	2763	2370	2670
Surface area, $m^2.kg^{-1}$	336	402	604	33300	13720	350	11230	28600

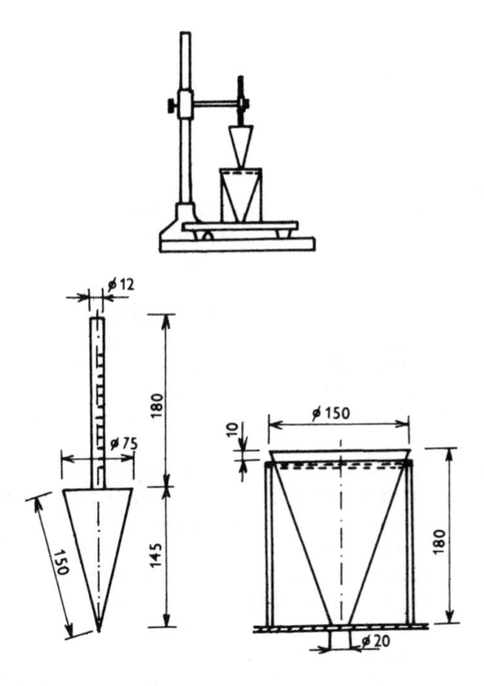

Fig. 1. A scheme of the workability test, in accordance with the Czecho-Slovak Standard ČSN 72 2441

Note : The workability in mm is expressed in terms of the immersion depth of standard cone into the fresh mortar

A combined DTA-DTG method was used for an estimation of free lime content in the plain mortar and mortars with different pozzolans. The amounts of some components (Ca^{2+}, K^{+}, Na^{+}) and pH were determined in the solutions, obtained by leaching of mortar specimens. 50g of ground mortars (passing 0.09 mm) was leached in 200 g of distilled water for a time of 24 hours before being filtrated and above-mentioned components analyzed.

An investigation of a deterioration rate of E-glass fibres in the matrix followed, making use a scanning electron microscope JOEL (type JSM 325) of Japan.

3 Test results and discussion

The test results of fresh mortars and physico-mechanical properties of water-cured mortar specimens are given in Tables 2-5. The free lime content and the amounts of components in the solution from mortar specimens are summarized in Table 6. A view of the original E-glass fibres is shown in Figure 2 and those embedded in the matrix in Figures 3 and 4.

3.1 Properties of fresh mixtures

From the test results given in Table 2, remarkable differences in the properties of fresh mixes with no fibres can be seen, depending on the type of pozzolana. From the additives tested, even in combination with superplasticizer in relatively high amount (1 wt% of dry solid) a strong negative effect on the workability was proved for diatomite and activated clay. An air content, in case of all the additives used, was smaller compared with plain OPC mortar (Table 2).

With E-glass fibres, the workability of fresh mortars decreased compared with no-GFRC mortars, except for the addition of silica fume (Mix No.12). In all cases, except for the metakaolinite (Mix No.13), the air content increased compared with no-GFRC mortars (Table 3).

The mortars with REZAL glass fibres (Table 4) are characterized by the decrease in the workability of fresh mixes. An increase in air content compared to above-mentioned mixtures can be seen, the most remarkable being in the case of condensed silica fume (Mix No.20) and activated clay (Mix No.24).

The properties of fresh mortars with Cem-FIL 1 glass fibres, in relation to the type of pozzolans, are comparable to those with REZAL glass fibres. An air content in the mixtures with individual additives was the highest from all the types of glass fibres used.

The test results clearly show the differences in the properties of fresh mortars, being influenced both by the type of glass fibres and the particular effect of pozzolana, other proportions and conditions being constant. While the effect of the physical state of glass fibres is supposed to be of a great importance in relation to the workability of fresh mortars, some topo-chemical reactions in the system may be the reason of some remarkable differences in the air content.

Table 2. Properties of fresh mortars and physico-mechanical properties of water-cured mortar specimens. No fibres used.

		Fresh mortars			Hardened mortars		
Mix No.	Type of pozzolana	Density ($kg.m^{-3}$)	Workability (mm)	Air cont. (%)	Dynamic modulus (GPa)	Compres. strength (MPa)	Tensile strength (MPa)
1	-	2025	>130	9.0	28.9	35.0	8.0
2	fly ash	2100	>130	4.2	28.4	40.2	8.3
3	zeolite	2070	100	5.5	38.8	41.1	6.3
4	silica fume	2108	125	4.2	30.8	41.8	8.9
5	m-kaolinite	2090	115	7.0	26.2	33.0	7.5
6	slag	2103	>130	3.8	29.4	36.2	7.3
7	diatomite	2090	55	4.2	29.7	33.7	6.4
8	act. clay	2050	85	8.2	21.7	29.6	6.6

Table 3. Properties of fresh mortars and physico-mechanical properties of water-cured mortar specimens with E-glass fibres (3 wt %).

Mix No.	Type of pozzolana	Fresh mortars Density ($kg.m^{-3}$)	Workability (mm)	Air cont. (%)	Hardened mortars Dynamic modulus (GPa)	Compres. strength (MPa)	Tensile strength (MPa)
9	-	2062	>130	10.8	29.7	33.3	8.2
10	fly ash	1970	120	10.6	28.8	40.6	10.5
11	zeolite	1990	85	9.5	28.4	44.9	10.6
12	silica fume	2097	130	5.5	30.8	51.6	11.1
13	m-kaolinite	2075	100	5.8	26.5	39.7	13.5
14	slag	2013	130	10.7	29.6	43.7	10.4
15	diatomite	2065	40	7.5	26.1	35.5	9.3
16	act. clay	1985	50	9.5	19.8	25.9	8.0

Table 4. Properties of fresh mortars and physico-mechanical properties of water-cured mortar specimens with REZAL 5 glass fibres (3 wt %)

Mix No.	Type of pozzolana	Fresh mortars Density ($kg.m^{-3}$)	Workability (mm)	Air cont. (%)	Hardened mortars Dynamic modulus (GPa)	Compres. strength (MPa)	Tensile strength (MPa)
17	-	2040	90	13.8	23.6	34.7	9.1
18	fly ash	1920	80	13.2	25.8	32.9	9.7
19	zeolite	2015	55	10.6	27.3	37.6	10.2
20	silica fume	1924	90	13.5	26.2	38.9	10.6
21	m-kaolinite	2052	50	8.5	26.0	35.8	11.8
22	slag	2065	60	9.0	27.1	36.9	11.8
23	diatomite	2049	45	6.7	26.0	23.2	10.9
24	act. clay	1915	50	13.0	20.7	24.1	9.7

Table 5. Properties of fresh mortars and physico-mechanical properties of water-cured mortar specimens with Cem-FIL 1 glass fibres (3 wt %)

Mix No.	Type of pozzolana	Fresh mortars: Density ($kg.m^{-3}$)	Workability (mm)	Air cont. (%)	Hardened mortars: Dynamic modulus (GPa)	Compres. strength (MPa)	Tensile strength (MPa)
25	-	1855	90	19.0	23.0	30.5	9.2
26	fly ash	1987	75	11.5	27.2	44.2	11.0
27	zeolite	2070	50	9.8	28.2	50.4	11.2
28	silica fume	1890	95	14.0	27.4	38.5	12.5
29	m-kaolinite	1930	70	9.0	25.4	34.6	12.8
30	slag	2053	60	8.5	27.9	49.4	12.2

3.2 Physico-mechanical properties

From the test results given in Table 2, the differences in pozzolanic activity of individual pozzolans can be seen. Only in the case of silica fume no decisive loss of compressive strength or tensile strength can be seen after 28 days' water curing. A remarkable decrease in the mechanical properties compared to plain mortar can be seen on the mortar specimens with activated clay. In this case, however, the amount of the material added is unusually high and serves only for comparison with other additives, mainly in relation to the durability of glass fibres in the matrix.

Using E-glass fibres, the positive effect on the compressive strength (compared to the mortars with no fibres- Table 2) can be seen, except for clay addition (Mix No.16). While silica fume proved the highest positive effect on compressive strength, the most remarkable effect of metakaolinite on tensile strength was observed (Table 3).

The combination of REZAL glass fibres proved beneficial in relation to the strength increase in case of silica fume and metakaolinite addition. The strength increase, however, was influenced markedly by an increase in air content of the fresh mortars. This is probably the main reason why, despite of increased alkali resistance - see for example Komloš at al. (1988), (1991) - REZAL glass fibres didn't show such a positive effect on relative tensile strength as non-alkali resistant E-glass fibres.

The same relations may be seen also in the case of Cem-FIL 1 glass fibres mortars (Table 5): although with individual pozzolans the strength increases compared to above-mentioned types of glass fibres, air entrainment of fresh mixes seems to be a decisive factor influencing the physico-mechanical properties of mortar specimens. It needs more detailed investigation and explanation, mainly in relation to the use of condensed silica fume (Mix No.28).

3.3 Free lime and alkali content

From the test results given in Table 6, the differences in the pozzolanic activity of individual pozzolans can be seen. Condensed silica fume proved to be an excellent pozzolanic additive when compared with other additives tested. The test results obtained are in general agreement with other experiments and accepted view in this field , e.g. Uchikawa (1986), Jambor (1963), Tenoutasse and Marion (1985), Madej et al. (1990).

The most remarkable decrease in alkali content was observed for the mortars with metakaolinite addition, however, the method used didn't allow to obtain the representative composition of pore solution.

3.4 Durability of E-glass fibres in the matrix

From the Figure 3 it is seen, that E-glass fibres deteriorated in the matrix of plain OPC mortar to high degree. On the other hand, no visible corrosion of the fibres was observed when combined with pozzolanic additives (Fig. 4). SEM observations thus confirmed high efficiency of the pozzolans tested, in relation to the protective effect of non-alkali resistant glass fibres. The observations are in agreement with the results of physico-mechanical properties of mortar specimens (Table 3) and also in relation to the other characteristics of cement matrix (Table 6). The differences between individual pozzolans, however, can be proved only after longer investigation.

Table 6. Amount of the components in the solutions from mortar specimens and free lime content (DTA-DTG method)

Mix No.	Type of pozzolana	Component				
		Ca^{2+} (as CaO) (wt %)	K^+ (mg/l)	Na+ (mg/l)	pH	$Ca(OH)_2$ (wt %)
1	-	921	410	252	12.48	11.56
2	fly ash	823	356	132	12.49	8.31
3	zeolite	739	384	160	12.44	8.60
4	silica fume	656	322	156	12.43	3.51
5	m-kaolinite	767	254	155	12.44	6.24
6	slag	795	326	147	12.45	7.40
7	diatomite	809	354	176	12.41	8.49
8	act. clay	725	314	292	12.40	8.31

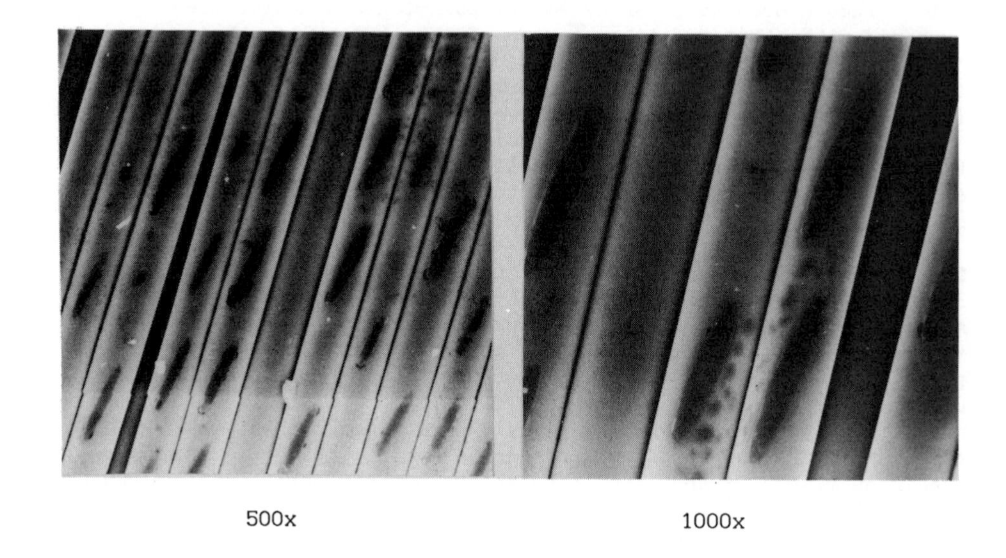

Fig.2. SEM of original E-glass fibres

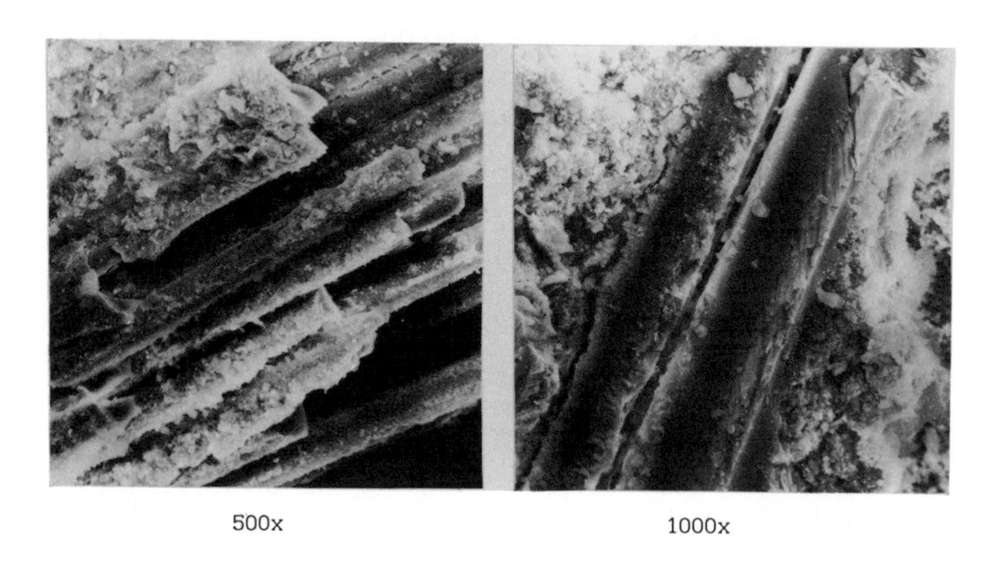

Fig.3. SEM of E-glass fibres in the plain cement matrix after 28 days of curing in water

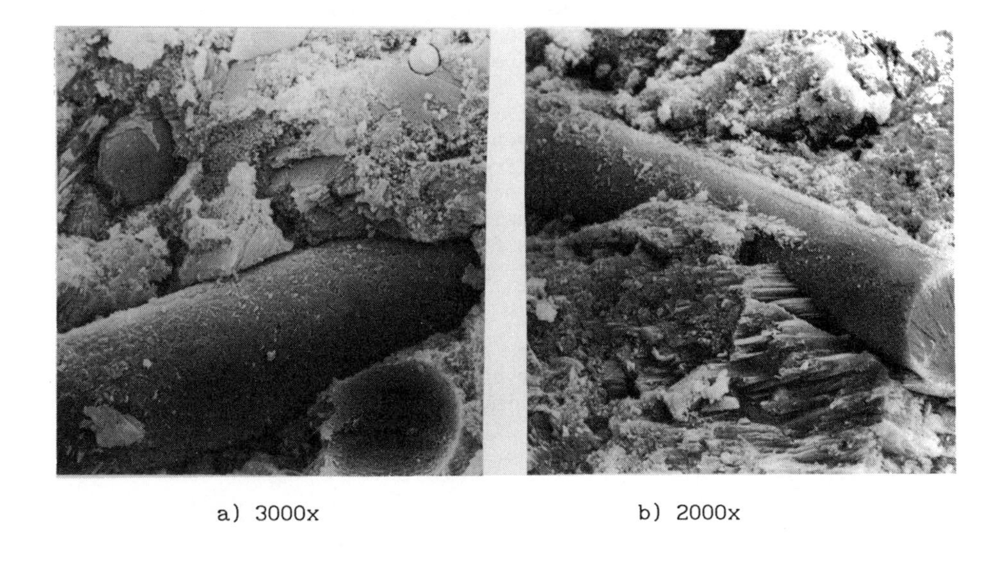

a) 3000x b) 2000x

c) 3000x d) 2000x

Fig.4. SEM of E-glass fibres in cement matrix after 28 days of curing in water: a-with fly ash; b-with zeolite; c-with metakaolinite; d-with condensed silica fume.

4 Conclusions

1. The properties of fresh mortars (workability, air content) are influenced by the type of glass fibres. A remarkable air entrainment of fresh mortars prepared with alkali resistant glass fibres (REZAL 5, Cem-FIL 1) compared to E-glass fibres was observed.

2. The pozzolanic activity of different types of pozzolanic additives, together with air entrainment of fresh mortars, plays an important role in relation to the physico-mechanical properties of mortar specimens. In relation to the tensile strength, a great positive influence was observed when glass fibres were combined with metakaolinite, condensed silica fume or granulated blast-furnace slag.

3. No deterioration of E-glass fibres embedded in water-cured mortar specimens after 28 days was observed by SEM in cement matrix with the additives studied. Long term investigation is necessary as to compare the protective effect of different types of pozzolanic additives.

5 References

Ambroise, N. and Marion, A.M. (1986) in **Proceedings of International Symposium on Durability of Glass Fiber Reinforced Concrete** (S.Diamond Ed.), PCI, Chicago, pp. 285-292.

Jambor, J. (1963) **Stavebnícky časopis,** XI, 1-2, 115-136.

Komloš, K. Vaniš, M. Babál, B. and Kozánková, J. (1988) Properties of glass reinforced cement composites, in **Proceedings of 13th Congress of IABSE**, Helsinky, pp.39-44.

Komloš, K. Vaniš, M. Babál, B. and Kozánková, J. (1991) Durability control of GRC composites, in **Proceedings of the Second International RILEM/CEB Symposium**, Ghent, pp. 302-312.

Leonard, S. and Bentur, A. (1984) Improvement of the durability of glass fiber reinforced cement using blended cement matrix, **Cement and Concrete Research**, 14, 717-728.

Malhotra, V.M. (Editor) (1989) **Proceedings of the 3rd International Conference on Fly Ash, Silica Fume, Slag and Natural Pozzolans in Concrete, ACI SP 114**. Trondheim, Norway.

Madej, J. Madejová, J. and Jakubeková, D. (1990) IR spectroscopic study of silica fume-modified cement pastes. **CERAMICS-Silikáty**, 34, 131-141.

Murat, M. and Al Cheikh, A (1989) Behavior of E-glass fiber in basic aqueous medium resulting from the dissolution of mineral binders containing metakaolinite, **Cement and Concrete Research**, 19, 16-24.

Tenoutasse, N. and Marion, A.M. (1985) The influence of silica fume on the hydration and microstructure of OPC pastes, in **Proceedings of the 2nd Int.Conf. on Use Fly Ash, Silica Fume, Slag and Mineral Admixtures in Concrete, ACI SP 91**, (Editor: V.M.Malhotra), Suppl. Paper, Vol III.

Uchikawa, H.(1986) Effect of blending component on hydration and structure formation, in **Proceedings of the 8th Int.Cong. on the Chemistry of Cement**, Rio de Janeiro, Principal Report, Vol.1, pp. 79-155.

Acknowledgements

The Author would like to thank to Karol Komloš, D.Sc., from the Institute of Building and Architecture of Slovak Academy of Sciences in Bratislava, for his useful comments to this work.

11 INFLUENCE OF FIBRE PARAMETERS ON COMPRESSIVE DEFORMATION OF STEEL FIBRE REINFORCED CEMENTS

A. E. S. ABDUL-MONEM* and R. S. BAGGOTT
University of Salford, UK

Synopsis
The compressive deformation of steel fibre reinforced cements containing up to 15% volume fraction of fibre is described, with particular reference to the influence of the geometry of the very fine melt overflow stainless steel fibres. Composites were prepared by slurry infiltration of a steel fibre bed and cured both at room temperature and under autoclave conditions. The two methods of curing enabled the influence of matrices of high strength and very high strength to be compared.

The experimental data is discussed with respect to the nature of the fibre contribution to deformation and strength and its dependency upon fibre concentration, length and diameter, and matrix strength.
Keywords: Compressive strength, Composites, Steel Fibres, Autoclave, Calcium Silicates, Infiltration.

1 Introduction

Steel fibre reinforced cementitious composites can be prepared with high volume fractions of fibre by simple infiltration methods. Slurry infiltrated fibre concrete, developed by Lankard (1984), has been used successfully in pavement and bridge deck overlays, security concrete, refractory concrete, explosive and seismic resistant structural applications as well as for the manufacture of precast concrete products (Lankard and Newall (1984), Balagurn and Kendzulak (1987)). A variant of this type of material with superior properties in some respects can be produced by autoclaving (Baggott and Sarandily (1986)).

The tensile strength and toughness of these materials are very considerably greater than normal steel fibre reinforced concrete (Naaman (1991), Baggott and Abdel-Monem (1991)). Very large improvements in compressive toughness are also obtained but only small changes in ultimate compressive strength.

---

[1] Present Address: University of Minia, Egypt

Fibre Reinforced Cement and Concrete. Edited by R. N. Swamy. © 1992 RILEM.
Published by E & FN Spon, 2-6 Boundary Row, London SE1 8HN. ISBN 0 419 18130 X.

In the case of compressive strength both positive and negative changes in ultimate strength have been reported in the literature but there does not appear to be an overall explanation for the variations.

The influence of different types of steel fibre on reinforcing behaviour in low volume fraction composites has been widely investigated but little data is available on the influence of fibre characteristics in high volume fraction composites. Of particular interest are the recently developed melt overflow steel fibres which are available in nominal diameters of around 100 μm, of irregular cross section (Fibre Technology Ltd). They provide very fine fibres with excellent bond properties and compatibility with high alkaline environments.

The paper presents experimental data on the compressive deformation and strength of autoclaved and room temperature cured composites fabricated by infiltration and containing up to 15% volume fraction of melt overflow fibres. The results are part of a comprehensive investigation of the mechanical properties of these composites (Abdel-Monem (1991)). The overall objective addressed in the paper is to consider the basis for the various relationships that have been observed between the fibre parameters, the matrix strength and the compressive modulus and strength of the composites. Three particular aspects are considered, the influence of fibre geometry, the influence of volume fraction and the influence of autoclaving.

2 Experimental Materials and Methods

The matrix consisted of 60 parts Ordinary Portland cement, 40 parts of silica flour (BM500) and a mixture of 27 parts of water and 3 parts of superplasticiser (Conplast 430) all by weight. These proportions produced a wet mix of very fluid consistency and a hardened material after autoclaving of very high strength.

Stainless steel melt overflow fibres (Microtex), of general appearance as shown in Figure 1 were used. Their essential features are variability of overall shape, a kidney shaped cross-section and a rough surface nature. Six types of fibre were tested, 10mm, 20 mm and 50mm long, each at 0.10 and 0.15 mm maximum dimension of the kidney shape (subsequently referred to as diameter for convenience).

Composites were fabricated with 5, 10 and 15% by volume of each type of fibre. Fabrication consisted of placing the required volume of fibres in a steel mould (50 x 330 x 50mm) by hand dispersion and subsequent pressing with 15% volume fraction arrays.

The slurry was prepared using a combination of blender and planetary mixer to ensure good dispersion of the silica flour. Slurry infiltration was achieved by direct gravity induced flow assisted by external vibration. The moulds were covered for 24h then stripped and allowed to cure for another 24h before cutting into cube specimens 50 x 50 x 50mm size.

The various specimens were either autoclaved with a cycle of 2h build-up

Figure 1 Microtex Fibres

time, 8h dwell-time at 180 °C and 3h blow-down time, or room temperature cured under water at 20 °C for 28 days.

Testing was carried out in a servo hydraulic testing machine with a constant rate of cross head movement of 0.3 mm/min. The parallel mould edge faces were placed in contact with the platens. Specimen deformation was measured by two linear variable differential transducers placed between the loading platens.

3 Results

Figures 2 and 3 illustrate typical load deformation curves for autoclaved and room temperature cured material. The essential features of the composite curves are a linear region extending to approximately 90% of the maximum load followed by a non-linear region identifying continued load bearing capability well beyond the maximum load.

Modulus of elasticity values of autoclaved material lay in the range 6850 to 9480 N/mm^2 with a mean value of 7930 N/mm^2, apparently independent of fibre content. In the case of room temperature cured material modulii values lay in the range 4100 to 5560 N/mm^2, with a mean value of 4540 N/mm^2, again without any dependence upon fibre content.

Deformation in the non-linear region was accompanied by the gradual development of cracks from the loaded faces predominantly in two planes.

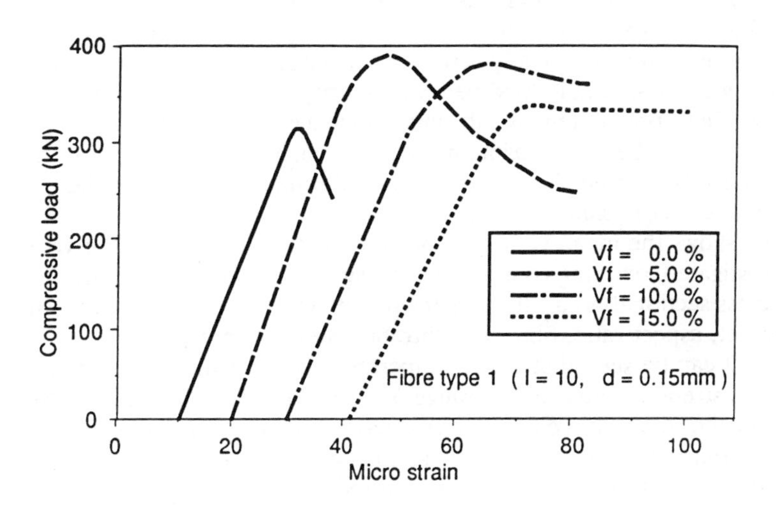

Fig. 2 Typical load - strain curves for autoclaved materials

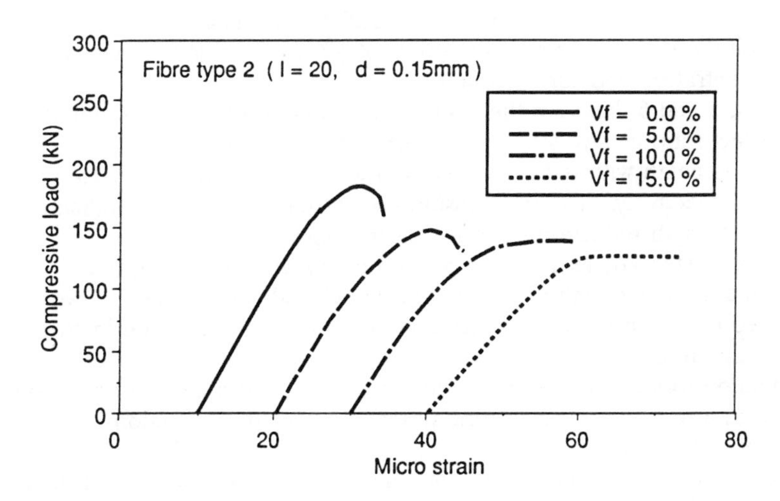

Fig. 3 Typical load - strain curves for room temperature cured materials

The angle of these major cracks was consistent in all composites and was aligned 15 to 20 degrees from the direction of compressive force. Eventually the cracks propagated completely across the specimen although even at this stage there was considerable load bearing capacity in the truncated cone remaining. The fibre content significantly influenced the shape of the deformation curves beyond maximum loads. At higher volume fractions there was a considerable further 'linear' region of deformation parallel to the deformation axis at maximum load.

Table 1 shows the mean values of the compressive strength for the various mixes investigated, defining compressive strength as maximum load divided by original cross-sectional area. It indicates the effect on strength of fibre length and diameter, aspect ratio, fibre concentration and curing condition.

Overall it can be seen that room-temperature cured composites had average compressive strength values in the range 51 to 63 N/mm^2 compared to 68 N/mm^2 for the un-reinforced matrix whereas autoclaved composite strengths were in the range 107 to 148 N/mm^2 compared to 120 N/mm^2 for the un-reinforced matrix. Fibre addition therefore reduced the strength of room temperature cured material regardless of fibre type whereas in the case of autoclaved material increases in strength were observed with the shortest fibres and either increases or reductions depending on fibre type and concentration with the longer fibres.

A significant dependence on length is evident in Table 1, the longer the fibre the weaker the composite with only one exception, ie the autoclaved composite containing 10% of 20mm long, 0.1mm diameter fibres. The dependency is very pronounced at 5% and 15% concentration in the case of autoclaved composites.

The range of diameter available for investigation was narrow hence only marginal effects were expected. The results indicate a slight increase in strength with the thicker fibres of the order of 5% observable with every length, concentration and curing condition.

Figures 4, 5 and 6 illustrate the combined effects of length and diameter by plotting aspect ratio against strength and generally show a reducing compressive strength with increasing aspect ration for 5%, 10%, and 15% composites respectively. The relationship is clearly non-linear and indicates diminishing strength reductions with increasing aspect ratio.

No overall systematic relationship was observed with increasing fibre volume content and strength in the case of autoclaved composites. However, the averaged results at each concentration did indicate an optimum in strength at 10% volume fraction.

The data on room temperature cured material indicates a slight reduction in strength as a result of incorporating fibres, however the reduction is independent of fibre concentration.

Comparisons between autoclaved data and room temperature cured data indicate greatly improved strength as a result of autoclaving, amounting to 75% for the un-reinforced matrix and up to 130% for certain composites.

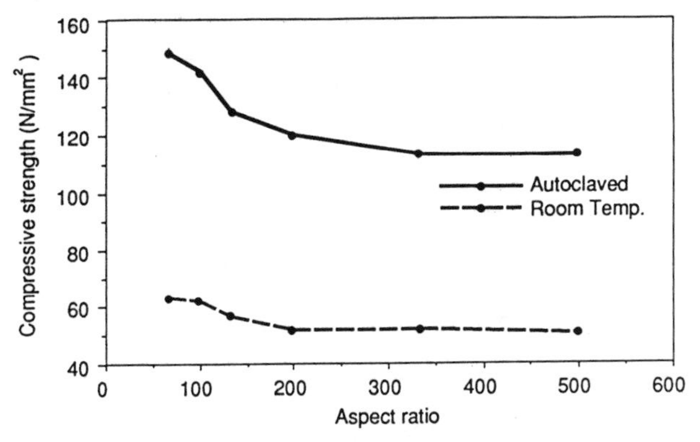

Fig. 4 Relationship between compressive strength and aspect ratio for composites with Vf = 5%

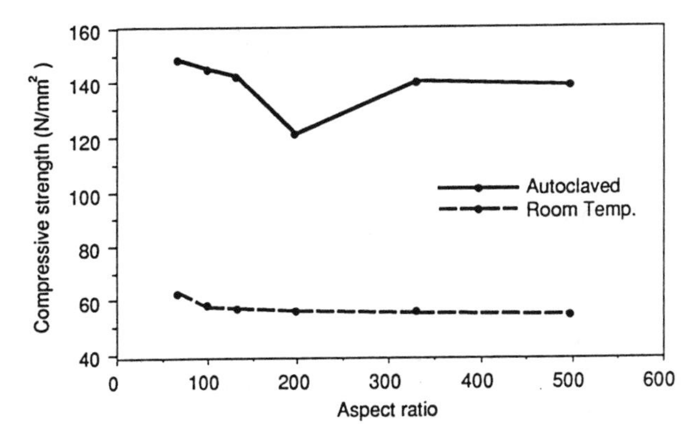

Fig.5 Relationship between compressive strength and aspect ratio for composites with Vf =10%

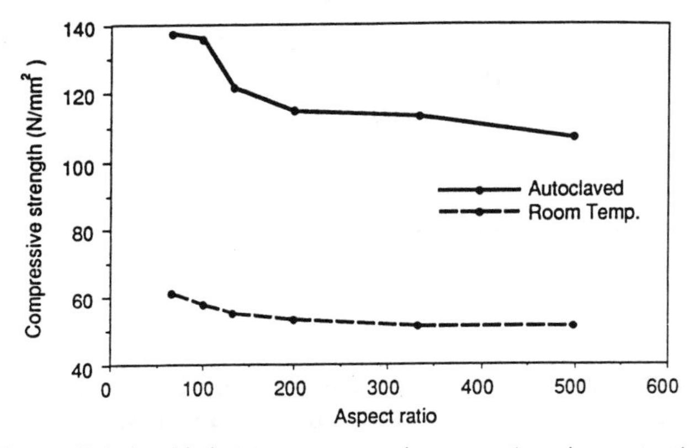

Fig. 6 Relationship between compressive strength and aspect ratio for composites with Vf =15 %

Table 1 **Compressive Strength Values**

Mix No	Fibre Properties			Vf%	Compressive Strength in N/mm²	
	l mm	d mm	l/d		Autoclaved	Room Temperature Cured
1	-	-	-	0.0	119.5	67.8
2	10		66.66		148.3	63.3
3	20	0.15	133.33		128.3	57.0
4	50		333.33		113.0	52.0
				5%		
5	10		100		142.3	62.4
6	20	0.10	200		120.0	52.2
7	50		500		112.8	50.7
9	10		66.66		148.4	63.0
10	20	0.15	133.33		142.2	56.8
11	50		333.33		140.4	55.0
				10%		
12	10		100		145.0	57.4
13	20	0.10	200		121.2	55.8
14	50		500		138.8	54.4
16	10		66.66		137.7	61.1
17	20	0.15	133.33		121.4	55.5
18	50		333.33		113.0	51.2
				15%		
19	10		100		136.0	58.2
20	20	0.10	200		114.6	53.0
21	50		500		106.5	51.2

4 Discussion

The overall results demonstrate that an extremely tough cement based material with very high compressive strength can be produced with very simple technology. A major attribute of the material is that it is not defect sensitive.

The contribution of the fibres is conventional in that it is primarily one of imparting post first crack load support. The higher the volume fraction of fibres the greater is the inelastic strain that can be accommodated without any load drop.

The minor changes to ultimate compressive strength values that are observed with lower strength matrices and coarser steel fibres are also observed with the very high strength matrices and very fine fibres. The following discussion is concerned with these secondary effects with a view to obtaining some insight into the apparent inconsistencies reported in the literature as to the influence of fibres on compressive strength.

In order to evaluate such effects it is useful to consider three somewhat paradoxical questions thrown up by the data, namely:

Why does adding up to 15% of a fibre 30 - 40 times stiffer than the matrix have no effect on modulus of elasticity?
Why does increasing the length of the fibre reduce the strength of all composites investigated in a systematic way whereas the effects of increasing quantity of fibre ranged from none at all in one matrix and variable with fibre type in the other?
Why does an identical fibre array increase the strength of one matrix and decrease it in another?

Considering first the modulus of elasticity.

The first point to make is that the results support the observations made in the literature on different matrices and steel fibre types. Naaman (1985) concluded that up to 5% addition did not affect the slope of the ascending portion of the deformation curve. A recent study by Homrich and Naaman (1987) on the properties of SIFCON in compression indicated that using up to 23% volume fraction of fibre, oriented either normal or parallel to the applied load did not affect the modulus of elasticity. The present results enable a direct comparison to be made of the effect of identical fibre arrays on matrices of significantly different moduli, 7930 N/mm^2 compared to 4540 N/mm^2.

According to the rule of mixtures a continuous aligned composite containing 15% of fibres of modulus 40 times that of the matrix should have a modulus almost x7 that of the un-reinforced matrix.

$$E_c = E_m V_m + E_f V_f \qquad (1$$

where E_c = composite modulus = 6.8, when E_m = matrix modulus = 1, V_f = volume fraction of fibre = 0.85, E_f = fibre modulus = 40 and V_f = volume fraction of fibre = 0.15

Simplistic theories for discontinuous random arrays reduce this factor to x2, (Lim et al 1986).

$$E_c = E_m V_m + E_f n_1 n_2 V_f \qquad (2$$

where n_1 = 0.4 a fibre orientation index and n_2 = a fibre length index

This level of increase was observed between the two basic matrices despite the experimental scatter and should therefore have been observed with fibre stiffening. In the case of the 23% Sifcon aligned discontinuous fibre composites the theoretical difference should have been much greater and its non-occurrence is even more significant. Three possible explanations can be made for the discrepancy: the effect is masked by experimental scatter despite the previous argument, the theoretical treatment of the effect of discontinuous fibres on the basic rule of mixtures is invalid or there is a modulus reduction mechanism compensating for the modulus increase due to increasing fibre content. Further work is necessary to resolve this issue.

The second question concerns the influence of the various fibre characteristics on strength. The pertinent observations are that a change of fibre in an array changes the composite strength to a greater extent than increasing the number of fibres in an array.

The apparent contradiction between the lack of effect of increasing fibre content and the significant influence of fibre length can be explained on the basis of fibres having three separate effects. These are to reduce the 'within composite' matrix strength indirectly, to reinforce via crack suppression and to eventually exacerbate matrix cracking. A composite strength either greater, equal or less than the un-reinforced matrix strength could therefore be produced depending upon the relative magnitude of the three effects.

It has been shown that the addition of fibres can increase the 'within-composite' matrix strength in tension, (Aveston, Mercer and Sillwood (1974), Aveston, Cooper and Kelly (1974), Johnson, Baggott and Abdel-Monem (1991)). Whether the modification is an increase or decrease depends upon the details of the actual composite system. In compression, strength reducing mechanisms are frequently associated with increased porosity. This is thought to be the case for the present results, ie it is probable that the reduction mechanism is one of reduced 'within-composite' matrix density because of less effective displacement of entrapped air during the infiltration of both higher density fibre arrays and arrays containing longer fibres. However, porosity data was much too variable to demonstrate such an effect.

Reinforcing mechanisms in compression can occur and significant strength increases have been reported, for example Oiter and Naaman (1986) obtained increases of up to 70% in concrete. The overall conceptual framework for reinforcement that is usually invoked is that the fibres provide lateral restraint or confinement of the developing tensile stresses perpendicular to the direction of compression. In other words conventional tensile crack suppression reinforcing mechanisms operate locally.

However, the present results indicate both strength increases and strength reductions for composites with identical entrapped air content and identical fibre arrays.

It follows that there must be further fibre matrix interactions which eventually override the crack suppression mechanism. Such negative reinforcement appears to increase with increasing fibre content and fibre aspect ratio. This is thought to be connected with the nature of the load transfer systems possible in compression. In addition to the shear stress transfer from matrix to fibre usually taken as the basis for tensile reinforcement, direct compressive loading can occur on all surfaces of the fibre and end loading in particular could make a significant contribution to overall fibre loading. Direct fibre to fibre contact, especially in the densely packed 15% volume arrays, also provides a means of distributing load throughout the fibre array. With increased fibre content combined with very effective fibre loading there is a greater likelihood of localised high stresses occurring because of, for example a particular fibre being end loaded sufficiently to develop buckling stresses or a geometric array of connecting fibres that could generate expansive forces.

Whether the nett result of positive and negative reinforcement is a strength increase will be dependent upon the geometry of the fibre, the nature of the fibre/matrix bond and the strength of the matrix.

This speculative model is able to explain qualitatively the various trends observed.

The lower strength with longer fibres, which is the reverse of the effect of length in tension, is attributed to the combination of a lower 'within-composite' matrix strength with the earlier development of negative reinforcement. It should be noted that the longer fibres were oriented to a greater extent perpendicular to the direction of loading than the shorter fibres. This, according to the confinement model, should have produced greater rather than lesser reinforcement.

The contribution of end loading of fibres in compression is thought to be responsible for the slight increase in strength with increasing fibre diameter. The thicker the fibre the less will be the buckling tendency and hence delay in the onset of negative reinforcement. This trend supports the observations concerning length and indicates that increasing fibre aspect ratio is detrimental to compressive strength.

The third question identified for consideration relates to the observation that certain identical fibre arrays can increase strength in one matrix but decrease it in another. A similar effect has been reported (Oiter and Naaman 1986) with hooked steel fibres which reinforced in concrete but had no effect in mortar. Invariably the increase in strength occurred with the stronger autoclaved matrix and with the shorter fibres. The influence of autoclaving is to upgrade the strengthening component by providing improved resistance to fibre induced matrix stressing via increased 'within-composite' matrix modulus and strength together with increased bond between fibre and matrix. These effects are particularly helpful with the shorter fibres.

Final comments are worth making regarding other factors that must contribute to the seemingly inconsistent nature of the results in the literature on compressive deformation. It is probable that all of the post first crack deformation characteristics are influenced by the shape and size of the test specimen. Clearly only empirical relationships can be expected from representing the internal forces inducing deformation by the equivalent uniaxial compressive stress.

5 Conclusions

Very high strength, very high volume fraction steel fibre reinforced cementitious composites exhibit the same type of behaviour in compression as low volume fraction normal strength composites.

The effects of fibre geometry and concentration are secondary compared to matrix properties in determining composite modulus and strength.

Strength changes of plus or minus 20% can be obtained depending upon appropriate selection of fibre type and concentration.

The relationships between strength and fibre parameters are the reverse of those observed in tension, ie low aspect ratios give greater strengths, increased fibre content reduces strength and better reinforcement is obtained in higher strength matrices.

The major benefit from fibre incorporation for compressive deformation is in changing the mode of failure from brittle to pseudo-ductile and in imparting toughness.

Suggestions were made of three broad types of qualitative fibre-matrix interaction necessary to explain the results but detailed quantitative mechanisms of deformation have still to be determined.

6 References

Abdel-Monem, A.E.S. (1991) The Strength and Fracture Characteristics of Autoclaved High Volume Fraction Steel-Fibre Reinforced Cementitious Composites, PhD Thesis, University of Salford, England.

Aveston, J., Cooper, G.A. and Kelly, A. (1971) Single and multiple fracture in the Properties of Fibrous Composites, Conference Proceedings National Physical Laboratory. IPC Science and Technology Press. Guildford, pp 15-26.

Aveston, J., Mercer, R. and Sillwood, J. (1974) Fibre Reinforced Cements. Scientific foundations for specifications. National Physical Laboratory Conference Proc., Composite-Standards Testing and Design, pp 93-103.

Baggott, R. and Abdel-Monem, A.E.S. (1991) Aspects of Bond in High Volume Fraction Steel Fibre Reinforced Calcium Silicates, ibid.

Baggott, R. and Sarandily, A. (1986) Very high strength steel fibre reinforced autoclaved mortars, in Rilem Symposium on Developments in fibre reinforced cement and concrete, paper 5.3.

Balagurn, P. and Kendzulak, J. (1987) Mechanical Properties of Slurry Infiltrated Concrete (SIFCON), in "Fibre Reinforced Concrete Properties and Application", ACI SP-105, Detroit, Shah et al (Editors), pp 247-267.

Fibre Technology Ltd, Brookhill Road, Pinxton, Nottingham, England.

Homrich, J. and Naaman, A. (1987) Stress-strain Properties of SIFCON in Compression, "Fibre Reinforced Concrete Properties and Applications", ACI SP-105, Detroit, Shah et al (Editors), pp 283-304.

Johnston, C. D. (1974) Steel Fibre Reinforced Mortar and Concrete, a Review of Mechanical Properties, in AC1 SP-44, pp 127-142.

Lankard, D.R. (1984) Slurry infiltrated Fibre Concrete (SIFCON): Properties and Applications, Vol 42, Materials Research Society Proceedings of Symposia, Boston, Massachusetts, USA, "Very High Strength Cement Based Materials" November 27-28.

Lankard, D.R. and Newall, J.K. (1984) Preparation of Highly Reinforced Steel Fibre Concrete Composites, in ACI SP-81, pp 287-306.

Lim, T.Y., Paramasivam, P., Mansar, M.A. and Lee, S.L. (1986) Tensile Behaviour of Steel Fibre Reinforced Composites, "FRC 86", Rilem Symposium, Swamy R M, Wagstaffe, R A & Oakley P R, (Editors).

Naaman, A.E. (1985) High Strength Fibre Reinforced Cement Composites, Materials Research Society, Symp. Proc. Vol. 42, Young J F, (Editor), pp 219-214.

Naaman, A.E. (1991) Tailored Properties for Structural Performance, "International Workshop on High Performance Fibre Reinforced Cement Composites" Mainz, Germany, June 24-26.

Oiter, D. and Naaman, A.E. (1986) Steel fibre reinforced Concrete under Static and Cyclic Compressive Loading, "FRC 86", Rilem Symposium, Swamy R M, Wagstaffe R A & Oakley P R, (Editors)

12 COMPRESSIVE STRENGTH AND MODULUS OF HIGH EARLY STRENGTH FIBER REINFORCED CONCRETE

F. M. ALKHAIRI and A. E. NAAMAN
The University of Michigan, Ann Arbor, MI, USA

Abstract
This study is part of an extensive on-going experimental investigation on the mechanical properties of HESFRC. HESFRC is defined here as achieving a minimum compressive strength of 5 ksi (35 MPa) at 24 hours. This paper discusses the compressive properties and elastic modulus of the hardened HESFRC composite with time. The properties of the fresh composite are not described here due to space limitations, but can be found elsewhere (Naaman and Alkhairi 1991). The effects of latex and silica fume were investigated and compared to plain HESFRC. Two fiber types were used, namely, hooked end steel and polypropylene fibers. For the hooked steel fibers, two aspect ratios were examined corresponding to 30/50 fibers (i.e., 30 mm long and 0.05 mm in diameter) and 50/50 fibers (i.e., 50 mm long and 0.05 mm in diameter). Several combinations of fiber types were examined using 1% and 2% fibers by volume of the concrete mix. Test results include the overall stress-strain response, the compressive strength, and the elastic modulus, all measured at 1, 3, 7, and 28 days. The requirement for HESFRC mixes of achieving a compressive strength of 5 ksi (35 MPa) or greater at 1 day was generally satisfied by all mixes except those containing polypropylene fibers and/or latex. In all, 16 different mixes were investigated comprising 220 specimens.
Keywords: Fiber Reinforced Concrete; Compressive Strength; Modulus of Elasticity; Latex; Silica Fume; Microsilica; High Early Strength; Polypropylene; High Performance Concrete; Stiffness

1 Objective and Scope

The main objective of this investigation was to study the effects of different types of fibers on the compressive properties of High Early Strength Fiber Reinforced Concrete (HESFRC) defined here as concrete achieving a compressive strength of 5 ksi (35 MPa) at 24 hours under normal moist curing conditions. The following three major objectives were sought: 1) to achieve a minimum compressive strength of 5 ksi (35 MPa) in 24 hours; 2) to obtain the complete experimental stress-strain curve of various HESFRC mixes tested at 24 hours, and then compare the results to tests performed at 3, 7, and 28 days; and 3) to compare the values of strength and modulus of elasticity of the composite at 24 hours and 28 days.

2 Test Program

The compression tests were subdivided into three major groups as shown in Fig. 1. For each group, two types of steel fibers and one type of polypropylene fibers were used. The steel fibers consisted of 30/50 and 50/50 hooked steel fibers (trade name:

Fibre Reinforced Cement and Concrete. Edited by R. N. Swamy. © 1992 RILEM.
Published by E & FN Spon, 2-6 Boundary Row, London SE1 8HN. ISBN 0 419 18130 X.

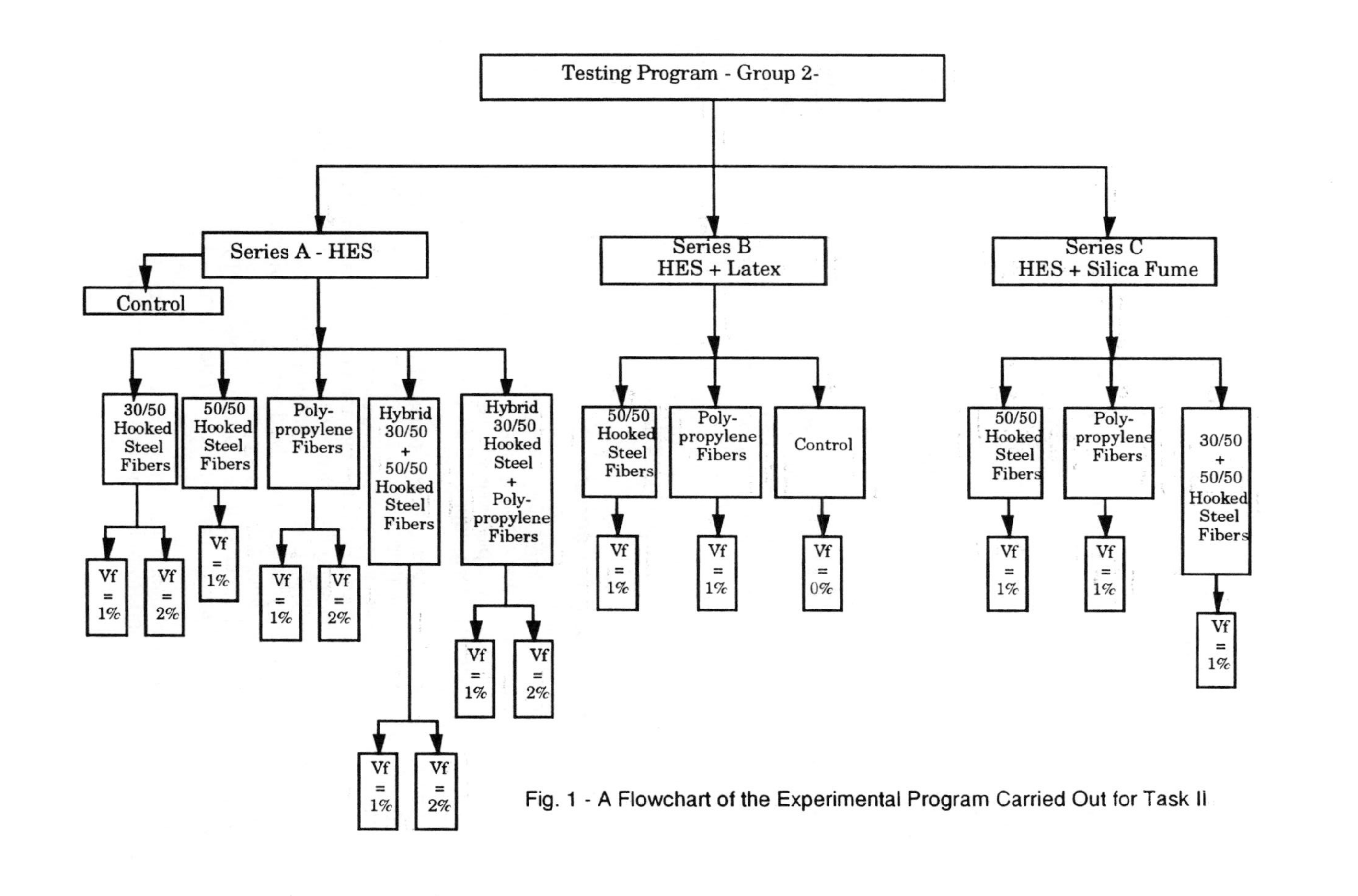

Fig. 1 - A Flowchart of the Experimental Program Carried Out for Task II

DRAMIX), to be designated from now on as 30/50 and 50/50, respectively. The diameter of the 30/50 and 50/50 fibers is 0.5 mm and their length is respectively 30 mm and 50 mm, leading to an aspect ratio of 60 and 100, respectively. Fig. 1 shows that two volume fractions of fibers (1% and 2%) were investigated for all mixes of series A, except for the mix containing 50/50 fibers for which only 1% fibers by volume was used due to difficulties encountered in mixing 2% by volume of fibers.

For each mix, standard cylindrical specimens were prepared for testing at 1, 3, 7, and 28 days. At least 3 specimens were tested for each parameter. Since the normal cylinder size used in this investigation was 4 x 8 inch, 2 cylinders of size 6 x 12 inch were also prepared and tested at 1 day to provide some correlation between the two sizes.

3 Mix Composition and Mix Proportion

The following materials were used to prepare all HESFRC mixes: Type III high early strength cement, sand (Type 2-NS), 3/4" (19 mm) max size washed crushed limestone, air entraining admixture (AEA - Vinsol Resin), corrosion inhibitor (DCI), superplasticizer (Melment), two types of hooked steel fibers, and one type of single filament polypropylene fibers. As shown in Table 1, two of the mixes of series A, called hybrid mixes, contained two different fiber lengths or two different fiber materials. It was anticipated that a hybrid mix may have some advantages, such as one fiber contributing to higher strength while the other to increased toughness or ductility.

The water/cement (w/c) ratio for series A was kept at 0.34. In designing the mixes of series B, it was assumed that the latex emulsion had 50% water content. Several trial mixes were tried but achieved a one day compressive strength less than 5 ksi. Finally, in order to achieve the required 1 day strength of 5 ksi, it was decided to reduce the w/c ratio of mix series B from 0.34 to 0.30 while maintaining the workability of the mix. Series C which contained silica fume had a water/cementitious (i.e., cement + solid microsilica) ratio of 0.32. Since silica fume was supplied as an emulsion, the amount of water contained was assumed equal to 50% and was accounted for in computing the water/cementitious ratio. Table 1 shows the mix proportions used to design mix series A, B, and C, while Fig. 4 describes the mix ID code.

4 Curing and Testing Procedure

Curing of all HESFRC specimens was carried out by covering the specimens with a plastic sheet while they were still inside the molds. The specimens were then removed from their molds and placed in plastic bags at room temperature until the time of testing. All cylinders were capped 2 to 3 hours before testing using a sulfur compound.

Each specimen was subjected to two types of tests. The first test was conducted on the 4" x 8" (100x200 mm) specimens only to measure the elastic modulus. Fig. 2 shows the test fixture used; it consists of two aluminum rings separated by temporary bracing. The bracing served to hold the top and bottom rings apart at exactly 4" (100 mm) gage length thus allowing the initial zeroing of the LVDT's attached to measure deformation. The second type of test was conducted on the 4" x 8" (100x200 mm) and 6" x 12" (150x300 mm) cylinders. It involved testing the cylinders up to failure while recording the entire load-deformation response (see Fig. 3). In each case, the load was measured by a load cell attached to a 300 tons capacity INSTRON universal testing machine, while the deformation was measured using three LVDT's (Linear Voltage Differential Transducers) placed at 120° around the specimen. Stress vs. strain response curves were obtained from three identical cylinders, and an average stress-strain curve was obtained (Naaman and Alkhairi 1991). Average curves were used to

Mix ID	[1]Ad. Type	C (pcf)	W (pcf)	Total w/c	S (pcf)	CA (pcf)	Mel. (pcf)	AEA (pcf)	DCI (pcf)	Ad. (pcf)
Control	---	850 (1)	235 (0.28)	0.34	1250 (1.47)	1550 (1.82)	29.75 (3.5%)	4.25 (0.5%)	78.5 (9.2%)	---
A1%S3	---	850 (1)	235 (0.28)	0.34	1250 (1.47)	1550 (1.82)	29.75 (3.5%)	4.25 (0.5%)	78.5 (9.2%)	---
A2%S3	---	850 (1)	235 (0.28)	0.34	1250 (1.47)	1550 (1.82)	29.75 (3.5%)	4.25 (0.5%)	78.5 (9.2%)	---
A1%S5	---	850 (1)	235 (0.28)	0.34	1250 (1.47)	1550 (1.82)	29.75 (3.5%)	4.25 (0.5%)	78.5 (9.2%)	---
A1%P0.75	---	850 (1)	235 (0.28)	0.34	1250 (1.47)	1550 (1.82)	29.75 (3.5%)	4.25 (0.5%)	78.5 (9.2%)	---
A2%P0.75	---	850 (1)	235 (0.28)	0.34	1250 (1.47)	1550 (1.82)	29.75 (3.5%)	4.25 (0.5%)	78.5 (9.2%)	---
A1%S3S5	---	850 (1)	235 (0.28)	0.34	1250 (1.47)	1550 (1.82)	29.75 (3.5%)	4.25 (0.5%)	78.5 (9.2%)	---
A2%S3S5	---	850 (1)	235 (0.28)	0.34	1250 (1.47)	1550 (1.82)	29.75 (3.5%)	4.25 (0.5%)	78.5 (9.2%)	---
A1%S3P0.5	---	850 (1)	235 (0.28)	0.34	1250 (1.47)	1550 (1.82)	29.75 (3.5%)	4.25 (0.5%)	78.5 (9.2%)	---
A2%S3P0.5	---	850 (1)	235 (0.28)	0.34	1250 (1.47)	1550 (1.82)	29.75 (3.5%)	4.25 (0.5%)	78.5 (9.2%)	---
B1%S5	Latex	900 (1)	98 (0.11)	.30	1250 (1.47)	1550 (1.82)	31.5 (3.5%)	2.7 (0.3%)	90 (9.2%)	190 (0.21)
B1%P0.5	Latex	900 (1)	98 (0.11)	.30	1250 (1.47)	1550 (1.82)	31.5 (3.5%)	2.7 (0.3%)	90 (9.2%)	190 (0.21)
B0%Con	Latex	900 (1)	98 (0.11)	.30	1250 (1.47)	1550 (1.82)	31.5 (3.5%)	2.7 (0.3%)	90 (9.2%)	190 (0.21)
C1%S5	Silica Fume	900 (1)	198 (0.22)	0.32	1250 (1.47)	1550 (1.82)	31.5 (3.5%)	2.7 (0.3%)	90 (9.2%)	100 (0.11)
C1%P0.5	Silica Fume	900 (1)	198 (0.22)	0.32	1250 (1.47)	1550 (1.82)	31.5 (3.5%)	2.7 (0.3%)	90 (9.2%)	100 (0.11)
C1%S3S5	Silica Fume	900 (1)	98 (0.22)	0.32	1250 (1.47)	1550 (1.82)	31.5 (3.5%)	2.7 (0.3%)	90 (9.2%)	100 (0.11)

Table 1 - Design Proportions for Mix Series A, B, and C

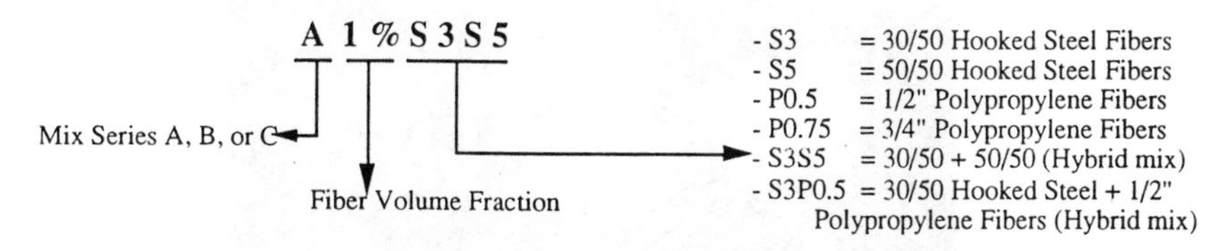

Fig. 4 Mix ID Code

[1] Ad. = Additive, C = Cement, W = Water, S = Sand, CA = Coarse Aggregates, Mel. = Melment

Fig. 2 - Test Set up for Measurement of Elastic Modulus

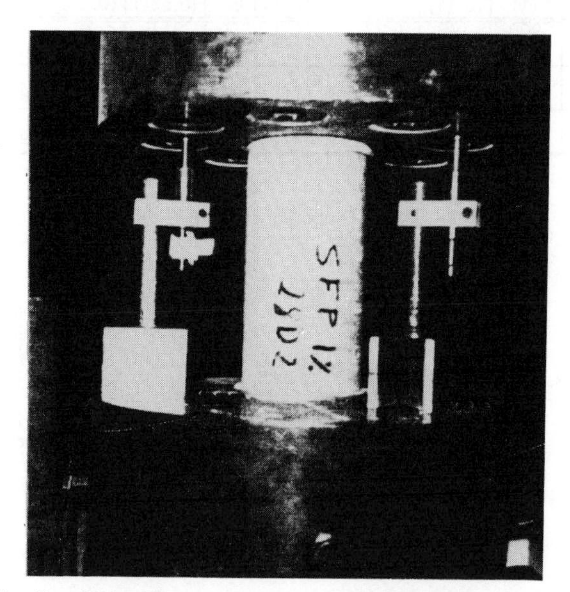

Fig. 3 - Test Set-up Used to Determine the Stress-Strain Response in Compression clarify the influence of study parameters such as time, specimen size, fiber content, fiber type, and the addition of either microsilica or latex.

5 Results and Discussion

5.1 Stress vs. Strain Response with Time

5.1.1 Series A - Plain HESFRC

Fig. 5 shows the results obtained from testing the control time series. The figure shows a sudden drop in the load carrying capacity beyond cracking. Fig. 6 shows the effect of adding 1% by volume of 30/50 hooked steel fibers, where it can be observed that in addition to the slight (17%) increase in f'_c at 1 day relative to the strength of the control mix, there was also a significant increase in the area under the curve, indicating a substantial increase in ductility. The figure also shows that the slope of the descending branch slightly decreases with an increase in the compressive strength. The increase in f'_c for the series tested at 3, 7, and 28 days was respectively 11.6%, 26.3%, and 28.5% in comparison to the series tested at 1 day. Fig. 7 compares the compressive strength as obtained from a 4" x 8" (100x200 mm) and 6" x 12" (150x300 mm) cylinders tested at 1 day. Generally, the 4" x 8" cylinders lead to a slightly higher compressive strength (of the order of 13%) than the 6" x 12" cylinders.

Fig. 8 shows the test results of the mixes containing 2% by volume of concrete of 30/50 hooked steel fibers. The trend observed in this figure follows the trend observed for the 1% fibers (Fig. 6).

Fig. 9 illustrates the results for the hybrid mix containing 30/50 + 50/50 hooked steel fibers at a total of 1% by volume. Several observations can be arrived at from Fig. 9: 1) the strength at 3, 7, and 28 days increases 14%, 25%, and 43%, respectively when compared to the 1 day strength; 2) the slope of the descending branch decreases with an increase in strength (age).

Fig. 10 presents the results of the mix containing 2% by volume of polypropylene fibers. The figure clearly shows two major drawbacks in using polypropylene fibers. The first is the significant drop in f'_c (-40%) at 1 day relative to the control mix. The reason for this sharp drop is not very clear and may be attributed to the low elastic modulus of the polypropylene fibers and their poor bonding properties in comparison to steel fibers. The second drawback is the steep descending branch of the 7 and 28 day curves. Unlike the 30/50 steel fiber mix at 1% and 2% volume fraction (Figs. 6 and 8, respectively), the polypropylene mix at 2% volume fraction shows a significantly lower ductility. This may also be explained by the low elastic modulus and poor bond properties of polypropylene fibers.

Fig. 11 describes the results of the hybrid mix containing 1% of 30/50 hooked steel fibers and 1% of polypropylene fibers (total volume fraction = 2%). It can be noted from this figure that replacing 1% by volume of polypropylene fibers (Fig. 10) by an equivalent amount of 30/50 hooked steel fibers significantly improves the behavior. Although Fig. 11 shows a slight decrease (8%) in the f'_c at 1 day relative to the 5 ksi target value, the slope of the descending branch does not change, indicating that ductility was maintained at 1, 3, 7, and 28 days.

5.1.2 Series B - HESFRC with Latex

Fig. 12 shows the stress-strain curves for the 50/50 hooked steel fiber mix at 1% volume fraction, modified with latex. It was observed during mixing that latex significantly improved the workability of the mix as if latex particles acted as a lubricant (Mason 1981, Bharyava 1981), while causing a significant reduction in the one day compressive strength. By testing few trial mixes at 1 day, it was decided to slightly reduce the w/c ratio of the latex-modified fiber reinforced concrete mixes from 0.34 to 0.30 so as to increase the 1 day strength. Despite the reduction in the w/c ratio, the compressive strength at 1 day was 10% less than the 5 ksi (35 MPa) target value.

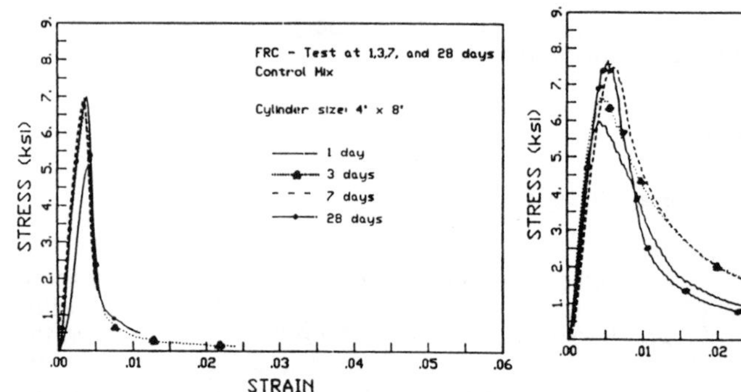

Fig. 5 - Stress vs. Strain Response of Control Mix With Time

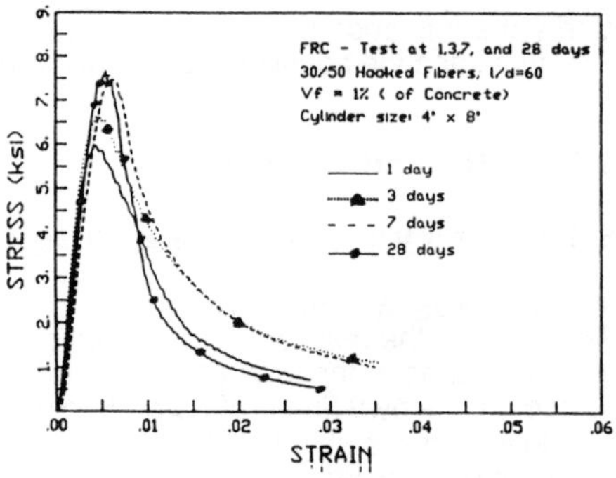

Fig. 6 - Stress vs. Strain Response of 30/50 Mix With Time (V_f = 1%)

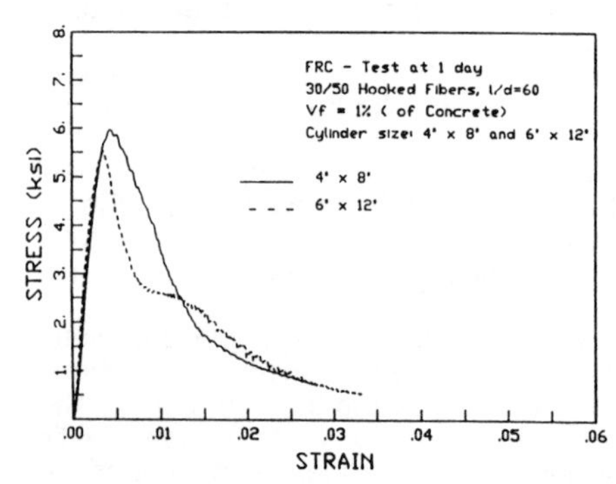

Fig. 7 - Effect of Cylinder Size on the Stress vs. Strain Response of the 30/50 Mix (V_f = 1%)

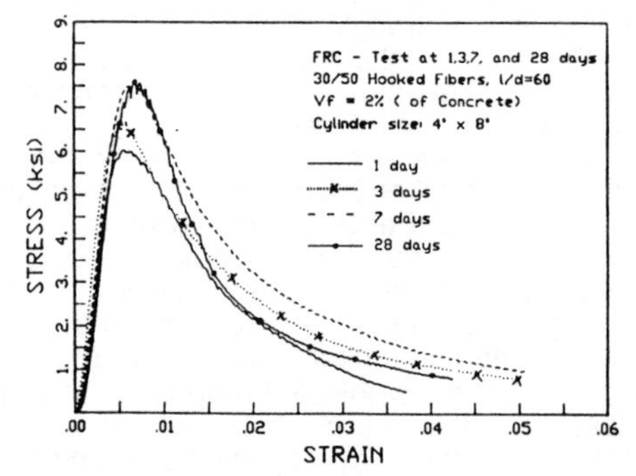

Fig. 8 - Stress vs. Strain Response of 30/50 Mix With Time (V_f = 2%)

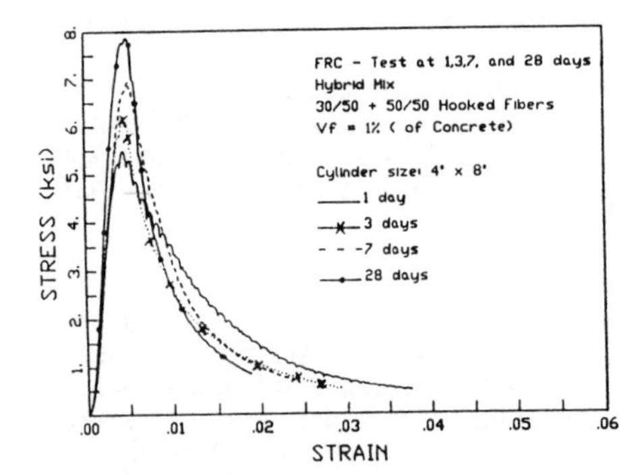

Fig. 9 - Stress vs. Strain Response of (30/50 + 50/50) Hybrid Mix With Time (V_f = 1%)

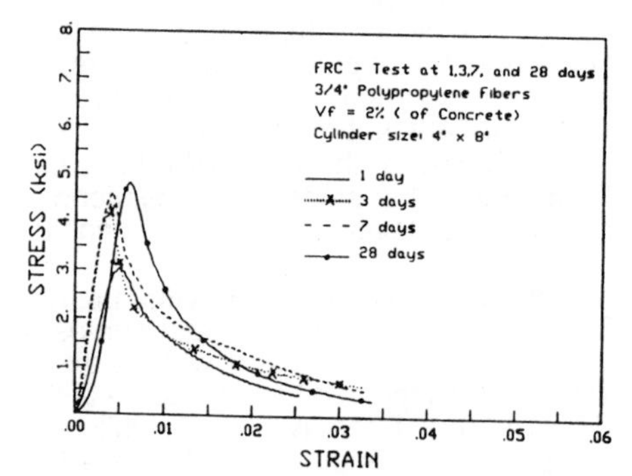

Fig. 10 - Stress vs. Strain Response of Polypropylene Mix With Time (V_f = 2%)

Fig. 12 also shows that the addition of latex enhances the ductility of FRC with time. A similar observation was reported (Soroushian, Aouadi, Naji 1991), where, using micrographs of pulled-out carbon fibers, it was observed that the failure mode in the presence of latex is partially shifted away from the interface, that is the fiber-cement interfacial bond strength was observed to be greater than the matrix shear strength.

5.1.3 Series C - HESFRC with Silica Fume

Fig. 13 describes the results of test series C containing silica fume in wet form used with 1% by volume of 30/50 + 50/50 hooked steel fibers. The w/c ratio of the mix had to be slightly adjusted due to the presence of microsilica as shown in Table 1. The figure shows that the strength at 1 day slightly decreases below the minimum 5 ksi (35 MPa) target value (4.7 ksi). The figure shows that the addition of silica fume increases the strength at 3, 7, and 28 days relative to the 1 day strength by 10%, 50%, and 54%, respectively. However, an increase in the slope of the descending branch was observed at 7 and 28 days, indicating a slight loss in ductility.

5.2 Stress vs. Strain Response - Comparative Evaluation of Different Mixes

Figs. 14 and 15 compare the 1 day and 28 day test results of the 30/50 and the polypropylene fiber mixes with $V_f = 2\%$. It can be observed that the response of the 30/50 fiber mix is superior to the polypropylene mix in terms of both compressive strength and ductility. Figs. 14 and 15 suggest that the interfacial bond properties of the steel fibers outperform those of the polypropylene fibers. They also suggest that the use of polypropylene fibers at $V_f = 2\%$ may cause a significant reduction in the compressive strength in comparison to the use of 1%, while the use of 30/50 fibers in the same volume fraction causes a significant improvement in the compressive strength. The above results may be due to the fact that it is very difficult to properly mix 2% polypropylene fibers by volume, which usually leads to a large amount of entrapped air.

Fig. 16 compares the results of the hybrid mixes of series A defined as containing one or two types of fibers at a volume fraction equal to 1%, all tested at 1 day. It is observed that: 1) the use of either 30/50, 50/50 or a combination of them, causes insignificant changes in the compressive strength and ductility, and 2) the substitution of 0.5% by volume of 30/50 fibers by an equivalent volume fraction of polypropylene fibers reduces the 1 day strength by almost 50%, thus causing a significant reduction in ductility. Therefore, the presence of polypropylene fibers at volume fractions as low as 0.5% is not desirable for compression properties.

Fig. 17 compares the results of series A at 1% volume fraction of fibers tested at 28 days. It can be observed that both the 30/50 and 30/50+50/50 fiber mixes gave the highest compressive strength and behaved virtually similarly. However, they showed a sharper increase in slope in the post-peak response when compared to the 50/50 fiber mix, which, in turn, showed a slightly lower compressive strength. The 30/50+polypropylene fiber mix led to the lowest compressive strength and ductility.

Fig. 18 compares the same mixes discussed in Fig. 16 tested at 1 day except that the volume fraction of fibers is increased from 1% to 2%. It is observed that the 30/50 fiber mix still outperforms the 30/50+polypropylene and the 30/50+50/50 hybrid mixes both in terms of strength and ductility. Contrary to the mixes shown in Fig. 16, the 30/50+polypropylene fiber mix containing 2% performed better than the 30/50+50/50 mix. The reason for that is not very clear.

Fig. 19 compares the 3 mixes of series A (see Fig. 18) tested at 28 days. It can be observed that the 30/50 fiber mix gives the best overall response. The performance of the 30/50+50/50 mix was similar to the 30/50 mix except for a slightly lower compressive strength. The response of the 30/50+polypropylene mix was acceptable but with noticeably lower compressive strength in comparison to the two other mixes.

Fig. 20 shows the effect of latex (series B) and silica fume (series C) on a mix containing 50/50 fibers at 1% volume fraction, all tested at 1 day. The figure shows

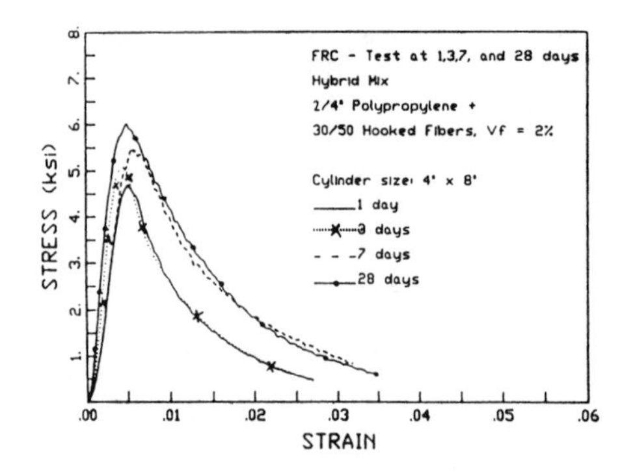

Fig. 11 - Stress vs. Strain Response of (30/50 + Polypropylene) Hybrid Mix With Time (V_f = 2%)

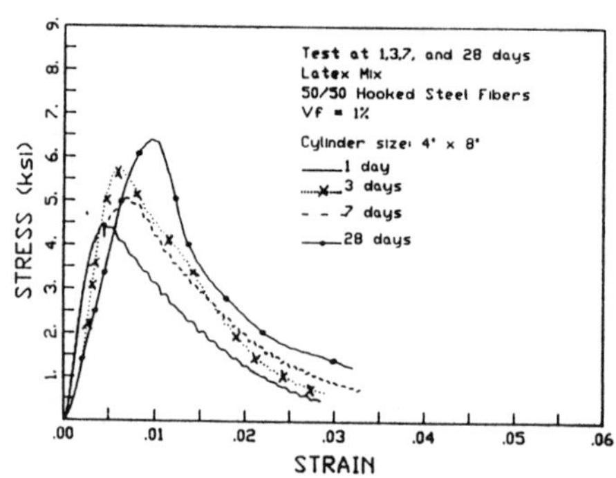

Fig. 12 - Effect of Latex on the Stress vs. Strain Response of 50/50 Mix With Time (V_f = 1%)

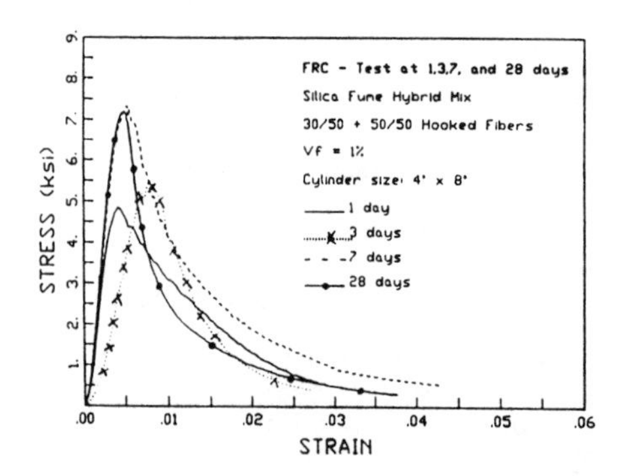

Fig. 13 - Effect of Silica Fume on the Stress-Strain Response of 30/50 + 50/50 Hybrid Mix With Time (V_f = 1%)

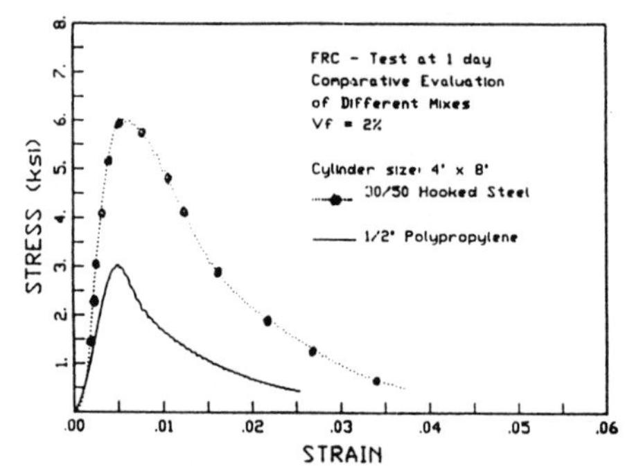

Fig. 14 - Comparative Evaluation of the Stress vs. Strain Response of the 30/50 and Polypropylene Mixes at 24 Hours (V_f = 2%)

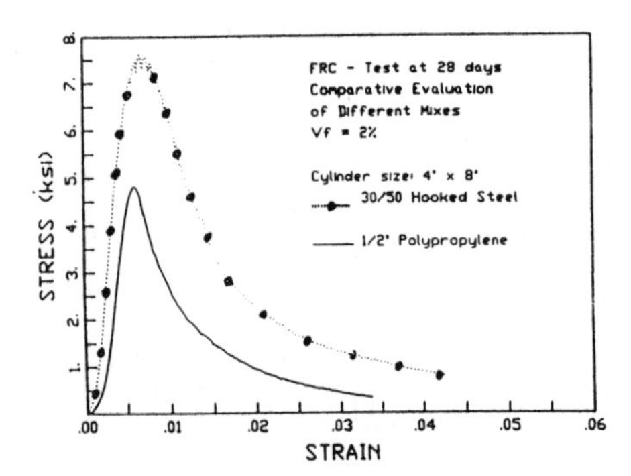

Fig. 15 - Comparative Evaluation of the Stress vs. Strain Response of the 30/50 and Polypropylene Mixes at 28 Days (V_f = 2%)

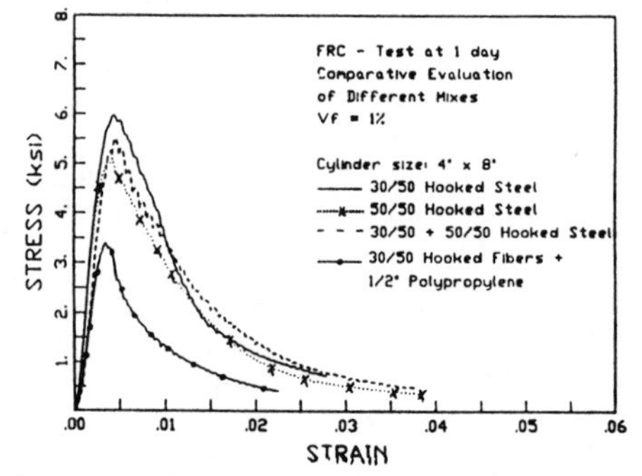

Fig. 16 - Comparative Evaluation of the Stress vs. Strain Response of Different Mixes at 24 Hours (V_f = 1%)

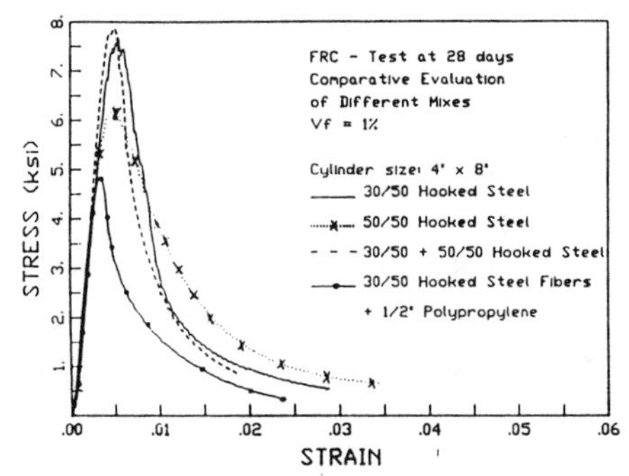

Fig 17- Comparative Evaluation of the Stress vs. Strain Response of Different Mixes at 28 Days (V_f = 1%)

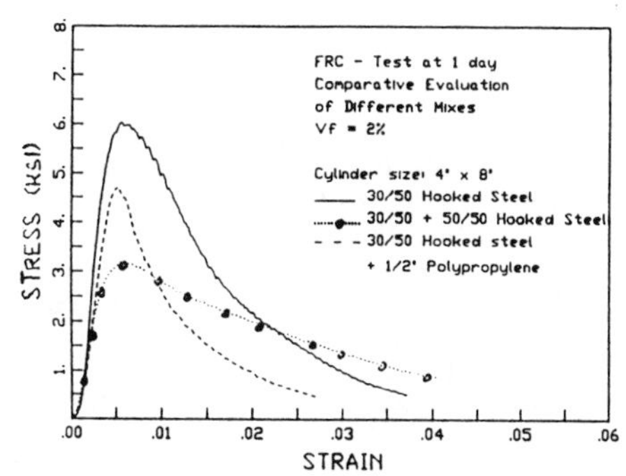

Fig 18 - Comparative Evaluation of the Stress vs. Strain Response of Different Mixes at 24 Hours (V_f = 2%)

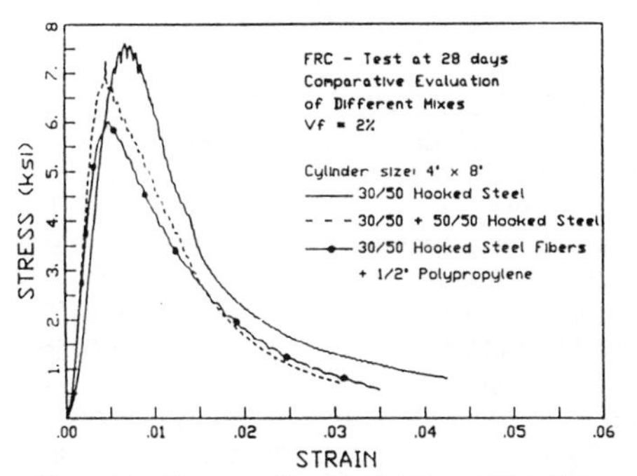

Fig. 19- Comparative Evaluation of the Stress vs. Strain Response of Different Mixes at 28 Days (V_f = 2%)

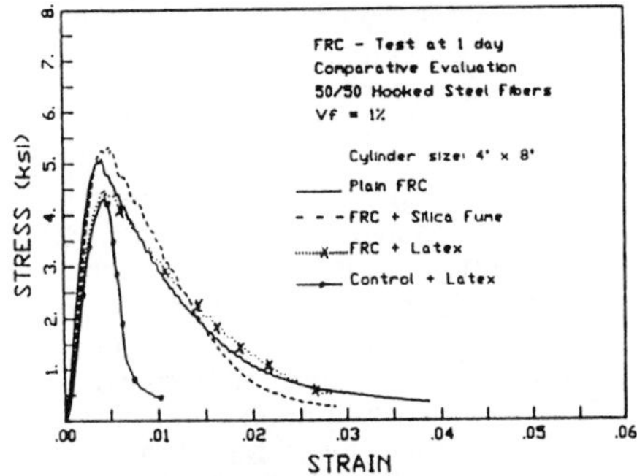

Fig. 20- Effect of Microsilica and Latex on the Stress-Strain Response of the 50/50 Mix at 24 Hours (V_f = 1%)

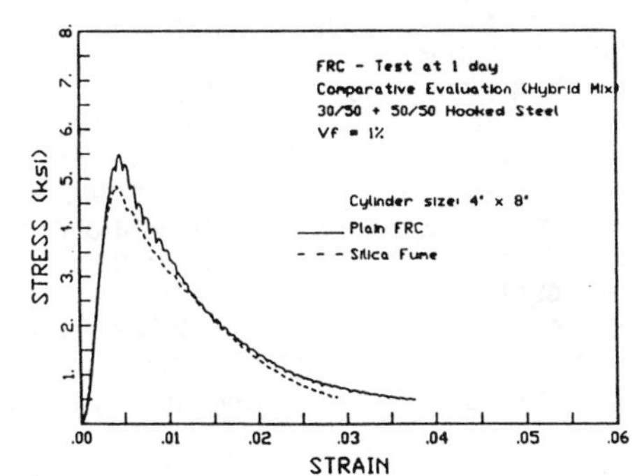

Fig. 21 - Effect of Microsilica on the Stress-Strain Response of the (30/50 + 50/50) Mix at 24 Hours (V_f = 1%)

unexpectedly that silica fume has little or no effect on the stress-strain response when compared to the plain fiber reinforced concrete mix of series A. This might be attributed to the small amounts of microsilica used in the current investigation (5% solids by weight of cement). The mix containing latex led to a slight decrease in the compressive strength despite the fact that the w/c ratio was decreased from 0.34 to 0.30 as show in Table 1. The water content included the water in the latex emulsion. Here, the compressive strength of the fiber reinforced and unreinforced concrete mix containing latex is the same; however, the latex mix with fibers shows a higher ductility compared to the unreinforced latex matrix.

Fig. 21 compares the effect of silica fume addition on the hybrid mix 30/50+50/50 containing 1% by volume of fibers and tested at 1 day. Insignificant changes in the response is observed, indicating that silica fume has no effect at early ages of FRC mixes.

5.3 Compressive Strength, f'_c

Fig. 22 compares the compressive strength, f'_c, of the 30/50 mix at 1% and 2% fiber content. The increase in strength of both mixes over the control mix at 1, 3, 7, and 28 days was approximately 17%, 2%, 10%, and 11%, respectively. The figure also shows that increasing the volume fraction of the 30/50 fibers from 1% to 2% does not significantly change f'_c.

Fig. 23 describes the results of the mix series A containing 2% fibers by volume. The figure shows that the 30/50 fiber mix still leads to an overall increase in f'_c over the control mix. The 30/50 and (polypropylene+30/50) fiber mixes achieved the target 5 ksi (35 MPa) compressive strength while other mixes did not meet this criterion. The 30/50 and (polypropylene+30/50) fiber mixes led respectively to a 20% increase and a 7% reduction in f'_c at 1 day over the control mix, while the (30/50+50/50) and polypropylene mixes led to a 33% reduction in f'_c over the control mix at 1 day. The presence of at least 1% by volume of 30/50 fibers in the polypropylene+30/50 fiber mix enhanced the overall behavior, even in the presence of 1% polypropylene fibers. The (30/50+50/50) hybrid fiber mix did not give satisfactory results. It is believed that this observation might be attributed to the presence of large entrapped air voids (8%) (Naaman and Alkhairi 1991). Fig. 23 also shows that the 28 day strength of the 30/50+50/50 hybrid fiber mix was approximately equal to that of the control mix. The polypropylene mix gave the lowest strength with an average 30% reduction over the control mix at all time intervals. Similar to the case of the 30/50+50/50 hybrid mix.

Fig. 24 illustrates the effect of silica fume on the compressive strength. The figure surprisingly shows that silica fume had little effect on the 1 day compressive strength for the 50/50 and 30/50+50/50 fiber mixes when compared to the control mix (Note: the control mix does not contain silica fume). However, the addition of microsilica slightly increased the strength at 7 and 28 days over that of the control mix. This increase was respectively 10% and 20% for the 50/50 fiber mix and, 6% and 6% for the (30/50+50/50) fiber mix. The polypropylene mix did not meet the target 5 ksi (35 MPa) 1 day strength and led to an average strength reduction over the control mix equal to 32%, 26%, 20%, and 16% at 1, 3, 7, and 28 days, respectively.

Fig. 25 shows the effect of silica fume on the compressive strength, f'_c, for the mix containing 1% by volume of 30/50+50/50 fibers. It is observed that the addition of microsilica at 5% solids by weight of cement has no significant effect in increasing f'_c at 1 and 3 days relative to the plain FRC mix, while slightly changing the strength at 7 and 28 days.

5.4 Modulus of Elasticity, E_c

The purpose of this section is to show the correlation between the measured values of modulus of elasticity, E_c, and compressive strength, f'_c, and their variation with time, for all HESFRC mixes. It should be noted that the scatter observed in the elastic modulus tests was much larger than that observed for the compressive strength tests.

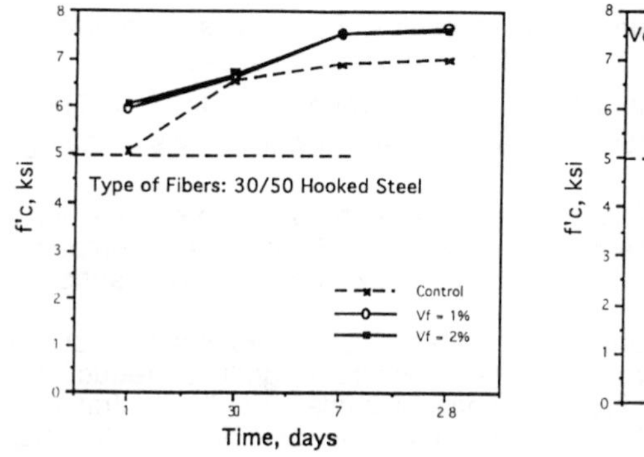

Fig. 22 - Compressive Strength, f'c vs. Time, days

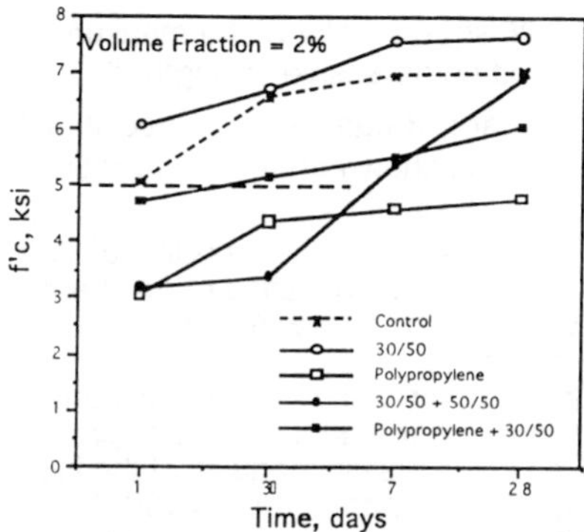

Fig. 23 - Compressive Strength, f'c vs. Time, days

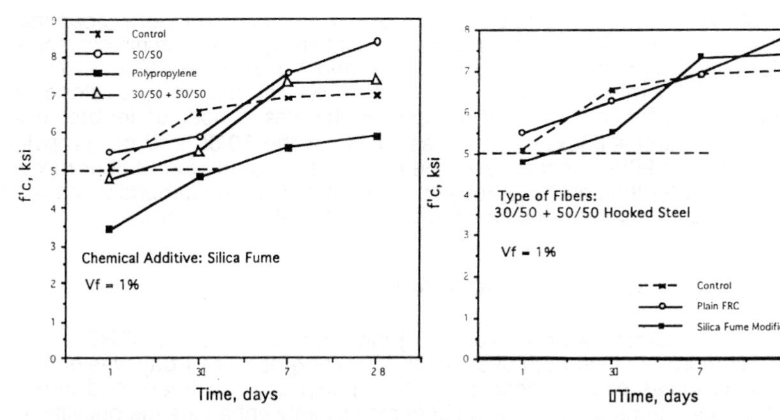

Fig. 24 - Compressive Strength, f'c vs. Time, days

Fig. 25 - Compressive Strength, f'c vs. Time, days

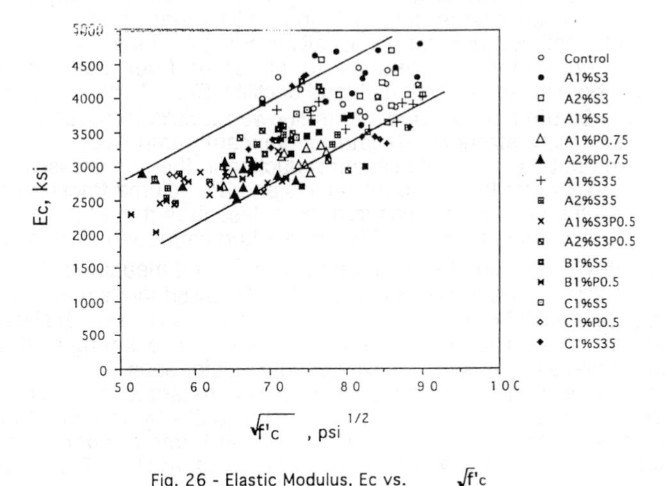

Fig. 26 - Elastic Modulus, Ec vs. $\sqrt{f'_c}$

Fig. 26 shows a plot of the measured elastic modulus, E_c vs. the square root of the measured compressive strength, f'_c for all cylinders tested. As expected, and except for few data, it can be observed that E_c is linearly proportional to $\sqrt{f'_c}$. The two straight lines shown on the figure represent the upper and lower bounds of the E_c vs. $\sqrt{f'_c}$ relationship.

Fig. 27 compares the modulus of elasticity results for the 30/50 mix at 1% and 2% fiber volume fractions. Little or no change is observed at 1, 3, and 7 days compared to the control mix. However, at 28 days, the 1% and 2% mixes show a substantial increase in E_c.

Fig. 28 compares the polypropylene mix at 1% and 2% fiber volume fractions. It is observed that the addition of polypropylene fibers causes a significant reduction (up to 30%) in the elastic modulus of the HESFRC composite relative to the control mix.

Fig. 29 compares the (30/50+50/50) hybrid fiber mix at 1% and 2% volume fractions. Here, both mixes showed elastic modulus values lower than the control mix. This reduction in stiffness is believed to be primarily caused by the relatively high volume of voids entrapped during mixing. The 1% and 2% fiber mixes had a 5% and 8% air content, respectively, compared to 4% air content for the control mix (Naaman and Alkhairi 1991).

Fig. 30 shows the effect of latex on the 50/50 and polypropylene fiber mixes (series B), and a comparison with plain concrete containing latex. The elastic modulus values relative to the (control + latex) mix at 1, 3, and 7 days were +40%, +10%, and -10% for the 50/50 mix and 0%, -25%, and -25% for the polypropylene mix, respectively. At 28 days, however, latex improved the elastic modulus for both mixes relative to the (control + latex) mix, increasing E_c for the 50/50 and polypropylene mixes by 100% and 90%, respectively. The performance of the 50/50 fiber mix was observed to be better than the polypropylene fiber mix. It can be generally concluded that latex significantly improves the long term elastic modulus of the mix.

6 Conclusions and Recommendations

The following conclusions were arrived at: 1) the requirement for HESFRC mixes of achieving a compressive strength of 5 ksi (35 MPa) or greater at 1 day was generally satisfied by all mixes except those containing polypropylene fibers and latex; 2) reinforcing a concrete matrix with steel fibers significantly enhances the ductility at all ages of testing; 3) the compressive strengths obtained at 1 day from the 4"x8" (100x200 mm) cylinders were slightly higher (13%) than those obtained from the 6"x12" (150x300 mm) cylinders for almost all mixes; 4) the response in compression of the polypropylene fiber mix relative to all other fiber concrete mixes was unsatisfactory, both in terms of strength and ductility; 5) the hybrid fiber mix containing 1% of (30/50 + 50/50) hooked steel fibers showed a 50% increase in f'_c at 28 days relative to the 1 day strength; 6) latex significantly improves the workability of HESFRC mixes but causes a significant reduction in the compressive properties at early ages; 7) except for the 30/50 mix at 1% and 2% volume fraction of steel fibers, the elastic modulus of all mixes was in general less than that of the control mix; this may be explained by the fact that all FRC mixes had entrapped air due to mixing; 8) a strong correlation was found between measured E_c and measured $\sqrt{f'_c}$; 9) the 30/50 mix at 1% and 2% volume fractions of steel fibers showed the highest elastic modulus at all times (1,3,7 and 28 days) among all mixes of series A, and 10) silica fume at 5% by volume had no significant effect on the compressive strength at 1 day, while causing an increase in the compressive strength at later ages.

The following recommendations can be made with regard to the optimizing HESFRC mixes in compressive behavior: 1) the use of 1% to 2% by volume of 30/50 hooked steel fibers gave optimum composite properties in terms of compressive strength, elastic modulus, and ductility when compared to all other HESFRC mixes; 2) next in

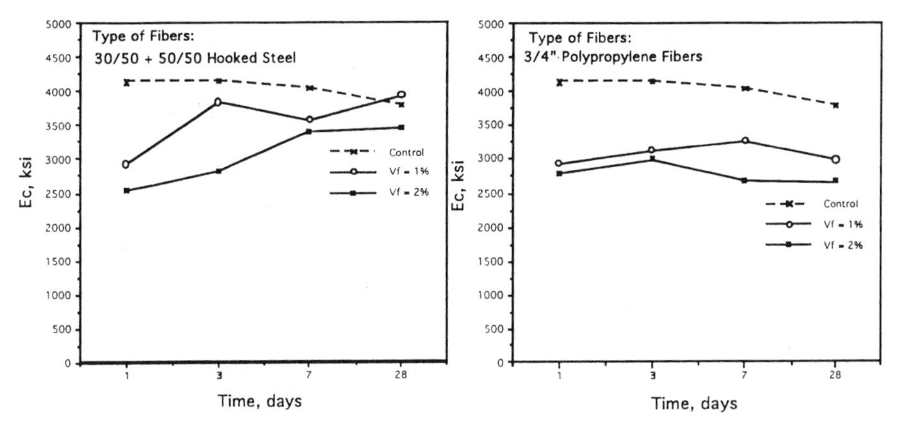

Fig. 27 - Elastic Modulus, Ec vs. Time, days

Fig. 28 - Elastic Modulus, Ec vs. Time, days

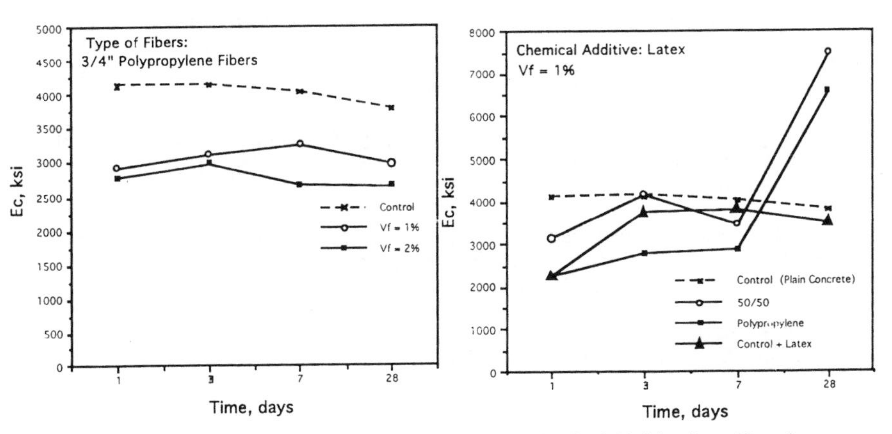

Fig. 29 - Elastic Modulus, Ec vs. Time, days

Fig. 30 - Elastic Modulus, Ec vs. Time, days

performance was the use of 1% by volume of either (30/50+50/50) steel fibers or 50/50 steel fibers; both mixes performed very similarly in terms of compressive strength, elastic modulus, and stress-strain response; 3) the use of polypropylene fibers alone to improve the compressive strength and elastic modulus properties is not desirable since mixes containing 1% to 2% by volume of polypropylene fibers showed reduction in compressive strength when compared to the control mix; 4) despite the fact that latex improved the workability of all HESFRC mixes, it is not desirable to use latex with HESFRC mixes for the purpose of improving early age compressive properties, however the use of latex for the purpose of improving the long term compressive stress-strain response is desirable; and 5) the addition of up to 5% solids by weight of cement of silica fume is not recommended to improve the 1 day compressive strength.

7 Acknowledgments

The research described herein was supported by the Strategic Highway Research Program (SHRP) under contract C-205, Mechanical Behavior of High Performance Concretes. SHRP is a unit of the National Research Council that was authorized by section 128 of the Surface Transportation and Uniform Relocation Assistance Act of 1987.

This research program is undertaken by a consortium of three Universities, namely: North Carolina State University (prime contractor) with P. Zia (Project Director), S. Ahmad, and M. Lemming; University of Michigan with A.E. Naaman (principle investigator), and University of Arkansas with R.P. Elliot and J.J. Schemmel. The program manager at SHRP is Inam Jawed.

The authors of this paper would like to acknowledge the help of several collaborators; namely, M. Harajli (visiting professor), and graduate students P. Strzyinski, I. Khayaat, and B. Campbell. Opinions expressed in this paper are those of the authors and do not necessarily reflect the views of SHRP.

8 References

Naaman, A.E. and Alkhairi, F.M. (1991) Fresh and Hardened Properties of High Early Strength Fiber Reinforced Concrete: Compressive Strength and Elastic Modulus with Time. **A Report submitted to the Strategic Highway Research Program (SHRP C-205), Washington D.C.**

Mason, J.A. (1981) Overview of Current Research on Polymer Concrete: Material and Future Needs. **Application of Polymer to Concrete, SP-69, American Concrete Institute, Detroit, pp. 1-20.**

Bharyava, J.K., Polymer-Modified Concrete for Overlays: Strength and Development Characteristics. **Application of Polymer to Concrete, SP-69, American Concrete Institute, Detroit 1981, pp. 205-218.**

Soroushian, P, Aouadi,F., and Naji,M. (1991) Latex-Modified Carbon Fiber Reinforced Mortar. **ACI Material Journal, V. 88, No. 1.**

13 FLEXURAL TESTING OF STEEL FIBRE REINFORCED REFRACTORY CONCRETE AT ELEVATED TEMPERATURES

P. J. ROBINS and S. A. AUSTIN
Department of Civil Engineering, University of Loughborough, UK

Abstract
Flexural testing of fibre reinforced refractory concrete was carried out at elevated teperature on beams which had previously been heated on one face only in an electric furnace. Specimens were kept at temperature during testing in an insulating box. Toughness, LOP and peak strengths reduced with increasing temperature. Generally the performance of the wire fibre mixes was better than that of the melt exract fibre mixes, but the difference between the two reduced as temperature increased.
Keywords: Fibre reinforcement, refractory, flexural strength, toughness, elevated temperature

1 Introduction

This paper describes a series of flexural tests carried out on steel fibre reinforced refractory beams at elevated temperature. The aim was to heat each specimen on one face, to simulate a refractory lining in-situ, and then to test it in flexure whilst keeping it hot using an insulating box. This was a technique that used previously by Austin et al (1987) and involved modification of some existing equipment, including an electric spalling furnace. Two fibre types were investigated, at three levels of reinforcement, to determine their effect on strength and toughness.

The two fibre types investigated were a hooked ended drawn wire and a melt extract fibre, both 25 mm long and made from 304 stainless steel. The refractory concrete was a typical medium duty calcium aluminate (HAC) cement castable. The castable was reinforced with 2, 3 and 4% by weight of each fibre and 12 beams were cast from each mix together with 100 mm cubes. A plain control mix was also tested.

Fibre Reinforced Cement and Concrete. Edited by R. N. Swamy. © 1992 RILEM.
Published by E & FN Spon, 2-6 Boundary Row, London SE1 8HN. ISBN 0 419 18130 X.

Three beams were tested in flexure at each test temperature of ambient, 550 C, 700 C and 850 C. Hot testing was achieved by preheating six specimens on one face in a furnace, and then placing one specimen at a time in the insulating box mounted on a testing machine where it was loaded to failure.

2 Materials and mix proportions

2.1 Refractory concrete

The castable was a standard medium duty calcium aluminate cement bonded castable. The chemical analysis quoted for the material was:

Al_2O_3	38.0%
SiO_2	48.0%
CaO	5.8%
Fe_2O_3	4.7%

2.2 Fibres

The wire fibre was an 18% Chromium/8% Nickel stainless steel hooked ended drawn wire fibre, 25 mm long by 0.4 mm diameter, manufactured by N V Bekaert S A, Belgium (Dramix ZL 25/40 304SS).

The melt fibre was an 18% Chromium/8% Nickel stainless steel melt extract fibre manufactured by Fibretech Ltd, UK (ME 304 25 mm).

2.3 Mix proportions

The castable was supplied in 14 bags each nominally of 25 kg. For each mix two bags were used and 20% by weight of water was added, based on the weight of the contents.

The fibres were added at 2, 3 or 4% by weight of the dry castable. Typically, the mix proportions were:

Castable	50 kg
Fibres	1, 1.5 or 2 kg
Water	10 kg

3 Test procedures

3.1 General approach

The test procedures adopted for mixing, casting, curing, conditioning and testing were based on appropriate British and US national standards wherever possible. Since the work involved a combination of refractory and

Figure 1. Beams in furnace

fibre reinforced concrete certain compromises had to be made. In particular the flexure specimen sizes for the two materials are quite different, so in this investigation we adopted a refractory style specimen, but applied toughness measurements developed for OPC fibre concretes.

3.2 Specimen preparation

Twelve beams 244 mm long by 64 mm high by 64 mm wide were cast from each mix together with three 100 mm cubes.

The specimens were prepared in accordance with BS 1902-703. This involved the following:

(1) ball in hand test for consistency;
(2) compaction on a vibrating table for > 1 min and < 5 min;
(3) covering samples with polythene and stripping after 18-48 hours;
(4) placing in a humidity cabinet at 20-25 C and 95% humidity up to a total of 48 hours;
(5) air drying for 24 hours; and
(6) oven drying at 110 C to constant weight.

The drying was achieved within 3 days, so that specimens were tested at an age of 7 days. This procedure is similar to that described in ASTM C860:77 and C862:82

During casting the beams for hot testing had two steel hooks, made from L shaped pieces of 6 mm reinforcing bar, cast in to enable them to be easily removed from the furnace. The hooks were located near the beam support positions so that they would not affect the strength or toughness of the beams.

3.3 Firing and heating

After drying, two sets of three beams were stacked vertically in the 384 mm wide by 244 mm high opening in the door of the furnace (Figure 1). The temperature was increased at a rate between 5 and 8 C/min up to the test temperature minus 50 C and then at 1 to 2 C/min up to the test temperature where it was held for at least 2 hours before the specimens were removed one by one for flexural testing. This procedure is in accordance with BS 1902 Part 706 Appendix A.

3.4 Cold crushing strength

Each cube was weighed after drying and tested at 7 days in accordance with BS 1902 Part 706, with the exception of the cube size (100 mm instead of 75 mm).

3.5 Ambient flexural strength

Flexural strength (modulus of rupture) was determined following drying at ambient temperature in accordance with BS 1902 Part 404, with the exception of the specimen dimensions which were 244 x 100 x 64 mm instead of 230 x 114 x 64 mm (brick size). The width was increased to 100 mm to obtain a realistic temperature gradient across the specimens and the length was 14 mm longer to fit our furnace more easily; the width increase will have the effect of producing marginally lower strengths compared to a BS specimen.

The specimens were tested on a 180 mm span in centre point loading with loading rollers able to accommodate twist. The loading rate specified in the standard is 9 N/mm2/min and this was converted to a deflection rate of 0.05 mm/min as the Instron 6025 testing machine was operated in displacement control mode.

The test procedure is similar to that of ASTM C133:1982a which requires a 230 x 114 x 65 mm specimen to be tested on a 178 mm span at a rate of 1.3 N/mm2/min or 1.33 mm/min.

3.6 Hot flexural strength

Each specimen to be tested at elevated temperature was removed from the furnace and placed in a purpose built insulating box which had slots to accommodate the loading rollers for testing in flexure (see Figure 2). Each beam had two steel hooks cast in to facilitate their removal when hot. The box consisted of a plywood shell lined with 38 mm thick insulating board on the top, bottom and ends, and 76 mm on the back which butted up to the hot face of the beam.

The box was then transferred to the Instron machine and the specimen tested immediately (Figure 3) using the same method adopted for the ambient tests. The peak load (used to calculate the modulus of rupture) was

Figure 2. Insulating box

Figure 3. Beam under test

reached approximately three minutes after removal of the beam from the furnace and the test stopped approximately seven minutes later (10 minutes total).

The heat loss during this period was assessed by instrumenting a dummy set of six beams with thermocouples. Each beam had four thermocouples embedded along its length with the welded joint flush with the hot face of the beam. The temperature at the cold face was measured with a hand held thermocouple. Typical temperatures for the middle third of a beam are given in Table 1 for each of the three test temperatures.

Table 1 Typical hot and cold face temperatures during testing

Time after	550 C		700 C		850 C	
removal (mins)	hot (C)	cold (C)	hot (C)	cold (C)	hot (C)	cold (C)
0	545	100	680	110	820	130
2	470	97	605	115	715	128
4	450	108	575	115	673	145
6	435	111	553	118	647	142
8	416	110	531	122	625	140
10	400	110	507	118	600	139

From the table it can be seen that the hot face temperature dropped by approximately 85 C, 90 C and 125 C respectively in the first three minutes, representing a 15% reduction for each of the three test temperatures. The drop by the end of the flexure test was typically 145 C, 170 C and 220 C, which are a 25% reduction.

3.7 Toughness

Toughness measurements were made from load deflection curves recorded during flexural testing of the beams. Two standard methods for measuring the toughness of fibre reinforced concrete are ASTM C1018 (1989) and JSCE SF4 (1984). Both these standards relate to OPC as opposed to refractory concrete and are based on a specimen and loading geometry taken from their related standards for flexural strength, namely 100 x 100 x 400 mm on a 300 mm span under four point loading. In this work the span to depth ratio is similar (180/64 = 2.81 compared with 3.0) but the loading was centre point as opposed to third point. Deflection of the underside of the specimen was measured with a transducer at mid span.

The toughness values reported in this paper have been calculated using the JSCE (1984) approach. This method defines toughness as the area under the load/deflection curve up to a fixed deflection of 1/150 of the span (1.2 mm here). Toughness values are dependent on specimen test geometry and can therefore only be compared with results obtained using the same geometry. However, this method of determining toughness has the advantage that the value obtained is not dependent on the limit of proportionality, which is influenced by fibre type and content and can be difficult to identify on the curve. A more thorough discussion of the relative merits of the JSCE and ASTM toughness measurement approaches is discussed by Robins and Austin (1992) elsewhere. This paper concludes that, for a fibre reinforced castable tested at temperature, the JSCE gives more meaningful values.

4 Density and compressive strength

The average dry (unfired) density and compressive strengths, determined from three cubes at 7 days are plotted against fibre content in Figure 4.

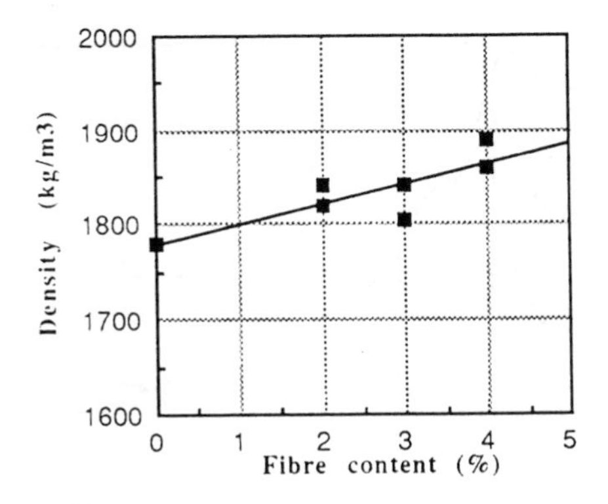

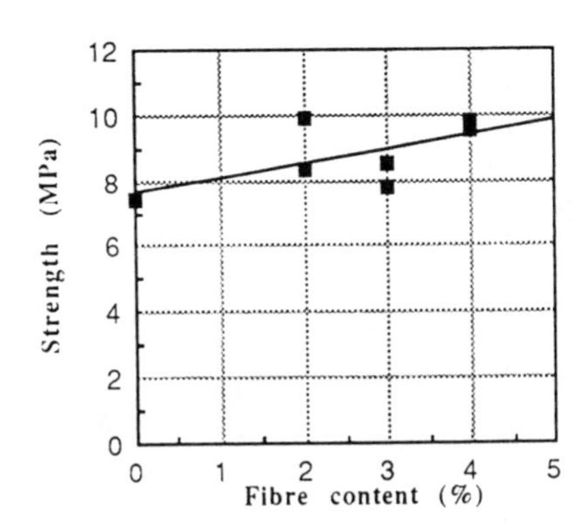

Figure 4. Density and compressive strength

These show strength increasing slightly with fibre content and this increase is in line with what is observed for steel fibres in an OPC (around 10 to 15% increase with a 3% fibre content). Both figures show that there was some variability in the mixes and indicate that the 2% wire and 4% melt mixes were stronger than expected and the 3% melt weaker. As a pre-blended material was used it is likely that the fluctuations in strength are largely the result of varying cement and fines contents from bag to bag.

These fluctuations in compressive strength will affect the observed flexural strength and toughness of the material as the tensile strength and bond of the matrix are both related to the compressive strength. These variations have been compensated for by normalising the flexural strength and toughness data using the best fit straight line for the compressive strength data of

Figure 4. Thus the strength and toughness values plotted in Figures 5 to 9 have been obtained by multiplying the observed value by the ratio:

$$\frac{\text{best fit compressive strength}}{\text{observed compressive strength}}$$

The curves plotted in Figures 5 to 9 are all least squares best fit, using either a straight line or a quadratic curve, whichever seemed more appropriate.

5 Flexural strengths

5.1 LOP strength

Limit of proportionality strengths were calculated in addition to peak strengths (modulus of rupture). The LOP strength and its relationship to the peak strength is of interest because it gives an indication of the effect of the fibres on the composite behaviour.

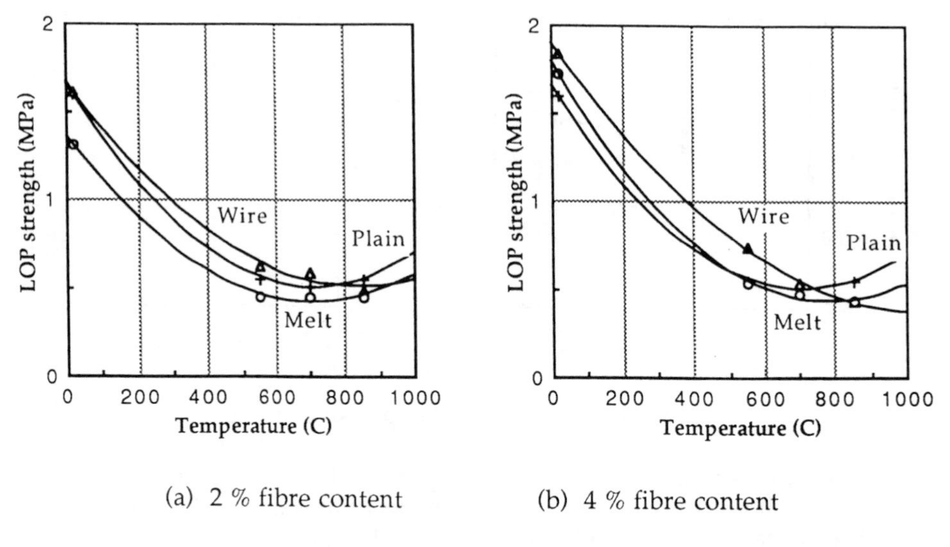

(a) 2 % fibre content (b) 4 % fibre content

Figure 5. LOP strengths

Figure 5 shows the effect of test temperature on normalised LOP strength forthe 2 % and 4 % fibre contents and the plain control. The graphs show that strength decreased with increasing temperature, but at a reducing rate. The wire fibres increased the LOP strength at 550 C, maintained or slightly increased strength at 700 C and resulted in a slight reduction at 850 C. In contrast, the melt fibres generally reduced LOP strengths but at ambient the 4 % mix produced an increase in LOP strength. The wire fibres produced higher strengths than the melt fibres, typically the difference being around 25 %.

The LOP strengths of the fibre mixes at elevated temperatures were similar to or slightly below that of the plain mix, with the exception of the wire fibres at 550 C which showed a moderate linear increase with fibre content.

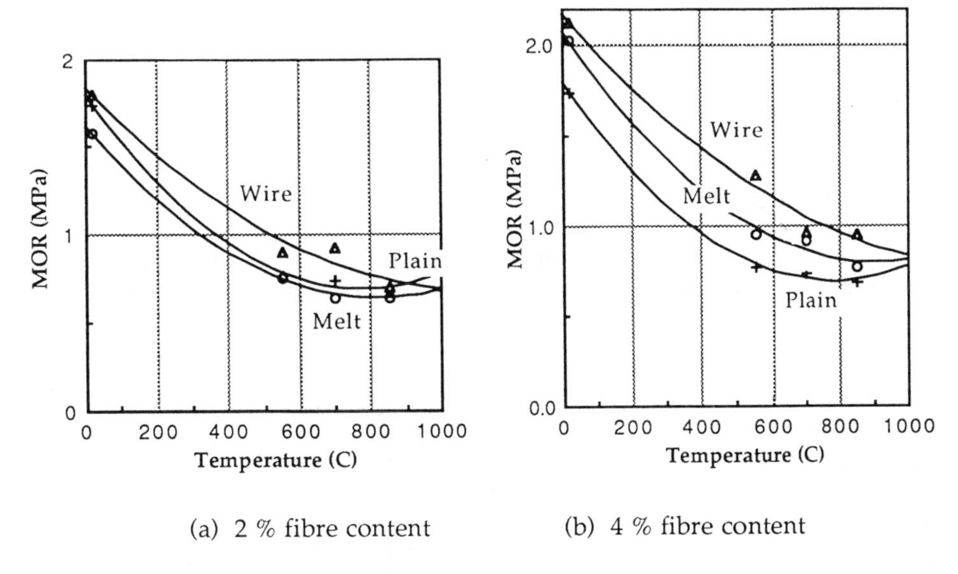

(a) 2 % fibre content (b) 4 % fibre content

Figure 6. Peak flexural strengths

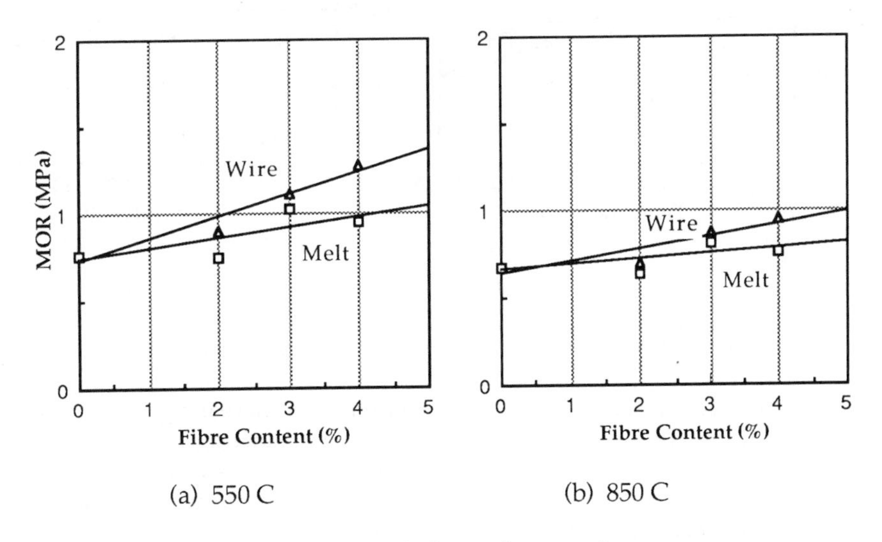

(a) 550 C (b) 850 C

Figure 7. Peak flexural strengths

4.3 Peak flexural strength (modulus of rupture)

Figure 6 shows the effect of test temperature on normalised MOR strength for the 2 % and 4 % fibre contents and the plain control. The graphs show that strength decreased with increasing temperature, the Dramix fibre reinforced concretes being superior to both the melt reinforced and plain concretes at all fibre contents.

Figure 7 shows the variation in normalised MOR strength with fibre content for the beams tested at 550 C and 850C. At all temperatures the wire fibres produced higher strengths than the melt fibres, typically the difference being around 20%.

6 Flexural toughness

The average normalised values of toughness, calculated as the area under the load/deflection curve up to 1.2 mm central deflection, are plotted for the 2 % and 4% fibre mixes against test temperature in Figure 8.

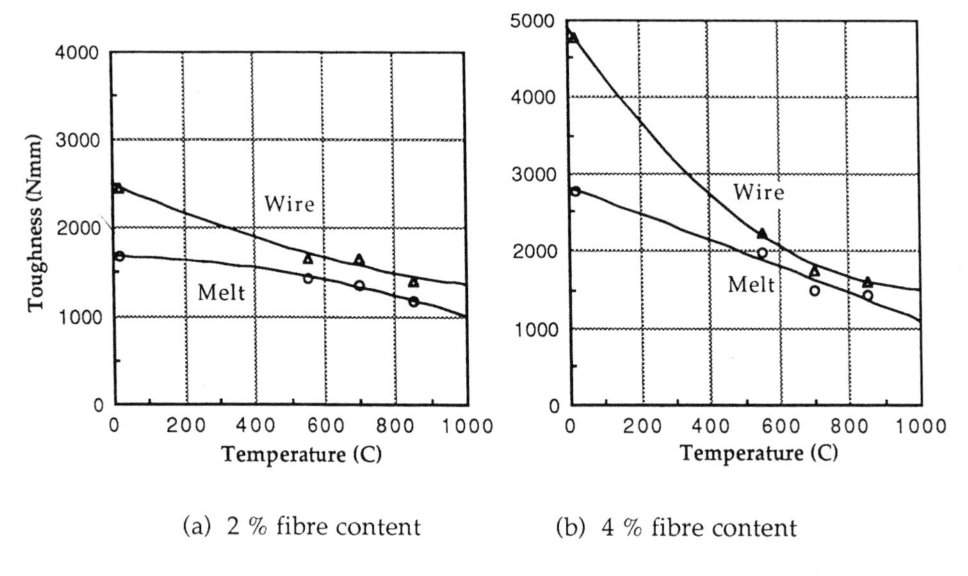

(a) 2 % fibre content (b) 4 % fibre content

Figure 8. Toughness at elevated temperatures

This figure shows that toughness decreased with increasing temperature. With wire fibres the relationship is clearly curved, starting from a very high toughness at ambient, whilst with melt extract fibres it is much flatter with a possible reverse in the curvature. Generally, as temperature increased the difference between the two fibres became smaller.

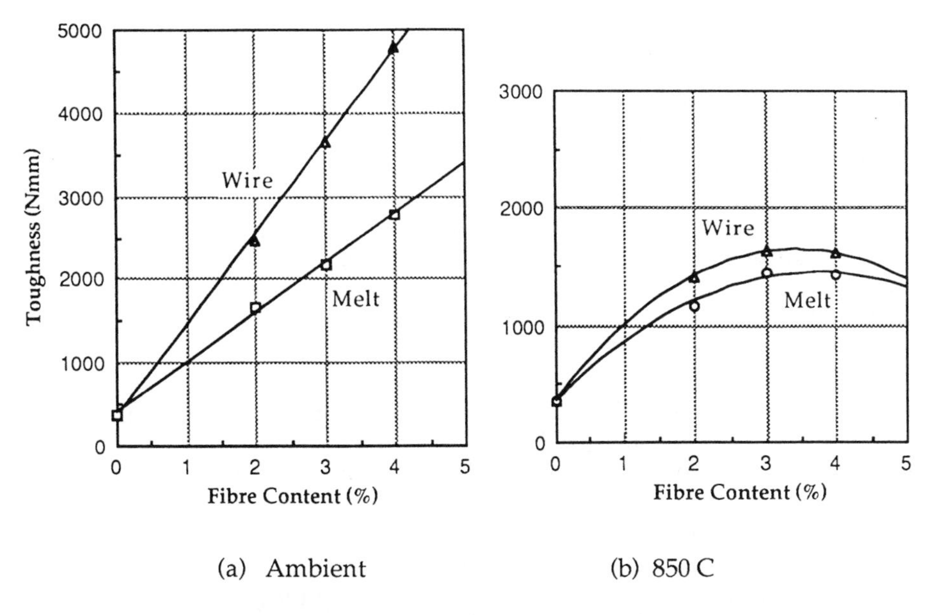

Figure 9. Variation in toughness with fibre content

Figure 9 shows the variation in normalised toughness with fibre content for the beams tested at ambient and 850 C. At all temperatures the wire fibres consistently produced a tougher material than the melt fibres, typically the increase in toughness being around 60% at ambient and 10% to 15% at temperatures of 550 C to 850 C. The effect of fibre content varied with test temperature. At ambient toughness increased linearly with increasing fibre content, but at temperature the rate of increase with fibre content decreased as temperature increased (ie the curves flatten off at higher temperatures).

7 Conclusions

7.1 Compressive strength and density

As expected, the density and compressive strength increased with fibre content. The scatter of the results, due to variations in the pre-blended castable from bag to bag, clearly affected the measurements of flexural strength and toughness and consequently the calculated values were normalised using the ratio of the expected and observed compressive strength for each mix. This normalisation enabled clearer trends of flexural strength and toughness against fibre content and test temperature to be discerned.

7.2 Flexural strength

As would be expected, the LOP and peak flexural strengths reduced with increasing test temperature, ceramic bond formation being unlikely to start below 900 to 1000 C for an ordinary CA cement. The wire mixes consistently out-performed the melt extract ones.

With the exception of the ambient results, the limit of proportionality and peak flexural strengths were found to vary linearly with fibre content. The Dramix fibres consistently produced higher strengths than the melt extract fibres, the difference being around 20% to 25%. The results indicate that, at elevated temperatures, to get an equivalent performance to 2% wire requires a melt extract fibre content of around 4% or more.

7.3 Toughness

The flexural toughness values reported were determined from beam load.deflection curves using JSCE approach, which calculates toughness as the area under the curve up to a fixed deflection of 1/150 of the span. Toughness increased with fibre content; at ambient temperatures the increase was linear, whereas at elevated temperatures it was non-linear, the rate of increase reducing with increasing fibre content.

The wire fibre reinforced castable was consistently tougher than the melt extract fibre reinforced castable, typically the difference in toughness being around 60% at ambient and 10% to 15% at temperatures of 550C to 850C. The results indicate that 3.5% by weight of melt fibres was required to get an equivalent performance to 2% wire at ambient, and 2.5% to 3% at elevated temperatures.

References

American Society for Testing Materials (1977) Standard recommended practices for determining consistency of refractory concrete, **ASTM C860:77**, American Society for Testing Materials, Philadelphia.

American Society for Testing Materials (1982a) Recommended practice for preparing refractory concrete specimens by casting, **ASTM C862:82**, American Society for Testing Materials, Philadelphia.

American Society for Testing Materials (1982b) Standard test methods for cold casting strength and modulus of rupture of refractory bricks and shapes, **ASTM C133:82**, American Society for Testing Materials, Philadelphia.

American Society for Testing Materials (1985) Flexural toughness and first crack measurement of fibre reinforced concrete, **ASTM C1018:85**, American Society for Testing Materials, Philadelphia.

Austin S A, Robins P J and Beddar M (1987) Influence of fibre geometry on the performance of steel fibre refractory concrete in **6th Int Conf of Composite Materials ICCM and ECCM**, Elsevier Applied Science, London, pp 2.80-2.89.

British Standards Institution (1984) Determination of modulus of rupture at ambient temperature, **BS1902 - 404:1984**, British Standards Institution, London.

British Standards Institution (1987a) Preparation of test pieces from dense castable by vibration. **BS1902 - 703** British Standards Institution, London.

British Standards Institution (1987b) Testing of materials as preformed pieces, **BS1902 - 706:1987**, British Standards Institution, London.

Japan Society of Civil Engineers (1984) Method of tests for flexural strength and flexural toughness of steel fibre reinforced concrete, **JSCE-SF4**, Concrete Library of Japan Society of Civil Engineers, No 3, June 1984, pp 58-61.

Robins P J and Austin S A (1992) Definition and measurement of toughness of fibre reinforced refractory concrete. Submitted to **Brit. Ceram. Trans. and J.** , January 1992.

14 CREEP PREDICTIONS FOR FIBER REINFORCED CEMENTITIOUS COMPOSITES BY THEORETICAL METHODS

C. H. YOUNG
Department of Construction Engineering, National Taiwan Institute of Technology, Taipei, Taiwan, ROC
J. C. CHERN
Department of Civil Engineering, National Taiwan University, Taiper, Taiwan, ROC

Abstract
This research establishes some theoretical formulations to describe the creep behavior of steel fiber reinforced concrete (SFRC), under various influencing factors including exposure environments, fiber contents, and applied stress states. The bond stress along the interface of fibers and matrices was found as the main reason why the fibers can resist the creep deformation of concretes. This fiber–matrix interfacial bond was evaluated theoretically and the formulations to calculate the creep strain of SFRC were derived. Creep strains of the steel fiber reinforced concrete can be obtained from these theoretical formulations based on the creep and shrinkage test data of the corresponding unreinforced plain concrete. Some comparisons between experiments and theoretical results of proposed model were performed to validate the theory. The results of the proposed model can fairly show that the restraint capability of fibers on creep of concretes is much more distinct under drying environment, tensile stress state and higher content of fibers.
Keywords: Creep, Fiber, Cementitious, Composites, Theoretical.

1 Introduction

Steel fiber reinforced concrete (SFRC) is a new construction material which provides several mechanical benefits to overcome many natural defects of concrete, such as brittleness, low tensile strength, low crack resistance, et al. Most literatures focused on the investigations of elastic behaviors of SFRC, but very few described the long–term behaviors such as creep of SFRC either on experiment or theory. Mangat and Azari (1985) performed a series of experiments to observe the creep behavior of SFRC under compression and developed some formulations to predict the creep strain [1]. According to their investigation the creep strain of SFRC were consistently lower than that of ordinary plain concrete (OPC) in the duration of applied load. However, their model can only be applied for specimens in the drying environment and under compressive load. Chern and Young (1989) set up an extensive series of experiments to study the compressive creep behavior of SFRC under various influencing factors, such as curing conditions, ages of concrete, fiber contents, applied loading states[2,5]. They concluded that the creep strains of SFRC were generally lower than that of OPC especially under drying environment and were slightly lower under moist environment. This study intends to establish a

Fibre Reinforced Cement and Concrete. Edited by R. N. Swamy. © 1992 RILEM.
Published by E & FN Spon, 2-6 Boundary Row, London SE1 8HN. ISBN 0 419 18130 X.

theoretical model to investigate how the fibers affect the creep behavior of SFRC. The bond stress along the fiber–matrix interface due to combined effect of Poisson's effect of the applied sustained stress and shrinkage deformation can share the stress, arising from external load, in the matrices and result in the creep recovery of matrices [3]. According to the superposition theory [4], one can calculate the final creep strain of SFRC.

Creep strains of steel fiber reinforced cementitious composites can be obtained from these theoretical formulations based on the given creep and shrinkage test data of corresponding unreinforced plain concrete. Some experimental data of creep of SFRC were used [2,5] to compare with the results of proposed model. The proposed model was also compared with that obtained by the Mangat's theoretical model [1].

2 Mechanism of creep in SFRC

Almost all the mechanical properties of SFRC are concerned with the interfacial mechanism between fibers and matrices. The bond on the interface also affects the creep behavior of SFRC. When a sustained load applies on this composite, the bond stress of fibers on the interface arises due to the relative lateral deformation between fiber and matrices owing to combined effect of Poisson's effect and shrinkage deformation. The bond stress then causes the reduction of stress in matrices and the creep recovery occurs gradually with the increase of time. This is the main reason to explain how the addition of fibers can reduce the creep strain of cementitious matrices in the SFRC composites.

3 Proposed creep model

As described previously, the existing analytical models for creep behavior of SFRC is very limited because of the complication for the problem. The fibers are randomly distributed with various orientations, an idealized model shown in Fig. 1 was used for the simplicity of analysis.

3.1 Basic assumptions and idealized model

The creep model for concrete matrices reinforced with randomly oriented short fibers of length l is based on the assumption that aligned fibers of equivalent length l_e and spacing s are all in the same direction as loading, as shown in Fig. 1. The spacing s in the idealized model could be obtained from following equation [6]:

$$s^3 + l_e s^2 - B^2 (H - l_e/2) \frac{l}{L} = 0 \qquad (1)$$

in which, s is the spacing of fibers, l_e is the equivalent length of fiber equals to $0.41l$, l is the fiber length, B is the equivalent breadth of specimen, H is the height of specimen, L is the equivalent length of a continuous fiber which equals $4v_f/\pi d^2$ and v_f is the volume content of fiber.

An sample element, shown in Fig. 1, was taken away from the idealized model as a typical representative unit to study the creep behavior of SFRC (see

Fig. 2). Each fiber in this unit having equivalent length l_e is assumed to be surrounded by a cylinder of concrete matrices with length l_e and diameter s.

3.2 Calculation of bond stress

The combined result due to the Poisson's effect of the sustained load and shrinkage in matrices induces the contact pressure on the surface of fiber (see Fig. 2(a)). This pressure, assumed to be uniformly distributed, could be derived easily from the geometric relationship [1]:

$$P_c = \frac{E_c(\epsilon_s - \epsilon_{\ell c})[(s/2)-(d/2)]}{(\frac{d}{2})\left[\left\{\frac{(s/2)^2+(d/2)^2}{(s/2)^2-(d/2)^2} + \nu_c\right\} + \frac{(1-\nu_s)}{E_s/E_c}\right]} \tag{2}$$

in which, p_c is the contact pressure radially applies on the lateral surface of fiber

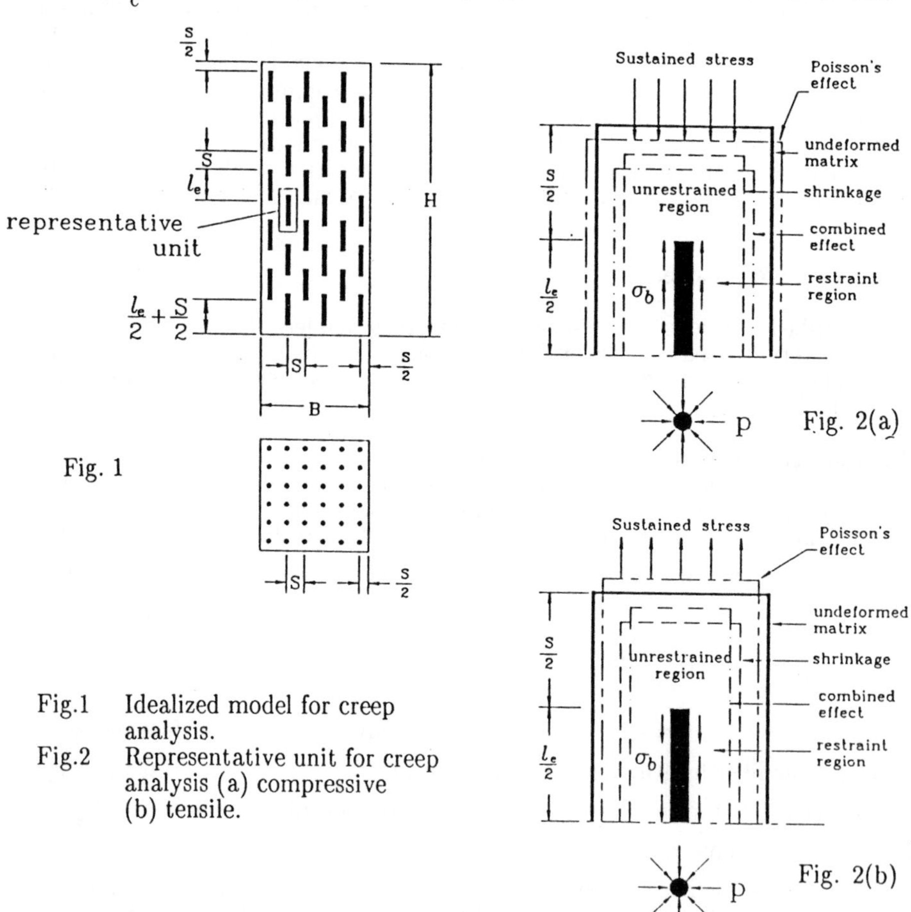

Fig.1 Idealized model for creep analysis.

Fig.2 Representative unit for creep analysis (a) compressive (b) tensile.

due to compressive sustained load, ϵ_s is the shrinkage of concrete matrices which is function of time of exposure in drying environment, ϵ_ℓ is lateral creep of the concrete matrices caused by the sustained external load due to Poisson's effect, s is the spacing of fibers in the idealized model, d is the diameter of fiber, E_c is the elastic modulus of concrete which is function of age of concrete while loading applies, ν_c is the Poisson's ratio of concrete matrices, ν_s is the Poisson's ratio of steel fiber and E_s is the elastic modulus of steel fiber.

The shrinkage of concrete matrices obtained from the experimental data of corresponding unreinforced plain concrete is concerned with time of exposure in the drying environment. The lateral strain ϵ_{ℓ_c} in the concrete matrices surrounding the fiber in the representative unit caused by the sustained external loading can be evaluated as below:

$$\epsilon_{\ell_c} = \nu_{cp}\,\epsilon_{oc} \tag{3}$$

where, ϵ_{oc} is the axial creep strain of the corresponding unreinforced concrete matrices and ν_{cp} is called creep Poisson's ratio. According to the available literatures [7,8] this value is more reliable in the range of 0.16 to 0.25.

If one considers the case of tensile sustained stress in Fig. 2(b), the contact pressure on the lateral surface of fibers will be higher than in compressive state shown in Fig. 2(a) due to the coincident direction of lateral deformation due to lateral creep and shrinkage. As compared with Eqn. (2), the contact pressure under tensile stress state can be derived as below:

$$P_t = \frac{E_c(\epsilon_s + \epsilon_{\ell_c})[(s/2)-(d/2)]}{\left(\frac{d}{2}\right)\left[\left\{\frac{(s/2)^2+(d/2)^2}{(s/2)^2-(d/2)^2} + \nu_c\right\} + \frac{(1-\nu_s)}{E_s/E_c}\right]} \tag{4}$$

According to the theory of friction, the frictional bond stress along the surface of fiber can be calculated as below:

$$\sigma_{fc} = \frac{\mu\, E_c(\epsilon_s - \epsilon_{\ell_c})[(s/2)-(d/2)]}{\left(\frac{d}{2}\right)\left[\left\{\frac{(s/2)^2+(d/2)^2}{(s/2)^2-(d/2)^2} + \nu_c\right\} + \frac{(1-\nu_s)}{E_s/E_c}\right]} \tag{5a}$$

$$\sigma_{ft} = \frac{\mu\, E_c(\epsilon_s + \epsilon_{\ell_c})[(s/2)-(d/2)]}{\left(\frac{d}{2}\right)\left[\left\{\frac{(s/2)^2+(d/2)^2}{(s/2)^2-(d/2)^2} + \nu_c\right\} + \frac{(1-\nu_s)}{E_s/E_c}\right]} \tag{5b}$$

in which, μ is the coefficient of friction between the fibers and matrices and σ_{fc} and σ_{ft} represent the frictional bond stress under compressive stress state and tensile stress state respectively.

The frictional bond stress arises due to the Poisson's effect and shrinkage deformation. The total bond stress must include another fraction of bond stress called initial bond strength σ_i, which exists natively on the interface of fibers and matrices and is not caused by contact pressure. This initial bond strength is very difficult to determine from experiments. Therefore, the authors proposed an approximate method to calculate this value.

In general, the order of creep strain of concrete lies between 10^{-4} to 10^{-3} from the experimental results of creep test under load duration from 1 day to several thousands days. During the occurrence of creep in cementitious matrices, the steel fibers do not creep under normal temperature, therefore, the values from 10^{-4} to 10^{-3} could be looked as the relative deformation between fibers and matrices. The corresponding value of bond stress due to the relative deformation in the fiber pull–out test is close to and slightly higher than the adhesive bond strength σ_{ad}. The authors recommended that the adhesive bond strength could be used in practical application instead of the initial bond strength. It includes two practical meanings: (1) The initial bond strength is difficult to determine from experiments as well as from theoretical calculations while the adhesive bond strength can be determined easily from pull–out test of fibers. (2) The adhesive bond strength used is the lower bound of the initial bond strength. This indicates the approximation underestimates the exact value and is on the safe and conservative side. The typical sketch of the adhesive bond strength in the bond–slip relationship of pull–out test is shown in Fig. 3.

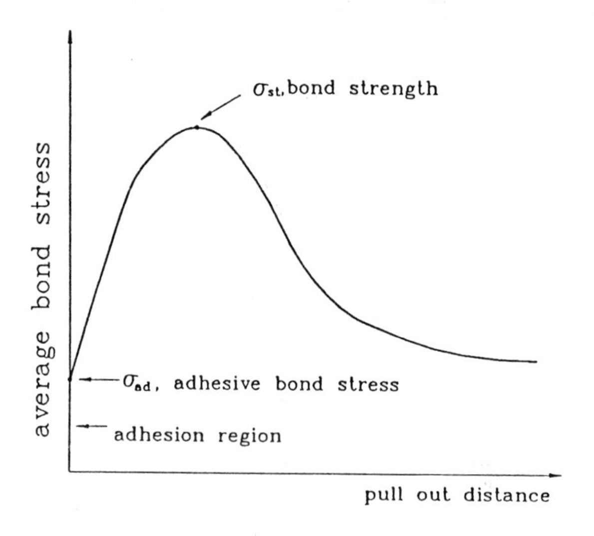

Fig.3 Adhesive bond strength in the pull–out test of fibers.

An empirical formulation, shown in below, to predict the adhesive bond strength can be obtained from the process of regression analysis of available pull–out test data of steel fibers. (see Fig. 4)

$$\sigma_{ad} = 0.027\, f_c \tag{6}$$

in which, f_c is the compressive strength of concrete at corresponding ages of concrete. The adhesive bond strength is dependent not only on compressive strength of cementitious matrices but also surface conditions of fibers. The empirical formulation proposed is suitable for smooth fibers only, i.e., another types of fibers will result in another constants other than 0.027. On their extensive test on several types of fibers, Young and Chern [5] also proposed the constant 0.019 for indented fibers and 0.013 for crimped fibers by using the same process of regression analysis from available test data.

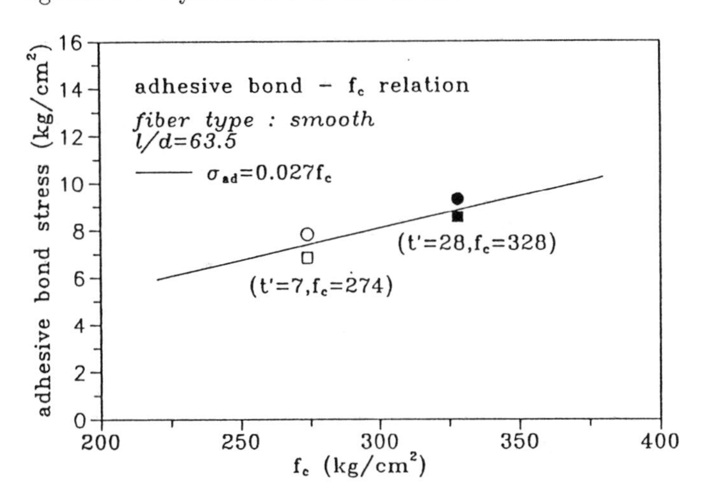

Fig.4 Regression analysis for the adhesive bond strength.

Rostasy and Keep (1982) [9] also recommended an approximate calculation formulation, $\sigma_{ad} = 0.033\ f_c$, of the adhesive bond strength for the steel reinforcing bars.

In Eqn. (5), the shrinkage strain ϵ_s and lateral creep strain ϵ_{ℓ_c} are all function of time after casting of concrete, t, and time t' at which the sustained load starts to apply on the specimen. Therefore, the frictional bond stress σ_f is also the function of t and t', which can be represented as $\sigma_f(t,t')$. As described previously, the adhesive bond stress is concerned with the compressive strength of concrete matrices, i.e., it is function of the curing age of concrete t', which could be expressed as $\sigma_{ad}(t')$. Consequently, the resultant bond stress comprises two parts, i.e. the frictional bond stress shown in Eqn. (5) and the adhesive bond strength given in Eqn. (6), is evaluated as below :

$$\sigma_{bc}(t,t') = \sigma_{fc}(t,t') + \sigma_{ad}(t') \quad (7a)$$

$$\sigma_{bt}(t,t') = \sigma_{ft}(t,t') + \sigma_{ad}(t') \quad (7b)$$

in which, σ_{bc} and σ_{bt} represent the bond stress on the fiber–matrix interface under compressive and tensile stress state respectively. The subscripts c and t denote compressive and tensile sustained stress state respectively.

In the early stage of time, the lateral creep strain is higher than shrinkage and only σ_{ad} exists because the friction bond stress vanishes in this stage (Eqn. 5). In the later stage the increase rate of shrinkage is faster than that of lateral creep strain, therefore, the friction bond stress behaves dominantly in this stage. It means that the increase of shrinkage results in higher contact pressure on the surface of fibers and then increases the bond stress.

3.3 Creep strain of SFRC

As shown in Fig. 2(a) and 2(b), the direction of bond stress is opposite to that of applied sustained load, therefore, the net stress due to the external load in the matrices reduces. The magnitude of the reduced stress could be derived in accordance with the equilibrium relations.

$$\Delta\sigma_c(t,t') = \pi\, d\,\frac{l_e}{2}\,\frac{\sigma_{bc}(t,t')}{A} \tag{8a}$$

$$\Delta\sigma_t(t,t') = \pi\, d\,\frac{l_e}{2}\,\frac{\sigma_{bt}(t,t')}{A} \tag{8b}$$

in which, A is the cross–sectional area of the representative unit. Then, the net stress due to applied sustained load is expressed by:

$$\sigma_c(t,t') = \sigma_{0c} - \Delta\sigma_c(t,t') \tag{9a}$$

$$\sigma_t(t,t') = \sigma_{0t} - \Delta\sigma_t(t,t') \tag{9b}$$

in which, σ_{oc} and σ_{ot} are the initial applied compressive and tensile sustained stress in the matrices respectively. The value of σ_c or σ_t is lower than the original applied sustained load σ_{oc} or σ_{ot} on the concrete matrices surrounding the fiber; consequently, the creep recovery occurs in the matrices and the final creep strain in the matrices will be reduced. This is the major reason why fibers can restrain the creep deformation of concrete and the lower creep strain of SFRC than that of OPC.

According to the principle of superposition [4], one can evaluate this final creep strain after creep recovery as an integration form:

$$\epsilon_{fc}(t,t') = \sigma_{0c}J_c(t,t') - \int_{t'}^{t} \frac{\Delta\sigma_c(\tau)}{\Delta\tau}\, J_c(t,\tau)\, d\tau \tag{10a}$$

$$\epsilon_{ft}(t,t') = \sigma_{0t}J_t(t,t') - \int_{t'}^{t} \frac{\Delta\sigma_t(\tau)}{\Delta\tau}\, J_t(t,\tau)\, d\tau \tag{10b}$$

in which, $J_c(t,t')$ or $J_t(t,t')$ is called compressive or tensile creep function of the material defined as strain at time t caused by a unit stress applied at time t'. $J_c(t,t')$ and $J_t(t,t')$ represent the magnitude of creep occurring in the corresponding unreinforced concrete under the applied sustained load σ_o. Eqn.

(10) is too complicated for calculation, one can evaluate it by a simpler and approximate expression similar to that used for reinforced concrete by Samra [3]:

$$\epsilon_{fc}(t,t')=\sigma_{0c}J_c(t,t')-\frac{\Delta\sigma_c(t,t')}{2E_c(t')}[E_c(t')J_c(t,t')+1] \tag{11a}$$

$$\epsilon_{ft}(t,t')=\sigma_{0t}J_t(t,t')-\frac{\Delta\sigma_t(t,t')}{2E_c(t')}[E_c(t')J_t(t,t')+1] \tag{11b}$$

The first term of Eqn. (11) equals creep strain of OPC and the second term represents the reduced creep strain due to fibers. One can easily observe from Eqn. (11) that the variation of creep strain between SFRC and OPC increases consistently with the increase of time because the bond stress σ_{bc} (or σ_{bt}) and the reduced stress $\Delta\sigma_c$ (or $\Delta\sigma_t$) in the matrices also increase due to the development of contact pressure on the lateral surface of fibers. Eqn. (11) expresses the creep strain in the matrices surrounding the fiber, i.e., the restraint region in the representative unit. The final creep strain of SFRC in the representative unit would be expressed as below:

$$\epsilon_{cc}=\frac{\epsilon_{fc}(\ell_e/2)+\epsilon_{oc}(s/2)}{(\ell_e/2)+(s/2)} \tag{12a}$$

$$\epsilon_{ct}=\frac{\epsilon_{ft}(\ell_e/2)+\epsilon_{ot}(s/2)}{(\ell_e/2)+(s/2)} \tag{12b}$$

in which, ϵ_{oc} and ϵ_{ot} are the creep strain of OPC in the unrestrained region above the fiber in the representative unit under compressive and tensile stress state respectively (see Fig. 2(a) and 2(b)).

4 Compared with test results and Mangat's model

In order to validate the proposed theoretical model, the experimental data for creep strain of SFRC obtained from Chern and Young (1989) [2] were used for comparison in fiber effect and environment effect; and other sets of test data obtained from Young and Chern (1990) [5] were used for comparison in the effect due to applied stress state. The corresponding shrinkage and creep test data of OPC in the same series of tests were used for substitution into Eqn. (5) and (7) to evaluate the bond stress. The compressive strength and Young's modulus at the corresponding age of concrete matrices obtained from test were also used in the computation. The specimen in the test is a cylinder with diameter of 15 cm and height of 30 cm. The coefficient of friction μ is 0.04 following the recommendation of Mangat and Azari [6] because the fibers in the test are all straight and smooth on shape and the creep Poisson's ratio were used by the value of 0.2 in accordance with the investigation of the literatures about the creep Poisson's ratio which was discussed previously. The value of elastic Poisson's ratio of concrete matrices and fiber are 0.2 and 0.3.

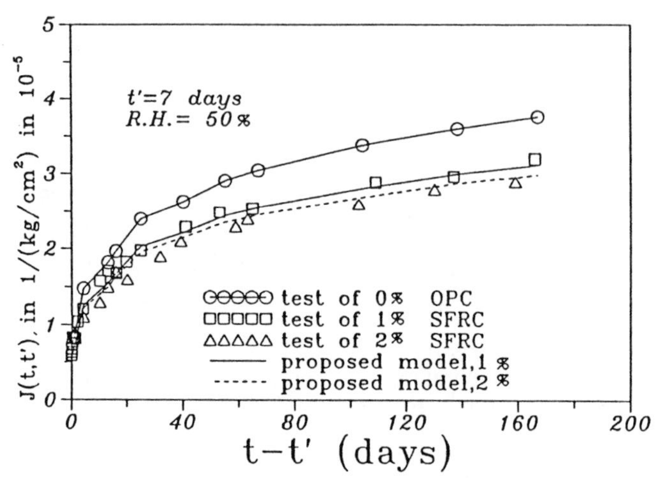

Fig.5 Comparison of experimental results and proposed model under various fiber contents.

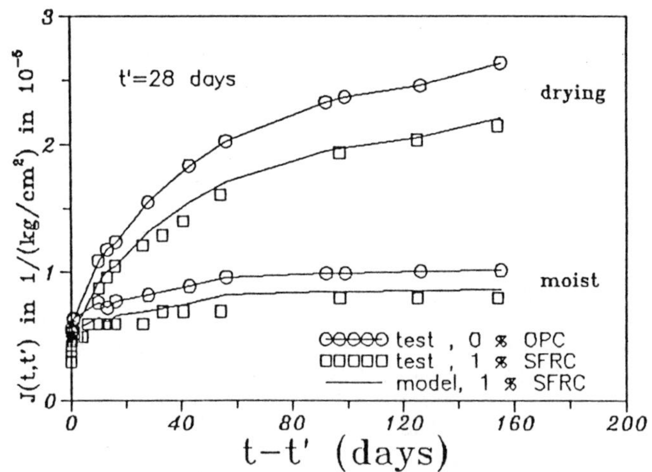

Fig.6 Comparison of experimental results and proposed model under various environments.

4.1 Effect of fiber content

The higher fiber content resulting in smaller spacing s between fibers (Eqn. 1) will reduce more stress in the matrices and thus lead to less creep strain (see Eqns. 5–12). Fig. 5 shows the theoretical results compared with the experimental results. One can easily observe that the higher fiber content results in lower creep strain in the matrices. The restraint capability of fibers does not linearly increase with the increase of fiber content, contrary, the increase of restraint capability becomes less obvious as the fiber content increases. This phenomenon can be fairly described in the proposed model as well.

4.2 Effect of curing environment

Eqn. (5), Fig. 2(a),2(b) show that the shrinkage has significant effect on the interfacial bond stress. One may see that the restraint capability of fibers on creep is more distinct in the drying condition, in which the shrinkage develops. If the specimen was tested in the moist environment, no shrinkage occurs and the frictional bond stress vanishes in this condition, i.e., only adhesion bond stress existed. It could be expected that the restraint capability of fibers on creep is not distinct in the moist environment. Fig. 6 shows the experimental and theoretical results of the creep strain of SFRC in these two environmental conditions, moist state of 100% R.H. and drying state of 50% R.H. It is clearly observed that the restraint capability of fibers on creep in the moist condition is less obvious than that in drying condition both found in the experimental results and the analysis by using the proposed model.

4.3 Effect of loading state

Fig. 2(b) shows that the lateral deformation owing to Poisson's effect under tensile sustained stress coincides with the deformation of shrinkage, therefore, the contact pressure is higher under tensile state than under compressive state. It results in higher frictional bond stress and lower creep strains of SFRC under tensile sustained stress state. (see Eqn. (5) and (11)). In addition, Fig. 7 shows that the creep strain of corresponding ordinary plain concrete is higher under tensile state than under compressive state from experiments. Consequently, owing to higher bond stress on the fiber–matrices boundary and higher creep strains of corresponding ordinary plain concrete under tensile stress state, the second term in Eqn. (11) which denotes the creep reduction due to addition of fibers in the case of tensile state is larger than compressive state. Fig. 7 shows the results of predictions by using proposed model and their comparisons with test results under 2% fiber content of SFRC and two different stress states. The proposed model could fairly describe the time–dependent material behavior of SFRC.

4.4 Comparison with Mangat's model

Mangat's model [1] neglects the adhesion bond stress so that it can not describe the creep behavior of SFRC in the moist environment condition; this is due to the absence of shrinkage and the vanish of frictional bond stress in this condition. In addition, they evaluated the creep strain in the reinforced matrices surrounding the fiber by using the linear proportional relationship among the original applied sustained stress σ_0, the corresponding creep strain of the unreinforced concrete, and the net applied stress in the reinforced matrices. The proposed model was derived using the phenomenon of creep recovery occurring in the reinforced matrices surrounding the fiber and evaluates the creep strain of reinforced concrete matrices by the integral type of superposition theory in the time domain.

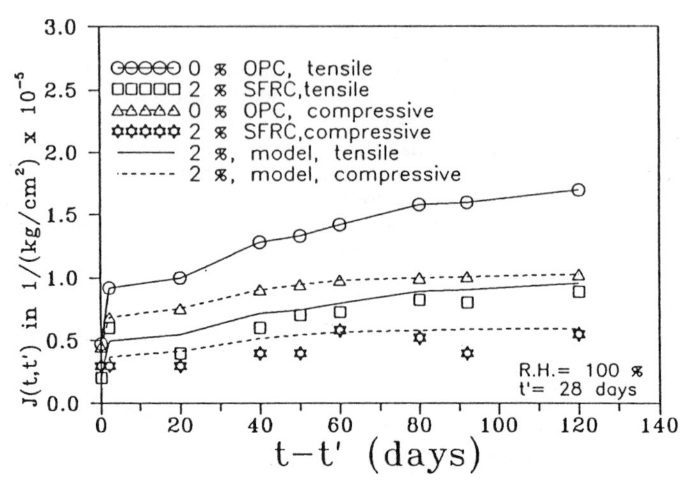

Fig.7 Comparison of experimental results and proposed model under various applied sustained stress states.

In addition, Mangat's model [1] discussed only with the compressive sustained stress, but the proposed model could be used under compressive stress state as well as under tensile stress state. Fig. 8 shows a numerical comparison between the proposed model and Mangat's model under drying environment and compressive sustained stress state. The proposed model was found to describe the creep behavior of SFRC more exactly than Mangat's model.

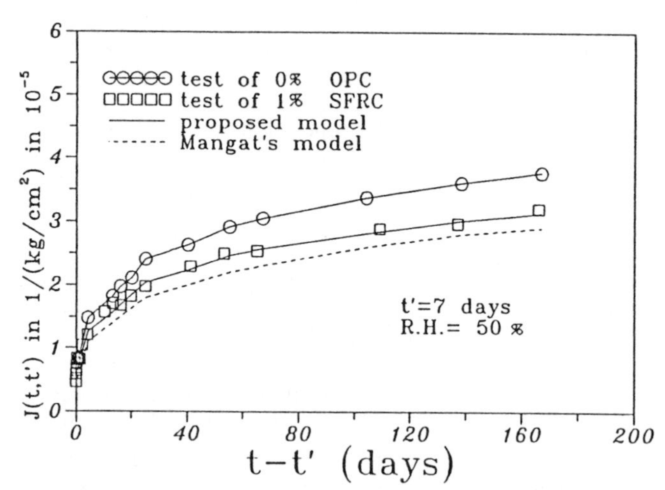

Fig.8 Comparison between proposed model and Mangat's model.

5 Conclusions

(1) The proposed model considers both the adhesive bond stress and the frictional bond stress on the interface between fibers and matrices so that it can fairly describe the creep behavior of SFRC not only under the drying environment but also under the moist environment.
(2) The proposed model can describe that restraint capability of fibers on creep is much more distinct in drying environment condition than that in moist condition.
(3) Higher fiber content results in lower creep of SFRC. But the restraint capability of fibers does not increase linearly with the increase of fiber content. The increase of restraint capability becomes less obvious while the content of fibers is increasing.
(4) The proposed model can describe that the restraint capability of fibers on creep is much more distinct under tensile sustained stress state than in compressive stress state.
(5) Compared with Mangat's model, the proposed model can be applied to wider range and give more exact predictions to the creep behavior of SFRC

6 Acknowledgement

Support from the National Science Council of Republic of China under Grant No. NSC78–0410–E002–07 to National Taiwan University is gratefully acknowledged.

7 References

1. Mangat, P.S. and Azari, M.M., "A theory for the creep of steel fiber reinforced cement matrices under compression". Journal of Materials Science, V 20, 1985, pp. 1119–1133.
2. Chern, J.C. and Young, C.H., "Compressive creep and shrinkage of steel fiber reinforced concrete". The International Journal of Cement Composites and Lightweight Concrete,V. 11, 1989, pp. 205–214.
3. Samra, R.M., "Creep model for reinforced concrete column". ACI Structural Journal, V. 86, 1989, pp. 77–82.
4. McHenry, D., "A new aspect of creep in concrete and its application to design". Proc. ASTM, V, 43, 1943, pp. 1069–1084.
5. Young, C.H., "The viscoelastic and fracture behavior of fiber reinforced cementitious composites". Ph.D. thesis, National Taiwan University, Taipei, Taiwan 1990. (in Chinese).
6. Mangat, P.S. and Azari, M.M., "A theory for the free shrinkage of steel fiber reinforced cement matrices". J. of Materials Science, V. 19, 1984, pp.2183–2194.
7. Meyer, H.G., "On the influence of water content and of drying conditions on lateral creep of plain concrete". Materials and Structures, Paris, V.2, 1969, pp.125–131.
8. Gopalakrishnan, K.S., Neville, A.M. and Ghali, A., "Creep Poisson's ratio of concrete under multiaxial compression". ACI Journal, V. 66, 1969, pp.1008–1020.
9. Rostasy, F.S. and Kepp, B., "Time–dependence of Bond, Bond in Concrete".Applied Science Publishers, London, England, 1982.

15 FLEXURAL STRENGTH OF STEEL WOOL REINFORCED MORTAR

J. R. AL-FEEL and N. K. AL-LAYLA
Civil Engineering Department, Mosul University, Iraq

Abstract
This investigation presents test results of a series of experiments to determine the flexural strength of cement mortar reinforced with steel wool. The strength of plain mortar is compared with the strength of mortars reinforced with different percentages of steel wool. The casting method used was impregnation of steel wool mat with a cementitious slurry. Different cement to sand ratios were used with a different volume percentage of steel wool. The steel wool has an irregular shape and can be quite long and thin, thus providing a reinforcing unit with a high aspect ratio. A regression analysis based on the test results was performed and an equation for the prediction of the flexural strength of mortar reinforced with steel wool was derived. It was found that the addition of steel wool enhanced the strength and toughness.
Keywords: Cement mortar, Cement paste, Flexural strength, Steel wool, Toughness.

1 Introduction

Steel fibres are used in cement matrices to improve some of the weak properties, such as tensile strength, ductility, and toughness. The volume of steel fibres is limited to about 3% due to difficulties in mixing, compacting, and finishing the fresh mix (ACI Committee 544, 1988). Composites with much higher steel fibre volume (SIFCON) were tried. The method used is based on the preparation of a prepacked steel fibre bed and subsequent infiltration of the spaces in the bed with a cement slurry (Lankard, 1984).

Another approach for reinforcing brittle cement matrices was tried (Bentur, 1987, 1988). This time steel wool was used as reinforcement. The steel wool mats are placed in the forms and slurry or mortar with fine sand is infiltrated into the fibre skeleton. Modulus of rupture values of 17 to 20 MPa could be obtained with steel wool volume contents of less than 2%, but for higher volume contents there was no strength enhancement.

Fibre Reinforced Cement and Concrete. Edited by R. N. Swamy. © 1992 RILEM.
Published by E & FN Spon, 2-6 Boundary Row, London SE1 8HN. ISBN 0 419 18130 X.

The steel wool consists of fibres with irregular shape and they can be quite long and thin, thus providing a reinforcing unit with a high aspect ratio that can be as large as 500.

The fibres in the wool have the potential for providing a good bond with the matrix due to their high aspect ratio and tortuous shape.

The aim of the present paper is to investigate the influence of low content of steel wool on the flexural strength and toughness of cement paste and mortars.

2 Experimental programme

Two series of tests with two sizes of sand, one passing No. 16 and the other passing No. 25 sieve, were used to study the effect of sand/cement ratio, and the percentage of steel wool by weight on the flexural strength and toughness index. Composites with five different steel wool contents were used (0, 0.5, 0.75, 1.0, 1.25% by weight), for each size of sand. The mixes are shown in Table 1.

Specimens of 20 x 20 x 300 mm were used. A thin layer of mortar was usually poured at the bottom of the mould to act as a bed for the composites then the steel wool was placed as a mat. After that the mould was filled with the slurry until the wool mat was fully impregnated. The water/cement ratio of the slurry was varied from 0.35 to 0.55 as the percentage of the fibre was changed from 0 to 1.25. Table vibration of the specimens was used, and the specimens were stripped after 24 hours and cured in water for 14 days. At the age of 14 days specimens were tested in flexure over a span of 250 mm, with the loads applied at the third points. The specimens were tested under a constant rate of movement at 0.16 mm/min. Deflections were measured at the midspan section. The results were the average of at least three specimens.

Table 1. Mix proportions

S*/C	W/C	Steel wool (% by weight)
0	0.35	0, 0.5, 0.75, 1.0, 1.25
0.5	0.4	0, 0.5. 0.75, 1.0, 1,25
1.0	0.45	0, 0.5. 0.75, 1.0, 1.25
1.5	0.5	0, 0.5, 0.75, 1.0, 1.25
2.0	0.55	0, 0.5, 0.75, 1.0, 1.25

* Two sizes of sand were used, passing No. 16 and 25 sieves.

3. Results and discussion

3.1 Effect of steel wool on flexural strength

The results of the flexure tests for specimens subjected to curing for 14 days are shown in Figs. 1 and 2. Points plotted in the figures represent the average of three specimens. The increase in steel wool content led to continuous improvement in the flexural strength of the composite as compared to the unreinforced matrix. But some of the mixes that had a high content of steel wool showed a decrease in flexural strength. This may be attributed to the method of preparing the specimens, and the chosen mix proportions.

The optimum fibre content depends on the mix proportions and the fineness of the sand.

The increase in the flexural strength of the cement paste mix was about 36% for steel wool content up to 1.25% by weight where the increase is about 19% for the mix with S/C of 1.5 and for sand passing the No. 16 sieve, the effect of sand/cement ratio on the flexural strength is shown in Table 2. As a result, the flexural strength for mixes containing sand sizes No. 16 and No. 25 decreases with the increase in sand content, as was obtained by Bentur (1989).

By using the present results and the results of other authors (Bentur, 1987, 1989), a regression equation based on 44 observations was obtained for the flexural strength of steel wool reinforced mortars as a function of the flexural strength of the unreinforced matrix and weight percentage of the used steel wool. The correlation coefficient of the

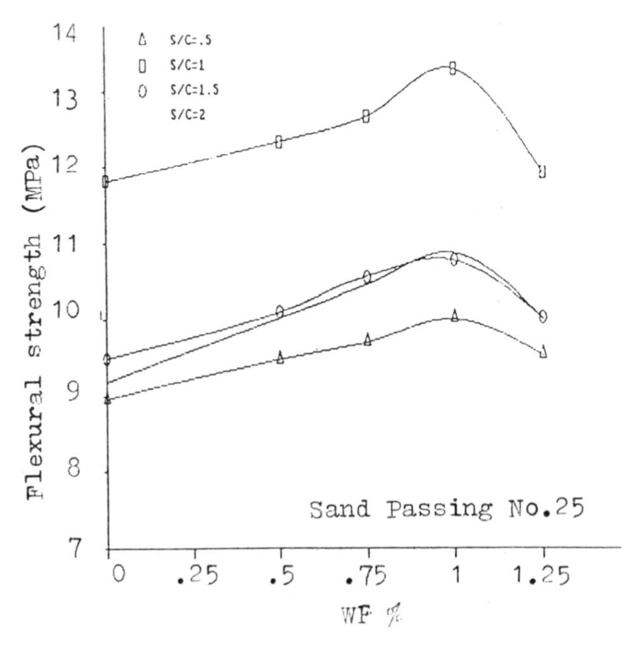

Fig. 1 Effect of steel wool content on the strength of the composite

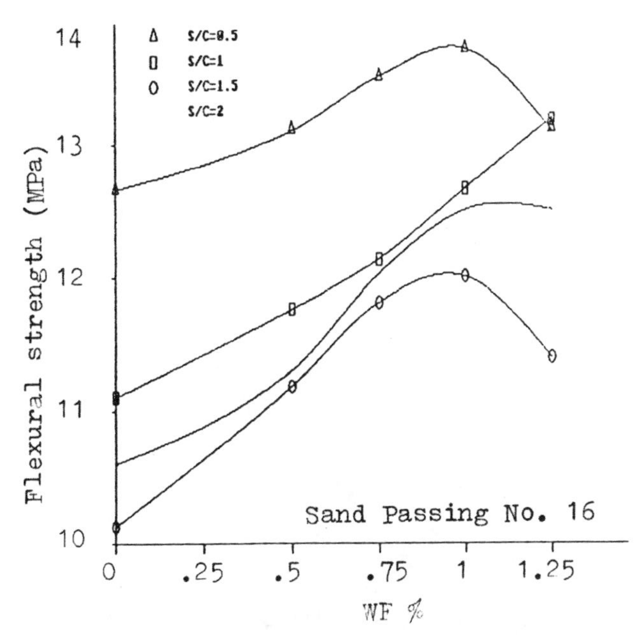

Fig. 2 Effect of steel wool content on the flexural strength of the composite

proposed equation is 0.987 which means an excellent degree of correlation between the investigated variables. The equation is of the following form:

$$f_c = 1.07f_m + 0.78w$$

where f_c = flexural strength of the composite;
f_m = flexural strength of the matrix; and
w = percentage weight of the steel wool.

3.2 Effect of steel wool on load-deflection curve

The load-deflection curves shown in Figs. 3 and 4 are for two mixes, C : S : W/C = 1 : 1 : 0.45 and cement paste for different steel wool contents. Toughness is a kind of measurement index of energy absorption capacity (Leslaw, 1989). For the present investigation the indices are shown in Figs. 5 and 6. These indices are used to characterize the fracture resistance of materials when they are subjected to static loads. Energy absorbed by the specimen is represented by the area under the load-deflection curve. The figure shows that as the steel wool content increased, the toughness index increased compared to the unreinforced matrix.

The value of toughness index depends on the steel wool content and size of sand. The maximum increase of toughness indices for two mixes C:S of 1:0 and 1:1 are 2.23 and 2.17 respectively compared to the unreinforced matrix.

The fracture mode of the beam without fibres was brittle as usual and failed suddenly at small deflections by separation into two pieces; whereas the specimens containing

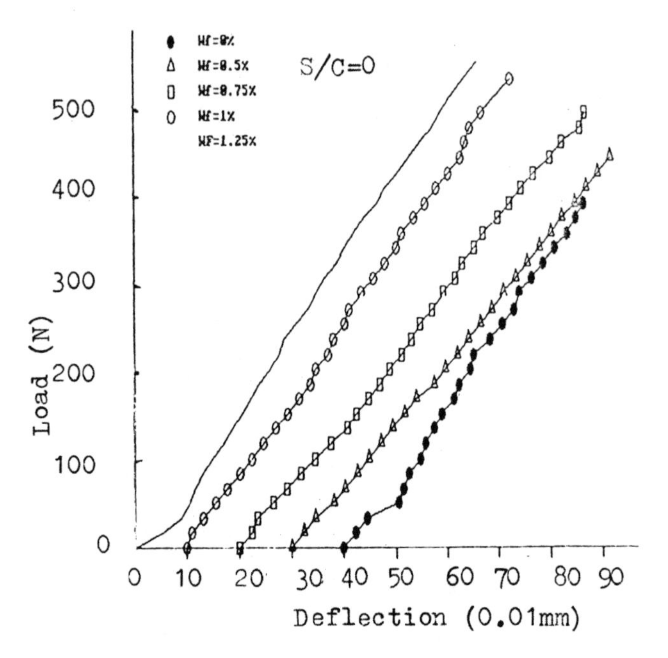

Fig 3 Effect of steel wool content on the load-deflection curve of the composite

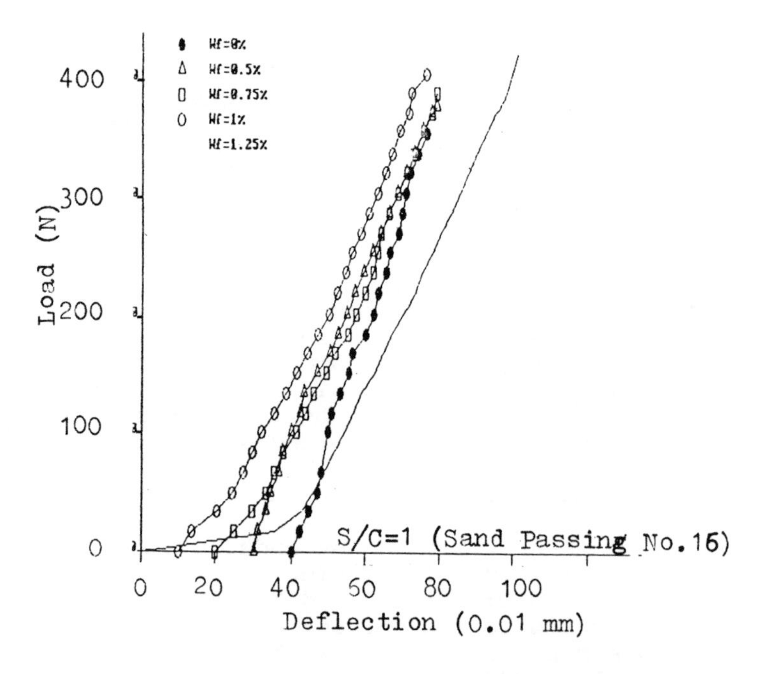

Fig. 4 Effect of steel wool content on the load-deflection curve of the composite

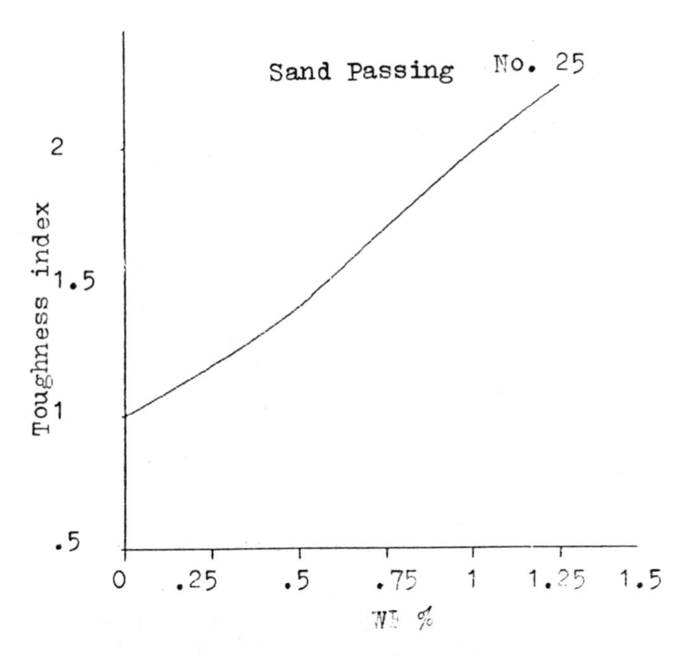

Fig. 5 Effect of steel wool content on the toughness index

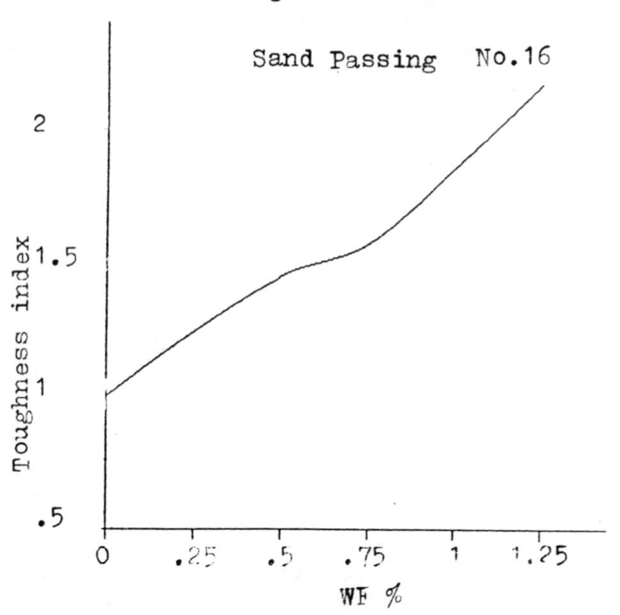

Fig. 6 Effect of steel wool content on the toughness index

steel wool failed by gradual development of single or multiple cracks, and the specimen retained its integrity even after the attainment of the maximum load.

Table 2. Effect of sand/cement ratio on flexural strength

S/C	W/C	Fibre (%)	Flexural strength (MPa)
0	0.35	1.25	17.24
0.5	0.4	1.0	13.1
1.0	0.45	0.75	12.13
1.5	0.5	0.5	11.1

4 Conclusions

Based on the results obtained in this investigation the following conclusions can be derived:

1. It is possible to use low contents of steel wool to produce cement composites by slurry infiltration to get flexural strength values up to 17 MPa.
2. The flexural strength of the mixes decreased with the increase in sand content.
3. A marked increase in the toughness was noted even when low contents of steel wool are used to reinforce the matrices.
4. Refined methods for preparing the specimens, placing the reinforcement, and proper mix proportions may result in higher flexural strength and toughness of composites.

5 References

ACI Committee 544 (1988) Design consideration for steel fibre reinforced concrete. J. ACI, 85, 563-580.

Bentur, A and Gree, R (1987) Cement reinforced with steel wool. J. Ltwt concrete, 9, 217-223.

Bentur, A (1989) Properties and reinforcing mechanisms in Steel wool reinforced cement, in **Fibre Reinforced Cement and Concretes** (eds R.N. Swamy and B. Barr), Applied Science, London. pp. 101-110.

Lankard, D.R. and Newell, S.K. (1984) Preparation of high reinforced steel fibre reinforced concrete composite in fibre reinforced concrete, in Fibre Reinforced Concrete, ACI SP-81, (ed. G.C. Hoff) Detroit, pp. 287-306.

Leslaw Hebda, Lech Rudzinski (1989) Influence of material structure of SFRC on toughness index, in Fibre Reinforced Cements and Concretes (eds R.N. Swamy and B. Barr), Elsevier Applied Science, London, pp. 388-400.

PART THREE

ENGINEERING PROPERTIES, DYNAMIC BEHAVIOUR

16 CHARACTERISTICS OF ACOUSTIC EMISSION IN STEEL FIBRE REINFORCED CONCRETE

YUE CHONGNIAN and DAI WENBING
Chongqing Institute of Architecture and Engineering,
Chongqing, China

Abstract
In this paper the characteristics of acoustics emission (AE) in steel fibre reinforced concrete (SFRC), such as: 1. Kaiser effect in SFRC; 2. The AE characteristics of initial crack in SFRC on flexural test; 3. The difference and the analogy between the AE parameters of concrete matrix cracking and that of steel fibre pulling out; 4. The distribution law of AE parameters in SFRC have been studied. Based on above research the formation and propagation of cracks in SFRC has been analyzed and the concept of split of SFRC has been discussed.
Key words: Acoustic emission, Kaiser effect, Steel fibre reinforced concrete, Initial crack point of SFRC

1 Introduction

Acoustic emission (AE), as a powerful test technology, has got a wide application in research on materials. At present it is mainly used for study of metals, alloys, rocks and composites, which consist of metal matrix and metal reinforcing fibre or plastics with plastic fibres. It is little that this technique has been used for test on composites with concrete matrix. However, some researchers have studied AE phenomena in pure concrete and concrete reinforced by steel bar. Almost nothing has been done in the field of steel fibre reinforced concrete (SFRC). For this reason authors have engaged in research in this field.

To determine the intial crack point of material is one of the imporant applications of AE, but this point is not always able to determined by AE. It depends on whether the change of microstructure in material under loading will manifest in the AE diagrams [1], [2], [3].

In order to investigate the AE characteristics of material effectively, it is essential to study the factors which effect the AE characteristics of material. Since the loading history is a basic factor for AE, the Kaiser effect of SFRC has firstly been investigated. In this paper authors tried to find the difference between AE characteristics of concrete cracking and that of fibre pulling out from the matrix. The distributions of AE parameters of SFRC with different fibre contents have been studied as well.

Fibre Reinforced Cement and Concrete. Edited by R. N. Swamy. © 1992 RILEM.
Published by E & FN Spon, 2-6 Boundary Row, London SE1 8HN. ISBN 0 419 18130 X.

2 Experiments

The concrete beams 430x100x50mm with different content of steel fibre have been prepared as specimens, the concrete consists of Portland blastfurnace slag cement 425#, river sand with fineness modulus M_k=2.87 and water. The mix proportion of specimens for fracture analysis on flexure is cement:sand:water=1:2.4:0.45, and that for Kaiser effect is 1:2.4:0.5. the steel fibre 0.35x0.60x25mm with equivalent diameter 0.517mm and L/D=48 was produced in Hangzhou Steel Fibre Factory by chopping technology. The steel fibre content for test on kaiser effect is 2.5% (by volume) and on other tests is 0%, 0.5%,1.0%,2.0%,2.5% and 3% respectively.

The specimens were loaded by three point bending with span 400mm.

All specimens have been tested on INSTRON -1346 test machine. For Kaiser effect the circular load was exerted. In each loading circle the load increased by about 15% of the maximum load before the maximum load and beyond that no constraint for it. The rate of loading was controlled by deformation in the middle of the beam (1/mm deflection/500 sec.)

During experiment all AE signals have been registered by AE test machine AET-5000A. Test parameters of AET-5000A were adopted as follows:

Sensors with resonance frequency 175KH
Pre-amplifier 60dB gain with band 125KH-1MH
Main amplifier (regulatable) 10-40dB gain
Threshold current 0.2-0.3A

On testing beams two AE sensors were fixed on one side of beam at distance of 300mm.

3 Results and discussions

1. The Kaiser effect of SFRC

By circular loading SFRC beams with 2.5% SF (by volume) the characteristics of AE event counts and relationship of deformation of beam to load exerted on it has been registered. As experiments show, in every loading circle before the maximum load there is almost no AE signal untill the deformation of loaded specimen reaches that of last circle. As soon as the deformation exceeds that of last circle, the AE events can be clearly observed, see Fig.1, Fig.2, Fig.3, Fig.4, Fig.5, Fig.6,Fig.7, Fig.8. In the AE event counts vs time (t) diagram the AE signals can also be registered at the early stage of unloading in whole loading circle, and at late stage AE is quiet. However, when beyond the maximum load (Pmax) the deformation of tested specimen does not got that of previous circle, also burst many AE events, which can be registered by test device.In that period when deformation of specimen reaches near the previous deformation of last loading circle the AE event counts are raising apparently, see Fig.1, Fig.3, Fig.4, Fig.5, Fig.6, Fig.7, Fig.8. By increase of the times of loading circles the AE event count raising tendency is weakening, at last it will disappear at all. From AE event counts vs time diagrams mentioned above during almost whole process of circular loading burst more or less AE events when the load beyond maximum load.

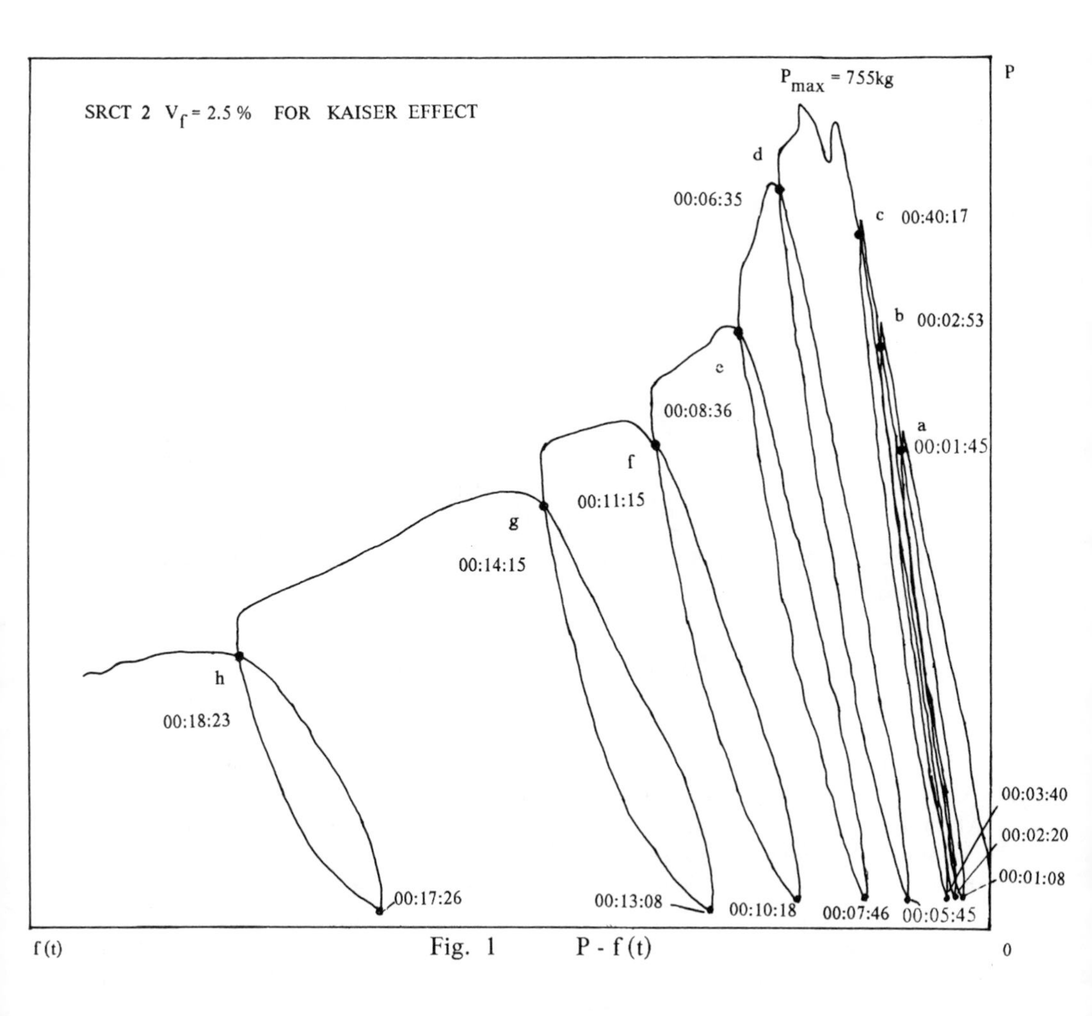

Fig. 1 P - f(t)

From those experiments it is found that the Kaiser effect of SFRC is clear before maximum load and is not apparent beyond it. This result coincides with Boyce's view on valid zone of Kaiser effect. By Boyce[4], the completely valid zone of Kaiser effect is only in the elastic zone or zone in which the crack can stablly propagate. Beyond that zone crack starts to develop unstablly and Kaiser effect will disappear completely, [5], [6].

2. The AE characteristics of initial crack in SFRC on flexural test

In order to study AE characteristics of SFRC when the matrix begins to crack, seven groups of specimens, which contain SF 0%, 0.5%,1.0%, 1.5%, 2.0%, 2.5%

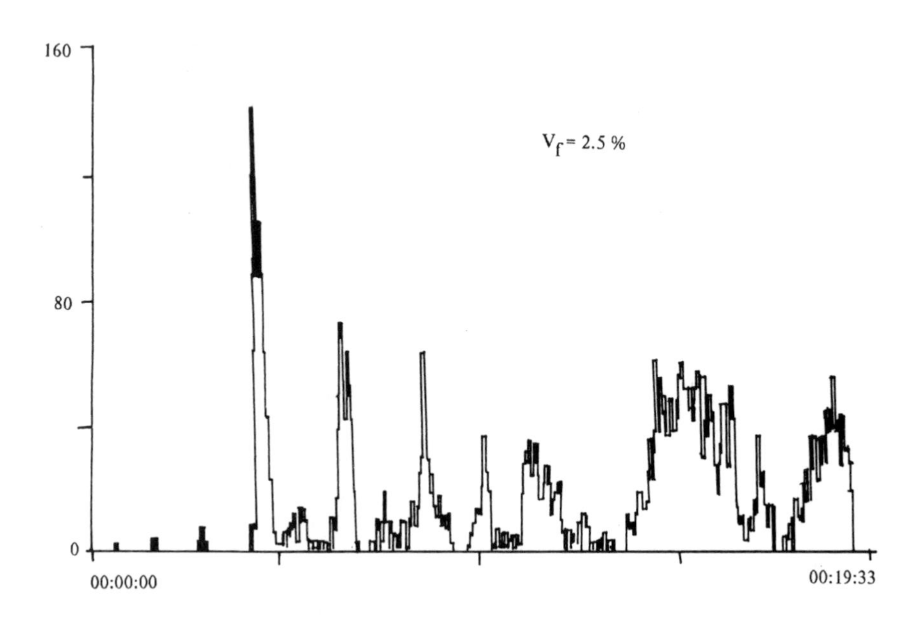

Fig. 2 Events VS time

and 3% (by volume) respectively, have been tested on flexure. During whole experimental process all AE characteristic diagrams have been registered. Since the AE apparatus can only register AE parameters with real time in association with computer, all diagrams were calibrated by the rate of loading. For study on crack propogation the AE event counts vs time diagram is adopted.

By observation and analysis of all AE event count vs time diagrams of loaded SFRC specimens, which contain seven kinds of fibre volumes, the AE diagrams can be divided into two categories. Category A is the diagrams of SFRC, which contain SF no more than 1.0%. In this category during testing before AE event appears suddenly, does not exist a continuously rising process of AE event counts, see Fig.9, Fig.10, Fig.11. Category B is for composites with SF more than 1.5%. In that category there is a continuously rising process of AE event almost in whole process, especially it may be clearly observed before a large amount of AE event appears, see Fig.12, Fig.13, Fig.14. In all Fig. mentioned above AE event counts vs time diagrams have been coodinated with load vs deflection (p-f) curves.

For category A the point on AE event counts vs time diagram, where AE event count increases abruptly, can be considered as point of initial crack in SFRC. Checking the load at this point, it can be found that this point coincides with the maximum load. Before this point the p-f curve is nearly a straight line. Since SFRC in this category contains fibre less than critical volume Vcr, fibre can not afford to

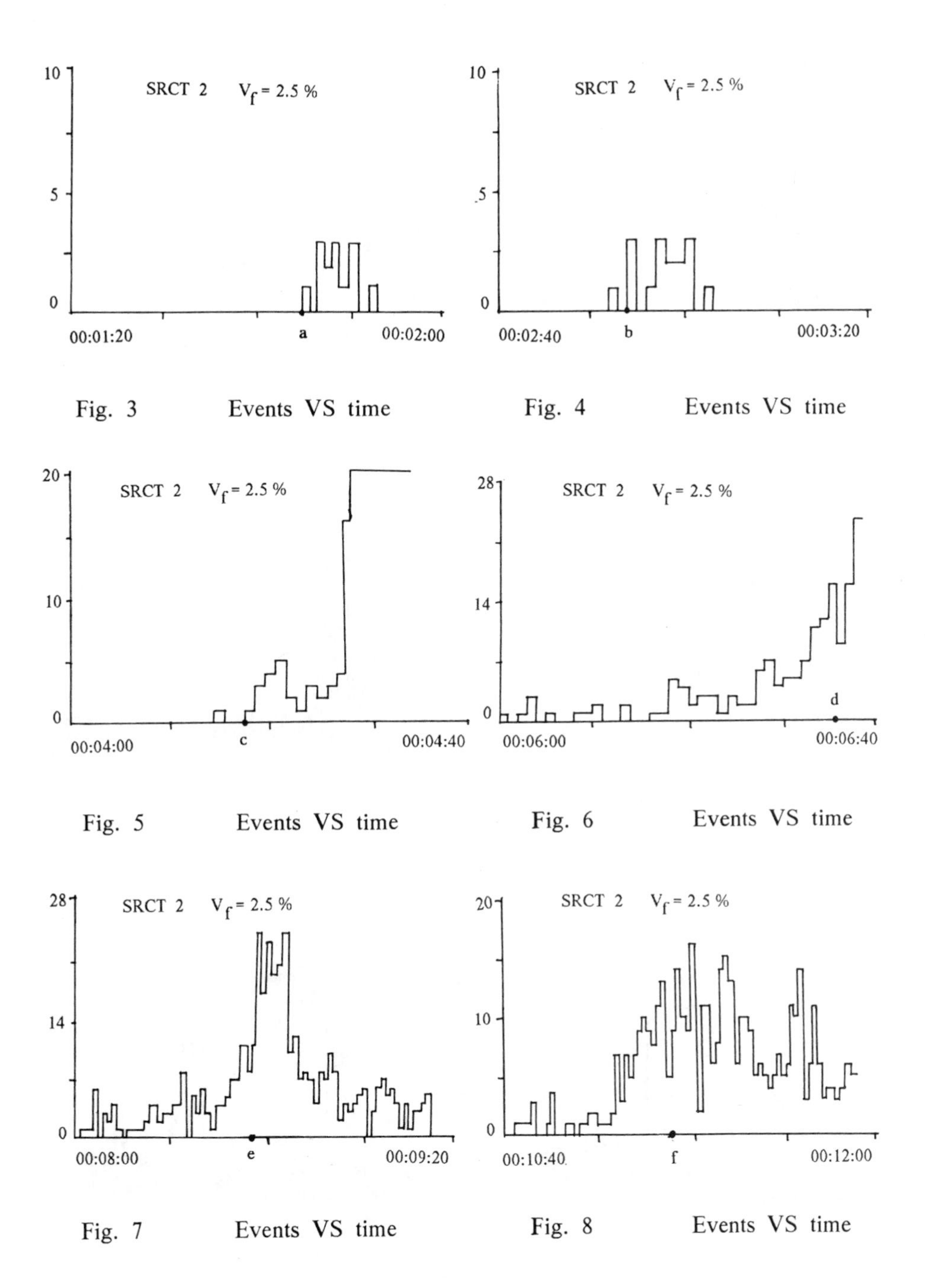
SRCT 2 V_f = 2.5 %
10
5
0
00:01:20
a
00:02:00
Fig. 3 Events VS time
SRCT 2 V_f = 2.5 %
10
5
0
00:02:40
b
00:03:20
Fig. 4 Events VS time
SRCT 2 V_f = 2.5 %
20
10
0
00:04:00
c
00:04:40
Fig. 5 Events VS time
SRCT 2 V_f = 2.5 %
28
14
0
d
00:06:00
00:06:40
Fig. 6 Events VS time
SRCT 2 V_f = 2.5 %
28
14
0
00:08:00
e
00:09:20
Fig. 7 Events VS time
SRCT 2 V_f = 2.5 %
20
10
0
00:10:40
f
00:12:00
Fig. 8 Events VS time

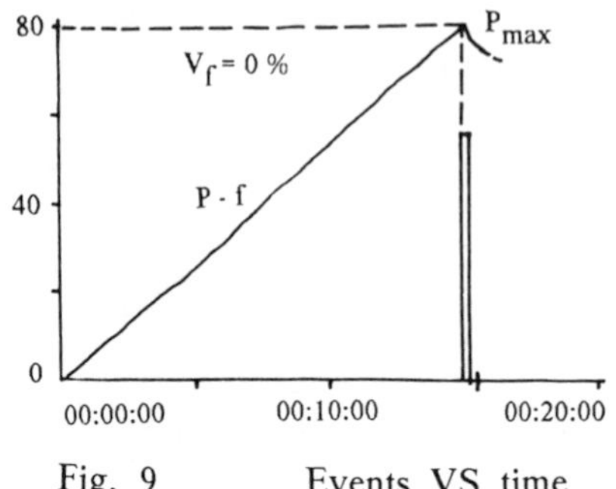

Fig. 9 Events VS time

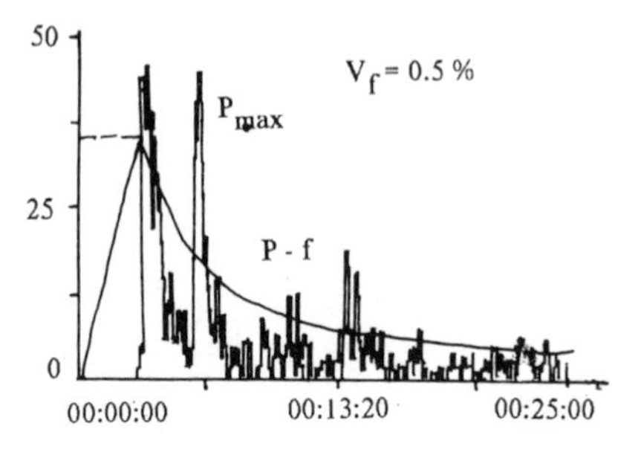

Fig. 10 Events VS time

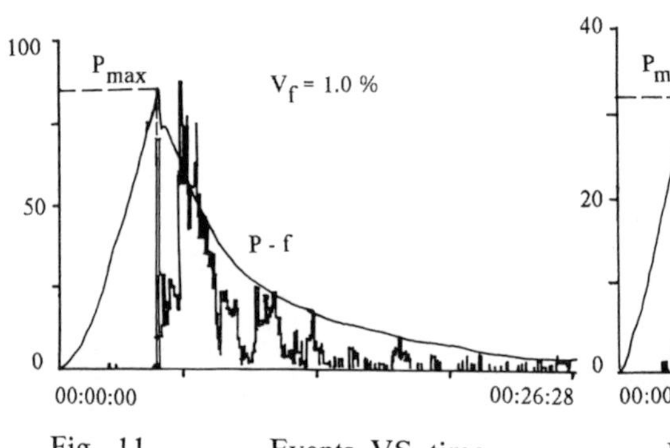

Fig. 11 Events VS time

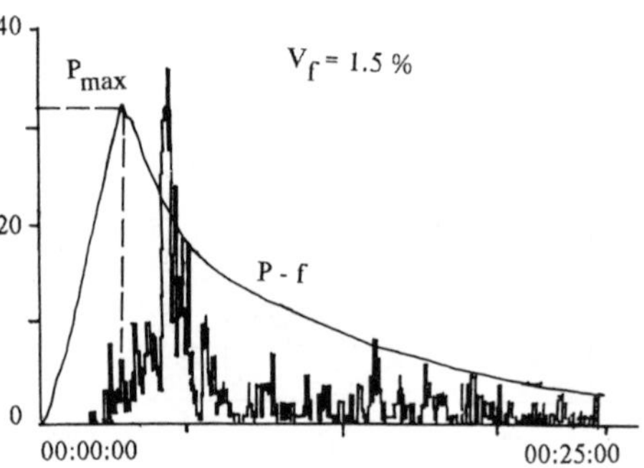

Fig. 12 Events VS time

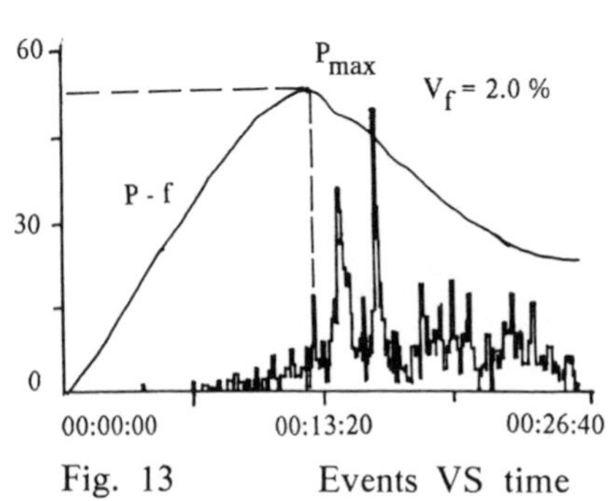

Fig. 13 Events VS time

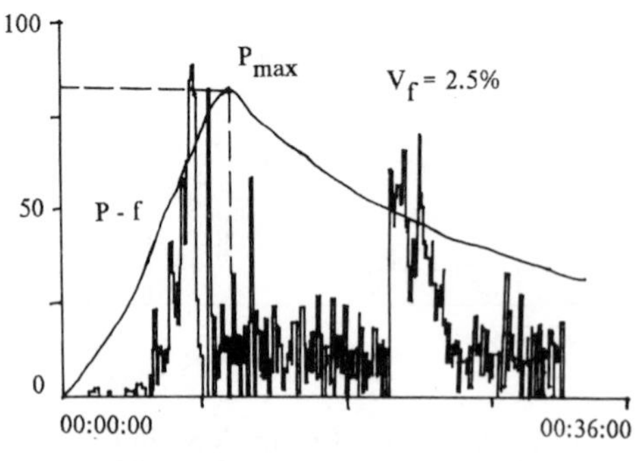

Fig. 14 Events VS time

stand the load after matrix cracks.

For category B there are two possibilities of initial crack point in the AE diagrams. The first one is the point, at which the first group of AE events bursts. Checking with the load-deflection curve, this point is just at about 50% load of the maximum. It is in the elastic stage composite. The second one is the point where AE event count increases considerablly, see Fig. 12, Fig.13, Fig.14. However, this point is far.beyond the maximum load. Investigating the location of AE signals by AE apparatus while the signals start to burst, it has been discovered that AE events distribute almost by whole length of the specimen and don't concentrated in the middle of beam, see Fig.15. In this case microcracks might appear along whole

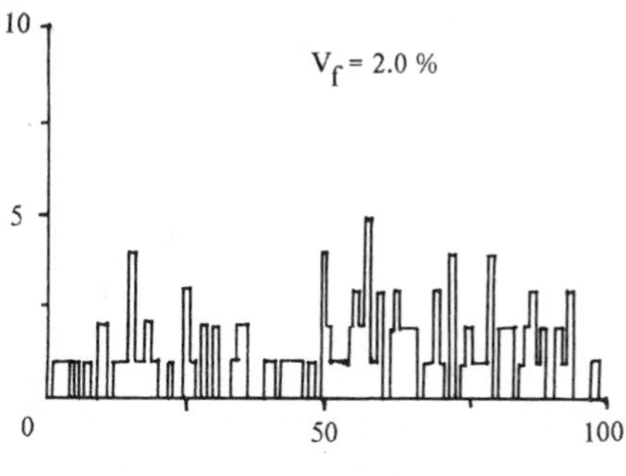

Fig. 15 Location of events

tensile zone of beam. Therefore, the start point of AE event can not be considered as the point of initial crack of main crack. Owning to difficulty to find initial crack point simply by AE event counts, then other AE parameters, such as Ringdown Count (RC), Event Duration (ED),Peak Amplitude (PA) and Energy with time (loading) were checked. But all those AE parameters have the same feature as the AE event count has, see Fig.16, Fig.17. By above excperiments the initial crack point of SFRC with V_f . 1.5% can not be simply determined by AE technique.

Combining AE characteristics of fracturing SFRC with feature of internal structure deformations of composite in fracture process the crack propagation has been analyzed. From AE event diagrams of two category of SFRC the AE characteristics of failfure of material are quite different. For specimens of category A there is almost no AE event untill the load reaches Pmax. It means that in early stage little microcrack exists in material, and no crack develops in it. As soon as the load gets Pmax the sudden increase of AE event means that there is unstable propagation of crack in form of failfure with single crack. Since SFRC in this category contains fibre less than Vcr, fibre in composite can not resists the development of crack effectively. When strain energy accumulated in material reaches certain value it can cause unstable cracking and bursts a lot of AE events, as show in AE diagrams. Composite of this category breaks nearly as the pure concrete. Then with pulling out fibre from concrete specimen deforms rapidly and AE event decreases gradually. For specimens of category B, before load reaches Pmax there will burst some AE events in material. Increase AE event means that a series of microcracks has been formed in the tensile zone of specimen due to load. Only load gets Pmax AE events start to increase suddenly. At this stage main crack has been formed from one of microcracks and begins to propagate unstablly. The AE noise expresses that process. Because fibre content in this kind of SFRC is more than Vcr. Fibre can arrest the microcracks and delay the unstable propagation of crack. When the fibre bridging crack can not bear the increasing load and starts to pull out from matrix, in composite bursts a large amount of AE events. Then AE decreases with increase of gradually pulled out fibre from matrix.

The AE event count vs time diagrams has 'uneven' characteristic. In every AE-t diagram several peaks and valleies can be observed. Uneven characteristic of AE event counts vs time of test expresses that the composite fails in form of uncontinuous by time crack propagation with ununiform rate. The tiny waves of AE event count show formation and development of microcracks or microslip of fibre from matrix and the existance of high peaks and deep valleies means that the main crack propagates and a great portion of energy releasees in form of AE events. From AE characteristic of tested composite the whole process of crack propagation, from formation of crack to failure of specimen, shows that the process of releasing energy repeats "high rate - low rate - high rate" pattern, corresponding to "crack propagation - energy accumulation - crack propagation". Especially in composite with Vf > Vcr, at early stage of loading formation of series of microcracks can be clearly observed, when the load exceeds 85% Pmax, main crack starts to form in SFRC and gradually propagates with pulling out of fibre from matrix.

As discussed above, by concept of fracture mechanism in, SFRC does not exist an initial crack point, as in traditional materials does. From internal structure changes the initial crack is a process in SFRC. So initial crack point can not be detectived by AE.

3. The difference and the analogy between the AE parameters of concrete matrix cracking and of steel fibre pulling out.

In order to determine AE characteristics the AE parameters of pure concrete specimens have been registered. For observing the AE parameters of fibre pulling out specimens with $V_f = 1.0\%$ and 2.0% were tested. For test on specimens was exerted 85% Pmax for 15 min. At that time it can be assumed that the matrix has completely cracked, then begins registeration of AE parameters of specimens, on which loading extends to failure. Using this procedure the registered AE parameters are mainly meeting to fibre pulling out.

Observing AE distributions, which were taken by above procedures, the feature and pattern of diagrams either with matrix cracking, or with fibre pulling out are very similar. However, for two cases the consituencies of AE parameters, such as RC, ED and RT, are quite different. For concrete cracking they range more widely, distribute more evenly and give relatively large average data. For fibre pulling out the AE distributions range less widely and give relatively small average data, see Tab.1.

Due to problem of AE apparatus, it can only receive the maximum amplitude as 77dB, instead of 115 dB normally. the Peak Amplitude and Energy, as a function of PA, are not right for analysis.

Table 1. AE Parameter

AE Parameter	RC (times)	ED (m sec)	RT (μ sec)	PA (db)	ENG
Cracking of conerete	1--283	0--3100	0--1020	33--77	35--112
Pulling out of fibres (V_f =1.0%)	1--68	0--440	0--200	28--77	44--107
Pulling out of fibres (V_f =2.0%)	1--64	0--430	0--180	30--77	35--112

RC---- Ring Down Count ED--- Event Duration RT---- Rising Time
PA---- Peak Amplitude ENG---Energy

From difference of fracture mechanism between concrete cracking and fibre pulling out, the difference of AE parameters can be explained. Since concrete matrix cracking is caused by tensile stress and failure is somehow explosive, the energy releasing per unit time is quite large, so average data of RC, ED and RT is large. But fibre pulling out is a gradual process, it expresses more gently. It is understandable that the average data of RC, ED and RT is relatively small.

4. The distribution law of AE parameters in SFRC

In loading process of SFRC AE parameters carry characteristics as follows: (1) AE distributions of ED, PA and energy is basicly by normal distribution, see Fig.18, Fig.19, Fig.20; (2) AE events with small RT is more than that with large ones, particularly most of events concentrete near zero point. It means AE events in SFRC have abrudt feature with many cases of sudden release of energy, see Fig.21; (3) Observing RC distributions in SFRC, linear distributions can be found. Small RC are more than large ones. It is interesting that for RC of all fibre content SFRC there is a limitation nearly equal to 68, see Fig.22, Fig.23, Fig.24, Fig.25, Fig.26, Fig.27; (4) From slope rates (K) of RC distributions it can be observed that the more V_f in SFRC, the less sloperate of RC distribution, see Fig.28. It means that the more V_f, the less average value of RC, the more gentle failure will be. And composite of this kind is less brittle and shows more quasi-elastic feature.

4 Conclusions

1. In SFRC clear Kaiser effect exists before maximum load. Beyond maximum load Kaiser effect is not distinct. With increase of circling times of load Kaiser effect will gradually disappear in SFRC.
2. AE technique can not be used to determine the initial crack **point** of SFRC.
3. In SFRC with Vf < Vcr before Pmax almost no AE event exists. SFRC of this category fails suddenly, and at that time emits a large amount of acoustic signals.

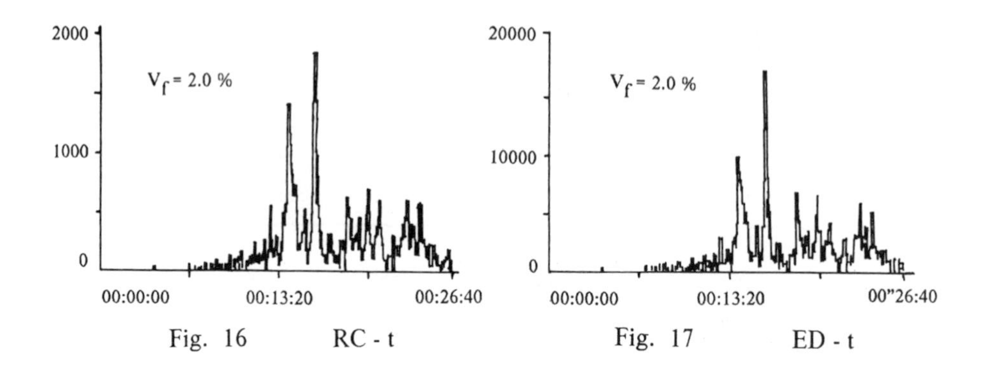

Fig. 16 RC - t

Fig. 17 ED - t

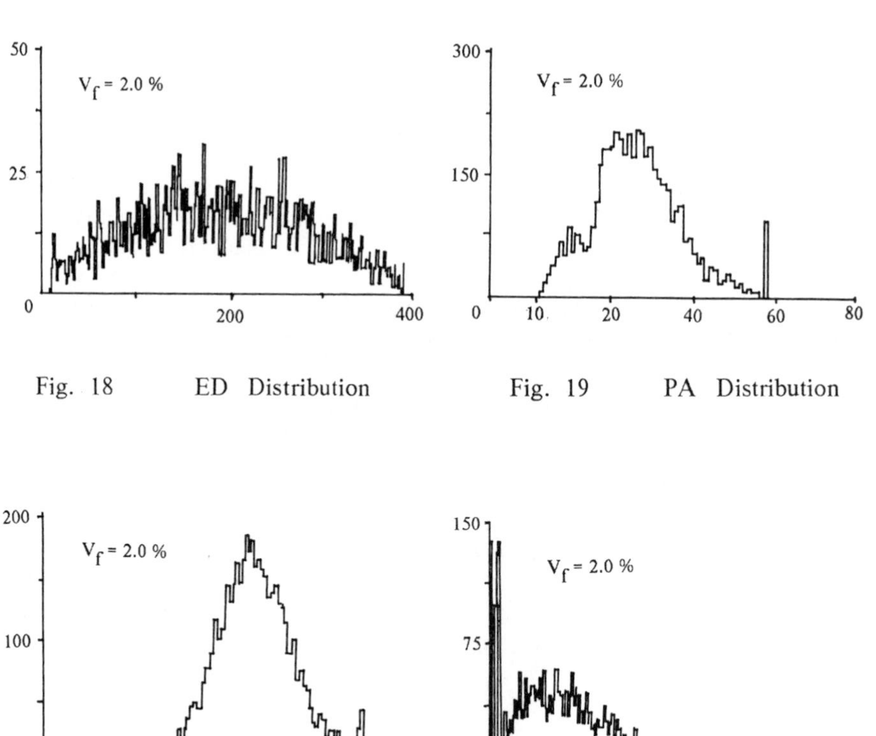

Fig. 18 ED Distribution

Fig. 19 PA Distribution

Fig. 20 Energy Distribution

Fig. 21 RC Distribution

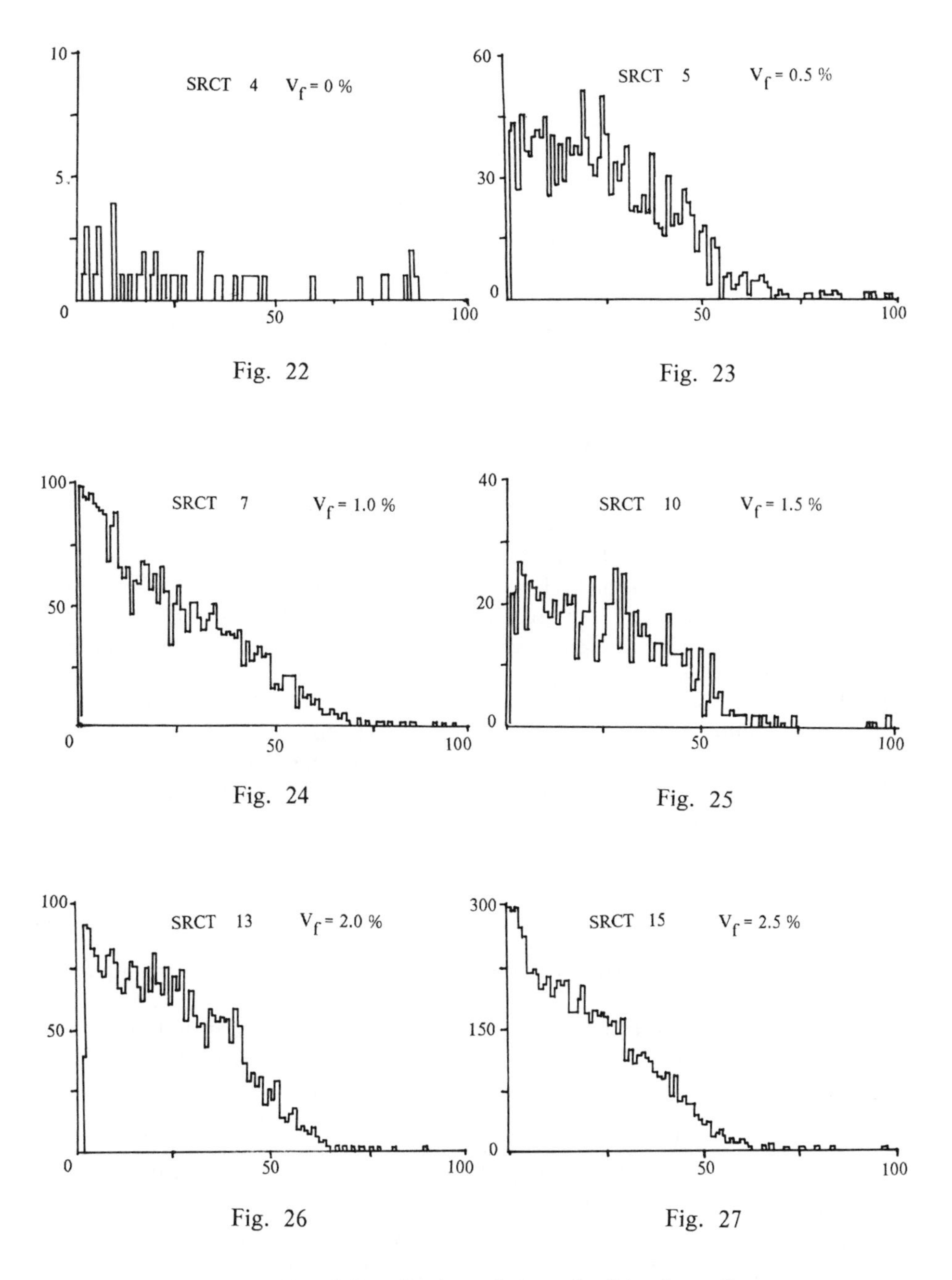

Fig. 22 -27 Distribution of events by Ringdown Counts

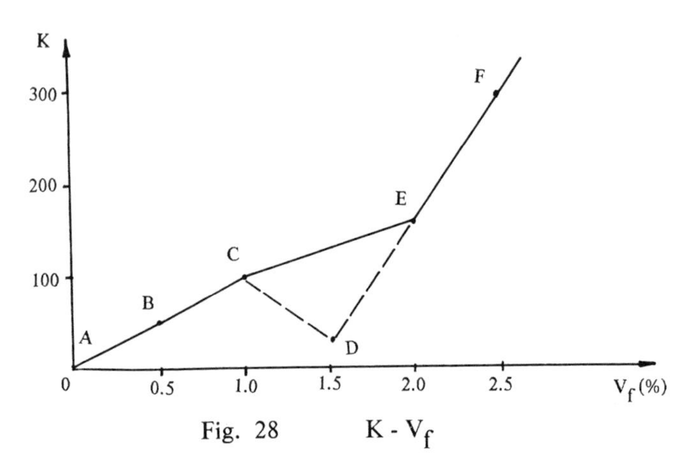

Fig. 28 K - V_f

AE events bursts before Pmax in SFRC with Vf > Vcr, and AE event counts raise with increase of load.

4. Concrete matrix cracking emits acoustic signals which have wider range of RC, ED and RT, as 1-283 times, 0-3100 m sec and 0-1020 μ sec, respectively. During steel fibre pulling out SFRC bursts AE signals with parameters, which have narrower range of RC, ED and RT, as 1-68 times, 0-440 m sec, 0-200 μ sec and 1-64 times, 0-430 m sec, 0-180 μ sec for Vf = 1.0% and 2.0% respectively. For steel fibre pulling out the average values of these AE parameters are lower than those for concrete matrix cracking.

5. In fracture process of SFRC composite distributions of AE characteristics have following features: (1) ED, PA and Energy of AE distribute basicly by normol distribution (2) AE events with small RT burst more than that with large RT; (3) RC distributions have linear characteristic; (4) SFRC with more Vf gives less slope rate of RC distribution.

5 References

1. Yuan Zengmin, Ma Yukuang, He Zenyun. Acoustic emission technique and applications, pp124-136

2. Susuki. at el, Classification and assessement of AE diagram of test on fracture toughness, Japen inspection on material failure, Vol. 30, No, 11, 1981, pp890

3. Chong Gang, Research on behaviors of steel fibre reinforced lime-sand concrete, M.S. Thesis, Chongqing Institute of Architecture and Engineering, pp29-41.

4. Boyce G.M.. A study of Acoustic Emission Response of Various Rock Types, M.S. Thesis, Drexel University

5. J.J. Mcetroy, R.M. Koerner, A.E. Lord, An Acoustic Jack to Asses in Situ Rock Behavior, Int. J. Rock Mech. Min. Science, 1985 Vol. 22, No. 1, pp21-29

6. Yoshikawa, at el, Kaiser effect of AE in rockinfluences of water and temperature, The 4th AE symposium, Tokyo, 1978, pp7/21-7/39

17 CRACK DEVELOPMENT IN PLAIN AND STEEL FIBRE CONCRETE DUE TO AN EXPANDING STEEL BAR

A. T. MOCZKO
Technical University of Wroclaw, Wroclaw, Poland
D. H. DALHUISEN and P. STROEVEN
Delft University of Technology, The Netherlands

Abstract
The paper reports on a study of cracking accompanying the expansion of a lineal steel element embedded in selected types of concrete. The tests involved the pulling of a cone-shaped steel bar through 50 mm thick sawn disks of plain and steel fibre concretes. Basic examinations, comprising acoustic emission and deformation measurements, have been carried out for young (7 days old) and standard (28 days old) specimens. Among other things, tests have shown the mechanisms of micro- and macrocrack formation in different concretes to be properly reflected by the acoustic emission parameters. Simultaneously, the significant role of fibres was established in controlling the cracking behaviour. Due to their ability of arresting microcracks and inhibiting their growth, a delay of the macrocrack formation has been observed. Availing oneself of this phenomenon could be essentially important for reducing the corrosion risks of reinforced concrete constructions under unfavourable conditions and thereby increasing their serviceability life.
Keywords: Bar Expansion, Cracking, Steel Fibre Concrete, Plain Concrete, Age of Concrete, Acoustic Emission.

1 Introduction

Corrosion of the steel reinforcement constitutes one of the major durability problems in concrete technology. Particularly under severe conditions - such as in marine applications - measures are required to prevent premature cracking of the concrete. Most commonly, high concrete qualities are used. Finite corrosion rates can in some cases, nevertheless, endanger the serviceability of the concrete construction within its planned life time. This is definitely so under marine conditions where chloride concentrations near the reinforcement will attain inadmissible levels after a relatively short period of time, de Wind et al. (1987).

In general, the corrosion effect could be considered as a problem of an expanding steel bar embedded in concrete. This provokes unfavourable tensile stresses in tangential direction around the lineal element and inevitably leads to concrete cracking. This phenomenon was analytically discussed by Stroeven (1991).

Fibre Reinforced Cement and Concrete. Edited by R. N. Swamy. © 1992 RILEM.
Published by E & FN Spon, 2-6 Boundary Row, London SE1 8HN. ISBN 0 419 18130 X.

It is obvious that the rate of cracking is significantly influenced by the structure of the concrete itself. Especially, an important role is played by the natural discontinuities and structural microdefects which constitute an immanent feature of concrete. These defects exist in the concrete structure from the beginning of its manufacturing and usually form at the contact zones between the aggregate and the hardening cement paste. They are the sources of the first microcrack formation under increasing loads. More detailed explanation of this phenomenon was given by Flaga et al. (1989).

According to Hoff (1987) durability resistance could be seriously reduced once cracks open up more than about 0.2 mm. Thus, the delaying of crack formation and of their further development appear as fundamental problems. Currently particular attention is given among others by Lovata (1989) and Shah (1990) to the application of different types of fibres to control the cracking behaviour of concrete.

Simultaneously the development of the advanced testing techniques is observed. For the recording of the cracking process the acoustic emission technique is particularly useful. Reference can be given to Ohtsu (1987), Leaird and Taylor (1989), Mobasher, Stang and Shah (1990) and Moczko (1991). In the present tests, acoustic measurements have been performed to evaluate the effects of age and of steel fibre additions on the cracking behaviour of concretes.

2 Experimental details

The tests were executed in the Stevin Laboratory of the Civil Engineering Faculty, Delft University of Technology. Two concrete mixes listed in Table 1 have been designed assuming equal volume of both compositions.

Table 1. Concrete mix compositions

MIX CODE	AGGREGATE GRADING (kg/m^3)						FIBRES (kg/m^3)	CEMENT (kg/m^3)	WATER (kg/m^3)
	0-0.25	0.25-0.5	0.5-1	1-2	2-4	4-8			
C	193	281	263	263	299	458	-	400	200
FC	189	274	258	258	292	446	118	400	200

A Portland Cement type A, satisfying the Netherlands Standard NEN 3550 was employed. Further use was made of a good quality river aggregate with a maximum grain size of 8 mm, tap water and 1.5% by volume of plain steel fibres with a length and diameter of 12.5 and 0.4 mm, respectively. The mixing, vibration and casting procedure was similar for all series and performed according to relevant Netherlands Standards.

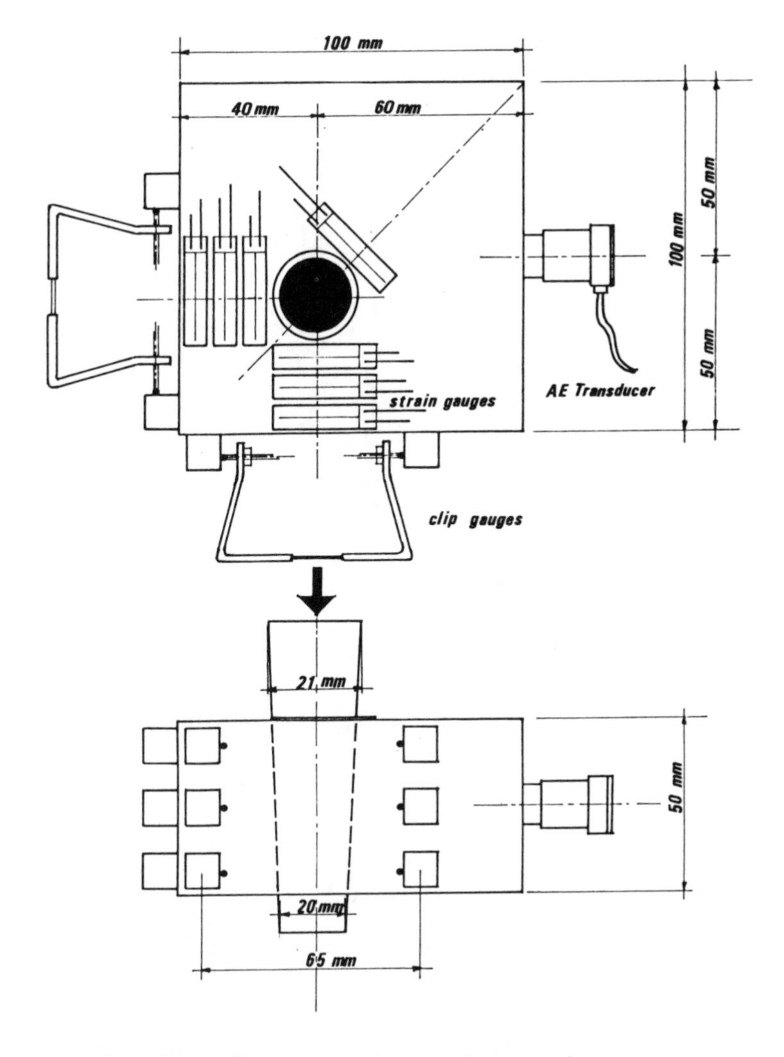

Fig. 1. View of test specimen and measuring system.

In the tests four groups of specimens were considered:

- series C7 - 7 days old plain concrete,
- series FC7 - 7 days old steel fibre concrete,
- series C28 - 28 days old plain concrete,
- series FC28 - 28 days old steel fibre concrete.

Each group consisted of 7 test specimens (Fig. 1.) and 12 standard cubes with linear dimensions of 150 mm. All specimens and cubes were after casting until the day of testing stored in a climatic chamber with an air temperature of 20^{0}C and a relative humidity of 95%.

The main element of the testing set up was constituted by a steel cone which is pushed with a constant rate of displacement through a similarly shaped hole in a concrete specimen. An example of a test specimen, provided with the measuring system is shown in Fig. 1. Additionally, the overall view of the experimental set up is presented in Fig. 2.

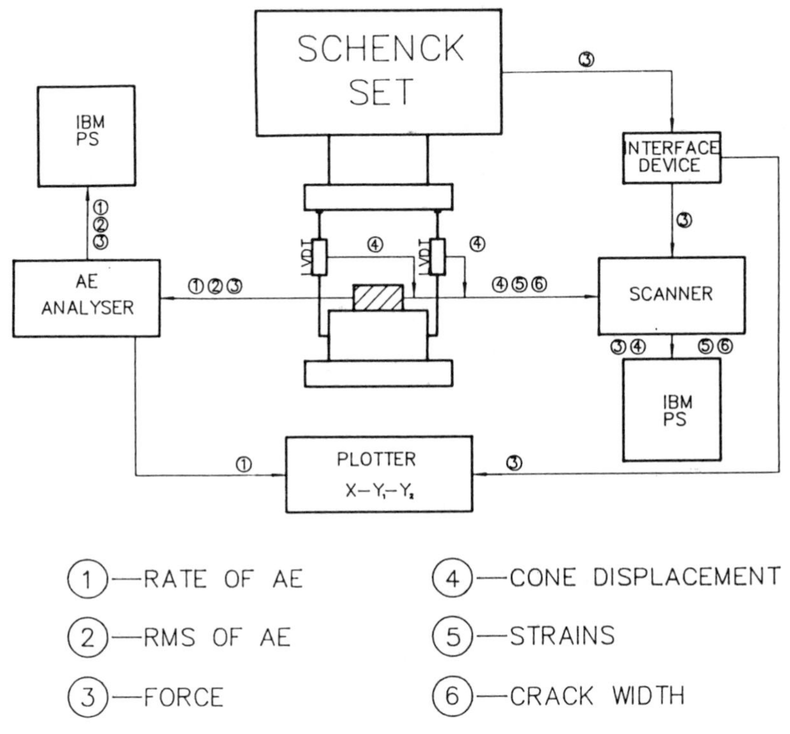

Fig. 2. Experimental set up.

The rate of applied cone displacement, amounting to 7.5 μm per second, was controlled by two LVDT's. Simultaneously, the surface strains were measured by strain gauges of 20 mm length. The width of the leading crack was registered by clip gauges. The acoustic emission response has been monitored additionally by means of a Polish acoustic emission analyser EA-3, designed and manufactured at the Institute of Fundamental Technological Research, Polish Academy of Sciences, Warsaw. The rate of AE counts and the RMS value of the analysed signals have been measured simultaneously using a piezoelectric transducer of a 200 kHz resonance frequency. The 'ring down' counting method with a specified rejection threshold of 1 V and a global signal amplification of 86 dB was used. In all tests a constant counting rate equal to 10 times per second was maintained. The friction between all contact zones was eliminated by polishing the specimens and greasing all surfaces with vaseline. All examinations were carried out at a temperature of 20°C and a relative humidity of 54%.

Independently, compressive and tensile splitting strengths were determined using standard cubes.

3 Results and discussion

The results obtained reflect the appreciable differences in cracking behaviour between plain and steel fibre concretes. It was manifested by post-peak differences in the load-displacement curves and in the acoustic emission response. Fig. 3. shows examples of two selected 28 days old specimens.

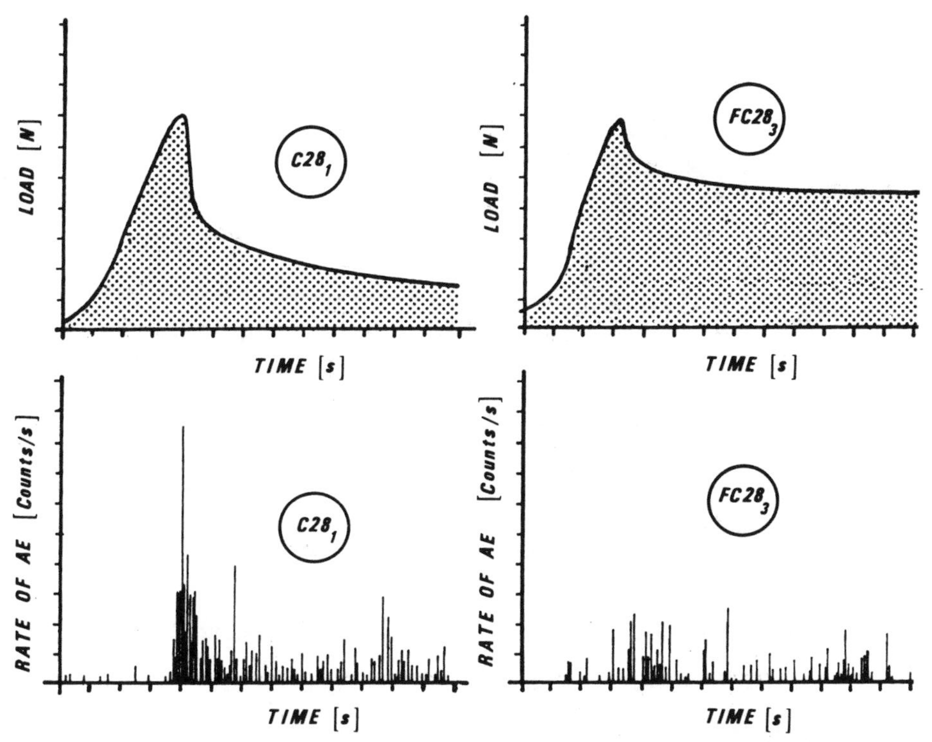

Fig. 3. Development of load and AE activity under controlled conditions of bar expansion. Examples of 28 days old specimens, plain concrete at the left and steel fibre concrete at the right, are presented.

It can be noted that the acoustic emission activity properly reflects the damage evolution process in the tested concrete composites. It is seen that the maximum of acoustic emission rate occurs around the BOP (Bend Over Point). Nevertheless, the character of this phenomenon is essentially different for plain and fibre reinforced concrete.

In plain concrete an impetuous acoustic discharge is associated with a considerable load drop following BOP. In case of steel fibre concrete, the post-peak load stabilizes after an insignificant decline. The acoustic emission activity in the post-peak range is characterized by the absence of a distinct peak. It can also be observed that the total number of AE counts registered over the post-peak cracking range is considerable less for the fibre concrete then for the plain one. This is illustrated in Fig. 4, giving data of four selected specimens, which are representative for the considered types of concrete composites. The significant influence is demonstrated of age and of steel fibre additions on the cumulative AE effect.

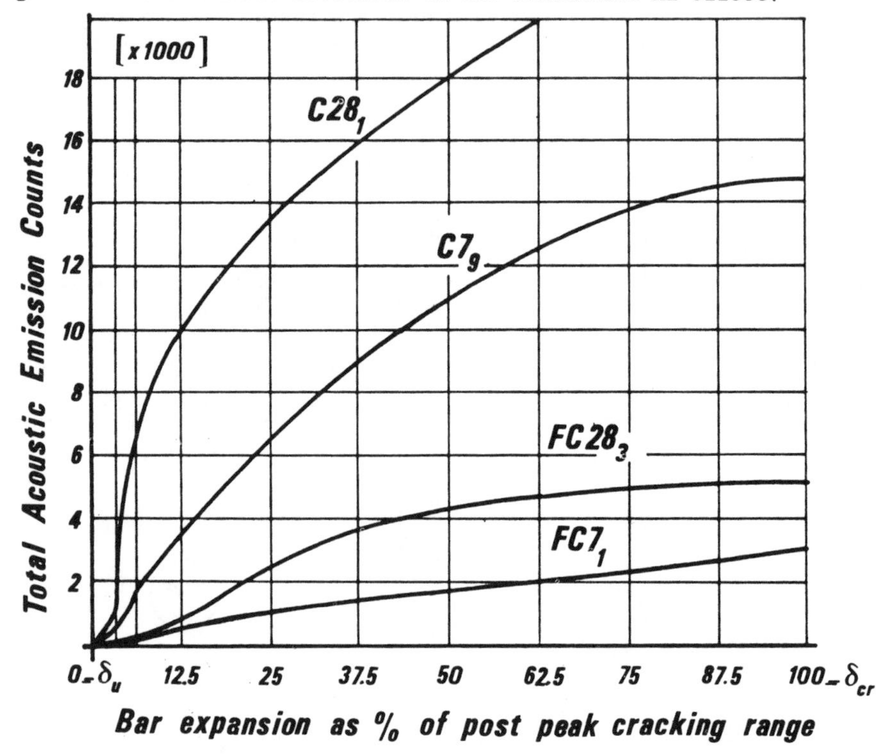

Fig. 4. Total AE counts recorded over the post-peak cracking range.

It is obvious that the lower emittivity of fibre concretes is due to the more controlled growth of the cracks, whereas the higher total number of AE counts for older concretes is the result of increased brittleness. For further analysis the idealization of the observed damage evolution has been given in Fig. 5.
The accepted code notation can be explained as follow:

- BOP - bend over point; notion which is usually attributed to the end of ascending branch of load-displacement relationship,

- F_u - ultimate load; load corresponding with bend over point,

- F_c - critical load; load corresponding with the critical width of the leading crack, assumed to be 0.2 mm,

- δ_u, δ_{cr} - ultimate, resp. critical bar expansion,

- $\Delta\delta$ - bar expansion over post-peak cracking range,

- δ_0 - fictitious point of zero bar expansion,

- E - cracking energy dissipated over post-peak cracking range.

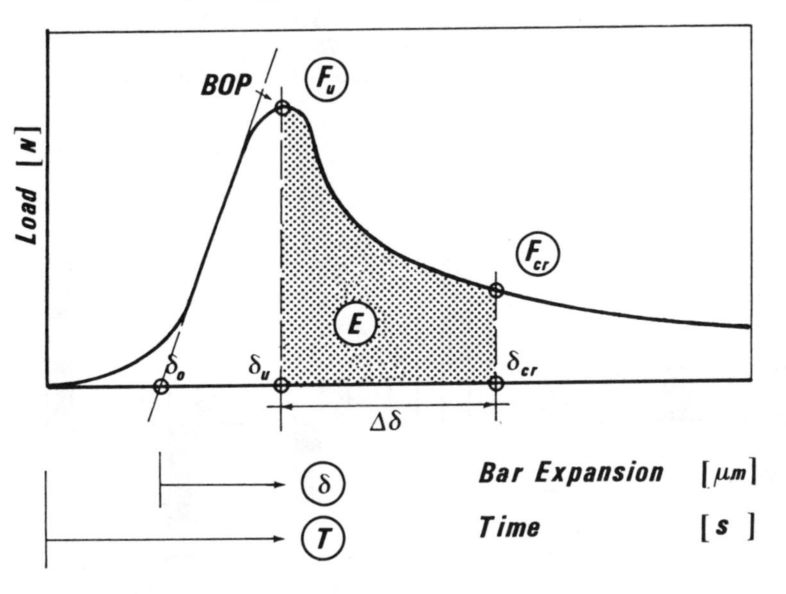

Fig. 5. Idealization of load - bar expansion curve.

To properly characterize material behaviour the following global evaluation parameters are used:

- $K = f_c/f_{ts}$ - brittleness factor,

- f_c, f_{ts} - compressive, resp. tensile strength,

- $\alpha = F_u/\delta_u$ - rate of pre-peak load increase,

- $\beta = F_{cr}/F_u \times 100\%$ - relative post-peak load drop,

- Σ_{AE} - total AE counts over post-peak cracking range,

- $\gamma_{AE} = \Sigma_{AE}/\Delta\delta$ - rate of AE activity in post-peak cracking range,

- $\xi_{AE} = \Sigma_{AE}/E$ - total AE counts per unit of cracking energy.

The determined average values of these parameters with a coefficient of variation of about 10-15% are listed in Table 2 and presented as bar chart in Fig. 6.

Table 2. Average values of selected evaluation parameters

SERIES CODE	K (%)	α $\frac{N}{\mu m}$	β (%)	$\Delta\delta$ (μm)	E (N x mm)	Σ_{AE} $\left(\begin{matrix}counts\\ x10^3\end{matrix}\right)$	γ_{AE} $\left(\frac{counts}{\mu m}\right)$	ξ_{AE} $\left(\frac{counts}{N \times mm}\right)$
C7	10.4	306	35	20.4	43.9	15.3	750	349
FC7	9.4		70	23.4	86.6	3.6	154	42
C28	12.4	362	30	14.5	27.2	25.3	1740	930
FC28	10.2		70	21.0	65.8	5.8	276	88

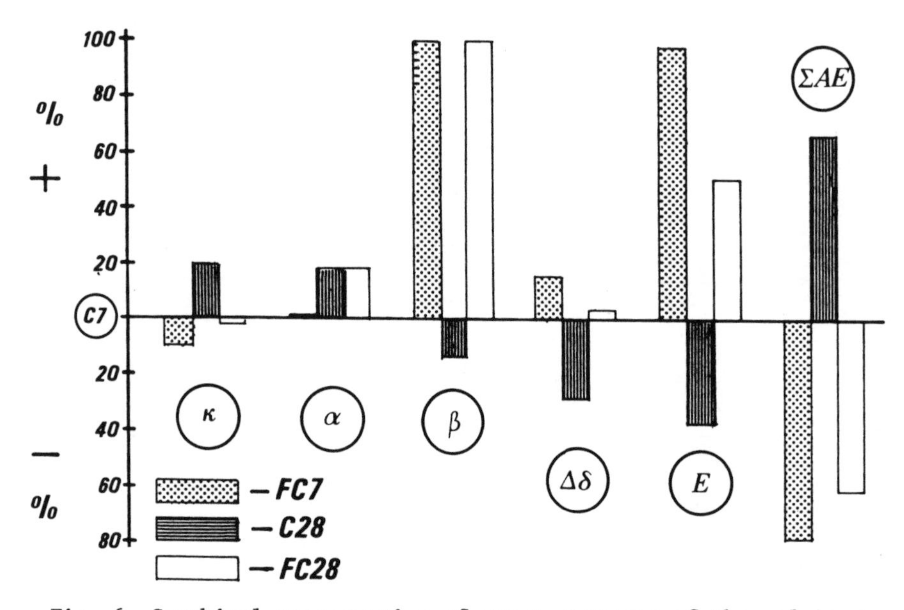

Fig. 6. Graphical presentation of group averages of the selected evaluation parameters normalized with respect to those of 7 days old concrete.

The factor, 'K' defined as the ratio of the compressive and tensile splitting strength, characterizes the 'brittleness' of the material. This parameter reveals two antagonistic tendencies. On the one hand, older concretes manifest an increase in brittleness, whereas fibre additions, on the other hand cause a reduction. The brittleness factor of 28 days old fibre concrete is as a result found to roughly equal the value obtained for 7 days old plain concrete. Generally, the favourable age effect on compressive strength in the nominator is partly compensated by the improved splitting tensile strength due to fibre addition in the denominator.

The parameter 'α' reflects concrete behaviour prior to BOP. In this loading range only the age effect on the curve's slope, connected with the the stiffness of testing set up, is found to be significant. Hence, influences of the fibre addition can be neglected.

In the post-peak range the fibre's influence on cracking behaviour is predominant. It can be seen that among the presented parameters - relative post-peak load drop (β), bar expansion over post-peak cracking range ($\Delta\delta$) and accompanying amount of dissipated energy (E), the last one seems to be most sensitive. Its values for composites of the same age but with or without fibres differ by a factor of two. Thus, fiber reinforced concretes display a considerably improved toughness which is due to fibre's ability of arresting microcracks and inhibiting their growth. Simultaneously the age effect is also revealed. Hence, it can be seen that older concretes require a smaller amount of energy to obtain the same presumably critical crack width. Since these two effects counteract each other, the significant improvement of the cracking resistance associated with fibre addition is gradually reduced by the age development. Nevertheless, it is obvious that fiber reinforcing appears essentially for delaying the crack coalescence instability.

These characteristics are confirmed by acoustic emission measurements. The relevant evaluation parameters are rate of acoustic emission activity in post-peak cracking range (γAE) and ratio of dissipated energy over the post-peak cracking range and total AE counts (ξAE). Particularly promising seems to be coefficient 'ξAE' which can be denoted as **Degree of Cracking Instability (DCI)**. This parameter expresses the relationship between the acoustic emission response and the external energy provoking this phenomenon. The obtained results indicate that fibres substantially decrease the rate of AE. It is resulting from the fibre's ability to control the growth of micro-and macrocracks. The acoustic emission response becomes more uniform, without the impetuous discharges. The rate of acoustic activity is reduced in spite of a larger number of cracks then in the plain concrete case. The strong age effect is also visible over this loading range. For older concretes a considerably larger number of AE counts is monitored then for younger composites.

4 Conclusions

Based on the obtained results the following conclusions can be drawn.

- The cracking process induced in the concrete composites by bar expansion could be properly monitored by strain and acoustic measurements.

- In the initial loading stage the stiffness of the composites was only significantly influenced by age.

- In the post-peak range even small amounts of steel fibres were found to considerably improve toughness by controlling crack development, despite increasing embrittlement due to age effects.

- The **Degree of Cracking Instability - DCI** (ξAE) is proposed as a relevant parameter for evaluation of structural integrity of concrete composites subjected to loading.

5 Acknowledgements

Tests reported here were carried out within the framework of a cooperation program between the Civil Engineering Faculties of the Technical University of Wroclaw and of Delft University of Technology. A fellowship position which was granted to the first author by the Netherlands Organization for Scientific Research (NWO) allowed to perform the tests in Delft. This particular support is greatly appreciated.

6 References

Flaga, K. Jabor, J. (1989) The role of interior stresses in the structural destruction of fibre reinforced concrete, in **Fibre Reinforced Cements and Concretes: Recent Developments** (eds. R.N.Swamy and B.Barr), Elsevier Applied Science , London and New York, pp. 219-228.

Hoff, G. (1987) Durability of fiber reinforced concrete in a severe marine environment, in **Concrete Durability** (ed. J.M.Scanlon), ACI Spec. Publ. SP-100, Detroit, pp. 997-1041.

Leaird, J.D. Taylor, M.A. (1989) Acoustic emission investigations into some concrete constructions problems. **J. of Acoustic Emission**, pp. 322-325.

Lovata, N.L. (1989) An analysis of post-peak loading conditions in fibrous concrete composites, in **Fibre Reinforced Cements and Concretes: Recent Developments** (eds. R.N. Swamy and B.Barr), Elsevier Applied Science , London and New York, pp. 513-522.

Mobasher, B. Stang, H. Shah, S.P. (1990) Microcracking in fiber reinforced concrete. **Cement and Concrete Research**, 20, pp. 665-676.

Moczko, A.T. (1991) The age effect in cracking behaviour of plain concrete, in **Brittle Matrix Composites 3** (eds. A.M. Brandt and I.H.Marshall), Elsevier Applied Science, London and New York, pp. 240-247.

Ohtsu, M. (1987) Acoustic emission characteristics in concrete and diagnostic applications. **J. of Acoustic Emission**, 6, pp. 99-108.

Shah, S.P. (1990) Toughening of cement-based materials with fiber reinforcement, in **Fiber-Reinforced Cementitious Materials** (eds. S.Mindess and J.Skalny), Materials Research Society, Pittsburgh, pp. 3-13.

Stroeven, P. (1991) Experimental and theoretical aspects of composite mechanical behaviour of concrete containing an expanding steel bar, in **Proc. of the 10th Congress on Material Testing** (ed. E.Czoboly), Budapest, pp. 424-429.

Wind, G. de, Stroeven, P. (1987) Chloride penetration into offshore concrete and corrosion risks, in **Concrete Durability** (ed. J.M.Scanlon), ACI Spec. Publ. SP-100, Detroit, pp. 1679-1690.

18 THE DEVELOPMENT OF AN INSTRUMENTED IMPACT TESTING APPARATUS

H. MAHJOUB-MOGHADDAS, N. J. S. GORST and B. I. G. BARR
University of Wales College of Cardiff, Cardiff, UK

Abstract
The paper describes the construction and instrumentation of a repeated drop-weight impact testing apparatus. The apparatus has two main variables - mass of drop-weight (ranging from 1 to 4 kg) and drop-height (up to 2.5m). The instrumentation comprises two load cells and a number of accelerometers. The two load cells are used to monitor the impact load and the reaction response at the support. The test results (which are recorded by means of a Signal Memory Recorder) include peak impact load, impulsive load, accelerations etc. The apparatus has been used to study flexure, shear and torsion strength of steel and polypropylene fibre reinforced concrete. Typical results are presented for a 2% (by weight) steel fibre reinforced concrete.
Keywords: Testing, Impact Testing, Fibre Reinforced Concrete, Repeated Drop-Weight Apparatus, Flexure Strength, Torsion Strength.

1 Introduction

Impact studies on fibre reinforced concrete (FRC) materials have been limited in the past - relative to studies on their static toughness properties. However, it is well-known that impact tests readily show the enhanced properties of FRC over plain concrete. The enhanced toughness of FRC materials can be observed in many ways including improved cracking resistance, improved post-cracking toughness and the ability to absorb large amounts of energy during the fracture process. Early impact tests were based on the development of expensive testing equipment and, unfortunately, the results generally showed high variability.

Impact testing has been reviewed by Banthia (1987). Two main types of impact tests have been developed. In the first type, test specimens are broken by a single blow and the energy absorbed during the test is measured. In the second type, test specimens are broken by a number of repeated blows of known energy. The two most popular types of testing machines are the pendulum-type machines (generally modified Izod and Charpy machines) and drop-weight machines. Many researchers have instrumented such testing machines to various levels of sophistication.

A repeated drop-weight impact testing apparatus was proposed by ACI Committee 544 (1978). The apparatus was designed to compare the

Fibre Reinforced Cement and Concrete. Edited by R. N. Swamy. © 1992 RILEM.
Published by E & FN Spon, 2-6 Boundary Row, London SE1 8HN. ISBN 0 419 18130 X.

relative merits of different fibre concrete mixes and to evaluate the improved performance of a fibre mix relative to a conventional plain concrete mix. Although this apparatus has not been widely used and has been much criticised (e.g. by Swamy and Jojagha (1982)), there is considerable merit in the basic ideas suggested by the ACI Committee regarding impact tests. The proposed test used a simple geometry and the apparatus could be readily constructed. The range of the number of blows to cause failure in the test specimens gave a reasonable scale within which the merits of different fibres and fibre concentrations could be determined. The same principles apply to the testing apparatus reported here.

The repeated drop-weight testing apparatus reported here was initially designed by Barr and Baghli (1988) for testing notched beams in bending (Mode I - type loading). After suitable modifications by Mahjoub-Moghaddas (1991), the apparatus was used for testing specimens in which the loading was primarily shear (Mode II - type loading). Currently the apparatus is being used by Gorst (1992) to test cylinders and cores in torsion (Mode III - type loading). The instrumentation of the apparatus has also been developed during the last four years.

2 Test apparatus

The impact testing apparatus is a very simply arrangement of three main components - a heavy supporting base, the drop-weight support and guide system and the impacting masses. Full details of these three components have been reported earlier by Barr and Baghli (1988). The base is constructed around a heavy steel channel with various holes and attachments to support a range of test specimens. The heavy base rests directly on the concrete floor in the laboratory. A framework attached to the base supports the drop-weight guide system and an associated safety cage. The guide system consists of two vertical steel rods 2.5m long along which the impacting masses drop freely. The guide system ensures that the impacting mass does not rotate during its descent and hence impact takes place perpendicular to the test specimens. The contact zone of each impacting mass is rounded - in order to create a line contact with the test specimen.

A schematic view of the test apparatus and test specimen geometries is shown in Fig.1. The apparatus itself has two main variables - the impacting masses and the drop-heights. To date masses of 1, 2 and 4 kg have been used in the experimental studies, although larger and smaller masses may be used in future work. The drop-height can be varied between 0.25m and 2.5m - in the tests carried out to date heights of 0.5, 1.0 and 2.0 m have been used. The apparatus allows various combinations of masses and drop-height to be used which result in the same impacting energy (i.e. mgh is kept constant).

The above facility allows the effects of loading rate to be considered. Impact testing is superficially very simple but great care is required in the interpretation of the test results. Reinhardt (1987) has reported that "a variety of testing methods has been used to show the rate effect. However, one should be cautious

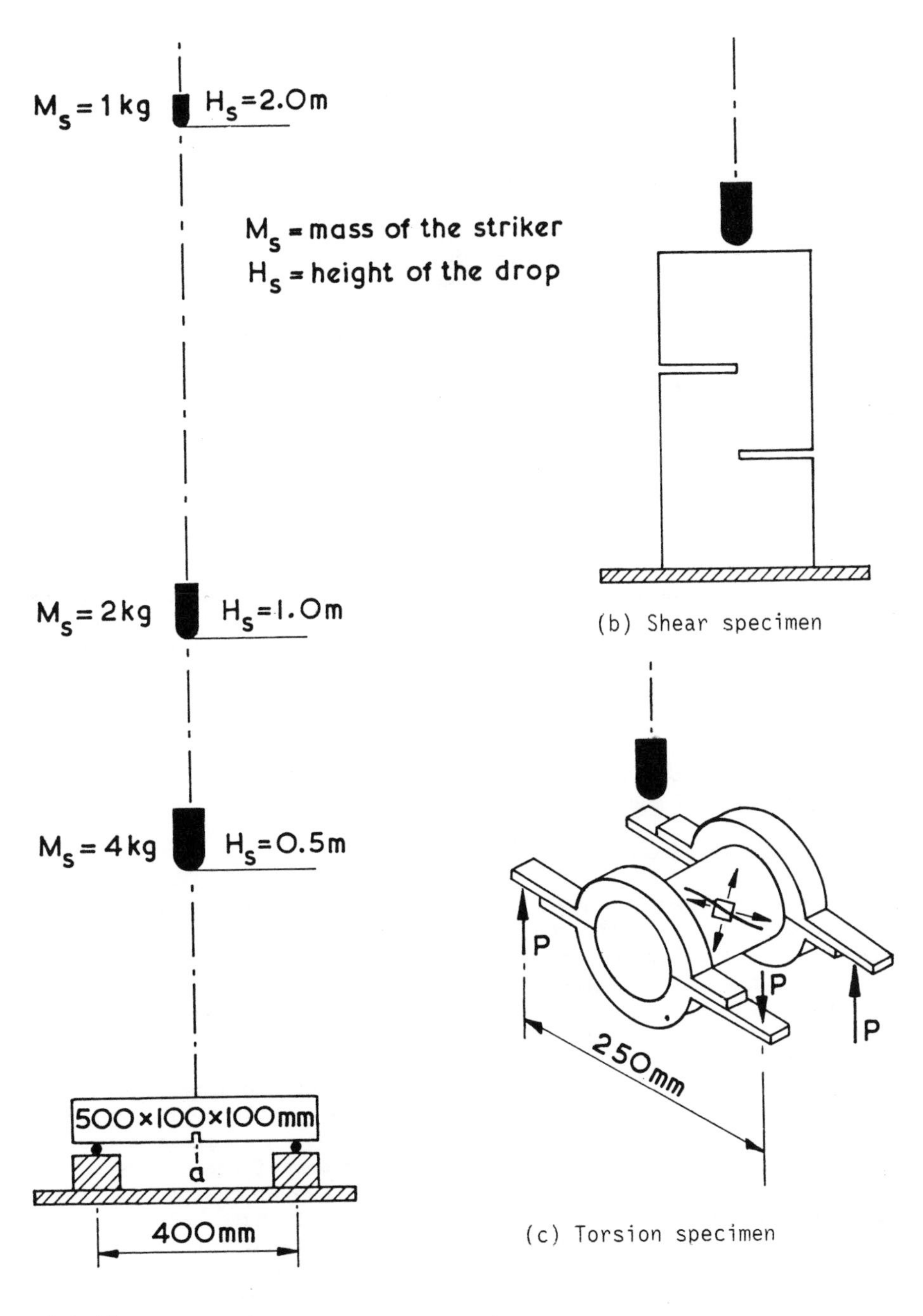

(b) Shear specimen

(c) Torsion specimen

(a) Flexure specimer

Fig.1. Schematic views of test variables and geometries

interpreting results since other effects may also have a significant influence on the results. Inertia effects, local damage, or stress wave reflection can be mentioned which should receive due attention". The apparatus illustrated in Fig.1(a) allows a study of some of these effects to be carried out.

Fig.1 also illustrates the three main test specimen geometries which have been studied. Fig.1(a) illustrates one of the flexure test specimens investigated by Barr and Baghli (1988) who also studied more compact flexure beams in which the support span was reduced from 400mm to 200mm. In both cases both notched and un-notched flexure specimens were investigated. Fig.1(b) illustrates the short-beam shear test geometry used by Mahjoub-Moghaddas (1991) to study the shear fracture characteristics of FRC materials. Further details regarding this and similar shear-type test specimens are presented in a companion paper by Mahjoub-Moghaddas and Barr (1992). The torsion test specimen, illustrated in Fig.1(c), has been used by Tokatly (1991) to study the torsional impact strength of steel and polypropylene FRC.

The repeated drop-weight impact test apparatus was developed initially to evaluate the impact resistance of FRC materials in terms of the number of blows required to cause failure. During the early days of the application of the testing system it was realised that the number of blows required to produce the first crack could also be determined. In its simplest form, the test apparatus offers a very simple method of evaluating and comparing toughness in terms of the number of blows to cause cracking or failure. However, the test apparatus has been continuously developed in terms of instrumentation. The current level of instrumentation includes two load cells and a number of accelerometers. The instrumentation allows a much better appreciation of the factors which influence impact results for FRC materials.

3 Instrumentation

The number of blows to cause cracking and/or fracture do not provide an appreciation of the fracture process. Therefore, various attempts have been made to learn more regarding the fracture process by instrumentation of the impact apparatus. The instrumentation varies with the test geometry under investigation - Fig.2 shows a schematic view of the instrumentation for the shear test geometry illustrated in Fig.1(b).

Similarly two load cells were used to monitor the impact tests using the flexure specimen illustrated in Fig.1(a). The first load cell was placed on top of the specimen at mid-span to monitor the impacting load and the second was located at one of the supports to monitor the reaction during impact. In the case of the torsion test specimens illustrated in Fig.1(c), the load cells have been replaced by instrumentation on the arms of the split collars. Strain gauges have been attached to the loading arms so that they also act as load cells. To date only one accelerometer has been used in the torsion tests - at the centre of the flat surface of the cylinder adjacent to the impacting load point.

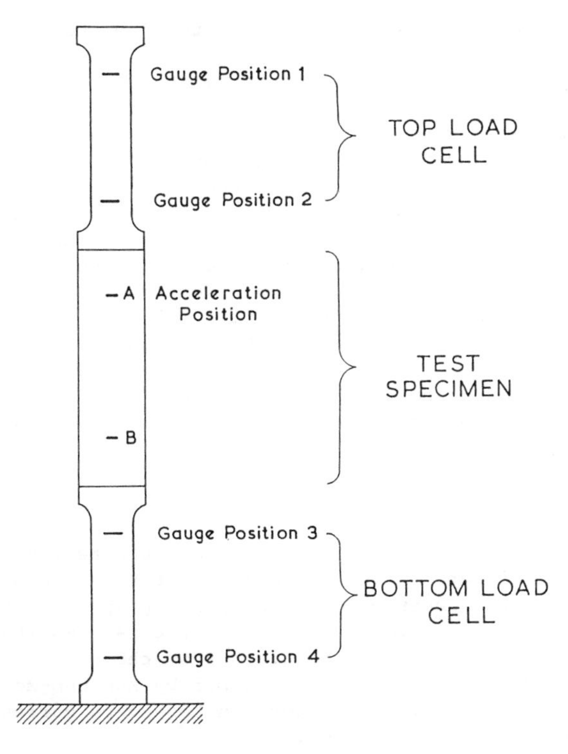

Fig. 2. Schematic view of instrumentation (for shear test specimen)

The two load cells shown in Fig.2 consist of a main central load measuring element together with two protective end fittings. The main central part is a hollow steel cylinder, 300mm long, with an external diameter of 80mm and a wall thickness of 8mm. The distance between the gauge positions (both load cells) was 200mm. The acceleration of the test specimen during the impact event was measured by means of a Kistler piezotron accelerometer. This type of accelerometer has a resolution of 0.1g and is capable of measuring up to 50,000g. In some cases the accelerometer was attached by means of a supporting frame and in other cases by adhesive.

The output signals from the load cells and accelerometers were amplified by means of a Fylde type FE 359 TA amplifier, capable of amplifying the output signed by x 5000. The output was recorded by means of a 16-channel Signal Memory Recorder (SMR) in conjunction with a display screen VDU. The image on the screen could be expanded or compressed horizontally and vertically as required. The test data was stored on a floppy disk during the tests and analysed later.

A typical set of load-time graphs obtained during one impact event is illustrated in Fig.3. These four graphs were obtained by the two load cells in an impact test on a shear test specimen in which the impacting mass was 4kg and the drop-height was 0.5m. The peak load is given by the maximum amplitude and the time to reach this peak load is given by the corresponding distance from the origin. The SMR is able to integrate the load-time signals to give the area under the graph (impulsive load). The results presented in Fig.3 also allows the time lapse between various events to be determined. The displacement of the three gauges (2, 3 and 4) along the time axis relative to Gauge No.1 give a good impression of the passage of the impact event through the testing arrangement.

A typical acceleration-time curve is shown in Fig.4. As in the case of the results shown in Fig.3, this graph was obtained in an impact test on a shear test specimen in which the impacting mass was 4kg and the drop-height was 0.5m. In the case of the acceleration-time graphs the following information was determined: maximum acceleration (measured in terms of g), time taken to reach maximum acceleration and the time difference between gauge No.2 (bottom of first load cell) and the acceleration position. Combining the results shown in Figs.3 and 4 allows the time taken for stress waves to travel from one monitoring position to another to be accurately determined.

4 Results

Results have been obtained from studies on all the test geometries illustrated in Fig.1 for various types of concrete - plain concrete, lightweight aggregate concrete, steel FRC, polypropylene FRC and polymer concrete. Only a limited number of typical results are presented here - typical results obtained from the shear-type of test specimens are presented in a companion paper by Mahjoub-Moghaddas and Barr (1992). The results from the torsion tests are currently being obtained and will be reported later.

A typical set of results for the peak load, P_{max}, and time to peak load, T, is presented in Table 1. These results were obtained from a double notched shear test specimen made of 2% (by weight) steel FRC and the impacting energy was 20 Nm. The top half of the table was obtained by a 2kg mass being dropped through 1m whereas the bottom half was obtained when a 4kg mass was dropped through 0.5m. The number of blows required to cause fracture in the test specimens were 11 and 10 respectively. (The corresponding number of blows required to cause the first crack were 6 and 5).

The corresponding data on acceleration is presented in Table 2. These results suggest that varying the impacting height and mass (but for the same impacting energy) has no significant effect on the acceleration values. This is an interesting result and may be due to the relative masses of the impacting mass and the testing arrangement.

The results presented here are examples drawn from an extensive study carried out by Mahjoub-Moghaddas (1991). The main conclusions reported by him include:

1. The peak load and the impulsive load were greater in the top load cell than in the bottom load cell.
2. Increasing impact energy results in an increase in P_{max} and the impulsive load but the increase is at a reduced rate.
3. Plain concrete and FRC give identical results when monitored by the load cells and by the accelerometer.

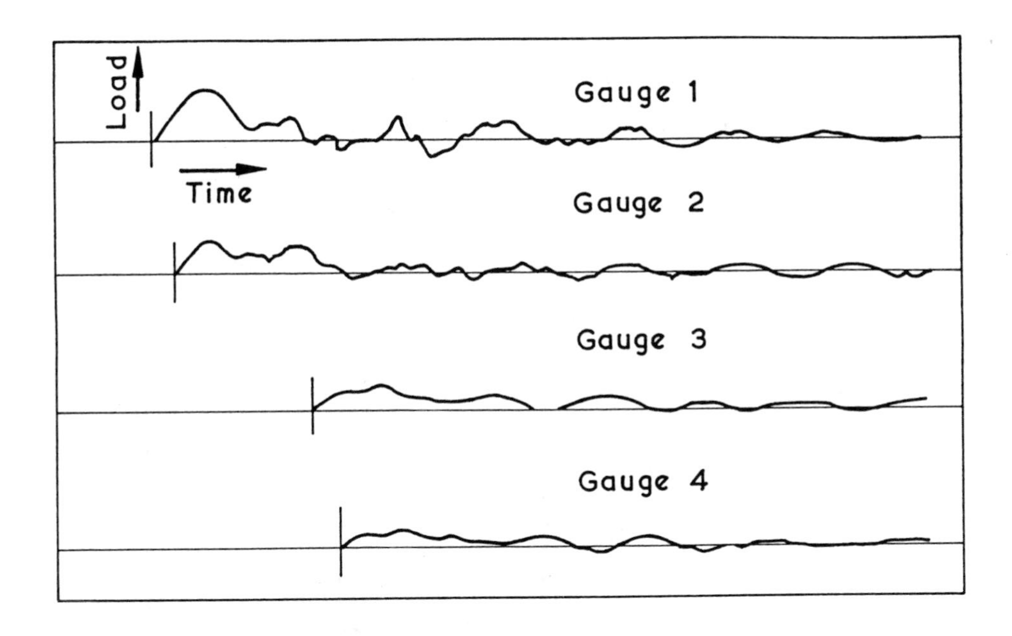

Fig.3. Typical load-time output from two load cells
(Shear test specimen - P = 4 kg and H = 0.5 m)

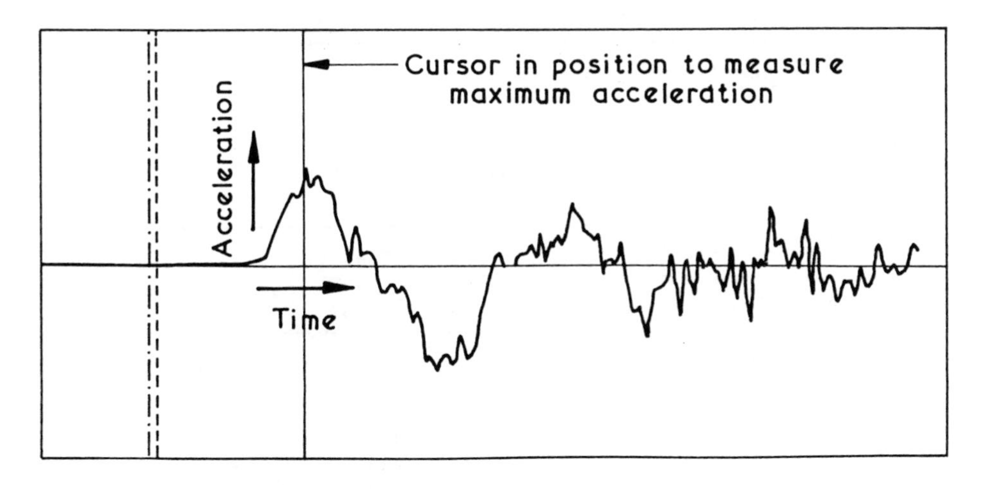

Fig.4. Typical acceleration-time output
(Shear test specimen - P = 4 kg and H = 0.5m)

Table 1. Peak load and peak load time results
(2% Steel FRC - Impacting Energy = 20 Nm)

Blow No.	Gauge 1 P_{max}(KN)	Gauge 1 T(s)	Gauge 2 P_{max}(KN)	Gauge 2 T(s)	Gauge 3 P_{max}(KN)	Gauge 3 T(s)	Gauge 4 P_{max}(KN)	Gauge 4 T(s)
1	91.2	122	55.2	60	38.4	148	24.0	172
2	97.2	116	60.0	52	58.8	86	32.4	84
3	99.6	128	58.8	72	67.2	68	44.4	76
4	99.6	114	70.8	148	81.6	98	60.0	76
5	99.6	122	69.6	154	61.2	70	43.0	88
*6	103.2	116	60.0	66	45.6	74	32.4	90
7	104.4	116	70.8	146	55.2	66	39.6	68
8	106.8	116	72.0	146	44.4	70	31.2	80
9	98.4	120	67.2	152	27.6	80	28.0	118
10	100.8	114	82.8	136	26.8	118	18.4	118
11	93.6	116	82.8	148	26.4	208	18.0	214
1	67.2	120	51.6	154	52.8	66	52.8	62
2	68.4	118	55.6	180	46.8	84	30.0	72
3	74.4	120	58.8	142	62.4	66	66.0	72
4	75.6	112	63.6	178	64.8	60	62.4	70
*5	74.4	120	62.4	186	53.6	72	52.0	82
6	74.4	114	67.2	178	54.8	86	52.4	70
7	64.8	130	66.0	194	47.6	110	41.2	80
8	74.4	118	57.6	184	38.8	126	32.4	108
9	76.8	106	56.4	104	29.2	204	13.2	142
10	75.6	110	54.0	104	16.8	282	9.6	152

*First crack appearance

Table 2. Acceleration and time results
(2% Steel FRC - Impacting Energy = 20 Nm)

Blow No.	P = 2 Kg, H = 1 m		P = 2 Kg, H = 1 m	
	Accel. (g)	T(s)	Accel. (g)	T(s)
1	2236	115	1669	84
2	2236	72	1701	114
3	2079	78	2047	116
4	2047	52	2142	90
5	2205	54	*2866	140
6	*1638	126	2394	126
7	2488	92	2394	150
8	1858	100	1827	132
9	2016	114	2047	70
10	1984	50	2598	64
11	2382	56	-	-

4. The acceleration of the bottom half of the shear-type test specimen is significantly less than the acceleration of the top half.

5 Conclusions

The paper reports on the development and, in particular, the instrumentation of an impact testing apparatus. Two load cells have been used to study impact tests on various fracture test specimens. The load cells, in conjunction with one or more accelerometer, have been used to accurately monitor impact events. The instrumentation allowed the peak load, impulsive load, acceleration and the time taken for stress waves to travel from one monitoring position to another to be accurately determined. This instrumentation allowed the full picture of each impact event to be recorded.

The instrumented repeated drop-weight testing apparatus reported here may be used at two levels. In the first case it may be used as a simple method of evaluating and comparing toughness in terms of the number of blows required to cause cracking and/or fracture. At the second level, the instrumentation allows a better understanding of the factors which influence the fracture process under impact loading to be developed.

6 References

ACI Committee 544 (1978) Measurement of properties of fibre reinforced concrete, Journal ACI Proceedings 75, No.7, pp283-289.

Banthia, N.P. (1987) Impact resistance of concrete, Ph.D. Thesis, The University of British Columbia, pp313.

Barr, B. and Baghli, A. (1988) A repeated drop-weight impact-testing apparatus. Magazine of Concrete Research 40, No. 144, pp167-176.

Gorst, N.J.S. (1992) Ph.D. Thesis (In preparation) University of Wales.

Mahjoub-Moghaddas, H. (1991) Tensile and shear impact strength of concrete and fibre reinforced concrete Ph.D. Thesis, University of Wales, pp409.

Mahjoub-Moghaddas and Barr, B. (1992) Impact shear strength of FRC materials, in proceedings of RILEM Symposium on Fibre Reinforced Cement and Concrete (Ed. R.N. Swamy) E & FN Spon, London, pp 234-44

Reinhardt, H.W. (1987) Loading rate effects, Draft Chapter D, RILEM Technical Committee 89-FMT, in Fracture Mechanics of Concrete Testing: Report No.25-87-16, Stevin Laboratory, Delft, The Netherlands.

Tokatly, Z.A.Y. (1991) Torsional and Mode III strength of concrete, Ph.D. Thesis, University of Wales, pp341.

19 PERMEABILITY AND RESISTANCE TO IMPACT AND ABRASION OF POLYPROPYLENE FIBER REINFORCED CONCRETE

P. SOROUSHIAN and F. MIRZA
Department of Civil and Environmental Engineering,
Michigan State University, East Lansing, MI, USA

Synopsis

The effects of collated fibrillated polypropylene fiber on the impact resistance, chloride permeability and abrasion resistance of concrete materials, incorporating different types of pozzolanic materials were assessed experimentally. A factorial design of experiments together with sufficient replications of tests were adopted in order to generate data for powerful statistical analyses based on which reliable conclusions could be derived.

Fibrillated polypropylene fibers were observed to provide concrete materials with improved impact resistance but did not change the permeability characteristics or abrasion resistance of Concrete. The effects of fiber reinforcement and pozzolan addition on impact resistance were found to interact; the combined effects of pozzolans and fibers were found to be more than additive. Fibers were more effective in increasing the impact resistance of concrete materials incorporating pozzolanic materials.

All the conclusions derived in this investigation are based on comprehensive statistical analyses of laboratory test results, accounting for random experimental errors. Levels of confidence are specified for all conclusions.

Keywords: Abrasion Resistance, Impact Resistance, Permeability, Polypropylene Fiber Reinforced Concrete, Pozzolanic Materials, Test.

1 Introduction

Collated fibrillated polypropylene fibers have gained popularity in the recent years for use in concrete at relatively low volume fractions, mainly to reduce cracking at early ages under the effects of restrained plastic shrinkage cracking. Polypropylene fibers are also expected to enhance certain aspects of hardened concrete properties, including impact resistance, permeability and abrasion resistance. In light of the variations in test results, considering the relatively small effects of fibers at low dosages, one may question the statistical reliability of conclusions based on limited test data regarding the effects of polypropylene fibers on the hardened concrete properties.

The main thrust of this research is to produce a comprehensive experimental data base for powerful statistical analyses which produce conclusions, through level of confidence,

Fibre Reinforced Cement and Concrete. Edited by R. N. Swamy. © 1992 RILEM.
Published by E & FN Spon, 2-6 Boundary Row, London SE1 8HN. ISBN 0 419 18130 X.

regarding the effects of low volume fractions of collated fibrillated polypropylene fibers on the impact resistance, permeability and abrasion resistance of concrete materials incorporating different levels of different types of pozzolanic materials.

2 Background

2.1 Permeability

There is a growing awareness of the importance of concrete permeability in regard to the long-term durability of concrete structures. If it could be possible to keep aggressive substances (Sulfates, Chloride ions, etc.) out of concrete by virtue of low permeability, then associated problems such as freeze-thaw deterioration, corrosion of reinforcement and formation of expansive components would be mitigated. Permeability of concrete can be determined by measuring the rate of permeation of liquids, gases or ions into concrete. The rapid chloride permeability test (adopted by AASHTO) has produced dependable test results for comparative investigation of the permeability of different concrete materials.

Concrete permeability is influenced by, among other factors, the strength, water/cement ratio, cement content, level of compaction and curing conditions of concrete. Also, careful attention should be given to aggregate size and grading, thermal and drying shrinkage strains, and avoiding premature or excessive loading in order to reduce the incidence of microcracking in the transition zones, which appear to be a major cause of increase permeability of concrete in practice. The utilization of polypropylene fibers in concrete enhances the drying shrinkage cracking characteristics of the material, and thus presents potentials for reducing permeability (Al-Tayyib and Al-Zahrani, 1990).

2.2 Impact resistance

Concrete materials are subjected to impact loading in various applications, including pile driving, hydraulic structures, airfield pavements, protective shelters and industrial floor overlays. Impact resistance represents the impact energy absorption capacity of concrete prior to failure. Since plain concrete is a brittle material, it has a relatively low energy absorption capacity under impact loads. Due to the fact that there is no standard test method for the measurement of the impact resistance of concrete, different test procedures have been developed and adopted by different investigators. Reports on earlier test result present in consistent conclusions in regard to the significance of the effects of polypropylene fibers on the impact resistance of concrete (Mindess and Vondran, 1988; Zollo, 1984; Malisch, 1986).

2.3 Abrasion resistance

Not much attention has been paid to concrete abrasion, despite the fact that poor abrasion resistance in highway concrete can accelerate pavement deterioration. Concrete surfaces (floors, highways and slabs) are subjected to abrasion/wear due to attrition by sliding, scraping or percussion. The ability of concrete to resist abrasion is one of its important characteristics in a variety of circumstances.

In general, the abrasion resistance of conventional concrete is a function of finishing conditions, aggregate type, and compressive strength of concrete. Finishing procedure plays an important role in deciding the abrasion resistance of concrete surfaces. Limited tests conducted on polypropylene fiber reinforced concrete indicated that abrasion resistance is improved with the use of collated fibrillated polypropylene fibers (Vondran, 1983).

3 Experimental Program

The effects of polypropylene fiber reinforcement on the impact resistance, permeability and abrosion resistance of concrete materials were investigated. The interactions of fibers and pozzolonic materials in deciding the impact resistance and permeability of concrete were also assessed.

Experiments were conducted following a 2^2 factorial design. The two variables of the study were polypropylene fiber volume fraction (0% and 0.1%) and pozzolan content (0% and 25% by weight of cement substituted with fly ash or slag, and 10% by weight of cement substituted with silica fume). Table 1 summarizes the experimental design used in this investigation. For each mix, 6 permeability test specimens, 12 impact specimens, and 30 abrasion specimens were prepared and tested in order to provide sufficient data for powerful statistical analyses and deriving reliable conclusions. In the case of the abrasion test, only the effects of polypropylene fiber reinforcement on concrete mixtures without pozzolans were investigated.

Table 1. Experimental Desgin.

Poz. Binder Ratio \ Vf (%)	0	0.1
0 %	*	*
b%	*	*

Poz. : Pozzolanic Matearial
b= 25 for Fly Ash and Slag, And 10 for Silica Fume.

The materials used in this experimental study are briefly introduced below:

Polypropylene Fibers: Collated fibrillated polypropylene fibers having a length of 19mm (0.75in). Table 2 present the physical properties of these fibers.

Cement: Type I cement with the chemical composition shown in Table 3 was used in this investigation.

Fly Ash: Class F fly ash with the chemical composition given in Table 3. The specific gravity of this fly ash was 2.245, and its fineness was 32.2% retained on # 325 sieve.

Slag: Ground granulated blast furnace slag was used in this investigation. The specific gravity of this slag was 2.9.

Silica Fume: Silica fume with the chemical properties presented in Table 3 was used. The specific gravity of silica fume was 2.3.

Coarse Aggregate: Natural river gravel with maximum size of 19 mm (0.75 in) was used. Table 4 presents the gradation of coarse aggregate.

Fine Aggregate: Natural sand with finesse modulus of 3.0 was used. The gradation is presented in Table 4.

Air Entraining Agent: A completely neutralized vinsol solution air entraining agent was used in this study.

Table 2. Physical Properties of Polypropylene Fibers.

Tensile Strength	550-760 MPa (70-110 Ksi)
Young's Modulus	3.5-4.7 GPa (500-700 Ksi)
Specific Gravity	0.9
Melting Point	160-170 °C (320-340°F)
Ultimate Elongation at Rapture	~ 10 %
Water Absorbtion	< 0.02 %
Chemical reactivity	Inert

Table 3. Chemical Properties of The Matearials.

Component	CaO	SiO_2	Al_2O_3	Fe_2O_3	SO_3	MgO	K_2O	C	Na_2O
Cement	63.24	21.14	5.76	2.93	2.46	2.06	0.79	---	---
Fly Ash	2.60	47.00	22.10	23.40	---	0.70	2.00	4.30	---
Silica Fume	---	96.50	0.15	0.15	---	0.20	0.04	1.4	0.2

Table 4. Aggrgate Gradation(% Passing)*.

Sieve Size mm (in.)	19.0 (3/4)	12.5 (1/2)	9.5 (3/8)	4.75 (No.4)	2.36 (No.8)	1.18 (No.16)	600μm (No.30)	300μm (No.50)	150μm (No.100)
Coarse Aggregate	100	94	70	11	5	----	----	----	----
Fine Aggregate	----	----	100	100	90	72	46	18	4

* The % Passing satisfied the ASTM C-33.

Superplasticizer: A naphtalene-based superplasticizer was used in this investigation.

The mix proportions for all the mixes of this investigation are presented in Table 5. The mixtures were designed to provide a slump of 89± 13 mm (3.5±0.5 in.) and air content of 7±1%; for this purpose adjustments were made in water/cement ratio and dosage of air entraining agent.

Table 5. Mix Proportions Kg/m³.*

Pozzolan ---------- Binder	V_f (%)	Cement	Coarse Aggregate	Fine Aggregate	Water	Pozzolan	Air Entraining Agent	Superplasticizer
0 %	0	401	1003	802	161	-----	0.201	-----
	0.1	395	987	789	177	-----	0.197	-----
25 %	0	292	975	780	175	97.5	0.643	0.205
FLY ASH	0.1	292	974	779	175	97.4	0.584	0.234
25 %	0	300	1000	829	160	100	0.390	-----
SLAG	0.1	299	996	825	164	99.6	0.299	-----
10 %	0	353	982	784	177	39.2	0.530	1.060
SILICA FUME	0.1	353	981	784	176	39.2	0.530	0.565

* lb/yd³ = 1.685 Kg/m³

4 CONSTRUCTION

The mixing procedure for plain concrete mixtures basically followed ASTM C-192. The mixer was first loaded with the coarse aggregate and a portion of water; the mixer was then started, and the fine aggregate, cement, pozzolan (when used), and the rest of water were added and mixed for 3 minutes. This was followed by 3 minutes of rest and then 2 minutes of final mixing. The fibers, in the case of fibrous mixtures, were added at the beginning of the mixing process.

All the specimens were demolded after 24 hours, and then moist cured at 23 $\pm$1.7 ^{0} C (73.4 $\pm$3 ^{0}F) and 100% relative humidity for three days. They were then exposed to interior laboratory conditions at 23.17 ^{0}C (73.4$\pm$3 ^{0}F) until the test age of 28 days.

Abrasion specimens were subjected to special finishing procedures that represent conventional slab finishing practices as follows:

1. Screeding (strike off): Screeding of the extra concrete on the surface of the specimen was performed immediately after casting using a straight-edge (wood) that was moved across with a sawing motion and advanced forward a short distance with each movement.

2. Bullfloat (Darby): Immediately after strike off, bullfloat was performed in a similar manner to eliminate high and low spots and embed large aggregate particles. Bullfloat was performed using aluminum bullfloat and completed before bleed water accumulated on the surface of the specimen.

3. Floating: After the bleed water evaporated floating was done using an aluminum float held flat on the surface and moved with sawing motion in a sweeping arc so that holes were filled, bumps were cut off, and ridges were smoothed. Float pressure was sustained throughout the process with 1.5 mm (0.08 in.) indentation.

4. Troweling: Troweling came after floating, using a steel trowel moved in the same manner as floating.

5 Test Procedures

5.1 Permeability

The chloride ion permeability test was performed following AASHTO T-277 (Rapid Determination of the chloride permeability of concrete). This test is based on a relationship between the electrical conductance and the resistance to Chloride penetration. A cylindrical specimen 102 mm (4.0 in.) in diameter and 51 mm (2.0 in.) high is used for this test. The sides of this specimen are sealed and it is dried under vacuum. The specimen is subsequently saturated by immersion in water and then connected to a cell with chloride and sodium solutions applied to negative and positive charge surfaces, respectively (see

Figure 1). After six hours, the total ampere-seconds (coulombs) of charge passed during the 6-hour test period is recorded. The test results are then evaluated using the qualitative classification of Table 6.

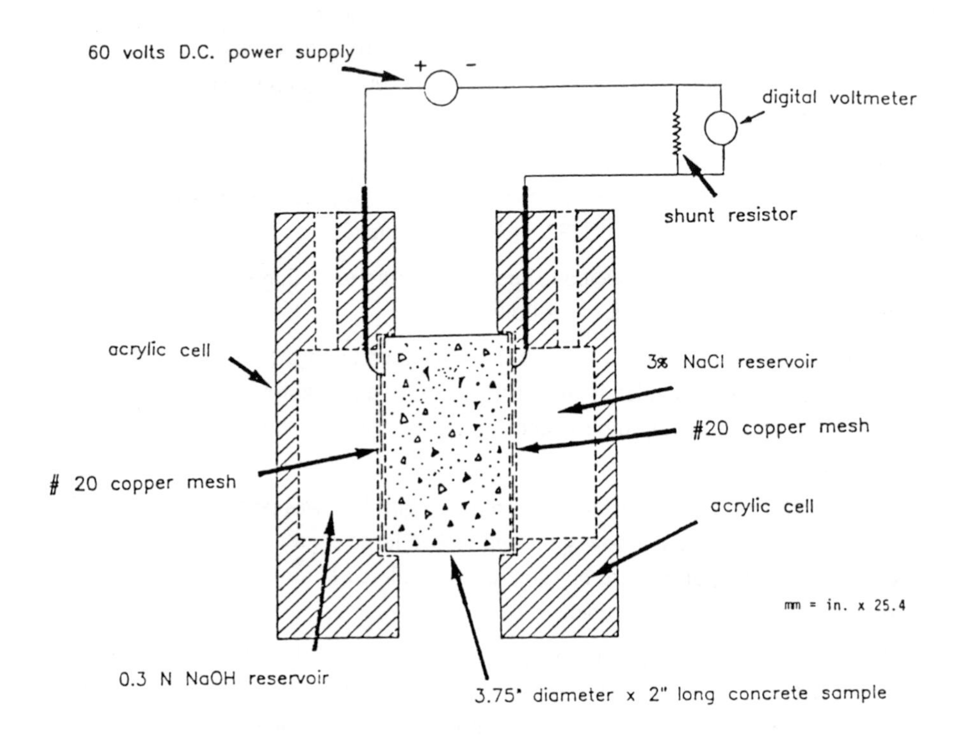

Fig. 1 Schematic of Rapid Permeability Test Apparatus.

Table 6. Chloride Permeability Based on Charge Passed (Whiting, 1981).

Charge Passed (Coulombs)	Chloride Permeability	Typical of
> 4,000	High	High water-cement ratio, conventional (>0.6) PCC
2,000-4,000	Moderate	Moderate h water-cement ratio, conventional (0.4-0.5) PCC
1,000-2,000	Low	Low h water-cement ratio, conventional (<0.4) PCC
100-1,000	Very Low	Latx-modified concrete, Internally sealed concrete
<100	Negligible	Polymer imprgnated concrete, polymer concrete

5.2 Impact

The impact test set-up is shown in Figure 2 (ACI Committee 544, 1989). The specimen is a cylinder 152 mm (6.0 in.) in diameter and 63.5 mm (2.5 in.) high. The test simply consists of repeatedly dropping a hammer on a steel ball supported by the specimen, while observing the formation of cracks and failure of the specimen. The number of blows required to cause the first visible crack on the top and the ultimate failure are both recorded.

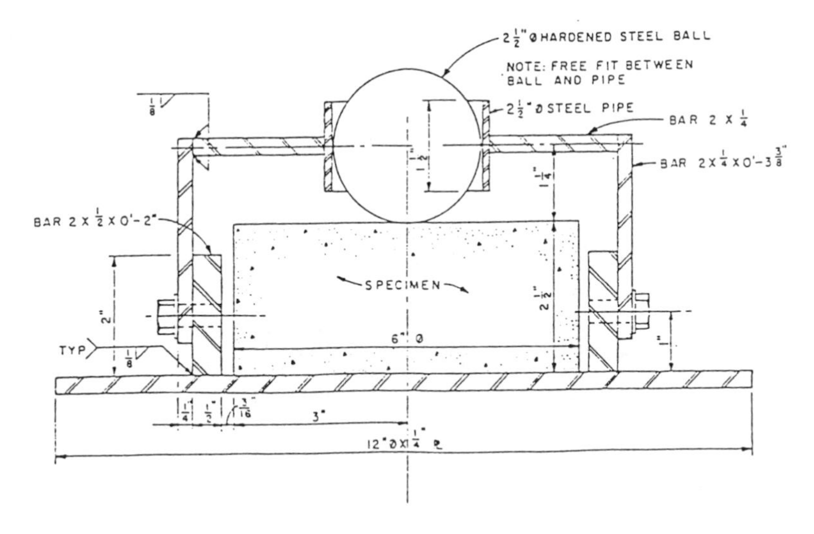

Fig. 2 Schematic of Impact Test Apparatus.

5.3 Abrasion

A drill press device was used in this test (ASTM C-944-80). The specimen was fastened in the device after its mass had been measured. Abrasion with a load of approximately 98 N (22 lb) was continued for 6 minutes after contact between cutter and the surface. At the end of the 6-minute abrasion period, the mass of the specimen was determined to the nearest 0.1g. Loss in mass was calculated and used as a measure of abrasion resistance.

6 Test Results and Discussion

The raw permeability, impact resistance and abrasion test data are presented in this section together with the results of statistical analyses data. The number of replication of different test were large enough for powerful statistical analysis at 95% level of confidence.

6.1 Permeability

The raw permeability test data are presented in Table 7. The average permeability test results together with the 95% confidence intervals are presented in Figure 3. Polypropy-

Table 7. Permeability Test Results (Coulombs).

Binder Type / V_f %	Cement	Fly Ash	Slag	Silica Fume
0	3770 3133 3437 5231 4962 4398	7338 5069 3727 6044 6413 7270	2064 2634 2329 2317 2583 2382	1164 1184 1492 1252 1072 1230
0.1	6609 6968 4085 4458 3743 5025	5078 6460 4821 7205 7185 6155	2404 3261 2942 2867 2785 2853	1274 1089 1971 1943 1462 1312

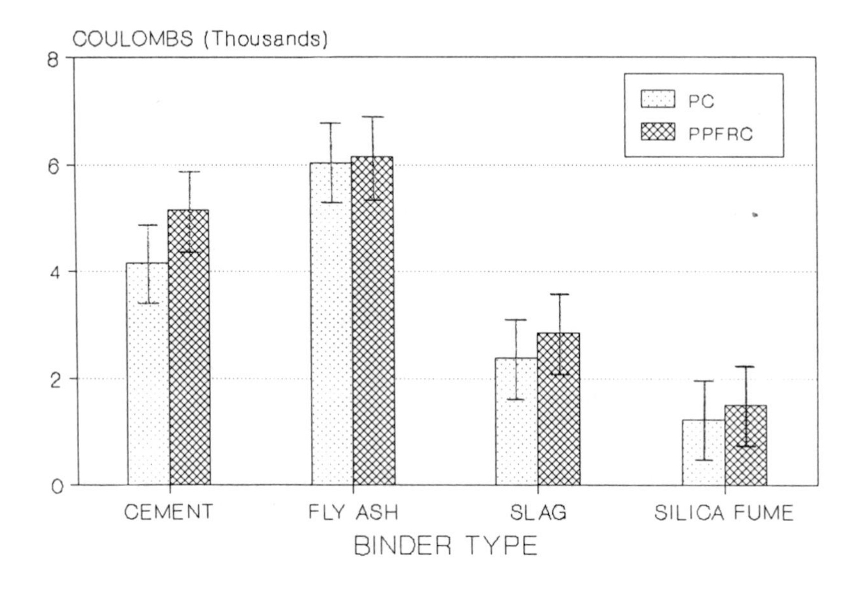

Fig. 3 Permeability Test results

lene fibers at 0.1% volume fraction, as may be seen in this figure, did not influence the permeability of concrete at 95% level of confidence. Pozzolanic materials (except for the fly ash used in this investigation) are observed in Figure 3 to effectively reduce concrete permeability. There was no interaction between pozzolans and polypropylene fibers in deciding concrete permeability, indicating that fibers do not change the generally positive effects of pozzolans on permeability. It is worth mentioning that, in conditions encouraging plastic shrinkage cracking, the arrest of shrinkage cracks in young concrete by fibers could lead to reduced permeability of polypropylene fiber reinforced concrete when compared with plain concrete. These conditions, however, did not exist in this investigation.

6.2 Impact

Table 8 presents the raw impact test results. The average number of blows to first crack and failure, together with the 95% confidence intervals, are presented in Figure 4, respectively. Analysis of variance of impact test results indicated that, at 95% level of confidence, pozzolans reduce the impact resistance while polypropylene fibers have positive effects on the impact resistance. A positive interaction was also found between the effects of fibers and pozzolans on impact resistance, in the sense that fibers produced a larger increase in the impact resistance of pozzolan concrete when compared with plain concrete (see Figure 5). On the average, plain pozzolan concrete has 40% less ultimate impact resistance than plain conventional concrete, and the percentage increase in the ultimate impact resistance of conventional and pozzolan concretes with the addition of polypropylene fibers were 50% and 100%, respectively.

6.3 Abrasion

Table 9 present the raw abrasion resistance test data. The average values of mass loss under abrasion together with the corresponding 95% confidence intervals are presented in Figure 6. Analysis of variance of the abrasion test data indicated that, at 95% level of confidence, polypropylene fiber reinforcement had not statistically significant effects on the abrasion resistance of concrete obtained this specific test procedure.

7 Summary and Conclusions

The effects of collated fibrillated polypropylene fibers, at 0.1% volume fraction, on the impact resistance, chloride permeability and abrasion resistance of concrete materials incorporating different levels of different types of pozzolans were investigated experimentally. Sufficient replicated test data were produced in order to confirm the validity of the following conclusions at 95% level of confidence:

1. While pozzolans generally reduce concrete permeability, polypropylene fibers have no statistically significant effects on the chloride permeability of concrete. Fibers also

Table 8. Impact Test Results (Number of blows).

(a) First Crack

V_f % / Binder Type	Cement	Fly Ash	Slag	Silica Fume
0	36 39 26 26 62 17 53 51 18 54	34 14 11 9 29 32 14 34 11	26 51 15 17 21 25 22 40 14 29 43 33	23 9 24 24 20 15 11 15 38 28
0.1	21 64 50 31 33 40 15 26 55 39	64 22 105 14 26 27 46 23 103 17 71	12 23 54 25 25 37 16 24 65 18 39	22 39 17 78 23 82 55 16 25 12

(b) Failure

V_f % / Binder Type	Cement	Fly Ash	Slag	Silica Fume
0	40 41 26 29 64 18 53 51 18 55	39 15 12 13 30 35 18 34 11	28 52 15 17 22 26 24 43 15 30 43 34	23 10 28 25 21 17 13 16 40 28
0.1	51 79 74 54 51 84 37 56 74 59	89 43 114 26 80 40 80 40 114 26 80	32 45 77 50 45 67 37 42 81 32 49	39 60 40 90 45 92 70 30 45 30

Table 9. Abrasion Test Results.

V_f %	Abrasion Resistance (Mass Loss, g) *
0	2.7 2.7 3.1 5.6 8.2 7.4 7.1 7.4 7.8 4.0 7.1 3.7 2.8 4.5 4.0 6.0 6.1 4.5 5.5 5.0 4.3 4.0 3.3 3.4 3.9 2.9 3.6 2.2 2.8 4.8
0.1	2.8 3.7 6.7 5.1 5.6 7.3 5.9 4.3 6.1 3.9 4.9 2.2 3.4 3.6 3.4 5.9 6.3 6.1 4.3 4.3 4.2 2.5 3.3 2.8 4.8 4.1 4.3 4.2 3.8 3.2

*1 g = 2.2×10^{-3} lb.

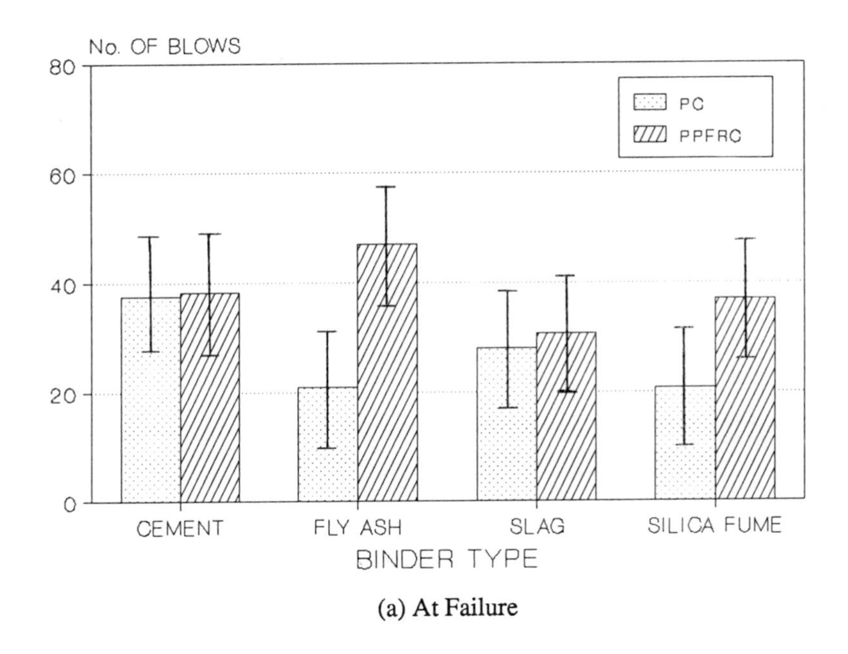

(a) At Failure

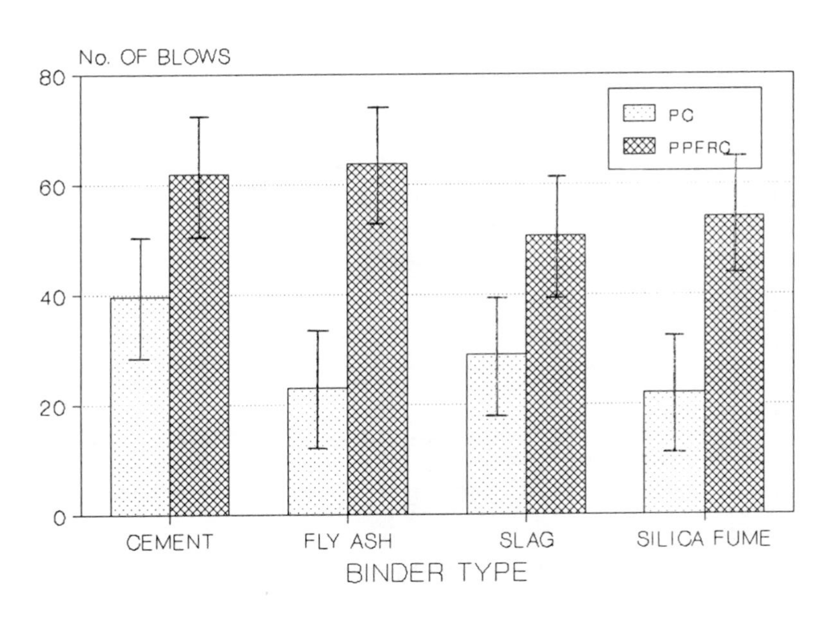

(b) First Crack

Fig. 4 Impact Resistance Test Results

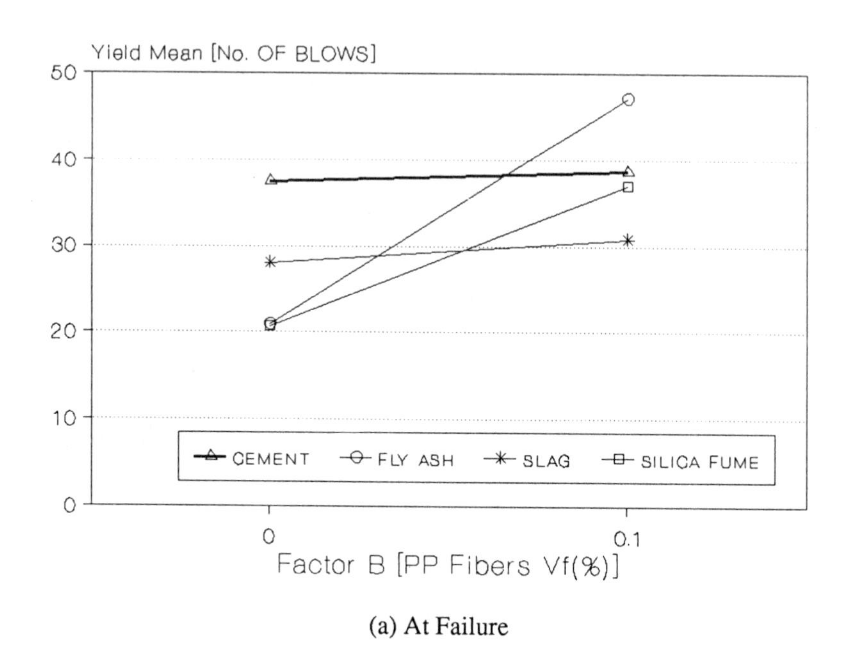

(a) At Failure

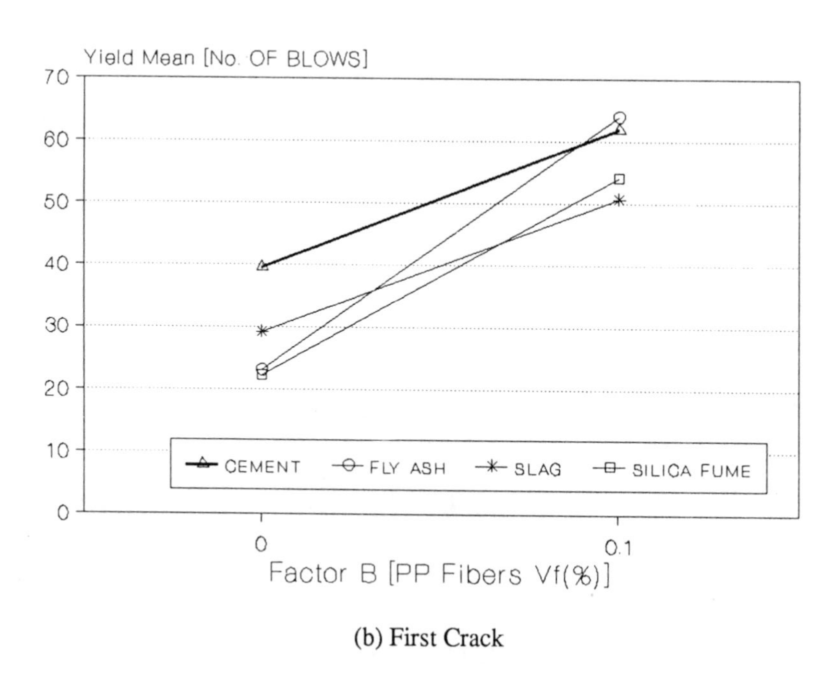

(b) First Crack

Fig.5 Interaction of Permeability Fibers and Pozzolans Effects on Impact Resistance of Concrete

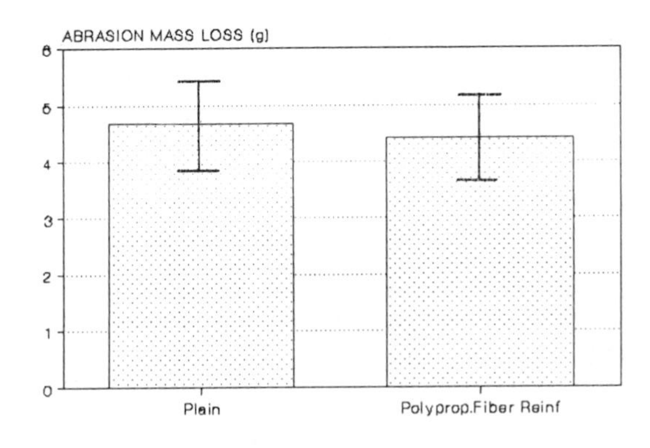

Fig.6 Abrasion Test Results

do not interact with pozzolans in deciding concrete permeability and this the generally positive effects of pozzolan on permeability would be evident to a similar extent in plain and polypropylene fiber reinforced concrete. It should be noted that in conditions encouraging plastic shrinkage cracking we may expect some positive effect of polypropylene fibers on permeability resulting from the crack-arrest action of fibers under these conditions.

2. Polypropylene fibers increase the impact resistance of concrete, on the average, by 50% and 100% in the cases without and with pozzolans, respectively. Pozzolanic materials reduce the impact resistance of concrete and there is a positive interaction between fibers and pozzolans in the sense that fibers are more effective in the presence of pozzolans in enhancing the impact resistance of concrete.

3. There is no statistically significant effects polypropylene fiber reinforcement on the abrasion resistance of concrete.

It is worth mentioning that the laboratory test procedures used in this investigation to assess the abrasion resistance, permeability and impact resistance of concrete may represent only some aspects of these complex properties of concrete. More laboratory and field test data are needed to fully understand the effects of polypropylene fibers on the abrasion resistance, permeability and impact resistance of concrete.

Acknowledgment

The research reported here was sponsored by the Research Excellence Fund of the State of Michigan. The materials used in this research were contributed by Fiber mesh Company (polypropylene fibers), W.R.Grace Company (air entrained agent and superplasticizer), Lansing Board of Water and Light (fly ash), Blue Circle company (ground granulated blast furnace slag), and Elkem Materials (silica fume). There contributions are gratefully acknowledged. The authors are also thankful to the composite materials and structures center and the Albert H. Case Center for the Computer-Aided Engineering at Michigan State University for there technical support of the project.

References

Al-Tayyib, A. and Al-Zahrahi, M.M. "Corrosion of Steel Reinforcement in Polypropylene Fiber Reinforced Concrete Structures". ACI Material Journal, Vol. 87, No. 2, March-April 1990, pp. 108-113.

Austin, S.and Robins, P. "Application of Polypropylene Fiber Reinforced Sprayed Concrete". Proceeding RILEM Symposium on Developments in Fiber Reinforced Cement and Concrete.

Bentur , A. Mindess, and Skalny, J. "Reinforcement of Normal and High Strength Concrete With Fibrillated Polypropylene Fibers". Cement Concrete Research, vol. 19, pp. 229-239.

Bentur , A., Mindess, S. and Vondran, G "Bonding in Polypropylene Fiber Reinforced Concrete". Cement composite and light weight concrete, vol. 11, No. 3, Aug. 1989.

Hannant, D.J. "Fiber Cement, and Fiber Concretes". John Wiley & sons, 1976.

Hannant, D.J. and Hughes, D.C. "Durability of Cement Sheets Reinforced with Layers of Continuous Networks of Fibrillated polypropylene Films". Third International RILEM Symposium on Developments in Fiber Reinforced Cement and Concrete, Volume 2, July 13-17, 1986.

Krenchel, H. and Hansen, A. "Durability of Polypropylene Fibers in Concrete". ACI, Special Publication, sp-81, 1987, pp. 626-643.

Krenchel, H. and Shah, S. "Application of Polypropylene Fibers in Scandinavian". Concrete International, March 1985, pp. 32-34.

Litvin, A. "Properties of Concrete Containing Polypropylene Fibers". Report to Wire Reinforce Institute. Construction Technology Laboratories, Portland Cement Association, Jan. 1985.

Malisch, W.R. "Polypropylene Fibers in Concrete, what do the test tell us". Concrete Construction, April 1986 pp. 363-368.

Measurement of Properties of Fiber Reinforced Concrete. ACI 544.2R-89, ACI Manual of Concrete Practice 1990, Part 5.

Mindess, S., Vondran, G. "Properties of Concrete Reinforced With Fibrillated Polypropylene Fibers Under Impact Loading". Cement and Concrete Research. Vol. 18, 1988, pp. 109-115.

Plante, P. and Biludean, A. "Rapid Chloride Ion Permeability Test: Data on Concrete incorporating Supplementary Cementing Materials". ACI, special publication, sp 114-30, pp. 626-643.

Standard Test Method for Abrasion Resistance of Concrete or Mortar Surfaces by the Rotating Cutter Method, Annual Book of ASTM Standards, ASTM C 944-90, 1991, pp. 483-485.

Standard method of test for rapid determination of the chloride permeability of concrete, AASHTO Designation: T277-83, AASHTO Materials Part II, Tests, 14th Edition 1986.

Standard Specification for Concrete Aggregates, Annual Book of ASTM Standards, ASTM C33-90, 1991, pp. 10-16.

Vondran, G "Making More Durable Concrete With Polymeric Fibers". American Concrete Institute, Detroit, MI, 1983, pp. 337-396.

Vondran, G. and Ebster, T. "The Relationship of Polypropylene Fiber reinforced Concrete to Permeability". ACI special publication, sp-108, 1985, pp. 85-97.

Webster, T. "Polypropylene Fibers May Protect Bridge Decks". Road and bridges magazine reprinted from june 1987, pp. 25-26.

Whiting, D., "Rapid Determination of the Chloride Permeability of Concrete". Report No. FHWA/RD-81/119, August 1981, P. 19.

Zollo, R. "Collated Fibrillated Polypropylene Fibers in FRC". ACI, special publication, sp-81, 1987, PP. 397-409.

20 IMPACT SHEAR STRENGTH OF FRC MATERIALS

H, MAHJOUB-MOGHADDAS and B. I. G. BARR
University of Wales College of Cardiff, Cardiff, UK

Abstract
Several test specimens have been used by researchers to determine the shear strength of plain and FRC concretes. Unfortunately almost all tests have been applied under static loading conditions, rather than by means of impact loading. This paper reports on impact tests carried out on three shear-type test specimen geometries - notched prisms, cylinders and cubes. The impact tests were carried out by means of an instrumented repeated drop weight testing apparatus. The main conclusion arising from this study is that fibres (steel and polypropylene) significantly enhance the shear strength of concrete under impact load conditions. Similar results were obtained by the prismatic and cylindrical specimens. The experimental study is supported by a numerical study of the test specimen geometries.
Keywords: Impact Testing, Fibre Reinforced Concrete, Shear Strength, Repeated Drop Weight Apparatus.

1 Introduction

As more fibre reinforced concrete is being used worldwide, the determination of the shear strength of concrete and FRC materials has become a subject of some considerable research interest. Barr (1987) has reviewed several test specimen geometries which have been developed to evaluate the shear strength of concrete and FRC materials. Unfortunately almost all the shear tests reported to date have been studied under static loading conditions rather than by means of dynamic loading. One of the main objectives of the work reported here was to compare three potential test specimen geometries for evaluating the shear strength of concrete and FRC under impact loading.

Test specimen geometries used for impact shear testing should have an overall shape which is suitable for use in the impact testing machine. (A number of impact testing machines have been developed and the test specimens will vary with the type of machine used). Moreover, the test specimens should be easy to prepare, should have a simple loading arrangement and give a predominantly shear type of failure. Three such test specimens are illustrated in Fig.1.

Most of the machines which have been developed to test FRC materials under impact conditions fall into one of two categories - machines of the pendulum type (modified Izod or Charpy machines) and

Fibre Reinforced Cement and Concrete. Edited by R. N. Swamy. © 1992 RILEM.
Published by E & FN Spon, 2-6 Boundary Row, London SE1 8HN. ISBN 0 419 18130 X.

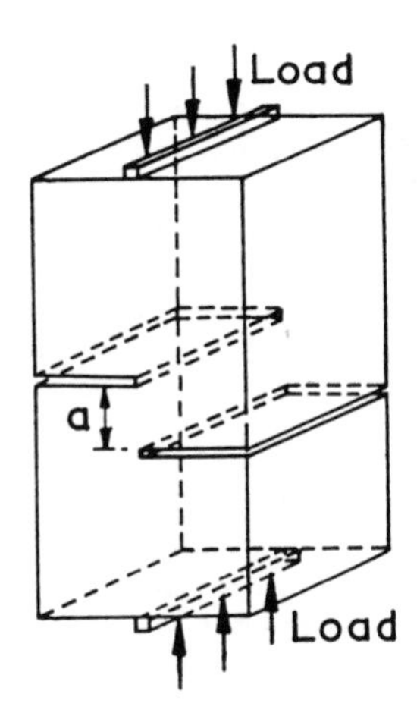

(a) Prism

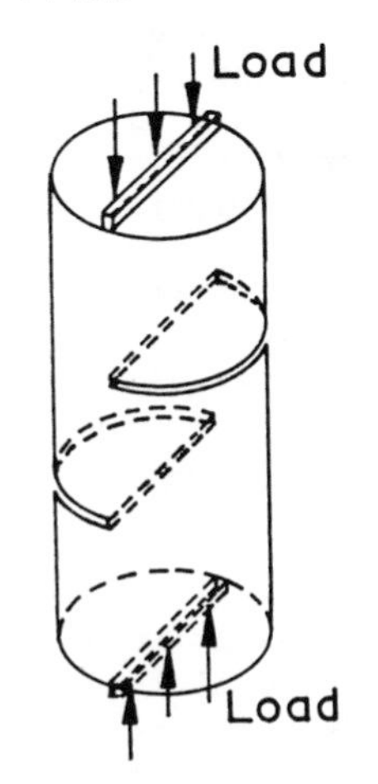

(b) Cylinder

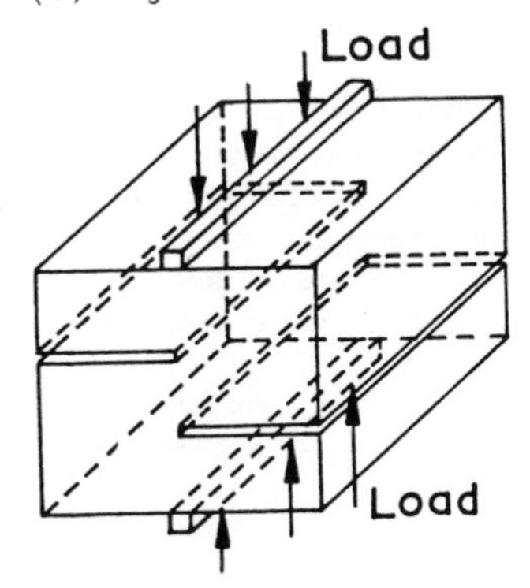

(c) Cube

Fig.1. Test geometries

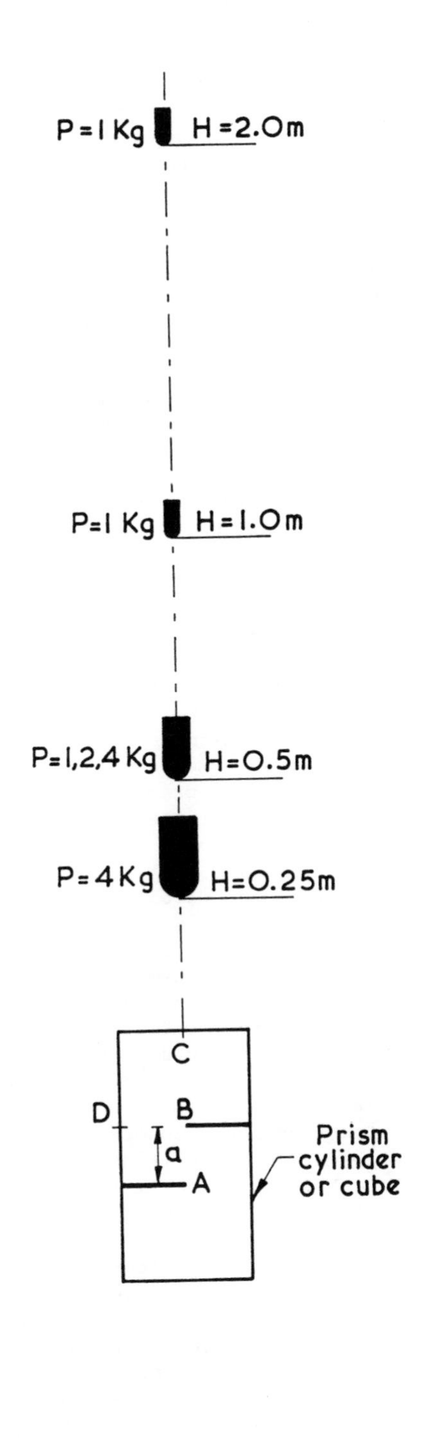

Fig.2. Impact system

machines of the drop weight type. Further details regarding impact testing machines are reported in a companion paper by the authors (1992). The apparatus used for the study reported here is illustrated in Fig.2. This repeated drop weight testing machine was initially designed by Barr and Baghli (1988) for testing notched beams in bending. After suitable modifications by Mahjoub-Moghaddas (1991), this machine has been used for testing various specimen geometries in which the loading was primarily shear. The impact testing machine is a very simple arrangement of three main components - a heavy supporting base, the drop weight guide system and the impacting masses.

A major objective of the work reported here was to investigate the shear strength of plain concrete and FRC under impact loading by means of the repeated drop weight impact testing machine. In particular, the authors were motivated to obtain a comparison between impact and static shear strength of FRC, since the static shear strength of FRC has been observed to be rather disappointing. The test specimens were struck repeatedly by a falling mass until failure occurred. The number of blows required to fracture the specimens was taken as a measure of the impact resistance. All impact events were instrumented and the details of the instrumentation and typical outputs are reported in a companion paper by the authors (1992). The impact test results presented here can be considered at two levels. In the first case, the impact resistance is given simply in terms of the number of blows to cause fracture which, in turn, can be converted to energy. Additionally, the instrumentation allows the investigators to determine with some degree of confidence the influence of the passage of the shock wave through the test specimens and the amount of energy dissipated in various processes, e.g. inertial effects.

2 Experimental details

The materials used for the basic concrete mix were as follows: Ordinary Portland Cement, sea-dredged sand (from Barry Island, South Wales) corresponding to Zone 3 and crushed limestone coarse aggregate (from Newport, Gwent) 10mm maximum size. The steel fibres were straight, 0.3mm in diameter and 30mm long (i.e. an aspect ratio of 100). The polypropylene fibres were of the fibrillated type (12000 denier) and of single length (50mm).

Only one basic concrete mix with the mix proportions of 1:1.8:2.8 representing cement:fine aggregate:coarse aggregate was used throughout the experimental work. The water/cement ratio was kept constant at 0.5. The fibres were used in percentages by weight of the total mix. Three different steel fibre contents were used (1%, 2% and 3%), whereas four polypropylene fibre contents were used (0.1%, 0.2% 0.3% and 0.4%).

The preparation of all mixes used in the experimental study was essentially similar. The dry fine aggregate, the coarse aggregate and the cement were mixed in a 2 cubic foot horizontal rotating pan mixer for about 3 minutes until a constant grey colour was obtained. Then the water was added gradually while the mixer was in motion. Then the mixer was stopped and all the unmixed dry materials on the pan surface was mixed in by means of a trowel and the mixing was continued for approximately another minute. When fibres were used, the basic mix was initially prepared as above and then the fibres were distributed into the mix in a random but uniform manner while the pan mixer was in motion. For the higher fibre contents i.e. 3% steel fibre and 0.3% and 0.4% polypropylene fibre, a trowel was used to help disperse the fibres.

From each mix of plain concrete, three control specimens (100mm cubes) were cast whereas for the fibre reinforced concrete mixes three control cubes of the basic mix and also three after the addition of fibre were manufactured. The test specimens and control specimens were compacted on a vibrating table. All specimens were demoulded after 24 hours and cured under water for a period of 28 days. After 28 days, the specimens were moved into an open laboratory i.e. curing continued in air until the specimens were tested at 90 days. During this time, the necessary work (notching etc) was carried out.

The first test specimen geometry which was used to study the shear failure was 200mm x 100mm x 100mm prisms with two opposite notches (Fig.1(a)). This type of specimen has been used (in static tests) by Liu et al (1985) to determine the Mode II stress intensity factor for plain and fibre reinforced concrete. The two opposite notches have a 50mm depth and the slot separation between the roots of the two notches was varied from 20mm to 50mm.

In some countries, the 200mm x 100mm cylinder is the standard test geometry for the evaluation of the compressive strength. Hence a double-notched cylindrical shear test specimen as shown in Fig.1(b) was also used to study the shear resistance of concrete and FRC under impact loading. In this case, the slot separation distance between the roots of the two 50mm deep opposite notches was varied from 20mm to only 30mm.

Compact cube shear test specimens have also been used to study the static shear strength and toughness FRC. This is the third type of test specimen geometry which was used in this experimental work (Fig.1(c)). This test specimen is a modified 100mm cube with two 50mm deep opposite notches. In the experimental study, the slot separation, between the roots of the two notches, was varied, from 15mm to 25mm. The greatest difficulty in the preparation of all three shear test specimens is to ensure that the ends of the two opposite notches are in alignment with the applied load.

A number of variables can be investigated using the impact testing apparatus shown in Fig.2. Various masses, ranging form 1 to 4 kg, were used in the study. (Masses outside this range could be used in future work). Various drop heights, ranging from 0.5 to 2m, were also used in the study. Various combinations of drop heights and masses can be used with this apparatus - for example, to provide the same impacting energy but covering a range of impacting velocities. In addition to the variables provided by the apparatus, the test specimens could also be varied - in terms of the slot separation distance and the fibre content. All these variables are included in the typical results presented in Tables 1 and 2. Space does not allow a full discussion here of all the variables investigated.

Table 1. Summary of results for number of blows to cause complete failure

Concrete type	Fibre content (%)	P (kg)	H (m)	Number of blows a=20mm	a=25mm	a=30mm
			0.5	9.5	7.17	11.0
Plain concrete	-	1	1	4.0	5.0	4.0
			2	1.83	1.5	2.0
		2	0.5	3.33	4.0	6.0
		4	0.25	3.17	3.0	3.33
			0.5	1.17	1.0	2.0
		1	1	11.67	13.5	11.17
Steel FRC	1	2	0.5	10.67	10.33	12.33
		4	0.25	8.67	10.33	13.17
		1	1	12.5	14.67	21.0
	2	2	0.5	8.67	19.17	18.5
		4	0.25	7.17	13.33	15.83
		1	1	20.83	19.17	21.6
	3	2	0.5	15.5	20.83	24.5
		4	0.25	16.17	16.33	24.83
		1	1	7.83	8.67	7.67
Polypropylene	0.1	2	0.5	5.67	7.33	8.0
FRC		4	0.25	4.83	5.5	8.5
		1	1	13.33	14.17	15.6
	0.2	2	0.5	10.83	12.67	14.67
		4	0.25	11.0	11.33	10.2
		1	1	12.17	10.5	14.17
	0.3	2	0.5	6.83	10.5	15.2
		4	0.25	5.67	6.67	12.17
		1	1	6.83	10.67	11.33
	0.4	2	0.5	5.83	6.83	10.17
		4	0.25	5.83	10.0	9.33

3 Results and Discussion

A summary of the test results obtained by means of the prismatic test geometry is given in Table 1. This table shows the effect of the variables investigated i.e. impacting mass, drop height, slot separation distance and also fibre type and content. Although six slot separation distances were investigated (ranging from 20mm to 50mm), the desired type of failure was only observed for slot separation distances of 20, 25 and 30mm.

It has been shown by Liu et al (1985) in earlier static tests that the ratio of slot separation to height of specimen can be in the range 0.1 to 0.25 for successful shear type of fracture to be developed. The impact tests reported here show that this ratio must be reduced to the range 0.1 to 0.15 for successful tests. Even for the 30mm slot separation test specimens, a small number of specimens fractured along the plane BC. This can be explained by means of a numerically analysis of the test geometry, which will be reported later.

The top third of Table 1 shows the plain concrete test results. The results for a 1kg mass dropped from three heights (0.5, 1 and 2m) are presented in the first three rows. The results show that when the energy per blow is increased by two times or by four times, the number of blows required for complete failure are reduced by more than two times or four times, respectively. A similar conclusion may be drawn from the results for the 2kg and 4kg impacting masses. These results show the greater efficiency of the heavier masses in producing failure (for the same energy per blow). These initial results suggest that larger masses travelling at a lower velocity (but with the same potential energy) have a higher impacting force, shown by the Force/Time curves, and that less energy is absorbed at the point of contact and in accelerating the test specimen compared with lighter masses travelling at higher velocities.

In the case of the steel FRC specimens, three impacting masses (1, 2 and 4kg) were used in conjunction with three corresponding impacting heights (1, 0.5 and 0.25m) to give a constant energy per blow (10N-m). The steel fibres result in an enhancement of the impact shear strength with the greatest enhancement being observed when the fibre content is increased from 0 to 1%. A similar combination of impacting masses and drop heights was used in the case of the polypropylene FRC specimens. The general trend of the polypropylene FRC test results is the same as for the steel FRC mixes. The two main conclusions to be drawn from Table 1 are that fibres improve the shear impact strength and that larger masses are more efficient than lighter masses in producing impact failure in the shear mode. The results presented in Table 1 also show that steel fibres are more efficient than polypropylene fibres in enhancing the impact strength. Furthermore, the results show that the 0.2% fibre content is the optimum polypropylene fibre content to achieve the best enhancement in the impact shear strength - this is probably due to the difficulties associated with good compaction in the case of the higher fibre concentrations.

A sample of the corresponding results for the double notched cylinder test specimens is presented in Table 2. Comparing like with

Table 2. Number of blows required for complete failure ; Cylindrical specimens

Concrete type	Impacting mass P (kg)	Impacting height H (m)	Slot separation a (mm)	Average No. of blows NB	Coeff. of variation V (%)
Plain			20	3.33	28.2
Concrete	2	0.5	25	3.83	27.9
			30	5.67	34.2
2% Steel			20	11.0	21.6
FRC	2	0.5	25	14.17	25.5
			30	20.33	19.8
0.3% Polypro.			20	8.5	22.3
FRC	2	0.5	25	10.0	25.8
			30	14.5	29.5

Table 3. Number of blows required for complete failure ; Cube specimens

Concrete type	Impacting mass P (kg)	Impacting height H (m)	Slot separation a (mm)	Average No. of blows NB	Coeff. of variation V (%)
Plain			15	5.83	33.5
Concrete	1	0.5	20	8.0	46.8
			25	5.0	-
		1	15	2.2	18.2
			20	2.5	20.0
	2	0.5	15	2.0	28.9
			20	2.0	-
2% Steel	1	1	15	7.3	34.0
FRC	2	0.5	15	5.5	32.8
0.3% Polypro.	1	1	15	5.5	22.9
FRC	2	0.5	15	5.0	32.6

like in Tables 1 and 2 shows that the test results in terms of the number of blows required to cause failure are very similar. Again, successful results were only obtained for slot separation distances of 20, 25 and 30mm. Unfortunately, the coefficients of variation were larger - generally between 20 and 30%.
Table 3 gives the corresponding results from the compact cube test specimen geometry. The results show that the desired plane of

Table 4. Energy used in fracture for different specimen geometries ; P = 2.0 kg, H = 0.5 m

Concrete type	Fibre content	Specimen geometry	Slot separation	Number of blows	Blows per unit-area	Average	Coeff. of variation	Total energy per unit-area
			a	NB	NB/A	(NB/A)	V	E/A
	(%)		(mm)		($1/mm^2$) $x10^{-3}$	($1/mm^2$) $x10^{-3}$	(%)	($N.m/mm^2$) $x10^{-2}$
		Short-beam	20	3.33	1.665			
			25	4.0	1.6	1.755	10.0	1.755
			30	6.0	2.0			
Plain concrete	-	Cylinder	20	3.33	1.665			
			25	3.83	1.532	1.696	8.7	1.696
			30	5.67	1.89			
		Cube	15	2.0	1.333	1.333	0.0	1.333
		Short-beam	20	8.67	4.335			
			25	19.17	7.668	6.057	22.5	6.057
			30	18.5	6.167			
Steel FRC	2	Cylinder	20	11.0	5.5			
			25	14.17	5.668	5.981	9.5	5.981
			30	5.67	1.89			
		Cube	15	5.5	3.667	3.667	0.0	3.667
		Short-beam	20	6.83	3.417			
			25	10.5	4.2	4.227	16.0	4.227
			30	15.2	5.067			
Polypropylene FRC	0.3	Cylinder	20	8.5	4.25			
			25	10.0	4.0	4.361	8.0	4.361
			30	14.5	4.833			
		Cube	15	5.0	3.333	3.333	0.0	3.333

failure was only achieved in the case of the 15mm slot separation distance. The coefficients of variation were very high for these test results - due to the small slot separation distance compared with the maximum aggregate size used in the concrete. This test geometry is not recommended for further use in such studies.

Table 4 gives a summary of the test results obtained in this study for plain concrete, 2% steel FRC and 0.3% polypropylene FRC. The number of blows to cause failure is given the fifth column and the corresponding number of blows per unit area in the sixth column. The results in the sixth column have been obtained on the assumption that the energy consumed in the fracture process is uniformly distributed across the slot separation area. Hence it has been assumed that the results are independent of the slot separation in producing the values in the seventh column. The coefficients of variation vary from 8 to 22%, which is acceptable for impact testing of FRC materials in the shear mode. The final column of Table 4 gives the total energy per unit area for the range of test variables. The final column of Table 4 shows clearly the similarity between the prismatic specimen test results and the notched cylinder test results. Furthermore, the enhanced impact resistance due to fibres is also shown by the results in Table 4.

Because of the problems encountered in the experimental work (e.g. wrong type of failure) a numerical analysis was carried out to study these experimental difficulties. A schematic view of the different failure modes is shown in Fig.3. For prismatic test specimens in which the slot separation distance exceeded 30mm, cracks were initiated at the top of the test specimens adjacent to the point of application of the impacting load. Thereafter these cracks propagated towards the nearest notch root i.e. along the plane CB. In the case of the 50mm slot separation specimens a third failure plane was observed - along the plane BD. The main objective of the numerical study was to try to explain the cause of the different failure modes obtained in the experimental work. A full description of the numerical study is beyond the scope of this paper. The analysis was carried out by means of the PAFEC package and typical results are shown in Figs.4 and 5. The loading was assumed to be a uniform line loading acting over the 100mm specimen thickness.

Fig.4 shows the normal and shear stress distributions along the plane between the roots of the two notches in the prismatic test specimen. It is observed that a small concentration of the shear stress occurs at the roots of the two notches. It is also observed that a compression stress concentration is developed at the roots of the notches, whereas an approximately uniform tensile stress

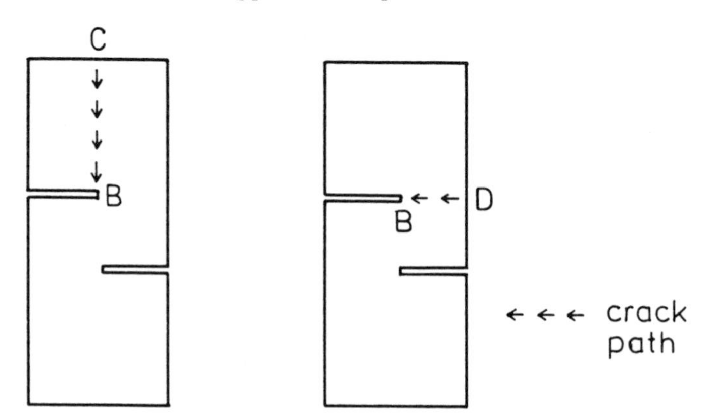

Fig.3. Schematic diagrams of various non-shear failure modes

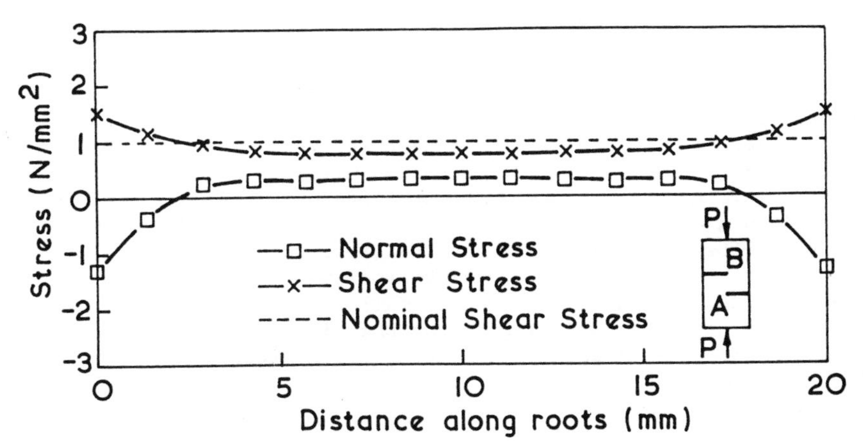

Fig.4. Stress distributions between roots of two notches (Prism with AB = 20 mm)

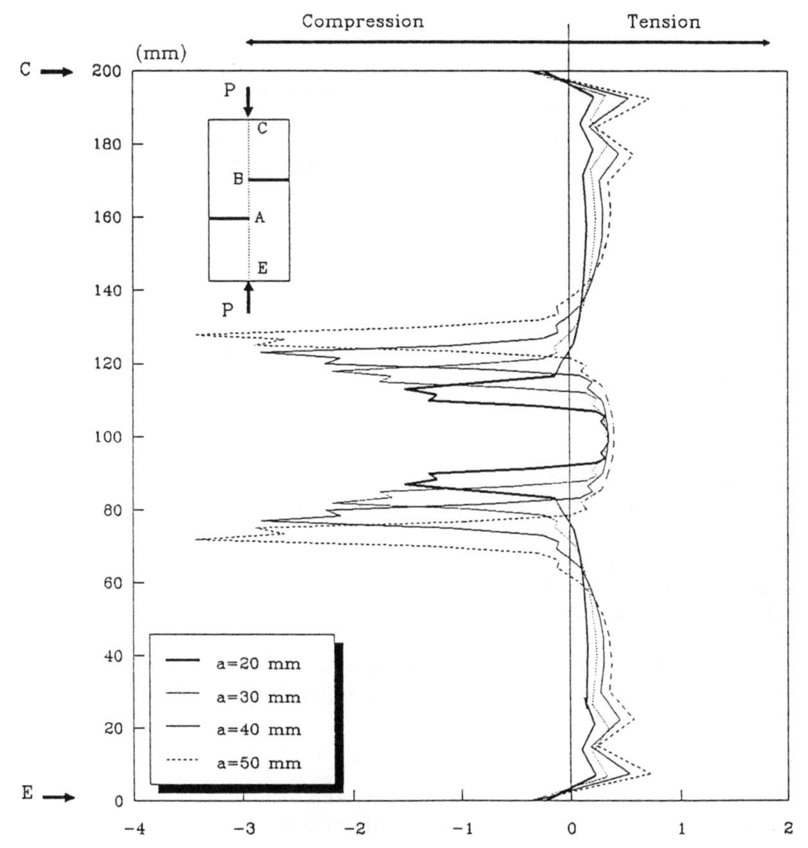

Fig.5. Stress distributions along the prism centerline

distribution occurs along the middle part of the slot separation distance. The results shown in Fig.4 are for a slot separation of 20mm. The complete range of slot separation distances was investigated numerically and the results showed that the maximum tensile stress was always less than 25% of the maximum shear stress.

Fig.5 shows the variation in the compressive and tensile stresses along the plane CE as the slot separation distance is varied from 20mm to 50mm. As the slot separation increases, the tensile stress increased in all three portions of the test specimen (BC, BA and AE). However, the tensile stresses along the planes BC and AE increased more than the increase along the slot separation plane, BA. Therefore, the possibility of tensile failure along the plane BC is more likely than tensile failure along the plane BA as the slot separation distance is increased - as observed experimentally.

4 Conclusions

The results show a significant difference in the mode of failure between static and impact shear tests using the same test specimen geometry. Increasing the impacting mass (for the same impact energy) and also increasing the energy at impact results in a more efficient impact event. The addition of polypropylene and steel fibres significantly increases the shear impact resistance of concrete. Numerical results have been used successfully to explain the occurrence of non-shear fracture in some of the test specimens. The results obtained by means of the prismatic and cylindrical test specimens were very similar. On the other hand, the results obtained from cubes showed a high coefficient of variation and such specimens are not recommended for further studies.

5 References

Barr, B. (1987) The fracture characteristics of FRC materials in shear, in Fibre Reinforced Concrete Properties and Applications (eds S.P. Shah and G.B. Batson) SP 105 American Concrete Institute, pp27-53.

Barr, B. and Baghli, A. (1988) A repeated drop-weight impact-testing apparatus. Magazine of Concrete Research 40, No. 144, pp167-176.

Liu, K., Barr, B. and Watkins, J. (1985) Mode II fracture of fibre reinforced concrete materials. International Journal of Cement Composites and Lightweight Concrete 7, No.2, pp93-101.

Mahjoub-Moghaddas, H. (1991) Tensile and shear impact strength of concrete and fibre reinforced concrete Ph.D. Thesis, University of Wales, pp409.

Mahjoub-Moghaddas, H., Gorst, N.J.S. and Barr, B. (1992) The development of an instrumented impact apparatus for testing concrete materials, in proceedings of RILEM Symposium on Fibre Reinforced Cement and Concrete (Ed. R.N. Swamy) E & FN Spon, London, pp 209-17

21 SIFCON SUBJECTED TO SHEAR: EFFECT OF MATERIAL ANISOTROPY ON STRENGTH AND STIFFNESS

J. G. M. van MIER and G. TIMMERS
Stevin Laboratory, Delft University of Technology,
The Netherlands

Abstract
The shear strength and stiffness of partially cracked Slurry Infiltrated Fibre Concrete are of interest in view of recently proposed structural applications. In the paper, results of displacement controlled biaxial tension/shear tests on double edge notched square SIFCON plates are presented. Because of the strong anisotropy of the SIFCON, the orientation of the fibre system with respect to the loading direction is the main variable. Two different steel fibres are used. The specimens are first pre-cracked in tension, whereafter shear is applied. The results show that both the tensile strength and the shear strength of the material depend to a large extent on the direction of the fibre system. The tensile and shear strengths are very moderate when fibres are oriented perpendicular to the tensile direction and parallel to the shear direction. The results indicate that the decrease in shear stiffness of the cracked SIFCON decreases less in comparison to plain concrete results. True shear fractures have been observed in the sense that an array of inclined tensile cracks develops in the weak planes between the fibre layers.
Keywords: SIFCON, Shear Strength, Shear Stiffness, Tensile Strength, Biaxial Testing, Anisotropy.

1 Introduction

In the last decades, many new concretes have been proposed for new applications. High performance materials such as high strength concrete, ordinary fibre concrete and slurry infiltrated fibre concrete have been developed. The application of such materials as a replacement for ordinary concrete will depend to a large extent on their performance. Often they are claimed to posses much better properties than conventional concretes, but it is the authors' opinion that much depends on which point of view is taken. A material may be improved in such a way that it performs very well under a specific physical or chemical loading, whereas under different loadings the material behaviour is rather moderate. For example, high strength concrete is developed, with main emphasis in increasing the compressive strength. At the same moment however, the ductility of the material may decrease, thereby limiting the range of applicability of the material. The future view presented by those emphasising only the extreme good properties of the newly developed material may therefore be somewhat overstated. This may have serious consequences, because after a disappointing experience with one of the new materials,

Fibre Reinforced Cement and Concrete. Edited by R. N. Swamy. © 1992 RILEM.
Published by E & FN Spon, 2-6 Boundary Row, London SE1 8HN. ISBN 0 419 18130 X.

consequences, because after a disappointing experience with one of the new materials, engineers may decide not to use the material all together. It seems better to take a more realistic point of view, and in fact the authors believe that attempts should be made to determine the weakest link in the new materials, because these will generally lead to a less favourable structural performance. This may sound rather negative, but it is felt that only through a realistic analysis of the properties of a new material, a good assessment of the application of the new material can be made. Maybe the term 'high performance' should be abandoned all together if it does not specify what the 'high performance' property is.

In the present study, one of the newly proposed 'high performance' materials is tested. Slurry Infiltrated Fibre Concrete (SIFCON) is subjected to combined tensile and shear states of stress. SIFCON was first proposed by Lankard & Newell (1984), and has been successfully applied in the repair of bridge decks. At present several researchers have proposed to use the material in structural applications, Naaman (1991). In view of these structural applications, the behaviour of SIFCON should be known under a variety of loading situations. For example, previous investigations, e.g. Homrich & Naaman (1988), have shown that the performance of SIFCON in uniaxial tension and compression is quite favourable. In this paper recent results of combined tension and shear experiments are described. The results were obtained with the sophisticated biaxial test-rig of the Stevin Laboratory, see Reinhardt et al. (1989), Van Mier et al. (1991). Main emphasis is on the shear strength and stiffness of partially cracked SIFCON with various inclined fibre systems.

2 Specimen preparation

For the biaxial experiments, square plates of size 100 x 100 mm^2 and 50 mm thickness are required. For manufacturing these specimens larger plates (300 x 300 x 50 mm^3) were casted in a battery mould containing four plates. The plates were casted in a vertical position. After sprinkling the fibres, the moulds were filled with a cement slurry. Two types of steel fibres were used in the experiments, viz. straight Dramix OL25/0.5 steel fibres and hooked Dramix ZL30/0.5 fibres. The numbers denote the length and diameter (in [mm]) of the fibres. Because the moulds were relatively narrow (50 mm), the fibres would allign preferentially in horizontal layers, Van Mier et al. (1991). Depending on the type of fibres, the mould could be filled with a smaller or larger amount of fibres: 11.47 % and 9.57 % (by volume) for the OL and ZL fibres respectively. The fibre volumes are the average quantities of four batches for each fibre type. The numbers indicate that - as expected - the hooked ZL fibres form a more entangled network than the straight OL fibres.

After the fibres were placed, the cement slurry was poured in the moulds as shown in Fig. 1. The cement slurry was an ordinary sand/cement mix containing 978.6 kg/m^3 Portland Cement Type B; a w/c-ratio of 0.39 (by wt.); 11.7 kg/m^3 Melment super plasticizer; and 822.4 kg/m^3 sand with a maximum aggregate size of 125 μm. When the slurry was poured along the sides of the moulds, the air could escape from the centre of the plates as shown in Fig. 1. This manufacturing method implies that the plates are filled from bottom to top. In an experiment with a mould where one of the sides was replaced by a plexiglass plate it was found that following this method all air was removed quite easily, thereby reducing the vibrational energy needed. During casting, the battery mould was placed on top of a vibrating table. The moulds were fixed to assure that the vibrational energy was transfered directly to the fibre mass in the moulds. This was achieved by clamping the mould on the vibrating table via supports that were placed directly on top of the fibre mass.

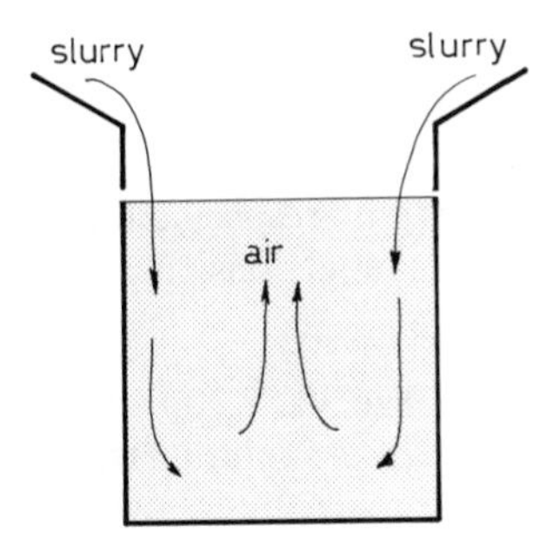

Fig. 1. Pouring the slurry in narrow moulds.

The specimens were demoulded after two days and placed in a fresh water basin in the laboratory (ambient temperature 17-18^{0}C). After 14 days the specimens were sawn. After sawing they were placed in fresh water again until an age of 28 days. At this moment, the specimens were removed from the basin and allowed to dry in the laboratory (50 % RH).

In Figure 2, two photographs of the fibre distribution in the specimens after pouring the slurry are shown. In Fig. 2a, a sawcut has been made parallel to the main fibre direction ($\alpha = 0^0$, see Figure 3), and in Fig. 2b, a section perpendicular to the main fibre direction is shown ($\alpha = 90^0$). The severe anisotropy of the fibre system will be obvious from these figures.

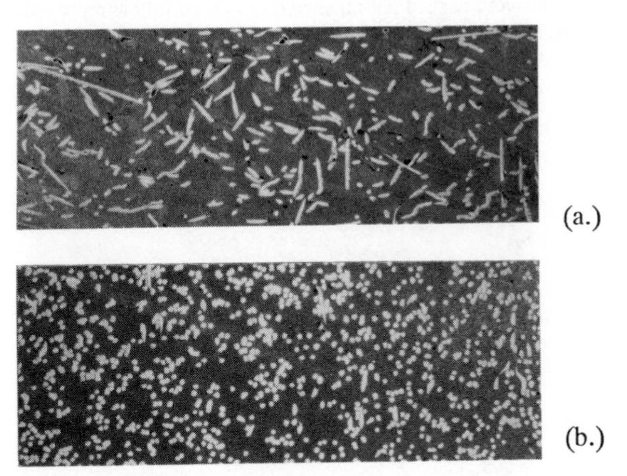

Fig.2. Fibre anisotropy in hardened SIFCON: (a) sawcut at $\alpha = 0^0$, and (b) at $\alpha = 90^0$.

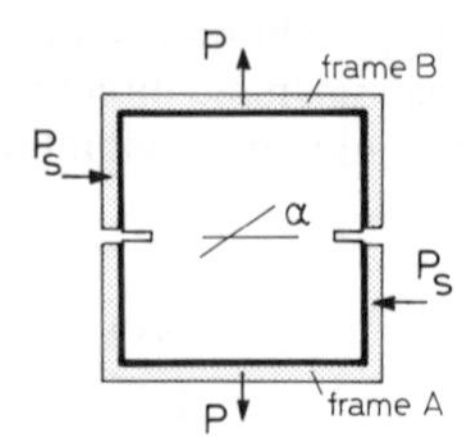

Fig. 3. Orientation of the fibres with respect to the tensile and shear loading directions

Each batch consisted of four plates of size 300 x 300 x 50 mm^3 and six 150 mm cubes. The average compressive strength f_{cc} and splitting tensile strength f_{spl} of the two SIFCON mixes were determined from the cubes (by following the procedure described in the Dutch codes) and the respective values are given in Table 1. In addition to the SIFCON batches, two castings of plain slurry were made. In Table 1, also the tensile strength f_t from the slurry and SIFCON mixes are included, but these results will be discussed in the next chapter.

Table 1. Strength results

Material	α [^{0}]	f_t[MPa]	f_s[MPa]	f_{cc}[MPa]	f_{spl}[MPa]
SIFCON ZL	0	3.09	5.87-8.01	70.5	6.38
SIFCON OL	0	2.86	6.01-8.64	68.2	6.27
Slurry*)	-	3.23	-	-	-
Slurry (2d)	-	2.20	-	-	-

*) The slurry tests were carried out on specimens cured following two distinct methods. Several tests were carried out where specimens were cured as described in paragraph 2. The slurry specimens designated 2d were kept under water until two days before testing.

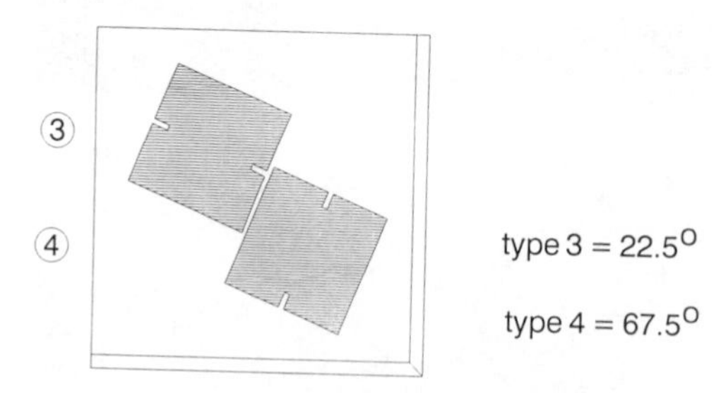

Fig. 4. Sawing double edge notched specimens (100 mm) with inclined fibre systems

The relatively large size of the SIFCON plates allowed the sawing of 100 mm specimens for the biaxial tests with inclined fibre systems as shown in Fig. 4. Two specimens were sawn with a rotating diamond saw from the larger plate. In the example of Fig. 4, a specimen with $\alpha = 22.5^0$ and a specimen with $\alpha = 67.5^0$ were sawn from the 300 mm plate. The specimens for the biaxial tests were double notched at half height in order to trigger crack growth at a known location. This was necessary because tests were carried out in displacement control.

3 Experiments

As mentioned before, the tension/shear experiments were carried out in the sophisticated biaxial test-rig of the Stevin Laboratory. In this apparatus, which is described in detail by Van Mier et al. (1991), a specimen can be subjected to any combination of tension and shear in displacement control. Square double edge notched specimens were used (see Fig. 4). The variables in the SIFCON experiments were the fibre type (two types: ZL30/0.5 and OL25/0.5), the inclination

Table 2. Overview of the SIFCON Tests

α[deg.]	0	22.5	45	67.5	90
δ_0 [μm]					
10	ZL/OL	ZL/OL	ZL/OL	ZL/OL 1802/1107*)	ZL/OL
25	ZL/OL	ZL/OL	ZL/OL 1902/1106	ZL/OL 1607/1007	ZL/OL
50	ZL/OL 1604/1202	ZL/OL 1801/1005	ZL/OL 1603/1006	ZL/OL 1606/1003	ZL/OL 1608/1108
100	ZL/OL 1702/1104	ZL/OL 1602/1103	ZL/OL 1605/1102 /1205	ZL/OL - /1206	?
200	ZL/OL 1903/1004	ZL/OL 1601/1001	ZL/OL 1901/1002 /1204	?	?
300	ZL/OL 1904/1201	ZL/OL 1701/1203	ZL/OL 1703/1105	?	?
400	ZL/OL	ZL/OL	ZL/OL 1803/1101	?	?

*) Specimen code: the first two numbers denote the batch number, the consecutive two numbers show the sequential specimen number for the specific batch.
Only those test combinations where specimen codes appear have been carried out.

of the fibre system defined through the angle α (Fig. 3) at five levels (α = 0, 22.5, 45, 67.5 and 90 degrees), and the axial crack opening before shearing (δ_0) at seven levels following the scheme of Table 2. Note that in order to study the full range of fibre orientations, also values of α between 90 and 180 degrees should be considered. For limiting the number of experiments in this first exploratory test series, specimens with fibre orientations between 0 and 90 degrees were investigated only.

3.1 Strength and ductility

In Fig. 5, characteristic tensile P-δ diagrams for the three different materials tested are shown. Included are test results for ZL and OL specimens ($\alpha = 0^0$), and a load-crack opening diagram obtained from a plain slurry specimen. The loading situation is clarified in the inset of Fig. 5. It is interesting to note that the tensile strength of the SIFCON mixes does not increase significantly. The strength of the SIFCON specimens is determined by the strength of the slurry in between of the fibre layers, at least when $\alpha = 0^0$. At larger fibre inclinations (see Fig. 9), the tensile strength of the SIFCON increases considerably. The same tendency was found for the OL and ZL specimens, see Van Mier et al. (1991).

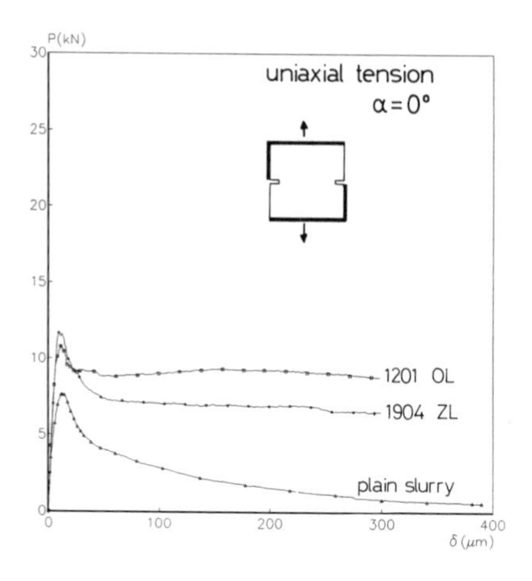

Fig. 5.Comparison of P-δ diagrams for SIFCON with OL and ZL fibres ($\alpha = 0^0$) and plain slurry.

In Fig. 6, P-δ and P_s-δ_s curves for OL specimens with $\alpha = 0^0$ are shown. Note that each of the specimens was loaded first in displacement control up to a prescribed axial crack opening δ_0. After unloading to P = 0, the specimen was fractured in shear while maintaining the load normal to the crack plane constant at P = 0. Crack opening due to sliding is then not restrained by external confinement, but through fibre bridging in the crack only. The P_s-δ_s curves show that the shear stiffness and maximum shear load decrease with increasing axial crack opening just before shearing. Note that the shear strength f_s for the specimens with $\alpha = 0^0$ is relatively low (see Table 1). For the OL specimens f_s is only 8-11 % of the compressive strength of the material, whereas the shear strength of the ZL specimens was approximately 9 to 13 % of the average

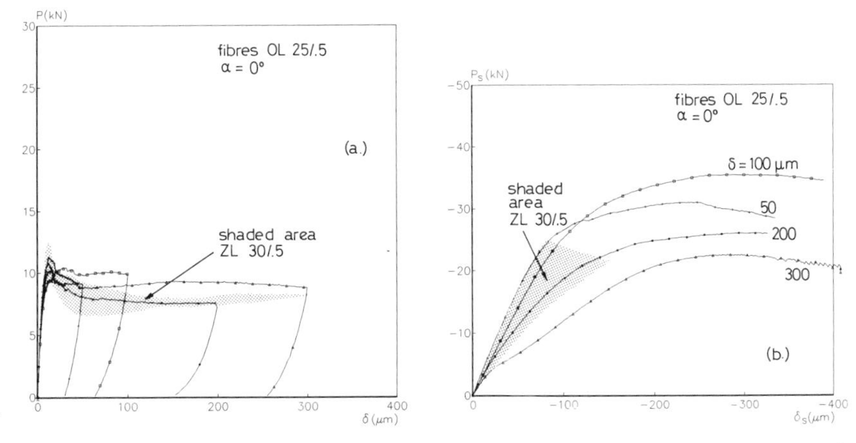

Fig. 6. P-δ plots (a) and P_s-δ_s plots (b) for experiments with OL fibres and $\alpha = 0^0$.

compressive strength. Needless to say that these values are very disappointing. The reason for the low shear capacity in specimens with 0^0 fibre systems is that weak planes exist between the fibre layers. Microscopic studies showed that true sliding failure may be obtained in these weak planes, see Van Mier (1990). Similar crack patterns were recently reported by Arslan et al. (1991) for ordinary FRC. In the latter case however, distributed cracking probably emerges due to the 'spreading' action of the fibres, and not because weak planes exist as in the case of SIFCON.

The effect of fibre type (OL or ZL) was negligibly small. These results are published before and are not repeated here, see Van Mier et al. (1991).

3.2 Effect of fibre anisotropy

The effect of fibre orientation is shown in Fig. 7 for OL specimens. In all cases, the specimens were first loaded to $\delta_0 = 50$ μm. The loading and unloading curves can be seen in Fig. 7a. The tensile strength of the material increases with increasing fibre orientation α. When $\alpha = 0^0$, the strength of the specimen is determined by the tensile strength of the weak planes between the fibre layers: only a relatively small number of fibres will intersect the fracture plane. It is very likely that the ductility at $\alpha = 0^0$ depends completely on the number of fibres intersecting the crack. Bridging caused by the slurry is extremely small, although the latter contribution may increase when crack growth is affected by shrinkage cracking, Van Mier et al. (1991). As soon as α increases, the number of fibres that intersect the potential fracture plane increases, leading to a delayed crack growth between the notches. At least this is what one should expect at first glance. However, the crack patterns shown in Fig. 8 indicate that for larger fibre inclinations, crack growth still occurs in the weak planes, which are now rotated with respect to the potential fracture plane between the notches. Surprisingly, crack growth is easier in these inclined weak planes, even though this is accompanied by an increase of axial load.

The P_s-δ_s diagrams of Fig. 7b show that shear fracture is not possible anymore when the fibres are slightly inclined to the shear direction. Already at $\alpha = 22.5^0$, the shear

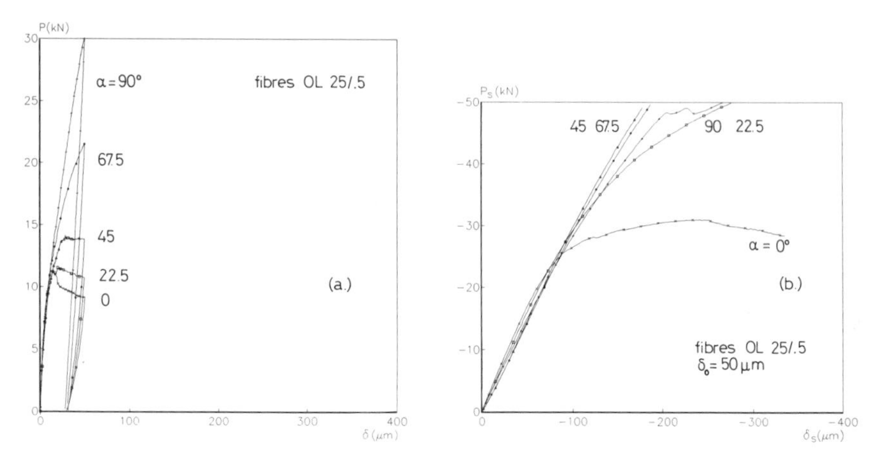

Fig. 7. Effect of fibre orientation on P-δ (a) and P_S-δ_S (b) response; OL fibres; axial unloading deformation δ_0 = 50 μm.

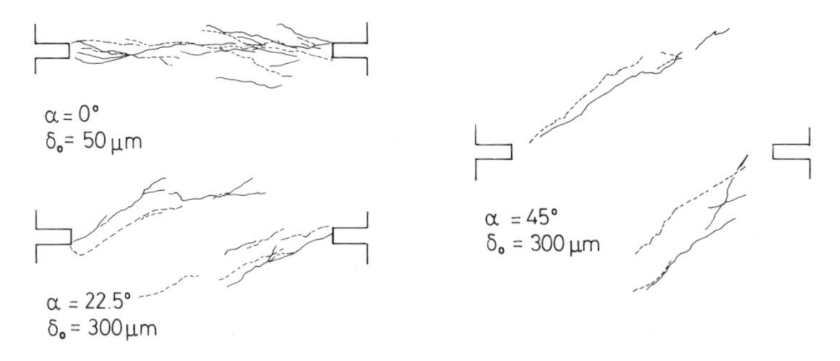

Fig. 8. Global crack patterns for specimens with different fibre orientations.

capacity of the specimen becomes so large that it cannot be fractured in the present test set-up. It should be mentioned that the maximum shear load in the biaxial machine is restricted to 50 kN. The slight curvature of the P_S-δ_S diagrams in the case of $\alpha = 22.5^0$ suggests however that cracking occured during shear. The other P_S-δ_S diagrams (45, 67.5 and 90^0) are more straight up to 50 kN, except that the 90^0 test suffered from cracking in the epoxy layer which was used to fix the specimen in the biaxial apparatus. This explains the 'peak' in the P_S-δ_S diagram for this test. The same behaviour as described above was found for the ZL experiments.

The growth of inclined cracks can be seen very clearly when the results of all 45^0 tests are compared. In Fig. 9 the P_S-δs diagrams for specimens with OL and ZL fibres at 45^0 are shown. The specimens were first pre-cracked up till different values of axial deformation. The initial crack openings are indicated in the figure. The "intial run" in the diagrams increases with increasing axial crack opening, which suggests that contact effects play a major role. This is no surprise in view of the orientation of the cracks in the specimens (see Fig. 8c). Indeed, recent microscopic studies revealed that crack closure occured during shear as shown in Fig. 10. Crack monitoring was done using a QUESTAR QM-100 Remote Measurement System, which combines a high resolution long distance microscope with a three dimensional motorized staging. Crack growth in a specimen can be studied from a relatively long distance. In Fig. 10, three stages of crack closure in specimen 1205 are shown (δ_0 = 100 µm). The crack patterns were taken just below the right notch. Clearly visible is the crack closure with increasing shear deformation δ_s.

The crack closure effect can likely be modelled using a static variant of Hertz's contact law $P_s = K * \delta_s^{3/2}$, where K is a constant mainly depending on the geometry of

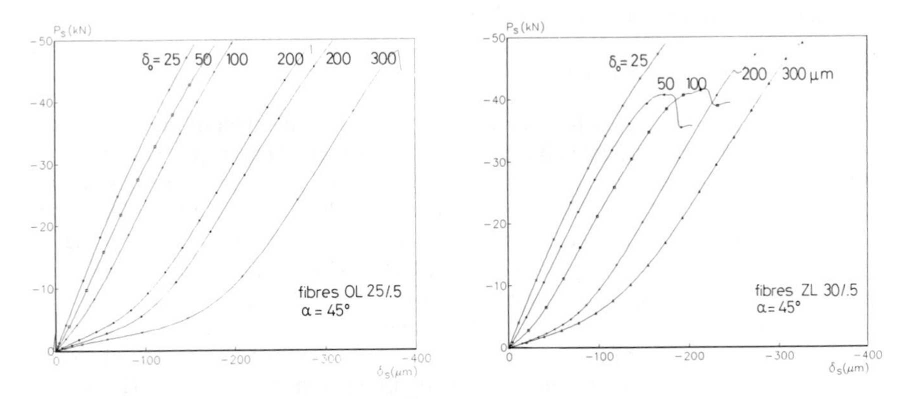

Fig. 9. P_s-δ_s diagrams for 45^0 OL (a) and ZL (b) specimens preloaded to different axial deformations.

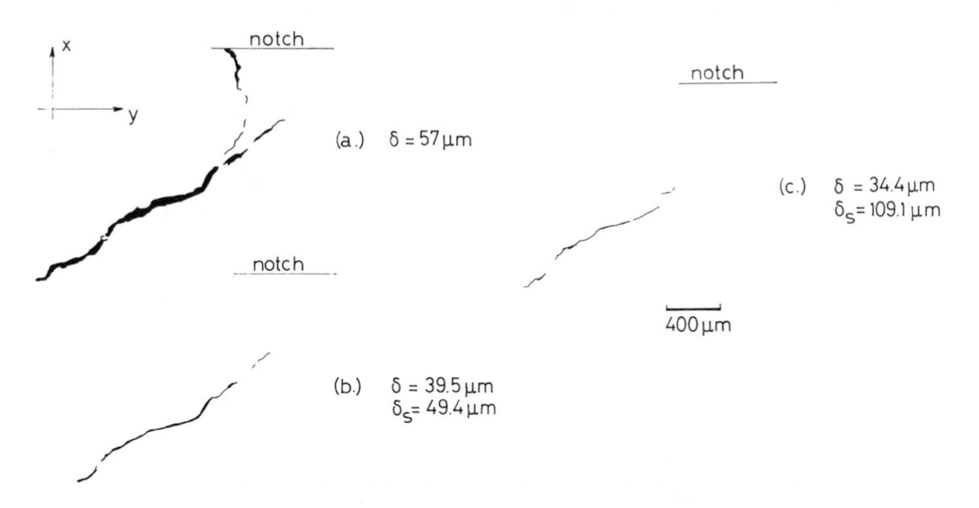

Fig. 10. Crack closure in a ZL specimen with 45^0 inclined fibres (specimen 1205, δ_0 = 100 μm) at three stages of loading.

the surfaces in contact, see Hughes & Speirs (1982). Such an approach is rather convenient for including the complex crack geometry. For the experiments of Fig. 9, values of K ranged between 1200 kN/mm$^{3/2}$ at small crack openings (δ_0 = 25 μm) to 200 kN/mm$^{3/2}$ for δ_0 = 300 μm. The values for the specimens with ZL fibres were slightly larger than those for the OL specimens, especially for the larger crack openings.

4 Shear stiffness

The P_s-δ_s diagrams of Fig. 6b indicate that the shear stiffness of the cracked SIFCON decreases with increasing initial crack opening. For the 0^0 and 22.5^0 experiments the shear stiffness has been determined, which is defined as

$$G = \tau/\gamma = (P_s * h)/(A * \delta_{sc}) \qquad ...(1),$$

where h is the width of the crack band and δ_{sc} the shear displacement of the crack zone. The shear displacement in the crack band is equal to the total shear displacement δ_s minus the elastic shear deformations outside the crack band. In the present analysis, the crack band width was taken h = 15 mm, which corresponds to the average width of the crack zone observed in $\alpha = 0^0$ specimens. Note that a continuum formulation for the decrease of shear stiffness seems possible for SIFCON, whereas this is not allowed in the case of plain concrete where deformations are localized in a single crack.

In Fig. 11, the results are presented for the 0^0 and 22.5^0 experiments. Also in the 22.5^0 specimens, the notion of a crack band still seemed to make some sense, whereas the specimens with larger fibre inclinations failed through diagonal cracking (Figs. 8 and 10). The bold dashed line in Fig. 11 connects the data points obtained from the preliminary tests, Van Mier (1990). The same tendency as observed before for plain concrete is found: after an initial rapid decrease of the shear stiffness at small crack openings, a moderate decrease is measured for larger crack openings. Although a direct

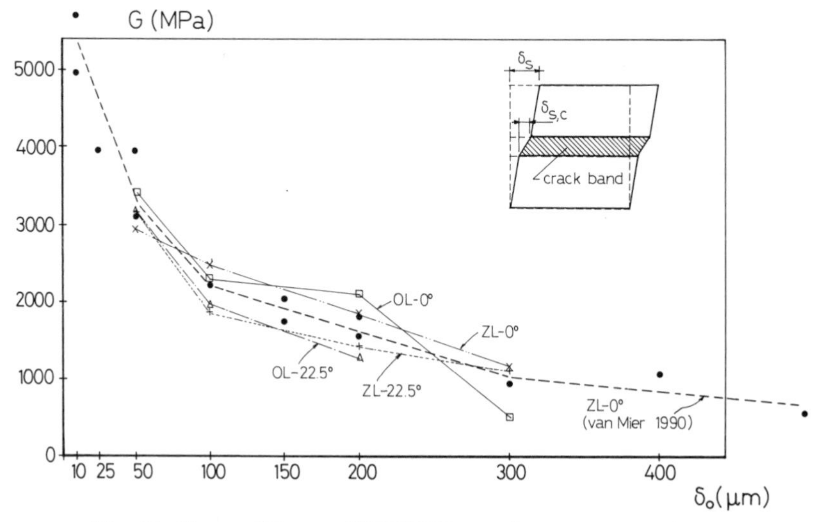

Fig. 11. Shear stiffness of SIFCON at various crack openings.

comparison with plain concrete results is difficult, mainly because the definition of a crack band width is almost imposible in plain concrete, the results seem to indicate that the shear stiffness increases when fibres are added, see Van Mier et al. (1991). Most likely this can be explained from the bridging action of the fibres that intersect the crack plane. The results of Fig. 11 show that the type of fibre had no significant influence on the shear stiffness of the SIFCON. Probably the type of fibre will be more important as soon as fibres are pulled out, which may occur in later stages of the fracture process.

5 Conclusions

Based on the results presented in this paper, and additional information published in Van Mier et al. (1991), the following conclusions can be drawn.

(1) True shear failure (defined as an array of short inclined tensile cracks) in SIFCON was observed. The shear fractures could develop because the highly anisotropic material structure contained oriented weak planes.

(2) When the orientation of the fibres is changed, no sliding failures could be obtained in the SIFCON, but rather a set of diagonal cracks would develop from the notches in the main fibre direction. In general specimens with inclined fibre systems could not be failed in shear in the present experimental set-up.

(3) A similar shear stiffness reduction as found for plain concrete was observed for SIFCON. A rapid decrease of shear stiffness is observed for narrow cracks, whereas the decrease diminishes at larger crack openings. The shear stiffness is somewhat higher than for the plain concrete, which may be explained from the fibre bridging.

(4) A significant effect of fibre orientation on the tensile strength and the shear strength of SIFCON was observed. Especially the shear strength is very low (viz. approximately 9-13% of f_{cc}) when fibres are oriented parallel to the shear direction.

(5) The SIFCON behaves considerably more ductile than plain slurry or plain concrete. The fibres can be regarded as an internal confinement, and the behaviour resembles the response of confined plain concrete. Similar findings were recently reported by Yin et al. (1990) for ordinary FRC subjected to biaxial loading.

(6) The results show that all future effort should be stored in optimizing the fibre distribution in the material. In order to avoid weak planes in the SIFCON, random fibre distributions are prefered.

6 References

Arslan, A., Hughes, T.G. and Barr, B. (1991), Mixed mode fracture - Including torsion in a new test specimen geometry, in *Fracture Processes in Concrete, Rock and Ceramics* (eds. J.G.M. van Mier, J.G. Rots and A. Bakker), Chapman & Hall, London/New York, pp. 737-746.

Homrich, J.R. and Naaman, A.E. (1988), Stress-strain properties of SIFCON in uniaxial compression and tension. Air Force Weapons Laboratory, *Final Report AFWL-TR-87-115*, August 1988.
Hughes, G. and Speirs, D.M. (1982), An investigation of the beam impact problem. *Technical Report 546*, Cement & Concrete Association, 117 pp.
Lankard, D.R. and Newell, J.K. (1984), Preparation of highly reinforced steel fibre reinforced concrete composites, in *Fibre Reinforced Concrete (ACI SP81)*, American Concrete Institute, Detroit, pp. 286-306.
Naaman, A.E. (1991), SIFCON: Tailored properties for structural performance, in Pre-Proceedings of the International Workshop on *High Performance Fiber Reinforced Cement Composites* (eds. H.W. Reinhardt and A.E. Naaman), Mainz, June 23-26, 1991, pp. 3-24.
Reinhardt, H.W., Cornelissen, H.A.W. and Hordijk, D.A. (1989), Mixed mode fracture tests on concrete, in *Fracture of Concrete and Rock* (eds. S.P. Shah and S.E. Swartz), Springer Verlag, New York, pp. 119-130.
Van Mier, J.G.M. (1990), Fracture of SIFCON under combined tensile and shearlike loading, in *Proceedings Symposium 'O' Fiber Reinforced Cementitious Materials* (eds. S. Mindess and J.P. Skalny), Mater. Res. Soc. Fall Meeting 1990, MRS Pittsburgh, Vol. 211, pp. 215-220.
Van Mier, J.G.M., Nooru-Mohamed, M.B. and Timmers, G. (1991), An experimental study of shear fracture and aggregate interlock in cementbased composites. *HERON*, Vol. 36, No. 4, pp. 1-104.
Yin,W.S., Su, E.C.M., Mansur, M.A. and Hsu, T.T.C. (1990), Fibre reinforced concrete under biaxial compression. *Engineering Fracture Mechanics*, Vol.35, No. 1/2/3, pp. 261-268.

22 POST-FATIGUE PROPERTIES OF STEEL FIBER REINFORCED CONCRETE

V. UKRAINCZYK and Z. RAK
University of Zagreb

Abstract
For the sake of choosing mix designs for investigation of railway sleepers, the fiber reinforced concretes were tested for fatigue stress and impact resistance. Experimentally have been investigated mechanical properties of steel fibre reinforced prisms before and after cyclic loadings. Fibers were dosaged in the three quantities: 0.9, 1.3 and 1.5 percent by volume. The maximum grain sizes of aggregate were 4, 8 and 16 mm. After two months curing the prisms were subjected to 10^4 cyclic flexural loads.The changes of ultra sonic pulse velocities in specimens and the residual bending strength and working diagrams have been tested. The results of fatigue tests prove that increase or decrease of residual bearing capacity,as measured by toughness indexes in comparison to the monotonic loaded prisms,correlated good with the quantities of fibres.
The impact properties were compared by means of falling weight test.
Keywords: *fiber reinforced concrete; steel fibers; fatigue test; impact test; residual strength; dynamic modulus.*

1 Introduction

The laboratory testing program was started to choose the concrete mix design for field investigation on railway sleepers. This is typical example of element subjected to large number of strong fatigue and impact loadings. Many sorts of existing sleepers made of wood or reinforced concrete or prestresed concrete or steel are not yet satisfactory,and in spite of higher investmenst,the long term costs of fiber reinforced concrete sleepers could be feasible solution.

In the literature exist reliable data on S-N relationships for fibre reinforced concrete beams, up to 2 milions cycles,based on flexural fatigue stress as the percentage of first crack strength,or in terms of actual applied stress versus number of loading cycles. They show that fiber content and aspect ratio are most important.

In our program the influence of three different quantities of drawn wire fibers from new production and three different maximum grain sizes of aggregate were studied. In these experiments, for the sake of shortening the time necessary for testing,the relations from S-N diagrams published by other autors /1 and 2/ were taken as the basis for selection of combination of high magnitude of cycling stresses at low number of

Fibre Reinforced Cement and Concrete. Edited by R. N. Swamy. © 1992 RILEM.
Published by E & FN Spon, 2-6 Boundary Row, London SE1 8HN. ISBN 0 419 18130 X.

cycles, to study the influence of constituents parameters on the fatigue properties of selected concrete mix designs. As the fatigue preloading of specimens only 10000 cycles were chosen with corespondingly high level of cycling stresses, calculated as 80 percent of the maximum static flexural strength. The residual properties of fatigue preloaded specimens were compared to the properties of only static loaded ones.

2 Experimental program

Mix proportions

The main characteristics of the concretes compositions of this experimental program have been (Table 1): three maximum grain size 4, 8 and 16 mm, combined with three quantities of drawn wire fibers 0.9, 1.3 and 1.5 Vol.%. Previonsly planned larger quantities of fibres (1.8 and more) were canceled from the program, because of tendency for balling. The steel fibres are 0.50 mm in diameter, 32 mm long and slightly bent at the ends. Ordinary portland cement, and river sand and crushed aggregate fractions 4 to 8 mm and 8 to 16 mm were used.

The quantities of water and plasticizer were adjusted to keep the adequate plastic consistency.

Testing methods

The examination of fatigue performance of concrete was performed on prisms 10x10x40 cm in the following steps:

- measuring the ultrasonic pulse velocity and calculating the dynamic modulus of elasticity (E_{dyn}),
- three point bending test (on the 300 mm span) with the purpose to determine the first crack stress (f_A), maximum stress (σ^s_{max}) and the working diagram P-y, where P is one point force and y is deflection,
- fatigue cyclic three point flexural loading on the 300 mm span, with the sinusoidal wave form, varying between 0.1 and approximately 0.8 of static flexural strength (maximum stress, σ^s_{max}),
- the number of cycles in all tests was 10000,with the rate of 1 cycle per second,
- measuring the ultrasonic pulse velocity and calculating the dynamic modulus of elasticity after the fatigue loading and then, static three point bending test to determine the residual first crack stress (f_A), maximum stress (σ^r_{max}) and the working diagram P-y.

The level and the number of cycling loadings was chosen on the basis of research works /1/ and /2/. All the cyclic loadings as well as the bending tests were performed on the strain controlled machine. The loading rate in static tests was 0.5 mm/min.

The impact resistance was tested in the drop-weight machine (Föppl hammer). The falling iron weight of 50 kg falls from a certain height on the tested piece, and the number of blows necessary to produce the first crack is the measure of impact resistance. The appearance of the first crack is followed by the sudden decrease in rebound of the weight.

Table 1. Composition and properties of fresh concrete

Constituents \ Code		MA 4-15	MA 4-13	MA 4-9	MA 8-15	MA 8-13	MA 8-9	MA 16-15	MA 16-13	MA 16-9
Cement	(kg/m^3)	480	480	480	450	450	450	420	420	420
Aggregate	(kg/m^3)	1502	1535	1529	1570	1600	1619	1689	1782	1784
0 - 4 mm		1502	1535	1529	843	858	869	570	602	603
4 - 8 mm		-	-	-	727	742	750	344	364	364
8 - 16 mm		-	-	-	-	-	-	775	816	817
Water	(l/m^3)	168	172	174	177	174	175	171	153	160
W/C		0.36	0.37	0.37	0.40	0.39	0.4	0.42	0.37	0.39
Admixtures:										
Plasticizer	(l/m^3)	4.8	4.8	4.8	4.5	4.5	4.5	4.2	4.2	4.2
AEA	(l/m^3)	0.19	0.19	0.19	0.2	0.2	0.18	0.25	0.22	0.22
Fibers	(kg/m^3)	120	100	70	120	100	70	120	100	70
	(Vol.-%)	1.5	1.3	0.9	1.5	1.3	0.9	1.5	1.3	0.9
Slump	(cm)	6	5	10	9	7	8.5	1.6	1.8	6
Vebe °		4	5	2	2	4	2	17	17	4
Pores	(%)	9.0	9.0	8.0	8.0	7.5	7.0	5.5	4.6	4.2
Volume density	(kg/m^3)	2275	2257	2258	2322	2327	2319	2404	2459	2438

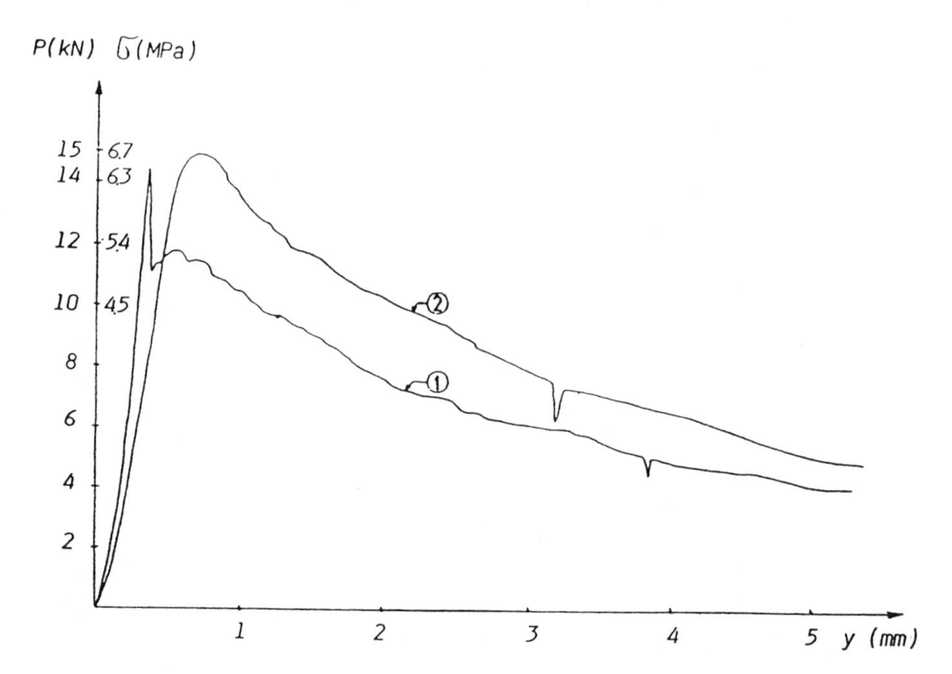

FIG 1. Characteristic P-y diagram (MA 4-13) for static flexural test
1 befor cycling loading   2 after cycling loading

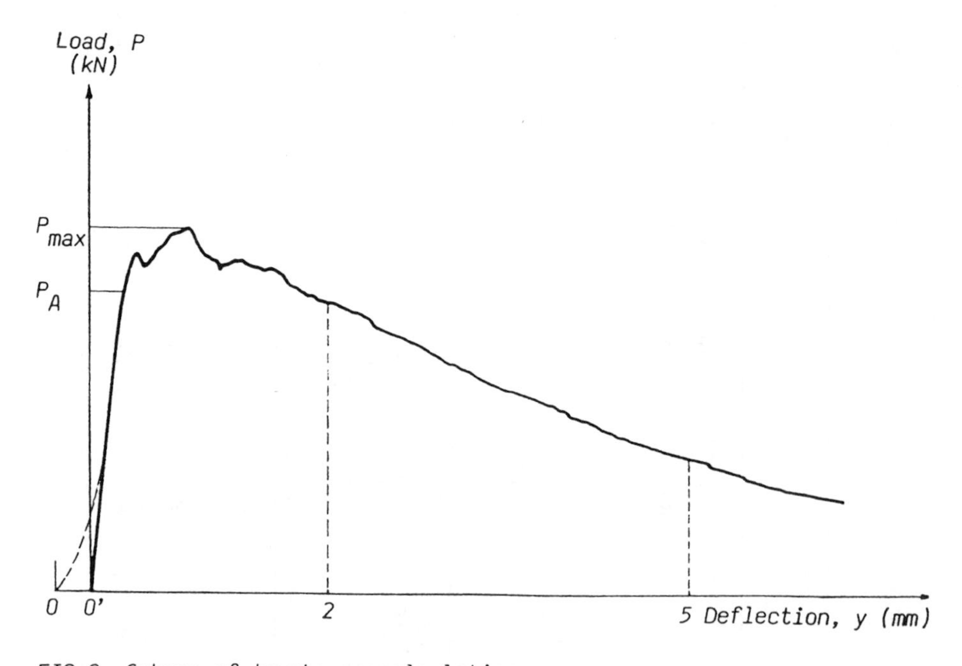

FIG 2. Scheme of toughness calculation

The height of drop was 50 mm, to enable the comparison of our results with other results /3/, using the energy equivalent per volume of the specimen. The testing of impact resistance was conducted on the cubes 10x10x10 cm obtained by cutting from the ends of prisms previously used in static bending tests.

3 Test results

In Fig. 1. are characteristic three point bending test P-y diagrams for the prisms casted of concrete mixture MA 4-13. The curve 1 is the working diagram before cyclic loading, and the curve 2 is obtained for prism of the same composition, but after cyclic loading. From such diagrams are determined the ultimate strengths, σ^s_{max}, and corresponding deflections. On the basis of these parameters were chosen the limits for fatigue cyclic loadings $\sigma_{max} \approx 0.8x\ \sigma^s_{max}$ and $\sigma_{min} = 0.1x\ \sigma^s_{max}$. The toughness of the specimen was calculated as the area under the P-y diagram till the deflection y=1/60=5 mm.

From the same diagram the toughness, T_{JCI} is calculated according to the proposal explained in /5/, as the area under the P-y diagram for deflection y=1/150, i.e. for 1=300 mm, y=2 mm.

In the Table 2 are summerized all measured test results and calculated parameters: first crack stress, ultimate strength, the area under the working diagram for the deflection y=5 mm and y=2 mm and the static and dynamic moduly of elasticity. The relations between the toughnesses are compared in Fig. 3. The toughnesses calculated for y=5 mm are compared because they better express the differences. It is visible that the residual toughness strongly depends on the fiber content, fatigue cyclic preloading and the relation between the maximum grain size and the fibre length. The optimum result is obtained for the mix composition coded MA 8-13 i.e. maximum grain size 8 mm with 1.3 Vol.% of fibres 32 mm long. This confirms the point of view obtained from static tests, that the optimum ratio between the fibre length and the maximum grain diameter is three or more. From these tests it could be concluded that increasing the fibre content above 1.3 Vol.% did not improve the fatigue resistance.

The results of impact resistance tests are reviewed in Table 3. For all combinations the impact resistance was about 4 times increased as compared to the plain concrete. However, the results for fibre reinforced specimens are in quite narrow limits, 202 till 260 blows as compared to 61 blow for specimens made of plain concrete. From these data could not be evaluated the relations between the impact resistance and the fibre contet or influence of the maximum grain size.

Table 2. Test results and calculated parameters

Code	Density	Compressive strength	Fiber content	Cycling flexural stress in % of σ_{max}	E_{stat}	E_{dyn}	Flexural strengths (MPa)		Toughness y=1/60 = =5 mm	T_{JCI} y=1/150 = =2 mm
							First crack f_A	Ultimate $\sigma^s_{max}/\sigma^r_{max}$		
	(t/m³)	(MPa)	(Vol.%)		(GPa)	(GPa)			(kN×mm)	(kN×mm)
					before cycling / after cycling					
MA 4-15	2.24	47.3	1.5	79	31.5	35.8	6.5	7.1	37.8	23.1
					30.0	33.1	7.0	7.7	45.0	25.4
MA 4-13	2.22	40.0	1.3	82	39.4	35.9	6.5	6.5	34.8	18.5
					23.1	33.1	5.8	6.8	42.5	22.3
MA 4- 9	2.24	46.0	0.9	81	28.2	38.3	6.3	6.4	23.5	15.2
					14.8	37.0	4.2	4.9	20.2	12.8
MA 8-15	2.31	40.4	1.5	82	28.8	41.1	6.5	6.5	39.9	21.1
					22.0	38.6	6.7	7.4	43.2	23.7
MA 8-13	2.28	41.0	1.3	83	32.1	38.7	5.9	7.2	47.0	24.8
					19.2	31.2	6.8	7.5	51.2	27.2
MA 8- 9	2.29	51.9	0.9	80	38.9	43.3	6.8	7.0	26.0	17.5
					25.3	42.6	6.5	6.7	25.3	16.8
MA 16-15	2.4	42.2	1.5	80	36.8	47.4	6.3	7.4	37.3	21.8
					35.1	41.6	6.0	6.9	36.0	20.9
MA 16-13	2.45	56.6	1.3	81	46.8	52.5	7.1	7.6	40.3	24.2
					43.6	48.7	6.7	7.2	37.5	23.0
MA 16- 9	2.42	57.0	0.0	79	50.5	54.8	7.2	7.7	22.1	16.9
					42.4	49.5	5.4	6.5	20.2	14.2

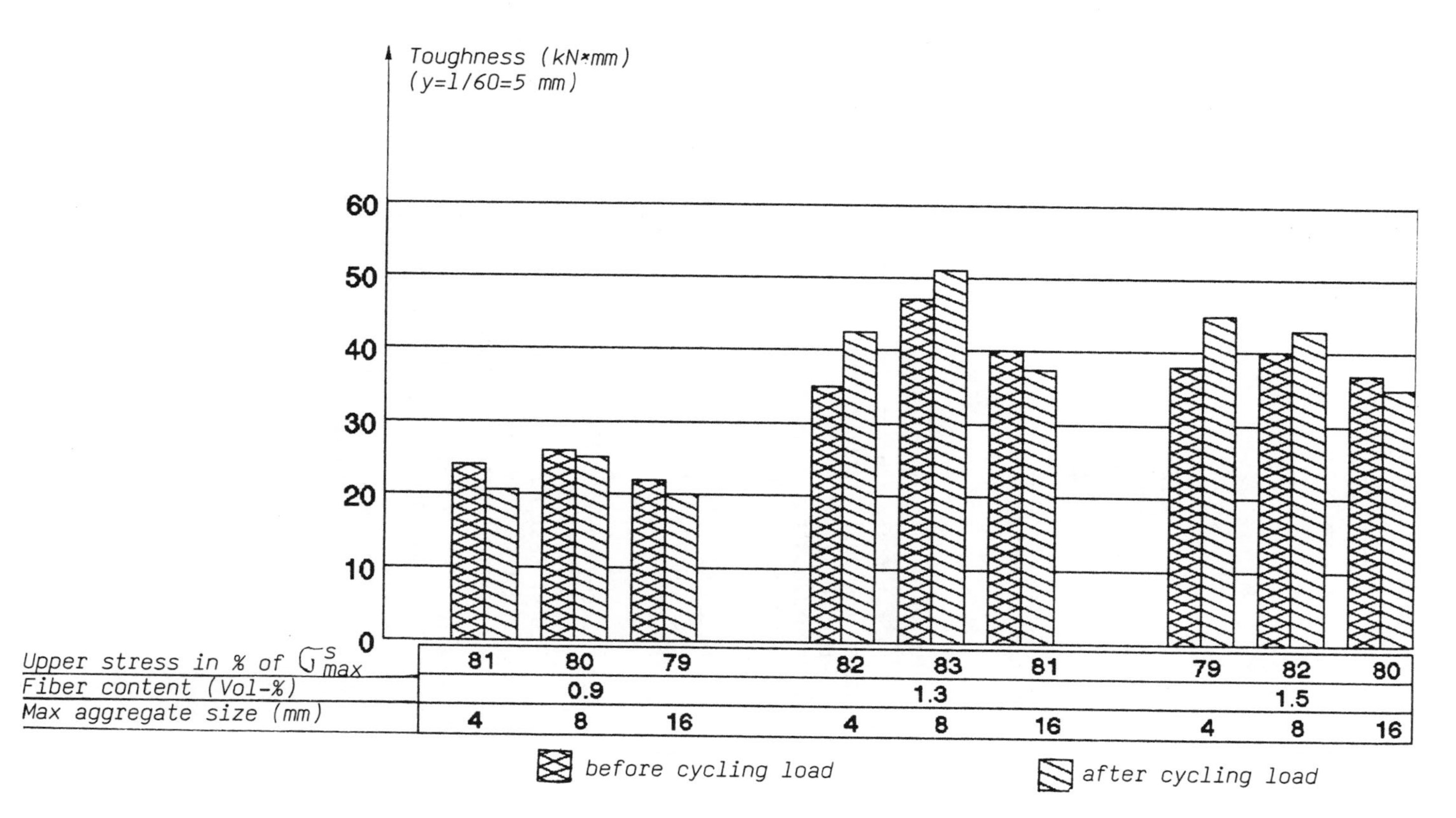

FIG 3. Toughness for monothonic static flexural test befor and after cycling

Table 3. Impact test results

Code	Max aggr. size (mm)	Fiber content (kg/m³)	Number of impacts
MA 4-15	4	120	206
MA 4-13	4	100	204
MA 4- 9	4	70	260
MA 8-15	8	120	207
MA 8-13	8	100	202
MA 8- 9	8	70	217
MA 16-15	16	120	212
MA 16-13	16	100	227
MA 16- 9	16	70	235
Plain Concrete	16	Ø	61

4 Comments

1) The static modulus of elasticity after cyclic loading was decreased.

2) In the cases of higher fiber contents, in spite of high stress in cyclic preloading, the postfatigue strength and residual toughness were increased, possiblly because of consolidation of the structural defects and tensile stresses caused by concrete shrinkage before loading. During the fatigue preloading the fibres take over the high stresses.

3) The ultrasonic pulse velocity and the calculated dynamic modulus of elasticity show weakening of the structure after the cyclic preloading, and from the static test results the differences are much greater.

4) The applied test procedure with the low number of cycles (10000), but high level of stress (0.8 x σ_{max}), has given significant differences in results of testing the specimens with various fiber content, with different maximum aggregate size and various ratio of fiber length to maximum aggregate grain diameter.

5) Impact resistance of steel fiber reinforced specimens was about 4 times greater than for plain concrete, irrespectively of the fibre content from 0.9 to 1.5 Vol.% and maximum grain size. This fact indicates that the chosen impact test technique is not valid to distinguish between very different fibre contents in concretes.

5 References

/1/ Butler,J.E., The performance of concrete containing high proportions of steel fibers with particular reference to rapid flexural and fatigue loadings, In: Fibre Reinforced Cements and Concretes. Recent Developments Edited by R.Swamy and B.Barr, Elsevier 1989.

/2/ Johnston,C.D., and Zemp,R.W., Flexural Fatigue Performance of Steel Fiber Reinforced Concrete - Influence of Fiber Content, Aspect Ratio, and Type, ACI Materials Journal,V.88, No.4, July--August 1991, pp. 374-383.

/3/ Jamrozy,Z., and Swamy,R.N., Use of Steel Fibre Reinforcement for Impact Resistance and Machinery Foundations, International Journal of Cement Composites, Vol.1, No.2, July 1979, pp. 65-76.

/4/ State-of-the-Art Report on Fibre Reinforced Concrete; Reported by ACI Commitete 544 (ACI 544.1R 82).

/5/ Gopalaratnam,V.S, et al, Fracture Toughness of Fiber Reinforced Concrete, ACI Materials Journal, V.88, No.4, July-August 1991, pp. 339-353.

23 STEEL FIBRE REINFORCED DRY-MIX SHOTCRETE: FIBRE ORIENTATION AND ITS EFFECT ON MECHANICAL PROPERTIES

H. ARMELIN
University of Sao Paulo, Sao Paulo, Brazil

Abstract
Steel-fibre reinforced shotcrete has a natural tendency for preferential fibre orientation. This 2-D random distribution of fibres becomes potentially dangerous to shotcrete (especially for tunnelling purposes) considering a natural tendency for a layering effect, which others believe may be, itself, responsible for up to 40% loss of compressive strength.

Cylinder cores (taken in both directions of fibre orientation) and cubes have been used to evaluate that possibility. Two different steel-fibres were employed at three different contents. Fibre influence on voids volume and material rebound have also been studied.
Keywords: Shotcrete, Fibre orientation, anisotropy.

1 INTRODUCTION

Steel-fibre reinforcement came to shotcrete in the early eighties, when large scale testes proved it to be at least as efficient as wire-mesh reinforcement (1)(2). Since then, more than twenty large scale applications of steel-fibre reinforced shotcrete (SFRS) have been made, mostly in Canada, Japan and Europe (1)(3).

Most of these applications of SFRS have been done in tunnelling and rock stabillization, where contractors have been able to significantly reduce excavation cycles (4) at a greater safety level. Also, other advantages such as less waste of concrete due to rebound and thinner linnings have often been mentioned (5).

2 FIBRE ORIENTATION

One very important aspect of fibre reinforced shotcrete is fibre orientation. Due to the spraying mechanism by which concrete is gradually deposited over the "aim", fibres tend to orient themselves in succesive planes, normal to the spraying direction (Figure 1).

Fibre Reinforced Cement and Concrete. Edited by R. N. Swamy. © 1992 RILEM.
Published by E & FN Spon, 2-6 Boundary Row, London SE1 8HN. ISBN 0 419 18130 X.

That bidimensional orientation effect has been previously described (2)(4) and becomes evident by visual observation of sawn SFRS (Figure 2).

That 2-D random distribution of fibres in shotcrete becomes a potentially dangerous influence on the mechanical properties if one considers two aspects concerning shotcrete in tunnel linnings:

a) In a tunnel linning structure, subject to external loads from the excavated ground, the main compression forces appear tangential to the linning, and therefore paralel to fibres' preferential direction (figure 3).

b) Plain (unreinforced) dry-mix shotcrete has a natural tendency for lack of homogeinety caused by a layering effect originated in the spraying/compaction process (Figure 4). That layering effect has been earlier described by Hills (7), who believes it could be responsible for a loss of compressive strength in the direction parallel to the layers of up to 40%.

Taking these two aspects into consideration, a test program was under taken in order to evaluate the possibility of the oriented fibres acting in a way as to enhance this layering effect, causing compressive strength loss in the direction normal to spraying (parallel to fibres).

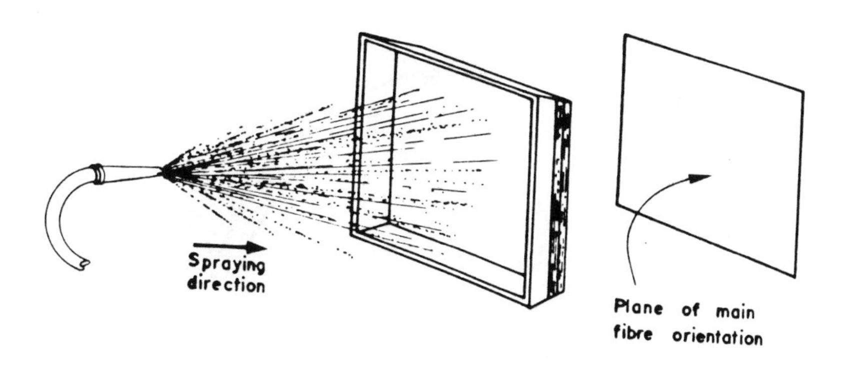

Fig.1. Due to the spraying mechanism, fibres tend to be oriented in planes normal to the spraying direction.

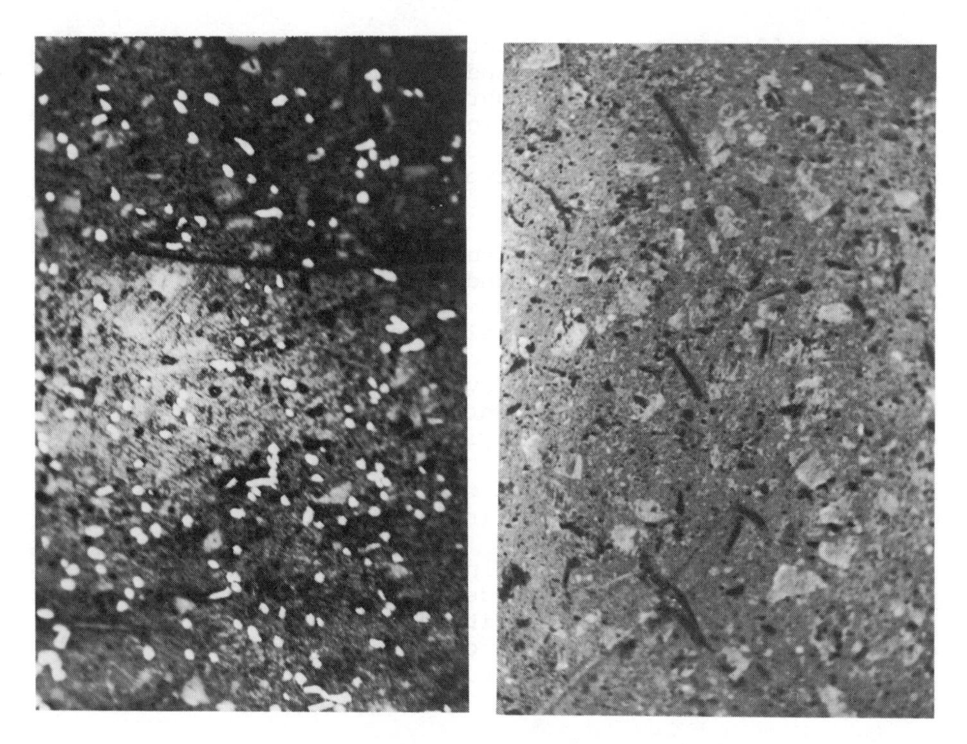

Fig.2. Fibre orientation in shotcrete: parallel to the spraying direction fibres appear as dark dots (a). Perpendicular to the spraying direction less fibres appear and their length becomes more evident (b).

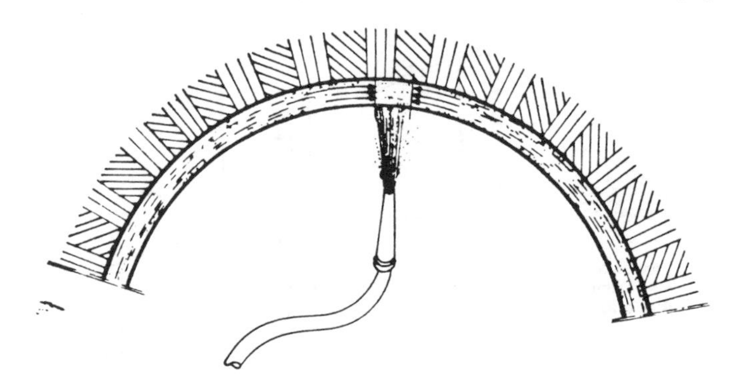

Fig.3. In a tunnel lining structure the main compression forces appear parallel to fibres preferential direction.

Fig.4. Cylinder core showing alternating layers of concrete ("layering effect").

3 TEST PROGRAM

The test program consisted basically of spraying plywood test pannels (50x50x16 cm) using dry-mix shotcrete with mix proportions commonly found in tunnelling sites (table 1). Two different types of steel-fibres available in the Brazilian market were added to this mix at three different percentages .

Table 1. Raw materials and proportions of mix

RAW MATERIALS

	CEMENT	SAND	AGG.	FIBRE		
Caract.	Ordinary portland Cement	2,4	max. 9,8 mm	(I) Straight (II) Curved		
Proportions (mass)	1	2,62	1,8	0,06 0,10	0,12 0,17	0,2 0,19

	Length (mm)	Diameter (mm)	Geometry
FIBRE TYPE (I)	24	0,35	
FIBRE TYPE (II)	24	0,40	

After spraying, the rebound material was collected from the ground and weighed after the spraying of each panel for an evaluation of a possible influence of the steel fibre content on the loss of concrete by rebound.

After three days of curing, test panels were subject to both the drilling of cylinder cores (7.5 x 15 cm) and sawing of 10 cm cubes (Figure 5). Speciemens were later capped and subject to compression test in parallel and perpendicular directions relative to fibres at the ages 10 and 11 days (respectively cylinders and cubes containing fibre type I) and 21 or 25 days (respectively cylinders and cubes containig fibre type II).

During shotcreting of the test panels the amount of water added to the mix was controlled in two ways in order to keep similar water:dry-materials proportion for all panels:

1) A manometer was installed at the water-pump and the water pressure was kept constant at 25 kgf/cm^2.

2) A 1kg sample of the shotcrete was taken from each of the test panels imediately after spraying for drying. Only those in which the water:dry-materials factor fell within the range of 10$\pm$0,5% were accepted.

After drying by heating of the previously mentioned sample of freshly sprayed shotcrete, the fibres present were collected with the aid of a magnet and weighed for an evaluation of the percentage of fibres that had been effectivelly incorporated to the shotcrete into the test panels.

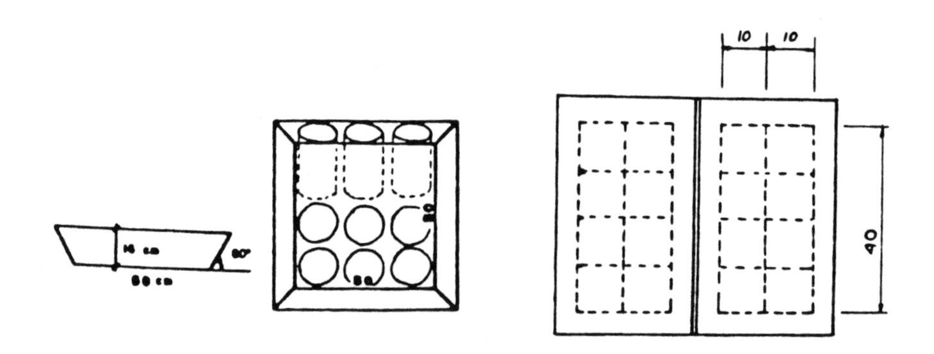

Fig.5. Position in which cylinder cores and cubes were taken from the test panels.

4 RESULTS

4.1 ACTUAL FIBRE CONTENT OF SHOTCRETE

Results for the content of steel-fibres in the shotcrete of the test panels are shown on figure 6, expressed as a function of the content of steel-fibres in the dry-mix. These results indicate, that the amount of fibres that actually become incorporated into shotcrete is a rather constant factor of the total ammount present in the dry-mix. This leads to believe that, due to differential rebound of fibres and concrete the true fibre content of the resulting shotcrete is only about 75% of that for which the dry-mix was conceived.

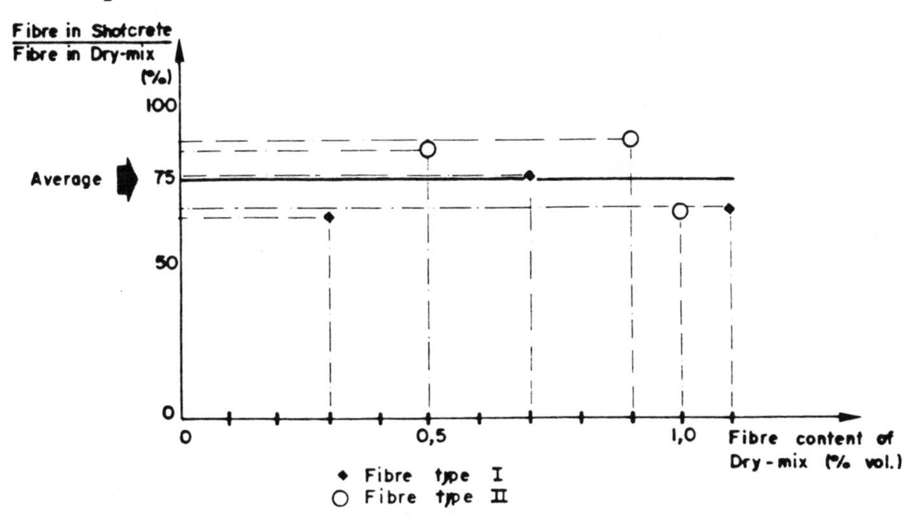

Fig.6. Actual fibre content of shotcrete: fibre content of shotcrete versus content of the dry-mix.

As for a possible influence of fibre geometry (straight or curved) on its ability to adhere to the shotcrete matrix during spraying, no appreciable difference is evident on figure 6, i.e. both fibres present similar results considering the erratic influence of material rebound on results of actual fibre content in shotcrete.

4.2 COMPRESSIVE STRENGTH

As for the influence of 2-D random distribution of fibres on compressive strength (tables 2 and 3) a statistical analysis of results shows no influence of fibre orientation when load was applied parallel or perpendicular relative to the direction of main fibre orientation. This result applies for cylinder and cube speciemens and for both fibres (types I and II).

An interpretation of these results can be made considering two aspects: First, the layering effect of shotcrete (described by Hills, 1980)(7) seems to be very dependent on the water content and efficiency of water mixing of the resulting mass, so that the dryer the sprayed concrete, the larger the tendency for inadequate compaction and therefore the appearence of the layering effect.

In this respect, the amount of water used for shotcreting the test panels for this study (water:dry-materials factor = 10 ± 0.5%) can be considered an usual value for tunnelling purposes, and seems to have fallen out of the "dry-range", subject to the layering effect.

Another aspect concerning the lack of evidence for anisotropic behaviour here is the rather low fibre content normally employed in shotcrete (usually around 1% by volume (1)(8)) which actually becomes even lower considering the values from figure 4: due to differential rebound of fibres, the actual content of dry-mix shotcrete is only about 75% of the dry-mix. Anisotropic behaviour due to fibre orientation in cast in place concrete is only likely to occur for higher fibre contents (above 1% - (9)(10)).

4.3 INFLUENCE OF FIBRES ON REBOUND

The possible influence of steel-fibres on rebound is of special interest to contractors who employ large amounts of shotcrete (eg. tunnelling) and for whom a possible increase in lost material by rebound may cause signifficant cost increases.

The results shown in Figure 7 do not indicate any clear influence of fibre content on rebound. A statistical analysis of these results would not be meaningfull since the influence of water content, air pressure, and size/position of test panels on rebound are not yet quantified by researchers working on shotcrete. However, as a general comment, one must take into consideration that fibres may tend to reduce rebound not for their physical presence, but by allowing the removal of traditional wire-mesh reinforcement (undoubtedly a significant barrier to spraying and therefore an influence on rebound).

4.4 VOIDS VOLUME

Permeable voids volume measurements were made since all Brazilian shotcrete specifications for dry-mix shotcrete limit this value to a maximum of 15% (usually, durability aspects are the main reason) and therefore any increase beyond that limit would make steel-fibre shotcrete improper for local applications.

The results obtained are expressed as the average of

Tables 2 and 3. Results of compression strength for cylinder cores and cubes (⊥ indicates stress normal to fibres main orientation and // indicates stress parallel).

FIBRE TYPE I

	Cylinder (AGE 10d)		Cube (AGE 11d)	
	Stress ⊥	Stress//	Stress ⊥	Stress//
Fibre Content 0,3% h = 9,7%	$\bar{x}$ = 40,1 n = 3 sd = 0,987	43,49 3 0,306	$\bar{x}$ = 43,49 n = 5 sd = 6,31	43,88 5 8,62
Fibre Content 0,7% h = 10,0%	$\bar{x}$ = 40,1 n = 3 sd = 1,0	39,7 3 1,08	$\bar{x}$ = 49,01 n = 7 sd = 1,30	45,46 5 4,35
Fibre Content 1,1% h = 10,0%	$\bar{x}$ = 38,7 n = 3 sd = 0,4	38,4 3 0,503	$\bar{x}$ = 47,21 n = 4 sd = 3,10	39,07 4 7,05

FIBRE TYPE II

	Cylinder (AGE 21d)		Cube (AGE 25d)	
	Stress ⊥	Stress//	Stress ⊥	Stress//
Fibre Content 0,5% h = 7,6%	$\bar{x}$ = 29,73 n = 3 sd = 5,62	32,57 3 9,46	$\bar{x}$ = 38,83 n = 4 sd = 5,65	50,19 4 4,31
Fibre Content 0,9% h = 9,4%	$\bar{x}$ = 52,60 n = 3 sd = 0,83	51,68 3 2,93	$\bar{x}$ = 56,87 n = 8 sd = 5,64	55,87 8 3,70
Fibre Content 1,0% h = 10,4%	$\bar{x}$ = 43,13 n = 3 sd = 0,31	41,80 3 1,31	$\bar{x}$ = 43,94 n = 4 sd = 4,64	41,77 4 8,4

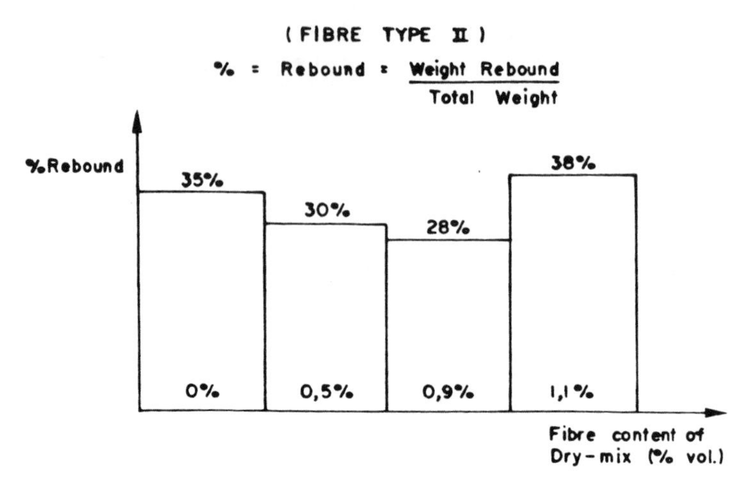

Fig.7. Rebound versus fibre content of the dry-mix (fibre type I only).

three 800g speciemens (Figure 8) and indicate no clear influence of either fibre geometry or content. Considerations on wether voids volume plays, in fact, a major role on durability would not be the objective of this work, but the main aspect of these results is that all values ranged close to the specified 15% maximum value, indicating that it is possible to produce steel-fibre reinforced dry-mix shotcrete that meets local standards.

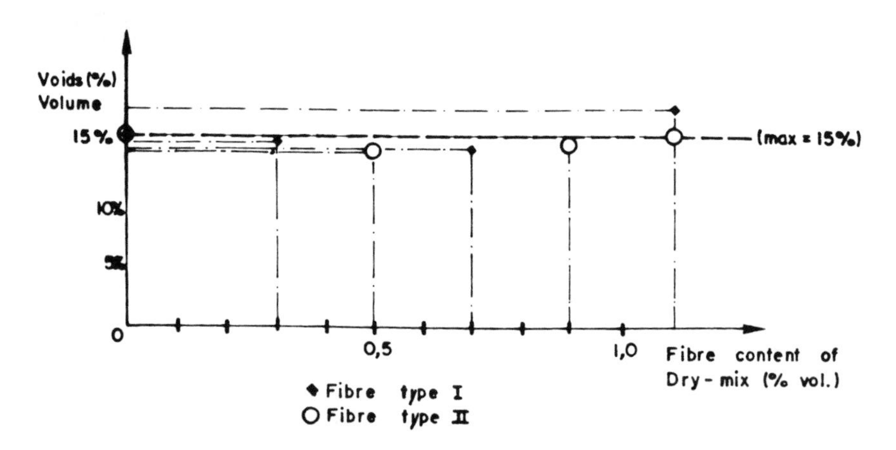

Fig.8. Voids volume (ASTM C 642) of hardened SFRS versus fibre content of the dry-mix.

5 FINAL CONSIDERATIONS

Steel-fibre shotcrete has a well-known, natural tendency, for fibre 2-D random orientation. That 2-D distribution of fibres could be a potentially dangerous aspect for dry-mix shotcrete (mainly for tunnelling purposes) considering its possible layering effect (described by Hills, 1980).

In contrast however, the results of this work do not indicate any tendency for drop of compressive strength parallel to fibres' main orientation. A possible explanation to these results is the relatively high water:dry-materials factor employed and the rather low actual fibre content of shotcrete (only about 75% of the original dry-mix).

As for a possible influence of steel-fibre content or geometry on rebound and voids volume, no clear tendency became evident, although the main aspect is that neither rebound nor voids volume were enhanced by the increase of fibre content.

REFERENCES

1. MORGAN, D.R. AND MOWAT, D.N. (1982) **A Comparative evaluation of Plain, Mesh and Fibre-reinforced Shotcrete.** ACI Int. Symp. on Fibre Reinforced Concrete, ACI SP 81, pp. 307-324.

2. OPSHAL, O.A. (1982) **Steel-fibre reinforced concrete for rock support.** Royal Norwegian Council for Scientiffic and Industrial Research (NTNF) pp. 121-139.

3. KOBAYASHI, K. (1983) **Development of fibre reinforced concrete in Japan.** Int. Journal of Cement and Composites, V.5, pp. 27-40.

4. MAIDL, B. (1984) **Handbuch des Tunnel un Stollenbau.** Konstrutionen und Verfahren, Verlag G1.

5. MORGAN, D.R. (1990) **Shotcrete support of underground openings in Canada.** Tunnelling Association of Canada, 8th annual meeting, Vancouver, B.C.

6. JAPAN TUNNELLING ASSOCIATION (1980) **Guide to the design and construction of steel-fibre concrete linnings**, Tokyo.

7. HILLS, D.L. (1980) **A review of sprayed concrete**, part 3. Concrete, pp. 32-34.

8. AUSTRIAN CONCRETE SOCIETY (1990) **Guideline on Shotcrete.** Vien, pp. 33.

9. EDGINTON, J. AND HANNANT, D. J. (1972) **Steel fibre reinforced concrete: The effect on fibre orientation of compaction by vibration**, Materiaux et Constructions, RILEM, Vol.5, No. 25, pp.41-44.

10. SWAMY, R.N. AND STAVRIDES, H. (1976) **Influence of the method of fabrication of strength properties of steel fibre concrete**, Materiaux et Constructions, RILEM, Vol. 9, No. 52, pp. 243-253.

11. ASTM C 642 - **Standard test method for specific gravity, absorbtion and voids volume in hardened concrete.**

AKNOWLEDGEMENTS

The author would like to aknowledge the financial support provided by the Companhia Brasileira de Projetos e Obras (CBPO) and the Scientiffic Council of the State of Såo Paulo (FAPESP). Also, the work and support of all researchers and technicians from the Instituto de Pesquisas Tecnologicas (IPT), where all the laboratory work was carried out, is very much appreciated.

24 STEEL FIBRE REINFORCED DRY-MIX SHOTCRETE: EFFECT OF FIBRE GEOMETRY ON FIBRE REBOUND AND MECHANICAL PROPERTIES

N. BANTHIA* and J. -F. TROTTIER
Dpt of Civil Engineering, Laval University, Ste-Foy, Quebec
D. WOOD
University of Toronto, Ontario, Canada
D. BEAUPRE
Department of Civil Engineering, Laval University, Ste-Foy, Quebec, Canada

Abstract
The paper will describe an experimental program in which five commercially available steel fibre geometries were investigated in dry-mix shotcrete. The fibres varied in geometry from end hooked to crimped, and their lengths varied between 25 and 30 mm. The intended fiber content was 60 kg/m^3. Twelve panels - two without fibres and ten with fibres--were shot. During shooting, appropriate tests were conducted to determine the fibre percent rebound. Once the panels had hardened, beams were sawn and cylinders were cored. While the cylinders were tested in compression, beams were tested in 4-point bending to obtain precise load-deflection curves. These curves were analyzed to obtain suitable toughness characterization parameters. In particular, the ASTM method of calculating "toughness indices" at multiples of first crack deflection and the Japanese Concrete Institute technique of obtaining "flexural toughness factors" from the energy absorbed to a certain midspan deflection, were used. It was noted that the rebound of steel fibres in shotcrete is, on a percent basis, far greater than the other ingredients and strongly dependent upon fibre geometry. Further, fibre percent rebound was found to be closely related to a fibre geometrical parameter called "*specific projected area*". In the hardened shotcrete, fibre geometry was found to have a decisive influence on flexural toughness. Both compressive and flexural strengths were found be marginally improved due to fibre presence.
Keywords:shotcrete, steel fibers, fiber geometry, rebound, toughness.

1 Introduction

The interest in fibre reinforced cement-based materials has grown steadily in the past few decades. In the case of conventionally placed (cast) concrete reinforced with fibres, significant research activity has taken place all over the world aimed at characterizing, optimizing and modeling the behaviour. Improvements in the

*Now at the University of British Columbia, Vancouver, B.C., V6T 1Z4, Canada

Fibre Reinforced Cement and Concrete. Edited by R. N. Swamy. © 1992 RILEM.
Published by E & FN Spon, 2-6 Boundary Row, London SE1 8HN. ISBN 0 419 18130 X.

physical and mechanical properties due to fibre reinforcement are well documented (1,2).

In the case of shotcrete, fibre reinforcement is believed to provide distinct placement advantages over plain or wire mesh reinforced shotcrete (3). The placement advantages coupled with improved material mechanical behaviour have led to a widespread use of fibre reinforced shotcrete in almost all the industrialized nations of the world (4-8). After its first experimental placement in 1971, the first practical application of steel fibre reinforced shotcrete (SFRS) was in 1973 when the U.S. Core of Engineers used it in a tunnel audit at Ririe Dam in Idaho (5). Since then fibre reinforced shotcrete has been successfully used in tunnel linings, slope stabilization works, culverts, sewers, dams, canals, swimming pools and reservoirs. Other major use of fibrous shotcrete is in repair and rehabilitation of marine, highway, railway and other structures. In spite of its widespread use, only a limited research activity has so far taken place to develop a fundamental understanding of fibre reinforced shotcrete. With its distinct production technique, the internal structure of shotcrete may be expected to be quite different form cast concrete, and as such, a great deal of existing information gathered from laboratory testing of fibre reinforced cast concrete may not be applied to fibre reinforced shotcrete. The major causes of limited research activity in the field of fibre reinforced shotcrete are the specialized nature of the equipment needed and the extensive preparations that usually precede a shooting.

Shotcrete may be divided into two broad categories: dry-mix and wet-mix. In the dry-mix process, the ingredients (cement, aggregates and admixtures) are pumped through a conveyer tube in the dry-state and water is added at the nozzle just before shooting on a formerly prepared surface. In the wet-mix process, on the other hand, all ingredients including water are premixed before shooting. Although, the use of the wet-mix process has its advantages, dry-mix process is more widely used.

From the materials behaviour point of view, fibres in shotcrete provide "toughness" or energy absorption capability leading to superior static, impact and fatigue resistance. Under flexure, while plain shotcrete fails essentially in a brittle manner at the occurrence of peak load, fibre reinforced shotcrete continues to support load well beyond matrix cracking up to large deflections. After matrix cracking, fibres bridge across cracks and undergo pull-out processes. Even at large displacements, therefore, the material maintains some residual strength and continues to absorb energy with increasing deformations. The extent to which steel fibre reinforced shotcrete can absorb energy beyond matrix cracking depends upon the pull-out resistance offered by the individual fibres.

Given the fact that straight, smooth fibres offer only a small resistance to their pull-out across a matrix crack (9), almost all steel fibres manufactured for use in shotcrete have deformed geometries. Deformed fibres develop a better anchorage bond with the surrounding matrix and offer a greater resistance to their pull-out. Deformed fibres, as a result, provide superior post-peak energy absorption capability or "toughness" than the straight, smooth fibres. However, even here there is a limit. If excessively deformed, the fibres develop too good an anchorage with the surrounding cementitious matrix and may fracture in the process of pull-out. This would provoke a brittle material response. Unfortunately, our current understanding of the fibre-matrix bond-slip characteristics for deformed fibres is insufficient, and it is not possible to determine what constitutes an optimized fibre geometry for a maximized composite toughness.

One of the major concerns, particularly with regards to dry-mix shotcrete, is in the extent of fibre rebound that can occur during the process of shooting. Experience and perception have both led to the belief that the geometry of the fibre is an important variable governing its rebound during shooting. Clearly, greater the fibre rebound, the lower the fibre volume fraction in the placed shotcrete and hence lower the observed improvements in the mechanical performance. In some earlier studies, deformed fibres were compared with straight, undeformed fibres (10), or various volume fraction of a deformed fibre were investigated (11). In a more recent study, Morgan et al (12) compared the physical, mechanical and rebound characteristics of two deformed fibre geometries. In the study reported here, a comprehensive investigation was undertaken to clearly understand the influence of fibre geometry on the rebound characteristics and hardened shotcrete properties.

2 Test Program

2.1 Preparation of Specimens

Five commercially available fibres were investigated. Their geometries, shown in Figure 1, varied from end-hooked to crimped to twisted. Fibre dimensions in Figure 1 were obtained by measurements on several fibres followed by averaging.

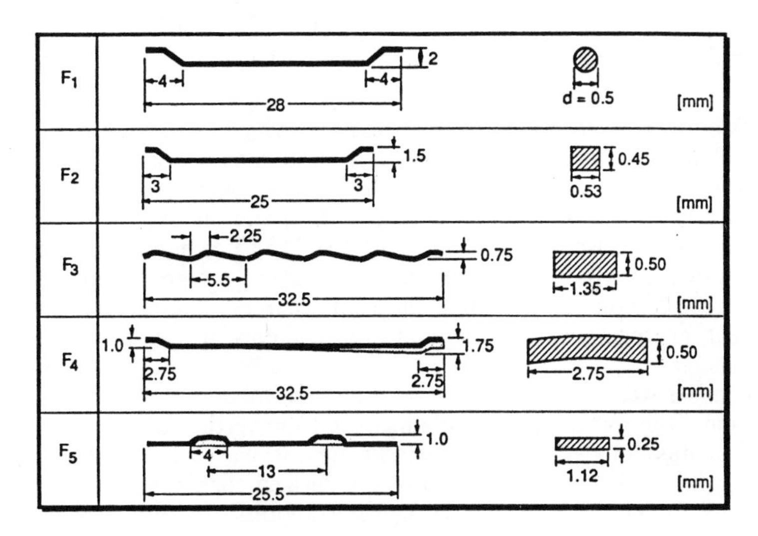

Fig. 1. Fibres Investigated

Individual fibre mass, volume, surface area, projected area and the calculated specific projected area (defined as the projected area per unit mass) are shown in Table 1. Average individual fibre mass and volume were determined by weighing a batch of fibres in air and then in water. Fibre projected area, defined here as the projected area on a plane parallel to the fibre plane, and fibre surface area were calculated from averaged dimensions.

Table 1. Characteristics of Fibres Investigated

Fiber	Fiber mass (m_f) mg	Fiber volume (v_f) mm^3	Fiber Total Surface Area (A_t) mm^2	Fiber Projected Area $(\bar{A}_p)$ mm^2	Specific Projected Area $(\bar{A}_{sp})$ $\bar{A}_{sp} = \frac{\bar{A}_p}{m_f}$ mm^2/g
F1	45.5	5.79	43.9	14.0	307
F2	59.1	7.52	49.0	13.2	224
F3	76.4	9.73	127.6	43.8	574
F4	181.8	23.16	211.2	44.6	245
F5	49.1	6.25	69.8	28.6	581

Shotcrete mixes were dry weight-batched and bagged in bulk-bin bags by a premix batching company. Six different bags of shotcrete premix - one control and five with fibres - each weighing about 1000 kg were transported to the shooting site. The compositions of the control (M0) and the five fibre reinforced mixes (MF1-MF5) are shown in Table 2. ASTM Type I cement, and silica fume obtained locally, were used. The fine and coarse aggregate gradation curves are shown in Figure 2. Note that within batch-plant accuracy, all mixes had about the same cement, aggregate, silica fume and fibre contents; it was only the type of fibre that changed.

The six mixes were shot on previously prepared wooden forms 800 mm x 800 mm x 200 mm. The forms were slightly inclined from vertical as shown in Figure 3. A N1-Modified model of Allentown Pneumatic was used as the shooting equipment. The equipment possesses two pressurized chambers, one for mixing and the other for pre-moisturizing. A Jaeger compressor with a capacity of 17 cubic meter per minute (600 cubic feet per minute) was used. The conveyer tube had an internal diameter of 38 mm with a Putzmeiser Nozzle at the end.

Table 2. Mix Compositions

Mix	Cement kg/m^3	Silica fume kg/m^3	Fine Aggregate kg/m^3	Coarse Aggregate kg/m^3	Steel Fibers kg/m^3	*Water Demand kg/m^3
M0	408.1	55.8	1278.3	435.0	-	142.0
MF1	417.1	55.8	1281.3	441.1	60.0	142.0
MF2	423.1	55.8	1284.3	441.1	60.0	142.0
MF3	429.1	55.8	1269.3	435.1	60.0	142.0
MF4	417.1	55.8	1278.3	438.1	60.0	142.0
MF5	414.1	55.8	1254.3	444.1	60.0	142.0

* Estimated

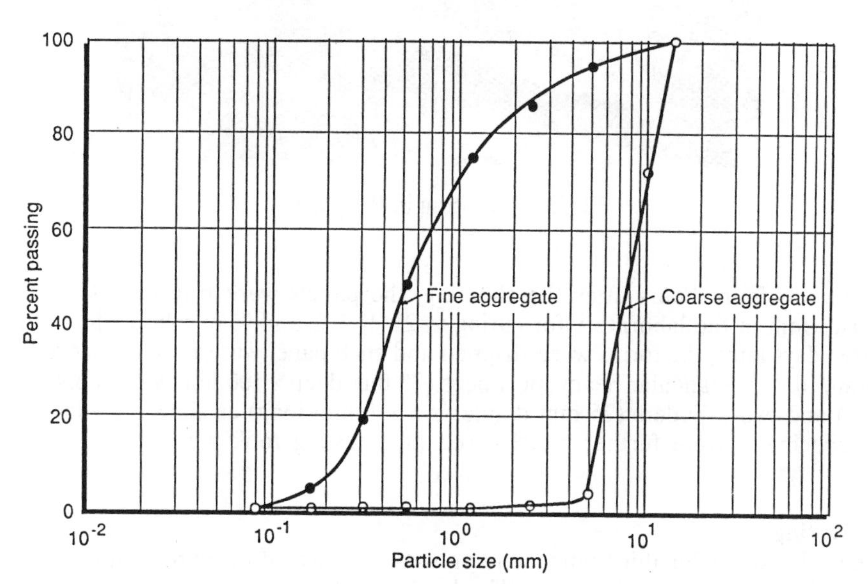

Fig. 2. Fine and Coarse Aggregate Gradation Curves

For the fibrous mixes, a 20 kg sample of in-place shotcrete was removed and "*wash-out*" tests were performed. The fibres were magnetically separated and weighed to determine the effective in-place fibre volume fraction. For the fibre rebound calculations, these in-place fibre volume fractions were subtracted from the original fibre volume fraction. After shooting, the forms were covered with a plastic sheet.

Fig. 3. Shooting in Progress

Twenty-four hours after the shooting, the panels were transferred to the moist room in the laboratory for curing at 23 $\pm$ 2°C and 95% relative humidity. Three days later, the forms were stripped and each panel was sawed and cored to obtain three rectangular beam specimens (75 mm deep x 100 mm wide x 350 mm long) and two cylinders (75 mm diameter x 150 mm long) as shown in Figure 4. All specimens were further moist cured until testing took place at the age of seven days.

2.2 Testing

Flexural tests under third point loading were performed on hardened shotcrete specimens as shown in Figure 5. The loading details are shown in Figure 6. A frame surrounding the specimen, similar to the one suggested in Japanese Concrete Institute Standard - called the "yoke" - was used to eliminate the extraneous deflections arising from support settlements. The other notable feature of the frame used was the joint flexibility that prevented imposing frame rigidity on the recorded deflections. Two linear variable differential transducers (LVDTs), one with a total travel of 1 mm (0.04 in) and the other with a total travel of 25 mm (1 in) were used. The 1 mm LVDT with its superior precision was used to record the deflections up to the occurrence of the peak load and the 25 mm LVDT was used for recording the large deflections after the peak load. Both the LVDTs were mounted on the same side of the beam and some minor twisting

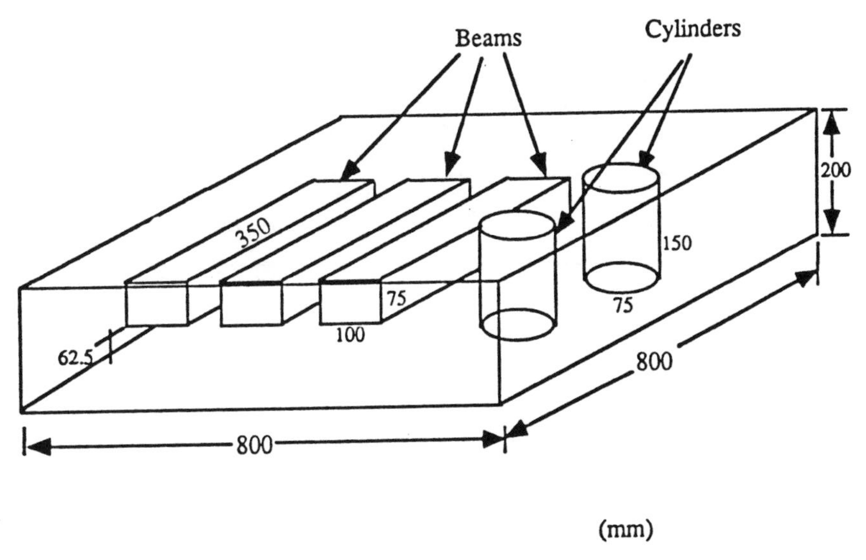

Fig. 4. Sawing and Coring Scheme

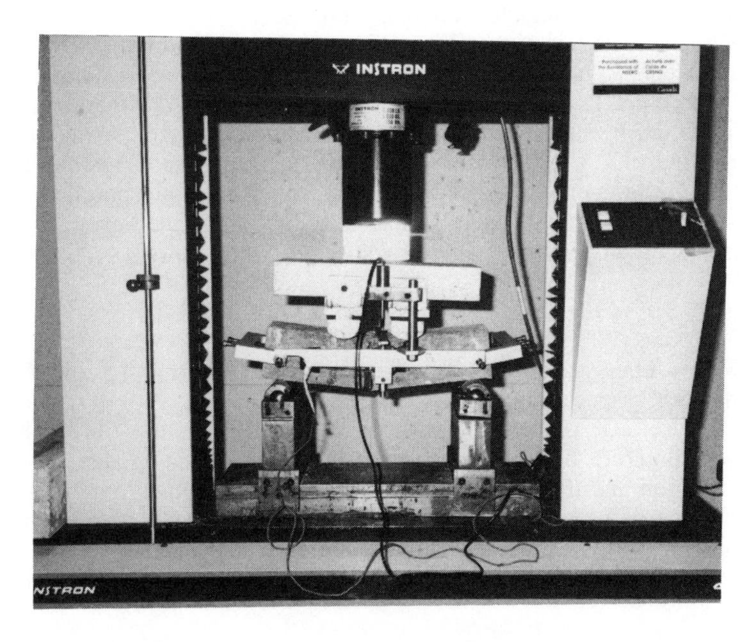

Fig. 5. Flexural Toughness Test Setup. A "yoke" surrounding the specimen helps eliminate spurious system deformations.

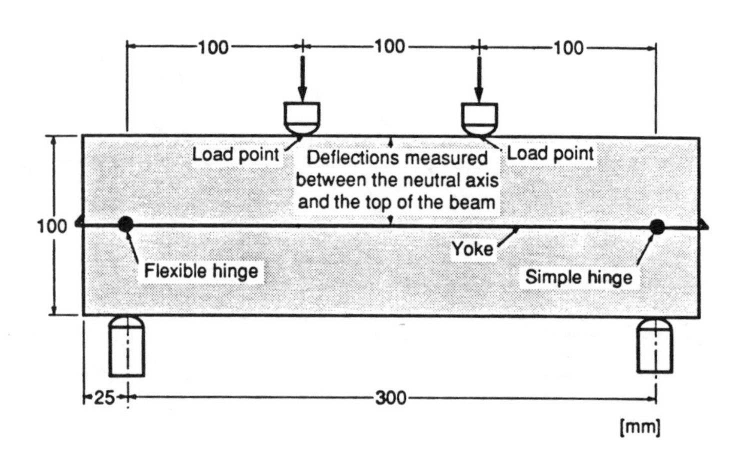

Fig. 6. Flexural Loading Details

twisting that may have occurred during testing was ignored. The applied load and the deflection data from the two LVDTs were continuously acquired by a data acquisition system based on a PC at a frequency of 30 Hz and the ensuing load-deflection plots were analyzed to characterize the material toughness. In particular, the American Society for Testing and Materials (ASTM) technique of determining the toughness indices and the Japanese Concrete Institute technique of determining flexural toughness factors, were used.

As per the ASTM technique (Figure 7), the energy absorbed by the flexural specimen to a certain multiple of first crack displacement is divided by the energy absorbed to first crack. The resulting quotients are termed "*toughness indices*" and they are qualified by subscripts 5, 10, 20, etc., using an elasto-plastic analogy as seen from Figure 7. On the other hand, in the Japanese Concrete Institute Standard SF-4, the energy absorbed by the specimen to a midspan deflection of $l/150$ (l = unsupported beam span) is normalized to obtain the flexural toughness factor (also called the 'equivalent flexural strength') as follows:

$$\text{Flexural Toughness Factor} = El/\delta\, bd^2 \qquad (1)$$

where, E is the energy absorbed, δ is the deflection ($l/150$) and b and d are the width and depth, respectively, of the specimen.

From Figure 7 it is clear that while characterizing steel fibre reinforced concrete or shotcrete by toughness indices, the test outcome depends largely upon the location and the way in which the beam deflections are measured. The deflections recorded by the machine cross-arm or an LVDT placed on top or bottom of the beam are subject to considerable experimental error due to support settlements and specimen rocking, especially for the small deflections

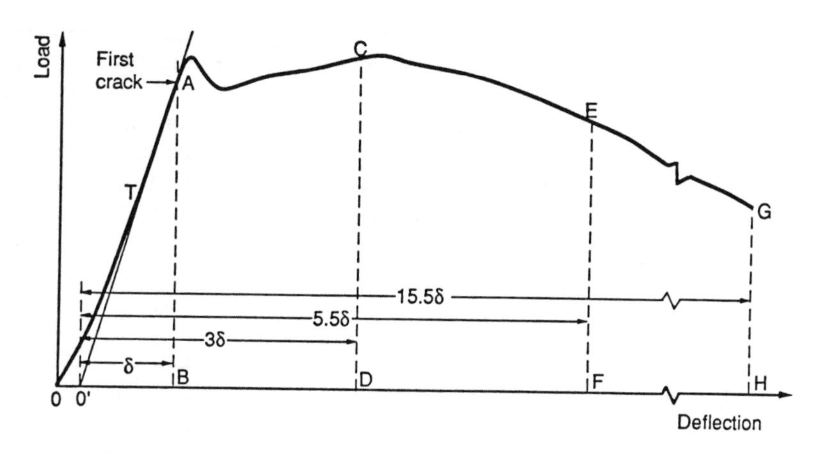

$$I_5 = \frac{\text{Area under the curve to } 3\delta}{\text{Area under the curve to } \delta}$$

$$I_{10} = \frac{\text{Area under the curve to } 5.5\delta}{\text{Area under the curve to } \delta}$$

$$I_{30} = \frac{\text{Area under the curve to } 15.5\delta}{\text{Area under the curve to } \delta}$$

For a perfectly elasto-plastic material like steel

$I_5 = 5.00$

$I_{10} = 10.00$

$I_{30} = 30.00$

Fig. 7. Calculation of Toughness Indices Using ASTM C1018-90.

that occur at the first crack load. Using elastic beam theory, the deflection of the beam, δ, under an applied load, P, is given by:

$$\delta = \frac{23P\ell^3}{1296EI}\left[1 + \frac{216\, d^2\,(1+\mu)}{115\,\ell^2}\right] \tag{2}$$

where, EI is the beam flexural rigidity, l is the beam span, μ is the Poisson's ratio and d is the depth of the beam section. Proper substitution indicates that the true beam deflections at the occurrence of first crack are substantially less than a millimeter. The true deflections being the deflections of the beam neutral axis, they may be more accurately obtained by installing a flexible "yoke" around the specimen (Figure 5).

Figure 8 shows three load vs deflection plots obtained on the same fibre reinforced specimen but with deflections measured differently (8). The three plots correspond to deflections measured by the "yoke", by an LVDT mounted directly on the top of the beam, and thirdly, by monitoring the cross-arm displacement. Figure 8 also shows the deflections at first crack, theoretical and measured. Notice that it was only with the "yoke" that the true first crack deflection could be measured and the deflections obtained from cross-arm and

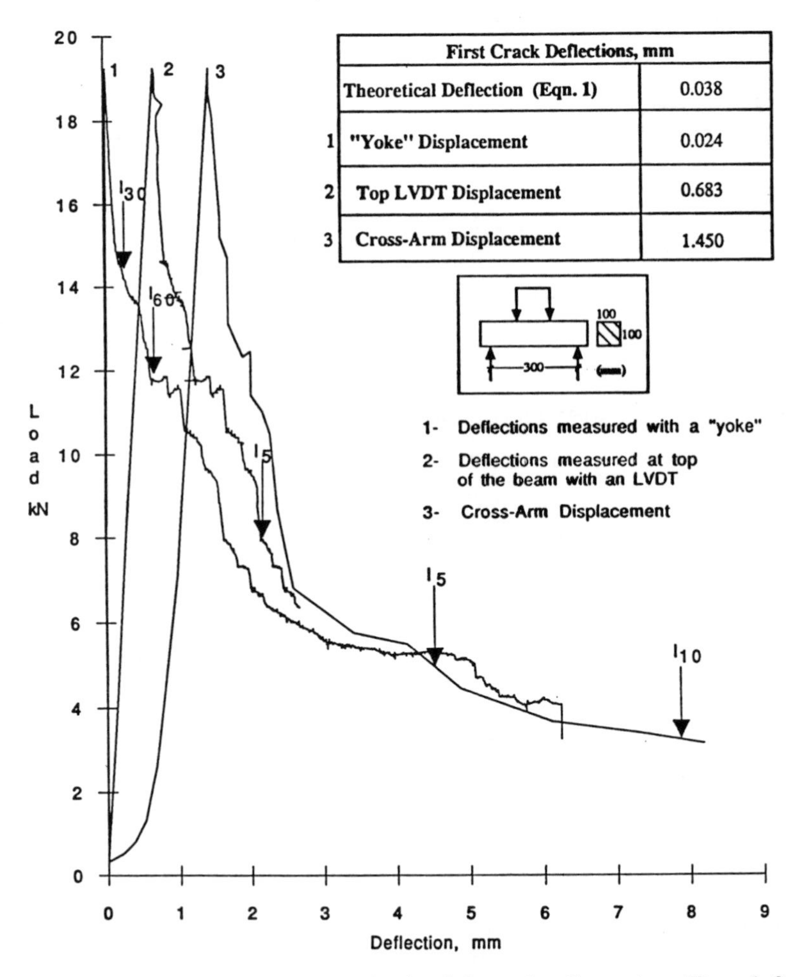

	First Crack Deflections, mm	
	Theoretical Deflection (Eqn. 1)	0.038
1	"Yoke" Displacement	0.024
2	Top LVDT Displacement	0.683
3	Cross-Arm Displacement	1.450

Fig. 8. Beam Displacements as Obtained from the Cross-Arm Travel, from a LVDT Placed on Beam-Top and from a "Yoke" Installed Around the Specimen.

LVDT on top were respectively, 38 and 18 times greater than the theoretical deflection. Needless to state, the toughness indices obtained using cross-arm or beam-top deflections will also be entirely different from the true values and hence misleading.

Also on the seventh day, cylinders (Figure 4) were sulfur-capped and tested for compressive strength.

3 Results

3.1 Fibre Rebound

In-place shotcrete samples in the plastic state were analyzed to obtain the in-place fibre volume fractions. These are shown in Table 3 along with the original intended fibre volume fraction and the calculated fibre rebound volume fractions. Notice that for all fibre types, the fibre rebound percentages are high in general, and significantly higher than the usually reported total rebound values for all

Table 3. Rebound Characteristics

Mix (1)	Original Fiber Volume Fraction (percent) (2)	Fiber Volume Fraction in In-Place Shotcrete (percent) (3)	Rebound Fiber Volume Fraction (2)-(3) (percent) (4)	Percent Rebound $\frac{(2)-(3)}{(2)} \times 100\%$ (5)
MF1	0.764	0.389	0.375	49.1
MF2	0.764	0.498	0.266	34.8
MF3	0.764	0.237	0.527	69.0
MF4	0.764	0.396	0.368	48.2
MF5	0.764	0.170	0.594	77.8

ingredients. The tendency for the fibres to rebound more than the other ingredients in dry-mix process has also been noted by Morgan et al (12) who reported up to 50% rebound. In the wet-mix process, however, significantly reduced fibre rebound may be expected.

While the importance of fibre geometry in governing the extent of fibre rebound has generally been recognized, no attempts have so far been made to relate a fibre "geometrical" parameter to the observed rebound. Hypothetically, fibre rebound, among other things including operator experience, could be related directly to the accelerations that a fibre experiences during the process of shooting. Under a given air pressure, fibre accelerations should be proportional to the projected surface area of the fibre and inversely proportional to the mass of the fibre. In turn, the quantity fibre projected area/fibre mass, termed here the specific projected area, $\bar{A}_{sp}$, may be related to fibre rebound. Figure 9 shows a plot of percent rebound (Table 3) as a function of the "specific projected area" for the various fibres (Table 1). Notice that the rebound is directly proportional to the specific projected fibre area. Although, the hypothesis is capable of explaining the higher fibre rebound of some fibres than others, much further

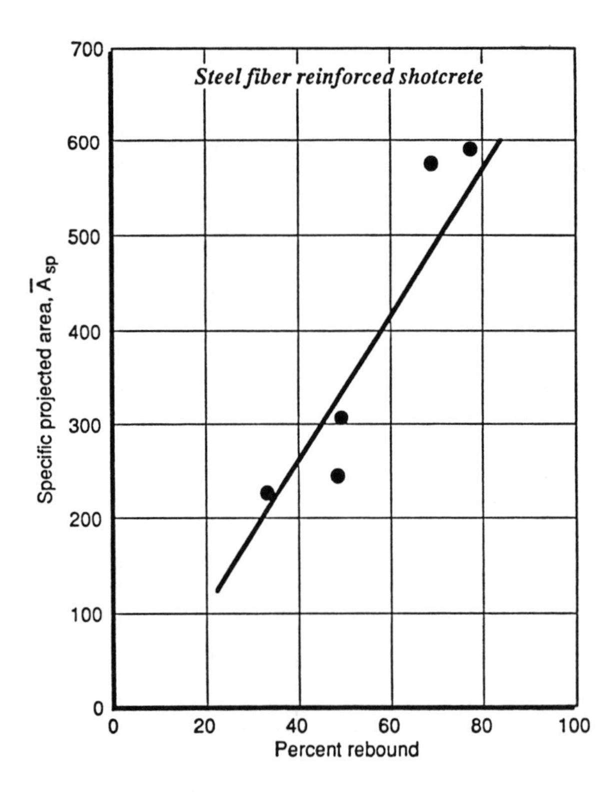

Fig. 9. Specific Projected Fibre Area as a Function of Fibre Rebound

investigation in this regard is needed. In this context, the use of high speed photography may be of much assistance.

3.2 Strengths

Table 4 shows the compressive and flexural strengths for the control and various fibre mixes from pairs of panels. The flexural strengths were obtained from the peak loads recorded in the flexural tests using elastic analysis. An acceptable inter-panel variability may be noted. Results indicate a 12 to 20% improvement in the compressive strength and a 13 to 24% improvement in the flexural strength in the fibrous mixes compared to the plain control mix. These strength improvements, however, do not appear to be related in any way to the volume fraction of the fibres left in the mix after the rebound. At the same time the geometry of the fibre does not appear to have any influence on compressive or flexural strength improvements.

Table 4. Compressive and Flexural Strengths

MIX	Compressive Strength		Flexural Strength (MOR)	
	Panel A (MPa)	Panel B (MPa)	Panel A (MPa)	Panel B (MPa)
MO	37.8 (1.73)*	31.6 (0.75)	4.61 (0.25)	4.03 (0.29)
MF1	37.6 (0.51)	36.9 (1.66)	5.18 (0.15)	4.66 (0.14)
MF2	34.2 (0.27)	40.3 (0.13)	4.92 (0.12)	5.03 (0.17)
MF3	37.1 (0.44)	42.6 (0.03)	5.50 (0.64)	5.28 (0.59)
MF4	40.3 (0.78)	39.5 (1.37)	4.58 (0.42)	5.22 (0.37)
MF5	38.9 (0.20)	37.6 (0.76)	5.20 (0.33)	4.51 (0.28)

* Numbers in parenthesis are standard deviations.

While the exact reasons for an improvement in the compressive and flexural strength due to fibres in shotcrete are not known, it may be due partly to the reduced water demand at the nozzle due to fibres. In retrospective, this could have been confirmed by an accurate assessment of the water:cement ratio in hardened shotcrete. Exact air content as well as the density of hardened shotcrete are essential to such a calculation.

3.3 Flexural Toughness

Representative load-deflection plots for the plain (M0), and the various fibrous shotcretes (MF1-MF5) are shown in Figure 10. Notice the unstable failure in plain shotcrete as opposed to a relatively stable failure in fibrous shotcrete. The flexural load-displacement curves were further analyzed as described before for standardized toughness parameters. The peak loads, the computed toughness indices and the flexural toughness factors are shown in Table 5. Also indicated in Table 5 are the average number of fibres crossing the section.

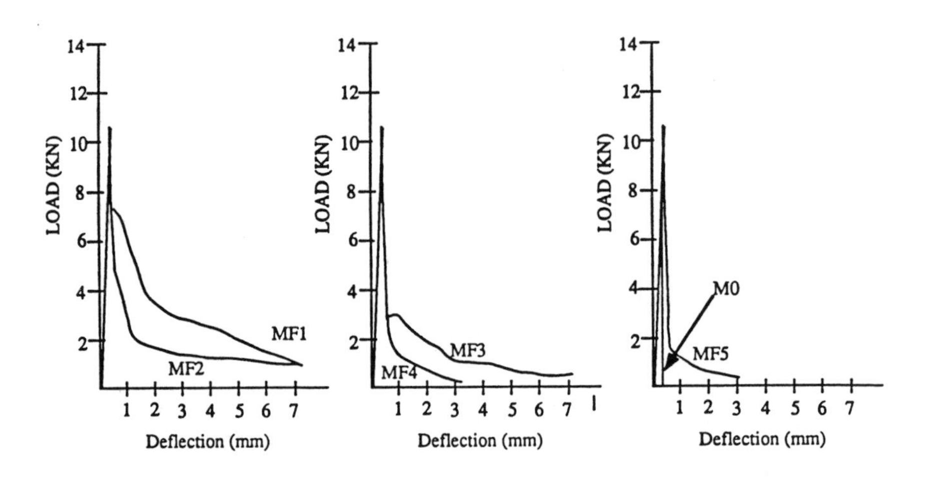

Fig. 10. Some Typical Load vs Deflection Plots

For the lower Indices, I_5 and I_{10}, the within batch variations were lower than for the higher Indices, I_{20} and I_{30}. Between the ASTM and the JCI analysis, the JCI method appears to yield a somewhat lower variability. The various toughness indices and the flexural toughness factors are shown as bar-charts in Figure 11. The toughness indices are plotted as a function of fibre volume fraction in Figure 12. Notice that both the lower Indices, I_5 and I_{10}, are not sensitive either to the fibre geometry or to the fibre volume fraction; it is only for the higher Indices, I_{20} and I_{30}, that the differences between fibres emerge. Fibre F1, even at a lower volume fraction, gave higher I_{20} and I_{30} Toughness Indices than Fibres F4 and F2. It is interesting to note that many current shotcrete specifications specify only I_5 and I_{10} values, which are not sensitive either to fibre geometry or to the fibre volume fraction when the deflections are measured off a "yoke" as used in this study. It has recently been shown (13) that the "yoke" alone is capable of measuring accurate displacements and the displacements at the peak load measured with the "yoke" match well with the theoretical values (Figure 8). While

Table 5. Flexural Toughness Results

Mix	Peak Load (kN)	First crack deflection (mm)	First crack energy (N·m)	Toughness Indices (ASTM)				Flexural Toughness Factor (JCI) (MPa)	Average Number of Fibers Crossing the Section.
				I_5	I_{10}	I_{20}	I_{30}		
M0	9.33 (1.18)*	0.024	0.22	1.0	1.0	1.0	1.0	0.05 (0.01)	---
MF1	10.50 (0.60)	0.030	0.32	4.28 (0.30)	7.52 (0.64)	13.26 (1.35)	18.34 (2.14)	2.070 (0.27)	67 (15)
MF2	10.49 (0.35)	0.032	0.38	4.37 (0.23)	7.38 (0.67)	11.69 (1.74)	14.50 (2.55)	1.420 (0.19)	36 (6)
MF3	11.21 (1.37)	0.031	0.35	4.51 (0.17)	7.55 (0.68)	11.26 (1.58)	13.76 (2.18)	1.294 (0.11)	26 (7)
MF4	10.84 (1.48)	0.029	0.25	4.46 (0.13)	7.39 (0.66)	10.72 (2.18)	12.87 (3.37)	0.895 (0.24)	11 (4)
MF5	10.22 (0.99)	0.033	0.34	4.28 (0.17)	6.50 (0.66)	8.62 (1.10)	9.90 (1.43)	0.807 (0.12)	25 (8)

* Numbers in parenthesis are standard deviations.

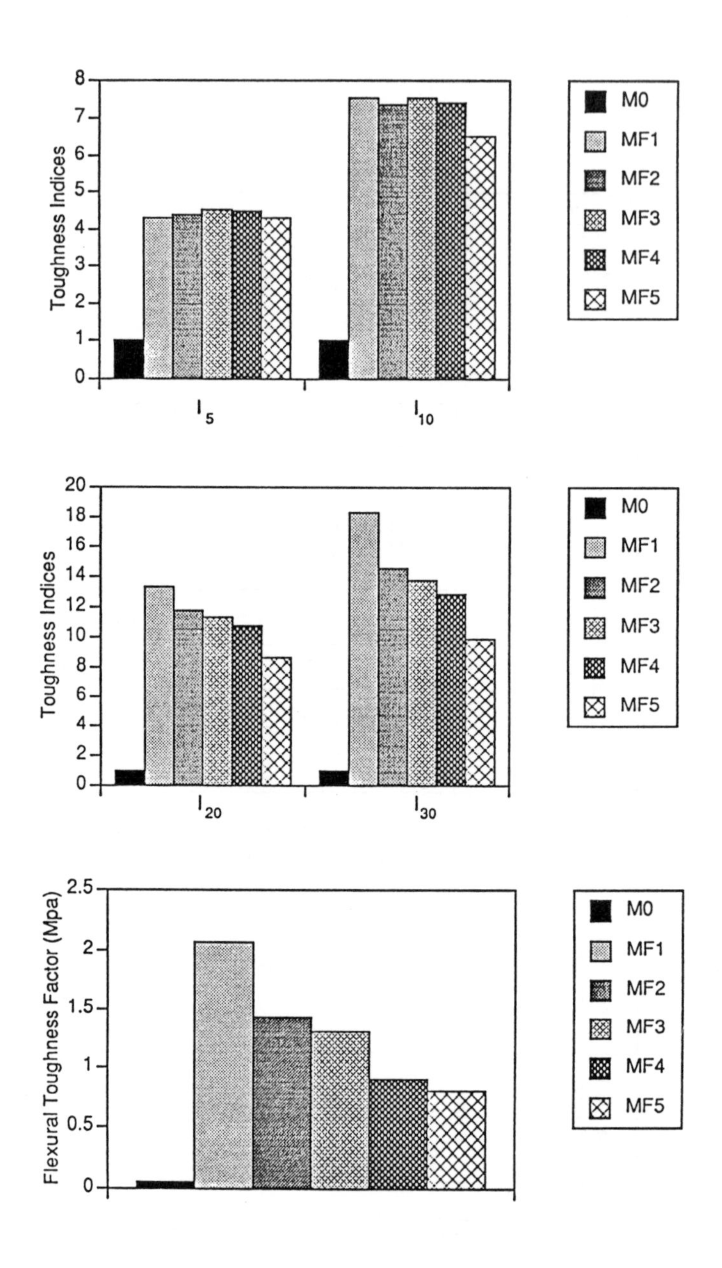

Fig. 11. Toughness Indices and Flexural Toughness Factors.

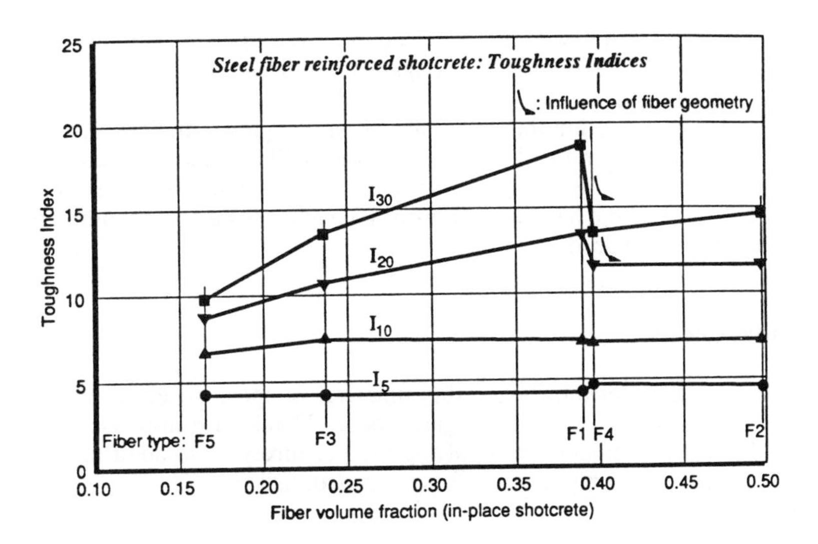

Fig. 12. Toughness Indices and Flexural Toughness Factors Plotted as a Function of Fibre Volume Fraction

characterizing fibre reinforced concrete or shotcrete using toughness indices, obtaining accurate displacement at the occurrence of peak load is of crucial importance.

The flexural toughness factors, that depend on the total energy absorbed to a certain midspan deflection after matrix cracking, are not sensitive to the characteristic of the curve at first crack. Consequently, small errors in the measurement of specimen displacement at first crack may be tolerated. This is, in the Authors' opinion, a distinct advantage over the current ASTM method of characterization. Flexural toughness factors are, as a single parameter, quite successful in representing the toughening capabilities of the various fibres (Figure 11).

Fibre F1 with the hooked ends, in spite of its greater rebound than Fibres F4 and F2, was found to yield the highest flexural toughness factors. This may be related directly to the large number of F1 fibres crossing a section at fracture. The small individual fibre mass for Fibre F1 may explain, to some extent, the large number of fibres crossing a matrix fracture.

4 Conclusions

1. The extent of fibre rebound during shotcreting appears to be directly related to the *specific projected fibre area* defined here as the fibre projected area per

unit fibre mass. There was, however, no relationship between the *specific projected fibre area* and ASTM Toughness Indices or the JCI Flexural Toughness Factors.

2. For the test system used in this study, the lower Toughness Indices, I_5 and I_{10}, are not sensitive either to fibre geometry or to fibre volume fraction. The differences between the various fibres emerge in the higher Toughness Indices, I_{20} and I_{30}, and also in the Flexural Toughness Factors recommended by the Japanese Standard.

5 Acknowledgments

The Authors wish to acknowledge the financial support received from the producers of the various fibres investigated here. Thanks are also due to King Packaged Materials Company for supplying the per-mixes. Continued support of Natural Sciences and Engineering Research Council of Canada is greatly appreciated.

6 References

1. Swamy, R.N. and Barr, B. (Eds.), Fibre Reinforced Cements and Concretes: Recent Developments, Elsevier Applied Science, 1989.
2. Bentur, A. and Mindess, S., Fibre Reinforced Cementitious Composites, Elsevier Applied Science, 1990.
3. Morgan, D.R. and Mowat, D.N., *A Comparative Evaluation of Plain, Mesh, and Steel Fibre Reinforced Shotcrete*, American Concrete Institute Special Publication, SP-81, 1984, pp. 307-324.
4. Ramakrishnan, V., *Steel Fibre Reinforced Shotcrete: A State-of the-Art Report*, in Steel Fibre Concrete, Proc. US-Sweden Joint Seminar (Eds. S.P. Shah and A. Skarendahl), 1985, pp. 7-24.
5. Henager, Charles, H., *Steel Fibrous Shotcrete: A Summary of the State of the Art*, Concrete International: Design and Construction, January 1981, pp. 55-58.
6. Morgan, D.R., *Dry-Mix Silica Fume Shotcrete in Western Canada*, Concrete International- Design and Construction, January, 1988, pp. 24-32.
7. Wood, D. F., *Application of Fibre Reinforced Shotcrete in Tunneling*, Proc. of First Canadian University-Industry Workshop on Fibre Reinforced Concrete (Ed. N. Banthia), Quebec City, Canada, 1991, pp. 183-196.
8. Morgan, D.R., *Use of Steel Fibre Reinforced Shotcrete in Canada*, Proc. of First Canadian University-Industry Workshop on Fibre Reinforced Concrete (Ed. N. Banthia), Quebec City, Canada, 1991, pp. 164-182.
9. Banthia, N., *A Study of Some factors Affecting the Fibre-Matrix Bond in Steel Fibre Reinforced Concrete*, The Canadian Journal of Civil Engineering, 17(4), 1990, pp. 610-620.

10. Ramakrishnan, V., Coyle, W.V. and Dahl, Linda Fowler and Schrader Earnest K., *A Comparative Evaluation of Fibre Shotcretes*, Concrete International-Design and Construction, January 1981, pp. 59-69.
11. Morgan, D.R., *Steel Fibre Shotcrete - A laboratory Study*, Concrete International - Design and Construction, January 1981, pp. 70-74.
12. Morgan, D.R., McAskill, N., Neill, J. and Duke, N.F., *Evaluation of Silica Fume Shotcretes*, Proc. CANMET/ACI International Workshop on Condensed Silica Fume in Concrete, Montreal, 1987, 34 pp.
13. Banthia, N. and Trottier, J.-F. *Discussion of Research Paper: Fibre-Type Effects on the Performance of Steel Fibre Reinforced Concrete* (by Soroushian and Bayasi in ACI Materials Journal, March-April 1991) to appear.

PART FOUR

FRACTURE BEHAVIOUR

25 TOUGHNESS BEHAVIOUR OF FIBRE REINFORCED CONCRETE

WEI-LING LIN
Department of Civil Engineering, National Chiao Tung University, Hsin Chu, Taiwan, Republic of China

Abstract
Four kinds of fibre material (iron wire, steel fibre, sisal fibre & glass fibre) were selected. The fibre reinforced concrete (FRC) specimens were prepared by using four kinds of water cement ratio (W/C) and five kinds of fibre content (F/C). The toughness behavior were tested and analyzed by using 245 KN. dynamic material testing machine. It is found that the best toughness behavior of FRC specimens shown is steel fibre, then in order of iron wire, glass fibre and sisal fibre etc. The toughness and toughness index both increase with fibre content and FRC curing age. While the toughness is decreasing with W/C, but the relationship between the toughness index and W/C isn't clear. The microstructural features of these four kinds of fibre materials were examined using scanning electron microscope, and the results could be related to their distinct toughness behaviors.
Keywords:Thoughness behavior, Fibre reinforced concrete, Iron wire, Steel fibre, Sisal fibre, Glass fibre, Water cement ratio, Fibre content.

1 Introduction

Fibre reinforced concrete (FRC) is considered improvement in the shortcomings of the behavior of low tensile strength and low tensile strain before fracture of plain concrete. It is much tougher and more resistant to impact, also flexibility in method of fabrication, so it can be an economic and useful construction new material, and has played an important role in the progress of modern concrete technology.

One of the Important properties of FRC is its toughness or energy absorption.It could reduce the risk of structural failure, especially in the areas where earthquake frequently occur. The toughness is represented by the area under the load-deflection curve. The following test which gives a toughness index may be useful to better quantify this property.

The toughness index is defined as a measure of the amount of energy required to deflect the 102x102x356 mm fibre concrete beam used in the modulus of rupture test a given amount compared to the energy required to bring the fibre beam to the point of first crack. Refer to Fig. 1. The toughness index is calculated as the area under the load-deflection curve out to a specified multiplies of the fist crack deflection (Dcr), divided by the area under the load-deflection (L-D) curve of the

Fibre Reinforced Cement and Concrete. Edited by R. N. Swamy. © 1992 RILEM.
Published by E & FN Spon, 2-6 Boundary Row, London SE1 8HN. ISBN 0 419 18130 X.

fibre beam up to the first crack strength (proportional limit, defined as first deviation from linearity). Then,

$$\text{Thoughness Index}(I_i) = \frac{\text{Area under L-D curve, to iDcr centre point deflection}}{\text{Area under L-D curve to first crack}}$$

The toughness index is influenced by the amount, length, configuration,strength and ductility of the fibres and other factors such as cement content and aggregate amounts, etc.

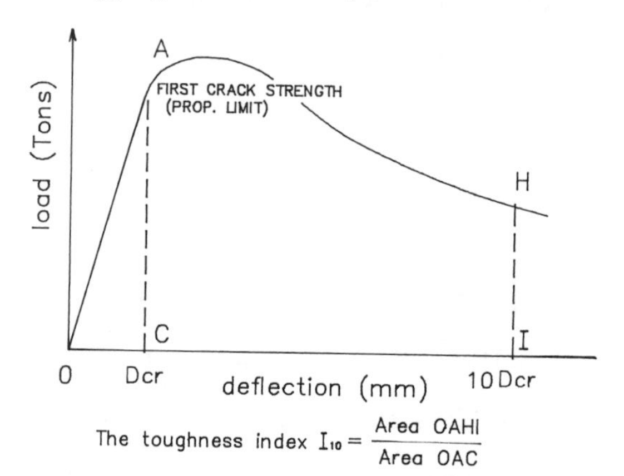

Fig.1 Calculation of toughness index from load–deflection curve

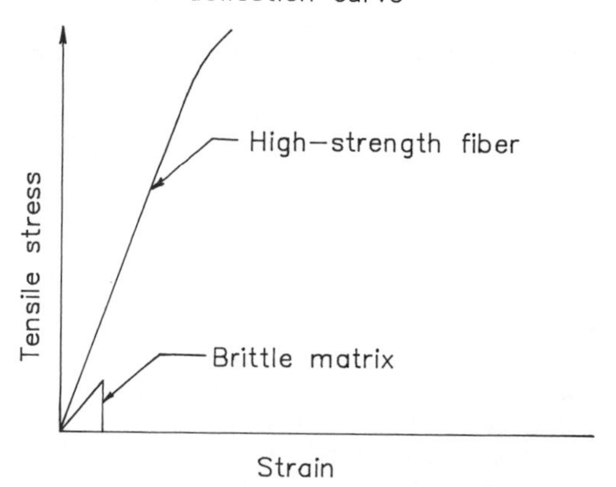

Fig.2 Tensile stress–strain curves for fibres with brittle matrices

The relative stress-strain curves in tension for a strong fibre (steel, glass, asbestos, polypropylene, sisal, kevlar, etc.) and a brittle matrix (Portland cement, gypsum, etc.) are shown in Fig. 2. The strain a fracture for the brittle matrix is considerably smaller (less than 1/50) than that for the fibre. As a result, when a fibre-reinforced brittle matrix composite is loaded, the matrix will crack long before the fibre can be fractured. Once the matrix is cracked,one of the following types of failure of the composite occurs:

(a) The composite fractures immediately after the matrix cracking.

This type of behavior is shown in Fig. 3a.i.e. polymeric fibres at low volume fraction.

(b) Although the maximum load on the composite is essentially the same as that of the matrix alone, the composite continues to carry decreasing load after the peak, as shown in Fig. 3b. i.e. steel fibres at low to moderate volume fraction. The post-cracking resistance is primarily provided by the pulling out of fibers from the cracked surfaces.Although no significant increase in the tensile strength of the composite is observed, a considerable increase in the toughness of the composite (area under the complete stress-strain curve) can be obtained.

(c) Even after the cracking of the matrix, the composite continues to carry increasing tensile stresses; the peak stress and the peak strain of the composite are greater than those of the matrix alone; and during the inelastic range (between the first cracking and the peak), multiple cracking of the matrix occurs (Fig. 3c). i. e. carbon fibres at moderate to high volume fraction.It is clear that this mode of failure results in the optimum performance of both the matrix and the fibers.Various aspects of this type of behavior are discussed in what follows (for composites subjected to uniaxial tension).

The composite will carry increasing loads after the first cracking of the matrix if the pull-out resistance of the fibre at the first crack is greater than the load at first cracking; while at the cracked section, the matrix does not resist any tension and the fibres carry the entire load taken by the composite. With an increasing load on the composite, the fibres will tend to transfer the additional stress to the matrix through

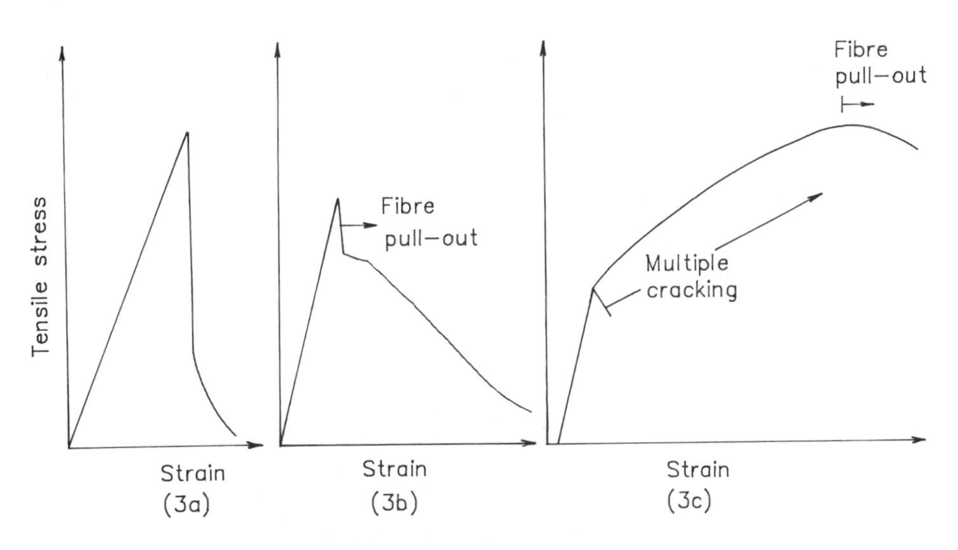

Fig.3 Three possible composite stress–strain curves for fibre–renforced brittle matrices

bond stresses. If these bond stresses do not exceed the bond strength, then there may be additional cracking in the matrix. The process of multiple cracking will continue until either fibres fail or the accumulated local debonding will lead to fibre pull-out.

The process of multiple cracking is a key element in assuring the maximum performance of strong fibre in a brittle matrix. With multiple cracking, it is possible to achieve substantially larger composite peak stress and especially peak strain as compared to those at first matrix cracking. Multiple cracking reduces crack widths, which is an important design consideration. The extent of multiple cracking can be increased by increasing the volume fraction of fibres, by improving the bond strength, as well as by increasing the aspect ratio of fibres.

Typical fibre types used in concrete could be classified as natural material i.e. sisal fibre, and processed by-products i.e. iron wire; steel fibre and glass fibre etc. The purpose of this paper is to evaluate the effect of various kinds of fibres properties to the essential properties of FRC,especially the toughness behavior. In particular, the influence of different water/cement ratio (W/C), fibre content (F/C) and curing age on the toughness behavior of FRC was investigated.

2 Experimental details

2.1 Material and mix preportion of fibre reinforced concrete

The cement used is ordinary Portland cement (ASTM Type 1), The aggregate used for making fibre reinforced concrete (FRC) were crushed gravel and river sand, bulk specific gravity of which were both equal to 2.62, The cumulative grain diameter distribution and their major physical properties of aggregates are shown in Table 1 and Table 2. Their grading and properties are as specifed by ASTM C33.

There are four kinds of fibre used in the testing. ie. iron wire, steel fibre, sisal fibre and glass fibre. Their physical properties are shown in Table 3. The microstructure morphology of these fibres are examined by using scanning electron microscope. The typical feature are shown in Fig. 4. The major character of these fibres are described briefly as follows:

Sisal fibre is a natural fibre material, soft; porous; high absorption of water, gradual accumulation of water results its volume swelling;Iron wire fibre were produced by cutting or chopping a uniform diameter wire.ductility, refined, low absorption of water;Steel fibres used were flat and crimped shape with a high tensile strength;Glass fibres used

2.2 Mix propotion design

Water/cement ratio is the major parameter in the mix proportion of FRC, the mix design of FRC is by the guideline of ACI 318. Four values of W/C were selected for test specimen. The mix proportion of FRC are shown in Table 4. The glass fibre has the property of high absorption of water, so the mix proportion of FRC contain glass fibres be revised and shown in Table 4.The required quantity of fibres and replaced coarse aggregate by fibres are shown in Table 5.

2.3 Testing equipment and procedure

The flexural toughness of FRC were tested and analyzed by using 245

Table 1. The cumulative grain diameter distribution of aggregate

unit:%

sieve no. / aggregate	3/4"	1/2"	3/8"	#4	#8	#16	#30	#50	#100	#200
coarse	100	98.6	62.8	6.6						
fine				98.1	86.4	75.4	55.5	28	6.2	

Table 2. The physical properties of aggregate

aggregate	specific weight	unit weight ($Tons/m^3$)	absorption (%)	abrasion (%)	fine moduous (F.M.)	sound loss (%)
coarse	2.62	1.515	1.35	22.48	—	0.06
fine	2.62	1.604	1.42	—	2.5	0.95

Table 3. The properties of fibre materials

fibre materials	length (mm)	diameter (mm)	aspect ratio (L/D)	specific weight ($Tons/m^3$)	load (Kg)	tensile strength (MPa)	elastic modulus (10^3MPa)
sisal	30	0.08-0.03	165	0.88	1.5	200	16
iron wire	40	0.75-0.80	50-53	7.85	22.0	441	195
steel	50	1.11	45	7.84	45.0	445	200
glass	25	0.007-0.02	1250-3571	2.65	52.0	512	70

KN. dynamic testing machine. It is automatic controlled and the load-deflection curves could be plotted in the recorder. The dimensions of test specimens were 10x10x35cm. Test specimens with different mix proportion of fibre material,W/C and F/C content etc. were installed in the machine for testing. The arrangement is shown in Fig. 5. It is same as that of a simple beam subjected to central load, and the test procedure is according to the specification of ASTM C1018. The toughness behavior of each specimen tested until its deflection is 1.9mm. This is the method of characterizing toughness, ACI committee 544 has adopted. In the ASTM C1018(1990) method, the toughness index is defined as Fig.1 shown.

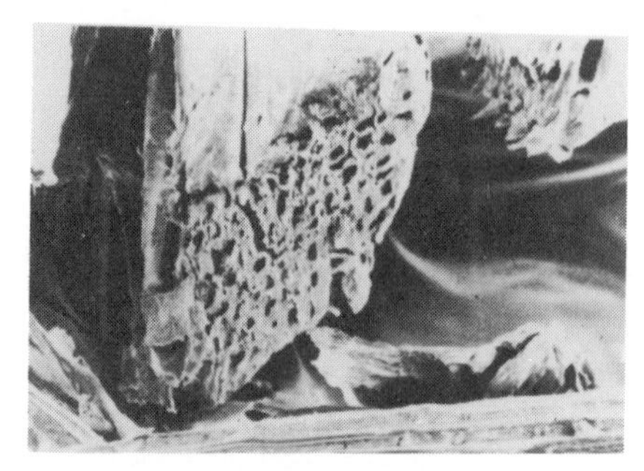

(4a) sisal fibre

(4c) steel fibre

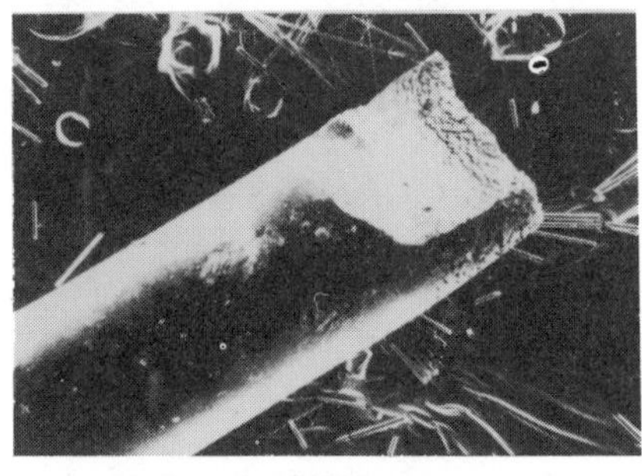

(4b) iron wire

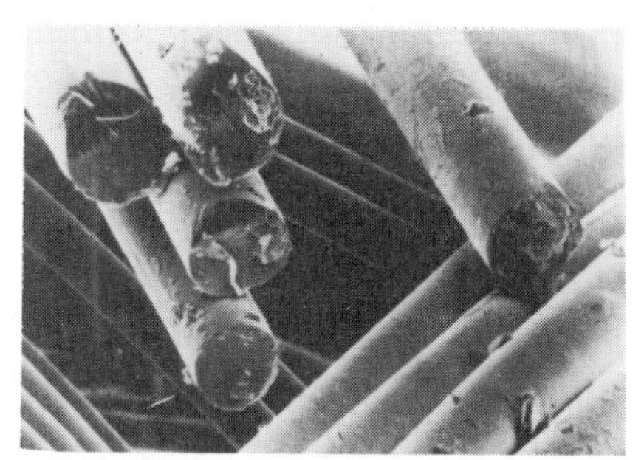

(4d) glass fibre

Fig.4.The fibre microstructure examined by using scanning electronic micrograph technique.

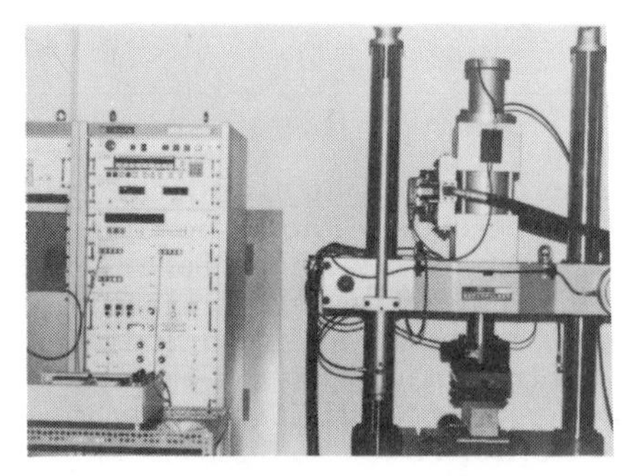

(5a) 25 tons dynamic material testing machine

(5b) Toughness behavior testing

Fig.5.The arrangement of toughness behavior testing

Table 4. The proportioning FRC mixes

unit: Kg/m^3

water/cement ratio (W/C)	0.48	0.55	0.62	0.68
strength* (Kg/cm^2)	365.9	320.4	253.0	225
water	230(250)	230(264)	230(275)	230(283)
cement	479(520)	428(480)	428(480)	338(416)
fine aggregate	663(1140)	714(1156)	714(1156)	781(1190)
coarse aggregate	890(200)	890(20)	890(250)	890(200)

( ) :Materials used in proportioning mixes for glass fibre reinforced concrete .
* Strength is the 28 days Compressive strength of plain concrete specimen.

Table 5. The proportioning mixes of FRC

unit: Kg/m^3

fiber content (F/C)	0.5%	1.0%	1.5%	2.0%	2.5%
sisal fibre	4.42	8.84	13.26	17.68	22.1
iron wire	34.27	68.54	102.81	137.08	171.35
steel fibre	39.20	78.40	117.6	156.8	196.0
glass fibre	13.25	26.50	39.75	53.0	66.25
replaced coarse aggregate	-13.1	-26.20	-39.30	-52.4	-65.5

3 Results and discussion

The data obtained could be discussed as follows:

3.1 The effect of fibre material to the toughness of FRC

Fig. 6 and Fig. 7 show the typical load-deflection curves of FRC specimen produced by four kinds of fibre and under the condition of constant water/cement ratio (W/C = 0.48) and two kinds of fibre content (F/C = 0.5% and F/C = 1.0%)respectively. And Table 6 shows the comparison of toughness index of FRC, with relation to the kinds of fibre material and fibre content. These data are under the condition of curing age of 28 days and W/C = 0.48. From the above Figures and Tables. It is shown the highest toughness and toughness index of FRC is that of steel fibre, the next is iron wire; sisal fibre and glass fibre are the lowest. Also with the increased of fibre content, the toughness and toughness index will be increased. All of these phenomena could be explained as follows:

1. Because the shape of steel fibre is deformed (crimped),resulting its higher bond strength, and the better characteristics of steel fibre i.e. high tensile strength and low absorption of water could be developed. Therefore its toughness behavior is the best.
2. Although the surface of iron wire fibre is smooth and not crimp, but it has high level of stiffness, so its toughness behavior is only next to that of steel fibre, but better than that of glass fibre and sisal fibre.
3. Both of glass fibre and sisal fibre have the properties of higher absorption of water, insufficient stiffness, easy to coil itself up into a mass during mixing, uneven distribution of fibre in the concrete etc. another factor is too light in specific weight of sisal fibre, will causing aggregate segregation easily, resulting their poor toughness.
4. The diameter of glass fibre is only 7-20 μ m, the fibre spacing is 1 -3 μ m, but the diameter of cement grain is about 30-60 μ m. Therefore the cement grain cann't enter the fibre space, the diameter of cement grain is about 3-4times of glass fibre and soft character of glass fibre etc. results bad bond strength between the glass fibre and cement, therefore its tensile strength isn't as good as that of steel fibre, and its toughness is poor.
5. Due to the fact of bad bonding strength between soft fibre material with matrix, causing the fibre could not take the additive load transmitted,resulting the fibre be pulled out and the stress dropping sharply,until the re-equilibrium condition between the external load of FRC and the bearing load of fibre is attained, then the decreasing rate of the load-deflection curve is changed to gentle.Form Fig. 6 and Fig. 7. It is found that the FRC maked by sisal fibre material, will be pulled out rapidly on the instant of microcrack just initiate, therefore it is quite obvious that the load -deflection curve is descending sharply.

3.2 Toughness behavior of iron wire fibre reinforced concrete

Fig. 8 shows the load-deflection curve of iron wire fibre.. It is obvious that the flexural toughness is increasing by higher fibre content;has inverse relationship with W/C and increasing by curing age. Also, refer to Table 6 and Table 7. It is obvious that the toughness index are gradually increasing by fibre content. But there are not any remarkable relationship with W/C. The above phenomena could be discussed as follows:

1. W/C could affect the bonding strength of fibre with matrix, also affect the developing of its strength, therefore reducing its value,

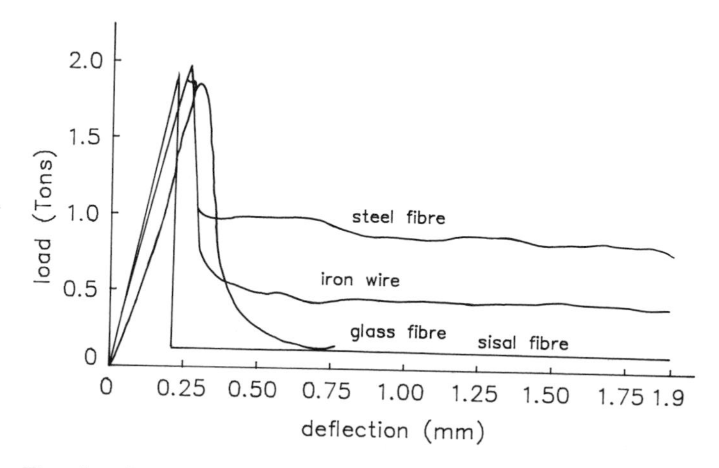

Fig 6 Typical load–deflection curve of fibre reinforced concrete(F/C=0.5%,W/C=0.48)

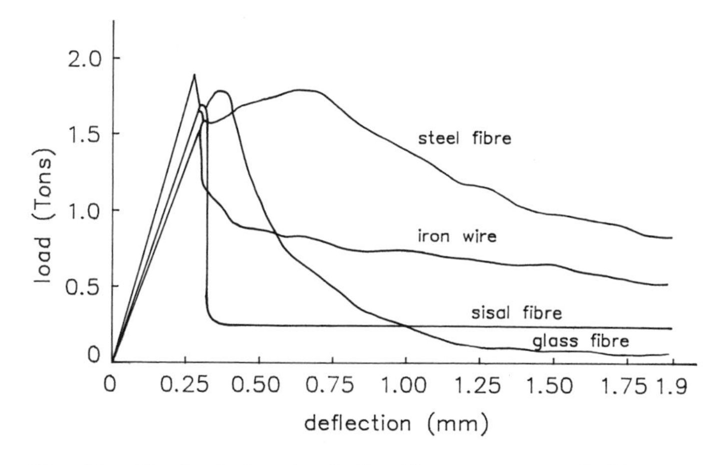

Fig 7 Typical ioad–deflection curve of fibre reinfoeced concrete(F/C=1%,W/C=0.48)

then the toughness of FRC will be increased. But no obvious effect to the toughness index,because the toughness under the first crack strength is also increasing relatively, thus the variation of toughness index is not so clear.

2. As the fibre content (F/C) increasing, the stiffness of F/C is increasing relatively, then the energy absorption is expanding and toughness is also relatively improving. While the toughness increment under the first crack strength is small, therefore its

toughness index is increasing clearly.

3. The stiffness of FRC, manufactured by using ductile fibre, is increasing by curing ages. Hence its toughness is also increasing relatively.

Table 6. The toughness indices of FRC

(W/C = 0.48, Curing age = 28 days)

fibre content (F/C)	0.5%	1.0%	1.5%	2%
iron wire	4.97	9.72	9.44	10.46
sisal fibre	2.12	4.55	5.29	—
steel fibre	7.02	9.96	15.03	15.43
glass fibre	2.54	3.44	5.01	—

Table 7. The toughness indices of SFRC

(Curing age: 28 days)

W/C \ F/C	0.5%	1%	1.5%	2%
0.48	7.02	9.96	15.03	15.43
0.55	7.14	9.75	11.55	10.41
0.62	8.51	10.70	10.05	12.86
0.68	6.05	6.58	17.73	20.39

3.3 Toughness behavior of steel fibre reinforced concrete

1. The relationship between fibre content and toughness of steel fibre reinforcing concrete (SFRC) is shown in Fig. 9 and Table 6. It is obvious that their toughness behavior is increasing by fibre content.
2. The relationship between water/cement ratio and toughness of SFRC is shown in Fig. 10 and Table 7. It is obvious that the lower W/C value, resulting the higher toughness. But the proportional relationship between W/C and toughness index is not clear.
3. The effect of curing age to the toughness behavior is shown on Fig. 11. It is found that toughness increasing by curing age. For making a comparision,the fibre content had changed from F/C = 0.5% to 1% or 2%, for W/C = 0.48 or W/C = 0.55% respectively. The test results shown that the toughness behavior relative to

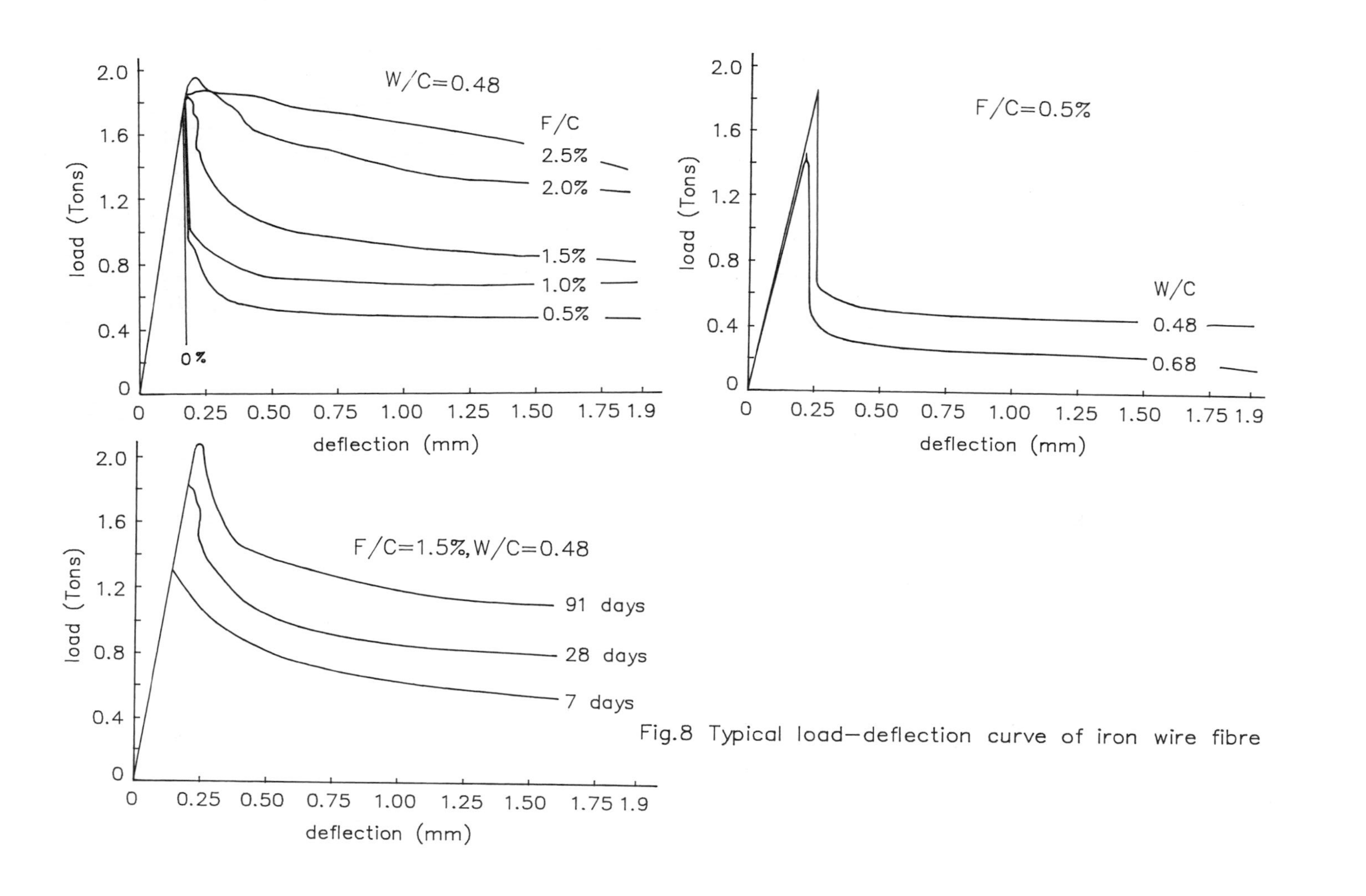

Fig.8 Typical load–deflection curve of iron wire fibre

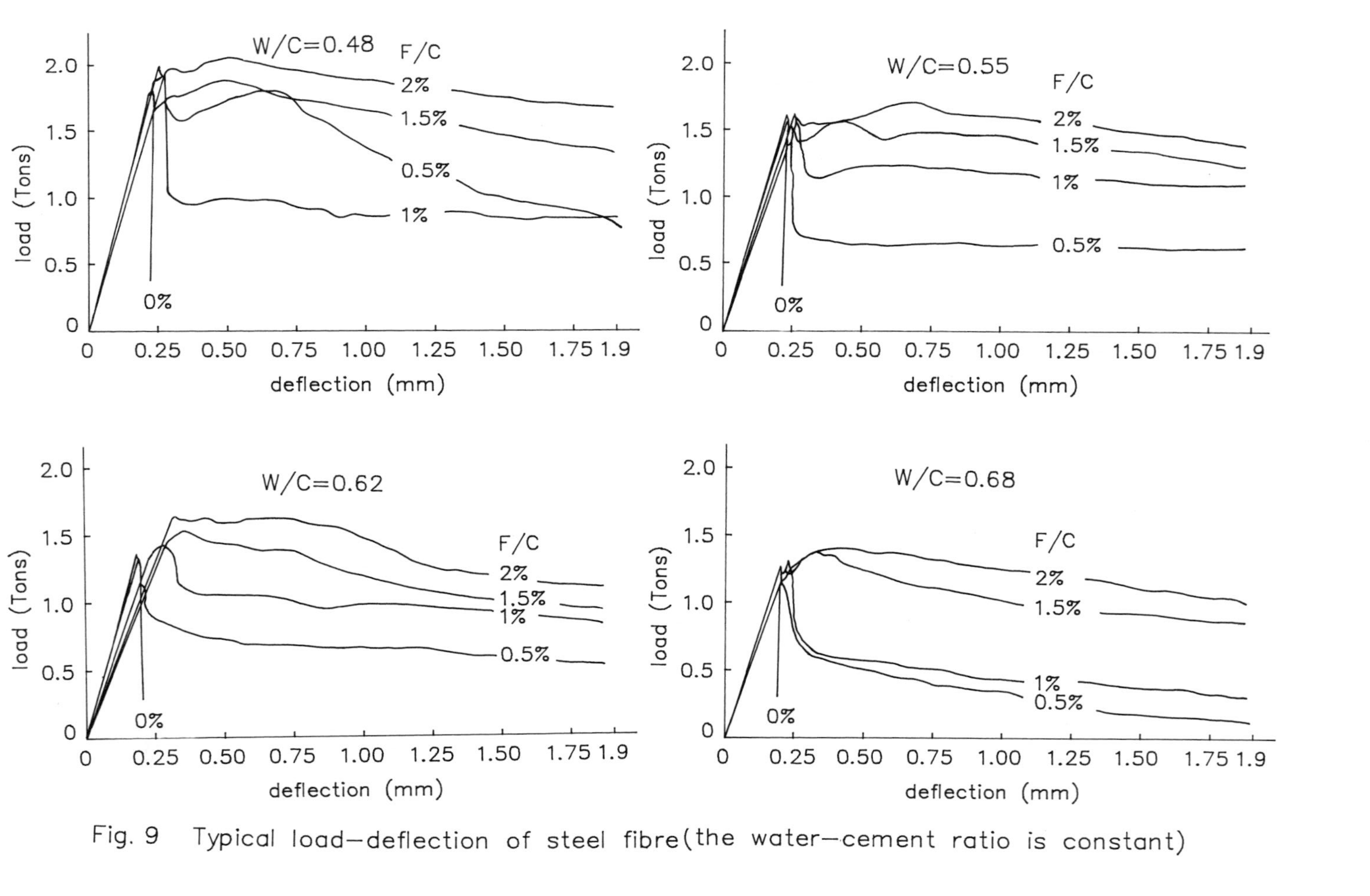

Fig. 9 Typical load–deflection of steel fibre(the water–cement ratio is constant)

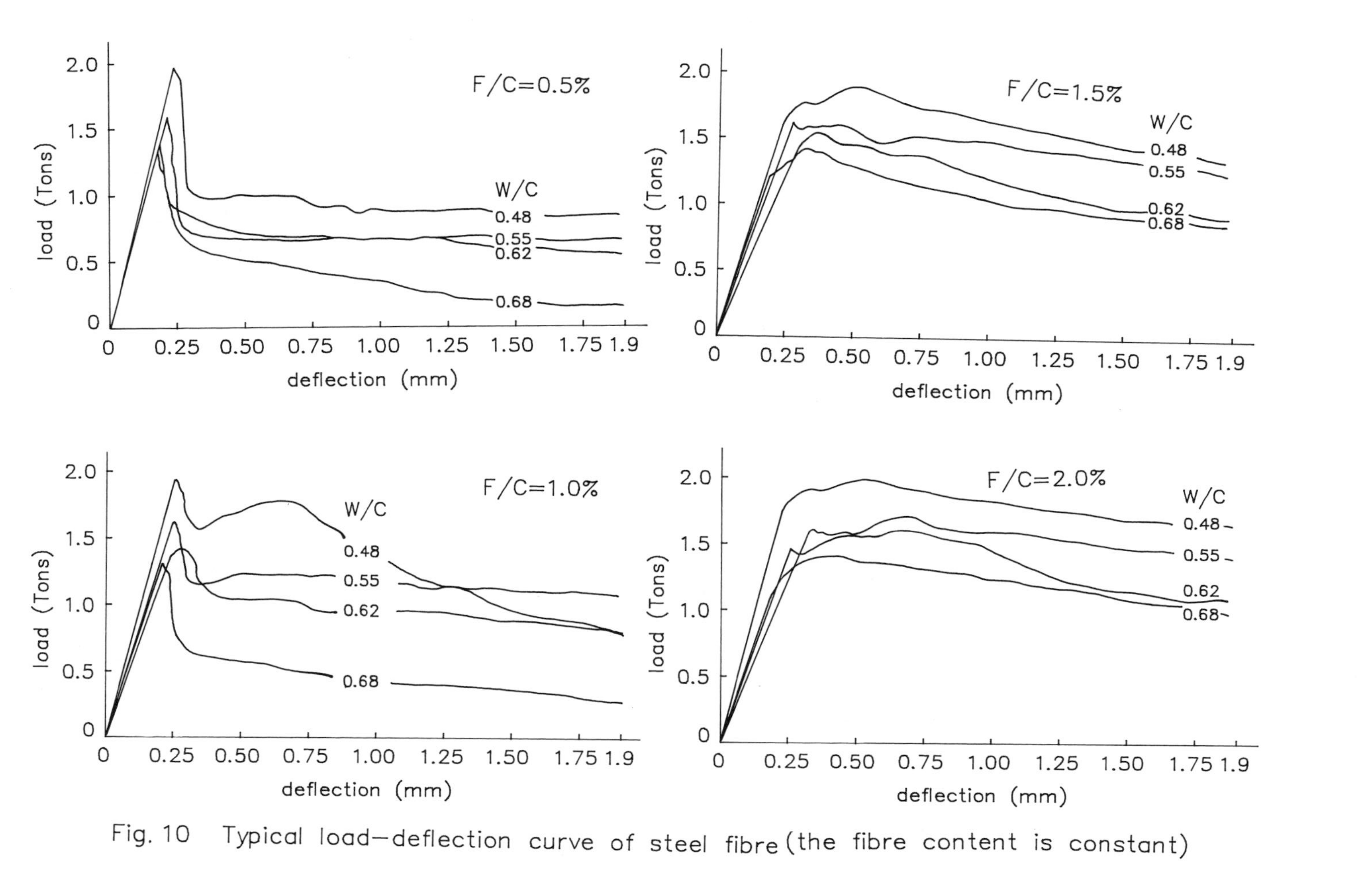

Fig. 10 Typical load–deflection curve of steel fibre (the fibre content is constant)

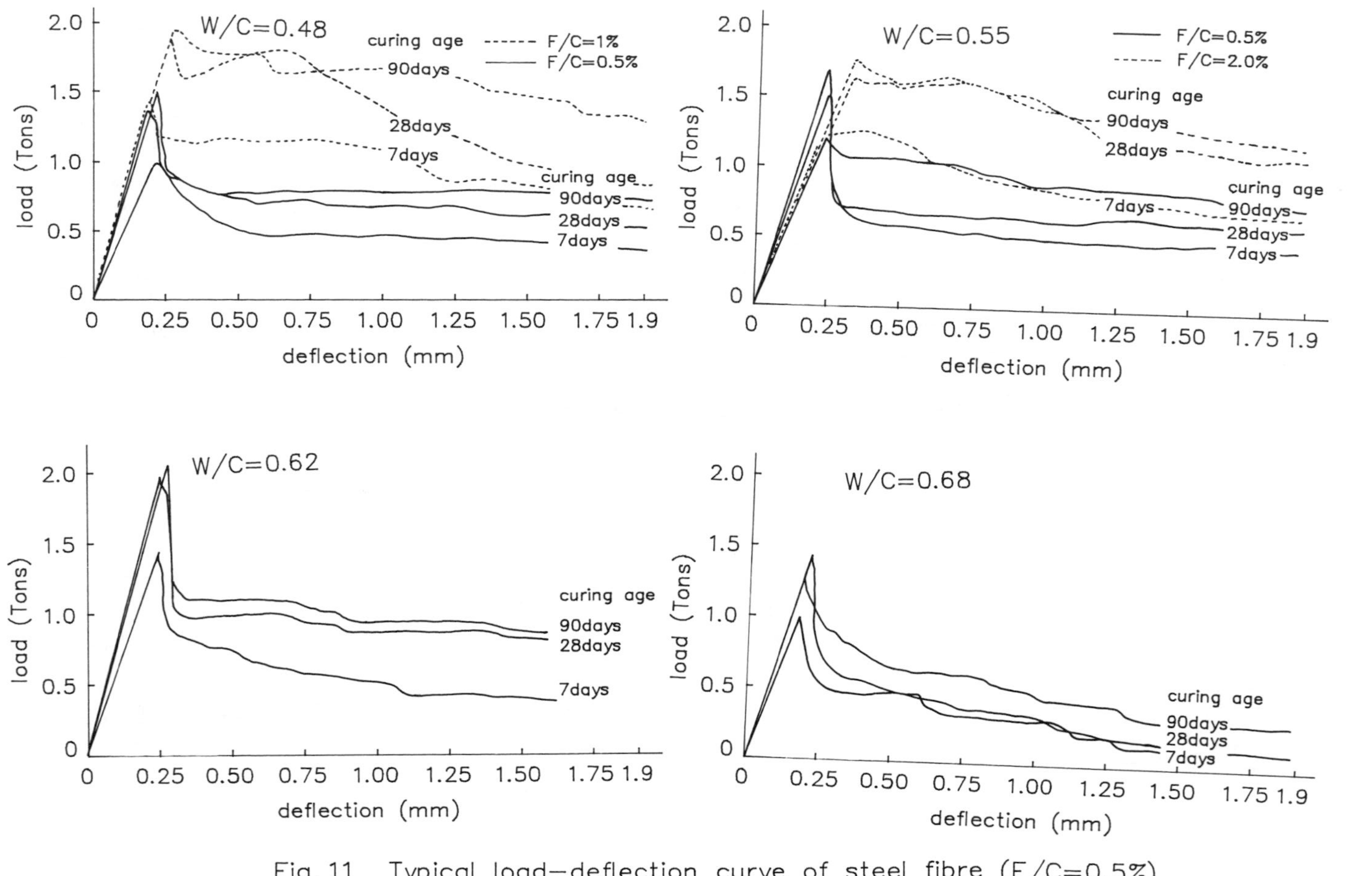

Fig. 11 Typical load–deflection curve of steel fibre (F/C=0.5%)

curing age is more apparent by the adding of fibre content.Also it is found that in the cases of W/C = 0.55, F/C = 2% and W/C = 0. 62,F/C = 0.5%, the increments of toughness and toughness index for curing age of 90 days and 28 days respectively are small. Therefore it could be concluded the effect of curing age to toughness of FRC after 28 days of curing age is gradually reduced.

3.4 Toughness behavior of glass fibre reinforced concrete
Fig. 12 shows the effect of fibre content to the flexural toughness of glass fibre reinforcing concrete (GFRC). Although their relationship is not clear, but this composite material could improving the first crack strength and ultimate strength of concrete, and also the flexural resistance of GFRC is better than that of plain concrete. It can improve the ability of crack expanding resistance and enlarge the flexural toughness of plain concrete is an obvious fact.

3.5 Toughness behavior of sisal fibre reinforced concrete
Fig. 13 is a load-deflection curve of sisal fibre reinforcing concrete.Because sisal fibre has the properties of light specific weight, high water absorption and large swelling volume etc. When this composite concrete is on continued drying. It is found that further loss of water, resulting considerable shrinkage and reducing the bonding strength between the fibre with matrix, therefore their strength is descending and flexural toughness behavior is poor.

4. Conclusion

Based upon the above experimental results, the following conclusion could be obtained:

1. The best toughness behavior of FRC specimens shown is that of steel fibre,then in order of iron wire, glass fibre and sisal fibre.
2. In proportion as the fibre content and curing age increase, the flexural toughness and toughness index will strengthen; The toughness behavior of FRC has inverse proportion to water/cement ratio, while the toughness index has not apparent relationship with water/cement ratio.
3. Experiments using soft fibre, i.e. glass or sisal, have been shown that it could improve the crack expanding resistance of plain concrete and enlarge its flexural toughness. In order to have a good workability, the fibre content and aspect ratio of additive fibre should be lower. It is recommended the upper limit condition of sisal fibre are aspect ratio $L/D \leqq 165$, fibre content $F/C < 2\%$ and length $L \leqq 30mm$, and for glass fibre of rich grading and $F/C \leqq 2.0$ etc.
4. The better reinforcing effect of steel fibre is recommended as aspect ratio $(L/D) \leqq 100$, length (L) = 20 - 50mm. Under optimum condition of workability,economics and strength development etc. the fibre content is recommended as 1-2%.
5. Improper mixing will affect the quality of FRC, therefore it is recommended that a better design of mix proportioning before application is necessary to get a good quality assurance. It is bet-

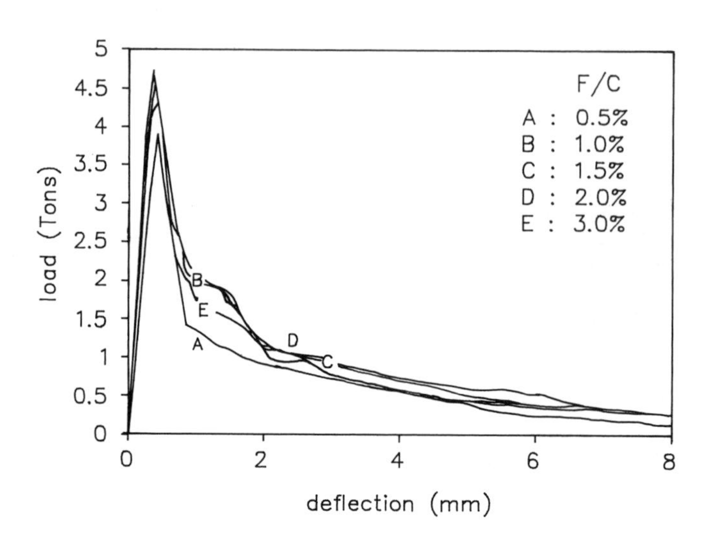

Fig. 12 Typical load–deflection curve of glass fibre

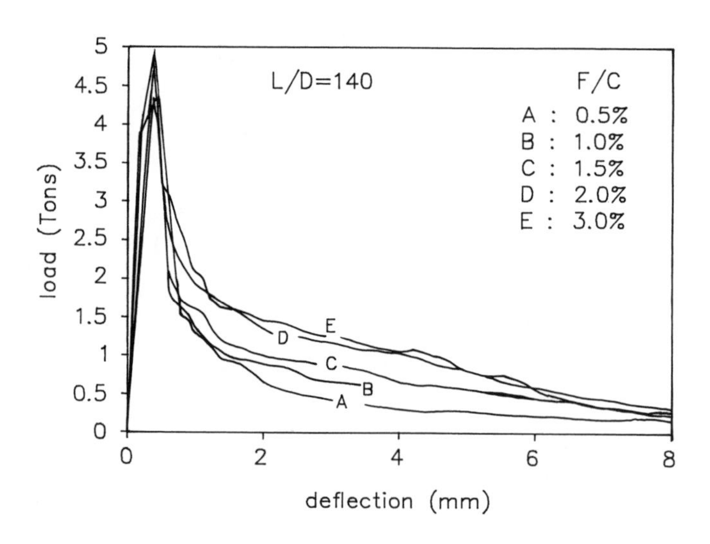

Fig. 13 Typical load–deflection curve of sisal fibre

ter to using water/cement ratio W/C = 0.48, the maximum diameter of aggregate D = 3/8" and fine/coarse aggregate ratio S/A = 5 0%. Also add concrete admixtures of natural pozzolan materials to FRC is helpful for its improvement of workability and reduction in permeability.

5. References

ASTM C 1018(1990), "Standard method of test for flexural toughness and first crack strength for FRC (using beam with third point loading)",**ASTM standards for Concrete and Aggregates,** Vol.04.0 2, pp.550-560.

Johnston, C.D. (1980), in **Progress in Concrete Technology,** (ed. V. M. Malhotra), CANMET.Ottawa, pp. 452-504.

Kormeling, H.A., Reinhardt, H.W. and Shah, S.P. (1980), Static and fatigue properties of concrete beams reinforced with continuous bars and with fibers, **J. Am.Concr. Inst., Proc.,** 77, No. 1, pp. 36-43.

Report AÇI 544. 1R-82 (1982) **Concrete Int.,** Vol. 4, No. 5, pp. 9-30.

Schrader E.K. (1987), "Formulating guideline for testing of fiber concrete in ACI Committee 544" RILEM symposium 1987, **Test and Test Methods of Fiber Cement Composites,** (Ed. R.N. Swamy), The Construction Press Ltd. Lancaster,England, pp. 17.

Shah, S.P. (1984), "Fiber Reinforced Concrete," in **Handbook of Structural Concrete,**(Eds. F.K. Kong. R.H. Evans, E. Cohen, and F. Roll), McGraw-Hill Book company, New York.

26 PRESSING OF PREMIXED GRC: INFLUENCE OF FIBER LENGTH ON TOUGHNESS

J. AMBROISE and J. PERA
Laboratoire des Matériaux Minéraux, INSA Lyon,
Villeurbanne, France

Abstract
In the first part of the work, the bending strength of premixed GRC composites manufactured by pressing was studied. The glass contents were 1.5 % and 2.5 % by weight, respectively. The fibre lengths were 4.5 mm, 12 mm and 25 mm.
In the second part, various mixes of fibres were investigated. The best results were obtained with the mix : 55 % 12 mm long fibres + 45 % 25 mm long fibres.
Keywords : Premixed GRC, Pressing, Flexural Strength, Toughness, Fibre Length, Mix of Fibres.

1 Introduction

Data on the properties of premixed GRC manufactured by pressing are very scarce in the literature. Hills (1975) studied the bending and impact strengths of premixed GRC composites varying the glass content from 1 % to 15 % by weight ; the fibre length was kept constant. So, it was interesting to study the influence of fibre length on the bending properties of such a composite material.

The fibre length is normally 12 mm because above this length the mix becomes difficult to work. A fibre of 25 mm is generally found to be the maximum useable. The second question investigated was to see if a mix of fibres can lead to good bending properties.

2 Experimental procedure

The matrix was composed of :

- 60 % pozzolanic binder containing 80 % white ordinary portland cement and 20 % metakaolin ; metakolin is a reactive pozzolan which improves the long-term strength properties of GRC. Its properties were investigated by Ambroise and al (1986-88),
- 30 % silica sand,
- 10 % limestone filler.

Fibre Reinforced Cement and Concrete. Edited by R. N. Swamy. © 1992 RILEM.
Published by E & FN Spon, 2-6 Boundary Row, London SE1 8HN. ISBN 0 419 18130 X.

AR CEMFIL 2 glass fibres were used in two proportions by weight : 1.5 % and 2.5 %. The fibre lengths were 4.5 mm, 12 mm and 25 mm, respectively.

The water solid ratio was 0.31 and a superplasticizer was used to produce a high quality slurry to achieve the necessary workability and allow for the uniform incorporation of fibres. The superplasticizer content was 1.3 % by weight of plain white cement.

The composite boards were fabricated in the form of rectangular sheets, 0.50 x 0.60 m and approximatively 20 mm thick. The sheets were pressed in order to obtain a 10 mm thickness ; the applied pressure was 0.67 MPa.

First, the proportions of AR glass fibres in all boards were 1.5 % to 2.5 % by weight and the fibre lengths were 4.5 mm, 12 mm and 25 mm, respectively. In the second part of the work, fibres of different lengths were mixed in the following proportions :

55 % 4.5 mm long fibres + 45 % 12 mm long fibres ;
55 % 4.5 mm long fibres + 45 % 25 mm long fibres ;
55 % 12 mm long fibres + 45 % 25 mm long fibres.

The mix contents were 1.5 % and 2.5 %.

All boards were given an initial cure of 24 hours in moist air at 30°C ; they were then cut into 70 mm x 280 mm specimens coupons and this was followed by 28 days'storage in air at 65 % HR and 20°C.

The specimens were then tested in 3 point bending on an unsupported span of 200 mm. The load was applied using a ADAMEL LHOMARGY testing machine with a cross head speed of 5 mm/min. while the load-deflection curve was recorded.

Flexural toughness is defined as the area under the flexural load-deflection curve and is an indication of the energy absorption capability of a material. The flexural toughness indices used in the data analysis presented in this report were developed by C.D. Johnston (1982) and ASTM Standard C1018-84. A flexural toughness index is defined as the area under the load-deflection curve up to a specified deflection criteria divided by the area up to the deflection at first cracking.

Flexural toughness indices have been determined for specific deflection criteria (Fig. 1). These criteria are the deflection up to the first cracking (σ), the deflection up to three times the first cracking deflection (3σ), the deflection up to 5.5 times the first cracking deflection (5.5σ), and the deflection up to 15.5 times the first cracking deflection (15.5σ).

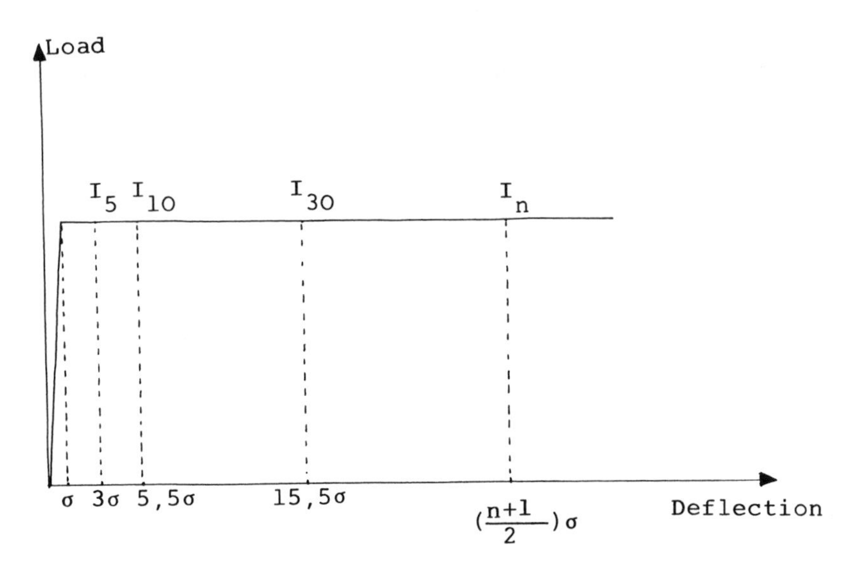

Fig. 1. Toughness idices according to ASTM Standard C1018-84.

Flexural toughness index I_5 is defined as the area under the load-deflection curve up to 3σ divided by the area up to σ. For a perfectly elastic-plastic material, the flexural toughness index I_5 would equal 5.0. Indices I_{10} and I_{30} respectively equal to 10.0 and 30.0 would indicate elastic-plastic behaviour.

A toughness index I_5 greater than 5.0 would indicate that the area under the load-deflection curve is greater than that for elastic-plastic behaviour. This could be an indication of increasing load carrying capacity beyond matrix cracking. The reverse could be said for a toughness index less than 5.0 This same line of reasoning may be used for values of toughness indices I_{10} and I_{30}.

In ASTM Standard C1018-89 specific reference to the I_{30} index was deleted and replaced by the option to determine I_{20} (end-point 10.5σ) because experience has shown that most load-deflection relationships reach a steady-state slope well before this end-point.

The strength retained after first crack is an other important facet of material performance from the point of view of short or long term structural safety or integrity Residual strength factors for the delection intervals associated with specific toughness indices are easily derived from the toughness indices, for example $R_{5,10} = 20(I_{10}-I_5)$ and $R_{10,20} = 10(I_{20}-I_{10})$ (ASTM Standard C1018-89). Such factors give a good indication of the shape of the load-deflection relationship after first crack.

3 Results and discussion

3.1. Influence of fibre length

The experimentally obtained load-deflection curves are shown in figures 2 to 4. It is seen that the fibre length has an influence on the peak load recorded during the flexural test. The best results were obtained for a 4.5 mm fibre length ; it is certainly due to the pressing of specimens which is easier when the fibres are shorter.

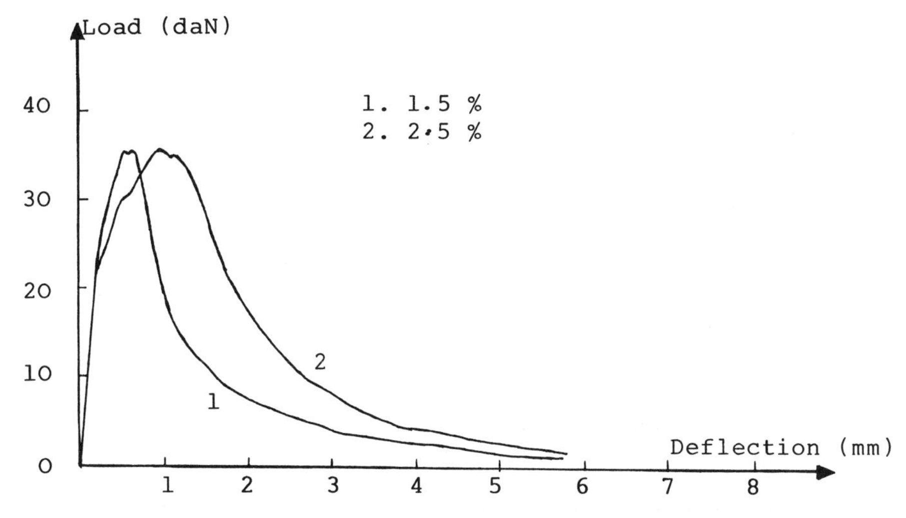

Fig. 2. Load-deflection curves - Fibre length : 4.5 mm .

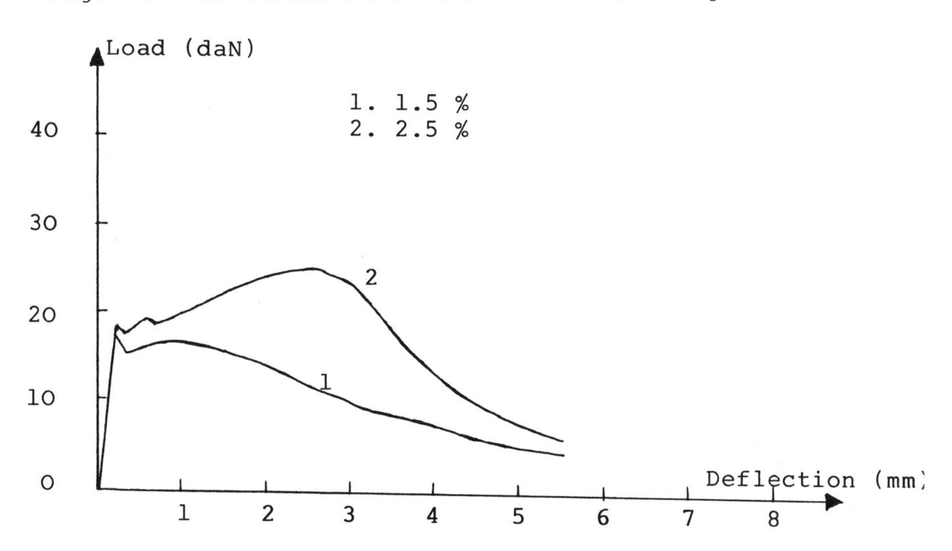

Fig. 3. Load-deflection curves - Fibre length : 12 mm.

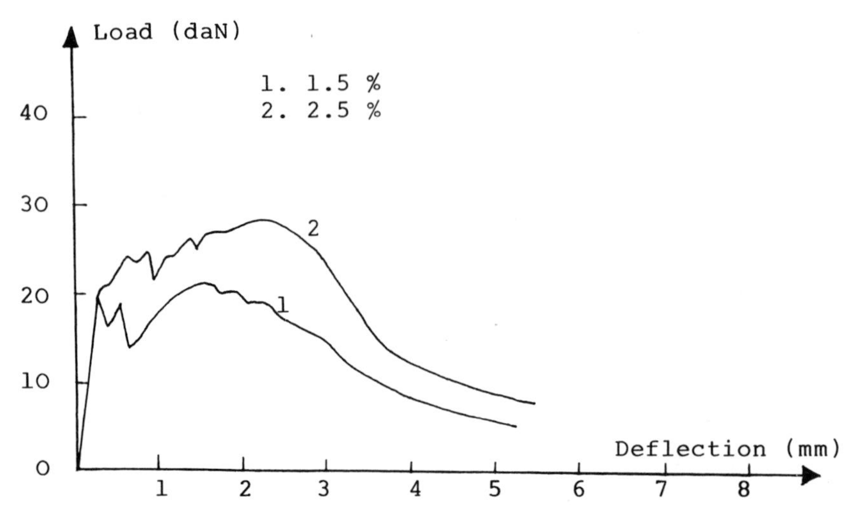

Fig. 4. Load deflection curves - Fibre length : 25 mm.

Average toughness indices are plotted in Table 1.

Table 1. Toughness indices

Fibre length (mm)	Proportion of fibres (%)	I_5	I_{10}	I_{20}	I_{30}
4.5	1.5	5.5	8.5	10.8	11.8
	2.5	6.0	12.5	18.5	20.8
12	1.5	4.7	9.3	17.9	24.6
	2.5	5.0	10.3	22.4	35.5
25	1.5	4.5	9.0	19.0	25.4
	2.5	5.5	11.9	15.8	36.2

Values of I_5 to I_{30} for composites containing 2.5 % 12 mm long fibres or 2.5 % 25 mm long fibres were all greater than the standard values. This indicates that out to 15.5 times the cracking deflection (σ), the post-cracking performance for these composites was better than elastic-plastic behaviour, as it can be seen from the residual strength factors.

The residual strength factors associated with I_5, I_{10}, and I_{20} are plotted in Table 2.

Table 2. Residual strength factors

Fibre length (mm)	Proportion of fibres (%)	$R_{5,10}=20(I_{10}-I_5)$	$R_{10,20}=10(I_{20}-I_{10})$
4.5	1.5	60	23
	2.5	130	60
12	1.5	92	86
	2.5	106	121
25	1.5	90	100
	2.5	128	139

Fully brittle behaviour corresponds to a residual strength factor of zero and perfectly plastic or yield-like behaviour to a factor of 100. When the amount of 12 mm or 25 mm long fibers equals 2.5 % by weight, values of $R_{5,10}$ and $R_{10,20}$ are greater than 100. Such values indicate a good ductility of composites.

Because of a higher MOR (Modulus of rupture) value (12.2 MPa instead of 10.9 MPa), the use of 25 mm long fibres is recommended.

3.2. Influence of fibre mixes

The load-deflection curves are shown in figure 5 ; the toughness indices are reported in Table 3 and the residual strength factors are plotted in Table 4.

1. 1.5 % - "4.5 mm + 12 mm"
2. 2.5 % - "4.5 mm + 12 mm"
3. 1.5 % - "4.5 mm + 25 mm"
4. 2.5 % - "4.5 mm + 25 mm"
5. 1.5 % - " 12 mm + 25 mm"
6. 2.5 % - " 12 mm + 25 mm"

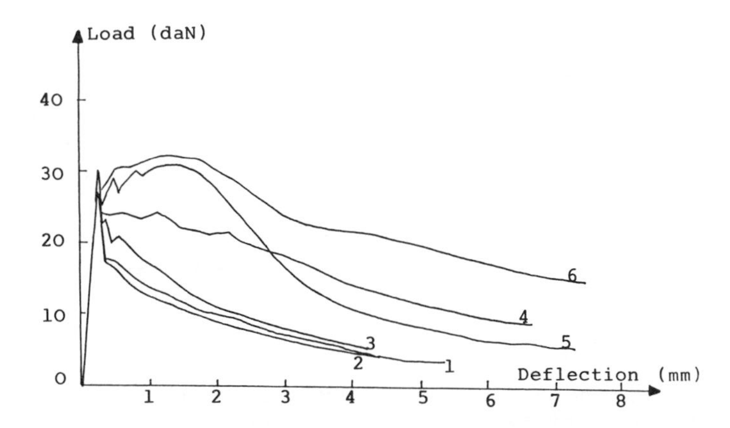

Fig. 5 Load-deflection curves - Mixes of fibres.

The bests results are obtained with the mix containing 55 % 12 mm long fibres and 45 % 25 mm long fibres, as well for MOR values as for toughness indices. The optimum glass content is 2.5 % : such composites behave as an elastic-plastic material out to 10.5 times the cracking deflection. The residual strength factors are lower than when 2.5 % 25 mm long fibres are used alone. The MOR value is slightly increased : 14.4 MPa instead of 12.2 MPa.

Table 3. Toughness indices - Mixes of fibres

Mix.	Proportion (%)	I_5	I_{10}	I_{20}	I_{30}
4.5 mm	1.5	3.6	5.8	9.2	11.4
+					
12 mm	2.5	3.7	6.2	10.0	13.1
4.5 mm	1.5	4.2	7.3	11.7	15.0
+					
25 mm	2.5	4.4	8.7	16.4	22.7
12 mm	1.5	4.9	10	19.0	24.2
+					
25 mm	2.5	5.1	10.4	20.3	28.1

Table 4. Residual strength factors - Mixes of fibres

Mix.	Proportion	$R_{5,10}$	$R_{10,20}$
4.5 mm	1.5	44	34
+			
12 mm	2.5	50	38
4.5 mm	1.5	62	44
+			
25 mm	2.5	86	77
12 mm	1.5	102	90
+			
25 mm	2.5	106	99

4 Conclusion

When manufacturing premixed GRC by pressing, the use of 2.5 % 25 mm long fibres leads to the best bending properties. A mix of fibres has only marginal effects on these properties.

References

Ambroise, J., Murat, M., and Péra, J. (1986). Durability of E-Glass Fiber in Portland Cement-Metakaolinite Matrix. Proceedings of the **International Symposium on the Durability of Glass-Fiber Reinforced Concrete**, Chicago, S. Diamond Editor, pp. 285-292.

Ambroise, J., Dejean, J., Foumbi, J., and Péra, J. (1987) Metakaolin blended cements : an efficient way to improve GRC durability and ductility. Proceedings on the **6th Biennal Congress of the Glass Reinforced Cement Association**, Edinburgh, Glass Reinforced Cement Association, (1988), pp. 19-27.

Ambroise, J., Dejean, J., and Péra, J. (1988). Study of fiber-matrix interfaces in metakaolin-blended cement GRC composites. Proceedings of the **Materials Research Society**, Boston, Vol. 114, pp. 175-180.

Hills, D.L., (1975) in Glass Fibre Reinforced Cement (eds A.J. Majumdar and V. Laws), Oxford BSP Professional Books, Oxford, (1991), pp. 86-87.

Johnston, C.D. (1982) ASTM Cement, Concrete and Aggregates, CCAGDP, Vol. 4, N° 2, pp. 53-60.

27 POLYPROPYLENE FRC: FIBER-MATRIX BOND STRENGTH

J. R. L. DYCZEK and M. A. PETRI
Technical University of Mining and Metallurgy, Cracow, Poland

Abstract
The strength of the fibre-hardened cement paste bond is a quality which determines to a considerable degree such properties of the FRC as: strength, modulus of rupture and fracture energy. This inter facial bond strength in theory should be equal to the shear strength of the weaker component, in this case, of the cement matrix. However, in practice this strength assumes markedly lower values. The reason for this is its specific layered structure resulting from the heterogeneous nucleation and crystallization of calcium hydroxite taking place on the fibre surface.
This phenomenon has been described for steel fibre reinforced concrete and for GRFC. In case of organic fibres, in particular those produced only in a cut with the surface rendered rough on purpose, there exists the problem of determining the real value of the strength of the fibre-matrix boundary using direct methods. In the present study an attempt has been made to estimate the value of τ basing on the post-cracking behavior of the material,i.e, the work of fracture and the fracture modulus. The investigations were carried out on FRC reinforced with short, space oriented polypropylene fibres. The Krenit Standard type of fibres, produced by Danaklon were used. The properties of fibers and its application for cement pastes and mortars were described by Krenchel [9] and Davies [10]. The cement matrix was modified with an addition of acrylene resin or silica fume.In case of autoclaved samples an addition of ground quartz sand was used
Keywords: fiber reinforced cement, polypropylene, flexure, work of fracture, modulus of rupture, interfacial bond strength.

1 Introduction

When analyzing the fracturing process of FRC and the crack development we can distinguish two extreme cases:
- the fibres are pulled out of a cement matrix,
- the fibres are broken and some of them are next pulled out.

1.1 Fracture energy

The process of energy absorption may thus be described in two ways:

Fibre Reinforced Cement and Concrete. Edited by R. N. Swamy. © 1992 RILEM.
Published by E & FN Spon, 2-6 Boundary Row, London SE1 8HN. ISBN 0 419 18130 X.

Case I - the fibres are pulled out. This phenomenon occurs when the composite is reinforced with short fibres of poor fibre-matrix bond strength with simultaneous relatively great capacity of the fibres to carry the load.
Three stages can be distinguished in this process: debonding of the fibre from the matrix, pulling out of the fibers, and secondary distribution of stresses. Hence the total energy absorbed in the cracking process of a composite of this type will be equal to the sum of the above mentioned components, enlarged by the fracture energy of the matrix

- increase of the fracture energy on the basis of the mechanism of fibre debonding from the matrix. This case has been described by Kelly [1] and by Outwater and Murphy [2] for composites of high fibre-matrix bond strength. The calculated values usually did not exceed 500 J/m^2.For materials of the FRC type, in particular for the case when fibers are being pulled out without prior tearing, any unambiguous interpretation of the phenomenon has not been put forward as yet. In this situation the authors have adopted a simplifying assumption $\tau_d=\tau_s$ which allows to neglect this effect in further calculations.
- pulling out of fibers. The mechanism has been described by Cottrell [3] and Kelly [4]. After the fibre has been torn off from the matrix it is pulled out. The resistance comes from the friction forces proportional to τ_s, the length of anchor of the fibre in the matrix, the geometry (length, diameter) and the number of the fibers.

$$\gamma_p I = \frac{Vf \cdot \tau \cdot l^2}{6 \cdot df} \qquad (1)$$

-after the fibers have been torn off or pulled out of the matrix, their ends are unable to absorb the excess of the elastic energy stored in the deformation process. This energy (γ_r) must be transferred to the matrix. In the case under consideration (pulling out) it is equal to:

$$\gamma_r I = \frac{Vf \cdot \tau^2 \cdot l^2}{6 \cdot E \cdot df} \qquad (2)$$

- the energy of the matrix fracture (γ_s) should be calculated taking into consideration the volume fraction of the fibers in the composite:

$$\gamma_s I = \gamma_m \cdot (1-Vf) \qquad (3)$$

The total expenditure of energy in the cracking process with dominating pull-out process will thus be equal to

$$\gamma_{tot} I = \gamma_p I + \gamma_r I + \gamma_s I \qquad (4)$$

Case II- the fibers break before pulling out of the matrix at a length lc/2 and smaller. Bearing in mind that the value of the critical length of the fibre is usually calculated as $lc=(\sigma f \cdot df)/(2 \cdot \tau)$, the formulae (1), (2) and (3) will take the forms:

$$\gamma_p II = \frac{Vf \cdot lc^2 \cdot \sigma f}{12 \cdot l} \qquad (5) \qquad\qquad \gamma_r II = \frac{Vf \cdot df \cdot \sigma f^3}{6 \cdot E \cdot \tau} \qquad (6)$$

$$\gamma_s II = Vf \cdot \left(\frac{l}{df} \right) \cdot \gamma_m + (1-Vf) \cdot \gamma_m \qquad (7)$$

In case of Eq.(7), i.e. the energy of forming new surfaces, the expenditure of energy connected with phase distribution has been taken into consideration as suggested by Masterson, Atkins and Feldbeck [5]. The total fracture energy is equal to:

$$\gamma_{tot} II = \gamma_p II + \gamma p_r II + \gamma_s II \qquad (8)$$

The above considerations refer to composites reinforced with continuous fibers, arranged parallel with respect to each other and to the applied load. When short, space-oriented fibers (3D) are used according to [6] we have only half the number of fibers in the cross-sectional area.
It follows from the above that the value of the fracture energy of the composite is a complex function, depending on the mechanical properties, geometry, number and arrangement of the reinforced fibers and the strength of the interphase fibre-matrix boundary.

Table 1. Properties of fibres, Danaklon, Krenit Standard

	Units	Producers data	Data for computation
Length	mm	6	$l = 6$
Thickness	μm	35-46	$a = 37.5$
Width	μm	100-250	$b = 175$
Density	g/cm^3	0.9	$\rho = 0.9$
Tensile strength	MPa	450-500	$\sigma f = 475$
E - modulus	GPa	13-18	$Ef = 15.5$
Ultimate strain	%	6-8	$\varepsilon f = 7$
Diameter	μm	-	$df^* = 61.7$

df^* - was calculated as: $df^* = 4 \cdot \frac{Af}{Pf}$ Eq. (22) in the text

where: Af - cross-sectional area of fibre
Pf- perimeter of fibre

Table 2. Calculated values of energy, consumed in fracture process of FRC reinforced with polypropylene DANAKLON fibres. Vf=0.02

l	τ MPa	γ_p [J/m^2]	γ_r [J/m^2]	γ_s [J/m^2]	γ_{tot} [J/m^2] (100%)
l=4 mm	1.5	648 (96.5)	4.1 (0.6)	19.8 (2.9)	671
	2.5	1080 (97.2)	11,4 (1.0)	19.8 (1.8)	1111
	4.0*	1331 (86.5)	178 (11.6)	31.5 (2.0)	1539
l=6 mm	1.5	1458 (97.7)	13.7 (0.9)	19.8 (1.4)	1491
	2.5*	2210 (87.5)	285 (11.0)	38.6 (1.5)	2593
	4.0*	886 (80.8)	178 (16.3)	31.5 (2.9)	1096
l=8 mm	1.5	2590 (98.0)	32.5(1.2)	19.8 (0.8)	2643
	2.5*	1703 (84.1)	285 (14.0)	38.6 (1.9)	2026
	4.0*	666 (76.1)	178 (20.3)	31.5 (3.6)	874

* rupture of fibers
( ) percentage part

For FRC reinforced with polypropylene fibers, the properties of which are listed in Table 1, the anticipated changes in the fracture energy have been calculated as functions of τ, l, Vf, on the basis of Eqs (4) and (8). The calculation results are shown on the graphs 1-5 and in Table 2.
In Fig 1. the maximum on the curve (γ_{tot} vs l) corresponds to a fibre length equal to the critical length calculated for the assumed τ=2.5 MPa. The above observation was made for the first time by Cottrel.The dependence of the fracture energy on the fibre-matrix strength for FRC differing in the length of the reinforced fibers is shown in Fig.2. The maximum points on the curves correspond to those values of τ at which the fibers of a given length may be break. The vertical sections A, B and C have been made for the following values of τ=1.5, 2.5 and 4 MPa. Figures 3-5 show the changes in γ_{tot} vs Vf for the above values of τ. When the fibre-matrix bond strength is equal to 1.5 MPa (Fig.3), in each of the presented cases of l, only the process of fibre pulling -out is to be expected.
An increase in the interphase strength up to 2.5 MPa (Fig.4) is responsible for the fact that in the the case of cracking of FRC reinforced with fibers 4 and 6 mm in length the fibers will be pulled out, and fibers 8 mm in length will be torn off. When τ is further increased up to the value of 4 MPa (Fig.5) all three types of fibers will be torn off in the cracking process. The greatest capacity for energy absorption is demonstrated by material reinforced with the shortest - 4 mm fibers. A detailed energy balance for the cases under consideration, broken down for γ_p, γ_r and γ_s, is shown in Table 2. It follows from the table that in the cracking process of FRC reinforced with polypropylene fibers, the process of pulling out of the fibers

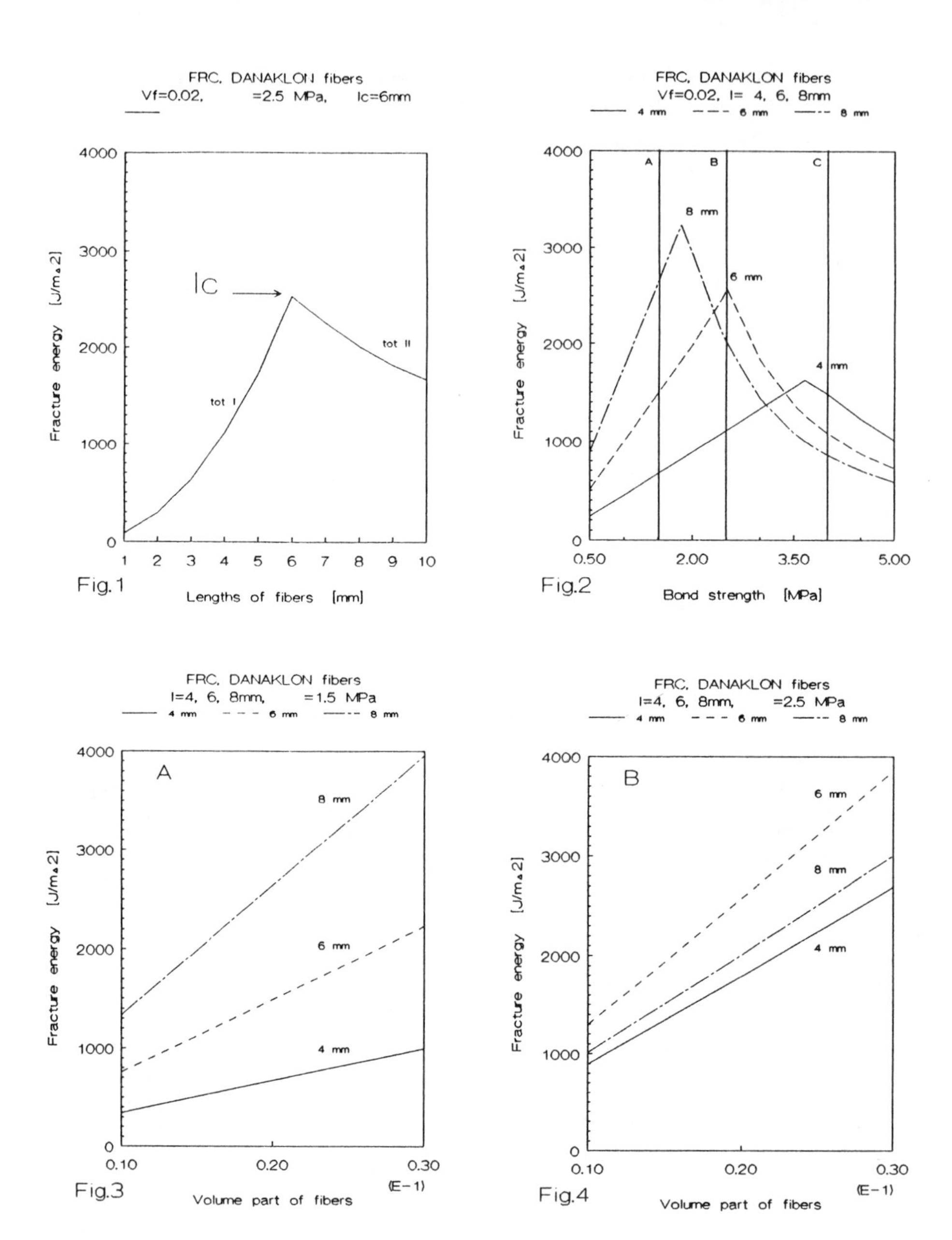

Fig.1

Fig.2

Fig.3

Fig.4

has the greatest share in the total fracture energy. It diminishes when the fibers are torn off, nevertheless it still dominat.
The value of the fracture energy is determined under conditions of a controlled process of crack propagation. In practice, the expenditure of energy connected with uncontrolled cracking of composites under conditions of dynamic loading (impact resistance) or static loading (work of fracture) is also measured. In the latter case, at the moment when after cracking properties are analyzed, the measured values of WOF are very close to the values of the fracture energy determined by the controlled process of crack propagation. Thus it may be assumed that in case when the cracking process takes place at a low velocity of deformation then $\gamma_{tot} \simeq$ WOF.

1.2 Modulus of rupture

The load capacity of the composite after cracking of a brittle matrix is usually defined as:

$$\sigma_c = \eta_1 \cdot \eta_2 \cdot Vf \cdot \sigma f \qquad (9)$$

where: η_1 - efficiency factor for fibre orientation,

η_2 - length efficiency factor (for short fibers).

For 3-D fibers taking into consideration in Eq (9) the efficiency factors given: by Laws [7] at the assumption $\tau_d = \tau_s$ we obtain the following equations:

$l < lc$

$$\text{MOR I} = \frac{14 \cdot l}{100 \cdot lc} \cdot Vf \cdot \sigma f \qquad (10)$$

$l > lc$

$$\text{MOR II} = \frac{1}{5} \cdot \left(1 - \frac{5 \cdot lc}{14 \cdot l}\right) \cdot Vf \cdot \sigma f \qquad (11)$$

These are correct solutions for pure tension. In case of bending the right hand sides of Eqs (10) and (11) should be multiplied by a constant equal to 2.44, Hannant [8].
Hannant analyzed the post-cracking flexural behaviors of FRC using a rectangular stress block in the tensile zone of the beam and the proposed the following formula for the space orientation (3-D) of the fibers in the composite:

$l < lc$

$$\text{MOR'I} = 1.22 \cdot Vf \cdot \tau \cdot \frac{l}{df} \qquad (12)$$

$l > lc$

$$\text{MOR'II} = 1.22 \cdot \left(1 - \frac{lc}{2 \cdot l}\right) \cdot Vf \cdot \sigma f \qquad (13)$$

The load capacities of the material described by the equations (10), (11) and (12), (13) for polypropylene fibers as functions of the length of the fibers used (Fig.6) and the strength of the fibre-matrix bonds (Fig.7) point to a considerable qualitative similarities and great quantitative differences depending on the proposed solution. In both cases the curves were plotted for bending samples with short, space-distributed fibers (3-D).

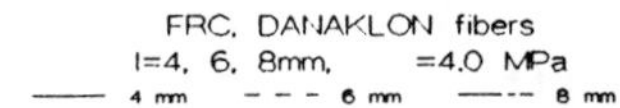

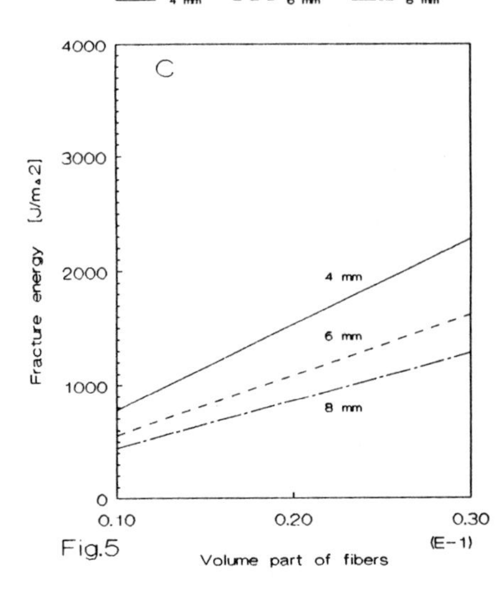

Fig.5

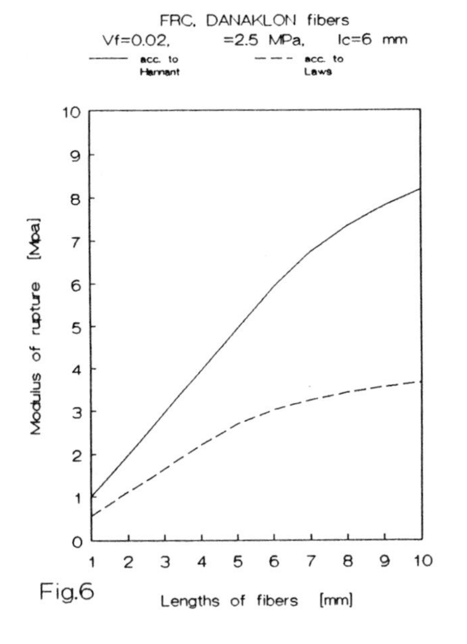

Fig.6

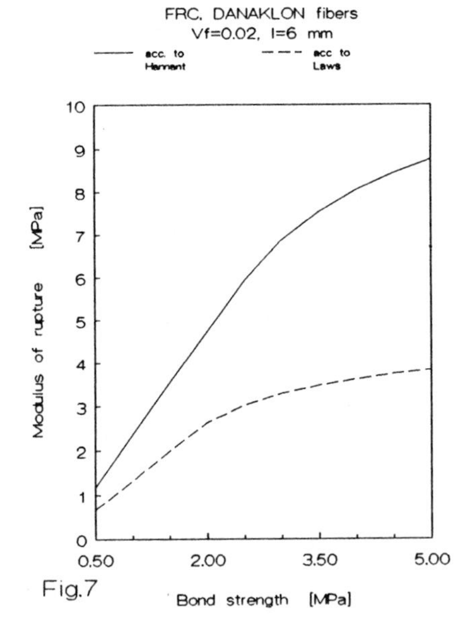

Fig.7

2 Experimental details

Beams 10 x 25 x 100 mm were obtained using the method of casting paste into steel moulds. Short, 6 mm polypropylene fibers, of Krenit Standard type, produced by Danaklon were added directly to the mixture. Before the mass was poured into the moulds it was subjected to the vacuum process and after pouring - to vibration. The properties of the fibres are listed in Table 1. The standard Portland cement 35 was employed. The quartz sand was ground to obtain the specific surface 280 m^2/kg. Acrylic resin produced by BASF (trade name Acronal A 603) was used in the form of its 50% water dispersion.The mix proportion of the FRC paste are given in Table 3. Samples from the series A, B, C, D, E were kept in water for28 days, and afterwards samples D, were dried for 12 h at 60 ^{o}C. Samples from the series F after 24 h storage in the atmosphere of 100% RH at 20 ^{o}C were autoclaved for 8 h at 160 ^{o}C in the atmosphere of saturated steam pressure.

Table 3. Mix proportions of matrix

	Cement	Silica fume	Quartz sand	Acrylic resin	w/c	w/ss	Cure
A	100	-	-	-	0.33		28d/water
B	90	10	-	-	0.33		28d/water
C	97	-	-	3	0.33		28d/water
D	97	-	-	3	0.33		28d/water 12h/60 ^{o}C
E	47.5	5	47.5		0.54	0.26	28d/water
F	47.5	5	47.5		0.54	0.26	autoclav. 8h/160 ^{o}C

2.1 Durability test

In order to determine the durability of the fibers at elevated temperature and pressure, with high pH, the following experiment was carried out. Samples of polypropylene fibers were placed in teflon containers with a water suspension of cement in such a way that being immersed in the solution they were simultaneously situated above the layer of the hydrated cement. The whole assembly was closed in steel autoclaves and subjected to hydrothermal treatment at 100, 160 and 180 ^{o}C for 8 h. The SEM results of the observation are shown in Figs 8-10. It has been found that at 180 ^{o} C the samples of polypropylene fibers became destroyed,whereas after autoclaving at 160 ^{o}C they retained their original habit.

2.2 Flexural strength test

FRC samples, 10 x 25 x 100 mm were loaded in a system of 3 point bending with the spacing of the supports equal to 80 mm. The deformation rate was 15.64 x 10^{-3} cm/min. The experiment was continued until complete separation of the halves of the fractured sample. The

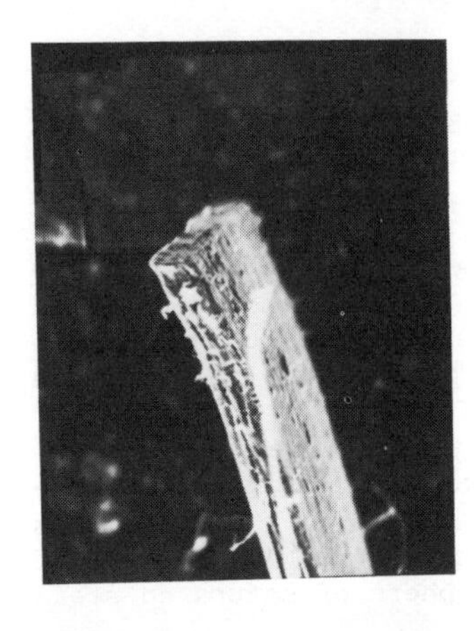

Fig.8
Polypropylene fibre cured 8 h at 100°C in hydrothermal conditions

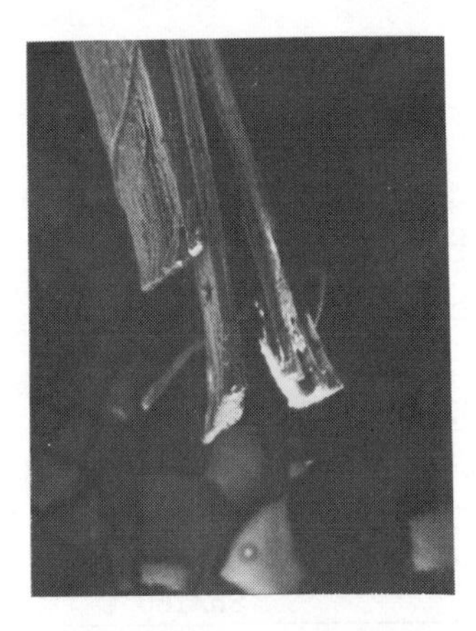

Fig.9
Polypropylene fibers cured 8 h at 160°C in hydrothermal conditions

Fig.10
Polypropylene fibers cured 8 h at 180°C in hydrothermal conditions

fracture work was calculated as the total area under the stress-strain curve.

Table 4. Measured values of work of fracture, WOF [J/m^2]

	Vf = 0.01		Vf = 0.02		Vf = 0.03	
A	1356	σ = 130	2214	σ = 308	3377	σ = 251
B	1736	σ = 104	2996	σ = 48	4104	σ = 256
C	821	σ = 157	1640	σ = 229	2423	σ = 279
D	1484	σ = 226	2531	σ = 313	3580	σ = 627
E	489	σ = 57	1345	σ = 28	2186	σ = 83
F	510	σ = 97	984	σ = 55	1213	σ = 88

σ - standard deviation

3 Results

The post-cracking properties of the tested materials are shown in Table 5 (modulus of rupture) and Table 4 (work of fracture) as the mean values derived from 6 measurements.

Table 5. Measured values of modulus of rupture, MOR [MPa]

	Vf = 0.01		Vf = 0.02		Vf = 0.03	
A	4.22	σ = 1.09	6.67	σ = 0.94	9.27	σ = 1.47
B	4.54	σ = 0.07	6.39	σ = 0.8	8.68	σ = 1.12
C	3.03	σ = 0.80	5.95	σ = 0.74	6.98	σ = 0.78
D	4.66	σ = 0.80	8.60	σ = 0.37	9.99	σ = 0.71
E	3.15	σ = 0.32	5.64	σ = 0.42	7.45	σ = 0.52
F	2.98	σ = 0.36	5.10	σ = 0.10	6.77	σ = 0.66

σ - standard deviation

Figures 11-16 illustrate the dependencies WOF vs Vf in correlation with the equations for the straight line:

$$WOF = A \cdot Vf + B \qquad (14)$$

The constant values of the equations are each time given in the proper figure.
Basing on the theoretical considerations presented earlier an attempt was made to assess the strength of the fibre-matrix inter facial boundary. Assuming that the values γ_{tot} determined under conditions of controlled development of the crack are close to the values of WOF obtained in the bending test of the sample at a small deformation

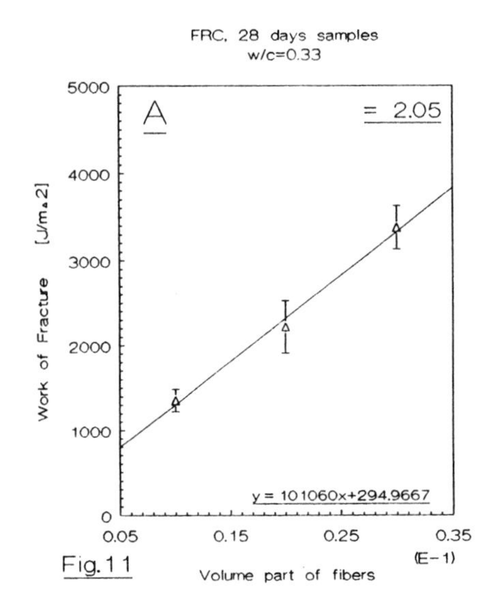

Fig.11

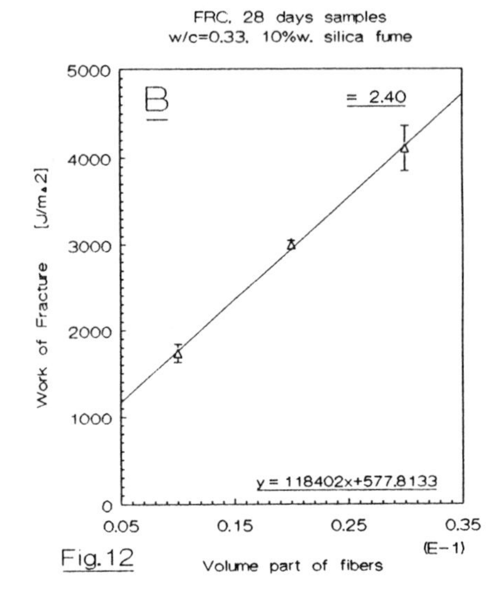

Fig.12

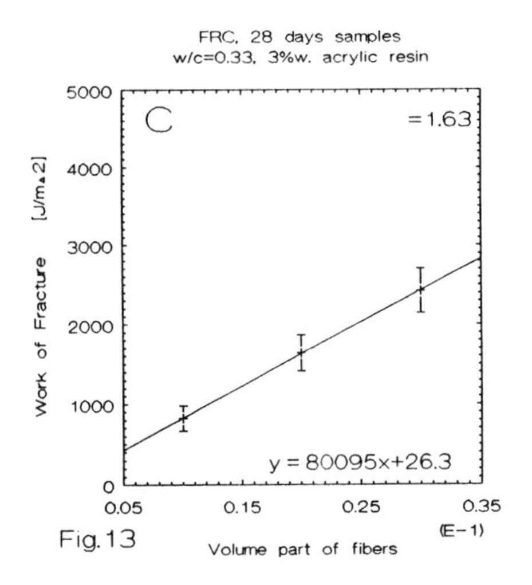

Fig.13

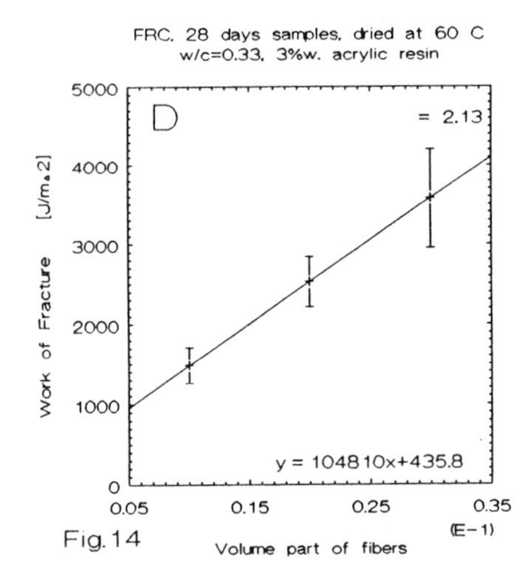

Fig.14

rate, Eqs (4) and (8) for 3-D random fibers, after little transformations may take the form:

$$\gamma_{tot}I = (\frac{\tau\cdot l^2}{12\cdot df} + \frac{\tau^2\cdot l^2}{12\cdot Ef\cdot df})\cdot Vf + (1 - Vf/2)\cdot\gamma_m \quad (15)$$

$$\gamma_{tot}II = (\frac{lc^2\cdot\sigma f}{24\cdot l} + \frac{\sigma f^3\cdot df}{12\cdot Ef\cdot\tau} + (\frac{lc}{2\cdot df})\cdot\gamma_m)\cdot Vf + (1 - Vf/2)\cdot\gamma_m \quad (16)$$

Since with the numbers of reinforced fibers (1-3%) used in our experiment the term (1 - Vf/2) changes within a small range it has been assumed that it is constant. Hence we may write:

$$A\ I = \frac{\tau\cdot l^2}{12\cdot df} + \frac{\tau^2\cdot l^2}{12\cdot Ef\cdot df} \quad (17)$$

$$A\ II = \frac{lc^2\cdot\sigma f}{24\cdot l} + \frac{\sigma f^3\cdot df}{12\cdot Ef\cdot\tau} + \frac{lc}{2\cdot df} \quad (18)$$

When knowing the value of A (directional coefficient of Eq.(14) it is possible using formula (17) or (18) to calculate τ, i.e. the value of the fibre-matrix strength.

All the above formulas, both for WOF and for MOR have been derived for fibers of circular cross section. The polypropylene fibers produced by Danaklon have a rectangular cross-section. Hence before starting the calculations it is necessary to determine the value of the equivalent mean df^*. It was determined on the basis of the following reasoning.

A fibre with the cross-section Af and the perimeter Pf is fastened in a matrix along a length equal to x. This fibre is being pulled out or break under the action of the applied force F which is counterbalanced by its "fastening". Thus we may write:

$$F = \tau\cdot Pf\cdot x \quad (19)$$

The unit stress in the fibre is equal to:

$$\sigma\Phi = (F/Af) = \tau\cdot x\cdot(Pf/Af) \quad (20)$$

A formula, analogous to formula (20), for fibers of circular cross-section equivalent to fibers of rectangular cross-section is:

$$\sigma f = (4\cdot\tau\cdot x)/df^* \quad (21)$$

By equating Eqs (20) and (21) we get: $df^* = 4\cdot(Af/Pf)$ (22)

When observing the cracking process in the tested materials it has been found that only in case of autoclaved samples (series F) the fibers were torn off. Thus in this case formula (18) was used in the calculations. In the other cases the fibers were rather pulled out and not break ,although it was not possible to establish this fact each time definitely. Thus, when calculating τ for the materials in the series A, B, C, D and E Eq.(17) was employed.

The values of the fibre-matrix interfacial boundary strength calculated on the basis of the dependence WOF vs Vf are listed in Table 5.
Figures 17-22 and Table 6 show the dependence MOR vs Vf, correlating then with the equation for a straight line:

$$MOR = A \cdot Vf + B \qquad (23)$$

In the figures there have been also plotted the courses of the dependence MOR vs Vf calculated on the basis of Laws equations (10) and (11) as well as those of Hannant (12) and (13). They were calculated using the values τ listed in Table 5.

4 Discussion

From the theoretical considerations referring to WOF and presented in the Introduction, the detailed solutions of which for the space oriented (3-D) fibres in the matrix are the Eqs (4) and (8), it follows that in the fracturing process of FRC maximum energy is absorbed when the length of the reinforcing fibres is equal to their critical length (Fig.1). However, as the critical length of the fibres is a composite function both of the strength and of the fibre geometry as well as of the strength of the fibre-matrix interfacial boundary, each case of FRC should be considered separately. From the considerations,the illustration of which are Eqs (15) and (16) it follows that the mechanical properties of the matrix have very small influence on the ability of the composite to absorb energy in the process of its fracture ($\gamma_{tot}I$ and $\gamma_{tot}II$ for Vf=0.02 is aprox. 1-3 kJ/m^2, γ_M for a cement paste matrix is only about 20 J/m^2). This conclusion is only apparently correct since the strength of the matrix which is a function of its micro structure, mineral composition etc. obviously affects the strength of the fibre-matrix bond, one must realize that in the fracture process when the fibre is separated from the matrix, the shearing takes place on the side of the weakest component (in this case-matrix), unless, obviously,this component is the interfacial boundary (which in case of FRC is very probable). If when designing the composite, we want to obtain one of great capacity for energy absorption it must be remembered that the application of a longer fibre does not necessarily produce this effect. As shown in Figs 2,3,4 and 5, only a deliberate decision based on the knowledge of both the properties of the employed fibres and the strength of the fibre-matrix bond will make possible the attaining of such result. In case the ability of FRC to carry load after the fracturing of the matrix is still retained (this quality being defined as the modulus of rupture -MOR) the problem is less complicated. The value of MOR increases proportionally with the increase of the length of the fibres used as well as with the strength of the fibre-matrix bond. Figs 6 and 7 illustrate these dependences for the polypropylene fibres produced by the firm Danaklon. The curves corresponding to the solutions by Laws and Hannant have been plotted on the curves. These curves are of the same

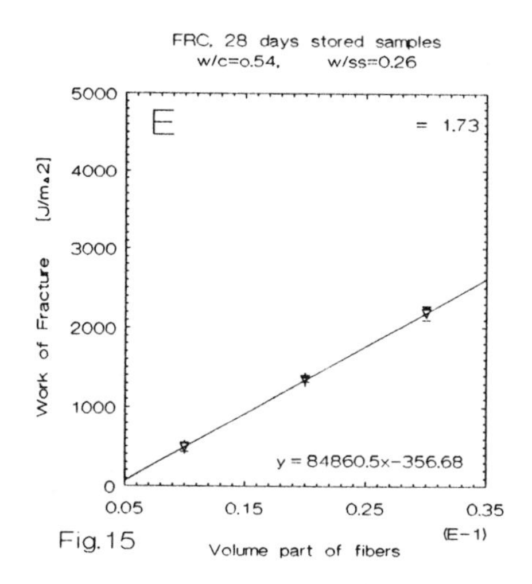

Fig.15

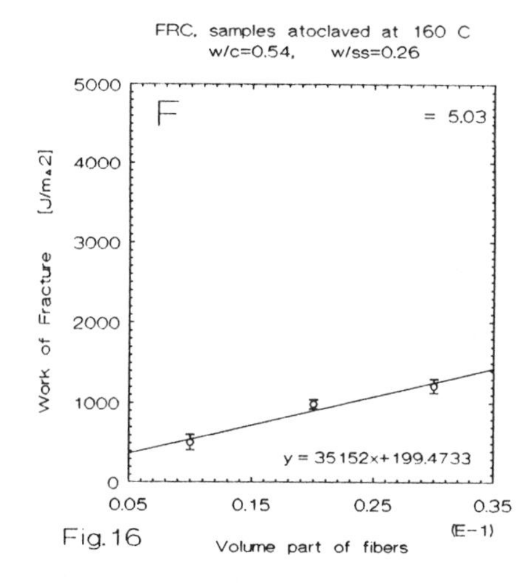

Fig.16

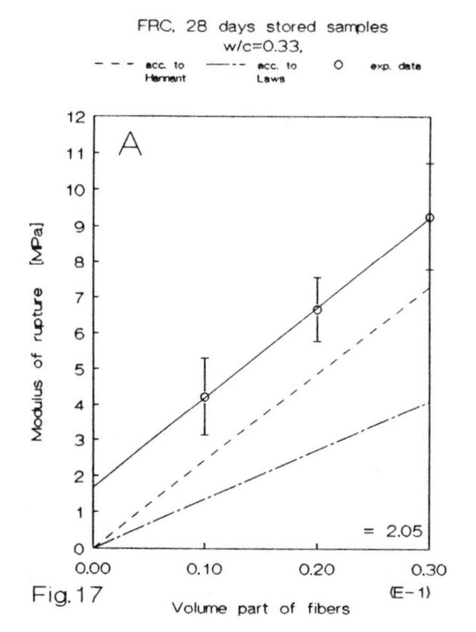

Fig.17

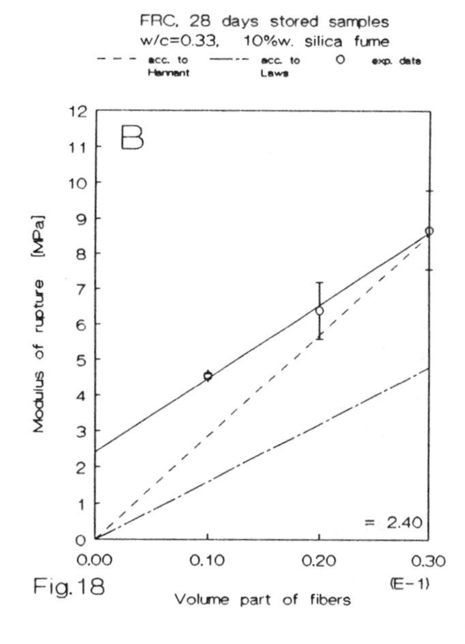

Fig.18

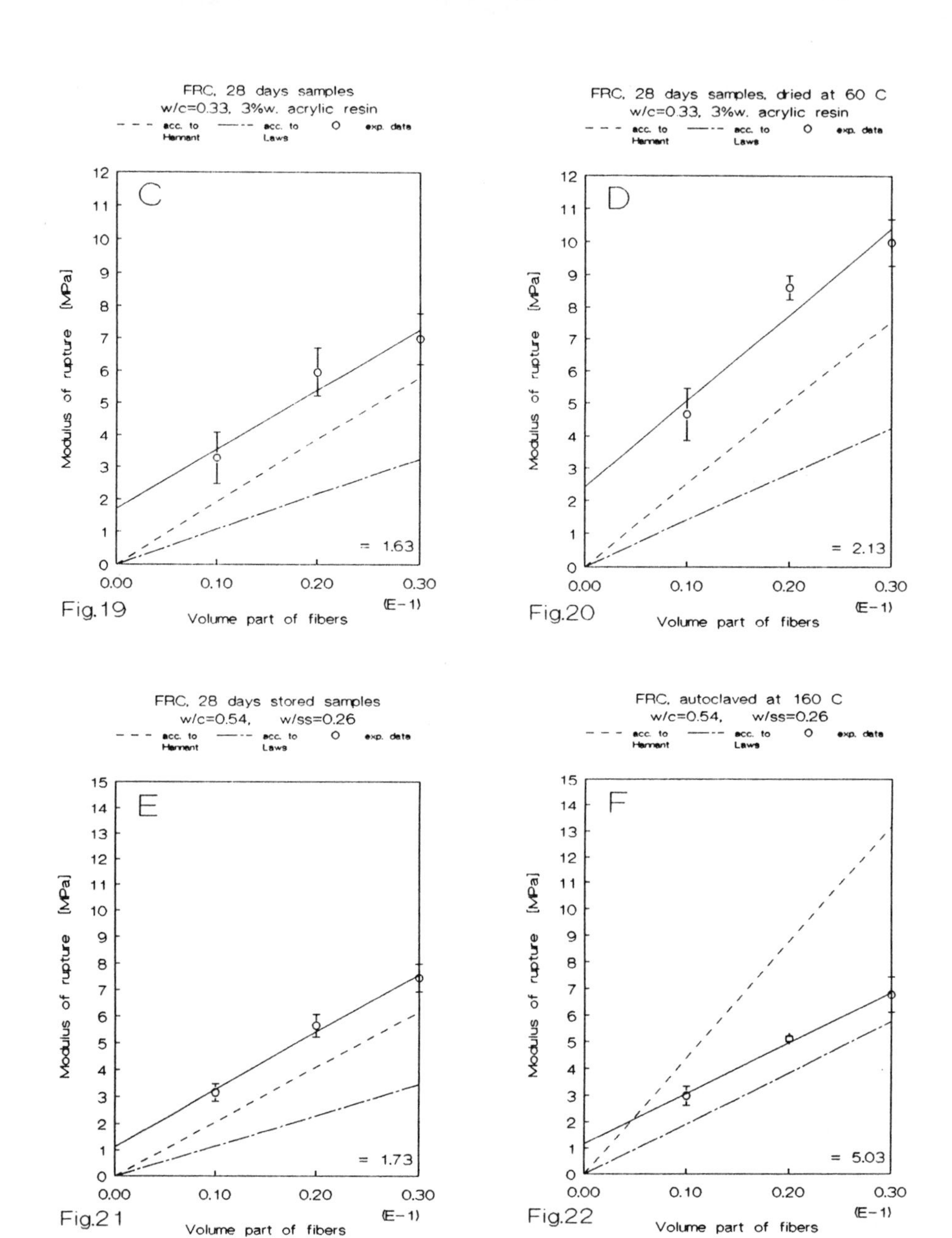

Fig.19

Fig.20

Fig.21

Fig.22

character in terms of quality, but differ in terms of quantity. The values of the energy connected with pulling out of the fibres γp, secondary distribution of stresses γr, energy of formation of new surfaces, as well as their sum γ_{tot}, calculated from Eqs (1),(2),(3) and (5),(6),(7) for polypropylene fibres, are listed in Table 2. These values have been calculated for Vf=0.02 for various fibre lengths (4,6 and 8 mm) and the strength of the fibre matrix boundary. It follows from these data that for selected variables in the fracture process of FRC, pulling out of the fibres absorbs most energy (76-98%), next comes the secondary distribution of stresses (0.6-20.3%), whereas the formation of new surfaces plays a marginal role only (0.8-3.6%). In case when the fibres are torn off, the share of the secondary distribution of stresses and of the energy of forming new surfaces is on the increase. The results of experimental studies are listed in Table 4 (WOF) and Table 6 (MOR) and in Figs 11-22. From the dependence WOF vs Vf, the values of the strength of the fibre-matrix bond have been calculated using formula (17) or (18). The values (τ),presented in Table 6, should be treated as estimated data,obtained by means of indirect measurements.

Table 6. Calculated values of τ from WOF vs Vf

	AI or AII	τ [MPa]
A	101060	2.05
B	118402	2.40
C	80095	1.63
D	104810	2.13
E	84860	1.73
F	35152*	5.03

* - AII

It follows from these data:

- an addition of silica fume amounting to 10% of the amount of cement, in the case of FRC reinforced with polypropylene fibres, caused an increase of work expenditure in the fracture process. Such a phenomenon takes place when τ increases with simultaneous satisfaction of the condition $l<lc$. The calculated values of τ presented in Table 5 are equal to, respectively:for regular FRC (A)-2.05 MPa, for FRC with 10% addition of silica fume (B)-2.40 MPa.

- an addition of acrylic resin in the case when the samples were examined immediately after taking them out of the water, caused a reduction of the determined values of WOF. After the samples had been dried at 60°C, this process causing polymerization of the acrylic resins, greater capacity of the material for energy absorption was observed as well as an increase of the value of τ. It must be added, however, that the addition of resin did not result in any considerable quantitative improvement of the properties of the material in comparison with the materials of the A series (without additions).

-samples of the series E and F were made of mixtures essentially differing from the others. Besides silica fume a relatively great quantity of ground quartz sand was added. The ratio water/cement which determines the mechanical properties of the matrix, was increased to 0.54 giving consideration to the workability of the mix. Samples of the E series were kept for 28 days immersed in water, those of series F were autoclaved at 160°C. A significant effect of autoclaving both on the mechanical properties of the materials and their fracture mechanism as well as on the calculated values of the strength of the interfacial boundary was observed. Material E, not autoclaved, showed relatively poor capacity for energy absorption (high w/c), and low calculated value τ = 1.73 MPa. The process of autoclaving, besides causing evident changes in the mineral composition and in the micro structure of the matrix, strengthened the fibre-matrix bond. It has been observed that during the fracturing process of the material, the fibres were torn off. A decrease in the value of WOF was observed. The calculated value of τ, at the assumption that the fibres are being torn off, was equal to 5.03 MPa. The authors believe this value to be overestimated, assuming that the fibres differ in the cross-sectional area and not all become divided into single elements. Hence it may be assumed that the fibres are not torn off in each case.

Figures 17-12 and Table 6 show the values of MOR vs Vf obtained experimentally. Against this background there are plotted the curves calculated from the equations proposed by Laws (10),(11) and Hannant (12), (13), these calculations being carried out on the basis of the value of τ obtained from the dependence WOF vs Vf listed in Table 5. From the presented results it follows that the highest values of MOR were obtained in case of FRC samples with 10% SF and an addition of acrylic resin (after its polymerization at 60°C). Contrary to earlier assumptions, autoclaved samples in spite of the calculated high values of τ = 5.03 MPa, did not exhibit any increase in the load carrying capacity (MOR). On the curves shown in Figs 17-22 one can see that the experimental data do not correlate with the equation MOR= A·Vf, which would be in agreement with the proposals of both Laws and Hannant, but rather with the equation of the type MOR=K+A·Vf. The values of the constant K in the analyzed cases are equal from 1 to 2.5 MPa. The directional coefficient of the straight line of the experimental data in case of FRC with low τ is dose to the model of Hannant, whereas in the case of a material with great strength of the fibre-matrix bond it is rather conforming to Lows formula. This is clearly seen in Figs 20 and 21 for the materials E and F.

5 Conclusion

1. The properties of FRC in the post-cracking conditions depend only on the mechanical properties of the reinforcing fibres and their adhesion to the matrix.
2. The possible existence of a correlation between the strength of

the matrix and that of the fibre-matrix interfacial boundary cannot be excluded.

3. The work of fracture (WOF) attains its highest value when the length of the reinforcing fibres is equal to their critical length.
4. Using the dependence WOF vs VΦ one can estimate the value of the strength of the the fibre-matrix bond.
5. An addition of silica fume or acrylic acid to the cement paste increases the strength of the interfacial boundary.
6. The process of autoclaving FRC increases the value of τ, this, however, is not accompanied by any improvement of the mechanical properties of FRC, at least with reference to polypropylene fibres.
7. The effect of τ on the modulus of fracture (MOR) is not unambiguous. With low values of the strength of the interfacial boundary MOR correlates with the equations proposed by Hannant whereas in case of high values-with those proposed by Laws.

6 References

1. Kelly A., (1970) **Proc.R.Soc.**, A 319, 95
2. Outwater J.D.,Jr., Murphy M.C., (1969) 26 Annual Conference, Society of Plastics Industry, **Reinforced Plastics-Composite** Division, Paper 11 C
3. Cottrell A.H., (1964) **Proc.R.Soc.**, A 282, 2
4. Kelly A., (1973) **Strong Solids**, Clarendon Press, Oxford
5. Masterson T.V., Atkins A.G., Feldbeck D.K., (1974) Interfacial Fracture Energy and the Toughness of Composites. **Jnl. Mater. Sci.**, No.9
6. Aveston J., Mercer R.A., Sillwood J.M., (1974) **Fibre-reinforced Cement**-Scientific Foundations for Specifications Standards, Testing and Design. Nat.Phys.Lab.Conf.Proc.,
7. Laws V., (1971) The Efficiency of Fibrous Reinforcement of Brittle Matrix. **Jnl. Physics** D: Applied Physics, No.4.
8. Hannant D.J., (1978) **Fibre Cements and Fibre Concretes**, John Wiley
9. Krenchel H., Jensen H.W.,(1980) **Organic reinforcing fibers for cement and concrete**, 'Fibrous Concrete. The Concrete Society', Procedings of the Symposium on Fibrous Concrete, London
10. Davies D.,(1990) Applications of Krenit Fibres for High Performance Mortars, **Nordisk Betong** 2-3

7 Appendix

The following symbols are used in the paper:

Vf - volume part of fibres
σf - fibre strength, or average pull-out stress in fibres
Ef - E -modulus of fibre
l - length of fibres
lc - critical length of fibres
df - diameter of fibre

MOR - modulus of rupture
WOF - work of fracture
γ_m - fracture energy of matrix
γ_p - fibre pull out energy
γ_r - redistribution of stresess energy
γ_s - forming of new surfaces energy

28 TENSILE PROPERTIES OF STEEL FIBRE REINFORCED CONCRETE

M. P. LUONG and H. LIU
CNRS-LMS, Ecole Polytechnique, Palaiseau, France
J. L. TRINH and T. P. TRAN
CEBTP, St Rémy lès Chevreuse, France

Abstract
Since concrete is generally weak and brittle in tension compared to its capacity in compression, steel fibre reinforcing is a practical means to improve the tensile performance as well as the tensile post-cracking behaviour of this cementitious material. Question still remains on a relevant evaluation of these properties in the current practice. This paper describes experimental results which have been obtained in our laboratory for Mode I fracture of steel fibre reinforced concrete, using special test methods for evaluating the fracture toughness, ductility and energy absorption capacity of the cementitious composite. The fracture toughness has been evaluated by various toughness indices reported in literature which are easily determined from the load-displacement curve of direct tension tests. The tensile cracking process has been detected by analysing the signal evolution of ultrasonic waves propagating through the specimen which is subjected to increasing static loads. A nonlinear analyser based on a multidimensional Fourier transform permits to separate the nonlinear part from the linear one of the response.
Keywords: Fracture Tests, Steel Fibre Reinforced Concrete, Mode I Test Specimen, Post-cracking, Toughness Index, Nonlinear Analyser.

1 Introduction

The tensile strength of concrete is important for concrete structures as a basic mechanical property and factor of good durability. Since concrete is generally weak and brittle in tension compared to its capacity in compression, steel fibre reinforcing is a practical means developed for a better control of the tensile performance as well as the tensile post-cracking and post-yield behaviour[1]. Steel fibres, distributed in the brittle material, can resist against the formation or propagation of cracks thanks to fibre bridging, - which develops owing to the fibre-matrix bond mechanism and fibre stiffness, - around the crack faces.

However debates still exist on the relevance of the usual test methods (for current concretes) capable to characterize properly these properties. Among other questions it is sometimes also quoted that the test specimens are not appropriate with respect to fibre length.

This paper describes special testing devices in use for measuring experimentally Mode I (Luong 1989) fracture resistance of steel fibre reinforced concrete. The proposed testing arrangements are very simple,

Fibre Reinforced Cement and Concrete. Edited by R. N. Swamy. © 1992 RILEM.
Published by E & FN Spon, 2-6 Boundary Row, London SE1 8HN. ISBN 0 419 18130 X.

practical and reliable as well as rigorous from the point of view of mechanics.

2 Uniaxial tensile strength

Many types of tensile tests have been devised, using equipment and techniques that range from the crude and empirical, with results that are almost impossible to interpret analytically, to the theoretically elegant, which are almost impossible to execute practically. But lack of confidence in test methods may be slowing acceptance, increasing costs.

The greatest difficulty in the direct test for the determination of tensile strength of concrete is the gripping of specimens. To achieve uniform tensile stress distribution and for easy gripping, especially prepared specimens are required which are difficult to machine. As a result, a great number of indirect methods have been developed for determining tensile strength of concrete and rock materials (Bernaix 1969, Hawkes & Mellor 1970, Roberts 1977, Lama & Vutukuri 1978, Arnaout et al 1990).

In spite of this need for an accurate knowledge of tensile strength, a complete satisfactory tensile test for building materials has not been developed so far. Several methods, more or less sophisticated, were later developed, among which the splitting test has received special attention and has been standardized in many countries. Unfortunately the tensile strength of concrete is found to be highly sensitive to the testing techniques and to the geometry of the specimens used. This fact has prevented a deep understanding of the behaviour of concrete under tension and, therefore, has made the practical application of the tensile strength values obtained from tests highly unreliable.

Most concrete materials are more or less brittle ; when unconfined, the test samples cannot yield plastically to relieve the stress concentrations that are produced at localized points around the specimens, where these are gripped to be pulled apart by the testing machine. Consequently premature failure originates at these points. Difficulties in ensuring truly axial loading also exist, so that the specimen is likely to be twisted or bent when gripped and pulled apart from either end.

Various direct and indirect testing methods have been developed, in attempt to resolve these difficulties and to determine the tensile strength of concrete and rock materials. These can, in general, be grouped as follows :

i. Bending tests : bending of prismatic and cylindrical specimens and bending of discs.
ii. Hydraulic expansion tests.
iii. Diametral compression of disc : brazilian test and ring test.
iv. Miscellaneous methods : diametral compression of cylinders, diametral compression of spheres, compression of square plates along a diameter and centrifugal tension.

The split cylinder brazilian indirect tension test (Carneiro 1947) has often been adopted and provides a simple means for obtaining the tensile strength of concrete as well as rocks. The popularity of this test stems from the fact that it is not only easy to perform, but it often uses the same cylindrical specimens and testing equipment that are used to determine unconfined compressive strength. Despite its popularity, however, it does not provide a precise measurement of the tensile strength. Past studies (Wright 1955) have shown that the tensile strength as determined by the split cylinder tests may overestimate the true tensile strength of brittle materials by as much as 50 %.

3 Proposed direct tensile test

Direct tension applied to a specimen is theoretically the simplest method of determining the tensile strength. It is both practical and theoretically meaningful. The direct test is of greater fundamental value because the stress field of an isotropic specimen is determined directly by the applied loading and the boundary conditions, irrespective of the material properties. Indirect tests have the inherent disadvantage that a stress-strain relationship must be assumed in order to obtain usable results (Habib et al 1964). The usual assumption of linear elasticity, with equal moduli in compression and tension, is invalid for some concretes and many rocks. However, there are major difficulties in gripping the specimens and applying a load parallel to the axis of the specimen.

The proposed test specimen (Luong 1986) is a cylindrical tube (Fig. 1). The test specimen is easily prepared, with two parallel flat ends and two inversed tubular coaxial borings. The external surface requires no particular preparation.

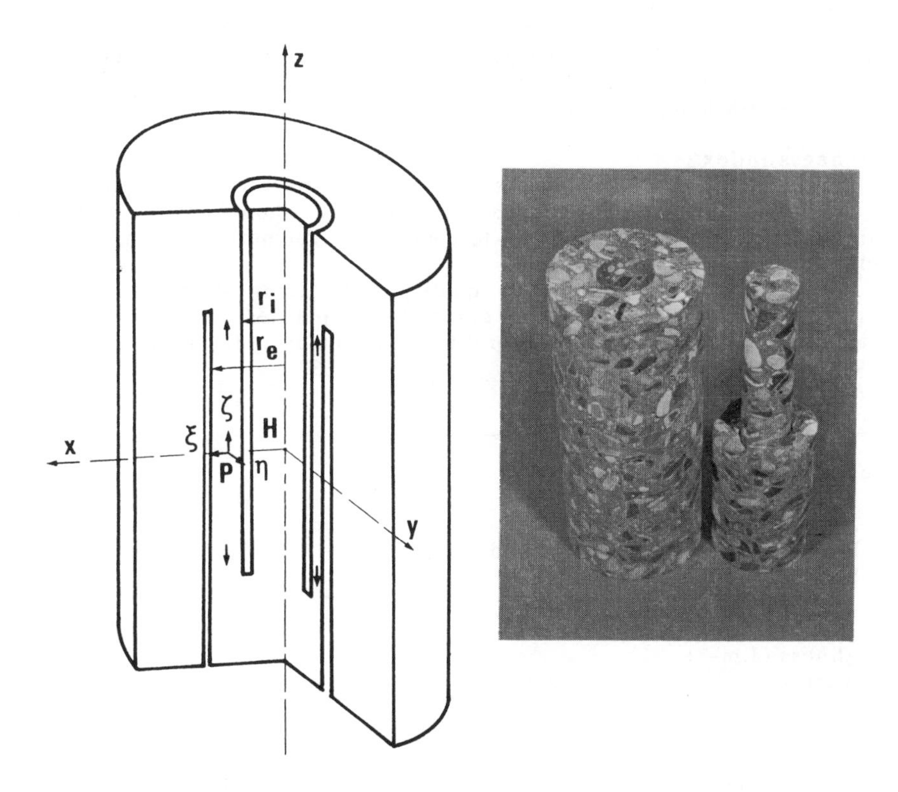

Fig.1. Proposed tensile test specimen.

Concrete mix

The mix used is (per m^3)

Aggregate size	12.3/20 mm :	485 kg
	8/12.5 mm :	359 kg
Sand	1.6/5 mm :	156 kg
	0/1.6 mm :	701 kg
Cement (CPA HP Ciment Lafarge Saint Vigor) :		385 kg
Superplasticizer Sikament HR 401 :	9.6 kg	
Mixing water :		137 litres
Dramix steel fibres	(60 mm long)	40 kg

The curing conditions provided during a fortnight period are : ambient temperature (20 °C) and relative humidity : > 95 % (mist conditioning room).

The compressive strength at 28 days old, measured on the standard cylinders (0.16 m in diameter and 0.32 m high), are 56.2 MPa, 57.3 MPa and 59.9 MPa.

Small test specimen

The small specimens were cylinders 120 mm long and 70 mm in diameter. Several tests with dimensions (r_e = 25 mm for external radius and r_i = 13 mm for internal radius) gave values of 2.185 MPa, 1.25 MPa, 1.41 MPa, 2.16 MPa and 3.07 MPa for the uniaxial tensile strength $f_t = F/[\pi(r^2_e - r^2_i)]$. These scattered results suggest a very strong scale effect due to the high influence of fibre distribution in the failure area (Zhan et al 1991), the fibre length being 60 mm.

Toughness index

A simple method of evaluating the post-cracking performance of materials is based on a toughness index that ideally should be a material characteristic, independent of the test specimen geometry, dimensions and loading conditions.

For ease of comparison and quality control, it should be readily obtained from a load-deflection curve of simple tests. Some toughness indices have been reported in literature (Barr et al 1982, Wang & Backer 1989).

Toughness indices	
ACI	5.14
Johnston (n = 3)	1.86
Barr et al.	0.375
Wang (n = 20)	3.07
Toughness ($J.m^{-2}$)	
Wang	260

Table 1. Toughness indices calculated for mode I failure on a steel fibre reinforced concrete specimen.

In order to ensure the feasibility of the tensile test, a finite element calculation - using an isotropic, homogeneous and linear elastic constitutive law - has optimized the geometry of the specimen (Luong 1990). Numerical results have shown that the tensile stresses are uniformly distributed in the central part of the intermediary tube subject to tension. In the vicinity of the notch ends, mixed mode fracture generally do not occur because of its higher resistance if compared to mode I strength. This has been shown by several series of experimental studies on concrete and rock materials (Luong 1989 and 1990). Usual imperfections such as lack of end surface parallelism less than 10 ° or eccentricities due to the borings less than 10 % have slight influence on the tensile strength obtained with this type of specimen.

The proposed configuration converts the applied load on the specimen into a tension, so that the usual compressive test machine can be used. The cylindrical symmetry permits self alignment of loads parallel to the axis of the specimen. The test requires no special device for specimen gripping and compressive loads for example can be applied without any precaution. There will be no tendency to cause bending so that unwanted stress concentrations are avoided. In particular, this test can be conducted in hostile environmental conditions : high temperature testing, long duration testing such as creep or relaxation experiment, immersed conditions, high confining pressure, under radioactive irradiation, shock loading and so on.

The main advantage of the proposed direct tensile test is evidenced by the uniformity of uniaxial tension all along the test specimen. Force-displacement curve can be readily obtained as shown on (Fig. 2). Thanks to its cylindrical symmetry, there is no bending or torsional stresses, no stress concentrations arising from geometrical irregularities of the sample and no end restraint effects perturbing the stress field : most of observed failures occur far from notch ends.

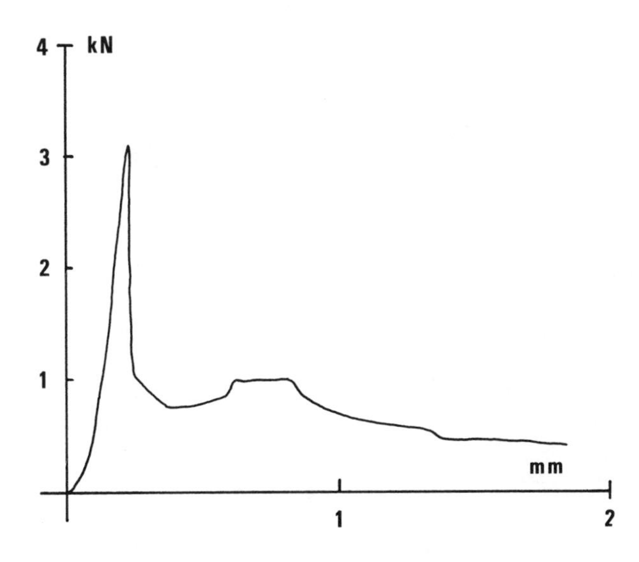

Fig.2. Force-displacement curve obtained from direct tension test.

The toughness index is a measure of the energy absorbed during fracture in the nonlinear portion of the load-deflection curve. It gives the numerical value to the toughness and is independent of the units used. The index is particularly useful in analysing the rheological behaviour of the material. It readily discriminates between high and low energy absorbing materials. Table 1 shows some values of the so-defined toughness indices based on the load-deflection curve.

Large test specimen

Some large specimens (10 cm x 50 cm x 50 cm) with 4000 cm³ of volume under tension have been prepared and maintained during testing by a steel frame (Fig. 3) in order to keep the main advantages of the proposed cylindrical test specimen. The tests were performed on a closed-loop electro-hydraulic loading machine of 50 kN capacity. Four electro-mechanical extensometers of 30 mm and 60 mm gauge length were glued on the specimens to measure the deformation at different locations. One LVDT measured the deformation over the crack zone.

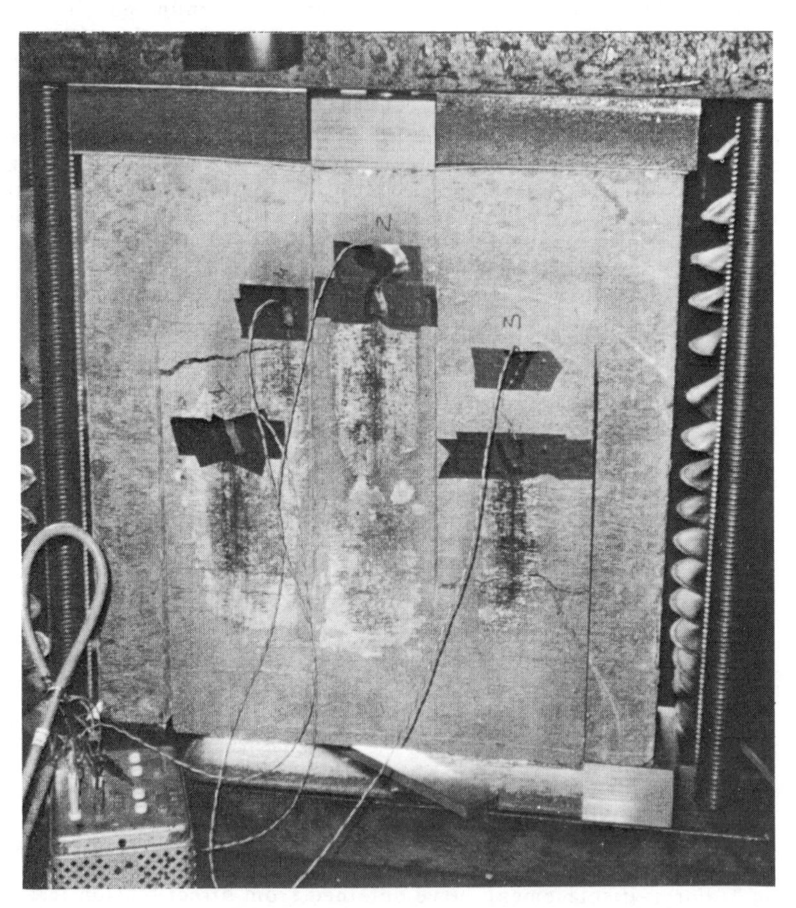

Fig.3. Large tensile test specimen equipped with strain gauges.

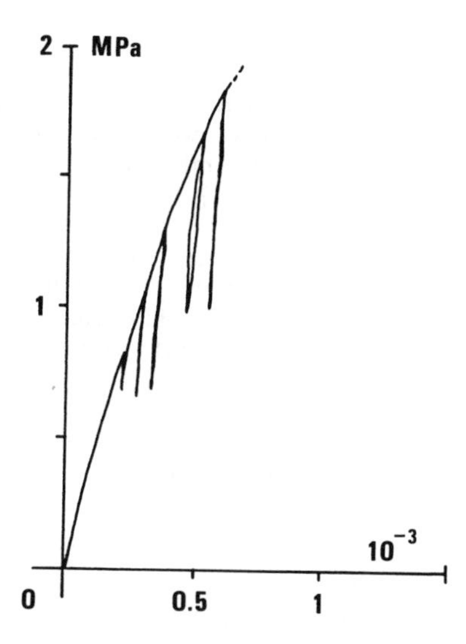

Fig.4. Stress-strain curve of steel fibre reinforced concrete subjected to increasing static tension.

A typical stress-strain curve using strain gauge is shown in Fig. 4. Fig. 5 gives the force-displacement curve. Using Hillerborg's technique (Hillerborg 1989), the characteristic length of the tested cementitious composite has been determined thanks to the relationship :

$$\ell_{ch} = EG_F / f^2{}_t = 248 \text{ mm}$$

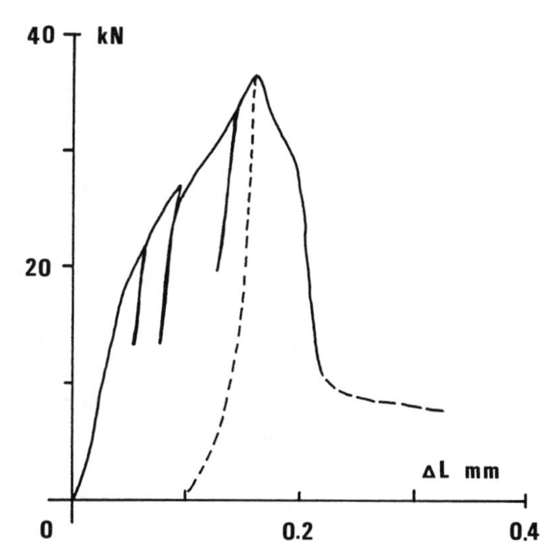

Fig.5. Tension-displacement curve of steel fibre reinforced concrete subjected to increasing static tension.

4 Nonlinear analyser

Linear analysers based on one-dimensional Fourier transforms (FTs) and modal analysis programmes are popular tools used by experimentalists in linear structural dynamics. In this experimental work, an input-output non-parametric approach (Liu & Vinh 1991) has been chosen to portray the nonlinear behaviour of reinforced concrete subject to tension.

A nonlinear functional Volterra series has been used to detect the occurrence of the nonlinear behaviour of steel fibre reinforced concrete subject to increasing tensile loading. In this functional, the total response of the system y(t) is decomposed into components of various orders.

(1) $$y(t) = y_1(t) + y_2(t) + \dots + y_n(t)$$

Each component is defined by a functional

(2) $$y_n(t) = \int_{-\infty}^{\infty} \dots \int_{-\infty}^{\infty} h_n(\tau_1,\tau_2,\dots,\tau_n) \prod_{k=1}^{n} x(t-\tau_k)\, d\tau_k$$

The first order component is described by linear convolution

(3) $$y_1(t) = \int_{-\infty}^{\infty} h_1(\tau)\, x(t-\tau)\, d\tau$$

where x(t) denotes the input function
$h_1(t)$ the first order impulse response that describes the linear behaviour of the system.

The other components request more than one time variables and multidimensional signal processing is needed. As an example, let us describe the second order (nonlinear) component of the response. From Eq.(2), we have the following expression :

(4) $$y_2(t_1,t_2) = \int_{-\infty}^{\infty}\int_{-\infty}^{\infty} h_2(\tau_1,\tau_2)\, x(t_1-\tau_1)\, x(t_2-\tau_2)\, d\tau_1 d\tau_2$$

where τ_1,τ_2 are the two time variables and $h_2(\tau_1,\tau_2)$ denotes the second order impulse response. Within this mathematical framework, two dimensional Fourier transform is appropriate for the study of the material behaviour in the frequency domain.

(5) $$\mathscr{F}_2\,[h_2(\tau_1,\tau_2)] = \int_{-\infty}^{\infty}\int_{-\infty}^{\infty} h_2(\tau_1,\tau_2)\, \exp(-j\omega_1\tau_1 - j\omega_2\tau_2)\, d\tau_1 d\tau_2$$

This expression defines the second order transfer function $\mathscr{H}_2$ with two circular frequency variables ω_1 and ω_2 . $\mathscr{F}$ is the Fourier transform operator.

(6) $$\mathscr{H}_2\,(\omega_1,\omega_2) = \mathscr{F}_2\,[h_2(\tau_1,\tau_2)]$$

In this application, a single impulse has been used as input

$$x(t) = a\,\delta(t-T)$$

Then from (1) and (2), we obtain :

$$(7)\quad y(t) = \sum_{i=1}^{n} y_i(t) = \sum_{i=1}^{n} \int_{-\infty}^{\infty} \cdots \int_{-\infty}^{\infty} h_i(\tau_1,\tau_2,\ldots\tau_i) \prod_{k=1}^{i} a\,\delta(t-T-\tau_k)\, d\tau_k$$

$$= \sum_{i=1}^{n} a^i\, h_i(t-T)$$

From the single impulse of m different test magnitudes, Eq.(7) gives :

$$(8)\qquad y^m(t) = \sum_{i=1}^{n} a_m^i\, h_i(t-T)$$

where m denotes the mth test.

Under matrix form, Eq.(8) is as follows :

$$(9)\qquad \{y\} = [a]\,\{h\}$$

where

$$(10)\qquad \{y\} = (y^1(t), y^2(t), \ldots, y^m(t))^T$$

$$\{h\} = (h_1(t-T), h_2(t-T), \ldots, h_n(t-T))^T$$

$$(11)\qquad [a] = \begin{bmatrix} a_1^1 & a_1^2 & \ldots & a_1^n \\ a_2^1 & a_2^2 & \ldots & a_2^n \\ \ldots & \ldots & \ldots & \ldots \\ a_m^1 & a_m^2 & \ldots & a_m^n \end{bmatrix}$$

or

$$(12)\qquad \{h\} = [a]^{-1}\,\{y\}$$

This equation gives the impulse response of various orders with only one time variable.

Test procedure and results

The pulse-transmission method using videoscan piezoelectric transducers V150 (0.25 MHz) have been applied as nondestructive testing, in conjunction with a pulser-receiver Panametrics 5052PR which provides high-energy broadband performance. The pulser section produces an electrical pulse to excite a piezoelectric transducer, which emits an ultrasonic pulse. The pulse travels through the specimen which is subjected to a given static tension, to a second transducer acting as a receiver. The transducer converts the pulse into an electric signal which is then amplified and conditioned by the receiver section and made available for nonlinear analysis.

Fig. 6 presents the evolution of the ultrasonic pulse travelling through the specimen, showing (1) the occurrence of tensile cracking which modifies the longitudinal part (the faster) of the pulse signal and (2) the effects of microcracking which affect its transversal part (the slower). Fig. 7 shows the change of dissipated energy when the static tensile load increases. It demonstrates the threshold of microcracking which generates nonlinear effects in wave propagation mechanisms.

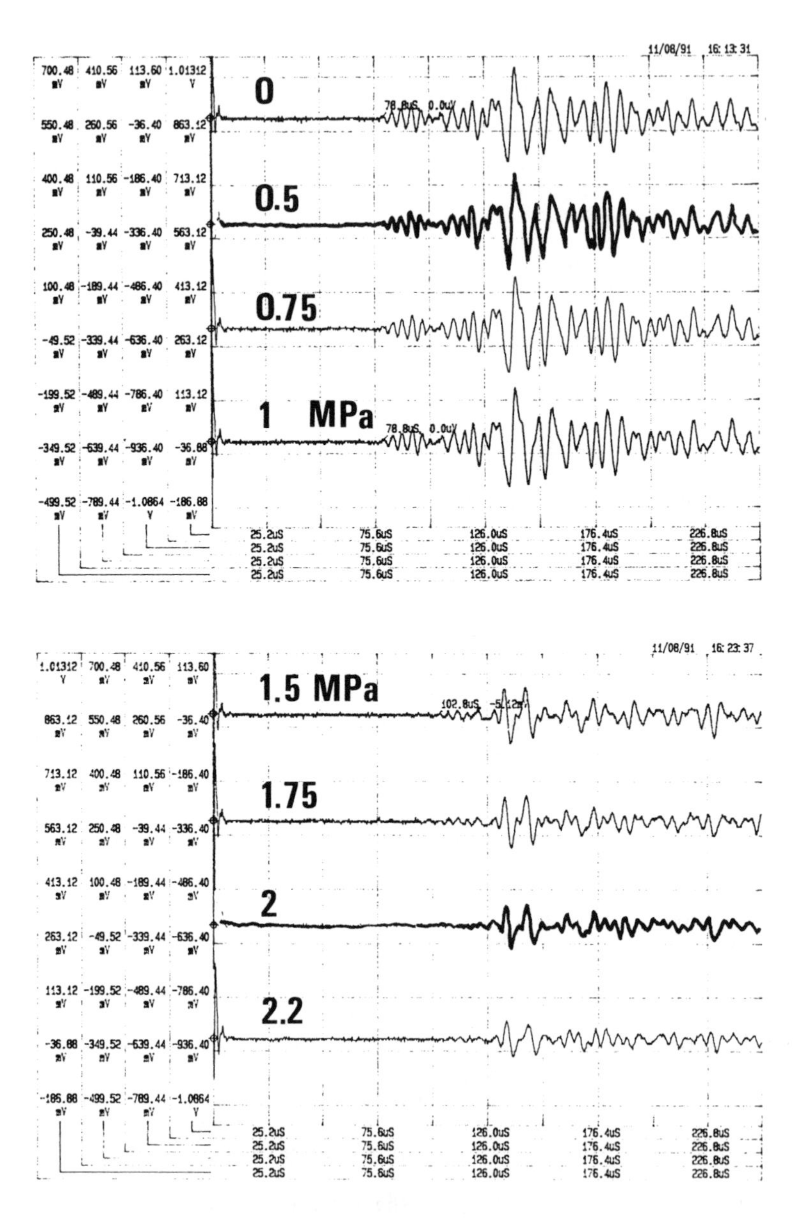

Fig.6. Evolution of the ultrasonic pulse travelling through the specimen under static tension loading.

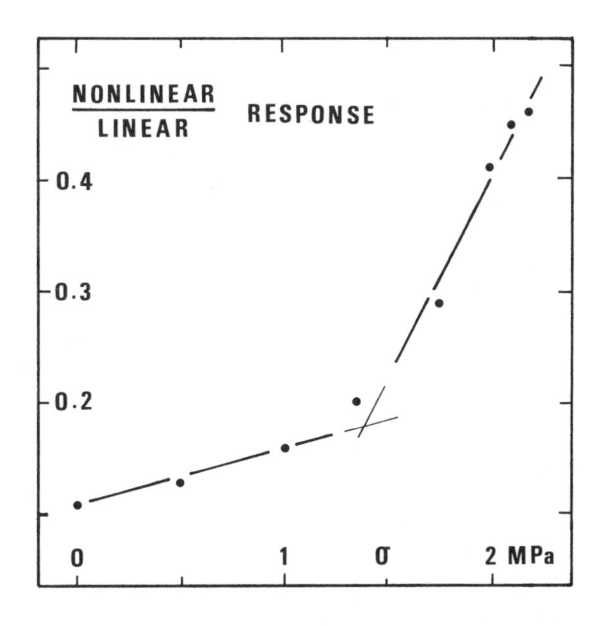

Fig.7. Variation of the ratio nonlinear/linear response of the ultrasonic pulse travelling through the specimen subject to static tension loading.

5 Concluding remarks

The unsuitability of the usual testing techniques for the determination of tensile strength has suggested these devices for evaluating the direct tensile properties of steel fibre reinforced concrete specimens which correspond respectively to mode I fracture resistance.

The proposed testing arrangements for direct tension strength measurements simplify the loading equipment by the use of a uniaxial test machine with the combined compression loading frames. They are practical and reliable. The method does not require special testing equipment and/or ancillary mechanical components.

In spite of a significant scattering of the experimental results obtained from the cylindrical tensile specimens, it can be thought that a larger size, approaching the characteristic length ℓ_{ch} , may be fully representative of the tensile behaviour of FRC materials.

They also facilitate the testing procedure when the concrete materials are subjected to severe and hostile environmental conditions. And finally they can all be used as routine tests more readily achieved than on the conventional triaxial tests in the cases of low mean stress level.

The nonlinear analyser reveals to be very useful for the detection of microcracking process announcing the occurrence of propagating damage in FRC material subject to tension. It can be used to monitor nondestructively and continously the whole fatigue damage process of FRC materials so that the damage mechanisms can be quantitatively explained.

6 References

Arnaout S., Pavlovic M.N. and Dougill J.W. (1990) A new method for the measurement of tensile properties of curved specimens. **Materials and Structures/Matériaux et Constructions**, 23, 296-304.

Barr B.I.G., Liu K. and Dowers R.C. (1982) A toughness index to measure the energy absorption of fibre reinforced concrete. **Int. J. of Cement Composites and Lightweight Concrete**, 4(4), 221-227.

Bernaix J. (1969) New laboratory methods of studying the mechanical properties of rocks. **Int. J. Rock Mech. Min. Sci. & Geomech. Abstr.**, 6(1), 43-90.

Carneiro F. (1947) Une nouvelle méthode d'essai pour déterminer la résistance à la traction du béton. **Réunion des Laboratoires d'Essais de Matériaux**.

Chern J.C., Young C.H. and Wu K.C. (1989) A nonlinear model for mode I fracture of fiber reinforced concrete. **Fracture Mechanics : Application to Concrete**, ACI SP-118, 91-112.

Habib P., Morlier P. et Radenkovic D. (1964) L'application de l'essai brésilien dans la mécanique des roches. **Symposium Rheology and Soil Mechanics**, Grenoble, France.

Hawkes I. and Mellor I. (1970) Uniaxial testing in rock mechanics laboratories. **Engineering Geology**, 4(3), 177-285.

Hillerborg A. (1989) Fracture mechanics and the concrete code. **Fracture Mechanics : Application to Concrete**, ACI SP-118, 157-169.

Lama R.D. and Vutukuri V.S. (1978) **Handbook on mechanical properties of rock**, Trans Tech Pub., 1.

Liu H. and Vinh T. (1991) Multidimensional signal processing for nonlinear structural dynamics. **Mechanical Systems and Signal Processing**, 5(1), 61-80.

Luong M.P. (1986) A new test for tensile strength measurements. **Revue Française de Géotechnique**, 34, 69-74.

Luong M.P. (1989) Fundamental modes of strength development for Fontainebleau sandstone. **Rock at Great Depth**, Maury & Fourmaintraux eds, Balkema, 157-163.

Luong M.P. (1990) Tensile and shear strengths of concrete and rock. **Engineering Fracture Mechanics**, 35, 127-135.

Roberts A. (1977) **Geotechnology - An introductory text for students and engineers**, Pergamon Press.

Wang Y. and Backer S. (1989) Toughness determination for fibre reinforced concrete, **Int. J. of Cement Composites and Lightweight Concrete**, 11(1), 11-19.

Wright P.J.F. (1955) Comments on an indirect tensile test on concrete cylinders. **Magazine of Concrete Research**, 7(27), 87-96.

Zhan Z.F., Fouré B. & Trinh, J.L. (1991) Characterizing tests in tension for fibre reinforced concrete, **Workshop "High Performance Fibre Reinforced Cement Composites"**, Mainz, June 24-26.

PART FIVE

MODELLING

29 OPTIMIZING THE COMPOSITION OF POLYPROPYLENE FIBRE REINFORCED CEMENTITIOUS COMPOSITES

L. HEBDA, L. RUDZINSKI and B. TURLEJ
Kielce University of Technology, Poland

Abstract
Optimizing the composition of polypropylene fibre cementitious composites is proposed. Two parameters of fibres were taken into account in this test: their length and volume content in matrix. Basing on seven points statistical Box design, authors presented the optimization procedure, where both the bending strength and toughness index I_5 were the criteria of optimization. This procedure is supported by statistical method of experiment design.

In this test the commercial polypropylene fibres were used and their length and volume changed from 5 mm to 30 mm and 0.1% to 3.0 % of matrix volume, respectively. The matrix was standard cement mortar using in Poland in cement strength testing and concrete with coarse aggregate 2 - 8 mm.

The results obtained are presented on the graphs in the form of relation: $y = f(x_1, x_2)$, where y was property measured (bending strength or toughness index) and x_1, x_2 were test factors considered (fibre volume and fibre length). Thus the obtained relations show the changes of mechanical properties of composites tested fibre reinforced mortar and concrete as the function two variables.

Basing on analysis of these relations the authors concluded that the optimum composition taking into account both fibre volume and fibre length may be differed in dependence on property considered.

1 Introduction

There is a broad range of fibre reinforced composites and within fibres of each material there is a great variety of characteristics, all of which are demanded to provide some advantage.It stands to reason that a lot of attempts have been made to optimize parameters of fibres in order to achieve profitable strength properties of composites.

The experiments were carried out at the Kielce University of Technology. Two parameters of polypropylene fibres applied were taken into account: their length and volume concentration. Both the bending strength and toughness index I_5 were the criteria of optimization. The three point bending test was performed using testing machine of INSTRON type. The deflection of each beam of 40x40x160 mm in

Fibre Reinforced Cement and Concrete. Edited by R. N. Swamy. © 1992 RILEM.
Published by E & FN Spon, 2-6 Boundary Row, London SE1 8HN. ISBN 0 419 18130 X.

the mid-span was recorded. The application of the experimental design gave the possibility to determine analytical relations between the variables with very restricted number of specimens.

2 Test procedure

The test was realized using two series of specimens. In the first series the cement mortar was the matrix in cement composite material with fibres. In the second series - the concrete was the matrix.

The tests were performed using the statistical methods, particularly the methods of the experimental design. These methods are very convenient and efficient, particularly in material and technological tests. Taking into account the known experimental designs it was decided to use 7-points (first series) and 10-points design of Box plan which are shown in Tables 1 and 2.

2.1 The First Series

Commercial ordinary Portland cement of quality "45" from Malogoszcz plant was used with natural quartz river sand. Mix proportions was as follows: cement:water:sand = 1:0.5:3.

The mortars were prepared in a forced horizontal concrete mixer with a vibrating sieve for introduction of fibres to prevent the phenomenon of balling. The sand was used in air-dry state. No admixtures were added to mortars tested.

Table 1. 7-points Design of Box Plan

Experiment number	Factors combination	
	x_1	x_2
1	0.866	0.500
2	0.000	0.000
3	0.000	- 1.000
4	0.000	1.000
5	0.866	- 0.500
6	- 0.866	- 0.500
7	- 0.866	0.500

On the basis of performed design it would be possible to estimate the coefficients of the function characterizing the examined object, which is a square equation as follows:

$$y = a_0 + a_1 x_1 + a_2 x_2 + a_{11} x_1^2 + a_{22} x_2^2 + a_{12} x_1 x_2 \qquad (1)$$

where:

y - measured properties,

a_i- coefficient of regressionfunction,

x_i- considered test factors.

The samples were cast as beams of 40x40x160 mm. After 48 hours the beams were demoulded and stored at a foggy room at constant temperature of +18°C and relative humidity of 95%. At the age of 28 days in the mid-span of each sample an initial notch of 5 mm depth was sawn. Then, the beams were stored at constant conditions of 60% RH and +18°C. The tested samples were loaded at the age of 90 days.

Table 2. 10-points Design of Box Plan

Experiment number	Factors combination	
	x_1	x_2
01	0.707	0.707
02	0.707	- 0.707
03	- 0.707	0.707
04	- 0.707	- 0.707
05	1.000	0.000
06	- 1.000	0.000
07	0.000	1.000
08	0.000	- 1.000
09	0.000	0.000
10	0.000	0.000

Two types of fibres were applied:
a) polypropylene fibres, cut from fissile polypropylene foil (producer of foil: Institute of Flax Industry in Lodz) with irregular cross-section, near to rectangle; the tensile strength of 75 MPa and Young's modulus of 1200 MPa; the length of fibres was changed in the test from 0.5 to 3.0 cm and fibre volume concentration was varied from 0.1 to 2.0%.
b) commercial polypropylene fibres looking like down, length of 0.6 cm; their volume content in matrix has changed from 0.1 to 1.5%.

2.2 The Second Series
The same type of cement was used like in the first series. The sand was taken from Sukow sand pit and was more fine than the natural river sand in the first series.

The crushed dolomitic limestone from Laskowa quarry was used as the coarse aggregate. This aggregate contained one comercial fraction size of 2-8 mm. Dolomitic limestone density was equal to 2850 kg/m^3 and b-ulk density was 2810 kg/m^3. Its porosity was 3.0% and full absorbabil-ity was no more than 2.5%. The concrete mix proportions were shown in Table 3.

Table 3. Concrete Mix Proportions - Second Series of Test

Cement	Sand	Coarse aggregate	Water	W/C
450 kg/m^3	540 kg/m^3	1250 kg/m^3	180 kg/m^3	0.40

The fibres of the type a) from the first series were used like reinforcement of concrete matrix.

The samples were cast as beams of 50x50x400 mm.The same initial notch was sawn in the mid-span of each sample like in first series and the conditionsof curing were the same too.

2.3 The Test Procedure

Basing on above-mentioned data the range of experiment was defined as follows (the same for both series):

Table 4. The Range of Experiment

Variable Factor	Unit	Symbol	Central Point	Range Of Change
Length of fibre	cm	x_1	1.75	±1.25
Fibre volume	%	x_2	1.05	±0.95

The constant factors were:
- mix proportion,
- conditions of samples preparation and conditions of samples curing,
- conditions of test realization.

The test program was shown in Table 5.

Table 5. Test program

The test number	Variable Factors			
	First Series		Second Series	
	x_1	x_2	x_1	x_2
01	2.83	1.52	2.63	1.72
02	1.75	1.05	2.63	0.38
03	1.75	0.10	0.87	1.72
04	1.75	2.00	0.87	0.38
05	2.83	0.57	3.00	1.05
06	0.67	0.57	0.10	1.05
07	0.67	1.52	1.75	2.00
08	-	-	1.75	0.10
09	-	-	1.75	1.05
10	-	-	1.75	1.05

The tests were carried out at the samples age of 90 days. The three points bending test (Fig. 1) was realized in testing machine of the "INSTRON" type with constant move of the piston which was equal to 0.05 mm per second. The course of the load and mid-span deflection were recorded. For each mix composition three samples were tested.

The toughness index I_5 was calculated according to ASTM C1018 definition (Fig. 2).

3 Results

The effects of fibres length and volume on bending strength and toughness index I_5 were shown in Fig. 3. and in Fig. 4. respectively for composite materials with cement mortar matrix and fibres of the a) type.

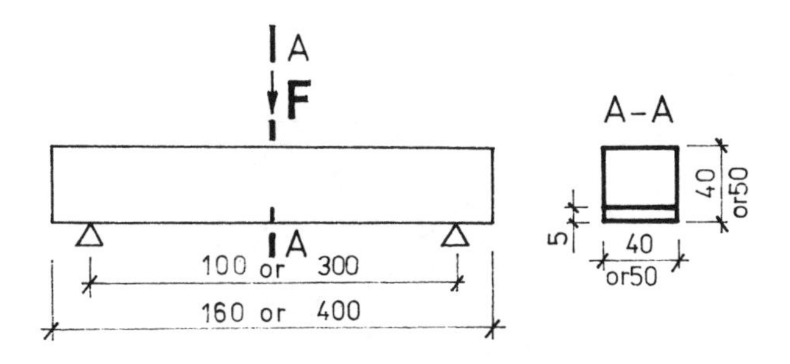

Fig. 1. Dimensions and shape of tested specimens

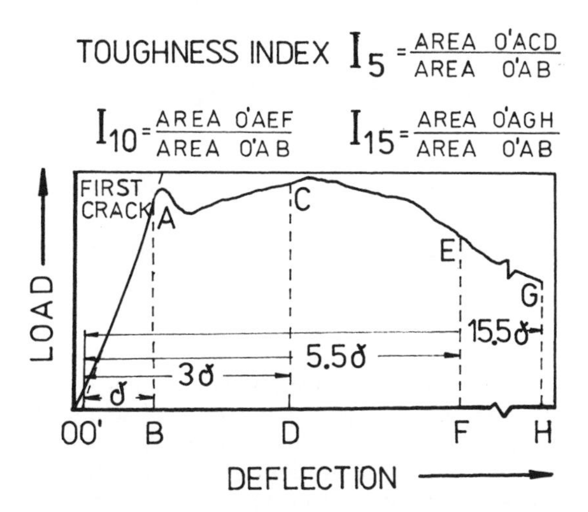

Fig. 2. Toughness index definition according to ASTM C1018

The influence of fibres volume concentration on bending strength and toughness index is shown in Figure 5 for composite with mortar matrix and fibres of the b) type.

The change of bending strength and toughness index I_5 magnitudes due to both different length and volume content of fibre of b) type in the case of composite materials with concrete matrix was presented in Fig. 6 and in Fig 7.

Under Fig. 3, Fig. 4, Fig. 6 and Fig.7, the functions which describe the graphic picture were recorded

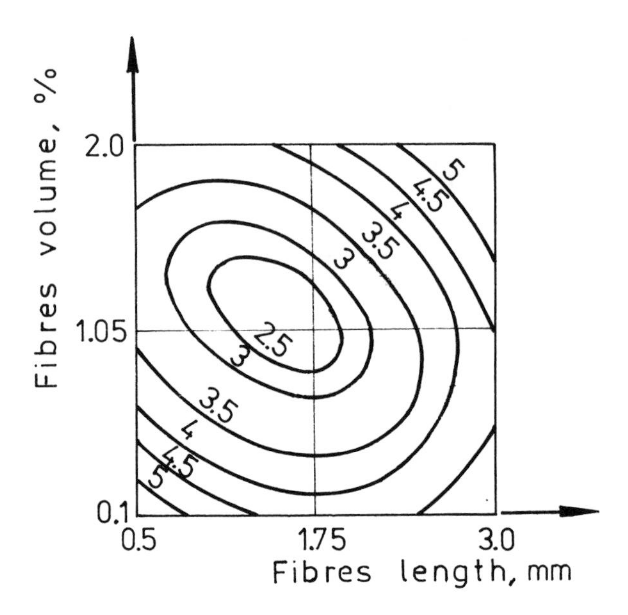

Fig. 3. The strength [MPa] distribution in composite materials with mortar matrix.

$$y = 5.43 + 1.17x_1 - 0.08x_2 + 2.52x_1^2 + 3.27x_2^2 + 1.60x_1x_2\,; \qquad R=0.99 \quad (2)$$

4 Discussion

As we can see from Fig. 3 there is the worst point of bending strength of composite with mortar matrix and fibres of the type a). It is situated near to the central point of the range of experiment. Probably, it is the result of the worst composition of fibres length and fibres volume concentration in this matrix. It is interesting that quicker increase of bending strength is observed in two directions from this worst point:
- direction to high volume of fibres and to longer fibres,
- direction to low volume and shorter fibres.

$$y = 3.67 + 0.35x_1 + 2.53x_2 - 0.19x_1^2 - 0.23x_2^2 + 1.99x_1x_2\,; \qquad R=0.98 \quad (3)$$

It follows from Fig.3 that in central region of the range of experiment the compositions of two factors (fibres volume and fibres length have not good influence on structure of matrix. The fibres are too short to have practical significance on bending parameters. At same time their high contents has dominant negative influence on composite structure.

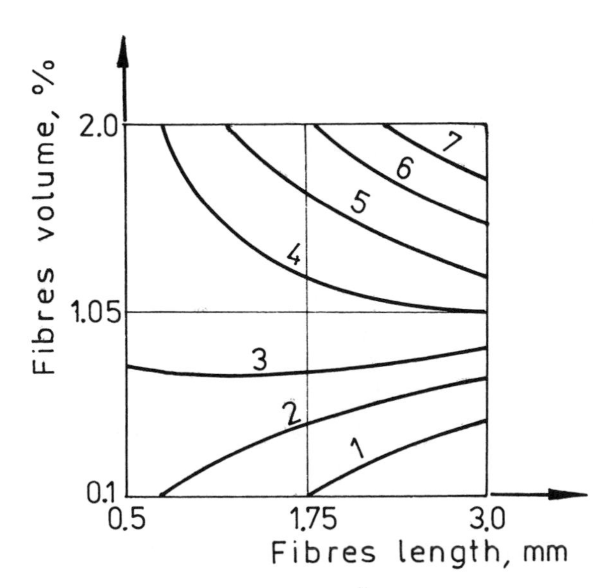

Fig. 4. The toughness index I5 distribution in composite materials with mortar matrix

The view of toughness index I5 distribution (Fig. 4) shows clearly that the direction to higher fibres volume and to longer fibres is the most profitable for bending properties of composite materials. In the case of increase in the fibres length their contact area with matrix has grown up and the number of fibres in the cross-section is bigger too.

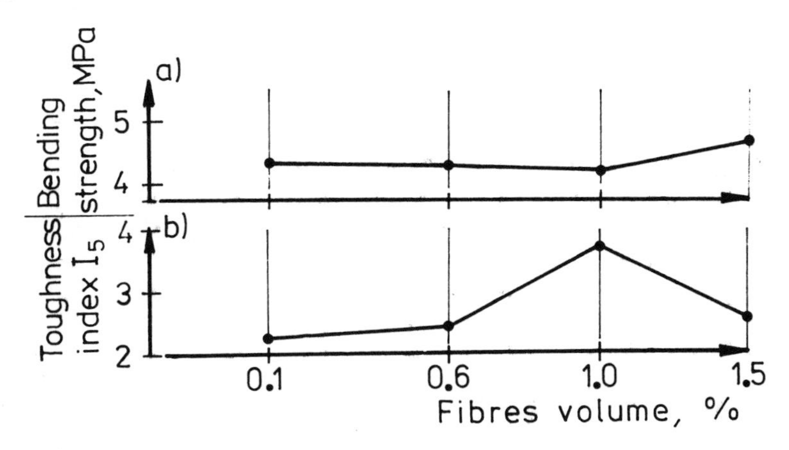

Fig. 5. Bending strength and toughness index I5 vs. fibre volume concentration in composite material with mortar matrix and down of polypropylene fibres.

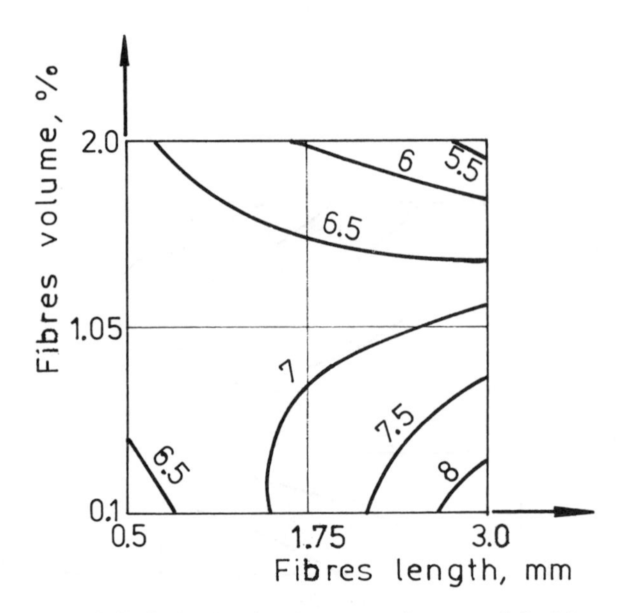

Fig. 6. The strength [MPa] distribution in composite material with concrete matrix.

$$y = 6.85 + 0.19x_1 + 0.63x_2 + 0.05x_1^2 - 0.25x_2^2 - 0.80x_1x_2\,; \qquad R=0.89 \quad (4)$$

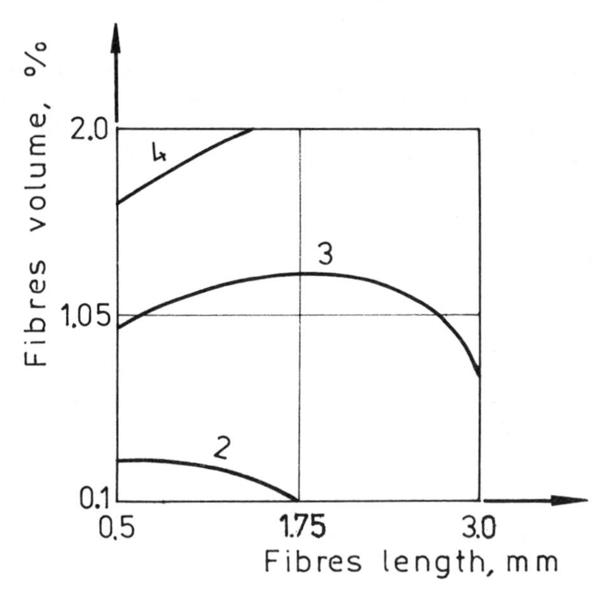

Fig. 7. The toughness index I_5 distribution in composite material with concrete matrix.

$$y = 2.78 + 0.03x_1 + 0.93x_2 + 0.31x_1^2 + 0.15x_2^2 - 0.49x_1x_2\,; \qquad R=0.97 \quad (5)$$

It follows from Fig. 5 how important is the optimal point of fibres contents. For the same polypropylene fibres but with different length and more down form we have obtained different results. From the energetic point of view, the optimal volume of fibre of the b)type on the level of 1% (toughness index I_5), but from bending strength point of view - the optimal volume of fibres of the b) type is on the level of 1.5% or higher.

The addition of coarse aggregate to cement mortar completely change distribution of fibres volume and fibres length influence on bending strength and toughness index I_5.

In the case of more complex concrete matrix, only long fibres have significant influence on bending strength and it is observed for low and high concentrations of fibres (Fig. 6). From the energetic point of view (toughness index), the content of fibres (Fig. 7) is the most important factor. The longer fibres have better contact zone with matrix (high surface of contact) and this is the base parameter which influences on bending strength. The bigger number of fibres in cross-section of a sample influences more significantly on toughness index.

5 Conclusions

The conclusions from the test presented above are as follows:

i) the optimum composition of fibres volume concentration and fibreslength depend on the form of fibres, the type of matrix (cement paste, cement mortar or concrete) and the kind of analyzed parameter (strength, durability, elasticity etc.);

ii) the criteria and the aim of the optimization should be defined before start of test procedure;it is very helpful at the stage of a testrange definition;

iii) the methods of the experimental design are the most effective optimization procedures in testing of new materials.

30 MODELIZATION OF THE TENSILE STRESS–STRAIN CURVE OF GLASSFIBRE REINFORCED CEMENTS

M. L. SANCHEZ PARADELA
Department of Construction, Faculty of Architecture of Madrid,
V. SANCHEZ GALVEZ
Department of Materials Science, Civil Engineering School of Madrid, Spain

Abstract
Glassfibre Reinforced Cement (GRC)is a composite material widely used for cladding panels. The knowledge of its tensile behaviour is a very important aspect for the design of members. There are lacks in the models of Fibre Reinforced Cements currently used, particularly in the stresses and strains for matrix crack initiation and growth as well as in the ultimate tensile stress and its corresponding strain. Therefore, a new model has been developed that predicts the tensile behaviour of GRC. Expressions for the Limit of Proportionality (LOP), Bend Over Point (BOP) and Ultimate Tensile Strength (UTS) have been derived. Theoretical stress-strain curves are compared to experimental results both from the literature and from a test programme carried out by the authors. Agreement between theory and experiments is pointed out.
Keywords: Glassfibre, Composite material, Fibre Reinforced Cement, Modelisation, Stress-strain curve, Tensile strength.

1 Introduction

GRC produced by the addition of small amounts of glass fibre to cement or mortar is a composite material that exhibits good strength and ductility. Usually, it is utilised for the production of cladding panels, using the spray-suction method, in which the fibre is introduced into a slurry previously produced by mixing the cement with the water, the additives and the sand. In this process, fibres cut at lengths in the range between 30 and 40 mm. are distributed at random in 2-D and GRC panels, extremely thin (around 1 cm. thick), are fullfilled with a polyestirene core,thus being fairly lightweight and highly fire resistant as well as low thermal conductor. For that reasons its use has increased dramatically in the last decades and there are a large number of examples of GRC utilisation for cladding panels (PCI,1981).

Fibre Reinforced Cement and Concrete. Edited by R. N. Swamy. © 1992 RILEM.
Published by E & FN Spon, 2-6 Boundary Row, London SE1 8HN. ISBN 0 419 18130 X.

GRC shows an ageing problem, becoming brittle with time, due to the attack of cement alkalis to the glass fibre. That problem lead to the development of AR glass fibres (alkali resistant) as well as to the utilisation of different procedures of fibre protection, either by additions to the cement paste or by the use of low alkalinity cements (Bentur and Mindess,1990). Nowadays, there are some companys stating to having solved the problem by means of special compositions (Thiery,1989).

Modelisation of the tensile behaviour of these composite materials is highly interesting for the designer, because it enables the analysis of the factors influencing such behaviour and even acting on the different components of the composite to achieve specific mechanical properties.

The majority of models of the tensile behaviour of fibre reinforced cements are based on the theory developed by Aveston, Cooper and Kelly (1971). It is widely assumed that the whole tensile stress-strain curve exhibits three regions (Nathan et al.,1977), elastic, cracking and post-cracking behaviour respectively, for which the majority of the authors assume linear relationships (Bentur and Mindess,1990).

Yet there are large discrepancies between the authors about stresses and strains for the passage from one region to another. With respect to the ultimate tensile strength of the composite material, it is usually accepted the following simple expression

$$\sigma_{RC}=V_f\sigma_{Rf}K \tag{1}$$

where σ_{RC} is the strength of the composite, σ_{Rf} the tensile strength of the fibre, V_f the fibre volume content and K a nondimensional parameter, called efficiency, dependent upon the length and orientation of the fibres and the shear strength of the bond between fibre and matrix. There are large discrepancies in the literature about the values of the strength efficiency (Oakley and Proctor,1975). With respect to the corresponding ultimate composite strain,simply lower and upper limits of it have been proposed (Bentur and Mindess,1990).

2 The model

2.1 Hypothesis

In the present model it is also assumed that the tensile stress-strain curve of glassfibre reinforced cement shows three regions clearly different, corresponding to three different mechanical behaviours:

I: The first region, corresponding to linear elastic behaviour, will be assumed to follow the rule of mixtures, in agreement with all preceding authors. Thus, the stress-strain curve in this region is given by the straight line

$$\sigma = E_c \, \epsilon \tag{2}$$

where E_c is Young's modulus of the composite, given by

$$E_c = E_m V_m + E_f V_f \eta_0 \eta_1 \tag{3}$$

where E_m and E_f are Young's modulus of matrix and fibre respectively, V_m and V_f are volume content of matrix and fibre and η_1 and η_0 are efficiency factors for length and orientation.

II: The second region, where the stress-strain behaviour is non linear. In this region, microcracks develop and grow in the matrix. Those microcracks do not lead to the failure of the composite because their propagation is arrested by the fibres. However, progressive microcracking of the matrix produce a marked reduction of the composite stiffness, thus a non linear stress-strain curve is observed.

III: The third region, starting after matrix cracking. It is necessary to include a higher fibre content than a critical value and the use of fibres long enough to achieve a real third region, which otherways could disappear, leading to the ultimate strength of the composite before the end of region II. In this region, load is carried by the fibres bridging the cracks and may even increase through fibre straining. This region ends when fibres break or are pulled out of the matrix, the corresponding stress being the ultimate strength of the composite.

The model developed approximates the tensile stress-strain curve of GRC for each one of those three abovementioned regions, as well as the stresses and strains separating one region from another.

2.2 Region I

In the linear elastic domain, the composite is assumed to be governed by the rule of mixtures (eqs. 2 and 3), where the orientation efficiency factor is 3/8 and η_1 is assumed to be

$$\eta_l = 1 - \frac{l_t}{2 l_f} \tag{4}$$

where l_f is the fibre length and l_t is the transmission length of the stress, which will be assumed constant.

The end of Region I is governed by the beginning of matrix microcracking. Matrix flaws start growing, with the shape of microcracks, which are arrested by the fibres avoiding the unstable propagation. The stress in the composite for microcrack initiation is called Limit of Proportionality (LOP) and is given by the following expression: (S.Paradela and S.Galvez, 1991a)

$$LOP = E_c \epsilon_1 \tag{5}$$

where

$$\epsilon_1 = \frac{\sigma_{mc}}{E_m} \tag{6}$$

and

$$\sigma_{mc} = \frac{K_{IC}}{2.1\sqrt{\pi \frac{\overline{s}}{2}}} \tag{7}$$

where symbols in eqs. 5 and 6 have usual meanings. K_{IC} is the Fracture Toughness of the matrix and $\overline{s}$ is the average fibre spacing, given by: (Krenchel, 1975)

$$\overline{s} = \sqrt{\frac{\pi^2 n r^2}{2 V_f}} \tag{8}$$

where n is the number of filaments in a strand, r is the radius of the filament and V_f is the fibre volume content.

2.3 Region II

Stress-strain curve in region II is obtained taking into account the decrease in specimen stiffness produced by the increase in the surface of microcracks. Equation 2 is still valid, but the matrix modulus E_m is no more a constant, thus a non linear equation is achieved. Details of the derivation of the stress-strain curve have been published elsewhere and will not be repeated here. (Valiente and S. Galvez, 1986). The following equation is obtained:

$$E_m = \frac{E}{1 + A\overline{s}} \tag{9}$$

where A is the cracked area per unit volume and E the Young's modulus of an ideal material, containing no flaws. A is not constant in region II, it increases with the strain ϵ according to: (Valiente and S.Galvez, 1986)

$$\epsilon^2 = 2\frac{G_c}{\overline{s}E}\left(l_n \frac{A}{A_0} + \overline{s}(A - A_0) \right) \tag{10}$$

where A_0 is the initial value of A and G_c is the fracture energy per unit area of the cement matrix, related to its fracture toughness by

$$G_c = \frac{K_{IC}^2}{E_m} \tag{11}$$

Region II ends with the beginning of matrix macrocracking. Fibres bridging the cracks have been impedding its propagation. For a stress level high enough, this equilibrium is lost and macrocracks start its propagation across the whole specimen. The corresponding composite stress is called Bend Over Point (BOP), and can be derived by using Fracture Mechanics criterion for the rupture of the matrix: (S.Paradela and S.Galvez,1991a)

$$K_I\ (ext) - K_I\ (fibres) = K_{IC} \tag{12}$$

where
$$K_I(ext) = \sigma_m \sqrt{\pi a_{max}} \tag{13}$$

and
$$a_{max} = \frac{K_{IC}^2}{\pi f_c^2} \tag{14}$$

$$K_I(fibres) = \gamma \frac{\sigma_m}{\sqrt{\pi a_{max}}} \sum_i \sqrt{\frac{2a_{max} - i\bar{s}}{i\bar{s}}} \tag{15}$$

where i= 1/2, 3/2.....up to $2a_{max}/\bar{s}$

$$\gamma = \frac{S_f}{\bar{s}} \left(\frac{E_f}{E_m} \eta_0 \eta_1 + \frac{V_m}{V_f} \right) \tag{16}$$

In equations 12 to 16, f_c denotes the ultimate tensile strength of the cement matrix, S_f is the area of the fibre cross section and all other symbols have usual meanings.

Equation 12 is an implicit expression in σ_m, the matrix stress for macrocracking initiation, which after being obtained permits the obtention of the stress in the fibres after matrix cracking σ'_f and finally the stress in the composite (BOP):

$$\sigma'_f = \sigma_m \left(\frac{E_f}{E_m} + \frac{V_m}{V_f \eta_0 \eta_1} \right) \tag{17}$$

$$BOP = \sigma'_f\ V_f\ \eta_0\ \eta_1 \tag{18}$$

2.4 Region III

For stresses higher than the BOP, Region III starts. In this region, firstly multiple cracking takes place, several macrocracks being propagated across the whole specimen width. After multiple cracking completion, the specimen is fractured into separate blocks, the external load being carried by the fibres bridging the macrocracks. Multiple cracking is completed when the spacing between cracks is shorter than l_t, the transmission length. The multiple cracking process takes place at slowly increasing stress, since succesive propagating cracks require higher stresses.

In the present model, however, for simplicity reasons, the process will be assumed to hapen at constant stress. At the end of the multiple cracking process, the cracks will be assumed to be equally spaced, the average crack spacing being $\bar{x} = 2/3\ l_t$ (See fig. 1).

At a cracked section, fibre strain in the x direction (direction of loading) is called ϵ'_f, ϵ_{f2} being the corresponding strain at the point in the middle between the cracks. Strain difference $\Delta\epsilon_f$ is equal to $2/3\ \Delta\epsilon_{fc}$, where $\Delta\epsilon_{fc}$ is such strain difference when the first crack occurs:

$$\Delta\epsilon_{fc}=\frac{BOP}{E_f V_f \eta_0 \eta_1}-\frac{BOP}{E_m V_m + E_f V_f \eta_0 \eta_1} \tag{19}$$

The average fibre strain $\bar{\epsilon}_f$ is proposed to be given by the following formula, which is similar to that included in the Model Code for modeling bond between concrete and reinforced bars: (CEB, 1990)

$$\overline{\epsilon_f}=\epsilon'_f-\beta_t \Delta\epsilon_{fc} \tag{20}$$

where β_t is a parameter around 0.4.

There are situations where multiple cracking does not occur along the extensometer gauge length, probably due to weakness in a particular section or difficulties to avoid small bending effects during tensile testing. In that case the general strain pattern shown in figure 1, must be corrected as depicted in figure 2, which is a sketch of the strain distribution along the x direction around a single

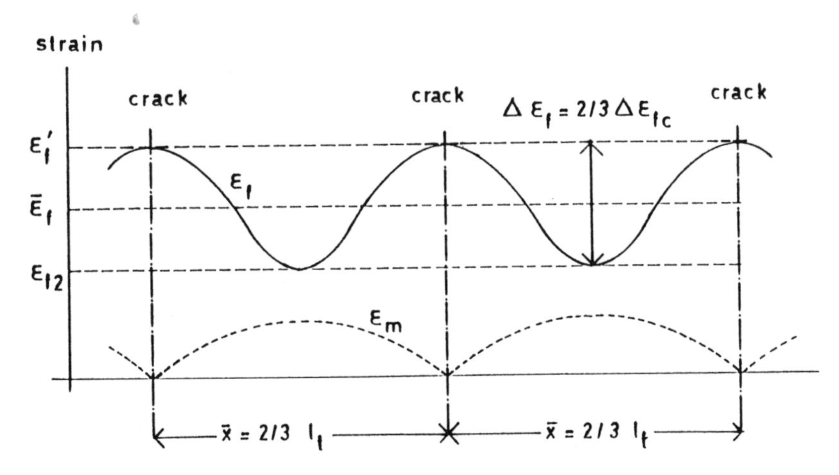

Fig. 1. Strain distribution along the x direction for a multiple crack pattern

crack. The average fibre strain along the extensometer gauge length is given by the following equation:

$$\overline{\epsilon_f}=(\epsilon'_f-\beta_t\Delta\epsilon_{fc})\frac{l_t}{l}+\epsilon_{f2}\left(1-\frac{l_t}{l}\right) \tag{21}$$

where ϵ_{f2} is the strain in the uncracked zone, still given by:

$$\epsilon_{f2}=\frac{\sigma}{E_mV_m+E_fV_f\eta_0\eta_1} \tag{22}$$

Stress-strain relationship in region III may then be derived simply by relating the external stress σ with the fibre strain ϵ'_f for fibres oriented in the direction of loading at a cracked section.

With that aim, let us consider a cracked section. At any angle θ with the direction of loading x, there are fibres bonded to the matrix, whose strain is $\epsilon'_f \cos^2\theta$ and which are thus carring a stress $E_f\ \epsilon'_f \cos^2\theta$ as well as fibres for which the embedded length at one side of the crack is shorter than half the transmission length l_t, fibres that according to the hypothesis of the model, will slip hereafter. For simplicity reasons, it will be assumed that slipping fibres do not carry any load. The probability of a crack cutting a fibre in the central zone, out from the two zones near the ends of the fibre, each $l_t/2$ of length is $1-l_t/l_f$, therefore the average stress for fibres oriented at an angle θ with the direction of loading is:

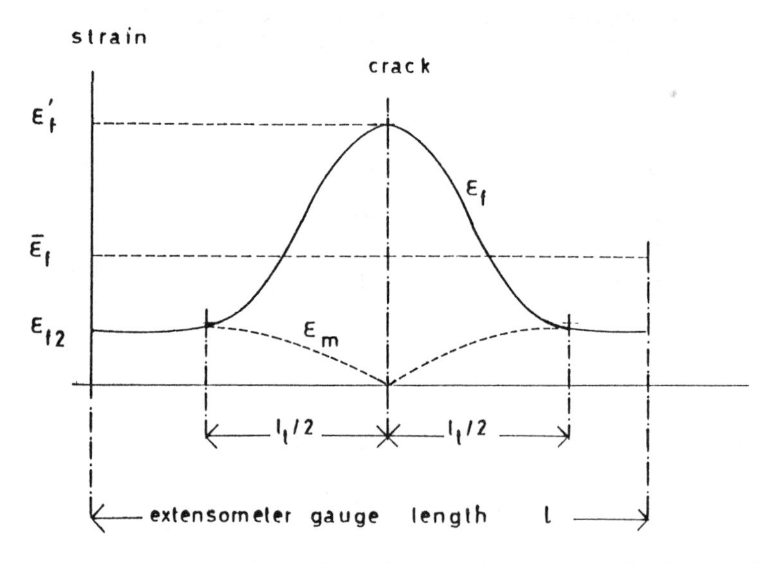

Fig. 2.- Strain distribution around a cracked section for a single crack pattern

$$\overline{\sigma_\theta}=\left(1-\frac{l_t}{l_f}\right)E_f\epsilon'_f\cos^2\theta \tag{23}$$

In the direction of loading, the average stress is obtained by adding the contributions for all possible fibre orientations. Assuming that the probability of fibre orientation is the same for any angle, the following expression is derived:

$$\sigma=\frac{V_f}{\pi}\int_{-\frac{\pi}{2}}^{+\frac{\pi}{2}}\overline{\sigma_\theta}\cos^2\theta d\theta=E_f V_f\epsilon'_f\left(1-\frac{l_t}{l_f}\right)\frac{3}{8} \tag{24}$$

Finally, subtituting ϵ'_f from eqs. 20 or 21, and calling ϵ to $\bar{\epsilon}_f$, the stress-strain relation in region III is obtained:

$$\sigma=E_f V_f\frac{3}{8}\left(1-\frac{l_t}{l_f}\right)(\epsilon+\beta_t\Delta\epsilon_{fc}) \tag{25}$$

$$\sigma=E_f V_f\frac{3}{8}\left(1-\frac{l_t}{l_f}\right)\left(\left(\epsilon-\frac{\sigma\left(1-\frac{l_t}{l_f}\right)}{E_m V_m+E_f V_f\eta_0\eta_l}\right)\frac{l}{l_t}+\beta_t\Delta\epsilon_{fc}\right) \tag{26}$$

which are linear relationships.

2.5 Rupture

The rupture of the composite can take place by fibre slipping if fibres are short ($l_t>l_f$). In this case, the stress-strain curve has not a true region III, the ultimate tensile strength being lower than the BOP.

For usual fibre lengths, ($l_f>l_t$), the rupture of the composite material occurs when the first fibre reaches its tensile strength. It will be assumed that the fracture of this fibre promotes succesively the fracture of all fibres bridging the same crack because the load is transfered from the fractured fibre to the other ones, which are then overstressed and reach their strength too. The highest fibre stress appears in fibres aligned to the external load direction, at a cracked section, where the strain is ϵ'_f and thus according to the assumptions mentioned above, fracture will occur when E_f ϵ'_f be equal to the fibre strength σ_{Rf}. Introducing this rupture condition into eqs. 20 to 22, 25 and 26, the ultimate tensile strength of the composite as well as the ultimate strain are achieved:

For the case of multiple cracking

$$\epsilon_{RC}=\frac{\sigma_{Rf}}{E_f}-\beta_t\Delta\epsilon_{fc} \tag{27}$$

$$\sigma_{RC}=\frac{3}{8}\left(1-\frac{l_t}{l_f}\right)V_f\sigma_{Rf} \tag{28}$$

When a single crack exists along the extensometer gauge length

$$\epsilon_{RC}=\left(\frac{\sigma_{Rf}}{E_f}-\beta_t\Delta\epsilon_{fc}\right)\frac{l_t}{l}+\frac{\sigma_{RC}}{E_mV_m+E_fV_f\eta_0\eta_l}\left(1-\frac{l_t}{l}\right) \tag{29}$$

σ_{RC} being given by eq. 28.

3 Examples

With the aim of checking the validity of the expressions achieved and to show with practical examples the way of using them, firstly two cases will be analyzed utilising the model. The two examples correspond to actual tests performed by Oakley and Proctor (1975).It is widely recognised the difficulty of performing valid direct tensile tests on GRC (Green et al.,1978), thus it is not frequent to find out tensile stress-strain curves in the literature. A good exception is the abovementioned work of Oakley and Proctor.

The model requires the use of several properties both of the fibres and the matrix.The two cases analyzed are a young GRC, after 28 days curing and an aged GRC, after 1 year natural ageing. For these materials, the values of the parameters required are listed in Table 1.

Table 1. Values of parameters used in modelisation of Oakley and Proctor tests

	Young GRC (28 days)	Aged GRC (1 year)
Fibre Young's modulus E_f (GPa)	70	70
Matrix Young's modulus E_m (GPa)	20	22
Fibre volume content V_f (%)	4.5	4.5
Fibre length l_f (mm)	38	38
Number of filaments in a strand n	204	204
Filament radius r (μm)	6.25	6.25
Transmission length l_t (mm)	6.6	7.4
Fibre perimeter p (mm)	2.83	2.83
Matrix fracture toughness K_{IC} (MPa$\sqrt{m}$)	0.413	0.413
Matrix tensile strength f_c (MPa)	3.5	3.2
Fibre tensile strength σ_{Rf} (MPa)	1300	900

Fibre properties used are equal to those described in Oakley and Proctor paper. With respect to the properties of the matrix, not indicated by those authors, have been assumed, values used being similar to those obtained experimentally by other authors (Sagar and Prat,1986; Granju and Maso,1984; Nair,1975; Majumdar,1975). Fibre strength after ageing is equal to that obtained experimentally by Larner, Speakman and Majumdar (1976), for AR fibres. Finally, the transmission length l_t used in the calculations is given by

$$l_t = \frac{2E_f S_f \Delta \epsilon_{fc}}{\tau_c p} \tag{30}$$

where p is the perimeter of the fibre and τ_c is the shear stress for fibre debonding,for which Oakley and Proctor (1975) propose 1.1 MPa for young GRC and 0.9 MPa for aged GRC.

Using the values of the parameters indicated, theoretical stress-strain curves of both GRC have been obtained, and have been plotted in figures 3 and 4, together with the experimental ones as were published by Oakley and Proctor (1975). As can be seen, there is an excellent agreement between experimental results and theoretical curves, the small discrepancies observed being lower than usual differencies between experimental results from one specimen to another.

For further checking of the model, theoretical curves have been obtained to be compared to some results of a wide experimental research programme, performed by the authors (S.Paradela and S.Galvez,1991b), including several direct tensile tests on GRC specimens. The details of tests performed are included in the reference. For comparison of predictions of the model to experimental results, the values of the parameters used are indicated in Table 2. Fibres properties are identical to those used in the preceding modelisation, except the length, which is 35 mm., and the volume content, 5%, both values being the actual ones used in our tests. Matrix properties were determined by testing cement paste specimens.

Finally, transmission lengths were increased with respect to the first modelisation, accordingly to the decrease in bond strength observed. In this case, region III is modelled using the expressions corresponding to a single crack pattern, since that was the situation observed and it is easily understood, taking into account the high values of the transmission length used.

Figures 5 and 6 show both theoretical curves as obtained from the model with the values of the parameters indicated and experimental curves (3 curves corresponding to 3 different specimens in the same conditions) for GRC after 28 days curing and GRC artificially aged by immersion in water

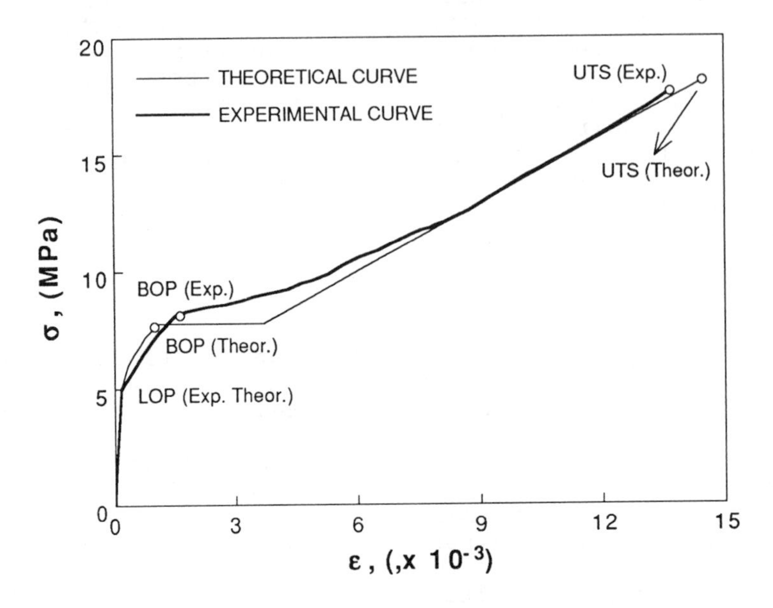

Fig. 3. Theoretical and experimental stress-strain curves for young GRC (28 days) tested by Oakley and Proctor

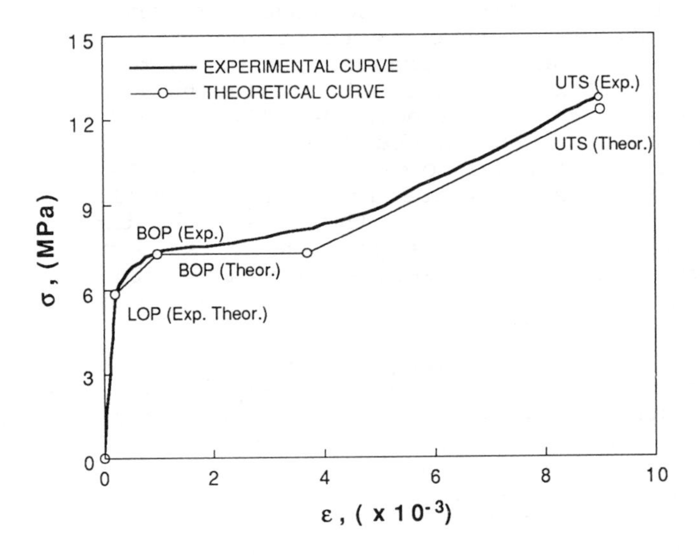

Fig. 4. Theoretical and experimental stress-strain curves for aged GRC (1 year) tested by Oakley and Proctor

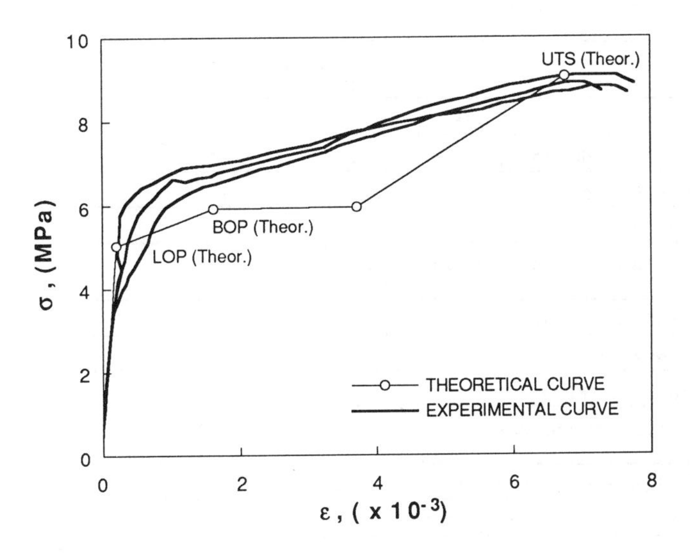

Fig. 5. Theoretical and experimental stress-strain curves for young GRC (28 days) tested by the authors

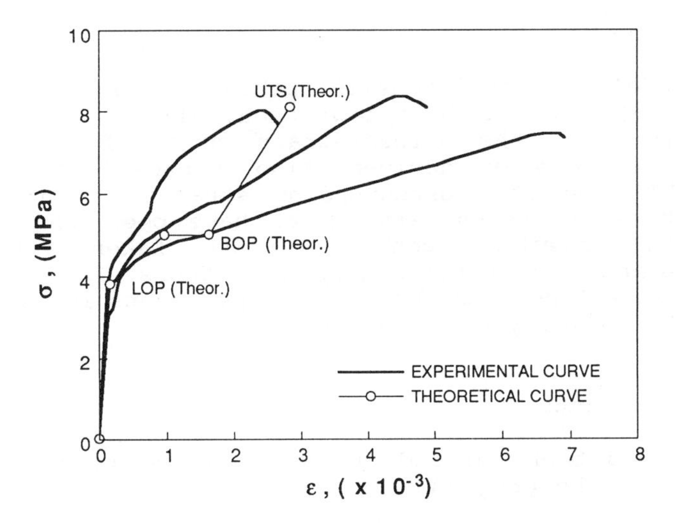

Fig. 6. Theoretical and experimental stress-strain curves for aged GRC (4 weeks at 50 C) tested by the authors

Table 2. Values of parameters used in modelisation of tests performed by the authors

	Young GRC (28 days)	Aged GRC (1 year)
Fibre Young's modulus E_f (GPa)	70	70
Matrix Young's modulus E_m (GPa)	20	22
Fibre volume content V_f (%)	5	5
Fibre length l_f mm	35	35
Number of filaments in a strand n	204	204
Filament radius r (μm)	6,25	6,25
Fibre perimeter p (mm)	2,83	2,83
Bond strength τ_c (MPa)	0,35	0,44
Matrix fracture toughness K_{IC} (MPa$\sqrt{m}$)	0,4	0,3
Matrix tensile strength f_c (MPa)	2,5	2,7
Fibre tensile strength σ_{Rf} (MPa)	1300	750

at 50 C for 4 weeks. Again, a good agreeement between theoretical curves and experimental results is observed, the scatter in the experimental curves being higher than discrepancies with the theoretical ones.

4 Conclusions

A new model of the tensile behaviour of Glassfibre Reinforced Cement has been developed. The model involves three regions of the stress-strain curve, for each one simple expressions are derived. Also, expressions for LOP, BOP and UTS and its corresponding strains are included.

Theoretical stress-strain curves have been compared to experimental results, both of tests taken from the literature and tests performed by the authors. In all cases, a good agreement between theory and experiments is observed, showing the validity of the model.

5 Acknowledgements

This work has been carried out with financial support by CICYT,through the programme PB87-0805.

6 References

Aveston,I. Cooper, G.A. and Kelly, A. (1971) Single and Multiple Fracture, in **The Properties of Fibre**

Composites,IPC Science and Technology Press, pp. 15-26.
Bentur, A. and Mindess, S. (1990) **Fibre Reinforced Cementitious Composites.**Elsevier Applied Science.
CEB (1990) **Model Code.**Chapter 3. General Models.
Granju, J.L. and Maso, J.C. (1984) Hardened Portland Cement Pastes, Modelisation of the microstructure and evolution laws of mechanical ptoperties. III Elastic Modulus. **Cement and Concrete Research.**,14, 539-545.
Green, M.F. Oakley, D.R. and Proctor, B.A. (1978) Tensile testing of glass reinforced cement sheet, in **RILEM Symposium on Testing and Test Methods of Fibre Cement Composites**,Sheffield,The Construction Press, Lancaster,pp. 439-449.
Krenchel, M. (1975) Fibre spacing and specific fibre surface, in **RILEM Symposium on Fibre Reinforced Cement and Concrete,** London, The Construction Press, pp. 69-79.
Larner, L.T. Speakman, K. and Majumdar, A.J. (1976) Chemical Interaction between Glass Fiber and Cement. **Journal of Noncristalline Solids.**,20, 43-74.
Majumdar, A.J. (1975) Properties of fibre cement composites, in **RILEM Symposium on Fibre Reinforced Cement and Concrete,** London, The Construction Press, pp. 279-313.
Nair, N.G. (1975) Mechanics of glass fibre reinforced cement, in **RILEM Symposium on Fibre Reinforced Cement and Concrete,** London, The Construction Press, pp. 81-93.
Nathan, G.K. Paramasivam, P. and Lee, S.L. (1977) Tensile behaviour of fibre reinforced cement paste. **Journal of Ferrocement.**,7, 59-79.
Oakley, D.R. and Proctor, B.A. (1975) Tensile Stress-Strain Behaviour of Glass Fibre Reinforced Cement Composites, in **RILEM Symposium on Fibre Reinforced Cement and Concrete,**London, The Construction Press, pp.347-359.
PCI Commitee on Glass Fiber Reinforced Concrete Panels (1981) Recommended Practice for Glass Fiber Reinforced Concrete Panels. **Journal of the PCI.**, 28-90.
Sagar, V. and Prat, P.L. (1986) Advances in Fracture Research, in **Fracture 84**, Pergamon Press, pp. 2809-2816.
Sanchez Paradela, M.L. and Sanchez Galvez, V. (1991a) Aplicacion de la Mecanica de la Fractura para la prediccion de la fisuracion de cementos reforzados con fibras de vidrio. **Anales de Mecanica Fractura.**,8,112-7.
Sanchez Paradela, M.L. and Sanchez Galvez, V. (1991b) Comportamiento a traccion de cementos reforzados con fibras de vidrio. **Informes Construccion.**,43,no.413,77-89.
Thiery, J. and Genis, A. (1989) High Durability Glass Cement Composites: New Vetrotex System, in **7th Biennial Congress of the Glassfibre Reinforced Cement Association**,Maastricht, GRCA, Newport, pp. 335-344.
Valiente, A. and Sanchez Galvez, V. (1986) A micromechanical model for the tensile stress-strain curve of fibre reinforced cements, in **RILEM Symposium on Developments in Fibre Reinforced Cement and Concrete**,Sheffield, (eds R.N. Swamy, R.L. Wagstaffe and D.R. Oakley), paper 1.1.

31 ENERGY DISSIPATION DURING STEEL FIBRE PULL-OUT

G. CHANVILLARD
Laboratoire Géomatériaux, ENTPE, Vaulx en Velin Cedex, France

Abstract
In the field of steel fibres reinforced concrete, there exists a large variety of fibre geometries governing the properties of fibre reinforced concrete.
It is now well accepted that when a wiredrawn non-straight steel fibre is being pulled out from a matrix, it straightens. This permits to develop a micro-mechanical model, based on the friction behaviour at the fibre-matrix interface and on the anchorage behaviour which provides plastic strains in the steel.
It is shown that both these basic phenomena interact. Sensitivity studies made with the theoretical model show that the anchorage component is the most important phenomenon which governs the pullout behaviour of the fibre.
From this analysis, we conclude that the critical size of the non-straight fibre cannot be given only in terms of its critical length any longer.
Finally, we evaluate the crack opening displacement at which the peak pullout load is theoretically obtained. It appears that the mobilization of the fibre is quite instantaneous, but, as a consequence of the crack plane surface effect, the behaviour of the fibre in concrete macrocracks is more complex.
Keywords : Steel Fibres, Modelling, Pull-out, Bond, Friction, Plastification, Energy, Geometry, Coupling Effect

1 Introduction

In the field of steel fibres, there exists a large variety of geometries governing the properties of fibre reinforced concrete.

With a view to optimizing the fibre-matrix association, the behaviour of one single steel fibre must be understood. With pullout tests, it has been shown that each wiredrawn non-straight steel fibre dissipates both plastic energy and friction energy [STROEVEN, P., 1979; BURAKIEWICZ, A., 1978; HUGHES, B.P.et al., 1975; MAAGE, M., 1977; NAAMAN, A.E. et al.,1989; CHANVILLARD, G., 1990].

So, a micro-mechanical model has been proposed to describe the behaviour of these fibres during the pull-out [CHANVILLARD, G., 1992; CHANVILLARD, G. et al., 1991; CHANVILLARD, G. et al., june 1991]. On the basis of this model, the part of each previous energy, which strongly interact, can be evaluated according to the fibre geometry.

At a second stage, the sensitivity of the fibre behaviour is studied, according to variation of the model parametres, which are the friction coefficient and the yield stress of the steel (YS).

Fibre Reinforced Cement and Concrete. Edited by R. N. Swamy. © 1992 RILEM.
Published by E & FN Spon, 2-6 Boundary Row, London SE1 8HN. ISBN 0 419 18130 X.

The case of the corrugated steel fibres commonly used, is analyzed in this paper from the rate of energy dissipation point of view and discussed from the crack size point of view. Recommendations are made on the concept of optimal fibre for fibre reinforced concrete.

Finally, the efficiency of the steel fibre is defined, which allows to argue on the possible properties of the real composite.

2 The micro-mechanical model

The pullout behaviour of wiredrawn non-straight steel fibre have been extensively studied. It is now well accepted that when such a fibre is being pulled out from a matrix, it straightens. In addition, the cut-out of the experimental specimens allows to notice that the internal damage of the matrix were of minor importance [CHANVILLARD, G., 1990; BANTHIA, N. et al., 1991]. As a result, the fibre slips into the path of its initial geometry.

After the complete debonding of the fibre, we can easily define a displacement field, based on the previous experimental observations. Assuming that the steel has a perfectly plastic rigid behaviour without hardening, we can use the balance of the mechanical energy to describe the behaviour of an infinitesimal element of the fibre [CHANVILLARD, G., 1992]. This leads to :

$$dP \cdot d\delta - dT \cdot d\delta = M \cdot dC \cdot ds \quad (1)$$

with
- dP increase in pullout load along ds
- $d\delta$ infinitesimal slip of the fibre element
- dT friction load along ds
- M plastification moment
- dC variation of the fibre element during $d\delta$
- ds length of the fibre element

We assume that the shear stress and the radial stress at the interface between the fibre and the matrix are governed by a Coulomb friction law, expressed in components. This equation means that there exists no residual cohesion after complete debonding of the fibre :

$$dT = f \cdot dN \quad (2)$$

with
- dT shear component
- dN radial component
- f friction coefficient

Combining geometrical equations with equilibrium conditions (2) and energies equation (1), we can write the model as :

$$dP = (f \cdot P \cdot C + M \cdot C')\, ds \quad (3)$$

with C' the first derivative of the curvature C.

This is a differential equation and it is obvious that there exists a strong interaction between the friction and the mechanical anchorage provided by the fibre geometry.

Ultimately, it may be asserted that the total behaviour of non-straight steel fibres can't be considered as the sum of a friction component and an anchorage component without paying much attention to the coupling effect.

The theoretical pullout load-slip curve is obtained by integrating the differential equation along the embedment length of the fibre, with the boundary condition P=0 at the embedment end of the fibre.

In previous papers, we have shown the agreement between the predictions of the micro-mechanical model and the experimental pull-out curves [CHANVILLARD, G., 1992; CHANVILLARD, G. et al., june 1991]. So, the conclusion to be drawn is that the elementary phenomena which governed the behaviour of non-straight fibres have been brought to the fore.

3 Sensitivity of the model

In order to study the effect of both friction and anchorage phenomena, we have made a sensibility study of the model. For this purpose, theoretical variations of the modelling parametres have been considered and the corresponding responses have been compared.

The two parametres investigated were the friction coefficient f and the yield stress (YS) of the fibre steel, which governs the plastification energy. Figures 1 and 2 give the theoretical load - slip curves for the studied values of both parametres.

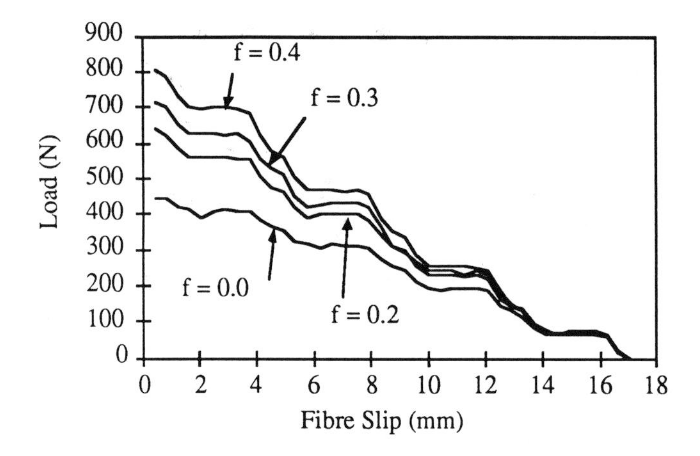

Fig.1. Effect of the friction coefficient on the load - slip curves.

It appears clearly, at first, that the higher the value of these parametres, the higher the pullout load.

We can evaluate the quantitative significance of each parametre variation with table 1. The relative variation is defined by the ratio of the pullout load variation to the parametre variation.

This is in accordance with experimental studies on the effect of the water/cement ratio (W/C) of the concrete matrix. It is well accepted that the bonding between concrete and steel is improved with decreasing W/C ratio. This results in a higher frequency of fibre failure for low W/C ratio [CHANVILLARD, G., 1992]. In the case of debonding fibre, the cohesion vanishes and only friction is mobilized. As a consequence, the W/C ratio is of less significance on the slipping behaviour [STROEVEN, P., 1979; GRAY, R.J. et al., 1984; CHANVILLARD, G., 1992]. Figure 3 shows the effect of this ratio variation on the pullout behaviour of a corrugated wiredrawn steel fibre [CHANVILLARD, G., 1992]. This effect is of the same order of magnitude, as the friction coefficient variation (fig. 1).

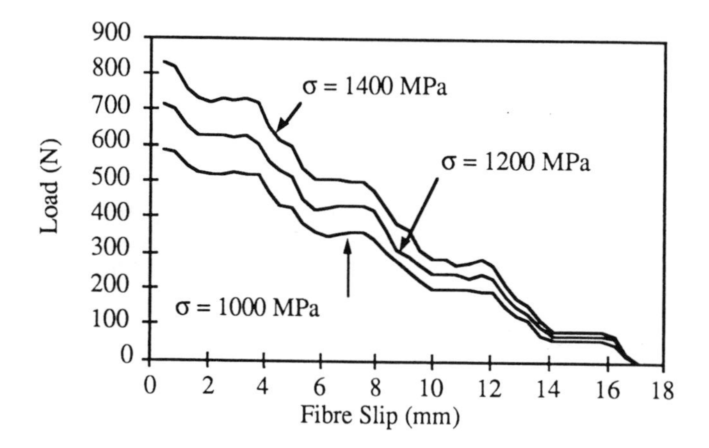

Fig.2. Effect of the YS on the load - slip curves.

Table 1 : Significance of the effect of each parametre.

Variation	Friction coefficient	Yield stress
parametre	25%	17%
pullout load	11%	17%
relative	0.44	1.0

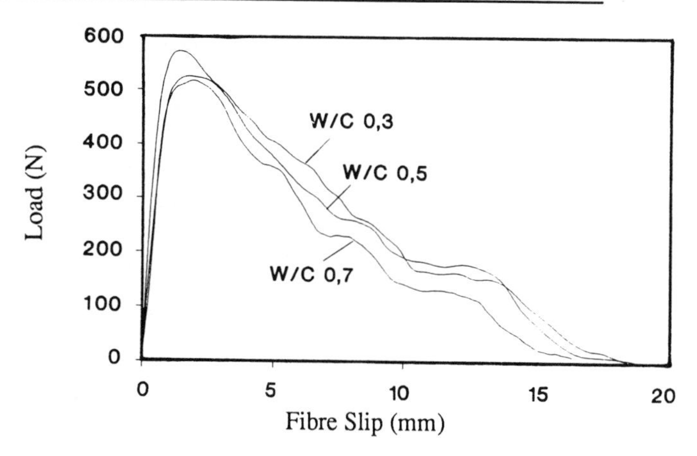

Fig.3. Effect of the W/C ratio on the load - slip curves.

About the yield stress (YS), we can point out an interesting paradox. We have seen that the higher the YS, the higher the pullout load resulting from a higher plastic energy.

At the same time, increasing the YS provides a higher capacity of the fibre in tension. Moreover, the pullout load varies with the cube of the radius resulting from plastic energy evaluation while the tension capacity varies with the square. So, an optimal radius of the fibre can be found which dissipates the maximum plastic energy without fibre break.

In fact, this approach is very complicated resulting from the coupling of the two basic phenomena. Finally, from the above table, we can conclude, on the basis of the relative variation, that the effect of the yield stress is more prominent than that of the friction coefficient.

4 Coupling effect

4.1 Theoretical evaluation

On the basis of the theoretical micro-mechanical model, the coupling effect has been studied. The friction load, provided by the friction energy, and the plastification load, provided by the plastic energy, have been theoretically separated. Each of these loads was evaluated with the following formulas :

friction : $dP_f = f \cdot P \cdot C \cdot ds$ (4)

plastification : $dP_p = M \cdot C' \cdot ds$ (5)

Both formulas are differential equations. Remember that the plastification moment M is governed by the interaction diagram with the pullout load P as normal load.

By integrating the previous equations along the embedment length of the fibre, the total pullout load is divided into two parts and given by the sum of these two parts, so :

$$P = P_f + P_p \quad (6)$$

A numerical example is given on figure 4 for the case of the corrugated steel fibre. For this geometry, the friction load represents in average on the complete behaviour only 17 % of the total load, against 83 % for the plastification load. These proportions are the following at the beginning of the pullout : 30% friction and 70% plastification. This is further evidence of the greater part played by the plasticication effect.

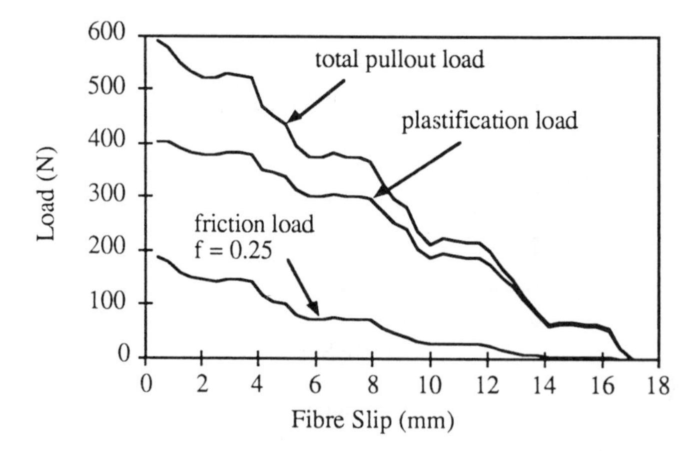

Fig.4. Separation of the pullout load in two part.

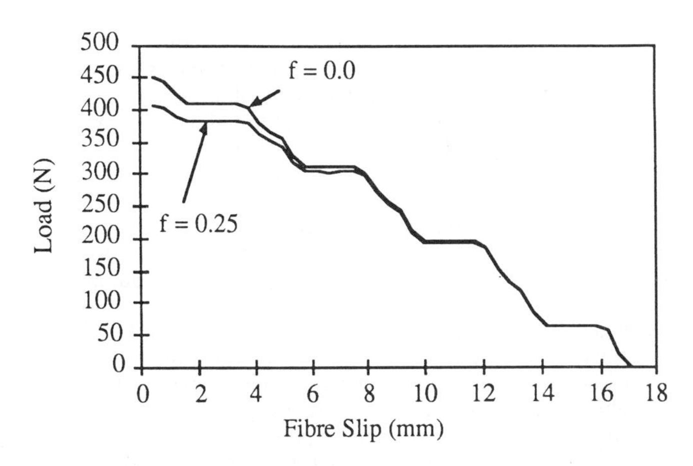

Fig.5. Coupling effect based on the plastification part of the load.

Thanks to the previous approach, we were able to analyze the coupling effect on the pull-out load - slip curve. Figure 5 shows the plastification part of the pull-out load for two cases of friction. First, the friction coefficient was taken as null and secondly, it takes the value of 0.25 obtained experimentally.

It is clear that due to the coupling effect, the higher the friction coefficient, the lower the plastic energy. This seams surprising at first, but it results directly from the decrease in the plastification moment on the interaction diagram, when the normal load increases.

4.2 Load distribution

To have another view of this coupling effect, we can analyze the distribution of these loads along the embedment length of the fibre (fig. 6).

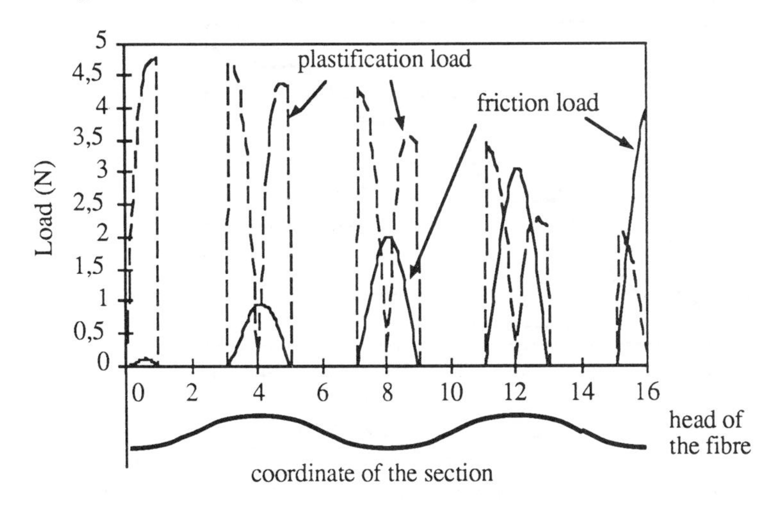

Fig.6. Distribution of the loads along the fibre.

First, it appears that along the straight part of the fibre geometry, no energy is dissipated. This comes directly from the no-cohesion hypothesis and from the constant curvature along these parts. The reality of the load distribution in the straight part of the fibre is probably affected by residual curvature from initial deforming. So, we assume that this distribution is valid after complete debonding (viz. after mobilization of each of the basic phenomena), but before large displacement (viz. before large residual strains and steel hardening).

The coupling effect appears very clearly. As the pullout load increases along the length, the plastification moment and consequently the plastification load decrease. Reversely, the radial component increase with the pullout load, resulting in an increasing friction load.

From this analysis, we conclude that the straight part of the fibre geometry is effective on the behaviour only in the mobilization stage by increasing the fibre length and so the bonding surface withe the matrix. After debonding, with respect to the model, the pullout load will be the same with or without these straight parts. This is an interesting conclusion from the optimization aim of the fibre geometry point of view.

Finally, figure 7 shows the cumulative loads from figure 6. The results are in agreement with the finite element results published by Banthia [BANTHIA, N. et al., 1991], and bear out our analysis.

5 Strains analysis

Knowing relatively precisely the stress states (or loads) in each section of the fibre, the longitudinal strain can be evaluated. We know that the validity of the present micro-mechanical model is limited only to the behaviour of the fibre after complete debonding.

However, the basic phenomena which govern the pull-out behaviour take place during the debonding and mobilization of the fibre. It follows that the higher pullout load, given by the model, is theoretically obtained exactly after the complete debonding, meaning that each fibre section undergoes a finite slip.

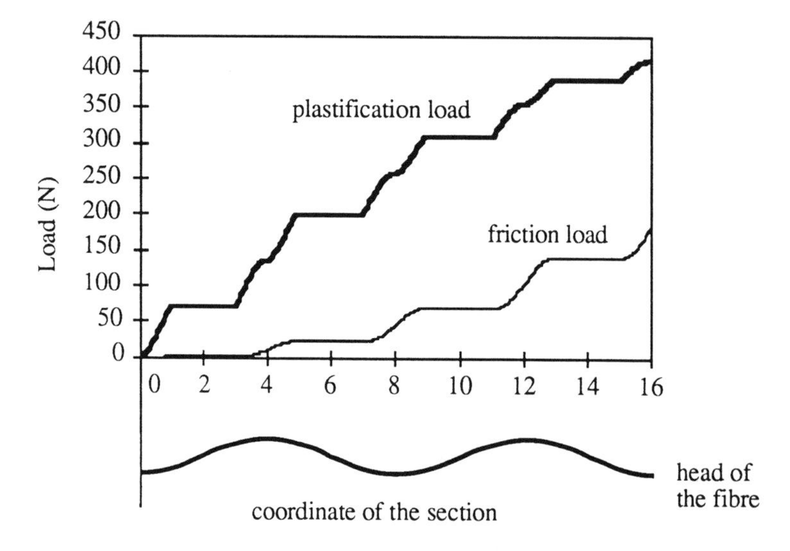

Figure 7 : Distribution of the cumulative loads along the fibre.

Assuming now that the fibre keeps an elastic behaviour axially, the strain in each section is given by :

$$\varepsilon = \frac{F}{S \cdot E} \tag{7}$$

with S section of the fibre
E elastic modulus of the steel
F axial load in each section with F=P (pullout load) at fibre head

By Integrating this equation 7 along the embedment length of the fibre, we can evaluate the displacement of the head of the fibre.

$$\Delta u = \int \varepsilon \cdot dl \tag{8}$$

The theoretical crack opening displacement, at which the peak pullout load is obtained, is about 0.04 mm with respect to the load distribution as shown in figure 7. This means that the fibre is theoretically able to restrain the crack opening quite instantaneously after its localization.

Such a behaviour would make the steel fibres able to play a prominent part as reinforcement for the concrete.

In fact, the single fibre pull-out and beam flexure experimental tests have shown the discrepancy between the theoretical behaviour and the experimental one. There are not in accordance at the time of cracking.

We propose the following simple explanation. The tridimensional stress state around the fibre produces surface effect in the crack plane, resulting in local crushing of the matrix. As a consequence, the mobilization of the fibre is really effective only after a larger crack opening displacement(meaning a few tenth millimeters), due to local staightening (fig. 8). It is interesting to point out that this local effect appears whatever the geometry of the fibre, even in the case of a straight fibre, simply due to the random orientation [NAAMAN, A.E. et al., 1976].

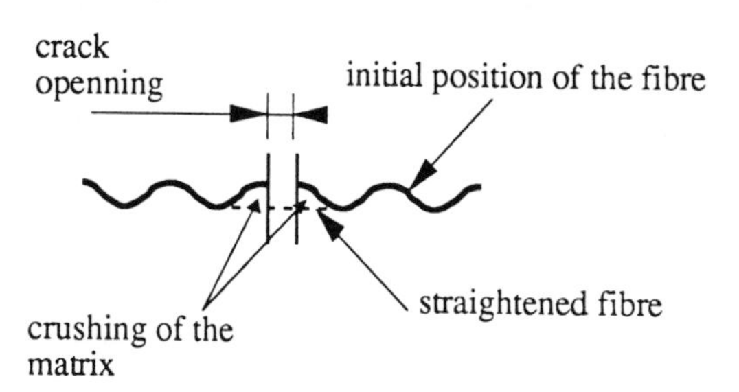

Fig.8. Local straightening of the fibre.

6 Discussions and conclusions

The behaviour of wiredrawn steel fibres results in a complex interaction between friction behaviour and anchorage behaviour. On the basis of the theoretical micro-mechanical model, we have studied the coupling effect between these two basic phenomena.

It appears clearly that the geometry of the fibre which provides the anchorage is of major importance in the overall behaviour . So, the critical size of the nonstraight fibre cannot be given in terms of critical length any longer (this concept has been based on friction considerations for straight fibre).

It is now necessary to specify the critical size regarding to the plastification capacity of the steel and according to the geometry of the fibre. The friction at the fibre-matrix interface only changes the pullout load, after debonding, by coupling effect.

The satisfactory understanding of the behaviour of the wiredrawn steel fibre draws us to the conclusion that the behaviour of structural concrete could now be more easily analyzed.

In fact, their is a critical step in this approach resulting from the mobilization of the steel fibres in structural macrocracks. We have shown that the complete debonding and so the corresponding peak pullout load is theoretically obtained for very small displacement of the fibre. At the same time, the local crushing of the matrix results in local straightening, and, as a consequence, a larger crack opening displacement is needed before mobilization of the fibre becomes effective.

In conclusion, the behaviour of the concrete depends on the behaviour of the pulling out fibre, but in addition, on the size effect provided by the structure itself.

7 References

BANTHIA, N., TROTTIER, J.-F., PIGEON, M., KRISHNADEV, M.-R. (1991) *Deformed Steel Fiber Pull-out : Material Characteristics and Metallurgical Processes,* International Workshop on High Performance Fiber Reinforced Cement Composites, Mainz, p.306-315

BURAKIEWICZ, A. (1978) *Testing of Fibre Bond Strength in Cement Matrix,* Proceeding, RILEM Symposium : Testing and Test Methods of Fibre Cement Composites, Sheffield, The Construction Press, p. 355-365.

CHANVILLARD, G. (1990) *Intéraction fibres d'acier-matrice : analyse expérimentale et modélisation,* Réunion Scientifique des groupements de recherches coordonnées, GRECO GEOMATERIAUX, CNRS, Toulon, France, 14 p.

CHANVILLARD, G. (1992) *Analyse expérimentale et modélisation micromécanique du comportement des fibres d'acier tréfilées, ancrées dans une matrice cimentaire,* Ph.D. thesis, University of Sherbrooke, Canada

CHANVILLARD, G., AITCIN, P.-C. (1991) *Micro-mechanical modeling of the pull-out behavior of corrugated wiredrawn steel fibers from cementitious matrices,* Fibre Reinforced Cementitious Materials, Materials Research Society, Symposium Proceedings, Ed. S. Mindess, Vol. 211, p. 197-202.

CHANVILLARD, G. , AITCIN, P.-C. (june 1991) *On the Modelling of the Pull-out Behaviour of Steel Fibres,* International Workshop on High Performance Fiber Reinforced Cement Composites, Mainz, p. 328-338

GRAY, R.J., JOHNSTON, C.D. (1984) *The Effect of Matrix Composition on Fibre/Matrix Interfacial Bond Shear Strength in Fibre-Reinforced Mortar,* Cement and Concrete Research, vol. 14, n° 2, p. 285-296.

HUGHES, B.P., FATTUHI, N.I. (1975) *Fibre Bond Strengths in Cement and Concrete,* Magazine of Concrete Research, vol. 27, n° 92, p. 161-166.

MAAGE, M. (1977) *Interaction Between Steel Fibers and Cement Based Matrixes,* RILEM, Matériaux et Constructions, vol. 10, n° 59, p. 297-301.

NAAMAN, A.E., NAMUR, G., NAJM, H., ALWAN, J. (1989) *Bond Mechanisms in Fiber Reinforced Cement Based Composites,t*University of Michigan, Ann Arbor, Report N° UMCE 89-9

NAAMAN, A.E., SHAH, S.P. (1976) *Pull-Out Mechanism in Steel Fiber-Reinforced Concrete*, Journal of the Structural Division, Proceeding of the American Society of Civil Engineers, vol. 102, n° 8, p. 1537-1549.
STROEVEN, P. (1979) *Micro- and Macromechanical Behaviour of Steel Fibre Reinforced Mortar in Tension*, Heron, vol. 24, n° 4, p. 7-40.

32 DEFORMED STEEL FIBER PULL-OUT MECHANICS: INFLUENCE OF STEEL PROPERTIES

M. R. KRISHNADEV and S. BERRADA
Department of Mining and Metallurgy, Laval University, Quebec,
N. BANTHIA and J.-F. FORTIER
Department of Civil Engineering, Laval University, Quebec,
Canada

Abstract
The paper presents some data on the influence of steel properties on the pull-out resistance of deformed steel fibers bonded in a cementitious matrix. Seven types of steels including plain carbon steels, martensite, stainless steels and HSLA (High strength low alloy) were investigated. Strength and ductility of these seven steels, were investigated. Strength and ductility of these steels covered a wide range. A decisive influence of the steel properties on the pull-out resistance was demonstrated.
Keywords: Steel fibre, concrete, strength, ductility, pull-out, metallography.

1. Introduction

It is well-known that addition of steel fibres to concrete brings about considerable improvements in its static, dynamic and fatigue properties (1-5). At low volume fractions, while the strength is not improved by the fibres, energy absorption capability or "toughness" is dramatically improved. Given its low strain capacity, the cement-based matrix cracks early in the loading process, and it is only after matrix cracking that fibers bridge the cracks and provide structural integrity. With further increase in the load, the fibers bridging the cracks undergo pull-out processes and fiber reinforced concrete exhibits post-matrix-cracking ductility which is entirely absent in concrete without fibers. Clearly, the extent to which fiber reinforced concrete can absorb energy in the post-matrix-cracking stage depends upon the energies absorbed by the individual fibers undergoing pull-out. A single fiber bridging a matrix crack and undergoing pull-out, therefore, is the fundamental "unit" of which the overall toughness of steel fiber reinforced concrete is composed.

Many factors play important role in governing the pull-out resistance of steel fibers bonded in cementious matrices. Note worthy are fiber geometry, matrix curing conditions, test temperature, stress-rate, etc. (6). However, one factor of major importance, that has surprisingly received very little attention, is the

Fibre Reinforced Cement and Concrete. Edited by R. N. Swamy. ©1992 RILEM.
Published by E & FN Spon, 2-6 Boundary Row, London SE1 8HN. ISBN 0 419 18130 X.

properties of steels employed for making the fibers. The role played by steel's chemical comosition, metallurgical microstructure, and the flow and fracture properties during pull-out is not yet fully understood. Also, there is no accepted standard for selecting steel for imparting a desired characteristic to SFRC. In this context, an in-depth study has been undertaken to understand the influence of metallurgical characteristics and flow properties of steel on the pull-out behavior and toughness of SFRC. In this paper, the effects of steel's chemistry, microstructure, and mechanical properties on the pull-out behavior of deformed fibers from cement-based matrices are presented. Toughness data will be reported in the near future. A variety of steels covering a broad range of compositions, microstructures, strength levels, and ductility were chosen and pull-out tests were conducted on end-deformed fibers embedded in cement matrices. Through metallographic and micro-hardness analyses of virgin and pulled-out fibres, an attempt has been made to relate the variations in pull-out performance to the variations in metallurgical characteristics.

2. Experimental

Materials and specimens

Composition of the seven steels used in the study are shown in Table 1. Steels A to C are normal carbon steels, steel D is a martensitic steel, steel E is an austenitic stainless steel, and steels F and G are HSLA (high strength low alloy) steels. Steel G is an experimental composition containing Cu, P, Ni, Si, and Cr for imparting enhanced atmospheric corrosion resistance. Both F and G contain microalloying elements (Nb, Ti) for grain refinement and precipitation strengthening. These types of steels are being used increasingly for critical structural applications requiring a good combination of high strength and low temperature toughness. The steels chosen cover microstructures varying from ferrite-pearlite to martensite. Their strength and ductility also cover a wide range. These properties are shown in Table 1. Some of the steels were tempered to various degrees for a fundamental look at fiber pull-out resistance as affected by heat treatment (Table 2).

Using these steels, end-deformed fibers (0.4 mm thick; 2.0 mm width) were produced by slitting cold-rolled sheets. Fibers were embedded in a concrete matrix with the following proportions: Cement: 373 kg/m^3, Water: 168 kg/m^3, Aggregates (20 mm): 390 kg/m^3, Aggregates (14 mm): 585 kg/m^3, Sand: 800 kg/m^3, Water Reducing Agent: 2 ml/kg: Air Entraining Agent: 0.10 ml/kg. This concrete had developed a compressive strength of 35 MPa after 28 days.

Testing

Pull-out tests were carried out in a floor mounted Instron at a cross-head speed of 0.5 mm/minute. In all tests, pull-out load versus crack opening displacement (C.O.D.) curves were digitally acquired. By measuring the area under these curves, pul-out energies were obtained.

Table 1. Chemical compositions (wt.%) and tensile properties of steels

Steel	C	Si	Mn	Cr	Ni	Mo	Nb	Cu	Ti	UTS(MPa)	Ef (%)
Steel A 1008	0.08	0.1	0.45							870	0.55
Steel B 1018	0.18	0.1	0.6							1026	0.54
Steel C 1015	0.45	0.2	0.6							1420	0.50
Steel D (Martensite)	0.17	0.03	0.45							1600	0.90
Steel E Stainless steel	0.05			18	8	2				850	3.5
Steel F (HSLA)	0.08	0.011	0.43	0.011	.014	.004	0.039	0.01		1400	0.88
Steel G (HSLA)	0.07	0.7	1.2	0.6	0.6		0.05	1.2	0.18	1450	0.60

Table 2. Tensile properties of heat treated steels

Steel	Treatment	Y.S. (MPa)	U.T.S. (MPa)	Sf (MPa)	Ef (%)
A	CR	776	870	870	0.55
	5 min. 600°C	716	810	585	2.3
B	CR	890	1026	1026	0.54
	5 min. 600°C	810	960	750	2.78
HSLA" G"	CR	1235	1400	1400	0.88
	5 min. 600°C	1061	1262	988	3.21

Note: Y.S. : Yield Strength; UTS: Ultimate Stress; Sf : Stress at fracture; Ef : Elongation at fracture; CR : Cold rolled

Standard metallographic techniques were used for preparing fiber specimens for microscopic observations. Normally, 2% nital was used for etching samples to reveal microstructural features. For observing the surface of the fiber after pull-out, a part of pulled-out fiber was mounted in plexiglass and finely polished. These samples were examined in a JEOL SEM II scanning electron microscope operating at 25 kV. To reveal dislocation substructure and copper precipitation, transmission electron microscopy was employed. Thin foils were prepared by electrochemical polishing method (7).

Microhardness measurements were conducted on fibers before and after the pull-out by using Leco (M-400 T) microhardness tester. A 25 gram-force indenter was employed and resulting Vickers hardness numbers were tabulated.

3. Results and Discussion

3.1 Pull-out test on steel fibers

The two parameters of interest in a pull-out test are the peak pull-out load and the total pull-out energy. Figures 1 and 2 show some representative load-deflection curves. The peak pull-out loads and the total pull-out energies for the various steels are tabulated in Table 3. Quantities in Table 3 are averages of four specimens with a standard deviation between 3 and 8%.

Peak pull-out loads

As seen in Figures 1 and 2, the peak pull-out loads are considerably influenced by the characteristics of the fibers themselves. The results may be discussed separately for plain carbon steels (Steels A, B and C) and the alloy steels (Steels D,E,F and G).

Plain Carbon Steels

For plain carbon steel fibers (Figure 1 and Table 3), the peak load varied proportionately with the percentage of carbon; higher the percentage of carbon, higher was the load registered. In addition, the deflection at the peak load was also found to decrease with an increase in the carbon content (i.e. with an increase in the pearlite content). This may be related to the fact that, although pearlite has virtually no effect on the yield stress, it increases the tensile strength and reduces ductility (8). Thus the increase in the pul-out peak load may be related to the improvement in the ultimate strength of steel itself increasing the pull-out resistance.

Alloy Steels

Martensite steel (Steel D), with a highly dislocated sub-structure and strong barriers to dislocation movement, has a very high strength but poor ductility. This leads to a higher peak pull-out load in pull-out than

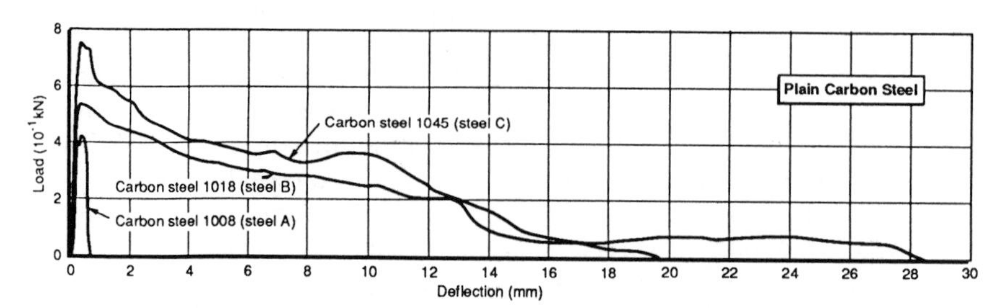

Figure 1. Pull-out load-displacement plots for plain carbon steel fibers

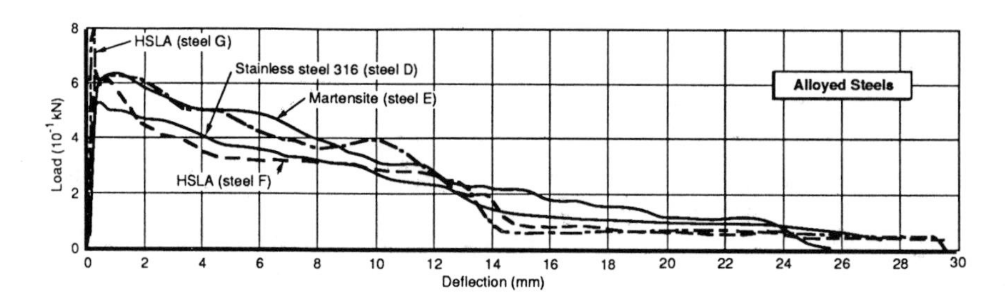

Figure 2. Pull-out load-displacement plots for alloyed steel fibers

Table 3. Pull-out test results

Steel	Peak load (kN)	Energy (N.m)	Deflection at peak load (mm)
Steel A	0.43903	0.4	0.71
Steel B	0.52716	4.2793	0.69
Steel C	1.01320	6.4983	0.44
Steel D	0.60186	5.7191	0.8
Steel E	0.50665	4.1252	0.74
Steel F	0.56786	3.8713	0.63
Steel G	0.88695	6.5377	0.39

stainless steel (Steel E) which had only a moderate strength but very good ductility (Figure 2, Table 1 and Table 4). So, it appears, as in the case of Carbon Steels above, that the strengfth of steel is a more dominating factor than its ductility.

In the case of the HSLA steels (Steels F and G), the steel with copper (Steel G) gave higher peaks loads and lower deflections at peak pull-out loads than did HSLA steel without copper (Steel F). This is due to the finer grain size brought by sufficient Nb addition (0.05%) for steel G. Transmission electron microscopy (TEM) was carried out to detect the precipitation of copper in steel G. Figure 3 shows extremely fine Cu-precipitates distributed uniformly in the matrix. Such a uniform distribution of precipitation permits unifrom distribution of slip thereby preventing strain localizations and possible cleavage fracture initiation near a stress concentrator(9) Matrix also had a high density of dislocations which gives additional strengthening. It appears that both grain refinement and precipitate strengthening are beneficial to pull-out resistance.

Energy absorbed
Once again, the absorbed energy values may be discussed separately for the plain carbon steels and the alloy steels.

Plain Carbon Steels
Based on the pull-out results in Figure 1 and Table 3, it may be noticed that steel fiber containing 0.08% carbon (steel A) absorbs less energy than that with 0.18% (steel B) or 0.45% carbon (steel C). Given the fact that steels containing high levels of carbon have higher strengths and lower ductilities, it may be deduced that high strength and low ductility in fiber steels leads not only to an improved peak pull-out load as seen before, but also to better pull-out energies. Steel A with low carbon also had lot more inclusions than other steels. This implies that even the purity of steels (i.e. inclusion level) is an important consideration while selecting steels. The tendency for cracking increases with an increase in the inclusion content as there are more sites for crack nucleation (10, 11)

Alloy steels
As evident from Figure 2 and Table 3, fibers made of martensite steel exhibited better pull-out energy absorption as compared to those made of stainless steel. Once again, a combination of low ductility and high strength in martensite gave a higher pull-out energy than a combination of high ductility and low strength in stainless steel. In the HSLA steels (Steels F and G), Steel G containing copper with additional precipitation strengthening gave better pull-out energies than steel F which did not have any copper.

Table 4. Microhardness of steel fibers before and after pull-out

Steel	Hardness Vickers (virgin)	Hardness Vickers (pull-out)
Steel A	224	224
Steel B	285	286
Steel C	378	378
Steel D	340	355
Steel E	296	296
Steel F	302	317
Steel G	452	460

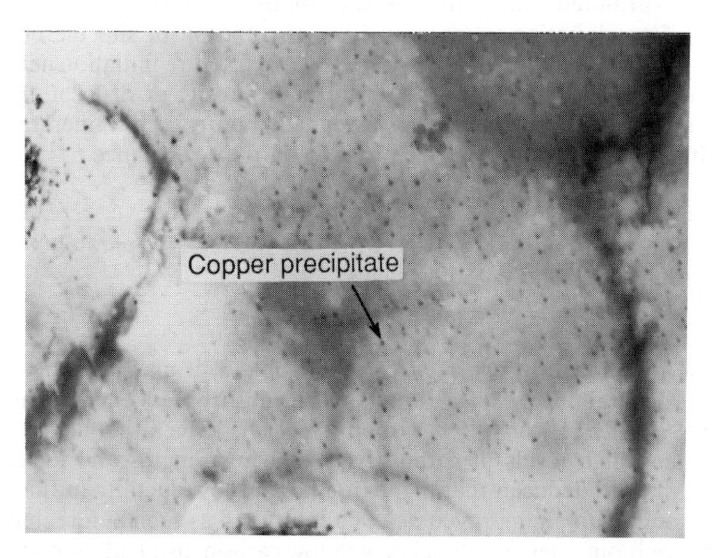

Figure 3. TEM, uniform precipitates of copper in steel G

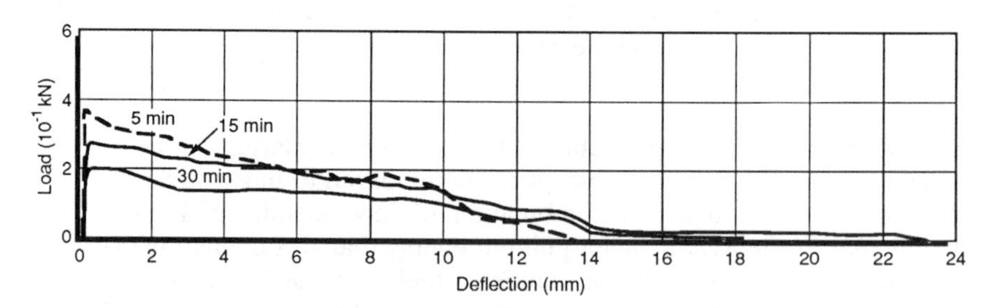

Figure 4. Effect of heat treatment on the pull-out load-displacement plots for HSLA steel (steel F) fibers

3.2 Effect of tempering on the pull-out behavior

Steels were heat treated at 600°C for various lengths of time for further modification of the properties. Some results obtained from heat treatment are shown in Table 2. In pull-out tests, the deflection at the peak load, the peak load itself, and the energy absorbed were found to decrease with increased tempering (Figure 4 and 5). This may be related to the microstructural changes that take place on tempering. In the case of light tempering (tempered for 5 minutes), there is only a partial recovery of dislocations and strength remains high and ductility remains low. But in the highly tempered sample (tempered for 15 minutes), recrystallization and grain growth takes place causing the dislocations to move easily as compared to the partially recovered sample. This results in softening of the matrix leading to a lower strength and high ductility. This confirms what has already been observed previously, i.e. the strength of the fiber plays a more significant role than ductility in fiber pull-out.

3.3 Microhardness measurements

Table 4 indicates the microhardness values for fibers before and after the pull-out. The numbers in Table 4 are average values over 25 mm length of the fiber. Notice that the microhardness after the pull-out is marginally higher than that before the pull-out. A much greater increase in the microhardness in the case of round fibers has been previously reported (12).

Among the steels studied, it was observed that the steels with the higher ultimate strengths, higher micro-hardness numbers, and low to moderate ductility had to a superior pull-out performance.

3.4 Scanning Electron Microscope Observations

In a pull-out test, the shear bond between the fiber and the matrix is lost much early in the process followed by fiber straightening and eventual pull-out under frictional forces. In a Scanning Electron Microscope it was observed that all the cracks on the surface of the fiber were perpendicular to the direction of the loading (Figure 6). All fibers except those from steel A were pulled-out during testing; fibres from Steel A fractured during the test. This behavior is a result of its low strength and high inclusion content. Figure 7-a shows that the cracks are continuous for Steel B compared to the HSLA Steel G where the cracks are mainly confined to the precipitates (Figure 7-b) and are blunted due to the intrinsic toughness of the steel matrix (i.e. extra fin-grained ferrite and bainite) and due to better compositional control and cleanliness. This may, to some extent, explain the better pull-out behavior of Steel G over Steel 1018. For all the steels studied, microscopic observations confirmed that the cracks start from the inclusions (Figure 7-a), indicating that it is important to have a minimum of impurities in fibers for SFRC.

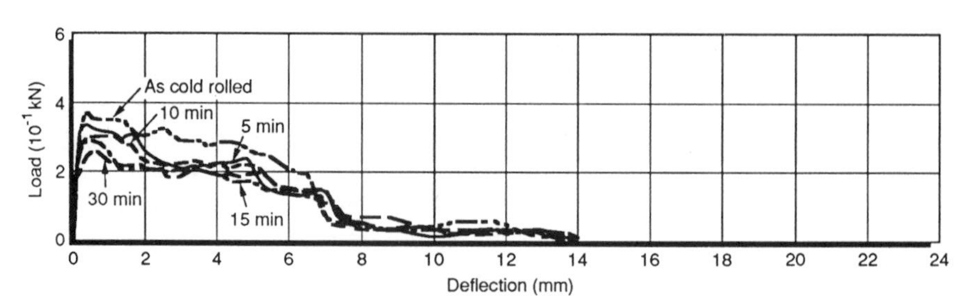

Figure 5. Effect of heat treatement on the pull-out load-displacement plots for plain carbon steel fibers (1018)

Figure 6. SEM photograph of pull-out steel fiber showing many surface cracks

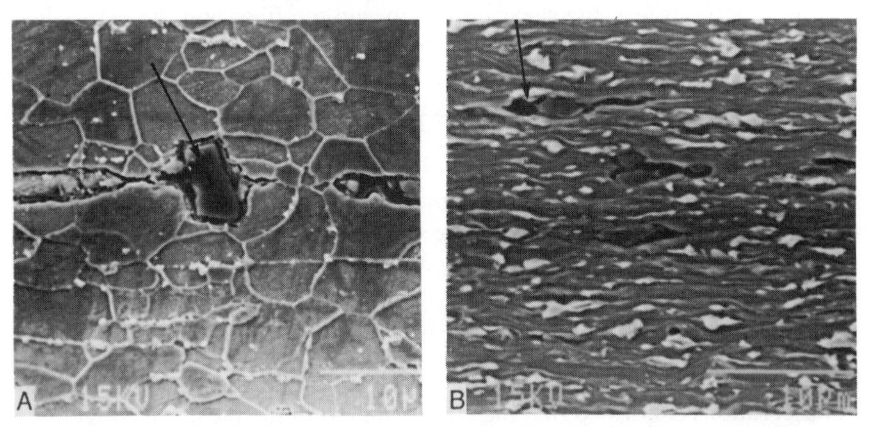

Figure 7. SEM photographs of pull-out steel fiber showing
A) continuous cracks and
B) cracks mainly confined to the precipitates

4. Conclusions

1 Although often ignored, steel's chemical composition and metallurgical microstructure have significant influence on the pull-out behavior of SFRC

2 The strength of the fiber is a more dominating factor than its ductility. For achieving high performance characteristics as indicated by pull-out tests, fibres need to have very high strengths and moderate ductility rather than very high ductility and moderate strengths. Pull-out performance of steel fibers appears also to be influenced by steel cleanliness and the existence of strengthening mechanisms like precipitation strengthening.

3 Considerable future research is necessary before a comprehensive understanding of the influence of steel metallurgical characteristics on the performance of SFRC may be developped.

5. References

Swamy, R.N., and Barr, B., (Editors), Fiber Reinforced Cements and Concretes: Recent Developments, Elsevier Applied Science 1989.

Shah, S.P. and Rangan, B.V., ACI Journal, Proc., 68(2), Feb. 1971, 126-134.

Ramakrishnan, V., Oberling, G. and Tatnall, P., ACI, SP-105, 1987, pp. 225-245.

Banthia, N., Midness, S. and Bentur, A., Materials and Structures (Paris). 20(119), 1987, pp.293-302

Suaris, W. and Shah, S.P., Journal of Structural Division, ASCE, 109(7), July 1983, pp. 1717-1741.

Banthia, N., Cdn J. of Civil Eng, 17 (4), 1990, pp. 610-620.

Krishnadev, M.R. et al, "Fracture-Microstructure-Mechanical Property Relations in an Advanced HSLA Steel", Microstructural Science, Vol. 14, 1986, pp. 189-203.

Pickering, F.B., "Physical Metallurgy and the Design of Steels", Applied Science Publishers, 1978, pp.60-69.

Cutler, L.R., Krishnadev, M.R., "Microstructure-Fracture-Touughness Relations in Copper Strengthened HSLA Steels", Microstructural Science, Vol.10, 1981, pp. 79-90.

Kozasu, I. and Tanaka, J., "Effects of Sulfide Inclusion on Toughness and Ductility of Structural Steels", Sulfide Inclusion in Steel, Vol. 6, ASM, American Society for Metals, 1975, pp. 286-308.

DeArdo, A.J and Hamburg, E.G., "Influence of Elongated Inclusions on the Mechanical Properties of High Strength Steel Plates", Sulfide Inclusion in Steel, Vol.6, ASM, American Society for Metals, 1975, pp. 286-308.

Banthia, N. and Trottier, J.-F., Pigeon, M. and Krishnadev, M.R., "Deformed Steel Fiber Pull-out: Material Characteristics and Metallurgical Process", Proc.Int.Workshop on Fiber Reinforced Concrete, Mainze (Germany), 1991.

33 MICROMECHANICS OF FIBER EFFECT ON THE UNIAXIAL COMPRESSIVE STRENGTH OF CEMENTITIOUS COMPOSITES

V. C. LI and D. K. MISHRA
Advanced Civil Engineering Materials Research Laboratory,
Department of Civil Engineering, University of Michigan,
Ann Arbor, MI, USA

Abstract
A micromechanical model is presented for the uniaxial compressive strength of fiber reinforced cementitious composites (FRCC). The model is based on the classical models of compressive failure of brittle solids containing sliding microcracks that induce wing-crack growth under compressive loading. The concepts of increased microcrack sliding resistance and wing-crack growth retardation associated with fiber bridging are exploited to produce a strengthening effect of fibers on composite strength. The concept of defect introduction associated with fiber volume fraction is included to produce a composite strength degradation. The combined effects result in a composite compressive strength which increases initially and subsequently drops with increasing fiber content, as has been observed in FRCs reinforced with a variety of fibers. This paper represents an extension of a preliminary study of the influence of fibers on compressive strength of FRCC by Li (1991). More accurate stress intensity factor calibrations for wing-crack growth and interaction are employed in the present paper. The predicted general trends of compressive strength change with fiber parameters remain unchanged from the original work.
Keywords: Fiber Reinforced Cementitious Composites, Concrete, Compressive Strength, Fiber Bridging, Micromechanics, Model.

1 INTRODUCTION

Early experimental studies of compressive strength of fiber reinforced cementitious composites using steel, glass and polypropylene fibers (e.g., Shah and Rangan, 1971; Fannela and Naaman, 1983) suggested that the influence of fibers on the compressive strength was insignificant at low volume fractions. Both increase and decrease of compressive strength with different fiber types have been experimentally observed (e.g., Akihama et al., 1986a; Ward and Li , 1990; Zhu, 1990). Even for the same material, there is mounting evidence that compressive strength may first rise followed by a drop with increasing fiber volume fraction. These observations suggest that the addition of fibers in a cement composite using conventional mixing procedure leads to a competing process of strength improvement as well as degradation. Some recent research, however, suggest that compressive strength can be enhanced substantially even at high fiber volume fraction, when special processing techniques, presumably leading to reduced matrix defects, are employed (Tjiptobroto, 1991; Naaman et al., 1991). Li (1991) proposed a simplified model based on micromechanics of fiber reinforcement to explain some of these effects. In this brief paper the previous model is modified by utilizing more accurate formulations for the stress intensity factors for initiation, propagation and interaction of sliding microcracks in the composite.

Fibre Reinforced Cement and Concrete. Edited by R. N. Swamy. © 1992 RILEM.
Published by E & FN Spon, 2-6 Boundary Row, London SE1 8HN. ISBN 0 419 18130 X.

The proposed model is fundamentally based on well known micromechanical models of compressive failure in brittle solids (Horii and Nemat-Nasser, 1986; Ashby and Hallam, 1986; Kemeny and Cook, 1991). The influence of fibers on microcrack sliding and extension is based on crack bridging studies carried out in recent years (Li and co-workers, 1991, 1992). The model is kept to be as simple as possible in order to obtain close form solutions which elucidate the micromechanical parameters controlling the strengthening and the weakening mechanisms. It is found that depending on the effectiveness and amount of fiber bridging, and the degree to which fibers introduce defects to the composite, both increase and decrease of compressive strength can be derived from increasing fiber content. Furthermore, it is identified that the fiber/matrix bond strength and a snubbing coefficient can lead to higher compressive strength for a given fiber type and aspect ratio, and for a given composite fabrication process. A summary of compressive strength change with fiber volume fraction is presented in Figure 1, the details of which can be found in Li (1991).

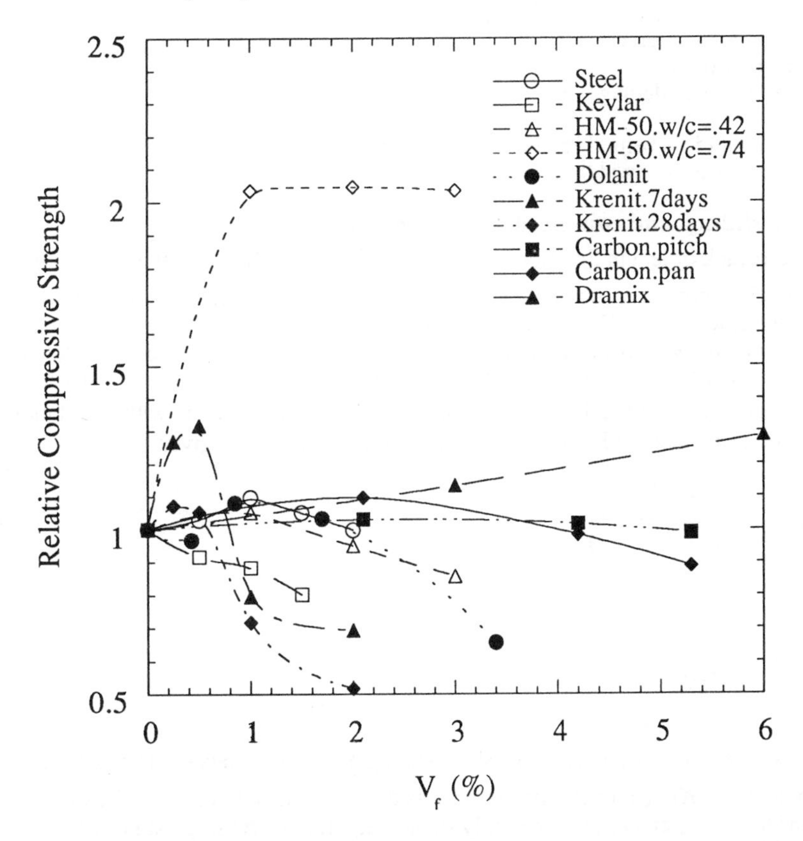

Figure 1. Compressive Strength of Various Fiber Reinforced Cementitious Composites Relative to the Matrix Compressive Strength, as a Function of Fiber Volume Fraction.

2 A MICROMECHANICAL MODEL

Unstable propagation of a critical tensile crack accounts for the failure of brittle solids under tension. However, under compressive loading the microcracks in the solid come under a local tensile field at their tips causing initiation of "wing-cracks". The extension of wing-cracks under such a local tension has been demonstrated to be unstable initially and becomes stable as the crack length increases. However presence of other microcracks and the interaction between them induces instability resulting in final failure. When fibers are present in such a body, they affect the crack propagation by increasing the resistance to sliding of the initial microcracks and opening of the wing cracks by crack-bridging. Thus the models used for fiber bridging developed for tensile strength can also be used in case of compressive strength. The proposed model can be divided into the following distinct parts that take into account different simultaneously occurring mechanisms:

(i) Crack propagation due to sliding and opening
(ii) Crack-crack interaction
(iii) Crack bridging
(iv) Fiber induced damage

The fundamental ideas behind each of the above concept has been discussed by Li (1991) and will only be summarized here.

2.1 Crack sliding and wing-crack propagation

For the microcrack of length $2a$ oriented at an angle β (Figure 2) in a brittle solid subjected to the uniaxial compressive load σ, a shear stress τ is generated which causes frictional sliding on the crack faces. The shear sliding in turn creates a singular stress field with tensile components on opposite quadrants, causing the initiation of tensile "wing-cracks". A simplified approximate expression for the stress intensity factor of straight wing-cracks of length ℓ parallel to the loading axis for the most critical sliding crack with orientation $\beta = 45°$ has been obtained by Horii and Nemat-Nasser (1986) and is given by:

$$K_I = \frac{2\tau\sqrt{\pi a}}{\pi\sqrt{2(\ell / a + 0.27)}} \tag{1}$$

where

$$\tau = \frac{1}{2}\sigma(1-\mu) \tag{2}$$

and μ is a coefficient of friction against shear sliding of the crack faces. The normalized crack driving force $K_I / (\sigma \sqrt{\pi a})$ drops rapidly as the wing-crack length (ℓ / a) extends. This means that wing-cracks are inherently stable, requiring increasing load to continue its growth. These wing-cracks lose their stability by crack-crack interaction, which raises the stress intensity factor and is a function of the crack density (i.e. how close the interacting cracks are to each other). Interaction between wing-cracks, therefore, is critical in leading to a critical load in compression -- the compressive strength of the solid with many microcracks.

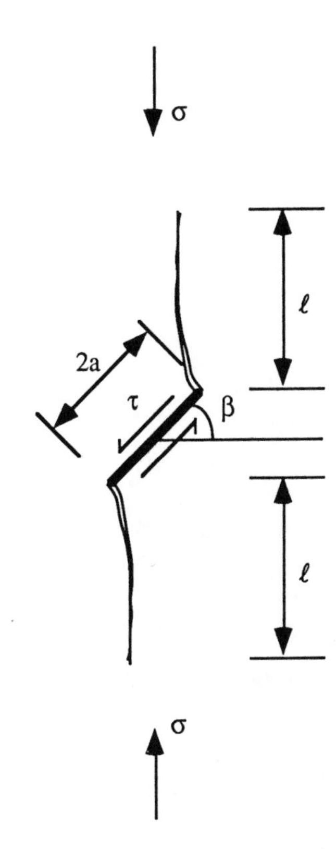

Figure 2. Wing-Crack Growth Induced by Sliding of Microcrack as the Basic Mechanism of Compressive Failure in Brittle Solids.

2.2 Crack interactions

Under overall compressive stress fields, Ashby and Hallam (1986) assumed crack interactions to result in the buckling of material columns formed by subparallel wing-crack growth. For this model, the stress intensity factor due to interaction between wing-cracks under uniaxial compression loading is given by:

$$K_I' = \frac{\sqrt{2}}{\pi}\sqrt{D_o}\left(\ell_o + \frac{1}{\sqrt{2}}\right)^{1/2} \sigma\sqrt{\pi a} \tag{3}$$

where

$$D_o = N_A \pi a^2 \tag{4}$$

and N_A is defined as the number of cracks per unit area. ℓ_o is the normalized wing-crack length ℓ / a. The intensity of crack interaction is dictated by D_o,which may therefore be regarded as an initial damage parameter of the solid in this model. For example, a poorly compacted concrete may be expected to have a large D_o value.

Assuming that linear elastic fracture mechanics holds on the scale of micro-defects, the condition for crack propagation is given by:

$$K_I + K_I' = K_{IC} \tag{5}$$

and $K_{IC} = K_m$ is the fracture toughness of the cementitious material without fibers. Inverting (5) the normalized compression load (σ_o) required to propagate the crack to length (ℓ / a) may be calculated:

$$\sigma_o \equiv \frac{\sigma\sqrt{\pi a}}{K_m} = \frac{1}{A[\ell_o, \mu, D_o]} \tag{6}$$

where

$$A[\ell_o, \mu, D_o] \equiv \frac{1-\mu}{\pi\sqrt{2(\ell_o + 0.27)}} + \frac{\sqrt{2}}{\pi}\sqrt{D_o}\left(\ell_o + \frac{1}{\sqrt{2}}\right)^{1/2}$$

This solution is illustrated in Figure 3 for different initial damage levels (D_o). The compressive strength is given by the peak of these curves, and is shown to decrease with the amount of initial damage. It is interesting to note that for tensile loading, ($\sigma\sqrt{\pi a}$) / K_m is of the order of unity. Since for cementitious materials, the compressive strength is typically one order of magnitude higher, Figure 3 suggests that the typical initial damage level must be of the range of 0.0001 to 0.0015. For the calculations to follow, we choose D_o= 0.0005 as the natural flaw density of typical cementitious materials without fibers.

3 STRENGTHENING EFFECT OF FIBER ADDITION

3.1 Resistance to crack-sliding

Based on frictional resistance to the pull out of randomly oriented fibers bridging across a crack, Li (1991) developed the following equation for reduction in shear stress acting on a sliding crack.

$$\tau_B = \frac{1}{2} s V_f \left(1 - \frac{2a\sigma}{L_f G}(1-\mu)(1-\nu)\right) \tag{7}$$

where G and ν are the composite shear moduli and Poisson's ratio respectively, V_f is the fiber volume fraction and L_f is the fiber length. The reinforcement index s is defined as

$$s = g\tau_f \left(\frac{L_f}{d_f}\right) \tag{8}$$

where d_f is the diameter of the fiber and τ_f is the interfacial shear strength. For typical values of the snubbing coefficient f (associated with inclined fiber pull-out, Li 1992) ranging from 0 to 1, the snubbing factor g ranges from 1 to 2.3.

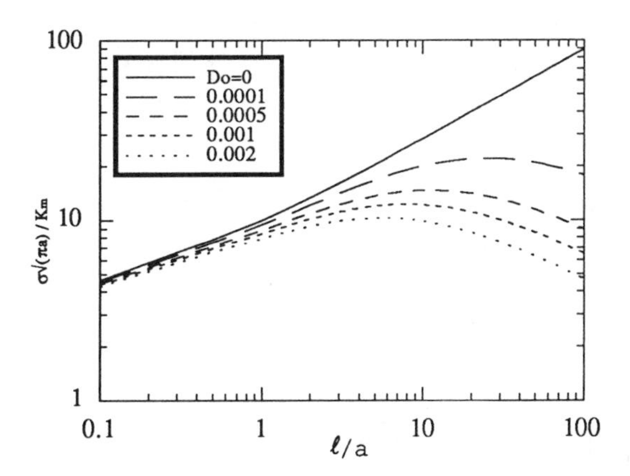

Figure 3. Normalized Compression Load Required to Drive a Wing-Crack of Length ℓ, for Four Different Initial Damage Level D_o. Calculated for $\mu = 0.5$.

The net shear stress acting on the sliding microcrack is therefore given by

$$\tau = \frac{1}{2}\sigma(1-\mu) - \tau_B \tag{9}$$

3.2 Resistance to wing-crack growth

The wing-crack, like the sliding microcrack, will also be bridged by fibers. As the wing-crack grows, increasing amount of bridging fibers will lead to an increase in crack closing pressure in an enlarging 'process zone'. Based on an R-curve concept associated with fiber bridging, Li (1991) developed an expression for the total toughness against which the wing-cracks have to propagate against:

$$K_{IC} = K_m + \sqrt{\frac{\ell}{\ell_o^*} E G_o} \tag{10}$$

where E and G_o are the composite Young's moduli and the composite fracture energy, respectively and ℓ^* defines the wing-crack length extension at which the linearized R-curve reaches the plateau value for the bridging fracture energy (Li, 1991). In the case of FRC, Li (1992) found that G_o can be related to the fiber and interfacial properties:

$$G_o = \frac{1}{12} s L_f V_f \tag{11}$$

3.3 Assessment of fiber strengthening effect

Combining eqns. (1), (3), (7) - (11) we obtain an expression relating the normalized compressive load σ_o required to maintain the normalized wing-crack length ℓ_o:

$$\sigma_o \equiv \frac{\sigma\sqrt{\pi a}}{K_m} = \left\{\frac{B[\ell_o,\mu,V_f,c]}{A[\ell_o,\mu,D_o]} + C[K_o,s_o,V_f]\right\}\frac{1}{D[s_o,V_f]} \tag{12}$$

where

$$B[\ell_o,\mu,V_f,c] \equiv (1-\mu)\left\{1+\sqrt{\frac{\bar{a}}{\ell_o^*}c\ell_o V_f}\right\};$$

$$C[K_o,s_o,V_f] \equiv \frac{s_o V_f}{K_o}; \text{ and}$$

$$D[s_o,V_f] \equiv (1-\mu)\left(1+4(1-\nu^2)s_o V_f\,\bar{a}\right)$$

and the non-dimensional parameters are defined as:

$$\ell_o \equiv \frac{\ell}{a};\ \ell_o^* \equiv \frac{\ell^*}{L_f};\ \bar{a} \equiv \frac{a}{L_f};\ s_o \equiv \frac{s}{2E};\ c \equiv \frac{1}{12}\frac{sL_f E}{K_m^2};\ K_o \equiv \frac{K_m}{E\sqrt{\pi a}}.$$

We are now in a position to access the positive influence of fibers on the load bearing capacity as the wing-cracks grow (Figure 4a). In this figure, the initial damage magnitude is fixed ($D_o = 0.0005$), and a family of curves is generated for various fiber volume fraction. (Other fixed parameters in this and subsequent calculations are $\ell_o^* = 20$; $\bar{a} = 0.1$; $c = 800$; $K_o = 0.0002$; $s_o = 0.01$. They are chosen to represent typical FRCs but can be adjusted for specific material systems.) Because σ_o scales linearly with V_f and s (eqn. 11), this family of curves may also be interpreted as a result of the influence of the interfacial bond strength or fiber aspect ratio. Figure 4b shows the monotonic increase in compressive strength σ_c with V_f. For the present set of parameters, compressive strength is shown to increase by more than 100 % for V_f up to 2%.

4 WEAKENING EFFECT OF FIBER ADDITION

As mentioned before, the addition of fibers beyond a certain optimal level may adversely affect the compressive strength due to introduction of additional defects and difficulties in processing. To account for this effect Li (1991) has suggested a simple modification in the initial flaw density or damage index parameter D_o following experimental observations. By introducing the fiber induced damage index k, we redefine D_o as given by equation (13) and replace the previous definition in equation (4).

$$D_o = (N_A \pi a^2)\, e^{kV_f} \tag{13}$$

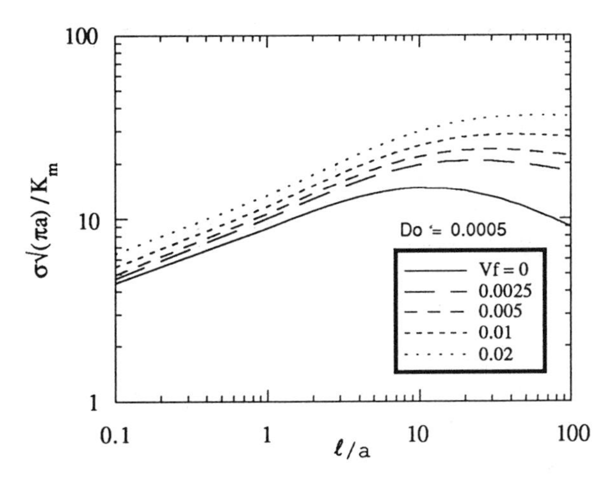

Figure 4a. Strengthening Effect of Fibers: Normalized Compression Load Required to Drive a Wing-Crack of Length ℓ, for Five Different Fiber Volume Fractions. Parametric Values Used are $\ell_o^* = 20$; $D_o = 0.0005$; $\bar{a} = 0.1$; $c = 800$; $K_o = 0.0002$; $s_o = 0.01$.

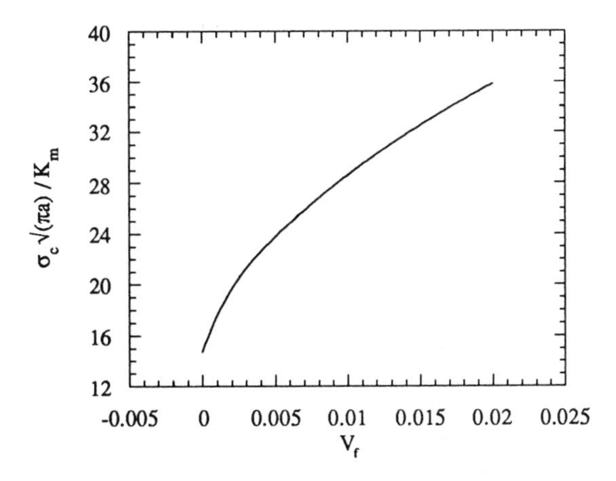

Figure 4b. Strengthening Effect of Fibers: Predicted Compressive Strength Increases with Fiber Volume Fraction, when No Fiber Induced Damage Effect is Included. Parametric Values Used are $\ell_o^* = 20$; $D_o = 0.0005$; $\bar{a} = 0.1$; $c = 800$; $K_o = 0.0002$; $s_o = 0.01$.

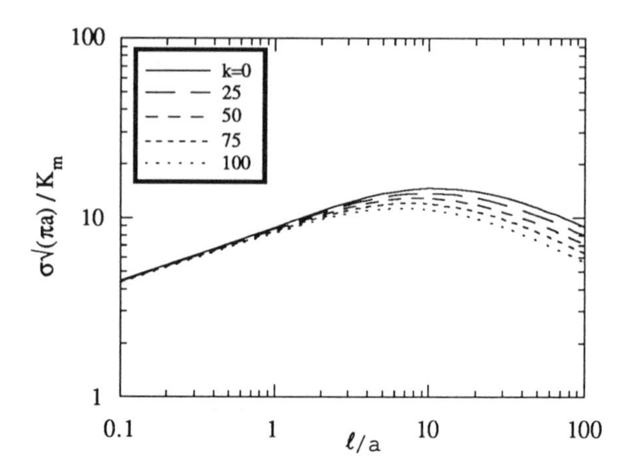

Figure 5a. Damage Effect of Fibers: Normalized Compression Load Required to Drive a Wing-Crack of Length ℓ, for Five Different Fiber Induced Damage Index k ($V_f = 0.01$). Parametric Values Used are $\ell_o^* = 20$; $D_o = 0.0005$; $\bar{a} = 0.1$; $c = 0$; $K_o = 0.0002$; $s_o = 0$.

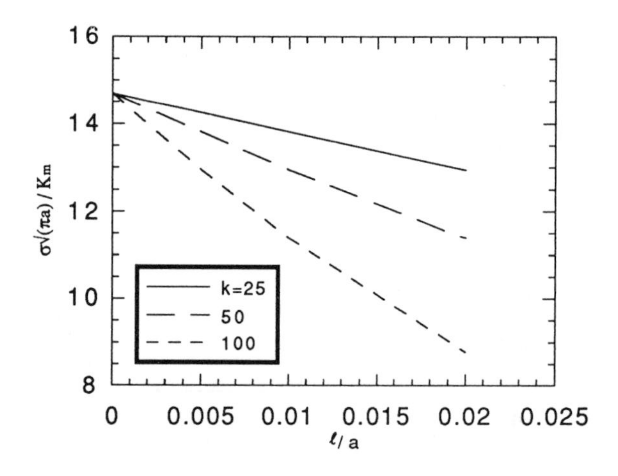

Figure 5b. Damage Effect of Fibers: Predicted Compressive Strength Decreases with Fiber Volume Fraction Due to Fiber Induced Damage Effect. Parametric Values Used are $\ell_o^* = 20$; $D_o = 0.0005$; $\bar{a} = 0.1$; $c = 0$; $K_o = 0.0002$; $s_o = 0$.

The parameter k is probably dependent upon the fiber type and processing techniques and has to be evaluated by experimental investigation for particular fiber-matrix systems.

Now we can evaluate the damage effect of the fibers. For the sake of clarity, we assume the fiber reinforcement index $s = 0$ (resulting in $c = s_o = 0$) in equation (12) so that the beneficial effects of fibers are suppressed. Figure 5a illustrates the damage effect of the fiber on the stress-crack length response for different values of k for a given volume fraction $V_f = 0.01$. As expected with increasing k, strong crack interaction reduces the compressive strength. Figure 5b indicates the negative influence of fiber addition on compressive strength.

5. COMBINED STRENGTHENING AND WEAKENING EFFECT OF FIBER ADDITION

Eqns. (12) and (13) may be used to study the effect of fiber on compressive strength in FRCs, when microcrack sliding resistance, wing-crack growth resistance, and damage introduction are operational simultaneously, as is suggested by experimental data such as that shown in Figure 1. Figure 6a shows the normalized compression load required to drive a wing-crack of length ℓ, for various fiber volume fractions. In Figure 6b, we show that the compressive strength may continue to rise even beyond 4% when the fiber damage index is small (e.g. $k = 25$), but rapidly drops beyond 0.4% when the fiber damage index is large (e.g. $k = 100$). In between these extremes, compressive strength is seen to rise initially with fiber volume fraction, and then decreases with additional amount of fibers. These predictions of fiber induced compressive strength changes are in qualitative agreement with the experimental data shown in Figure 1.

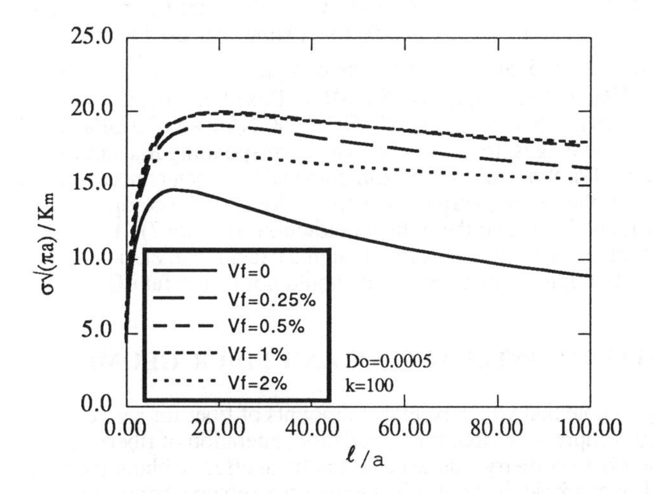

Figure 6a. Combined Strengthening and Damage Effect of Fibers: Normalized Compression Load Required to Drive a Wing-Crack of Length ℓ, for Five Different Fiber Volume Fractions. Parametric Values Used are $\ell_o^* = 20$; $D_o = 0.0005$; $\bar{a} = 0.1$; $c = 800$; $K_o = 0.0002$; $s_o = 0.01$, and k=100.

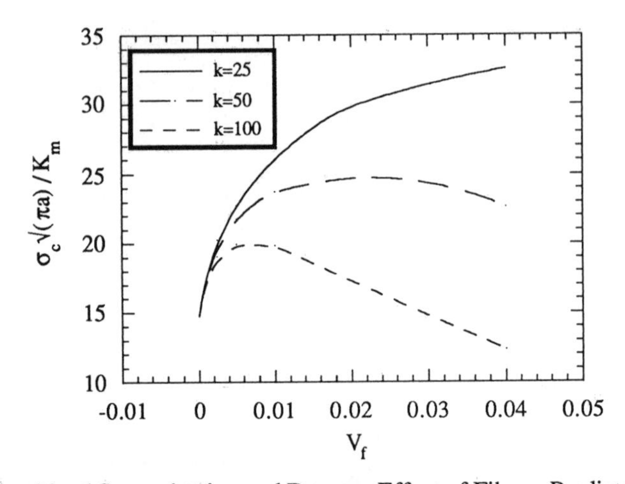

Figure 6b. Combined Strengthening and Damage Effect of Fibers: Predicted Compressive Strength Change with Fiber Volume Fractions, for Different Fiber Induced Damage Index k. Parametric Values Used are $\ell_o^* = 20$; $D_o = 0.0005$; $\bar{a} = 0.1$; $c = 800$; $K_o = 0.0002$; $s_o = 0.01$.

Figure 7 shows model predictions for the Krenit (a polypropylene) fiber reinforced concrete data (also shown in Figure 1). The original compressive strength data shows a higher value for the 28 days composite in comparison to the 7 days composite. When the data is normalized with respect to the plain concrete ($V_f = 0$) compressive strength, the relative strength for the 7 days composite lies above that of the 28 day composite. Common parametric values chosen for both sets of data are: $\ell_o^* = 40$; $D_o = 0.002$; a = 6mm; L_f/d_f = 12mm/.08mm = 150 (an effective diameter is used for the thin film-like fiber); $\tau = 2$ MPa, $\mu = 0.5$, and k = 250. The compressive strength of the plain concrete at 7 days is 23.1 MPa, and at 28 days is 38.8 MPa. Based on this, the elastic modulus and matrix fracture energy are estimated at 20 GPa, 150 N/m, and 30 GPa, 200 N/m for the composites of the two ages, respectively. The corresponding K_m values are then 1.75 MPa $\sqrt{m}$ and 2.5 MPa $\sqrt{m}$. All the non-dimensional parameters needed as model input can be calculated from this set of parameteric values. Reasonable comparisons can be found between experimental data and theoretical predictions (Figure 7). However, it should be mentioned that there is plenty of uncertainty in the exact parameteric values (since they are not measured), although the numbers used should not be too far off.

6. FIBER/MATRIX INTERACTION AND FIBER GEOMETRY EFFECT

It is interesting to note that while excessive amounts of fiber for a given fabrication process can lead to compressive strength degradation, alteration of fiber/matrix interaction property, or the fiber geometry, can lead to beneficial effect without the attendant damage introduction. For example, it is conceivable that fiber/matrix bond strength or the snubbing factor could be increased without causing a rise in the initial amount of damage. This is in fact one of the assumptions behind eqn. (12), and we illustrate this idea with Figure

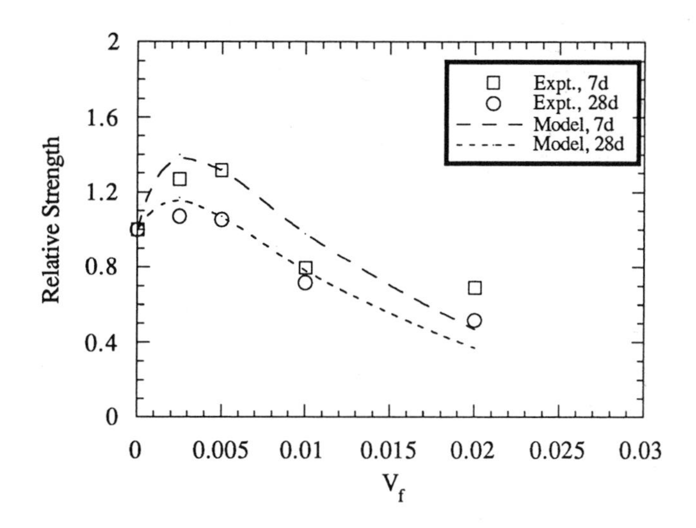

Figure 7. Model predictions for the Krenit fiber reinforced composite.

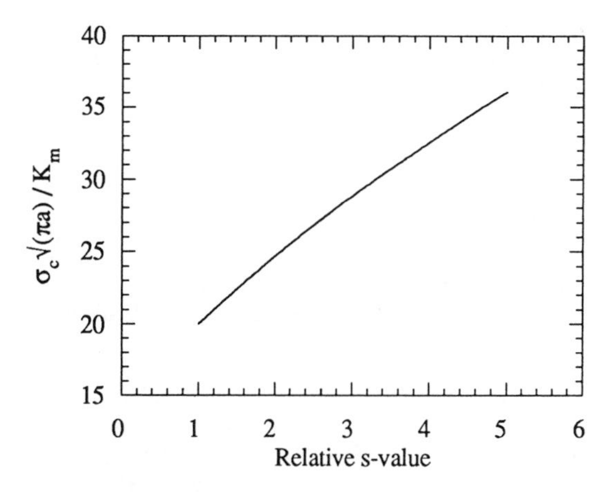

Figure 8. Compressive Strength Increase with Fiber Reinforcement Index s. The s-values have been normalized by the reference magnitude of s such that $s_o = 0.01$ and $c = 800$ as used in all the preceding calculations. Other Parametric Values Used are $\ell_o^* = 20$; $D_o = 0.0005$; $\bar{a} = 0.1$; $c = 800$; $K_o = 0.0002$; $s_o = 0.01$.

8, which shows the relationship between compressive strength σ_c and the reinforcement index s, at a fixed fiber volume fraction and fiber damage index ($V_f = 0.01$ and $k = 100$).

Note that the reinforcement index s as defined in eqn. (8), is directly proportional to g, τ_f, and L_f/d_f. Therefore variation in s may be interpreted as variation in any one of these parameters, with the others fixed. Figure 8 indicates a monotonic increase in the compressive strength with the s - value, suggesting the importance of these parameters in controlling the compressive strength in FRCs. However, it should be pointed out that the amount of initial damage may be expected to increase with fiber aspect ratio, even though this notion has not been incorporated in the present model.

7 FURTHER DISCUSSIONS AND CONCLUSIONS

Although the microcrack sliding model of compressive strength in brittle material has been discussed (eqn. (6)) in the context of uniaxial loading in the present paper, extensive studies (Horii and Nemat-Nasser, 1986; Ashby and Hallam, 1986; Kemeny and Cook, 1991) have shown that the compressive strength is very sensitive to confining stresses. This can be seen in the sensitivity of the stress intensity factor of the wing-cracks to normal compressive load. This notion is in accord with experience in cementitious materials, for which confinements are general prescriptions to derive higher compressive load bearing capacity. The present model of compressive strength for FRC shows that fibers can be exploited to increase the compressive strength and may therefore act as a passive confining pressure. This passive confinement idea was first proposed by Yin et al, (1990), who discovered this beneficial effect of fiber in a series of biaxial steel FRC tests.

The positive effect of fibers on the elastic modulus of composites has received extensive attention (e.g. Tandon and Weng, 1986; Wakashima and Tsukamoto, 1991). However, it is also well known, at least for some cementitious composites, that fiber can degrade the composite modulus to below that of the matrix modulus (e.g. Naaman et al, 1991). It is likely that fibers would induce a competing process of modulus improvement and degradation in cementitious composites, in the same manner that it influences compressive strength. In fact, many of the basic elements of the present work could be applied to analyze FRC elastic modulus.

The present work represents a preliminary look at how fibers in FRC contributes or degrades composite mechanical properties. The results based on the present model appear to capture much of what has been experimentally observed in compressive strength change due to fiber addition. These modelling results (particularly Figures 4 and 8) indicate that fibers can significantly improve compressive strength of FRCC if the weakening effect of fiber is controlled via novel processing routes. A difficulty in applying the current model, however, lies in the lack of knowledge in some micromechanisms and micromechanical parameters. These include, for example, the detail micromechanisms in the way fiber resist microcrack sliding, and the general unavailability of parameteric values of a, D_o, and k. Additional research is required to tackle these issues. The present work provides a framework for which these future research should be organized.

ACKNOWLEDGEMENTS

Research at the ACE-MRL have been supported by research grants from the National Science Foundation (Program manager: Dr. K. Chong) and from the Air Force Office of Sponsored Research (Program manager: Dr. Spencer Wu) to the University of Michigan, Ann Arbor. Helpful discussions with H. Horii, J. Huang, N.N. Jakobsen, J. Kemeny, and H. Stang are gratefully acknowledged.

REFERENCES

Akihama, S., Nakagawa, H., Takada, T. and Yamaguchi, M., Experimental study on aramid fiber reinforced cement composites "AFRC" mechanical properties of AFRC with short fibers. In RILEM Symposium on Developments in Fiber Reinforced Cement and Concrete, FRC86, Vol. 1, Swamy, R.N, Wagstaffe, R.L. and Oakley, D.R. (ed.), Paper 2.5, 1986a.

Ashby, M.F. and Hallam, S.D., The failure of brittle solids containing small cracks under compressive stress states. *Acta Metall.* 34 No. 3, (1986), 497-510.

Fannela, D. A., and Naaman, A. E., Stress-Strain Properties of Fiber Reinforced Concrete in Compression, *J. of ACI, Proceedings*, Vol. 82, No. 4, (1983), 475-483.

Horii, H. and Nemat-Nasser, S., Brittle failure in compression: splitting, faulting, and brittle-ductile transition. *Phil. Trans. Royal Soc. London*, 319, (1986), 337-374.

Kemeny, J. M., and Cook, N.G.W., Micromechanics of deformation in rocks. In *Toughening Mechanisms in Quasi-Brittle Materials*, S.P. Shah (ed.), Kluwer Academic Publishers, (1991), 155-188.

Li, V.C., A simplified micromechanical model of compressive strength of fiber reinforced cementitious composites. Accepted for publication in the *J. of Cement and Concrete Composites*, 1991.

Li, V.C., Post-crack scaling relations for fiber reinforced cementitious composites. ASCE *J. of Materials in Civil Engineering*, Vol. 4, No. 1, (1992), 41-57.

Li, V.C. and Leung, C., Tensile failure modes of random discontinuous fiber reinforced brittle matrix composites. In *Fracture Processes in Concrete, Rock and Ceramics,* J.G.M.Van Mier, J.G. Rots and A. Bakker (eds.), publisher: Chapman and Hall, (1991), 285-294.

Li, V.C., Wang, Y., and Backer S., A micromechanical model of tension-softening and bridging toughening of short random fiber reinforced brittle matrix composites. *J. Mechanics and Physics of Solids*, V. 39, No. 5, (1991), 607-625.

Li, V.C., and Wu, H.C., Pseudo Strain-Hardening Design in Cementitious Composites. To appear in *High Performance Fiber Reinforced Cement Composites*, H. Reinhardt and A. Naaman (eds.), Chapman and Hall, 1991.

Naaman, A. Otter, D. and Najim, H., Elastic modulus of SIFCON in tension and compression. ACI *Materials Journal*, Vol. 88, No. 6, Nov.-Dec., (1991), 603-612.

Rooke, D.P., and Cartwright , D.J., *Compendium of Stress Intensity Factors*, The Hillingdon Press, Middx, 1976.

Sammis, C.G. and Ashby, M.F., The failure of brittle porous solids under compressive stress states, *Acta Metall.* V. 34, (1986), 511-526.

Shah, S. P., and Rangan, B. V., Fiber Reinforced Concrete Properties, *J. of ACI, Proceedings*, Vol. 68, No. 2, (1971), 126-135.

Tandon, G.P. and Weng, G.J., Average stress in the matrix and effective moduli of randomly oriented composites, *Composites Science and Technology* 27, (1986), 111-132.

Tjiptbroto, P., Tensile strain hardening of high performance fiber reinforced cement based composites. *Ph.D. Thesis*. Department of Civil Engineering, University of Michigan, 1991.

Wakashima K. and Tsukamoto, H., Mean-field micromechanics model and its application to the analysis of thermomechanical behavior of composite materials. In press in *Materials Science and Engineering A.*, (1991).

Ward, R., Yamanobe, K., Li, V.C., and Backer, S., Fracture resistance of acrylic fiber reinforced mortar in shear and flexure, in *Fracture Mechanics: Application to Concrete*, Eds. V. Li and Z. Bazant, ACI SP-118, (1989), 17-68.

Yin, W. S., Su, C. M., Mansur, M.A., and Hsu, T.T.C., Fiber Reinforced Concrete Under Biaxial Compression, *Engineering Fracture Mechanics*, Vol. 35, No. 1/2/3, (1990), 261-268.

Zhu, B.Y., Behavior of concrete with synthetic organic fibers, in *Darmstadt Concrete*, Vol. 5, (1990), 249-255.

34 A THEORETICAL MODEL OF HYBRID FIBRE REINFORCED CONCRETE

YUE CHANGNIAN and CHEN YAOLIONG

Chongqing Institute of Architecture and Engineering, Chongqing, China

Abstract

In order to predict the critical moment of unstable crack propagation of hybrid fibre (steel fibre and glass fibre) reinforced concrete, a theoretical model is presented. Based on this model and linear elastic fracture mechanics, the methods of calculations of the length of effective crack, the critical stress intensity factor and critical crack opening displacement have been proposed. The calculated results have been checked with data of tests on flexural and tensile notched specimens.

Key Words: Hybrid fibre, Fracture behavior, Hybrid fibre reinforced composite, Model for fracture behavior

1 Introduction

By Nair [1] the glass fibre (GF) is distributed in concrete in form of bundlles of glass filament. Then GF forms a kind of cell with product of cement hydration. A kind of composite is formed as concrete reinforced by the cells. The reinforced mechanism of this composite is different from that of concrete reinforced with steel fibre (SF). The propagation of microcrack in this composite is resisted by the cells when the microcracks extend. Due to large size, with the same volume of fibre, the effect of SF on resistance of microcracks is not so evident. SF only resists macrocracks effectively. So it can be considered that these two kinds of fibre have different levels for reinforcing matrix. Sun Wei [2] and K. Kobayashi and R. Cho [3] studied the properties of composite with hybrid fibre (HF) after or before it cracks. Since then the fracture behavior of this kind of composite has not been researched yet. It is well known that there is a zone will influence the stable and unstable crack propagation extremely. A theoretical model is proposed by authors for analysis of fracture behavior. In this model whole crack (main crack and microcrack) is replaced by an equivalent crack. By this model, K_{Ic}, J_{Ic} and critical crack opening displacement (COD) can be determined, COD and J_{Ic} are considered as fracture toughness indices of composite with HF.

Fibre Reinforced Cement and Concrete. Edited by R. N. Swamy. © 1992 RILEM.
Published by E & FN Spon, 2-6 Boundary Row, London SE1 8HN. ISBN 0 419 18130 X.

2 Model for fracture behavior of composion with hybrid fibre
Three assumptions:

1. Linear profile of crack;
2. fibre closing pressure (FCP) is normal to the crack plane;
3. matrix with short GF is linear elastic. Assumption 3 is approved by study of other researchers [4,5] and test of authors.

Model:

With above assumptions a model has been proposed in Fig 1. In this Fig. A_0 is the length of initial crack. At unstable crack propagation, the whole length of crack is expressed by A_0+L_f. Lp shows a microcrack zone in front of the tip of main

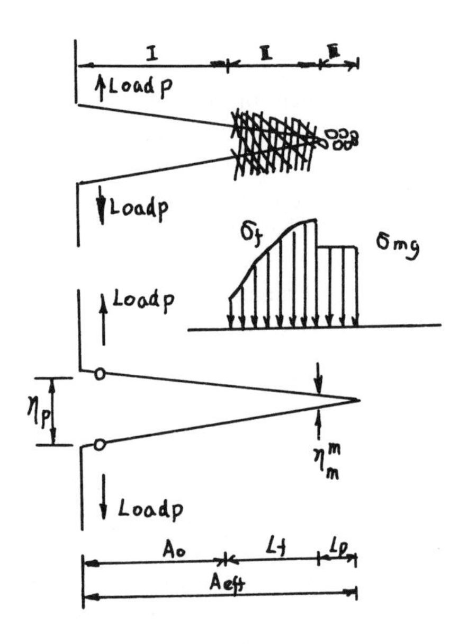

Fig. 1 Concept of proposed model

crack. Whole crack can be replaced by an equivalent crack, which can be divided into three zones (I,II,III, in Fig 1.). Crack in I zone is prepared previously. In this zone there is no fibre across the crack, and no FCP for matrix. This zone is called as fibre traction free zone. In II zone fibre has been pulled and begins to slip relatively to matrix. It exerts force to close the crack. II zone is called as fibre bridging zone. III zone is a zone of microcracks. Thus, in composite relationship between FCP and COD can be expressed as follow [6]:

$$\sigma_f = \sigma_{max}(1 - \eta/\eta_m^f)^2 \qquad (1)$$

where σ_f , η --- FCP and COD under certain load.

σ_{max}, η_m^f --- maximum stress and COD at maximum stress, when the fibre can bridge the crack effectively.

Stress in III zone is guite complicated. Because GF is thiner and much more highly distributed than SF, so GF can contribute more to resiste microcrack than SF can. Therefore force in III zone for resisting micropcrack is mainly exerted by GF. Lp is very short. The bond stress in this zone exerting to matrix can be considered as a constant. In II zone short GF has been pulled out, σ_f is only exerted by SF, $\sigma_f = \sigma_{fs}$.

The proposed model differs from the model of Wecharctang and Shah [6] by two aspects: firstly, in proposed model, due to resistance of GF to microcrack in mortar, stress in III zone must be considered. In their model, there was no GF, so in that zone stress can be ignored; secondly, for analysis by their model, the length must be measured. During crack propagation, at the tip of main crack there are several brenches and curve cracks. It is difficult to identify the exact position of tip of main crack. By proposed model, according to P -$\eta_p$curve the effective length of crack, A_{eff}, K_{Ic} and COD can be evaluated.

Calculation of test on flexural and tensile specimens:

Crack mouth displacement ($\eta_p$or CMD) and load (p) can be measured by test. In Fig. 2,$\eta_m^m$is COD for matrix. It can be measured by test on pure mortar. It is sure that when the width of main crack tip (Y = Ao + L_f) reaches$\eta_m^m$the crack begins to extend

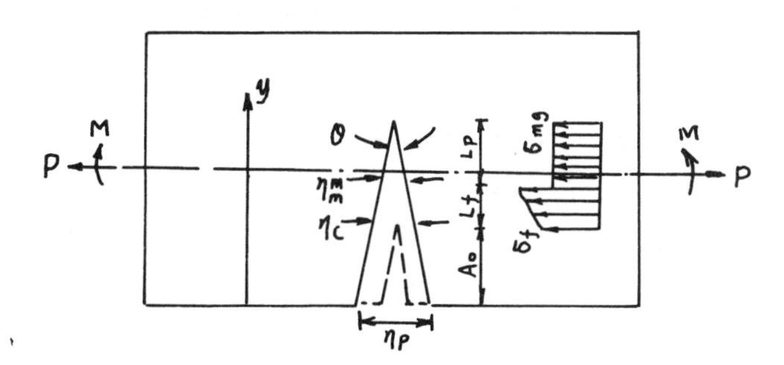

Fig. 2 Proposed model used in test

unstably. However, only knowing η_p and η_m^m, the crack opening displacement at arbitrary section of crack can not be determined, thus other parameters of fracture mechanics can not be determined. In order to solve this problem angle θ must be ensured. By Okamura [7]:

$$\theta = \Delta\lambda_{pm} \cdot P + \Delta\lambda_m \cdot M \tag{2}$$

where $\Delta\lambda$ pm, $\Delta\lambda$ m --- increments of the complians of members caused by the existance of a crack; P---axial load; M---moment.

From definition of $\Delta\lambda_{pm}$ and $\Delta\lambda_m$ they can be written as follows:

$$\Delta\lambda_{pm} = 2(1-\mu^2)(EB)^{-1}(6/W)G_{pm}(a/W) \tag{3}$$

$$\Delta\lambda_m = 2(1-\mu^2)(EB)^{-1}(6/W)G_m(a/W) \quad (4)$$

where: μ ---Poisson ratio; E---Young's modulus;

B,w,a---geometry dimensions, see Fig. 3.

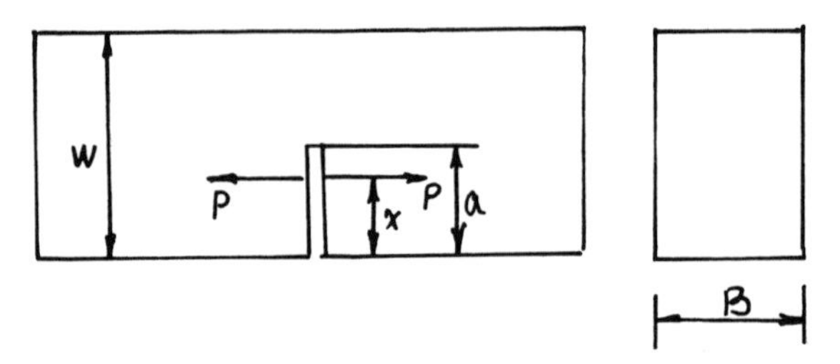

Fig. 3 Scheme for calculation of η_c

If stress intensity factors caused by axial load and moment respectively are following:

$$K_p = \sigma_p(\pi \cdot a)^{1/2} Y_p(a/W) \qquad (\sigma_p = P/B.W) \quad (5)$$

$$K_m = \sigma_m(\pi \cdot a)^{1/2} Y_m(a/W) \qquad (\sigma_m = 6M/B.W^2) \quad (6)$$

then:

$$G_{pm}(a/W) = \int_0^{a/W} \xi \cdot Y_p(\xi) Y_m(\xi)\, d \quad (7)$$

$$G_m(a/W) = \int_0^{a/W} \xi \cdot Y_p^2(\xi)\, d \quad (8)$$

where:

$$Y_p(\xi) = 1.12 - 0.231\xi + 10.55\xi^2 + 21.72\xi^3 + 30.39\xi^4 \quad (9)$$

$$Y_m(\xi) = 1.112 - 1.40\xi + 7.33\xi^2 + 13.08\xi^3 + 14.0\xi^4 \quad (10)$$

Since crack opening displacement depends on external load, geometry and size of specimen and fibre clossing pressure, and fibre clossing pressure is caused by crack opening displacement itself, in order to determin θ the iterative method has to be used. At first an initial η_c has to be given. Then by η_p and η_m^m which have been measured by experiment, η can be determined geometricly. According to Eq. (2) can be found, thus to calculate η_p' and $\eta_m'^m$. If $\eta_p' = \eta_p$ and $\eta_m'^m = \eta_m^m$, the assumed η_c is reasonable. Otherwise assume another η_c and repeat above calculation, untill $\eta_p' = \eta_p$ and $\eta_m'^m = \eta_m^m$. Then the A_{eff} and other parameters of fracture mechanics can be obtained.

The basic procedure to solve problem by proposed model is shown as follows: If at section $Y=A_0$ the width of crack is η_c, apparently it must be $\eta_m^m < \eta_c < \eta_p$. Then

$$A_{eff} = [\eta_p / (\eta_p - \eta_c)] \cdot A_o \quad (11)$$

Where: $A_{eff} = A_o + L_f + L_p$

$$L_p = [\eta_m{}^m / (\eta_p - \eta_c)] \cdot A_o \quad (12)$$

$$L_f = [(\eta_c - \eta_m) / (\eta_p - \eta_c)] \cdot A_o \quad (13)$$

Simultaneously, assume, must be: $\eta_c < (1 - \eta_p / W) \cdot A_o$ (14)

And if: $C = 1 - \eta_p / \eta^f_m$; $D = \eta_p / \eta^f_m \cdot 1 / A_{eff}$ (15)

then Eq. (1) will be:

$$\sigma_f = \sigma_m (C + DY)^2 \quad (16)$$

The stress intensity factor caused by load as shown in Fig. 3:

$$K_I = F(x/a) \cdot 2P(\pi \cdot a)^{-1/2} \quad (17)$$

where:

$$F(x/a) = 3.52(1 - x/a)(1 - a/W)^{-3/2} - (4.35 - 5.28\, x/a)(1 - a/W)^{-1/2} + \{[1.30 - 0.30(x/a)^{3/2}][1 - (x/a)^2]^{-1} + 0.83 - 1.6\, x/a\}[1 - (1 - x/a)\, a/W] \quad (18)$$

In fibre bridging zone assume:

$$F_1 = \int_{-1}^{1} F\{[1/2\, L_f(x + 1) + A_o]\, A_{eff}^{-1}\}\{C + D[1/2\, L_f(x + 1) + A_o]\}^2 dx \quad (19)$$

then:

$$K_{p1} = B F_1 L_f \sigma_m (\pi \cdot A_{eff})^{-1/2} \quad (20)$$

In microcracking zone assume:

$$F_2 = \int_{-1}^{1} F\{[1/2\, L_p(x + 1) + A_o + L_f]\, A_{eff}^{-1}\}\, dx \quad (21)$$

then

$$K_{p2} = B F_2 L_p \sigma_{mg} (\pi \cdot A_{eff})^{-1/2} \quad (22)$$

Therefore, the stress intersity factor K_{pI} caused by fibre bridging force:

$$K_{pI} = K_{p1} + K_{p2} \quad (23)$$

Then from Eq(5) an equivalent force (p) can replace fibre clossing pressure:

$$P = K_{pI}\, B\, W[(\pi \cdot A_{eff})^{1/2}\, Y_p(A_{eff}/W)]^{-1} \quad (24)$$

For specimen on flexure:

$$\theta = \Delta\lambda_m M - \Delta\lambda_{pm} P \quad (25)$$

$$K_1 = Y_m (A_{eff}/W)\, 6M\, (\pi \cdot A_{eff})^{1/2} (B\, W^2)^{-1} \quad (26)$$

For specimen on tension:

$$\theta = \Delta\lambda_{pm} (P' - P) \quad (27)$$

$$K_I = Y_p (A_{eff} / W)\, p' (\pi \cdot A_{eff})^{1/2} (B\, W)^{-1} \quad (28)$$

where: P'---external loading tensile force

The η'^{m}_{m} and η'_p can be calculated by following Eq.:

$$\eta'^{m}_{m} = 2 L_p \operatorname{tg} (\theta / 2) \quad (29)$$

$$\eta'_p = 2 A_{eff} \operatorname{tg} (\theta / 2) \quad (30)$$

3 Test and Analysis

In order to research fracture behavior of mortar with HF, using proposed model, flexural and tensile specimens were prepared.

Flexural specimens are single-edge notched beams 416x100x50 mm with initial crack A_0 = 50 mm in the middle, and loaded as in Fig. 4. In specimens mortar is reinforced by SF (25 mm long, with equivalent diameter 0.48 mm) and GF (30 mm long, with average diameter 12μ). Total fibre content (by volume) in every group of specimens is constant (2%). The ratio of two kinds of fibres (λ=GF/SF) varies from 0.0 to 0.5 by every 0.05. All fibre distribute randomly in mortar.

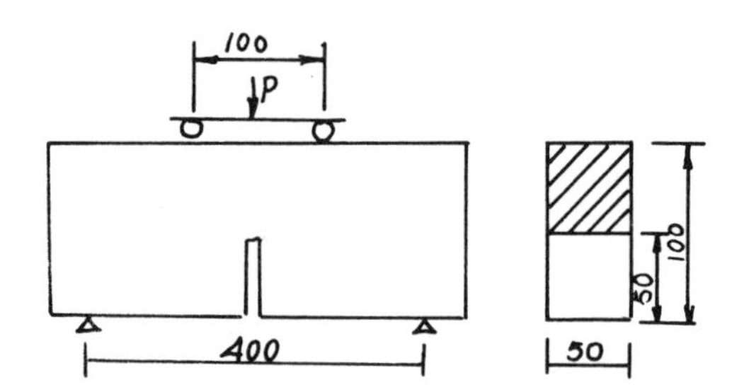

Fig. 4 Loading state of flexural specimen

Tensile specimens have lengthened '8' shape with cross section in the middle 30x25mm. the initial crack is 12.5mm deep in the middle of specimen. GF content (by volume) is 0.75%, SF---0.484%. GF distributes randomly and SF is parallel to the loading axis.

During test all P---$\eta_p$curves were registered. Before test on P---$\eta_p$curves basic parameters of materials have been measured and calculated as in Tab. 1. A computer program was used for above procedure.

Then all flexural and tensile test results have been evaluated by a computer program of iterative method calculation. The results of calculation are shown in Tab. 2.

Tab. 1 Basic parameters

Compressive Strength of Mortar (Mpa)	Young's Modulus of Mortar (Gpa)	Fracture Strength Mortar (Mpa)	Tensile Strength of Mortar with Short GF (Mpa)	Fracture Strength of Mortar with Short GF (Mpa)	of SF from Mortar (Mpa)
47.3	23.3	0.118	2.827	0.579	2.82

Tab. 2 The Results of caleulation with proposed modle*

specimen Designation	COC (mm)	CMD(mm)	Load(KN)	A_{eff} (mm)	K_{IC} ($Nm^{3/2}$)	J_{Ic} (Jm^{-2})
B00	6.24E-3	0.25	3.5	51.4	1.67E+6	47.4
B05	7.01E-3	0.25	3.5	51.4	1.67E+6	61.7
B10	7.66E-3	0.24	3.6	51.6	1.75E+6	39.2
B15	7.33E-3	0.21	3.3	51.8	1.61E+6	48.5
B20	8.01E-3	0.25	3.8	51.6	1.85E+6	61.6
B25	8.92E-3	0.27	3.1	51.7	1.76E+6	77.0
B30	9.57E-3	0.35	3.5	51.4	1.69E+6	75.3
B35	9.21E-3	0.15	3.3	53.1	1.68E+6	74.3
B40	8.91E-3	0.35	3.8	51.3	1.83E+6	61.5
B45	9.33E-3	0.36	3.4	51.3	1.62E+6	53.8
B50	8.76E-3	0.37	4.5	51.2	2.16E+6	77.4
NSG	0.0102	0.22	1.58	13.1	1.31E+6	

* B----- ---Specimen for flexural test

NDG-----Specimen for tensile test, 05 ---- = 0.05

From Tab. 2, both COD and J_{Ic} are raised to second power with evidently. Crack opening displacement and fibre clossing pressure influence on each other. In the stage of stable crack propagation crack opening displacement increases continously, and fibre clossing pressure changes corelatively. Then it resistes the crack opening. Of cause fibre clossing pressure depends on λ . Due to influence between COD and FCP, it is reasonable to consider that influence on COD as a squared function of λ . If in the stage of stable crack propagation the relationship between COD and FCP is quasielastic as shown by tests, J_{Ic} will be equivalent to COD.

COD calculated by proposed model and J_{Ic} can be considered as fracture toughness indices of composite with HF, K_{Ic} is only a parameter of material at moment of unstable crack propagation. It can not express whole process of crack propagation and has not any clear relation with λ. So K_{Ic} is unsuitable as an index of fracture toughness for composite with HF.

4 Conclusions

1. Both J_{Ic} and COD can be considered as indices of fracture toughness for mortar reinforced by HF. They are related to λ as a squared function. Due to no evident relationship between K_{Ic} and, K_{Ic} is unsuitable for assesment of fracture toughness of mortar reinforced by HF.

2. It is approved by experiments that proposed model can express fracture behavior of mortar with HF.

3. As shown by results of calculation, COD in mortar with short GF (5mm) is much larger than that in mortar with long (30mm). It may be described by that the short GF resists microcrack more effectively.

5 References

1. N.G. Nair, Mechanics of glass fibre reinforced cement, U.K., RILEM SYMPOSIUM 1975, FIBRE REINFORCED CEMENT AND CONCRETE, General Editor Adam Neville in association with D.J. Hannant, A.J. Majumdar, C.D. Pomeroy and R.N. Swamy. pp 81-83

2. Sun Wei and Zhan Bin-Gen, Crack arrest effect of fibre reinforced high strength cement matrix, P.R. China, 2nd International Symposium on cement, 1989, pp 125-132

3. K. Kobayashi and R. Cho, Flexural characteristics of steel fibre and polyethylene fibre hybrid reinforced concrete, UK, Composite, Apr. 1982, pp 164-168

4. A.C. Jaras; K.L. Litherland, Microstrctural features in glass fibre reinforced cement composites, UK, RILEM SYMPOSIUM 1975, FIBRE REINFORCED CEMENT AND CONCRETE, General Editor Adam Neville in association with D.J. Hannant, A.J. Majumdar, C.D. Pomeroy and R. N Swamy, pp 327-334

5. D.R. Oakley; B.A. Proctor, Tensile stress-strain behavior of glass fibre reinforced cement composites, UK, RILEM SYMPOSIM 1975 FIBRE REINFORCED CEMENT AND CONCRETE, General Editor Adam Neville in association with D.J. Hannant, A.J. Majumdar, C.D. Pomeroy and R. N. Swamy, pp 347-359

6. M. Wecharatang; S.P. Shah, A model for predicting fracture resistance of fibre reinforced concrete, UK, Cement and concrete research, Vol. 13, 1983, pp 819-829

7. Hiroyuki Okamura; K. Watanbe; T. Takno, Deformation and strength of crack member under bending moment and axial force, Engineering frauture Mechanics, Vol. 7, 1975, pp 531-539

35 FINITE ELEMENT APPLICATION OF A CONSTITUTIVE MODEL FOR FIBER REINFORCED CONCRETE/MORTAR

D. LIU and D. J. STEVENS
Department of Civil and Environmental Engineering,
Clarkson University, Potsdam, USA

Abstract
The addition of strong, ductile, discontinuous fibers with high aspect ratio to concrete or mortar results in Fiber Reinforced Concrete (FRC) or Fiber Reinforced Mortar (FRM), which, when compared to the plain material, exhibits superior ductility, failure toughness, strength, impact resistance, and fatigue strength. Laboratory tests show these improved properties are a result of the randomly distributed fibers which act as crack arrestors. Currently, an adequate numerical approach does not exist for the analysis of structures composed of FRC or FRM. In this paper, a recently developed constitutive model for the pre-peak response of FRC/FRM is presented and then extended to the post-peak regime using a nonlocal approach. The existing, local model is based loosely within the theory of mixtures; for the plain concrete or mortar, an anisotropic, strain-based, continuum damage/plasticity model is used. To represent the effect of the fibers, a simplified model that accounts for both the resistance of the fibers and the enhanced resistance of the matrix is employed. In the Finite Element implementation of the model, the nonlocal approach is shown to lessen mesh sensitivity and to improve the predictions of energy dissipation. The results of the Finite Element Model are shown to compare well with the experimental results from two separate test programs.
Keywords: Fiber Reinforced Concrete, Fiber Reinforced Mortar, Mixture Theory, Finite Elements, Constitutive Model, Nonlocal Approach.

1 Introduction

The incorporation of steel fibers into concrete or mortar modifies a large number of its mechanical properties, particularly when fibers that develop a good mechanical bond are used. As observed experimentally, fibers that are randomly distributed act as microcrack arrestors, improving the ductility, failure toughness, postcracking strength, impact resistance, and fatigue strength, while reducing the number of cracks and the mean crack width. Typically, the amount of steel fibers varies between 0.5 and 2.0 percent, by volume. Examples of the application of steel FRC include slabs, bridge decks, pavement overlays, blast resistant structures, tunnel reinforcement, and precast bridge segments, spillways and bridge piers. In addition, a number of experimental programs have studied FRC's potential for use in seismic resistant structures.

A number of approaches for modeling FRC/FRM have been attempted in the

Fibre Reinforced Cement and Concrete. Edited by R. N. Swamy. © 1992 RILEM.
Published by E & FN Spon, 2-6 Boundary Row, London SE1 8HN. ISBN 0 419 18130 X.

past, including: fracture mechanics, mixture models, fictitious crack models, and continuum damage mechanics. The majority of models have taken either a fracture mechanics or law of mixtures approach. In mixture models, the response of the composite is taken as a volume weighted sum of the stresses in the matrix and the fiber. The interaction between the two materials, which is influenced by the fiber spacing, fiber orientation, fiber properties, and the fiber/matrix interface properties, is usually accounted for with an 'efficiency' factor. Typically, mixture models are limited to predicting the uniaxial response, but a multi-axial model using plasticity theory was developed by Tanigawa et al. '83 in which the debonding behavior and orientation of steel fiber are characterized, and the post-failure behavior is obtained by introducing the damage index in terms of strain.

Other examples of FRC/FRM models include Naaman et al. '73, who combined a statistical composite model with fracture mechanics criteria to model the ductile and brittle modes of failure and Swamy '73, who proposed a crack control-composite mechanics approach to develop equations for the prediction of first crack and ultimate flexural strength. The fictitious crack model was also applied by Petersson '80 to the modeling of FRC; in this approach, a stress-strain relationship is used up to peak stress; thereafter, in the strain softening region, a stress-crack width curve is used, since stress-strain laws are nonunique when the deformation localizes. Lastly, the theory of continuum damage mechanics was applied to the modeling of FRC in uniaxial tension and compression by Fanella and Krajcinovic '85 who used a parallel bar model coupled with damage laws for the composite and matrix.

In this paper, a multi-axial constitutive model, previously developed by the authors for the pre-peak response of FRC and FRM, is briefly reviewed and then extended to the softening region, using a nonlocal approach. The nonlocal constitutive model is then implemented into a general purpose Finite Element code and the test specimens from two laboratory studies are modeled. The comparisons between the model and experimental data are shown to agree well.

2 Constitutive law for FRC/FRM

The existing constitutive model for fiber reinforced cementitious materials was initially developed for FRC; however, it also works well for FRM, since concrete and mortar are somewhat similar materials. A nonlocal version of the model is then presented and the Finite Element implementation is discussed.

2.1 Review of mixture model for FRC/FRM

As presented by Stevens and Liu '91, the constitutive model for FRC is developed loosely within the theory of mixtures as follows

$$\boldsymbol{\sigma} = \boldsymbol{\sigma}^c + \boldsymbol{\sigma}^f \tag{1}$$

where $\boldsymbol{\sigma}^c$ and $\boldsymbol{\sigma}^f$ are the stresses in the concrete and fiber, respectively. Unlike typical mixture approaches, the summation is not weighted by the volume percentage, since the fiber contribution reflects **both the fiber and (fiber-enhanced) matrix response.** A strain-based, continuum damage/plasticity model is employed for the plain concrete or mortar, and an *effective* fiber model is used to account for the fiber resistance and the fiber-enhanced resistance of the matrix.

2.2 Continuum damage/plasticity model for concrete (or mortar)

The model for the matrix (concrete or mortar) is based upon the theories of continuum damage mechanics and plasticity. It is assumed that the majority of the large scale response of concrete or mortar is determined by two small scale phenomena: the initiation, growth and coalescence of microcracks and the pressure-dependent (frictional) tangential movement of the microcracks surfaces. Damage is created by the growth and coalescence of the microcracks, while permanent deformations are created through frictional slip across the microcracks; thus, a combination of continuum damage and plasticity theories is used.

The stress and strain are decomposed as

$$\boldsymbol{\sigma} = \boldsymbol{\sigma}^e - \boldsymbol{\sigma}^p \tag{2}$$

$$\boldsymbol{\epsilon} = \boldsymbol{\epsilon}^e + \boldsymbol{\epsilon}^p \tag{3}$$

With the fourth order secant stiffness tensor, $\mathbf{C}$, these decompositions lead to the following stress-strain relations:

$$\boldsymbol{\sigma} = \mathbf{C} : \boldsymbol{\epsilon}^e \tag{4}$$

$$\boldsymbol{\sigma}^e = \mathbf{C} : \boldsymbol{\epsilon}, \qquad \boldsymbol{\sigma}^p = \mathbf{C} : \boldsymbol{\epsilon}^p \tag{5}$$

In this model, the fourth order secant stiffness tensor is decomposed to reflect the changes due to direct tensile Mode I cracking and the compressive mode of splitting:

$$\mathbf{C} = \mathbf{C}^o - \mathbf{C}^+ - \mathbf{C}^- \tag{6}$$

$\mathbf{C}^o$ is the original elastic stiffness. $\mathbf{C}^+$ and $\mathbf{C}^-$ are defined as the changes in the positive (tensile) and negative (splitting/compressive) effective stiffness. To preserve consistency and retain simplicity, the evolution of damage and permanent deformations is determined by the same elastic strain-based surface. To model the compressive mode of damage and deformation, an 'inelastic', isotropic surface, g^- is used. To account for the strong directionality of the Mode I and wing cracks, a tensile kinematic 'inelastic' surface, g^+, is used:

$$g^+(\boldsymbol{\epsilon}^e, \boldsymbol{\alpha}^+) = \tau^+ - r_o^+ = 0, \quad g^-(\boldsymbol{\epsilon}^e) = \tau^- - r^- = 0 \tag{7}$$

where

$$\tau^+ = \sqrt{< \boldsymbol{\epsilon}^{e+} - \boldsymbol{\alpha}^+ >: \mathbf{C}^o :< \boldsymbol{\epsilon}^{e+} - \boldsymbol{\alpha}^+ >}, \qquad \tau^- = \sqrt{\boldsymbol{\epsilon}^{e-} : \mathbf{C}^o : \boldsymbol{\epsilon}^{e-}} \tag{8}$$

$$\boldsymbol{\epsilon}^{e+} = \sum_{a=1}^{3} < \epsilon^e_{(a)} > \underline{e}^e_{(a)} \otimes \underline{e}^e_{(a)}, \qquad \boldsymbol{\epsilon}^{e-} = \boldsymbol{\epsilon}^e - \boldsymbol{\epsilon}^{e+} \tag{9}$$

$$< \boldsymbol{\epsilon}^{e+} - \boldsymbol{\alpha}^+ >_{ij} = < \epsilon^{e+}_{ij} - \alpha^+_{ij} > \tag{10}$$

τ^+ and τ^- are the energy norms of the shifted positive elastic strain tensor and negative elastic strain tensor. r_o^+ and r^- are the energy thresholds for further damage/permanent deformation; $\boldsymbol{\alpha}^+$ is a kinematic 'back strain' that defines the current origin of g^+. $\epsilon^e_{(a)}$ and $\underline{e}^e_{(a)}$ are the eigenvalues and normalized eigenvectors

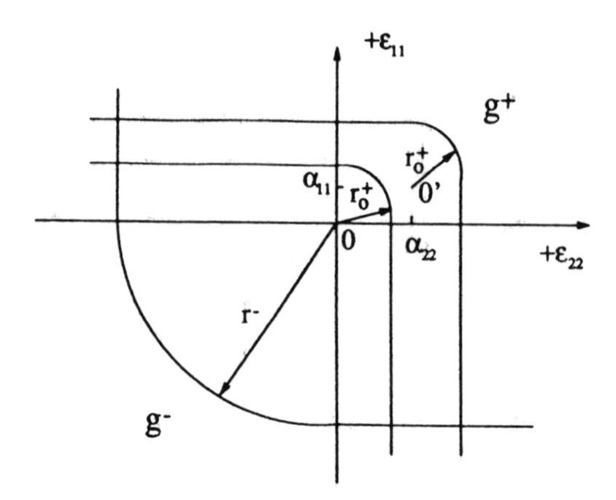

Fig.1. 'Inelastic' surfaces for a plane strain condition, with $\mathbf{C} = \mathbf{I}$ (for plotting purposes).

of $\boldsymbol{\epsilon}^e$, respectively and $< x >$ is the McAuley bracket. These surfaces are shown in Fig. 1 for a plane strain state, with $\mathbf{C} = \mathbf{I}$ for plotting purposes.

The inelastic surfaces in Eq. 7 are combined with associated flow rules to define the total elastic degradation, $\mathbf{C}_t^+$ and $\mathbf{C}_t^-$, in which all cracks are assumed to be open and active. To account for crack closure, **effective** stiffness, $\mathbf{C}^+$ and $\mathbf{C}^-$, are used in the constitutive model; these stiffnesses are defined later. The rate equations for the **total** stiffnesses are:

$$\dot{\mathbf{C}}_t^+ = \dot{\mu}^+ \frac{\partial g^+}{\partial \boldsymbol{\epsilon}^e \otimes \boldsymbol{\epsilon}^e}, \quad \dot{\mathbf{C}}_t^- = \dot{\mu}^- \hat{H}(tr\boldsymbol{\epsilon}^{e+}) \frac{\partial g^-}{\partial \boldsymbol{\epsilon}^e \otimes \boldsymbol{\epsilon}^e} \tag{11}$$

where $\dot{\mathbf{C}}_t^+$ and $\dot{\mathbf{C}}_t^-$ are the positive and negative total stiffness rates, respectively. $\dot{\mu}^+$, $\dot{\mu}^-$ are the consistency parameters. The Heaviside function, $\hat{H}(*)$, is used to accommodate pressure effects.

The kinematic evolution (translation) of the tensile surface is governed by an associated flow rule:

$$\dot{\boldsymbol{\alpha}}^+ = \dot{\mu}^+ \frac{1}{H^+} \frac{\partial g^+}{\partial \boldsymbol{\epsilon}^e} \tag{12}$$

$$H^+ = \frac{1}{\tau^+ + \tau^-} \gamma_1 \hat{H}(\lambda_{min} - \epsilon_{ii}^{e-}) + \gamma_2 \hat{H}(\epsilon_{ii}^{e+} - \lambda_{max}) + \gamma_3 \hat{H}(\epsilon_{ii}^{e}) \tag{13}$$

H^+ is the 'hardening/softening' function, which is determined from laboratory data. In addition, $\epsilon_{ii}^{e-} = tr(\boldsymbol{\epsilon}^{e-})$, $\epsilon_{ii}^{e+} = tr(\boldsymbol{\epsilon}^{e+})$, and λ_{min} and λ_{max} are the minimum and maximum eigenvalues of $\boldsymbol{\epsilon}^e$. γ_1, γ_2, and γ_3 are material constants.

The isotropic evolution of the compressive surface is taken as:

$$\dot{r}^- = \frac{\dot{\mu}^-}{2\tau^- H^-} \tag{14}$$

$$H^- = \frac{A^- - B^-\tau^-}{\tau_o^-}\alpha_c \tag{15}$$

where H^- is the 'hardening function', determined from laboratory data through curve fitting. A^- and B^- are material constants, as is α_c which is used to account for confinement:

$$\alpha_c = \begin{cases} \gamma_4 & \text{if } tr(\epsilon^{e+}) = 0 \text{ and } tr(|\mathbf{E}^{e-}|) \neq 0; \\ 1 & \text{otherwise.} \end{cases} \tag{16}$$

$\mathbf{E}^{e-} = \epsilon^{e-} - tr(\epsilon^{e-})\delta/3$ is the deviatoric tensor of the compressive elastic strain tensor. γ_4 is a material constant.

In the development of the elastic degradation, it is assumed that all of the cracks are active and open. To capture stiffness recovery under reversed loading, Ortiz '85 proposed that the effective stiffnesses, $\mathbf{C}^+$ and $\mathbf{C}^-$, be related to the total stiffnesses, $\mathbf{C}_t^+$ and $\mathbf{C}_t^-$ as follows:

$$\mathbf{C}^+ = \mathbf{P}^+ : \mathbf{C}_t^+ : \mathbf{P}^+, \qquad \mathbf{C}^- = \mathbf{P}^- : \mathbf{C}_t^- : \mathbf{P}^- \tag{17}$$

where

$$P_{ijkl}^+ = Q_{ik}^+ Q_{jl}^+, \qquad \mathbf{P}^- = \mathbf{I} - \mathbf{P}^+ \tag{18}$$

$\mathbf{P}^+$, $\mathbf{P}^-$ are the positive and negative orthogonal projection operators; $\mathbf{I}$ is the fourth order identity tensor. The second order tensor $\mathbf{Q}^+$ is defined as:

$$\mathbf{Q}^+ = \sum_{a=1}^{3} \hat{H}(\epsilon_{(a)}^e) \underline{e}_{(a)}^e \otimes \underline{e}_{(a)}^e \tag{19}$$

The development of permanent deformation is also signaled by the inelastic surfaces; associated flow rules are used to define the plastic stress evolution:

$$\dot{\sigma}^p = \dot{\mu}^+ \frac{\partial g^+}{\partial \epsilon^e} + \dot{\mu}^- \frac{\partial g^-}{\partial \epsilon^e} \tag{20}$$

2.3 *Effective* fiber model

Fibers that develop sufficient bond with the concrete or mortar will resist crack opening (tensile and splitting) through debond, friction- and pressure-dependent pullout, mechanical interlock, yielding and dowel action. In addition, the fibers act as local stress reducers, lowering the stress concentrations in the vicinity of cracks or fields of cracks. All of these mechanisms contribute to the improved mechanical response of FRC or FRM, relative to the plain material. In the model of Stevens and Liu '91, a simple technique is used to account for both the crack-closing effects of the fibers across open or 'active' cracks and the enhancement of the matrix response when fibers are present.

In the *effective* fiber model, the fiber response is assumed to augment the resistance of FRC/FRM in the directions of tensile strain only. Assuming that the fiber distribution across the 'active' cracks is uniformly random, the relationship between normal tensile strain and the resulting effective fiber stress is approximated as elastic/perfectly plastic/linear softening as shown in Fig. 2. The pre-peak region reflects the elastic contribution of the fibers and the enhanced tensile response

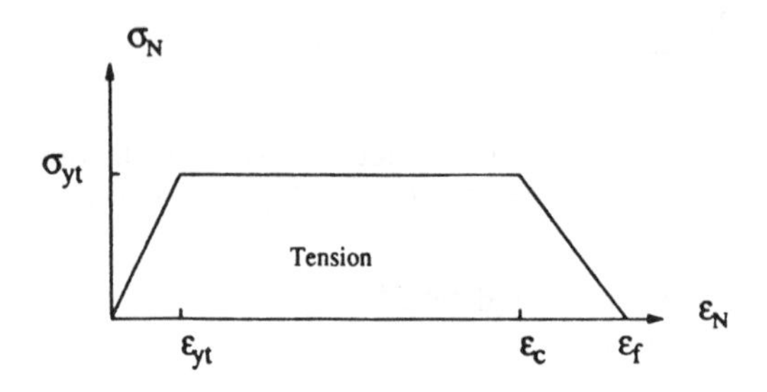

Fig.2. *Effective* fiber stress-strain curve.

of the matrix. The plastic plateau is due to fiber pullout, debond, and yielding of the fibers. The linear softening region represents the pullout of the fibers and final exhaustion of the matrix tensile resistance. The 'yield strain' ϵ_{yt}, 'yield stress' σ_{yt}, critical strain ϵ_c, and final strain ϵ_f, may be extracted from experimental tensile stress-crack width curves through energy equivalence arguments. The contribution of fibers in the direction of compressive strains is taken as zero.

The experimental results of Yin et al. '89 imply that increased confinement will increase the tensile resistance of the fiber. To incorporate this effect, the 'yield stress' σ_{yt} is modified as:

$$\sigma_{yt} = \sigma_{yt}^{o}[1 + \gamma_5 \frac{|\epsilon_{ii}^{-}| + min(\epsilon_{11}, \epsilon_{22}, \epsilon_{33})}{|\epsilon_{ii}^{-}|} - \frac{|\epsilon_{ii}^{+}| - max(\epsilon_{11}, \epsilon_{22}, \epsilon_{33})}{|\epsilon_{ii}^{-}|}] \tag{21}$$

where σ_{yt}^{o} is the 'yield stress' under zero confinement; γ_5 is a constant.

Assuming that the fiber distribution is uniformly random, the magnitude of the fiber stress vector $\underset{\sim}{\sigma}_N$ on the active tensile plane can be calculated and then converted to the stress tensor $\boldsymbol{\sigma}^f$:

$$\boldsymbol{\sigma}^f = \sum_{N=1}^{m} |\underset{\sim}{\sigma}_N| \underset{\sim}{e}^{(N)} \otimes \underset{\sim}{e}^{(N)} \tag{22}$$

where m is the number of active cracks (maximum 3). The capabilities of this constitutive model for FRC were investigated and reported by Stevens and Liu '91. The model's predictions were compared to uniaxial and biaxial test results from 3 different test programs. Very good correlations between the model and the test data were seen.

3 Nonlocal finite element analysis

As with plain concrete, FRC exhibits strain-softening in tension and in compression. Tanigawa et al. '80 reported that the strain-softening portion of uniaxial compressive stress-strain curves of FRC is size dependent and that brittleness increases with the slenderness of the specimen. Due to the heterogeneity created by coalesced microcracks, strain softening is a 'structural' phenomenon, not a material characteristic. Since strain-softening is size dependent, the post-peak region must be considered within the framework of boundary value problems, not just as one aspect of the material model. In the solution of problems with strain-softening materials, the constitutive model or numerical technique must be modified to remove the typically observed mesh sensitivity and to predict the correct energy dissipation.

A continuum approach with nonlocal damage has recently been shown to be an effective approach for the analysis of strain-softening structures (Bazant and Lin '88; Pijaudier-Cabot and Bazant '88; Stevens and Krauthammer '89). In this approach, the variables that govern strain softening in a material are taken as nonlocal. In the present model, the positive hardening function H^+ determines the rate at which the concrete or mortar softens; therefore, a nonlocal form of H^+ is used. The nonlocal variable $\bar{H}^+$ represents the weighted spatial average of H^+ over a representative volume surrounding each point $\mathbf{x}$ and is given by:

$$\bar{H}^+ = \frac{1}{V_r(\mathbf{x})} \int_V H^+ \alpha(\mathbf{s} - \mathbf{x}) dV \tag{23}$$

in which

$$V_r(\mathbf{x}) = \int_V \alpha(\mathbf{s} - \mathbf{x}) dV \tag{24}$$

and V is the volume of the structure, $V_r(\mathbf{x})$ is the representative volume at point $\mathbf{x}$, and $\alpha(\mathbf{s} - \mathbf{x})$ is a weight function that was proposed by Pijaudier-Cabot and Bazant '88 as:

$$\alpha(\mathbf{s} - \mathbf{x}) = exp[-(k|\mathbf{s} - \mathbf{x}|/l)^2] \tag{25}$$

k is a constant and is equal to 2 for the case of two dimensions; l is the characteristic length of the material which must be determined from laboratory data.

Strain softening in FRC also occurs when the fibers pull out of the matrix. In the present model, it is assumed that the softening due to concrete and mortar damage is more significant than the fiber pull-out softening; therefore, the fiber softening is treated as a local phenomenon.

The nonlocal constitutive model for FRC was implemented into FEAP (Finite Element Analysis Program), developed by Taylor '87. The structure of FEAP is described by Zienkiewicz and Taylor '89. FEAP is designed for solving linear and non-linear problems, operating in a UNIX environment. The four-node, isoparametric, quadrilateral elements with two degrees of freedom are used for the computations reported later.

4 Results

Results from two test programs were used to verify the proposed approach. In the first study by Shah '87, **notched** 76.1 mm x 279.2 mm x 28.6 mm (depth x span x thickness) beams composed of Fiber Reinforced Mortar were loaded in 3 point bending. The fibers were 25.4 mm long brass coated smooth steel with aspect ratio of 63.5. The volume percentages evaluated were 0.6% and 1.38%. The mortar strength was not reported but the mix composition would suggest that it was a weak material. In the second study, Ward and Li '90 tested **unnotched** 114 mm x 342 mm x 62.7 mm specimens in 3 point bending. The fibers were 25.4 mm long crimped steel with aspect ratio of 28.5.

Table I. Material values

	Shah '87		Ward and Li '90	
E (MPa)	15,862		41,379	
ν	0.22		0.2	
r_o^+ $(MPa)^{1/2}$	3.32×10^{-3}		9.97×10^{-3}	
r_o^- $(MPa)^{1/2}$	2.49×10^{-2}		2.49×10^{-2}	
A^- $(MPa)^{-1/2}$	0.36		0.36	
B^- $(MPa)^{-1}$	0.725		0.725	
γ_1 (MPa)	8.28×10^{-3}		8.28×10^{-3}	
γ_2 (MPa)	3.45×10^{-3}		3.45×10^{-3}	
γ_3 (MPa)	1.72		2.41	
γ_4	6.0		6.0	
	$V = 0.6\%$	$V = 1.38\%$	$V = 0.5\%$	$V = 1.0\%$
γ_5	1.1	1.1	1.1	1.1
σ_{yt}^o (MPa)	1.03	1.72	2.21	2.90
ϵ_{yt} $\times 10^{-3}$	8.0	5.0	1.0	1.0
ϵ_c $\times 10^{-1}$	5.0	5.0	8.0	8.0
ϵ_f $\times 10^{-1}$	6.0	6.0	9.0	9.0
	Plain	FRM	Plain	FRM
l (mm)	12.7	25.4	12.7	38.1

The material properties that were used in the analysis are shown in Table I. Since comprehensive test data are not available for the plain mortar response and for the fibers used in the two programs, these material values were adopted (with small modification) from a previous study, in which the pre-peak response of the local version of the model was verified (Stevens and Liu '90). The characteristic lengths used for the specimen results presented by Shah '87 were taken as 12.7 mm for plain mortar and 25.4 mm for FRM, and for Ward and Li '90, as 12.7 mm for plain mortar and 38.1 mm for FRM. These values for plain mortar are determined by $l \approx 2.8d_a$ in which d_a = the maximum size of the aggregate in mortar, proposed by Bazant and Pijaudier-Cabot '89. The maximum aggregate size used in the Shah '87 tests equals 4.76 mm. The characteristic lengths for FRM were chosen by trial and error. This should be measured experimentally in the future. Note that the characteristic length used for Ward and Li's specimens is greater than that used for Shah's specimens; Ward and Li '90 used crimped fibers which distribute and resist the damage better than the straight fibers of Shah's specimens.

Fig. 3 shows good agreement between the predicted and experimental values of Shah '87; in this case deflection was measured at the loading point. The model used a 4x6 mesh with equal size elements. It can be seen that the model predicts the peak loads for both the plain beams and FRM beams quite well and also shows the great improvement of energy dissipation of FRM beams. Because the *effective* fiber model is a local model, some instabilities (or oscillations) of the FRM beams are seen after the peak load; this will be corrected in further development of the model.

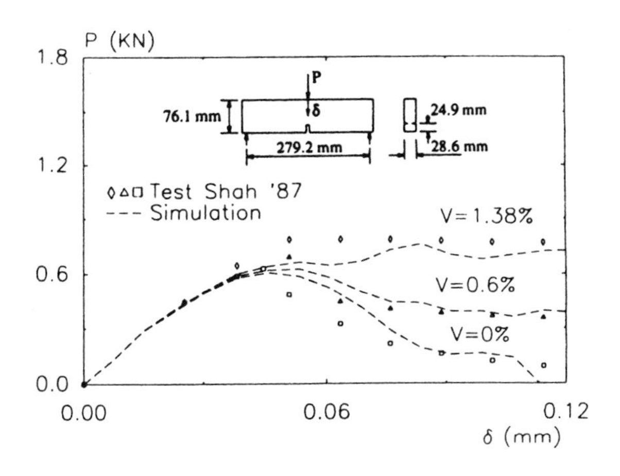

Fig.3. Predicted and measured load-deflection curves for Shah '87.

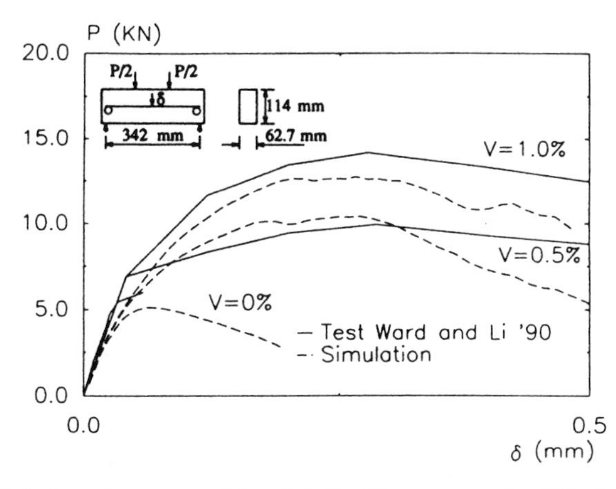

Fig.4. Predicted and measured load-deflection curves for Ward and Li '90.

Fig. 4 presents the model's results and test data of Ward and Li '90. The deflection at midspan was measured relative to the supports; therefore, the deformation due to crushing at the supports is not reflected in the deflection measurements. A 4x12 mesh was used in the model. The uniaxial compressive strength of the plain mortar for the test was 57 MPa; the constitutive model, without fibers, predicts a peak strength of 62.7 MPa which is reasonably close. Again, the model satisfactorily predicts the peak loads and demonstrates the brittle behavior of plain mortar beams and the ductile behavior of FRM beams. With regards to these two examples, it appears that the improvement of peak loads of the **notched** FRM beams is less significant in comparison to the improvement of energy dissipation, but, in the **unnotched** FRM beams, the peak loads and energy dissipation are both increased greatly.

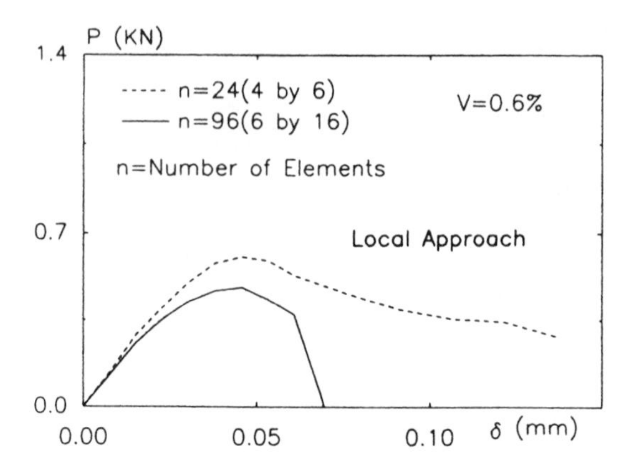

Fig.5. Local analysis of three-point bending beam.

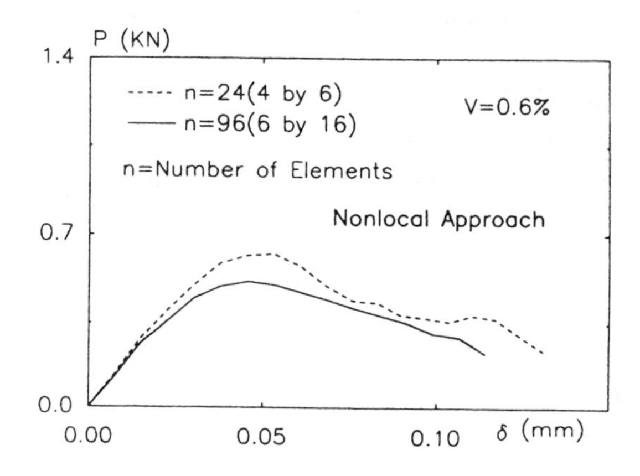

Fig.6. Nonlocal analysis of three-point bending beam.

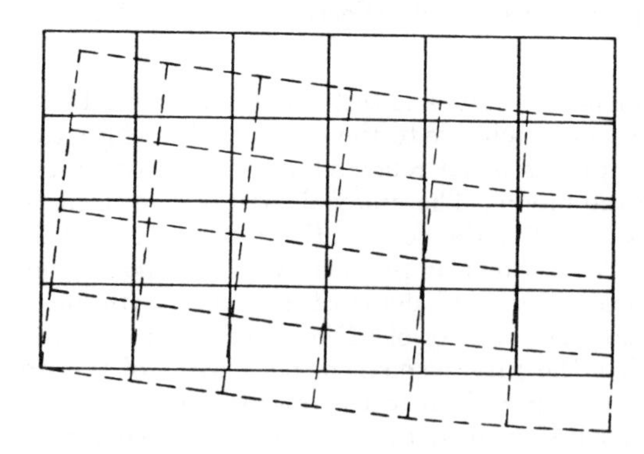

Fig.7. Undeformed and deformed mesh patterns for Shah '87

Figs. 5 and 6 show a comparison of local and nonlocal finite element solutions for different numbers n of finite elements. In the local approach, as the mesh is refined, the post-peak curve becomes steeper and more brittle. If the mesh were further refined, the response would become even more brittle and the predicted amount of energy dissipation would decrease. Thus a local model would predict failure in the FRM structure without predicting a reasonable amount of energy dissipation.

Fig. 7 shows the deformed mesh (magnified) superimposed upon the undeformed mesh. This figure points out one weakness of the nonlocal approach: the localization that occurs at the tensile crack cannot be reproduced unless a highly refined mesh is used. The model, with a relatively crude mesh, does not allow a sharp discontinuity to form across the Mode I type crack. In future development of the model, a special finite element with an embedded localization shape function will be implemented into FEAP to overcome this. While this is a shortcoming, it is important to note that the correct macro-response (load versus deflection) curve was predicted. The nonlocal method assures that the correct amount of energy is dissipated on the structural level, even though the sharp discontinuities in stress and strain at the crack are not predicted.

5 Conclusion

The finite element implementation of the constitutive model for FRC/FRM satisfactorily reproduces the load-deflection curves of notched and unnotched beams composed of plain mortar and and FRM. The nonlocal approach is shown to be simple and effective method for removing mesh sensitivity and improving the predictions of energy dissipation in the analysis of strain-softening structures.

6 References

Bazant, Z.P. and Lin, F-B (1988) Nonlocal smeared cracking model for concrete fracture. **J. Str. Eng.**, ASCE, 114(11), 2493-2510.

Bazant, Z.P. and Pijaudier-Cabot, G. (1989) Measurement of characteristic length of nonlocal continuum. **J. Engrg. Mech.**, ASCE, 115(4), 755-767.

Fanella, D. and Krajcinovic, D. (1985) Continuum damage mechanics of fiber reinforced concrete. **J. Eng. Mech.**, ASCE, 111(8), 995-1009.

Jenq, Y.S. and Shah, S.P. (1986) Crack propagation in fiber-reinforced concrete. **J. of Structural Engrg.**, ASCE, Vol. 112, No. 1, 19-34.

Naaman, A. Argon, A. and Moavenzadeh, F. (1973) A fracture model for fiber reinforced cementitious materials. **Cem. and Con. Res.**, 3(4).

Ortiz, M. (1985) A constitutive theory for the inelastic behavior of concrete. **Mech. of Mat.**, 4, 67-93.

Petersson, P.E. (1980) Fracture mechanics calculations and tests for fiber reinforced cementitious materials. **Advances in Cement Matrix Composites, Materials Research Society Symposium**, Boston, pp. 95-106.

Pijaudier-Cabot, G. and Bazant, Z.P. (1987) Nonlocal damage theory. **J. Eng. Mech.**, ASCE, 113(10), 1512-1533.

Ramakrishnan, V. and Kumar, M.S. (1987) Constitutive relations and modeling for concrete fiber composites, a state-of-the-art report. **Int. Sym. on Fibre Reinforced Concrete**, Madras, India, Vol I, 1.21- 1.56.

Shah, S.P. (1987) Strength evaluation and failure mechanisms of fiber reinforced concrete. **Int. Sym. on Fibre Reinf. Conc.**, Madras, India, Vol. I, 1.3-1.19.

Stevens, D.J. and Krauthammer, T. (1989) Nonlocal continuum damage/plasticity model for impulse loaded RC beams. **J. Str. Eng.**, ASCE, 115(9), 2329-2347.

Stevens, D.J. and Liu, D. (1992) Strain-based continuum damage/plasticity model with mixed evolution rules for concrete. To be Published June, 1992 in the **Journal of Engineering Mechanics**, ASCE.

Stevens, D.J. and Liu, D. (1991) Constitutive modeling of fiber reinforced concrete. Tentatively Accepted for Publication in **Symposium for New Developments in Fiber Reinforced Concrete for the 21st Century**, ACI.

Tanigawa, Y. Yamada, M. and Hatanaka, S. (1980) Inelastic behavior of steel fiber reinforced concrete under compression. **Int. Sym. Mat. Research Society: Advances in Cement-Matrix Composites**, Boston, pp. 107-118.

Tanigawa, Y. Yamada, K. Hatanaka, S. and Mori, H. (1983) A simple constitutive model of steel fiber reinforced concrete. **Int. J. of Cem. Comp. and Lightweight Con.**, 5(2), 87-96.

Taylor, R. (1987) **FEAP and PCFEAP, Users-Manual.** University of California, Berkeley.

Ward, J.R. and Li, C.V. (1990) Dependence of flexural behavior of fiber reinforced mortar on material fracture resistance and beam size. **ACI Material Journal**, v. 87, No. 6, 627-637.

Yin, W.S. Su, E.C.M. Mansur, M.A. and Hsu, T.T.C. (1989) biaxial tests of plain and fiber concrete. **ACI Mat. J.**, 86(3), 236-243.

Zienkiewicz, O.C. and Taylor, R. (1989) **The Finite Element Method.** McGraw-Hill Book Company, Fourth Edition, Volume I.

36 NON-LINEAR FINITE ELEMENT ANALYSIS OF STEEL FIBRE REINFORCED CONCRETE MEMBERS

S. A. AL-TAAN and N. A. EZZADEEN
Civil Engineering Department, Mosul University, Mosul, Iraq

Abstract
A numerical procedure based on the finite element method is developed for the nonlinear analysis of reinforced fibrous concrete members subjected to monotonic loads. The proposed method is capable of tracing the response of these structures up to their ultimate load ranges. The predicted results include displacements, strains, curvatures, slopes, stresses, and member end actions. An iterative scheme based on Newton-Raphson's method is employed for the nonlinear solution algorithm. A frame element with a composite layer system is used to model the structure. The constitutive models of the nonlinear material behaviour are presented to take into account the nonlinear stress-strain relationships, cracking, crushing of concrete, debonding and pull-out of the steel fibres, and yielding of the reinforcement. The analytical solution of steel fibre reinforced concrete short columns and beams are compared with published experimental test results to verify the accuracy of the proposed procedure, and to demonstrate the validity of the adopted and developed material constitutive models.
Kewords: Beams, Columns, Finite Element, Flexural Strength, Nonlinear (Analysis), Reinforced Concrete, Steel Fibres.

1 Introduction

Since the early sixties, extensive researches and developments have been carried out to study the effects of steel fibres on the mechanical properties of concrete and cement mortar. Some of these studies were based on empirical data derived from laboratory experiments. Model testing can give an insight into the structural behavior, but it is very costly and time-consuming procedure. Therefore, several numerical methods have been developed to predict the flexural strength of reinforced fibrous concrete members (Henager, 1977, Swamy and Al-Taan, 1981) and to estimate their response under monotonic loads (Craig, 1987, Soroushian and Reklaoui, 1989).

Fibre Reinforced Cement and Concrete. Edited by R. N. Swamy. © 1992 RILEM.
Published by E & FN Spon, 2-6 Boundary Row, London SE1 8HN. ISBN 0 419 18130 X.

In this study a numerical procedure, based on the finite element method, is presented to analyze reinforced fibrous concrete members. A computer program was developed to demonstrate the validity of the procedure and to make it useful for the analysis of real frames. It also helps in predicting the effects of steel fibers on the reinforced concrete members economically and with a minimum efforts.

2 Materials constitutive relationships

The constitutive model of the nonlinear material behaviour should be in such a form that it can be easily incorporated into a numerical analysis procedure to simulate the structural behaviour. In the subsequent sections, mathematical modelling of the materials, used in this work, are given.

2.1 Fibrous concrete

2.1.1 Compressive constitutive model

In compression, an empirical model suggested by Soroushian and Lee (1989) is adopted for this study (Fig. 1). The model represents the compressive stress-strain relationship of steel fibre reinforced concrete as a function of the matrix strength f'_c and the fibre reinforcement index $(V_f \cdot l_f / d_f)$.

The model consists of a curvilinear (parabola) ascending portion,

$$\sigma = - f'_{c_f} \frac{\varepsilon}{\varepsilon_{Pf}} \left(\frac{\varepsilon}{\varepsilon_{Pf}} - 2 \right) \quad \text{for} \quad \varepsilon \leq \varepsilon_{Pf} \tag{1}$$

and a bilinear descending branch,

$$\sigma = z \left(\varepsilon - \varepsilon_{Pf} \right) + f'_{c_f} \geq f_o \quad \text{for } \varepsilon > \varepsilon_{Pf} \tag{2}$$

where,

σ = concrete compressive stress;
ε = concrete compressive strain;
f'_{c_f} = compressive strength of SFRC;
ε_{Pf} = compressive strain at peak stress;
z = slope of the descending branch; and
f_o = residual stress.

The tangent modulus E_t in the ascending portion can be obtained by differentiating Eq. 1,

$$E_t = \frac{2 f'_{c_f}}{\varepsilon_{Pf}} \left(1 - \frac{\varepsilon}{\varepsilon_{Pf}} \right) \tag{3}$$

and it is assumed to be zero in the descending portion to avoid numerical difficulty in using negative stiffness.

The variables $(f_o, f'_{c_f}, z, \varepsilon_{Pf})$ in this model were

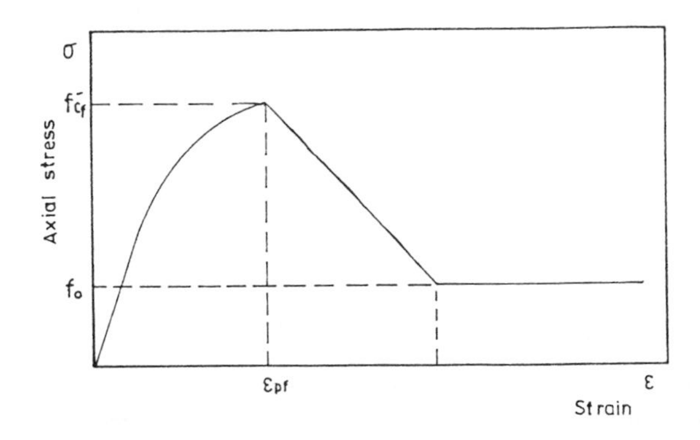

Fig. 1. Compressive constitutive model of SFRC.

represented by the following empirical expressions :

$$f'_{c_f} = f'_c + 3.6\,\frac{Vf \cdot lf}{df} \quad \text{(MPa)} \qquad (4.a)$$

$$f_o = 0.12\, f'_{c_f} + 14.8\,\frac{Vf \cdot lf}{df} \quad \text{(MPa)} \qquad (4.b)$$

$$z = -\,343 \cdot f'_c\,[1 - 0.66\,(\frac{Vf \cdot lf}{df})^{1/2}\,] \le 0 \quad \text{(MPa)} \qquad (4.c)$$

$$\varepsilon_{pf} = 0.0007 \cdot Vf \cdot \frac{lf}{df} + 0.0021 \qquad (4.d)$$

where,

f'_c = matrix compressive strength (MPa);
Vf = volume fraction of the fibres;
lf, df = length and equivalent diameter of the fibre respectively (mm).

It can be seen from Eq. 4 that fibres improve the post-peak ductility, energy obsorption capacity and, to some extent, the strength of concrete.

2.1.2 Tensile constitutive model

The ascending part of the tensile stress-strain curve, up to the first cracking, is similar to that of plain concrete. Then it tends to deviate from linearity due to the process of microcrack propagation in the matrix prior to the formation of a continuous crack system at peak stress. This part can be modelled by a bilinear curve, as shown in Fig. 2 (Soroushian and Lee, 1989), where the tensile strength f_{tf} and its corresponding strain ε_{tf} are:

$$f_{tf} = f'_t\,(1 + 0.016 \cdot Nf^{1/3} + 0.05 \cdot \pi \cdot df \cdot lf \cdot Nf\,) \qquad (5)$$

$$\varepsilon_{tf} = \varepsilon_t\,(1 + 0.35 \cdot Nf \cdot df \cdot lf) \qquad (6)$$

where,

f'_t=tensile strength of matrix (MPa);

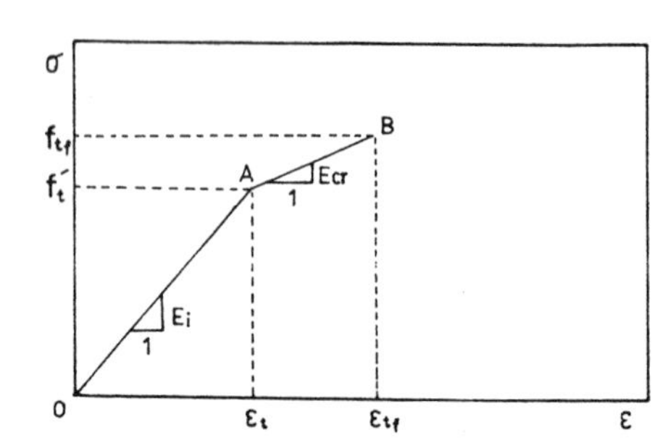

Fig.2. Pre-peak tensile constitutive model of SFRC.

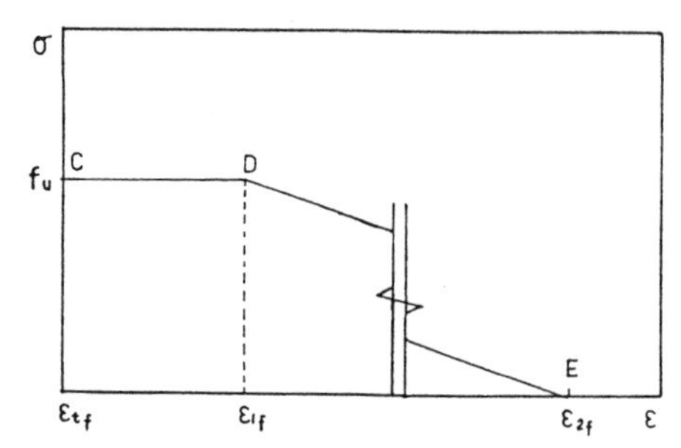

Fig.3. Post-peak tensile constitutive model of SFRC.

Nf = number of fibres per unit cross section area;
= $\beta \cdot (4 \cdot Vf / \pi \cdot df)$
β = orientation factor (actual β is roughly average of 2 and 3 dimensional conditions); and
ε_t = matrix cracking strain (mm/mm);

In this study, orientation factor equal to (0.5) is used. Thus, the number of fibres per unit cross sectional area can be rewritten as,

$$Nf = 2\ Vf / \pi \cdot df^2 \tag{7}$$

The post-peak tensile behaviour of steel fibre reinforced concrete is determined by pull-out action of fibres. The peak pull-out resistance of fibres is reached directly after the composite material reaches its peak tensile strength. Therefore, the bilinear stress-strain curve, shown in Fig. 3, is used to simulate the post-peak region of the tensile constitutive relationship of SFRC.

Assuming a uniform distribution of fibre length across the crack, a mean embedment length of lf/ 4 is obtained. Thus, the average tensile stress fu, can be derived by multiplying the average bond stress τu by the interfacial area of all fibres crossing the cracked section,

$$fu = 0.5 \cdot Vf \cdot \tau u \cdot lf / df \tag{8}$$

The average characteristic bond strength τu can be predicted using the empirical expression derived by Soroushian and Lee (1989), including the effects of the number of fibres per unit area (Naaman and McGarry, 1974), and the effect of fibre shape (Henager, 1977):

$$\tau u = (\ 2.62 - 0.0036\ Nf)\ kf \tag{9}$$

where, kf = bond efficiency factor.

In Fig. 3, ε_{1f} is the strain at which slippage of fibres is assumed to begin. When the pull-out commence, the stress carried by one fibre due to the interfacial bond is:

$$\varepsilon_{1f} \cdot Ef = \frac{\pi \cdot df \cdot \tau u}{\pi \cdot df^2 / 4} \cdot \frac{lf}{4} = \frac{lf \cdot \tau u}{df}$$

where, Ef is modulus of elasticity of the steel fibre; thus,

$$\varepsilon_{1f} = \frac{lf \cdot \tau u}{df \cdot Ef} \tag{10}$$

As assumed by Sakai et al (1986) ε_{2f} is taken as 0.1 mm/mm.

The form of this curve (post-peak region) is similar to those assumed by various investigators (Lim et al, 1986, Craig, 1989).

2.2 Reinforcing steel

The superior resistance of fibre concrete to cracking and crack propagation, allows the use of high strength steels in practice without the risk of exceeding the limit state of crack width and deflection. A multilinear stress (σ)-strain (ε) model is used in this study for high strength steel having a specified yield strength greater than 414 MPa (Fig. 4).

For such materials a yield point may be defined as the stress at which a specified total strain εy (0.35% according to the ACI-code,1989) or a specified offset (0.2 %) is attained.

In the absence of test data; the ultimate strength fsu is assumed to be approximately 1.55 times fy (Chen, 1982), and Esh equal to 0.25 times Esy.

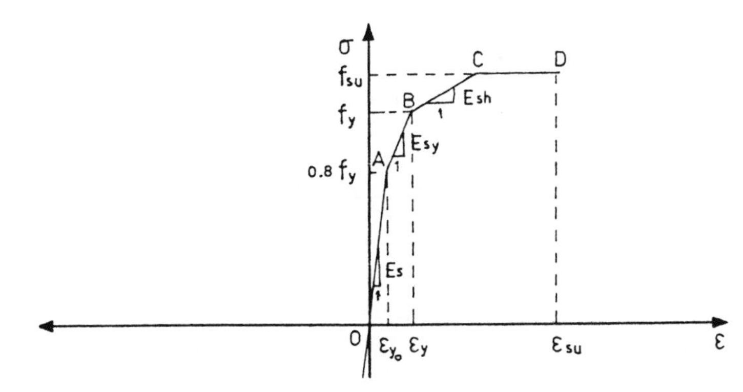

Fig. 4. Multilinear stress-strain model for high strength steel.

3 Finite element idealization

A typical reinforced concrete element is shown in Fig 5. A frame element with a composite layer system is used to model the structure. The x-axis needs not coincide with the centroidal axis of the element. The reference plane of the element is defined as the plane containing the x-axis and is also perpendicular to x-y plane. Each element is divided into a discrete number of concrete and steel layers.

Each node has three degrees of freedom with an additional axial displacement degree of freedom u_n, which is eliminated by static condensation. The internal degree of freedom u_n is introduced in the calculation in order that the axial strain should match the linear strain variation from bending.

A parabolic interpolation function for the axial displacement between the nodes $\underline{N}_a$, and a cubic interpolation function (Zienkiewicz, 1977) for the other displacements $\underline{N}_b$ are used. These two functions can be written as follows:

$$\underline{N}a = \left[\frac{1}{2}(1-\xi) \ , \ \frac{1}{2}(1+\xi) \ , \ 1-\xi^2 \right] \tag{11}$$

$$\underline{N}_b = \left[\frac{1}{4}(2-3\xi+\xi^3) \ , \ \frac{-L}{8}(1-\xi)(1-\xi^2) \ , \ \frac{1}{4}(2+3\xi-\xi^3) \ , \ \frac{L}{8}(1+\xi)(1-\xi^2) \right] \tag{12}$$

where, ξ is the local coordinate = $2(x-x_c)/L$; L is the element length, and $x_c = L/2$.

The incremental strain $\Delta\varepsilon$ at any point within an element is represented by the following equation:

$$\Delta\varepsilon = [\ B\ ]\{\Delta r\} \tag{13}$$

where,

$[\ B\ ] = [\ \underline{N}_{a,x} \ , \ -y\ \underline{N}_{b,xx}\ ]$;

$\{\Delta r\}$ = the incremental nodal displacement vector.

The tangential equilibrium equation can be written as:

$$\{dR\ \}^j = (\int_v [\ B\ ]^T E_t [\ B\ ] dv\ )\{dr\} \tag{14}$$

or

$$\{dR^j\} = [\ K_t\ ]\ \{dr\} \tag{15}$$

where, {dR } is the external nodal forces.

The tangent stiffness matrix $[K_t]$ in Eq. 15 is of order (6×6), since the column and row those containing the displacement degree of freedom u_n, has been eliminated using static condensation process (Cook, 1981).

When the equilibrium correction is performed on the solution, the exact tangent stiffness needs not be formed (Kang and Scordelis, 1980). Therefore, $[K_t]$ is evaluated at the mid-length of the frame element by the layer integration. Unlike tangent stiffness, the internal resisting loads have to be evaluated as accurate as possible for the unbalanced load iteration procedure. Thus, these loads are evaluated by three Gaussian quadrature points along the length combined with the layer integration through the depth of the element.

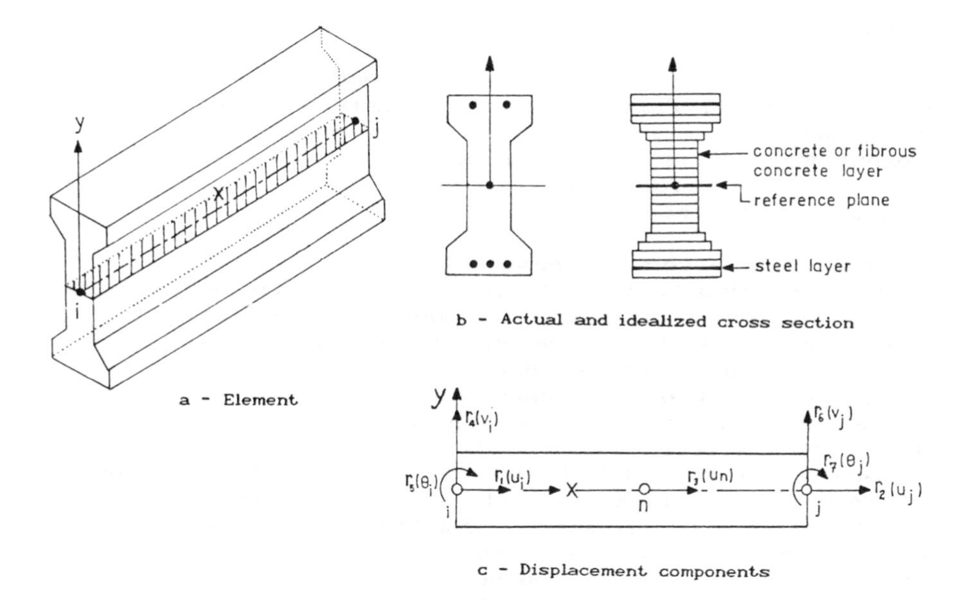

Fig. 5. Element and cross section.

4 Nomlinear solution algorithm

An incremental-iterative technique is employed for the non-linear solution, where the total load is divided into load increments {ΔR}. Within each increment, successive iterations are performed to get through more accurate results as follows:

a) Calculate tangent stiffness matrix for each element and transform them into global coordinate. Assemble the structure tangent stiffness matrix.
b) Solve $[Kt]\{\Delta r\}=\{\Delta R\}$ for displacement increments and transform them into local coordinates to obtain element end displacement increments.
c) Calculate the strain increments. The total strain ε is then obtained by adding $\Delta\varepsilon$ to the previous total.
d) Compute the stress σ by the nonlinear stress-strain low.
e) Calculate element end forces in local coordinates. Transform them to global coordinates to obtain the internal resisting load $\{R^i\}$.
f) Subtract the internal resisting load $\{R^i\}$ from the total nodal forces $\{R^j\}$ to obtain the unbalanced forces $\{R^u\}$.
g) Set $\{\Delta R\} = \{R^u\}$, then go to step (a).
h) Continue the procedure (a-f) until the displacement increment $\{\Delta r\}$ are within allowable tolerance. Then, the same procedure (a-g) is performed with the next load increment.

5 Results

Two examples are presented in this section. In the first example, a steel fibre reinforced concrete columns (without steel bars) are analyzed to test the accuracy of the present method for predicting the ultimate load capacity of such structures. The later example consists of reinforced fibrous concrete beams.

5.1 Fibre reinforced concrete columns

A number of plain and steel fibre reinforced concrete short columns subjected to axial and eccentric loads, tested by Niyogi and Chawla (1982), are analyzed in this section.

All the columns have identical geometry. The loads were applied from both ends of each column with eccentricities (e) varied from 0 to 50 mm with respect to the central axis of the columns (Fig. 6).

Each column is modelled by 12 equal length elements, and the cross section is divided into 20 concrete layers. The eccentricities are simulated by shifting the reference plane to coincide with the points of application of the loads, as shown in Fig. 6. The cube compressive strength of concrete f_{cu}, illustrated in Table 1, are related to

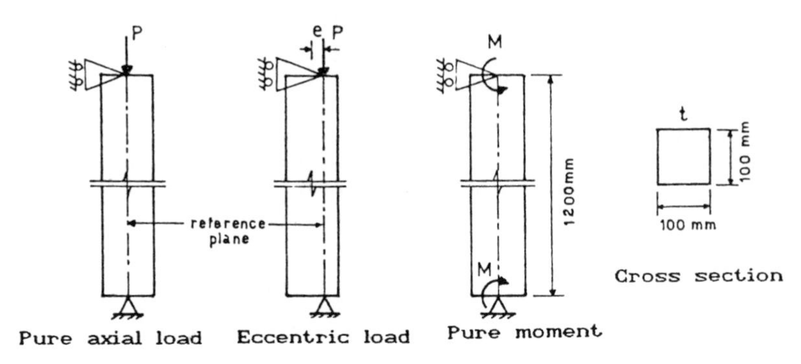

Fig. 6. Geometry and loading.

Table 1. Material properties of the columns

e/t	f_{cu} (MPa)		f'_t (MPa)		Fibre index
	Plain	Fibre	Plain	Fibre	
0.0	25.42	27.70	1.68	2.97	V_f = 2%
0.063	20.88	24.56	1.50	2.63	
0.125	20.80	25.10	1.52	2.72	l_f = 50 mm
0.250	21.53	24.67	1.47	2.47	d_f = 1.0 mm
0.375	23.05	25.56	1.60	2.77	
0.500	24.53	27.59	1.51	2.69	k_f = 1.0
Pure moment	23.80	27.27	1.58	2.99	E_f = 200×10^3 MPa

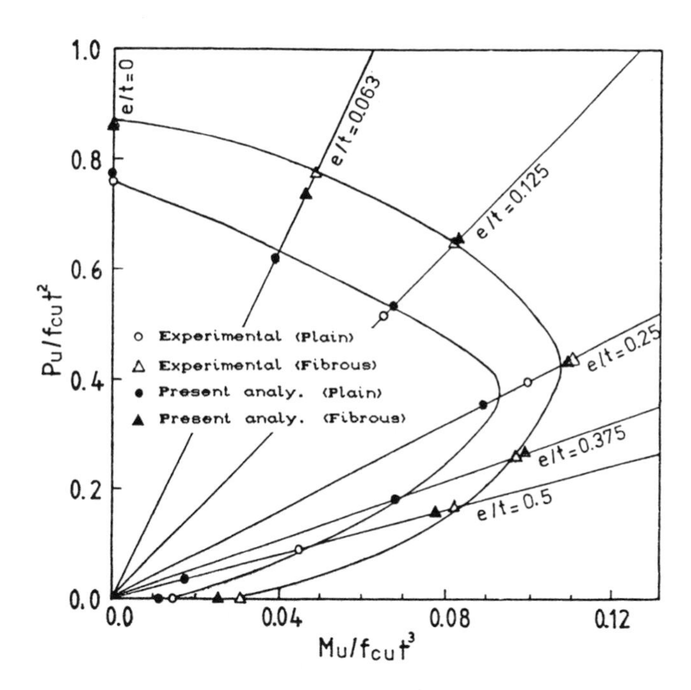

Fig. 7. Load-moment interaction diagram.

that of the standard cylinders (Neville, 1977).

Fig. 7 shows the load-moment interaction of the plain and fibrous concrete columns. Good agreement is observed between the present and the experimental results. Some differences can be seen in this comparison, especially in the case of plain concrete columns. They might be attributed to the small size of the specimens and to the absence of conventional reinforcements, which make the tests very sensitive to locations of the applied loading and the initial cracks.

Further, the test results of the columns, subjected to pure moments, was obtained by loading them as beams at the third points. While, the present results are calculated by applying pure moments at the ends of the columns. For this reason, some difference is observed between the two methods due to the difference in the resultant moment distribution along the length of the columns.

5.2 Reinforced fibrous concrete beams

Two simply supported reinforced fibrous concrete beams, tested by Swamy and Al-Ta'an (1981), is selected for the analysis. The geometry and material properties of the test beams are shown in Fig. 8 and Table 2 respectively.

Due to symmetry, one half of each beam is modelled by 10 equal length elements. The cross section of each element is divided into 15 concrete layers and 2 steel layers.

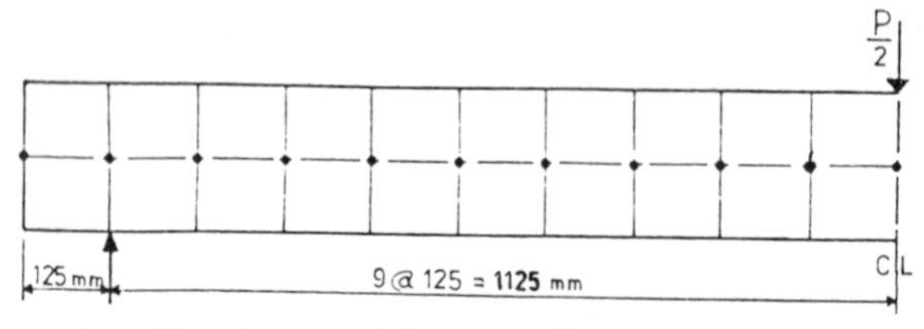

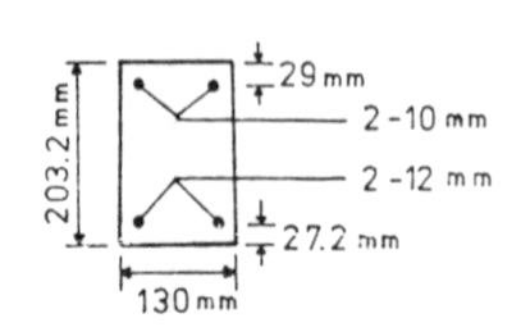

Fig. 8. Geometry and details of the beam.

Table 2. Details of test beam

Beam No.		DR10	DR12	Remarks
V_f	(%)	0.0	1.0	Crimped steel
$l_f \times d_f$	(mm)	0.5×50	0.5×50	E_f = 210 kN/mm^2
f_{cu}	(MPa)	36.97	40.0	
E_i	(kN/mm^2)	28.13	28.35	
f_y	(MPa)	460	460	0.2 % offset

Very good agreements can be observed between the present and the experimental results, represented by the load-deflection curves and the load-steel strain curves at mid-span of the beams, as shown in Figs. 9 and 10.

The slight underestimation near the ultimate loads can be attributed to the approximation in modeling the stress-

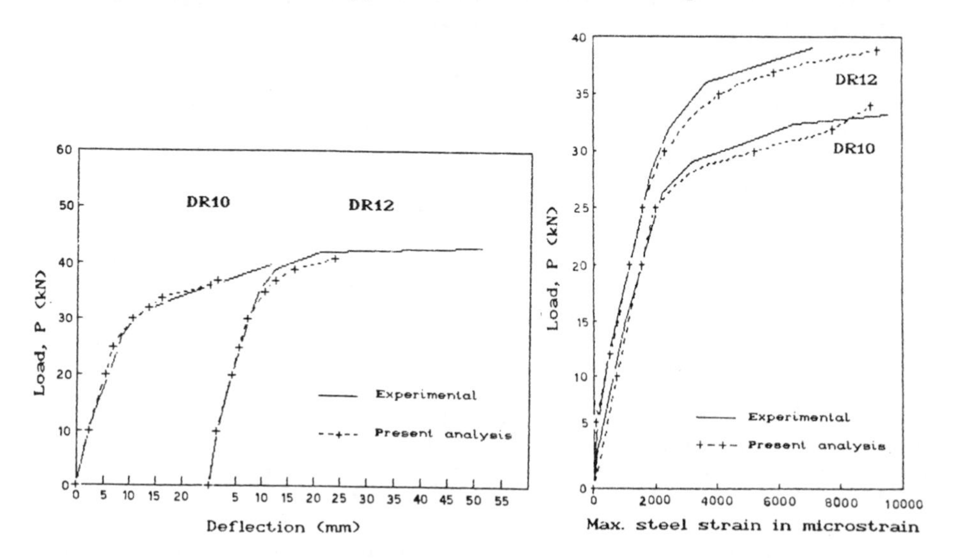

Fig. 9. Comparison of mid-span deflection.

Fig. 10. Comparison of steel strain.

strain curve of the high strength reinforcing steel, where it is approximated as a multilinear curve. Also, the experimental difficulties in tracing the response of the structure at its final stages, due to the strong material nonlinearity, may cause such differences.

Another correlation is presented in Fig. 11 for the crack propagation as traced in the test beam (Al-Taan, 1978) and in the analytical model. Good agreement is observed between the present and the test results. As it is expected, a slight increases in the length of the cracks are indicated by the analytical model. This is because of the small crack width at the ends of the crack patterns, which can not be detected visually or by magnifying glass, especially in the case of fibrous concrete. While, the propagation of these cracks can simply be traced numerically.

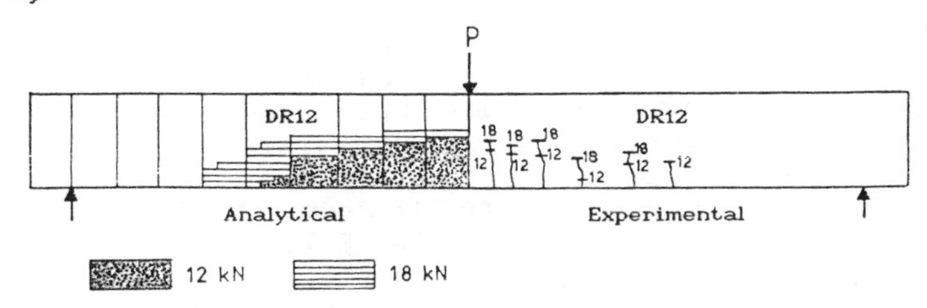

Fig. 11. Comparison of crack propagation.

6 Conclusions

Based on the numerical results obtained in this work, the following can be concluded:
(1) The proposed method of the finite element analysis appears to be valid and powerful tool for the nonlinear analysis of fibre reinforced concrete members, and capable of predicting the response of such structures under monotonic loads.
(2) The adopted and developed material constitutive relationships proved to give satisfactory results at the service and ultimate load stages.
(3) Better results can be obtained if the complete experimental data of the stress-strain diagram of the reinforcing steel, and a refined stress-strain curves for concrete in tension and compression are available.

7 References

ACI Committee 318 (1989) **Building Code Requirements for Reinforced Concrete.** ACI 318-89, Drtroit.
Al-Ta'an, S.A. (1978) **Structural Behaviour of Conventionally Reinforced Concrete Beams With Steel Fibres.**

Ph.D. Thesis, University of Sheffield, Sheffield.
Chen, W.F. (1982) **Plasticity in Reinforced Concrete.** McGraw-Hill Co., N.Y.
Cook, R.D. (1981) **Concepts and Applications of Finite Element Analysis.** John, Wiley and Sons Inc., USA.
Craig, R.J. (1987) Flexural behavior and design of reinforced fibre concrete members, in **Fibre Reinforced Concrete Properties and Applications** (eds S.P. Shah and G.B. Batson), ACI SP-105-28, Detroid, pp. 517-563.
Henager, C.H. (1977) Ultimate strength of rinforced seel fibrous Concrete Beams, in **Fibre Reinforced Materials: Design and Engineering Applications,** (ICE), London, pp. 151-160.
Kang, Y.J. and Scordelis, A.C. (1980) Nonlinear analysis of prestressed concrete frames. **J. Structural Div.**, 106, ST2, 445-462.
Lim, T.Y., Paramasivam, P., Mansur, M.A. and Lee, S.L. (1986) Tensile behaviour of steel fibre reinforced cement composites, in **RILEM Symp. on Developments in Fibre Reinforced Cement and Concrete** (eds R.N. Swamy, R.L. Wagstaffe and D.R. Oakley), University of Sheffield Press, Sheffield, pp. 7-15.
Naaman, A.E. and McGarry, F.J. (1974) Probabilistic analysis of fibre reinforced concrete. **J. Enging Mechanics,** 100, EMZ, 397-413.
Neville, A.M. (1977) **Properties of Concrete.** Pitman Publishing Ltd., London.
Niyogi, S.K. and Chawla, A.P. (1982) Fibre reinforced concrete short columns under uniaxially eccentric loads. **J. Civil Engrs.**, 73, Part 2, 199-206.
Sakai, M. and Nakamura, N. (1986) Analysis of flexural behaviour of steel fibre reinforced concrete, in **RILEM Symp. on Developments in Fibre Reinforced Cement and Concrete** (eds R.N. Swamy, and R.L. Wagstaffe and D.R. Oakley), University of Sheffield Press, Sheffield, pp. 27-34.
Soroushian, P., and Lee, C.D. (1989) Constitutive modeling of steel fibre reinforced concrete under direct tension and compression, in **Recent Developments in Fibre Reinforced Cements and Concretes** (eds R.N. Swamy and B. Barr), Elsevier Science Publishers Ltd., Essex, pp. 363-377.
Soroushian, P., and Reklaoui, A. (1989) Flexural design of reinforced concrete beams incorporating steel fibres, in **Recent Developments in Fibre Reinforced Cements and Concretes** (eds R.N. Swamy and B. Barr), Elsevier Science Publishers Ltd., Essex, pp. 454-466.
Swamy, R.N. and Al-Ta'an, S.A. (1981) Deformation and ultimate strength in flexure of reinforced concrete beams made with steel fibre concrete. **J. ACI,** 78, 5, 395-405.
Zienkiewicz, O.C. (1977) **The Finite Element Method.** McGraw-Hill Co., Maidenhead.

37 SHEAR RESPONSE OF REINFORCED FIBROUS CONCRETE BEAMS USING FINITE ELEMENT METHOD

K. MURUGAPPAN, P. PARAMASIVAM and K. H. TAN
Department of Civil Engineering, National University of Singapore

Abstract
A study on the use of a finite element formulation for the analysis of reinforced fibrous concrete (RFC) beams under predominant shear is presented. The formulation treats the cracked reinforced fibre concrete as an orthotropic non-linear elastic material, based on a smeared rotating crack model. Approporiate constitutive relations for the cracked RFC, accounting for the post cracking tensile strength and softening in compression are used in the principal stress/strain directions to define the secant moduli. The crack model along with a four noded quadrilateral finite element has been used to predict the behaviour of reinforced fibrous concrete beams. The analytically obtained results show good correlation with the test results of six reinforced steel fibre concrete I beams, and also some of the test results available in the literature.
Keywords: Finite element anlysis, Fibre concrete, Stress-strain relations, Tensile strength.

1 Introduction

The inclusion of short discrete steel fibres in a reinforced concrete member increases the strength, stiffness and improves the ductility (1-5). Recent research works indicate that the use of steel fibres as shear reinforcement can lead to a significant increase in the shear strength of RC beams. Various semi-empirical relations have been proposed to

Fibre Reinforced Cement and Concrete. Edited by R. N. Swamy. © 1992 RILEM.
Published by E & FN Spon, 2-6 Boundary Row, London SE1 8HN. ISBN 0 419 18130 X.

determine the ultimate shear capacity of RFC beams (1-5). Swamy et al. (2) utilise the flexural strength, whereas Narayanan et al (4) suggest the use of the split cylinder strength of fibre concrete cylinders. Also, there are studies by Lim et al (5) indicating that for ultimate load predictions, the effect of fibres can be expressed in terms of the equivalent shear reinforcement in the form of stirrups. Swamy et al. (6) have reported the finite element modelling of the behaviour of RFC beams using the eight noded isoparametric hexahedral elements. The present study is aimed at the modelling of the shear behaviour of the RFC beams using a plane stress finite element formulation. In this paper, a finite element formulation, treating the cracked RFC as an orthotropic non-linear elastic material, based on a smeared rotating crack model is presented. Constitutive relations for the reinforced fibrous concrete in a biaxial state of stress, incorporating the post-cracking tensile strength and also accounting for the softening of fibre concrete under compression have been used in this study. The results of this finite element formulation are compared with the test results of the present experimental investigation and also that of Lim et al (5).

2 Problem statement

The response of reinforced fibre concrete beams with different shear span to effective depth ratio (a/d) and varying volume fraction of the fibres are investigated in this study using a four noded quadrilateral finite element. The range of (a/d) in the present analysis is between 1.5 and 3.5. The volume fraction of the fibres varied from 0 to 1.0%, and in all the beams reported in this study, hook ended steel fibres, 30 mm long and 0.5 mm diameter were used. All the beams were subjected to two point loading. Since the beams are symmetrically loaded, only one half needs to be modelled for the analysis. Figure 1a shows a typical mesh used for the analysis of the beams, along with the load and support conditions.

The formulation is based on the following assumptions :

(1) The cracks are uniformly distributed over the element(Smeared crack representation) and the orientation of the cracks is defined by the orientation of the principal strains.

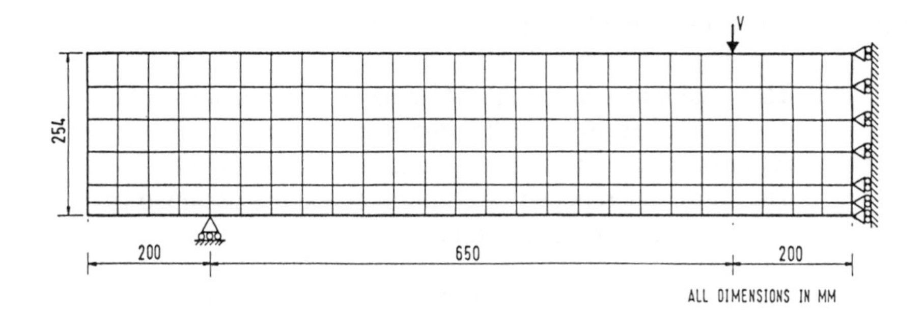

Fig. 1a A typical finite element mesh used

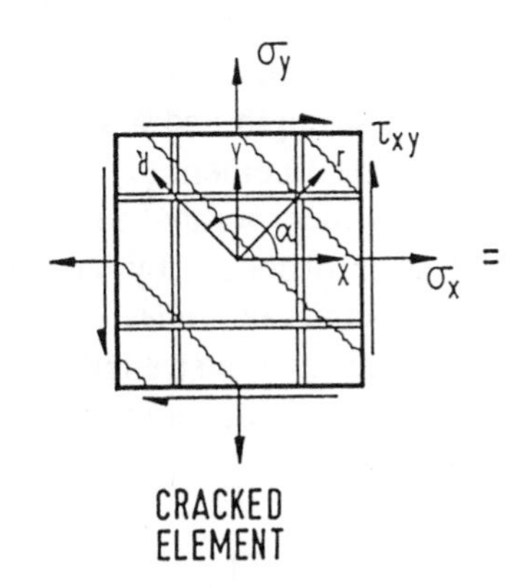

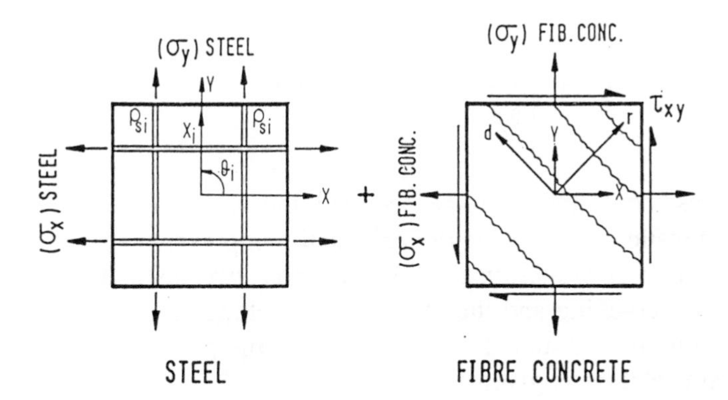

Fig. 1b State of stress in a cracked element

(2) The reinforcing bars are assumed to be uniformly distributed over the element.

(3) The axial stress in the reinforcements is dependent only on the axial strains and that the shear resistance of the reinforcements normal to their axis (x_i) is negligible.

(4) The fibre concrete and the reinforcement are assumed to be perfectly bonded. ie., there is no slip.

The following section describes the finite element formulation and the solution procedure based on the above assumptions, using a secant stiffness approach, in which the principal stress-strain relations are used to define the secant moduli.

3 Principal tensile stress-strain relations for fibre concrete

The inclusion of steel fibres in ordinary concrete results in a slightly enhanced cylinder compressive strength and a significantly improved strain at peak stress due to the confinement (7) provided by the fibres after cracking. This influence is incorporated into the principal compressive stress-strain relation proposed by Vecchio and Collins (8) accounting for the softening of concrete under compression , with f_c' and ε_0 taken as the cylinder compressive strength and strain at peak stress of fibre concrete cylinders. The average principal compressive stress-strain relation used is (Fig. 2):

$$\sigma_d = f_{c2max}\left[2\left(\frac{\varepsilon_d}{\varepsilon_0}\right) - \left(\frac{\varepsilon_d}{\varepsilon_0}\right)^2\right] \tag{1}$$

$$where \quad \frac{f_{c2max}}{f_c'} = \left(\frac{1.0}{0.8 - 0.34\,\dfrac{\varepsilon_r}{\varepsilon_0}}\right) \leq 1.0$$

In the above expression ε_d and ε_r are respectively the principal compressive and tensile strains. σ_d is the principal compressive stress.

The principal tensile stress-strain relation used has two distinct portions, one before cracking and the other after cracking. Before cracking, the behaviour is assumed to be elastic and the principal tensile stress strain relation is given as (Fig. 3):

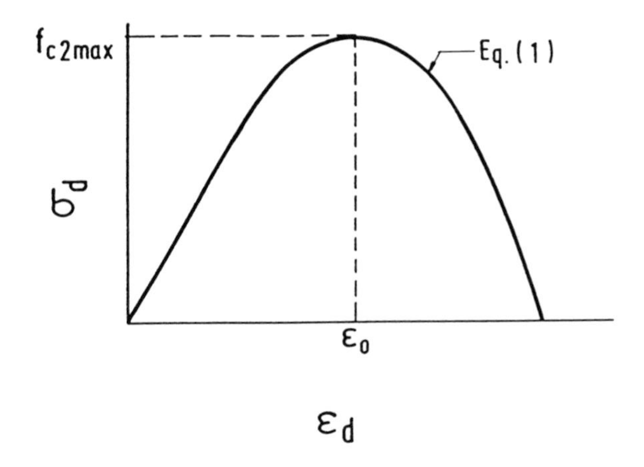

Fig. 2 Principal compressive stress-strain relation

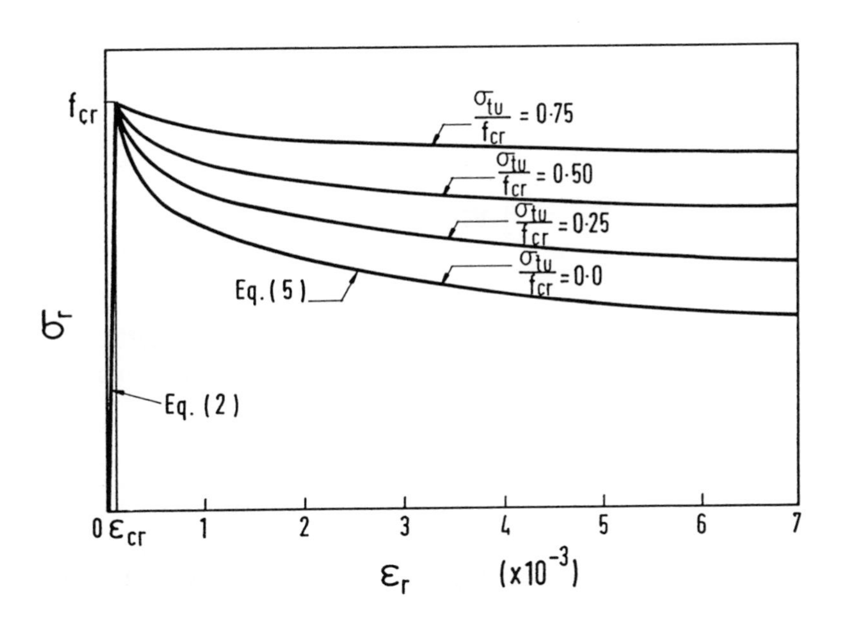

Fig. 3 Principal tensile stress-strain relation

$$\sigma_r = \varepsilon_r E_c \quad for \quad \varepsilon_r \leq \varepsilon_{cr} \tag{2}$$

where, E_c is the Young's modulus and ε_{cr} is the cracking strain of fibre concrete. The cracking strain ε_{cr} is computed as $\varepsilon_{cr} = f_{cr}/E_c$, with f_{cr} and E_c taken as :

$$f_{cr} = 0.33 \sqrt{f_c'} \tag{3}$$

$$E_c = 2 f_c' / \varepsilon_0 \tag{4}$$

In the above equations, f_c' and ε_0 are respectively the compressive strength and strain at peak stress for the fibre concrete cylinders.

After cracking, there exists a significant transfer of stress across the cracks in the fibre concrete (9). This stress often defined as the post-cracking strength is designated here as σ_{tu}. To account for this post cracking tensile strength, the average principal tensile stress-strain relation used is :

$$\sigma_r = \frac{f_{cr} + (\beta\, \sigma_{tu})}{1 + \beta} \quad for \quad \varepsilon_r > \varepsilon_{cr}$$

$$where\ \beta = \sqrt{\frac{\varepsilon_r - \varepsilon_{cr}}{0.005}} \tag{5}$$

The value of σ_{tu} depends on the orientation factor η_0', the length efficiency factor η_l , the volume fraction V_f, and the ultimate bond strength τ_u of the fibres and can be taken as (9):

$$\sigma_{tu} = \frac{\eta_l\, \eta_0'\, V_f\, l_f\, \tau_u}{2\, r'} \tag{6}$$

l_f is the length of the fibres and r' is the ratio of the cross section to the circumference of the fibres.

4 Finite element formulation

The state of stress in the cracked RFC element is as shown in Fig. 1b. X and Y are the global co-ordinates and the principal directions of the cracked fibre concrete are denoted by d and r. The direction of the reinforcement component is indicated as x_i. In developing the element stiffness matrix for the finite element analysis, the material stiffness matrix [D] relates the stresses to the strains as (10):

$$\{f\} = [D]\{\varepsilon\} \tag{7}$$

For a plane stress analysis as in this case, $\{f\} = [\ \sigma_x\ ,\ \sigma_y\ ,\ \tau_{xy}\ ]$ and $\{\varepsilon\} = [\ \varepsilon_x\ ,\ \varepsilon_y\ ,\ \gamma_{xy}\ ]$, where σ_x , σ_y , τ_{xy} are the normal and shear stresses in the global co-ordinate system. ε_x , ε_y , and γ_{xy} are the corresponding strains. The material stiffness matrix [D] for the conventional elastic analysis consists of the elastic constants and can be found from a standard text (10). However, the material stiffness for the cracked element has to be computed as the sum of the stiffness of the cracked RFC and the stiffness of the steel reinforcements. So, the material stiffness for the cracked RFC and the steel components are first computed in their respective principal directions as $[\ D_c'\ ]$ and $[D_s']_i$ (i=1,n) respectively. Then they are transformed to the global (X-Y) co-ordinate system using the corresponding transformation matrices. The composite material stiffness matrix is then obtained as the sum of the stiffness matrices $[D_c]$ and $[\ D_s\ ]$. The computations involved are summarised as :

$$\begin{aligned} [D] &= [D_c] + \sum [D_s]_i \\ [D_c] &= [T_c]^T\ [D_c']\ \ [T_c] \\ [D_s]_i &= [T_s]_i^T\ [D_s']_i\ [T_s]_i \end{aligned} \tag{8}$$

The transformation matrix for the concrete component $[\ T_c\ ]$ is computed as given in Eq. 9 using $\theta = \alpha$ (Fig. 1b). The transformation matrix $[\ T_s\ ]_i$ for the steel component i is also computed as given in Eq. 9 with $\theta = \theta_i$ (Fig. 1b). Here α and θ_i are respectively the principal strain directions for the cracked RFC and the reinforcement component i respectively.

$$[T] = \begin{bmatrix} \cos^2\theta & \sin^2\theta & \cos\theta \sin\theta \\ \sin^2\theta & \cos^2\theta & -\cos\theta \sin\theta \\ -2\cos\theta \sin\theta & 2\cos\theta \sin\theta & \cos^2\theta - \sin^2\theta \end{bmatrix} \tag{9}$$

Since a secant modulus approach has been adopted in this study, the secant moduli are used to define the material stiffness of the cracked RFC and the steel components. The evaluation of the material stiffness matrices in the principal directions system are explained in the following paragraphs.

The cracked fibre concrete is modelled as an orthotropic material, with its principal directions along d and r. Assuming that the Poisson's effect is negligible after cracking, the material stiffness for concrete in the principal directions is given by (11):

$$[D_c'] = \begin{bmatrix} \overline{E}_d & 0 & 0 \\ 0 & \overline{E}_r & 0 \\ 0 & 0 & \overline{G} \end{bmatrix} \tag{10}$$

In the above expressions, $\overline{E}_r$, $\overline{E}_d$, and $\overline{G}$ are the secant modulii and can be computed using the constitutive relations (Eqns. 1-6) as :

$$\overline{E}_r = \frac{\sigma_r}{\varepsilon_r}, \quad \overline{E}_d = \frac{\sigma_d}{\varepsilon_d}$$

$$\text{and} \quad \overline{G} = \frac{\overline{E}_r \overline{E}_d}{\overline{E}_r + \overline{E}_d} \tag{11}$$

For each reinforcement component i in an element, the material stiffness matrix in the local co-ordinate system along the reinforcement is computed using the reinforcement ratio ρ_{si} and the secant modulus of the steel $\overline{E}_{si}$ as:

$$[D_{si}'] = \begin{bmatrix} \rho_{si}\overline{E}_{si} & 0 & 0 \\ 0 & 0 & 0 \\ 0 & 0 & 0 \end{bmatrix} \tag{12}$$

In this study the behaviour of steel is assumed to be elastic-perfectly plastic, and the secant modulus for the steel components are computed using this idealisation.

Once the material stiffness matrix [D] is obtained, the element stiffness matrix for the four noded quadrilateral element can be evaluated using the standard procedures as $[k] = \int_v [B]^T [D] [B] \, dV$, [B] being the strain shape function (12). The details of the computations involved are presented in Appendix I.

The solution procedure adopted for obtaining the response of a beam for a given loading condition is shown in the flow chart (Fig.4).

5 Convergence characteristics and numerical stability

From the solution procedure (Fig. 4), it could be seen that, for a given load, the response of the beam is obtained using an iterative technique. The first iteration begins with the material stiffness defined using the elastic constants of concrete before cracking, and the elastic solution is obtained. The subsequent iterations utilize the secant stiffness obtained using the state of strain in the preceding iteration. Hence it is found that the stresses and strains for a given element change significantly for the first few iterations, when the secant moduli change from the uncracked to that of cracked. The final solution is obtained when the secant modulii converge. It is observed that the local failure of an element does not affect the convergence and hence the numerical stability of the solutions. Redistribution to adjacent elements occurs in this case and a numerically stable solution is obtained till the structure (here the beam) attains its ultimate load. The convergence of the displacements and stresses for the Beam-4 tested by Lim et al (5) (Table 1) are given in Fig. 5, when ten elements were used across the depth of the beam. The load deflection response for the beams is given in Figs. 6a - 6f. The convergence is faster at lower loads compared to higher load levels, essentially because only a few elements are cracked. Normally it is found that around 20 to 25 iterations are required for the convergence of the displacements and stresses.

Also, it is observed generally that, a less stiffer response is obtained with mesh refinement. This was observed in almost all the beams analysed. Fig. 1a shows a typical mesh with six elements across the depth of the beam used in the present study. Fig. 6e shows the load Vs

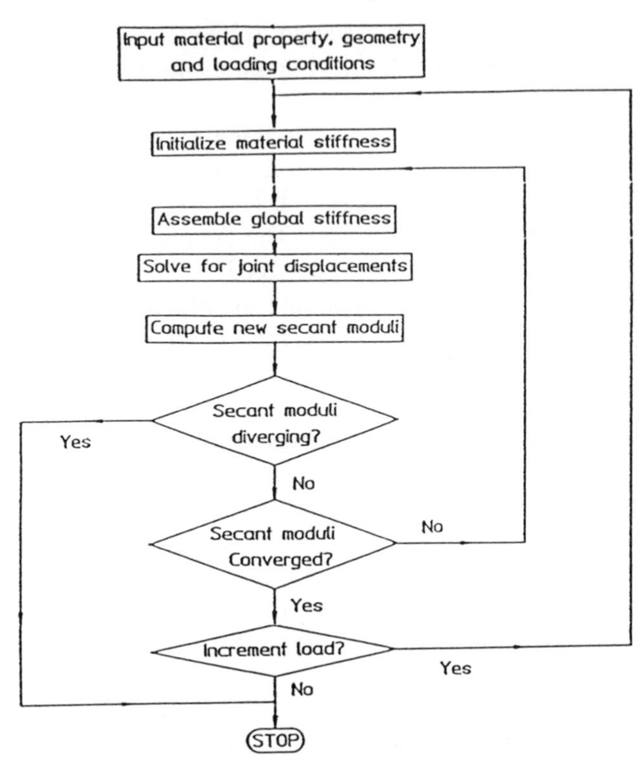

Fig. 4 Solution procedure

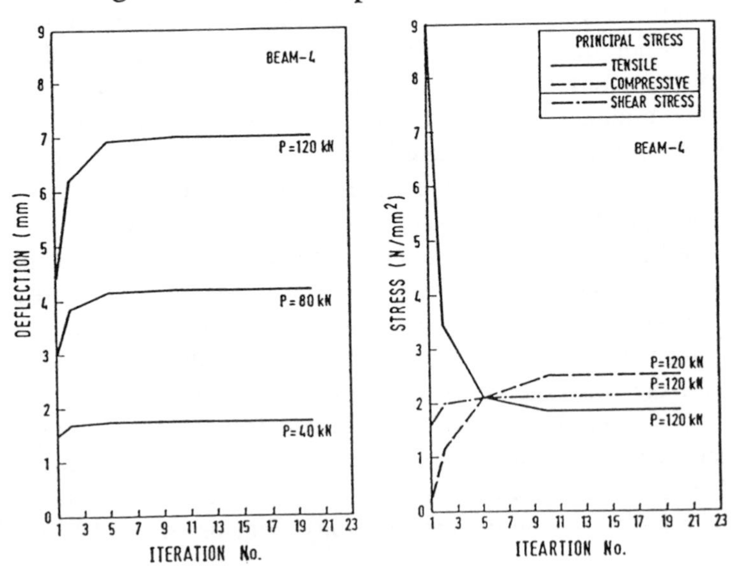

Fig. 5 Convergence characteristics

Table 1. Details of the beams tested in ref. 5

Designation	a_1/d	V_f %	σ_{tu} N/mm^2	Ult. load kN (Experiment)
Beam-1	1.5	1.00	1.68	295.0
Beam-2	1.5	0.50	0.84	270.0
Beam-3	2.5	1.00	1.68	165.2
Beam-4	2.5	0.50	0.84	127.4
Beam-5	3.5	1.00	1.68	134.8
Beam-6	3.5	0.50	0.84	98.8

Note : a_1 is the clear distance between support and load point.

Table 2. Details of the beams tested in the present study

Designation	a/d	V_f %	σ_{tu} N/mm^2	Ult. load kN (Experiment)
B-1	2.0	0.00	1.68	126.3
B-2	2.0	0.50	0.84	218.0
B-3	2.0	0.75	1.68	180.9
B-4	2.0	1.00	0.84	210.3
B-5	2.5	1.00	1.68	154.2
B-6	1.5	1.00	0.84	307.0

midpoint displacement response for the beams tested by Lim et al (5), obtained using both six and ten elements across the depth of the beam. All the other solutions presented in this paper are obtained using the mesh with ten elements across the depth. The solution was obtained using a main frame computer and the program was coded in FORTRAN 77. The solution for one problem roughly took 1200 seconds of cpu time, varying with the number of load steps used.

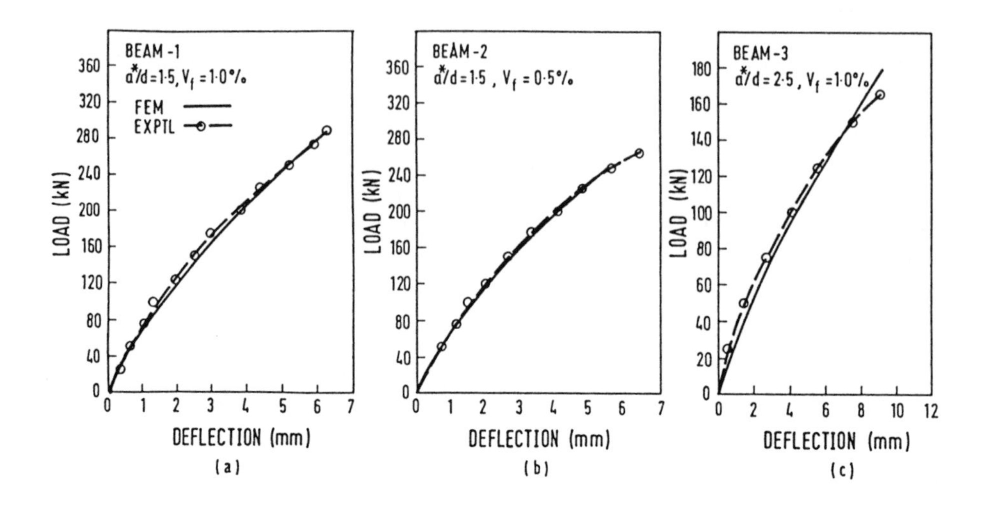

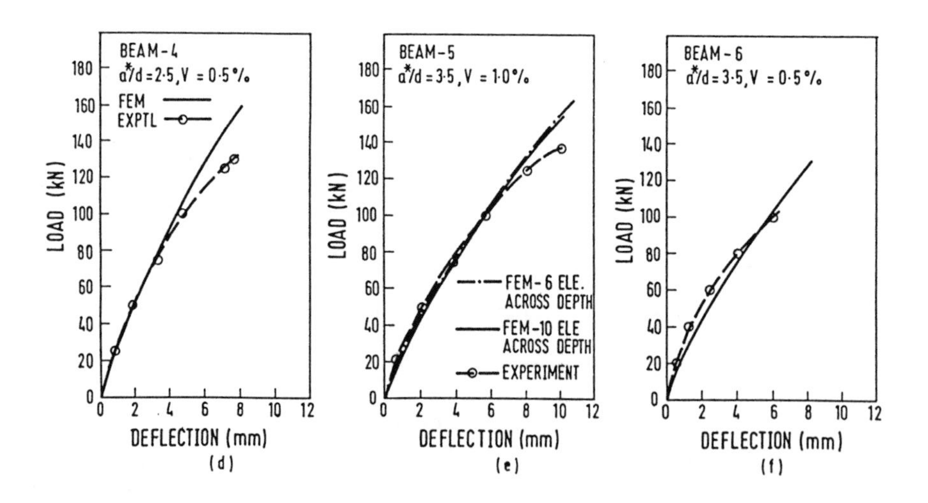

Fig. 6 Comparison of load Vs mid-point deflection for beams tested by Lim et al. (5)

6 Comparison with experimental results

The finite element formulation described in this paper was used to obtain the response of two series of beams, one tested by Lim et al (5), and the other tested during the present study. The sectional and other details of the beam are given in Table-1 and Table-2. The beams tested by Lim et al (5) were of rectangular cross section and were reinforced with four 16 mm bars, each with an yield strength of 90 kN. Figs 6a to 6f show the predicted and the experimental load deflection response of the beams. It is found that the finite element formulation could model the stiffness before and after cracking and hence predict the complete load-deflection behaviour well.

Fig. 7 shows the cross section of the beams tested during the present study and also the schematic diagram of the loading arrangement for a shear span to effective depth ratio of 2.0. Figs. 8a to 8f show the comparison of the steel strains in the longitudinal reinforcing bars in the shear span at a distance of 'd' from the load point. It is found that the finite element formulation could model well the diagonal web cracking and hence predict the steel strains reasonably well. Also the formulation could estimate the influence of the fibres on the load deformation response and also the steel strains. It could be observed from Figs. 7 and 8 that the analytical response predictes was slightly stiffer than the experimental results. This is probably due to the difficulty in modelling the localised cracking in the concrete, which analytically has been modelled to be smeared.

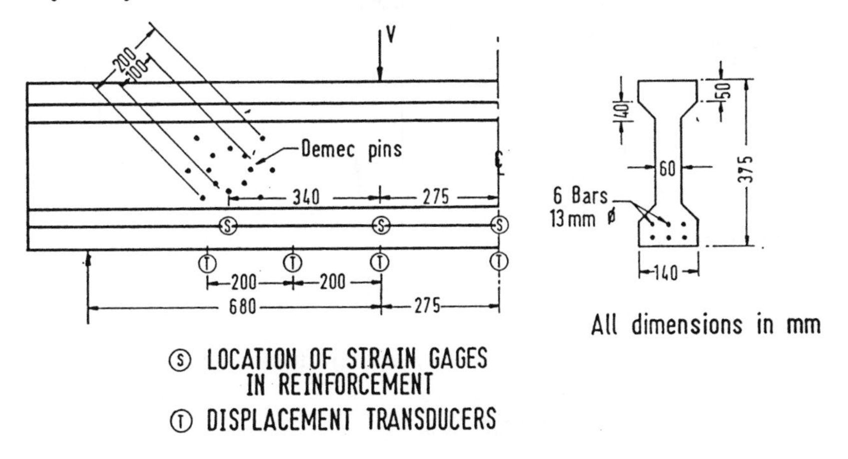

Fig. 7 Loading arrangement and instrumentation

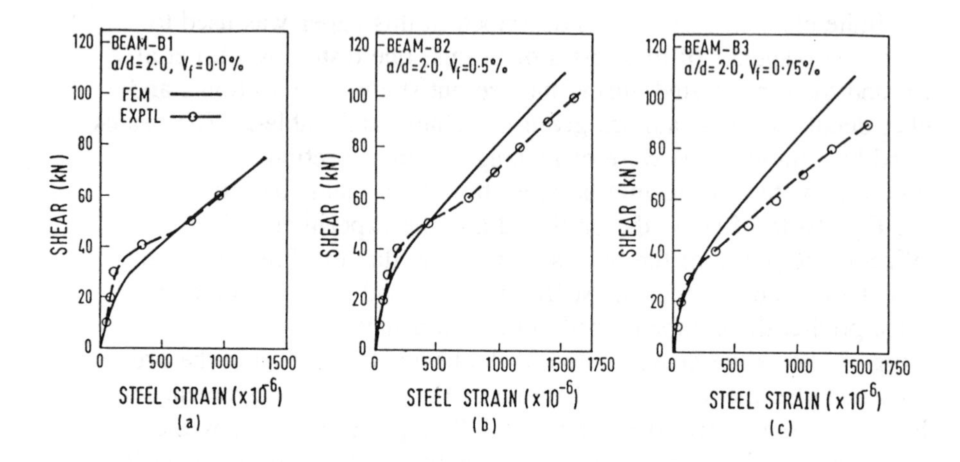

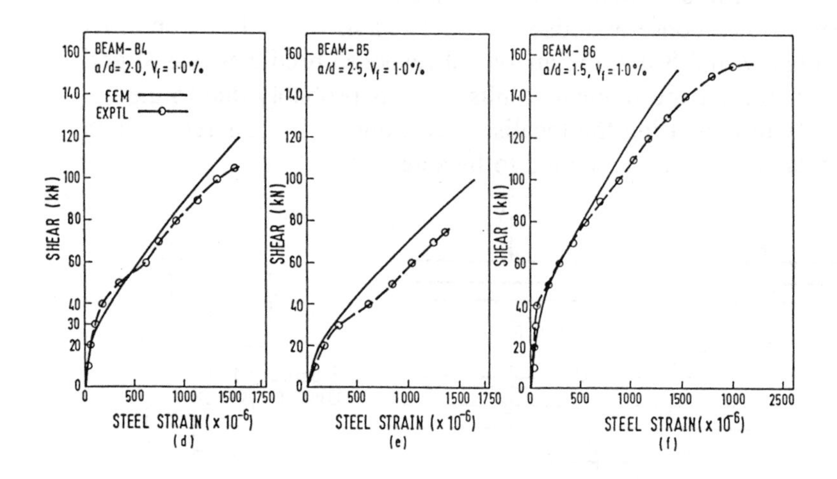

Fig. 8 Comparison of steel strains for beams tested in the present study

7 Conclusions

A finite element formulation for the analysis of the RFC beams based on a total equilibrium approach, applicable for obtaining the path independent response is presented. The cracked reinforced fibre concrete is modelled as an orthotropic material. The secant moduli defining the material properties are computed using the principal stress-strain relations proposed for the fibre concrete, based on the concept of average strain and stress. It is observed that the solution procedure was numerically stable with good convergence characteristics. The response of the beams obtained using the formulation presented in this paper compared well with the experimental results of the present study and some results available in the literature, thus confirming its validity.

8 Acknowledgement

The work upon which the paper is based was supported by National University of Singapore research grant RP880646.

9 References

1. Mansur, M.A., and Paramasivam, P., " Fiber reinforced concrete beams in torsion, bending and shear",ACI Journal,Proceedings, Vol.82, No.1, Jan - Feb 1985, pp 33-39.
2. Swamy, R.N., and Bahia, H.M., "Effectiveness of steel fibers as shear reinforcement", Concrete International; Design and construction,Vol. 7, No. 3, Mar. 1985, pp 35-40.
3. Mansur, M.A., Ong, K.C.G., and Paramasivam, P., "Shear strength of fibrous concrete beams without stirrups", Proceedings ASCE, Journal of structural Division, Vol. 112, N0. 9, Sep 1986, pp 2066-2079.
4. Narayanan, R., and Darwish, I.Y.S., "Use of steel fiber as shear reinforcement", ACI Structural Journal, Vol. 84, No.3, May-Jun 1987, pp 216-227.
5. Lim, T.Y., Paramasivam, P., and Lee, S.L., "Shear and moment capacity of reinforced steel fiber concrete beams", Magazine of Concrete Research, Vol. 39, No. 140, Sep 1987,

pp 148 - 160.

6. Swamy, R.N., Jones, R., Chiam, T.P., "Modelling structural behaviour of reinforced concrete beams with steel fibres", Proc. International symposium on fibre reinforced concrete, Madras, Dec. 1987, Vol. 1, pp. 1.219-1.232.
7. Fanella, D.A., and Naaman, A.E., "Stress-strain properties of fiber reinforced mortar in compression", ACI Journal, Vol. 82, No. 4, Jul-Aug 1985, pp 475-483.
8. Vecchio, F.J., and Collins, M.P., " Response of reinforced concrete to in-plane shear & normal stresses",Publication No. 82 -03, Dept. of Civil Engg.,Univ. of Toronto,Mar 1982,pp 78-84.
9. Lim, T.Y., Paramasivam, P., and Lee, S.L., " Bending behaviour of Steel fiber concrete beams", ACI Structural Journal, Vol. 84, No. 6, Nov - Dec 1987, pp 524-536.
10. Zienkiewicz, O.C., " The Finite Element Method", Third Edition, Tata McGraw Hill Publishing Co. Ltd., 1986, New Delhi, pp 98-101.
11. Vecchio, F.J., "Reinforced concrete membarane element formulations", Proceedings, ASCE, Journal of the Structural Division, Vol. 116, No. 3, March 1990, pp 730-750.
12. Chandrupatala, T.R., and Belegundu, A.D., " Introduction to finite elements in engineering", Prentice Hall, Englewood Cliff, New Jersy, 1991, pp. 194-200.

10 Appendix I: Element stiffness for 4 node quadrilateral element

The element displacement vector { q } is given as :

$$\{ q \} = [\; u_1 \;\; v_1 \;\; u_2 \;\; v_2 \;\; u_3 \;\; v_3 \;\; u_4 \;\; v_4 \;]$$

The matrix [J] relates the derivatives of local co-cordinates (η,ξ) to the derivatives of global co-ordinates (X,Y) and is defined as (Fig 1b):

$$[\, J \,] = \begin{bmatrix} J_{11} & J_{12} \\ J_{21} & J_{22} \end{bmatrix} = \begin{bmatrix} \frac{\partial x}{\partial \xi} & \frac{\partial y}{\partial \xi} \\ \frac{\partial x}{\partial \eta} & \frac{\partial y}{\partial \eta} \end{bmatrix}$$

where,

$$J_{11} = \frac{1}{4}\left(-(1-\eta)\,x_1 + (1-\eta)\,x_2 + (1+\eta)\,x_3 - (1+\eta)\,x_4 \right)$$

$$J_{12} = \frac{1}{4}\left(-(1-\eta)\,y_1 + (1-\eta)\,y_2 + (1+\eta)\,y_3 - (1+\eta)\,y_4 \right)$$

$$J_{21} = \frac{1}{4}\left(-(1-\xi)\,x_1 - (1+\xi)\,x_2 + (1+\xi)\,x_3 + (1-\xi)\,x_4 \right)$$

$$J_{22} = \frac{1}{4}\left(-(1-\xi)\,y_1 - (1+\xi)\,y_2 + (1+\xi)\,y_3 + (1-\xi)\,y_4 \right)$$

The strain $\{\varepsilon\}$ is related to the nodal displacement matrix {q} as: $\{ \varepsilon \} = [\,B\,]\,\{\,q\,\}$. [B] is computed as : $[\,B\,] = [\,A\,]\ [\,G\,]$ with [A] and [G] computed as follows:

$$[\,A\,] = \frac{1.0}{\det[\,J\,]} \begin{bmatrix} J_{22} & -J_{12} & 0 & 0 \\ 0 & 0 & -J_{21} & J_{11} \\ -J_{21} & J_{11} & J_{22} & -J_{21} \end{bmatrix}$$

$$[\,G\,] = \frac{1}{4} \begin{bmatrix} -(1-\eta) & 0 & (1-\eta) & 0 & (1+\eta) & 0 & -(1+\eta) & 0 \\ -(1-\xi) & 0 & -(1+\xi) & 0 & (1+\xi) & 0 & (1-\xi) & 0 \\ 0 & -(1-\eta) & 0 & (1-\eta) & 0 & (1+\eta) & 0 & -(1+\eta) \\ 0 & -(1-\xi) & 0 & -(1+\xi) & 0 & (1+\xi) & 0 & (1-\xi) \end{bmatrix}$$

The element stiffness matrix is then computed as :

$$[\,k\,]^e = t_e \int_{-1}^{1} \int_{-1}^{1} [\,B\,]^T\ [\,D\,]\ [\,B\,]\ \det[\,J\,]\ d\xi\ d\eta$$

with t_e taken as the thickness of the element.

PART SIX
STRUCTURAL BEHAVIOUR

38 ULTIMATE FLEXURAL STRENGTH OF REINFORCED CONCRETE BEAMS WITH LARGE VOLUMES OF SHORT RANDOMLY ORIENTED STEEL FIBRES

H. I. AHMED and R. P. PAMA
Asian Institute of Technology, Bangkok, Thailand

Abstract
The paper describes the results of an investigation conducted at the Asian Institute of Technology to determine the influence of high fiber volumes on the ultimate flexural strength of reinforced concrete beams. Tests were carried out on twenty-four isolated simply supported beams having conventional steel reinforcement with steel fiber concrete. The fiber volume was taken as high as 4%. This amount is twice as high as previously reported by other research workers.

A method for predicting the behaviour in flexure of reinforced concrete beams with high volume of fiber reinforcement is also presented.
Keywords: Fiber reinforced concrete, Flexural strength, Steel fibers, Beams, Stress-strain relationship

List of symbols

a = effective shear span
A_c = area of concrete
A_s = area of longitudinal steel
A'_s = area of compression steel
C_c = compressive force in concrete
C_s = compressive force in steel
d = effective depth
d_f = fiber diameter
d' = distance from top fiber of the beam to the centroid of the compression steel
d_l' = distance from extreme fiber in tension to the centroid of tension steel
F_{ct} = post-cracking strength
k_l , β_1 = factors associated with simplified stress block
l_c = critical length of the fiber
l_f = length of the fiber
M_u = ultimate flexural strength of the section
M_y = yield strength of the section
r = ratio of fiber cross-sectional area to its perimeter

Fibre Reinforced Cement and Concrete. Edited by R. N. Swamy. © 1992 RILEM.
Published by E & FN Spon, 2-6 Boundary Row, London SE1 8HN. ISBN 0 419 18130 X.

T = flexural toughness
V_f = volume fraction of the fiber
V_m = volume fraction of the matrix
x = depth of the neutral axis
y = distance of centroid of compressive area from neutral axis
ε = strain
ε_t = strain in tension reinforcement
ε'_s = strain in compression reinforcement
ε_y = yield strain of longitudinal reinforcement
ε'_y = yield strain of compression reinforcement
ε_{cu} = ultimate compressive strain in concrete
ε_{cr} = strain at cracking in composite
σ_c = stress in composite
σ_{cr} = cracking stress of the composite
σ_b = flexural strength of the composite
σ_{fu} = ultimate flexural strength of the fiber
σ_m = stress in the matrix
τ_b = average bond stress
η_l = length efficiency factor
η_b = bond efficiency factor

1 Research background

Considerable research has already been done in investigating the various properties, influencing parameters, method of preparation and testing of fiber reinforced concrete [Bentur (1986), Guan and Zhao (1986), Henager (1980), Kukreja et al. (1980), Nathan et al. (1977) and Pakotiprapha (1973)]. Of particular interest has been the use of short randomly oriented steel fibers to enhance the structural behaviour of concrete beams in shear and flexure [Pakotiprapha (1973), Rajagopalan et al. (1974) and Shah and Rangan (1970)]. However, few studies have been made so far to investigate the flexural behaviour of SFRC beams with conventional reinforcement [Henager and Doherty (1976), Lim et al. (1987), Rajagopalan et al. (1974), Swamy and Sa'ad (1981) and Uomoto and Weeraratnee (1986)]. Most of the studies, beside having various restrictions have one aspect in common that they all have practically been restricted in scope to the investigation of low fiber volumes. This is due to the problems encountered in preparation of SFRC mixes with high fiber volumes, i.e. decrease in workability and the phenomenon of "balling or "clumping" in which the fibers bunch together, resulting in an unworkable and segregated mix. However it is very well established that significant changes in the behaviour of SFRC occur with increasing fiber content. Paradoxically, a general weakening effect has been reported along with a decrease in ductility when fiber volumes are increased excessively [Agustin (1989) and Hughes and Fattuhi (1976)]. Although most studies related to the strength in flexure of SFRC beams follow the assumption of a linear relationship between fiber volume fraction (V_f) and post-cracking strength (Fct), the phenomenon mentioned above leads to a non-linear relation between the two beyond a certain value of V_f [Kobayashi et al. (1977)]. Only a few studies are available investigating the behaviour of SFRC or SFRM (steel fiber reinforced mortar) for higher volume fractions of fibers [Josifek and Lankard (1987) and Kobayashi et al. (1977)]. For concrete the effects of a higher V_f was studied by Kobayashi et.al (1977), however, the study is limited to the relationship between V_f and F_{ct} and is not extended to the strength behaviour of full scale beams. The present study investigates the behaviour of SFRC beams reinforced with high volumes of steel fibers in addition to conventional bar reinforcement.

2 Experimental Investigation

The experimental investigation comprises of casting control specimens for tests for compressive strength, flexural strength and tensile strength as well as casting of full scale SFRC beams.

A constant mix proportion of cement: fine aggregate :coarse aggregate :w/c of 1:2:2: 0.5 was used. The maximum size of coarse aggregate was 10 mm and fine aggregate was natural sand passing ASTM #7 sieve. The steel fibers were galvanized steel wire of No.24 gauge (0.5 mm dia) cut into lengths of 30 mm giving an aspect ratio of 60. The properties of the steel reinforcement are shown in Table 1.

The details of loading arrangement and reinforcement details are shown in Fig.1. A total of 24 beams were tested in the program.

A consistent mixing procedure as suggested by Agustin (1989) and Weeraratnee (1985) was adopted for mixing of SFRC. Wet mixing followed dry mixing for 45 seconds, which continued till the mix became uniform. A layer of fibers was sprinkled and mixing was started again for 5 seconds. The mixer was stopped again and another layer of fibers was added. The process continued till all the fibers were incorporated in the mix . Special care was taken to prevent the fibers falling on top of each other in small clusters, as it inevitably leads to balling. Small fiber masses tending to show the signs of developing into interlocked fiber balls were broken apart manually by using a three pronged garden fork in such a way that the fibers did not bend. For each specimen and beam, the casting was done in three layers . Each layer was thoroughly compacted using external vibrations.Control specimens comprised of 152 mm cubes, 100 x 100 x 500 mm prisms and double tensile strength specimens for compression , flexural strength and strength in direct tension respectively of the composite. The double tensile strength test used was as suggested by Uomoto (1986). Fig.2 shows typical details of

Table 1. Description and mechanical properties of steel reinforcement

Type	Nominal dia (mm)	Yield strength (MPa)	Yield strain (microns)	Tensile strength (MPa)	Young's modulus (MPa)
Deformed	16	393	2160	582	1.82×10^5
Round	9	220	1100	277	1.96×10^5
Galvanized wire	0.5	319	--	483	--

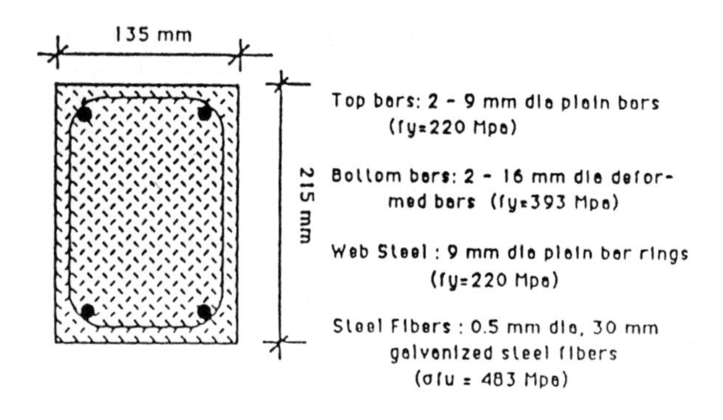

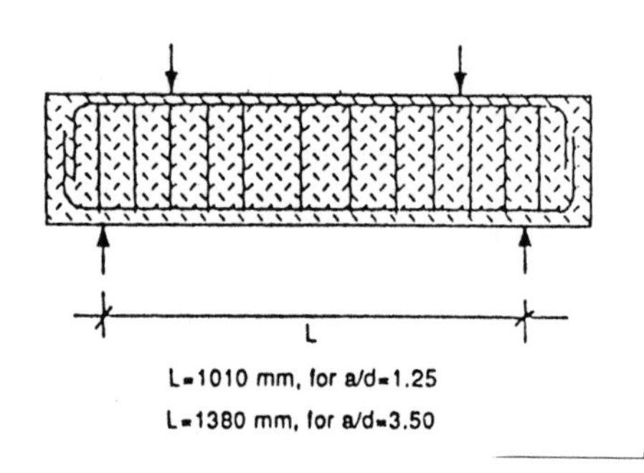

Fig.1. Typical details of test beam's span and cross section.

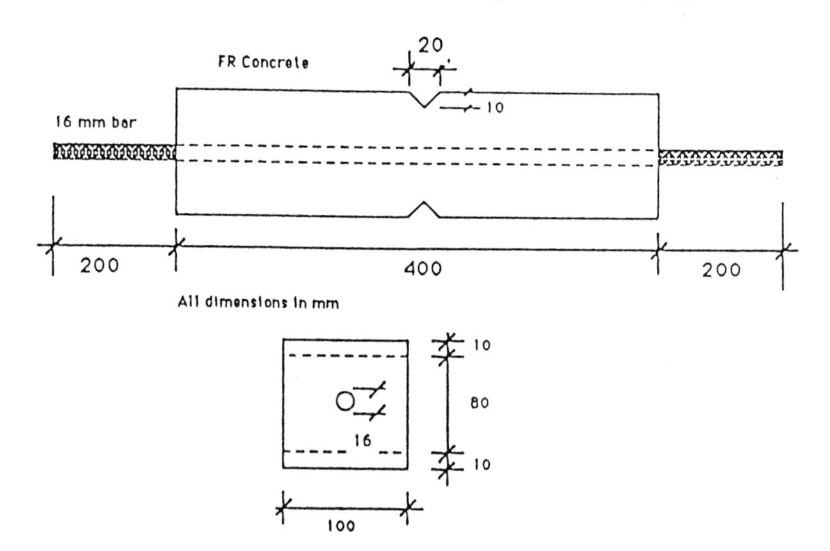

Fig.2. Details of double tensile strength specimen.

such a specimen. All prisms were tested under third point loading. All the specimens and beams were cured for 7 days and tested at the age of 21 days.

3 Test results

3.1 Effect on tensile strength

Fig.3 and 4 show the typical behavior of the tensile specimens at varying fiber percentages. It is obvious that for plain concrete, with increasing tensile load, the tensile stress

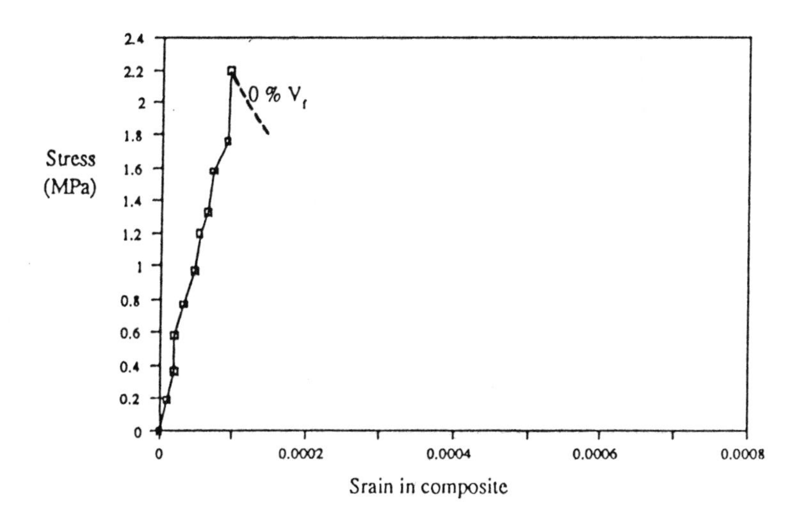

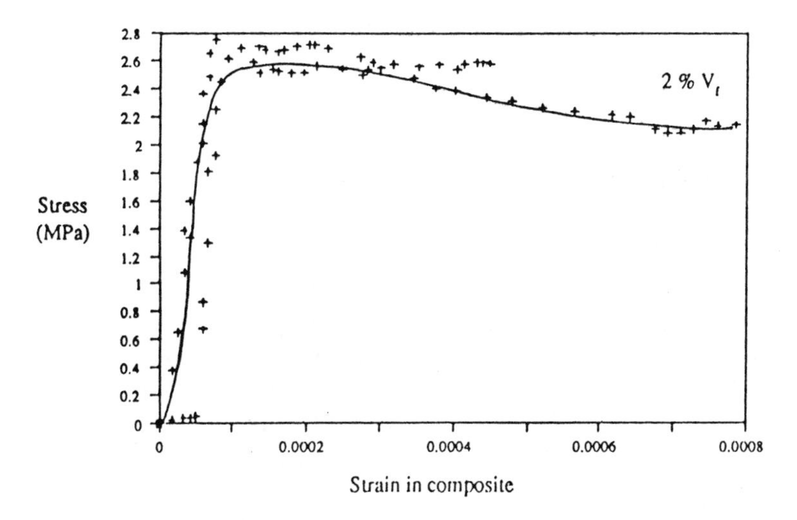

Fig.3. Average tensile stress-strain curves at 0% and 2% V_f.

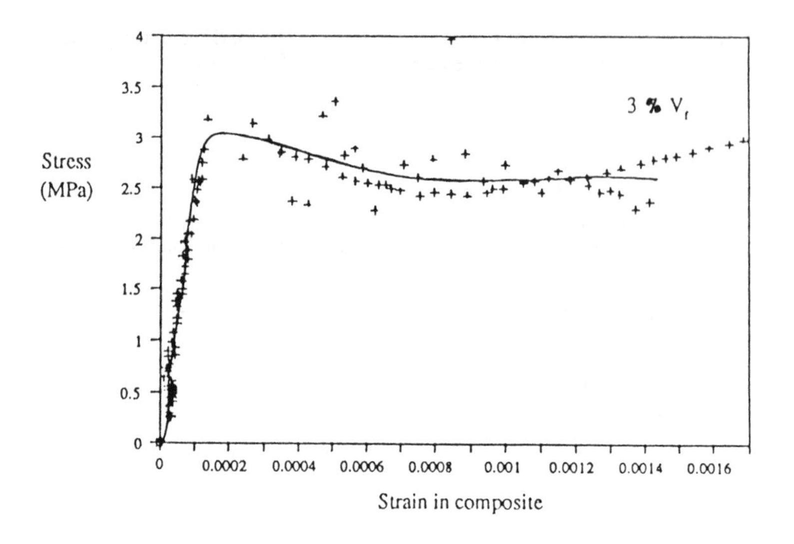

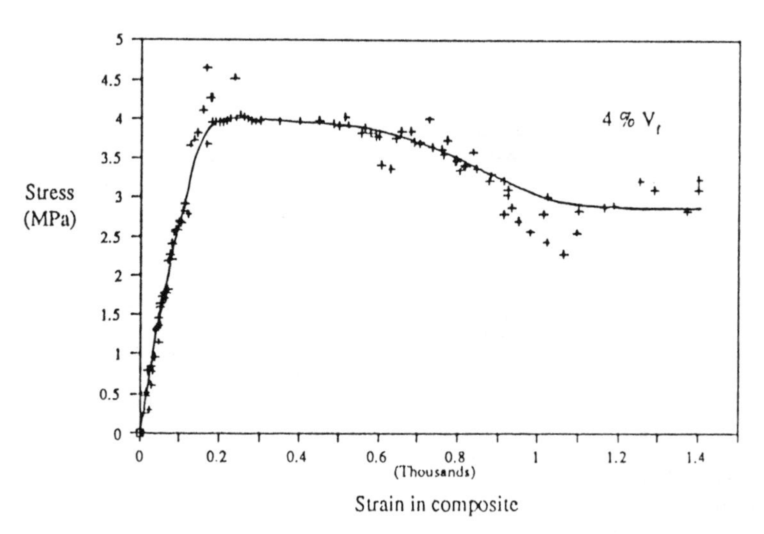

Fig.4. Average tensile stress-strain curves at 3% and 4% V_f.

in the matrix increased until the cracking stress is reached. Beyond the cracking stress, the matrix cracked and no longer carried any stress. In the case of SFRC specimens, the composite continued to carry a certain amount of stress even after cracking of the matrix has occurred. The behaviour of SFRC specimens in double tensile test in the present study may be clearly categorized into three stages. In the first stage, a linear elastic response was observed till the matrix cracked. In this first elastic stage of response, the slope of the curve may be defined by the law of mixture. As soon as cracking occurred in the composite, in the second stage of response, the load is transferred to the fibers and a gradual drop in stress is observed along with large increase in the composite's strain compared with the first stage of response. The gradual drop in stress continued until sufficiently large strains in the composite was attained and the stress dropped to a constant level. It should be noted that for lower fiber volumes, typically below or close to 1%, a sudden drop in stress is reported by many researchers [Swamy and Sa'ad (1981), Lim et al. (1987), Uomoto and Weeraratnee (1986) and Agustin (1989)]. However, in the present study, at high fiber volumes, a similar trend has not been observed. Instead the slope of the curve in the second stage was found to be milder indicating a slowly progressing matrix cracking phenomenon as evidensed by the non-linear portion of the curve at the junction of the first and second stage of the response, resulting in a gradual transfer of load from the matrix to the fibers. The observation which emerges is the fact that increase in the fiber content did not result in increase in F_{ct} (post- cracking strength, at the level of constant stress in the composite) in linear proportion, as Fig.3 and 4 clearly show that doubling the V_f from 2 to 4% did not result in twice the F_{ct} value at 2 %. Thus, it can be concluded that the relationship between fiber content and F_{ct} is not linear. It is generally believed that at lower fiber volumes typically below 1%, this relationship is almost a straight line, but with increasing fiber volume the signs of non-linearity begin to show. Kobayashi et al. (1977) reported the same effect and attributed the onset of non-linearity to increased fiber volume or aspect ratio.

Within the scope of this study, an attempt is made to quantify the non-linear behaviour of SFRC at higher fiber content.

3.2 Effect on compressive strength

Several studies have reported a marginal increase in the compressive strength of concrete with the inclusion of fibers [Swamy and Sa'ad (1981), Agustin (1989), Ramachandran et al. (1981), Naaman (1985) and Edgington (1978)]. Pakotiprapha (1973) reports the compressive strength of SFRC to be even lower than that of plain concrete. Within the scope of this study, from Table 2 it can be seen that compressive specimens tested at various fiber volumes displayed the compressive strength to be about the same as that of plain concrete even upto 4% V_f. A similar observation has also been reported by Sabapathi et.al. [21]. However, a distinct change of failure pattern of the specimens was observed. Fig. 5 shows large increase in ductility with increasing fiber volumes. It

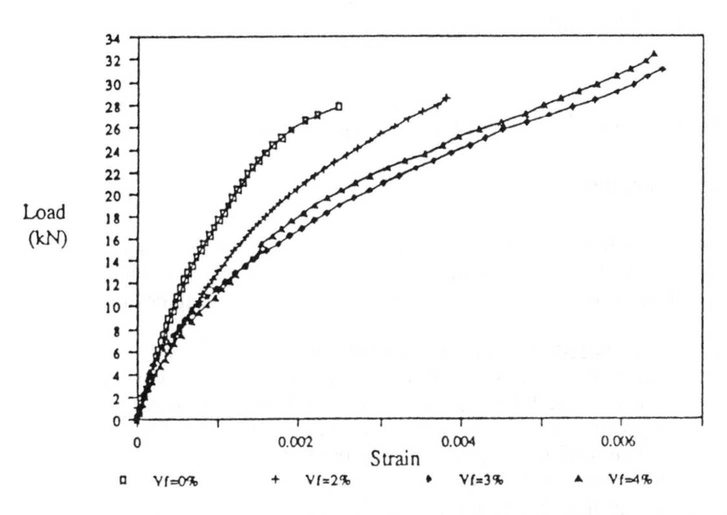

Fig.5. Comparative plots of stress- strain behaviour in compression at varying V_f.

Table 2. Details and designation of beam specimen

Series	a/d	Observed strength (MPa)			Ult. failure load (kN)	Failure mode	Specimen designation	Beam series
		f'_c	F_{ct}	σ_b				
V_f=0%	1.25	32.25	--	4.84	247.80	M/S	A1	Aa
					235.44	M/S	A1A	
					244.56	F	A2	Ab
					245.25	F	A2A	
					206.60 *	F	A3	
	3.5	32.25	--	4.84	87.99	F	A4	Ac
					78.48	F	A4A	
					82.99	F	A2A	
					79.46	F	A5A	
V_f=2%	1.25	31.95	2.22	7.66	271.93	F	B1	Ba
					259.40	F	B2	Bb
					235.24 *	F	B3	
	3.50	31.95	2.22	7.66	94.86	F	B4	Bc
					96.43	F	B5	
V_f=3%	1.25	32.56	2.59	9.36	289.10	F	C1	Ca
					196.20 *	F	C2	Cb
					278.60	F	C3	
	3.5	32.56	2.59	9.36	103.00	F	C4	Cc
					100.06	F	C5	
V_f=4%	1.25	32.55	3.07	9.94	264.87	F	D1	Da
					255.06	F	D2	Db
					259.97	S	D3	
	3.50	32.55	3.07	9.94	110.85	F	D4	Dc
					105.90	F	D5	

*Defective specimen

should be noted that no noticeable increase in the peak stress was observed indicating that actually a decrease in compressive stiffness has occurred. The large strains indicate that the inclusion of fibers results in general improvement in compressive toughness.

3.3 Effect on flexural strength and toughness

For plain concrete specimens failure took place without any warning, thus indicating a brittle failure as expected. SFRC specimens exhibited a slow ductile mode of failure without any sign of crushing or spalling of concrete at the top region of the prism in compression. The load-deflection curve remained linear upto failure for 0% V_f specimens. In SFRC prisms a ductile behaviour was observed. Close inspection of the crack at failure revealed that failure took place by cracking in the matrix followed by the gradual pullout and debonding of the fibers rather than fiber fracture. Thus, large

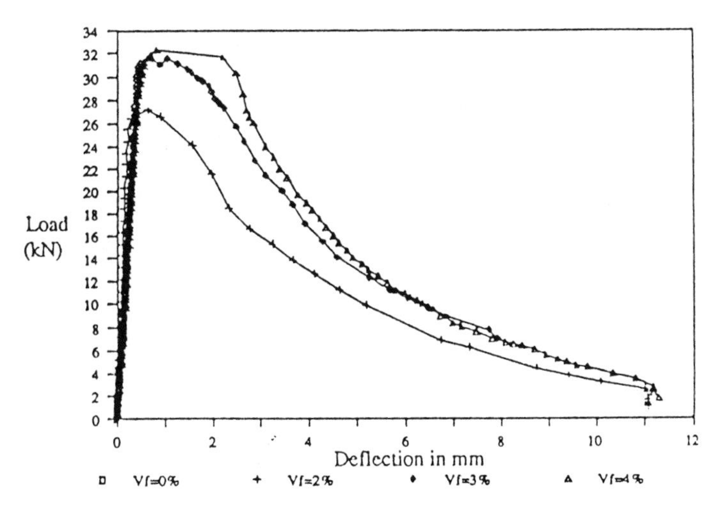

Fig.6. Comparative plots of load-deflection curves at varying V_f.

ductility may be attributed to large energy absorption in debonding and stretching of the fibers. Table 2 shows the observed values of flexural strength of the prisms at varying V_f. However, it can be seen that the rate of increase in flexural strength with increasing V_f decreased at higher fiber volumes. From Fig.6 it is obvious that the rate of increase in the flexural strength is not substantial as V_f increased from 3% to 4%. Once again, the reduction in the rate of increase of flexural strength with increasing V_f may be due to the problems of poor compaction and weakening effect as cited before. It can be observed that the area under the load-deflection curve increased substantially with increasing V_f thus indicating a greater improvement in flexural toughness by the inclusion of fibers. The present study attempts to form a linear relationship between F_{ct} and flexural strength σ_b.

4 Behaviour of SFRC beams at higher fiber volumes in flexure

A total of 24 beams were tested in the experimental program under simply supported condition subjected to third point loading. The location of strain and dial gages for the

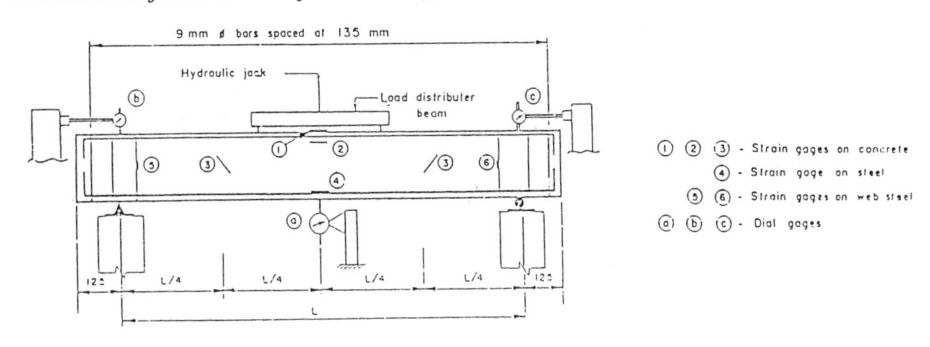

Fig.7. Test setup location of recording instruments on test beams.

test beams are shown in Fig.7. All data was obtained using a portable automatic data aquisition system TDS-301 hooked to a micro-computer.

4.1 Strength of beams and failure modes

Table 2 shows the failure loads and failure modes of the test beams. It has been observed that all the beams except D3 , failed in flexure irrespective of a/d ratio as flexural failure is independent of this parameter. The mode of failure for plain beams A1 and A1A was in moment-shear, thus, the failure was initiated with the yielding of steel but before the ultimate strain in the concrete in compression was reached, a sudden shear failure occurred. This indicated that the moment and shear strengths of the members were close enough to initiate M-S type of failure. However, Table 2 indicates that for SFRC beams of fiber content 2% onwards, beams without web reinforcement failed in flexure, indicating that the fibers significantly increased the shear strength to change the mode of M-S failure to flexure failure. It is observed that specimen D3 having 4 % fiber volume abnormally failed in shear demonstrating diagonal tension failure. The failure of beam D3 in shear probably occurred as the test specimen showed large voids on its surface, indicating a probable presence of similarly large voids within the internal structure.

Table 3 shows the percentage increase in flexural strength the increasing fiber percentage. Although, the increase of 11% in flexural strength at 2 % V_f as compared with that of plain concrete is not very substantial, an increase of 18% and 27% for 3 and

Table 3. Percentage increase in flexural strength

V_f (%)	M_u (kN-m)	% increase in M_u
0	26.88	
2	30.00	11.64
3	31.95	18.86
4	29.38	9.3
	34.15	27

4% fiber volumes may be regarded as substantial. The increase in the yield moment of the SFRC beams was also observed indicating the effectiveness of fibers in sharing the tensile stresses with conventional bar reinforcement.

The wide cracks at failure showed fiber pullout across the cracks. No sign of fiber fracture was observed. For all SFRC beams, at ultimate load, the inclusion of fibers in compression zone prevented disintegration of the compression concrete. Unlike ordinary concrete, there was no breaking up of the compression concrete and no falling of debris.

4.2 Load-deflection curves

Fig.8 shows the ultimate deflections of SFRC beams to be substantially large as compared with those of plain concrete beams. The increase in ultimate deflections varied from 2 times for series B beams having 2% V_f to almost 3 times for series D beams having 4% fiber volume. The load-deflection curves indicate that SFRC beams possess enormous ductility. Thus, it can be concluded that fiber reinforced beams have higher toughness. It can be seen that the fibers are able to reduce the deflections at initial stages of loading and have increased the deflection towards failure, thus exhibiting

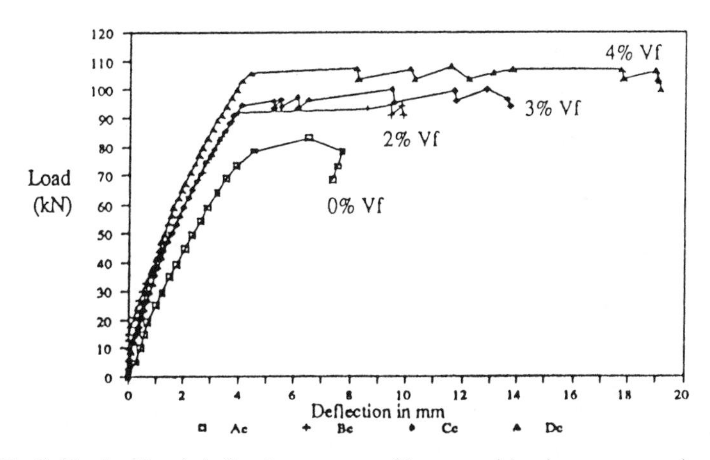

Fig.8. Typical load-deflection curves of beams with stirrups at varying V_f.

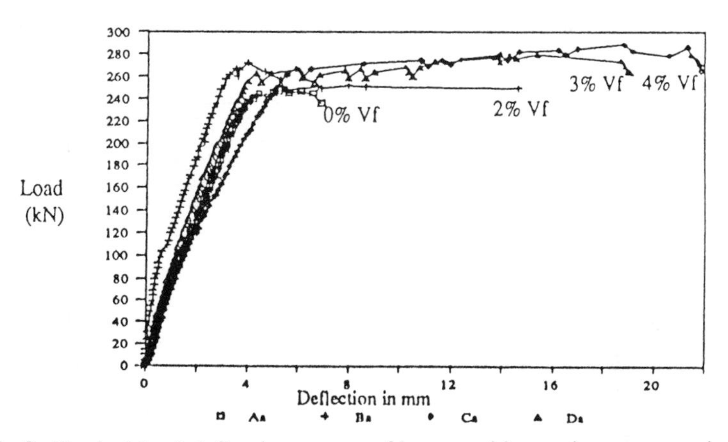

Fig.9. Typical load-deflection curves of beams without stirrups at varying V_f.

beneficial behaviour both at service conditions and ultimate state.

Considering the first crack to occur at the load at which load-deflection curve deviates from linearity [Swamy and Sa'ad (1981)] , Figs.8 and 9 show no marked difference in the values of first crack strength at varying fiber volumes. This observation is in accordance with the observations made by the other researchers [Rajagopalan et al. (1974), Shah (1987)].

4.3 Load-strain curves

Figs. 10 to 11 show typical load-strain curves of test beams for longitudinal reinforcement. The plot of stress-strain indicates that the inclusion of the fibers lead to reduced strains in longitudinal reinforcement in comparison with plain concrete beams for the same level of loading. At 3 % V_f the lowering of strain was more as compared to 2% Vf beams. Similarly for 4% V_f beams the lowering of strain was greater as compared with 3% V_f beams. This shows the effectiveness of fibrous concrete in sharing the tensile load with the main steel of the beam. This may be regarded as a beneficial effect in controlling crackwidths as they are directly related to steel stresses.

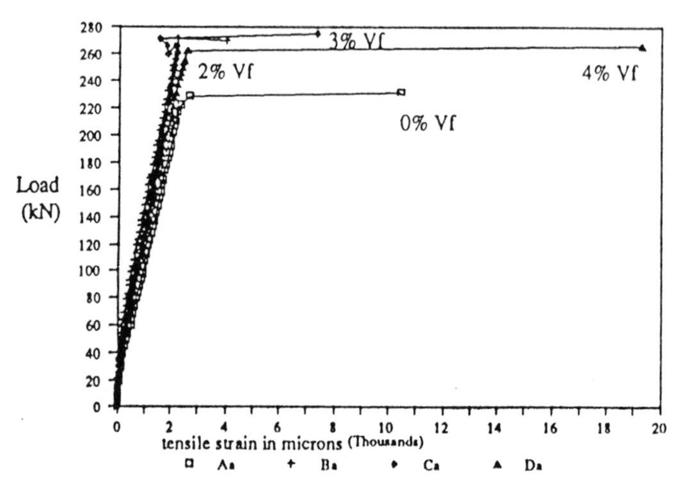

Fig.10. Typical load-main steel strain curves of beams without stirrups at varying V_f.

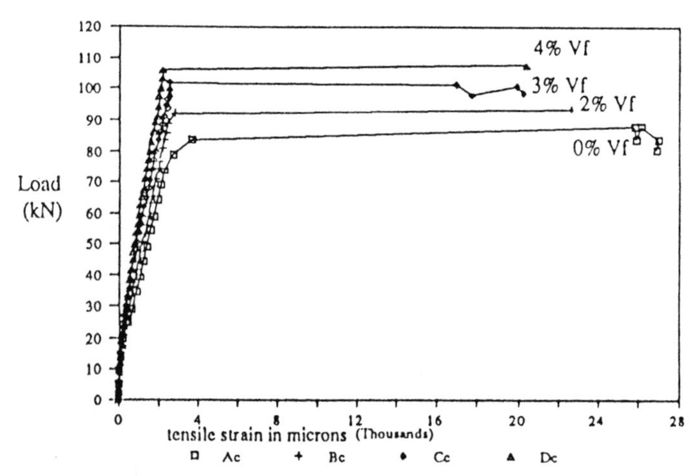

Fig.11. Typical load-main steel strain curves of beams with stirrups at varying V_f.

A slight kink in the elastic range of the curves (Figs.8-12) show the cracking of the matrix. Once again it is obvious that the deviation of load-strain curve from initial tangent for different fiber volumes occurs at the same load level as plain concrete, confirming the observation made from the load-deflection curves that cracking load is insensitive to the variations in fiber volume.

Load-compression strain curves for the test beams are presented in Figs.12. This shows that the strains corresponding to the same load are higher in SFRC beam series Ba, Ca, and Da as compared to plain concrete beams. The increased strains show a higher degree of compressibility and thus shows a greater compressive toughness and ability to undergo plastic deformation.

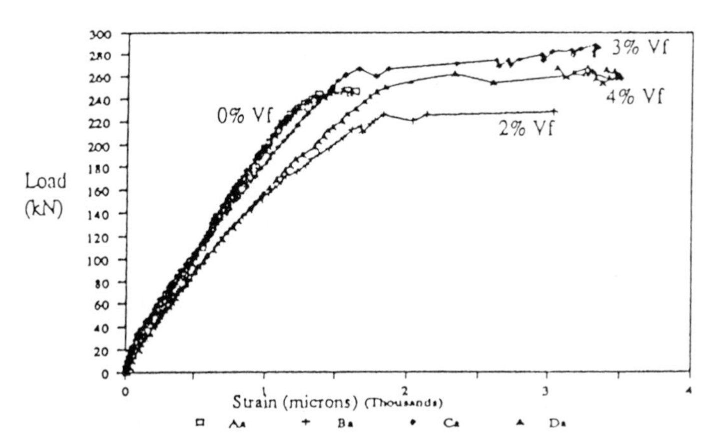

Fig.12. Typical load-compressive strain at extreme fiber curves of beams without stirrups at varying V_f.

5 Theoretical considerations

5.1 Idealization of tensile stress block for concrete

Based upon the observations in the study, a three stage idealized stress-strain relationship is adopted for the stress block in tension as compared with the popular idealization of rectangular stress block for low fiber volumes [Lim et al. (1987), Uomoto and Weeraratnee (1986), Agustin (1989) and Pakotiprapha et al. (1983)]. The adopted tensile stress block is shown in Fig.13.

Based on the law of mixture, neglecting the contribution of the matrix in carrying any stress and applying the corrections for orientation, bond efficiency, and length efficiency factor, the most common expression in use is given by,

$$F_{ct} = \eta_o \eta_l \eta_b V_f 2\, \tau_f \frac{l_f}{d_f} \tag{1.1}$$

In this study equation (1.1) has been modified to account for non-linear behaviour of the composite. Data from the present study as well as data from other references

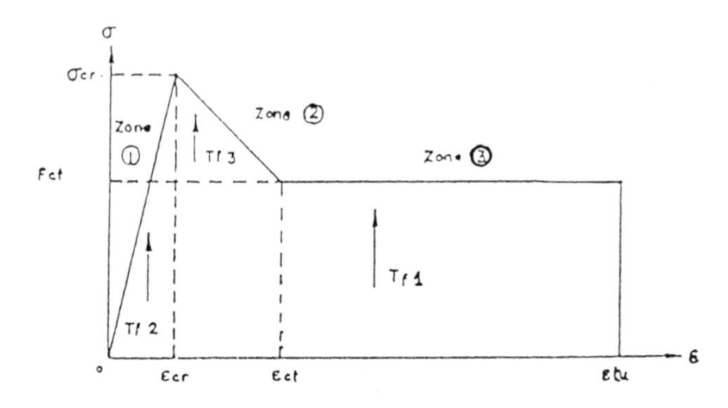

Fig.13. Proposed idealized tensile stress-strain relationship for SFRC at higher fiber volumes in concrete.

[Nathan et al. (1977), Agustin (1989) and Mansur et al. (1986)] etc. have been used to evaluate the phenomenological effects which contributes towards non-linear behaviour. Treating the extensive data with non-linear regression analysis equation (1.1) has been modified to,

$$F_{ct} = 0.4430 \left| \eta_o \eta_l \eta_b 2 \tau_f \frac{l_f}{d_f} V_f \right|^a \quad (1.2)$$

Existing expressions for " η_0 ", " η_1 " and " η_b " derived elsewhere [Pakotiprapha (1973), Pakotiprapha et al. (1983) and Mansur et al. (1986)] may be used with equation 1.2, where

a = phenomenological factor having a value 0.796.

The statistical co-efficient R^2 of 0.984 means a very strong correlation.

5.2 *Fct* and flexural strength relationship

When the data of flexural strength from the present study along with a large set of data from other sources [Uomoto and Weeraratnee (1986) andMansur et al. (1986)] was plotted against F_{ct}, a strong linear relationship emerged. Fig.14 shows a linear relationship between F_{ct} and σ_b. Using linear regression analysis a simplified equation for F_{ct} in terms of flexural strength σ_b is obtained by the line of best fit,

$$F_{ct} = 0.3753\, \sigma_b - 0.806 \quad (1.3)$$

Using equation (1.3) F_{ct} can be predicted by simply testing 100 mmx100 m x500 mm standard prisms. Table 5 shows the comparative results of the conventional and the proposed equations.

5.3 Development of an idealized compressive stress block

Stress-strain relationship for compression as proposed by Sabapathi and Achyutha (1989) has been adopted with minor modifications to formulate an equivalent compressive stress block. Thus for SFRC,

$$f = \frac{A\varepsilon}{\left(1 + B\varepsilon + C\varepsilon^2\right)} \quad (1.4)$$

where A, B, and C are the constants which are to be computed using the boundary conditions. Sabapathi and Achyutha (1989) gives the boundary conditions to be satisfied:

i) $\varepsilon = 0$, $f = 0$ ii) $\varepsilon = 0$, $df / d\varepsilon = E_c$

iii) $\varepsilon = \varepsilon_o$, $f = f'_c$ iv) $\varepsilon = \varepsilon_{0.85}$, $f = 0.85 f'_c$

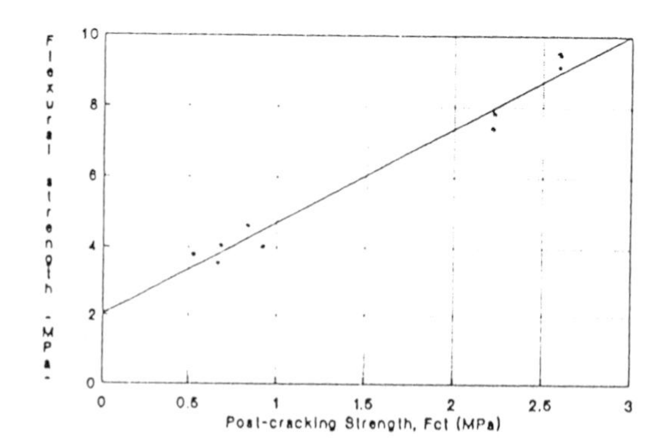

Fig.14. Linear relationship between F_{ct} and flexural strength of points.

Table 4. Comparison of observed and computed flexural strengths using observed and computed F_{ct} values.

V_f (%)	a/d	Beam series	Ult. flexural strength (kN-m) M_u	M_u^*	M_u'	Yield strength, M_y (kN-m) Observed	Computed	M_y Computed / Observed	M_u^* / M_u	M_u' / M_u
0	1.25	Aa, Ab	27.73	26.40	--	--	--		0.952	--
	3.50	Ac	25.90						0.978	
2	1.25	Ba, Bb	29.88	30.90	30.10	26.69	27.66	1.036	1.034	1.007
	3.50	Bc	30.13			27.92		.990	1.025	0.999
3	1.25	Ca, Cb	31.93	31.74	31.58	29.31	28.98	.989	0.994	0.989
	3.50	Cc	31.97			28.65		1.011	0.993	0.987
4	1.25	Da, Db	29.25	32.82	32.82	27.90	29.26	1.048	1.117	1.122
	3.50	Dc	34.15			31.76		.921	0.956	0.956

M_u= observed, M_u^*=from proposed model with observed F_{ct}.
M_u'=using proposed model with computed F_{ct} by Eq. (1.2)

Table 5. Comparative results of observed and computed values of post-cracking strength (F_{ct})

Reference	V_f	Observed (1)	Eq. (1.1) (2)	Eq. (1.2) (3)	Eq. (1.3) (4)	(2)/(1)	(3)/(1)	(4)/(1)
Mansur et al. (1986)	0.50	0.53	0.369	0.481	0.612	0.696	0.908	1.155
	0.75	0.68	0.554	0.664	0.706	0.815	0.976	1.038
	1.0	0.83	0.738	0.835	0.913	0.889	1.006	1.100
Agustin (1989)	1.0	0.915	0.949	1.075	--	1.037	1.17	--
	2.0	1.415	1.887	1.866	--	1.340	1.32	--
Nathan et al. (1977)	2.0	1.792	1.994	1.959	--	1.113	1.093	--
	3.0	2.608	2.991	2.706	--	1.147	1.037	--
	4.0	3.850	9.980	3.400	--	1.034	0.883	--
Present study	2.0	2.220	1.845	1.813	2.070	0.831	0.816	0.932
	3.0	2.590	2.770	2.504	2.700	1.070	0.967	1.042
	4.0	3.070	3.690	3.150	2.842	1.202	1.026	0.925

*All results in N/mm^2

$$\varepsilon_o = \left(\left|10.3\frac{l_f}{d_f} + 492.0\,V_f - 5.3\frac{l_f}{d_f}V_f - 560.0\right|\right)10^{-5} \tag{1.5}$$

$$\varepsilon_{0.85} = \frac{5.0}{\sqrt{f_c'}}\left(492 + 12.4\frac{l_f}{d_f}V_f - 8.0\frac{l_f}{d_f} - 380.0\,V_f\right)10^{-5} \tag{1.6}$$

Based on equation (1.4), the parameters of the equivalent compressive stress block are as given as,

$$k_1 = \frac{\frac{1}{\varepsilon_{cu}}\frac{A}{C}\left\{\left[\frac{1}{2}\ln\left|\frac{(\varepsilon_{cu}+a)^2+b^2}{(a^2+b^2)}\right|\right]-\frac{a}{b}\left[\tan^{-1}\left(\frac{\varepsilon_{cu}+a}{b}\right)-\tan^{-1}\left(\frac{a}{b}\right)\right]\right\}b_1}{\beta_1 f_c} \tag{1.7}$$

$$\beta_1 = 2.\left[1-\frac{\frac{1}{\varepsilon_{cu}}\left[(\varepsilon_{cu}+a)-a\ln\left|\frac{(\varepsilon_{cu}+a)+b^2}{b^2}\right|+\frac{(a^2-b^2)}{b}\tan^{-1}\left(\frac{\varepsilon_{cu}+a}{b}\right)\right]}{\left\{\frac{1}{2}\ln\left|\frac{(\varepsilon_{cu}+a)^2+b^2}{(a^2+b^2)}\right|-\frac{a}{b}\left[\tan^{-1}\left(\frac{\varepsilon_{cu}+a}{b}\right)-\tan^{-1}\left(\frac{a}{b}\right)\right]\right\}}\right] \tag{1.8}$$

$$\frac{B}{C}=\alpha \qquad \frac{1}{C}=\beta \qquad and \qquad a=\frac{\alpha}{2} \qquad b=\sqrt{\left(\beta-\frac{\alpha^2}{4}\right)}$$

Using the expressions for k_1 & β_1, the equivalent compressive stress block taking into account the large ductility of SFRC in compression can be constructed.

With the completion of compressive & tensile stress blocks, the process for flexural analysis of the beams can be carried out using the principles of force equilibrium and strain compatibility. The general expression will be of the following form,

$$M=C_c(d-\bar{Y})-T_{f1}\left(\frac{X_1}{2}-d\right)-T_{f2}\left[(X_1-d)-\frac{X_2}{2}\right]-T_{f3}\left[(X_1-d)+\frac{X_3}{3}\right]+C_s(d-d') \tag{1.9}$$

The compressive and tensile forces and their respective distances from the neutral axis are illustrated in Fig.15.

6. Conclusions

(a) The properties of hardened concrete vary with the inclusion of the steel fibers.
(b) The behavior of SFRC beams in flexure at higher fiber volumes follows similar trends as observed for lower fiber volume fractions, provided it is ensured that the difficulties of clustering , balling and compaction at mixing and placing of the mix are effectively controlled. If proper compaction could not be achieved then it is expected that the strength would progressively reduce after showing an increase at lower fiber volumes, or members may fail prematurely in shear before developing their full flexural capacity.
(c) The effect of steel fibers in increasing the ultimate strength of beams having conventional bar reinforcement is found to be limited upto fiber volumes as high as 2 %. Still higher fiber volumes results in reasonable increase in flexural strength although it may not be most economical. Thus, the inclusion of fibers may be intended basically to improve the shear strength when conventional bar reinforcement is used and the increase in flexural strength may be regarded as a secondary reinforcement.
(d) Fiber concrete beams show significant inelastic deformations and substantially high ductility at failure. Also, the fibers form an effective crack arrest mechanism, thus the fibers may be incorporated for controlling cracking and deformations and for higher ductility and toughness at all stages of loading rather than with the main objective of substantial gain in flexural strength.

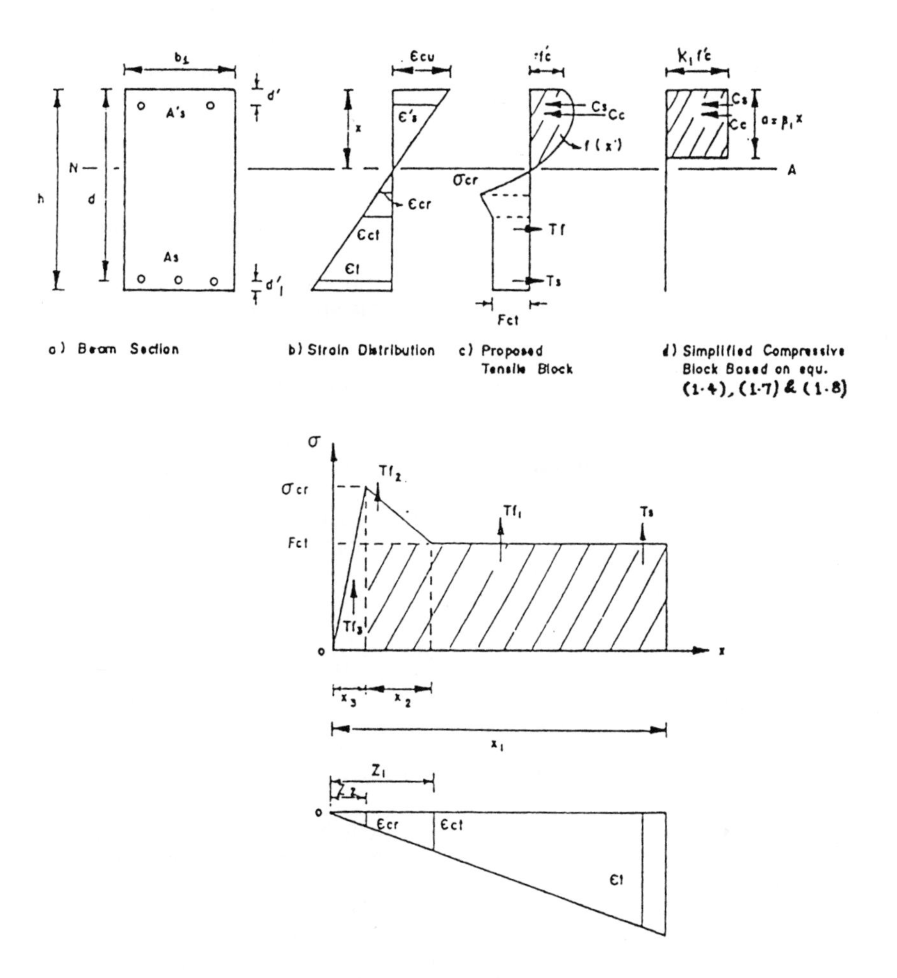

Fig. 15 Forces and strain distribution in SFRC beam X-section using proposed stress blocks.

(e) The most important parameter influencing the flexural strength of SFRC beams is found to be post- cracking strength F_{ct}. An improved empirical equation is proposed to predict the post-cracking strength within the scope of this study taking into account the non-linearity of the relationship at higher V_f. Furthermore, a linear relationship is established linking F_{ct} to flexural strength of SFRC prisms as an alternative to the conventional approach.

(f) Using the principles of force equilibrium and strain compatibility, the flexural strength of SFRC beams may be predicted using the above equations and SFRC beams may be analyzed for any given loading condition.

7 References

Agustin, R. (1989) **Shear Behaviour of Fiber Reinforced Concrete Beams**, M. Eng. Thesis, Asian Institute of Technology Bangkok, Thailand.

Al-Ta'an S. A. and Al-Feel, J. R. (1986) Role of steel fiber reinforcement in flexural failure, **Development in Fiber reinforced Cements and Concretes**, FRC-86, Rilem Symposium, England, Vol.2., pp.10

Bentur, A. Gray, R. J. and Mindess, S. (1986) Cracking and pullout, processes in fiber reinforced cementitious materials, **Developments in Fiber Reinforced Cement and Concrete, Rilem Symposium**, England, pp.7

Edgington, J. Hannat, D. J. and Williams, R.I.T. (1978) Steel fibre reinforced concrete.**Fibre Reinforced Materials**, BRE Building Research Series, Vol. 2. pp.112-128.

Guan Liqui and Zhao Gaufan (1986) A study of the mechanism of fibre reinforcment in short steel fibrous Concrete, **Developments in Fiber Reinforced Cement and Concrete, Rilem Symposium,** England, pp.10.

Henager, C. H. (1980) Steel fibrous concrete-a review of testing procedures, fibrous concrete. **Concrete International**, 7(3)16-28.

Henager, C.H. and Doherty, T.J. (1976) Analysis of Reinforced Fibrous Concrete Beams. **Journal of the Structural Division**, ASCE Vol. 102, No. ST.1, 177-188.

Hughes, B.P. and Fattuhi, N.I. (1976) The workability of steel fibre reinforced concrete. **Magazine of Concrete Research**, 28(96), 157-161.

Josifek, C. and Lankard, D.R. (1987) SIFCON: Slurry Infiltrated Fibrous Concrete, **Proceedings of the International Symposium on Fiber Rein forced Concrete**, Madras, pp. 7.15-7.23.

Kazusuke, Kobayashi and Ryo, Uche Cho (1977) Strength and deformation of steel fiber reinforced concrete in uniaxial tension (in Japanese), **Proceedings of the Japan Society of Civil Engineers**, 527, pp 22-27.

Kukreja, C. B. Kaushik, S. K. Kanchi, M. B. and Jain, O. P. (1980) Tensile strength of steel reinforced concrete. **Indian Concrete Journal**, 54(7), 184-188.

Lim, T. Y. Paramasivam, P. and Lee, S. L. (1987) Shear and moment capacity of reinforced steel fibre concrete beams. **Magazine of Concrete Research**, 39(140), pp.148-160.

Mansur, M. A. Ong, K.C.G. and Paramasivam, P. (1986) Shear strength of fibrous concrete beams without stirrups, **Journal of Structural Engineering**, ASCE, 112(9), pp.2066-2079.

Naaman, A. E. (1985) Fiber reinforcement for concrete. **Concrete International**, 7 (3), pp. 21- 25.

Nathan, G.K. Pramasivam, P. and Lee, S.L. (1977) Tensile behavior of fiber reinforced cement paste. **Journal of Ferrocement**, 7(2), 59-79.

Pakotiprapha, B. (1973) **Mechanical Properties of Cement Mortar with Randomly Oriented Short Steel Fibers**, M. Eng. Thesis, Asian Institute of Technology,

Thailand.
Pakotiprapha, B. Pama, R.P. and Lee, S.L. Analysis of a bambo fiber-cement paste composite. **Journal of Ferrocement**, 1983. Vol.13, No.2. pp.141-159.
Rajagopalan, K. Parameswaran V. S. and Ramaswamy, G. S. (1974) Strength of steel fiber reinforced concrete beams. **Indian Concrete Journal**, 48(1), 17-25.
Ramachandran, V.S. R.F. Field, Beavdoin, J.J. (1981) **Concrete Science**, Heydon & Son Ltd. pp. 169-223.
Sabapathi, P. and Achyutha, H. (1989) Stress - strain characteristics of steel fibre rein forced concrete in compression. **Indian Concrete Journal**, 1989. Vol. 69, PTCI 4. pp.257-261.
Shah, S.P. and Rangan, B.V. (1970) Effects of reinforcement on ductility of concrete, **Proceedings of ASCE**, Vol.96, No.ST.6, pp.1167-1184.
Shah, Surendra P. (1987) Strength evaluation and failure mechanisms of fiber rein forced concrete, **Proceedings of the International Symposium on Fiber Reinforced Concrete**, , Madras, India. pp. 1.3-1.19.
Swamy, R. N. and Sa'ad A. Al-Taan (1981) Deformation and ultimate strength in flexure of reinforced concrete beams made with steel fiber concrete, **ACI Journal**, Technical Paper, Title No. 78-36, 395-405.
Uomoto, T. and Weeraratnee, R. K. (1986) Flexural and shear capacities of reinforced concrete beams using steel fiber reinforced concrete, **The First East Asian Conference on Structural Engineering and Construction**, Bangkok, pp. 634-645.
Weeraratnee, R.K. (1985) Shear Behaviour of Singly Reinforced concrete Beams Cast with Fiber Concretes, Phd. Dissertation, University of Tokyo, Japan, pp.162.

39 POLYPROPYLENE FIBRE CONCRETE BEAMS IN FLEXURE

S. GHOSH and A. ROY
Department of Civil Engineering, Jadavpur University, Calcutta, India

Abstract
Twenty PFRC beams were tested to destruction to study their overall behaviour in pure flexure. Some inferences are made on flexural capacity, crack arresting property, overall flexural stiffness etc. of PFRC beams. An expression of ultimate moment of resistance has been developed, where different stress block and post yield parameters are considered which are established from the analysis of experimental data and previous references. Theoretically predicted ultimate moment of resistance values are compared with experimental results and good agreements have been observed.
Keywords: Concrete, Beam, Flexure, Fibre, Composite, Cement.

1 Introduction

The idea of using fibre to reinforce weak and brittle matrices is not new and it exists in early history of civilization. The use of straw in sun-dried mud blocks and horse hair in gypsum plaster has been known since ancient times. The present development of fibre reinforced cement concrete is only about three decades old. The fibres used are asbestos, carbon, jute, glass, nylon, polypropylene, polythene, steel etc. The selection of such fibres are primarily based on their availability, basic properties, compatibility with cement, cost effectiveness and durability of fibres in the composites.

Fibres may be used either in long aligned wire forms or as short discrete randomly distributed fibres. The use of aligned continuous straight fibres is as old as reinforced concrete and the properties of such composite material are well established. In the current development of fibre cement composites the use of short discrete fibres offers more exciting prospect. Fibre reinforced concrete mix with fibre reinforcement in the form of short discontinuous discrete fibres which act effectively as a rigid inclusion in the concrete matrix and shows substantial improvements in the static and dynamic strength properties. Polypropylene fibre

Fibre Reinforced Cement and Concrete. Edited by R. N. Swamy. © 1992 RILEM.
Published by E & FN Spon, 2-6 Boundary Row, London SE1 8HN. ISBN 0 419 18130 X.

is found to be a most-effective synthetic fibre which can be used with cement concrete.

For the last twenty years researchers have given attention to develop steel fibre reinforced concrete. The design procedures have been developed by considering the additional tensile resistance available from steel fibres. The design procedure explained in ACI Special Publications SP44, SP8 and SP105. The guidance regarding proportion and testing workability of steel fibre reinforced concrete has been given in ACI-544-3R. The main parameters which effect the structural behaviour are (a) Aspect ratio, (b) Shape and (c) Orientation of fibres. The volume of steel fibres generally lies between 0.5 to 1.5% of total volume of concrete. By various tests it has been observed that the increase in compressive strength of steel fibre reinforced concrete varies between 0 to 23% compared to conventional Cement Concrete. Due to large frictional force and development of large fibre bending energy during fibre pullout more number of narrower cracks appear and eventually give more toughness. The shear compressive, flexural and tensile capacities increase approximately 3 to 10 times as compared to the conventional concrete due to addition of fibre in concrete. The total absorption of energy could be increased by 40 to 100 times of any unreinforced beam depending on different other parameters. The increase in shear strength has been observed about 20 to 150% with steel fibres having deformed and crimped end.

Recently the adoption of carbon fibre in case of some light-weight concrete shows low shrinkage, more resistance against freezing and thawing and more durable in hot environment. The length and diameter usually adopted lies between 3 to 10 mm and 15 to 20 micrometers respectively.

Asbestos fibre may give pollution problem in the surrounding atmosphere. Glass fibre may give deterioration at early stage in highly alkaline atmosphere. Natural fibre may give decay. Considering the above points and also cost benefit aspect more attention has been put in development of synthetic fibres reinforced concrete. The present investigation is primarily intended to understand the behaviour of polypropylene fibre reinforced concrete beams in pure bending. Cube test, indirect tensile test i.e. split cylinder test and flexural (prism) tests have also been performed. A theoretical analysis has been developed to predict ultimate moment of resistance of the section. The theoretical values are compared with experimental results and a good agreement has been observed.

Polypropylene fibres are synthetic fibres having isotactic configuration and circular cross-section and available in two forms, monofilaments fibres and film fibres. The raw material polypropylene derived from the monomeric C_3H_6 is a pure hydrocarbon.

The manufacturing process is similar to the production of nylon and rayon and by extrusion of synthetic polymers into fibres by spinneret. The extruder is fitted with a die to produce a tubular or flat film which is then slit into tapes and monoaxially stretched. Due to stretching a molecular orientation takes place which gives the higher tensile strength of fibres.

The types of fibre are charactered by expressing as length in metres/kg or by old textile designation of the denier, i.e. the weight in grams of 9,000 m of yarn. In general the fibres are supplied in spool form and generally chopped in the length of 25 mm and 75 mm.

The basic characteristics of the polypropylene fibres may be mentioned as follows

a. High melting point (approximately 165°C), which allows to use concrete at higher temperature.
b. Resistant to most chemicals due to its inertness.
c. The water demand for the fibres is nil due to its hydrophobic surface.
d. Higher tensile strength.

There are, of course, certain disadvantages of these fibres when the same shall be used in concrete. Those are :

a. In case of fire the additional porosity shall be obtained due to its combustibility. The porosity will be equal to the volume of fibre in concrete.
b. For monofilament fibres a poor bond strength with concrete.
c. Sunlight and oxygen may affect the quality. Incorporation of pigment and stabilizers would provide the resistance.

2 Theoretical expression to predict ultimate moment of resistance

Following assumptions were made for developing the ultimate moment capacity of a PFRC beam section.

a. Plane section remain plane after bending.
b. Tensile strength of fibre concrete has been taken into account.
c. Maximum strain at extreme fibre in compression is 0.0035.
d. Maximum strain in steel $(f_y/E_s) + 0.002$.
e. The maximum stress in concrete 0.68 times the characteristic compressive strength of concrete (f_{ck}).
f. The maximum stress in tension reinforcement is taken as the stress corresponding 0.2% of proof stress/yield stress.

The strain and stress diagrams are shown in Fig. 1.

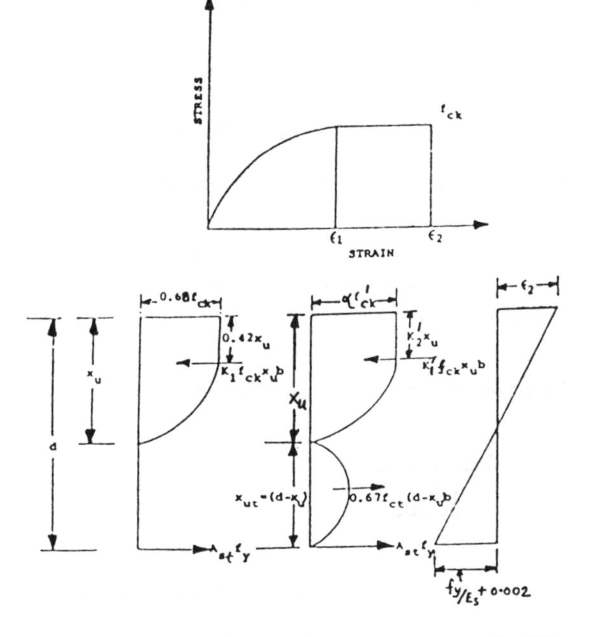

Fig. 1. Stress and strain pattern of fibre concrete.

The area of compression stress block

$$= \frac{2}{3}\left(\frac{\epsilon_1}{\epsilon_2}\right)(x_u)(\alpha f'_{ck}) + \frac{(\epsilon_2 - \epsilon_1)}{\epsilon_2}(x_u)(\alpha f'_{ck})$$

$$= K'_1 f'_{ck} x_u$$

Depth of c.g. of stress block from extreme compressive fibre :

$$\frac{\left(\frac{2}{3}\right)\left(\frac{\epsilon_1}{\epsilon_2}\right)(x_u)(\alpha f'_{ck})\left(x_u - \frac{5}{8}\frac{\epsilon_1}{\epsilon_2}x_u\right) + \left(\frac{1}{2}\right)\left(\frac{\epsilon_1-\epsilon_2}{\epsilon_2}\right)^2 (x_u)(\alpha f'_{ck})}{\left(\frac{2}{3}\right)\left(\frac{\epsilon_1}{\epsilon_2}\right)(x_u)(\alpha f'_{ck}) + \left(\frac{\epsilon_2-\epsilon_1}{\epsilon_2}\right)(x_u)(\alpha f'_{ck})}$$

Tensile force on tensile steel

$$= A_{st} f_y$$

Tensile force due to fibre concrete below neutral axis

$$= \frac{2}{3} f_{ct} (d - x_u) b$$

Equating compressive and tensile forces,

$$K_1' f_{ck}' x_u b = (A_{st} f_y) + \frac{2}{3} f_{ct} (d - x_u) b$$

$$\text{or,} \quad x_u = \frac{(A_{st} f_y) + (\frac{2}{3} f_{ct} bd)}{\left\{(K_1' f_{ck}') + (\frac{2}{3} f_{ct})\right\} b} \tag{1}$$

From strain diagram critical neutral axis

$$x_{uc} = \frac{\epsilon_2 d}{(\frac{f_y}{E_s}) + 0.002 + \epsilon_2} \tag{2}$$

For under reinforced beams, the ultimate moment

$$M_{uf} = A_{st} f_y (d - K_2' x_u) + \frac{2}{3}(d - x_u)(f_{ct} - b) \left\{ d - \frac{d - x_u}{2} - K_2' x_u \right\} \tag{3}$$

3 Experimental programme

3.1 Casting of specimens

The experimental program was undertaken to test standard cubes, cylinders and prisms for studying strengths in compression, indirect tension and flexure. The samples were prepared without any fibre and with polypropylene fibres of quantity 1 Kg, 2 Kg, 3 Kg and 4 Kg per m^3 of concrete volume. Concrete used are of average grade M15 and M20.

The concrete mix was made using ordinary portland cement conforming to IS Hand Book (SP23), well graded 20 mm down stone chips, 2.36 mm down river sand and potable quality of water. The mixer was a tilting type with capacity $0{\cdot}2\ m^3$. $0{\cdot}1\ m^3$ volume was mixed at a time. The casting and testing of cubes, cylinders and prisms were done following relevant IS codes. Twenty eight days under water curing was done for all the samples. Testing was performed between 1 month to 1½ months after casting. A total of three or more numbers of samples tested and the average values taken for study and presentation. To study the PFRC beams in pure flexure, a total number of twenty beams were casted. In these beams the quantity of polypropylene fibre was varied as 1 Kg, 2 Kg,

3 Kg and 4 Kg per m^3 of concrete volume. The mix proportion was adopted same as in the case of cube, cylinder and prisms. The steels used for flexural reinforcements had yield stress 320 N/mm^2 and modulus of elasticity 2×10^5 N/mm^2. In case of HYSD bars 0.2% proof stress was 410 N/mm^2. In all the cases the specification of polypropylene fibres was 15 denier (15 gms. weight of 9,000 m long fibre). These had specific gravity as 0.9 and tensile strength 400 N/mm^2.

3.2 Testing and test results

The testing of specimens was done in universal testing machines of 100 MT (980 KN) capacity for cubes, cylinders and prisms and 10 MT (98 KN) capacity for beams and the rate of increase of load was 400 Kg (3,923 N) per minute. The increment of load was 4-6 KN for beams tested in pure flexure i.e. under two point symmetrical loading system. The strains along depth were measured at load points and mid span. Sufficient strain and deflections were measured at every stage of loading in order to draw M-Ø diagram of beams and stress-strain diagram of fibre concrete. The tests of cubes, cylinders and prisms were done following IS:516. The average results were taken for each case for study. Table 3 shows the average results from the tests of cubes, cylinders and prisms having no fibre as well as with fibres of quantities 1 Kg, 2 Kg, 3 Kg and 4 Kg per cubic metre of concrete. Detail of test specimen of beams are given in Tables 1 and 2.

3.3 Interpretation of test results

Some inferences can be made from the test results of compression, indirect tensile and prism test.

a. Increase of compressive strength is maximum (46.19%) for concrete with 2 Kg/m^3 fibres as compared to the plain concrete.
b. Increase of compressive strength reduces in case of the higher volume of fibre.
c. Increase of tensile strength is maximum (137.16%) for concrete with fibres of 3 Kg/m^3 concrete volume as compared to the plain concrete.
d. Increase of flexural strength is maximum (50.17%) in case the fibre content is 3 Kg/m^3 of concrete volume.

The ultimate moment capacity of different beams have been recorded and the average percentage increase values against equivalent beam without fibres have been compared and shown in Table 4. It is seen that at a fibre content of 3 Kg/m^3 concrete volume the increase of ultimate value is the maximum (25.6%).

Fig. 3 shows an indicative representation of load-deflection response at central point for beams without fibre and with fibre of 2 Kg/m^3 volume of concrete. This shows an improvement of ductility of fibre reinforced beams.

The curvatures of the beam at different loading cases

Table 1. Detail of beam specimens

Span of the beam : 1900 mm ;
Top reinforcement : Nil (Nominal black wire) ;
Bottom reinforcement: 3-8 ⏀ HYSD bars ;
Stirrup : 2^{L}6 Ø m.s. @ 150 c/c ; Beam section :150 x 200 mm
Proof stress of main steel reinforcement : 410 N/mm^2 ;
Yield stress of stirrup reinforcement : 320 N/mm^2.

Beam Mkd.	Average grade of concrete	Fibre content (Kg/m^3 of concrete)	$\frac{w}{c}$ ratio	Mix proportion
B1	M15	0	0.55	1:2:4
B2	"	1	"	"
B3	"	2	"	"
B4	"	3	"	"
B5	"	4	"	"
B6	M20	0	0.48	1:1.8:3.6
B7	"	1	"	"
B8	"	2	"	"
B9	"	3	"	"
B10	"	4	"	"

Table 2. Detail of beam specimens

Span of the beam : 1400 mm ;
Top reinforcement : Nil (Nominal black wire) ;
Bottom reinforcement : 3-8 Ø m.s. bars ;
Stirrup : 2^{L}6ϕm.s. @ 100 c/c ; Beam section : 125 x 150 mm ;
Proof stress of main steel and
stirrup steel reinforcement : 320 N/mm^2.

Beam Mkd.	Average grade of concrete	Fibre content (Kg/m^3 of concrete)	$\frac{w}{c}$ ratio	Mix proportion
B1A	M15	0	0.6	1:2:4
B2A	"	1	"	"
B3A	"	2	"	"
B4A	"	3	"	"
B5A	"	4	"	"
B6A	M20	0	0.5	1:1.8:3.6
B7A	"	1	"	"
B8A	"	2	"	"
B9A	"	3	"	"
B10A	"	4	"	"

Table 3

3.1 Compressive strength (by Cube test)

Type of specimen	Quantity of fibre added in Kg/m^3 of concrete	Average compressive strength (N/mm^2)	% of increase in strength of P.F.C.over O.C.
OC	0	15.50	-
PFC	1	22.44	+44.77
PFC	2	22.66	+46.19
PFC	3	16.44	+06.06
PFC	4	15.95	+02.90

3.2 Tensile strength (by Split cylinder test)

Type of Specimen	Quantity of fibre added in Kg/m^3 of concrete	Average tensile strength (N/mm^2)	% of increase in strength of P.F.C.over O.C.
OC	0	1.13	-
PFC	1	1.44	+27.43
PFC	2	1.69	+49.55
PFC	3	2.68	+137.16
PFC	4	1.48	+30.97

3.3 Flexural strength (by Prism test)

Type of specimen	Quantity of fibre added in Kg/m^3 of concrete	Average value of modulus of rupture(N/mm^2)	% of increase in strength of P.F.C.over O.C.
OC	0	2.93	-
PFC	1	3.20	+09.21
PFC	2	4.12	+40.61
PFC	3	4.40	+50.17
PFC	4	3.90	+33.10

OC - ordinary concrete(M15); PFC - polypropylene fibre concrete.

Table 4. Increase in ultimate moment of resistance for fibre concrete beams over conventional reinforced concrete

Beam Mkd.	Quantity of fibre added in Kg/m^3 of concrete	Grade of concrete (Average)	% Increase over conventional reinforced concrete beams
B1	0	M15	-
B2	1	"	+ 6.54%
B3	2	"	+13.10%
B4	3	"	+19.64%
B5	4	"	+13.10%
B6	0	M20	-
B7	1	"	+ 8.30%
B8	2	"	+18.00%
B9	3	"	+25.60%
B10	4	"	+14.00%

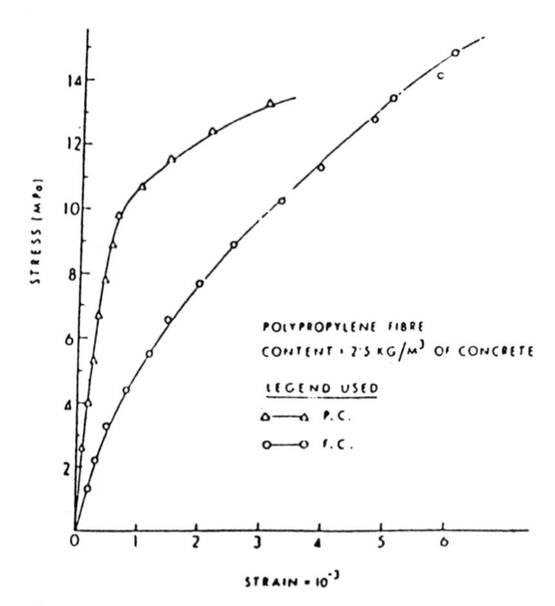

Fig. 2. Stress vs. strain plot for plain and fibre reinforced concrete cubes.

were calculated based on the measured strains at distances of 20, 40, 100 and 130 mm from top of beams. The strains at different levels as mentioned were taken for different section (entire right half of the beam) at an interval of 150 mm c/c in order to get M-Ø diagrams i.e. flexural stiffness at different sections and also to note the change of flexural stiffness available along the length of the beam. The moment-curvature relationship at mid span has been shown in Fig. 4 . Stress-strain plot of fibre concrete is shown in Figs. 2.

3.4 Background analysis of experimental data

The experimental data of ten beams, cubes, cylinders and prisms were analysed to establish beam stiffness, stress block parameters etc. From the strain data of different sections, M-Ø diagram were drawn and flexural stiffness (EI) and stress block parameters of different sections were estimated. Assuming parabolic distribution of flexural stiffness (EI) along the length of the beam, quarter span and mid-span deflections were computed theoretically and checked against actual experimental deflection data. The process was repeated at different stages of loading and accordingly stress block parameters etc. were established. The beams were analysed using proposed theoretical expression and compared with companion ten beams (Ref. Table 5).

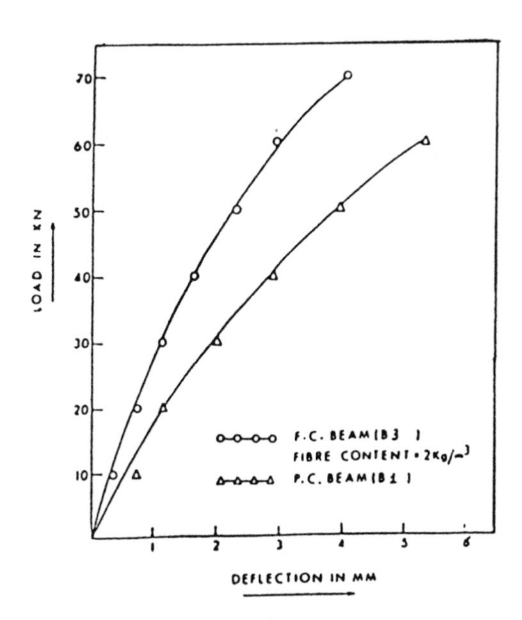

Fig. 3. Load vs. central deflection plot.

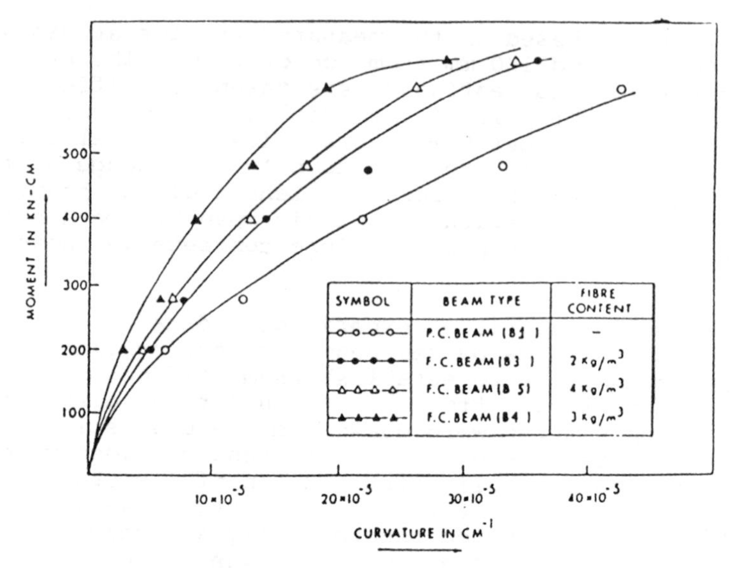

Fig. 4. Moment - curvature curve.

Table 5. Comparison of theoretical results with experimental results

Beam Mkd.	Quantity of fibre added in Kg/m3 of concrete	Grade of concrete (Average)	Variation between M_{uF}theoretical and M_{uF}experimental
B2	1	M15	+ 3.82%
B3	2	"	+ 5.21%
B4	3	"	+ 6.75%
B5	4	"	+16.62%
B7	1	M20	+ 5.10%
B8	2	"	+10.2%
B9	3	"	+ 8.60%
B10	4	"	+12.30%

4 Conclusions

The present investigation shows the improvement of strength parameters due to adoptation of polypropylene fibres. The strength parameters increases with the increase of fibre content upto a certain value and then decrease even with increase of fibre content. The useful range observed as 2 to

3 Kg per m^3 of concrete volume. The ductility of the beams improved due to fibre insertion which are reflected from load-deflection and moment-curvature curves. The theoretical predicted values of ultimate moment of resistance shows good agreement with the experimental results.

5 Future recommendations

The present investigation shows the tremendous possibility of future studies in this field. The improvement of structural behaviours should be properly utilised on some suitable fields of application, such as circular pipes, thin-webbed sections, slab-column connections, etc.

6 Notations

d = effective depth of beam ;
f_y = yield stress in steel reinforcement ;
E_s = modulus of elasticity of steel ;
f_{ck} = characteristic compressive strength of ordinary concrete ;
f'_{ck} = characteristic compressive strength of fibre concrete ;
$\alpha f'_{ck}$ = stress at the outer most compression fibre ;
ϵ_1 = strain in fibre conc. beyond which plastic strain is considered to occur ;
ϵ_2 = maximum strain in fibre concrete at the outer most compression fibre ;
x_u = neutral axis depth from top of beam ;
K_1 = average stress coefficient for ordinary concrete;
K'_1 = average stress coefficient for fibre concrete ;
K_2 = coefficient of depth of compressive force for ordinary concrete ;
K'_2 = coefficient of depth of compressive force for fibre concrete ;
f_{ct} = tensile stress of fibre concrete in bending ;
M_{uf} = ultimate moment of resistance of the fibre reinforced concrete action ;
M_u = ultimate moment of resistance of reinforced concrete section ;
A_{st} = area of steel reinforcement in tension ;
a_f = distance between the c.g. of the compressive stress block and the c.g. of the tensile steel reinforcements.

7 References

Batson, G. Ball, C. Bailey, L. Landers, E. and Hooks, J. (1972) Flexural fatigue strength of steel fibre reinforced concrete beams. Journal of ACI, (november 1972), 69 (11), 673-677.

Batson, G. Jenkins, E. and Spatney, E. (1972) Steel fibres as shear reinforcements in beams. Journal of ACI, October 1972, 69 (10), 640-644.

Hannant, D.J. (1978) Fibre cements and fibre concretes, John Wiley & Sons, 1978, 241 p.

Kukreja, C.B. Kaushik, S.K Kanchi, M.B. and Jain, O.P. (1980) Tensile strength of steel fibre reinforced concrete. Indian Concrete Journal, July 1980, 54(7), 184-188.

Narayan, R. and Darwish, I.V.S. (1987) Use of steel fibres as shear reinforcement, ACI Structural Journal, May-June 1987, 84(3), 216-227.

Ray, Arunachal and Ghosh, Somnath (1990) A study on natural and synthetic fibre concrete. National Seminar on Innovative Technology for Low Cost Rural Housing by ASCE-IS and RHDC, Calcutta, 27th April 1990, AR-1 to AR-8.

Romualdi, J.P. and Batson, G.B. (1963) Mechanics of crack arrest in concrete, ACSE (Mech. Engg. Division), June 1963, 89 (EM3), 147-168.

Romualdi, J.P. and Mandel, J.A. (1964) Tensile strength of concrete affected by uniformly distributed closely spaced short lengths of wire reinforcements. Journal of ACI, June 1964, 61(6), 657-671

Swamy, R.N. and Mangat, P.S. (1975) The onset of cracking ductility of steel fibre concrete, Cement and Concrete Research, January 1975, 5(1), 37-53.

40 FIBRE REINFORCED CONCRETE CHANNELS AS SURFACE REINFORCEMENT FOR FLEXURAL MEMBERS

A. S. PARULEKAR, I. I. PANDYA and S. K. DAMLE
Applied Mechanics Department, Faculty of Technology and Engineering, M. S. University of Baroda, Baroda, India

Abstract
This paper describes experimental investigations carried out on nine beams consisting of plain and fibre reinforced concrete (FRC) channels to serve as permanent formwork for beams. The channels provide skin reinforcement or surface reinforcement for flexural members.
A modified theory, based on load-slip curves obtained from the pullout tests on fibres, is presented for the analysis of such FRC covered plain and fibre reinforced concrete beams. The experimental and theoretically estimated values of ultimate loads show good agreement. The results also indicate the improved load carrying capacity with reduction in deflection and crack-width.
Keywords: Beams, Ductility; Fibre Reinforced Concrete, Load Tests, Steel Fibres, Surface Reinforcement.

1 Introduction

Concrete is inherently weak material in tension. This drawback leads one to reinforced concrete with steel in which tension will be taken by steel for flexural members. Another method for strengthening weak and brittle matrices such as concrete is to incorporate strong fibres in it. The inclusion of steel fibres is more beneficial in improving the flexural strength, crack resistance and energy absorption capacity of concrete [4,6,7]. Use of fibre-reinforcement throughout the length of the member gives good performance characteristics at working as well as at ultimate load. In this case cost of construction is relatively high due to the high cost of fibres compared to the cost of other materials. Hence, it is advisable to use the fibres in the most critical tensile region only. In case of flexural members, a fair guess of the most critical region can be arrived at. Hence, fibre reinforced concrete is used as skin reinforcement in this

Fibre Reinforced Cement and Concrete. Edited by R. N. Swamy. ©1992 RILEM.
Published by E & FN Spon, 2-6 Boundary Row, London SE1 8HN. ISBN 0 419 18130 X.

investigation [4]. This skin reinforcement can be provided in the form of FRC thin sheets and channels in cases of slabs and beams respectively. The present study reported here is an experimental attempt to verify the feasibility of such 'Skin Reinforced' structural members.

2 Experimental Investigations

Casting of FRC covered plain concrete beams was done in two stages. In the first stage, the outer FRC channels were cast. In the second stage, these FRC channels act as direct formwork for casting of composite beams as shown in Fig.1.

Static testing was done on nine beams consisting of three series, each series consisting of three beams subjected to two point loading. The beams were simply supported on a span of 500 mm. Steel fibres, having diameter of 1.11 mm and aspect ratio of 80 with 0%, 0.75% and 1.5% by volume, were added to concrete having proportion of 1:2:4 with w/c = 0.5. Three channels of 30 mm thickness with 1.5% fibres were also tested in different positions to check various loading conditions during transportation, erection as well as placing of in-situ concrete. The beams B1 to B3 were cast and tested to determine ultimate load and compared with the theoretical value calculated from the presented theory for FRC covered plain concrete beams. The beams B4 to B9 were cast and tested to determine the increase in ultimate strength compared to FRC covered plain concrete beams by providing fibres in core concrete in different percentages. The fully fibre reinforced beams B10 to B12 were tested to determine the ultimate strength of the beams. These results were compared with beams B7 to B9 to check whether there was any slip between the core concrete and channels. The fibre reinforced concrete channels C1 to C4 were cast and tested to know their load carrying capacity, load-deflection characteristics and modes of failure. See Fig.2.

All the beams and channels were tested on a specially provided bending bench of Universal Testing Machine of 500 kN capacity under two point loading. Midspan deflections were recorded with the help of dial gauges at an interval of 2 kN.

3 Theoretical Analysis

A flexure thoery for the analysis of steel fibre reinforced concrete beams in the post cracking range, based on the available load-slip curves obtained from the pull-out tests on fibres was proposed by P. Sabapathi and H. Achyutha [5] as shown in Fig.3 and Fig.4. The same

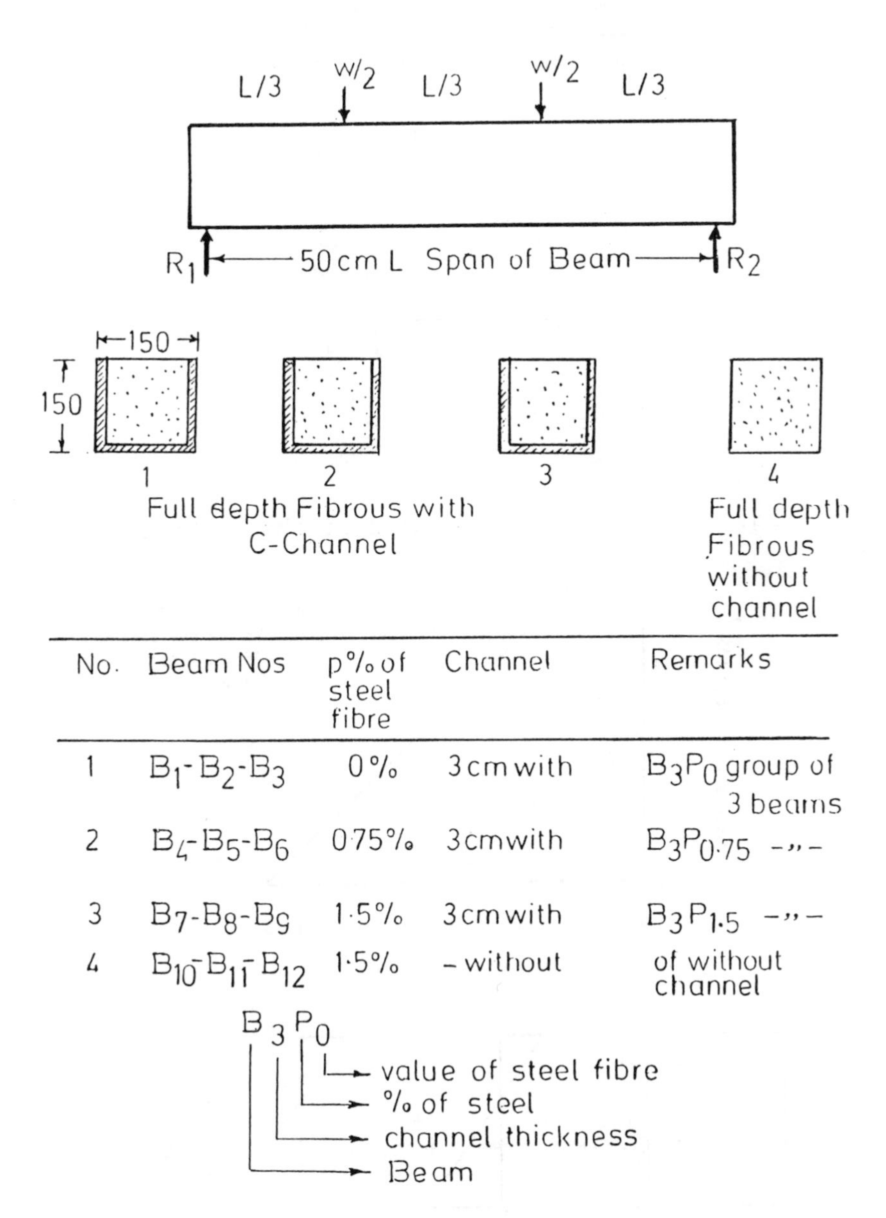

No.	Beam Nos	p% of steel fibre	Channel	Remarks
1	B_1-B_2-B_3	0%	3 cm with	B_3P_0 group of 3 beams
2	B_4-B_5-B_6	0.75%	3 cm with	$B_3P_{0.75}$ -"-
3	B_7-B_8-B_9	1.5%	3 cm with	$B_3P_{1.5}$ -"-
4	B_{10}-B_{11}-B_{12}	1.5%	- without	of without channel

Fig. 1. Typical loading arrangement for experiment.

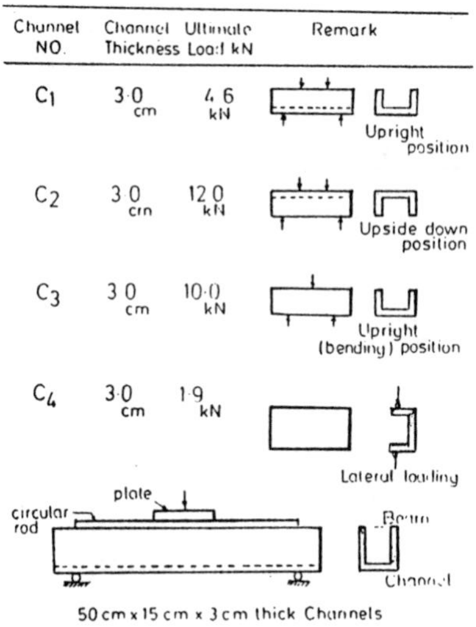

Channel NO.
Channel Thickness
Ultimate Load kN
Remark
C1
3·0 cm
4·6 kN
Upright position
C2
3·0 cm
12·0 kN
Upside down position
C3
3·0 cm
10·0 kN
Upright (bending) position
C4
3·0 cm
1·9 kN
Lateral loading
plate
circular rod
Beam
Channel
50 cm x 15 cm x 3 cm thick Channels

Fig. 2. Experimental results of channel testing.

a Fibre length 0·5 on both sides
b Fibre length 0·25 on one side and 0·75 on other side of the crack
Stress in fibre, f / fo
Strain in beam at fibre location, ε / εo
0
0·25
0·50
0·75
1·00
1·0
2·0
3·0
4·0
5·0
a
b

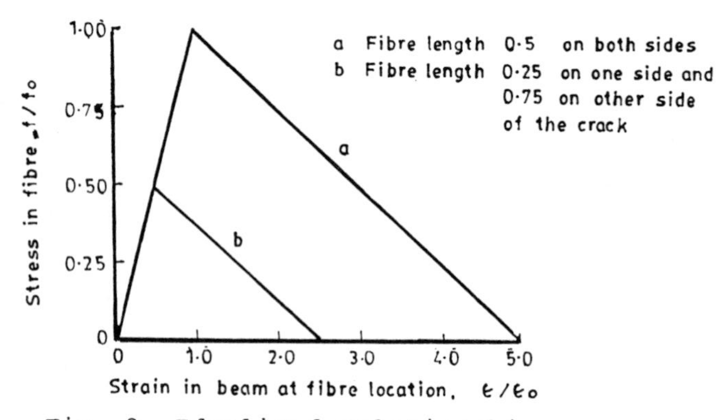

Fig. 3. Idealized relationship.

N
A
kh
(1 - k) h
b
εc
x
dx
αεo
εc = Fc/Ec
(1-k) h / α
A
εo
B
αεo = Fc/Ec (1-k)/k
(a)
(b)

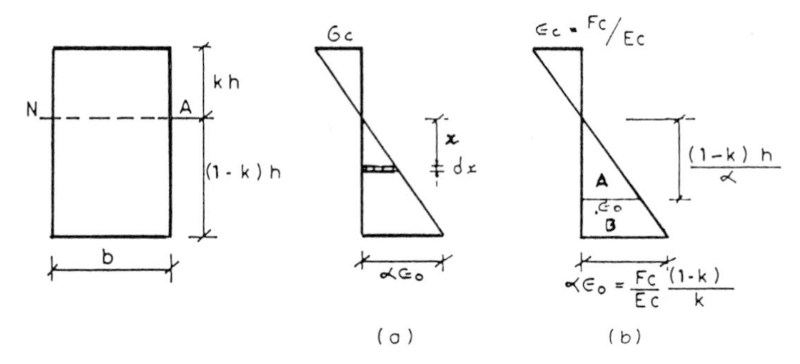

Fig. 4. Strain diagram for fibre reinforced beam.

theory was modified to take into account the effect of 'skin reinforcement' [4].

The formulae for various sectional parameters for the 'skin reinforced' section were worked out as shown below. See Fig.5 and Fig.6.

k = Neutral axis factor $= 1.0 \Big/ \left\{ 1 + \sqrt{\left[\dfrac{\alpha}{2m\, V_{f_1}\, \gamma} \right]} \right\}$

j = Lever arm factor $= \frac{2}{3}k + Y_T (1-k)$

M_R = Moment of resistance $= \gamma (1-k)\, j\, f_o\, V_{f_1}\, b h^2$

where, $= R\, f_o\, V_{f_1}\, b h^2$

α = Ratio of extreme layer tensile strain to the strain in the fibre at its maximum stress.

n = Modular ratio.

V_{f1} = Volume fraction of fibres in channels.

γ = Coefficient for the average tensile stress for all the fibres in the tensile zone.

Y_T = Ratio of distance of the centre of total tensile force from the neutral axis to the depth of the tensile zone.

f_o = Maximum stress attained in the fibre having equal length on either side of the crack.

The parameters γ and Y_T were determined in rigorous analysis for the section [1].

The parameter α was found out directly with the help of curve given in [2] for corresponding value of volume fraction of fibres as equal to 1.2. Design curves were

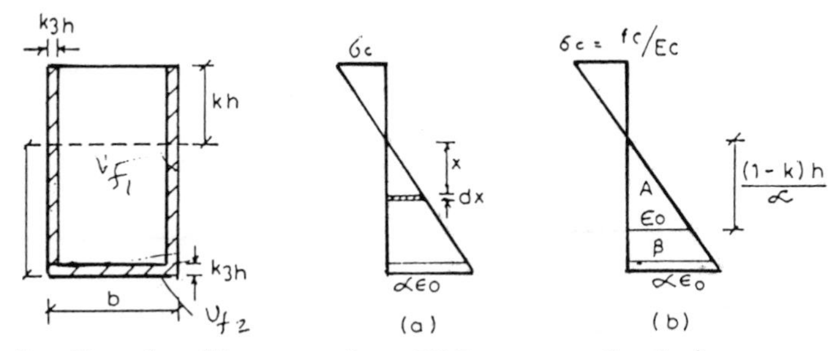

Fig. 5. Strain diagram for FRC covered plain concrete beam.

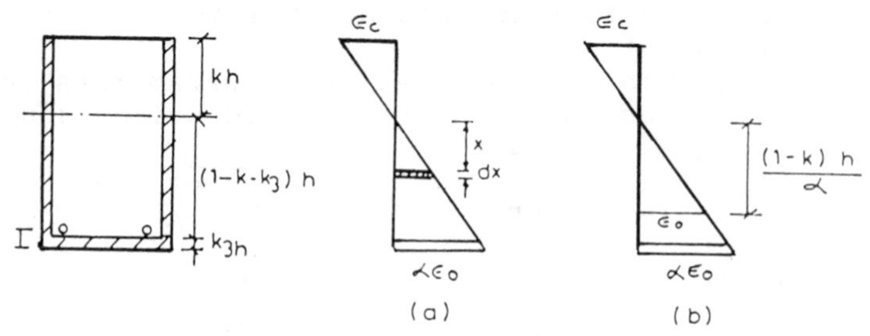

Fig .6. STRAIN-DIAGRAM FOR FRC COVERED RCC BEAM

plotted using the above formulae which are shown in Fig.7. Using these curves for FRC covered plain concrete beams,

From the strain diagram of fibre-reinforced concrete

$$\epsilon = \frac{x\,\alpha\,\epsilon_0}{(1-k)\,h}$$

Stress in the fibre at this layer will be $x \alpha f_0 (1-k) h$

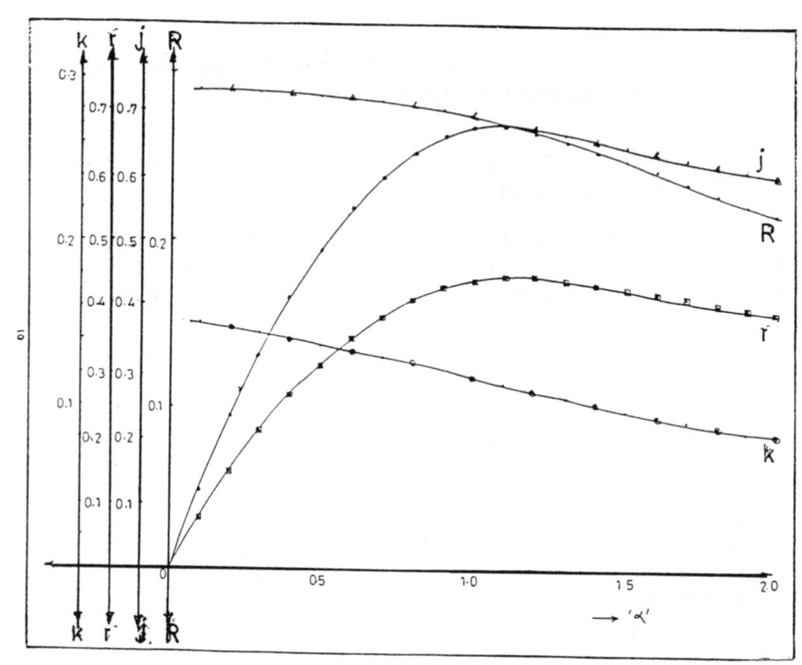

Fig. 7. Design curves for FRC beams.

Average stress will be $x \propto f_o [2(1-K)h]$

Total tensile force will be given by

$$\int_0^{(1-K)h} V_f \left[\frac{1 - x\propto}{(1-K)h}\right] b\,dx \left[\frac{x \propto f_o}{(1-K)h}\right] + V_f \left[\frac{x\propto(\beta-1)}{(1-K)h\beta}\right] b\,dx \left[\frac{x \propto f_o}{2(1-K)h}\right]$$

$$= \int_0^{(1-K)h} f_o V_f \, b\,dx \{r'\}$$

Equating strain compatibility condition

$$\frac{\propto \epsilon_o}{n(1-K)h} = \frac{\epsilon_c}{Kh} \quad \therefore \epsilon_o = \frac{n\epsilon_c(1-K)}{\propto K}$$

$$f_o = \frac{n f_c (1-K)}{\propto K} \qquad (A)$$

Equation tensile and compressive forces

$$\frac{f_c}{2} \cdot K = (1-K)\, r\, V_f f_o$$

$$\therefore f_o = \frac{f_c K}{[2(1-K) r V_f]} \qquad (B)$$

Equating equation (A) and (B), we get

$$K = \frac{1}{1 + \sqrt{\left[\frac{\propto}{2n V_f + r}\right]}}$$

And $\quad j = \frac{2}{3}K + Y_t(1-K)$

From the above parameter M i.e. Moment of Resistance factor R

$R = r(1-K)j \quad$ where $M = R f_o V_f b h^2$

$r = 0.4438$

$j = 0.6765$

$R = 0.2675$

f_o is calculated with the theory given in [7].

$f_o = 807.6$ N/sq.mm.

Moment of Resistance is given by

$M_R = R \cdot f_o \cdot V_{f_1} \cdot b \cdot h^2$

$= 0.2675 * 807.6 * 0.004584 * 60 * 150 * 150$

$= 1.33 * E{+}06$ Nmm

$= 1.33$ kNm.

Actual average Moment of Resistance for FRC covered plain concrete beams is

$M_R = 21.20/2 * 0.16 = 1.6960$ KNm.

4 Test Results

The test results for beams B1 to B12 are tabulated in Table 1 along with estimated theoretical ultimate loads for the beams. All beams recorded flexural failure with single dominating crack in the middle region. The failure was ductile in nature because of fibre reinforcement.

Table 1: EXPERIMENTAL RESULTS

Beam No No	% Fibre in Core Concrete	Experimental Ult.Load kN P_{EXP}	Theoretical Ult.Load kN P_{THE}	Beam Group with Ave. Deflection	$\frac{P_{EXP}}{P_{THE}}$
B1	0.0	20.5	16.715		1.22
B2	0.0	22.0	16.715	B3PO	1.32
B3	0.0	21.0	16.715	.17 mm	1.25
B4	0.75	21.5	21.095		1.02
B5	0.75	27.0	21.095	B3P.75	1.27
B6	0.75	24.0	21.095	.15mm	1.13
B7	1.5	23.5	25.348		0.93
B8	1.5	26.0	25.348	B3P1.5	1.02
B9	1.5	24.0	25.348	.13mm	0.95
B10	1.5	29.0	25.348		1.14
B11	1.5	26.4	25.348	BoF	1.04
B12	1.5	30.0	25.348	.24mm	1.18

5 Discussions and Conclusions

From Table 1, it is clear that the ultimate moment of resistance obtained by the presented flexural theory is about 10% to 20% less than the results obtained experimentally. This oversafety in the design can be attributed to the value of equivalent mean bond stress which was calculated with the help of empirical relationship [2]. This relationship was derived for homogeneous distribution of fibres in the section which is not the case here. The empirical formula needs to be modified to take into account this effect.

There is only marginal increase in ultimate load, when additional fibres were introduced in core region. This further strengthens the fact that the fibres in critical tensile region give maximum benefit.

The beams B7 to B12 were specifically tested to check whether there was any slip between the earlier cast channels and later cast concrete at ultimate load. It is seen that the difference in ultimate load values for the two sets viz. B7 to B9 and B10 to B12 was about 10% although there were no visible signs of any separation of channels. This marginal difference in strength can be attributed to maximum size of aggregate which was very less in channel section as well as more homogeneous distribution of fibres in fully fibre reinforced beams.

Plain concrete is a highly brittle material and failure occurs almost suddenly. On the other hand, in case of FRC covered plain concrete beams, with only 0.45% fibres by volume as a surface reinforcement, the change in type of failure is significant. The beams resisted about 50% more load after the first major crack was visible. At ultimate load, the beams did not break into two halves like plain concrete but showed higher deflections at constant loads. This is one of the major advantages of fibre reinforced concrete. This ductility is introduced in the concrete by the fibres due to the effect of fibre bridging cracks and thus acting as crack arrest mechanisms.

From the test results on channels C1 to C4, it was observed that the channels can safely withstand the weight of cast-in-situ concrete for normal formwork spans. When checked for concentrated loads, it was seen that one man can safely work on it. Channels were very weak to resist lateral loads due to improper bond between channel bottoms and sides. Some channels even cracked at this critical junction before in-situ concreting; but this weakness did not get reflected in test results of beams cast from these channels. This may be due to the effective grouting ofthe cracks during in-situ concreting.

6 Acknowledgement

The study was carried out in the Applied Mechanics Department, Faculty of Technology and Engineering, M.S. University of Baroda, Baroda, India. The authors thank the authorities for providing necessary help and laboratory facilities.

7 References

1 A State of Art Report in Fibre Reinforced Concrete. Proceedings, ACI, Vol.70, Nov.1973.

2 Dave, N.J. (1987) Repair and/or Strengthening of concrete structural members by bonding a FRC layer as external reinforcement. Proceedings, international symposium on fibre reinforced concrete, Madras, pp 454-461.

3 Ithape, S.B. (1980) Skin fibre reinforced slabs and beams. M. Tech. Dissertation, Indian Institute of Technology, Bombay, India.

4 Parulekar, A.S. (1990) Study of skin fibre reinforced concrete structural members. M.E. Dissertation, Maharaja Sayajirao University of Baroda, Baroda, India.

5 Sabapathi, P. and Achyutha, H. (1989) Analysis of steel fibre reinforced concrete beams. Indian Concrete Journal, pp 257-261.

6 Shah, S.P. and Rangan, B.V. (1971) Fibre reinforced concrete properties, Proceedings, ACI Journal, Vol.68, No.2, pp 126-134.

7 Swamy, R.N. (1974) The Technology of steel fibre reinforced concrete for practical applications, Proceedings, The Institute of Civil Engineers, Part-I, Vol.56.

41 FIBRE TYPE EFFECTS ON THE CURVATURE IN STEEL FIBRE REINFORCED CONCRETE RINGS

J. HÁJEK, K. KOMLOŠ, T. NÜRNBERGEROVÁ and B. BABÁL
Institute of Construction and Architecture, Slovak Academy of Sciences, Bratislava, ČSFR

Abstract
The influence of the type and size of fibers on the strength and ductility of steel fibre reinforced concrete on 21 rings subjected to the two edge load was investigated. The test results have enabled a quantitative evaluation of the effect of different shape and surface treatment of fibres on the flexural strength and quasi-plastic behaviour of FRC rings in comparison with those made of plain concrete. A special attention is paid to the variation of curvature with increasing load.
Keywords: Plain concrete, Steel fibre reinforced concrete, Fibre type, Deflection, Curvature.

1 Introduction

One of the suitable possibilities of application of steel fibre reinforced concrete are sewer pipes. A ring representing the cross section of the pipe acts under two edge loading as a hyperstatic circular closed frame. Tests carried out previously on plain concrete pipes have shown that the total collapse could take place only when cracking limit in the lateral cross sections was reached. The ratio between the load at the formation of the first crack on the side and the load at the formation of the first crack in the top or bottom cross section varied from 1.10 to 1.20. An explanation of such a behaviour could be looked for in the influence of descending branch of the stress-strain diagram of concrete in tension at the top and bottom cross sections on the gradual increase of rotations of these sections. The undamaged lateral parts of the ring prevent the abrupt failure. It could be expected that because of the reduced slope of descending branch the randomly dispersed reinforcement prevents the sudden enlargement of the curvature in the critical section. The undamaged parts of the hyperstatic element contribute to the retardation of failure deformation, as well.

2 Test programme

The rings with an inner diameter of 500 mm, wall thickness of 75 mm and width of 150 mm were chosen as test specimens. Tests were carried

Fibre Reinforced Cement and Concrete. Edited by R. N. Swamy. © 1992 RILEM.
Published by E & FN Spon, 2-6 Boundary Row, London SE1 8HN. ISBN 0 419 18130 X.

out on seven series each consisting of three rings (21 elements altogether). One of the series was made of plain concrete, the other six differed from each other in the type of dispersed reinforcement and in the diameter of fibers and their length, respectively. The tests procedure with vertical deflection (the change of the ring inner diameter laying in the direction of the loading force) rate control was applied (Fig. 1).

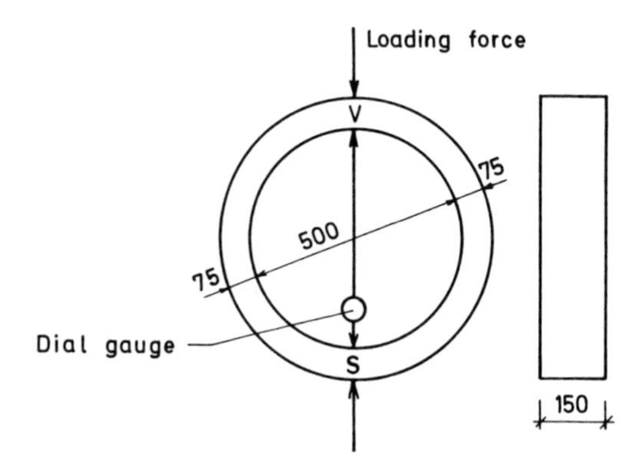

Fig. 1. Size of rings and scheme of loading.

Portland cement type PC 400 in the amount of 450 kg.m^{-3} was used throughout the experiments. The applied aggregate grading was in accordance with the curve B8 after DIN 1045. The fibre weight fraction was 3 % i.e. 72 kg.m^{-3}. The steel fibres of the diameter D = 0.50 mm and D = 0.375 mm with aspect ratio L/D = 75 and of diameter D = 0.40 mm with aspect ratios L/D = 75 and L/D = 100 were used. The designation of the series is given in Tab. 1. The values of compression strength, modulus of rupture, and dynamic modulus of elasticity tested on prisms 100x100x400 mm (mean value from three samples) are in Tab. 2.

Table 1. Designation of testing series and type of applied steel fibre

Designation of series	Fibre Length L (mm)	Fibre Diameter D (mm)	L/D	Type of steel fibre
S01	-	-	-	-
S02	37.5	0.50	75	black wire, annealed
S03	23.5	0.315	75	Bohumín
S04	30.0	0.40	75	black wire non-annealed
S05	40.0	0.40	100	Hlohovec
S06	40.0	0.40	100	Duoform
S07	40.0	0.40	100	Dramix

Table 2. Mechanical properties of concrete used for individual testing series

Designation of series	Modulus of rupture (MPa)	Compression strength (MPa)	E_{dyn} (GPa)
S01	3.87	32.9	34.3
S02	4.49	34.4	35.2
S03	4.48	35.5	34.6
S04	4.92	36.0	35.2
S05	5.34	36.7	35.6
S06	5.92	39.9	36.3
S07	5.48	37.1	35.4

The tests of rings under the two edge loading were carried out with mean rate of the vertical deflection of 0.015 $mm.min^{-1}$. The test was considered as finished when the vertical deflection reached a value of 1.5 mm (in the case of S06 and S07 series a value of 2 mm). In addition to vertical deflection, the relative strains at the top, bottom and both lateral sections of the ring were measured using electrical resistance strain gauges. The loading force was registrated through potentiometrical sensing element. More detailed information see Hájek et al. (1982).

3 Test results

The tests were carried out with an intent to serve as a basis for future comparison with theoretical analysis. For this purpose it is suitable to replace the obtained experimental curves by analytical functions in order to have the possibility to determine integrals or derivations. Therefore we used for comparison of load carrying capacity of rings the relationship between the loading force F versus the vertical deflection w which the loading procedure was controlled by. The experimental values were fitted with the aid of polynomial of the 6 degree in the form

$$V = V_m . \sum_{i=1}^{6} \alpha_i (w/d)^i, \qquad (1)$$

where V is the two edge load per width of the ring ($kN.m^{-1}$), V_m is the maximal value defined as local extreme of regression polynomial, w is the vertical deflection, d is the diameter of the center line (determined by measuring of each ring particularly), α_i (i=1,..6) are dimensionless coeficients.

The relationships between the loading force F and the vertical deflection w derived on the basis of this regression are illustrated in Fig. 2. The peaks of the regression curves that are considered to

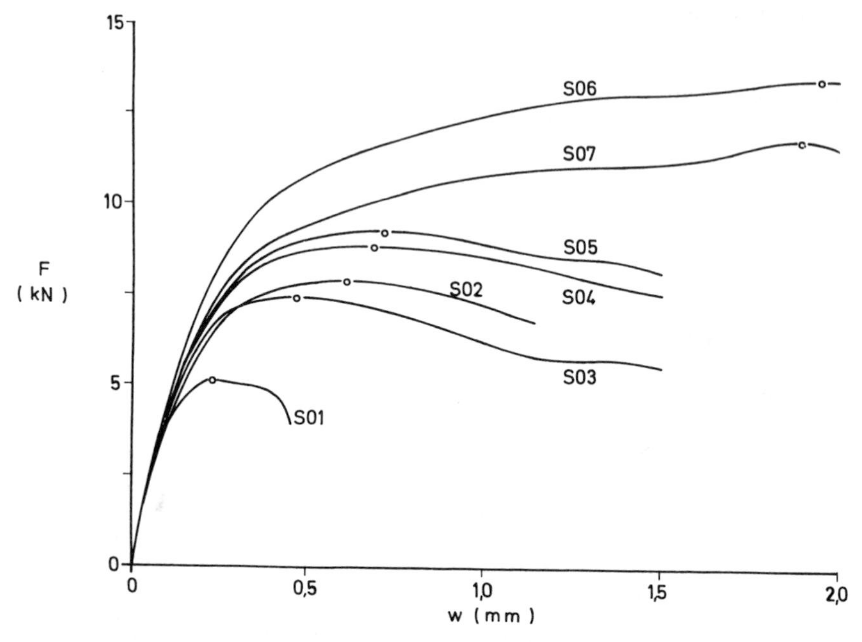

Fig. 2. The two edge loading force F and the vertical deflection w relationship.

be the load carrying capacity of the rings are circled. It can be seen on the Fig. 2 how the load bearing capacity of the ring is influenced by the fiber type and its surface treatment. The numerical values are presented in Tab. 3. It follows from Tab. 3 that 2.61 multiple of load

Table 3. Maximal force F_m and corresponding vertical deflection w_m and strain energy needed to reaching the load bearing capacity W_m

Designation of series	Maximal force		Deflection at the peak		Strain energy at max. force	
	F_m (kN)	Ratio to S01	w_m (mm)	Ratio to S01	W_m (J)	Ratio to S01
S01	5.12	1.00	0.228	1.00	0.80	1.00
S02	7.89	1.54	0.607	2.66	3.73	4.66
S03	7.44	1.45	0.463	2.03	2.65	3.31
S04	8.87	1.73	0.688	3.02	4.81	6.14
S05	9.27	1.81	0.719	3.15	5.26	6.58
S06	13.48	2.63	1.954	8.57	22.04	27.55
S07	11.77	2.29	1.898	8.32	18.47	23.09

carrying capacity of plain concrete rings was achieved at series S06, while the corresponding limit increment of the vertical diameter increased as much as 8.56 times.

The relationship between the loading force and the vertical deflection in the form of regression function (1) allows to determinate the strain energy at given deflection by its integration. In Fig. 3 the strain energy versus the vertical deflection

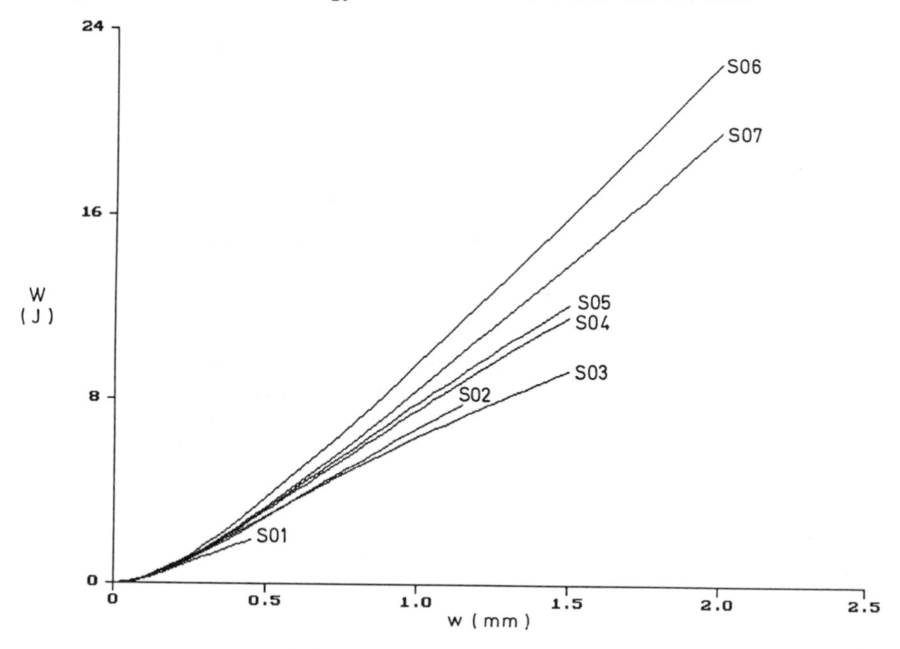

Fig. 3. The strain energy W and the vertical deflection w relationship.

relationships for all tested series are plotted. The values of the energy required for achievement of the peak of stress-strain diagram as well as the values of energy absorbed by steel fibre reinforced and plain concrete rings are given in Tab. 3. As it is known the load carrying capacity as well as the ductility can be characterized through the strain energy (see e.g. ACI Committee 544 (1988)).

Examining the cracks development we proceeded from strains at tension surfaces at the top and bottom as they were registered by electrical resistance strain gauges during the test. For reciprocal comparison of individual test series we choose the strain value of 150 $\mu m.m^{-1}$ which corresponds to the crack formation in bending and that of 800 $\mu m.m^{-1}$ which corresponds to the crack width of 0.05 mm. The two edge load at these strains is related to the modulus of rupture f_r, as it is plotted in Fig. 4. It can be seen in Fig. 4 that the difference between strains at the top and bottom is not significant. Replacing experimental values by quadratic parabola we obtained the formula for strains of 150 $\mu m.m^{-1}$

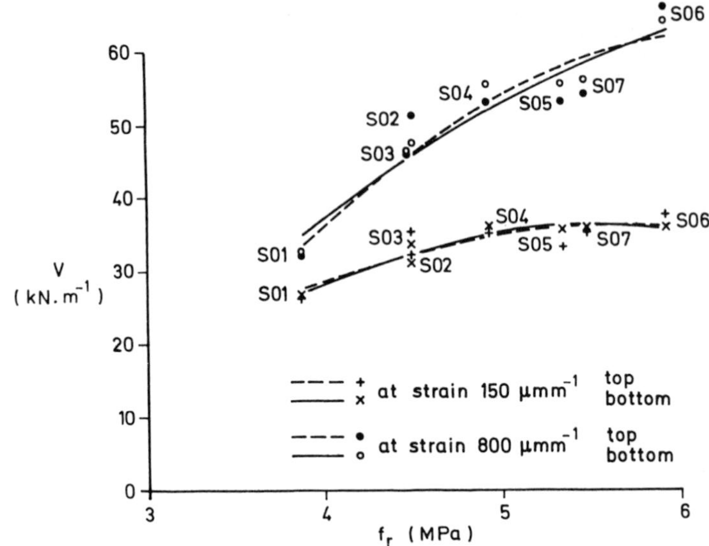

Fig. 4. The relationship between the modulus of rupture f_r versus the two edge load V at strain 150 μm.m^{-1} and 800 μm.m^{-1} of the tensile surface at the top and bottom of rings.

$$\frac{V_{150}}{V_{0,150}} = -2.1896 + 4.8802 \frac{f_r}{f_{r0}} - 1.6905 \left(\frac{f_r}{f_{r0}} \right)^2, \tag{2}$$

whith $V_{0,150}$ = 27.3 kN.m^{-1}, and for strains of 800 μm.m^{-1}

$$\frac{V_{800}}{V_{0,800}} = -3.0285 + 5.6709 \frac{f_r}{f_{r0}} - 1.6424 \left(\frac{f_r}{f_{r0}} \right)^2, \tag{3}$$

with $V_{0,800}$ = 34,50 kN.m^{-1}. The curves in Fig. 4 show that at the level of the first defect (the level defined by the the strain of 150 μm.m^{-1}) the influence of the different fibre types of reinforcement is not expressive. But at the level corresponding to visible cracks (when the strain of 800 μm.m^{-1} is reached) that influence is considerable.

The measured strains at the tensile and compressive surface of the ring enabled evaluation of the curvatures at the top cross section. The experimental couples of data (w, κ), i.e. the vertical deflection and the curvature were fitted by polynomial of the 4 degree in the form

$$w = \sum_{i=1}^{4} a_i \kappa^i, \tag{4}$$

where w is the vertical deflection, κ is the curvature at the top cross section, a_i $(i=1,...4)$ are parameters of the regression polynomial. From the values of w obtained in this way, we determined the loading force F versus the curvature κ relationships with the aid of equation (1). These relationships are plotted in Fig. 5 for all series of rings tested. The curvature of $\kappa = 3\ km^{-1}$ and that of $\kappa = 15\ km^{-1}$ approximately corresponds to the tensile strain of 150 $\mu m.m^{-1}$ and that of 800 $\mu m.m^{-1}$, respectively. It can be seen in Fig. 5 that in

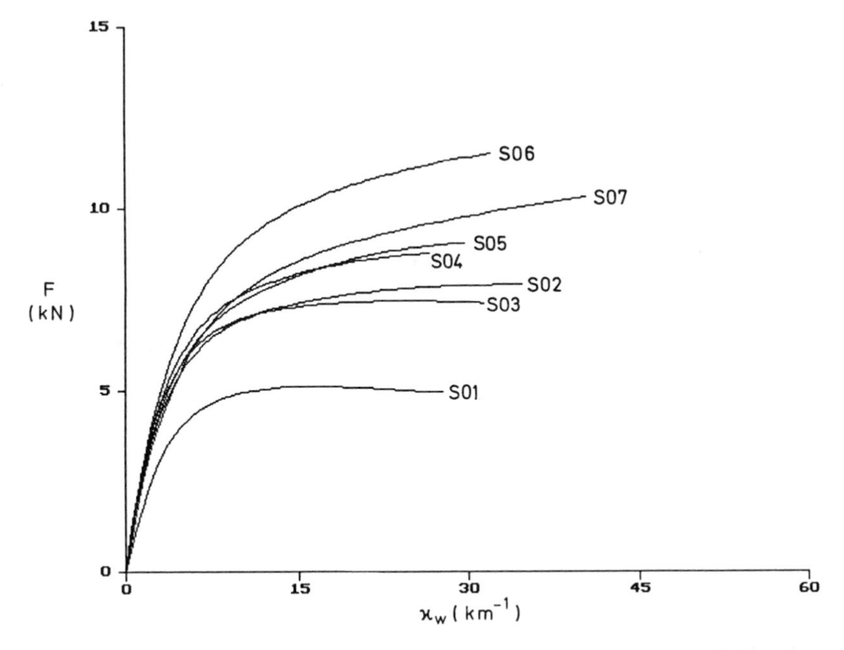

Fig. 5. The two edge loading force F and the curvature κ of the top cross section.

the first case the loading force of the reinforced rings is approximately the same and it is ca 1.3 multiple of the loading force of plain concrete ring, while in the second case ($\kappa = 15\ km^{-1}$) the influence of applied fibre manifested itself expressively and e.g. the force of the rings of the series S06 increased as much as 1.8 multiple for those made of plain concrete.

These results indicate that applying steel fibre reinforced concrete it is possibble to consider the increase of two edge load at the limit of serviceability, including partly the raise of the modulus of rupture partly the raise of ductility.

4 Conclusions

Based on the results reported in this paper, the following conclusions can be drawn:

- on the basis of the strain energy absorbed by the rings to reach the failure the increase of load carrying capacity and that of ductility was determined depending on the type of fibre reinforcement;

- the load carrying capacity ratio of fibre reinforced to plain concrete rings can reach a value of 2 and the applicability limit a value of 2.3;
- the best results were obtained with steel fibres used for casting rings of the series S06;
- the obtained results are presented in a form of fitted expressions (instead of discrete values) which can serve for comparison with analytical solutions;
- the results served as a basis for the provision of the Czechoslovak code ČSN 72 3149 (1985) which thus enabled the calculation of the pipes made of steel fibre reinforced concrete, too.

5 References

ACI Committee 544. (1988) Measurement of Properties of Fiber Reinforced Concrete. **ACI Materials Journal,** 6, 583-593

ČSN 72 3149 (1985) **Design of concrete pipes.** Praha ÚNM, ČSFR

Hájek, J. Komloš, K. Nürnbergerová, T. and Šveda, M. (1982) Concrete rings with randomely dispersed steel fibres, in **Concretes reinforced by dispersed fibres and their application in mining, geotechnic and engineering,** Ostrava, ČSFR, pp. 418-423

42 USE OF STEEL FIBERS AS SHEAR REINFORCEMENT IN HIGH STRENGTH CONCRETE BEAMS

S. A. ASHOUR and F. F. WAFA
King Abdulaziz University, Jeddah, Saudi Arabia

Abstract
The use of steel fibers as shear reinforcement in high-strength concrete beams is investigated in this study. Sixteen high-strength fiber reinforced concrete beams were tested by two point loading. The variables were the steel fiber content (0, 0.5, 1.0 and 1.5 %) and the shear-span/depth (a/d) ratio(1,2,4 and 6). Concrete with average compressive strength of about 96 MPa (14,000 psi) and one type of hooked steel fiber were used in this investigation. Fiber addition noticeably improved the shear strength of the tested high-strength concrete beams and this strength enhancement significantly increases as the a/d ratio decreases. By modelling of the shear strength due to the fibers as vertical stirrups, the two equivalent factors corresponding to shear reinforcement ratio and spacing are determined. Based on the test results, an empirical equation is proposed to predict the extra strength provided by the steel fibers.
Keywords: compressive strength; beams (supports); cracking; deflection; diagonal tension; high-strength concrete; high-strength fiber-reinforced concrete; shear strength; shear-span/depth ratio; steel fibers; shear reinforcement; stirrups.

1 Introduction

High-strength concrete is considered as a relatively brittle material, and the post-peak portion of its stress-strain diagram almost vanishes and descends steeply with the increase in compressive strength (Swamy 1986,1987 , ACI 1984, Naaman and Homrich 1985 , and Wafa et al. 1991). This inverse relation between strength and ductility is a serious drawback in the use of high-strength concrete. A compromise between strength and ductility can be obtained by using discontinuous fibers. Addition of fibers to concrete makes it a homogeneous and isotropic material and converts brittleness into a ductile

Fibre Reinforced Cement and Concrete. Edited by R. N. Swamy. © 1992 RILEM.
Published by E & FN Spon, 2-6 Boundary Row, London SE1 8HN. ISBN 0 419 18130 X.

behavior (ACI 1982 , Rumualdi and Batson 1983, Swamy et al. 1974) . When concrete cracks, the randomly oriented fibers start functioning, arresting both micro-cracking and its propagation, thus improving strength and ductility. Addition of fibers only slightly influences the ascending portion of the stress-strain curve but leads to a noticeable increase in the peak strain (strain at peak stress) and a significant increase in ductility (Naaman and Homrich 1985, Wafa et al. 1991).

The behavior of steel fiber-reinforced normal-strength concrete beams subjected to predominant shear has already been thoroughly investigated (Jindal 1984, Narayanan and Darwish 1987, Swamy and Bahia 1985). Test results have shown that due to the crack-arresting mechanism of the fibers the first-crack shear strength increased significantly, the improvements in ultimate shear strength of fiber-reinforced concrete (FRC) beams were of the same order as that obtained from conventional stirrups even at a fiber volume fraction of 1.0 %, and that the steel fibers were effective in supplementing or even replacing the stirrups in beams.

2 Experimental Program

2.1 Test Specimens

Sixteen high-strength fiber reinforced concrete (HSFRC) beams were tested in this investigation. All beams were singly reinforced and without shear reinforcement. The variables were: (a) shear-span/depth (a/d) ratio, and (b) steel fiber content (V_f). The cylinder strength of the concrete matrix used was about 96 MPa (14,000 psi) and cross-section of all the beams was kept constant at 125 x 250 mm (5x10 in.). Fig. 1a presents the detailed testing program. Each beam is designated in a way to indicate shear-span/depth ratio, and steel fiber content. Thus specimen B-2.0-1.5 represents a beam with a/d = 2.0, and V_f of 1.5 %.

2.2 Materials

In the testing program, 22 mm grade 60 deformed steel bars having 460 MPa (66,700 psi) yield strength were used as flexural reinforcements. The concrete mix proportion was 1:0.25:2.5 (cement:sand:coarse aggregate) to produce concrete with compression strength of about 96 MPa (14,000 psi). Ordinary portland cement (type I) and a natural desert sand with a fineness modulus of 3.1, and a specific gravity of 2.69 were used. Coarse aggregate (crushed basalt) of maximum size 10 mm (3/8 in.), having a crushing strength of 7.7 %, a specific gravity of 2.84 and an impact value of 3.8 % was used. Light gray densified

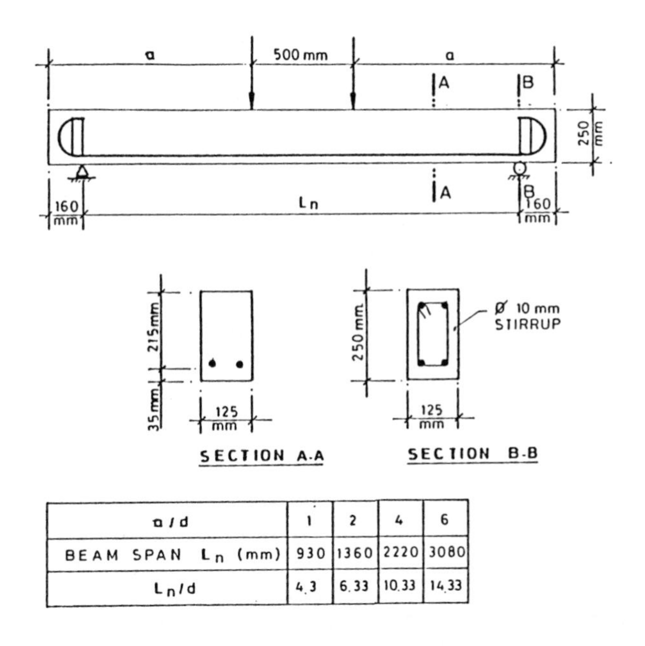

a/d	1	2	4	6
BEAM SPAN L_n (mm)	930	1360	2220	3080
L_n/d	4.3	6.33	10.33	14.33

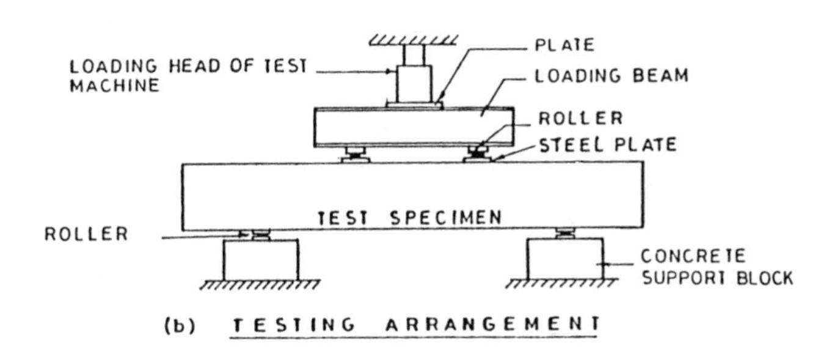

Fig. 1 : Details of test beams and testing arrangements.

microsilica (20 % by weight of cement) with a specific gravity of about 2.2, a bulk density of 6.0 kN/m^3 (37.4 lb/ft^3) and a specific surface of 23 m^2/g was used. Hooked-ends mild carbon steel fibers were used in this investigation. The average length of fibers is 60 mm (2.36 in.), the nominal diameter is 0.8 mm (0.03 in.), the aspect ratio is 75 and the yield strength is 260 MPa (37,500 psi). The bundles are in groups of about 30 steel fibers of average width 24 mm (0.94 in.). The collating fibers create an artificial aspect ratio of approximately 15. A superplasticizer was used and the mixing time was increased to produce uniform mixing of concrete without any segregation.

The measured concrete strengths were based on an average value of 3 specimens. Three 150 x 300 mm (6 x 12 in.) cylinders were cast to determine the concrete compressive strength, three identical cylinders were used to determine the splitting tensile strength. Additionally three 150 x 150x530 mm (6 x 6 x 21 in.) beams were tested to find the modulus of rupture of the concrete used. The concrete was placed in three layers and was internally and externally vibrated immediately afterward. All beams and control specimens were cast and cured under similar conditions. The specimens were kept covered with polyethylene sheets until 24 hours before testing (28 days) to prevent the loss of moisture from the specimens.

2.3 Test Procedure

The beams were simply supported and were subjected to a two-point load as shown in Fig. 1b. The distance between the two point loads was kept constant at 500 mm(20 in.). Special bearing assemblies (roller, guide plates, etc.) were designed to facilitate the application of load to the test specimens. The beam mid-span deflection and the end rotation were measured with the help of transducers. Two internal strain gauges were glued to the main reinforcement (one on each bar) to measure the tensile strain of the reinforcing bars at mid-span, and one external strain gauge glued to the top surface of the concrete to measure the compressive strain. The two-point load was applied to the beam by means of a 400 kN hydraulic testing machine. The load was applied in fifteen to twenty-five increments up to failure. At the end of each load increment, observations and measurements were recorded for mid-span deflection, rotation, strain gauge readings, curvature, and crack development and propagation on the beam surfaces.

3 Experimental Results

Test results from the sixteen HSFRC beams are presented in Tables 1 and 2. These tables include the material properties and the experimental shear strengths of the tested beams.

3.1 Load-Midspan Deflection

Fig. 2 presents the effect of the fiber addition on the load-midspan deflection curves for the tested HSFRC beams with a/d = 1 and 2. The addition of steel fibers not only enhanced the ultimate shear strength of the tested beams but also increased the stiffness at all stages and hence reduced the deflection for a given load. Fig. 2 shows that the beam ductility increases as V_f increases and is

Table 1 : Material Properties of the Tested Beams.

Beam (1)	a/d (2)	V_f (%) (3)	f'_c (MPa) (4)	f_r (MPa) (5)	f'_{sp} (MPa) (6)
B-1-0.0	1	0	94.80	9.87	6.73
B-2-0.0	2	0	94.90	9.83	6.70
B-4-0.0	4	0	92.30	10.33	6.07
B-6-0.0	6	0	92.00	9.99	6.30
B-1-0.5	1	0.5	99.00	11.33	8.00
B-2-0.5	2	0.5	99.10	11.37	8.00
B-4-0.5	4	0.5	95.40	11.27	9.10
B-6-0.5	6	0.5	95.83	11.70	8.47
B-1-1.0	1	1.0	95.30	14.13	10.93
B-2-1.0	2	1.0	95.30	14.13	10.93
B-4-1.0	4	1.0	97.53	15.53	9.70
B-6-1.0	6	1.0	100.50	15.17	9.70
B-1-1.5	1	1.5	96.40	16.77	10.20
B-2-1.5	2	1.5	96.60	16.87	10.23
B-4-1.5	4	1.5	97.10	16.20	9.93
B-6-1.5	6	1.5	101.32	17.12	9.80

Table 2 : Experimental Shear Strength and Corresponding Equivalent Shear Reinforcement Parameters.

Beam (1)	Experimental Shear Strength V_{cf} (MPa) (2)	Strength Enhancement due to Fibers $v_{cf} - v_c$ (MPa) (3)	Area of Equivalent Stirrups Av (mm2) (4)	Spacing S (mm) (5)	Reinforcement Ratio ρ_v (%) (6)	Mode of Failure* (7)
B-1-0.0	7.098	-	-	-	-	A-R
B-2-0.0	3.480	-	-	-	-	A-R
B-4-0.0	1.897	-	-	-	-	F-S
B-6-0.0	1.494	-	-	-	-	F-S
B-1-0.5	9.090	1.992	58.19	d/2	0.4330	A-R
B-2-0.5	4.820	1.340	39.14	d/2	0.2913	A-R
B-4-0.5	2.270	0.373	10.90	d/2	0.0811	F-S
B-6-0.5	1.950	0.456	13.32	d/2	0.0991	F
B-1-1.0	12.740	5.642	82.41	d/4	1.2266	A-R
B-2-1.0	6.060	2.580	75.36	d/2	0.5608	F-S
B-4-1.0	3.170	1.273	37.19	d/2	0.2767	F-S
B-6-1.0	1.960	0.466	13.61	d/2	0.1013	F
B-1-1.5	13.950	6.852	100.08	d/4	1.4895	A-R
B-2-1.5	7.210	3.730	54.48	d/4	0.8108	F-S
B-4-1.5	3.510	1.613	47.12	d/2	0.3507	F-S
B-6-1.5	1.980	0.486	14.20	d/2	0.1057	F

* A-R= Arch=Rib failure, F-S= Flexure-Shear failure,F=Flexure failure

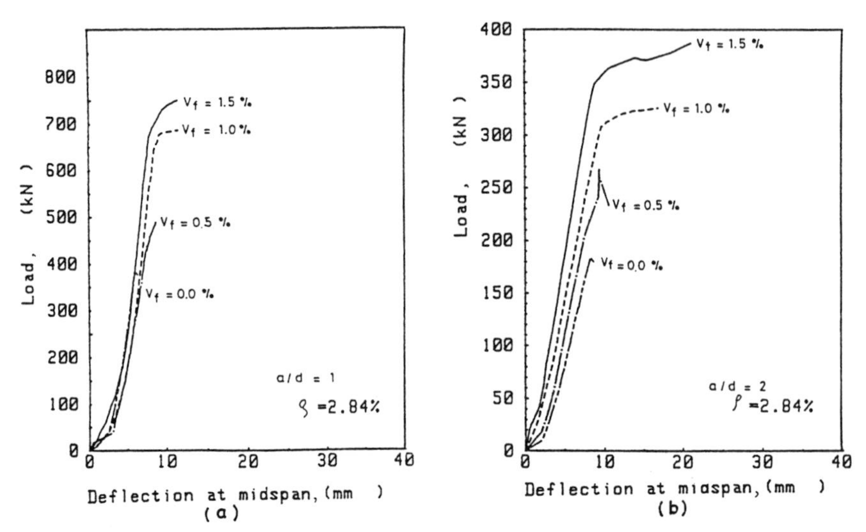

Fig. 2 : Load-deflection curves for beams with (a) a/d = 1, and (b) a/d = 2.

more pronounced as a/d becomes larger.

3.2 Modes of Failure:

High-strength singly reinforced concrete beams without any stirrups or fibers tested in this investigation exhibited four distinct modes of failure. Slender beam (a/d>6) exhibited flexural failure (mode I). Beam with very low reinforcement ratio failed in mode I irrespective of the a/d values. Beams with somewhat smaller value of a/d (4<a/d<6) exhibited either flexure-shear or diagonal tension failures (mode II). Short beams (2.5<a/d<4) failed in either shear-tension or shear compression (mode III). Finally, deep beams (a/d < 2.5) failed in mode IV. The beams can fail in a number of ways, but the crushing of the arch-rib failure was the predominant type.

HSFRC beams exhibited mode of failures generally similar to that of identical beams without fiber, but the presence of high percentage of steel fibers transformed the mode of failure into a more ductile one as indicated in Table 2. Beams with no fibers and with moderate length (3<a/d<6) failed soon after the formation of the first diagonal cracks, while for HSFRC beams with relatively high percentage of fibers several diagonal cracks were observed, thus indicating the redistribution of stress beyond cracking. HSFRC beams with low fiber content (less than 1.0%) exhibited sudden failures at the ultimate stages, although these were far less catastrophic compared with that of the beams without fibers. The steel fibers become more effective after the formation of shear crack and continue to resist the principal tensile stresses until the complete pullout of all fibers occurr at one critical crack.

3.3 Experimental Shear Strength:
Table 2 and Fig. 3 show the effects of both the shear-span/depth (a/d) ratio and the fiber content on the experimental shear strength. Test results show that by

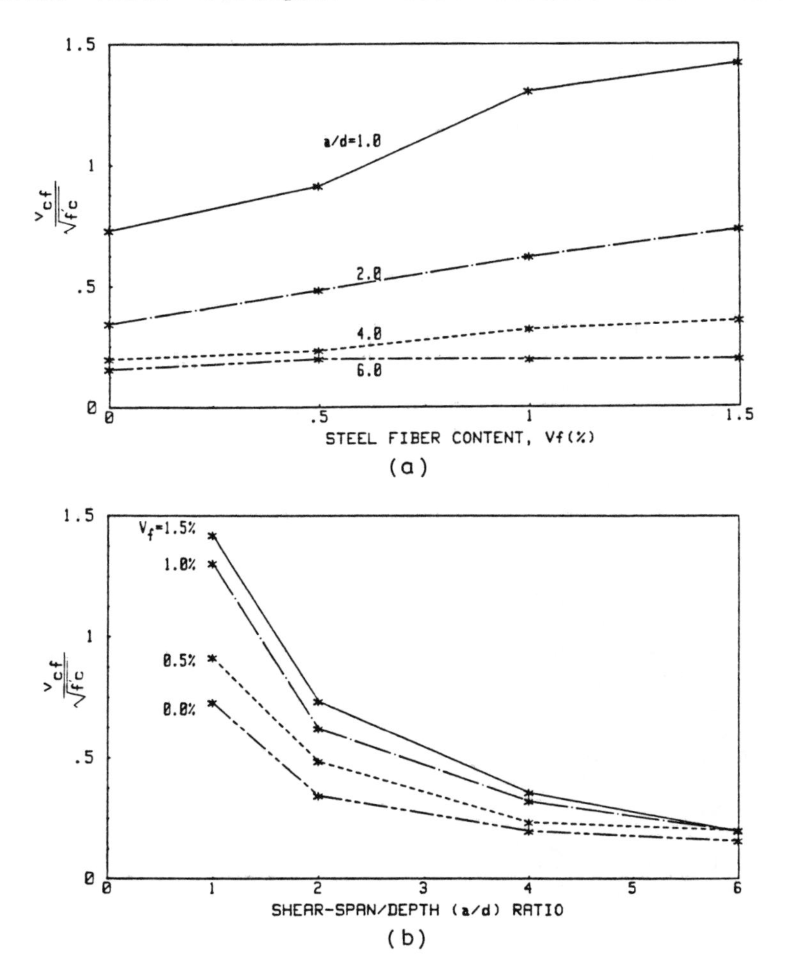

Fig. 3 : Effect of (a) steel fiber content (V_f), and (b) shear-span/depth (a/d) ratio on the experimental shear strength.

increasing the fiber content the shear strength improves irrespective of the a/d values, but the rate of increase is higher as the a/d becomes smaller. Increasing the fiber content from zero to 1.5% causes an increase of the shear strength equal to 96.6 and 32.2 % for a/d = 1 and 6, respectively.

3.4 Equivalent Shear Reinforcement Ratio:

Reinforced concrete beams, without shear reinforcement and fibers, resist shear mostly resisted by the concrete upto the cracking load (ACI-ASCE 1973), then shear is resisted by the aggregate interlock, the dowel action of the main reinforcement, and the resistance of the still uncracked concrete at the top of the beam. The ACI Code (1989) shear design provides equations to estimate the shear strength of plain concrete beams. Wafa et al. (1991) found that the ACI Code (1989) always gives a lower bound estimate of the shear strength of high-strength plain concrete beams, and proposed empirical equations which are a modification of Zsutty's equations (1968):

For a/d > 2.5

$$v_c = 2.1\,(f'_c\,\rho\,\frac{d}{a})^{0.33} \quad \text{(MPa)} \qquad (1a)$$

For a/d < 2.5

$$v_c = (\frac{2.5}{a/d})\ [\text{Eq. 1.a}] \quad \text{(MPa)} \qquad (1b)$$

These equations were examined using other researcher works on high-strength concrete beams and were found to provide good prediction (Wafa et al. 1991).

When fibers are added to the concrete matrix, an extra resisting shear component (v_f) contributed by fibers is added to the shear strength of plain concrete (v_c). Thus, the total internal action (v_{cf}) of FRC beam that resists the applied external shear is given as:

$$v_{cf} = v_c + v_f \qquad (2a)$$

Thus, the additional shear strength contributed by the fibers (v_f) is given as:

$$v_f = v_{cf} - v_c \qquad (2b)$$

The fiber contribution (v_f) is thought as the shear resistance provided by equivalent vertical stirrups having a shear reinforcement ratio of ρ_v and can be modeled as:

$$\rho_v = \frac{v_f}{f_{vy}} = \frac{A_v}{b.S} \qquad (3)$$

Where:

A_v = area of the stirrups (mm^2),
b = width of the beam (mm),
f_{vy}= yield strength of stirrups (MPa).
S = spacing of stirrups (mm),

The enhancement in shear strength, contributed by the vertical component of the fiber pullout forces along the inclined shear cracks, can be theoritically estimated as (Narayanan and Darwish 1987):

$$v_{ft} = 0.41\ \tau\ \frac{L_f}{D_f}\ d_f\ V_f \qquad (4a)$$

where:

v_{ft} = theoritically predicted shear strength contributed by fibers, (MPa),
τ = ultimate bond strength of fibers (MPa),
L_f = length of fiber (mm),
D_f = diameter of fiber (mm),
d_f = bond factor that accounts for differing bond characteristic of the fiber, and equals 0.5 for round fiber, 0.75 for crimped fibers and 1.00 for indented fibers.
V_f = fiber content.

However, Fig. 3 shows that the extra shear strength provided by the fibers is a function of the shear-span/depth (a/d) ratio, and for a given fiber content, a higher shear strength was obtained for smaller a/d ratio. A regression analysis of the test results showed the theoretically predicted contribution to the enhancement of the shear strength by fibers can be modeled as:

$$v_{ft} = 0.41\ c\ \tau(L_f/D_f)\ d_f\ V_f^{m}\ (a/d)^{n} \qquad (4b)$$

where c, m and n are numerical values and equal to 0.0265, 1.2 and 1.1, respectively. Substituting the estimated fiber contribution to the shear strength enhancement (Eq. 4b) for v_f in Eq. 2 yields:

$$\varrho_v = \frac{12.17\ (0.41\ \tau\ d_f\ \frac{L_f}{D_f})\ V_f^{1.2}\ (\frac{d}{a})^{1.1}}{f_{vy}} \qquad (5)$$

The enhancement in shear strengths (v_s) contributed by the equivalent shear reinforcement which is equal to v_{ft} calculated as (ACI 1989):

$$v_s = \frac{A_v}{S.b} f_{vv} \quad (6)$$

Where:

A_v is calculated from Eq.3 and ρ_v is calculated from Eq. 5.

Thus, the total predicted shear strength of the equivalent beam (v_{eq}) is the sum of the strength of the plain concrete and the strength can be given by the equivalent stirrups (Eq. 6).

$$v_{eq} = v_c + v_s \quad (7)$$

The proposed equations (Eqs. 1, 5 to 7) are used to model the tested HSFRC beams as equivalent stirrupped beams. The corresponding ρ_v is calculated for each beam assuming no fibers and the associated area of the stirrups is obtained considering the ACI Code (1989) provisions regarding the maximum stresses in the stirrups and the maximum spacings. These shear reinforcing ratios, areas and spacings are presented in Table 2.
Since only one type of fiber was used in this investigation, the term($0.41\ \tau\ d_f\ L_f/D_f$) in Eq. 5 is a constant value and equal to 87 and the value of the ultimate bond strength of the fiber (τ) is 3.77 MPa (Wafa et al. 1991). Thus, for the type of fiber used in this investigation Eq. 5 can be simplified as:

$$\rho_v = \frac{1058\ V_f^{1.2}}{f_{vv}} (d/a)^{1.1} \quad (8)$$

Based on the proposed model, the shear strength of the equivalent stirrupped beams (Eq.7) have been estimated and compared with the experimental results of the tested HSFRC beams. Fig. 4 shows that the equivalent stirrupped beams follow closely the experimental results. Fig. 5 presents comparison between the experimental equivalent ρ_v estimated using Eq. 3 and the proposed equivalent ρ_v estimated using Eq. 5 as a function of a/d and V_f. A close agreements is observed for all a/d and V_f values. However, it should be noticed that for a/d = 6, the beam behavior is primarily flexural and the results obtained using Eq. 7 are not directly applicable.

4 CONCLUSION

Based on the test results of sixteen high-strength fiber-reinforced concrete beams presented in this paper, the

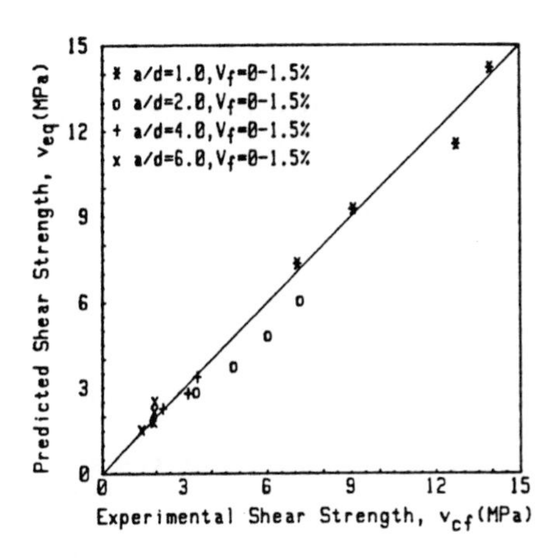

Fig. 4 : Comparison of the experimental and predicted shear strength.

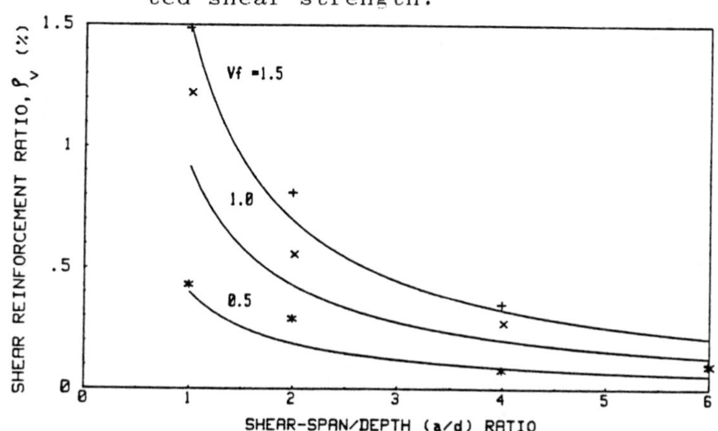

Fig. 5 : Comparison of the experimental and predicted shear reinforcement ratio ρ_v.

following conclusions are drawn:

1. The modes of failure of the tested beams changed from brittle failures for beams without fibers to more ductile failures for beams with fibers.
2. The shear strength of the tested beams increases with an increase in the fiber content, and the increase was inversely related to shear-span/depth ratio.
3. The hooked-end steel fibers acted very effectively as reinforcement against shear failure.
4. A methodology has been suggested for estimating the efficiency of the hooked-end steel fibers as a replacement of equivalent vertical stirrups.

5 ACKNOWLEDGEMENT

The experimental work was carried out in the Structures Laboratory of the Civil Engineering Department of King Abdulaziz University, Jeddah, Saudi Arabia. The financial support under research grant No. 069/410 by the Scientific Research Administration of King Abdulaziz University is gratefully acknowledged.

6 REFERENCES

ACI-ASCE Committee 426, (1973) "The Shear Strength of Reinforced Concrete Members," Journal of the Structural Division, ASCE, Vol. 99, No. ST6, pp. 1091-1187.

ACI Committee 318,(1989) "Building Code Requirements for Reinforced Concrete (ACI 318-89)," American Concrete Institutes, Detroit, 353 p.

ACI Committee 363, (1984) "State of the Art Report on High-Strength Concrete,(ACI 363R-84)," American Concrete Institute, Detroit, 18 p.

ACI Committee 544, (1982) "State of the Art Report of Fiber Reinforced Concrete," ACI Concrete International, V. 4, N.5, pp. 9-30.

Jindal, R.L., (1984) "Shear and Moment Capacities of Steel Fiber Reinforced Concretre Beams," Fiber Reinforced Concrete, SP-81, American Concrete Institute, pp. 1-16.

Naaman, A.E., and Homrich, J.R., (1985) "Properties of High-Strength Fiber Reinforced Concrete," High-Strength Concrete, SP-87, American Concrete Institute, Detroit, pp. 233-249.

Narayanan, R.,and Darwish, I.Y.S.,(1987)"Use of Steel Fibers as Shear Reinforcement," ACI Structural Journal Proceedings V. 84, No. 3, pp. 216-227.

Rumualdi, J.P., and Batson, G.B., (1983)"The Mechanics of Crack Arrest in Concrete," Journal of the Engineering Mechanics Division, ASCE, Vol. 89, pp. 147-168.

Swamy, R. N., (1987) "High-Strength Concrete-Material Properties and Structural Behavior," High-Strength Concrete, SP-87, American Concrete Institute, pp. 110-146.

Swamy, R.N., (1986) "Properties of High-Strength Con-Concrete," Cement, Concrete, and Aggregates, Vol. 8, No. 1, pp. 33-41.

Swamy, R.N., and Bahia, H.M.,(1985) "The Effectiveness of Steel Fibers as Shear Reinforcement," ACI Concrete International, Vol. 7, No. 3, pp. 35-38.

Swamy, R.N., Mangat, P.S., and Rao,C.V., (1974) "The Mechanics of Fiber Reinforcement of Cement Matrices," Fiber Reinforced Concrete, SP-44, American Concrete Institute,Detroit, pp. 1-28.

Wafa, F.F., Ashour, S.A., and Hasanain, G.S.,(1991)

"Flexural and Shear Behavior of High-Strength Fiber Reinforced Concrete Beams." Final Report, Research No. 410/069, College of Engineering, King Abdulaziz University, Jeddah, Saudi Arabia, 107 p.
Zsutty, T.C.,(1968) "Beam Shear Strength Prediction by Analysis of Existing Data," ACI Journal, Proceedings V. 65, No. 11, pp. 943-951.

43 POLYPROPYLENE FIBRE CONCRETE BEAMS IN SHEAR–FLEXURE

S. GHOSH and A. ROY
Department of Civil Engineering, Jadavpur University, Calcutta, India

Abstract
Thirty-five Polypropylene fibre reinforced concrete (PFRC) beams were tested to destruction to study their behaviour in shear-flexure mode. The contribution of concrete in compression, concrete in tension, longitudinal reinforcement and fibres in particular were estimated from experimental results. An expression for ultimate shear capacity has been developed involving parameters such as fibre content, a/d ratio, grade of concrete, p_c/p_t ratio, p_t etc. A good agreement between theoretical and experimental results were observed.
Keywords: Concrete, Shear, Flexure, Fibre, Composite, Cement.

1 Introduction

The present development of fibre reinforced cement and concrete is only about three decades old. The fibres included are asbestos (Chrysotile, Crocidolite), carbon, cellulose, Glasscem-Fil filament, 204 filament strand, kevlar, Nylon, Polypropylene (Monofilament, fibrillated) polyethylene, steel, Akwara, Alumina, coconut fibres, perlon, rock wool, sisal.

Polypropylene fibres were suggested as an admixture to concrete in 1965 by Goldfein for construction of blast-resistant buildings for the U.S. Corps of Engineers. The development of Polypropylene in a new form, the isotactic configuration, and commercial production in the 1960s, offered the textile industry a potentially low-priced polymer capable of being converted into useful textile fibre. Polypropylene fibres then became available in two forms, monofilaments (or spinneret) fibres and film fibres. These fibres are normally circular in cross-section. Fibrillated films twisted into the form of fibres have a softer handle than spinneret fibres and were mainly developed for use in rope and twine, but proved to be useful in concrete as well. The types of fibre are characterised by figures expressing the length in metres per kilogram. The fibres are supplied in spool form for cutting on site, or are chopped by the

Fibre Reinforced Cement and Concrete. Edited by R. N. Swamy. ©1992 RILEM.
Published by E & FN Spon, 2-6 Boundary Row, London SE1 8HN. ISBN 0 419 18130 X.

manufacturer, usually in staple lengths between 25 and 75 mm. Purchasing the fibres on spools and cutting to the required length in the precast works can lead to considerable savings. Spools are also easier to handle in transport and require less storage space.

The present investigation is undertaken to study the behaviour of polypropylene fibre reinforced concrete beams in shear-flexure mode. Ordinary concrete is very weak in tension. This deficiency is normally compensated by using reinforcing steel. However by this technique the basic tensile strength of concrete cannot be improved. The basic tensile strength of ordinary concrete can be improved by adding fibres of different nature. Fibre addition to concrete often is a convenient means of achieving improvements in many engineering properties of concrete such as fracture toughness, fatigue resistance, impact resistance, shear strength and flexural strength. But most of the developments are centred to steel fibre concrete. Polypropylene fibre reinforced concrete may be very useful in connection to low cost housing technology. A study on PFRC beams under shear-flexure mode may be useful in understanding the subject in a better way.

An experimental investigation has been carried out to study the overall behaviour of polypropylene fibre reinforced concrete beams in shear-flexure mode. The experimental investigation consisted of testing on thirty-five beams out of which some beams are conventional RC beams with and without stirrups. The complete details are shown in Table 1. The beams were tested under two-point loading with shear-span to depth ratio that varied from 1 to 2.5.

2 Experimental programme

2.1 Casting of specimens

The experimental programme was undertaken to test standard cubes, cylinders and prisms for studying strengths in compression, indirect tension and shear. The samples were prepared without any fibre and with polypropylene fibres of intensities 1, 2, 3 and 4 Kg/m^3 of concrete.

The concrete mix was made using ordinary portland cement conforming to IS Hand Book (SP23), well graded 20 mm down stone chips, 2.36 mm down river sand and potable water. The mix proportion was 1:2:4 by volume and water cement ratio 0.6. The mixer was a tilting type with capacity 0.2 m^3. 0.1 m^3 was mixed at a time. The casting and testing of cubes, cylinders and prisms were done following relevant IS codes. Twenty eight days under water curing was done for all the samples. Testing was performed between 1 month to 1½ months after casting. A total of at least three samples tested and the average values were taken for analysis and presentation.

In this investigation a series of tests on polypropylene fibre reinforced concrete beams were performed and also on ordinary reinforced concrete beams for comparison. The total of thirty-five beams were divided into three groups. The detail of beam specimens tested were detailed in Tables 1 & 2. In all the cases the specification of polypropylene fibres was 15 denier (15 gms. weight of 9,000 m long fibre). These had specific gravity as 0.9 and tensile strength 400 N/mm^2.

Table 1. Detail of PFRC beams and testing parameters (first phase of testing)
Loading scheme : Two point symmetrical loading

Beam Mkd.	Fibre content (Kg/m^3 of concrete)	$\frac{a}{d}$	Other details
B1	1	1.4	Average grade of concrete = M15
B2	2	1.4	Beam section = 125 mm x 150 mm
B3	3	1.4	a = 175 mm ; Span = 1400 mm
B4	4	1.4	A_{st} = 3 - 8ϕ m.s. bars
B5	0	1.4	A_{sc} = 2 - 6ϕ m.s. bars
B6	1	1.4	Average grade of concrete = M20
B7	2	1.4	Beam section = 125 mm x 150 mm
B8	3	1.4	a = 175 mm ; Span = 1400 mm
B9	4	1.4	A_{st} = 3 - 8ϕ m.s. bars
B10	0	1.4	A_{sc} = 2 - 6ϕ m.s. bars
B11(PC)	0	2	Average grade of concrete = M15
B12(PCS)	0	2	Beam section = 150 mm x 250 mm
B13(FC)	2.5	2	a = 450 mm ; Span = 1900 mm
B14(FCS)	2.5	2	A_{st} = 2 - 12ϕ HYSD bars
B15(FPC)	2.5	2	A_{sc} = 2 - 6ϕ m.s. bars
B16	2.5	1	Average grade of concrete = M15
B17	2.5	2	Beam section = 150 mm x 200 mm
B18	2.5	2.5	Span = 1900 mm
B19	0	2.5	A_{st} = 2 - 12ϕ HYSD bars
			A_{sc} = 2 - 10ϕ HYSD bars
B20	0	1	Average grade of concrete = M20
B21	0	2	Beam section = 150 mm x 200 mm
B22	0	2.5	Span = 1900 mm
B23	2.5	1	A_{st} = 2 - 12ϕ HYSD bars
B24	2.5	2	A_{sc} = 2 - 10ϕ HYSD bars
B25	2.5	2.5	

* Yield stress of m.s. bars = 320 N/mm^2 ;
0.2% proof stress of HYSD bars = 410 N/mm^2.

Table 2. Detail of PFRC beams and testing parameters (second phase of testing)
Loading scheme : two-point symmetrical loading
Yield stress of m.s. bars : 320 N/mm^2
0.2% proof stress of HYSD bars : 410 N/mm^2

Beam Mkd.	Fibre content (Kg/m^3 of concrete)	$\frac{a}{d}$	Other details
B26	1.75	1.4	Average grade of concrete = M15
B27	1.75	2.25	Beam section = 150 mm x 200 mm
B28	2.25	1.4	Span = 1900 mm
B29	2.25	2.25	A_{st} = 3 - 12 ϕ HYSD bars
			A_{sc} = 2 - 10 ϕ HYSD bars
B30	1.75	1.4	Average grade of concrete = M20
B31	1.75	2.25	Beam section = 150 mm x 200 mm
B32	2.25	1.4	Span = 1900 mm
B33	2.25	2.25	A_{st} = 3 - 12 ϕ HYSD bars
			A_{sc} = 2 - 10 ϕ HYSD bars
B34	2	1.4	Average grade of concrete = M15
B35	2	2.25	Beam section = 150 mm x 250 mm
			Span = 1900 mm
			A_{st} = 2 - 12 ϕ HYSD bars
			A_{sc} = 2 - 6 ϕ m.s. bars

Some beams were provided with no shear reinforcements. Some beams were cast with plain concrete and was marked as P.C. Beams cast with fibre reinforced concrete were marked as F.C. The beams cast with both fibre reinforced concrete and plain concrete for upper half and lower half of the beam respectively were marked as F.P.C. The beams cast with plain concrete with stirrup were marked as P.C.S. and beams cast with fibre reinforced concrete with stirrup were marked as F.C.S. The beams were tested under two-point symmetrical loading adjusting different a/d ratio.

2.2 Testing of specimens

The testing of specimens was done with universal testing machines of 100 MT (980 KN) capacity for cubes, cylinders and prisms and 10 MT (98 KN) capacity for beams and the rate of increase of load was 400 Kg (3923 N) per minute.

3 Theoretical expression for shear

Emperical formula suggested by Krefeld and Thurston for ultimate shear resistance is given as :

$$\frac{V_{ult}}{b\,D} = 0.35\sqrt{\sigma_{cu}} + \frac{300\,p_t}{\frac{M}{Vd}} \quad \text{(in metric system)}$$

where,

V_{ult} = ultimate shear force at failure ;

b = width of the beam ; D = overall depth of the beam ;

σ_{cu} = cube compressive strength of fibre concrete ;

$\frac{a}{d} = \frac{M}{Vd}$ = shear span to effective depth ratio (M=Pa) ;

p_t = percentage of tensile reinforcement.

The above formula do not include the influence of compression steel.

In the experiments[3], it was found that a considerable increase in the percentage of shear resistance occurs with the increase of percentage of compression steel (p_c/p_t) with a fixed percentage of tension steel and also when the percentage of tension steel is increased by varying the ratio of p_c/p_t as in former case.

Considering the contribution of compression steel modified emperical expression[3] has been established on the basis of experimental results[3] of forty-six beams. The experimental data of beams marked B1 to B25 and the previous results[3] were analysed and an expression for ultimate shear capacity of PFRC beam sections was established. The proposed equation is basically an extended form of Krefield and Thurston's work based on dimension analysis. The value of 'K' is to be found out by using Figs 1, 2 and 3 and an expression $K = (K_1 K_2 K_3)^{1/3}$.

Proposed equations for predicting ultimate shear capacity and cracking shear are as follows

$$\frac{V_{ult}}{b\,D} = \frac{0.35\sqrt{\sigma_{cu}}}{\sqrt{\frac{a}{d}}} + \frac{300\,p_t}{\frac{M}{Vd}} \times \frac{1}{K(1 + \frac{p_c}{p_t} + \frac{V_f}{V_c})}$$

and

$$\frac{V_{crack}}{b\,D} = \frac{0.20\sqrt{\sigma_{cu}}}{\sqrt{\frac{a}{d}}} + \frac{270\,p_t}{\frac{M}{Vd}} \times \frac{1}{K(1 + \frac{p_c}{p_t} + \frac{V_f}{V_c})}$$

where,

p_c = percentage of compression steel ;

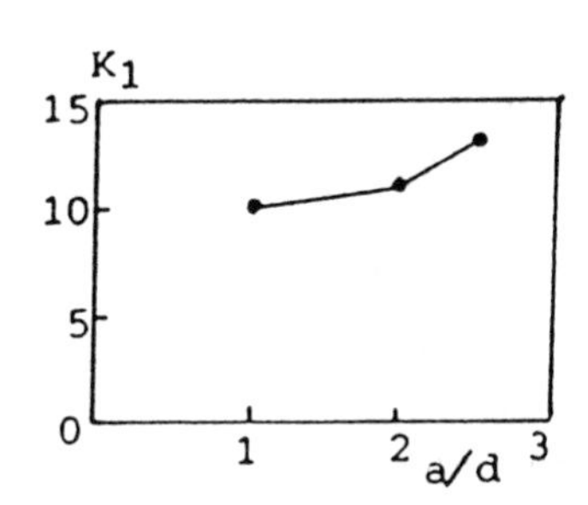

Fig.1. a/d vs. K_1 plot

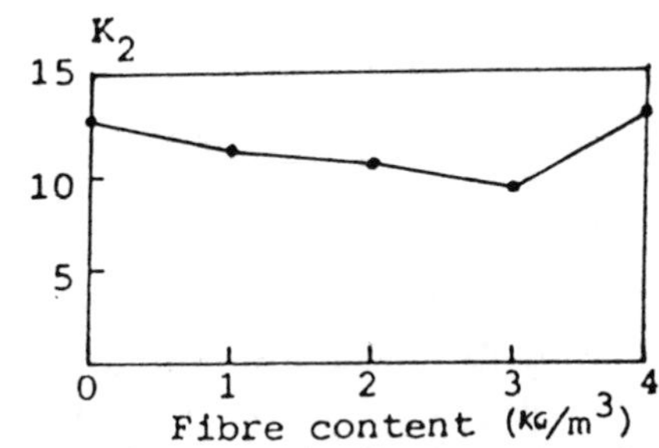

Fig.2. K_2 vs. Fibre content plot.

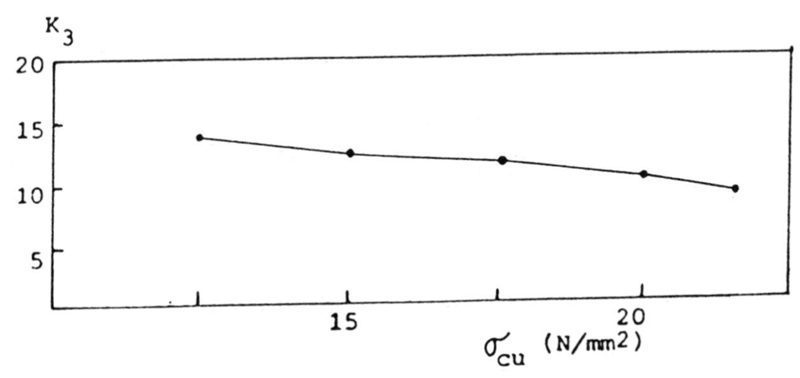

Fig. 3. σ_{cu} vs. K_3 plot.

K = a factor involving unknown effects of combined stresses and varies with percentage of compression steel and fibre content ;

V_f = total volume of fibre in the beam ;

V_c = volume of beam ;

M = average moment in shear span = $\frac{P.a}{2}$.

In the 2nd phase of investigation on ten beams (complete details are shown in Table 2) were tested to destruction and experimental results were compared with theoretical results using proposed equations and a reasonable agreement was observed. Tables 3 to 6 and Figs. 4 to 7 show some interesting results.

Table 3. Comparison of results

Central deflection (δ_c) in mm	Load on beams in tonnes					Ratio of loads					
	P C	F C	P C S	F C S	F PC	$\frac{FC}{PC}$	$\frac{FCS}{PCS}$	$\frac{PCS}{PC}$	$\frac{FCS}{FC}$	$\frac{FC}{FPC}$	$\frac{FPC}{PC}$
1.675	3.85	4.45	5.3	5.45	5	1.15	1.03	1.38	1.22	0.89	1.30
4.875	7.5	8.15	10.7	11.0	8.8	1.09	1.03	1.43	1.35	1.93	1.17
5.025	7.95	9.05	11.65	12.45	9.2	1.14	1.07	1.46	1.37	0.98	1.16
5.8625	8.35	9.95	12.2	13.95	9.3	1.19	1.14	1.46	1.40	1.07	1.11
6.7	8.5	10.6	12.3	15.2	9.4	1.25	1.24	1.45	1.43	1.13	1.10
At failure	8.5	11.75	12.3	17	9.4	1.38	1.38	1.45	1.45	1.25	1.10

* The above values are taken from experimental graphs.

Table 4. Ratio of Ultimate load to Cracking load

Sl. No.	Beam Mkd.	Cracking load (ton) W_{cr}	Ultimate load (ton) W_u	W_u/W_{cr}
1	P C	7.2	8.5	1.18
2	F C	8.6	11.75	1.37
3	P C S	7.8	12.3	1.57
4	F C S	10.1	17.0	1.69
5	F PC	7.2	9.4	1.30

Table 5. Compressive Tensile and Flexural strength of FRC and PC

Type of concrete	Av. cube comp. strength, in N/mm^2	Av. split cylinder direct tensile strength in N/mm^2	Flexural tensile strength from prism test in N/mm^2
F RC	15.55	2.087	3.93
P C	14.55	1.7	2.72

Table 6. Increase in ultimate shear stress for fibre concrete beams over conventional reinforced beams

Quantity of fibre added in Kg/m^3 of concrete	Ultimate shear stress in (N/mm^2)	% increase over conventional reinforced concrete beams
4	2.30	-2.54%
O	2.36	-
3	2.88	+22.03%
2	2.88	+22.03%
2	3.20	+35.6%
1	2.68	+13.56%

* Average grade of concrete M15.

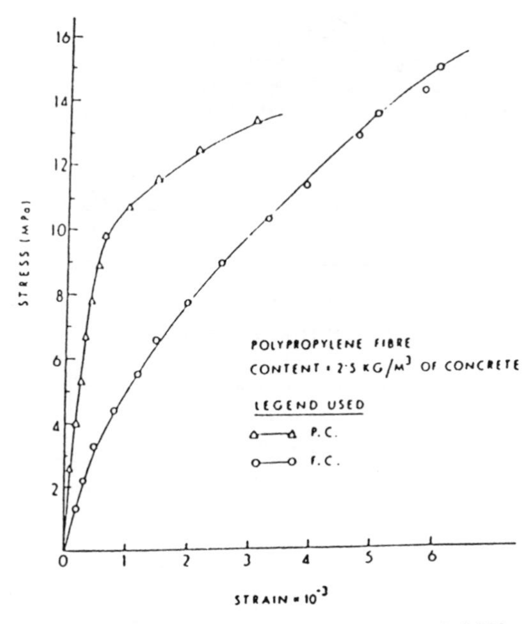

FIG.4 STRESS vs. STRAIN PLOT FOR PLAIN AND FIBRE REINFORCED CONCRETE CUBES.

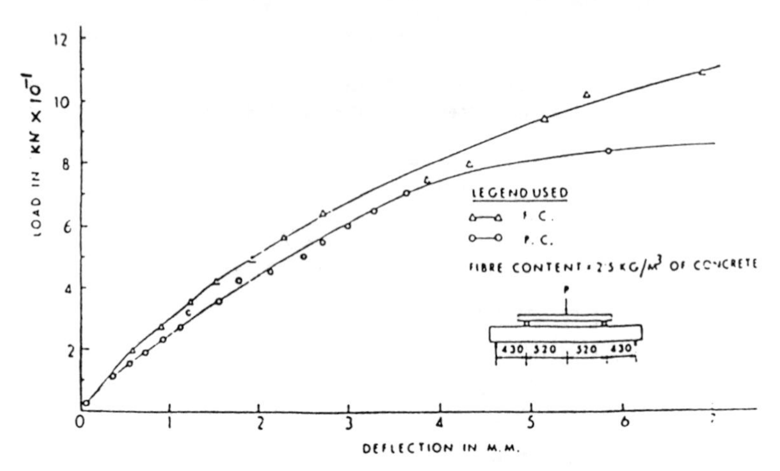

FIG.5 LOAD vs. CENTRAL DEFLECTION FOR THE BEAMS MKD F.C. AND P.C.

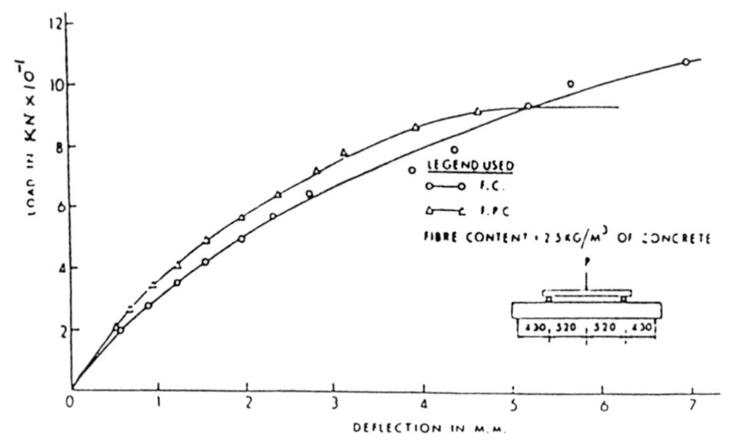

FIG. 6 LOAD vs CENTRAL DEFLECTION FOR THE BEAMS MKD F.P.C AND F.C.

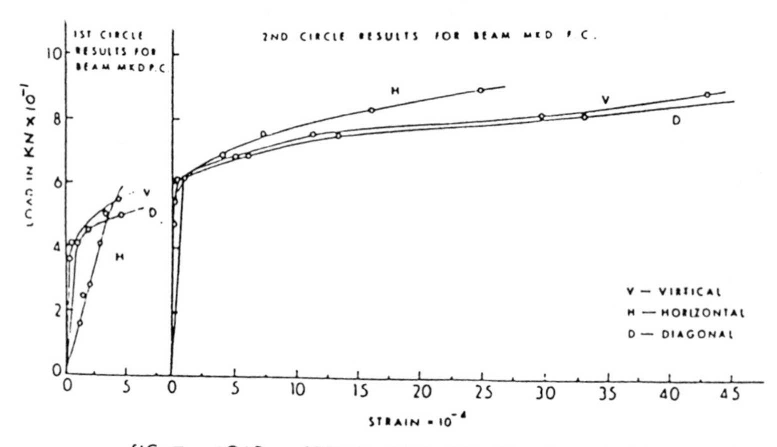

FIG 7 . LOAD vs. STRAIN PLOT FOR BEAMS MKD F.C. AND P.C.

4 Conclusion

The present investigation shows the improvement of strength parameters due to adoption of polypropylene fibres. The strength parameters increases by increase of fibre content upto a certain value and then decrease even with increase of fibre content. The useful range observed as 2 to 3 Kg per m^3 of concrete volume. The ductility of the beams improved due to fibre insertion which are reflected from load deflection and moment-curvature relationships. The theoretical predicted values of shear strengths show a good agreement with the experimental results (TABLE 7).

Table 7. Comparison of theoretical results with experimental results

Beam Mkd.	Variation of theoretical results from experimental results in %	
	V_{crack}	V_{ult}
B26	10.2	12.5
B27	11.6	13.7
B28	10.0	15.6
B29	8.6	5.7
B30	9.6	7.8
B31	10.6	15.2
B32	3.2	6.8
B33	5.6	7.9
B34	11.0	9.0
B35	14.0	10.6

5 Notations

- a = distance from the support to the nearest load i.e. shear span ;
- b = width of beam ; d = effective depth of beam ;
- D = overall depth of beam ; V = shear force ;
- V_c = volume of beam ;
- V_f = total volume of fibre in the beam ;
- σ'_{cu} = characteristic cube compressive strength of concrete at 28 days IN N/mm^2;
- K = a factor involving unknown effects of combined stresses and varies with percentage of compression steel and fibre content ;
- p = applied concentrated load ;
- p_c = percentage of compression steel ;

p_t = percentage of tension steel ;
V_{crack} = effective cracking shear force ;
V_{ult} = effective ultimate shear force ;
M = average moment in shear span.

6 References

Batson, G. Ball, C. Bailey, L. Landers, E. and Hooks, J. (1972) Flexural fatigue strength of steel fibre reinforced concrete beams, **Journal of ACI**, November 1972,69(11), 673-677.

Batson, G. Jenkins, E. and Spatney, E. (1972) Steel fibres as shear reinforcements in beams, **Journal of ACI**, October 1972, 69 (10), 640-644.

Ghosh, D. Dutta, A. Kundu, D. and Sengupta, B. (1990) Contribution of compression steel towards shear-flexure failure of reinforced concrete beams. **Journal of Institution of Engineers (India)**, Vol. 71, September 1990, 65-67.

Hannant, D.J. (1978) **Fibre cements and fibre concretes**, John Wiley & Sons, 241 p.

Krefield, M.J. and Thurston, C.W. (1966) Contribution of longitudinal steel to shear resistance of reinforced concrete beams. **Journal of the American Concrete Institution**, Vol. 63, No. 3, March 1966, 325 p.

Kukreja, C.B. Kaushik, S.K. Kanchi, M.B. and Jain, O.P. (1980) Tensile strength of steel fibre reinforced concrete. **Indian Concrete Journal**, July 1980, 54(7),187-188.

Ray, Arunachal and Ghosh, Somnath (1990) A study on natural and synthetic fibre concrete. **National Seminar on Innovative Technology for Low Cost Rural Housing by ASCE-IS and RHDC**, Calcutta, 27th April 1990, AR-1 to AR-8.

Ray, Arunachal and Ghosh, Somnath (1991) Polypropylene fibre reinforced concrete as construction material. **Third NCB International Seminar on Cement and Building Materials**, January 1991, New Delhi.

Narayan, R. and Darwish, I.V.S. (1987) Use of steel fibres as shear reinforcement, ACI Structural Journal, May-June 1987, 84(3), 216-227.

Romualdi, J.P. and Batson, G.B. (1963) Mechanics of crack arrest in concrete, **ACSE (Mech. Engg. division)**, June 1963, 89(EM3), 147-168.

Romualdi, J.P. and Mandel, J.A. (1964) Tensile strength of concrete affected by uniformly distributed closely spaced short lengths of wire reinforcements. Journal of ACI, June 1964, 61(6), 657-671.

Swamy, R.N. and Mangat, P.S. (1975) The onset of cracking ductility of steel fibre concrete, **Cement and Concrete Research**, January 1975, 5(1), 37-53.

44 CONTRIBUTION OF FIBRES IN SFRC BEAMS FAILING IN SHEAR

A. CHERIYAN and S. KRISHNAN
College of Engineering, Trivandrum, India

Abstract
The experimental study on influence of fibre reinforcement on shear-flexure interaction for longitudinally reinforced concrete beams without web reinforcement, is being dealt with in this paper. The influence of fibre reinforcement in the area of shear is well established. The authors propose a mathematical model in line with that of Gaetano et al[3] to quantify effect of fibre contribution ver conventional concrete beams.
Key words: Analysis, Beams, Contribution of fibres, Fibre Volume, Flexural, Capacity, Longitudinal reinforcement, Shear span of depth ratio.

1 Introduction

Even though concrete has served the construction industry as the leading building material of this decade, its widespread application in certain specialised fields has been somewhat restricted in the past due to its low tensile strength, susceptibility to cracking and lack of ductility. These shortcomings have been sought to be corrected by a number of innovative modifications of which, perhaps the most promising has been the development of Steel Fibre Reinforced Concrete composites. The highly salubrious effect that inclusion of steel fibres produces in improving the above mentioned properties of all types of concretes has persuaded research workers to study the effect of inclusion of fibres in several other areas also in which concrete performance has been found wanting.

Concrete subjected to shear has always been such a troubled area. Lack of strength in shear coupled with the violent nature of shear failures has forced designers to adopt measures which would eliminate the possibility of shear failures completely and reinforce adequately with a significant quantity of web reinforcement. Imperfect understanding of shear failures has also encouraged such an overcautious approach. However, the fact that some of factors that reduced the capacity of beams failing in shear (related to low tensile strength of concrete and

Fibre Reinforced Cement and Concrete. Edited by R. N. Swamy. © 1992 RILEM.
Published by E & FN Spon, 2-6 Boundary Row, London SE1 8HN. ISBN 0 419 18130 X.

lack of ductility), were precisely those which could be improved by SFRC and hence a number of such studies have been undertaken recently.[1,2,3] Although significant contributions and higher load carrying capacities have been noted in these tests due to the inclusion of fibre reinforcements, no clear picture has emerged regarding either the action of fibres or their exact contribution in enhancing shear capacity of reinforced concrete beams under different loading conditions.

One of the major difficulties in quantifying the contribution of steel fibres in reinforced concrete beams resisting shear has been, indeed, the lack of clear understanding of the shear problem itself. Recent work, notably by Gaetano et al[3], redefined the so-called'shear failures' as failures at reduced flexural capacity due to the effect of shear. A combined beam and arch action model proposed in Gaetano's[3] work has succedded in predicting the detrimental effect of shear on flexural capacity, the effect of the shear span to depth ratio (a/d) and the 'critical' a/d ratio at which flexural capacity is a minimum.

The approach outlines above has been used in this paper to evaluate the results of a series of SFRC beams under various flexure -shear combinations and has provided a basis for evaluating and predicting the contributions of fibres in RC beams. Only members without web reinforcement have been analysed and a mathematical model to assess the contribution of fibres separately in SFRC beams has been suggested.

2 Test set up and experimental results

The results of tests on 12 reinforced concrete beams (each test result being the average for two identical specimens) have been presented in this paper with fibre content and a/d ratio as the main variables. Beams tested are of uniform rectangular cross section having dimensions 120 mm x 200 mm with effective depth of 175 mm reinforced longitudinally with two bars 10 mm diameter. Three fibre contents were used viz. 0, 0.75 and 1.2% by volume of conventional concrete using 10 mm aggregate, river sand, and ordinary portland cement. Concrete strengths were established by 150 mm cube control specimens, a minimum of three in each case. Steel fibres used were ordinary tying wire type, 0.52 mm diameter cut to average aspect ratio (l/d)80-90.

Material properties and specimen details are given in Tables 1 and 2 Test results of various specimens are given in Tables 3 and 4.

Table 1 Properties of materils used

10 mm crushed aggregate	FM	5.93
	Sp.Gravity	2.68
River sand	FM	2.9
	Sp.Gravity	2.60
Ordinary portland cement		
7 day compressive strength		18 N/mm^2
10 mm ϕ torsteel f_y		380 N/mm^2
Steel fibres		
diameter		0.51
aspect ratio		80.90
	f_y	480 N/mm^2

Table 2

Specimen details	
Mix	M20
Size of beam	120 x 200 x 2500 mm
Longi. reinforcement	2 Nos. 10 mm Torsteel
Fibre Reinforcement by volume percentage	0, 1, 1.5
Shear span to depth ratio (a/d) tested	1, 2, 3, 4

Table 3 Test Results

Beam Desig nation	a/d	V_f %volume	Matrix strength N/mm^2		Moment KNM		Mode of failure
			σ_c	σ_{sp}	M_{cr}	M_u	
A1	1	0.0	30.1	2.7	6.04	7.48	SP
A2	2	0.0	30.0	2.72	9.18	11.22	ST
A3	3	0.0	30.0	2.71	8.93	11.48	ST
A4	4	0.0	31.0	2.8	10.88	12.58	F
B1	1	0.75	34.0	3.55	7.82	9.18	SP
B2	2	0.75	32.0	3.5	11.73	13.86	F
B3	3	0.75	34.0	3.2	10.71	12.62	SF
B4	4	0.75	34.0	3.54	11.22	13.26	F
C1	1	1.2	33.0	3.4	9.18	10.97	SP
C2	2	1.2	26.8	3.6	10.29	11.31	ST
C3	3	1.2	29.0	3.4	11.22	14.03	F
C4	4	1.2	33.0	3.3	12.24	12.92	F

F flexure failure SF shearflexure
ST shear tention SP shear proper.

Table 4 Moment Characteristics

Beam	V_f	a/d	Enhanced moment over conventional in KNM M_{cr}	M_u	$\frac{M_{cr}}{M_u}$	$\frac{M_u}{M_{fl}}$	$\frac{M_{cr}}{M_{fl}}$	Mode of failure
A1		1			0.81	0.596	0.48	SP
A2		2			0.82	0.89	0.73	ST
A3		3			0.77	0.912	0.7	ST
A4		4			0.86	1.0	0.86	F
B1	0.75	1	1.78	1.7	0.85	0.69	0.59	SP
B2	0.75	2	2.55	2.64	0.801	1.0	0.885	F
B3	0.75	3	1.78	1.14	0.85	0.95	0.8	SF
B4	0.75	4	0.34	0.68	0.85	1.0	0.85	F
C1	1.2	1	3.14	3.49	0.84	0.85	0.71	SP
C2	1.2	2	1.11		0.91	0.875	0.79	ST
C3	1.2	3	2.29	2.55	0.8	1.0	0.87	F
C4	1.2	4	1.36	0.34	0.95	1.0	0.95	F

M_{cr} cracking movement. M_u ultimate movement.

M_{fl} moment of flexure failure.

3 Gaetano's model

The importance of the study reported by Gaetano et al[3] is that it recognises for the first time the irrationality in separating shear resistance and flexural capacity of an RC beam in which shear and flexure are interconnected by equilibrium considerations. The flexural capacity of a beam as calculated from consideration of the section properties of midspan section alone is also of dubious value as shearflexure interaction is bound to limit the final moment of resistance of the member at failure in which properties of several other cross sections of the member, the disposition of the reinforcing steel, the diameters of the aggregate used and particularly the a/d ratios play an important part [4,5,6]

The model assumes that both beam action and arch action as exemplified in the ACI building code equation for nominal shear stress viz. $\upsilon_c = 1.9\ (f_c')^{\frac{1}{2}} + 2500\ V_u d/M_u$ (valid in fps units only)(1) occur simultaneously and contribute directly to the shear resistance of the beam and indirectly to the limiting flexural capacity to which it is related through the a/d ratio (M/Vd).

The equation proposed by Gaetano for the shear capacity of RC beam is

$$\frac{M_u}{M_{fl}} = \xi \frac{0.83\rho^{1/3}\, fc'^{\,1/2}\, a/d + 206.9\rho^{5/6}\,(a/d)^{-3/2}}{\rho fy\,(1 - f_y/1.7fc')} \qquad (2)$$

where

ξ = function taking aggregate size effect into account
ρ = tensile reinforcement ratio A_s/bd
fc' = compressive concrete strength (cylinder)

a/d = shear span depth ratio

f_y = yield strength of tensile reinforcement

M_u = moment capacity including shear influence, and

M_{fl} = moment capacity in pure flexure

This model not only shows good agreement with test results on several RC beams failing in shear and flexure under various a/d ratios but is also able to identify a critical a/d ratio at which flexural capacity is a minimum which is also an experimentally evident fact. In this work the Gaetano's model is suitably modified and is able to quantify the fibre contribution in SFRC beams.

4 Proposed model

An analysis of the test results clearly defines the extent of contribution made by the fibres in SFRC under various flexure - shear combinations. The proposed model is based on the combined beam and arch action model for relative flexural capacity as proposed by Gaetano et al.

Application of the Gaetano model directly does not show agreement with test results thereby indicating that:

(i) there is a significant contribution to strength by the fibres which has to be accounted for and that
(ii) since the influence of concrete strength in Gaetano's model appears in the beam action component as fc',evidently depicting the tensile rather than the compressive strength of concrete, there is good reason to replace this factor in the analysis of SFRC beams by the corresponding improved tensile strength of concrete obtained from test results, viz the σsp value.

The Gaetano's model modified as in (ii) above shows better agreement with test results but still needs further refinement to include the effect of the fibres themselves. The increase in strength due to the inclusion of fibres is exhibited by test beams both at cracking stage as also at ultimate stage by almost the same value (Figs. 1 & 2). It is therefore, proposed to account for the fibre contribution by a single equation including all parameters that affect both beam action and arch action viz. tensile strength (σ_{sp}), fibre volume ratio ρ' and a/d, with the quotient for each parameter chosen bearing in mind the corresponding quotient in the Gaetano's equation.

The proposed equation for additional moment capacity due to the inclusion of fibres is given by.

$$M_{fibre} = K. \sigma_{sp} \, bd^2 \, (\rho')^{1/3} \, (a/d)^{-3/2}$$ where

σ_{sp} = split tensile strength, directly measured cylinder specimen of SFRC.

ρ' = fibre volume expressed as %age usually kept at < 2% from economical considerations and also to form matrix controlled composite.

a/d = shear span - depth ratio which determines the extent of shear-flexure interaction and

κ = A constant determined from experimental results in these case = 2/3

5 Discussions

Shear-flexure interaction of beams reinforced longitudinally, without web reinforcement, is analysed based on a modified version of the very recent mathematical model[3] suggested by Gaetano Russo et al. The model where the combined effect of shear and flexure is taken care of by the arch action and beam action agrees well with the experimental results for plain concrete specimen for most of the shear span to depth ratios.

The mathematical model of reference gives a measure of flexural capacity for various shear span to depth ratios as shown in Table 5 and Figs.3, 4 & 5. It is seen that the moment capacity agrees well in a conservative fashion when the flexural influence is defined by the split tensile strength of the concrete used.

The agreement of the model for plain concrete members in most of the loading situations (defined by M/Vd or a/d) enabled the authors to use the same model for SFRC members. The results are tabulated and compared with the experimental results in Table 6 & 7.

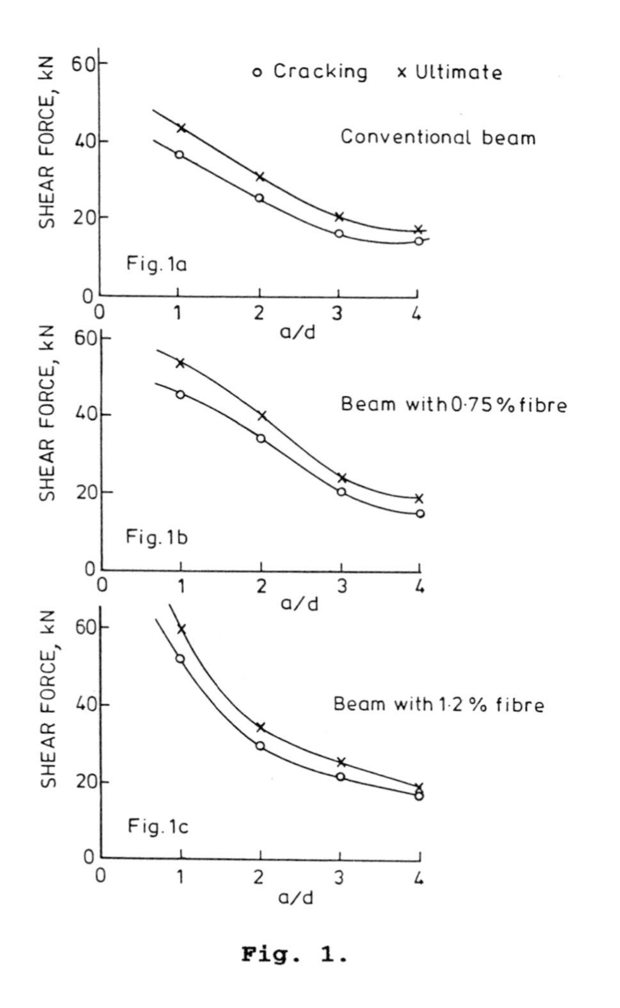

o Cracking x Ultimate
Conventional beam
Beam with 0·75% fibre
Beam with 1·2 % fibre
SHEAR FORCE, kN
a/d
Fig. 1a
Fig. 1b
Fig. 1c

Fig. 1.

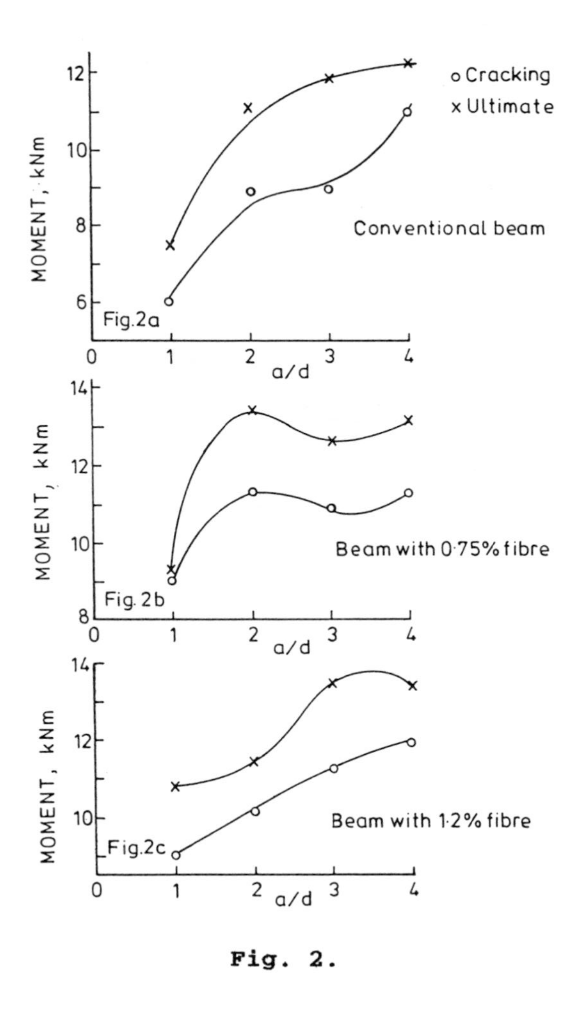

o Cracking
x Ultimate
Conventional beam
Beam with 0·75% fibre
Beam with 1·2% fibre
MOMENT, kNm
a/d
Fig.2a
Fig.2b
Fig.2c

Fig. 2.

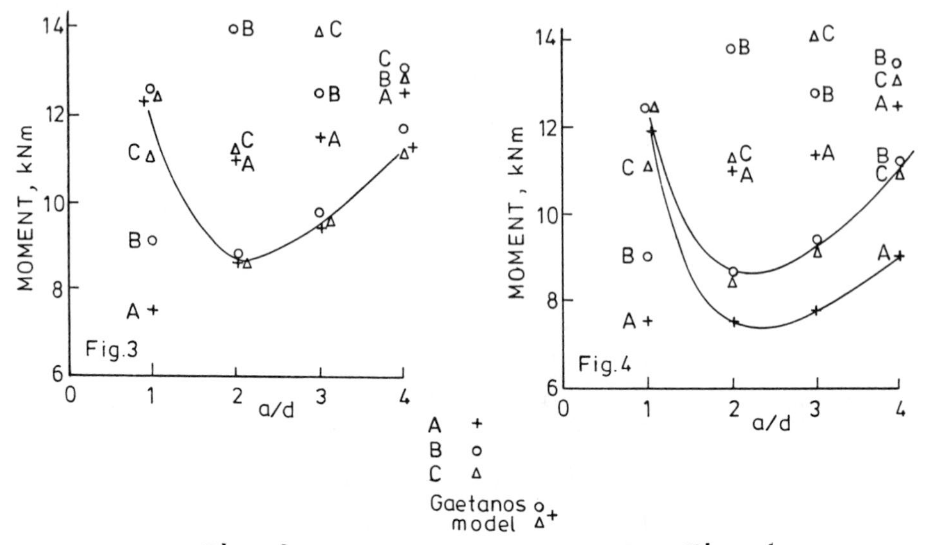

Fig. 3. **Fig. 4.**

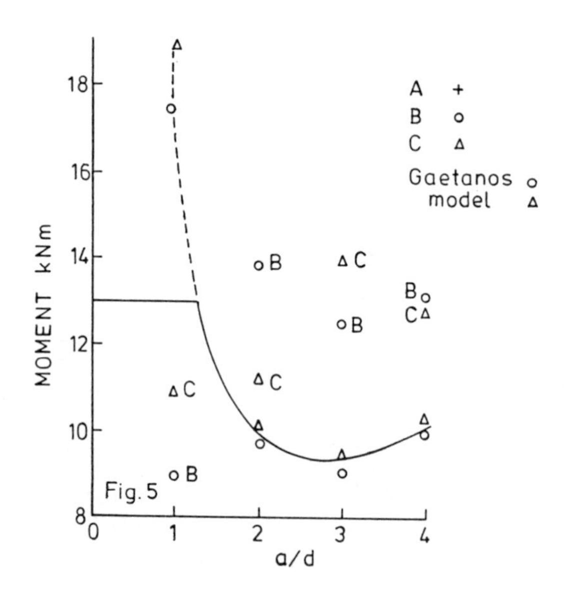

Fig. 5.

Table 5 Gaetano's Model

Beam	a/d	M_{ub}	M_{ua}	M_{uG}	$M_{u\sigma sp}$	M_u Expt.	M_u/M_{uG}	$M_u/M_{u\sigma sp}$
1	2	3	4	5	6	7	8	9
Conventional beam:								
A1	1	2.52	9.92	12.44	11.86	7.48	0.6	0.63
A2	2	5.04	3.5	8.54	7.4	11.22	1.31	1.52
A3	3	7.5	1.909	9.409	7.75	11.48	1.22	1.48
A4	4	10.09	1.24	11.334	9.04	12.58	1.11	1.39
SFRC beam:								
B1	1	2.65	9.92	12.57	12.44	9.18	0.730	0.737
B2	2	5.3	3.5	8.8I	8.54	13.86	1.5	1.6
B3	3	7.89	1.909	9.8	9.47	12.62	1.28	1.33
B4	4	10.6	1.24	11.84	11.32	13.26	1.11	1.17
C1	1	2.53	9.92	12.46	12.36	10.97	0.88	0.887
C2	2	5.06	3.5	8.56	8.39	11.31	1.32	1.34
c3	3	7.59	1.909	9.499	9.244	14.03	1.47	1.5
C4	4	10.12	1.24	11.34	11.02	12.92	1.13	1.17

M_{ub} beam action contribution.
M_{ua} arch action contribution.
M_{uG} ultimate moment due to Gaetano et al.
$M_{u\sigma sp}$ ultimate moment when σ_{sp} is incorporated in M_{ub}.

M_u expt " " experimental value.

Table 6 Proposed Model

Beam	M_{uG}	$M_{cr} = 0.8\ M_{uG}$	M_{cr} Expt	M_{cr} Expt/M_{cr}
Conventional concrete beams:				
A1	12.44	9.95	6.04	0.61
A2	8.54	6.8	9.18	1.35
A3	9.41	7.5	8.93	1.19
A4	11.334	9.06	10.88	1.2

Table 7

Beam	M_{uG} C.C.Beam	M_{cr}(CC) (o.8 M_{uG})	M_f	M_u (M_{cr}+M_f)	M_u Expt	M_{uE}/M_u
Fibrous concrete beams:						
B1	12.44	9.952	7.78	17.732	9.18	0.52
B2	8.54	6.832	2.75	9.582	13.86	1.44
B3	9.409	7.528	1.49	9.018	12.62	1.40
B4	11.334	9.06	0.972	10.03	13.26	1.32
C1		9.95	8.85	18.8	10.97	0.58
C2		6.83	3.28	10.11	11.31	1.12
C3		7.53	1.7	9.23	14.03	1.52
C4		9.06	1.07	10.13	12.92	1.27

M_f Fibre contribution = $K\sigma_{sp}\, bd^2\, (\rho')^{1/3}\, (a/d)^{3/2}$

M_{cr} 0.8 M_{uG}

M_{uG} M_u Gaetano

CC Conventional Concrete.
Note: Moments are expressed in Kilo Newton Metres.

The proposed model relates the enhancement of flexural capacity influenced by shear, over that of the conventional concrete member to the matrix characteristics and the loading conditions defined by split tensile strength and the shear span to depth ratio respectively; in other words it expresses the enhancement due to fibre inclusion in terms of σ_{sp} and a/d ratio.

The model incorporates the variables, which take care of the overall performance of the member, which control the shear, moment and their interaction. In the formulation of the model, in addition to the conventional assumptions two more are added viz. the effect of fibre reinforcement (ρ') is the same as that of the longitudinal reinforcement and hence the term $(\rho')^{1/3}$ and the fibre effect influences the cracking and ultimate moment capacities to the same order of magnitude as is given in figs. 1 and 2.

The greater influence of shear in shorter shear spans is taken care of in the model by the term $(a/d)^{3/2}$ and the overall performance in flexure by σ_{sp}.

The important observation that can be made from the test results is that the enhancement in flexural capacity (and correspondingly in shear capacity also) is alsmost the same at cracking stage as well as at ultimate stage. This means that the fibres are in action throughout the loading range and their contribution is the same at and after cracking. This may suggest that having delayed the appearance of cracks, the fibres are relatively passive in the post-cracking stage except for increased tensile strength contribution to beam actions and improvement in some of the other failure characteristics discussed later.

The choice of the quotients for ρ' and a/d is also of interest, the influence of the fibres in modifying matrix behavious has already been accounted for by incorporating the increased split tensile strength in place of $(fc')^{\frac{1}{2}}$ in the equation for beam action in the Gaetano's model. However in members failing in shear the fibres across the shear crack are seen to have pulled out and not broken. Therefore, these fibres have acted like reinforcements across the crack and their contribution in the resistance against shear has to be of the same form as longitudinal reinforcements. Hence the term $(\rho')^{1/3}$ in the proposed model is chosen similar to $(\rho)^{1/3}$ in the original model.

Since the contribution of the fibres is seen to be obtained in one stage prior to cracking, the proposed model also consists of single factor to be added to the calculated value as per the modified Gaetano's model. In the gaetano's model the a/d contribution is included both in the beam action component and in the arch action component. Since σ_{sp} has already been included in M fibre in line with beam action, the a/d component in the proposed equation is given a quotient -3/2 as in the original model.

Thus the proposed model is nothing but an extension of the Gaetano's model and therefore reflects many of the important observed characteristics of the flexureshear interaction phenomena which have formed the basis of theat model and which have been drawn from several published shear test results.

The calculated values as per the model agree well with test results except for a/d = 1. And the agreement is to a comparitively conservative order when split tensile strength is allowed to take care of the performance as can be seen from Tables (5,6,& 7).

The comparative statements in Tables 5,6,& 7 indicate that the quantification of the fibre contribution by the proposed model is significant and in line with test results in the shear- flexure interaction regions.

Another observation from tests which cannot be quantified but which definitely deserves mention here is that even though inclusion of fibres in RC beams has not in certain cases caused enough enhancement of shear strength as to cause a change in mode of failure from shear to flexure, it has definitely helped to reduce the violence and suddenness of the resulting failure. The beam is also perceptibly less damaged after failures and there is, therefore, good justification in using SFRC in place of ordinary RCC to reduce the bad effects of unexpected shear failures.

Conclusion

Within the limited scope of the experiments conducted and analysed for RC & SFRC members, the following conclusions can be drawn.

1. Fibre reinforcement enhances tensile strength and hence the shear strength which is considerable in shorter shear span depth ratios.
2. Fibres smeared in the matrix delays cracking, and crack propagation and retains structural integrity for longer duration which leads to pleasant ductile failure.
3. Influence of shear in flexure in reducing the flexural capacity reduces with the inclusion of fibre. Shear failure without attaining flexural capacity reduces to a smaller range than in conventional RC beams.
4. Action of fibres is felt quantitatively upto cracking and thereafter, the fibres improve the quality of the failure. The proposed model is able to predict the contribution of the fibres relatively accurately for a wide range of a/d ratios.
5. The approach described herein follows the basic concept of regarding shear in RC beams as a phenomenon limiting flexural capacity through destructive interaction. The concept was first outlined by Gaetano Russo and others and is borne out by the observations on SFRC beam tests reported herein.
6. The approach outlined here is based on a modification and extension of the Gaetano's model. As this model has been derived from the equation for nominal shear stress given in the ACI Building regulations the proposal made herein for SFRC beams are not also likely to conflict with current design specifications.

Acknowledgement:

Financial assistance to this research work by centre for Applied and Fundamental Research, College of Engineering, Trivandrum is gratefully acknowledged.

References

1. Gordon B. Batson,(1985) "Use of Steel Feibres for Shear Reinforcement and Ductility," U.S. Sweden Joint Seminar (NSF-STU) Stockholm. June 3 - 5, 1985. 2.Jindal, R.L.(1984) Shear and Moment Capacities of Steel Fibre Reinforced Concrete Beams. ACI SP81, 1984. 3. Gaetano Russo, Gaetano Zingone, and Giovanni Puleri, "Flecture-Shear Interaction Model for Longitudinally Reinforced Beams," ACI Structual Journal, V.88, No.1, January-February 1992. 4. "The Shear Strength of Reinforced Concrete Members," ACI 426R-74 (Reaffirmed 1980) ASCE, V.99, No.ST6, June 1973, August 1974. 5. James G. Mac Gregor, "Reinforced Concrete - Mechanics and Design," Prentice Hall Englewood Cliffs, New Jersey. 6. Michel D. Kotsovos, "Mechanism of Shear Failure," Magazine of Concrete Research V.35, No.123, June 1983.

45 EXPERIMENTAL INVESTIGATION ON FIBER REINFORCED PRESTRESSED CONCRETE BEAMS UNDER SHEAR

R. S. RAJAGOPAL
SERC, CSIR Campus, Taramani, Madras, India
S. SIDDAPPA
Bangalore University, Bangalore, India

Abstract

Shear failure of conventional reinforced and prestressed concrete beams usually occurs by tensile failure of concrete in the shear span. Shear failure is brittle in general. To prevent this type of failure and to increase the shear strength, shear reinforcement in the form of stirrups is incorporated. Steel fibres act as substitutes for shear reinforcement and especially in case of thin webs where bending and fixing of bars is difficult. In addition, because of the uniform distribution of fibres and its random orientation through out the volume of concrete uniform effective reinforcement is automatically established, thereby providing a substantial improvement in the resistance to the formation and growth of cracks besides enhanced ductility and better spalling resistance.

The paper presents the results of investigations undertaken on the effect of steel fibres on the behaviour and strength of fibre reinforced prestressed concrete beams (with Para-bolic cables) under shear. Tests were carried out on eight post tensioned beams, including two companion bonded and unbonded prestressed beams without fibres. Both bonded and unbonded beams studied at different contents ranging from 0% to 0.75%. The fibrous prestressed beams showed marked improvement in first crack load and ultimate load over that of plain prestressed concrete beams.

The theoretical prediction of ultimate shear strength from a knowledge of the fiber properties, the cube strength, amount of tensile reinforcement and the applied prestress has been compared with experimental results. The predictions obtained by the theoretical method were found to be satisfactory.

Key words: Prestressed Reinforced Concrete, Steel Fibres, Shear Strength, Cracking Shear, Ultimate Shear, Deflection.

Fibre Reinforced Cement and Concrete. Edited by R. N. Swamy. © 1992 RILEM.
Published by E & FN Spon, 2-6 Boundary Row, London SE1 8HN. ISBN 0 419 18130 X.

1 INTRODUCTION

Whenever principal tensile stresses with in the shear span of a prestressed concrete or post-tensioned concrete beam exceed the tensile strength of concrete, diagonal cracks will develop in the webs along the shear span, followed by sudden collapse of the beam.

Addition of fibers, its uniform distribution and corresponding random orientation throughout the volume, will convert brittle failure to a ductile one. Another significant consequence of fiber addition is the improvemnt in the resistance to the formation and growth of cracks and a substantial improvement in tensile strength.

Studies by several investigations on the shear strength of fibre concrete have confirmed the effectiveness of steel fibres in improving the ductility, spalling resistance, dowel resistance, arch action, first crack shear strength and ultimate shear strength. The tests reported in this paper were carried out on prestressed concrete beams reinforced by two 12mm diameter High Strength Deformed Bars (HYSD) of grade Fe 415 and steel fibres of 0.3mm dia. The beams were designed to fail in the shear mode. The main objectives of the study was to examine the combined effect of fibre reinforcement and of prestressing.

The results of 8 prestressed concrete beams tested under shear, are discussed. Based on the CP110 of practice and studies carried out by other investigators, analytical prediction of the shear capacity of prestressed concrete beams has been made by including the effect of fibre incorporation and these values have been compared with experimental data.

2 REVIEW OF EARLIER WORKS

Shear strength improvement is greatly dependent on good workable mix and the fibre volume fraction of the mix, as well as resistance to the formation of a crack and its propagation offered by the fibers. Fibers having large aspect ratio (L/D) have been found to exhibit greater pull-out strength than the fiber, having smaller L/D values. By changing the surface texture by crimping of fibres and by indenting, improves the bond strength characteristics compared to round straight fibers. Therefore the three main factors that influence the strength of a fiber concrete are fiber volume fraction, fiber aspect ratio L/D, and the extent of bond between the fiber and the concrete. The influence of these three factors has been incorporated into a combined parameter called the fiber factor F given by

$$F = (L/D).\rho_f.d_f \qquad (1)$$

where d_f is the bond factor and accounts for differing bond characteristics of the fiber. Based on large series of pullout tests, d_f was assigned a relative value of 0.5 for round fibres, 0.75 for crimped fibres, and 1.0 for indented or duoform fibres.

2.1 Split Cylinder Strength of Fibre Concrete (f_{spf})

Direct tension test, modulus of rupture tests, and split cylinder tests have been employed to measure the tensile strength of fibre reinforced concrete. Narayanan and Kareem (1983) have been shown from their studies using modulus of rupture and the split cylinder tests, that the strengths obtained from split tests on fibre concrete cylinders can be employed for computational purposes. A relationship connecting the ultimate split cylinder strength of steel fibre concrete f_{spf} with its compressive strength and fibre factor was obtained by them as

$$f_{spf} = (f_{cuf}/A) + B + C\sqrt{F} \qquad (2)$$

where A = a non dimensional constant = $(20 - f^{1/2})$
B = a dimensional constant = 0.7 MPa
C = a dimensional constant = 1 MPa

2.2 Ultimate Shear Strength of Fibre Reinforced Concrete Beam (V_u)

The shear resistance of conventional reinforced concrete beam without web reinforcement is influenced by the shear force in the uncracked compression zone, the aggregate interlock force in the cracked tensile concrete, and the dowel force in the main steel.

The aggregate interlock force decreases with crack width and therefore can be neglected. The dowel force is uncertain in nature and probably of minor value. All the three, however, are dependent on concrete strength and reinforcement ratio. A semi-empirical design equation based on 91 test results carried out by several investigators, has been proposed by Narayanan and Darwish (1987b) for predicting the ultimate shear strength (V_u)of fibre reinforced concrete beams. The general form of this equation given below, covers the different terms such as split cylinder strength f_{spf}, dowel action provided by the amount of tensile reinforcement $\rho = A_s/bd$, and the shear span ratio a/d. The last term V_b takes in to account the contribution of the fibre pullout force along the inclined crack.

$$V_u = e \left[A' f_{spf} B' \rho (d/a) \right] + V_b \qquad (3)$$

(All units are in N and mm)

where e = a non-dimensional factor which takes in to account the effect of arch action and is given by
= 1.0 when $a/d > 2.8$
= 2.8 (d/a) when $a/d \leq 2.8$
A' = a non dimensional constant = 0.24
B' = a dimensional constant = 80 N/sq.mm
f_{spf} = split cylinder strength of fibre concrete calculated from equation (2) in N/sq.mm
ρ = the amount of tensile reinforcement (A_s/bd)
a = the shear span of the beam, mm
d = the effective depth of the beam, mm
V_b = is the vertical fibre pullout stresses along the inclined crack (N/sq.mm)

$$V_b = 0.41\ \tau\ F \qquad (4)$$

Where F is the fibre factor given by equation 1 and τ is the average fibre matrix interfacial bond and is taken to be 4.15 N/sq.mm

2.3.1 Ultimate Shear Strength of Fibre Reinforced Prestressed Concrete Beams

To determine the strength capacity of the conventional prestressed concrete sections, it is necessary to consider separately the regions of a member uncracked in flexure and regions cracked in flexure ASCE-ACI Task Committee-426 (1973), Reynolds (1974) and Ramaswamy (1976).

2.3.2 Sections Uncracked in Flexure

In regions of member uncracked in flexure, Vco can be obtained using the following equation from CP110 Narayanan and Darwish (1987).

$$V_{co} = 0.67bh\left[f_t^2 + 0.8\ f_t\ f_{cp}\right]^{1/2} \qquad (5)$$

Where f_t is the tensile strength of concrete limited to $f_t = 0.24\sqrt{(f_{cu})}$ which after making an allowance of 1.5 towards partial safety factor would become $0.36\sqrt{(f_{cu})}$. f_{cp} is the compressive stress due to prestressing centroidal axis. Factor 0.8 incorporated in the code are removed and rewritten as

$$V_{co} = 0.67bh\left[f_t^2 + f_t\ f_{cp}\right]^{1/2} \qquad (6)$$

In the case of prestressed concrete beams containing fibres, the above equation f_t has been approximated to f_{spf} by equation (2) and rewritten as

$$V_{co} = 0.67bh\left[f_{spf}^2 + f_{spf}\ f_{cp}\right]^{1/2} \qquad (7)$$

2.3.2 Sections Cracked in Flexure

In a similar manner, the empirical equation given in CP110 (1972) for the shear carried by the concrete in regions of a member cracked in flexure may be modified for a pre-stressed fibre concrete beam, as fallows

$$V_{cr} = (1 - 0.55\, f_{pe}/f_{pu})\, Vu\, bd + M_o\, V/M \qquad (8)$$

where d is the distance from the extreme compression fibre to the centroid of the tendons at the section considered.

M_o is the moment necessary to produce zero stress in the concrete at depth d, $M_o = 0.8\, f_{pt}\, I/Y$ where

f_{pt} is the stress due to prestress only at depth d and distance Y from the centroid of the concrete section which has second moment of area I.

f_{pu} is the characteristic strength of the prestressing wires, N/sq mm.

f_{pe} is the effective prestress after all losses have occurred. For the purposes of this equation f_{pe} shall not be put greater than 0.6 f_{pu}.

V & M are the shear force and bending moment respectively at the section considered due to ultimate loads

V_{cr} should be taken as not less than $0.1bd\sqrt{(f_{cu})}$.

V_u is the ultimate shear stress of fibre concrete beams given by equation 3.

3 PRESENT INVESTIGATIONS

3.1 Details of Test Specimens

The experimental investigation consisted of testing 8 beams with identical rectangular cross section of 100mm x 200mm with effective span 1200mm.

The beams were classified as series 1,2,3 and 4. There were two beams in each series 1A, 1B, 2A, 2B, 3A, 3B, and 4A, 4B. The beams 1A,2A,3A and 4A were unbonded post tensioned beams and the beams 1B, 2B, 3B and 4B were bonded post tensioned beams. All were prestressed by parabolic cable with an effective prestressing force of 54 kN.

The beams in the test series were all intended to fail in shear, therefore it became necessary to provide untensioned steel in order to obviate flexural failure and ensure shear failure.

The beams were cast from a mix with proportion of 1:1.4:2.2 by weight of cement, sand and aggregate, by weight. The water-cement ratio was 0.44. Ordinary portland cement, natural river sand with a fineness modulus of 3.3 and specific gravity 2.6 and coarse aggregate with fineness modulus of 6.6 and maximum size 10mm were used. The mix was designed to give a cube strength of about 35

N/sq mm at 28 days.

Round steel fibres with an aspect ratio (L/D) of 100, and bond factor d_f = 0.50 were used in preparing all the fiber concrete beams. The fibres had an ultimate strength of 1980 N/sq mm. Cold drawn stress relieved high tensile wires of 4 mm dia were used to prestress the beams. Three specimens of high tensile wires of length 600 mm with gauge length 40 mm were tested in an universal testing machine. The wires had an ultimate tensile strength of 1980 N/sq mm (f_{pu}). Two 12 mm high yield strength deformed bars were provided to improve the flexural capacity of the members and to ensure failure by shear in all cases. The ultimate tensile strength was found to be 420 N/sq mm.

Form work for casting the rectangular beams consisted of two steel channels and two wooden planks of the same cross section having 30 mm diameter hole at the center, placed at both the ends of a form work. Two metallic pipes were bent to parabolic shape with a central rise (eccentricity) equal to 37.5 mm for providing the duct in prestressed members. All the ingredients were weighed separately. The fiber was first mixed with coarse aggregate by sprinkling in dry condition and taking care to avoid balling and interlocking of the fibres. Fine aggregate and cement were mixed together in dry condition. Water was added gradually by sprinkling and simultaneously mixing was continued manually to get a uniform mix.

Bottom cover concrete was put first in to the mold and well compacted manually by using a tamping plate. The untensioned reinforcement was then placed in position. The parabolic shaped metallic pipe was inserted through the holes in the wooden planks along with 4mm dia mild steel anchorage reinforcement. Two mild steel rods of 10mm dia and 1000mm length were inserted in the holes of metal pipe at two sides to facilitate grouting. The concrete was poured in two layers and each layer was well compacted by a needle vibrator to ensure the uniform distribution of fibers between the steel pipes and anchorage reinforcement and to prevent the balling of fibers. Care was taken while vibrating the concrete so as not to displaced the pipes from the original profile and eccentricity. The vertical clips provided at the ends restrained the longitudinal movement of the pipe. The top surface was smoothly finished using trowels. Control specimens of standard size consisting of 3 cubes and 3 cylinders were cast correspondingly to each fibre volume fraction considered for the study to find characteristic strength and split tensile strength of fiber concrete.

The grout rods and metal pipe were kept rotating and moved slightly forward and backward after initial setting time to prevent adhesion of green concrete to their surface and to facilitate their subsequent removal. Then these rods and pipe were taken out after final setting time of concrete.

The beam was demoulded after 24 hours and cured for 28 days along with control specimens by covering with wet gunny bags. Cubes and cylinders were tested on the day of testing of beams. The beams were all prestressed (post-tensioning) after 28 days by using Killick Nixon hand operated jack. Mild Steel bearing plates of size 100mm x 100mm x 8mm were used on the end of the beams to serve as anchorage plates. HT wires, after being stressed to the required prestressing force were simultaneously anchored on the MS plate by using 4mm wedges and barrels. After prestressing the cable, duct of bonded beams was grouted to ensure good bond between H T wires and concrete. The grouting was done by injecting cement slurry with a w/c ratio 0.5, through the grout holes using a standard Killick Nixon grout pump. Then the beams were cured for 7 days.

4 TEST PROCEDURE

The beams were simply supported over a span of 1.20m and tested under two-point loading. **Fig 1** gives the schematic diagram of the beam test setup and cross section details of the beam. Dial gauges were used to measure the deflection at load point and at mid span, while a DEMEC gauge of 150 mm gauge length was used to measure longitudinal strains under the loads and at midspan and transverse strains in the shear span. The beams were tested by applying transverse load through a hand operated hydraulic Jack of 50 kN capacity. The load increment was 10 kN. Dial gauge and demec gauge measurements corresponding to each load increment were recorded. Load at first visible shear crack, ultimate load and crack width and its propagation pattern were recorded. Crushing and split cylinder tests were carried out on control specimens on same day

5 TEST RESULTS

Table 1 shows the details of test beams with different fiber volume and test results. **Fig 3** shows the moment-curvature relationships for all the beams with varying volume percentages of fibres. Mid span deflection for all the beams, both bonded and unbonded beams are shown in **Fig 2**. **Fig 4** shows the load Vs crack width at shear span. **Fig 5** shows the crack pattern for all the beams.

6 DISCUSSION OF EXPERIMENTAL RESULTS

All the beams exhibited similar linear behaviour from the initial loading and up to the occurrence of the first hair line inclined crack. Beams 1A, 1B (V_f=0%) failed soon

Table 1 Properties of Beams, Test Results and Comparison of Results

Beam series	Beam no	Fiber Vol %	Fiber factor	Crushing strength N/sq.mm	Splitting strength** N/sq.mm	Cracking load kN	Ultimate load kN [S]	% increase in cracking load
1	1A	0.00	0.000	48.66	3.13	41.20	91.20	-
	1B	0.00	0.000	48.66	3.13	61.20	101.20	-
2	2A	0.25	0.125	50.66	3.63	46.20	86.20	13.44
	2B	0.25	0.125	50.66	3.63	67.20	106.20	23.30
3	3A	0.50	0.250	53.11	3.92	51.20	81.20	26.90
	3B	0.50	0.250	53.11	3.92	74.20	121.20	30.70
4	4A	0.75	0.375	55.11	4.15	61.20	96.20	54.00
	4B	0.75	0.375	55.11	4.15	101.20	116.20	70.70

1A,2A,3A and 4A Unbonded beams

1B,2B,3B and 4B Bonded beams

** By formulae 2

S Shear Failure

$b = 100$ mm $d = 200$ mm

d_e = Effective depth at mid span 173.9 mm

= At quarter span 155.5 mm

$a = 300$, $a/d = 1.6$, $L/D = 100$

$As = 226$ sq mm $nA_{ps} = 37.71$ sq mm

$Pe = 54$ kN $Y = 37.5$ mm

Table 2 Calculated Ultimate Observed and Predicted Shear Strength from Equations 7 and 8 for PSFRC Beams

Beam series	Beam no	P_v kN	V_u MPa	V_{co} MPa	V_{cr} MPa	V_{uo} MPa	V_{up} MPa	V_{uo}/V_{up}	% increase in shear strength
1	1A	37.01	2.16	67.98	52.23	2.38	3.35	0.70	-
	1B	57.01	2.16	67.98	52.23	3.65	3.35	1.09	-
2	2A	42.01	2.52	74.20	56.84	2.70	3.65	0.73	13.44
	2B	63.01	2.52	74.20	56.84	4.05	3.65	1.10	23.30
3	3A	47.01	2.92	79.00	61.90	3.02	3.98	0.75	26.90
	3B	70.01	2.92	79.00	61.90	4.77	3.98	1.19	30.70
4	4A	57.01	3.20	84.51	65.34	3.66	4.20	0.87	54.00
	4B	97.01	3.20	84.51	65.34	6.23	4.20	1.48	70.70

Y = Eccentricity of the parabolic prestress at quarter span is 31 mm

O = Slope of the cable = 0.0775

Vertical component of shear force due to parabolic prestress = 4185 N

Net shear force at quarter span for the beam 1A = 41200 - 4185 = 37015 N

f_{pe} = 1431.90 MPa f_{pe}/f_{pu} = 0.34 f_{cp} = 2.7 MPa $V_{uo} = V_o/bd$ $V_{up} = V_{cr}/bd$

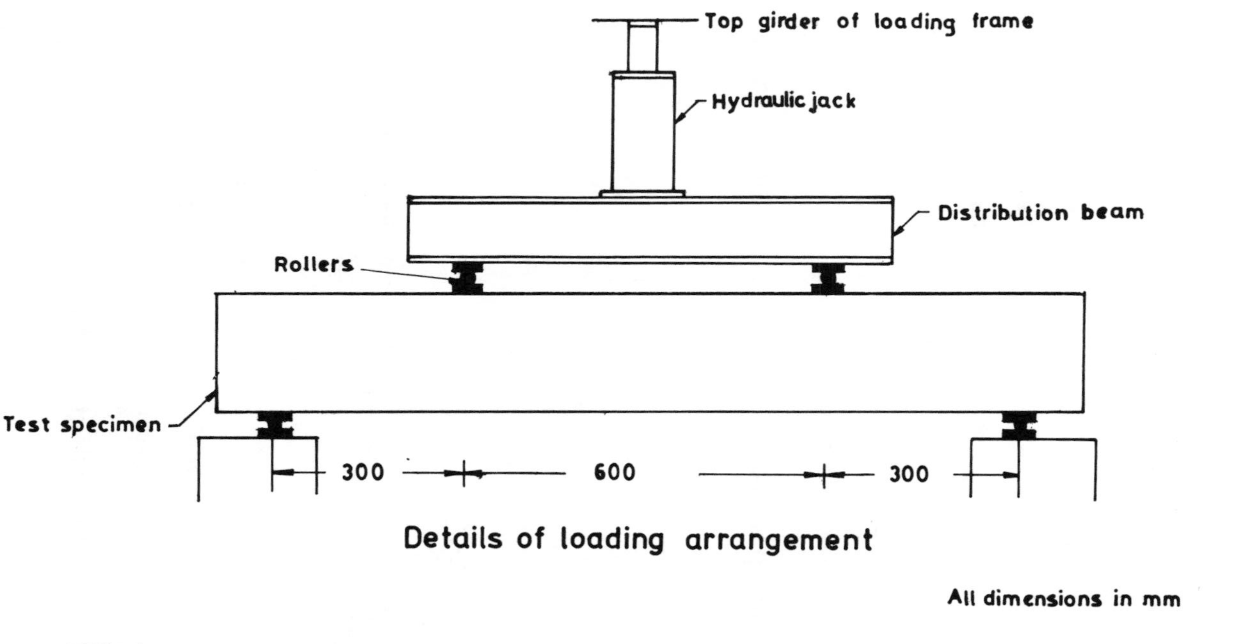

Details of loading arrangement

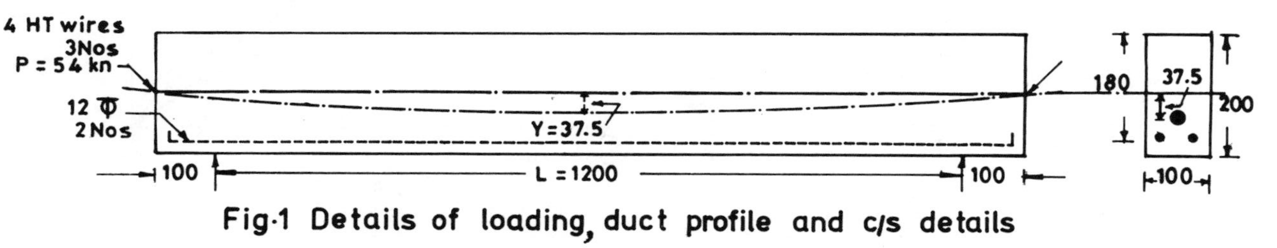

Fig·1 Details of loading, duct profile and c/s details

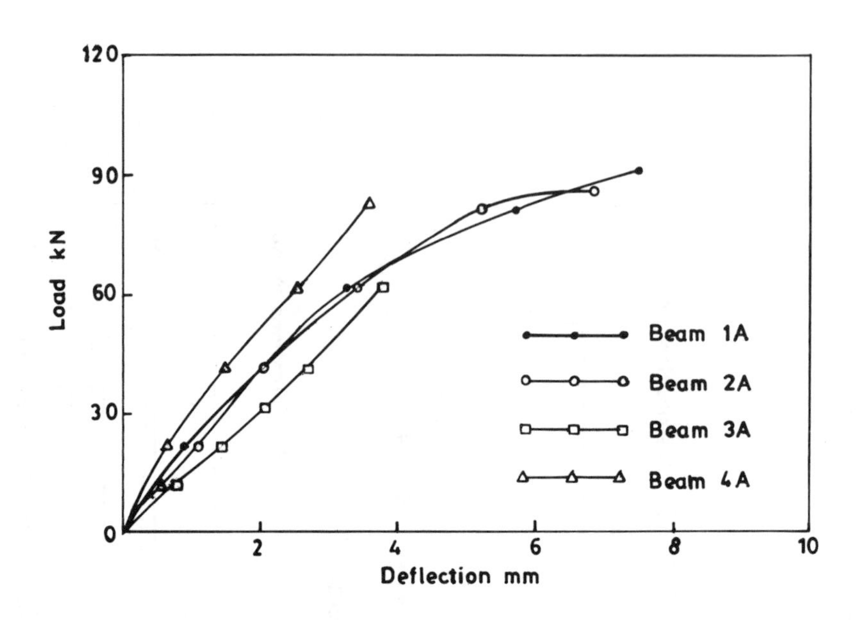

Fig 2(a) Mid span deflection for unbonded beams

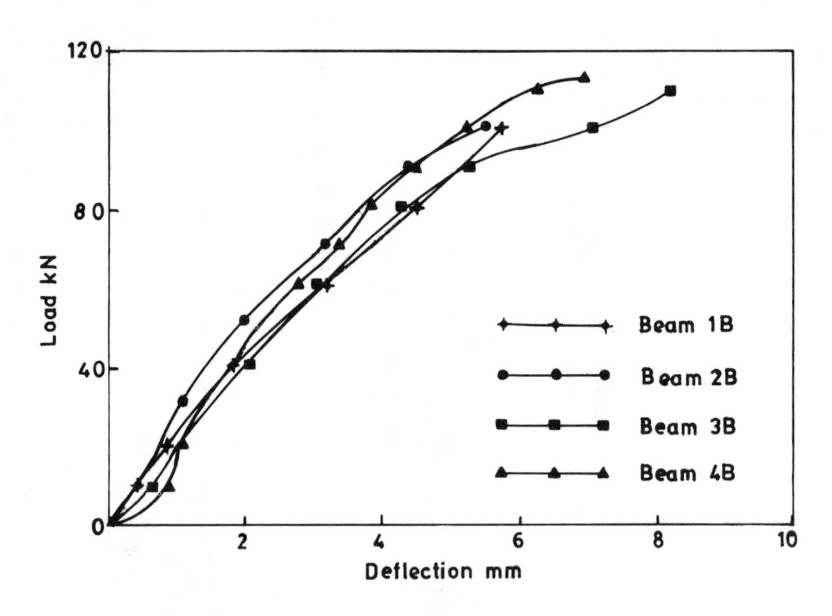

Fig 2(b) Mid span deflection for bonded beams

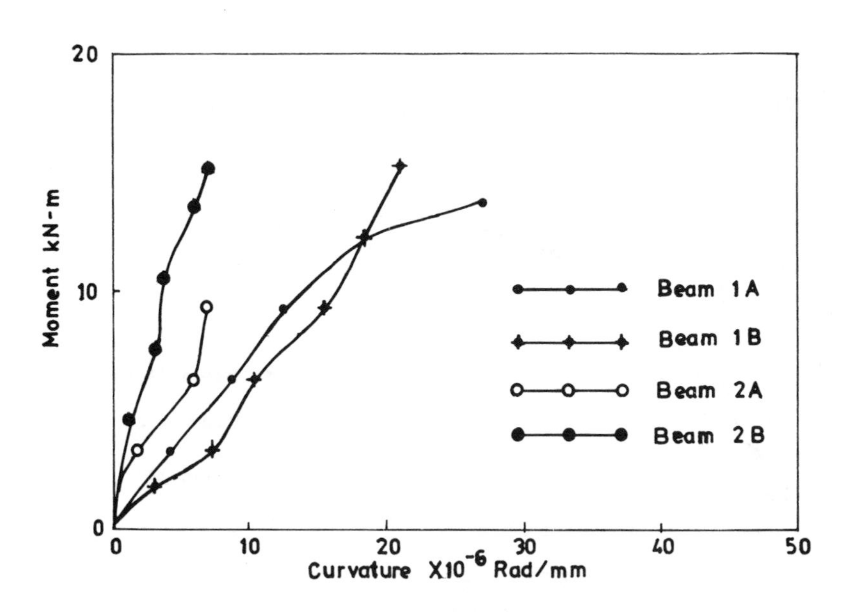

Fig ·3(a) Moment curvature relation for beams

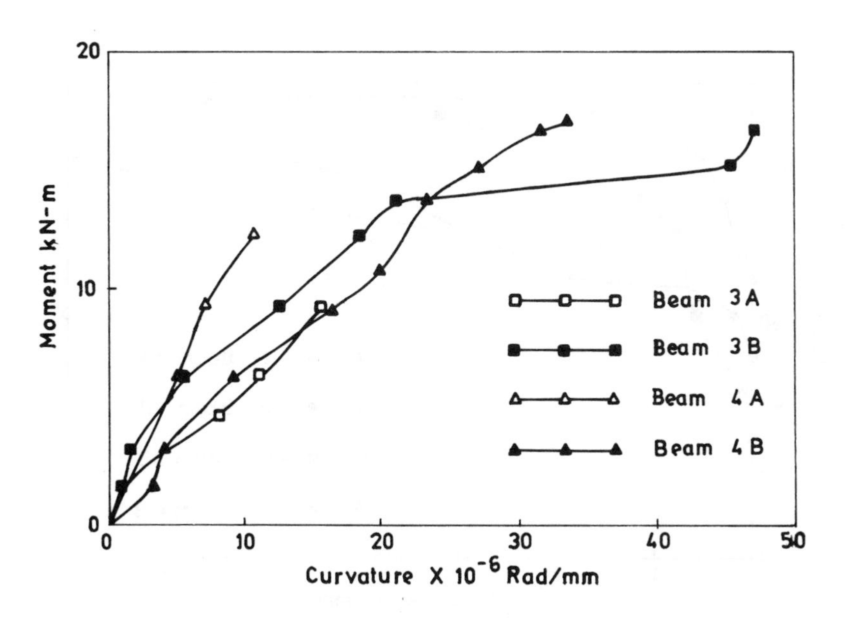

Fig· 3(b) Moment curvature relation for beams

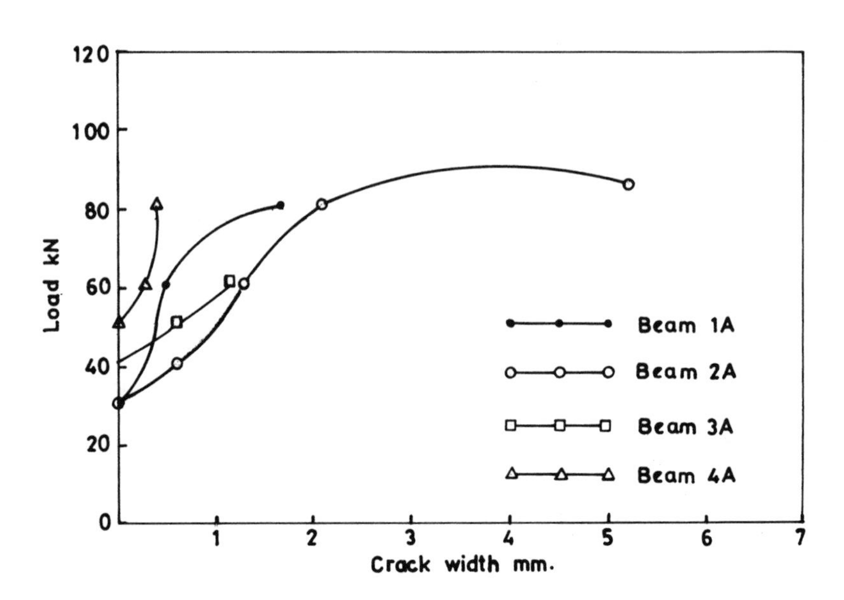

Fig 4(a) Variation of crack width for unbonded beams

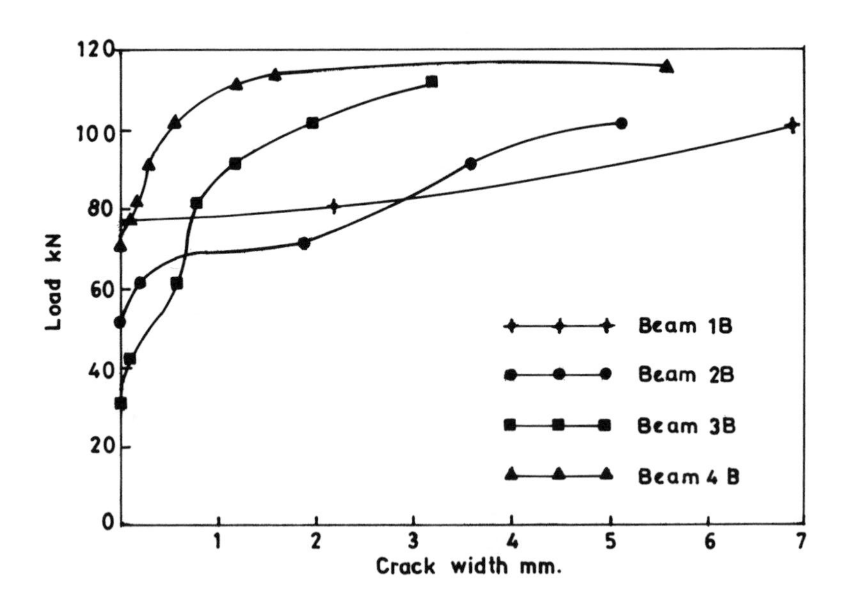

Fig 4(b) Variation of crack width for bonded beams

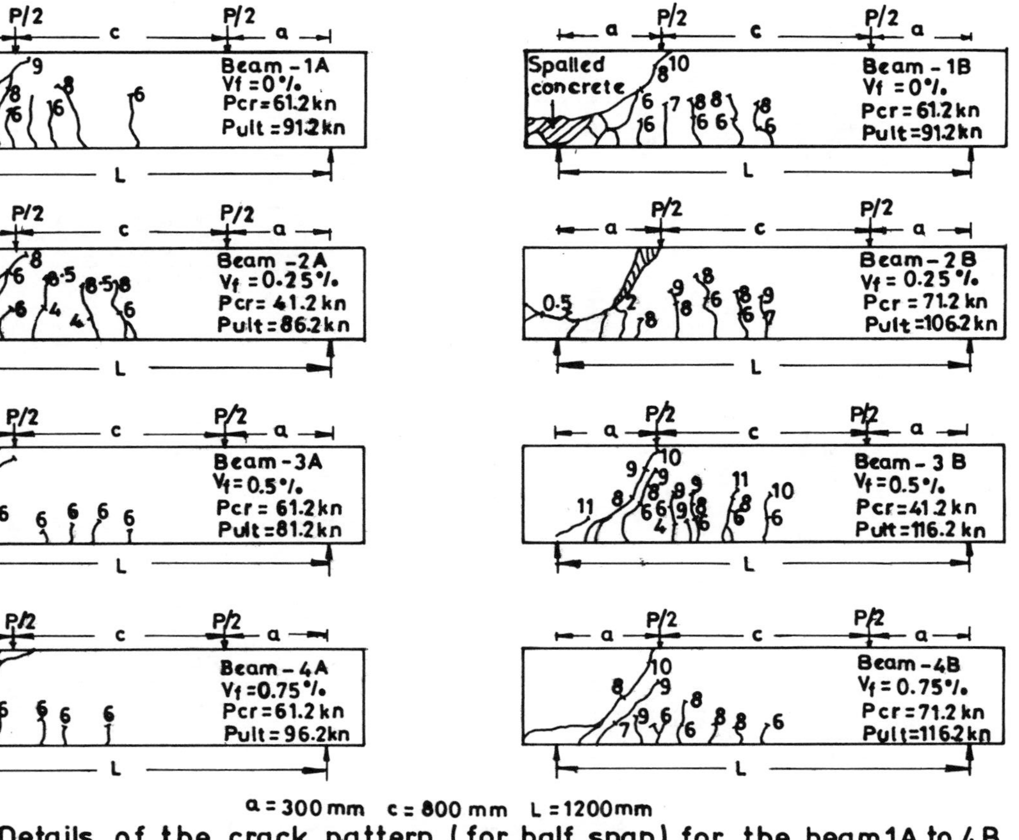

a = 300 mm c = 800 mm L = 1200 mm

Fig.5 Details of the crack pattern (for half span) for the beam 1A to 4B after the formation of the first diagonal crack. Beams with fibers continued to resist higher loads even after the formation of first shear crack. In many cases beams did not fail suddenly, thus indicating a considerable improvement in the ductility of beams.

Several diagonal cracks with smaller crack width were observed in the beams with relatively high percentage of fibres, owing to the redistribution of stresses after first cracking. The fibres may be thought of as becoming effective after shear cracking and they continued to resist the principal tensile stresses until the complete pullout of all fibres occurred at one critical crack width.

It was observed that the inclusion of steel fibres eliminated the spalling of concrete. Fibres, in addition to prestressing, imparted to the beams, the capacity to hold parts of the concrete together in the post cracking stages, thus preventing spalling even near advanced stages of failure.

The effect of fibre volume is shown in **Table 2** where both the cracking load and ultimate load are found to increase progressively with increase in fiber volume percent. Percentage increase in cracking load of a fibrous beam compared with a beams without fibres is as large as 54% in unbonded beams and 71% in bonded beams. This is probably due to the crack arrest mechanism exhibited by the fibres and due to the elastic fracture mechanism. Percentage increase in ultimate shear strength is shown in **Table2**. **Table 1** shows that there is substantial increase in crushing strength and splitting tensile strength with increase in the volume of fibre. The crushing strength and splitting tensile strength increased by about 13% and 35% respectively for 0.75% fiber volume. Load-deflection curves for all the beams are shown in **Fig 2**. It is clear from the curves that, for the same amount of load increment, the deflection is smaller in bonded fiber reinforced prestressed concrete (FRPSC) beams than the unbonded FRPSC beams with higher volume percentages of fibres.

Fig 3 shows the moment curvature characteristics of the beams at mid span. The curvature ($\emptyset$) was calculated by using equation $\emptyset=8\nabla/L^2$ where ∇ is the differential deflection between load point and mid point of the beam. It was observed that the magnitude of prestressing force and fiber volume affect the curvature characteristics. Higher fiber volume and high magnitude of prestressing force results in flatter curves. This indicates prestressed bonded beams with steel fibres are more effective in resisting transverse loads. **Fig 4** details the variation of average crack width with load. In all the beams flexural cracks appeared first, shear cracks observed in the later stage of loading. Shear cracks were found to propagate rapidly than flexural cracks, and finally all the beams failed in shear.The general crack pattern of the beams can be observed in the **Fig 5**. The fiber content and effect of parabolic prestress has influence over the number of cracks and the crack width. It was observed that the number of cracks in the beams increased with the increase in fiber volume percent. While loading the beams, the cracking of the beams was found to occur with audible noise due to pulling out of fibres. After the failure of the beam and while unloading, the cracks obtained were found to close again with the same audible noise due to the effect of parabolic prestress and fiber interfacial bond strength. The maximum crack width observed was 2mm in beams containing no fibres, and 5mm for beams containing fibres.

6.1 COMPARISON OF THEORETICALLY PREDICTED SHEAR WITH OBSERVED VALUES

The ultimate shear strength, calculated using equations 3,7,8 are reported in **Table 2** . By observing observed and predictive shear strength ratios, the theoretical predictions for beams 1B,2B,3B and 4B are conservative. The mean value of the observed ultimate shear strength/ predicted ultimate shear strength for 8 prestressed concrete beams were found to range from between 0.7 to 1.48. These comparisons shows that equations 7 and 8 can be employed to obtain satisfactory prediction of the ultimate shear of prestressed concrete beams containing steel fibres.

7 CONCLUSIONS

Based on the experimental investigations conducted in this study the following conclusions can be drawn:

1. Shear strength of rectangular PSFRC beams depends mostly on the concrete strength, reinforcement ratio, fibre properties and shear span to depth ratio.
2. The failure observed was sudden in the case of beam without fibres than in beam containing fibres, which was failed gradually. This is because of the crack arrest mechanism exhibited by steel fibre and due to interfacial bond strength of steel fibres.
3. Addition of increased volume of steel fibres increases the ultimate shear strength capacity.
4. Fibres control the propagation and widening of the cracks and render the beams more ductile.
5. Fibre reinforced post-tensioned bonded beams are more effective in load carrying, resisting deformation and in control of crack-width.
6. Moment-curvature relationship shows a decrease in curvature with the increase in fibre volume, thus indicating the stiffening of the beams with the inclusion of fibres. Also ductility improves with increasing volume percentage of fibres.
7. The equations presented by Narayanan and Darwish for

shear capacity of sections cracked in flexure and uncracked in flexure for fibre reinforced prestressed concrete beams were found to give reasonable agreement with test results.

8 ACKNOWLEDGMENT

The authors would like to express their thanks to the Chairman, Department Council, Department of Civil Engineering Bangalore University, for the facility provided to carryout experiment. The authors would also like to express their deep sense of gratitude to Sri: N.V. Raman, Director SERC, and Dr:V.S.Parameswaran, Deputy Director SERC for the encouragement given for publishing their paper.

9 REFERENCES

British Standards Institution, CP110:1972 **The Structural Use of Concrete, Part 1, Design, Materials and Workmanship,** London.

EI-Niema, E.I. (1991) Reinforced Concrete Beams with Steel Fibers under Shear. **ACI Structural Journal,** 178-183.

Joint ASCE-ACI Task Committee 426 (1973) The Shear Strength of Reinforced Concrete Members, **Journal of the Structural Division, ASCE,** 99, ST6, 1091-187

Narayanan, R. and Kareem-Palanjian, A.S. (1983) Steel Fibre Reinforced Concrete Beams in Torsion. **The International Journal of Cement Composites and Lightweight Concrete,** 5, No.4, 235-46.

Narayanan,R. and Darwish,I.Y.S. (1987a) Shear in Prestressed Concrete Beams Containing Steel Fibres. **The International Journal of Cement Composites and Lightweight Concrete,** 9, No.2, 81-90

Narayanan, R., and Darwish, I.Y.S. (1987b) Use of Steel Fibers as Shear Reinforcement. **ACI Structural Journal,** 216-227.

Ramaswamy, G.S. (1976) Modern Prestressed Concrete Design. Pitman Publishing Ltd, London,

IS-1343-1980 **Code of Practice for Prestressed Concrete.**

Reynolds,G.C. (1974) Shear Provisions for Prestressed Concrete in the United Code, CP110: 1972. **Cement and Concrete Asocciation,** Technical Report 42.500.

Sharma,A.K. (1986) Shear Strength of Steel Fibre Concrete Beams. **ACI Journal,** 83, 4, 624-628.

46 FIBER REINFORCED CONCRETE COLUMNS USING DOLOMITE AS COARSE AGGREGATE

H. H. BAHNASAWY, F. E. EL-REFAI and M. M. KAMAL
GOHBPR, Cairo, Egypt

Abstract

The deformation behaviour and strength of axially loaded fiber reinforced concrete columns casted with dolomite as coarse aggregate were investigated with special attention to their longitudinal deformations, cracking, ultimate load and mode of failure. Reinforced concrete columns casted with gravel were also tested for comparison. The main variables taken into consideration in this research were the longitudinal reinforcement, aggregates type and fiber content. Dolomite reinforced concrete columns showed higher initial cracking loads and ultimate strength than those casted with gravel.
Keywords: Dolomite and Gravel, Fibrous Concrete Mixes, Longitudinal Strains, Ultimate Strength, Mode of Failure.

1 Introduction

Extensive studies have been carried out to determine the potentiality of using different types of aggregates in concrete industry [(1) to (6)]. Dolomite is one of the most promising types of Egyptian aggregates in producing high strength concrete[(4),(5)]. Here in Egypt especially in Suez Canal area, there are many dolomite quarries. High strength concrete is the subject of many research work carried out recently around the world. On the other hand steel fibers enhanced greatly the mechanical properties of concrete[(7)to(10)]. Research has shown that the use of steel fibers in reinforced concrete columns has many benefits such as, increase its shear capacity, ductility and crack control[(12),(13)]. Usage of both dolomite as coarse aggregate and steel fibers in casting columns was the purpose of this research.

Fibre Reinforced Cement and Concrete. Edited by R. N. Swamy. © 1992 RILEM.
Published by E & FN Spon, 2-6 Boundary Row, London SE1 8HN. ISBN 0 419 18130 X.

2 Experimental Work

Eighteen concrete mixes were firstly investigated to determine the effeciency of using dolomite as coarse aggregate instead of gravel. Also to evaluate the enclusion of steel fibers on the consistency of the fresh concrete as shown in Table1.

The concrete mixes were prepared of Egyptian portland cement, sand and gravel from Pyramids quarries and dolomite from quarries at Suez Canal area. The properties of both gravel and dolomite are shown in Table 2. Two cement contents of 300 and 400 kg/m^3 of concrete were taken into consideration. Superplasticizer with high range of water reducing agent was used. The properties of the used superplasticizer are complying with ASTM C-494 requirements. Triangular twisted rugged surface steel fibers of 2 mm. breadth, 0.25 mm. thickness and 32 mm. length were used. Two fibers content of 1.5 and 2% by volume of concrete were also used in some tested mixes. Control mixes without fibers were casted and tested for comparison.

Compressive strength was carried out on 150x150x150 mm. cubes and four points flexural test was carried out on beams 100x100x500 mm. Bond test was carried out using concrete cylinders of 150x300 mm. with central rod 16mm. diameter. Modulus of elasticity test was carried out on cylinders Of 150x300 mm. dimensions. All the test specimens were cured in the laboratory atmosphere of 25°C temperature and 50% relative humidity until testing at 28 days age.

Twenty four reinforced concrete columns with 200x200x1300 mm. dimensions were casted and tested in this research. These columns represent two sets, one of them was reinforced with 4ø 8mm longitudinal mild steel bars (24/36), while the other set was reinforced with 4ø 13mm. longitudinal high strength steel bars (40/60). Tension tests carried out on the steel bars indicated that the yield stresses of the mild steel bars and the high strength steel bars were 280 and 500 N/mm^2 respectively. Concrete mixes No. 7,8,9 and 16,17,18 were used to cast gravel and dolomite columns respectively. The columns were casted horizontally in the forms to achieve the most compaction and uniformity of the concrete strength of the column length. Concrete control specimens were also casted. The columns and the control specimens were cured under wet burlap for 14 days after which they were allowed to air until time of testing at 28 days age.

Table 1. Fiber Reinforced Concrete Mixes.

Mix No.	Type of aggregate	Cement content kg/m³	% fiber by volume of concrete	Water cement ratio by weight	% Superplasticizer by cement weight	Slump (mm)
1	Gravel	300	——	0.50	——	100
2			——	0.43	1.0	90
3			1.5	0.43	2.0	80
4			2.0	0.43	2.0	75
5	Gravel	400	——	0.46	——	80
6			——	0.40	1.0	70
7			——	0.34	2.0	80
8			1.5	0.34	2.0	75
9			2.0	0.34	2.0	70
10	Dolomite	300	——	0.50	——	70
11			——	0.43	1.0	70
12			1.5	0.43	2.0	62
13			2.0	0.43	2.0	60
14	Dolomite	400	——	0.46	——	56
15			——	0.40	1.0	40
16			——	0.34	2.0	45
17			1.5	0.34	2.0	40
18			2.0	0.34	2.0	40

* Fine to coarse aggregate =1:2

Table 2. Prperties of coarse aggregates

Property	Dolomite	Gravel
Maximum nominal size (mm)	20.0	20.0
Percentage of fines	2.0	3.0
Unit Weight	2.5	2.4
Mass density (kg/m³)	1540.0	1720.0
Percentage of voids	38.4	28.3
Percentage of water absorption	3.5	2.5
Aggregate crushing value	17.4	15.1
Impact resistance value	16.22	14.8
Abrasion resistance value	30.0	22.0

The full dimensions and details of reinforcement of the columns and the used concrete mixes are shown in table 3 and Figs. 1 and 2.

One day before testing, the demec points were fixed in position by an adhesive with gauge length of 200 mm. The columns were carefully adjusted vertically with testing machine and all measurments were taken to apply the load axially. The loading was performed by a 500 ton hydraulic testing machine. The columns were tested to failure in about 10 increments of loading. Strains were measured at different positions along the middle cross-section and cracks were observed. The initial cracking loads as well as the ultimate loads were recorded.

Table 3. Scheme of reinforced concrete columns tested under axial loading

Set No.	Column No.	Mix No.	Type of Aggregate	Fiber content by volume of concrete	Longitudinal reinforcement	Lateral reinforcement
	C_1	7		0.0		
	C_2	8	Gravel	1.5	4 ø 8	ø 6 mm
	C_3	9		2.0		
1					Mild	
	C_4	16		0.0		
	C_5	17	Dolomite	1.5	steel	Stirrups
	C_6	18		2.0		
						about
	C_7	7		0.0		
	C_8	8	Gravel	1.5	4 ø 13	
	C_9	9		2.0		100mm
2					High	
	C_{10}	16		0.0	tensile	
	C_{11}	17	Dololmite	1.5	steel	
	C_{12}	18		2.0		

3 Analysis and Discussion of Test Results

3.1 Properties of concrete mixes

Table 4 and Figures 3 and 4 show the properties of the different concrete mixes taken into consideration in this

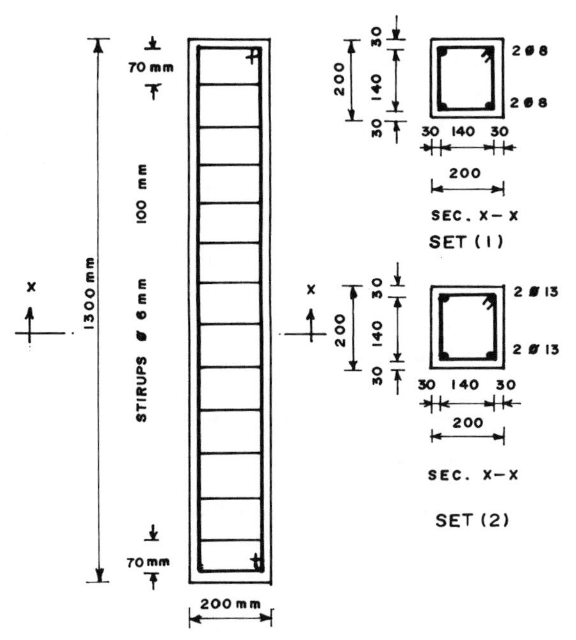

FIG.(1) : DETAILS OF COLUMN REINFORCEMENT

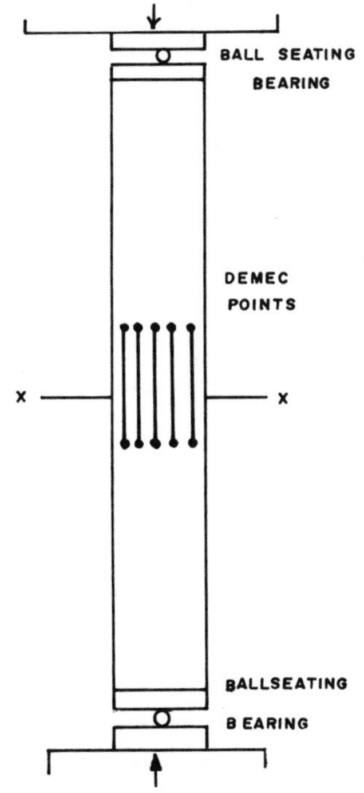

FIG. (2) : TEST OF COLUMNS UNDER AXIAL LOADING

investigation. Out of the slump test carried out on the fresh concrete mixes it was clear that the concrete mixes, casted with dolomite were of less consistency than that of those mixes casted with gravel for the same w/c ratio. The used superplasticizer (SP) increased the slump and facilitate the compaction of the tested mixes. Within the range of the (SP) percentages used in this study it was found that the higher the (SP) dose the higher the slump achieved and the higher the mechanical properties gained.

Table 4. Mechanical properties of tested mixes

Mix No.	Compressive Strength N/mm^2	Flexural strength N/mm^2	Bond strength N/mm^2	Modulus of elasticity kN/mm^2
1	22	3.4	2.5	17.0
2	30	5.0	3.2	19.0
3	35	7.0	4.0	22.0
4	40	9.0	6.6	24.0
5	35	5.0	4.0	26.0
6	40	6.5	5.0	21.0
7	45	7.0	6.0	22.0
8	52	10.0	8.0	24.0
9	57	12.0	9.0	26.0
10	33	5.8	4.0	19.0
11	40	7.0	5.0	20.0
12	50	12.0	8.0	23.0
13	56	14.0	9.0	25.0
14	45	7.0	5.0	21.0
15	55	9.0	7.0	22.0
16	63	16.0	8.0	23.0
17	67	16.0	10.0	25.0
18	70	18.0	12.0	27.0

*Results are average of 3 test specimens at 28 days age.

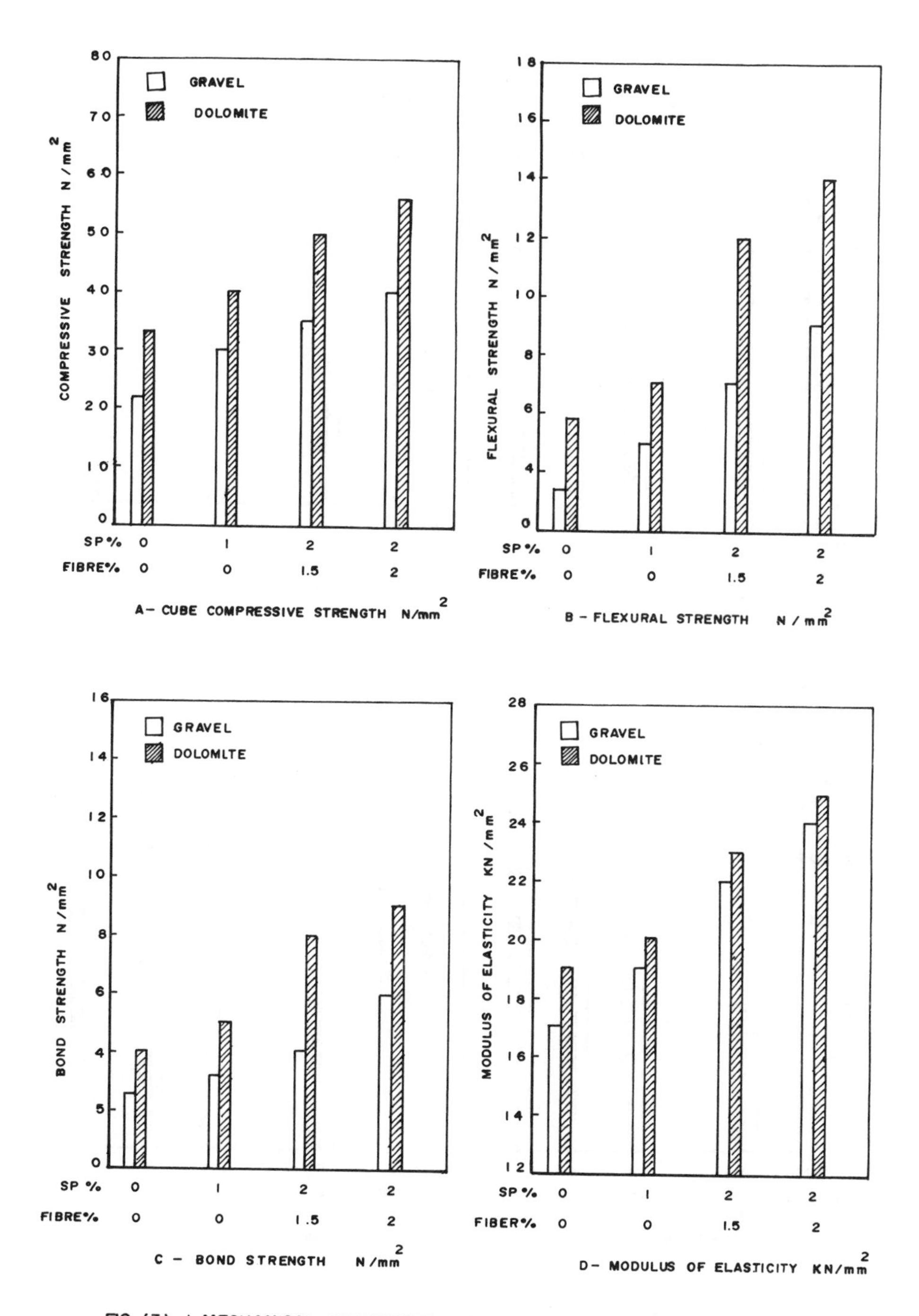

FIG. (3) : MECHANICAL PROPERTIES OF CONCRETE MIXES WITH CEMENT CONTENT = 300 kg / m^3

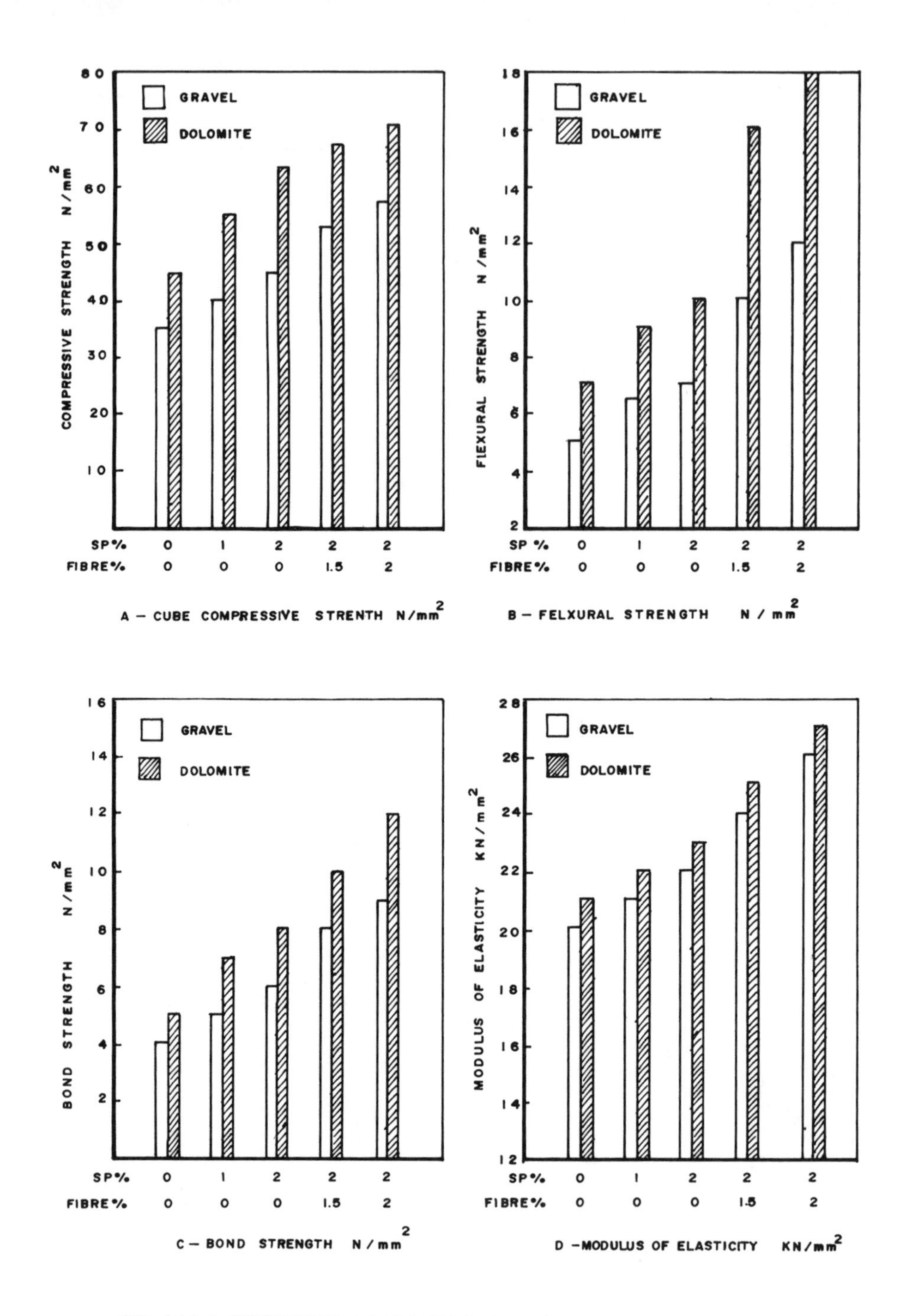

FIG. (4) : MECHANICAL PROPERTIES OF CONCRETE MIXES WITH CEMENT CONTENT = 400 kg/m^3

For the two concrete mixes casted with the cement content of 300 and 400kg/m^3, those mixes casted with dolomite showed higher mechanical properties than those casted with gravel for the same (SP) percentage and steel fiber percentagec. The concrete mix No. 18 casted with dolomite, cement content of 400 kg/m^3, (SP)/cement ratio of 2% and steel fiber content of 2% by volume has shown superior mechanical properties than the traditional concrete of concrete mix casted with gravel No. 5. The compressive strength, flexural strength, bond strength and modulus of elsticity of mix No. 18 were 200%, 360%, 300% and 104% of that of mix No. 5 respectively. These values represent a wide step forward in the concrete industry in Egypt.

3.2 Behaviour and strength of reinforced concrete columns.

3.2.1 Longitudinal strains

Figs 5 and 6 show a typical strain distribution of the longitudinal strains at the mid sections of the tested columns. It is quiet clear that although all the necessary precautions were made for the axial loading of the tested columns, small eccentricity was observed from the experimental results. The maximum compressive strains recorded for the tested columns are also shown in the same Figs. 5 and 6 at different stages of loading for those columns casted with gravel or dolomite respectively.

Generally speaking, higher strains were recorded for the columns casted with gravel in comparison with those columns casted with dolomite for the same concrete mix propartions, longitudinal reinforcement and stage of loading.

It was found that the higher the percentage of steel fibers inclusion, the lower the longitudinal strains at the same stage of loading. This phenomena was also recorded with regarding to the longitudinal reinforcement. Higher strains were recorded for the columns casted without fibers and reinforced with 4ø13 longitudinal bars and those reinforced with 4ø8 longitudinal bars than those casted of the concrete mix contained 2% steel fibers.

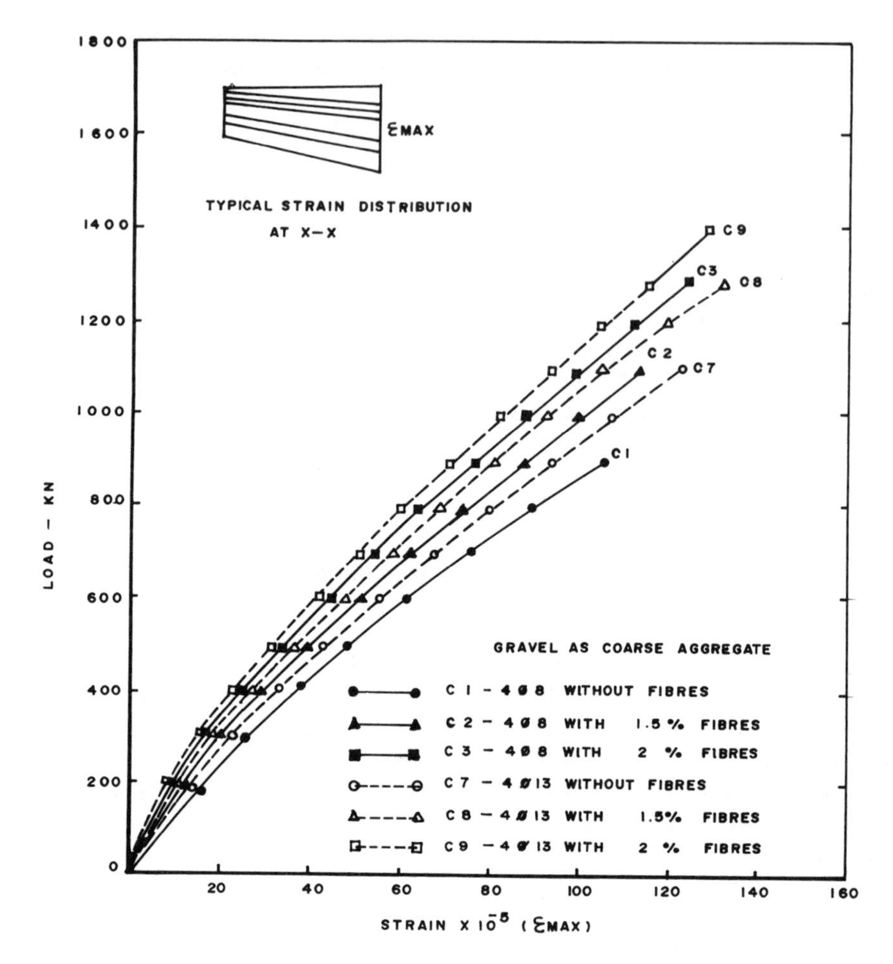

FIG. (5) : MAXIMUM COMPRESSIVE STRANS IN REINFORCED CONCRETE COLUMNS ON THE CENTRAL SECTION X – X (GRAVEL)

3.2.2 Ultimate loads

Fig. 7 and table 5 show the ultimate failure loads obtained experimentaly and calculated theoretically. Generally speaking, for the same longitudinal reinforcement and percentage of steel fibers, concrete columns casted with dolomite exhibited higher carrying capacity than those casted with gravel. For the concrete casted with gravel or dolomite as coarse aggregate, higher ultimate loads were recorded for those columns reinforced with higher

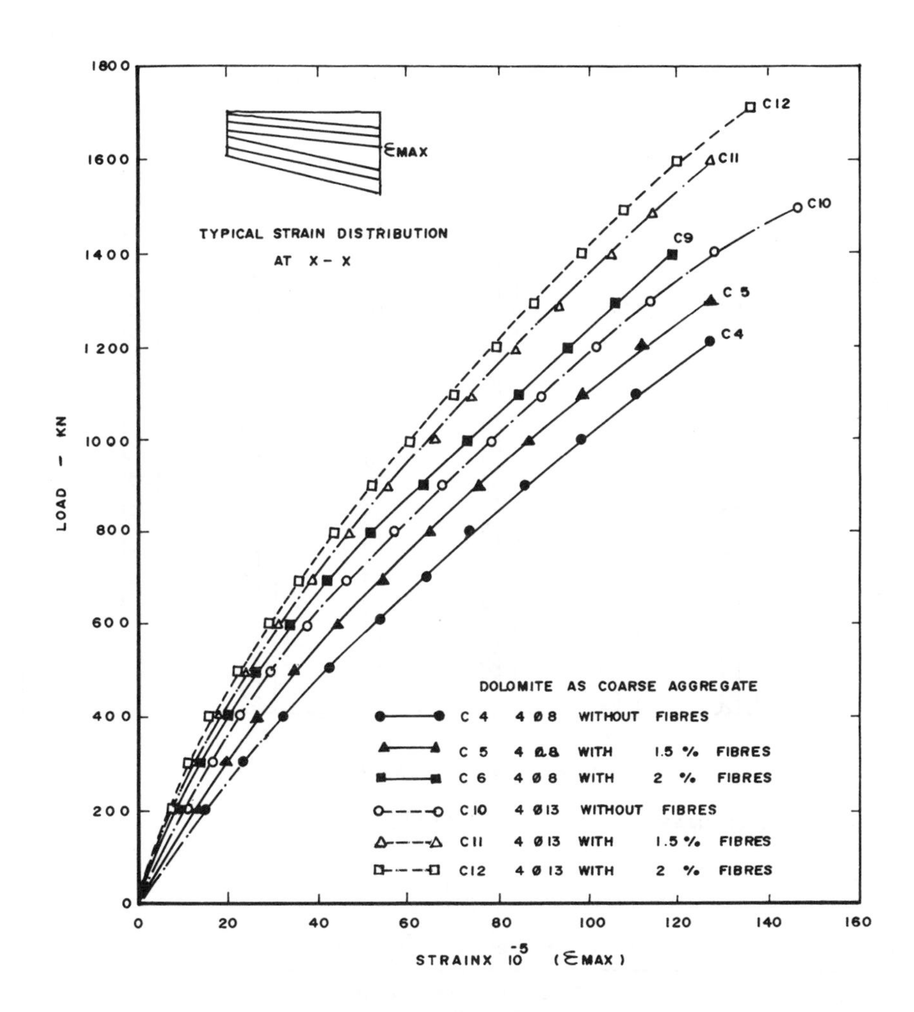

FIG. (6) : MAXIMUM COMPRESSIVE STRAINS IN REINFORCED CONCRETE COLUMNS ON THE CENTRAL SECTION X – X (DOLOMITE)

percentage of longitudinal reinforcement. The carrying capacity of the reinforced concrete columns increased with the increase of the steel fibers percentage. Those columns casted with dolomite, steel fibers of 2% by volume of concrete and reinforced longitudinally by 4ø8 or 4ø13 mm. bars showed ultimate carrying capacities of 143% and 146% respectively than those casted with gravel without fibers. Table 5 shows that the experimental test results of the ultimate loads of the reinforced concrete columns were about 87% in average of the ultimate loads calculated theoretically which confirms the existance of small eccentricity during loading.

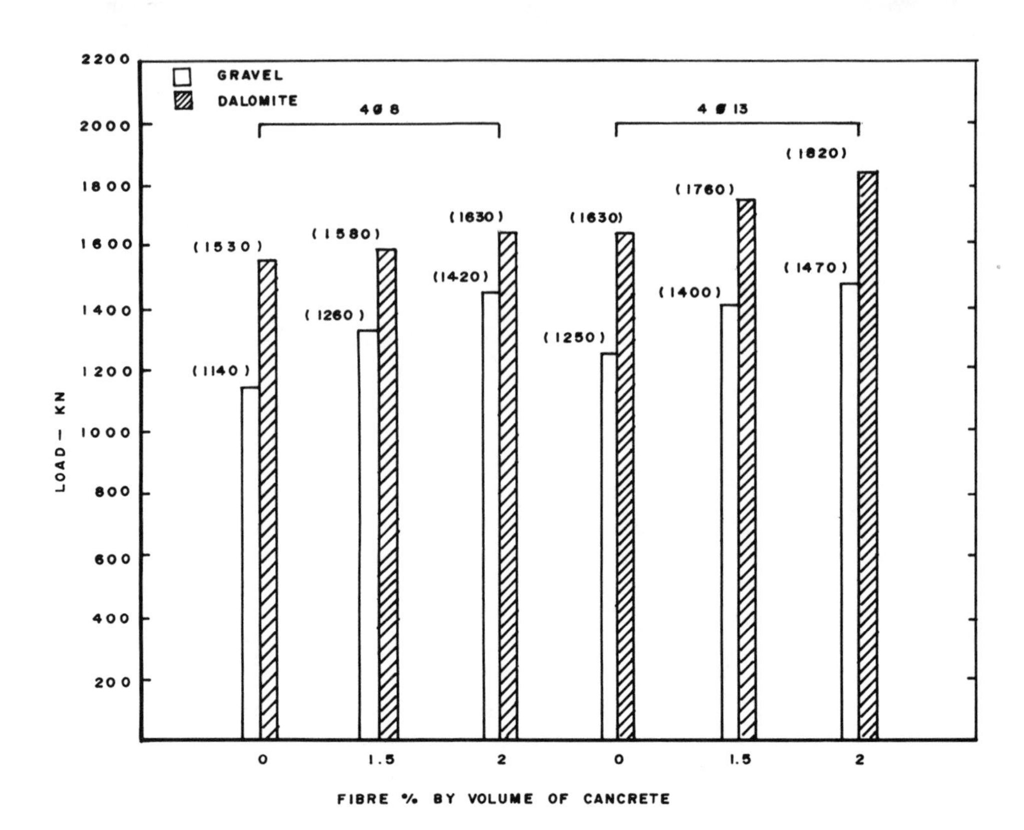

FIG. (7) : ULTIMATE LOADS OF THE TESTED COLUMNS

Table 5. Theoretical and experimental ultimate loads

Columns No.	Experimental Results Cracking Load (kN)	Experimental Results Ultimate Load (kN)	Theoretical* ultimate load	cracking load / Ultimate load	Exp. Ult. ld. / Theor. ult. Ld. %
C_1	1025	1140	1260	90.0	91.0
C_2	1070	1260	1440	85.0	87.5
C_3	1170	1420	1500	82.0	90.0
C_4	1345	1530	1730	88.0	88.0
C_5	1360	1580	1840	86.0	86.0
C_6	1370	1630	1920	84.0	85.0
C_7	1110	1250	1450	89.0	86.0
C_8	1220	1400	1640	87.0	85.0
C_9	1220	1470	1770	83.0	83.0
C_{10}	1400	1630	1930	86.0	84.0
C_{11}	1500	1760	2040	85.0	86.0
C_{12}	1510	1820	2120	83.0	86.0

* $P_u = 0.67\ C_{cu}\ (A_c - A_s) + F_y\ A_s$

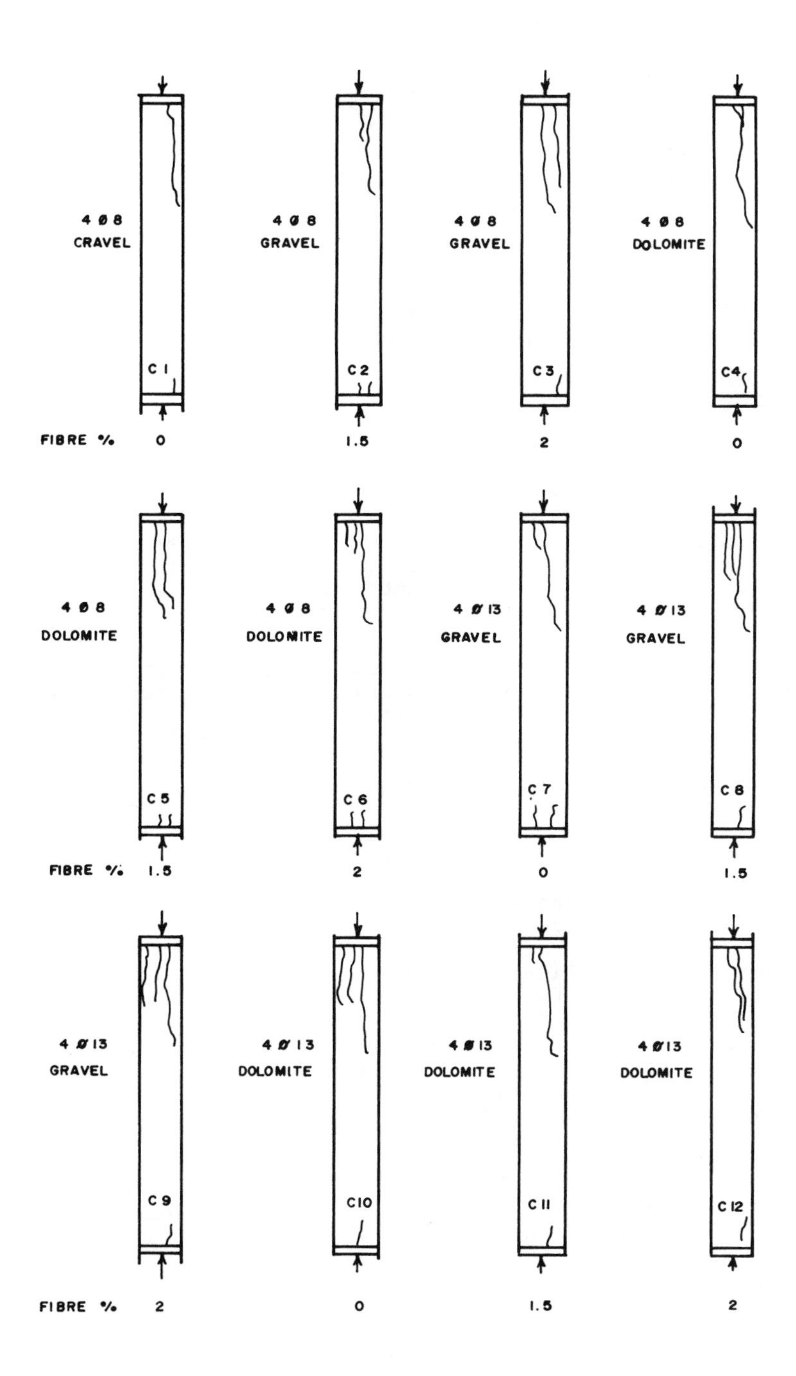

FIG. (8) CRACK PATTERN OF TESTED COLUMNS

3.2.3 Cracking and mode of failure

Out of the test results it was found that the enclusion of steel fibres increased the intial cracking loads as shown in Table 5. However, the ratio between the initial cracking load to the ultimate load decreased with the increase of the steel fibre percentage which gave more announcement before complete failure. All the columns were failed in compression with cracks concentrated more in one side of the columns rather than the other as shown in Fig. 8.

Conclusions

Out of this research study the following conclusions could be derived:-

1 Higher mechanical properties could be obtained by using dolomite as coarse aggregate instead of gravel.
2 Superior mechanical properties could be achieved by the inclusion of steel fibers in the dolomite concrete mix. Plasticizers are needed to enhance the workability which greatly affects the concrete mixing and compaction.
3 Using high strength concrete with dolomite as coarse aggregates and steel fibers enhanced greatly the deformation behaviour of axially loaded columns and increased their ultimate carrying capacity.
4 The longitudinal reinforcing bars in reinforced concrete columns could be reduced on the account of adding steel fibers and using dolomite as coarse aggregates.

5 References

1 Neville, A.M.,"Properties of Concrete",(1983), Pitman Publishing, London.
2 Shetty, M.S.,"Concrete Technology", (1982),S.SHAND Compand Ltd.,Ram Nagar, New Delhi.
3 Smith, R.C., "Materials of Construction", (1973), McGraw-Hill Book Company.
4 Kamal, M.M.,"(1985) Dolomite as Coarse Aggregate in Concrete Manufacture", First Egyptian Structural Engineering Conference, Cairo University.
5 Bahnasawy, H.H., and El-Refai F.E. (1990),"High Strength Concrete Using Dolomite as Coarse Aggregate", The Protection of Concrete Conference, Dundee, England.
6 Aiticin, P.C., (1985),"How to produce High Strength Concrete", Concrete Construction, Vol. 25, No, 3.
7 El-Refai, F.E. and Kamal, M.M.,"Behaviour of Polymer Fibre Reinforced Concrete Slabs",First Egyptian Structural Engineering Conference, Cairo University.
8 El-Refai, F.E. and Kamal, M.M.,(1986)"Behaviour of Polymer Fibre Reinforced Concrete Beams", Third International Symposium on Development in Fibre Reinforced cement and concrete, RILEM Technical committee, 49 TFR, Sheffield, England.

9 Kamal, M.M., Tawfik, S.Y. and Nosseir, M.H.(1987),"Polyester mortar", Jornal of Applied Polymer Science, Vol.33.
10 El-Refai, F.E. and Kamal, M.M.,(1987) Impact Resistance of Fibre Reinforced Concrete", First International Symposium on Housing Technology Production and Transfer from Research into Practice, Brazil.
11 El-Refai, F.E. and Kamal, M.M.,(1987),"Durability of Fibre Reinforced Concrete", Fourth International Conference on Durability of Building Materials and Components, Singapore.
12 El-Refai, F.E. and Kamal, M.M.,(1987)" Behaviour of Reinforced Concrete Columns", Journal of the Egyptian Society of Engineers, Vol. 26, No. 2.
13 Craig, R.J. Mc Connell, J. Germann, H., Dib, N., and Kashani, F., (1984), "Behaviour of Reinforced Concrete Columns", ACI SP-81,pp.69-105.

47 MECHANICAL BEHAVIOR OF COMPOSITE RC COLUMNS WITH THIN SECTION PRECAST GRC PANELS

E. MAKITANI
Department of Architecture, Kanto-Gakuin University, Japan
N. YANAGISAWA
Institute of Research of JDC Corporation, Japan
M. HAYASHI
Nihon Electric Glass Co, Japan
I. UCHIDA
Chichibu Cement Co., Japan

Abstract

The mechanical property of GRC with new lower alkali cement was investigated by flexural test for the length of chopped fiber, mesh type fibers and water/cement ratio, where the best combination of these factors was selected for the concrete form by the closed thin-section GRC precast panel. The shear test by repeated lateral load was carried out on the composite reinforced concrete column with closed thin-section GRC panel. The fact was recognized that the GRC web-panel is expected as not only the concrete form but the structural component.

On the basis of these results, the thin-section GRC precast panel was applied to the permanent concrete form of reinforced concrete column in four -story buildings in Japan.

Keywords: chopped and mesh glass fibers, low alkali cement, limit of proportionality, modulus of rupture, composite RC column with GRC precast panel, shear resistance, structural component, construction, laternal pressure, mechanical joint.

1 INTRODUCTION

Glass fiber reinfored cement and concrete, abbreviated by GRC, are generally made by ordinary portland cement as the matrix. Even if the alkali -resistant glass fiber were used for GRC, the deterioration of strength is not able to be fully restrained. For this reason, it is seemed that the problem of durability of the glass fiber in cement matrix prevent a remarkable development of glass fiber reinforced concrete.

The lower alkali cement for GRC was recently developed by Chichibu Cement Manufacturing Company in Japan. Using this new cement for GRC, the durability of GRC is enhanced and otherwise the mechanical properties and shrinkage by drying of cement can be improved, comparing with those of conventional GRC[1],[2].

For the purpose of applying the new GRC, using this lower alkali cement and the alkali-resistant glass fiber, to the permanent formwork of structures in

Fibre Reinforced Cement and Concrete. Edited by R. N. Swamy. © 1992 RILEM.
Published by E & FN Spon, 2-6 Boundary Row, London SE1 8HN. ISBN 0 419 18130 X.

building, tests were performed for investigating quantitatively the mixture of materials such as cement and water admixture, as well as glass fiber and the mechanical property. The most effective reinforcement method of glass fiber for flexural strength and ductility was selected from these tests. This reinforcement was applied to the concrete of the concrete column.

When the GRC panel is used for the reinforced concrete column as the concrete form, the composite column with closed thin-section GRC and reinforced concrete is constituted. Therefore, the GRC panel is expected to contribute to the enhancement of structural performance besides a role of concrete form in concrete members. Hence, for investigating the effect of the GRC panel as a structural component, the test by lateral cyclic loading was conducted on the composite column with a closed thin-section GRC and reinforced concrete. The contribution of the GRC panel to shear resistance was discovered from this test.

Furthermore, the GRC panel was applied to the concrete form of reinforced concrete columns in four-story buildings in Japan. The demonstration of construction is described in the present paper.

2 MECHANICAL PROPERTIES OF GRC

The mechanical property of the GRC is mainly influenced by the contents of admixture, types and content of glass fiber and water/cement ratio. Therefore, the tests were conducted for investigating how these factors affected the flexural properties of GRC. The results obtained from flexural tests of the GRC were herein indicated concerning the contents of admixture and water/cement ratio and reinforcement of glass fiber.

2.1 Test Specimen and Loading

The new GRC cement is constituted from higher alumina and lower lime to reduce the alkali and to suppress the deterioration of glass fiber and the shrinkage of concrete. River sand with a maximum size of 1.2mm was used as fine aggregate. The content of cement and sand is even by weight. The superplasticizer of 1% for cement by weight was mixed into mortar for enhancement of workability.

Table.1-Combinations of reinforcement by glass fiber and these contents by weight

unit:(%)

Kind of fibers	Location of mesh fibers				No mesh fiber
	5mm grid		10mm grid		
	Center	Bottom	Center	Bottom	
Chopped fiber 6mm length	2.77	0.23	2.74	0.26	3.00
Chopped fiber 6mm and 25mm length	1.39 1.39	0.23	1.37 1.37	0.26	1.50 1.50
Chopped fiber 25mm length	2.77	0.23	2.74	0.26	3.00
No chopped fiber	0.23	0.23	0.26	0.26	0

The chopped and mesh types of alkali-resistant glass fibers were used in the hybrid form, where they has a length of 6 and 25mm respectively. The whole content of glass fiber in the GRC occupies 3 percent by weight.

The mixing of new GRC cement, sand, glass fiber and superplasticizer, as shown in Table.1, was conducted by omni mixer in accordance with the manufacturing procedure which was indicated in Fig.1. The wet curing was taken for GRC panel in the constant temparature laboratory of 20 degrees centigrade. The testwas carried out on forty fifth day after placing of mortar.

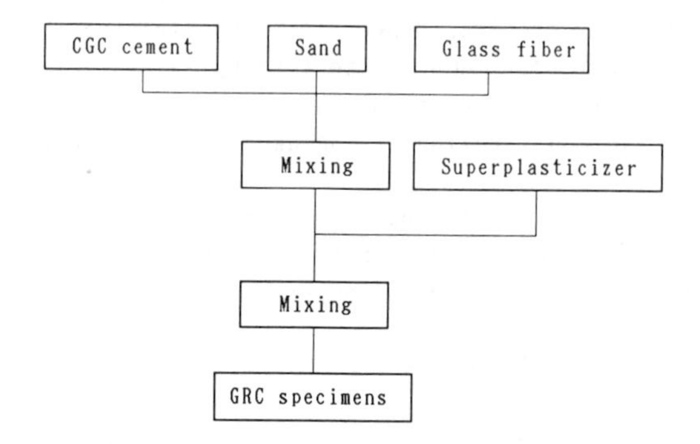

Fig.1-Manufacturing procedure of mixing for GRC specimens

The twenty four specimens were cut off from a flat plate to depth x width x length =10 x 50 x 300 mm for each type. The reinforcement for specimens was taken by only chopped and mesh fibers as a well as hybrid form of chopped and mesh fibers, of which the place was arranged either at the plane of center or bottom.

The monotonic loading was given by a concentrated force at the center of the specimen, which was supported at a span of 250mm.

The central deflection of specimen was measured by cantilever-type displacement meter.

2. 2 Flexural property of GRC

The mean flexural strengths of the limit of proportionality (LOP) and modulus of rupture (MOR) are plotted in Fig.2, when the mortars are reinforced by 25mm chopped glass fiber which possessed a content of 3 percent by weight and made by changing the contents from 1 to 3 percent of superplasticizer and the water/cement ratio from 35 to 40 percent. They were herein evaluated by twenty four specimens for each type where their coefficients of variation existed in the values of 0.08～0.17. The mean flexural strengths are also shown in Fig.2 for the GRC made by ordinary portland cement. It is found from these figures that both values of LOP and MOR for new GRC cement is greater than those of GRC used OPC, and the MOR value is greatest in the case when water/cement ratio and superplasticizer/cement ratio are equal to 40 and1 percents, respectively. This is the reason why new GRC and alkali resistant glass fiber excels conventional GRC in property of adhesive strength, since the calcium silicate-$C_4A_3\overline{S}$-$C\overline{S}$-Slag type low alkaline cement developed aiming for improving reinforcing effect of GRC in long term.

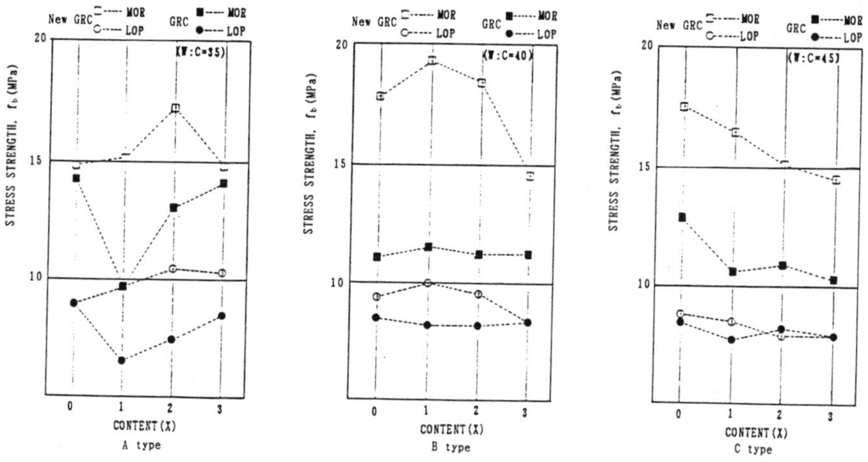

Fig. 2-Relationship between flexural strength and content of superplasticizer

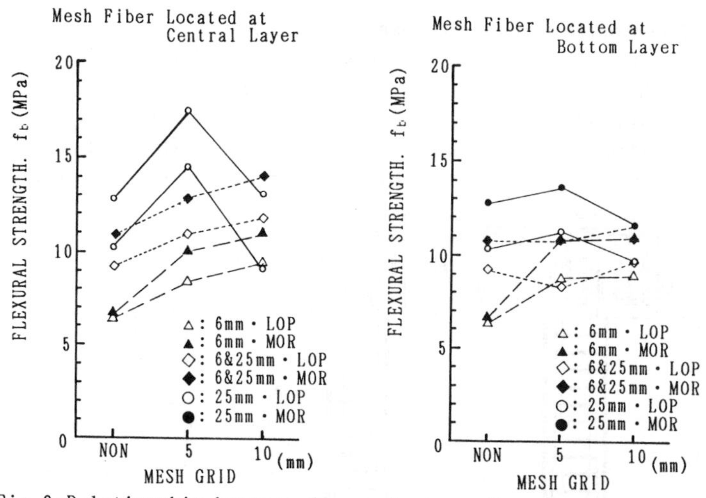

Fig. 3-Relationship between flexural strength and mesh grid

Using the mixture of GRC by which the flexural strengths become maximum, the flexural strengths of LOP and MOR are plotted in Fig. 3 for thin concrete plate reinforced by the chopped, mesh and hybrid fibers where the mesh is arranged either at the plane of center or bottom. The following facts can be recognized from these figures;

(1) Both values of LOP and MOR are maximum when GRC constitutes the hybrid

form of the chopped fiber with 25mm length and mesh fiber with 5mm grid whose place is arranged at the central layer.
(2) The values for mesh fiber with central layer are greater than those of mesh fiber with a bottom layer.

This is a reason why as the chopped fiber under the central mesh is restrained so as to distribute in the two-dimentional directions, its flexural effect may be enhanced.

3 COMPOSITE RC COLUMN WITH THIN-SECTION GRC PANEL

The optimal mixture and reinforcement for strength and ductility have been determined on the basis of the the fundamental data concerning the mechanical property of GRC which were previously obtained. This mixture and reinforcement was applied to the closed thin-section GRC inside of which the reinforcement of column and mechanical joint are previously installed before. The composite concrete structure is formed by monolithic integrity of the thin GRC panel and reinforced concrete. The test of the composite concrete column was conducted in order to certify whether its thin GRC webs panels constitute the shear -resistant component.

3.1 Test of Composite Concrete Column

The specimens are short columns with a square section of width x depth x length=200x200x600mm, of which longitudinal and shear reinforcements are shown in Fig. 4. The longitudinal reinforcement of the column is connected by a coupler for thread-deformed reinforcing at its bottom in order to install these reinforcements inside of the closed thin-section GRC precast panel.

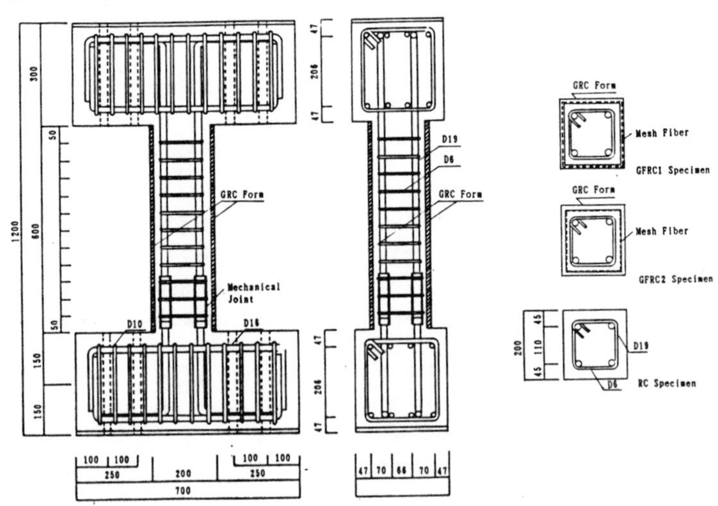

Fig. 4-Shape and reinforcement of specimen

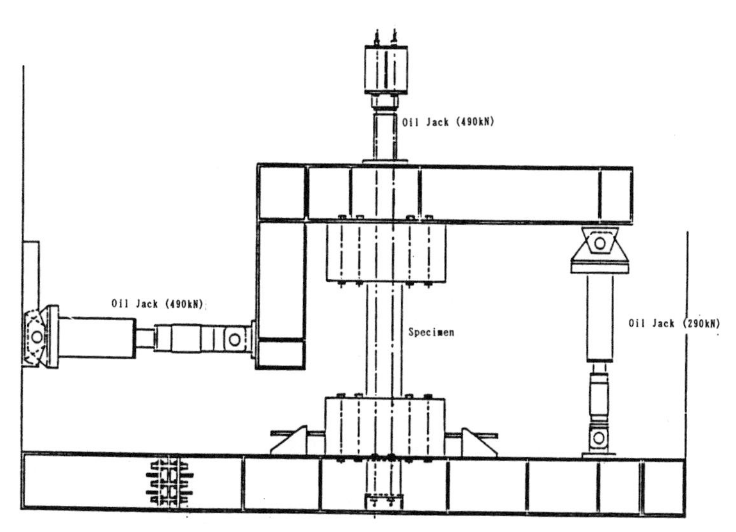

Fig. 5-Apparatus for lateral loading

The thin GRC panel with 10mm depth is reinforced by the hybrid fibers of 6 and 25mm length and 5mm grid mesh, of which the whole content is 3% by weight. The displacement of mesh fiber is arranged either at the center or the inside layers of the GRC panel. Hence, two composite concrete columns with GRC panel and one reinforced concrete column for comparison were made.

The lateral cyclic loadings were given by a 490kN oil jack used for push-off and/or pull-out. They were given by the control of lateral displacement at the top of column so that the angles of the column axis were equal to 1, 2, 4, 6, 8, 10, 12, 16, 20, 24×10^{-3} rad each one cycle. The axial loading of the column was given by a 490kN oil jack so that the compressive stress of 3 MPa kept a constant value in the section of the column. The apparatus for lateral loading is shown in Fig. 5.

3. 2 Results of Test

Multiple shear and flexural cracks were generated in the web and flange surfaces of RC column but these cracks were few in the GFRC columns. The shear failure occurs finally in RC and GFRC columns after the yield of longitudinal reinforcement. The GRC panels were separated from the concrete column at failure when the mesh fiber was located at the central layer. But they were not separated from the column at failure when the mesh was located inside it, because the GRC panel is monolithic with the concrete column by bonding with the mesh fiber. The cracks at failure are depicted in Fig. 6.

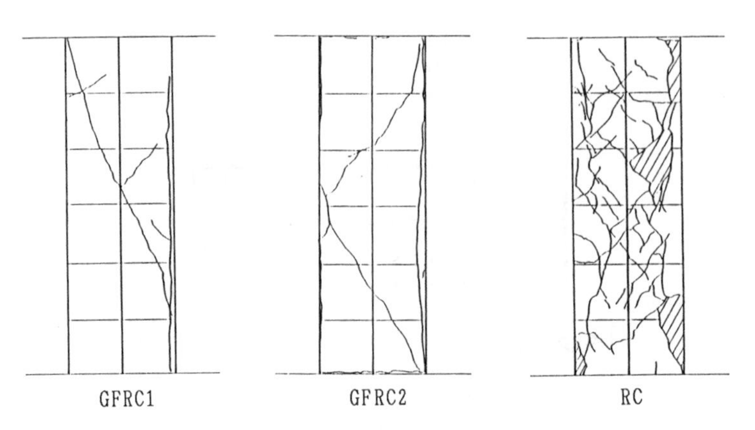

Fig. 6-Sketch of crackes at failure

The cyclic curves for load-lateral displacement at the top of column and their envelope curves are shown in Fig. 7. The cyclic curve for the GFRC1 column represents a difference in the positive and negative axis of X diection . This is due to formation of the shear failure due to inclined crack which occurred from the top to the bottom of the column. The maximum load for GFRC2 column is the greatest of three columns.

This is due to the contribution to shear resistance of the mesh fiber located at the central layer of GRC web-panel. The deterioration of load for the GFRC2 column is found after maximum load, because the mesh fiber has broken.

3. 3 Shear resistance of Columns

The shear failure mechanism of the GRC web panel was assumed as shown in Fig. 8, to investigate how the GRC web-panel might contribute to the shear resistance of the column. The tensile strengths of chopped fiber reinforced concrete and mesh fiber have to be evaluated in order to calculate the shear strength of the GRC web-panel. They can be obtained from direct tensile test, but they can be also estimated approximately by evaluating the stress block equivalent to tensile stress distribution under the neutral axis of specimen from the results of flexural test of GRC[6]. The equation of shear strength of the GRC web-panel has been derived from the above assumption.

3. 4 Tensile Strength of Material

The flexural behaviors of the thin GRC panel with chopped and mesh fibers are classified into four states of stress as shown in Fig. 9; (1) 0～LOP, (2) LOP～MOR, (3) Maximum load (MOR), (4) MOR～ultimate strength. If the depth of the neutral axis from the compression edge is assumed 0.2D at the maximum load[5], the moment M_{fu} by tensile force of chopped fiber reinforced concrete T_{fc} and mesh fiber T_{fm} are given about the place of the compressive resultant of

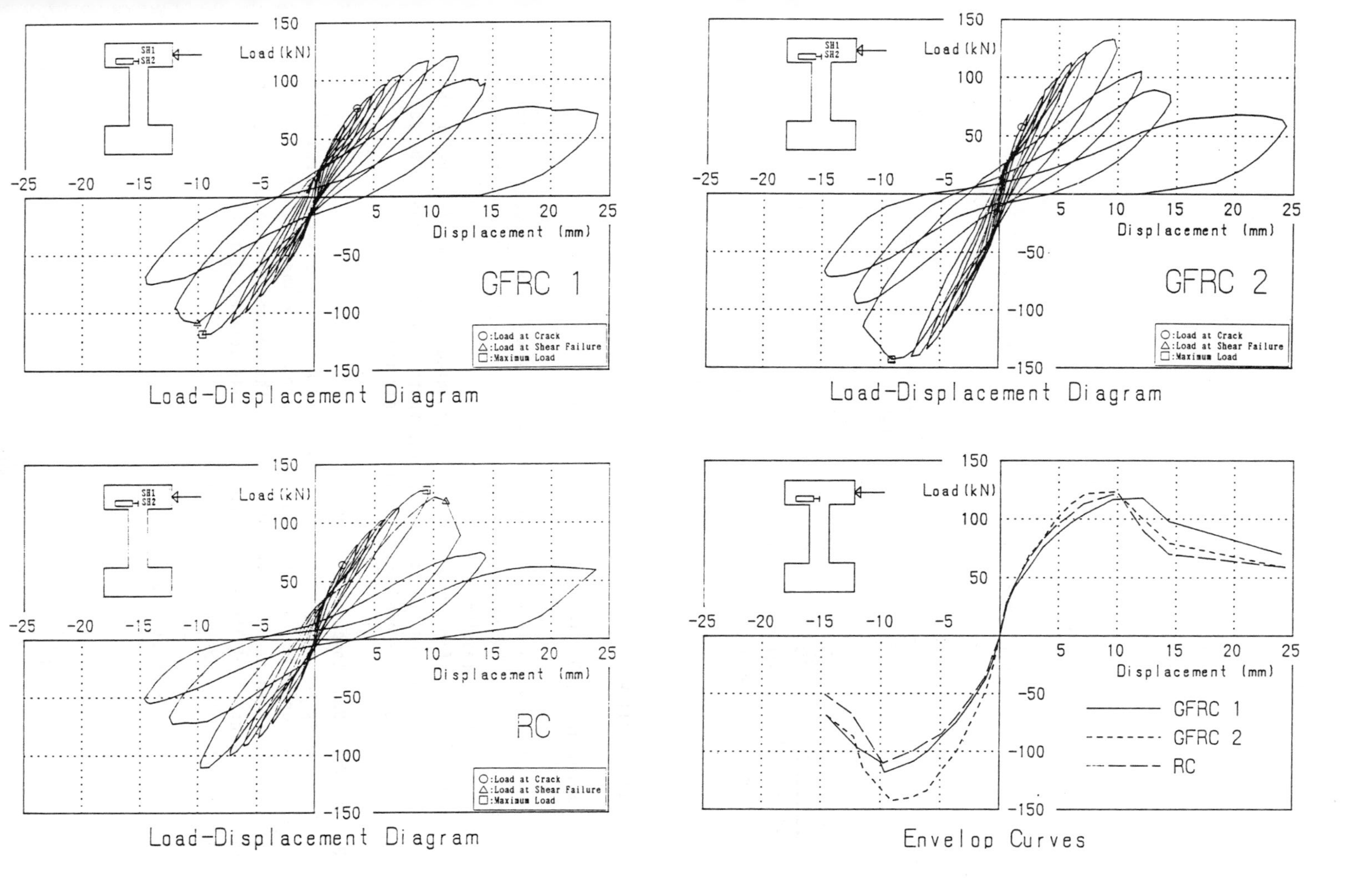

Fig. 7-Cyclic and evelope curves for load - lateral displacement

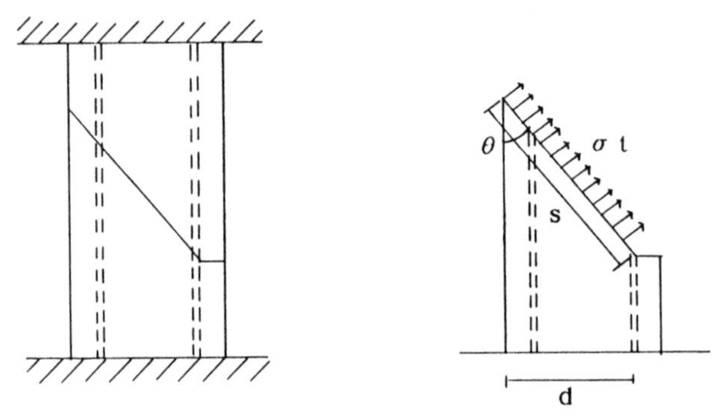

Fig. 8-Mechanism at shear failure of GRC web panel

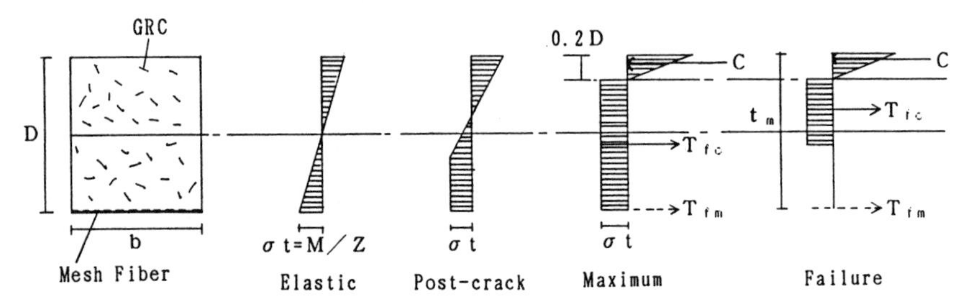

Fig. 9-Flexural behaviors thin GRC panel reinforced by chopped and mesh fiber

concrete as follows:

$$M_{fu}=0.426\sigma_t bD^2+T_{fm}(t_m-0.067D) \qquad (1)$$

where b and D are the width and depth of GRC panel, σ_t is the apparent tensile strength of chopped fiber reinforced concrete, t_m is the distance of mesh fiber from the compression edge.

The mean values of σ_t and T_{fm} can be calculated by equation (1), using the results of flexural test for the GRC panel, where they were evaluated by forty eight specimens and by confidence limit of eigty five percents.

3. 5 Shear Resistance of GRC Panel

The shear resistance of the GRC web-panel is generated by the chopped fiber reinforced concrete and the mesh fiber aross the shear crack with the inclination θ to the longitudinal axis of the column. Under the assumption that the length S of the shear crack is approximately equal to $d/\sin\theta$, it is represented by the following equation:

$$Q_{fc}=2(\sigma_t\cdot D\cdot S\cdot\cos\theta+T_{fm}\cdot d) \qquad (2a)$$

where d is the effective depth of the RC column. The angle θ can be expressed by angle ϕ of friction of concrete on the basis of the fracture standard of modified Mohr-Coulomb of concrete[3]:

$$\theta = \pi/4 - \phi/2$$

If ϕ is taken to be 37° corresponding to the angle of friction of concrete, θ is equal to 26.5°. Substituting this value into equation (2a), the following equation is obtained:

$$Q_{fc} = 2d(2\sigma_t \cdot D + T_{fm}) \qquad (2b)$$

3.6 Shear Resistance of GRC Column

The shear bearing capacity of the RC column is given by sum of those due to concrete Q_c, web reinforcement Q_w and axial force Q_a according to the Building Standard Law of Japan[4].

$$Q_{su} = Q_c + Q_w + Q_a$$

$$= \left\{\frac{0.053Pt^{0.23}(F_c+180)}{M/QD+0.12} + 2.7\sqrt{P_w \sigma_{wy}} + 0.1\sigma_0\right\} bj \qquad (3)$$

where P_t is the tension steel radio, F_c is the compressive strength of concrete (kgf/cm^2), M/Qd is the web span ratio, P_w is the shear reinforcement ratio, σ_{wy} is the yield strength of web reinforcement (kgf/cm^2), b is the width of the column and j is the distance between resultants of tension and compression.

Therefore, the shear resistances GRC column are given by sum of equation (2b) and (3).

$$Q_u = Q_{su} + Q_{fc}$$

$$= \left\{\frac{0.053P_t^{0.23}(F_c+180)}{M/Qd+0.12} + 2.7\sqrt{P_w \sigma_{wy}} + 0.1\sigma_0\right\} bj + 2d(2\sigma_t D + T_{fm}) \qquad (4)$$

3.7 Shear Capacity of GRC Column

The shear capacity of the GFRC and RC columns, using equation(4), is calculated by separating the shear resistance component such as concrete, web reinforcement and the GRC web-panel. The results of calculation are shown in Table.2. It is found from Table.2 that the ratio of the GRC web-panel to the composite column for shear resistance is approximately equal to 30 %. Therefore, the thin GRC web panel contributes considerably to shear resistance as a structural component besides being a concrete form.

The shear capacities of the GRC columns, which were calculated by equation (4), indicate such a good estimate that they is given by the ratios of 0.97 and 1.13 for the results of the test.

Table. 2-Shear resistance for structural components

Specimen	Maximum load	Structural components								Shear force	
		Concrete		web		Axial force		GRC			
	Qu (kN)	Qc (kN)	$\frac{Qc}{Qsu}$	Qw (kN)	$\frac{Qw}{Qsu}$	Qa (kN)	$\frac{Qa}{Qsu}$	Qfc (kN)	$\frac{Qfc}{Qsu}$	Qsu (kN)	$\frac{Qu}{Qsu}$
GFRC1	120.8	49.6	0.40	30.8	0.25	8.1	0.07	35.8	0.29	124.3	0.97
GFRC2	140.3	48.0	0.39	30.8	0.25	8.1	0.07	37.3	0.30	124.2	1.13
RC	122.0	64.0	0.58	37.9	0.34	8.1	0.07	—	—	110.0	1.11

4 APPLICATION OF GRC FORM TO CONCRETE COLUMN

The rational construction of building is required to reduce the works on the site by means of the prefabrication of parts. The precast GRC form for parts in building is one among such constructions. This construction has the following main merits:

(1) Treatment in the site of construction is very easy, as the GRC panel is thin and light.

(2) When the finish of the column is formed in fabrication of the GRC panel, the work at the site can be reduced.

(3) The durability enhances since the crack in the GRC panel is restrained by glass fiber.

The concrete forms by the GRC panel have been applied to all the concrete columns of four-story buildings in Japan. The structure is constituted from the frames in X and Y directions. The hight of the story is 3.4m in the first and fourth stories and 2.75m in the second and third stories. The columns in each story have the square section of 65cm for a side length.

It was made in the shape of a box by binding two GRC channels with the inside length of 32.5x65cm and the width of 2cm by adhesive. Also, it was reinforced by chopped glass fiber with a length of 25mm and 5% by weight, as well as mesh fiber located at the central layer, and was formed by the direct spray method.

As it is subjected to the lateral pressure agaist the GRC panel by the placing of ready mixed concrete, the cramps were arranged at a distance of 50cm.

To measure this lateral pressure, a pressure meter for was installed inside of its bottom, as shown in Fig. 10. The strain gauges were stuck on the top of, middle and bottom surfaces to evaluate the flexural stress which generates in the GRC panel by the lateral pressure due to ready mixed concrete. The measured internal pressure and strains in lateral and longitudinal directions were plotted for change of time in Fig. 11. It is recognized from these measurements that the strain increases from top to bottom and the lateral pressure against the GRC panel decreases by concrete hardening as time goes on. The results of measurement indicated that the maximum lateral pressure and flexural stress were equal to 44 kN/m^2 and 5.5 MPa, respectively. As the flexural strengths of LOP and MOR for GRC panels have been obtained by 13.8 MPa, 28.9 MPa, respectively, the GRC panel possesses the full safety for lateral pressure.

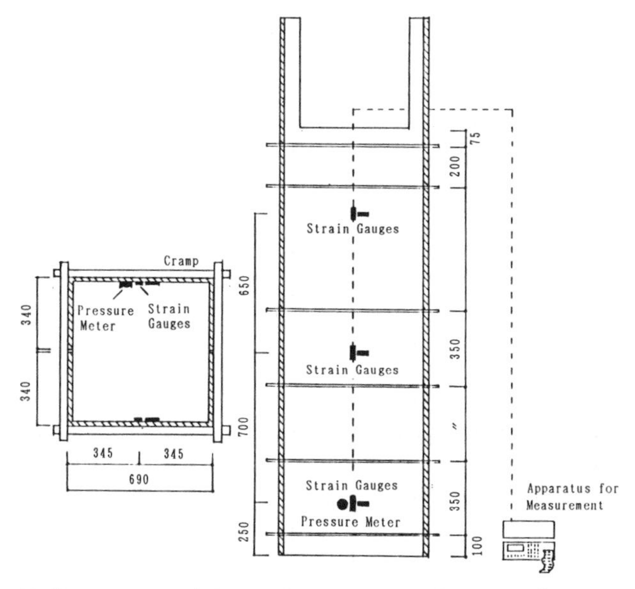

Fig.10-Measurment of lateral pressure and strain GRC panel

The photograph show the appearance of concrete form column by GRC panel under construction.

Photo.1-Appearance of concrete form of column by GRC panel under construction

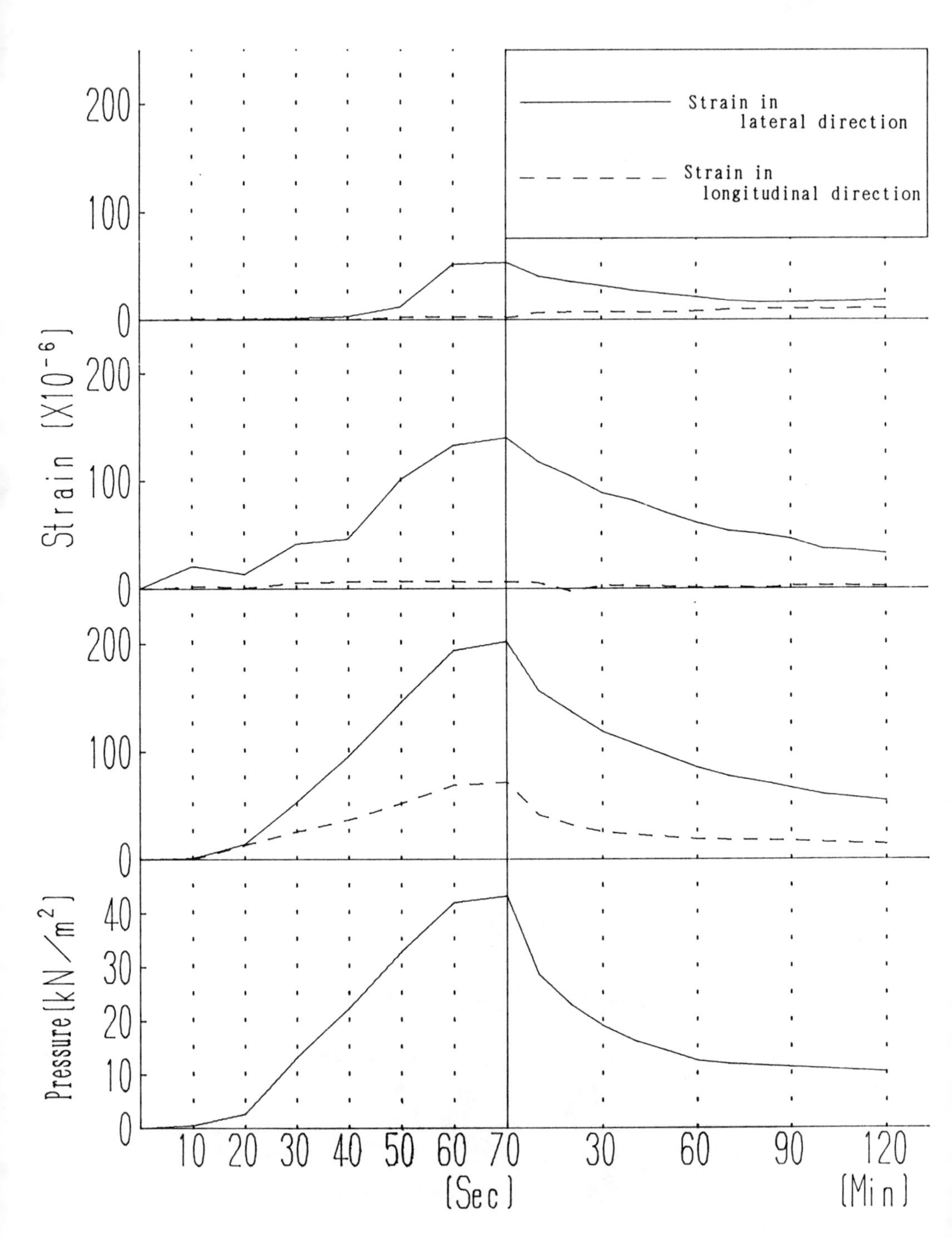

Fig.11-Strain and lateral pressure - time relationships at GRC panel

5 SUMMARY AND CONCLUSIONS

The properties of flexural strength of the GRC using by new low-alkali cement and reinforced by chopped and mesh glass fibers possesses the following merits in comparision with GRC by ordinary portland cement:

(1) The values of LOP and MOR are enhanced when the water/cement radio is from 35% to 40% and the content of the superplasticizer is from 1% to 2%.

(2) The maximum strengths are obtained if the hybrid type is of chopped and mesh glass fibers with a 25mm length and a 5mm grid, respectively, where the mesh fiber is located at the central layer.

On the basis of these results, the thin concrete form by the GRC panel was applied to the model of the RC column. The following facts concerning shear behavior were found from the model test:

(3) The crack strength increase remarkably by the crack restraint effect of the fiber.

(4) The GRC web-panel is expected as a structural component for shear resistance besides having a role as a concrete form. Therefore, it is possible to enhance the structural performance of the RC column.

By producting the concrete form by GRC at the precast concrete factory, it was applied to all the RC columns at four-story buildings in Japan as a permanent concrete form. The following facts were certified from this application to actual building:

(5) As the works at the site for fabrication and removal of form such as plywood panel are not required, the term of construction work can be reduced and otherwise it contributes to protection of wood resources.

(6) The safety for labor under construction made higher as the concrete form by GRC panel is very lighter compared to full precast concrete parts.

This application is the most common construction work as the scale of concrete form by GRC in Japan. As the panel is used as a concrete form in the building, there is no problem for fire resistant performance. But when it is evaluated as a structural component in design, it does not satisfy the fire resistance performance so long as the special specification is not used. In such a case the carbon fiber possessing the fire resistance is desired for use in the concrete form. Taking this opportunity, it is expected that the concrete form by fiber reinforced concrete might be applied to many buildings in future and be developed as a rational construction method.

6 References

1, I. Uchida, et al, : Development of Low Alkali Cement for new GRC, International Meeting on Advanced Materials, Tokyo, Japan, 1988

2, I. Uchida, M. Hayashi, et al , : Application Rearch on GFRC-CGC Cladding

panels, 7th Biennial Congress of the Glassfiber Reinforced Cement Association, Netherlands, 1989
3, W.F.Chen, Plasticity in Reinforced Concrete, McGraw-Hill company, 1982.
4, Building Standard Laws of Japan, The Building Center of Japan,1990.
5, D.J.Hannant, Fiber Cements and Concrete, John Wiley & Son,1978.
6, A.J.Majumdar and v.Laws, Glass Fiber Reinforced Concrete, BSP Professional Books, 1991.

48 BEHAVIOR OF CONFINED CONCRETE WITH STEEL FIBRES

A. H. AL-SHAIKH
College of Engineering, King Saud University, Riyadh, Saudi Arabia

Abstract
The behavior of laterally confined concrete prisms with steel fibers has been experimentally investigated. Prisms were loaded monotonically in the main (axial) direction until failure. Lateral confinement was provided by steel stirrups. One fiber volume and aspect ratio of 1% and 75, respectively, have been used. The main variable considered in this study was the method of curing. Identical specimens were cured either in water or under atmospheric conditions or in the environmental room under constant temperature of 60 °C. In addition, two stirrup diameters and two spacings were also used. Results were discussed regarding general behavior, strength, ductility and energy absorption and dissipation capacity. A noticeable enhancement in both strength and ductility was observed with some of the test specimens, while others showed much lower values, depending on the test variable used.
Keywords: Steel Fibers - Confinement - Curing - Stress-Strain

1 Introduction

It is well established that lateral confinement of a compression member improves its ability to withstand compressive loads and enhances its ductility. This improvement due to lateral confinement is obtained by utilizing the well-recognized fact that concrete compressed in more than one principal direction is able to undergo more deformation, and to achieve higher strength, than when compressed in only one direction. However, the behaviour of the confined concrete is affected by a number of factors including concrete strength; amount and distribution of longitudinal reinforcement; amount, spacing and configuration of transverse

Fibre Reinforced Cement and Concrete. Edited by R. N. Swamy. © 1992 RILEM.
Published by E & FN Spon, 2-6 Boundary Row, London SE1 8HN. ISBN 0 419 18130 X.

reinforcement; size and shape of confined concrete; ratio of confined area to gross area as well as others. The addition of steel fibers to the plain and confined concrete introduces another factor and, therefore, several researchers have studied its effect on the concrete characteristics and member behaviour (Swamy et al. (1974), Williamson (1977), Ramakrishnan et al. (1981), Fenella and Naaman (1985)). Both strength and ductility of the confined concrete was reported to increase by the addition of steel fibers (Craig et al. (1984), Mangat and Motamedi (1985), Ganesan and Murthy (1990)). The aim of this paper is to quantify the effect of the curing method on the compressive behaviour of confined steel fiber concrete. In addition to water curing, the effect of an atmospheric and high temperature curing were investigated. These two curing methods were chosen to represent possible situations due to bad construction practices and/or during summer and hot weather concreting.

2 Experimental program

2.1 Specimen and material details

The overall dimensions of the prisms were 150 mm x 150 mm x 450 mm. Lateral confinement was provided by overlapping steel stirrups with two bar diameters (4 and 6 mm) and two stirrup spacing (40 and 60 mm). Stirrups were held in position using four longitudinal straight steel bars (6 mm diameter) placed at the corners of each prism. Steel fibers used in this investigation were hooked with nominal fiber length of 50 mm and crimps at each end measuring 5 mm in length; and nominal fiber diameter of 0.8 mm (aspect ratio of 75). The fibers were made from low carbon steel wires with a density of 7800 Kg/m^3, and were produced in sheets of as many as 30 fibers glued side by side. A total of 24 prism were used in this study and cast using the mix proportions listed in Table 1. Plain standard cubes and cylinders were also cast and tested at 7 and 28 days. The constituents of the mix were carefully batched and mixed thoroughly in a mechanical mixer. Steel fiber were then slowly added and mixed for about 3 more minutes until uniform dispersion of fibers is obtained. Concrete was then placed in vertical steel moulds in three layers, each layer being vibrated for about 1 minute and a half. Specimens were cured for a day under damp hessian; after which they were demoulded, divided into three equal groups, and placed either in a water tank, outside the laboratory, under atmospheric conditions

Table 1. Mix proportions for one cubic meter

Coarse Aggregate (10mm+20mm)	1020	kg
Fine Aggregate	680	kg
Water	214	kg
Cement	486	kg
Steel Fiber (V_f=1%)	78	kg
Superplasticizer	7.69	liter

(temperature ranges between 20-25 °C) or in the laboratory environmental room at a constant temperature of 60 °C for 28 days before testing.

2.2 Loading and instrumentation details

All prisms were tested in compression in an Amsler 2000 kN capacity testing machine at a constant strain rate of 7 microstrains per second. Axial deformation of each prism were measured using 4 LVDT's placed opposite to the prism faces. All load and displacement measurements were recorded and analyzed using a mini computer and data logging system (Fig. 1).

Fig. 1 Test setup

3 Results and discussion

A summary of the test results showing the maximum stress and strain obtained for the various specimens is presented in Table 2. The first number for specimen designations in this table refers to the stirrups diameter, the second number refers to its spacing and the letters W, A and T refers to water, atmospheric and temperature curing, respectively. The effect of the different test variables on the prism's stress-strain behaviour is presented graphically in Figs. 2 through 10.

3.1 Behaviour of prisms

The behaviour and response of each prism to the applied load was observed carefully throughout each test, especially the manner in which cracks propagated until failure. In general, a similar behaviour under load was observed for all the steel fiber specimens, regardless of the method of curing used. The applied load rose rapidly in the initial stages of loading, while the corresponding measured longitudinal displacement were relatively small until the first visible crack was formed. Subsequently, the rate of longitudinal strain increased accompanied by a steady load increase until the maximum load, P_{ult}, was reached. The initial cracking load value varied between 50% - 75% of P_{ult} and, as loading continued beyond the first cracking load, fine surface crack propagated steadily, however, very little surface spalling was observed. Cracks were first observed in the prism middle region and had no regular orientation.

3.2 Effect of stirrup diameter

The effect of increasing the stirrups diameter, d_s, from 4 to 6 mm for the same spacing and method of curing is presented in figures 2 through 4. As expected, increasing the lateral confinement ratio resulted in more confinement to the core and, hence, a higher load carrying capacity. All three figures indicate an increase in the maximum stress, σ_{max}, reached using the 6 mm diameter stirrups, however, it can also be seen that this strength enhancement varied for the different curing methods. Although the water cured specimens recorded the highest compressive strength, a higher strength gain (about 22%) was observed for high temperature cured specimens compared to atmospheric (11%) and water curing (18%). This strength gain was calculated as the percentage

Table 2. Summary of principal results

Serial No.	Specimen Designation	P_{ult} (KN)	σ_{max} (MPa)	ε_{max} ($\mu\varepsilon$)
1	44W	722	32.1	4795
2	64W	859	38.2	5315
3	66W	637	28.3	6450
4	46W	585	26.0	6005
5	44A	674	29.9	7163
6	64A	740	32.8	7562
7	66A	639	28.4	7630
8	46A	637	28.3	5197
9	44T	614	27.3	7624
10	64T	750	33.7	7732
11	66T	641	28.5	7540
12	46T	559	24.8	6715

- f'_c= 43.1 MPa

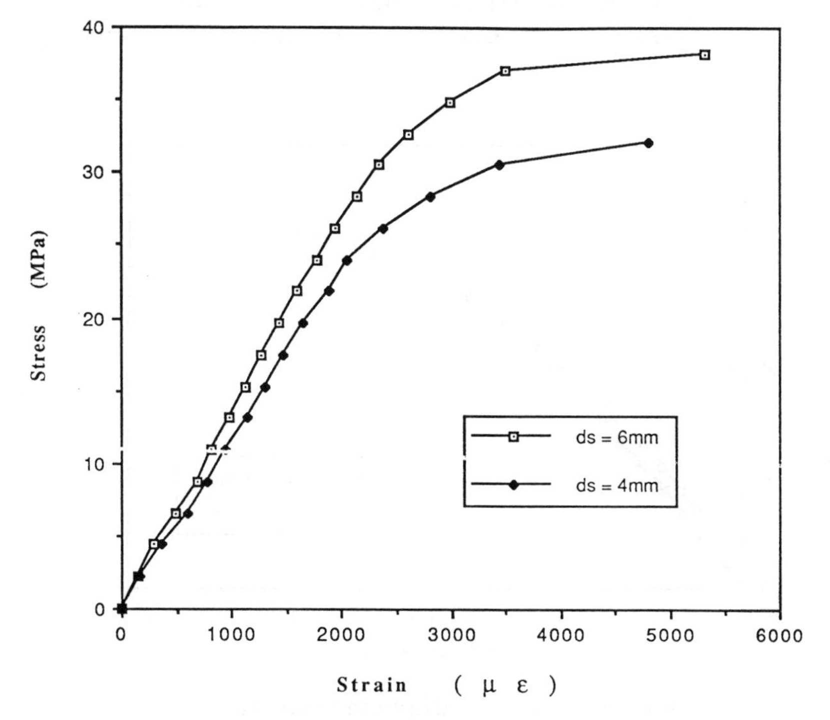

Fig. 2 Effect of stirrup diameter on the stress-strain diagram for water curing.

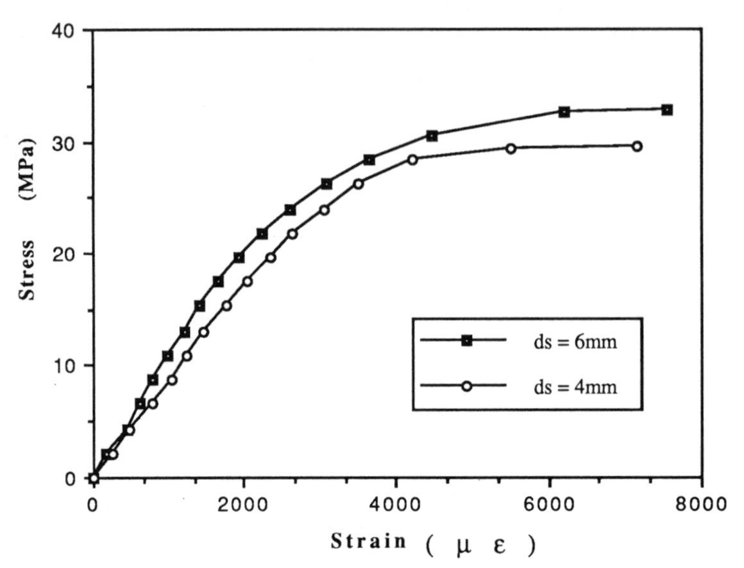

Fig. 3 Effect of stirrup diameter on the stress-strain diagram for atmospheric curing.

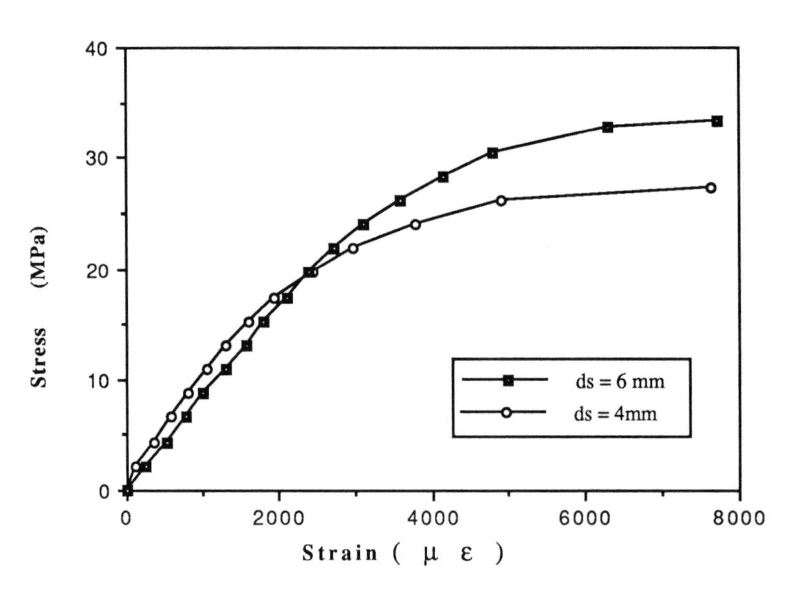

Fig. 4 Effect of stirrup diameter on the stress-strain diagram for temperature curing.

increase in the maximum strength due to the increase of d_s from 4 to 6 mm.

On the other hand, the maximum strain, ε_{max}, corresponding to σ_{max} showed no significant difference due to the increase of d_s, however, a higher strain was recorded for the atmospheric and temperature cured specimens as can be seen in Figs. 2 through 4. The initial modulus was also slightly increased by increasing d_s for water and atmospheric curing but an opposite result was observed for temperature curing.

3.3 Effect of stirrup spacing

The effect of the lateral confinement spacing, s, on the prisms stress-strain behaviour is also presented in figures 5 through 7. As can be seen from these figures decreasing the stirrups spacing from 60 mm to 40 mm resulted in an increase in the prisms maximum stress, σ_{max}, for all three curing methods. Here, again, the atmospheric cured specimens recorded the lowest strength gain (about 4%) as a result of the reduction in the stirrup spacing. The water and temperature cured specimens recorded a strength gain of 22% and 10%, respectively. The maximum strain, ε_{max}, also did not appear to be largely affect by the reduction in the confinement spacing.

3.4 Effect of the curing method

A comparison of the effect of the three curing methods used in this study is presented in Figs. 8 through 10. It can be clearly seen that, in all cases, water curing resulted in a higher ultimate strength while the temperature curing produced the lowest strength. As discussed earlier, the best results in terms of strength gain is expected with the combination of s = 40 mm and d_s = 6 mm. These results are presented in Figure 8 from which a maximum stresses of 34.5, 32.8 and 30.5 MPa for water, atmospheric and temperature curing, respectively, were obtained. Hence, a strength loss of about 5% and 11% is obtained due to atmospheric and temperature curing, respectively, as compared to water curing. This strength loss increased to about 8% and 15% for s = 40 mm and d_s = 4 mm, and to about 9% and 20% for s = 60 mm and d_s = 4 mm as can be seen from Figures 9 and 10. The ultimate strain, however, showed no consistent trend with the change in the method of curing but a strain in excess of 6000 microstrains was generally obtained. This large values of strain also reflected a high

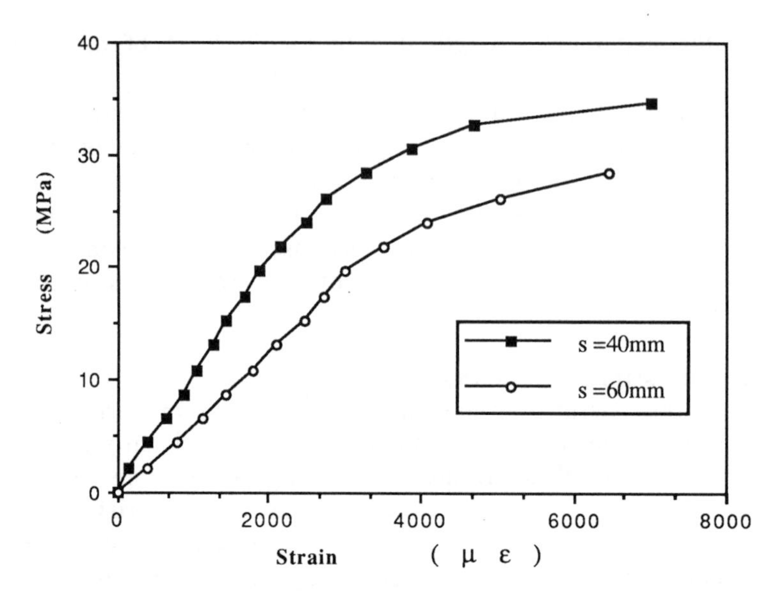

Fig. 5 Effect of stirrup spacing on the stress-strain diagram for water curing.

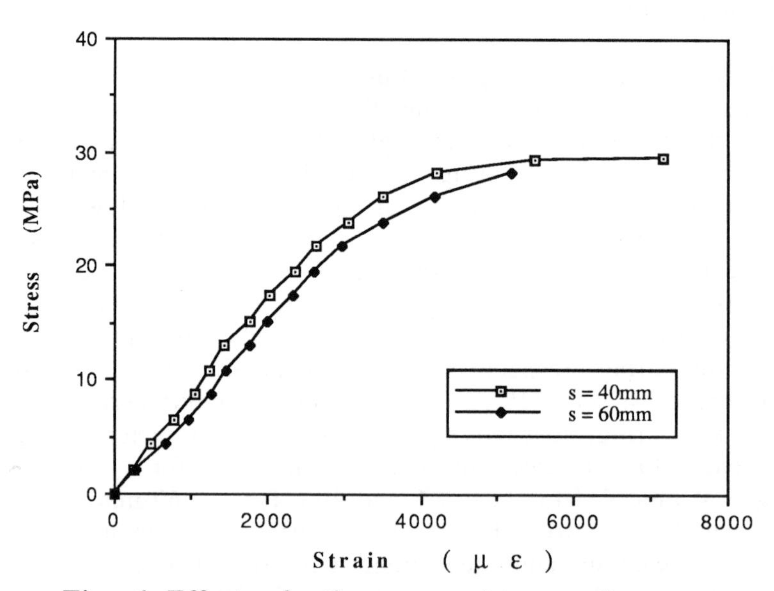

Fig. 6 Effect of stirrup spacing on the stress-strain diagram for atmospheric curing.

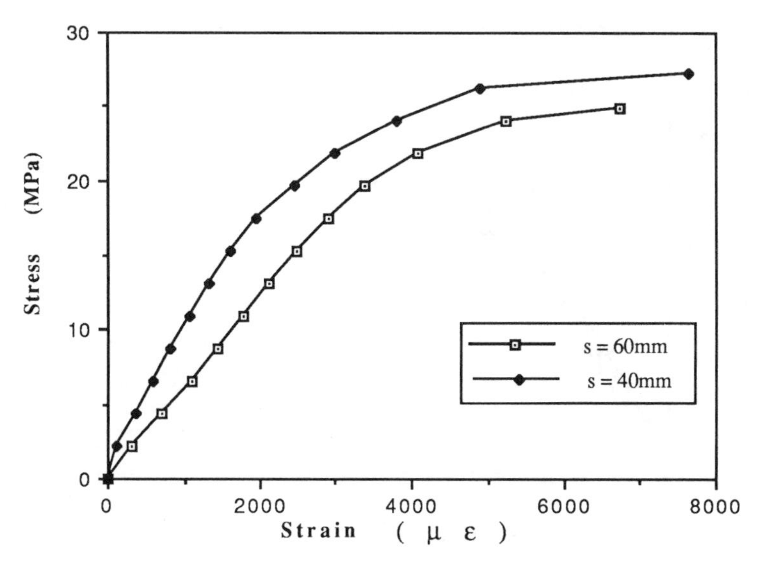

Fig. 7 Effect of stirrup spacing on the stress-strain diagram for temperature curing.

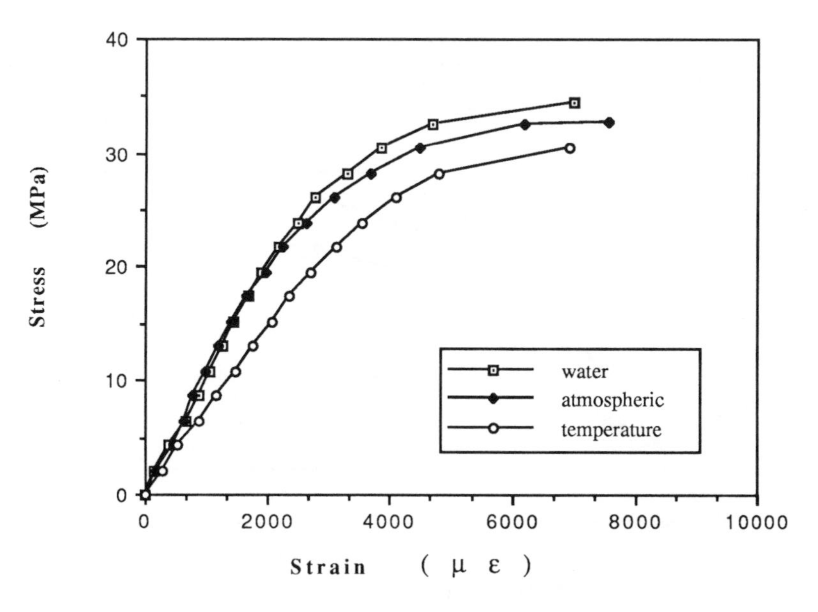

Fig. 8 Comparison of the various curing methods for s=40mm and ds=6mm

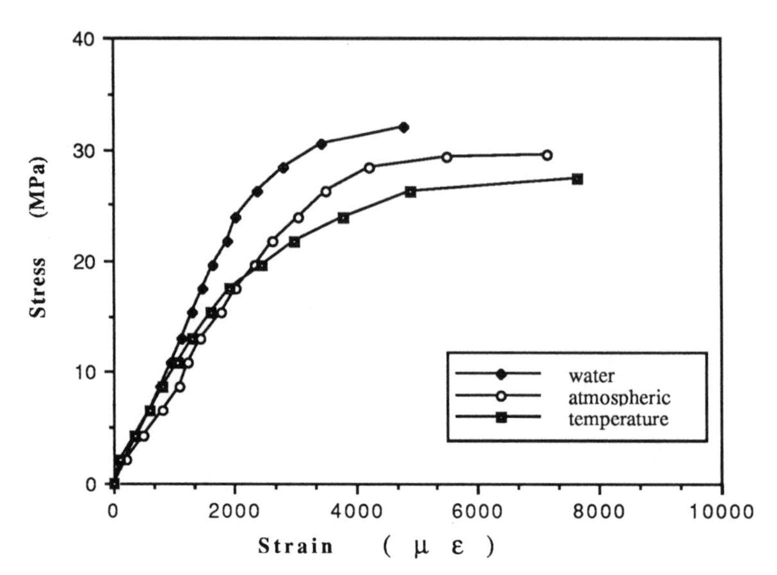

Fig. 9 Comparison of the various curing methods for s=40mm and ds=4mm.

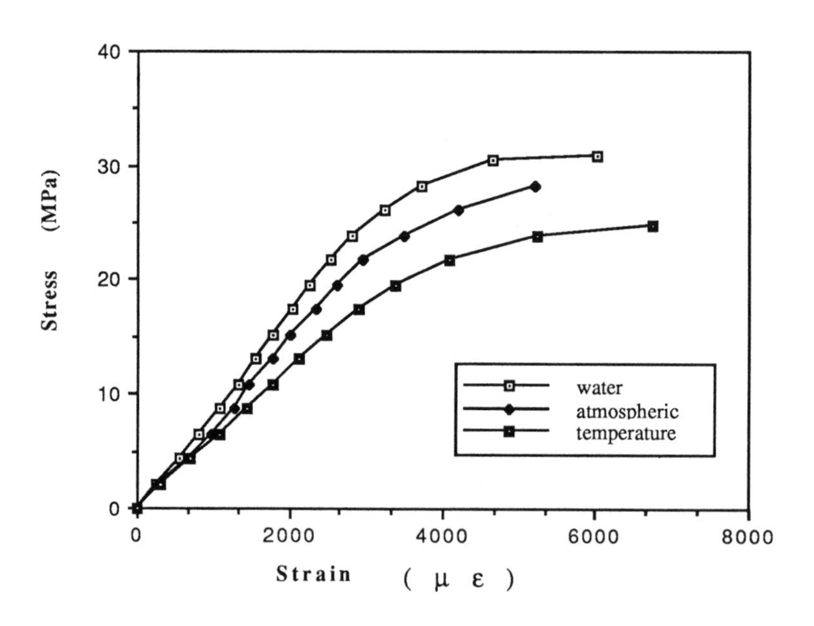

Fig. 10 Comparison of the various curing methods for s=60mm and ds=4mm.

energy absorption capacity of the tested prisms as can be depicted from the large area under the stress-strain curve. Water cured specimens also exhibited the largest energy absorption capacity; however, a noticeably high values were obtained under 60 °C temperature curing. Examining the specimens after failure suggested that the energy absorbed was largely dissipated in damaging the specimen. Furthermore, the initial modulus of elasticity appeared to be very slightly affected by the curing method used, especially between water and atmospheric curing as can be seen in the same figures.

4 Concluding remarks

The effect of three different curing methods on the behaviour of laterally confined concrete prisms with added steel fibers has been discussed in this paper. The results showed that although water curing produced the best results regarding strength and ductility, the use of steel fibers seems to help in reducing the strength loss due to other extreme curing conditions such as high temperature and, hence, improving the prism's general response. The use of steel fiber for hot weather concreting and curing may thus be suggested; however, more data and test results are needed.

5 References

Craig, R.J.; McConnell, J.; Germann, H.; Dib, N.; and Kashani, F. (1984) Behavior of reinforced fibrous concrete, **SP-81, ACI**, 69-103.

Fenella, D. A., and Naaman, A. E. (1985) Stress-strain properties of fiber reinforced mortar in Compression. **ACI Journal**, V. 82, No.4, 475-483.

Ganesan, N., and Ramana Murthy (1990) Strength and behavior of confined steel fiber reinforced concrete columns. **ACI Material Journal**, V. 87, No. 3, 221-227.

Mangat, P.S., and Motamedi, A. M. (1985) Influence of steel fiber and stirrup reinforcement on the properties of concrete in compression members. **International Journal of Cement composites Light Weight Concrete**, Vol. 7, No. 3, 183-192.

Ramakrishnan, V.; Coyle, W.V.; Kulandaisamy, V.: and Scharder, E.K. (1981) Performance characteristics of fiber reinforced concrete with Low fiber contents. **ACI Journal**, V. 78, No. 5, 388-394.

Swamy, R. N.; Mangat, P.S.; and Rao, C. V. S. K. (1974) The mechanics of fiber reinforcement of cement matrices. **SP-44, ACI,** 1-28.

Williamson, G. R. (1977) The effect of steel fiber on the compressive strength of concrete. **SP-44, ACI**, 195-207.

49 USE OF STEEL FIBRE CONCRETE IN SEISMIC DESIGN

B. KATZENSTEINER and S. MINDESS
Dept of Civil Engineering, University of BC, Vancouver, Canada
A. FILIATRAULT
Dept Civil Engineering, Ecole Polytechnique, Montreal, Canada
N. D. NATHAN
Dept of Civil Engineering, University of BC, Vancouver, Canada
N. BANTHIA
Dept of Civil Engineering, Universite Laval, Quebec, Canada

Abstract
An experimental research project was carried out at the University of British Columbia to test the effectiveness of steel fibre concrete in the seismic design of reinforced concrete structures. Two sets of parallel two-bay, two-storey plane frames were tested on an earthquake simulator. Both sets of frames were identical in geometry and were detailed to resist seismic loads according to the CAN3-A23.3-M84 Canadian concrete code. However, the second set of frames was constructed with steel fibre-reinforced concrete and had a substantial number of confining hoops removed from the joint regions. The steel fibres used were 30 mm-long x 0.50 mm-diameter collated hooked-end Dramix (Bekaert) fibres, and were added to the concrete at a loading of 60 kg/m^3 (0.76% by volume). Each set of frames was subjected to the same sequence of earthquake tests. Comparison of the data from the two sets of frames showed the steel fibre concrete to be effective in maintaining the integrity of the joint regions and in providing additional post-cracking energy absorption during the higher-magnitude earthquakes. The potential of steel fibre concrete in the seismic design of concrete structures is clearly demonstrated.
Keywords: Cracking, Ductility, Earthquake simulator, Energy dissipation, reinforced concrete, Seismic-resistant structures, SFRC.

1 Introduction

Concrete is used extensively in the construction of buildings. However, other materials must be added to the concrete to supplement its tensile capacity, most notably steel reinforcing bars of varying diameters. In concrete structures designed to resist reversing seismic loads, these steel bars are required in the top and bottom faces of beams and both sides of columns. In interior joints, the bars are usually continuous through the joint region, while in exterior joints, the bars are carried to the opposite face of the joint and anchored with a 90° hook. All of these longitudinal bars make for somewhat congested joint regions, particularly when there is a large percentage of steel in both the beams and the columns framing into a joint. The congestion in the joint region is further

Fibre Reinforced Cement and Concrete. Edited by R. N. Swamy. © 1992 RILEM.
Published by E & FN Spon, 2-6 Boundary Row, London SE1 8HN. ISBN 0 419 18130 X.

compounded by the presence of beam and column hoops, the latter of which are also placed within the joints themselves. These hoops provide resistance to shear forces, prevent buckling of the longitudinal bars under high compressive forces, and help to confine the concrete in the core of a member as it is cycled back and forth during an earthquake, thereby increasing the member's strain capacity and ductility. As can be expected, placing all of this steel is labour intensive, and therefore costly.

In an attempt to simplify the fabrication of these reinforcing cages, researchers have proposed removing some of the beam and column hoops from the joint regions, and using fibre-reinforced concrete instead of plain concrete to achieve the same confinement and ductility. The results of quasi-static tests done by Henager (1974) on beam-column joints that were modified in the above manner showed the modified joint to be stronger and more damage tolerant than a similar beam-column joint detailed for seismic forces in the conventional manner. These and other encouraging findings provided the incentive for performing some dynamic tests of large-scale fibre-reinforced concrete frames at the University of British Columbia's earthquake testing facility. For these tests, two structures consisting of two parallel, one-bay, two-storey frames were built. The first set was detailed for lateral forces using the Canadian CAN3-A23.2-M84 (1985) concrete code, while the second set was a modified version of the first and constructed with steel fibre-reinforced concrete (SFRC) instead of plain concrete. This paper describes the tests that were done on the two sets of frames, and discusses the significance of some of the results.

2 Concrete Frames

2.1 Design Constraints

There were several factors which influenced the design of the concrete frames. These included the physical limitations of the earthquake simulator itself, the available headroom in the earthquake engineering laboratory, and the concrete weights available for use as dead load on the structures. The earthquake simulator is an aluminum table 3.05 m x 3.05 m in plan, and can support a maximum vertical load of 160 kN. Given these specifications and other physical constraints in the laboratory, the dimensions finally chosen were a frame height of 2.7 m, a column spacing of 2.7 m, and a transverse spacing of the two frames of 1.2 m c/c.

2.2 Analysis and Design of Frames

Keeping in mind these constraints, the design of the frames was carried out according to the National Building Code of Canada (1985). For determining tributary loads, the frames were assumed to be part of a small office building located in Vancouver, Canada. The structural loads for which the frames were designed are given in Table 1. For determining the member forces due to the various load combinations, a static analysis was performed in which the imposed vertical loads were applied to the frames as point loads at the two locations where the 1.8m x 1.0m x 0.4m, 1700kg concrete blocks would rest on each beam.

Table 1. Structural Loads per Frame

Load Type	Description	Magnitude
Dead Loads	* Weight of frame	8.5 kN
	* Partition dead load of 1.0 kPa over lower beam 2.7m x 2.7m tributary area	7.3
	* 100 mm-thick slab @ 23.5 kN/m^3 over upper beam and lower beam 2.7m x 2.7m tributary area	17.1
Live Loads	* Office live load of 2.4 kPa over lower beam 2.7m x 2.7m tributary area	17.5
	* Snow load of 1.52 kPa over upper beam 2.7m x 2.7m tributary area	11.1
Wind Loads	* Lateral wind load at upper beam of 0.64 kPa over 2.7m x 2.7m tributary area	1.2
	* Lateral wind load at lower beam of 0.64 kPa 2.7m x 1.4m tributary area	2.4
Seismic Loads	* Equivalent lateral static force at upper beam due to seismic loading	1.3
	* Equivalent lateral static force at lower beam due to seismic loading	2.2

Using the member forces from the analysis, the first set of frames was detailed according to the CAN3-A23.3-M84 (1985) Canadian concrete code using a design concrete compressive strength of 30 MPa. As set out in the code, the "strong columns-weak beams" philosophy was used in designing the longitudinal beam and column reinforcement so that hinging of the beams would precede hinging of the columns during an earthquake. All of the longitudinal beam and column steel was carried through the joints and anchored in the confined joint core with standard 90° hooks. To facilitate fabrication, the longitudinal beam bars were spliced in the central portion of the beams. In the regions of the beams and columns adjacent to joints, closely spaced hoops were required for confinement of the core concrete. The final reinforcing details are shown in Figure 1.

For the second set of frames, the same concrete strength and longitudinal steel arrangement was used as for the first set of frames. However, some of the transverse steel was removed and steel fibres added to the entire concrete matrix. In particular, approximately every second beam and column hoop near the joints was removed, as were all of the confining hoops in the joints themselves. These modifications gave a uniform hoop spacing of 70 mm throughout the columns and beams, as shown in Figure 2. The fibres that were

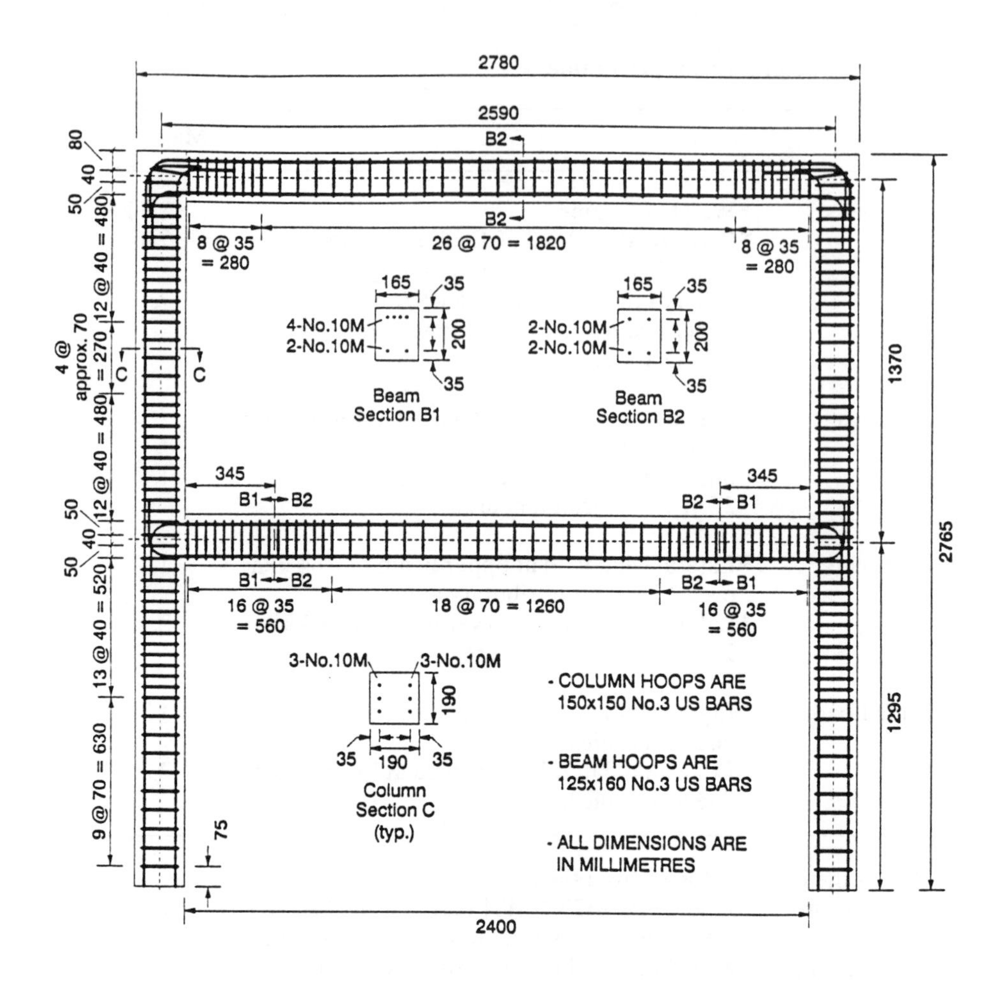

Fig. 1. Reinforcing Steel Layout of Structure #1 (Seismically-Detailed) Frames

added to the concrete of the second frames were ZP30/0.50 collated Dramix (Bekaert) steel fibres, which are 30 mm long x 0.50 mm in diameter (aspect ratio of 60), have a minimum tensile yield strength of 1170 MPa, and are hooked at the ends to increase the pullout resistance of the fibre. A 30 mm-long fibre was chosen so that the steel fibre concrete could be easily accommodated between the transverse reinforcing hoops. The fibres were added to the concrete at a loading of 60 kg/m^3 (0.76% by volume), which provided an economical and workable mix.

2.3 Instrumentation

A number of measuring devices, both internal and external, were used to record the performance of the frames during the earthquake tests.

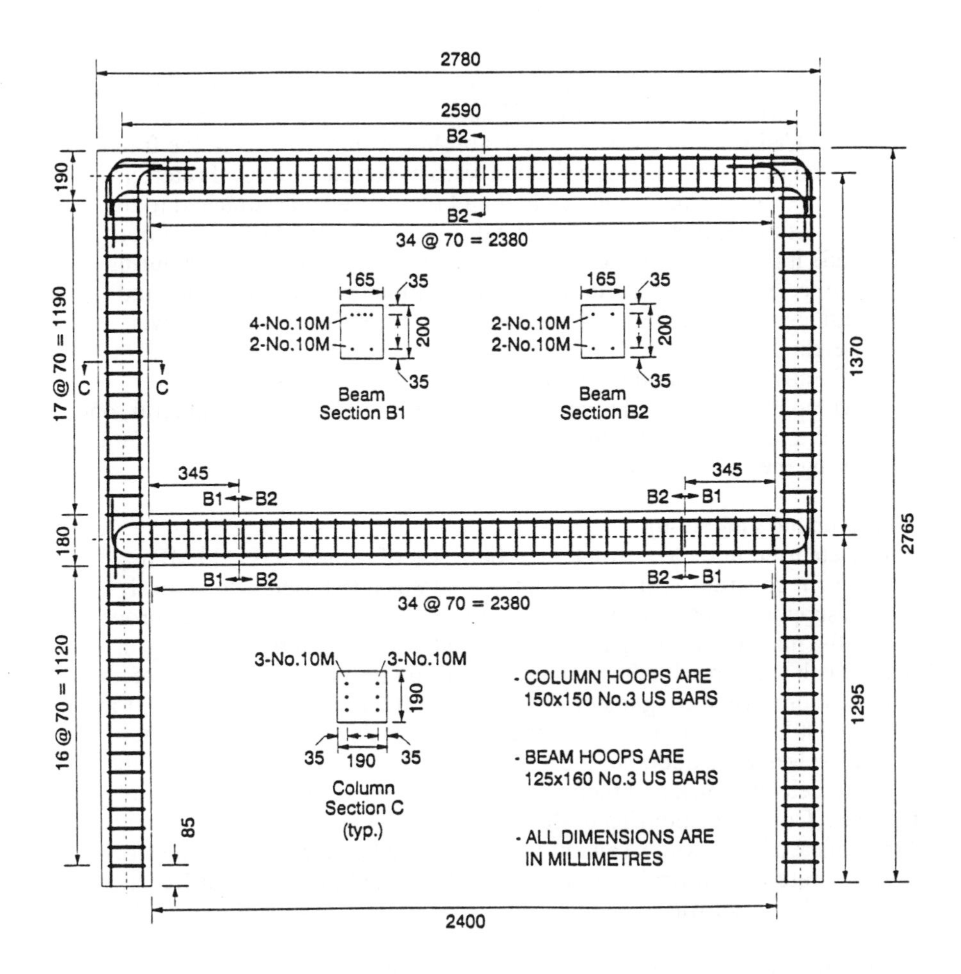

Fig. 2. Reinforcing Steel Layout of Structure #2 (SFRC) Frames

Internal measuring devices consisted of strain gauges mounted on the top and bottom longitudinal reinforcement of the beams at the column face and at a distance of d (the effective depth of the beams) from the column face. The strains from these gauges were used to measure the curvature of the beam in potential plastic hinge locations, which were then used to calculate sectional ductility demands. Both ends of the upper and lower beam of one frame were instrumented in this manner.

External measuring devices consisted of LVDTs (Linear Variable Differential Transformers), potentiometers, and accelerometers. A rigid aluminum frame to which two LVDTs were attached was mounted vertically on the beams at a distance of d from the column face. The LVDTs measured the top and bottom displacement of the aluminum frame relative to the column, and these displacements were then used to obtain the rotational- and curvature-time histories of the beams. A similar arrangement had been used earlier in earthquake tests of concrete frames by Hidalgo and Clough (1974) at the University of California at Berkeley. The absolute displacements of the frames were recorded by potentiometer units that were mounted on a fixed reference wall beside the earthquake simulator that had their wire leads attached to the outer faces of three joints. Recording the displacement time histories of both frames made it possible to detect any torsional motion of the structure during the tests. Accelerometers measured the acceleration levels at the upper and lower beam levels of one frame and at the centroids of the upper and lower stacked weights.

In total, thirty data channels, all of which were tied into a central IBM personal computer, were used to monitor the frames. Three video cameras, one of which was a high-speed system, were also used to record the response of the frames during the tests.

2.4 Construction and Set-up

The concrete used in the construction of both sets of frames was a 30 MPa normal-weight concrete with 90 mm slump, and was purchased from a local ready-mix supplier. In order to achieve optimal properties of the steel fibrous concrete, the maximum aggregate size was limited to 10 mm for the 60 kg/m^3 (0.76% by volume) fibre loading, which is based on the acceptable range of proportions for normal-weight steel fibrous concrete as described by Bentur and Mindess (1990). The composition of the concrete is shown in Table 2. The fibres themselves were added to the ready-mix concrete via the hopper on the back of the delivery truck and then mixed into the concrete for approximately 15 minutes. During the pouring of the first set of frames, it became apparent that the concrete's low water-cement ratio would make placement of the steel fibrous concrete very difficult, so to improve the workability of the SFRC, a liquid superplasticizer was added to the mix just prior to casting the second set frames.

Due to the limited space in the earthquake laboratory, the frames were cast horizontally in sets of two. Figure 3 shows the reinforcing cages and formwork

Table 2. Ready-Mix Concrete Composition

Ingredient	Amount per m^3	Vol. % per m^3
*Cement, Type 10 (CSA A5-M83)	270 kg	8.5%
*Flyash	70 kg	3.1
*10 mm Maximum-Sized Aggregate	1145 kg	41.9
*Sand	748 kg	27.8
*Water	147 kg	14.7
*Air-Entraining Agent	--	4.0
*Pozzolan 325N	1190 mls	--

Fig. 3. Reinforcing Cages and Formwork Arrangement

arrangement. Once the concrete had gained sufficient strength, the frames were lifted onto the simulator and connected with steel-angle cross-braces. The angles were sized so that the transverse natural frequency of the structure was roughly three times the longitudinal natural frequency, thereby minimizing the possibility of torsional motion during the earthquake tests. The column bottoms were fastened to the simulator via hinge units that allowed longitudinal rotation only. After fastening the frames, the concrete blocks and lead weights were placed on the beams and secured to prevent shifting during the tests. The weights represented the tributary dead load for both the upper and lower beams plus 25% of the snow load for the upper beams, resulting in a total mass of 11,060 kg for the test structure. Figure 4 shows the completed structure ready for testing.

3 Material Tests of Concrete

To determine the properties of the plain and steel fibrous concretes, 150 mm x 300 mm cylinders and 100 mm x 100 mm x 350 mm beams were cast during the pouring of the frames. Compression tests of the cylinders showed the average unconfined compressive strength to be 34.5 MPa for the plain concrete and 45.0 MPa for the steel fibrous concrete, both of which were greater than the 30 MPa specified to the ready-mix supplier. From loading the plain concrete beams in four-point bending according to ASTM C78 (1985), the average

Fig. 4. Completed Structure Prior to Testing

flexural strength of the plain concrete beams was found to be 4.2 MPa. The toughness of the steel fibre concrete was determined by loading the SFRC beams in four-point bending according to ASTM C1018 (1985) with a span length of 300 mm. A typical plot of total applied load versus midspan deflection from one of the tests is shown in Figure 5. The average values for the I_5, I_{10}, and I_{30}, toughness indices were found to be 4.4, 8.2, and 19.5, respectively, while the average ratios of toughness indices were 1.8 for I_{10}/I_5 and 2.4 for I_{30}/I_{10}. The average first-crack strength of these SFRC beams was 5.5 MPa.

4 Dynamic Tests

4.1 Experimental Programme

The dynamic tests performed on the two-frame structures were broken into three phases.

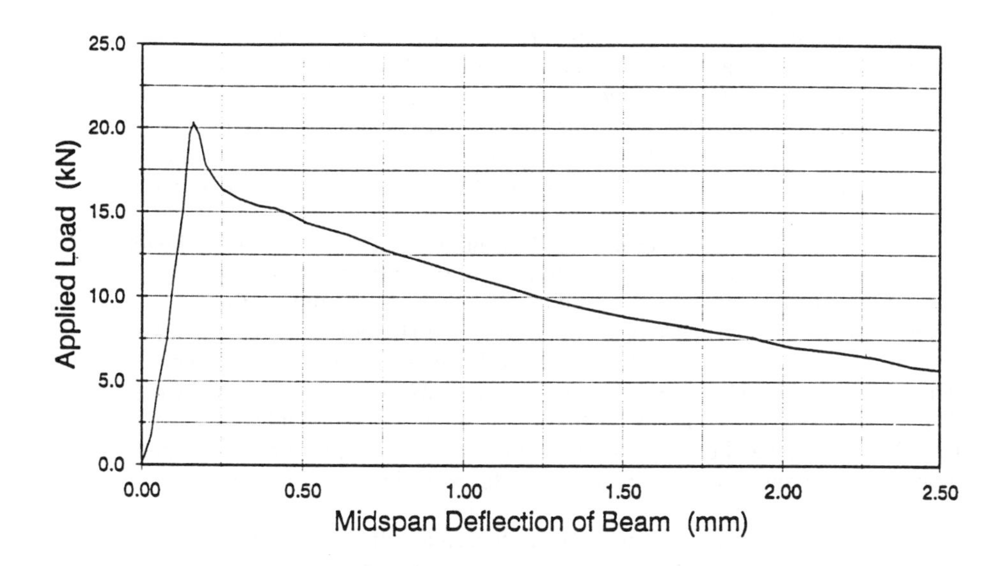

Fig. 5. Typical Plot from SFRC Flexural Beam Tests (ASTM C1018)

In the first phase, the dynamic properties (natural frequencies and damping ratios) of each loaded structure were determined by subjecting the frames to various forms of low-level excitation and recording the resulting structural response. The methods of excitation included impacting the frames with a battering ram and applying sinusoidal ground motions of varying frequency to the base of the columns via the earthquake simulator.

The second phase consisted of subjecting the structures to a total of eight earthquake acceleration records of increasing magnitude, and again recording the structural response. The first five tests used the Newmark-Blume-Kapur (1973) artificial earthquake acceleration record scaled to a maximum ground acceleration of 0.02 g, 0.04 g, 0.08 g, 0.16 g, and 0.32 g, respectively (g is the acceleration due to gravity). The sixth test consisted of an additional Newmark-Blume-Kapur earthquake scaled to a maximum ground acceleration of 0.32 g. For the seventh test, the acceleration record measured at the Oakland, California outer harbour wharf during the 1989 Loma Prieta earthquake was used, scaled to a maximum ground acceleration of 0.27 g. The eighth and final test simulated the expected major subduction-type earthquake for the Vancouver region, scaled to a peak acceleration of 0.30 g.

In the third and final phase of the experimental programme, some of the dynamic properties were again measured for comparison with the initial values. The variation of the properties reflected the amount of damage sustained during the earthquake tests, and thus served as another comparison of the two sets of frames.

4.2 Dynamic Properties of Frames

As mentioned in the previous section, low-level excitations were applied to each structure before and after the earthquake tests to determine the dynamic properties of the frames. The most important of these properties were the fundamental natural frequency of the structure (the resonant frequency at which the structure would vibrate during free-vibration response) and the damping of the structure (the ability of the structural materials to dissipate energy).

The natural frequency was obtained by inducing lateral vibrations of each structure and then measuring the frequency of the resulting free-vibration oscillations. In this manner, the fundamental natural frequency of the seismically-detailed structure was found to be 3.9 Hz before the earthquake tests and 1.8 Hz afterwards, while the SFRC structure's natural frequency decreased from 3.5 Hz before the earthquake tests to 1.6 Hz after the last test. The drop in frequencies indicates that the structures were roughly twice as flexible after the earthquake tests as before the tests, and reflects the large amount of damage sustained by both structures during the higher magnitude earthquakes.

The equivalent viscous damping of each damage-free structure was measured by first applying a sinusoidal input motion at the previously determined natural frequency and then suddenly stopping the input motion and recording the subsequent free-vibration acceleration response. Using the decaying amplitudes of this response, the damping was calculated by the logarithmic decrement method (described by Clough and Penzien (1975), for example) to be 1.7% of critical for the seismically-detailed frames and 3.0% of critical for the SFRC frames (critical damping is the lowest amount of damping that prevents an oscillatory-type of free-vibration response). These values are consistent with the typical 2% to 5% damping of most concrete structures.

4.3 Earthquake Test Results

For the purposes of this paper, only the results of the seventh earthquake test on each structure will be presented, which is the test in which the structures were subjected to the ground accelerations recorded at the Oakland outer harbour wharf during the 1989 Loma Prieta earthquake. The reason for this is that the Oakland earthquake acceleration record gradually builds up to the maximum ground acceleration of 0.27 g and then slowly dies away again (see Figure 6), which allows the maxima of the structural response to be more clearly illustrated than the maxima of other earthquake tests.

Using the displacements of the rigid aluminum frames that were mounted on the beams at a distance of d from the column face, the curvature-time history was determined for both the top and bottom beams. The maximum curvature of the bottom beams was 0.00012 radians/mm for the seismically-detailed structure and 0.00031 radians/mm for the SFRC structure, while for the top beams, the maximum curvature was 0.00004 radians/mm for Structure #1 and 0.00009 radians/mm for Structure #2. Taking the ratio of these curvatures to the actual yield curvatures (i.e. the yield curvatures based on the actual strength of the concrete and reinforcing bars) gave the maximum curvature ductility μ

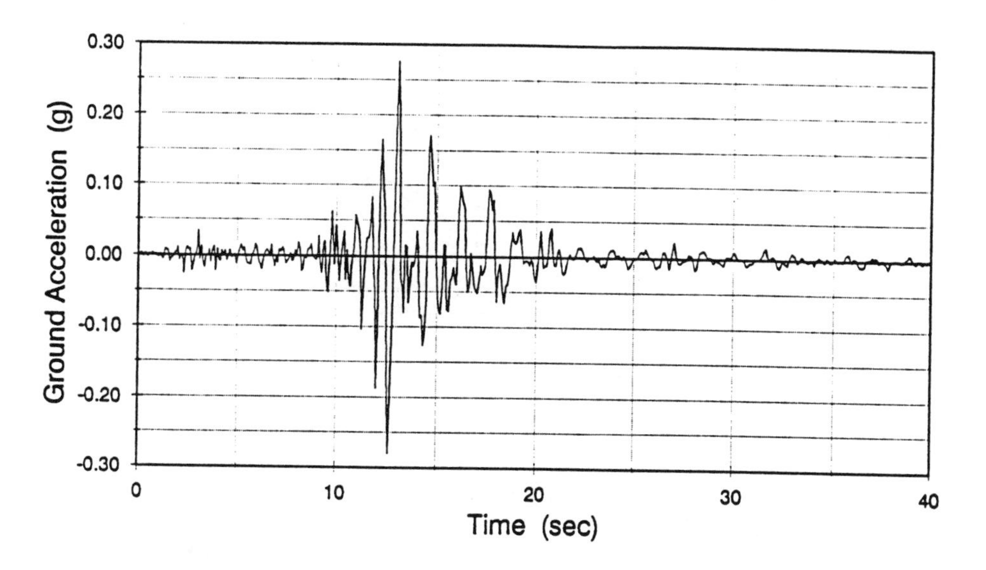

Fig. 6. Oakland Ground Acceleration Record from 1989 Loma Prieta Earthquake

attained at each particular beam section. For the bottom beams, μ was 5.4 for the seismically-detailed frames and 13.3 for the SFRC frames, while for the top beams, μ was 2.1 for Structure #1 and 4.4 for Structure #2. It should be noted that the maximum curvatures were attained during downward bending of both the bottom and top beams.

The overall energy-absorption capabilities of the two structures were compared by constructing plots of the structural base shear versus the lateral displacement of the top joint relative to the ground, where the base shear is calculated from the top and bottom joint acceleration records. These hysteretic plots are shown in Figures 7 and 8 for the strong-motion portion of the response starting at 10 seconds and ending at 20 seconds. This is the portion in which the ground acceleration reaches its maximum (see Figure 6), and in which the maximum responses are attained. The maximum relative displacement of the top joint was about 100 mm for the seismically-detailed frames and 125 mm for the SFRC frames, while the maximum base shear of both structures was about 70 kN. The hysteretic loops of the SFRC structure were more rounded than those of the seismically-detailed structure, resulting in approximately 40% more energy being dissipated by the SFRC structure per dynamic cycle (i.e. the area enclosed by one force-versus-displacement loop).

4.4 Visual Damage

Photographs that show the extent of cracking in one of the bottom joints after all eight earthquake tests are shown in Figures 9 and 10. The crack patterns and extent of cracking are roughly the same for both structures. Heavy damage

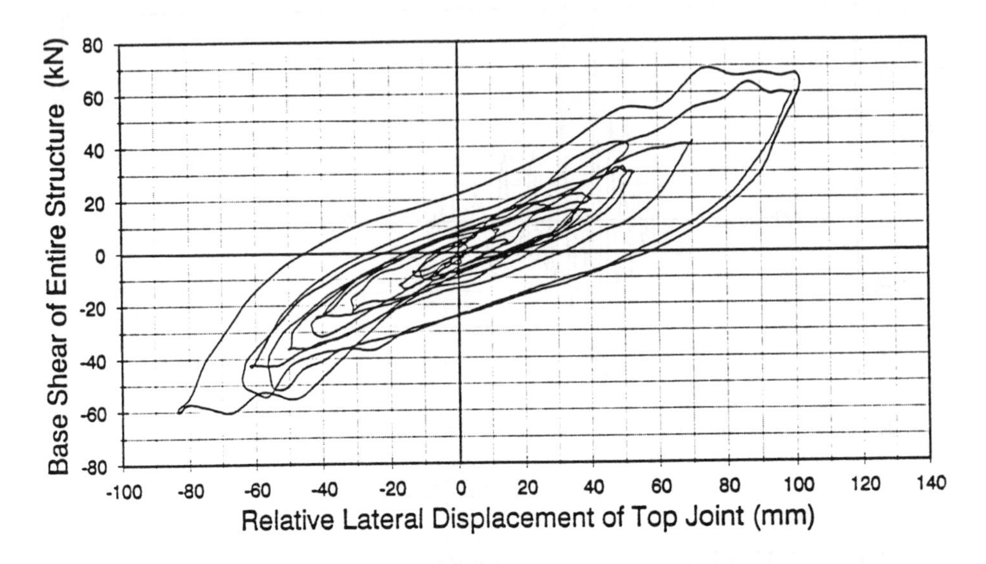

Fig. 7. Base Shear vs Lateral Displacement of Structure #1 (Seismically-Detailed) Frames

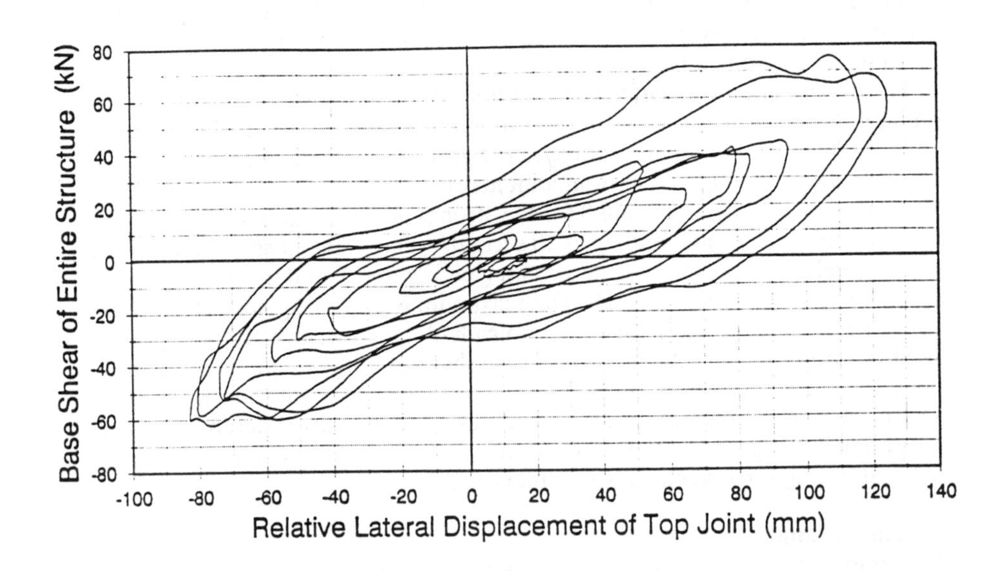

Fig. 8. Base Shear vs Lateral Displacement of Structure #2 (SFRC) Frames

Fig. 9. Structure #1 (Seismically-Detailed) Bottom Joint
After Last Earthquake Test

Fig. 10. Structure #2 (SFRC) Bottom Joint After Last Earthquake Test

occurred in the joint itself, with several large 45° shear cracks opening up during the higher-magnitude earthquakes. Flexural cracks of the beams at the column faces also became wider during the later stages of the testing programme.

After all the lead weights and concrete blocks were removed from the seismically-detailed structure, the spalled concrete cover was removed from the damaged joint regions. Figures 11 shows the same joint as in Figure 9, but with the loose cover removed. However, the spalled cover of the SFRC structure could not be removed by hand since the steel fibres were still holding the cracked concrete together.

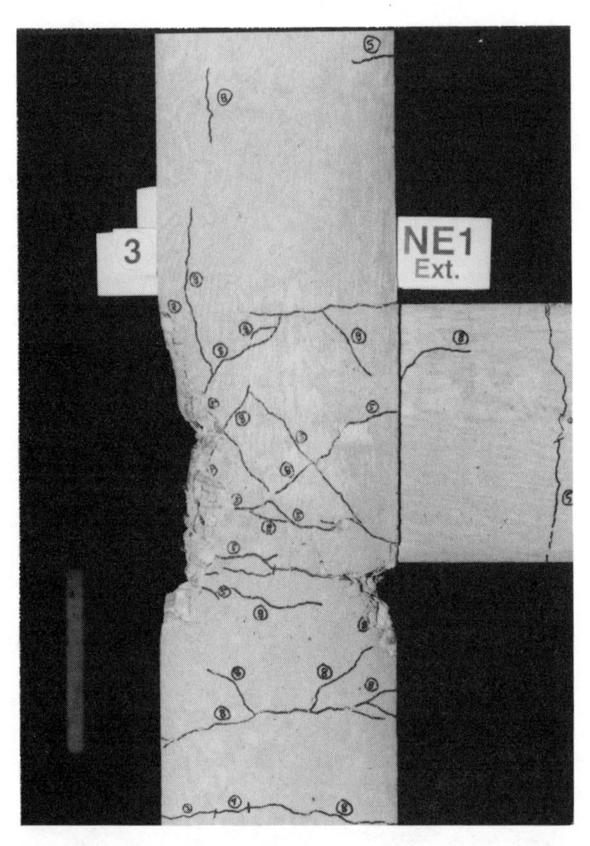

Fig. 11. Structure #1 (Seismically-Detailed) Bottom Joint After Last Earthquake Test and With Spalled Cover Removed

5 Discussion of Results

From the results presented in the previous sections, the effectiveness of steel fibre-reinforced concrete in the seismic design of structures has been demonstrated. During the Loma Prieta (Oakland) earthquake, the maximum ductilities attained by the SFRC frames were higher than the maximum ductilities achieved by the seismically-detailed frames. A higher ductility implies that more energy was absorbed inelastically by the SFRC frames than by the seismically-detailed frames. This difference was confirmed by the size of the base shear-versus-lateral displacement loops of the two structures. Since all structures are expected to yield during large earthquakes, the amount of energy that the structure can absorb inelastically plays a major role in the structure's ability to remain standing. Thus, the higher energy absorption is an important feature of the SFRC frames.

One other point to note concerning the use of SFRC in seismic design is its ability to maintain the integrity of the joint regions. As was mentioned in the last section, none of the cracked concrete cover could be removed from the damaged joints of the SFRC frames since the fibres were effectively holding the concrete together. Although the amount of force that this cracked SFRC could withstand during an earthquake may be small and hard to quantify, the energy dissipated by the cracked concrete of several SFRC joints may be significant. Furthermore, since the fibres maintain the integrity of the cracked concrete cover, they must also be effective in confining the concrete inside the core of the members, which again is reflected by the force-displacement behaviour of the SFRC frames.

6 Conclusions

The major conclusions of this paper are as follows:

1. A comparison of two sets of concrete frames subjected to an identical sequence of earthquake tests showed that the SFRC frames, constructed with fewer confining hoops, performed at least as well as the conventional set of seismically-detailed frames with regard to ductility and strength.
2. The base shear-versus-lateral displacement loops of the SFRC structure were more rounded than those of the seismically-detailed structure, resulting in a larger amount of energy being dissipated by the SFRC frames per dynamic cycle.
3. The steel fibres effectively maintained the integrity of the joint regions, both inside and outside the concrete core.

7 Acknowledgements

This research project was jointly funded by the Civil Engineering Department at the University of British Columbia and the Bekaert Corporation. Support was also provided by the Natural Sciences and Engineering Research Council, through the Network of Centres of Excellence on High Performance Concrete and through operating grants. The assistance of the Civil Engineering Department's technicians in the construction and testing of the frames is also gratefully acknowledged.

8 References

ASTM C78-85, (1985) Standard test method for flexural strength of concrete (using beam with third-point loading), Vol. 04.02: Concrete and Mineral Aggregates, American Society for Testing and Materials, Philadelphia, pp. 40-42.

ASTM C1018-85, (1985) Standard test method for flexural toughness and first-crack strength of fiber-reinforced concrete (using beam with third-point loading), Vol. 04.02: Concrete and Mineral Aggregates, American Society for Testing and Materials, Philadelphia, pp. 637-644.

Bentur, A. and Mindess, S., (1990) Fibre Reinforced Cementitious Composites, Elsevier Science Publishers Ltd., New York, pg. 183.

CAN3-A23.3-M84, (1985) Design of Concrete Structures for Buildings, Canadian Standards Association, Toronto.

Clough, R.W. and Penzien, J., (1975) Dynamics of Structures, McGraw-Hill, New York, pg. 70.

Henager, C.H., (1974) Steel fibrous ductile concrete joint for seismic-resistant structures, ACI Publication SP-53: Reinforced Concrete Structures in Seismic Zones, American Concrete Institute, Detroit, pp. 371-386.

Hidalgo, P. and Clough, R.W., (1974) Earthquake simulator study of a reinforced concrete frame, Report No. EERC 74-13, Earthquake Engineering Research Center, University of California, Berkeley, December, pg. 51.

National Building Code of Canada, Part 4, (1985) National Research Council of Canada, Ottawa.

Newmark, N.M., Blume, J.A. and Kapur, K.K., (1973) Seismic design spectra for nuclear power plants", Journal of the Power Division, American Society for Civil Engineering, Vol. 99, No. PO2, November, pp. 287-303.

50 STRENGTH AND BEHAVIOUR OF STEEL FIBRE REINFORCED CONCRETE SLABS SUBJECTED TO IMPACT LOADING

M. A. AL-AUSI, S. A-SALIH and A. L. K. ALDOURI
Baghdad, Iraq

Abstract
Twenty two simply supported concrete slabs reinforced with nominal steel bars and steel fibres were subjected to either static loading or repeated impacts by a hemispherical nose falling mass. The dimensions of the slabs were (1000x1000x50 mm). The test variables were the steel fibre content, falling mass and the height of drop. A special test rig was designed and fabricated to carry out static and impact tests. In each static test the load-deflection curve was determined in addition to the mode of slab failure. Load causing first crack and failure load were also recorded. For impact tests, maximum transient and residual deflections and the crater depth were measured with each impact blow. Number of impacts to cause first crack, initial and complete scabbing and slab failure were also recorded.

Test results have shown that slabs reinforced with steel fibres exhibited considerable resistance to impact loading by producing smaller deformations and sustaining more impacts to failure than slab without fibres (plain concrete slab). The results also indicated that scabbing is remarkably reduced in the slabs reinforced with steel fibres compared to plain ones.

1 Introduction

It is generally agreed that the most significant property needed by a material subjected to impact loading is its energy absorbing capability or toughness. It has also been found that adding fibres to a concrete matrix generally leads to a significant increase in its toughness, hence improved impact resistance (9). Until recently there has been no simple and practical test method to measure the impact strength of fibre reinforced concrete and to evaluate the effectiveness of the different types and volume fractions of fibres in improving the impact strength of fibre concrete (10). More recently ACI

Fibre Reinforced Cement and Concrete. Edited by R. N. Swamy. © 1992 RILEM.
Published by E & FN Spon, 2-6 Boundary Row, London SE1 8HN. ISBN 0 419 18130 X.

committee 544 (2) has developed such a method, and this test is currently being considered for adoption as an ASTM standard.

The experimental investigations of the impact resistance of fibre reinforced concrete (FRC) are of fundamental importance.The impact - load effects on FRC elements are very complicated and no exact theoretical solutions are available as yet. Impact resistance is generally determined by experimental studies carried out on specimens of FRC as well as on structural elements of various dimentions and shapes. Moreover, there are several methods of impact testing (3). For these reasons, the available experimental test results are not easily comparable (8). In spite of these difficulties, impact tests can broadly be classified into those which are more fundamental, and give an insight into the fracture process under impact, enabling quantification of energy absorption, and some understanding of the mechanism of energy absorption (10). In practice, both single and repeating impact loads are important. Impact from falling objects, aircraft or from motor vehicles come into the first category while machinery foundations fall into the second category.

2 Experimental Work

2.1 Test slab

Two way spanning simply supported slabs of dimensions (1000x1000x50 mm) were used in this investigation. Each slab was reinforced with 10 mm deformed bars, distributed across the section in two perpendicular directions at the bottom with a spacing of 250 mm and a clear cover of 10 mm. The same percentage (minimum reiforcement according to ACI - 83 requirements) (1) of longitudinal reinforcement was used for all tested slabs.

2.2 Control specimens

With each slab three 150 mm cubes and three 100x300 mm cylinders were cast to measure the compressive and indirect tensile strength of the concrete. Also three 100x100x500 mm prisms were cast, to measure the modulus of rupture.

2.3 Materials and mix proportion

Deformed steel bars of 10 mm diameter with avarage yield strength f_y of 470 Mpa were used.

High strength steel fibers with hooked ends, "Dramix" type were used with a nominal ultimate tensile strength of 1177 Mpa.

The length and diameter of the fibers were 60 mm and 0.8 mm respectively with an aspect ratio of 80.

Ordinary Portland cement and washed natural sand conforming to B. S. 882 (6) falling in the zone specified (overall limits) were used.

The coarse aggregate was rounded gravel conforming with B. S. 882 with a maximim size of 20 mm. The water was ordinary tap water.

A mix proportion of 1:1.85: 2.15: 0.48 (Cement sand: gravel: water) by weight was adopted throughout this study.

2.4 The Test Rig

A special test rig was designed and fabricated to meet the requirements of the impact and static tests so that the boundary conditions of the specimens were the same in both tests. The test rig consisted mainly of four parts as shown in Figure (1).

A clamping frame made of 100x45x5 mm steel channels was used to prevent any bouncing of the slab due to the impact. A vertical guide for the falling mass was used to insure impact at mid span.

Three drop hammers of 100 mm diameter and hemispherical nose were fabricated and used for the impact test. A special hammer with the same diameter and nose shape was used for the static tests.

3 Test Variables And Experimental Program

The main test variables in this research were the steel fibre content, the weight of the falling mass, and the height of the drop. The experimental program was divided into four groups according to the steel fibre volume fractions (Vf). Each slab was simply supported along its four edges. One specimen from each group was tested under static load and the others were subjected to repeated impacts.

Throught the impact tests either the falling mass or the height of the drop were varied (i.e. kinetic energy was varied) Table (1) shows the full detail of the experimental program.

The falling heights were selected to provide different impact energies and to insure that each slab tested under the hieghest impact energy (in the program) will fail in more than one impact blow. Six trials were made on a plain and fibre concrete slabs using different masses and falling heights for this purpose.

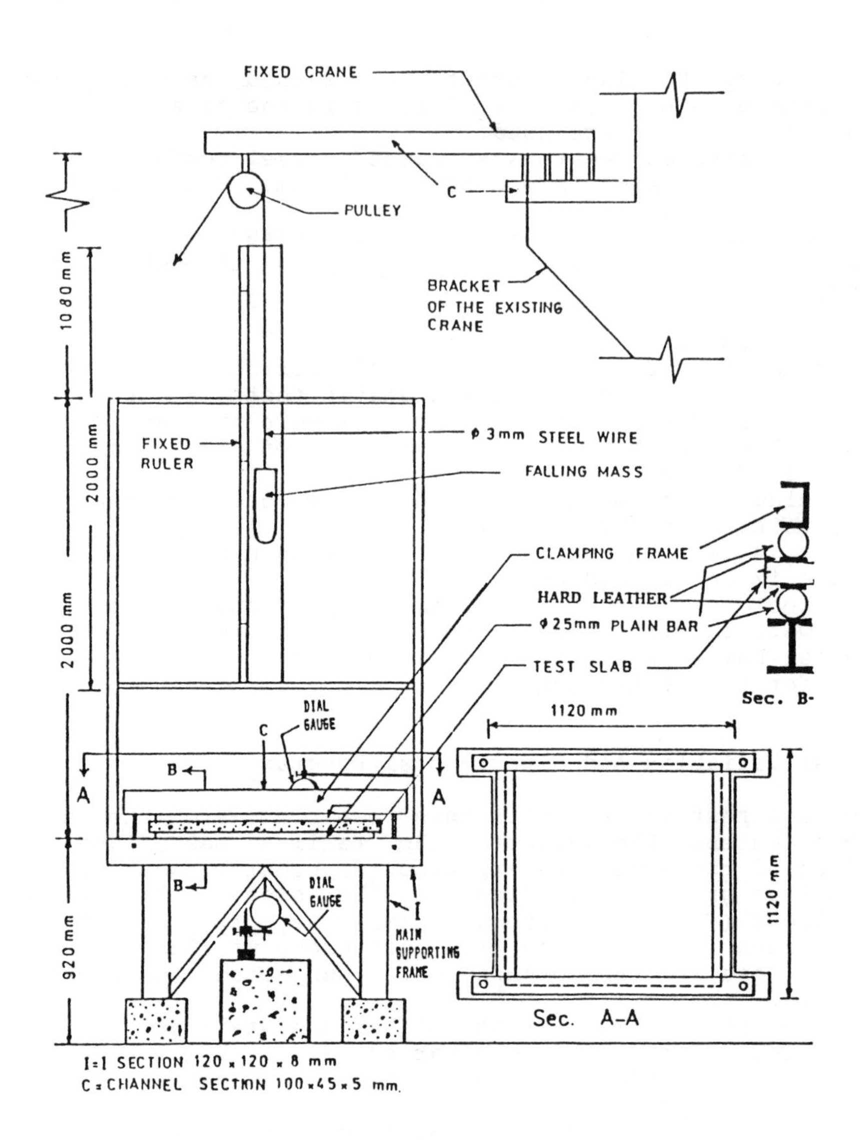

Fig. 1. The test rig.

Table 1. Experimental Test Program

Group No.	Slab No.	Fibre Content Vf%	Type of Test		Falling Mass (kg)	Falling Height (mm)
			Static	Impact		
1	S11	0.0	*			
	D12	0.0		*	11.2	1500
	D13	0.0		*	16.8	1500
	D14	0.0		*	22.4	1500
	D15	0.0		*	16.8	750
	D16	0.0		*	16.8	2250
2	S21	0.5	*			
	D22	0.5		*	11.2	1500
	D23	0.5		*	16.8	1500
	D24	0.5		*	22.4	1500
	D25	0.5		*	16.8	750
	D26	0.5		*	16.8	2250
3	S31	1.0	*			
	D32	1.0		*	11.2	1500
	D33	1.0		*	16.8	1500
	D34	1.0		*	22.4	1500
	D35	1.0		*	16.8	750
	D36	1.0		*	16.8	2250
4	S41	1.2	*			
	D42	1.2		*	11.2	1500
	D43	1.2		*	16.8	1500
	D44	1.2		*	22.4	1500

4 Experimental Procedure

The static tests were carried out using a hydraulic power controlled machine. The load was applied to the centre of the specimen, then load was increased at increments of 2.45 KN (250 Kg). At each load increment the central deflection was measured using 0.01 mm dial guage. First crack load was recorded and the test continued until failure. Then the failure mode and crack pattern were noted and recorded.

The impact tests were carried out using a drop hammer with a hemispherical nose. The impact test conditions for each group of slabs are given in Table (1). The impact energy ratios of the impact blows were 1, 1.5, 2, and 3 with respect to the lowest one. After the specimen had been fixed into position, the steel drop hammer was lifted by a winch up to the required height (between 750-2250 mm) and then released. Rebound of the drop hammer was prevented manualy.

With each impact blow, maximum transient and residual central deflection, and crater depth at impact zone were measured. The number of impacts to cause first crack, initial scabbing, complete scabbing and perforation were noted and recorded. Impact blows were continued until failure (defined as formation of performation which is visually apparent), then the diameter of the upper hole and scabbing area were recorded.

5 Static Test For The Deformation Constant

The well known Hertz contact law which describes the static compression of two elastic bodies is given by equation (1):

$$F = k.a^{3/2} \quad (1)$$

where F = The compressive force at the point of contact.
a = The relative approach of the striking bodies.
k = The Hertz (deformation) constant which depends on the elastic mechanical properties of the two bodies and on the shape of the two bodies at the contact zone.

The Hertz law had been used to characterize the impact zone (5, 6) i.e.

$$F(t) = k.\ a(t)^{3/2} \quad (2)$$

where F(t) is the impact force at any time (t) within the duration of impact.

If the Hertz contact law is to be followed to characterize the impact zone, then it is necessary to determine the deformation constant (k) which depends on the impact velocity and the impact zone mechanical properties (7). To study the effects of these factors, a dynamic test should be carried out. However as the special equipment for dynamic test were not available, a static test was carried out to determine the deformation constant neglecting the effect of the impact velocity. A hydraulic testing machine with a hemispherical head was used to apply a static load to a concrete specimen. The deformation was measured by a dial guage at each load increment.

For each group four specimens of dimension (250x250x50 mm), cut out from the tested slabs and three 150 mm cubes were tested.

6 Test Results

Table (2) shows the test resuts for control specimens which were cast with each slab.

Table 2. Test Results for Control Speciens

Group No.	Slab No.	Steel Fibre Content Vf%	Comp. Strength (Mpa)	Splitting Strength (Mpa)		Modulus of Rept. (Mpa)	
				Ist. crack	failure	Ist. crack	failure
1	S11	0.0	37.88	3.79	3.79	3.75	3.75
	D12	0.0	40.70	3.67	3.67	3.82	3.82
	D13	0.0	41.12	3.78	3.78	3.67	3.67
	D14	0.0	38.31	3.87	3.87	3.56	3.56
	D15	0.0	40.15	4.09	4.09	3.42	3.42
	D16	0.0	40.32	3.69	3.69	3.86	3.86
2	S21	0.5	39.36	3.82	5.23	3.78	5.25
	D22	0.5	40.92	4.02	5.69	3.46	5.16
	D23	0.5	41.60	4.30	5.49	3.34	4.44
	D24	0.5	37.68	3.56	5.16	3.24	4.60
	D25	0.5	38.89	3.70	5.29	3.75	5.97
	D26	0.5	39.10	3.66	5.11	3.82	5.89
3	S31	1.0	37.67	3.85	6.39	3.74	6.83
	D32	1.0	37.46	3.62	6.05	3.68	6.33
	D33	1.0	40.46	4.06	7.23	3.83	6.52
	D34	1.0	37.51	3.73	6.49	3.77	6.79
	D35	1.0	40.24	4.21	6.98	4.12	7.65
	D36	1.0	37.82	3.72	6.31	3.83	7.06
4	S41	1.2	37.22	3.96	3.36	3.81	7.32
	D42	1.2	39.33	4.27	7.47	3.85	7.35
	D43	1.2	40.40	4.16	7.45	4.22	7.06
	D44	1.2	37.91	4.18	7.22	3.94	7.40

For comparison purpose one slab from each group was tested under static load. During this test, visible cracks were initially observed at the slab centre propagating diagonally towards the corners. As the applied load increased, these cracks were extended, and new cracks developed with different orientations. No cracks were noticed on the compression face of the slabs at any stage of loading before failure.

Until the formation of the first crack the behaviour of all the tested plain and fibre concrete slabs were almost the same, and characterized by relatively small deformations.

Table (3) shows the static test result of the four slabs. All other slabs were subjected to repeated impact blows by a hemispherical nose hammer droped from a predetermined height. The impact test conditions are listed in Table (1). Crater depth made by the successive impacts were recorded. Figure 2 shows the relationship between number of impacts and accumulative crater depth, for the slabs. With each of of the impact blows the maximum transient and residual central deflections were recorded. The results are illustrated in Figures 3 and 4 respectively.

Table 3. Static Test Results

Slab No.	Steel Fibre Content Vf%	1st. Crack Load (kn)	1st. Crack Deflection (mm)	Ultimate Load (kn)	Ultimate Deflection (mm)	1st. Crack Load / Ult. Load (%)
S11	0.0	8.339	1.0	23.054	13.10	36.17
S21	0.5	9.320	1.30	26.487	20.20	35.19
S31	1.0	9.320	1.25	35.071	24.20	26.57
S41	1.2	9.320	1.12	38.750	26.00	24.05

Table (4) shows the details of impact test results. A significant influence of fibre content on the number of impacts to cause initial scabbing, complete scabbing and failure can be seen from this Table.

The failure mode was punching shear for all slabs with associated concrete scabbing on the lower faces of the slabs. In case of slabs with steel fibres it was observed that formation of a cone - shaped plug of material in the vicinitiy of impact is followed by its movement and crushing as a consequence of repeated impacts.

7 Static Test Results For Deformation Constant

As described befor, for each group of slabs four specimens of dimensions (250x250x50 mm) were cut out from the tested slabs, and three 150 mm cubes were tested statically to determine the Hertz (deformation) constant (k) of the impact zone.

During the test, the penetration was recorded at each load increment until specimen failure, then the average deformation of all the tested specimens was taken for each load increment. It was found that there is no significant difference between the results of plates and cubes.

The test results show that the presence of steel fibres in the mixes, affected the deformation constant significantly.

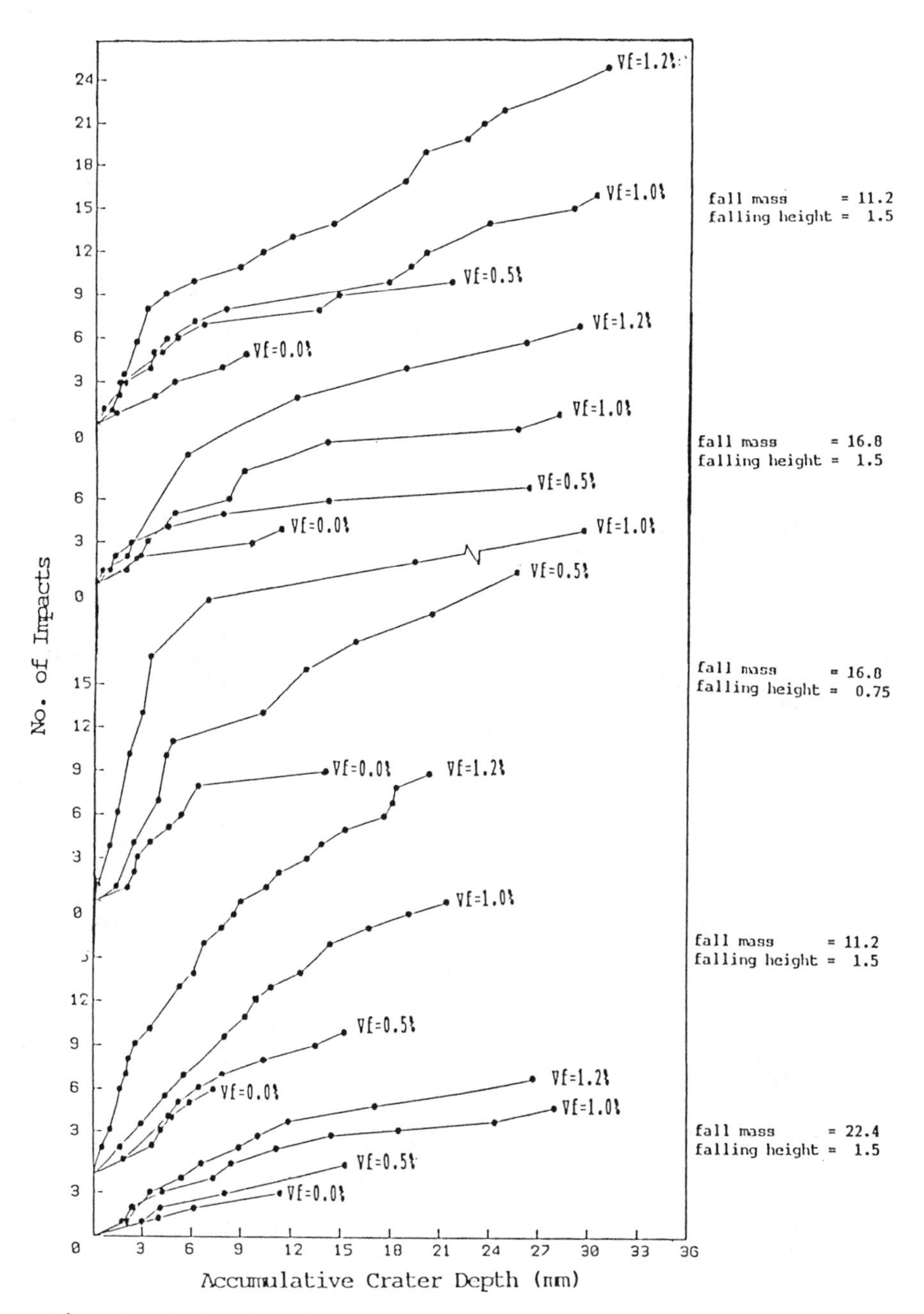

Fig. 2. Relationship between number of impacts and accumulative crater depth.

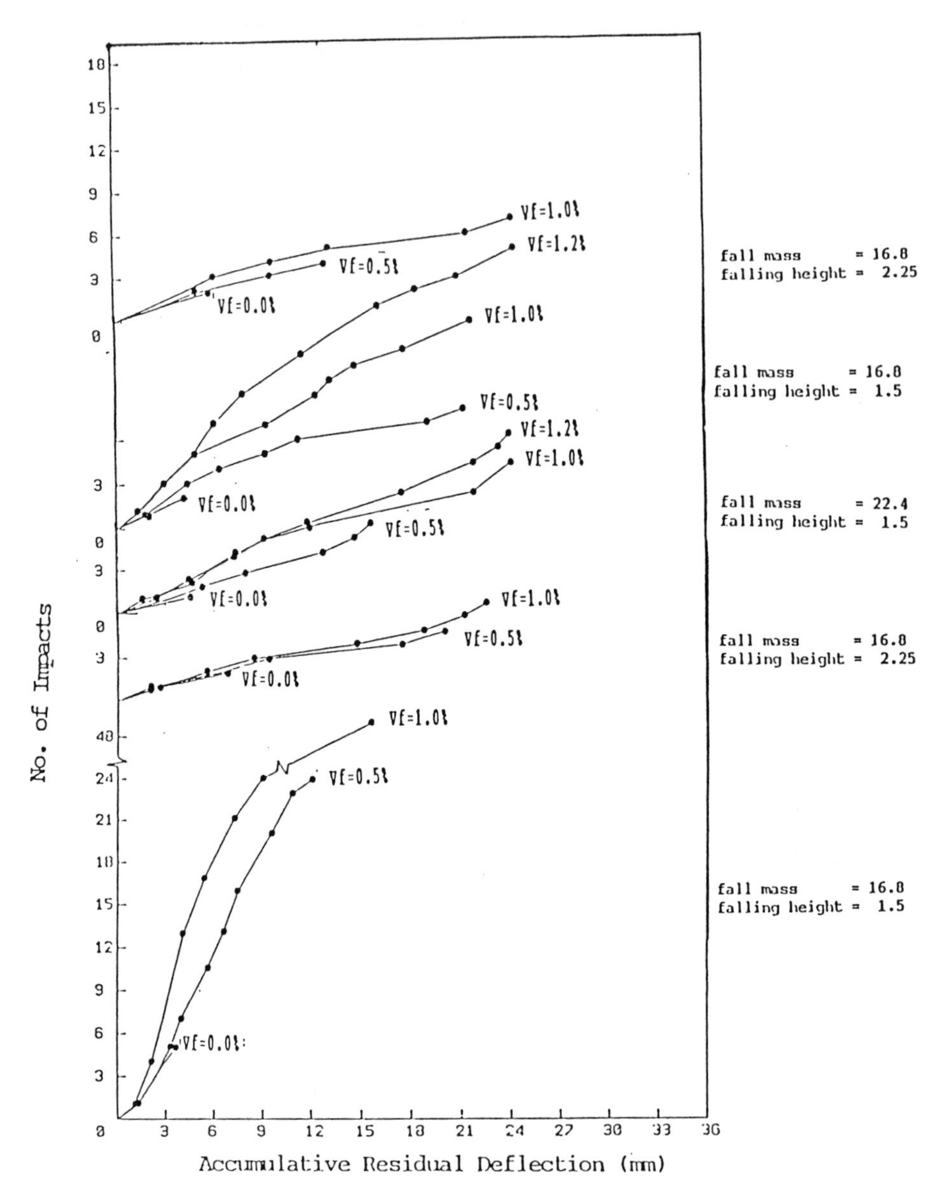

Fig. 3. Relationship between number of impacts and accumulative residual deflection.

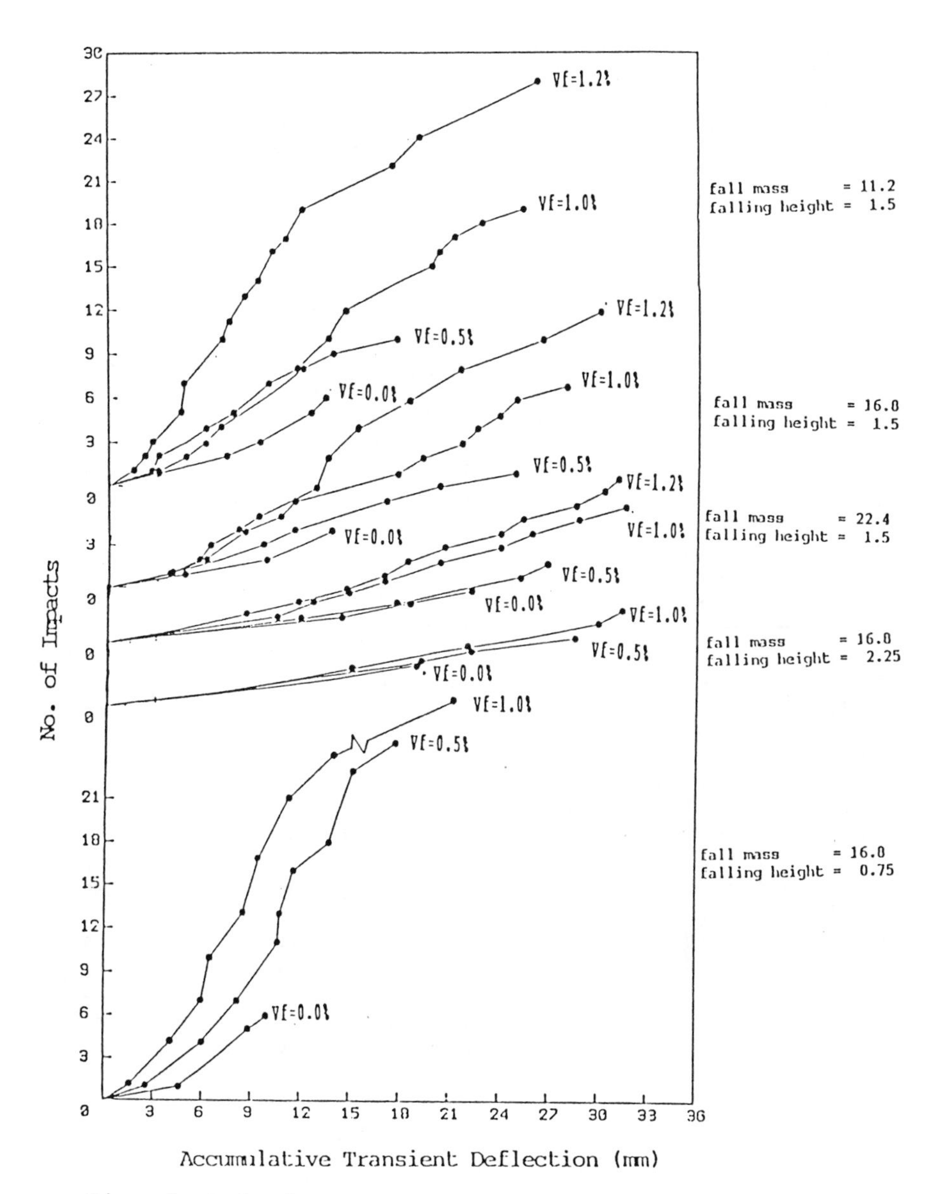

Fig. 4. Relationship between number of impacts and accumulative transient deflection.

The best fitting curves for the load - deformation relationship are shown in Figure (5).

A theoretical analysis is carried out based on the deformation constant to determine the force time history applied by the first impact blow to the specimen. The results will be published under a seperate paper.

Table 4. Impact Test Results

Group No.	Slab No.	Steel Fibre Content Vf%	Falling Mass (kg)	Falling Height (mm)	No. of impact to Cause				Diameter of Crater (mm)	Diameter of Scabbing Area (mm)
					1st. Crack	Initial Scabbing	Complete Scabbing	Complete Perforation		
1	D12	0.0	11.2	1500	1	2	3	6	71	230
	D13	0.0	16.8	1500	1	2	3	5	69	270
	D14	0.0	22.4	1500	1	2	2	4	55	120
	D15	0.0	16.8	750	1	4	6	9	71	280
	D16	0.0	16.8	2250	1	2	2	3	75	330
2	D22	0.5	11.2	1500	1	3	6	10	76	240
	D23	0.5	16.8	1500	1	3	5	8	75	240
	D24	0.5	22.4	1500	1	2	3	6	87	280
	D25	0.5	16.8	750	1	6	9	24	90	300
	D26	0.5	16.8	2250	1	2	3	5	88	280
3	D32	1.0	11.2	1500	1	5	15	19	97	260
	D33	1.0	16.8	1500	1	4	9	14	96	200
	D34	1.0	22.4	1500	1	3	6	10	94	250
	D35	1.0	16.8	750	1	9	38	49	100	265
	D36	1.0	16.8	2250	1	3	4	8	--	260
4	D42	1.2	11.2	1500	1	11	21	28	96	210
	D43	1.2	16.8	1500	1	6	13	19	97	230
	D44	1.2	22.4	1500	1	4	10	12	96	210

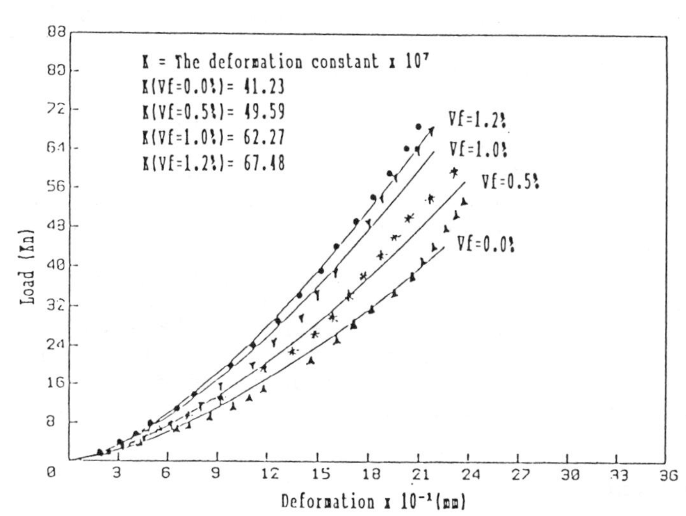

Fig. 5. Best fit curves for deformation constant results.

8 Conclusions

A Static Tests

1 The presence of steel fibre in the slabs delay the formation of the first crack, and controls the development of the cracks at the tension side. The cracks were much finer and greater in number compared to the plain concrete slab.
2 The inclusion of steel fibres produced a gradual punching failure, while slabs without fibres fail in sudden punching.
3 Fibre concrete slabs produced smaller deformations than plain concrete slab at all stages of loading.
4 The presence of steel fibres in the slabs increased the ultimate punching shear load of the corresponding plain slab. This increase was about 13% , 34% and 40% for slabs with 0.5, 1.0 and 1.2 per cent steel fibres respectively.

B Impact Tests

1 The mode of failure for all the slabs were in punching shear and the crack patterns were nearly the same as that for static slabs.
2 Slabs reinforced with steel fibres produced smaller deflections than plain concrete ones, at any number of impacts.

3. The residual and transient deflection are decreased, as the fibre fraction in the slabs increased under the conditions of varying number of impacts. This effect was decreased when the striker mass and velocity increased.
4. For the same applied kinetic energy, the relatively higher falling mass produced larger transient and residual central deflections and the higher impact velocity produce larger transient and residual deflections. This effect is more pronaunced when the steel fibre content increased.
5. The presence of steel fibres increased substantially the total number of impact to failure as compared to those for plain concrete slabs. This increase was about 67, 217 and 367 per cent for slabs with 0.5, 1.0 and 1.2 pre cent of steel fibres subjected to 11.2 kg falling mass, respectively.

9 References

ACI-318 (1983) Building code requirements for reinforced concrete, ACI Committee 318.

ACI COMMITTEE 544 (1978) Measurement of properties of fibre reinforced concrete. J. of Amer. Conc. Inst. Proc., Vol. 75, No. 7, PP. 253-290.

BANTHIA, N. and OHAMA, Y. (1989) Dynamic tensile fracture of carbon fibre reinforced cements. Fibre reinforced cement and concrete, recent development. International conference. School of engineering, University of Wales College of Cardiff.

BRITISH STANDARDS INSTITUTE (1973) Specification for aggregates from natural sources for concrete, metric units, B.S. 982, 1201, Part 2.

ERINGEN, A.C., AND LAFLYETTE, I. (1953) Transverse impact on beams and plates. J. of Applied Mechanics.

HUGHES, G., BEEBY, A.W. (1982) Investigation of the effect of impact loading on concrete beams. The Structural Engineer Vol. 608.

HUGHES, B.P., AND WATSON, A.J. (1978) Compressive strength and ultimate strain of concrete under impact loading. Magazine of Concrete Research, Vol. 30, No. 105.

RADOMSKI, W. (1981) Application of the rotating impact machine for testing fibre reinforced concrete. The Int. J. Cement Composites and Lightweight Concrete, Vol.3, No.1, PP 3-12.

SWAMY, R.N. (1984) New reinforcement concretes. Surrey University press, 2nd Edition.

SWAMY, R.N. and, JOJAGHA, A.H. (1982) Impact resistance of steel fibre reinforced lihtweight concrete. The international J. of Cement Composites and Lightweight Concrete, Vol. 4, PP 209-220.

51 STRENGTH AND BEHAVIOR OF VERTICALLY LOADED FIBER REINFORCED CONCRETE WALLS

A. A. SHAHEEN, M. M. E. NASSEF, A. ABDUL-RAHMAN
and A. H. ELZANATY
Cairo University, Egypt

Abstract
Test results on nine fiber reinforced concrete walls with and without openings are introduced and discussed. Different fiber contents were examined to study their effect on both the behavior and strength of the walls. Results indicated that the presence of fibers improves the serviceability, ductility and strength of the walls. Empirical equations were developed to predict the wall ultimate strength.
Keywords: Fiber reinforced concrete, Walls, Wall Openings, Steel fibers, Vertical Loads

1 Introduction

The objective of the current work was to investigate the effect of steel fibers on the Structural behavior and strength of concrete walls without traditional reinforcement.

An experimental study was conducted on 9-model walls. The main parameters investigated were :

1- Effect of percentage of fiber reinforcement on the behavior and strength of walls with or without openings.

2-Effect of openings on behavior and strength of plain and fiber reinforced concrete walls.

Test results are discussed in terms of initial cracking load, crack pattern, strain distribution along critical sections, ultimate capacity and mode of failure. Simplified empirical equations are proposed to predict the ultimate strength of such walls considering the dimensions of openings and the percentage of fiber content.

2 Test specimens, test procedure, and instrumentation

Table 1 describes schematically the test specimens, and Fig. 1 shows the dimensions, and the test arrangement. Mechanical strain gauges of 15 cm. length with accuracy of 1.33×10^{-5} mm/mm. were used to record strains at critical sections. Dial gauges with accuracy of 0.01 mm. were also used to record the deflection of opening edges. All walls were loaded incrementally using 500 tons compression Amsler machine' At low loads, each load increment was 5.0. tons. After cracking, the load increment was reduced. At each load increment, strain and deflection measurements were

Fibre Reinforced Cement and Concrete. Edited by R. N. Swamy. © 1992 RILEM.
Published by E & FN Spon, 2-6 Boundary Row, London SE1 8HN. ISBN 0 419 18130 X.

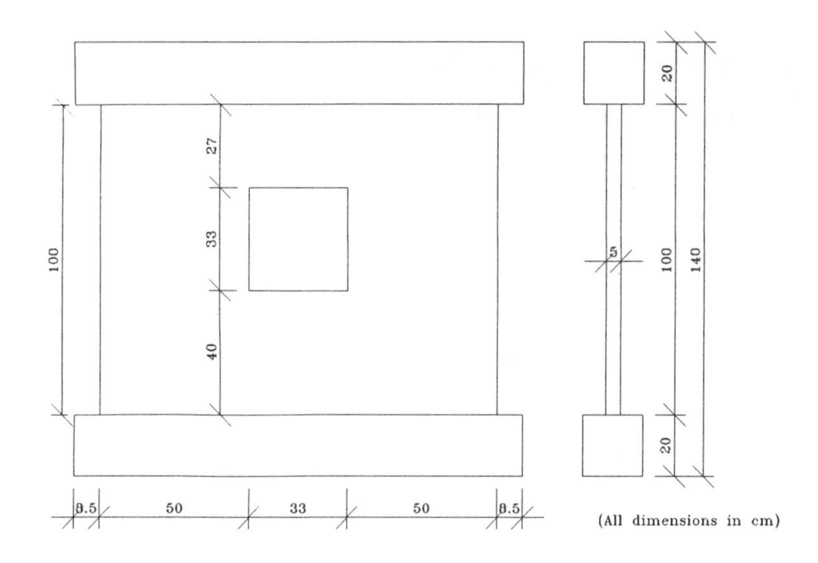

Vertically loaded wall model

Fig. 1. Dimensions of specimens and testing set up.

Table 1

Scheme of Vertically Loaded Tested Walls

Group	Wall	Identification	Scheme
I	SP	Solid plain concrete wall.	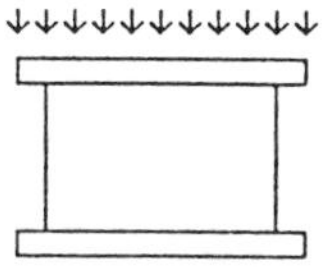
	SFI	Solid 0.5% fiber reinforced concrete wall.	
	SFII	Solid 1.0% fiber reinforced concrete wall.	
II	WP	Window opening plain concrete wall.	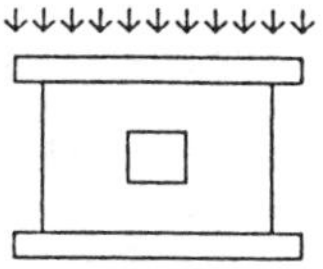
	WFI	Window opening 0.5% fiber reinforced concrete wall.	
	WFII	Window opening 1.0% fiber reinforced concrete wall.	
III	DR	Door opening plain concrete wall.	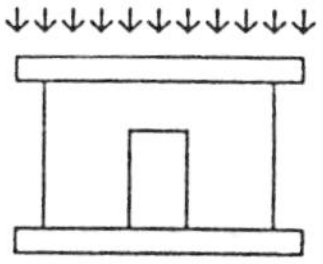
	DFI	Door opening 0.5% fiber reinforced concrete wall.	
	DFII	Door opening 1.0% fiber reinforced concrete wall.	

recorded and cracks, if any, were marked. The interval between each two successive load increments was from 15 to 30 mins. It took 3 to 4 hours to complete a test.

3 Behavior during testing

For solid walls; linear elastic behavior was observed up to cracking. All walls (SP,SFI, and SFII) showed maximum value of horizontal strain ϵ_x - as a measure of the splitting tensile stress- at the vertical center line of the wall and decreased gradually toward its edges, Fig.2. The measured maximum ϵ_x, at the same load, decreased with the increase of fiber content. With the increase of loading, cracking was observed. For plain concrete wall (SP), a single wide vertical crack occurred suddenly and failure followed immediately after cracking showing a somewhat explosive brittle failure at a load of about 95 tons/m`. For wall (SFI), at cracking load of about 105 tons/m` short discrete cracks were noticed. With the increase of loading more cracks formed and pre-existing cracks propagated and widened causing failure at a load of 120 tons/m`. A ductile failure was noticed with ample warning prior to it. In case of wall (SFII), no cracks occurred before failure. A sudden crushing compressive failure was observed at a load of 139 tons/m`. However, the integrity of the wall after failure was obviously preserved.

For walls with window opening; maximum tensile strain values were recorded in the horizontal dircetion (x direction), at mid point of upper and lower edges of the opening. These values decreased gradually toward the vertical edges of wall. Also relatively, high values were recorded at vertical lines tangential to the right and left edges of opening, Fig.3.

The part of the wall above the opening appeared to act as a deep beam transferring the vertical load acting on it to the supporting vertical parts beside the opening. The maximum vertical strain values were recorded at vertical lines tangential to the right and left edge of opening and decreased gradually toward the edges of wall Fig. 4. Minimum values were recorded directly above and below the openings indicating that the two vertical parts directly right and left of the opening act as columns. For WP, (without fibers) the first crack appeared at a load equal to 22.5 tons/m`. It was vertical and initiated at the center of the wall directly above the opening. With the increase of loading this crack propagated and widened while another crack directly below the opening started to propagate toward the lower edge of the wall. At higher stages of loading, a group of cracks propagated from corners of opening and extended vertically toward the upper and lower edges of wall causing crushing failure at the upper left corner of opening at a line load of 72 tons/m`, Fig.5.

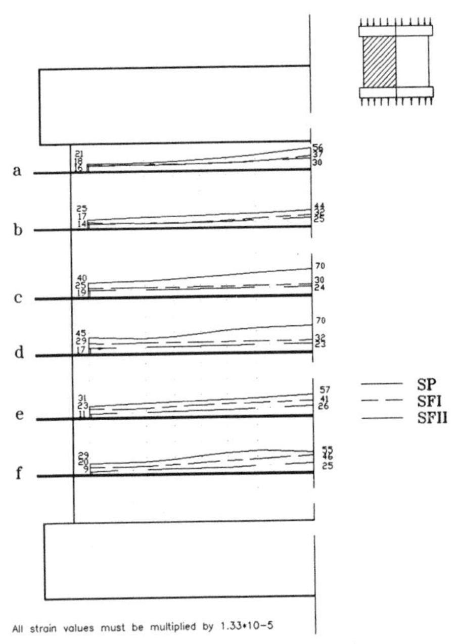

Fig. 2. Horizontal strain distribution at 64 t/m', walls SP, SFI and SFII.

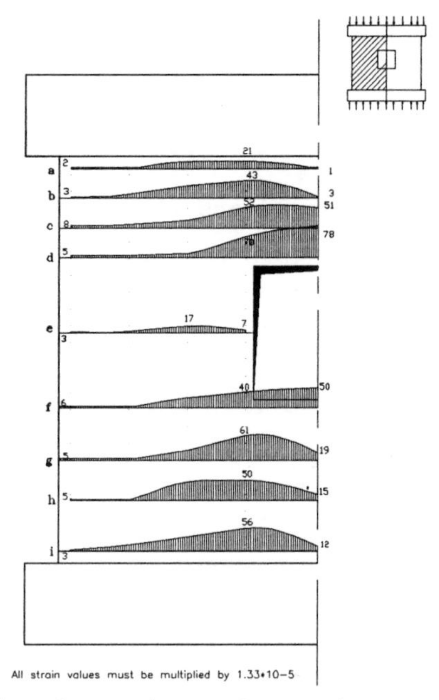

Fig. 3. Horizontal strain distribution at load = 64 t/m', WFII-Wall.

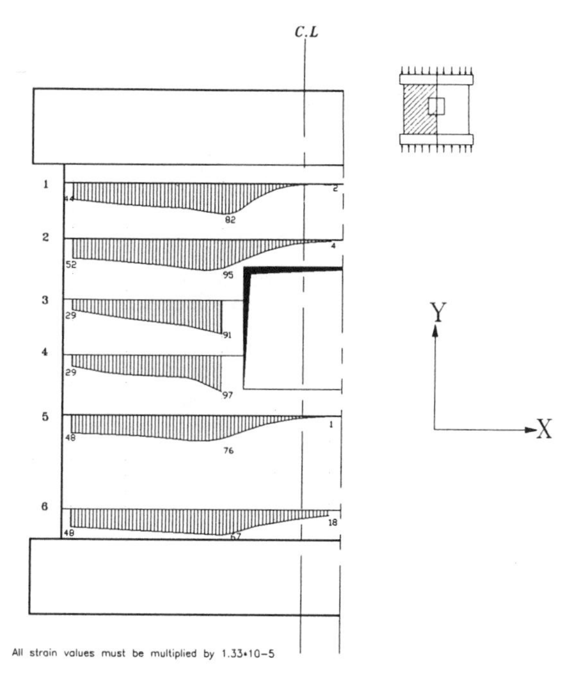

Fig. 4. Vertical strain distribution at loads = 64 t/m', WFII-Wall.

Fig. 5. Mode of failure of WP-Wall.

For wall WFI, (having 0.5% fiber content) an inclined crack was initiated above the window opening at a line load of 26 tons/m`. At higher stages of loading, this crack showed limited extension upward. Other cracks formed at the four corners of opening and extended vertically toward upper and lower edges of wall, with the increase of loading, causing failure at a line load of 79 tons/m`, Fig. 6. No crushing failure was observed.

For wall WFII, (having 1% fiber content), the first crack was initiated at a line load of 26 tons/m` at both upper and lower edges of opening at the same time. At higher stage of loading, these cracks showed limited propagation. Other cracks adjacent to the first crack upper and lower the opening propagated from the corners of opening and extended toward the upper and lower edge of the wall, Fig. 7. No crushing failure was observed and failure occurred at a line load of 90 tons/m`.

For walls with door opening; the strain distribution showed clearly the deep beam action of the part of the wall above the opening as explained earlier for the walls with window opening,

For DP wall (having no fibers), the first crack was initiated above the door opening at a line load of 15 tons/m`. With the increase of loading this crack propagated and widened and other cracks were observed at both right and left corners of the door opening and extended rapidly toward the upper edge of the wall causing failure at a line load of 48 tons/m`.

For wall DFI (having 0.5% fiber content), the first crack appeared at line load 27.5 tons/m` at the same location as that of DP. with the increase of loading other discrete short cracks initiated to the right and left of the first crack near corners

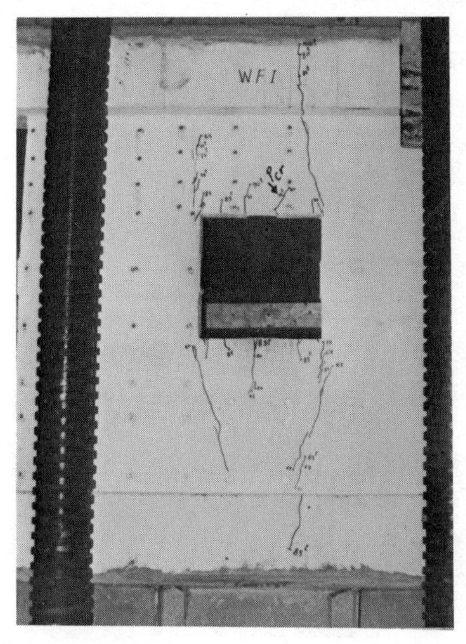

Fig. 6. Pattern of cracks and mode of failure of WFI-Wall, (0.5%).

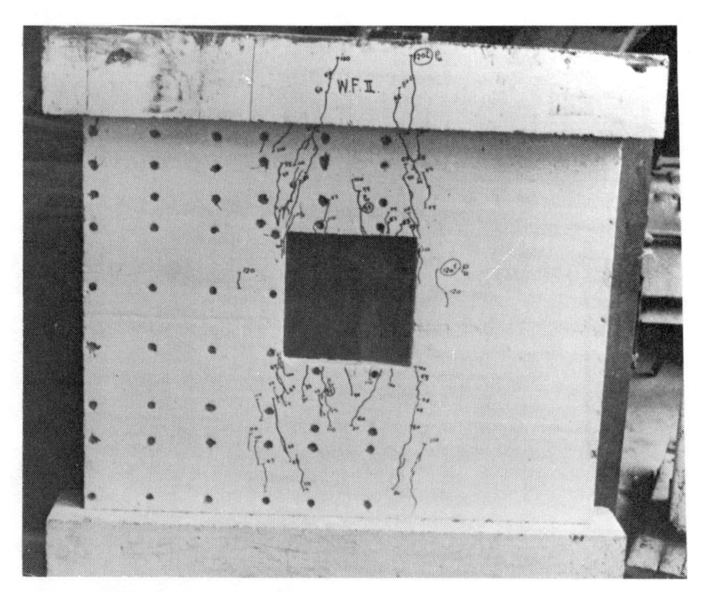

Fig. 7. Pattern of cracks and mode of failure of WFII-Wall, (1.0%).

of opening, and propagated vertically toward the upper edge of the wall. Failure was due to widening of cracks at a load of 64 tons/m`, Fig.9.

For wall DFII (having 1.0% fiber content), same behavior as that of wall DFI was observed, except that first crack appeared at a line load of 41.0 tons/m` while failure occurred at a line load of 82 tons/m`.

4 Effect of fiber content

In general fibers had a pronounced effect on improving behavior and strength of the model walls. They increased rigidity and stiffness of walls by increasing the modulus of elasticity of the concrete mix. Measured strain values were reduced in walls containing fibers. The reduction increased with the increase of fiber content.

Table 2 gives the cracking and ultimate loads of the three solid wall models. For wall SFI having 0.5% fiber content, the cracking load increased by 10.5% compared to SP wall. It was also clear that the fibers were effective in controlling cracking and minimizing crack width. In case of SP wall, the failure load was almost the cracking load causing explosive brittle failure, while in wall SFI the failure load was 14% higher than the cracking load showing somewhat ductile failure giving ample warning prior to it. In case of SFII wall, no cracks occurred before failure. A sudden crushing compressive failure was observed at a load equal to 139 tons/m`, which was 46% higher than ultimate load for SP wall. the integrity of the wall after failure was obviously

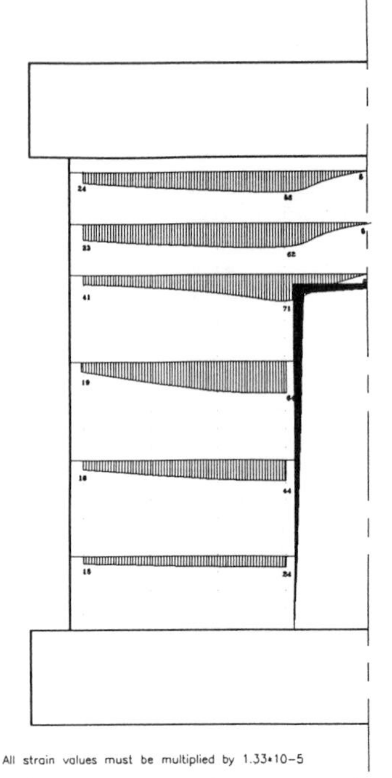

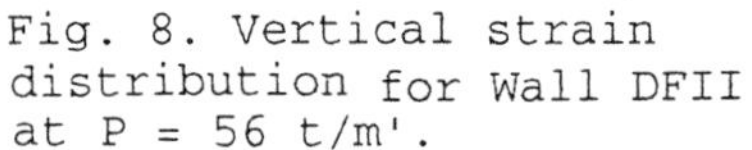

Fig. 8. Vertical strain distribution for Wall DFII at P = 56 t/m'.

Fig. 9. Pattern of cracks and mode of failure of DFI-Wall, (0.5%).

preserved. Knowing that the addition of fibers to concrete mix increases its tensile strength by a higher percentage than increasing its compressive strength. It can be concluded that adding 1% fiber content was sufficient to change the mode of failure from splitting tension failure in case of Sp and SFI walls to crushing compressive failure in case of SFII wall.

For wall WFI having 0.5% fiber content the cracking load was higher by 16% (i.e. at line load of 26 tons/m`) compared to WP wall, Table 3. However, comparing the cracking patterns it can be noticed that fibers had a pronounced effect on controlling

Table 2
Effect of fiber reinforcement on Cracking and ultimate Loads : Solid Walls

Wall	Cracking load (P_{cr}) (t/m')	%Increase	Ultimate load (P_u) (t/m')	%Increase
SP (0.0)	95	-	95	-
SFI (0.5)	105	10.5%	120	23.0%
SFII (1.0)	139	46.0%	139	46.0%

Table 3
Effect of Fiber Reinforcement on Cracking and ultimate Loads: Walls With Windows

Wall	Cracking load (P_{cr}) (t/m')	%Increase	Ultimate load (P_u) (t/m')	%Increase
WP (0.0)	22.5	-	72	-
WFI (0.5)	26	16%	79	9%
WFII (1.0)	26	16%	90	25%

cracking by increasing the number of cracks and hence decreasing the length and width. No crushing failure occurred, the failure was due to spreading of cracks in a conical shape above and below the opening. Also the ultimate load increased by 9% compared to WP wall (i.e. it was at line load of 79 tons/m`.

For wall WFII having 1.0% fiber content the cracking load increased by 16% while the ultimate load increased by 25% compared to WP wall. However, at the same load, after initial cracking the length and width of the cracks were less while the total number of cracks were more in case of wall WFII compared to WFI. The fibers bridging across the cracks help in controlling crack propagation and reducing crack width.

It can be concluded that the first crack load is controlled by the flexural tensile strength of the concrete mix. The strain measurements above the opening give a strain distribution similar to that of deep beam transferring the vertical load acting on it to the supporting vertical legs beside the opening. Comparing between Figs. 5,6,7 which indicate the mode of failure and final crack pattern of window opening wall group, it can be noticed that, the use of steel fibers in concrete walls can substantially improve the serviceability conditions by controlling crack formation and propagation and improving the associated deformation characteristics. The effectiveness of fibers in providing a ductile failure was undoubtedly clear during testing. For wall WP, the ultimate load was accompanied by a compression failure and the upper left corner of opening was shattered into pieces while for walls WFI and WFII a more ductile failure was observed and the integrity of the wall was preserved.

Also for door opening walls, Table 4 indicates an increase of cracking load by 50% and 100% and an increase of the ultimate load by 33% and 72% due to increase of fiber content from 0.0% to 0.5% and 1.0% respectively.

Comparison between the crack patterns of the three walls indicates that the number of cracks was less and rate of propagation was much higher in case of wall without fibers. Also at the same load, the cracks were longer and wider in case of the plain concrete wall DP. The failure load was associated with a more ductile failure in case of walls with fibers and the ductility increased with the increase of fiber content.

Table 4
Effect of fiber Reinforcement on Cracking and ultimate Loads: Walls With Doors

Wall	Cracking load (P_{cr}) (t/m')	%Increase	Ultimate load (P_u) (t/m')	% Increase
DP	15	-	48	-
DFI	22.5	50%	64	33%
DFII	30	100%	82.5	72%

5 Effect of Openings

It can be observed that the initial cracking load was reduced by 76%, 75% and 81% due to window opening and reduced by 84%, 79% and 78% due to door opening for cases of 0.0%, 0.5% and 1.0% fiber content respectively. Ultimate loads were also reduced by 24%, 34% and 35% due to window openings, and 49%, 47% and 41% due to door opening, for cases of 0.0%, 0.5% and 1.0% fiber content respectively.

The mode of failure changed from splitting type in solid walls to spreading of cracks around openings in walls with window or door openings.

Table 5
Effect of Openings on initial Cracking and ultimate Loads

V_f %		0.0%		0.5%		1.0%	
Type	Load (ton)	P_{cr}	P_u	P_{cr}	P_u	P_{cr}	P_u
Solid walls		95	95	105	120	139	139
Opening wall	Window	22.5	72	26	79	26	90
	% of reduction	76%	24%	75%	34%	81%	35%
	Door	15	48	22.5	64	30	82.5
	% of reduction	84%	49%	79%	47%	78%	41%

6 Prediction of Ultimate loads of Vertically Loaded Walls

Based on the test results of this research, the proposed equations by Madina and Desayi[2,3] and the ACI procedure for designing walls with the resultant inside the middle third, were modified to predict the ultimate strength of vertically loaded walls having fibers and containing opening as follows :

$$P_{us} = 0.7\ f\acute{c}\ Ag\left[1-(\frac{h}{32t})^2\right]\left[1.2-(\frac{h}{10L})\right]\gamma \quad (1)$$

$$P_{uo} = \alpha\ P_{us} \quad (2)$$

where
- P_{us} = ultimate strength of solid wall.
- fc' = Cylinder compressive strength.
- A_g = Gross area of wall section.
- L = Length of wall panel
- t = Thickness of wall panel.
- h = Hight of wall panel.
- γ = Fiber magnification factor to be determined according to the following equation:

$$\gamma = 1+0.36\ V_f - 0.16\ V_f^2 \quad (3)$$

where
- V_f = volume percentage of fibers.
- P_{uo} = Ultimate strength of walls containing opening.
- α = Reduction factor to take into account the effect of opening as follows:

$$\alpha = \left(1-K_1\frac{A_o}{A}\right)\left(1-K_2\frac{R_o}{R}\right) \quad for\ \frac{R_o}{R} \leq 0.541 \quad (4)$$

where
- A = cross sectional area of wall panel in plan.
- A_o = cross sectional area of opening in plan.
- R = Area of wall in front view, Fig.10.
- R_o = Area of opening in front view.
- K_1, and K_2 are constants : $K_1 = 0.75$, $K_2 = 1.85$

Table 6 gives summary of the computed and measured ultimate loads for the nine model walls. It can be noticed that predicted values are in good agreement with corresponding experimental values.

Table 6
Summary of the Computed and Experimental Ultimate Loads

P_u (ton)	Solid Walls			Window Opening Walls			Door Opening Walls		
	SP	SFI	SFII	WP	WFI	WFII	DP	DFI	DFII
P_u exp.	95	120	139	72	79	90	48	64	82.5
$P_{u_{th.}}$	96.7	121	137	65.7	82.4	93.4	52.2	65.5	74.1
$\frac{P_{th.}}{P_{exp.}}$	1.01	1.01	0.99	0.91	1.04	1.03	1.08	1.02	0.90

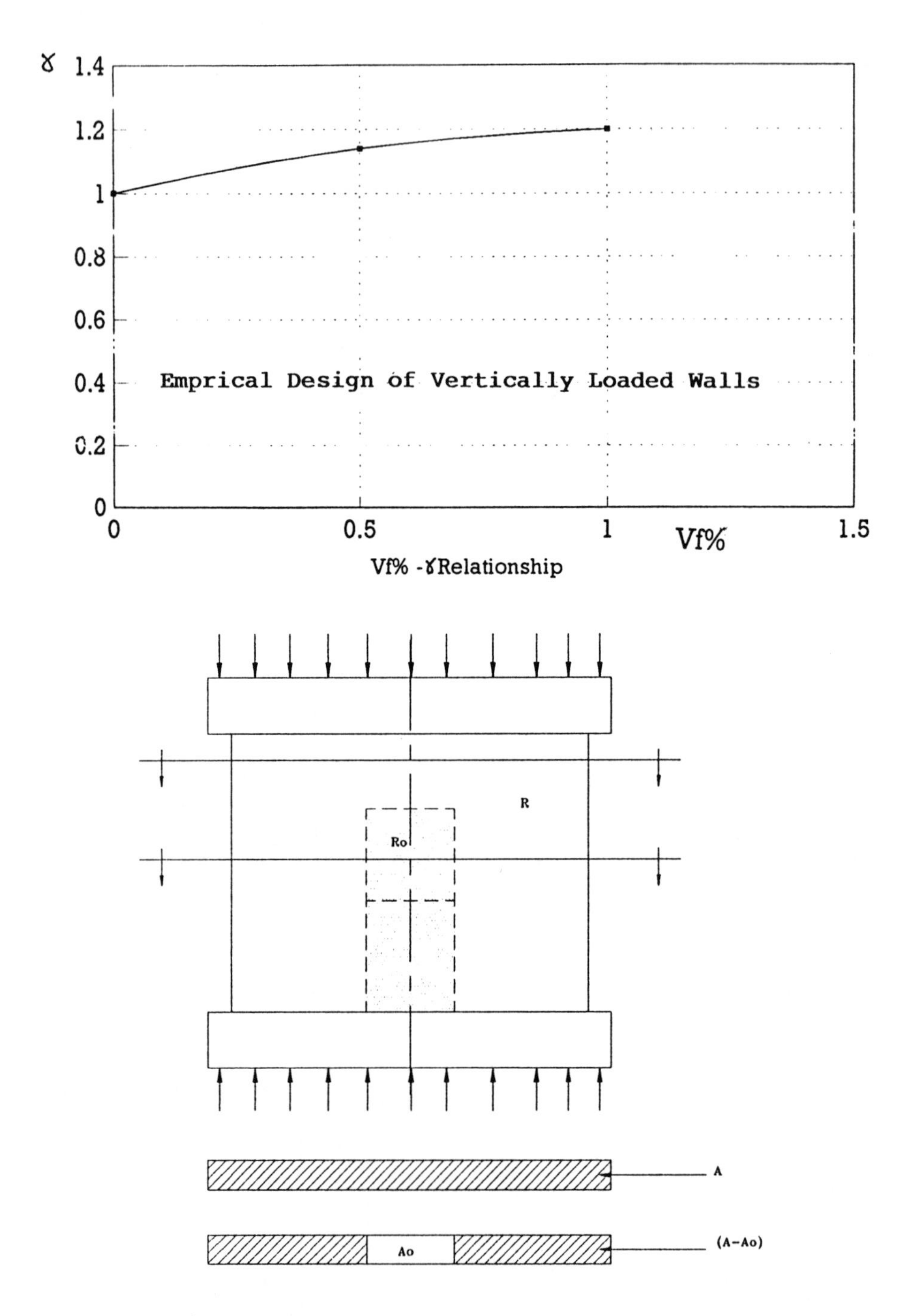

Fig. 10. Suggested empirical design method for vertically loaded walls.

7 Conclusions

1- The test results of this research, clearly, demonstrate the effect of steel fibers on enhancing ultimate strength and improving serviceability and ductility of concrete walls vertically loaded.
2- Equations for predicting ultimate strength of Fiber Reinforced concrete walls containing opening are proposed. The predicted values using these equations were in good agreement with the measured experimental values.

8 References

1 "Building Code Requirements for Reinforced Concrete", American Concrete Institute, 1990.
2 Madina and Desayi, "Ultimate Strength of R.C. Wall Panels in Two Way In-Plane Action", ASCE, May 1990, PP. 1384 - 1402.
3 Madina and Desayi, "Ultimate Strength of R.C. Wall Panels with Openings", ASCE, June 1990, PP. 1565 - 1578.

PART SEVEN
APPLICATIONS

52 FASTENING TECHNOLOGY IN FIBRE REINFORCED CONCRETE

E. WALTER and W. J. AMMANN
Hilti AG, Corporate Research, Schaan, Principality of Liechtenstein

Abstract:
In 404 tests, force and deformation controlled expansion anchors, chemical anchors and powder actuated bolts and nails with two dimensions were investigated in two kinds of steel fibre and polypropylen fibre reinforced concrete. The aim was to determine the kind and amount of changes in their force-deformation characteristics due to the presence of different fibres in concrete. The correct setting procedure was checked and the ultimate failure load, the failure mode and the force-deformation characteristics were measured. It was found that the recommended design loads for anchors and powder actuated elements can be used on fibre reinforced concrete as well. Neither steel fibres nor polypropylene fibres lead to an increase of the ultimate failure load of the tested fastening elements and the corresponding deformation of the anchors.
Keywords: Anchors, Bolts, Failure mode, Fastening elements, Fibre reinforced concrete, Force-deformation characteristics, Nails, Ultimate failure load

1. Introduction

Fastening is a technology by which different construction elements - structural and/or non structural - are joined together. There are three components of a complete fastening system - (1) the fastened construction element, (2) the fastening itself and (3) the base material. Together they all form a unity. However, the individual properties of each of the components of this system strongly influence the overall behaviour of the whole fastening system. The properties of the base material govern to a great extent the application, quality and behaviour of the different fastening systems.

Fibre reinforcement in concrete leads to a change of the basic properties. The increase in strength, ultimate strain and fracture toughness influences the whole fastening assembly.

A number of tests with different fastening systems were performed to determine the kind and amount of changes due to fibre reinforcement. Of primary

Fibre Reinforced Cement and Concrete. Edited by R. N. Swamy. © 1992 RILEM.
Published by E & FN Spon, 2-6 Boundary Row, London SE1 8HN. ISBN 0 419 18130 X.

importance was to study possible influences on the force-deformation behaviour of fastening elements.

2. Fastening technology

2.1 Type of fastening elements

The fastening elements used in construction are mainly mechanical and chemical anchors as well as powder actuated fasteners (for specific information see [1], [2], [3], [4]). A classification is possible according to Fig.1 and ref. [2] into:

Type A: Force (Torque) controlled expansion anchors
Torque controlled expansion anchors are cone anchors where a predetermined torque is applied to the bolt or nut. Thus, the cone at the lower end of the anchor is pulled into the slit anchor shell (sleeve). During this procedure the anchor shell is expanded and highly pressed against the borehole-wall. The anchor holds by friction. When an external tensile force is applied and exceeds a certain level the cone is pulled further into the shell and the expansion thus slowly increases up to the maximum possible value. The sleeve is not displaced longitudinally during this procedure (see e.g. [5]).

Type B: Deformation controlled expansion anchors
Hammer-set anchors are anchored by driving a cone (or plug) to a certain extent into a steel shell. By this procedure a radial expansion is produced and the steel shell is pressed against the concrete borehole-wall. The anchor holds by friction. A special type is the self drilling anchor. Self-drilling anchors have a cutting face and can therefore be used as a drill bit for making the anchor hole. The expansion operation is similar to Type B anchors.

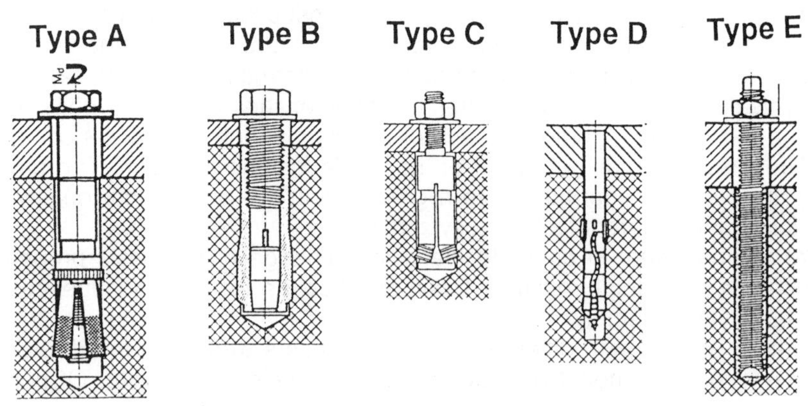

Fig 1: Different types of fastening elements

Type C: Undercut anchors
Undercut anchors require a locally widened hole or "undercut" to accept the expansion elements of the anchor. Due to the mechanical interlock (keying) for load transfer, the radial expansion forces are smaller than in Type A or B anchors.

Type D: Plastic anchors
Plastic anchors are used in a broad variety, mainly for non-structural applications with only minor requirements to integral safety.

Type E: Chemical Anchors
Frequently used in construction are adhesive anchors where a threaded rod is grouted in by means of a glas capsule containing epoxy acrylate resin, hardener and quartz granules. The anchor holds by bond between the threaded rod and the borehole-wall. Another type of chemical anchors use a two-component modified epoxy acrylate resin, pressed into the borehole.

Powder-actuated fasteners:
Powder-actuated fasteners are bolts or nails set with specific tools and with cartridge energy into concrete or even steel. During the penetration action into the base material, the concrete is displaced and highly compressed. Pressure and temperature rise due to friction leads to local sintering effects of the conrete and the metallic surface of the fastener. To avoid spalling of the concrete at the surface the fastening element can be set in a short predrilled hole (see fig. 2).

2.2 Working principles of fasteners
External forces acting on an anchor must be transmitted to the base material. This transmission is based either on friction, keying or bonding (see Fig. 3). To activate friction an expansion force has to be produced (type A and B anchors). The keying principle, explained in Fig. 3 shows a supporting force acting on the locally widened hole-walls by means of the expansion elements. If a fastening element works on the bonding principle, a synthetic resin adhesive mortar produces the bond between the threaded rod and the borehole-wall. With powder

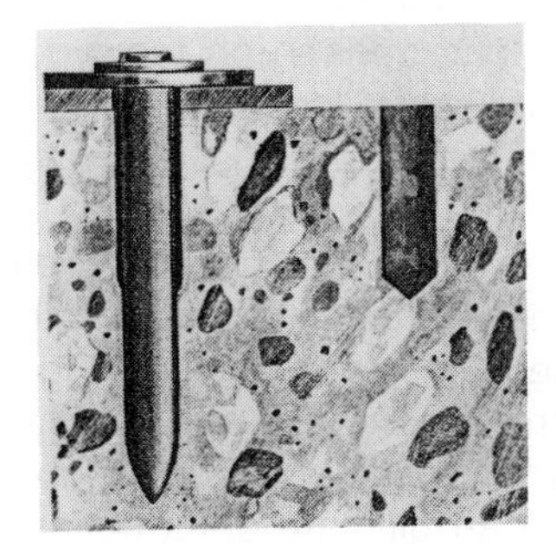

Fig 2: Reliable powder actuated fastening with a nail set into a predrilled hole

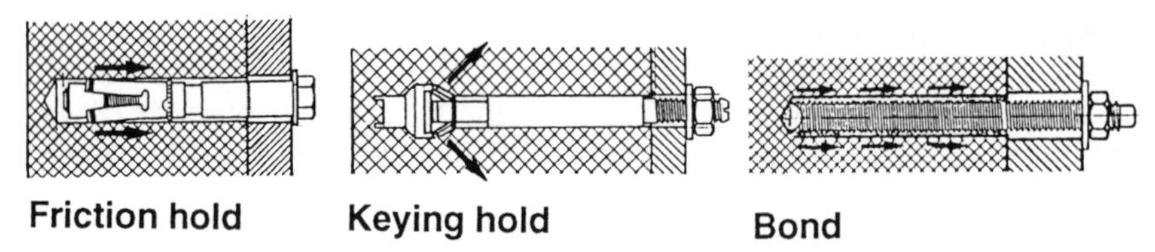

Fig 3: Working principles of fastening elements

actuated fastening the sintering effect provides bond between nail and subground. The holding power of many anchors comes from a combination of the previously mentioned working principles (see also [1], [2], [3],[10]).

2.3 Main factors influencing the fastening quality

Each component of a fastening system - the base material, the fastened construction element and the fastener - influences itself, and in interaction with the others, the quality of a fastening. Besides the anchor characteristic other factors like the type and direction of the applied force, the setting operation, the bore hole diameter, the geometry of embedment, the characteristics of the base material etc. are also important (see [1], [2], [6]). Dominant is the quality of the base material. Increasing the strength of the base material leads to an increase of the ultimate load. The increase is proportional to the tensile strength of the base material. Because, as a standard, only the compressive strength of concrete is tested most of the empirical formulas to determine the ultimate load use this parameter.

On the other hand the ultimate load of a powder actuated fastening is hardly dependent on the strength of the base material. However, the properties of the base material greatly influence the setting procedure and its quality.

2.4 Mechanical behaviour of fastening elements in tension

The mechanical behaviour of a fastening element can be characterized with the force-deformation characteristics and with the failure modes.

The force-deformation characteristics:

Figure 4 shows typical force-deformation curves for different anchor types. The displacement shown respresents the possible slip of the anchors in the borehole, the deformation of the concrete and of the anchor. The resulting deformations under service load and ultimate load depend on the type of fastening element. The overall deformation of force controlled expansion anchors is increased by the relative deformation between rod and sleeve through the follow-up expansion process.

Deformation controlled mechanical anchors show the smallest deformation. Due to the high friction forces, the slip of the anchor on the borehole is small. Self drilling anchors show more pronounced deformations (see Fig. 4). The load deformation diagram of an adhesive anchor is nearly elastic up to the ultimate load. Powder actuated fastening elements show a very steep force-deformation curve up to the ultimate load with an ultimate deformation between 0.3 and 0.4 mm.

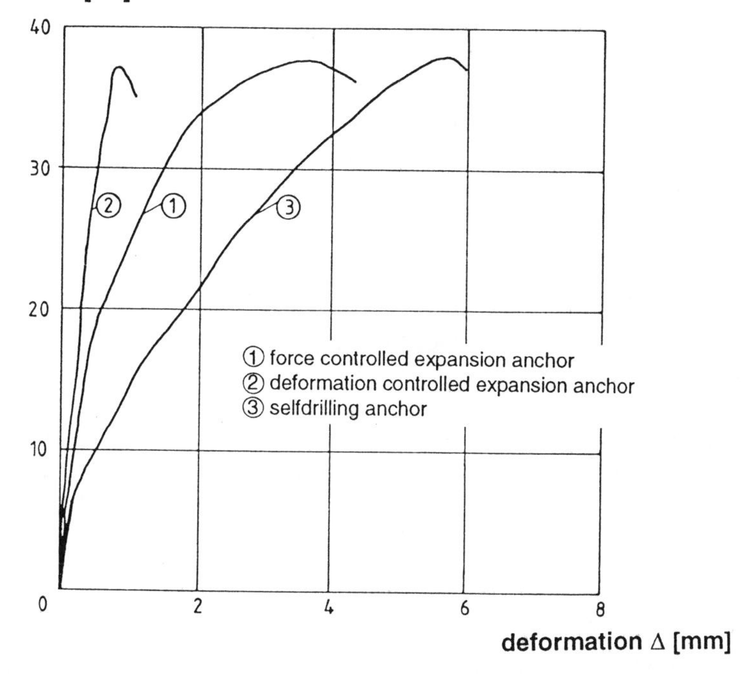

Fig 4: Typical force-deformation curves of fastening elements (from [1]), diametre M16, empedment depth h_v = 65 mm and concrete cube strength f_c = 25 N/mm2

Failure modes:

Figure 5 shows the possible failure modes of fastening elements:

a) The anchor developes a concrete failure cone. The concrete strength is fully activated.

b) If the fastening elements are spaced too close to each other, or placed too close to the edge, a common cone failure or/and edge failure, respectively, may occur at correspondingly reduced ultimate loads.

c) The conrete is split by the fastening element. This failure mode will occur only, if the geometric dimensions of the subground are too small, the fastening elements are placed too close to the edge or too close to each other, or the expansion forces are too high. The failure load is usually smaller than in mode a).

d) The anchor is pulled out of the borehole without significant concrete damage.

e) The bolt or anchor rod fails. For given material properties and fastening element dimensions this case defines the upper limit for the anchor ultimate failure load [7].

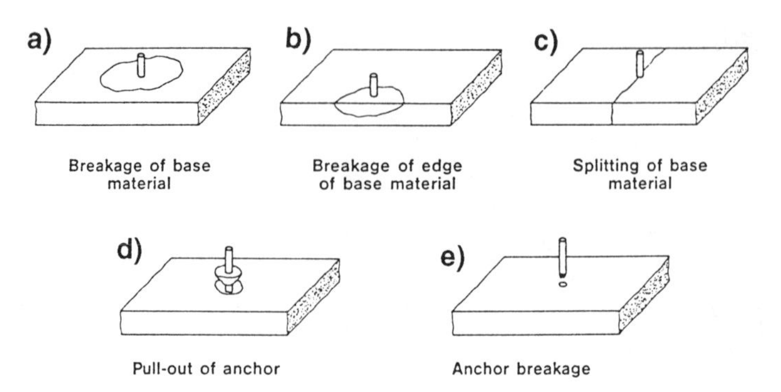

Fig 5: Failure modes (from[4])

Ultimate failure load:
The ultimate failure load is determined experimentally but there exist a number of empirical formulas for an estimation. For example for forced controlled mechanical anchors the ultimate load in uncracked concrete can be expressed as (see [6]):

$$F_u = 13{,}5 * h_v^{1.5} * f_c^{0.5} \qquad [N]$$

with h_v = depth of embedment [mm]
f_c = concrete cube strength [N/mm^2]

For cracked concrete the ultimate load is reduced depending very much on the type of anchor (see [1], [2]).

3. Behaviour of anchors in fibre reinforced concrete

The aim of these test series was to determine the probable change in the ultimate load and also the failure mode of different anchors in fibre reinforced concrete. Anchors of type A, B and E and two kinds of powder actuated fastenings of different dimensions were tested in static tension in uncracked concrete reinforced with three different types of fibre. In a second step the influence of fibre reinforcement on the force-deformation characteristics of the anchors was investigated.

3.1 Test set up

All anchor tests were performed using a hydraulic testing system, developped at the Hilti Corporate Research Centre. The hydraulic jack has a capacitiy of 80 kN. The tests are controlled and recorded on a HP 9826 computer. To represent the fastened construction element, a 10 mm thick washer was used, which at the same time served as the fixation point of the applied force.

During the test programme, the tests were run until the force-deformation curve reached the declining part; in the second programme the deformation controlled tests were run until zero applicable force. The deformation speed during the tests was 20.0 mm/minute. Figure 6 shows the overall test setup.

The testing of the powder actuated fastening elements was carried out with a mobile hydraulic tensile facility with the deformation speed of 330 N/sec. In this tests only the ultimate load was recorded. The bolts were set in a grid of 150x150 mm.

The anchors were installed in holes drilled with a rotary impact drilling machine type Hilti TE22 in a vertical down position. The anchors were set according to the recommended setting instructions. The experiences during the setting procedure in the fibre reinforced concrete were recorded.

The powder actuated fastening elements were set with the new method of predrilling a hole - so called Hilti DX-KWIK technology- of 23 mm depth and 5mm diameter before setting the bolt with a Hilti-DX M36 or Hilti-DX650 setting tool. The cartridge used was size red 6,8/11 M.

3.2 Test parameters

The anchors were set in fibre reinforced concrete slabs with the dimension of 4270 x 2550 x 200 mm. Concrete compression,splitting and bending tests were performed after 28 days for each specimen and compression tests at the day of the anchor tests. A concrete quality of 40 N/mm^2 after 28 days was planned. A maximum grain size of 16 mm was specified. No plasticizier was added.

The following fibres were used

- steel-fiber Dramix ZC 60/80
- steel-fibre Harex SF 01-32
- polypropylene fibre Forta Fibre D15.

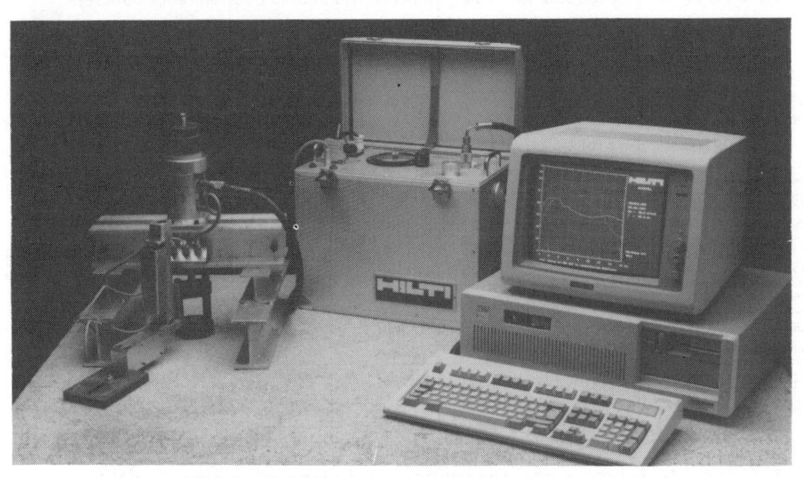

Fig 6: Test set-up of the anchor tests

The fibre contents were determined in accordance to the mix design for standard applications used in Europe. SIFCON-type contents between 4 % - 12 % of fibre reinforcement as reported in [9], were not tested. Figure 7 shows a summary of the fibre reinforced concrete mixtures and the results of the compression, splitting and bending tests. For reference purposes the strength values of a non-fibre-reinforced concrete are added (in Fig. 14-16 characterized with "NORMALB")

The following anchors were used (see [3]):

Type A:
heavy duty force controlled expansion anchor: HSL M12 and HSL M16
medium duty force controlled expansion anchor: HSA M12 and HSA M16

Type B:
deformation controlled expansion anchor: HKD M12 and HKD M16

Type E:
adhesive anchors: HVA M12 and HVA M16

The anchors were set one day in advance of testing. The concrete age and the concrete strength at that time are given in Fig. 8. Five and five specimens of each anchor diameter were tested.

The tested powder actuated fastening elements were the DNH 37-P8 S15, NPH-42 L15 and M8M-15-37 P8 (see [9]). The latter element is a threaded bolt with dimension M8. The drive pins DNH 37-P8 S15 and NPH-42 L15 serve to fasten metal strips etc. to concrete. The elements were set immediately before testing. Each element type was tested with 50 specimens. The concrete age varied between 30 and 70 days.

steel fibres ZC 60/80		steel fibres SF 01-32		polypropylen fibres D15		reference concrete		
sand 0/4	765	sand 0/4	765	sand 0/4	765	sand 0/4	765	kg/m3
gravel 4/8	478	gravel 4/8	478	gravel 4/8	478	gravel 4/8	478	kg/m3
gravel 8/16	669	gravel 8/16	669	gravel 8/16	669	gravel 8/16	669	kg/m3
cement HPC	300	cement HPC	300	cement HPC	300	cement HPC	300	kg/m3
w/c ratio	0.6	w/c ratio	0.6	w/c ratio	0.6	w/c ratio	0.6	
ZC 60/80	50	SF 01-32	50	FF D-15	50			kg/m3
compression strength								
28 days	44.21	28 days	49.97	32 days	47.15	31 days	41.86	N/mm2
189 days	49.35	202 days	61.18	73 days	50.1	60 days	48.73	N/mm2
532 days	53.54	538 days	59.79	510 days	51.18	600 days	55.12	N/mm2
splitting strength								
28 days	2.29	28 days	3.06	32 days	2.64	31 days	2.36	N/mm2
189 days	2.89	202 days	3.56	73 days	2.78	60 days	2.71	N/mm2
bending strength								
28 days	5.01	28 days	6.42	32 days	4.55	31 days	4.92	N/mm2
189 days	6.3	202 days	7.69	73 days	6.11	60 days	4.88	N/mm2

Fig 7: Concrete mixes used for testing and strength gained at different ages

Dramix ZC 60/80								
test number	type	number	size	conc. strength (N/mm2)	embedment depth (mm)	ultimate load (KN)	standard deviation (KN)	failure mode
HL12-90	HSL	5	M12	49.35	80	63.49	5.18	5*A/B
HLG12-91	HSLG	7	M12	53.54	80	71.82	8.72	1*A, 6*B
HL16-90	HSL	5	M16	49.35	105	111.93	3.53	5*A/B
HLG16-91	HSLG	5	M16	53.54	105	116.38	20.65	2*D, 3*B
HA12-90	HVA	5	M12	49.35	110	49.21	0.2	5*D
HA12-91	HVA	5	M12	53.54	110	75.45	6.02	5*B
HA16-90	HVA	5	M16	49.35	125	89.78	0.67	5*D
HA16-91	HVA	5	M16	53.54	125	121.79	9.04	1*A/B, 4*A
HSA12-90	HSA	5	M12	49.35	80	39.38	2.16	3*A, 2*D
HSA16-90	HSA	5	M16	49.35	100	59.64	2.27	5*A
HSA16-91	HSA	5	M16	53.54	100	80.52	7.25	3*A,2*D
HD12-90	HKD	5	M12	49.35	50	30.58	1.67	5*A/B
HD16-90	HKD	5	M16	49.35	65	49.88	2.3	5*A/B

Harex SF 01-32								
test number	type	number	size	conc. strength (N/mm2)	embedment depth (mm)	ultimate load (KN)	standard deviation (KN)	failure mode
DL12-90	HSL	5	M12	59.79	80	66.84	7.01	1*A/B, 2*B;2*D
DLG12-91	HSLG	5	M12	61.18	80	68.49	3.47	5*B
DL16-90	HSL	5	M16	59.79	105	111.1	8.49	2*B, 3* A/B
DLG16-91	HSLG	5	M16	61.18	105	137.85	14.34	5*B
DA12-90	HVA	5	M12	59.79	110	49.47	0.57	5*D
DA12-91	HVA	5	M12	61.18	110	90.77	9.24	4*A, 1*B
DA16-90	HVA	5	M16	59.79	125	88.63	5.37	4*D, 1*D/B
DA16-91	HVA	5	M16	61.18	125	133.92	16.27	5*A
DSA12-90	HSA	5	M12	59.79	80	38.84	1.76	4*A, 1*D
DSA16-90	HSA	5	M16	59.79	100	61.51	3.65	5*Z
DSA16-91	HSA	5	M16	61.18	100	83.35	3.88	3*A, 2*D
DD12-90	HKD	5	M12	59.79	50	31.12	1.81	5*A/B
DD16-90	HKD	5	M16	59.79	65	54.69	3.27	5*A/B

Forta Fibre D-15								
test number	type	number	size	conc. strength (N/mm2)	embedment depth (mm)	ultimate load (KN)	standard deviation (KN)	failure mode
FL12-90	HSL	5	M12	50.1	80	66.84	3.95	4*A,1*A/B
FLG12-91	HSLG	5	M12	51.18	80	72.1	5.01	5*B
FL16-90	HSL	5	M16	50.1	105	106.89	5.61	5*B
FLG16-91	HSLG	5	M16	51.18	105	132.03	5.51	5*B
FA12-90	HVA	3	M12	50.1	110	51.13	1.48	3*D
FA12-91	HVA	5	M12	51.18	110	85.65	7.66	5*A
FA16-90	HVA	3	M16	50.1	125	89.16	0.45	5*D
FA16-91	HVA	5	M16	51.18	125	120.14	7.01	3*B, 2*A
FSA12-90	HSA	5	M12	50.1	80	38.21	1.78	5*A
FSA16-90	HSA	5	M16	50.1	100	57.39	3.01	5*A
FSA16-91	HSA	5	M16	51.18	100	80.49	5.58	2*B, 2*A, 1*D
FD12-90	HKD	5	M12	50.1	50	32.79	2.83	2*B,1*D, 2*A/B
FD16-90	HKD	5	M16	51.18	65	49.53	2.76	5*B

reference concrete								
test number	type	number	size	conc. strength (N/mm2)	embedment depth (mm)	ultimate load (KN)	standard deviation (KN)	failure mode
RL12-90	HSL	5	M12	48.73	80	64.42	5.04	3*A/B, 1*D, 1*A
RLG12-91	HSLG	5	M12	55.12	80	76.77	4.56	5*B
RL16-90	HSL	5	M16	48.73	105	112.58	2.89	5*B
RLG16-91	HSLG	10	M16	55.12	105	124.82	17.27	8*B, 2*A
RA12-90	HVA	3	M12	48.73	110	50.15	2.19	3*D
RA12-91	HVA	5	M12	55.12	110	83.98	12.54	5*A
RA16-90	HVA	3	M16	48.73	125	92.81	1.86	3*D
RA16-91	HVA	5	M16	55.12	125	102.04	12.2	5*A
RSA12-90	HSA	5	M12	48.73	80	34.82	2.3	4*A, 1*A/B
RSA16-90	HSA	5	M16	48.73	100	60.75	3.21	5*A
RD12-90	HKD	5	M12	48.73	50	30.95	3.37	5*A/B
RD16-90	HKD	5	M16	48.73	65	46.01	1.94	5*A/B

Fig 8: Test results of the anchor pull out tests

3.3 Test results

Drilling operations and setting procedure:

No problems occured during the drilling process in the predefined fibre reinforced concretes. An optical control of the structure and geometry of the borehole showed that depending on the position and the angle of the steel fibres hit during the drilling operation, they were cut through or pressed against the borehole-wall. Tests to evaluate the correct setting procedure of the anchors were performed by pushing a gauge in the borehole after certain drilling distances (Drilling distance: 25 m, 50 m, 75 m, 100 m). This showed that after a longer drilling use in steel fibre reinforced concrete the easiness of the setting operation of an anchor decreased. This is due to the steel fibres sticking in the borehole.

The drilling speed (mm/min) on concrete with steel fibre reinforcement was only 5 % lower after a total drilling distance of 100 m.

No difficulties occured to apply the correct torque moment with force controlled mechanical anchors. There were no problems recorded in the setting and penetration procedure of powder actuated fastening elements in all tests.

Anchor tests:

The failure modes observed during testing fell into the following categories (see fig. 5):

a) concrete failure cone (base material)
b) anchor pull out (base material)
c) bolt failure (fastener material)

A bolt failure happened to the chemical anchors type HVA with the dimension of M12 and to some of the force controlled expansion anchors type HSA and HSL. The most frequent failure mode was a concrete failure cone alone or in combination with an anchor pull out. During a base material failure, radial cracks spread outwards from the anchor on the loaded surface, which either supports a pull out behaviour of the anchor or leads to a concrete failure cone immediately after the appearance of the cracks. Figure 8 summarized the failure modes in the three different fibre reinforced concrete specimens and in the reference concrete (column 9) (A = anchor pull out, B = concrete failure cone, D = bolt failure). An important observation was that with fibre reinforcement of the type Harex SF 01-32 some of the concrete cones broke out as one piece, compared to several cracked pieces clinged together in other tests. Figure 9 shows two typical failure cones of the above mentioned modes.

Results of the tensile ultimate strength (static) tests are summarized in Fig. 8. Depending on the different fibre reinforcement used, the load bearing capacity of each anchor type is shown. Column 5 lists the actual compression strength of 200 mm cubes at the time of testing; column 7 the ultimate load, and column 8 the standard deviation in relation to the number of tests executed.

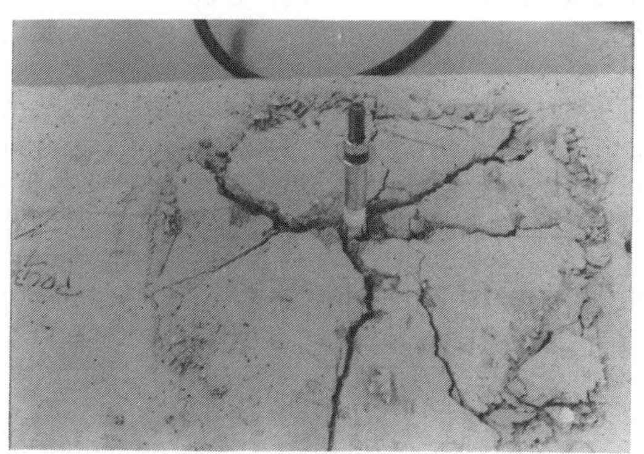

Fig 9: Typical failure cones of a HSL anchor in Harex SF 01-32 (left) and Dramix ZC 60/80 fibre reinforced concrete (right)

The force-deformation characteristics of the five specimens of force controlled expansion anchor type HSL M12 in fibre reinforced concrete (Harex SF 01/32) are shown in Fig. 10. The mean values of the force deformation curves of the other different types of anchors in concrete with different fibre reinforce-

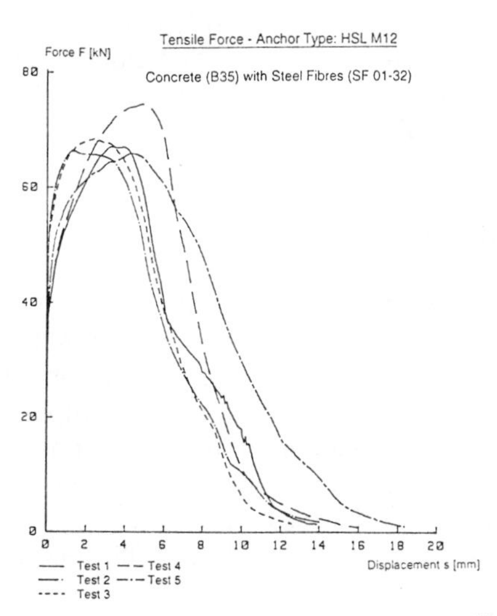

Fig 10: Force-deformation characteristics of a force controlled expansian anchor HSL M12 in steel fibre reinforced concrete (Harex SF 01/32)

ment are shown in Fig. 14-16. In the post failure behaviour - especially of anchors with the dimension M16 - a very jerky pull-out behaviour was observed under the deformation controlled testing.

The deformation controlled expansion anchor had no follow up expansion so the load deformation curve showed a linear behaviour up to 80-90% of the ultimate load. The bend in the force deformation curve signalizes that the entire anchor is pulled out which is followed in general by a concrete cone failure (see Fig. 15).

The load deformation characteristics of chemical anchors is very stiff until the ultimate load is reached. Depending on the failure mode either the force deformation curve stops suddenly due to bolt failure or, if the anchor is pulled out, the force decreases in a hyperbolic shape.

For an estimation of the absorbed energy the force deformation curves of the different fastening elements for all different fibre reinforced and reference concretes were integrated. Figure 11 shows in column 5 the deformation corresponding to the ultimate load and in column 6 the energy based on the force deformation curve up to the ultimate load. Column 7 gives the percentage of the used energy in relation to the maximum possible energy (= ultimate load x corresponding deformation) and in column 7 to 9 the same data for the complete load deformation curves are summarized. All data are mean values of the executed tests. Column 10 shows the percentage of the used energy up to the ultimate load in relation to the overall energy.

Powder actuated fastener tests

The failure modes of the powder actuated fastening elements in fibre reinforced concretes were either a pull-out failure or a small concrete cone failure after pull-out of the fastener. No bolt failure occurred. The results are summarized in Fig. 12. It shows the ultimate load of each type of fastening elements, the 5 %-fractile according to a chosen log-normal distributions (see Fig. 17), the actual concrete compression strength with 200 mm cubes at the time of testing and the penetration geometry of the nail or bolt.

4. Discussion of the results

4.1 Anchors in fibre reinforced concrete

Compared to the reference concrete without fibres the three types of fibre reinforcement had no significant influence on the setting procedure of the different kinds of anchors tested. The torque, for example, could easily be raised to the prescribed level. The influence of the fibres on the drilling procedure was only observable with steel fibres; but only on a lifetime test basis after a total of 100 m drilling distance. The drilling output is slightly lower with the preuse of the drill bit in steel fibre reinforcement, but which has no noteworthy influence on the single hole.

A comparison of the ultimate load of the anchors is shown in Fig. 13. Only the concrete strength and the kind of fibre reinforcement are the parameters

test number	type of anchor	fibres	ultimate load	deformation at ultimate load	energy at ultimate load	energy quota (col 6)	maximum deformation	overall energy	energy quota (col 9)	energy quota (col 6/col 9)
			(KN)	(mm)	(KNmm)	(%)	(mm)	(KNmm)	(%)	(%)
HLG12-91	HSLG M12	ZC 60/60	71.82	3.55	228	89	29.37	1243	59	18
HLG16-91	HSLG M16	ZC 60/60	109.28	10.85	831	70	30.44	1848	56	45
HA12-91	HVA M12	ZC 60/60	75.45	4.73	314	88	26.81	960	47	33
HA16-91	HVA M16	ZC 60/60	121.81	5.77	636	90	35.46	2188	51	29
HSA16-91	HSA M16	ZC 60/60	80.52	16.29	862	66	36.4	1723	59	50
DLG12-91	HSLG M12	SF 01-32	68.49	3.29	201	89	14.96	519	51	39
DLG16-91	HSLG M16	SF 01-32	137.85	3.45	371	78	26.97	2345	63	16
DA12-91	HVA M12	SF 01-32	90.77	3.56	276	85	31.79	1070	37	26
DA16-91	HVA M16	SF 01-32	133.92	5.32	630	88	33.38	2026	45	31
DSA16-91	HSA M16	SF 01-32	83.35	15.87	871	66	31.11	1819	70	48
FLG12-91	HSLG M12	FF D15	72.11	2.93	185	88	17.32	639	51	29
FLG16-91	HSLG M16	FF D15	132.03	3.83	377	75	26.34	1813	52	21
FA12-91	HVA M12	FF D15	85.65	4.51	331	86	33.81	1130	39	29
FA16-91	HVA M16	FF D15	120.14	4.47	480	89	36.35	2369	54	20
FSA16-91	HSA M16	FF D15	80.49	15.81	816	64	32.25	1553	60	53
RLG12-91	HSLG M12	-	76.77	3.51	233	86	17.53	698	52	33
RLG16-91	HSLG M16	-	124.82	5.96	509	68	42.86	3055	57	17
RA12-91	HVA M12	-	83.98	3.58	263	87	34.77	1159	40	23
RA16-91	HVA M16	-	102.04	2.78	184	65	27.65	1221	43	15

Fig 11: Force deformation data of the tests

bolts	type of fibres	compression strength	projection	penetration depth	ultimate tensile load	5% fractile
		(N/mm2)	(mm)	(mm)	(KN)	(KN)
DNH 37 P8 S1	ZC 60/80	51.93	7.61	13.39	8.25	6.83
M8H-15-37 P8	ZC 60/80	51.93	14.91	17.09	13.71	11.72
M8H-15-37P8	SF 01-32	49.97	16.07	15.92	13.55	11.33
NPH3-42 L15	SF 01-32	49.97	6.33	14.67	13.97	11.69
M8H-15-37P8	FF D15	50.56	15.85	16.14	17.93	12.91
DNH 37 P8 S1	FF D15	50.56	7.42	13.58	9.17	7.62

Fig 12 : Test results of the powder actuated fastening elements

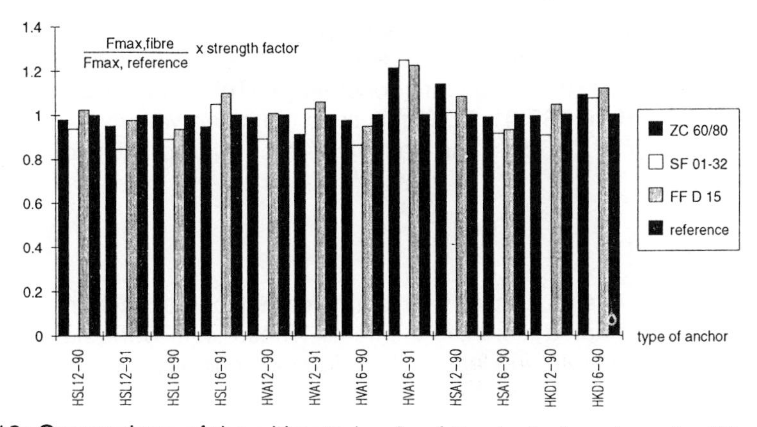

Fig 13: Comparison of the ultimate loads of the tested anchors in different fibre reinforced concrete

varied. Referring to empirical formulas mentioned in section 2.4 the data were standardized according to the square root of the concrete strength with fibre reinforcement in relation to the strength of the reference concrete. It is clearly evident that for all three fibre reinforced concretes the ultimate load varies only within the dispersion of results in concrete without fibres. In reference to the test results, there is no dependence on the type and dimension of the anchors. The failure modes were similar for all types of concrete.

The force deformation curves of each type of anchor in different fibre reinforced concretes is shown in Fig. 14-16. Up to the ultimate load, the shape of the curves does not differ very much. The follow up expansion behaviour of the force controlled expansion anchor is not influenced by the kind of fibres or the fibres. The force deflection curve of the chemical anchors are up to the ultimate load almost identical. This statement corresponds well with the results of the calculation of energy of the load deflection curves in different fibre reinforced concretes (see Fig. 11).

4.2 Powder actuated fastening elements

The kind of steel fibres and polypropylen fibres used in the tests did not influence the setting operation of the powder actuated fastenings. The penetration procedure was not impaired due to the presence of fibres. Figure 17 shows a comparison of the statistical interpretation of the ultimate failure load in different fibre reinforced concretes. This was pictured in a gaussian probability curve, whereby the failure propability P_f was assigned to the increasing size of events.

$$P_f = \frac{m - 0,5}{n}$$

m .. ordinal number

n .. number of tests

In concrete with fibre reinforcement type Dramix ZC 60/80 the standard deviation is for both tested specimens less than the standard deviation determined from former tests on standard concrete [9]. This is also true with fibre reinforcement type Harex SF 01-32. In this case the mean value was a little bit higher too. Forta fibre D15 had the same effect on the variation of the test results.

5. Conclusions

A total of 404 tests were performed with four different types of anchors and with 3 kinds of powder actuated fastening elements. These tests were static tensile tests in uncracked concrete and lead to the following conclusions.

Setting procedure:

- The setting procedure of an anchor by drilling the hole, putting it in (and in the case of force controlled expansion anchor, raising the torque moment) is not influenced by the presence of fibres in concrete. The steel fibres slightly influence the drilling speed.

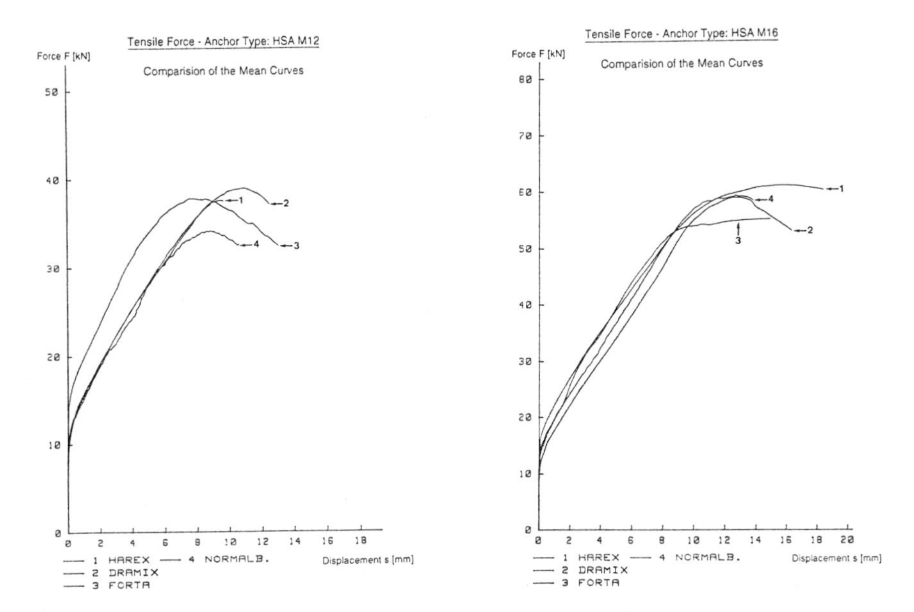

Fig 14: Comparison of the force-deformation curves of an force controlled expansion anchor in different fibre reinforced concretes

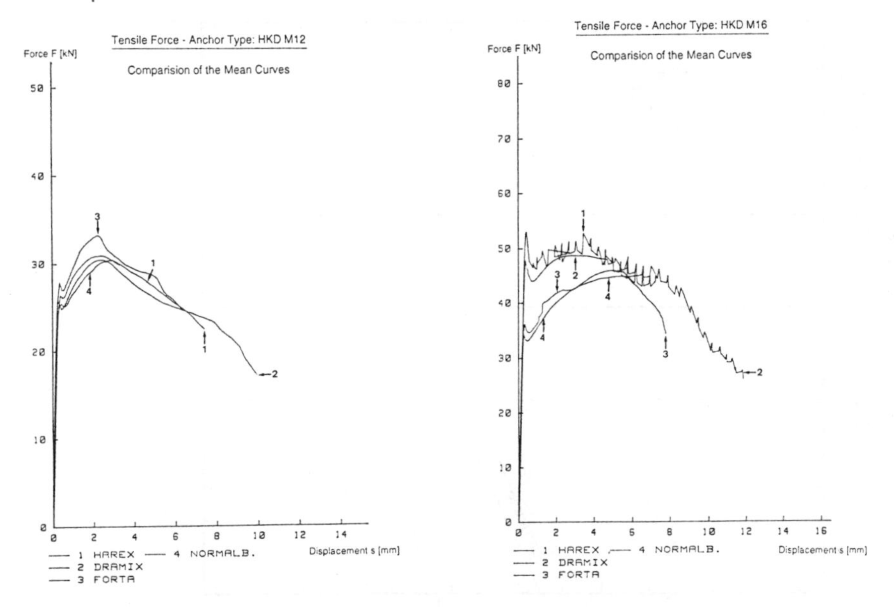

Fig 15: Comparison of the force-deformation curves of an deformation controlled expansion anchor in different fibre reinforced concretes

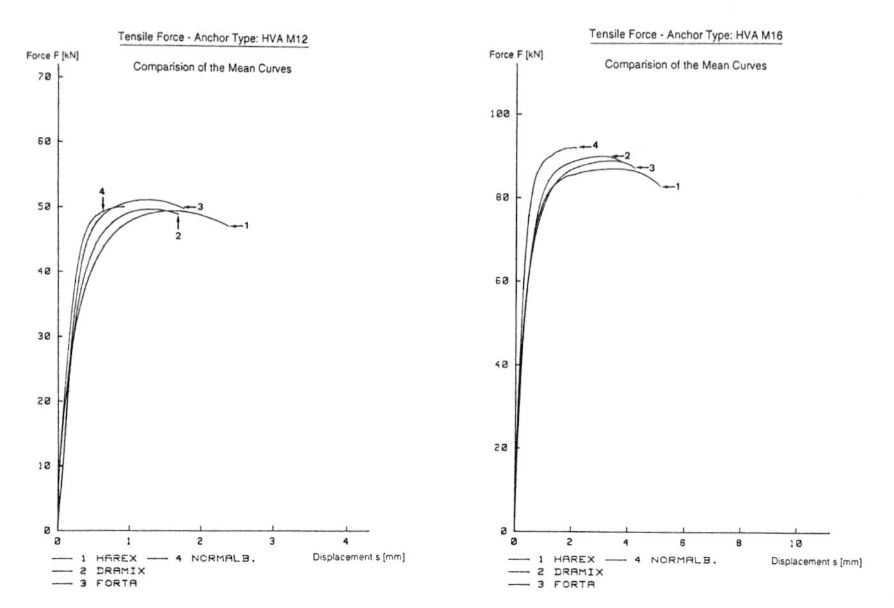

Fig 16: Comparison of the force-deformation curves of an chemical anchor in different fibre reinforced concretes

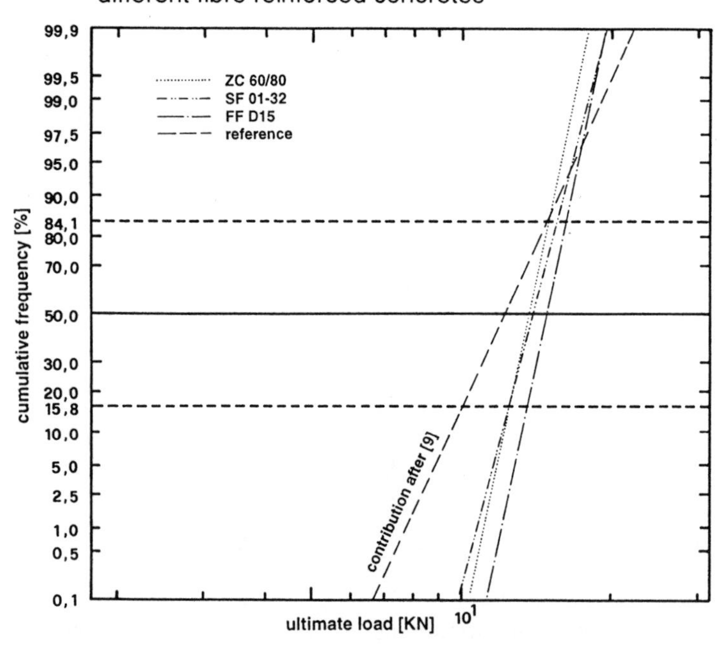

Fig 17: Comparison of the log- normal distribution of the ultimate load of powder actuated fastening elements in different fibre reinforced concretes

- Steel fibres or polypropylene fibres do not deviate the penetration procedure of the powder actuated fastening elements. A correct setting is achievable in fibre reinforced concrete.

Anchor tests:

- Neither steel fibres nor polypropylene fibres lead to an increase of the ultimate force and the corresponding deformations of the anchors. The recommended design forces for anchors can be used on fibre reinforced concrete as well.
- The follow up expansion procedure with expansion anchors is not disturbed due to the presence of steel fibres or polypropylene fibres.
- The force deformation characteristics of anchors in fibre reinforced concrete up to the ultimate force is comparable to normal concrete.

Powder actuated fastening elements:

- The mean value of the ultimate force of powder actuated bolts or nails in steel fibre or polypropylene fibre reinforced concrete is identical to normal concrete. The standard deviation, however is influenced positively.

References:

[1] Rehm G., Eligehausen R.; Mallée R.; "Fastening Technology" ("Befestigungstechnik") Betonkalender 1988, Verlag für Architektur und technische Wissenschaften, Berlin, 1988 (in german)

[2] CEB: "Fastenings to Reinforced Concrete and Masonry Structures", Bulletin d'Information Nr. 206, Part I, CEB Lausanne, 1991

[3] Hilti AG; "Fastening Manual Anchoring", Issue Aug. 1991, Hilti Corporation Fastening Systems, Schaan 1991

[4] Wisniewsky GH; "Fastening Technology"; Verlag moderne Industrie AG & Co, Landberg/Lech, 1988

[5] Wagner-Grey U.; "Theoretical Considerations on Loadbearing Behaviour of Expansion Anchors", in: Anchorage to Concrete, American Concrete Institute, ACI SP 103, Detroit, 1987

[6] Ammann W.: "Komponenten des Befestigungssystems", in: Sicherheit und Dauerhaftigkeit von Befestigungssystemen, Dokumentation SIA D055, Schweizerischer Ingenieur- und Architektenverein Zürich, 1990

[7] Eligehausen R.; "Anchorage to Concrete by Metallic Expansion Anchors", in: Anchorage to Concrete, American Concrete Institute, ACI SP 102, Detroit, 1987

[8] Naaman A. F.; Homrich J. R.; " Tensile Stress-Strain Properties of SIFCON", ACI Materials Journal, May-June, 1989, p. 244-251.

[9] Hilti AG; " Powder Actuated Fastening Technology", Issue 1986, Hilti Corporation Fastening Systems,, Schaan, 1986

[10] ACI, Special Publication SP 130, "Anchors in Concrete: Design and Behaviour", Editors: G.Senkiew, H. Lancelot, American Concrete Institute, 1992

53 THIN REPAIRS WITH METALLIC GLASS FIBRE REINFORCED CONCRETE: LENGTH CHANGES DURING THE FIRST 24 HOURS

J-L. GRANJU
Laboratoire Matériaux et Durabilité des Constructions, INSA - UPS Génie Civile, Toulouse, France
F. GRANDHAIE
Centre de Recherches de Pont à Mousson, and INSA - UPS Génie Civile, Toulouse, France

Abstract
The work reported here is a part of a general programme aimed at quantifying a ıd, if possible, modelizing the contribution of metallıc glass fibres to the durability of thin concrete repairs (about 50 mm) cast on horizontal concrete bases.

This paper presents a systematic study of the length changes of both the fresh repair concrete and the base of old concrete during the first 24 hours. The results show that :

The length changes of the fresh overlay concrete is a composition of its own length changes and of those of the base concrete (which swells, absorbing part of the water content of the fresh overlay).

The reinforcement by metallic glass fibres, even in proportion as low as vf = 0.3 %, reduces the swelling of the fresh overlay, whether its origin is autogenous or due to stretching by adhesion to a swelling support.

Keywords : Repair, Fibre Reinforced Concrete, Shrinkage

1 Introduction

The work reported here is part of a general programme started in late 1989 (Granju and al, 1991). Its aim is to quantify and, if possible, to modelize the contribution of metallic glass fibres to the durability of thin concrete repairs (about 50 mm thick) cast on horizontal concrete bases. The repairs of roads, highways and slabs are concerned. Their durability may be affected by the differential dimensional variations of, on the one hand the repair layer and on the other hand the base of old concrete.

This paper focuses on the study of the length changes in fresh repairs, during the first 24 hours.

It will be shown that, if the addtion of metallic glass fibres does not affect the shrinkage of the fresh repair concrete, it significantly reduces its length increases,

Fibre Reinforced Cement and Concrete. Edited by R. N. Swamy. © 1992 RILEM.
Published by E & FN Spon, 2-6 Boundary Row, London SE1 8HN. ISBN 0 419 18130 X.

whether they are autogenous swelling or stretching by adhesion to a swelling support.

The work has been carried out at LMDC (construction materials and durability laboratory) in Toulouse, France, with the cooperation of CRPAM (Pont à Mousson research center) in Pont à Mousson, France.

2 Fibres used

The fibres considered for this study are FIBRALEX fibres, produced by SEVA, France (De Guillebon, Sohm, 1986), a company of Saint Gobain Group. They are obtained by "melt spinning" : a thin flow of melted metal falls onto a cold spinning wheel at the contact of which it is cooled very quickly to harden into amorphous structure. The fibres obtained are shiny ribbons, a few micrometres thick, 1 to 2 millimetres in width and available in different lengths from 15 to 45 millimetres. They are flexible, have a high tensile strength as well as a strong bond with the concrete matrix and, above all, they are corrosion resistant ; this quality makes them especially appropriate for repairs subjected to chemical attacks (for instance by deicing salts or by sea water or sea mist). We have chosen to carry out this work with 30 mm long fibres, which are the most common ones. Their properties are presented in table 1.

Up to now, for repairs, these fibres have been mainly used in tunnel rehabilitations, particularly sewers, the fibre reinforced concrete being placed by shotcreting (Schacher, 1989). They have also been used for some slab repairs.

Table 1 - Fibre characteristics (metallic glass fibre)

Name :	Fibraflex
Nature :	Amorphous metallic alloy, $(Fe, Cr)_{80}$ $(P,C,Si)_{20}$
Length :	30 mm
Width :	1.6 mm
Thickness :	30 µm
Specific surface :	10 m^2/kg
Density :	7.2
Tensile strength :	≈ 2000 MPa
Elastic modulus :	140 000 MPa
Corrosion resistance :	rustproof

3 Experimental

The tests were carried out on two types of specimens : simple specimens to characterize the behaviour of the repair layer and composite specimens representing a real repair with its different components (a concrete overlay cast onto an old concrete base).

Two overlay concretes were considered, a plain one and a fibre reinforced one. Their compositions were chosen in such a way as to be representative of typical field mixtures used for thin repairs in France. They contain 400 kg/m^3 of Portland cement, have a water to cement ratio of 0.45, a maximum coarse aggregate size of about 10 mm and contain a superplasticizer. In both cases, the fine to coarse aggregate ratio was optimized using the Baron Lesage method (Baron, Lesage, 1969), and the dosage of the superplasticizer was adjusted to obtain the desired workability. The fibre content was fixed at a single value of 20 kg/m^3 (0.3% in volume). The characteristics of the mixtures are described in Table 2.

Table 2 - Characteristics of concrete mixtures

Materials
Cement : CPA HP from Boussens (France)
65 % C_3S, 8 % C_2S, 9 % C_3A, 8 % C_4AF
Blaine specific surface 4200 cm^2/g
Superplasticizer : Resin GT produced by Chryso (melamine)
Aggregates : river siliceous aggregates, sand 0-5 mm and gravel 5-12.5 mm
Fibres : Fibraflex 30 mm

Mixing sequence
1 mm : dry components
introduction of water, then 2 min of mixing
introduction of superplasticizer, then 2 min of mixing

Composition per m^3 concrete	Plain concrete	Fibre
Cement (kg)	400	400
Water (l)	180	180
Sand (kg)	655	839
Gravel (kg)	1116	932
Fibre (kg)	0	20
Superplasticizer (l)	2.7	4.2

Workability
The same for both mixes. Measured while vibrating (with a "Maniabililimètre LCL" (Baron, Lesage, 1969)), it corresponds for plain concrete to 120 mm slump.

Compressive strength at 28 days
Measured on cylinders kept at 100 % RH
Plain concrete : f_c=51 MPa   Fibre concrete : f_c=52 MPa

The specimens used are illustrated in Figure 1. The simple specimens were 50 mm thick (as the expected real repair overlays). The composite specimens consisted of an old concrete base on which the 50 mm thick repair overlay was cast without any bonding agent.

The bases (50 mm thick) were made with concrete of the same composition as the plain mixture, and were obtained by sawing in two parts 500 mm x 100 mm x 100 mm prisms. These prisms were wet cured for 28 days at 20°C, sawn and kept during 70 days at 20°C and 60 % R.H. The repair overlay was applied on their dry sawn surface after this 70 day's period. Examination of these sawn surfaces revealed that carbonation had penetrated to a depth of 0.5 mm, and very few microcracks caused by the sawing action could be seen using the replica technique(Ollivier, 1986).

As the repair concrete was fresh, the samples were tested in their molds (with smooth oiled PVC sides) and the ambient conditions were set at 20°C and 60 % R.H. As the dimensional variations appeared to be particularly sensitive to the ventilation conditions, over and around the specimens, all tests were performed under a no wind condition.

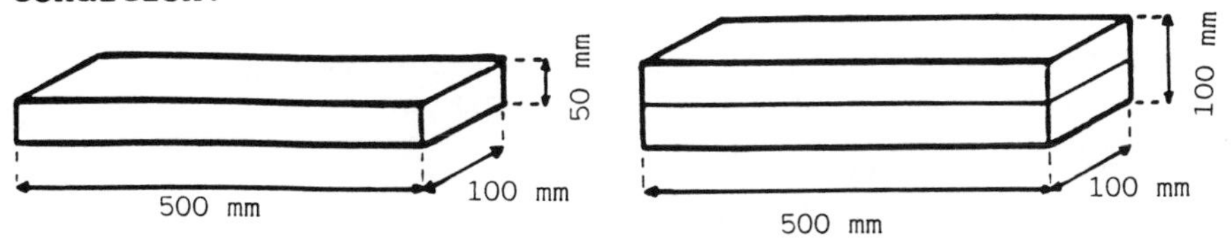

Simple specimen Composite specimen

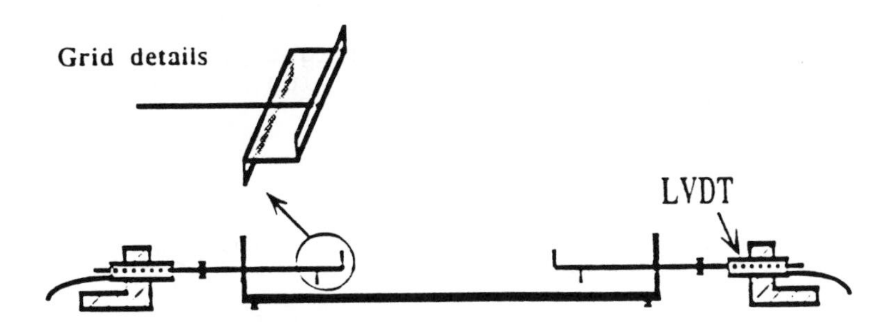

Device to measure the length variations of the fresh repair layer

Fig. 1. Experimentals

Three curing modes were considered :

With free evaporation through the top surface of the repair overlay.

Without free evaporation, the fresh overlay being capped with a polythene sheet which leaves a small air volume (= 2 mm thick) above the overlay surface.

With no evaporation, the polythene sheet being tightly applied onto the overlay fresh concrete with no visible trapped air.

The length variations of the repair layer as well as of the old concrete base were monitored continuously. The method used was the one developed by Detriché (1978) to study the shrinkage of mortar coatings. For the overlay measurements, grids anchored in the fresh concrete were each connected to an LVDT sensor as shown in figure 1. The LVDT supports were completely separated from the mold in order to avoid its dimensional variations. When required, the length changes of the old concrete base were measured with a similar device, the LVDT being then connected to two rods stuck in each end face of the base.

In all cases, thermocouples were also used to measure and follow the temperature of the repair overlay.

4 Results and discussion

Each curve or value presented here is the average of the test results from at least three specimens.

4.1 Simple specimens.

4.1.1 With free evaporation (figure 2)

The addition of fibres (20 kg/m^3 or 0.3 % in volume) has no visible influence on the measured length variations which are essentially shrinkage. The obtained curves show the classic shape described by Baron (1982) : a "primary shrinkage", followed by a period of slight swelling, and a "secondary shrinkage" (also called "drying, or hardening shrinkage" (Buil, 1982)).

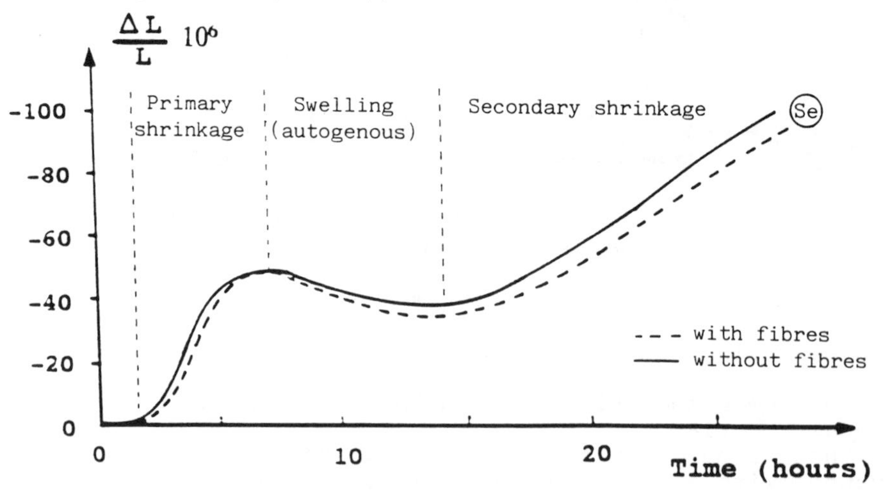

Fig. 2. Typical length variations of simple specimens with free evaporation

4.1.2 Whithout free evaporation and with no evaporation (figure 3)

In both cases, the addition of 20 kg/m³ of fibres has no visible effect on the measured shrinkage, but it significantly reduces the swelling of the fresh concrete layer. This swelling is autogenous and it is associated with the hydration process. The higher is its amplitude, the greater is its reduction by the fibres. The curves in figure 4, reporting the temperature variations, prove that this reduced swelling is not due to of a lower temperature rise of the fibre concrete layer. In fact, it exhibits a higher temperature rise than plain concrete (probably due to its higher superplasticizer dosage (Bensted, 1987)).

In the case of the specimens without free evaporation, the air volume beneath the polythene film capping allows evaporation from the fresh concrete, witnessed by a thick dew on the underface of the polythene sheet. Then the overlay shows a significant "primary shrinkage".

In the case of no evaporation, the obtained curves are in conformity with the general knowledge : as the "primary shrinkage" phase is reduced considerably (even can it disappear), the swelling phase is emphasized.

These results of simple specimens already show that metallic glass fibres, in a proportion of 20 kg/m³ (0.3 % in volume), significantly reduce the autogenous swelling of fresh concrete.

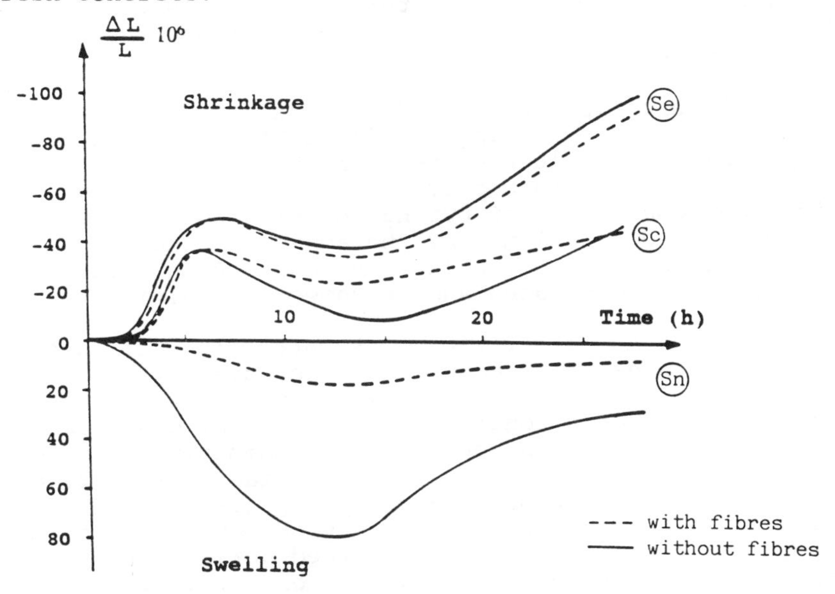

Se : free evaporation
Sc : without free evaporation
Sn : with no evaporation

Fig. 3. Simple specimens : length variations

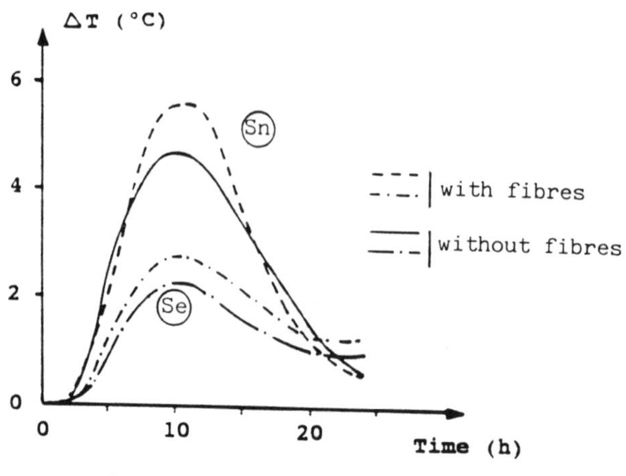

Fig.4. Simple specimens : temperature variations

4.2 Composite specimens
The curing mode with no evaporation,which is not realistic in the case of real repairs, was not considered here. So, only the two following cases were considered : with free evaporation and without free evaporation. The length variations of both, the fresh repair overlay and the old concrete base were measured.

4.2.1 Length variations of the old concrete base (figure 5)
The results show that, during the first 24 hours, the length changes are influenced neither by the nature of the overlay concrete (reinforced or not with fibres) nor by the two different conditions of curing considered here (with and without evaporation) nor by the adhesion conditions between the base and the overlay. The curve B in figure 5 is true for all these cases and it shows a significant swelling.

This swelling is attributed to the absorption of part of the water of the fresh concret in the eoverlay by the base concrete. This is confirmed by the study of the water absorption by the base concrete. As a prism of concrete identical to those used as bases in the composite specimens is positioned, its sawn face down, in contact with a free surface of water, it absorbs water by capillarity and swells. Its length change (reported by the curve Bw in figure 5) and the mass of water absorbed during the first 24 hours are similar to those measured on the bases of composite specimens.

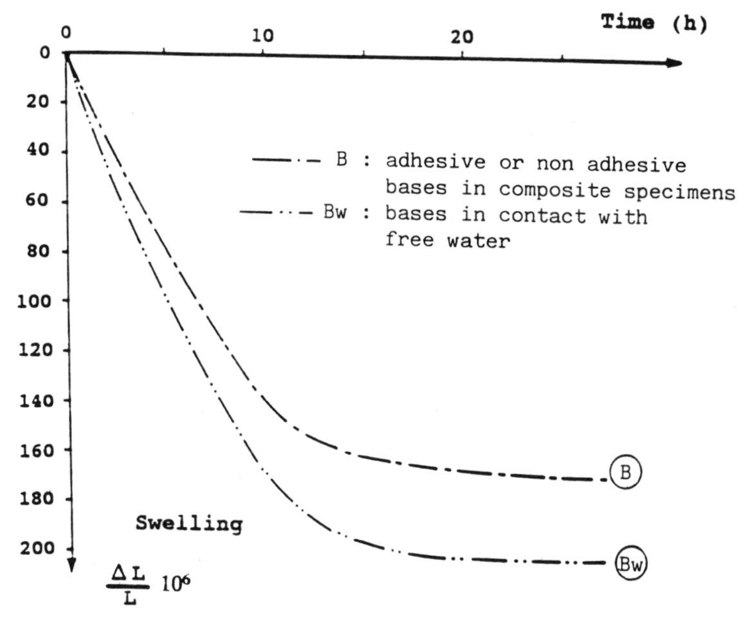

Fig. 5. Bases of old concrete : length variations

4.2.2 Length variations of the repair overlay (figure 5, 6 and 7)

For both curing modes, the curves show an initial phase of fast dilatation, followed by an evolution reminding the one noted with the simple specimens. The addition of fibres to the repair concrete reduces this initial dilatation.

The analysis of the results leads to the following conclusion : the length change of the repair tends to be the algebric sum of the own length change of the overlay and of the swelling of the old concrete base. This trend is modified by the addition of fibres in the overlay concrete and by the strengthening of the overlay.

Specimens with free evaporation

In order to determine the influence of adhesion between the repair overlay and the old concrete base, specimens were prepared in which a porous film (filter paper) had been interposed, prior to casting, between the base and the overlay of fresh concrete. This technique supresses the adhesion but keeps the water exchanges operative.

The obtained curves (curves Cn in figure 6) have the same shape as the ones obtained on simple specimens (Se). But the amplitude of the "primary shrinkage" is emphasised by the suction of part of the water content of the overlay by the base old concrete (Detriché and al, 1981, 1983).

To check to what extent the actual length variation of the overlay is equal to the algebric sum of its own length

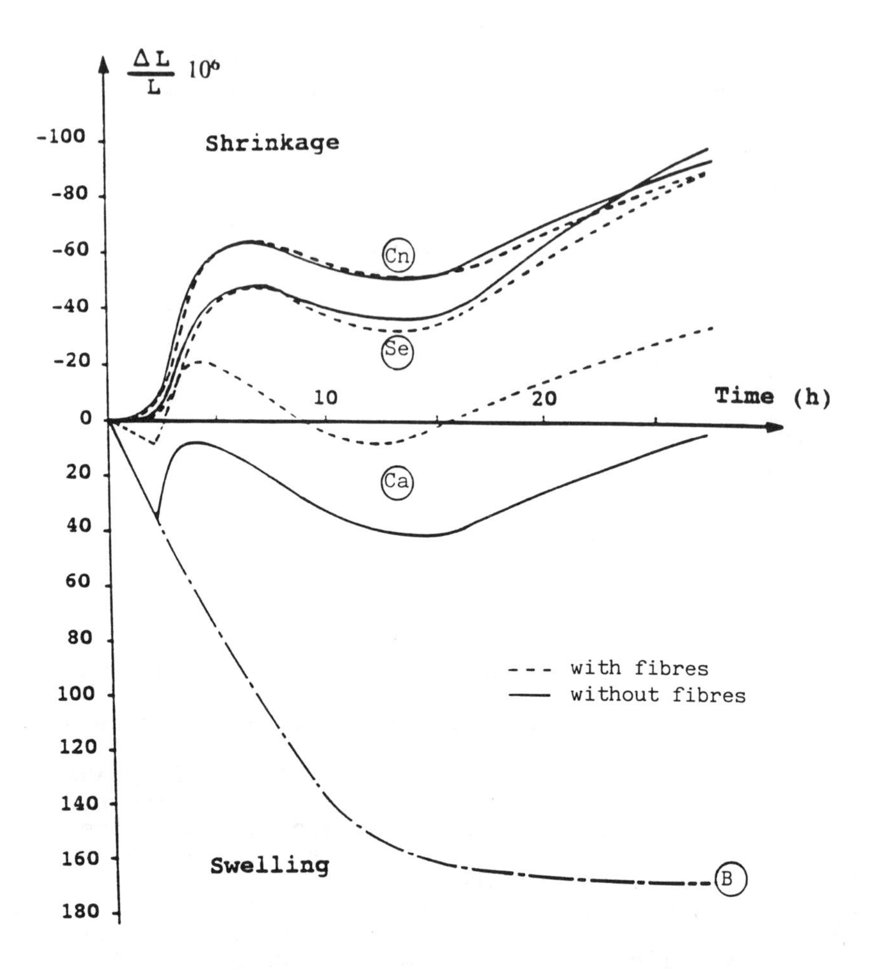

Se : simple specimens (reminded)
Ca : adhesive overlay in composite specimens
Cn : non adhesive overlay in composite specimens
B : bases in composite specimens

Fig.6. Composite specimens : length variations with free evaporation

change and of the swelling of the base, the actual value of this swelling, noted B, was compared with the calculated value Bc = Ca - Cn, representing the substraction : actual length change of the overlay minus the length change of the non adhesive overlay. The results of this comparison are reported in figure 7. During the first 4 hours, the values of Bc and B are in good agreement. Past this time a gap appears between them. It increases with the hardening and the strengthening of the overlay. It is clear that the fast initial dilatation of the fresh overlay is caused by the swelling of the base concrete. The former, which has not yet begun to stiffen nor to shrink, follows the length changes of the latter ; it is stretched by the swelling of the base. This phase ends when the "primary shrinkage" of the overlay concrete begins (but the stretching is still going on).

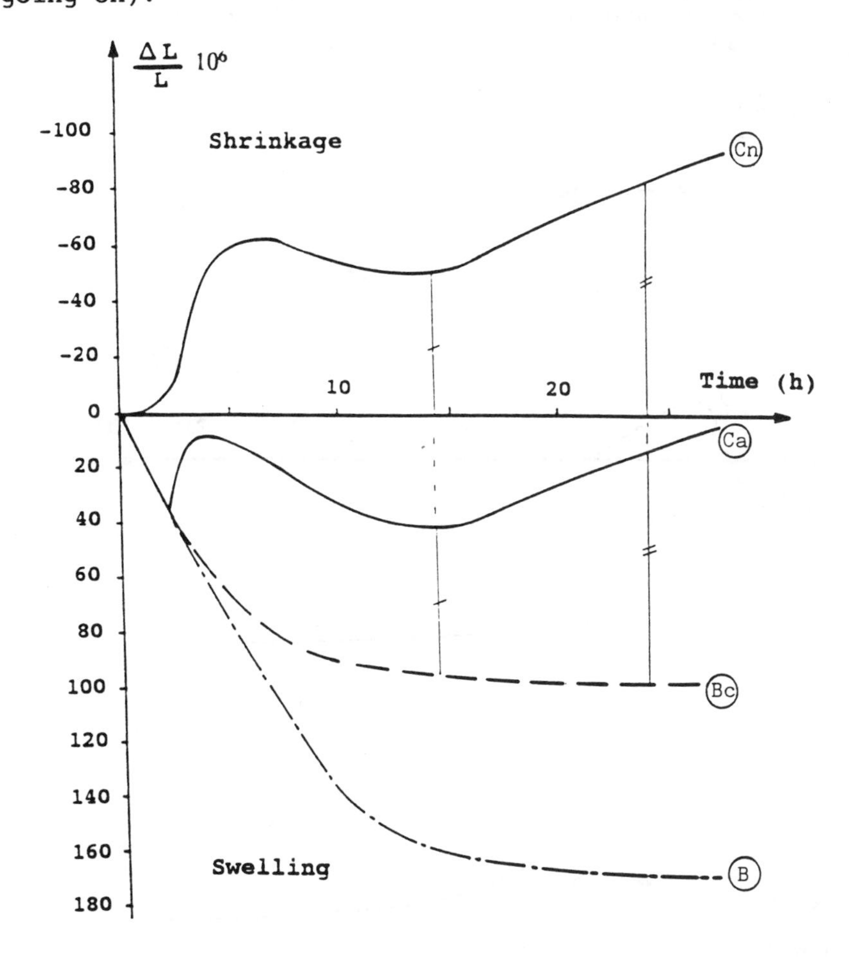

Ca : adhesive overlay in composite specimens
Cn : non adhesive overlay in composite specimens
B : bases in composite specimens
Bc : calculated length variations of the bases

Fig.7. Composite specimens with plain concrete overlay, with free evaporation : comparison between actual and calculated base length variations

The case of fibre reinforced repairs is presented in figure 6 and detailed in figure 8. One finds the same elements as with plain concrete repairs, but, in this case, even in the very first minutes, the fresh overlay never strictly follows the swelling of the base. The fibres provide a small but significant shear strength to the plastic concrete and so limits its stretching by the swelling base.

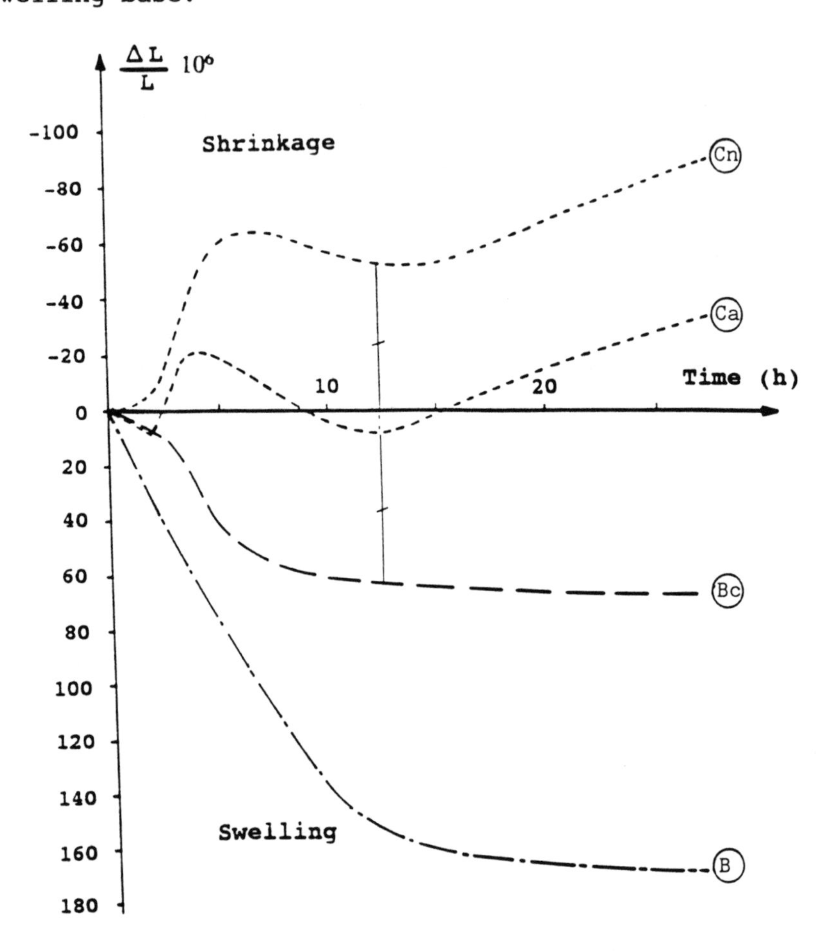

Fig.8. Composite specimens with fibre reinforced concrete overlay, with free evaporation : comparison between actual and calculated base length variations

Specimens without free evaporation (figure 9)

They behave as expected according to the previous findings.

In all cases the addition of fibres to the repair concrete reduces its length increases.

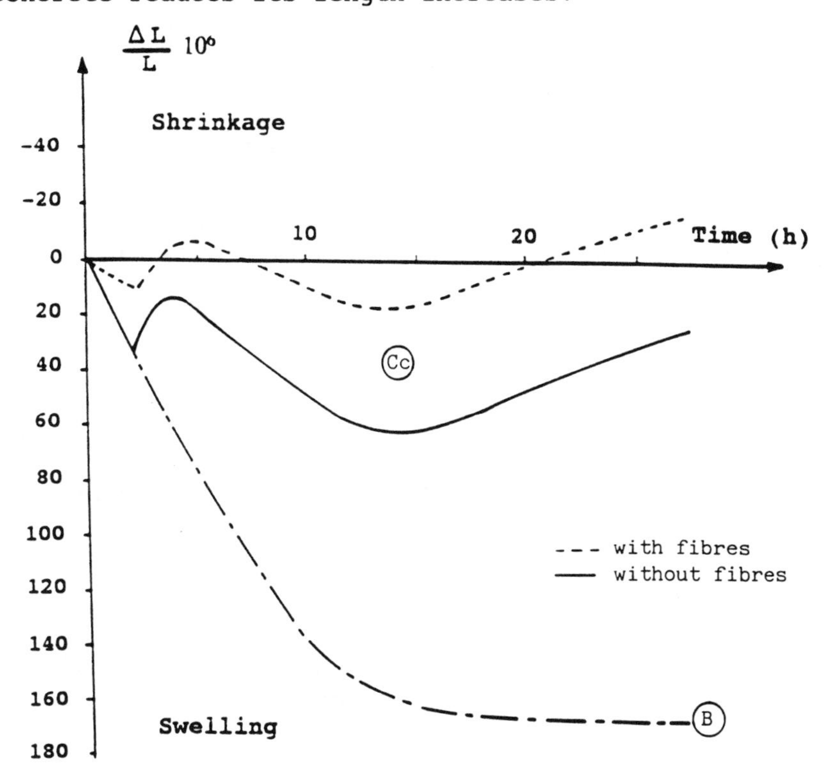

Cc : adhesive overlay in composite specimens
B : bases in composite specimens

Fig.9. Composite specimens : length variations without free evaporation

5 Conclusion

When an overlay of fresh concrete is cast over an old concrete base, this one absorbs part of the water content of the fresh concrete overlay and swells. The fresh overlay being adhesive to the base is stretched by the swelling of the latter. Its length changes are a composition of its own length changes and of those of the swelling old concrete base.

The reinforcement of the repair concrete by metallic glass fibres, in proportion as low as 0.3 % in volume (20 kg/m^3), has a significant effect on the length changes of the fresh concrete overlay.

First, it reduces the initial stretching of the overlay by the swelling of the base concrete.

Second, it reduces the own, autogenous, swelling of the overlay concrete ; the more covered is the overlay (to prevent evaporation), the higher is its autogenous swelling and the greater is the reduction of this swelling by the fibres.

It follows a significant reduction of the straining of the fresh overlay which could have positive effects on the durability of the repair.

6 Acknowledgments

The authors acknowledge the support of Pont à Mousson Research Center (CR PAM), the French National Association of Technical Research (ANRT), the National French Project "Voies Nouvelles du Matériau Béton", and the France-Québec cooperation institutions.

7 References

BARON, J. (1982) Le retrait de la pâte de ciment, chapter 27, in "Le Béton Hydraulique, connaissance et pratique", pp 485-501, **Presses de l'Ecole Nationale des Ponts et Chaussées.**

BARON, J. LESAGE, R. (1969) Etude expérimentale de la mise en place des bétons frais. **Rapport de recherche n° 37 du Laboratoire Central des Ponts et Chaussées**, Paris.

BENSTED, J. (1987) Some applications of conduction calorimetry to cement hydratation. **Advances in Cement Research**, 1, 1, october.

BUIL, M. (1982) Contribution à l'étude du retrait de la pâte de ciment durcissante. **Rapport de recherche n° 92 du Laboratoire Central des Ponts et Chaussées**, Paris.

DE GUILLEBON, B. SOHM, J-M. (1986)Metallic Glass Ribbons : a new Fibre for Concrete Reinforcement. **3rd International Symposium on Developments in Fibre Reinforced Cement and Concrete, Sheffield, U.K.**

DETRICHE, C.H. GRANDET,J. (1981) Influence de la succion des supports poreux sur la prise et la résistance au cisaillement des mortiers moulés à leur contact, **Matériaux et Constructions (RILEM)**, 14, 80.

DETRICHE, C.H. GRANDET,J. MASO, J.C. (1983) Dessiccation des mortiers d'enduit. **Matériaux et Constructions (RILEM)**, 16, 94.

DETRICHE,C.H. (1978) Analyse expérimentale du retrait de couches minces de mortier, mesuré depuis le moulage. **Matériaux et Constructions (RILEM)**, 11, 84, pp 247-259.

GRANJU, J.L. PIGEON, M. GRANDHAIE, F. BANTHIA, N. (1991) Pavement Repairs with Metallic Glass Fibre Reinforced Concrete, Laboratorary and Field Studies of Durability. **ACI International Conference on Evaluation and Rehabilitation of Concrete Structures and Innovations in Design, Hong Kong.**

OLLIVIER, J.P. (1986) A Non Destructive Procedure to Observe the Microcracks of Concrete by Scanning Electronic Microscopy. **Cem. Conc. Res.**, 15, pp 1055-1060.

SCHACHER, B. (1989) Béton renforcé de fibres de fonte FIBRAFLEX pour réseaux visitables", **Techniques Sciences Méthodes**.

54 OBTAINING GENERAL QUALIFICATION APPROVAL IN GERMANY FOR POLYACRYLO-NITRILE-FIBRE CONCRETE

H. HÄHNE
Hoechst AG, Werk Kelheim, Germany
H. TECHEN
König & Heunisch, Beratende Ingenieure, Frankfurt, Germany
J. -D. WÖRNER
Institut fur Massivbau, Technical University Darmstadt, Germany

1 Introduction

The behaviour of polyacrylonitrile-fibre concrete has been explained in several papers, for instance, Refs 1, 6, 11, 16, 19. For its application in Germany, fibre concrete needs either an approval for an individual case (Zustimmung im Einzelfall) or a general qualification approval (allgemeine bauaufsichtliche Zulassung). If the general qualification approval is given, it is possible to use the material without delay for corresponding authorization. The approval for an individual case will be given only for one special application.

Based on numerous experiments and experience with applications of Dolanit acrylonitrile fibre, general qualification approval was proposed for its use as an additional component for concrete. The necessary investigations for this general qualification approval were conducted at the Institut fur Massivbau of the Technical University in Darmstadt.

This paper is focused on the description of the material which was given to the authorities for attaining the general qualification approval rather than a report on newly obtained results.

2 Overview of investigations

To attain the general approval a variety of experiments to determine the behaviour of fresh and hardened concrete as well as the long-term behaviour were carried out. As far as the properties of fresh and hardened concrete are concerned, a parametric study was performed which included more than 200 mixtures. From the results of the parametric study, the following properties could be determined:

Fresh concrete:	workability	Hardened concrete:	density
	air content		compressive strength
	density		tensile strength
			flexural strength
			Young's modulus
			energy dispersion capacity (ductility)

Fibre Reinforced Cement and Concrete. Edited by R. N. Swamy. © 1992 RILEM.
Published by E & FN Spon, 2-6 Boundary Row, London SE1 8HN. ISBN 0 419 18130 X.

In addition, experiments were conducted on the pull out behaviour of single fibres, bond of ordinary reinforcement, shrinkage cracking, carbonation, water penetration, tightness, freeze-thaw/deicing-salt resistance and freeze-thaw durability.

Besides the information about the behaviour of the fibre concrete the material properties of the fibres had to be defined. This includes health aspects and the durability of the fibres in alkaline environments. Long-term investigations of roof tiles made with Dolanit fibre cement were extended by SIC tests such as are usually performed for glass fibres.

The results of the tests are briefly described in the next section without explaining the details of the test methods. An interpretation of the different results is briefly reported in the conclusion.

3 Results of the parametric study

3.1 Air content

The air content increases with increasing fibre amount (Fig. 1). The diameter of the fibres has only a minor influence on the air content (Fig. 2). Figure 3 shows the effect of different fibre lengths on the air content. The influence of cement content and of aggregate size can be seen from Figures 4 and 5.

3.2 Density

Figure 6 shows that the density of fresh concrete decreases with increasing fibre amount. No influence of the fibre length on density could not be found (Fig. 7). Thinner fibres lead to lower densities.

3.3 Compressive strength

In the framework of the parametric study, compressive strength was determined with 150 mm cubes and with cylinders (diameter 15 cm, length 30 cm). Figures 9 - 11 show the cube compressive strength for the varied parameters. Only the fibre diameter has a notable influence.

3.4 Flexural strength

Flexural strength was determined with two types of prism, with dimensions of 4 x 4 x 16 cm and span 10 cm, and 10 x 10 x 50 cm and span 40 cm, respectively. Maximum strengths were obtained for fibre amounts of 2.5 volume % (Fig. 12). Higher fibre amounts result in lower flexural strengths, the ductility however increases further. From Figure 13 it can be stated that an influence of the aggregates on the flexural strength can be observed for coarse aggregates smaller than 8 mm. Cement content (Fig. 14) plays an unimportant role on the flexural strength of fibre concrete. Additional investigation showed that the length of the fibres had no great influence on the flexural strength (Fig. 15). An increase in strength occurred with increasing fibre diameter (Fig. 16).

3.5 Young's modulus

From the comparison of the stress-strain relationship of fibre-free and fibre-reinforced concretes it can be stated that the Young's modulus is not changed. The determination

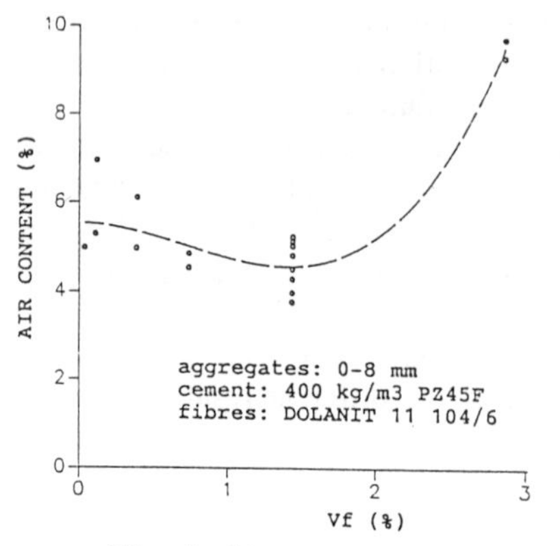

Fig. 1. Air content at different fibre contents.

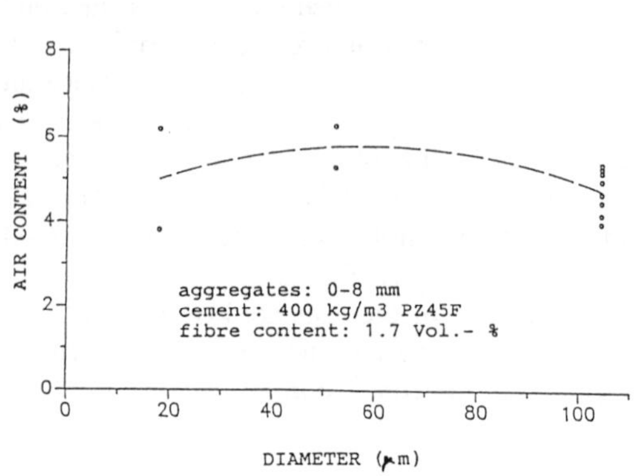

Fig. 2. Air content at different fibre diameters.

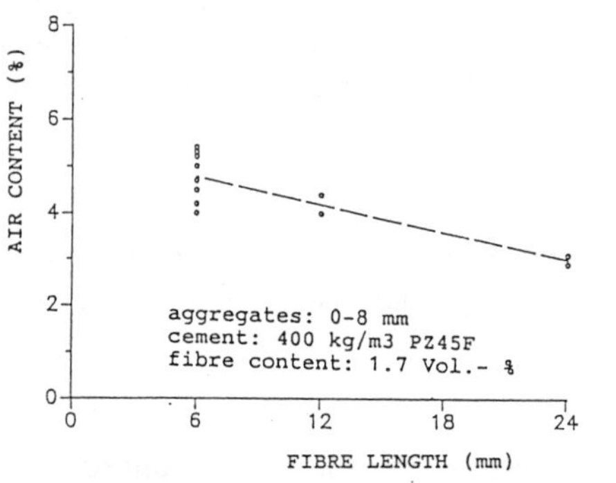

Fig. 3. Air content at different fibre lengths.

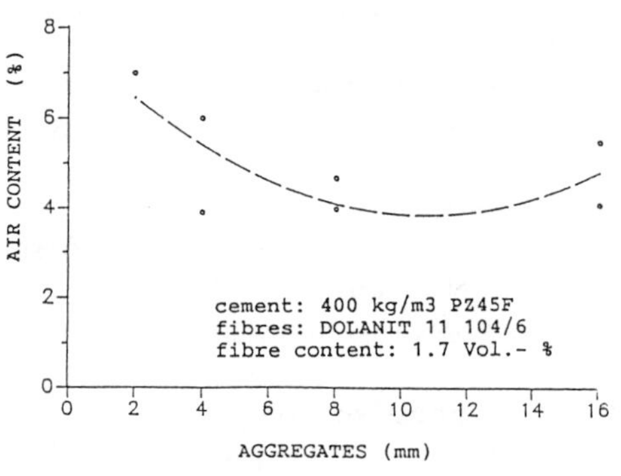

Fig. 4. Air content at different aggregates.

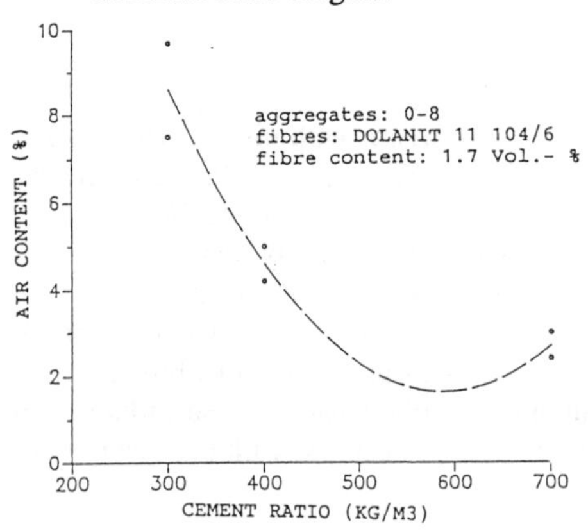

Fig. 5. Density at different cement contents.

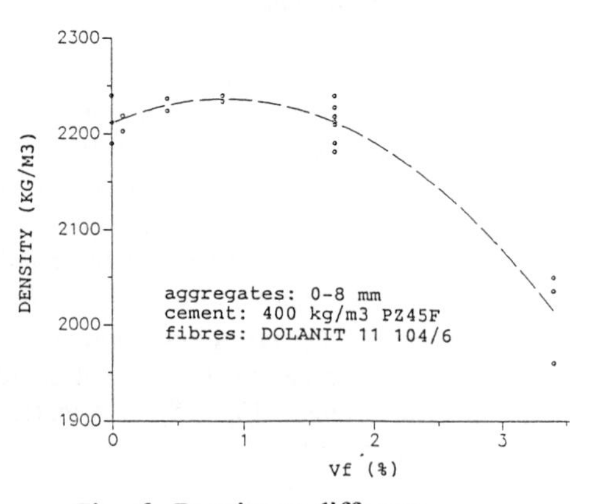

Fig. 6. Density at different fibre contents.

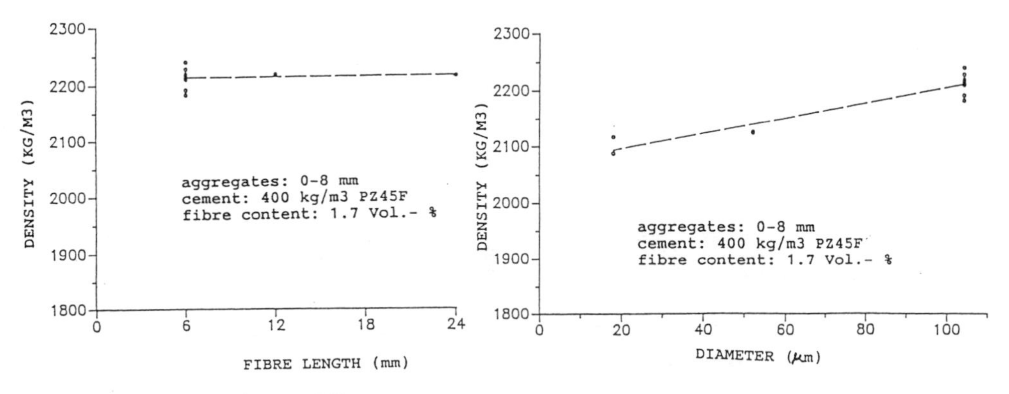

Fig. 7. Density at different fibre lengths.

Fig. 8. Density at different fibre diameters.

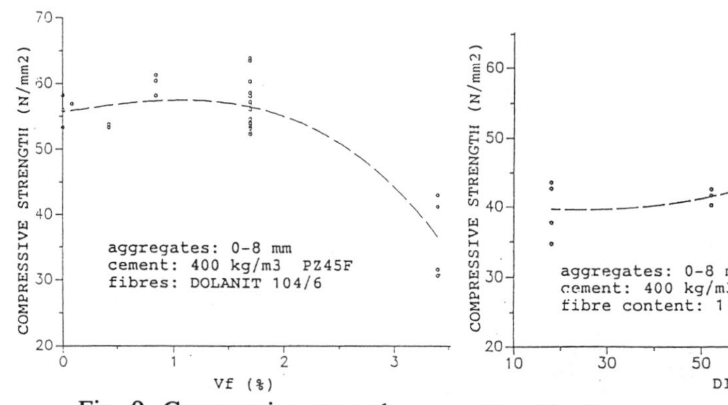

Fig. 9. Compressive strength at different fibre contents.

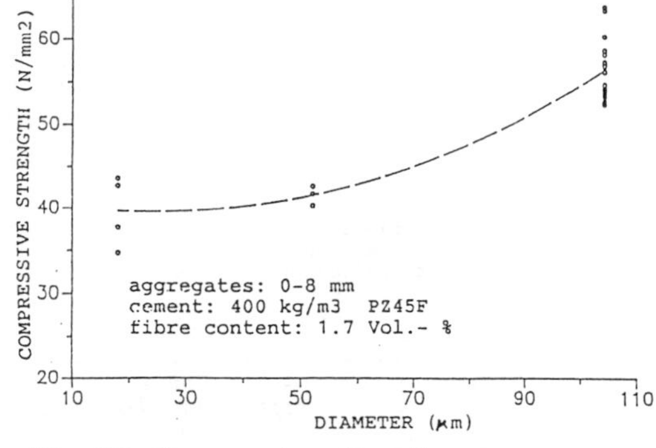

Fig. 10. Compressive strength at different fibre diameters.

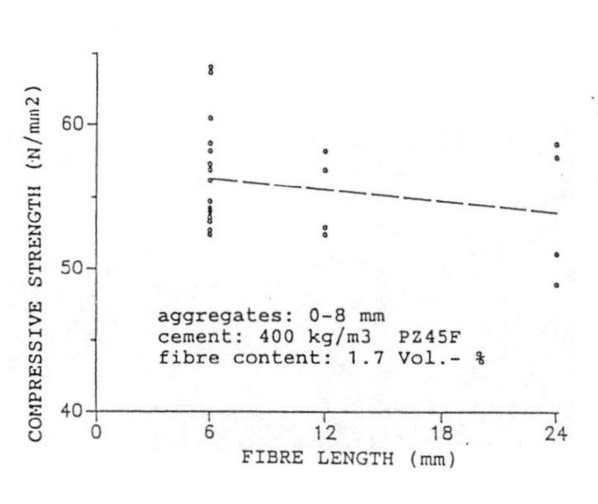

Fig. 11. Compressive strength at different fibre lengths.

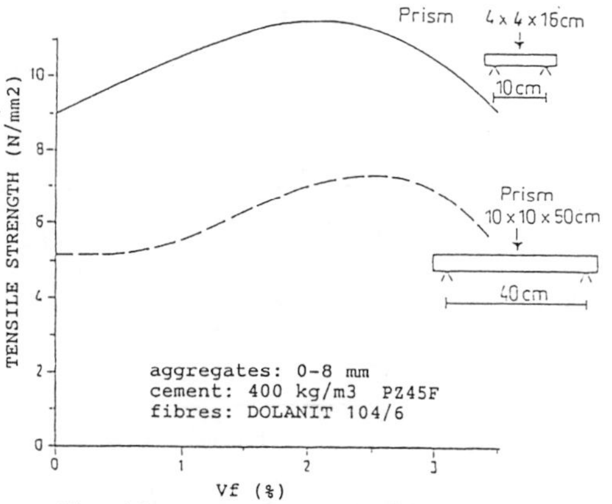

Fig. 12. Compressive strength at different fibre contents.

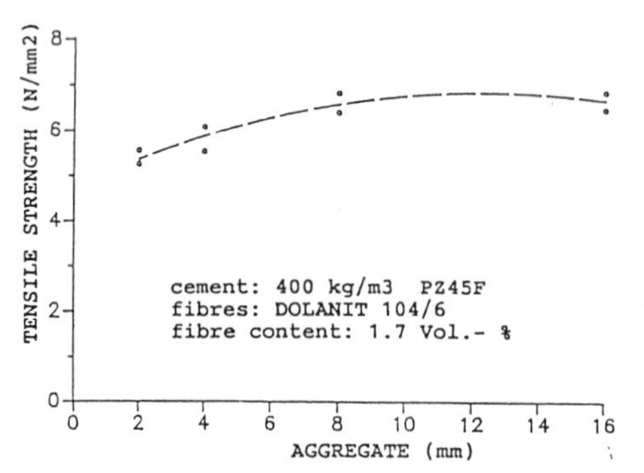

Fig. 13. Tensile strength with different aggregates.

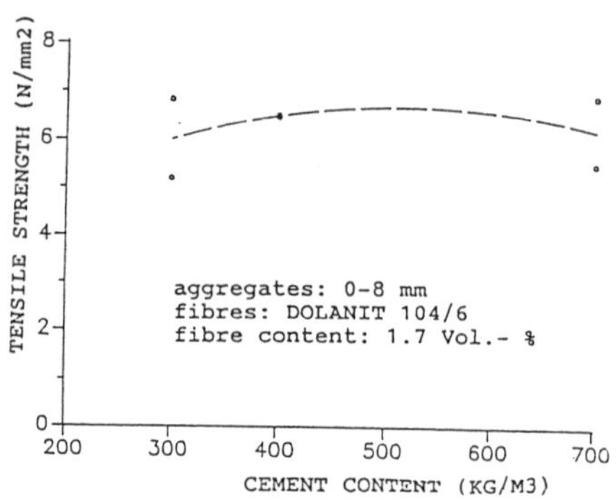

Fig. 14. Tensile strength at different cement contents.

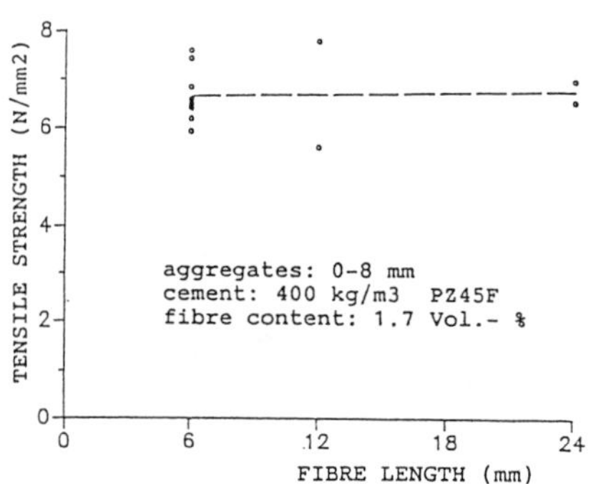

Fig. 15. Tensile strength at different fibre lengths.

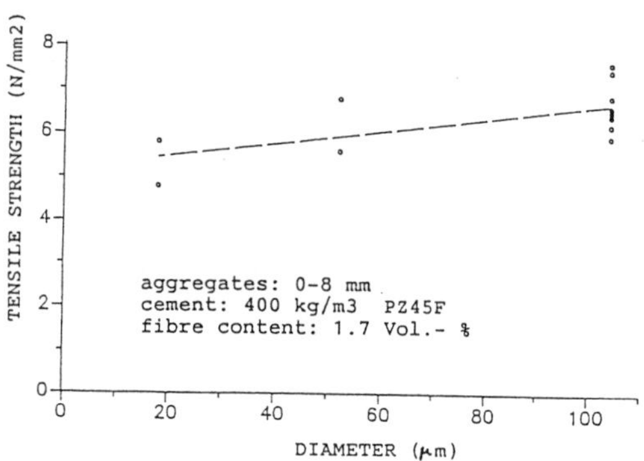

Fig. 16. Tensile strength at different fibre diameters.

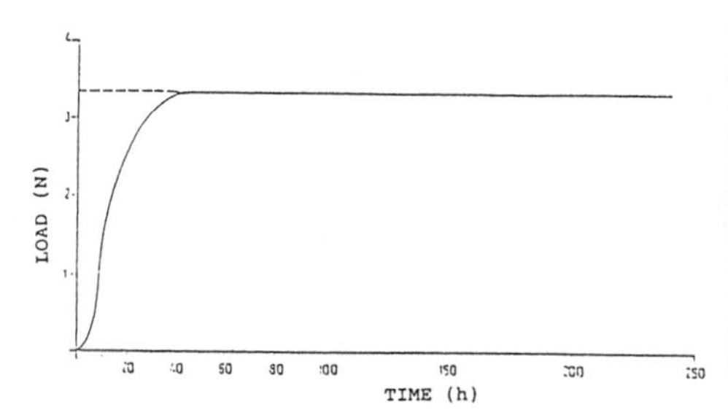

Fig. 17. Pull out behaviour dependant on the time.

Fig. 18. Carbonation behaviour.

of the Young's modulus from the bending experiments showed similar results. Up to a fibre content 2 volume % no significant change was found. Fibre amounts higher than 3.5 volume % lead to a smaller Young's modulus.

3.6 Fibre pull out

The bond behaviour of single fibres was determined by pull out experiments. The results given in Fig. 17 show that the bond is already built up after 2 days. The fibre strength is fully activated.

3.7 Carbonation

Experiments on carbonation took about 1.5 years. From the results a statement about the influence of the fibres on durability can be formulated. Fig. 18 shows that there is no significant difference between the carbonation depth of unreinforced and fibre-reinforced specimens.

3.8 Freeze-thaw/deicing-salt durability

In the framework of these experiments specimens with different fibre amounts and with air entraining agent were tested. The specimens were stored in salty water (3% solution). All specimens were tested in temperature cycles of 15 h at - 18°C and 9 h at +20°C (Fig. 19), which was repeated several times. After 7 cycles the loosened material was weighed. These tests showed that Dolanit has no influence on the freeze-thaw/deicing-salt durability.

3.9 Freeze-thaw durability

In contrast with the tests with deicing salt, no air entraining agent was used. The temperature cycles were similar to those described above. As in the case of the freeze-thaw/deicing-salt durability no significant influence of the fibres could be found.

3.10 Alkaline resistance of the fibres

The alkaline resistance is especially important to ensure the long-term effectiveness of the fibres. In addition to previously reported investigations with fibre cement sheets, SIC tests were performed. Figures 20 and 21 show the preparation of the specimens where the fibres were embedded in cement paste. After hardening, the specimens were stored in 50°C water to accelerate the chemical attack. Ten specimens were taken out at different times and the tensile strength of the fibres was determined. It was found that the tensile strength of Dolanit was reduced by about 8% after 28 days, a value which is within the standard deviation of the tensile strength of new fibres. Photographs with a scanning electron microscope showed no significant change of the fibre surface (Figures 22 and 23).

4 Conclusions

The experimental results, the most important of which are reported here, confirmed the ideas of using polyacrylonitrile fibres in concrete. The parametric study gave information about the influence of different parameters, such as fibre length, fibre quantity, fibre diameter, matrix composition, on the most interesting concrete

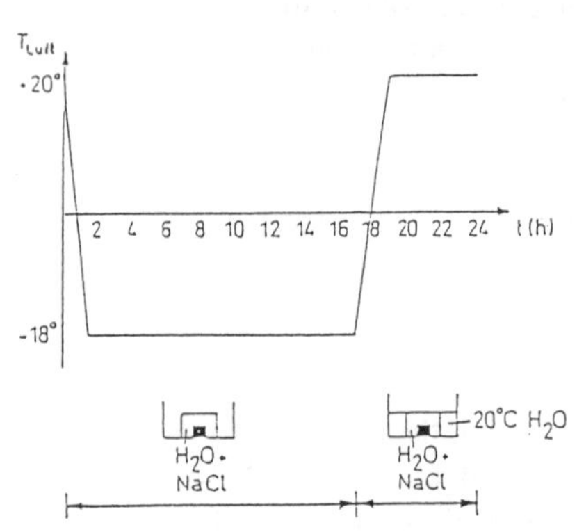

Fig. 19. Freeze-thaw/deicing-salt durability test.

Fig. 20. Fibres coated with an epoxy resin.

Fig. 21. Specimens prepared for the SIC test.

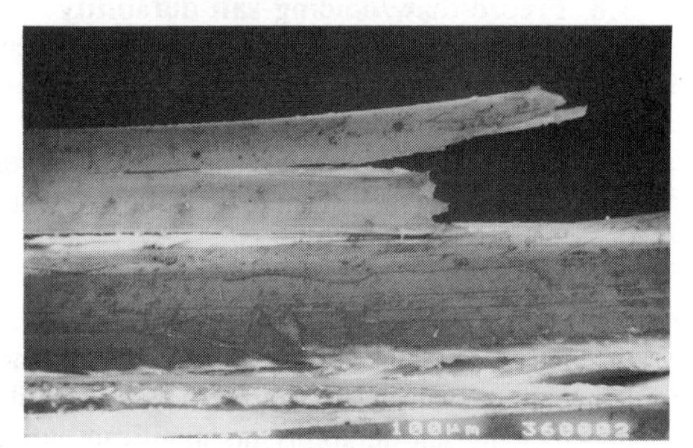

Fig. 22. Fibre surface after 1 day stored in hot water.

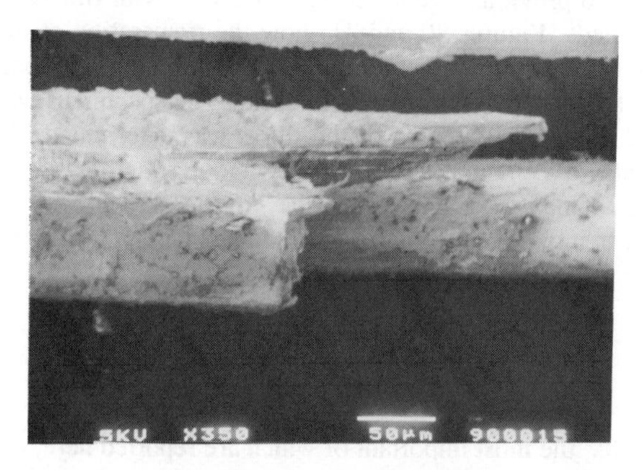

Fig. 23. Fibre surface after 28 days stored in hot water.

characteristics. Minimum effective fibre quantities as well as reasonable maximum quantities could be determined. A large proportion fibres leads for example to a smaller compressive strength due to the higher amount of water necessary for satisfactory workability. Although positive effects of the fibres were shown on flexural strength, durability and shrinkage cracking, the general approval will be limited to the statement that the concrete quality is not influenced negatively. As reported earlier, fibre concrete with polyacrylonitrile fibres has interesting characteristics which can be used in structures with special demands:

Flexural strength
- facade elements
- precast elements

Ductility
- impact resistant structures

Shrinkage cracking
- environmental protective structures
- edge beams of bridges

As soon as the general approval is given, further attempts for the focused use of specific characteristics for special demands will be made.

Meanwhile, a complete work in the field of design of fibre concrete in tension and bending has been formulated which allows the consequent design of structures (20).

References

[1] Hähne, H.: Hochfeste Acrylfasern für zementgebundene Verbundwerkstoffe, Lehrgang Faserbeton, Technische Akademie Esslingen, 1988

[2] Hähne, H.: Hochfeste Acrylfasern für Verbundwerkstoffe, Internationale Chemiefasertagung, Dornbirn 1986

[3] Pott, F.: Staub - Reinhalt. Luft 38 (1978)

[4] Hoechst AG: DOLANIT, Hochfeste Acrylfasern, Typenprogramm, 1988

[5] Hähne, H., S. Karl, J.D. Wörner: Properties of Polyacrylonitrile Fibre Reinforced Concrete, ACI, SP-105, 1987

[6] König, G., H. Hähne, J.D. Wörner: Behaviour and Applications of Polyacrylonitrile-Fibre-Concrete, Rilem, From Materials

Science to Construction Materials Engineering, Vol. 2, 1987

[7] Hähne, H., H. Schinkel, P. Huussen: DOLANIT-Fasermörtel - ein neues Baumaterial, Betonwerk + Fertigteiltechnik, Heft 8, August 1988

[9] Karl, S., J.D. Wörner: Experimental Investigations of Polyacrylonitrile Fibre Reinforced Concrete, Darmstadt Concrete Vol. 2, 1987, Annual Report of Institut für Massivbau Technical University Darmstadt, S. 137-148

[10] Wörner, J.D.: Anwendungsbeispiele für zementgebundene Verbundwerkstoffe mit hochfesten Acrylfasern, Faserbeton-Seminar Technische Akademie Esslingen 1988

[11] Hähne, H., G. König, J.D. Wörner: Behaviour, Design and Application of Polyacrylonitrile Fibre Concrete, Internationale Faserbetonkonferenz Cardiff, 1989

[12] Hähne, H., G. König, J.D. Wörner: Applications and Design of Polyacrylonitrile Fibre Concrete, FRC-Rilem, Cardiff, 1989

[13] Tschiedel, M.: Auszugsverhalten von Dolanit-Fasern, Vertieferarbeit, Darmstadt 1990, unveröffentlicht

[14] König, G., J.D. Wörner, B.Y. Zhu: Neuer Baustoff - Polyacryl (Dolanit) verstärkter Faserbeton, Forschungsbericht, Darmstadt 1990, unveröffentlicht

[15] Wörner, J.D.: Dauerhaftigkeit von Dolanit in Faserzementprodukten, Studie Frankfurt 1987, unveröffentlicht

[16] Zhu, B.: Verhalten von Faserbeton mit synthetischen organischen Fasern, Darmstädter Massivbau-Seminar, Bd. 3, Darmstadt 1990

[17] Tsukamoto, M.: Spezielle Eigenschaften von Faserbeton, Darmstädter Massivbau-Seminar, Bd. 3, Darmstadt 1990

[18] Unterlagen zum Antrag auf bauaufsichtliche Zulassung, Darmstadt 1990, unveröffentlicht

[19] Hähne, H., H. Techen, J.D. Wörner: Eigenschaften von Polyacrylnitrilfaserverstärktem Beton, Beton- und Stahlbeton, 1992

[20] Müller, M.: Ein Berechnungsverfahren für Faserbeton für Biegung und Normalkraft, Dissertation, Darmstadt 1992

55 DEVELOPMENT AND APPLICATION OF A GRC LIGHTWEIGHT PRESTRESSED SLAB FOR NETWORKED ROOFS

WEI JIN
China Building Materials Academy, Beijing, China

Abstract
The purpose of this paper is to discuss research and application of a new load–carrying roof slab – GRC slab for net–worked roof which combine GRC material with prestressed steel reinforced concrete. The conventional GRC material can't be used as load carrying member, so we carried our technical method on the basis of reinforcing low alkalinity cement with alkali–resistant glass fiber to improve the durability of GRC considerably and using prestressed steel reinforced concrete as rib for GRC slab to decrease volume around the rib and deadweight of slab. Comparing with the existing roof slab produced by conventional material, the GRC slab are excellent not only in physio–mechanical properties but also in technical economy. Furthermore, the paper have introduced the current applications of this slab in various large constructive engineerings.
Keywords: GRC (glass fiber reinforced cement); Net–worked roof; Slab; Prestressed Steel Reinforced Concrete.

1 Introduction

In recent years, the application area of glassfiber reinforced cement (GRC) have been enlarged continuously. But there are some existing problems when using alkali–resistant glass fiber to reinforcing ordinary cement in other countries, their GRC products are limited in non–loading structures. We carried our technical method on the basis of reinforcing low–alkalinity cement with alkali–resistant glass fiber in our country. The durability of GRC are considerably improved. In 1988, we began our research on load carrying member–GRC slab for net–worked roof according to the policy of improving material performances and ensure the durability. There are three kinds of roof slab for net–worked roof in China. A, metal–molded slab: These slabs have low deadweight and are convenient for transportation. But they have been used in small number of constructions because of their poor properties on warm–insulating and high cost in anticorrosion. B, steel reinforced concrete and steel mesh cement slab for net–worked roof. This slab has high volume density,

Fibre Reinforced Cement and Concrete. Edited by R. N. Swamy. © 1992 RILEM.
Published by E & FN Spon, 2-6 Boundary Row, London SE1 8HN. ISBN 0 419 18130 X.

low strength and poor crack resistance. These made them difficult for transportation. In addition to mentioned above, these slabs have long manufacturing period and complicated technical process. C, straw slab. These slabs have low deadweight and good thermal performance, but their strength are low. They need a great deal of steel purlins to bear load. Furthermore, water resistance of these slabs are poor, so they are not suitable for moist area. In view of these above, we began our research about net–worked roof slab with the size of 3M × 3M. The research results show that lightweight prestressed slab for net–worked roof have excellent physio–mechanical properties and good social economic benefits.

2 Research Process

2.1 Raw materials and mixing proportion

2.1.1 Cement

Use 525 strength grade* of sulphoaluminate low–alkalinity early strength cement. The physio–mechanical properties are in Table 1.

Table 1. Physio–mechanical Properties of Low–alkalinity Cement

Specific area (cm^2/g)	Setting time		1:3 Mortar tensile strength (MPa)			1:3 Mortar compressive strength (MPa)		
	Inital	Final	1 D	7 D	28 D	1 D	7 D	28 D
5011	0:50	3:30	2.5	2.8	3.2	39.6	48.8	60.3

2.1.2 Fiber

Alkali–resistant glassfiber roving. Its physio–mechanical properties are in Table 2.

2.1.3 Sand

Ordinary river sand, fineness modulus: 1.92~2.38, involved mud: <2%

2.1.4 Gravel

5~15mm pebble, involved mud: < 1%

---

*This is the cement grade determining by China Standarded Testing Method. 525 strength grade of cement means that the 28–days compressive strength of cement sample is 525 kg/cm^2 (51.5 MPa) and the 3–day, 7–day compressive strength and 3–day, 7–day, 28–day bending strength must meet the corresponding requirements of China Standard.

Table 2. Physio–mechanical Properties of Alkali–resistant Glassfiber

Diameter (μ)	Density (g / cm^2)	Elastic modulus MPa × 10^4	Tensile strength MPa	Poission ratio	Specific elongation of tensile limit (%)
13–14	2.78	6.26–7	1700	0.2	4

2.1.5 Admixture

Add in proper amount of organic admixture, cement / sand ratio: 1:0.7~ 0.8, water / cement ratio: 0.38~ 0.42, added fiber amount: 4~ 5% (weight percent to cement).

3. Researching and Manufacturing Process

3.1 Research process

There are not any integrate design methods and information about using GRC as load carrying members inside and outside our country. We carry our research to solve following problems. A, First, we research on physio–mechanical properties of load carrying member when loaded. Then according to the results from mechanical test and the half–life of GRC strength, select the rational thickness of slab surface. The physio–mechanical properties of GRC are in Table 3.

Table 3. Physio–mechanical Properties of GRC

Compressive strength (N / mm^2)	Vertical to fiber layer Parallel to fiber layer	76.0 51.0
Bending strength	Limit of proportionatity (N / mm^2) Damage (N / mm^2) Elastic modulus ($N / mm^2 \times 10^4$) Limit of deformation ($\mu\varepsilon$)	6.2 20.9 2.4 13900
Tensile strength	Limit of proportionality (N / mm^2) Damage (N / mm^2) Elastic modlus ($N / mm^2 \times 10^4$) Limit of deformation ($\mu\varepsilon$)	3.1 7.7 2.2 6770 .
Impact strength ($N \cdot mm / mm^2$)		22
Volume density (g / cm^3)		1.88
Air–dried density (g / cm^3)		1.95
Thermal coefficient (kcal / mh℃)		0.59
Molding process		Spraying method

B, Research structural design of net–worked roof slab. We use GRC slab and rib of prestressed steel reinforced concrete to design our structure. They are moulded in one time. Our design absorbs the advantages in both GRC and prestressed steel reinforced concrete. The designed member has high crack–resistance. Using prestressed steel reinforced concrete as rib for GRC slab can adopt high grade concrete to decrease volume around rib, furthermore, to decrease slab deadweight. When using this GRC slab as surface material, the slab thickness are considerably decreased, which can simplify moulding process and at the same time, this slab are excellent in both crack resistance (impact strength: 15∼20 kg–cm / cm^2) and strength. We have carried our research work after solving the two problems above. The technical process is as follows:

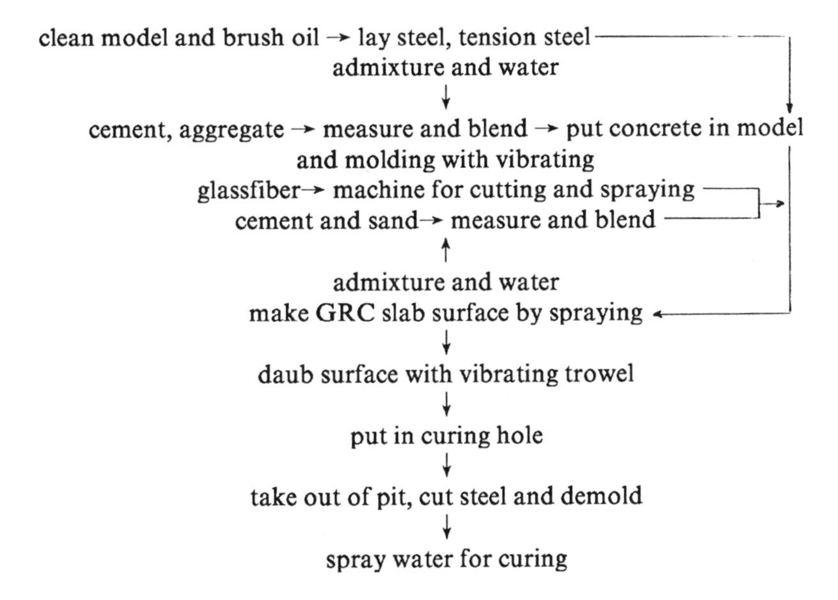

4. Properties of GRC Lightweight Prestressed Slab for Net Worked Roof

After finishing our research work, we entrusted Quality Testing Department, Xuzhou Standardization Bureau to test the properties of the slab by four–point bending method. The results are in Table 4. All the performance of this GRC net–wroked roof slab can match or superior to that of prestressed steel reinforced slab.

5. Application Results

This GRC lightweight prestressed slab is a kind of lightweight roof slab as a acces-

Table 4. Report of Structure Properties for Precast Member

Name and Type: GRC slab for net-worked roof, Manufacturing Date: Oct. 24, 1988 Test Date: Nov. 14, 1988

Item	Dimensions (mm)	Thickness of provitive layer (mm)	Number and type of main bar	Concrete grade	standard load (N / cm^2)	Testing index			
						Strength K value	Deflection (mm)	Crack resistance	Length of crack (mm)
Deisgn	2970 × 2970	15	2ΦL12	300#	2150	1.875	11.52	0.7	0.2
Meas-ure	2975 × 2972	15	2ΦL12	297#	according to 2150	2.01	Main rib: 3.25; auxiliary rib: 4.12; center rib: 6.15; Middle of diagonal: 8.43	1.49	

Diagram of loading, meter location and number

Crack information and damage feature: When loading to $3200 N / m^2$, there are micro crackers observated in auxiliary rib direction with max. cracking length of 0.02mm. To $4315 N / m^2$, there are 15 to 20 crackers respectively occuring on the sides of two auxiliary ribs with max. cracking length of 0.1 mm. These cracks uniformly disperse from the middle to the ends of span. When finishing test, the max. deflection of main bar and auxiliary bar are 1 cm and 1.4 cm respecitvely at middle point of span.

Conclusion	$K^s > \beta K$ $\quad f^s < [fd]$ $\quad Kf^s > [Kf]$ $\quad \delta af^s \max < [\delta df^s \max]$

sary for net–worked roof, one net with one slab. The preset iron elements on four corners of slab can be directly welded to the bearing load on the joints on upper chord of net frame: These slabs have low deadweight and are excellent in impact resistance and strength, which make them easy to be hoisted and transported. They have been widely used in Jiangsu, Beijing, Shandong, Liaoning and Jilin etc. provinces for about 30,000 M^2 in recent years, in which there are civil constructions, and industrial constructions, and have got good technical effects and economic benefits.

Fig.1 is the slab structure. Fig.2 to 10 are some of practical applications for this kind of GRC net–worked roof slab. GRC net–worked roof slab has low deadweight and high strength. Fig.2 is the loading test. The size of GRC net–worked roof slab is $3 \times 3m^2$. 1300 red bricks, total weight of more than 3 tons, are loaded on four–point bending slab. Safe coefficient $k > 2$; crack resistance > 1.15, depth of cracker < 0.1 mm. The impact strength of GRC net–worked roof slab is 15 times higher than that of ordinary concrete slab. Transportation of containerization can decrease the damaged rate to lowest. Fig.3 is the unloading site after 1500 KM transportation. GRC net–worked roof slab is very conventient to be hoisted and not necessary to use large hoisting equipment. As shown in Fig.4, this 4m × 4m / 2 slab can be hoisted only by ordianry headframe. Fig.5 shows the positioning of 3m × 3m GRC net–wroked roof slab. By four–angle three–point welding operation and one net frame with one slab, only five minutes are enough for six people to set one slab, so working time can be saved 1 / 3 and hoisting period are also shortened.Fig.6 is the construction of test hall of Nanjing Hydraulic Academy. The surface of GRC net–worked roof slab are moulded by high pressure spraying. The surface is close and smooth. The levelling blanket is only 15 mm thick. Operation can adopt normal method.GRC net–worked roof slab are worldwide advanced in durability. They have the advantages of low deadweight, high loading bearing capability, high strength, self–waterproof and fireproof etc. Their price is lower than that of other net–worked roof slabs. Using this slab, the roof structure can save steel 3 to 5 kg / m^2. Because the longer the span is, the higher the economic benefits are, they have got excellent reputation and active practices from many design and construction departments. Fig.7 is a ethene construction of 16,000 square meters in Fushun City.Fig.8 is the construction of Palace of Peace and Happiness, Beijing Friendship Hotel. This construction use more than 81 specifications of slabs, fully displaying their special performances. The picture in Fig.9 is the net–shell–like roof slab of Beijing East District Children's Palace. It has the advantages of self–waterproof, low price, good durability and easy treatment on roof construction. The net–shell–like roof slab is a kind of GRC roof slab especially assembled with the framework of net ball and curved surface. They can be shaped as triangular pyramid, rectangular pyramid and sphere etc. with beautiful appearance and strong solid sense. It is an ideal roof material for curved surface net–frame. GRC net–worked roof slab not only can be applied to industrial building, sports facilities, storehouse, hall for airport or railway station and exhibition hall but also can be used for old building reformation, because these slabs can satisfy the requirements on design and can provide the customers with the slabs of various specifications. Fig.10 is a reforming construction for the old building of Hubei Huanshi Cotton Mill.

Compared with steel reinforced concrete slab for net–worked roof, GRC lightweigth prestressed slab is 60~ 65 kg / m^2 lighter for each deadweight and 15 times higher in impact strength which ensure their low damage percentage when hoisting and transporting. This light GRC slabs do not need large lifting installation, so constructing period can be shortened 40%. In addition to above, GRC slab for net–worked roof can save steel 3 kg / m^2 compared with reinforced concrete slab. And the steel amount for net frame can be saved 2 kg / m^2 when the frame is designed in term of GRC lightweight prestress slab. Now GRC net–worked roof slab are used about 500000 M^2 per year in China. According to this used amount, the steel bar in slab and steel for frame can be saved 3000 tons and 2000 tons respectively per year. To sum up, the cost for total construction can be decr / eased. There are not any problems occuring in construction in recent years, so this kind of roof slab are admired by many disigning institutes and using companies. GRC lightweight prestressed slab for net–worked roof can't meet market demand now and will has a vast new world in developing and application area.

6. Discussion

This structural design of combining GRC with prestressed reinforced concrete has its special advantages which exist in both GRC and prestressed reinforced concrete. Part of the load on slab surface are carried by GRC and transfer to prestressed reinforced rib, then reach to net frame. This structural design has been patented in China. We think that this structure have its advantages not only in net–worked roof but also in large roof slab. And because GRC are spraying–molded, they can be made to net–worked roof slab with insulating structure for cold area (set insulating material on the middle of two layers of GRC, finish by one time and do not need any other construction for warm–keeping), we still have a lot of work to do on this research area. But no doubt, the construction field will benefit from this structure.

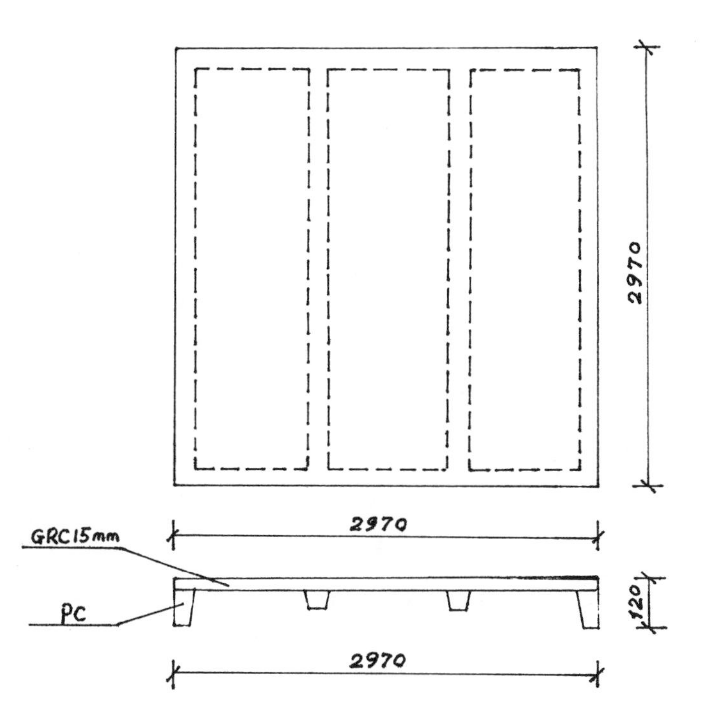

Fig.1 Slab structure

Fig.2 Loading test of GRC net–worked slab

Fig.3 The unloading site after 1500 KM transportation

Fig.4 Hoisting site of GRC net–worked roof slab

Fig.5 The positioning of 3m × 3m GRC net–worked roof slab

Fig.6 The construction of test hall of Nanjing Hydraulic Academy

Fig.7 A ethene construction of using GRC net–worked roof slab in Fushun

Fig.8 The construction of Palace and Happiness, Beijing Friendship Hotel

Fig.9 The net–shell–like roof slab of Beijing East District Children's Palace

Fig.10 A reforming construction for old building

References

Bill Blaha. (1985) Ramade Renaissance Hotel Sets Records for GFRC Usage, Concrete Products, January, pp.18–20.
Design Guide: Glassfibre Reinforced Cement (1984), Pilkington Brother Ltd. U.K.
Housing Project Sprouts up in Trinidad, Concrete International, March, (1985) pp.45–46.
Jin Wei (1989) Effect of Zeolite on Durability of GRC, Proceeding of the International Congress of Recent Developmen of Fiber Reinforced Cements and Increted, Sep., pp.121–125. U.K.
Mircea Halmagin (1981) The Development of GRC in Romania, Proceedings of the Internaional Congress on Glass Fiber Reinforced Cement, France.
M.W. Fordyce (1983) GRC and Buildings, Butterworth Co. (Publishers) Ltd.

56 FIBRE REINFORCED CONCRETE CONTAINERS: FROM CONCEPT TO MANUFACTURING

R. PECH
SOGEFIBRE, Saint-Quentin-en-Yvelines, France
B. SCHACHER
SEVA, Chalon-sur-Saône, France

Introduction
Radioactive waste conditioning is a major step in the process implemented in nuclear installations. The objective is to contain the radioactive material in nuclear waste as satisfactorily as possible for man and the environment; containment integrity has to be guaranteed over very long periods.
Medium-level (ML) and even very low-level (LL) waste is no exception to this rule. Cogema conducted research for many years and developed a new process to condition nuclear waste in containers reinforced with metal fibres, called fibre concrete containers.
This process has been welcomed by Andra, the French radioactive waste management agency and, consequently, the French Safety Authorities, as the most satisfactory up to now. Sogefibre, a subsidiary of SGN and Compagnie Générale des Eaux, has been manufacturing FRC containers in industrial quantities since July 1990.

1 Fibre reinforced concrete concept

1.1 Cogema need
Low-level and medium-level radioactive waste management in France requires an "Institutional Control Period" of the disposal facilities of 300 years. For the waste producer, two ways exist after having immobilized the waste:
- either to use a non-durable container and consequently, to improve storage conditions and/or waste immobilization material,
- or to use a durable container able by itself to guarantee a durable integrity; requirements on storage conditions and waste immobilization material become less severe.

Of course the specification reference [1] issued by Andra is more severe for durable container than non-durable container. Andra specification for durable container specifies several mechanical, physical and containment

Fibre Reinforced Cement and Concrete. Edited by R. N. Swamy. © 1992 RILEM.
Published by E & FN Spon, 2-6 Boundary Row, London SE1 8HN. ISBN 0 419 18130 X.

characteristics to be reached on the material and the container itself (see table 1).

Table 1. Andra specification content

TEST	NATURE OF SPECIMEN	STANDARD OR SPECIFICATION	ANDRA CRITERION
Compression	Material after 28 days	NFP 18.406	≥ 50 MPa
Splitting	Material after 28 days	NFP 18.408	≥ 4.5 MPa
Shrinkage	Material after 28 days	NFP 15.433	≤ 300 μm/m
Mass loss	Material after 28 days	NFP 15.433	≤ 35 kg/m^3
Permeability to nitrogen	Material after 28 days	ANDRA 322 ET 09.04	≤ 5.10^{-18} m^2
Resistance to gamma irradiation	Material after 28 days	ANDRA 330 ET 09.09	Strength variation: ≤ 20%
Diffusion of radionuclides in an aqueous medium	Material after 28 days	ANDRA 330 ET 09.07	Tritiated water: < 1.5 10^{-3} cm^2/d Caesium: < 10^{-3} cm^2/d
Permeability to water	Material after 28 days	ANDRA 322 ET 09.02	-
Density	Material after 28 days	ANDRA 330 ET 09.10	-
Porosity	Material after 28 days	ANDRA 330 ET 09.11	-
Water tightness	Container body	-	No trace of oozing
Load behaviour	Finished package	ANDRA 330 ET 09.12	Deformation ≤ 3%
Drop test	Finished package	ANDRA 330 ET 09.13	Containment not altered
Homogeneity of the encapsulation barrier	Finished package	ANDRA 330 ET 09.02	Heterogeneity < 20%
Immersion test	Finished package	ANDRA 330 ET 09.17	Containment not altered
Resistance to thermal cycling	Finished package	ANDRA 330 ET 09.04	Strength variation: ≤ 20 %

In addition to these tests, a durability assessment over 300 years must be performed to get the label "high performance".

Cogema, one of the main radioactive waste producers in France, decided in 1985 to condition parts of its waste in high performance containers. As, at that time, no existing container could be qualified as "high performance", a research and development programme was undertaken to find a concrete formulation compatible with the following main requirements:

- good mechanical strength;
- good resistance of microcracking allowing a good ability in containing the radioactivity;
- a substantial life (at least 300 years).

1.2 FRC research and development programme

1.2.1 Introduction
Many researchers from various countries have been performing works on mechanical characteristics of fibre reinforced concrete from several years. A considerable amount of data have been published. The purpose of this paper is not to re-issue the results of these research works, but, among their conclusions, there was a general agreement on the following FRC advantages compared to armoured or usual concrete:

- the matrix is homogeneously reinforced;
- the material is less fragile and more ductile;
- the development of microcracks during concrete life time is stopped inducing higher containment properties.

To define the FRC for Cogema purpose, two ways have been investigated simultaneously:

- selection of metallic fibre to be incorporated to the concrete;
- improvement of the concrete formulation.

1.2.2 Metallic fibre selection
The selected fibre called Fibraflex, developed and produced by Seva, a company of the Saint-Gobain group, has the geometrical and mechanical characteristics presented in table 2.

Table 2. Fibraflex features

Size	Length around 30 mm Width : 1 to 2 mm Thickness : 30 µm
Specific surface	10 m^2/kg (1 m^2/kg for steel fibre of 0.5 mm diameter)
Tensile strength	$\simeq$ 2000 Mpa
Modulus of elasticity	$\simeq$ 140,000 Mpa

It is produced on an industrial scale by quenching a liquid metal jet on a cooled wheel rotating at high speed. The high quenching rate, of about one million degrees Celcius per second, solidifies the liquid metal in an amorphous state (a non-cristalline state) which gives Fibraflex high flexibility (giving ease of use) and strong mechanical features.

Comparative studies have shown how the ribbon-shape of these fibres improves their mechanical effectiveness, compared to other steel fibres such as cylindrical shape or hooked ends fibres, for a given fibre length and content:

- adding to a mortar up to 2% enlarged or hooked ends fibres creates a reinforcing effect equivalent to the one obtained with only 0.6% of Fibraflex [2];
- a higher residual strength after cracking is obtained with a Fibraflex fibre content 2 times lower than the cylindrical steel fibre content (0.5 mm diameter; 30 mm length) for crack opening below 0.3 mm which prevents cracks growing during concrete life time [3].

Moreover, with the presence of chromium in the alloy, which contributes to the spontaneous creation of a very efficient passivation film, the amorphous state induces a very high resistance to corrosion. This resistance has been checked by tests performed in a liquid medium much more aggressive than the concrete interstitial liquid and compared to conventional carbon steel used for armoured concrete. Results are presented in table 3.

Table 3. Resistance to corrosion of fibre (Fibraflex)

	TEST DURATION	CORROSION SPEED μm/year	CURRENT OF CORROSION 10^{-6} A/cm²	POTENTIAL CORROSION mV/ecs
Carbon steel rebars	240 h	413 ± 15	36	- 386
Fibraflex	720 h	13 ± 6	1.1	- 40

(Solution of pH = 11.5 + chlorides and sulfates; temperature 50°C)

Fibraflex is shown in figure 1.

Fig. 1. Fibraflex

1.2.3 Concrete formulation
The incorporation of fibre into the concrete matrix has required an adaptation of the formulation. The purpose in this adaptation was to improve the packing efficiency and the workability of the fresh concrete. It has mainly consisted in increasing the sand/gravel ratio and the cement paste content.

To improve the containment properties, a very low granulometric sand has been incorporated to fill the interstices between the cement paste and the usual sand.

1.3 FRC characterization programme
To demonstrate the compliance of the FRC with Andra specification [1], a characterization programme has been carried out. The main results are given in tables 4 to 6.

Table 4. FRC material: mechanical and physical properties (after 28 days)

	INDUSTRIAL RESULTS (average after one year operation)	ANDRA CRITERION
- Specific density	2.4	
- Compressive strength	60 to 70 MPa	> 50 MPa
- Tensile strength	4.5 to 5.5 MPa	> 4.5 MPa
- Shrinkage	about 290 μm/m	< 300 μm/m
- Weight loss	0.7%	

Table 5. FRC material: containment properties after 28 days

	TEST CONDITIONS	TEST RESULTS	ANDRA CRITERION
Effective diffusion factor	$\simeq$ 1 year		
. Tritiated water	2 cm pellet	< 7.1 10^{-5} cm²/d	< 1.5 10^{-3} cm²/d
. Caesium	1 cm pellet	< 2.4 10^{-5} cm²/d	< 10^{-3} cm²/d
Water permeability	-	< 1.5 10^{-20} m²	-
Nitrogen permeability	-	3 10^{-20} m²	< 5 10^{-18} m²

Table 6. FRC material: irradiation strength

- Dose integrated	10^6 Gy
- Dose rate	1.2 to 1.4 10^3 Gy h^{-1}
- Sample weight	5 kg
<u>RESULTS</u>:	
- Volume of H_2 generated	0.220 l
- Radiolytic return ratio G (H_2)	0.017
- Volume of 0_2 consumed	1.14 l
- Radiolytic return ratio G (0_2)	- 0.09
- No cracks	

On the basis of the results of the characterization programme, Andra gave the approval to two types of container:

- a cylindrical container (CBF-C1) in January 1991;
- a cubical container (CBF-K) in July 1991.

These containers are described in the paragraph 2.1. Sogefibre FRC is prepared according to the French patents 88.16337 (December 12, 1988) and 89.08050 (June 16, 1989).

The typical fibre distribution inside the concrete is shown in figure 2.

Fig. 2. FRC cross-section

To check the integrity of Sogefibre FRC containers for a long period of time, simulations of the radionuclide migration taking account of the container surface degradation have been carried out. Although some assessments are still to be completed, it is clear today that the lifetime is consistent with the "Institutional Control Period" for low-and medium-level waste French disposal facilities, i.e. 300 years, if the FRC thickness at conditioning time is over 100 mm. This conclusion is the result of the following process:

- evaluation of the minimum FRC thickness for mechanical stability of the disposal facility;
- evaluation of the FRC degradated thickness by water-leaching after 300 years;
- evaluation of the minimum FRC thickness for containment;
- determination of the FRC overall thickness required at conditioning time.

Thus, Sogefibre got the label "high performance" for waste immobilized:

- in FRC in case of cylindrical containers C1 and C2;
- in mortar in case of cubical container K.

2 Fibre reinforced concrete container manufacturing

2.1 Existing containers

For Cogema need, three FRC containers have been designed. Their characteristics are given in table 7.

Table 7. Sogefibre existing FRC container

CONTAINER		OVERALL DIMENSIONS (mm)	USEFULL VOLUME (litre)	MINIMUM WALL THICKNESS (mm)
SHAPE	REFERENCE			
CYLINDRICAL	CBF - C1	h: 1200 dia.: 840	330	74
	CBF - C2	h: 1500 dia.: 1000	700	74
CUBICAL	CBF - K	1700 x 1700 x 1700	3000	100

A CBF-C1 container is shown in figure 3 and a CBF-K in figure 4.
FRC containers can be tailored to the specific needs of each customer, as regards:

- geometry (cylindrical, cubical, etc.);
- dimensions;
- wall thickness;
- closure system;
- gripping system;
- surface finish of outer walls.

As an example of the FRC manufacturing possibilities, figure 3 shows the very elaborated shape of the anti-float covers of the cylindrical FRC containers. The purpose of this cover is to avoid the floating of the waste drum when filling the container.

2.2 Manufacturing factory

The number of containers required by Cogema led to the construction of a FRC dedicated factory in Valognes, in Normandy (France), which entered into service in July 1990 with a rated capacity of 12,000 containers per year. This factory is shown in figure 5.
The manufacturing specifications imposed by Cogema, to guarantee that any FRC container (body and cover) delivered by Sogefibre fulfills Andra specification on material and container, are very stringent. Quality assurance procedures have been implemented (French standard AQ2) mainly consisting in:

- very stringent control for incoming raw materials (cement, sand, aggregates, fibres, etc.);
- control of process parameters (proportioning of

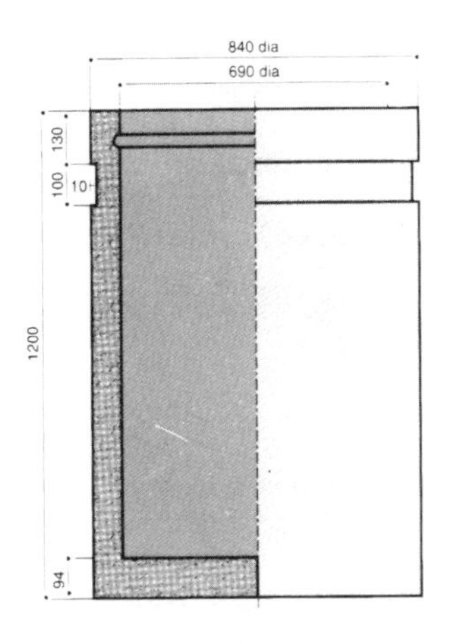

Fig. 3. CBF-C1 container with an anti-float cover

Fig. 4. CBF-K container

Fig. 5. Valognes FRC container manufacturing factory

concrete ingredients, mixing time, etc.);
- systematic measurement of concrete shrinkage and mechanical strength on test samples;
- dimensional and visual inspection of products;
- product traceability.

The manufacturing process applied in the Valognes factory is the following:
- fresh FRC preparation by mixing;
- fresh FRC pouring into vibrating moulds;
- 24 hours setting;
- removal out of moulds;
- 28 days cure time.

Most operations are remote controlled operations, managed by computer, allowing a high production ratio with reduced staff, eliminating the hard tasks.

The Valognes FRC manufacturing factory concept has been deeply influenced by nuclear industry high quality and high technology standards, and thus can be considered as an advanced factory in the concrete industry.

3 Conclusion

After an around 5 years conceptual phase, FRC has been sucessfully entering in an industrial manufacturing phase since July 1990. To provide other clients in nuclear or non-nuclear fields with FRC products is today the next goal for Sogefibre. Such a technological and industrial success should find applications in other sectors of the industry (prefabrication, thin concrete walls, environmental protections, etc).

4 References

[1] Andra specification STE 119.581S, Technical specifications for packages holding immobilized heterogeneous waste delivered in durable containers and intended for disposal in a near-surface facility.

[2] KASPERKIEWICZ J., and SKARENDAL A., -- Testing the influence of steel fibre parameters on toughness and cracking of fibre concrete. 2nd International Symposium on Brittle Matrix Composites. September 20-22, 1988, Cedzyna.

[3] ROSSI P., HARROUCHE N., and LE MAOU F., -- Comportement mécanique des bétons de fibres métalliques utilisés dans les structures en béton armé et précontraint. Annales de l'ITBTP, n° 479 bis, pp. 167-182, 1989.

57 DESIGN OF FIBRE REINFORCED CONCRETE FLOORS

Ö. PETERSSON
Swedish Cement and Concrete Research Institute, Stockholm, Sweden

Abstract
The use of fibre reinforced concrete industrial floors in Sweden is expanding. Rules for design of such floors have to be made. This project aim to give recommendation for design and testing of fibre reinforced concrete for industrial floors on ground. Full-scale testing of shrinkage has been made with the use of a steel frame of 2 * 5 meter. Different fibres and fibre content has been tested. Also full-scale testing of slabs subjected to ultimate load has been performed. Beams with six types of fibre and different fibre content with two types of concrete mix has been tested for flexural strength and toughness.
Keywords: Industrial floors, Fibre concrete, Bending, Toughness, Shrinkage.

1 Background

Concrete slabs on ground are generally reinforced, resting directly on ground, on a more or less defined sub-base. The problem with this type of concrete structure is the difficulty of controlling `wild' cracking, resulting in a shorter service life of the structure. The normal reinforcement work is heavy, time consuming and does not always achieve the expected results. The use of fibre-reinforced concrete would result in considerable improvement in work procedures and also the work environment.

The general aim of the project is to prepare recommendations for the design and construction of concrete slabs on ground using steel fibres as the only reinforcement, as well as studying the effects of the fibre reinforcement on structures subject to constraining forces.

Fibre Reinforced Cement and Concrete. Edited by R. N. Swamy. © 1992 RILEM.
Published by E & FN Spon, 2-6 Boundary Row, London SE1 8HN. ISBN 0 419 18130 X.

2 Laboratory testing of beams

In the study, a total of 60 beams were tested as regards to their cracking stress and toughness. Shrinkage testing was also performed in accordance with a standard procedure. A total of six fibre types produced by four steel-fibre manufacturers were tested. Table 1 shows the different fibre types, strength classes and fibre quantities tested. As shown in Table 1, K30 and K45 concretes were used (28-day compressive strengths 30 and 45 MPa respectively). The aim of the two strength classes was to test two classes of industrial concrete floor, one with a high wearing resistance and one with lower requirements on the wearing resistance.

Table 1. Fibre-reinforced concrete beams tested

				K30			K45		
Fibertype	**l**	**d**	**l/d**	**30**	**50**	**70**	**30**	**50**	**70**
	mm	mm		kg/m³			kg/m³		
Eurosteel	60	1,0	60	2	2	2	2	--	2
Twincone	54	1,0	54	2	2	2	2	--	2
Hörle	70	0,7	100	2	2	2	2	--	2
Dramix	60	0,8	75	2	2	2	2	--	2
Xorex	50	0,9	55	2	2	2	2	--	2
Euro-fiber	50	0,9	56	2	2	2	2	--	2

The testing of beams was essentially based on the ASTM C 1018-89 method, `Flexural Toughness and first-Crack Strength of Fibre-Reinforced Concrete (using beam with third-point loading)`\1\. The load and net deflection were recorded continually. The dimensions of the beams were: 800 mm long, 150 mm width and 100 mm depth. During testing, the following values were determined, cracking stress (f_{spr}), toughness index (I_5, I_{10} and I_{20}) and the residual stress ($R_{5,10}$ and $R_{10,20}$). To facilitate comparison between the efficiencies of different fibres, diagrams were drawn showing the residual-stress factor $R_{10,20}$ (see Figure 1). Each bar represents the average value of two tests).

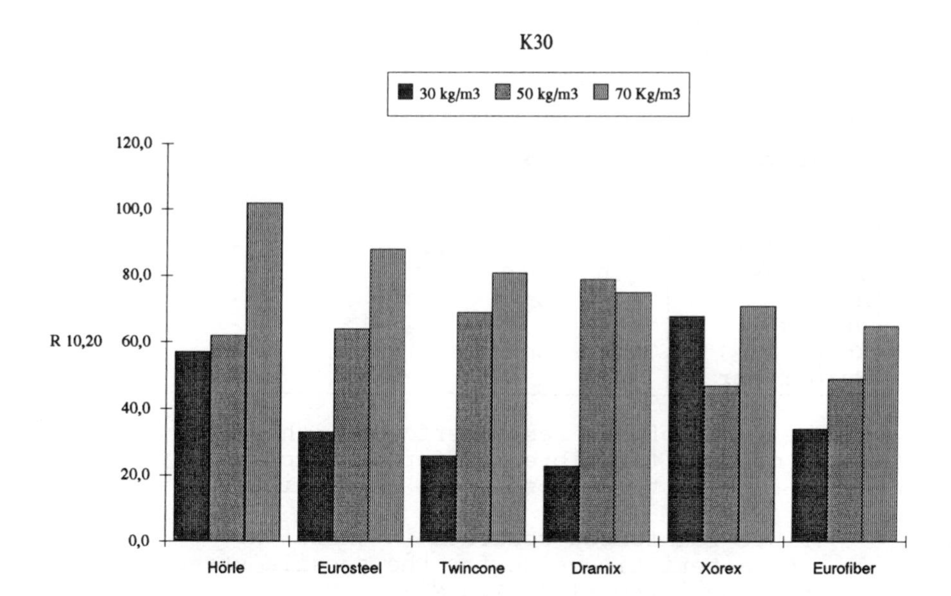

Figure 1. Theresidual-stress factor for the various fibres in K30 concrete.

Figure 1 shows how the residual-stress factor varies for the different fibres. The highest value for a K30 concrete was obtained using Hörle fibres. The work curves made during testing varied somewhat due to the difficulty of obtaining a homogeneous mix. The number of fibres was counted in the failure surface and was relatively uniform for the two tests. For all tested beams the average cracking stress, for all fibre contents, was 4.3 MPa for K30 and 5.7 MPa for K45 concrete. The cracking stress was also the maximum stress for all beams (K30 and K45) except for Hörle fibre with K30 and 70 kg/m^3.

One way of evaluating fibre-reinforced concrete beams is to apply special strength classes for fibre-reinforced concrete in which the purpose of testing is to check that a satisfactory level of toughness is achieved. A certain strength for fibre-reinforced concrete could contain a term that describes the effect of the fibre-reinforced concrete's toughness. A suitable criterion is the residual-stress factor. Reasonable values for the `imaginary` tensile strength in bending of the fibre-reinforced concrete may be assessed on the basis of the residual-stress factor multiplied by the cracking stress, see Table 2.

Table 2. Average values from tested beams for residual-stress factor ($R_{10,20}$) multiplied by the cracking stress, MPa.

	K30			K45		
Fibertype	**30**	**50**	**70**	**30**	**50**	**70**
	kg/m^3			kg/m^3		
Eurosteel	1.2	2.7	4.0	2.0	---	4.2
Twincone	1.0	2.9	3.7	1.7	---	3.9
Hörle	2.6	2.5	4.3	2.0	---	4.6
Dramix	1.0	3.6	3.4	3.1	---	4.9
Xorex	2.7	2.0	3.1	1.1	---	3.1
Euro-fiber	1.5	2.0	2.9	1.3	---	3.2

On the basis of the tests performed, the following strength classes for fibre-reinforced concrete may be suitable for the tensile strength in bending.

F1 = Fibre-reinforced concrete in which the characteristic value for the tensile strength in bending is equal to 1.0 MPa
F1.5 = As above, but with a value for the tensile strength in bending of 1.5 MPa. The classes continue up to 4.5 MPa in 0.5 MPa steps.

It is interesting to compare the proposed strength classes with what is commonly used or values proposed by fibre manufacturers. Compared to the beam test results it seems that the strengths proposed by manufactures are somewhat high \2\, \3\ see also Table 3. Compared to values used for shortcreating \4\ it can, on the other side, be somewhat low.

Table 3. Values used by Eurosteel. Concrete 25 MPa and a rupture safety factor of 1.7

Content fibre 60-100	20	25	30	35	40	kg/m3
Allowable flexural strengths	2.4	2.8	3.2	3.5	3.8	MPa

3 Full-scale load testing

Two concrete slabs measuring 3 x 3 metres were loaded to failure. A point load of up to 50 tonnes could be employed, using a hydraulic jack and a counterforce provided by a large steel beam. The steel beam was anchored to the underlying rock, using rock bolts see Fig 2.

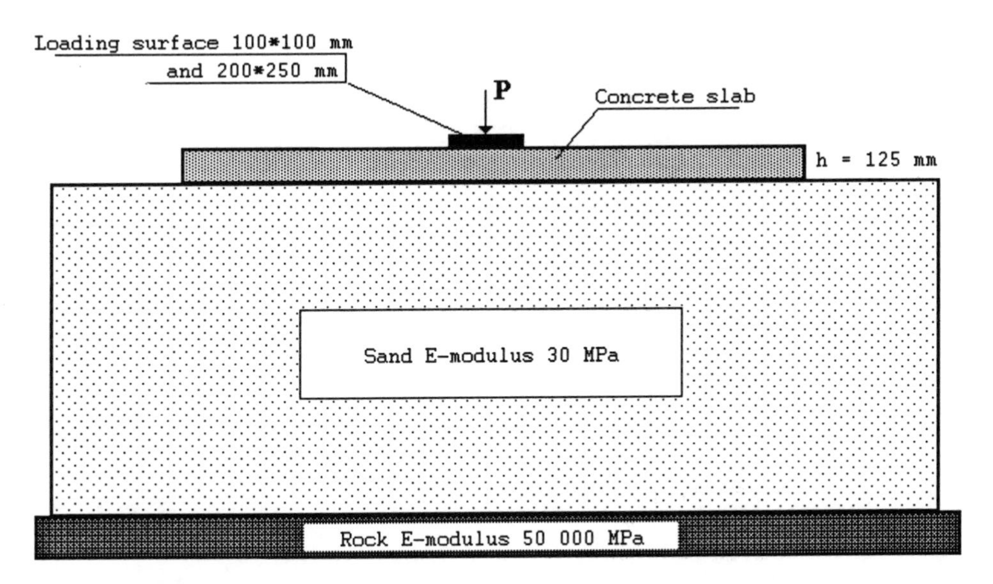

Figure 2. Test configuration

Before the load test was started, the modulus of elasticity of the sand was determined using plate-loading equipment. The test was performed in accordance with DIN 18134. The test was performed by subjecting the sand to a load by means of a 300 mm diameter plate. After the stress and associated deformation are read, the modulus of elasticity of the sand may be calculated. The equipment is completely computer controlled and the calculation is performed directly in the equipment's computer.

For loading the conventionally reinforced slab, ϕ8 s 150 in 2directions, a loading surface measuring 100x100 mm was used. When the load reached 298 kN, punching failure occurred. To avoid punching for the fibre-reinforced slab, 40 kg/m^3 Dramix 60/80, the size of the loading surface was increased to 250x200 mm. A pure tensile stress in bending was obtained in the fibre-reinforced slab at a load of 223 kN. The deformation of both slabs was recorded at a few points and at different load stages. The registrations were made at the loading surface, +300 mm, -300 mm, +700 mm, +1100 mm and 1400 mm from the loading point, i.e. at five points. Figure 3 shows the relationship between the deformation measured at the center of the slab (the loading point) under different loads.

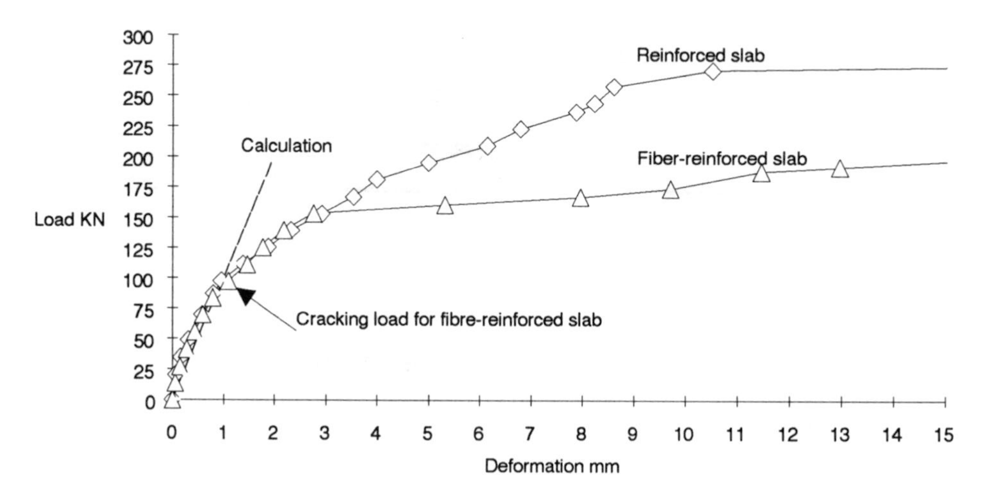

- Reinforced 3.3 cm^2/m (in 2directions) in the underside
- Fibre-reinforced slab, 40 kg/m^3 Dramix 60/80

Figure 3. Relationship between deformation and load on slabs.

Calculations with a software program \5\ is also shown in the figure. The marked point represents about 7 MPa in cracking stress. Cracking was observed at 150 KN for the fibre reinforced slab. No cracking was observed for the conventionally reinforced slab before punching failure. From the load tests performed in England \6\ and the tests, it can be seen that a considerable reserve of force remains after the cracking load has been reached. This shows that it is possible to use the properties of the fibre-reinforced concrete in design.

4 Full-scale shrinkage testing

To obtain relevant values of the fibre-reinforced concrete's ability to distribute cracking under restrained-shrinkage conditions, a number of full-scale tests were performed on concrete slabs, see Table 4. The so-called `ring tests', with restrained shrinkage, were not considered to provide a representative picture of the crack-distributing ability of fibre types and quantities. The test configuration for measuring restrained shrinkage involved using a very strong steel frame round a concrete slab. The slab could move freely along the sides of the 5 x 2 m slab. The ends were anchored to the steel frame using a number of reinforcing bars welded to the frame and cast into the slab. It was not possible to prevent movement of the concrete slab completely, but using the steel frame, the movements could be kept to a minimum and this arrangement permitted comparison of the crack-distributing ability of slabs containing different types and amounts of fibres. The tests were performed indoors at a relatively constant temperature of about 20 degrees Celsius and a relative humidity of about 60%. During testing, the temperature and relative humidity were recorded continually. Figure 4 shows the test configuration.

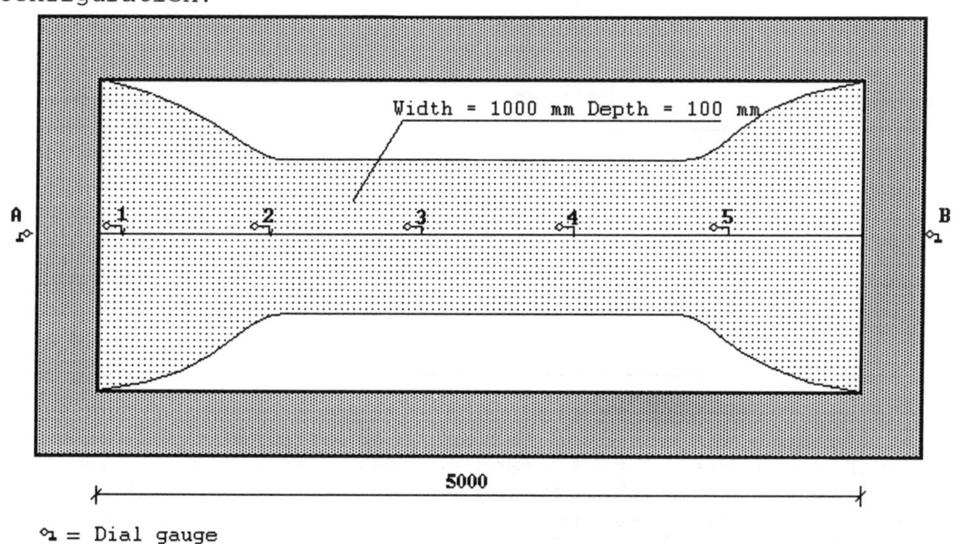

Figure 4. Test configuration for restrained shrinkage testing.

A total of six slabs were cast. A K45 concrete was used (compressive strength 45 MPa). Of these, one was unreinforced, four were steel-fibre reinforced and one was reinforced in the traditional way, see Table 4. The last slab, the one reinforced in the traditional way, was included to demonstrate the ability of the test configuration to imitate restrained shrinkage. To control the movements and measure the size of any cracks, the movement of the frame was measured by placing two dial gauges at the outside edges of the ends and fixing them to the surrounding floor. During casting of the slabs, studs were placed in the concrete at each metre. Movements were measured using a number of digital dial gauges attached to a beam fixed at one end of the steel frame and located at respective studs, to measure the movement at each metre point and also the total contraction of the frame see Fig 4. The results show that the time at which cracking of the slab started varied, despite relatively similar conditions during testing. The reason is probably to be found in variation in the binding agent from test to test.

Table 4. Data of restrained-shrinkage tests and days to start of cracking.

Sample No.	Fibre type	Qty, kg/m^3	Splitting strength MPa	Mean compr. strength MPa	No of days to start of cracking
1	Unreinforced	0	-	-	23
2	Dramix	30	-	-	21
3	Dramix	52	4,8	46,3	10
4	Twinecone	37	4,8	43,2	17
5	Dramix	17	4,2	42,6	7
6	Reinforced ϕ10 s 100 in one direction	8,6 cm2	4,3	42,5	42 Crack 1 50 Crack 2

During testing of restrained shrinkage, no significant difference was observed between the performance of end-anchored fibres (Dramix) and conically-anchored fibres (Twinecone). But the casting properties of the conically-anchored fibres were better, probably because they are smoother and thicker. The tests indicated in the performance of the slabs depending on the various quantities and types of reinforcement. None of the fibre-reinforced slabs cracked in more than one place. In this case the fibers had no ability to distribute cracks. The width of the crack increased with time and continued to increase for a long time without any new crack appearing see Figure 5. The crack appeared in bay 3-4 and the width after more than 30 days where approximate 0.7 mm.

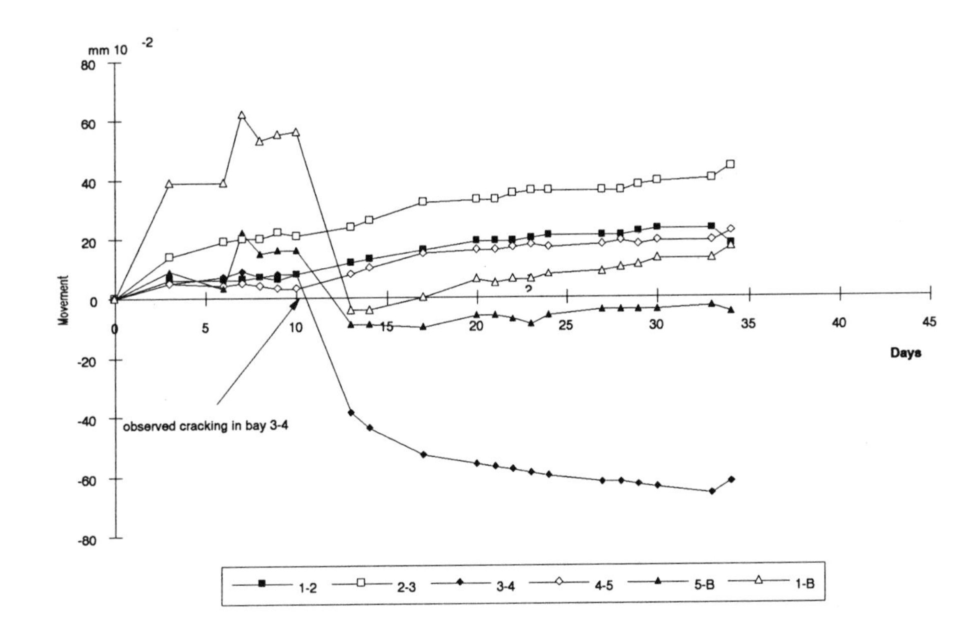

Figure 5. Movements at various points on the slab made using Dramix 60/80 fibres with 52 kg of fibres per m^3 of concrete.

In the conventionally reinforced slab, first one crack appeared and then a second one appeared after a couple of weeks. The reinforcement in this case distributed the cracking, which it did not do in the case of the fibre-reinforced slabs. Figure 6 shows a compilation of the shrinkage in the `bay' that cracked. Movements were measured over a length of 1000 mm.

5 Acknowledgement

Finacial support has been supplied by SBUF, Development Fund of the Swedish Construction Industri.

6 References

\1\ ASTM (1989) Standard Test Method for Flexural Toughness and First-Crack Strength of Fiber-reinforced Concrete (Using Beam With Third-Point Loading). ASTM C 1018-89

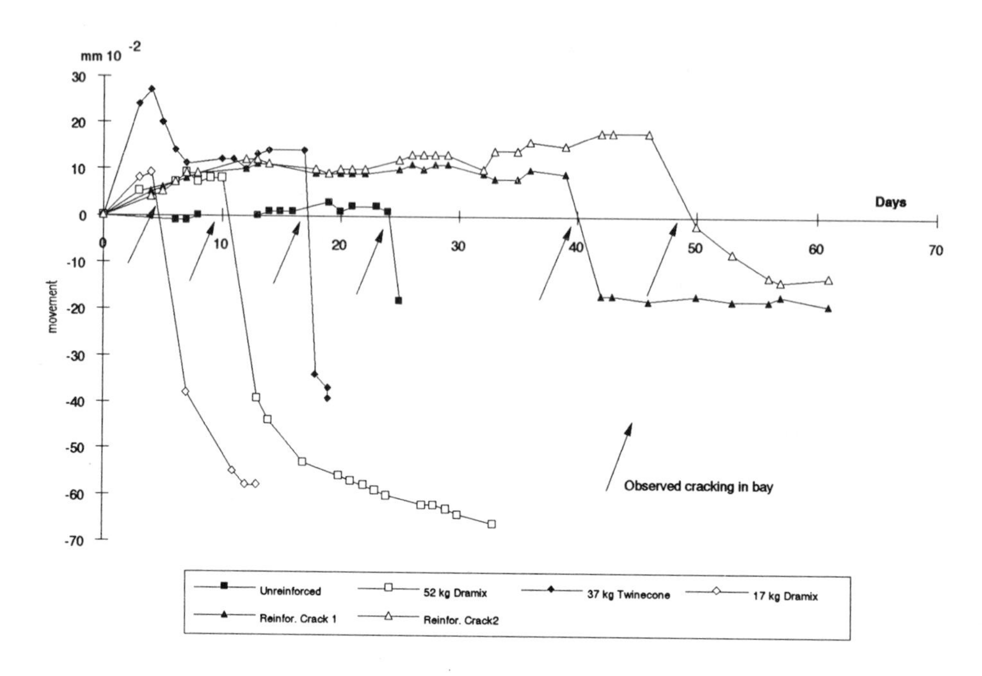

Figure 6. Movements over 1000 mm, measured in the `bay' in which the crack occurred in respective slabs.

\2\ J.Moens D.Nemegeer, Designing fiber reinforced concrete based on toughness characteristics, Concrete International nov 1991

\3\ X. Destree, Steelfiber reinforced concrete industrial floors: Design and construction for guaranteed performance, ACI Spring Convention 1990 Toronto.

\4\ Å. Skarendahl, Improving performance of steel fibre shotcrete (The Swedish experience), Rilem International Workshop, High Performance Fiber Reinforced Cement Composites, Stuttgart june 1991 (To be published).

\5\ F. van Cauwelaert, Five-layer program for concrete pavements (FLIP), Brussels 1991.

\6\ D. Beckett, Comparative tests on plain, fabric reinforced & steel fibre reinforced concrete ground slabs, Thames Polytechnic, Dartford, Kent Report No. TP/B/1 1989.

58 APPLICATION OF SFRC AS PROTECTIVE COVERING OF STRUCTURES SUBJECTED TO SUBSTANTIAL STRAINS AND DYNAMIC LOADING

R. GRAMATIKOVA
Bulgarian Acedemy of Sciences, Sofia, Bulgaria
S. SASISDEK and M. SLIWINSKI
Technological University, Krakow, Poland

Abstract
In construction and maintenance of bridges and overbridges, which are subjected to substantial strains and dynamic loads, the protective coverings are conventionally made of reinforced concrete. Application of SFRC instead of RC material with at least the same features considering its strength and durability but technologically more simple and lighter, was considered. It would allow the execution of works easier and diminish the weight of protective covering and thus the weight of a structure as a whole. On the basis of laboratory tests the authors proposed to substitute RC, used as a protective covering, with SFRC. The protective covering made of SFRC with "Harex" type fibres was applied on a railway bridge with a span of 223.8 m. In the paper the laboratory tests results of SFRC carried out prior to its application on the bridge are given. The observations on execution of works and preliminary test results on structural performance of SFRC in the bridge are also presented.

Keywords: Protective covering, dynamic loading, deformability, abrasive resistance, fatigue strength, impact strength.

1 Introduction

It is a frequent case that in engineering structures we come across the minor elements which do not play purely constructional role but nevertheless are subjected to substantial deformations and dynamic loading. These are usually protective elements contributing also to load bearing capacity of the structure e.g. bedding and protective covering of prefab ceilings, decks in bridges, boards on slender columns, cooling towers and chimneys, subgrading of railway and tramway lines, etc.

It is a common building practice that in such cases RC elements cast in situ or small size RC prefab slabs are used. This solution has disadvantageous features such as:

- difficulties in fabrication of thin RC elements, which are usually unnecessarily over-reinforced and therefore costly;

Fibre Reinforced Cement and Concrete. Edited by R. N. Swamy. © 1992 RILEM.
Published by E & FN Spon, 2-6 Boundary Row, London SE1 8HN. ISBN 0 419 18130 X.

- formation of cracks caused by excessive shrinkage or deflections;
- low resistance to impact and fatigue loads;
- unequal deflections of elements in prefab structures causing cracking of RC covering which requires introduction of fine net of tight dilations to stop it;
- difficulties in fabrication of slabs with thickness below 50 mm

The common cases of structures which are subjected particularly to substantial deformations and dynamic loadings are decks of steel bridges and overbridges. In such structures RC is being used for the top layer which serves as drainage, watertight corrosion protective cover for steel. The RC covering is usually fabricated unnecessarily thick and over reinforced, which increases the overall weight of the structure. The authors are of the opinion that the above disadvantageous features of RC coverings can be removed by using SFRC.

Basic advantageous features of SFRC such as:

- high resistance to dynamic loading
- substantial deformability under load
- low shrinkage
- high abrasive resistance

are specifically required when the protective covering serves as subgrading for railway lines.

The results of laboratory tests which support the authors thesis about the beneficial effect of substituting RC with SFRC in fabrication of different types of coverings are presented i n this paper. The selection of "Harex" type fibres to be used in SFRC as protecting covering on steel railway bridge is substantiated in the following sections.

The laboratory test results allowed the authors to recommend the application of SFRC in fabrication of the protective covering on steel railway bridge which was recently executed in Krakow, Poland.

2 Laboratory test results

2.1 Selection of fibres

SFRC with "Harex" type fibre of small aspect ratio has been used in concreting of airstrip pavements and floor finishings of industrial buildings with heavy traffic duty. The relatively low price of "Harex" type fibre and prompt delivery are also of significance.

The compositions of SFRC which have been prepared throughout the laboratory tests were as follows:

- type of fibre "Harex" SF 01-32 or "Dramix loose"
- fibre content by volume 0, 0.5, 1.0, 1.5%
- fibre content by weight 0 to 120 kg/m^3
- sand 0/2 mm 500 kg/m^3
- granite 2/8 mm 1090 kg/m^3
- portland cement 35 400 kg/m^3
- water 200 l/m^3

Slump measured on the slump test table was 30 to 35 mm.

The preparation of SFRC trial mixes containing the maximum 1.5% "Harex" type fibre and their testing confirmed very good workability of the mixes. It was also found out that if necessary the amount of 1.5% fibres can be exceeded without adversely affecting workability of fresh SFRC. In the case of "Dramix loose" fibre for mixes with 1.5% fibre content there were signs of the mixes being on the verge of acceptable, limiting workability of fresh concrete.

Series of SFRC laboratory test samples with above given compositions were made. Series of SFRC samples with "Dramix" type fibres were also made for comparison purposes.

2.2 Limited fatigue strength of SFRC with "Harex" type fibre

Based on results of tests carried out by Karpuchin [1], and Murdock [2], the following constant parameters have been assumed in the fatigue strength tests:

- characteristic of the cycle ρ_1 =0.1
- limiting number of cycles N_G=2x10^6 cycles

The following elements were used for the investigations, based on theoretical and experimental analysis:

- compression: blocks 5 x 5 x 5 cm
- bending: beams 5 x 5 x 16 cm

All specimens were cut out of big size plates.

Fatigue tests were carried out using the staircase method. Twenty samples were tested for each type of material and each loading condition. The above method was invented by Dixon and Mood [3] and for the first time was used in determination of limited fatigue strength by Ranson and Mehl [4]. The idea of the method is based on preliminary test of a sample which is subjected to loading close to the expected limited fatigue strength of the material. The range of expected dispersion of test results is divided into several levels of intensity of stress. Depending on whether the sample crushes in the first test or outlasts the assumed number of cycles, the following tests were carried out assuming lower or higher intensity of stress. The mean value and standard deviation is computed using the formulae given by Dixon and Mood:

$$\bar{x} = x_o + d\left(\frac{\Sigma i \times f_i}{F^2} \pm \frac{1}{2}\right)$$

where

$\bar{x}$ - mean value
x_0 - value of the lowest level of less frequently occurring event
d - distance between stress levels
i - chronological number of stress levels
f_i - frequency of event occurring on particular stress level
F - sum of events
s - standard deviation

The "+" sign is used here in case calculations are carried out taking into consideration events which did not take place.

Because the details of relations between the fatigue strength and frequency of vibrations, size of samples and the ρ parameter of the cycle are well known qualitatively and quantitatively for different types of concretes [5], the parameter of frequency was limited to one and one parameter of loading with two static schemes (bending and compression) was adopted in tests.

Table 1 Comprehensive list of mean values obtained in fatigue strength tests

Composite	Limited fatigue strength						Static strength	
	compression			bending				
	f_c^f Mpa	$\frac{f_c^f}{f_c}$	S Mpa	f_c^f MPa	$\frac{f_b^f}{f_b}$	S MPa	f_c MPa	f_b Mpa
C	18.4	0.58	1.8	2.95	0.59	0.26	31.8	5.0
SFRC	24.2	0.62	3.8	6.00	0.69	0.61	39.0	8.7

C - plain concrete, SF - "Harex"

The results obtained from the tests of limited fatigue strength are presented in Table 1 which provide evidence of the beneficial effect of using steel fibres, especially in bending.

2.3 Impact strength of SFRC with "Harex" type fibres

Complying with the suggestions of numerous authors concerning standardization of the determination of impact strength of SFRC and other concrete composites, it was decided to adopt the method with utilization of the results of the studies to practical purposes.

In compliance with Polish standards concerning dimensions and shape of specimens for the strength of concrete, the specimens in the form of rectangular blocks 15x15x6 cm were adopted. The change of the sample shape with respect to the ACI recommendations [6] and high strength of the materials studied compelled the authors to modify the drop device. A series comprising 15 specimens for each type of the composite was prepared. The samples were kept in steel moulds during the first 24 hours, then for 7 days at a temperature of 20°C and relative humidity of 95% and subsequently in laboratory conditions. All tests were performed after 90 days from the time of preparation of the samples.

Table 2 Results of studies on impact resistance of concrete composites

Composite	Weight of beater	Drop height	Appearance of first crack			Destruction		
			number of blows	standard deviation	first crack energy	number of blows	standard deviation	destruction energy
	G(N)	h(dm)	n	SD(n)	(J)	n		(J)
C	49	50	10	6.8	244	11	6.8	277
SFRC	49	50	12	3.5	297	46	18.1	443

C - plain concrete, SF - "Harex"

The results of studies on impact resistance of concrete composites are presented in Table 2 as the mean values obtained for each series of 15 specimens. The values correspond to appearance of the first crack as well as to the complete destruction.

The energy of destruction was calculated from the formula:

$$E = G \times h \times n$$

where:

G = weight of hammer,
h= drop height
and n = number of blows

The effect of contribution of steel fibres to the matrix is best seen when comparing the first crack and failure energy of steel fibre reinforced concrete and plain concrete. It can be distinctly seen that the fibres carried the load mainly after cracking of the concrete matrix.

The observations of depth and diameter of impact blow of the steel ball allows to estimate deformability of the material and its hardness. It should be observed, that at the moment of failure, the diameter and depth of the impact crater for specimens with addition of steel fibres were considerably larger than those in the plain concrete. As a result of the above the number of blows resisted prior to the failure are exerted on a specimen of smaller thickness but at the same time on a considerably greater area. One could then estimate the ultimate energy required for cracking or failure of the specimen as a function of the changing specimen thickness and the area of the impact crater (the area of action of the impact force on the specimen).

The characteristics picture of specimen's destruction in impact test in shown in Fig. 1.

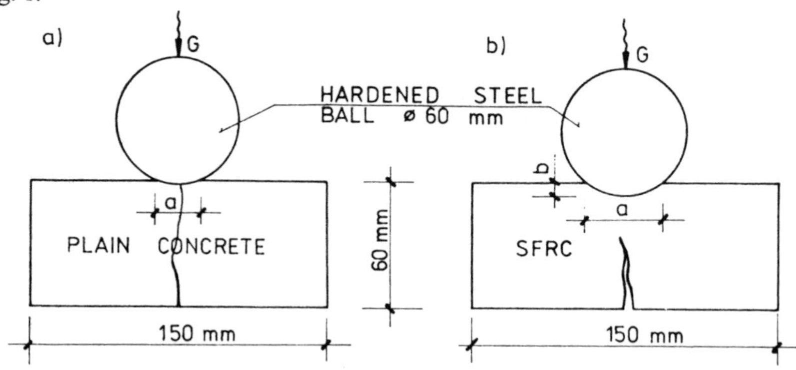

Fig. 1 Characteristic pictures of specimen's destruction in impact tests
a - diameter of impact crater made by ball at cracking
b - depth of an impact crater of a ball at cracking

This comparison should be made at any rate after the same number of impacts i.e. at the levels of the same supplied energy. The values of the diameter and depth of the impact crater measured at the moment of the first crack are given in Table 3. Each value is the mean of three measurements.

Table 3 Deformation of the specimen surface under impact load

Composite	First crack energy	Specimen deformations			
	(J)	diameter of crater (mm)	depth of a crater (b) (mm)	volume of crater (mm^3)	surface area of crater (mm^2)
C	244	19	1.55	222	574
SFRC	297	25	2.7	666	1004

2.3 Bending performance of SFRC

Bending performance investigations of SFRC were carried out on beams with dimensions 100x150x700 mm. The cube compressive strength of the plain concrete at 28 days was 39.4 MPa. SFRC specimens were reinforced in the whole volume with steel fibres which were randomly oriented in the matrix. Notations used in the tests carried out were as follows:

Fibre type	Volume content	Notation (element)
"Harex"	0.5%	1
	1.0%	2
	1.5%	3
"Dramix"	0.5%	4
	1.0%	5
	1.5%	6

The element made of plain concrete was marked with 0.

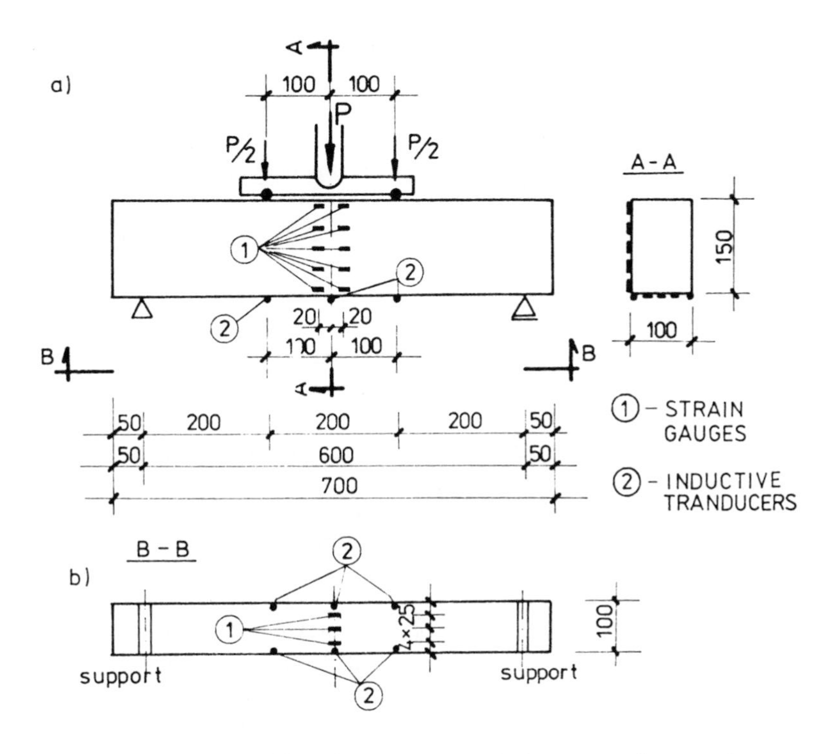

Fig. 2 Scheme of loading of beams and placement of measuring
a - measurings on front face of beams
b- measuring on bottom surface of beams

The beams were tested at 9 months, in pure bending as shown in Fig. 2. The loading was applied in stages, at intervals of 1 kN, till the complete destruction of the beam. The deformations along the height of the middle cross-section of the specimen were measured with the help of 10 strain gauges, while the deflections in the middle and at the third points of the beam span were measured by six inductive transducers (Fig. 2). The strain gauges on the bottom surface of the beams and the inductive transducers were used for recording the first crack formation. Scanning strain-gauge measuring bridge recorded the signals, coming from the gauges, for every stage of loading. The results were mathematically treated by computer "Hewlett Packard", connected to the bridge.

As a result of the experiment, the following average values were obtained for each element:

P_{cr} - the load at the moment of the first crack formation

P_n - the ultimate load

ε - relative deformations along the height of the middle cross-section for each load level

f - the deflection in the middle cross-section and at the third points for each load level

ε_c - maximum compressive relative deformation at the moment of the first crack formation

ε_t - maximum tensile relative deformation at the moment of the first crack formation

ε_{c1} - maximum compressive relative deformation at the destruction of the beam

f_u - deflection at destruction of the beam

On Fig. 3 distribution of relative deformations at P_{cr} and at P_u are shown.

Coming out from the received data, the graphs P-ε_c and P-f can be obtained for every element, as well as the distribution of the relative deformations along the height of the middle cross-section of the beams at P_{cr} and P_u. For the specimens, reinforced with fibres "Harex", these graphs are represented on Fig. 3 while the others, reinforced with fibres "Dramix", are shown on Fig. 4. Analysing these results, the following conclusions can be made:

(a) The fibre reinforcement increases the load level, at which the first crack appears. The relative increase of P_{cr} of the reinforced elements, in comparison with P_{cr} of the plain concrete element, is more considerable for specimens 4, 5 and 6 (fibre "Dramix") where it is 50%, 92% and 116% respectively, while for the element 1, 2 and 3 (fibres "Harex") it is 42%, 34% and 67% respectively. For the both types of reinforcement the increase of the volume content of the fibres results in increase of the relevant values of P_{cr}. This tendency is more clearly expressed when reinforcing with fibres "Dramix".

Table 4 Average values of deformations and deflections of tested beams

Element No	P_{cr} (kN)	ε_c x10^{-6}	ε_t x10^{-6}	f_{cr} (mm)	P_u (kN)	ε_{c1} x10^{-6}	f_u (mm)
0	12	102	70	0.23	19	170	0.36
1	17	140	112	0.29	21	180	0.38
2	16	108	128	0.21	23	216	0.45
3	20	100	148	0.18	26	260	0.55
4	18	132	120	0.35	26	220	0.60
5	23	124	152	0.38	32	320	0.80
6	26	100	180	0.84	37	420	1.03

(b) The fibre reinforcement increases the flexural tensile strength of the composite. This statement is very well confirmed by the results for reinforcement with fibres "Dramix". In this case, the increase of the volume content of the fibres from 0.5% to 1.0% and 1.5% leads to considerable increase of the flexural tensile strength of the composite, in comparison with plain concrete, and was 37%, 68% and 95% respectively. In the case of reinforcement with fibres "Harex", this positive influence of the fibres upon the flexural tensile strength is also registered, but to the smaller extent: the increase of the volume content of the fibres from 0.5% to 1.0% and 1.5% results in the increase of the flexural tensile strength of 11%, 21% and 37%, respectively.

(c) The fibre reinforcement of the both types improves very much the deformability of the elements. Comparison between the graph P-ε_c and P-f of the reinforced specimen from one side and of the specimen, made of plain concrete, from the other, shows that at one and the same load level, the relative deformations ε_c and the deflections f of the reinforced elements are smaller than those of the plain concrete element; in the reinforced specimens, the appearance of the first crack occurs at greater load values and the elastic zones of the graphs P-ε_c and P-f of these elements are longer than they are for the plain concrete specimen; the destruction of elements 1, 2, 3, 4, 5 and 6 occurs at greater values of the deformations ε_c and f_u than they are at the moment of failure of the plain concrete element; the increase of the volume content of the fibres result in increase of their effectiveness concerning deformability.

Although SFRC with "Dramix" fibre has shown more advantageous features concerning the bending performance, compared to SFRC with "Harex" fibre, the better workability of the fresh mix of the latter is to be stressed, compared to the workability of SFRC with "Dramix" fibres. Both types of SFRC have shown advantageous bending performance compared to plain concrete.

In further recommendations to use SFRC with "Harex" fibre in fabrication of the protective covering on the bridge the above mentioned two factors, namely, good bending performance features and very good workability of the fresh mix were taken into account.

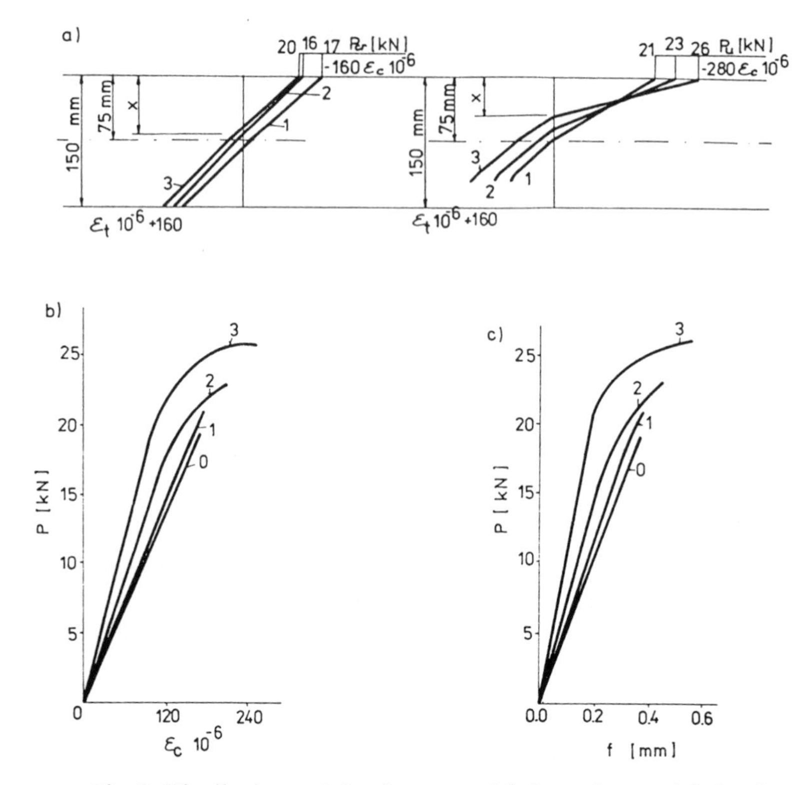

Fig. 3 Distribution and development of deformations and deflections in SFRC beams with "Harex" fibres
a - deformations ε at P_{cr} and P_n
b - P/ε_c graphs
c - P/f graphs

3 Application of SFRC as protective covering on steel railway bridge

The protective covering made of SFRC with "Harex" type fibres was fabricated on steel railway bridge with the cross section of a single lane as shown in Fig. 5. Span of bridge is L=6 x 37.3 = 223.8 m and it has two lanes. The anticipated deflection of a single span of the bridge can be up to f=100 mm. Total square area covered with SFRC on the bridge was 2000 m^2, volume of SFRC used was 120 m^3.

The thickness of protective covering made of SFRC varied from 80 mm to 40 mm to allow for drainage of water from within the bridge girder. Concrete was prepared as a ready mix concrete and transported to the site in mixing truck where it was placed and compacted by surface vibrators. The consistency of the fresh mix was K_2: the spread measured on the slump test table was 30 to 35 cm. Curing of SFRC covering in subsequent 7 days was done by intensive sprinkling of water on it.

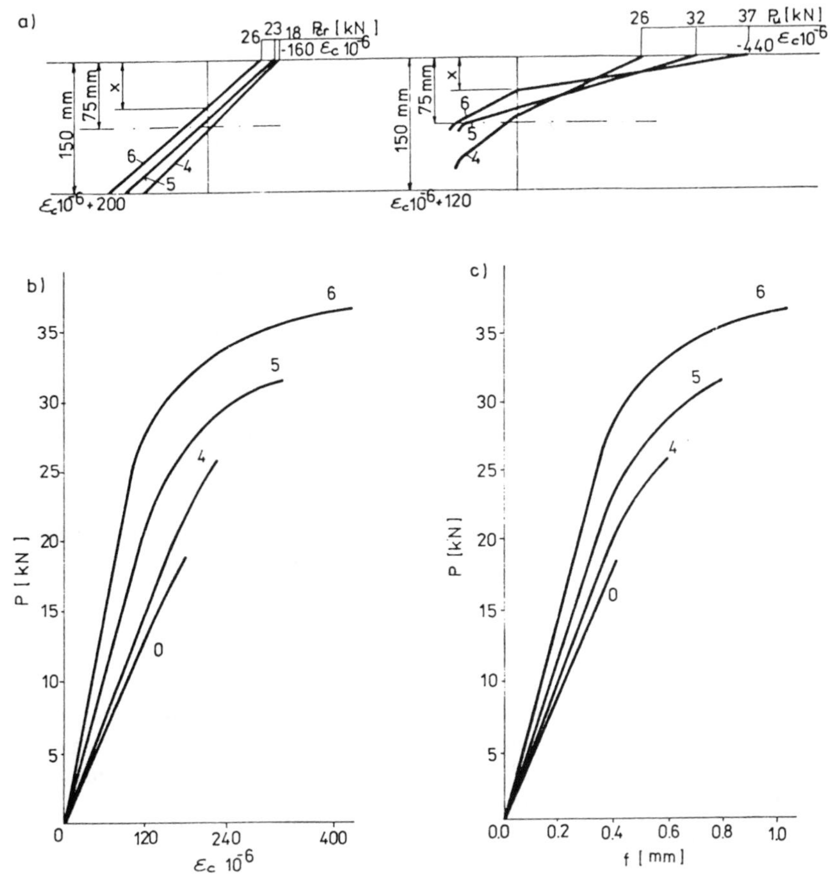

Fig. 4 Distribution and development of deformations and deflections in SFRC beams with "Dramix" fibres
a - deformations ε at P_{cr} and P_n
b - P/ε_c graphs
c - P/f graphs

The strength tests of SFRC carried on sample which were kept on the bridge, at 28 days time, have shown that SFRC reached assumed strengths:

$f_c = 37$ MPa $\qquad f_b = 8.5$ MPa

After 28 days from the time of placement of the protective covering, the surface was cleaned and it was impregnated with the polymer composition. At the same time the joints, which were spaced at 5 m distance, were filled up with the polymer mastic.

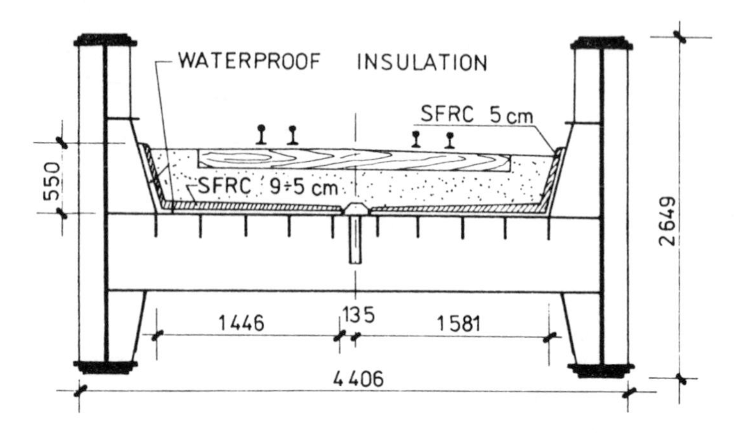

Fig. 5 Cross section of a single lane of the bridge

The observations carried out after 60 days time of placement did not reveal any defects in the SFRC covering e.g. cracks. The protective covering was commissioned for further works i.e. placement of the railway subgrading.

4 Conclusions

Results of tests on SFRC presented in the paper, proved the beneficial influence of "Harex" type fibre reinforcement on plain concrete. The limited fatigue strength and impact strength of SFRC are considerably higher compared to the strength of the plain concrete. Bending tests have also shown the advantageous performance of SFRC compared to the bending performance of plain concrete.

Although SFRC with "Dramix" type fibre performs better in bending compared to the performance of SFRC with "Harex" type fibre, good workability and stability of the mix, irrespective of fibre content of the mix, was the decisive factor in recommending the latter for fabrication of the protective covering of the bridge girders. Based on laboratory test results and on the observations on execution of works on the bridge the authors consider the application of SFRC as a reliable and universal material solution in fabrication of protective coverings of different types.

5 References

1 Karpuchin, N S, Issliedowani je wynosliwosti bietona pod wozdjejestwiem mnigokratno prilozenoj nagruzki Transzeldoizdat, Moskwa, 162.
2 Murdock, J W, The mechanism of fatigue failure in concrete, PhD Thesis, University of Illinois, 1960.
3 Dixon W J and Mood, A M, Journal of Amer. Statistic Assoc, no. 40, 1948
4 Ranson, J T and Mehl, R F, Symposium on fatigue with emphasis on statistical approach, II ASTM STP, no, 137, 3, Philadelphia, 1953.

5 Sasiadek, S, Fatigue strength of concrete with limestone aggregate, Proc. of International Symposium on Brittle Matrix Composites -BMC 3, eds. A M Brandt, I H Marshall, Elsevier Applied Science, 1991, pp 148-153.

6 Measurement of properties of fibre-reinforced concrete, Subcommittee Report, ACI Committee 544 Fibre-reinforced Concrete, in Testing and Test Methods of Fibre Cement Composite, RILEM Symposium 1978, ed R N Swamy, The Construction Press, pp 9-21.

7 Hibbert, A B and Hannat, D J, The design of an instrumented impact test machine of fibre concrete. In Testing and Test Methods of Fibre Cement Composites, RILEM Symposium, 1978, ed R N Swamy, The Construction Press, pp 99-105.

8 Gramatikova, R and Yamboliew, K, Experimental procedure for the establishment of the influence of different types of reinforcement in SFRC upon its bending performance, Proc. of the Sixth National Conference on Mechanics and Technology of Composite Materials, Bulgarian Academy of Sciences, Sofia 1991, pp 363.366.

PART EIGHT

AGEING AND DURABILITY

59 EFFECT OF HIGH TEMPERATURE ON THE MECHANICAL PROPERTIES OF FIBRE REINFORCED CONCRETE CAST USING SUPERPLASTICIZER

F. E. EL-REFAI, M. M. KAMAL and H. H. BAHNASAWY
GOHBPR, Cairo, Egypt

Abstract
This research was conducted to determine the effect of high temperature on the mechanical properties of steel fiber reinforced concrete casted using superplasticizer. The main variables taken into consideration were temperature of exposure, the duration of exposure and method of treatment after exposure to high temperature. The tested specimens were divided into two groups. The first group cooled down in air. The second group was quenched in water to cool down, then left in the laboratory atmosphere for one day before testing. Visual inspection was carried out with special attention to change in colour of concrete and occurance of hair cracks. Tests for compressive strength, splitting tensile strength, flexural strength and modulus of elasticity were carried out on all specimens.
Keywords: Fiber Reinforced Concrete, High Temperature, Duration of Exposure, Visual Inspection, Mechanical Properties, Cracking Pattern.

1 Introduction

The influence of elevated temperature on the mechanical properties of concrete is important for fire-resistance studies and also for understanding the behaviour of containment vessels, such as nuclear reactor pressure vessels, and ovens concrete lining at both service and ultimate conditions.

In recent years, the use of fiber reinforced concrete has become increasingly faster[1]. It is technically and economically feasible to produce ready mixed high strength fiber reinforced concrete using plain steel fibers with concrete materials. As the use of fiber reinforced concrete becomes more common, the probability of exposing it to high temperature also increases. Considerable research data are

Fibre Reinforced Cement and Concrete. Edited by R. N. Swamy. © 1992 RILEM.
Published by E & FN Spon, 2-6 Boundary Row, London SE1 8HN. ISBN 0 419 18130 X.

available on the mechanical properties of steel fiber reinforced concrete[2],[10]. Also research data are available on the effect of high temperature on the properties of ordinary concrete[11],[18]. This research aims to study the effect of high temperature on mechanical properties of steel fiber reinforced concrete after exposure to both different temperatures and durations.

2 EXPERIMENTAL DETAILS

2.1 Materials Used

Ordinary portland cement was used. The physical, mechanical, and chemical properties of this cement are in compliance with the limits of the Egyptian Standard Specifications 373, 1984.

Natural dry siliceous gravel of 20 mm N.M.S from El-Yahmoum quarries at Cairo was used as coarse aggregate. Sand from Pyramids zone was used as fine aggregate. The aggregate properties are in complete compliance with the limits of the Egyptian Standard Specifications 1109, 1973.

Superplasticizer with high range of water reducing agent was used. The properties of the used superplasticizer are complying with ASTM C-494 requirements.

Plain galvanized fibers of 25mm length and 0.25mm diameter with aspect ratio (l/d) 100 was used. The specific gravity of galvanized steel fiber was 7.6.

2.2 Steel Fiber Concrete Mix

Steel fiber reinforced concrete mix was prepared with the following constant mix proportions all over this research as shown in Table 1.

Table 1 Fiber Reinforced concrete mix proportions by weight

Cement	350 Kg/m^3
Fine aggregate	617.80 Kg/m^3
Coarse aggregate	1226.80 Kg/m^3
Water/cement ratio	0.46
Superplasticizer (% of cement weight)	1.00 %
Fiber (2% of concrete volume)	152.00 Kg/m^3

2.3 Properties of Fiber Reinforced Fresh Concrete

For all batches of SFRC mixes slump was ranged between 25–35 mm. Slump tests were used as an indicator of relative consistency for steel fiber reinforced concrete mixes.

2.4 Mixing, Casting and curing of test specimens

The mixing of fiber concrete was accomplished carefully by mechanical mixer with special steel fiber dispencer to have a uniform dispersion of the fibers and prevent the segregation or balling of the fibers during mixing.

Test specimens were casted, for compressive strength tests, 150 mm x 150 mm x 150 mm cubes and 150 mm x 300 mm cylinders were used. For tensile and modulus of elasticity tests 150 mm x 300 mm cylinders were used. All cylinders were casted, capped and cured according to ASTM C 192-81 recommendations. Also beams of 100 mm x 100 mm x 350 mm were casted for modulus of rupture tests. All beams were casted and cured according to ASTM C 78-75.

All specimens were cured for 28 days before exposing to different temperatures and different exposure times. A total number of 57 cubes, 171 cylinders and 57 beams were prepared for heating and testing.

Control specimens for different tests were left at laboratory temperature ≅ 25 °c and tested without heating. Groups of three specimens were exposed to each heat level for each duration time then cured by cooling in air or quenched in water then left, for 24 hours in laboratory atmosphere before testing.

Compressive strength, splitting tensile, modulus of elasticity and flexural tests were done according to ASTM
C 496- 81 and ASTM C 78-75.

2.5 Heating and Testing

An electric furnace was used for specimens heating. The furnace temperature ranged between 0 - 1200 °c, it was controlled by an electronic controller. Three specimens were heated each time in the furnace. All specimens were dried at about 60 °c before heating. The temperatures were continuously recorded by a thermocouple positioned at the center of the furnace. The approximate temperature - time curves for 200 °c and 600 °c are shown in Fig 1.

The following test procedure was followed:

1-Three temperature of exposure levels were chosen 200, 400, 600 °c.

2-The rate of heating was increased as shown in Fig 1.

Three specimens were heated in each time.
3-The duration of each heating period were 2, 4 and 6 hours for each level of exposed temperature.
4-After heating each group of specimens was treated once in air and the other by quenching in water and then left for 24 hours before testing. Scheme of experimental work is shown in Fig 2.

3 Test results and Discussion

The measured mechanical properties of steel fiber reinforced concrete mix for different levels of heating, different exposure times and different methods of treatment after heating are shown in Table 2. Test results are interpreted in Figs. 3 to 9 on which the discussion will be given.

3.1 Visual Inspection

Through careful observation, it was possible to notice that specimens colour changed to darker down after exposure to temperature of 400, for 4 and 6 hours duration time. A network of hair cracks was observed at 600°C which become deeper when the specimens treated by quenching in water. For ordinary concrete mixes, hair cracks appeared at 400°C (12).

3.2 Compressive Strength

The compressive Strength of cubes at different temperatures, different times of exposure and cooled in air or quenched in water is shown in Fig 3. Each point in the figure represents an average of the crushing compressive strength of three specimens (fc) compared with respect to the average crushing compressive strength of control specimens (fc'). The change in the compressive strength of fiber reinforced concrete heated specimens appears to follow a common trend. Initially, as the temperature increased to 200°C, the strength decreased compared to the control specimen. The temperature range over which the fiber reinforced concrete showed reduced strength increased with the time of exposure and method of treatment. With further increasing temperature to 400°C, the reduction of compressive strength increased reaching 0.5 and 0.74 of control specimens strength for both specimens cooled in air and quenched in water respectively. In the temperature range of 400 to 600 °C, the strength for both cases of cooling dropped sharply reaching a lower limits at six hours exposure time to heat. The reduction of compressive

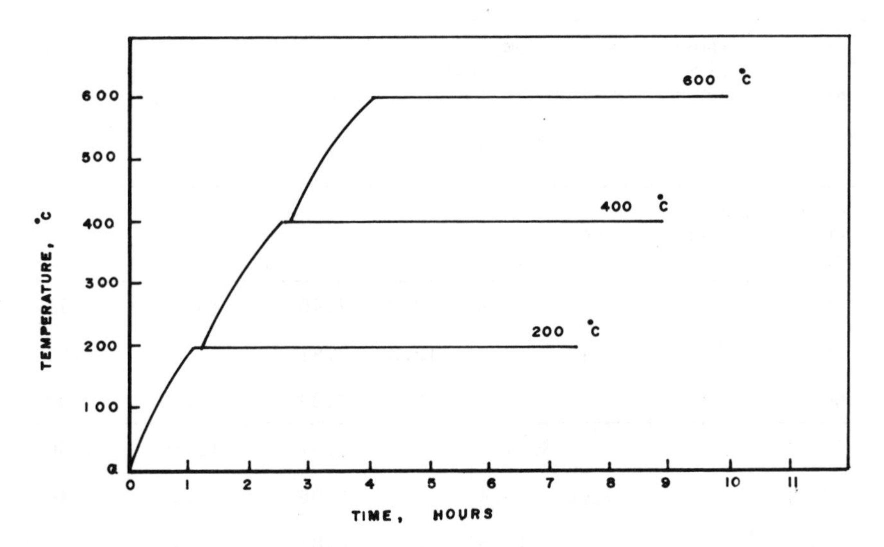

FIG. 1 : TEMPERATURE_TIME CURVES USED FOR HEATING OF CONCRETE SPECIMENS. (APPROXIMATE OVEN TEMPERATURE)

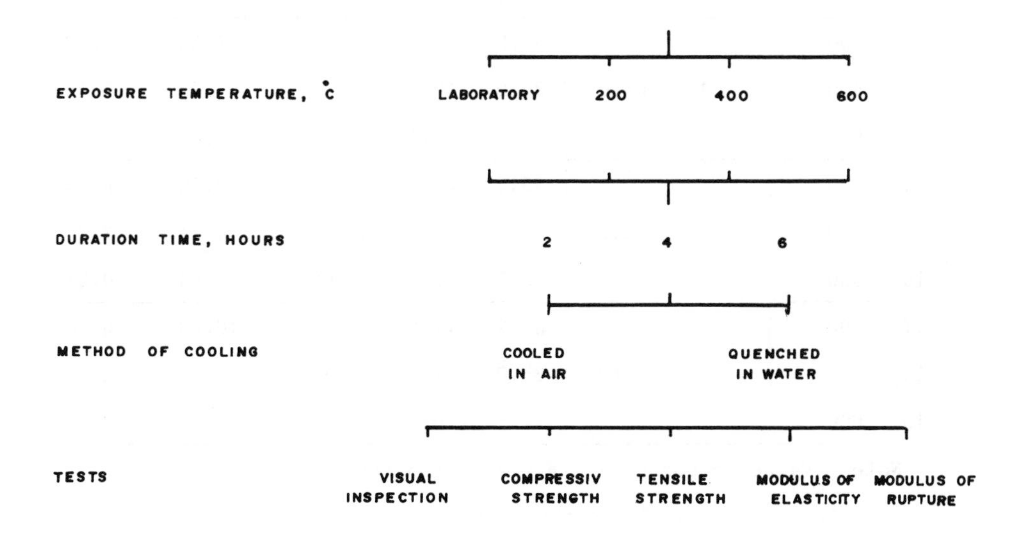

FIG. 2 : SCHEME OF EXPERIMENTAL WORK

Table 2. Experimental details and test reults.

Set No.	Oven Temper-ature °c	Exposed Time (hours)	Method of cooling	Compressive strength N/ mm^2		Tensile Strength N/ mm^2	Modulus of Elasticity KN/ mm^2	Modulus of rupture N/ mm^2
				Cubes	Cylin-ders			
1	control specimens at laboratory temperature about 25 °c			50.4	34.3	5.10	15.18	7.88
2	200			49.8	28.9	4.40	13.42	7.41
3	400	2	air	43.2	19.8	3.81	6.27	4.00
4	600			33.6	10.9	1.39	1.69	2.03
5	200			37.0	20.5	3.67	11.90	7.00
6	400	2	water	28.4	14.4	3.06	5.48	3.46
7	600			14.8	5.6	0.81	1.44	1.17
8	200			47.0	27.6	4.20	12.68	7.00
9	400	4	air	37.8	17.7	3.36	5.38	3.94
10	600			28.1	10.1	1.10	1.01	1.34
11	200			34.9	18.9	3.55	11.36	6.78
12	400	4	water	26.0	10.8	2.70	4.07	3.15
13	600			14.3	5.1	0.62	0.83	1.10
14	200			45.3	25.5	4.10	12.10	6.53
15	400	6	air	37.2	12.3	2.96	4.24	2.70
16	600			25.7	9.1	0.90	0.94	0.81
17	200			31.9	17.0	3.40	10.53	5.40
18	400	6	water	25.0	7.6	2.24	3.63	2.04
19	600			12.7	3.2	0.45	0.56	0.15

Note : Values reported are average of 3 tests

strength reached 0.25 and 0.41 of control specimens strength for both specimens cooled by air and quenched in water respectively.

In the temperature range of 25 to 400°C, the moisture content has a significant bearing on the strength of concrete. It is believed that absorbed water in concrete softens the cement gel or attenuates the surface forces between gel particles, thus reducing the strength [19]. Reduction in compressive strength of concrete at 200°C, has also been attributed to the triaxial state of stress apparently existing when the paste pores are filled with water [15].

At temperature above 400°C the percentage drop in strength was sharp. At these temperatures, the dehydration of the cement paste results in its gradual disintegration. Since the paste tends to shrink and the aggregate expands at high temperatures, the bond between the aggregate and the paste is weakened, thus gradually reducing the concrete strength [16].

The compressive strength of cylinders at different temperatures, different times of exposure and treated by both air and water is shown in Fig 4. Each point in the figure represents an average of the maximum compressive strength of three specimens (fcy) compared with respect to the maximum compressive strength. (fcy') at laboratory temperature. The reduction in compressive strength of heated cylinders appears to follow the trend of cubes with higher values. From the results, it is obvious that a decrease in the compressive strength of all tested cylinders started at 200 °C and progressed sharply until 600 °C. The reduction in compressive strength of fiber reinforced concrete cylinders increases with 15 to 40% than the reduction in cube compressive strength for all levels of heating, exposure times and method of treatment. It may be related to that the surface area of cylinders exposed to temperature is greater than cubes by about 30 %, thus the core of cylinder reaches to the thermal steady state condition before the cube.

3.3 Tensile Strength

The splitting tensile strength of cylinders at different temperatures, different times of exposure and cooled in air or quenched in water is shown in Fig.5. Each point in the figure represents an average of the maximum tensile strength of three specimens (ft) compared with the average maximum tensile strength (ft') at laboratory temperature. Initially as the temperature increased to 200 °C, the

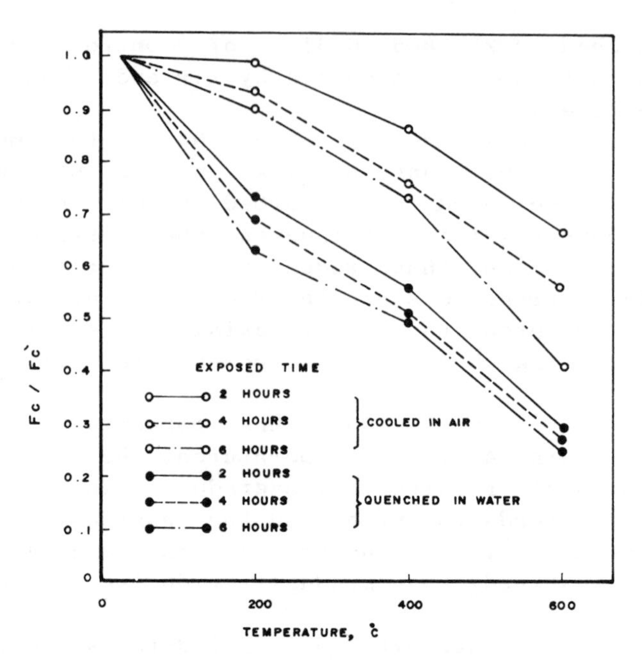

FIG. 3 : VARIATION OF CUBE COMPRESSIVE STRENGTH WITH TEMPERATURE

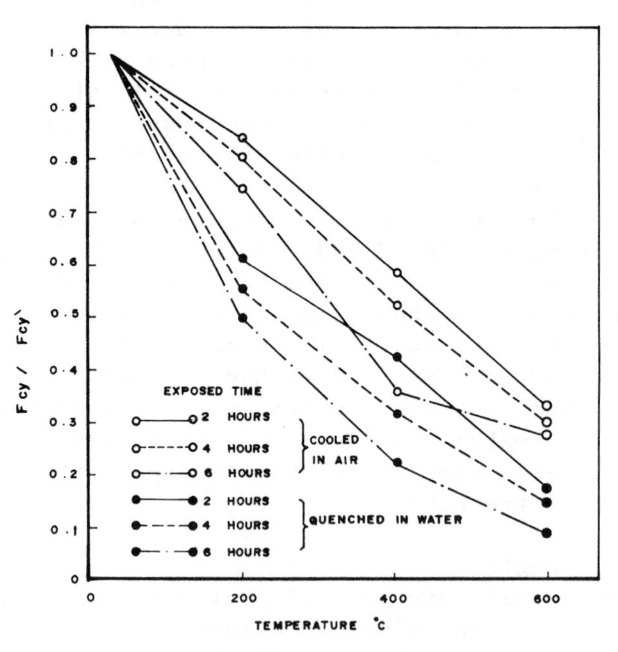

FIG. 4 : VARIATION OF CYLINDER COMPRESSIVE STRENGTH WITH TEMPERATURE

tensile strength decreased compared to the laboratory temperature strength. The temperature range over which the fiber reinforced concrete showed reduced strength increased with time of exposure. It reached 80 and 67 percent for specimens cooled in air and quenched in water respectively.

With further increase in temperature to 400°C, the reduction of tensile strength increased reaching 58 and 43 percent below laboratory temperature strength for both specimens cooled in air and quenched in water respectively. In the temperature range of 400 to 600°C, the tensile strength for both cases of cooling dropped sharply reaching a lower limits. The reduction of tensile strength reached 18 and 8 percent below laboratory temperature strength for both specimens cooled by air and quenched in water respectively.

3.4 Modulus of Elasticity

The static modulus of elasticity of cylinders at different temperatures, different times of exposure and cooled by air or quenched in water is shown in Fig 6. Each point in the figure represents an average of the maximum modulus of elasticity (E) of three specimens compared with respect to the average maximum modulus of elasticity (E') at laboratory temperature. The effect of temperature between 25 to 200 °C, on modulus of elasticity of both specimens cooled in air or quenched in water is similar to other mechanical properties of fiber reinforced concrete discussed before. The modulus of elasticity decreased sharply between 200 and 400 °C and also between 400 and 600°C. The fiber reinforced concrete modulus of elasticity was found to be minimum at 600°C. It reached about 5 percent of the modulus of elasticity of specimens at laboratory temperature. This sharp decrease in modulus of elasticity may be due to the increase of thermal coefficient of expansion of steel fibers than concrete which has an effect on decreasing the stress and increasing the strain of fibers reinforced concrete cylinders.

3.5 Modulus of rupture

The modulus of rupture of fiber reinforced concrete beams at different temperatures, different times of exposure and cooled by air and water is shown in Fig. 7. Each point in the figure represents an average of the maximum modulus of rupture (fr) of three specimens compared with the laboratory average maximum modulus of rupture (fr') at laboratory temperature. In the temperature range of 25 to 200°C, the modulus decreased slightly. At temperature above

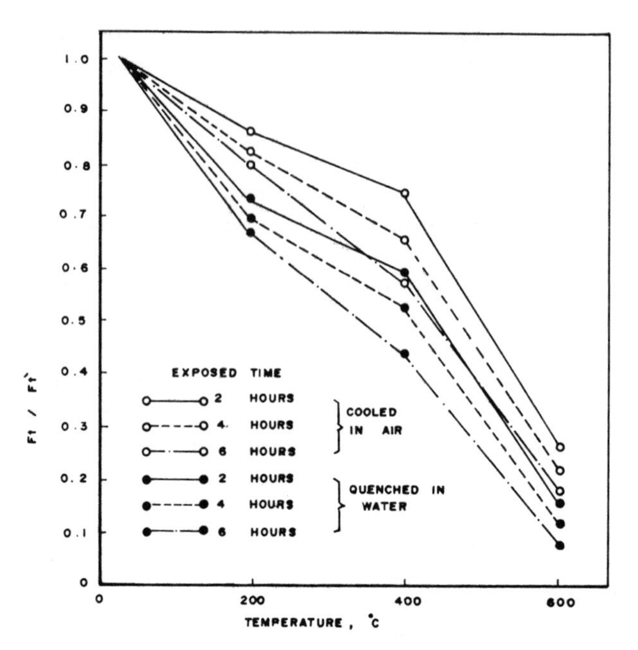

FIG. 5 : VARIATION OF SPLITTING TENSILE STRENGTH WITH TEMPERATURE

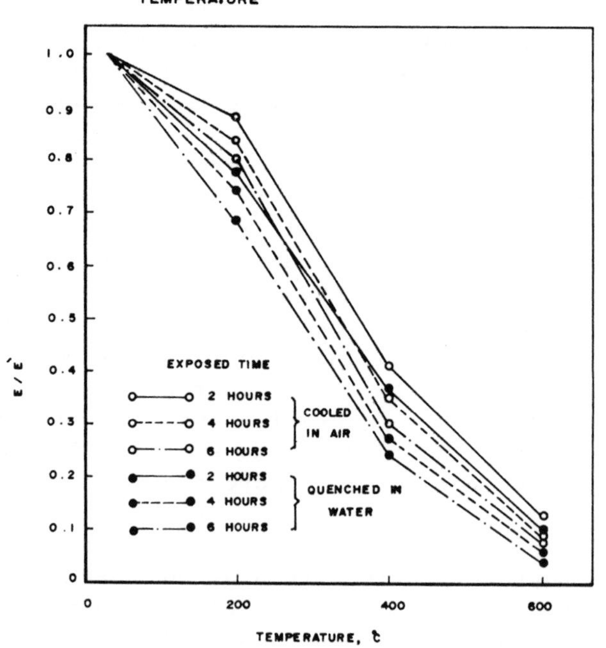

FIG. 6 : VARIATION OF STATIC MODULUS OF ELASTICITY WITH TEMPERATURE

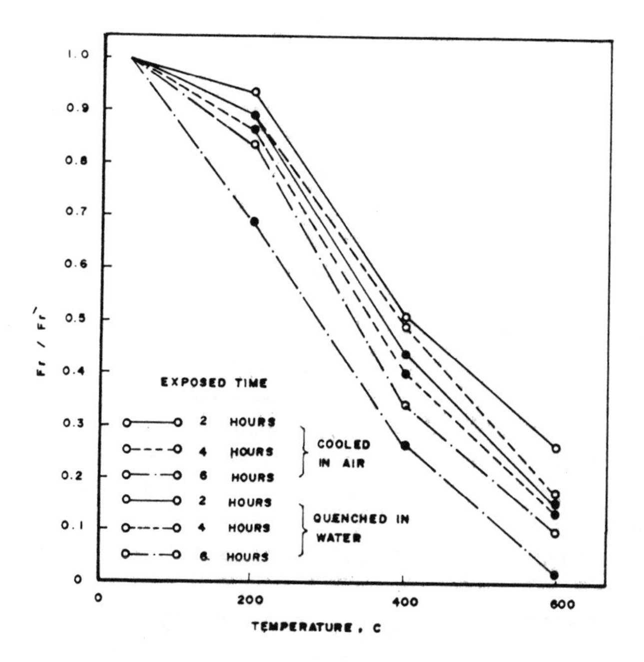

FIG. 7 : VARIATION OF MODULUS OF RUPTURE WITH TEMPERATURE

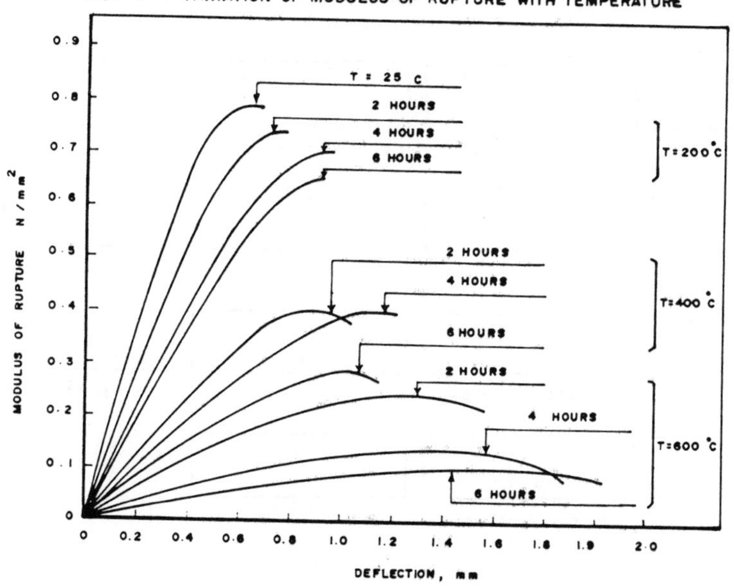

FIG. 8 : MODULUS OF RUPTURE AGAINST DEFLECTION AT HIGH TEMPERATURES COOLED IN AIR UNTIL FIRST CRACK

400 °C as the bond between fibers and concrete was gradually lost, the modulus of rupture decreased to about 10 and 2 percent of the value at laboratory temperature for air and water cooled specimens.

The relation between modulus of rupture and deflection at mid span of fiber reinforced concrete beams tested at three points loading at different temperatures, different exposure times and cooled in air is shown in Fig 8. The deflection at peak load did not very significantly within the temperature range at 25 to 200 °C. The deflection increased slightly between 2 to 6 hours exposed time. At 600°C, the deflection at peak load was about three times the deflection at laboratory temperature.

Crack pattern and failure mode of fiber reinforced concrete beams at different temperatures, different exposed times and cooled in air and water are shown in Fig. 9. The specimens exposed to a temperature of 400 to 600°C and quenched in water failed in brittle manner soon after reaching their peak strengths. The cracks were deep through the whole depth of the beams.

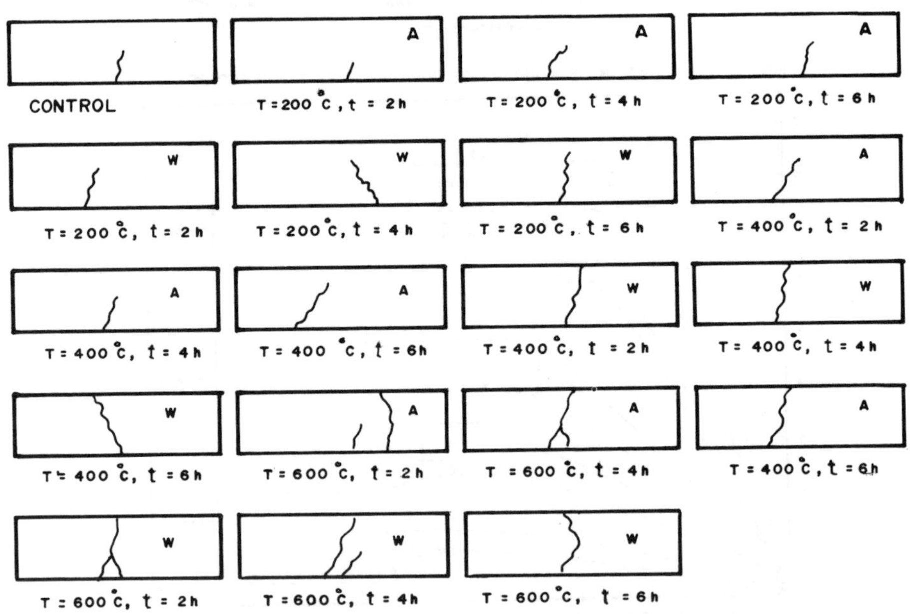

FIG. 9 : CRACKS AND FAILURE MODE OF BEAMS EXPOSED TO DIFFERENT TEMPERATURES (T), DURATION TIME (t) AND COOLED IN AIR OR QUENCHED IN WATER (W)

3.6 Influence of fiber on the mecchanical properties of concrete at high temperatures

Comparing the test results of this study with test results of plain concrete at high temperature[20], it was found that generally, a percentage of 2% of fibers improves the resistance of concrete at high temperature in the range of 5 to 28% approximately for the different duration times and method of treatment in this study.

4 Conclusions

Based on the results of this study, the following conclusions can be deduced.

1 When fiber reinforced concrete exposed to temperature in the range from 25 to 200°C, it showed a 1 to 27 % loss of compressive strength, 14 to 33% loss of tensile strength, 6 to 31 % of modulus of rupture and 12 to 31 % loss of modulus of elasticity. The range of loss depend on time of exposure to high temperature and cooling of specimens in air or quenched in water.

2 At temperature 400 °C, the fiber reinforced concrete progressively lost its strength which dropped to the range from 14 to 50% loss of compressive strength, 15 to 56 % of tensile strength, 49, to 74% of modulus of rupture and 59 to 76% of modulus of elasticity.

3 At temperature 600°C, the fiber reinforced concrete showed a loss in strength in the range from 33 to 75 % of compressive strength, 73, to 92% of tensile strength, 74 to 98 % of modulus of rupture and 89 to 96 % of modulus of elasticity.

4 The main factors affecting the strength loss of fiber reinforced concrete respectively are: High temperature, quenching heated specimens in water, increasing exposure time, and dimensions of specimens.

5 Fiber reinforced concrete resist high temperature more than plain concrete.

5 References

1 Fiber Reinforced Concrete (1948)- International symposium, Sp.81 American Concrete Institute, Detroit, PP 460.

2 Gopalaratnam, V.S. and Shah S.P., 1987, "Tensile Failure of steel Fiber Reinforced Mortar, Journal of

Engineering Mechanics, ASCE, V. 113,No.5,PP.635-652.
3 Johnston, C.D., (1980) "Properties of Steel Fiber Reinforced Mortar and Concrete", Proceedings, International Symposium on Fibrous Concrete (CI-80), Construction Press, Lancaster, PP. 29-47.
4 Johnston, Colin D.,(1982), "Steel Fiber reinforced and Plain concrete :, Factors influencing Flexural strength Measurement", ACI Journal, Proceedings V.79, No. 2, PP.131- 138.
5 P. "Complete Stress- Strain - Strain Curves for Steel Fiber Reinforced concrete in Uniaxial Tension and Compression", Testing and test Methods of Fibre Cement Composites,. RILEM Symposium 1978, construction Press, Lancaster,PP 399- 408.
6 Swamy, R.N., Al- Ta-an, S.A, and Ali, Sami A.R., (1987), " Steel Fibers for controlling cracking and Deflection", Concrete International : Design and Construction V.1, No.8, PP. 41-49.
7 Williamson, Gilbert R.,(1974) "The effect of steel Fibers on the Compressive Strength of Concrete". Fiber Reinforced Concrete, SP-44, ACI Detroit, PP. 195- 207.
8 Hannant, D.J., (1984), "Fiber Reinforced Cement and Concrete", Part 2. Practical Composites. Concrete (London), V.18, No 3, PP 21-22.
9 Snyder, M.Jack, and Lankard,David R., (1972) "Factors Affecting the Flexural strength of Steel Fibrous Concrete", ACI Journal, Proceedings V. 69, No.2, PP. 96-100.
10 Nanni. (1988) "Splitting- Tension Test for Fiber Reinforced Concrete", ACI Journal, Proceeding V.85 , No.4, PP. 229-233.
11 Malharta, H.L, (1956)," The Effect of Temperature on the Compressive Strength of Concrete," Magazine of Concrete Research (London), V.8, No.23, PP. 85-94.
12 Abrams, M.S., (1971) "Compressive Strength of Concrete at Temperatures to 1600F," Temperature and Concrete, SP-25 ACI, Detroit, PP. 33-58.
13 Saemann, J.C., and Washa, G.W., (1957),"Variation of Mortar and Concrete Properties with Temperature", ACI, Proceedings, V.54, No.5, PP. 385-395.
14 Hannant, D.J., (1964)", Effects of Heat on concrcte Strength", Engineering (London), V.203, No.21 PP. 302.
15 Davis, H.S, (1967) ,"Effects of High Temperature Exposure on Concrete", Material Research and Standards, V.7, No.10 PP. 452-459.

16 Castillo, C. and Durrani, A.J., (1990), "Effect of Transient High Temperature on High-Strength Concrete". ACI, Proceedings, V.87, No.1,PP 47-53.

17 Dias, W.P.S, Khoury, G.A., and Sullivan, P.J.E., (1990), "Mechanical Properties of Hardened Cement Paste Exposed to Temperature up to 700°C (1292 F)", ACI, Proceedings, V.87, No.2 PP. 160-165.

18 Papayianni, J., and Valiasis, T., (1991),"Residual Mechanical Properties of Heated Concrete Incorporating Different Pozzolanic Materials", RILEM, V.24 No. 140 PP. 115- 121.

19 Lankard, D.R., Birkimer, D.L., Fondfriest, F.F. and snyder, M.J., (1971), "Effects of Moisture content on the Structural Properties of Portland cement concrete Exposed to Temperature up to 500 F",Temperature and Concrete, SP-25, ACI, Detroit, PP. 95-102.

20 Kamal, M.M.,etal, 1990," The Effect of High Temperature on The Mechanical Prperties of Concrete", C.E.R.M, Vol 12, No 6, PP 86-99.

60 BEHAVIOUR OF FIBRE REINFORCED CONCRETE BEAMS EXPOSED TO FIRE

M. M. KAMAL, H. H. BAHNASAWY and F. E. EL-REFAI
GOHBPR, Cairo, Egypt

Abstract
The behaviour and strength of steel reinforced concrete beams cast with fibrous concrete and exposed to fire are included. The main variables taken into consideration were the type of the used steel fibres, fibres content, the temperature of exposure, the temperature duration and the method of treatment after exposure to fire. The deflection, longitudinal strains, cracking, ultimate loads and mode of failure were recoreded. Reinforced concrete beams cast with ordinary concrete and exposed to the same conditions as the steel fibre reinforced concrete were also tested for comparison .
Keywords : Steel fibrous reinforced concrete beams, Fire, Deflection, Longitudinal strains, Cracking, Ultimate load.

1 Introduction

During the last few decades extensive research work have been carried out to elucidate the potentiality and feasability of using fibrous concrete as a structural material. Most of these research dealt with the mechanical properties and durability of this non-traditinal material. Behaviour and strength of the different structural members made of fibre reinforced concrete have also been investigated (1 to 8). Most of these research were carried out in the laboratory or field environments. The effect of high temperature and fire on different types of building materials have been widely investigated (9 to 11). Among the many causes behinid fire accidents are the absence of the proper protection and maintenance of the gas or electrical systems and usage of synthetic materials as floor or wall coverings and furniture. The research in hand was conducted to study the behaviour and strength of reinforced concrete beams cast with different types of steel fibres and exposed to high temperature due to fire .

2 Experimental work

Five concrete mixes were firstly investigated to determine the effect of exposure to high temperature due to fire on the mechanical properties of plain and fibrous reinforced concrete specimens as shown in Table 1 .

The concrete mixes used were made of ordinary Portlanl cement, sand and gravel obtained from El-Yahmoum quarries near Cairo. The nominal maximum size of the used coarse aggregates was 20 mm. The mixes proportions by weight were 1 : 2 : 4 and the water cement ratio was o.5 by weight. The cement content used was 350 kg/m^3 of concrete. Galvanized plain round steel fibres of 0.5 mm diameter and 30 mm length were used in some mixes. Triangular twisted rugged surface steel fibres of 2 mm breadth, 0.25 mm thickness and 32 mm length were used with some other mixes. For each type of the used steel fibres two concrete mixes with two fibre contents of 1.5 % and 2 % by volume of concrete were tested to determine their mechanical properties after exposure to high temperature due to fire. A concrete mix without fibres was also tested for comparison. The fibres were dispersed uniformaly throughout the mix before adding the mixing water. Compression test was carried out on 150 x 150 mm cubes, four point bending test was carried out on beams of 100 x 100 x 500 mm to determine the concrete flexural strength and modulus of elasticty test was carried out using prisms of 100 x 100 x 500 mm dimensions. Test specimens were air cured in the laboratory atmosphere before exposure to fire at the age of 28 days after casting.

Twenty four reinforced concrete beams represent three different sets were cast and tested in this research as shown in Table 2 . Eight beams were cast of ordinary cement concrete, eight beams of plain

Fibre Reinforced Cement and Concrete. Edited by R. N. Swamy. © 1992 RILEM.
Published by E & FN Spon, 2-6 Boundary Row, London SE1 8HN. ISBN 0 419 18130 X.

Table 1 . Scheme of testing of fibre reinforced concrete specimens after exposure to fire

Mix oN	Type of steel fibres	Fibre content by vol . of	Temperature of exposure	Duration (hrs)	Treatment after exposure	Tested property
M1	——	——	25	-	air	Compressive strength Flexural strength Modulus of elasticity
M2			600	2	air	
M3			600	4	air	
M4			600	2	Water	
M5	Plain fibres	1.5	25	-	air	
M6			600	2	air	
M7			600	4	air	
M8			600	2	Water	
M9		2	25	-	air	
M10			600	2	air	
M11			600	4	air	
M12			600	2	water	
M13	Triangular twisted fibres	1.5	25	-	air	
M14			600	2	air	
M15			600	4	air	
M16			600	2	Water	
M17		2	25	-	air	
M18			600	2	air	
M19			600	4	air	
M20			600	2	Water	

For each test 3 Specimens were cast and testedConcrete mixes containing 1.5% fibre content were used in casting concrete beams

steel fibres concrete and eight beams of triangular twisted rugged surface steel fibres concrete. All the tested beams were of the same dimensions with 120x200 mm cross-section, a total length of 1800 mm and an effective span of 1600 mm. Each beam was reinforced longitudinally with 2 ϕ 13 mm mild steel bars, stirrups ϕ 6 mm with spacing 60 mm and 2 ϕ 8 mm stirrup hangers . The full dimensions and the details of the reinforcement of the beams are shown in Fig. 1 . Fibre reinforced concrete mixes with fibre content of 1.5 % by volume of concrete were used for each of the two types of steel used in this research as Shown in Table 2 . The concrete beams and the concrete control specimens were cast in layers in the moulds and compacted using electrical vibrator and rod tamping. They were cured in the laboratory atmosphere (25 ^oC and 50 % R. H.) for 28 days. For each set of beams two of them were tested without exposure to fire, two beams were exposed to fire at 600 ^oC for 2 hours and two beams were exposed to fire for four hours. These groups were left to cool down in the laboratory atmosphere and then tested. Another group consisted of two beams in each set was exposed to fire at 600 ^oC for 2 hours and then exposed to stream of water to cool down and then treated. Figure 2 shows the locations of Demec points fixed in positions to measure the longitudinal strains and the dial gauges for deflection recording. Beams were tested by the four point loading method with two equal concentrated loads at quarter points and spacing at 800 mm apart.

Table 2 . Scheme of testing of fibre reinforced concrete beams after exposure to fire

Mix N°	Type of steel fibres	Fibre content by vol . of concrete%	Temperature of exposure c°	Duration (hrs)	Treatment after exposure to fire	Tested property
B1	Without	———	25	-	air	Deflection Longitudinal strains Cracking Ultimate load Mode of failure
B2			600	2	air	
B3			600	4	air	
B4			600	2	Water	
B5	Plain fibres	1.5	25	-	air	
B6			600	2	air	
B7			600	4	air	
B8			600	4	Water	
B9	Triangular twisted fibres	1.5	25	-	air	
B10			600	2	air	
B11			600	4	air	
B12			600	2	water	

Each Symbol represents two beams

Loads were applied on successive increments. After each load increment longitudinal strains and deflections were recorded up till beam failure. Initial cracking load and cracking pattern were recorded with each load increment.

The fire room used was of a movable roof built specially for the sake of this research. The dimensions of that room were 2400 mm length, 900 mm width and 950 mm height. The room had internal steped walls at different heights to achieve different spans of beams. It was built with sand bricks and covered from inside with thermal bricks as shown in Fig. 3 . The fire room was provided with two gas sources which achieved uniform fire intensity along the room. The level of this fire sources could be changed up and down along the room height to achieve the required temperature of expousre to concrete beams which were layed on two supports 1600 mm apart during the fire. The concrete control specimens were layed on steel mesh during the fire. A steel plate was used as a roof cover to the room during the fire. The temperature of exposure was measured very close to the beam bottom and the control specimens usings a thermocouple connected to an apparatus for measuring temperature.

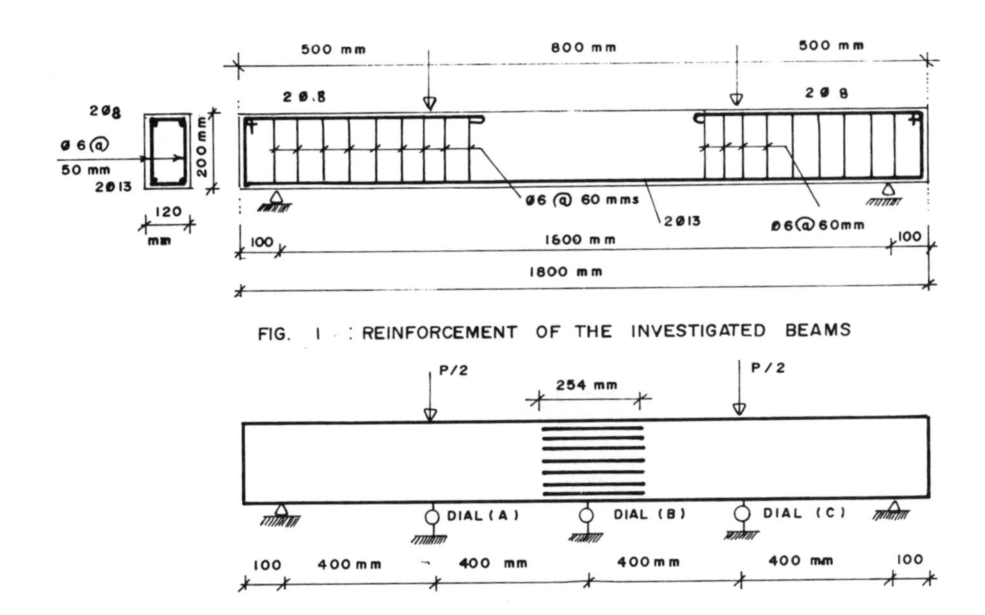

FIG. 1 : REINFORCEMENT OF THE INVESTIGATED BEAMS

FIG. 2 : POINTS OF DEFLECTION AND STRAIN MEASUREMENTS ON BEAMS

3 Analysis and discussion of test results

3.1 Effect of fire on mechanical properties of steel fiber reinforced concrete

Figure 4 shows the effect of exposure to high temperature due to fire on the compressive strength, flexural strength and modulus of elasticity of steel fibre reinforced concrete. It is clear that the inclusion of steel fibres in the concrete mix increased the mechanical properties in the laboratory atmosphere. Within the range of steel fibre content used in this investigation increasing the steel fibre content improved the mechanical properties. However for fiber content higher than 1.5% the rate of improvement the mechanical properties deceased.

Exposure to fire at 600 ^{0}C for 2 hours followed by air cooling led to distinguished decrease in the mechanical properties. The reduction in the concrete mechanical properties increased with the increase of the duration to high temperature. The rate of decrease in the mechanical properties due to exposure to high temperature was higher for the concrete specimens cast without fibres. It was found that the higher the fibre content in the concrete mix the higher the resistance to fire destructive effect as shown in Fig. 4 .

Treatment with water after exposure to fire at 600 ^{0}C for 2 hours led to a very high destructive effect. Cracking and high reduction in the mechanical properties of concrete were recorded for plain concrete specimens. Less destructive effect was found with those specimens cast with steel fibres. Insignificant differences were recorded in the mechanical properties for those specimens contained different percentages of steel fibres exposed to fire and treated with water.

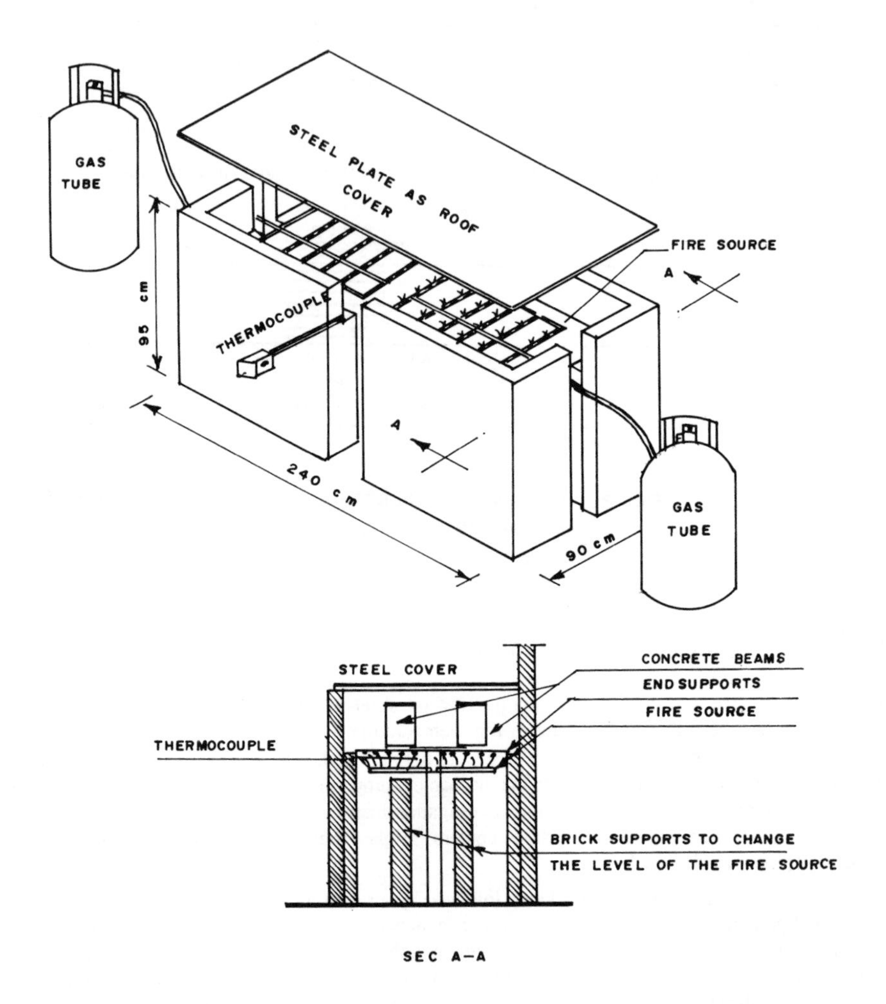

FIG. 3 : FIRE ROOM.

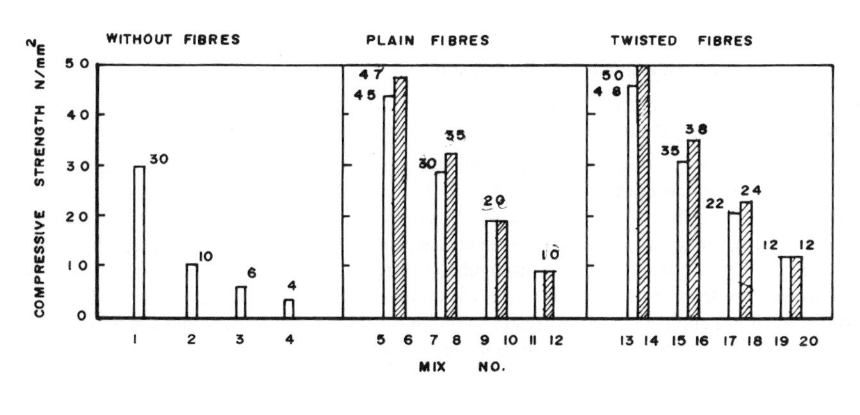

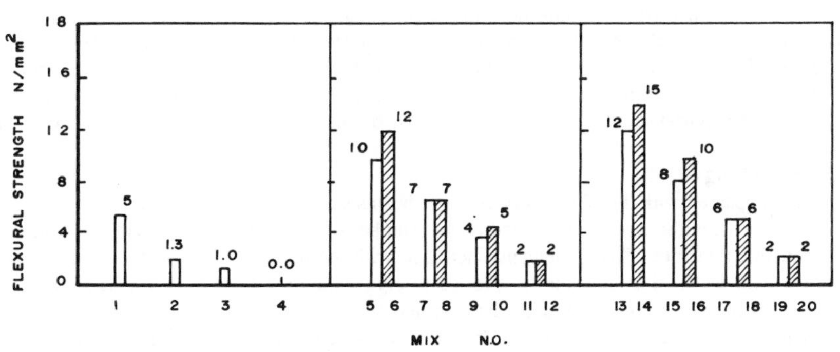

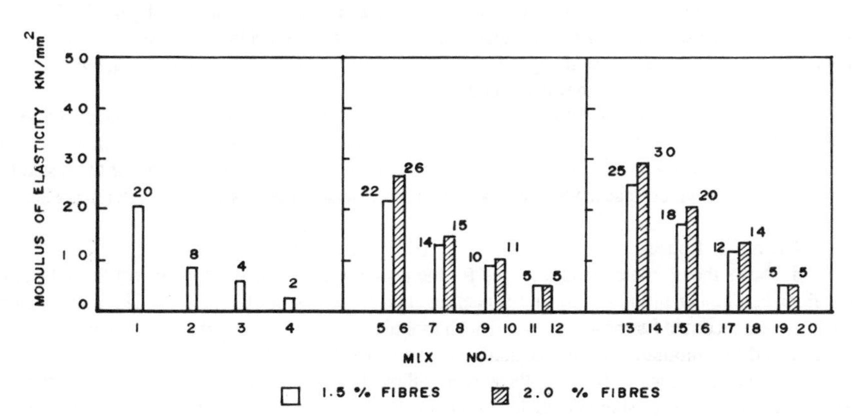

FIG. 4 : EFFECT OF HIGH TEMPERATURE ON THE MECHANICAL PROPERTIES OF STEEL FIBRE REINFORCED CONCRETE

3.2 Behaviour and strength of steel fibre reinforced concrete beams after exposure to fire

3.2.1 Deflection

Figure 5 shows the maximum deflection recorded at mid span of the beams under investigation under different stages of loading up till failure. At laboratory atmosphere (20 ^{0}C) those beams cast with triangular twisted rugged surface steel fibres exhibited the highest flexural rigidity and those cast without fibres, the lowest flexural rigidity. It was found that exposure to fire with 600 ^{0}C decreased the flexural rigidity of all the tested beams. The reduction in the flexural rigidity increased with the increase of the exposure duration to fire. Treatment of the beams with water after exposure to fire greatly decreased their flexural rigidity. Inclosure of steel fibres decreased the reduction in the flexural rigidity due to high temperature. Those beams cast with triangular twisted rugged surface steel fibres exhibited higher resistance to fire than those cast with plain round steel fibres which might be due to the higher bond between the fibres and the concrete .Treatment of reinforced concrete beams with water after exposure to high temperature led to tremendous decrease in their flexural rigidity.

3.2.2 Longitudinal strains

Figures 6 and 7 show the maximum compressive and tensile strains recorded at the extreme fibres of the tested beams respectively. At laboratory atmosphere those beams cast with triangular twisted rugged surface steel fibres exhibited the lowest strains at different stages of loading while those cast without fibers exhibited the highest strains. The longitudinal strains increased with the increased of the exposure temperature. The highest strains were recorded for those beams treated with water after exposure to fire.

3.2.3 Cracking

Hair cracks appeared with discolouring of concrete beams surfaces after exposure to fire before loading. Those beams cast without fibres exhibited bigger number of cracks with wider widthes than those cast with steel fibre concrete. This phenomena increased with the increase of the exposure duration to fire. Treatment with water after exposure to fire led to bigger number of cracks with wider widthes. Less cracking was observed with those beams cast with steel fibres.

Figure 8 shows the initial cracking loads for the tested beams after exposure to fire and cooling down in air or treatment with water. Concrete mixes cast with steel fibres showed higher initial cracking loads than those cast without fibres. The highest values of the initial cracking loads were recorded for those beams cast with trianguler twisted rugged surface steel fibres. The lowest initial cracking loads were recorded for the beams treated with water after exposure to fire.

Figure 9 shows the crack pattern of the tested beams at different stages of loading up till failure. It could be observed that those beams cast without fibres exhibited higher number of cracks with wider widthes than those cast with steel fibres. Cracks were concentrated in the middle part of the tested beams exposed to fire. The widest cracks were observed with those beams treated with water after exposure to fire.

3.2.4 Ultimate Loads

Figure 8 shows the ultimate loads recorded for the tested beams. It is quite clear that those beams cast with fibres had higher ultimate carrying capacities than those cast without fibres. The highest carrying capacities were recorded for those beams cast with tringular twisted rugged surface steel fibres. Exposure to fire decreased the ultimate carrying capacities of the tested beams. Those beams cast with steel fibers exhibited higher resistance to fire than those cast without fibres and consequently higher initial cracking loads and ultimate loads as shown in table 3. Treatment with water of those beams exposed to fire led to high reduction in their ultimate capacities.

4 Conclusions

1 Exposure to fire has a destructive effect on concrete and consequently a reduction in the mechanical properties which increases with the increase of the temperature of exposure and the fire duration.
2 Treatment with water of those concrete specimens exposed to fire led to dramatic decrease in their

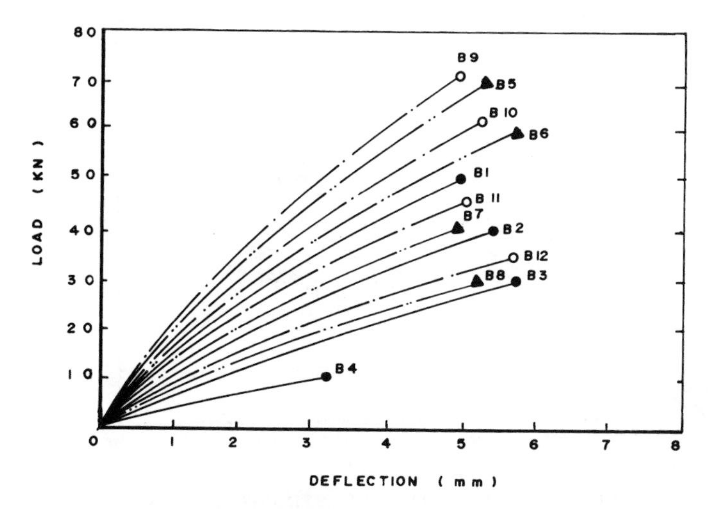

FIG. 5 : MAXIMUM DEFLECTION AT DIFFERENT STAGES OF LOADING OF TESTED BEAMS

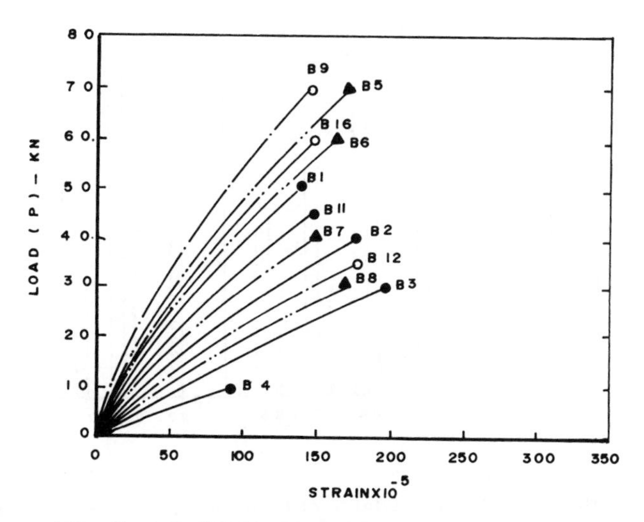

FIG. 6 : MAXIMUM COMPRESSIVE STRAIN AT EXTREME FIBRE OF TESTED BEAMS

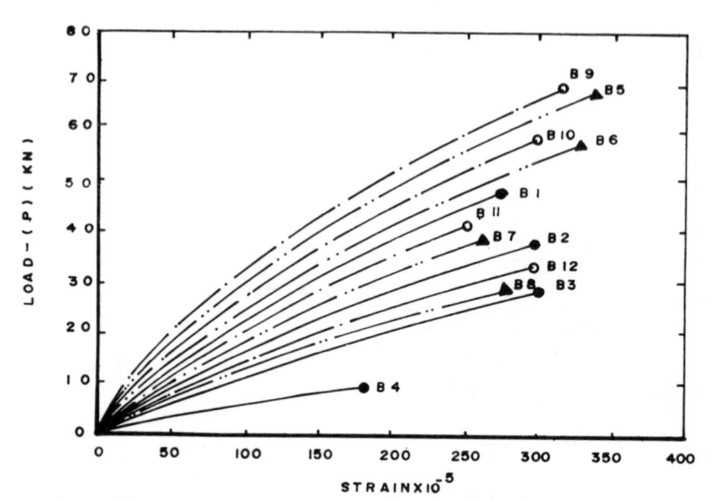

FIG. 7 : EXTREME FIBRE TENSILE STRAINS OF TESTED BEAMS

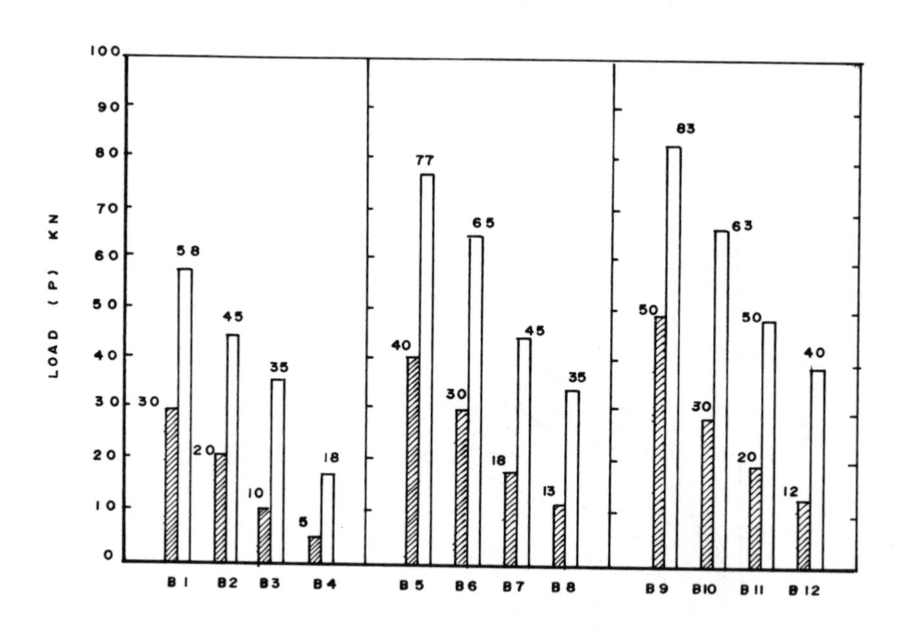

FIG. 8 : INITIAL CRACKING LOADS AND ULTIMATE LOADS OF TESTED BEAMS

Table 3. Relationships between cracking loads and ultimate loads of the tested beams

Beam N°	Cracking Load / Ultimate load %	Ult.ld.of beam exposed to fire / ult. ld. of reference beam B1 %
B1	51.7	100
B2	44.4	77.6
B3	28.6	60.3
B4	27.7	31
B5	52	132.8
B6	46.2	112.0
B7	40	77.6
B8	37.1	60.3
B9	60.2	143
B10	47.6	108.6
B11	40	86.0
B12	30	69

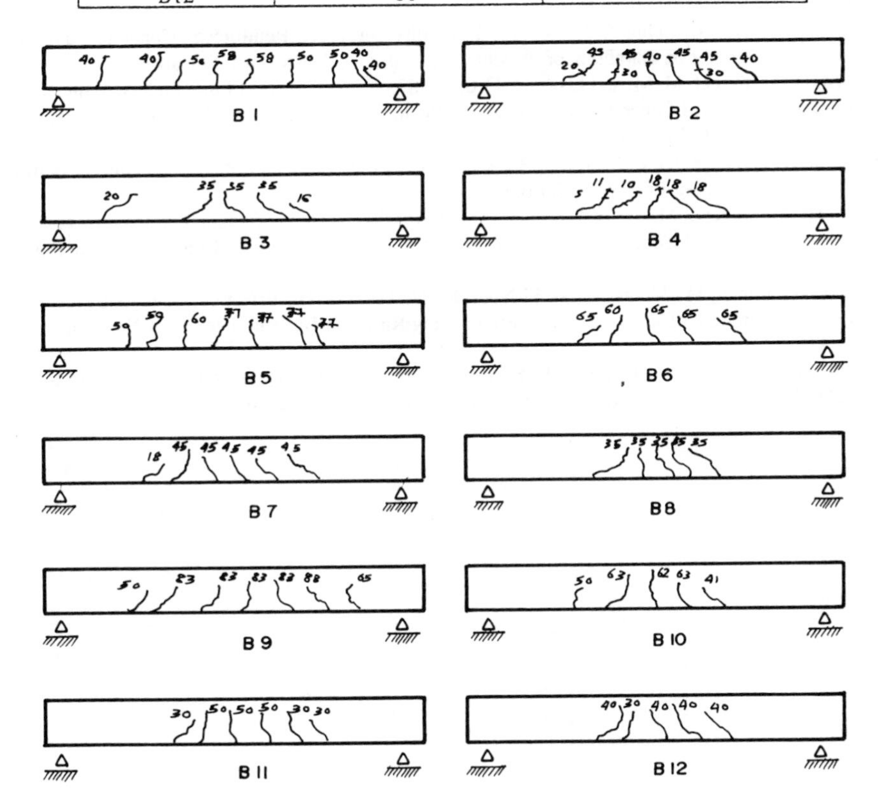

FIG. 9 : CRACK PATTERN FOR THE TESTED BEAMS

rigidity and strength.

3 Steel fibre reinforced concrete beams exhibited higher flexural rigidity, initial cracking loads and ultimate loads than those cast without steel fibres. The enhancement was greater when triangular twisted rugged surface steel fibres were used in comparison with the plain round fibres.
4 Extention of this research work will be carried out on structural members exposed to fire under different levels of loadings. Repair and strengthening of such members will also be investigated.

5 References

1 Swamy, R. N. and Al-Noori, (1975), "Flexural Behaviour of Fibre Concrete with Conventional Steel Reinforcement", Proc. RILEM Symp. Fibre Reinforced Cement and Concrete, Lancaster, U. K.

2 Rahimi, M. M. and Kesler, C. E., (1979), "Partially Steel-Fibre Reinforced Mortar", Jour. of the Struct. Div., ASCE, Vol. 105, No. ST1.

3 Samarai, M. A. (1980), "Fibrous Concrete : Its Effect on Crack Control in Reinforced Concrete Slabs", First Egyptian Structural Engineering Conference, Cairo University, Egypt.

4 El-Refai., F. E. and Kamal, M. M., (1986), "Behaviour of Polymer Fibre Reinforced Concrete Beams", Third International Symposium on Development in Fibre Reinforced Cement and Concrete, Rilem Technical Committee, 49-TFR, Sheffield, England.

5 El- Refai, F. E., Kamal, M. M., (1987), "Impact Resistance of Fibre Reinforced Concrete", First International Symposium on Housing Technology Production and Transfer from Research Into Practice, BRAZIL.

6 Kamal, M. M. and El-Refai, F. E., (1987), "Durability of Fibre Reinforced Concrete", Fourth International Conference on Durability of Building Materials and Components, Singapore.

7 El-Sayed, H. A., Kamal, M. M. and El-Refai, F. E., (1987), "Effect of Steel Fibre Inclusion on some Engineering Properties of Concrete and Corrosion of the Main Reinforcement", Journal of the Egyptian Society of Engineers, Cairo, Egypt.

8 Kamal, M. M. and El-Refai, F.E., (1987), Behaviour of Fibre Reinforced Concrete Columns", Journal of the Egyptian Society of Engineers, Cairo, Egypt.

9 Kamal, M. M., Metawi, M. A. and Kamel, M. (1990), "The Effect of High Temperature on the Mechanical Properties of Concrete", Civil Eng. Research Magazine, Faculty of Eng., Al-Azhar Univ., Cairo.

10 Tawfik, S. T. Kamal, M. M. and Abd El-Nour, K. N., (1991), "The Effect of Temperature on the Behaviour and Strength of Epoxy Mortar", Engineering Research Bulletin, Faculty of Eng., Univ. of Helwan, Cairo.

11 Latif, H. M., Ragab, A. and El-Refai, F. E., (1991), "Flexural Behaviour of R. C. Concrete Slabs Exposed to Fire", Fourth Arab Structural Engineering Conference, Faculty of Eng., cairo Univ.

61 MECHANICAL PROPERTIES OF FIBRE REINFORCED MORTAR EXPOSED TO GAMMA RADIATION

R. H. MILLS
Department of Civil Engineering, University of Toronto, Canada
C. SENI
AECL-CANDU, Mississauga, Canada

Abstract
Fibre reinforced mortar (FRM) modulus of rupture beams were exposed to 42 and 142 Mrad of Gamma radiation before testing. The effect of radiation was to increase strength and, except for steel reinforced FRM, reduce fracture toughness.
Keywords: Fibre Reinforced Mortar, Nuclear Shielding Radiation, Strength, Embrittlement, Toughness.

1. Introduction

These tests were designed to measure the influence of gamma radiation on flexural strength, elastic modulus, flexural rigidity and toughness index of Fibre Reinforced Mortar (FRM) beams. The specimens were subjected to two levels of gamma radiation, viz: (a) 42 Mrad, equivalent to 100 years service in a CANDU 6 concrete containment building; and (b) 142 Mrad, equivalent to 100 years service in a CANDU 6 concrete containment building, plus a major accident. For each level of exposure the performance was to be compared with companion specimens which were not exposed to radiation. Five different fibres were used in the concentrations shown in Table 1.

Table 1 Fibre Content and Characteristics of FRM's

Code	Name	Fibre Material	V_f % A	B	Length (mm)	Aspect Ratio
SFRM	Dramix	Steel	1.5,	2.5	30.0	75.0
GFRM	Cem-Fil	AR Glass	0.8,	1.6	25.0	1838.0
PFRM	Monofil	Polypropylene	1.5,	2.25	25.0	125.0
CFRM	Dialead	Carbon	0.5,	1.5	18.0	1059.0
AFRM	Kevlar 49	Aramid	0.75,	1.5	12.5	800.0

Fibre Reinforced Cement and Concrete. Edited by R. N. Swamy. © 1992 RILEM.
Published by E & FN Spon, 2-6 Boundary Row, London SE1 8HN. ISBN 0 419 18130 X.

2. Experimental Programme

2.1 Mortar Mix

The mortar mix was designed to give adequate workability with a water/cement ratio = 0.42. This necessitated the use of Lomar D Superplasticiser. The mix was modified by the addition of a Latex monomer to the fresh mix.

Mix details are given in Table 2. The mortar was mixed in a paddle mixer for two minutes before the fibres were added gradually by hand taking care to obtain uniform dispersion.

Table 2 Mix proportions of mortar matrix

	Mass kg/m^3
Rapid hardening cement (ASTM Type)	618
Quartz sand	1374
Water	259
Rhoplex latex emulsion	36
Lomar-D powder	24

2.2 Specimens

For each of the five fibre types, and for two fibre concentrations, a slab of FRM measuring 1 x 0.3 x 0.025 m was cast. After 14 days moist curing these specimens were cut into modulus of rupture beams measuring 300 x 60 x 25 mm. After a further 14 days moist curing the specimens were irradiated at dose rates varying from 1.54 to 1.79 Mrad/hour.

Each slab was saw-cut to provide 9 beams for each fibre and each fibre concentration. These were divided into 3 groups of 3 replicates. One group was tested after exposures ranging from 42 to 49 Mrad and a second after exposures ranging from 142 to 165 Mrad. The third group was left unexposed to radiation.

The 99 modulus of rupture beams included the following variables:

- 5 different fibres in the FRM
- 2 different concentrations of fibres in the FRM
- 3 specimens to be exposed to 42 Mrad i.e. 30 FRM + 3 plain = 33 in all
- 3 specimens to be exposed to 142 Mrad i.e. 30 FRM + 3 plain = 33 in all
- 3 specimens to be unexposed i.e. 30 FRM + 3 plain = 33 in all

2.3 Treatment

Before the 66 specimens were dispatched for irradiation, they were moist cured for 28 days. After irradiation at an average rate of 167 Mrad/hr, the beams were stored at 50% relative humidity until they were tested at 56 days. The control specimens were maintained in an atmosphere of 100 per cent R.H for 56 days until tested.

2.4 Flexural Tests

The specimens were tested in 3-point bending with auto-recording of load and deflection. The effectiveness of the fibre reinforcing, before and after radiation, was assessed according to the following parameters which are also illustrated in Figures 1 and 2.

- σ_e Flexural strength at the elastic limit also known as First Crack Strength (1).
- σ_c Flexural strength at maximum load.
- E Modulus of Elasticity calculated from the linear part of the load-deflection curve (ASTM C1018).
- TI The area under the load-deflection curve up to a specified end point divided by the area under the linear elastic part of the curve.
- I_n The end point, corresponding to deflection = $(n + 1)\delta/2$, which gives TI = n. In this investigation values I_5 and I_{10} corresponding to deflections 3δ and 5.5δ were adopted for comparison (Johnson 1991).
- E/20 The ratio of the area under the load-deflection curve in the post-cracking region divided by the area under the linear elastic curve. In this case the end point was defined by the intersection on the curve of a line drawn from the origin with 1/20 the initial slope.

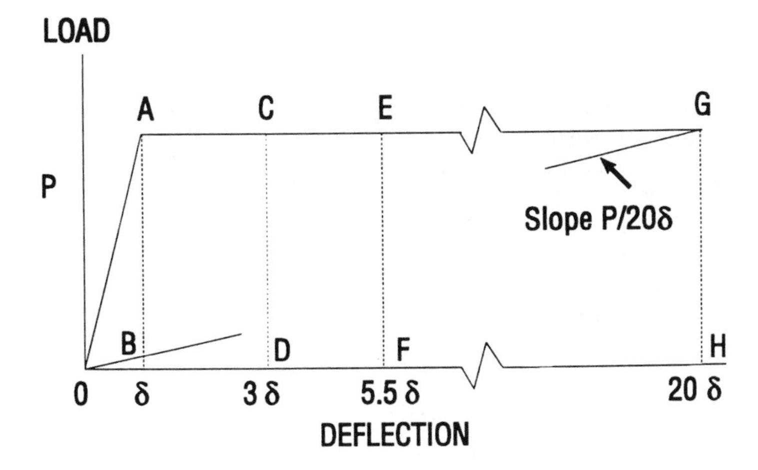

Fig. 1 Linear Elastic - Perfectly Plastic behaviour of a beam in flexure. The toughness index I_5 = Area OACD divided by Area OAB = 5. Similarly I_{10} = Area OAEF ÷ Area OAB. E/20 = Area BAGH ÷ Area OAB. Residual strength over the range CE = $R_{5,10} = 20\ (I_{10} - I_5)$ = 100 per cent.

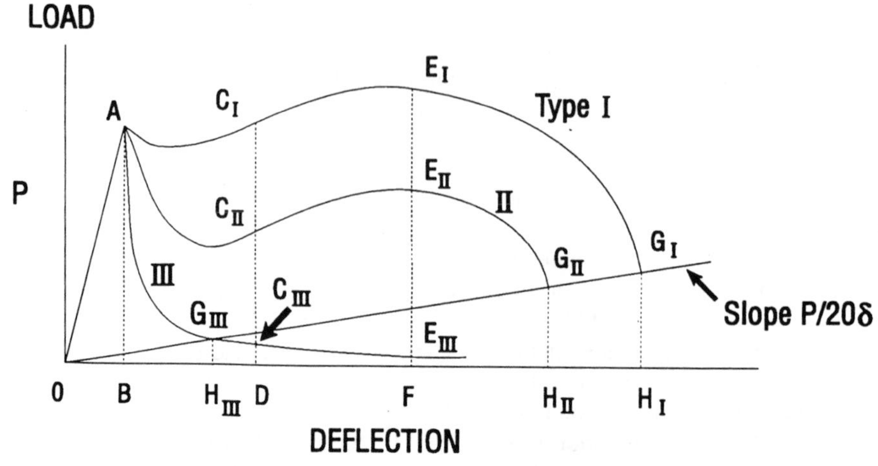

Fig. 2 Typical load-deflection curves for fibre reinforced mortar. Curve I shows quasi strain hardening in the post-cracking region: $I_5 > 5$ and $I_{10} > 10$. The residual strength $R_{5,10} = 20\ (I_{10} - I_5) > 100\%$. Curve II shows $I_5 < 5$, $I_{10} < 10$ and $R_{5,10} = 20\ (I_{10} - I_5) < 100\%$. This behaviour shows substantial ductility and good post-cracking characteristics though somewhat inferior to the Linear Elastic-Perfectly Plastic case. Curve III shows totally inadequate post-cracking performance with $I_5 << 5$, $I_{10} << 10$ and $R_{5,10} << 100\%$.

3. Results

The results of the flexural tests are summarised in Table 3. Typical curves showing the effects of radiation are given in figures 3, 4 and 5.

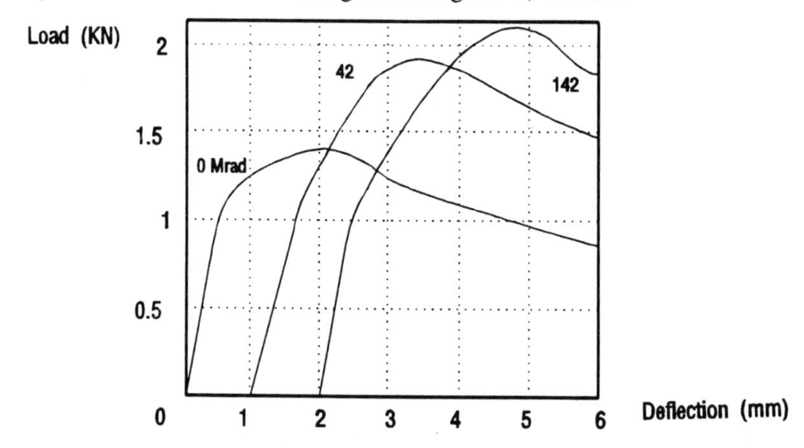

Fig. 3 Type I - load-deflection curves for specimens reinforced with 2.5 volume per cent steel fibres. Note strength increase with radiation.

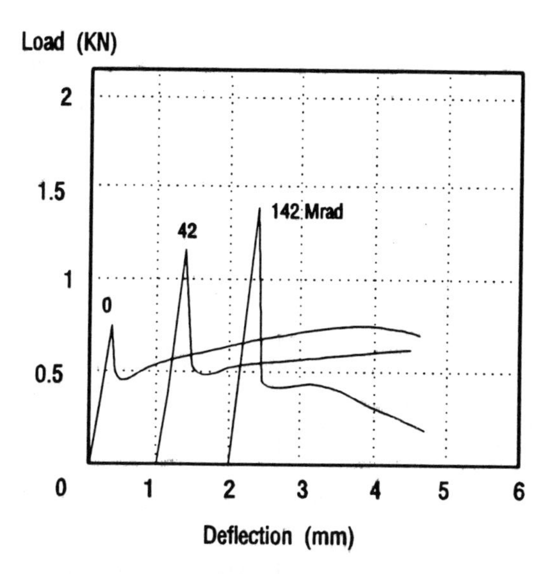

Fig. 4 Type II - load-deflection curves for specimens reinforced with 2.25 volume per cent of Polypropylene fibres.

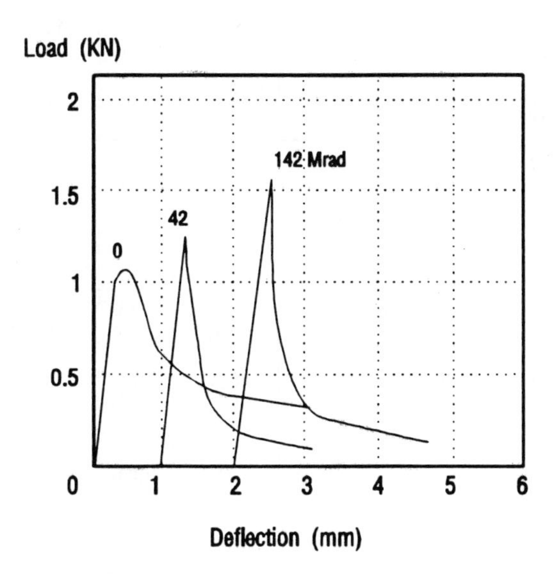

Fig. 5 Type III - load-deflection curves for specimens reinforced with 1.5 volume per cent of Dialead Carbon Fibres.

Table 3 Summary of Flexural Test Results

		Stresses				Toughness Indices		
		Elastic Limit σ_e (MPa)	Maximum Strength σ_{max} (MPa)	Residual Strength R5,10%	Elastic Modulus E (GPa)	I_5	I_{10}	E/20
Unreinforced	N-0	6.11 (0.05)	6.11 (0.05)	6	8.53 (0.29)	0.82 (0.13)	1.12 (0.33)	0.72 (0.11)
	N-42	10.10 (0.44)	10.10 0.44)	6	14.10 (1.54)	0.60 (0.06)	0.92 (0.07)	0.45 (0.00)
	N-142	11.84 (0.57)	11.84 (0.57)	7	14.82 (0.73)	0.56 (0.15)	0.92 (0.23)	0.38 (0.18)
Steel	SA-0	8.12 (0.72)	8.50 (0.93)	88	5.48 (2.07)	3.20 (0.67)	7.61 (0.33)	12.04 (3.63)
	SB-0	9.23 (0.16)	11.77 (1.21)	207	10.11 (2.39)	5.87 (0.46)	16.23 (2.74)	37.10 (5.80)
	SA-42	8.25 (0.61)	9.32 (0.87)	154	13.64 (0.41)	5.18 (0.84)	12.87 (2.55)	39.20 (8.20)
	SB-42	7.05 (0.40)	8.93 (0.07)	182	12.31 (0.06)	6.97 (0.47)	16.08 (1.64)	77.29 (15.06)
	SA-142	6.02 (0.16)	8.62 (0.07)	316	9.91 (1.69)	6.16 (0.41)	14.3 (0.8)	54.80 (12.93)
	SB-142	7.87 (0.35)	9.73 (0.52)	152	11.12 (1.86)	6.04 (2.39)	13.62 (5.19)	87.70 (36.6)
Glass	GA-0	5.77 (0.69)	5.92 (1.98)	99	11.93 (0.85)	4.34 (0.95)	9.30 (0.96)	10.04 (1.49)
	GB-0	8.40 (0.24)	8.71 (0.75)	76	6.77 (0.56)	4.24 (1.03)	8.04 (2.50)	9.84 (3.08)
	GA-42	9.82 (0.38)	7.03 (0.75)	58	12.16 (0.57)	3.15 (0.30)	6.03 (0.89)	6.66 (1.16)
	GB-42	11.36 (0.85)	12.26 (1.13)	20	8.32 (1.63)	2.4 4 (0.36)	3.45 (0.54)	2.93 (0.51)
	GA-142	10.60 (1.93)	6.84 (0.32)	57	12.26 (1.13)	3.31 (1.89)	6.16 (3.64)	7.63 (6.03)
	GB-142	11.82 (0.73)	8.64 (5.64)	30	11.97 (1.73)	2.97 (0.80)	4.45 (1.54)	4.22 (1.90)

Table 3 Summary of Flexural Test Results cont'd

		Stresses				Toughness Indices		
		Elastic Limit σ_e (MPa)	Maximum Strength σ_{max} (MPa)	Residual Strength R5,10%	Elastic Modulus E (GPa)	I_5	I_{10}	E/20
Polypropylene	PA-0	5.97 (0.59)	6.28 (0.86)	94	6.91 (1.75)	5.03 (2.55)	9.75 (5.47)	27.13 (11.43)
	PB-0	4.92 (0.59)	6.04 (0.04)	110	10.54 (0.92)	4.38 (0.92)	9.88 (1.83)	63.61 (30.51)
	PA-4.2	7.44 (0.58)	6.69 (1.04)	71	8.95 (0.94)	2.55 (0.15)	6.08 (0.90)	24.89 (15.31)
	PB-4.2	10.23 (0.01)	10.23 (0.01)	48	13.65 (0.11)	1.29 (0.42)	3.68 (0.07)	8.68 (1.61)
	PA-14.2	11.48 (0.38)	11.48 (0.38)	25	13.01 (2.10)	1.36 (0.46)	2.52 (0.46)	2.31 (0.67)
	PB-14.2	11.15 (0.55)	11.15 (0.55)	18	12.85 (0.03)	1.04 (0.19)	1.92 (0.25)	1.47 (0.48)
Kevlar	KA-0	5.73 (0.32)	5.98 (0.03)	16	6.38 (0.73)	1.61 (0.38)	2.42 (0.60)	1.86 (0.58)
	KB-0	5.64 (0.57)	6.16 (0.55)	30	6.63 (1.02)	2.44 (0.42)	3.94 (0.99)	3.66 (1.30)
	KA-4.2	9.48 (0.28)	9.31 (0.07)	11	8.08 (0.70)	1.10 (0.32)	1.63 (0.38)	0.84 (0.16)
	KB-4.2	8.43 (1.82)	7.53 (0.26)	18	7.58 (0.03)	2.73 (0.26)	3.63 (0.33)	3.84 (0.20)
	KA-14.2	9.95 (0.25)	9.95 (0.25)	14	9.56 (1.38)	1.44 (0.20)	2.12 (0.25)	1.55 (0.27)
	KB-14.2	9.27 (0.33)	9.27 (0.33)	13	10.23 (0.25)	1.25 (0.04)	1.89 (0.00)	1.30 (0.07)
Carbon	CA-0	6.82 (0.61)	7.27 (0.67)	30	9.52 (2.07)	2.91 (0.31)	4.40 (0.67)	4.89 (1.34)
	CB-0	6.38 (0.04)	6.80 (0.21)	49	8.86 (1.59)	4.50 (0.60)	6.95 (2.06)	8.05 (3.06)
	CA-4.2	8.94 (0.12)	9.18 (0.29)	17	10.20 (0.74)	1.91 (0.45)	2.74 (0.54)	2.41 (0.67)
	CB-4.2	9.51 (0.31)	9.51 (0.31)	35	10.60 (0.54)	2.21 (0.35)	3.98 (1.32)	3.02 (1.23)
	CA-14.2	10.85 (0.94)	10.85 (0.94)	10	11.80 (1.73)	1.25 (0.10)	1.74 (0.28)	1.35 (0.06)
	CB-14.2	10.37 (0.52)	10.37 (0.52)	13	10.40 (0.22)	1.31 (0.18)	1.96 (0.16)	1.49 (0.04)

3.1 The Effect of Radiation on Cracking Strength, σ_e.

In this context cracking strength, σ_e, is defined as the flexural stress corresponding to the first deviation from linearity of the load-deflection curve (ASTM 1018).

After exposure to radiation, all the FRM beams except those reinforced with steel fibres showed strength increases ranging from 40% (Glass B) to 128% (Poly B) as shown in Table 4. The unreinforced control groups showed 65% increase in strength and the Steel A group were weakened by 25 per cent. It should be emphasized that σ_e represents the end of linear elastic behaviour and should not be regarded as a disabling stress except in unreinforced FRM; or where the fibre reinforcement is ineffective.

Table 4 Effects of radiation on cracking strength. Figures give cracking strength σ_e MPa

Radiation MRad		0	42	142
Plain		6.1*	10.1	11.8
Steel	A	8.1	8.3	6.0
	B	9.2	7.1	7.9
Glass	A	5.8	9.8	10.6
	B	8.4	11.4	11.8
Poly	A	6.0	7.4	11.5
	B	4.9	10.2	11.2
Kevlar	A	5.7	9.5	10.0
	B	5.6	8.4	9.3
Carbon	A	6.8	8.9	10.9
	B	6.4	9.5	10.4

* This means 6.1 MPa

3.2 The Effect of Radiation on Residual Strength

The desired characteristics of fibre reinforced materials include the capacity to support tensile stresses and absorb energy after the matrix has cracked. The residual strength is defined as follows:

Let I_P = the toughness index corresponding to deflection $P\delta$; and
I_Q = that corresponding to deflection $Q\delta$; where
δ = the initial elastic deflection

The residual strength R_{PQ} is the area under the load-deflection curve between $P\delta$ and $Q\delta$ divided by the difference in deflection ($Q\delta$ - $P\delta$). Thus $R_{PQ} = (I_Q - I_P)/$

($Q\delta$ - $P\delta$). For perfectly plastic behaviour as shown in Fig. 1, $R_{P,Q}$ say < 10 per cent indicates extensive damage and severely diminished structural capacity. Values of R_{PQ} above 100 per cent indicate superior structural performance.

The results tabulated in Table 5 indicate that steel gave superior performance after radiation while Glass A, and Polypropylene A gave fair to good performance while Glass B, Polypropylene B and the two organic fibres, Kevlar and Carbon gave unsatisfactory performances after similar exposure.

Table 5 Effect of radiation on residual strength over the range 3.0 to 5.5 times the elastic deflection. The Residual strength is given as a percentage of the cracking strength.

Radiation MRad		0	42	142	Classification Type
Plain		6	6	7	III
Steel	A	88*	154	316	I
	B	207	182	152	
Glass	A	99	58	57	II
	B	76	20	30	
Polypropylene	A	94	71	25	II
	B	110	48	18	III
Kevlar	A	16	11	14	III
	B	30	18	13	
Carbon	A	30	17	10	III
	B	49	35	13	

* This means that the residual strength $R_{5,10}$ is 88 per cent of the cracking strength.

3.3 Toughness

Two alternative measures of toughness were used: (a) Toughness index I_n = Area under the load-deflection (P-δ) curve from $\delta = 0$ to $\delta = (n + 1)\delta^*/2$ divided by the area under the linear-elastic part up to the cracking load P^*; where P^* and δ^* refer to the first deviation from linearity; and (b) the ratio of the area under the curve after first crack to the intercept of a second line with slope = slope of the elastic curve ÷ 20, divided by the area under the elastic curve (Mills, 1981). In the case of I_n, n represents the toughness index for the linear elastic - ideal plastic case shown in Fig. 1 and the performance of FRM is judged by the extent to which behaviour of the material under test approaches or exceeds the value.

For example performance may be valued as follows for I_R:

$I_R \geq n$	Superior performance
$I_R \geq 0.6$ n	Satisfactory performance
$I_R \geq 0.3$ n	Fair performance
$I_R < 0.2$ n	Unsatisfactory performance

where n = the toughness index of a linear elastic - perfectly plastic material.

The results recorded in Table 5 indicate superior performance for Steel fibres, satisfactory performance for Glass A and marginal performance for poly A. Poly B and the Kevlar and Carbon fibre reinforced specimens gave unsatisfactory performances.

In the case of the intercept method, the intercept defines a displacement OH which varies with the type of load-displacement curve (Mills, 1981). For linear elastic - perfectly plastic behaviour this method would correspond to a deflection = 20δ or I_{39}. If OH > $(n + 1)\delta^*/2$, the characterisation of load-deflection behaviour will be similar to that defined by I_n. As the load-deflection curve approaches Type III (Fig. 2) the area reduces to small values so that TI = E/20 < I_n. The structural value of the FRM reduces with the value of TI = E/20. When the E/20 value of TI < I_5, the material is deemed to have no useful performance in the post-cracking region. Examples falling into this category are:

All unreinforced mortar specimens

Kevlar A - 42 Mrad

Kevlar B - 142 Mrad

3.4 Modulus of Elasticity

The apparent modulus of elasticity was calculated from simple beam theory and the slope P^*/δ^* as

$$E = \frac{P*}{\delta *} \cdot \frac{l^3}{48} \cdot \frac{1}{I} = \frac{P*}{\delta *} \cdot \frac{1}{I} \text{ if } \frac{P*}{\delta *} \text{ is in } \frac{N}{mm}$$

where I = the moment of inertia determined from the cross-section dimensions of each beam;

l = the span of the beam.

Linear regression of pooled results for each FRM and the unreinforced control gave empirical equations relating strength at the elastic limit, σ_e, and the apparent Elastic modulus, E.

3.5 Tensile Strain at Cracking

The data of Table 3 show considerable variability in cracking strength and apparent Modulus of Elasticity. This was probably due to micro-cracking resulting from drying during irradiation and handling between the laboratory and the site of radiation some hundreds of kilometers away. A comparison of flexural strain at first crack is given in Table 7.

Table 6 Effect of radiation on Toughness Indices (TI) I_5 = the actual toughness index at a deflection which would give TI* = 5 for the Linear Elastic - Ideal Plastic Case. I_{10} corresponds to TI* = 10 on the same basis. The Roman superscripts identify the performance class.

Radiation MRad		I_5			I_{10}		
		0	42	142	0	42	142
Plain		0.8^{III}	0.6^{III}	0.6^{III}	1.1^{III}	0.9^{III}	0.9^{III}
Steel	A	3.2^{II}	5.2^{I}	6.2^{I}	7.6^{II}	12.9^{I}	14.3^{I}
	B	5.9^{I}	7.0^{I}	6.0^{I}	16.2^{I}	16.0^{I}	13.6^{I}
Glass	A	4.3^{II}	3.2^{II}	3.3^{II}	9.3^{II}	6.0^{II}	6.2^{II}
	B	4.2^{II}	2.4^{III}	3.0^{II}	8.0^{II}	3.5^{II}	4.5^{II}
Polypropylene	A	5.0^{I}	2.6^{III}	1.4^{III}	9.8^{II}	6.0^{II}	2.5^{III}
	B	4.4^{II}	1.3^{III}	1.0^{III}	9.9^{I}	3.7^{II}	1.9^{III}
Kevlar	A	1.6^{III}	1.1^{III}	1.4^{III}	2.4^{II}	1.6^{III}	2.1^{III}
	B	2.4^{III}	2.7^{III}	1.3^{III}	3.9^{II}	3.6^{III}	1.9^{III}
Carbon	A	2.9^{III}	1.9^{III}	1.3^{III}	4.4^{II}	2.7^{II}	1.7^{III}
	B	4.5^{III}	2.2^{III}	1.3^{III}	7.0^{II}	4.0^{II}	2.0^{III}

Table 7 $E = a + b\sigma_e$ (MPa)

Group	a	b	R^2
Unreinforced	270	1300	0.99
Steel	1060	1180	0.61
Glass	2140	840	0.59
Polypropylene	1500	1080	0.86
Kevlar	540	920	0.93
Carbon	1020	1030	0.93

R^2 = Coefficient of determination.

Table 8 Influence of Radiation on Cracking Strain Apparent cracking strain - 10^{-6}

Radiation	Zero	Pooled 42 MRad and 142 MRad
Unreinforced	716	757
Steel	1196	623 (58)
Glass	862	960 (164)
Polypropylene	665	832 (60)
Kevlar	874	1056 (113)
Carbon	718	922 (52)
Pooled all FRM	863 (311)	878 (174)

Figures in parenthesis are standard deviations

4. Discussion

Early tests on the effect of radiation on strength of concrete (4) indicated that both strength and stiffness were reduced. In the present series, the cracking strength of Steel B reduced with radiation and also for Steel A at the higher dosage but in all other cases it improved with radiation.

Like normal reinforced concrete, the desired structural performance of fibre reinforced concrete relates to the post-cracking region. The stress at first crack is of importance for functional performance of a nuclear containment.

The three classes of load-deflection curve of Fig. 2 are illustrated by the typical curves of Figures 3, 4 and 5. Fig. 3 shows the excellent Class I characteristics of Steel B and also the significant improvement in post-cracking performance resulting from irradiation. Fig. 4 shows intermediate Class II behaviour with satisfactory strain-hardening of the non-irradiated FRM and progressive increase in cracking strength and deterioration of post-cracking performance resulting from irradiation of Polypropylene-reinforced specimens. Fig. 5 shows unsatisfactory Class III performance of Carbon-reinforced FRM.

The effect of irradiation is also reflected in Tables 5 and 6, where it is clear that only steel reinforced FRM retained good class I behaviour following exposure.

The American Concrete Institute (1989) and Canadian CSA 23.3 (1977) give the Elastic modulus of concrete as

$$E = 5000 \sqrt{f_c'} \text{ MPa}$$

Also the modulus of rupture strength

$$f_t = 0.6 \sqrt{f'_c} \text{ MPa.}$$

Combining these equations gives

$$E = 8333 f_t.$$

The formulae in Table 7 give values of E much less than values estimated from CSA 23.3 (1977). The CSA estimate is based on conventional concrete containing about 68 volume per cent of high-modulus coarse aggregate with relatively low flexural strength. The heavily polymer modified mortar of these experiments contained only 26 volume per cent of high modulus aggregate and its tensile strength was enhanced compared with normal structured concrete.

Table 7 generally shows increases of mean tensile strain at first crack resulting from irradiation, compared with the corresponding strains of unexposed controls. Many authorities, for instance Neville (1981), give cracking strains for concrete ranging from 300 to 1000 microstrain. The values of Table 7, as expected, tend towards the high side of this range.

In summary, all of the FRM's showed satisfactory strengths after irradiation but all except steel showed serious embrittlement, which was most severe in the case of the three organic materials.

Acknowledgement

This work was sponsored by the Candu Owners Group (COG) whose support is gratefully acknowledged.

References

ASTM C 1018 (1989) Standard test method for flexural toughness and first-crack strength of fibre-reinforced concrete.

Johnson, C.D. (1991) Methods of evaluating the performance of fiber reinforced concrete. **MRS Symposium Proceedings**, 211, 15-24, 1991.

Mills, R.H. (1981) Age-embrittlement of glass reinforced concrete containing blastfurnace slag. **Cement and Concrete Research**, 11, 3, May, 421-428.

Batten, R.W.C. (1960) Effect of radiation on the strength of concrete. **UK Atomic Energy Authority Report**, Harwell, 1960, 10.

ACI Committee 318 (1989) Building code requirements for reinforced concrete (ACI 318-89), **American Concrete Institute**, Detroit, 1989, 360.

Canadian Standards Association (1977) Concrete materials and methods of construction CSA A23.3-M77.

Neville, A.M. (1981) Properties of concrete. 281-287, **Pitman Publishing Ltd.**, London.

62 DURABILITY OF FIBRE REINFORCED CONCRETE RELATED TO TYPE OF CEMENT AND ADMIXTURES

O. A. SALAH ELDIN and S. N. ELIBIARI
Building Research Center, GOHBPR, Cairo, Egypt

Abstract

The influence of two different types of admixtures on the durability of steel fibre concrete made with ordinary portland cement and sulphate resistant cement is studied. Durability is assessed by measuring the water absorption, the compressive strength and the action of sulphate ions at elevated temperature of 100% humidity. Good results have been achieved by using fibres for durability properties. The performance of concrete made with sulphate resistant cement gives some differences compared with that made of ordinary portland cement. A noticeable improvement in the durability of fibre reinforced concrete was found when cured in elevated temperature of 100% humidity.

Keywords: Durability, Fibres, Ordinary Portland Cement, Sulphate Resistant Cement, Concrete, Mortar, Admixtures.

1 Introduction

Deterioration of concrete structures due to corrosion of steel reinforcement either as bars or fibres has often been attributed to the permeability of concrete which leads to the attack of chloride ions or the carbonation of concrete cover. Many researches (1-5) have been carried out on the effect of these two main factors on reinforced concrete. It seems that not only chloride ion attack and carbonation, but also some other factors such as sulphates contribute to the initiation and rapid propagation of corrosion specially in high temperatures and humidity conditions.

The goal of this investigation is to study the effect of

Fibre Reinforced Cement and Concrete. Edited by R. N. Swamy. © 1992 RILEM.
Published by E & FN Spon, 2-6 Boundary Row, London SE1 8HN. ISBN 0 419 18130 X.

using two different admixtures on the porosity of steel fibre reinforced concrete with ordinary portland cement or sulphate resistant cement. It also aims to study the effect of sulphate ions of concentration 2% $Mg\ SO_4$ in the mixing water of fibre reinforced mortar, using two admixtures and two types of cement.

2 Materials Used

The concrete mix used is made from siliceous sand and gravel of maximum size 20mm obtained from quarries near Cairo. These aggregates are in compliance with the Egyptian Standard Specifications (6). Two types of cement were used, i.e. ordinary portland cement and sulphate resistant cement which comply with the Egyptian Standard Specifications (7,8). The water used in mixing is drinking water.

The admixtures used are Z-10 and Z-12 of type B according to ASTM C 494. The specific gravity of the used steel fibres is 7.6 and their size is 0.3mm diameter and 200mm length.

The concrete mix proportions was taken cement: sand: gravel: water equal to 1.0: 1.79: 3.48: 0.5 by weight of cement. The cement content was 350 kg/m^3. The percentage of fibres is 1.5% of concrete volume which is equal to 114 kg/m^3 (≅ 5% by weight of concrete/m^3). The percentage of the admixtures for either Z-10 or Z-12 is 3% of the cement weight.

For the mortar mix, the ratio of cement to sand is 1:3 by weight and the water is 10% by weight of both cement and sand (W/C=4). The percentage of fibres is 1.5% by volume of concrete while the percentage of the added admixture was 3% of the cement weight for either Z-10 or Z-12.

3 Scheme of Experimental Work

Two sets of tests were carried out. The first set was carried out on concrete specimens to determine the water absorption of different concrete mixes, as recommended in BS 1881 (9). The tested cylinder cores were of 100mm diameter. The compressive strength of these cores was also determined.

The second set consisted of mortar specimens, to study

the effect of adding 2% $Mg\ So_4.7H_2O$ to the mixing water using ordinary portland cement. The specimens were cubes of dimensions 7x7x7 cms, cured for 7 days by submerging in water and then were kept at $50^{\circ}C$ and 100% relative humidity for three weeks and then tested at an age of 28 days. Reference specimens were submerged in water for 28 days and then tested for comparison.

Table 1 represents the tested concrete specimens while Tables 2,3 and 4 give the scheme of the mortar tested specimens of groups (1),(2) and (3) respectively. Group (1) represents mortar control specimens with no sulphates or admixtures, which are cured by submerging in water. Group (2) represents mortar specimens containing either fibre or admixtures or both without adding any sulphates, while group (3) represents mortar specimens containing sulphates in the mix. Each identification mark in these tables represents three specimens.

4 Tests carried out

4.1 Water absorption

One of the tests to evaluate the durability of the concrete is by measuring its water absorption. This test is carried out in accordance to the B S 1881 (9), with the exception that the cylinder cores used are of diameter 100mm instead of 75mm as recommended in the standard. The results given are average of three specimens.

4.2 Compressive strength

The previous concrete cylinder cores of 100mm diameter were tested in compression and the concrete compressive strength for the different mixes is determined according to B S 1881 (10).

4.3 Sulphate ion resistance

Mortar cubes 7x7x7 cm containing $Mg\ SO_4.7H_2O$ were cured for 7days by submerging in water and then exposed to $50^{\circ}C$ and 100% relative humidity for 21 days to accelerate the reaction. After 28 days, the compressive strength of these cubes were recorded. Control specimens without sulphate ions and cured by submerging in water were also tested at the same age.

Table 1. Concrete Specimens of Different Mixes

Cement type	Ordinary Portland						Sulphate Resistant	
Admixture type	—		Z-10		Z-12		———	
Fibre* content	0.0	1.5	0.0	1.5	0.0	1.5	0.0	1.5
Identification mark	C_1	C_2	C_3	C_4	C_5	C_6	C_7	C_8

* % of concrete volume

Table 2 Mortar Specimens of Group (1)-Control Specimens

Cement type	Ordinary Portland		Sulphate Resistant	
Fibre* content	0.0	1.5	0.0	1.5
Identification mark	R_1	R_2	R_3	R_4

* % of concrete volume

Table 3. Mortar Specimens of Group (2)- Without Sulphates

Cement type	Ordinary Portland						Sulphate Resistant	
Admixture type	—		Z-10		Z-12		———	
Fibre* content	0.0	1.5	0.0	1.5	0.0	1.5	0.0	1.5
Identification mark	M_1	M_2	M_3	M_4	M_5	M_6	M_7	M_8

* % of concrete volume

Table 4. Mortar Specimens Of Group (3)- Containing Sulphates

Cement type	Ordinary Portland						Sulphate Resistant	
Admixture type	——		Z-10		Z-12		————	
Fibre* content	0.0	1.5	0.0	1.5	0.0	1.5	0.0	1.5
Identification mark	L_1	L_2	L_3	L_4	L_5	L_6	L_7	L_8

* % of concrete volume

5 Test results and discussion

5.1 Concrete Specimens

Table 5 and Figures 1 and 2 give the percentage of absorption and the compressive strength of the tested concrete specimens. It may be observed from these results that the addition of either fibres or admixtures to the concrete, made of ordinary portland cement or sulphate resistant cement, decreases the water absorption . From the above results, it is found that the compressive strength of the tested concrete specimens increases and their absorption decreases when compared with control specimens, which are the main factors that affect durability.

When comparing the control mixes of concrete specimens having no fibres or admixtures, i.e C_1 made of ordinary portland cement and C_7 made of sulphate resistant cement, it is found that the absorption of the concrete made of sulphate resistant cement decreases to 91% than that of the concrete made of ordinary portland cement and its compressive strength increases by about 20%.

In this test program, admixtures are not used with the sulphate resistant cement as it includes a certain composition against aggressive enviroments. From the above results, it may be observed that the best results are achieved when using fibres with sulphate resistant cement (C_8), where it gives the best percentage of absorption (81%) and the highest compressive strength (236 kg/cm^2).

When using ordinary portland cement, it is recommended to use fibres and admixture Z-12, as it gives the highest compressive strength with ordinary portland cement and the percentage of water absorption is 88.6%.

5.2 Mortar Specimens

Tables 6 and 7 give the compressive strength of the mortar specimens of group (2) without sulphates and group (3) containing sulphates in addition to the control specimens of group (1).

Figure 3 shows the effect of fibres on the mortar compressive strength. It may be observed that for all the tested specimens, the addition of fibres increases the mortar compressive strength, except for the mortar specimens M_5, M_6 in group (2) made of ordinary portland cement and having the admixture Z-12, it is found that the

Table 5. Percentage of Water Absorption and Cylindr Compressive Strength of Concrete Specimens.

Cement Type	Indentification Mark	Type of Concrete	$(\frac{W_1-W_2}{W_2} \cdot \phi)$ *	%of Water absorption from control mix	Cylinder compressive strength kg/cm²
Ordinary Portland (O.P.C)	C_1	Control Mix	3.42	100	142
	C_2	Control Mix+Fibre	3.2	93.6	146
	C_3	Control Mix+Admix. Z-10	3.37	89.5	149
	C_4	Control Mix+Admix. Z-10 +Fibre	2.94	86	147
	C_5	Control Mix+Admix. Z-12	3.16	92.4	155
	C_6	Control Mix+Admix. Z-12 +Fibre	3.03	88.6	200
Sulphate Resistant (S.R.C)	C_7	Control Mix	3.11	91	170
	C_8	Control Mix+Fibre	2.79	81.5	236

* W_1 = Weight of wet specimens

* W_2 = Weight of dry specimens

* ϕ = Correction factor as given in BS 1881, Part 5:1970.

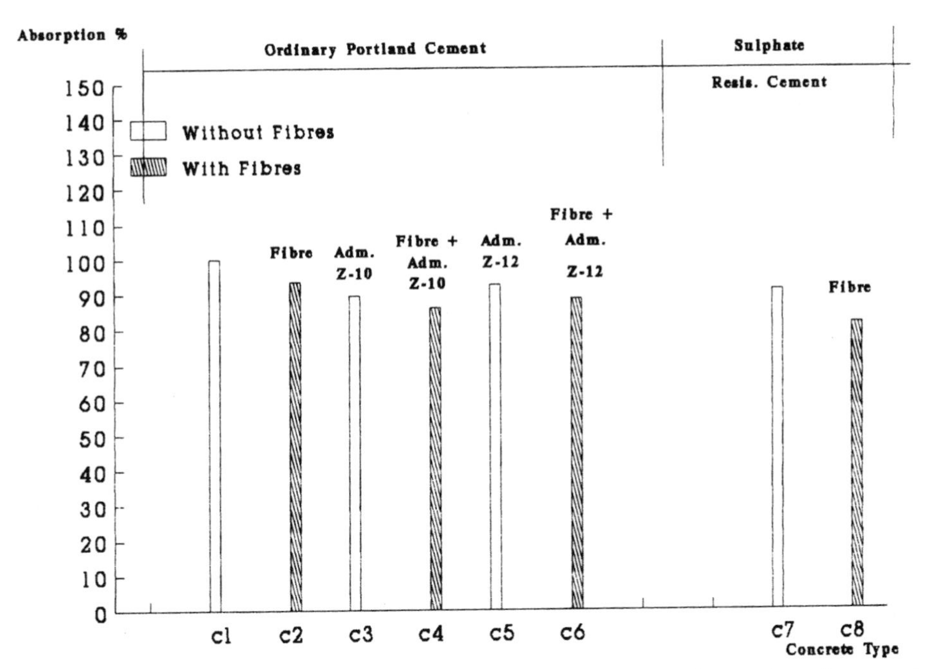

Fig. 1. Effect of fibres and admixtures on water absorption of concrete

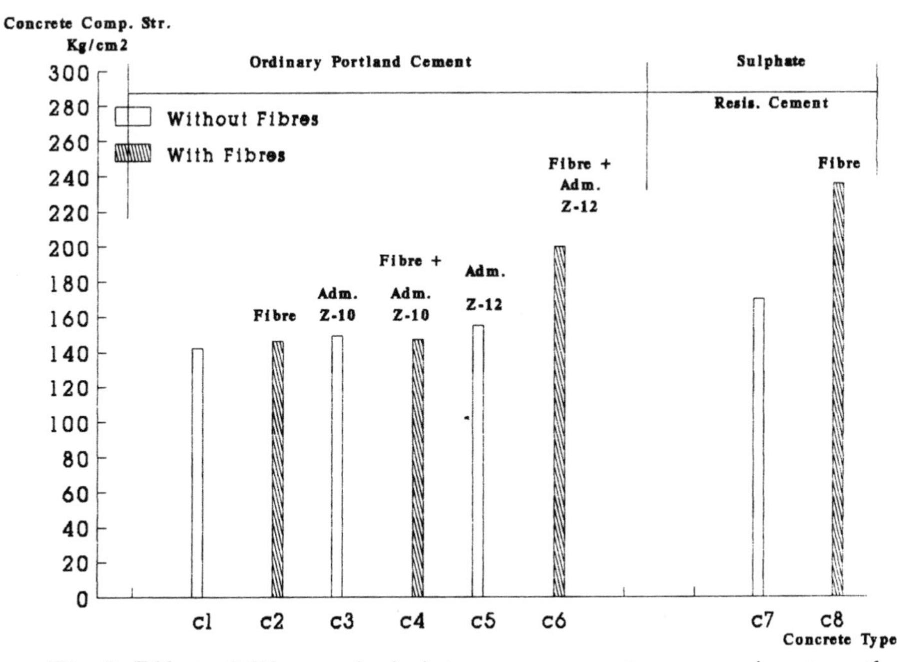

Fig. 2. Effect of fibres and admixtures on concrete compressive strength

Table 6. Compressive Strength of Mortar Specimens of Group (1) Control Specimens and Group (2) Without Sulphates.

Curing Conditions	Cement Type	Admixture Type	Fibre+ Content %	Identification Mark	Compressive* Strength kg/cm²
Submerged in Water	Ordinary Portland (OPC)	—	0.0	R_1	384
			1.5	R_2	473
	Sulphate Resistant (SRC)	—	0.0	R_3	292
			1.5	R_4	431
50°C Temperature and 100% relative humidity	Ordinary Portland (OPC)	—	0.0	M_1	400
		—	1.5	M_2	497
		Z-10	0.0	M_3	292
		Z-10	1.5	M_4	355
		Z-12	0.0	M_5	370
		Z-12	1.5	M_6	367
	Sulphate Resistant (SRC)	—	0.0	M_7	318
		—	1.5	M_8	387

* Each reading is average of 3 cubes

+ % of concrete volume

Table 7. Compressive Strength of Mortar Specimens of Group (1) Control Specimens and Group (3) Containing Sulphates.

Curing Conditions	Cement Type	Admixture Type	Fibre[+] Content %	Sulphates % by weight of Cement %	Identification Mark	Compressive[*] Strength kg/cm^2
Submerged in Water	Ordinary Portland (OPC)	—	0.0	—	R_1	384
			1.5	—	R_2	473
	Sulphate Resistant (SRC)	—	0.0	—	R_3	292
			1.5	—	R_4	431
50°C Temperature and 100% relative humidity	Ordinary Portland (OPC)	—	0.0	2	L_1	447
		—	1.5	2	L_2	527
		Z-10	0.0	2	L_3	422
		Z-10	1.5	2	L_4	466
		Z-12	0.0	2	L_5	401
		Z-12	1.5	2	L_6	501
	Sulphate Resistant (SRC)	—	0.0	2	L_7	394
		—	1.5	2	L_8	439

* Each reading is average of 3 cubes

+ % of concrete volume

compressive strength of the mortar with no fibres (370 kg/cm^2) is nearly the same as that of the mortar having fibres (367 kg/cm^2). This may bedue to unexpected experimental error.

The increase in the mortar compressive strength due to the use of fibres ranges between 10% to 50% for all the tested specimens.

Figure 4 shows the effect of curing conditions and admixtures on the mortar compressive strength. In general, it may be said that for all the tested mortar specimens either of group (2) (without sulphates) or group (3) (containing sulphates), the max. compressive strength is reached for the specimens having no admixtures and cured at 50°C temperature and 100% relative humidity. From Figure 4, it is clear that the addition of admixtures to the specimens cured in 50°C temperature and 100% relative humidity,decreases their compressive strength than that of specimens having no admixtures for all the groups either containing fibres or not. In order to analyse this fact, more chemical studies are needed to be done on the used admixtures.

The effect of the sulphates on the mortar compressive strength is shown in Figure 5. It is observed that the compressive strength of the specimens of group (2) without sulphates is lower than that of the specimens of group (3) containing sulphates. This fact is attributed to the curing conditions as the specimens are exposed to high temperature and 100% relative humidity. Under these curing conditions, the sulphates may react with the cement components producing new components of variable sizes such as mono and / or tricalcium sulpho aluminate (C_3AS) which may occupy the pores and consequently increases the compressive strength of the mortar specimens containing sulphates than the like specimens without sulphates. The produced components (C_3AS) have the ability to absorb water and become velominous by time which causes a lateral force and accordingly cracks the mortar .These cracks cause more moisture absorption. The presence of moisture in the surrounding enviroment causes decrease in the mortar compressive strength with time.

It may be noticed that control mortar specimens with sulphate resistant cement (Table 6) are less beneficial compared to that of ordinary portland cement which is unlike to the results of the concrete specimens (Table 5).

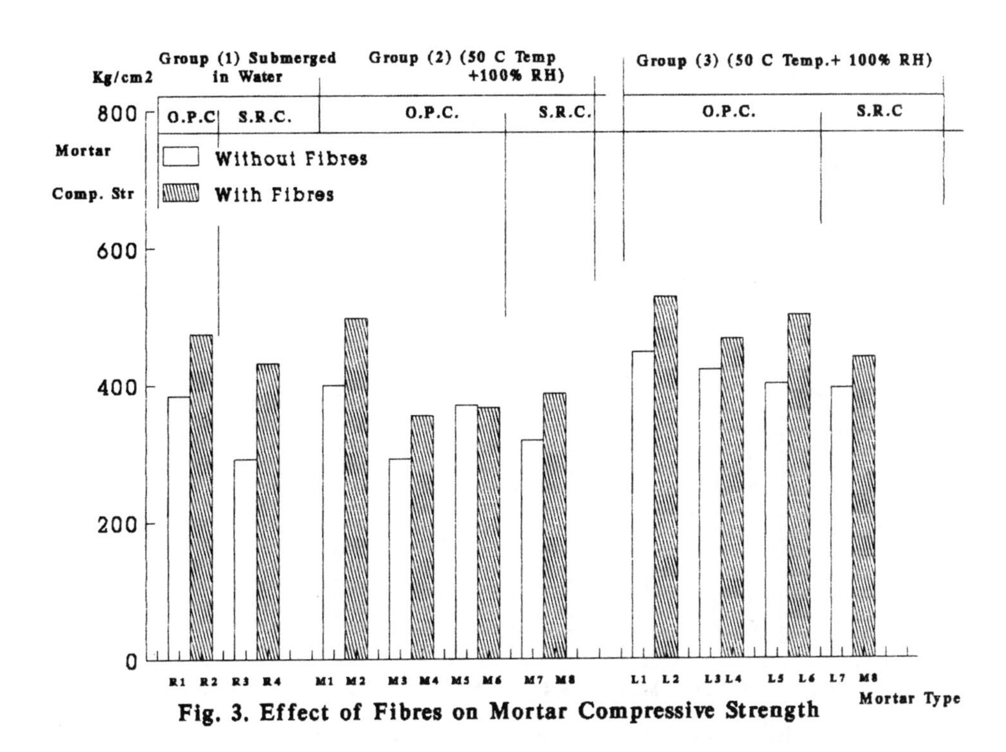

Fig. 3. Effect of Fibres on Mortar Compressive Strength

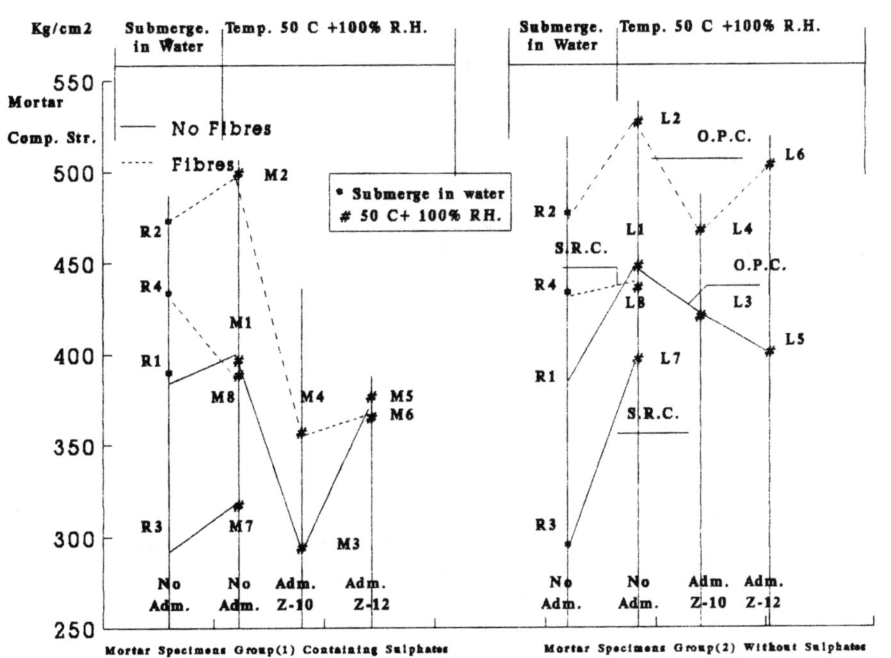

Fig. 4. Effect of Curing Conditions and Admixtures on Mortar Compressive Strength

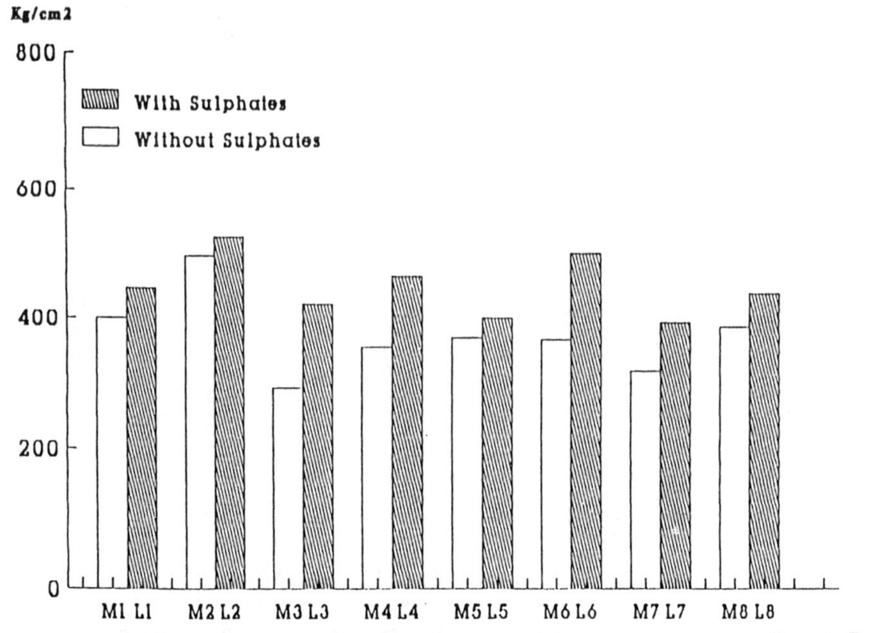

Fig. 5. Effect of Sulphates on Mortar Compressive Strength

This phenomena may be due to the disturbance which occured during the cutting of the concrete cores.

6 Conclusions

1 It is recommended to use steel fibres without the admixtures Z-10 and Z-12 of type B in concrete made of either ordinary portland cement or sulphate resistant cement as the fibres improve both strength and water absorption.

2 The best strength and water absorption was reached in concrete when using steel fibres and sulphate resistant cement without admixtures. This case gives the least percentage of absorption (81.5%) and the maximum compressive strength (236 kg/cm^2). This compressive strength is about 65% higher than that of the control mix made of ordinary portland cement and having no fibres or admixtures.

3 The use of steel fibres increases the compressive strength of mortars made of either ordinary portland cement or sulphate resistant cement. The mortar compressive strength for specimens containing steel fibres increases up to 1.5 times those having no fibres.

4 A noticeable increase in the compressive strength of the mortar specimens cured in $50^{\circ}C$ temperature and 100% relative humidity without using admixtures was touched compared with those cured by submerging in water.

5 It is recommended not to cure the mortar specimens containing sulphates in high temperatures and 100% relative humidity as it may give false high compressive strength at early ages .

References

1 ACI Committee 222, "Corrosion of Metal in Concrete", ACI Journal Vol. 82 No.1 January - February 1985, pp 3-31.

2 Scanlon, John M. (ed.) Concrete Durability- Kathrine and Bryant Mather International Conference, ACI SP 100, American Concrete Institute, Detroit, Michigan, 1987.

3 Slater, J.E. (ed.), Corrosion of metals in Association with Concrete, ASTM STP 818, American Society for Testing and Materials, Philadelphia, 1983.

4 Crane, Alan P. (ed.), Corrosion of Reiforcement in Concrete Construction, Society of Chemical Industry, Ellis, Horwood Ltd. London, 1983.

5 Malhotra, V.M (ed.), Performance of Concrete in Marine Environment, ACI, SP65, American Concrete Institute, Detroit, Michigan, 1980.

6 Egyptian Standard Specification ESS 1109-1973, "Aggregates from Natural Sources".

7 Egyptian Standard Specification ESS 373-1984, "Ordinary Portland Cement and Rapid Hardening Portland Cement".

8 Egyptian Standard Specification ESS 583-1986, "Sulphate Resistance Cement".

9 British Standard Institution BS 1881, Part 5:1970, "Methods of Testing Concrete for other than Strength".

10 British Standards, BS 1881 Part 120 1983, "Method for Determination of the Compressive Strength of Concrete Cores".

63 DURABILITY OF GLASS FIBRE REINFORCED POLYMER MORTAR COMPOSITES

M. NEELAMEGAN and V. S. PARAMESWARAN
Structural Engineering Research Centre, Madras, India

Abstract
Polyester resin mortar has proved itself to be a more durable and high strength material than ordinary cement concrete / mortar. In order to improve its ductility , glass fibres are generally used to reinforce the matrix. Investigations were carried out by the authors to evaluate the properties of this mortar, particularly, to find out its suitability for applications in aggressive environment. Tests to evaluate the Chemical resistance of resin mortars containing different percentage of chopped and E glass fibres were also carried out. Nine types of chemicals were used in these tests. The paper describes the details of laboratory investigations carried out on a number of Glass Fibre Reinforced Resin Mortar (GFRRM)specimens and highlights the relative superiority of GFRRM over conventional cement mortar in resisting aggressive chemical environs.
Keywords: E-Glass fibres, Polyester resin,Physical properties, Polymerisation, Filler, Reagent, Chemical resistance, Laboratory tests, Applications

1 Introduction

Durability of building materials is a very important factor that plays as vital a role as the strength and stiffness characteristics in the design of structures. In general , concrete and reinforced concrete deteriorate most intensively from exposure to acids in the form of aqueous solution or acidic gases which form acids when dissolved in water, and also by salt solutions and alkalis. The susceptibility of reinforced concrete structures to corrosion and their poor durability characteristics are amply evident from the pathetic service condition of a number of existing structures built in aggressive environment as shown in Fig.1(Mani and Neelamegam [1984]).

Fibre Reinforced Cement and Concrete. Edited by R. N. Swamy. © 1992 RILEM.
Published by E & FN Spon, 2-6 Boundary Row, London SE1 8HN. ISBN 0 419 18130 X.

Fig. 1. Damage of RCC structures due to aggressive agents (Mani and Neelamegan, 1984)

Since the last couple of decades, increased attention has been focused throughout the world on the durability of RC structures and development of new concrete composites which possess good mechanical and durability characteristics. One promising solution that has emerged as a culmination of pains-taking research work over the last few decades is the use of polymers for improving the durability of conventional cement concrete in addition to enhancing its strength. Extensive research work has been reported on polymer concrete composites since 1960 and this novel material has already found wide applications (Okada and Ohama[1987]) for construction and rehabilitation of several outstanding structures in Japan, USA, UK, Germany, and China. The Structural Engineering Research Centre, Madras has also carried out extensive investigations on several types of polymer concrete composites and identified their suitability and applications. In this paper, the authors have presented the results of studies on durability properties of polymer mortar/concrete composites with and without glass fibre reinforcement.

2 Polymer concrete/mortar composites

There are three types of polymer concrete composites in vogue today viz.,

(i) **Polymer cement composites (PCC)** in which a polymeric additive (latex or pre-polymer) is added to normal cement composites during the mixing stage itself.

(ii) **Polymer or resin mortar/concrete** in which polymer is the sole binder of aggregate in lieu of cement-water binder of cement composites.

(iii) **Polymer impregnated concrete/ mortar** in which monomers are impregnated into the pore system of hardened cement composites and polymerized, thereby filling the pores of cement composites and making them impermeable.

Table 1 gives the typical mechanical properties of all the three types of polymer mortar composites based on

Table 1. Properties of different types of mortars

Type of test	(OCM)	(PIM)	(REM)	(PCM)
Compressive strngth (MPa)	32.15	126.32	117.91	29.09
75 x 150 mm cylinder	35.53	126.80	115.02	30.87
	34.20	134.79	114.58	29.80
Split Tensile strength (MPa)	2.39	9.72	10.16	3.05
75 x 150 mm cylinder	2.82	10.27	8.05	3.53
	3.05	7.99	9.71	3.64
Direct Tensile strength (MPa)	2.50	10.74	11.03	2.95
	2.35	12.06	10.87	3.54
Flexural Strength (MPa)	6.62	23.91	27.22	10.12
40 mm x 40 mm x 160 mm beam	6.71	23.18	30.35	8.46
	6.62	26.49	29.80	9.20
Central deflection of	2.00	4.00	4.50	2.30
flexural beam (mm)	1.70	4.20	4.20	2.30
40 mm x 40 mm x 160 mm beam	2.10	4.25	6.00	2.30
Density (kN/m^3)	21.04	21.45	20.68	19.70

OCM - Ordinary cement mortar
PIM - Polymer impregnated mortar
REM - Resin mortar
PCM - Polymer cement mortar or latex modified cement mortar

the studies(Neelamegam and Ohama[1983]) carried out by the authors.

Amongst the the three types of polymer mortar composites mentioned above, resin mortar/concrete composites have proved to be the most popular due to several advantages like simple process technology, capability of rapid production, and excellent mechanical and durability properties. By virtue of these positive features,these composites have found a wide range of applications. However, the inherent brittleness of the material is a serious handicap for its ready acceptance in many applications. The addition of glass fibres to reinforce the polymer mortar and modify its structural behaviour has been accepted (Kurimoto[1980])as one of the techniques for alleviating this problem.

3 Durability properties of resin mortar/concrete composites with and without glass fibres

The ability of cement concrete composites to resist the aggressiveness of the environment depends primarily on the properties of its ingredients and especially that of cement mortar. Similarly, in case of glass fibre reinforced resin mortar composites, the durability characteristics are governed to a great extent by the properties of resin mortar. In order to assess the durability characteristics of any material, the following tests are necessary:

(a) Water absorption, (b) Water resistance, (c) Weatherability, (d) Freeze-thaw, (e) Strength under thermal cycling, (f) Strength reduction in soil, and (g) Chemical resistance.

4 Review of previous investigations

4.1 Work carried out in Japan

Chuo kouzai Co. Ltd., Japan produces various resin concrete components like man-hole frames and covers for communication conduits, hydrant joint boxes, pickling tanks, industrial waste solution processing tanks requiring high resistance to chemicals, high tension power cable ducts, etc. Some of the durability tests carried out by this firm on resin concrete composites are discussed below(Chuo Kouzai[1990]).

Chemical resistance

Polyester resin concrete composites were immersed for 80 days in sulphuric acid with a pH value of 2.0 and sodium hydroxide (NaOH) with a pH value of 12. The solutions were heated to three temperatures 40 deg.C, 60 deg.C, and 80 deg.C. The specimens were subjected to flexural and compression tests. As evident from **Fig. 2**, there was hardly any reduction in strength at 60 deg.C but about 10% reduction in strength was observed at 80 deg.C. At 40 deg.C, the strength was little affected by both the solutions.

Water resistance
Since resin concrete products may be exposed to underground water in service, test specimens were

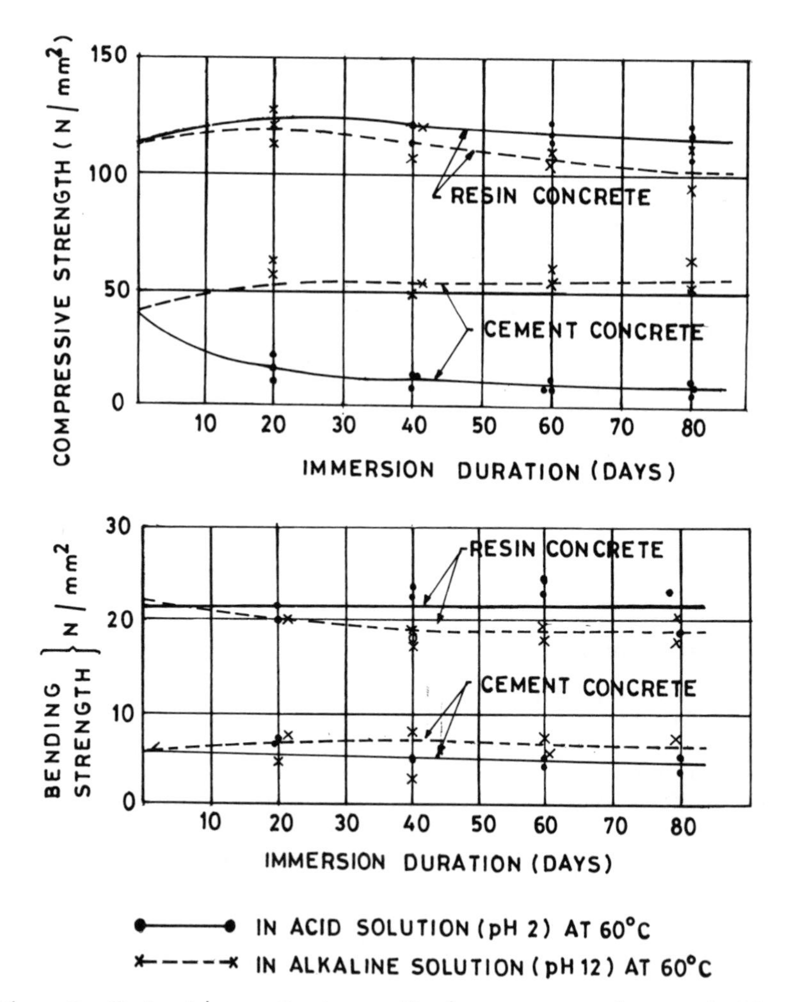

Fig. 2. Reduction of strength in compression and flexure due to chemical attack (Chuo Kozai, 1990)

tested for endurance by immersion in hot water which is the most severe among the water resistance degradation-acceleration tests, since it is estimated that 1500 hours of immersion in hot boiling water is equivalent to 60 years of immersion in water at normal temperature. It is seen from the test results presented in **Fig. 3** that for an immersion period of 1500 hours, there is hardly any perceptible change in bending strength at a temperature of 60 deg.C, whereas the strength is reduced by 25% for the same duration at a temperature of 100 deg.C. However, the bending strength was as high as 15 MPa even in the latter case.

Water absorption

Normal resin concrete/mortar composites showed zero water absorption at room temperature and normal atmospheric pressure. Application of 1 and 2 MPa of pressure for 48 hours also resulted in no water absorption. Increasing the water temperature to 100 deg.C caused a water absorption of 2.1% by weight after 200 hours of immersion(**Fig. 4**). Therefore, the water absorption of resin concrete can be almost taken as nil for normal underground water pressure and temperature conditions.

Freeze-thaw test

Resin concrete specimens were subjected to freeze-thaw cycles of 3.5 hours duration in accordance with ASTM specifications (ASTM C-666-84 [1984]) . After 210 cycles of testing, it was observed that the weight loss was a meager 0.1%.

Weatherability

The effect of ultraviolet rays on the flexural and compressive strength of resin concrete was tested by exposing the test specimens to radiation from 2 carbon arc lamps for a duration of 2000 hours. The deterioration in strength ranged from 1.1 to 1.7% while the colour became slightly yellowish.

Strength reduction in soil

Resin concrete specimens were buried in two types of soil viz., acidic soil of pH 1.5-2.3 at a temperature of 30-50 deg.C and alkaline soil of pH 8.5-9.5 at a temperature of 63-67 deg.C. Samples of buried specimens were periodically tested in flexure and compression at the end of every year. It was found that loss of strength was about 10% in acidic soil and 15% in alkaline soil at the end of 10 years as shown in **Fig. 5**.

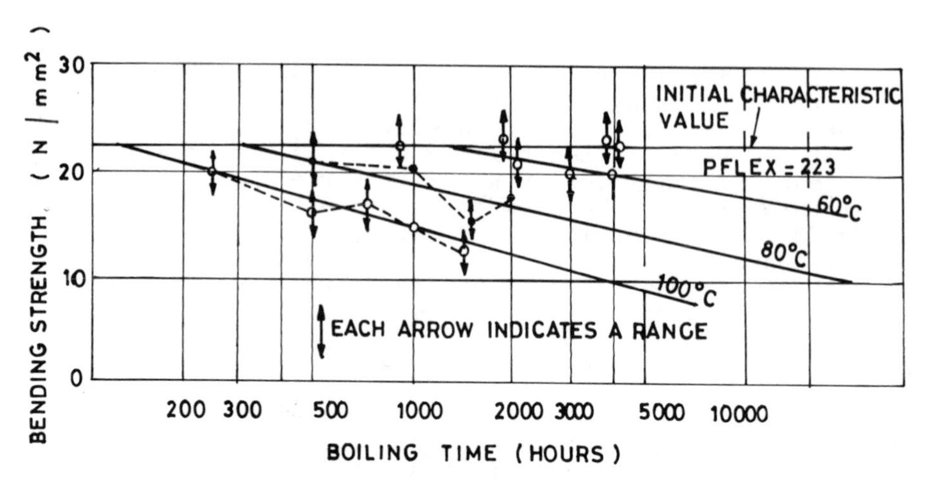

Fig. 3. Reduction in bending strength of resin concrete in boiling water with time for different immersion periods and at different temperatures (Chuo Kozai, 1990)

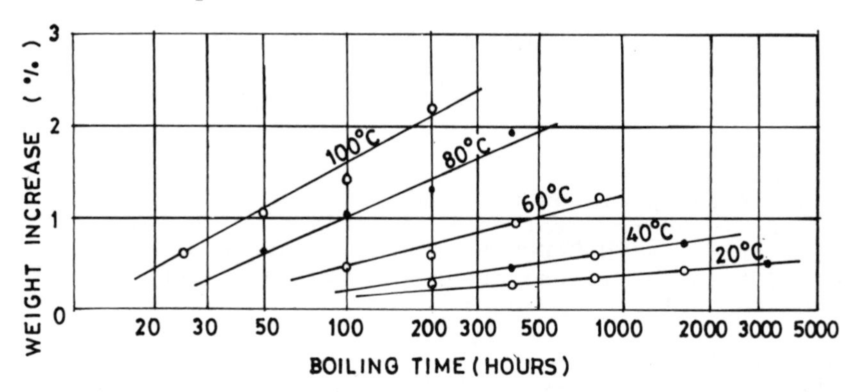

Fig. 4. Weight increase of resin concrete in boiling test (Chuo Kozai, 1990)

Effect of thermal cycles

Test pieces were subjected to cyclic thermal tests for 15, 30, 45, and 60 cycles, respectively, each cycle extending over a period of 4 hours at 65 deg.C and 4 hours at -20 deg.C. No change in strength was observed.

4.2 ACI Committee 515 findings

The recommendations given by this committee (ACI 515 [1966]) cover the durability of concrete in different types of chemicals .Suitable materials are also suggested as coating or protective barrier for concrete and reinforced concrete structures. Twenty five types of chemicals (acids,alkalis, and other chemicals) considered for their studies have been reported to

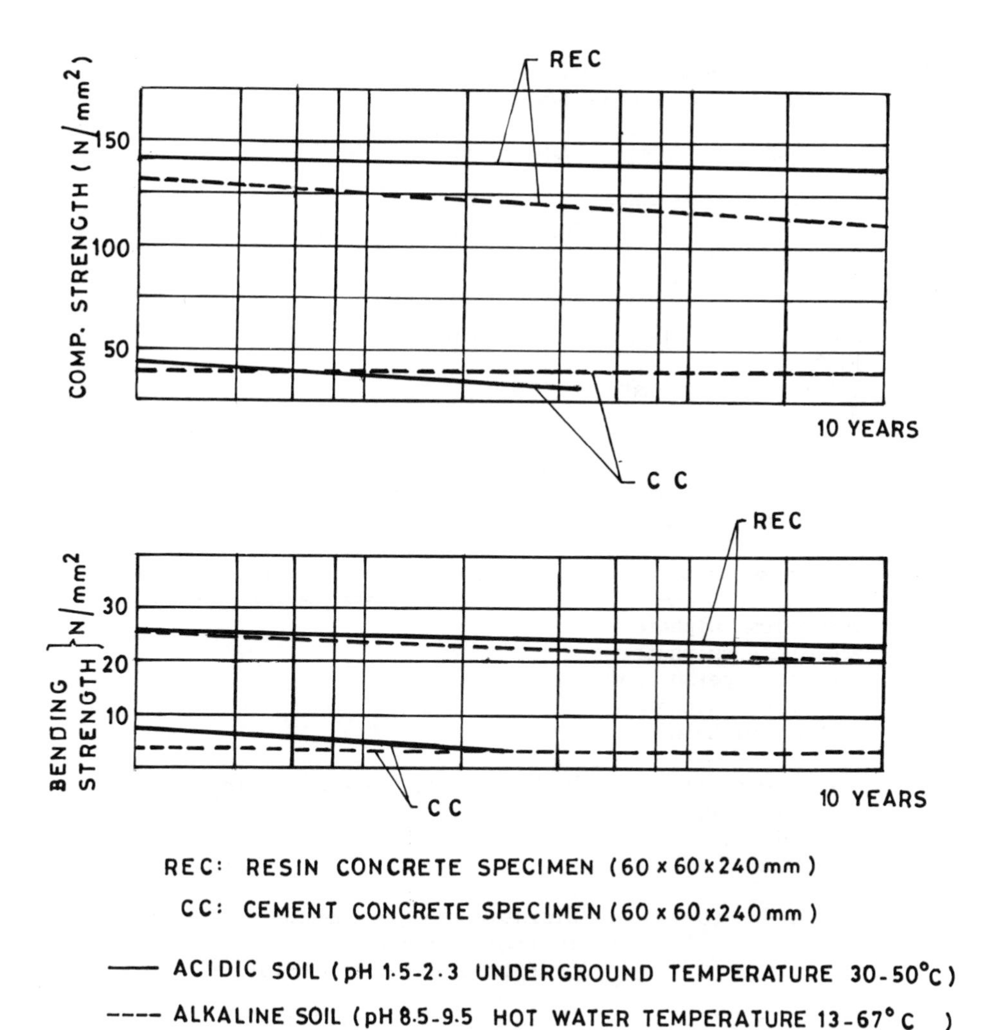

Fig. 5. Reduction in flexural and compressive strength of resin concrete buried in acidic and alkaline soil with time (Chuo Kozai, 1990)

cause medium and rapid disintegration of concrete structures. These are listed in **Table 2**. Polyester resin mortar/concrete is recommended by the committee as a protective barrier for concrete to avoid such disintegration brought about by the various types of aggressive agents listed in the **Table 2**.

Table 2 Performance of concrete against various agressive agents.

Sl.no.	Chemical	Action on concrete
1	Nitric acid 2%+5% Sulphuric acid Sulphurous acid	Disintegrates rapidly
2	Ammonium sulphite Fish liquor Potassium dichromate Sulphite liquor	Disintegrates.
3	Aluminium chloride Aluminium sulphate Ammonium nitrate	Disintegrates. In porous or cracked concrete, attacks steel.
4	Cobalt sulphate Ferrous sulphate Magnesium sulphate Manganese sulphate Potassium persulphate Potassium sulphate Sodium sulphate	Disintegrates concrete of inadequate sulphate resistance.
5	Cocoa bean oil, cocoa butter, coconut oil. Cotton seed oil	Disintegrates, especially in presence of air.
6	Bromine	Disintegrates in gaseous bromine & in liquid bromine if it contains hydrobromic acid and moisture.
7	Castor oil	Disintegrates, especially in presence of air.
8	Ferric sulphate	Disintegrates concrete of inadequate quality.
9	Sea water	Disintegrates concrete of inadequate sulphate resistance. Attacks steel in porous or cracked concrete.

Polyester resin mortar is recommended by the ACI committee as protective barrier for concrete to avoid disintegration by the attack of various aggressive agents listed in the above Table.

5 Present investigations

5.1 Raw materials used

The authors have carried out extensive studies on glass fibre reinforced resin mortar composites(GFRRM) prepared from commercially available isophthalic polyester resin (commercially marketed under the name CALPOL 68-210). Cobalt octate, methyl ethyl ketone peroxide (MEKP), and dimethylaniline (DMA) were used as promoters for polymerization reaction. The properties of the resin in liquid and solid states are given in **Table 3**. The GFRRM specimens were reinforced with E-glass fibres of lengths 6mm, 12.5mm, and 20mm. **Table 4** gives the physical and mechanical properties of glass fibres used. Five values of fibre content, 0, 1%, 2%, 3%, and 4% by weight were considered corresponding to each value of fibre length.

Heavy grade calcium carbonate and ordinary river sand passing 1.18mm sieve were used as filler and fine aggregate for all the mortar mixes.

5.2 Mix proportions and preparation of test specimens

Based on the earlier studies on resin concrete/mortars carried out at SERC, Madras, the following resin mortar proportion was selected for this investigation.

TABLE 3. Properties of CALPOL 68-210 resin (Isophthalic polyester*)

Sl. no	Physical properties in liquid state	Physical properties after curing
1.	Viscosity(Cps)at 25°C 850-1050	1. Flexural strength 100MPa
2.	Specific gravity at 25°C 1.11	2. Flexural modulus 3800MPa
3.	Acid number mg KOH/gm 10-15	3. Tensile strength 50MPa
4.	Styrene content 37+1%	4. Tensile modulus 4000MPa
5.	Shelf life at 20°C 6 months	5. Elongation at break 1.5%
6.	Flash point 32°C	6. Heat distortion temp.130°C
		7. Impact strength 4MJ/mm^2
		8. Hardness 40-50

* Data taken from technical brochures supplied by manufacturers

Table 4. Physical and chemical properties of glass fibres

Physical properties		Chemical analysis	
		Oxide	**%by weight**
Diameter of filament	9 - 12.7	SiO_2	52.4 - 54.0
Number of filaments	200	Al_2O_3	13.9 - 14.4
Tensile strength of glass fibre	2850-3164 MPa	B_2O_3	10.6 - 11.0
		TiO	0.02
Young's modulus of glass fibre	74000 MPa	MgO	3.4 - 4.6
		CaO	17.3
Specific gravity	2.69	Na_2O	0.08 - 0.1
		K_2O	0.03
Exterior surface area of strand	$0.62mm^2/mm$	Fe_2O_3	0.1

Resin and filler - 1 part each (by weight)
Fine aggregate (sand) - 4 parts (by weight)
Glass fibre - 0%, 1%, 2%, 3%, and 4% by weight of total mix
Fibre length - 6mm, 12.5mm, and 20mm

Normal mixing and compacting procedure was adopted in making the mortars. Different categories of test specimens were prepared from the mortars for use in the tests. These included 75mm diameter x 150mm long cylinders, 180mm x 180mm x 40mm plates, 70mm x 70mm x 40mm plates, 50mm x 50mm x 50mm cubes, 60mm x 60mm size dog-bone shaped tensile specimens, and 50mm x 50mm x 250mm beams. The details of the mixes used in casting the above specimens are given in **Table 5**.

5.3 Test procedure

a) Tests for mechanical properties

The properties of GFRRM mixes used in the investigations such as their strengths in compression, tension, flexure, and shear as well as their resistance to impact and abrasion were obtained by subjecting the corresponding test specimens to the relevant tests prescribed in the codes of practice. These properties are given in **Table 6**.

b) Tests for durability

All the test specimens were first subjected to water absorption and permeability tests at a pressure of 0.3 MPa. In all the cases, the specimens were found to be impermeable and non-water-absorbent.

Nine types of chemical solutions viz., 15% dil

Table 5. Mix proportions and details of mixes used for GFRRM

Sl. no.	Type of Mix	Mix proportion (parts by wt) Resin	Filler	Sand	% fiber by wt. (length in mm given in brackets)	Remarks
1	CM (Cement : Sand = 1:1.5, W/C = 0.43)					Very good cohesive mix
2	L	1.0	1.0	4.0	0.0	Cohesive mix
3	LF_2	1.0	1.0	4.0	2.0 (6.0)	Good mix
4	LF_4	1.0	1.0	4.0	4.0 (6.0)	More fiber content
5	MF_1	1.0	1.0	4.0	1.0 (12.5)	Very good cohesive mix
6	MF_2	1.0	1.0	4.0	1.0 (12.5)	Cohesive mix
7	MF_3	1.0	1.0	4.0	3.0 (12.5)	Good mix
8	MF_4	1.0	1.0	4.0	4.0 (12.5)	Higher fiber content
9	NF_1	1.0	1.0	4.0	1.0 (20.0)	Very good cohesive mix
10	NF_2	1.0	1.0	4.0	2.0 (20.0)	Cohesive mix
11	NF_3	1.0	1.0	4.0	3.0 (20.0)	Good mix for working
12	NF_4	1.0	1.0	4.0	4.0 (20.0)	Higher fiber content slightly low matrix content

TABLE 6. Properties of glass fibre reinforced resin mortars

Sl No.	Type of mix	Glass fibre content (wt %)	Length of the fibre (mm)	Compressive strength (MPa)	Split tensile strength (MPA)	Direct tensile strength (MPa)	Flexural strength (MPa)	Impact resistance No. of drops	Abrasion loss (mm)	Shear strength (MPa)
1	CM	0.0	-	32.5	3.10	-	3.0	1	3.00	-
2	L	0.0	-	64.0	8.35	9.19	20.1	1	2.90	12.5
3	LF2	2.0	6.0	69.4	9.33	8.53	20.9	1	2.59	14.5
4	LF4	4.0	6.0	56.6	9.04	8.26	19.1	4	2.91	16.1
5	MF1	1.0	12.5	61.0	8.46	7.74	21.5	1	2.28	14.1
6	MF2	2.0	12.5	49.4	8.09	9.64	19.6	2	-	16.1
7	MF3	3.0	12.5	43.0	8.36	9.53	20.3	6	1.62	18.1
8	MF4	4.0	12.5	52.1	9.98	8.04	19.73	14	1.96	13.6
9	NF1	1.0	25.0	58.0	9.10	-	20.8	3	-	13.4
10	NF2	2.0	25.0	69.11	9.37	8.13	20.7	4	1.75	16.1
11	NF3	3.0	25.0	62.1	10.73	-	20.3	36	1.29	18.7
12	NF4	4.0	25.0	56.9	10.51	7.24	-	31	1.60	14.6

Note: Resin - 1 part, Filler - 1 part, Sand - 4 parts by weight in all mixes
Values are average of three identical specimens

sulphuric acid (pH 0.64), 15% dil hydrochloric acid (pH 0.36), 15% dil nitric acid (pH 0.08), saturated sodium hydroxide (pH 7.25), saturated sodium chloride, saturated sodium sulphate (pH 7.45), 25% ammonia solution, acetone, and diesel were used for chemical attack tests. The chemicals were prepared and stored in separate plastic containers. GFRRM specimens were weighed and immersed in respective solutions(**Fig. 6**). The specimens were taken out and weighed periodically at intervals of 2, 6, 11, 14, 17, 22, and 26 days.

6 Test results and discussions

Figs. 7a and 7b show the percentage weight change in ordinary cement mortars and GFRRM specimens immersed in different types of chemicals at the end of 26 days. The

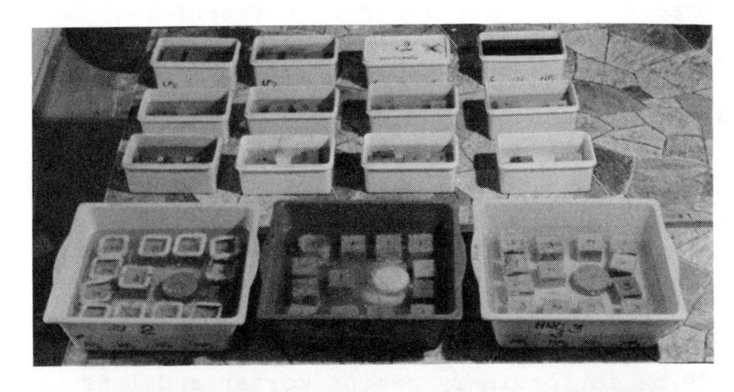

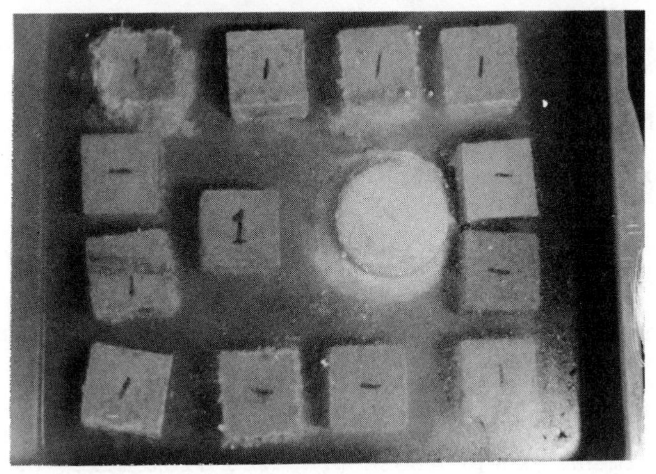

Fig. 6. A view of the GFFRM specimens under chemical attack test

relative performance of GFRRM and companion cement mortar specimens is discussed below.

6.1 Dilute sulphuric acid (pH 0.64)
Ordinary cement mortar specimens showed a weight loss of 72% after immersion for 26 days. In the same period, the weight loss recorded for all GFRRM specimens was less than 2.5% except for specimen NF4. Since the acidic concentration used in the tests was much more than that prevailing under normal aggressive type environments, GFRRM specimens can be expected to perform even better under such natural conditions. The loss in compressive strength of polyester resin mortar specimens without glass fibres was found to be around 18.7% of the initial strength when tested after 26 days of chemical attack. However, specimens from other types of mixes showed comparatively greater reduction in strength as shown in **Figs. 7a and 7b**. The values shown within brackets in these figures represent the change in compressive strength after 26 days immersion in the chemicals expressed as a percentage.

6.2 Dilute. hydrochloric acid (pH 0.35)
Ordinary cement mortar specimens suffered a loss in weight of 20% when subjected to attack by dilute. HCl at the end of 26 days. The corresponding figure for GFRRM specimens was less than 2.5%. The loss in compressive strength suffered by different types of mixes is again depicted in **Fig. 7a** .

6.3 Dilute nitric acid (pH 0.08)
It is observed that the loss in weight in this case is around 20.6% for ordinary cement mortar and less than 7.34% for GFRRM specimens. When the specimens were tested for compressive strength at the end of 26 days, they were found to exhibit enhancement in strength in contrast to loss in strength observed in case of dilute sulphuric and hydrochloric acids (**Fig. 7a**).

6.4 Saturated sodium sulphate (pH 7.45)
Under the attack of saturated sodium sulphate solution, the test specimens showed an insignificant change in weight ranging from -1% to +1% and a slight improvement in compressive strength was also noticed (**Fig. 7b**).

6.5 Saturated sodium hydroxide (pH 7.25) and saturated sodium chloride(pH 8.1)
Saturated sodium hydroxide caused a weight change ranging from +0.83% to -1.53% and an improvement in compressive strength after 26 days in case of all the test specimens except those prepared from mixes MF2 and NF1 (**Fig. 7b**). Only a limited number of test specimens

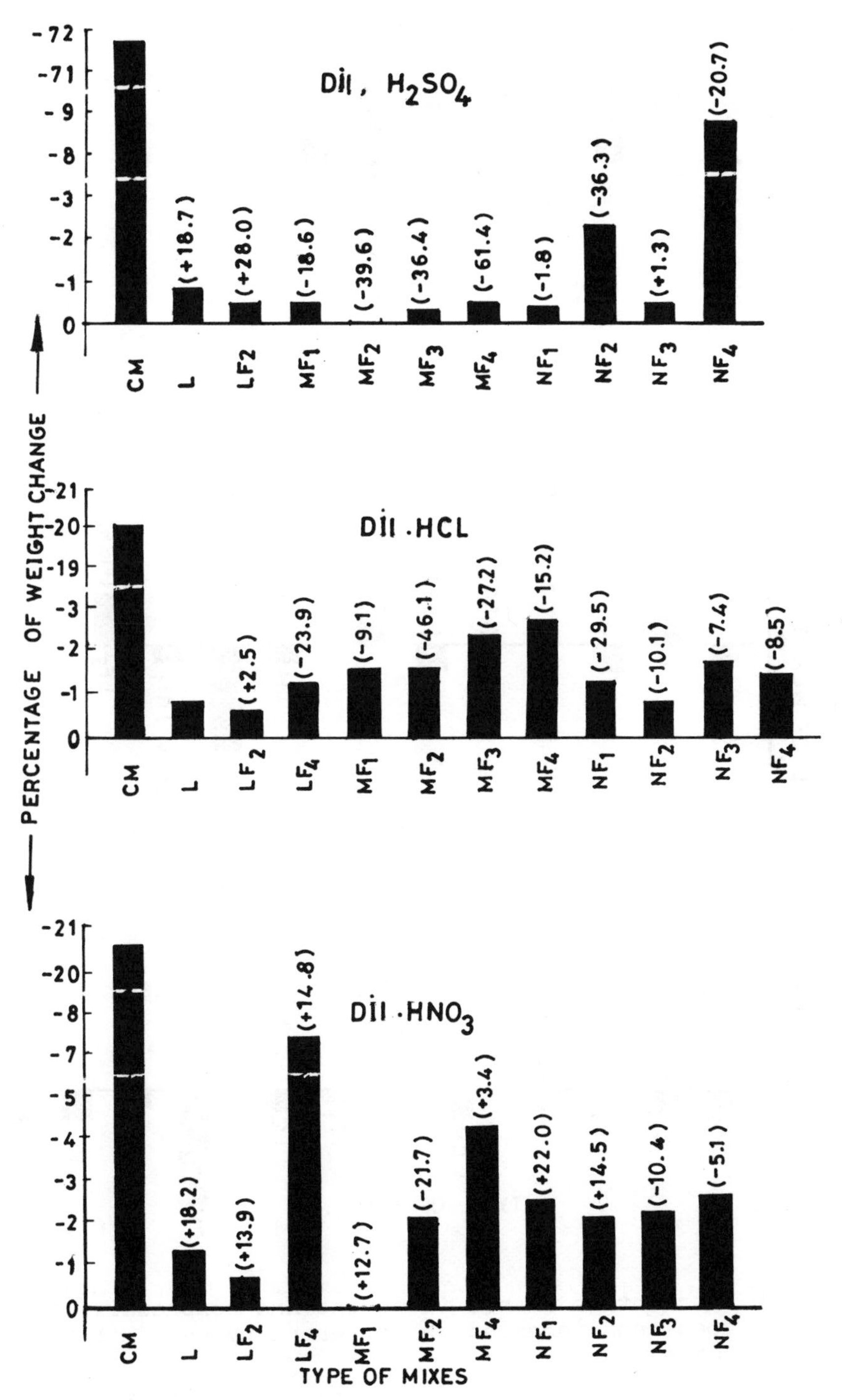

Fig. 7a. Percentage change in weight and compressive strength of test specimens immersed in different types of chemical after 26 days

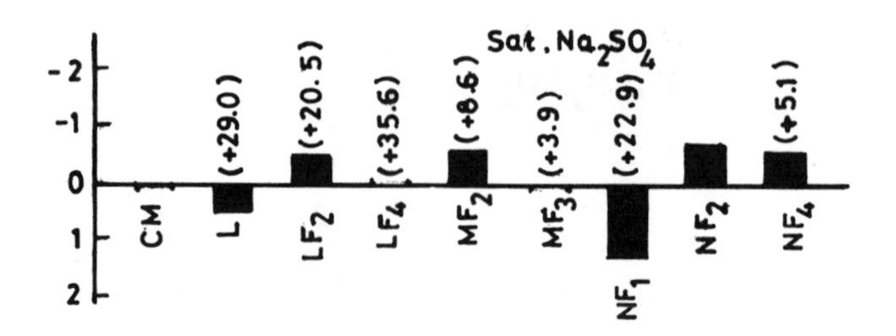

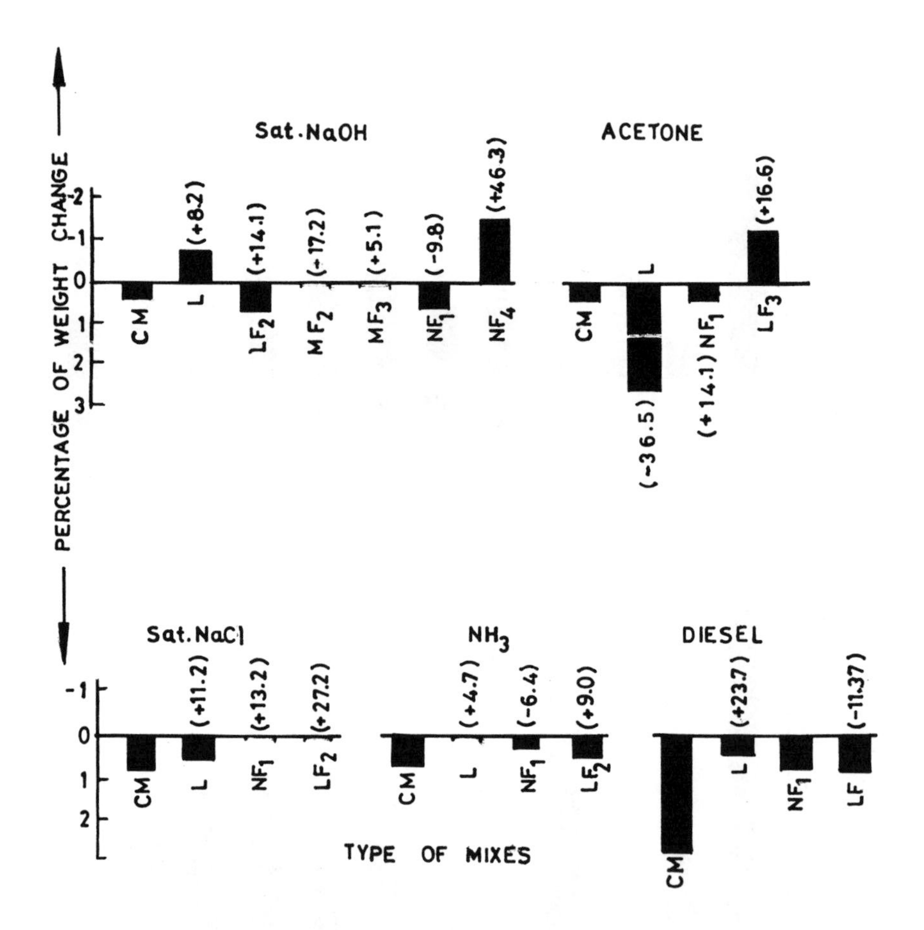

Fig. 7b. Percentage change in weight and compressive strength of test specimens immersed in different types of chemical after 26 days

were subjected to chemical attack by saturated sodium chloride solution and in all cases, there was a gain in weight accompanied by a corresponding rise in strength.

6.6 Acetone, ammonia, and diesel

Four types of mixes were investigated for their endurance under the attack of acetone, ammonia, and diesel. As evident from **Fig. 7b**, all the test specimens except those prepared from mix L in case of acetone and NF1 in ammonia, showed improvement in compressive strength and gain in weight.

7 Suggested applications

Based on the experimental data reported in this paper, GFRRM mixes can be recommended for application under the following situations:

(a) As protective barrier in underground RCC structures to surfaces exposed to chemical attack from soil.
(b) As "lost-forms" for concrete structures located in aggressive environment.
(c) In chemical factories, the existing structures can be given a protective cover of GFRRM mortar. Alternatively, GFRRM plates can be bonded to exposed surfaces using suitable gluing material.
(d) GFRRM products can be used in chemical factories as storage tanks for chemical processing, ducts for transporting chemical effluents, tiles for light- and heavy-duty factory floorings, etc.

8 Conclusions

From the investigations reported here, the following conclusions are drawn:

(1) Resin concrete specimens showed zero water absorption and near impermeability when immersed in water even under pressure .
(2) GFRRM specimens showed some reduction in compressive strength only in case of attack by sulphuric acid and hydrochloric acid solutions. However, the acid concentrations used in the study were rather high and a much better performance can be obtained under conditions prevalent in actual field environments.
(3) In case of all the salt solutions considered for the present study and in case of ammonia, acetone, and diesel, the percentage loss in weight was observed to be less than 1.5% in case

of GFRRM specimens. Also except for a few mixes, an improvement in compressive strength was noticed after subjecting the test specimens to chemical attack. However, the percentage gain in strength was rather insignificant. Similar results have been reported from the experimental work carried out by Chuo Kouzai Co., Ltd., Japan (Chuo Kouzai[1990]). ACI Committee 515 also recommends that polyester resin mortar/concrete can be used as a protective barrier for concrete to avoid disintegration by the attack of various aggressive chemicals.

9 Acknowledgement

The paper is published with the kind permission of Director, Structural Engineering Research Centre, Madras. The authors are grateful to the technical staff of the Concrete Composites Laboratory and particularly to Mr. M. Sadasivam and Mr. K.K Easow for the assistance rendered by them in carrying out the experimental studies reported in this paper. Thanks are also due to Mr. R. Bhaskaran and Mr. R. Veeraswami for the help provided by them in the preparation of drawings.

References

ACI Committee 515, (1966), "Guide for protection of concrete against chemical attack by means of coatings and other corrosion resistant materials", **Jnl American Concrete Institute**, pp1305-1392

ASTM C-666-84(1984), "Test methods for resistance of concrete to rapid freezing and thawing"

Chuo Kouzai Co., Ltd.., (1990), "Resin Concrete ", **Technical Data**, 890 Oazer Tambama, Hidaka-Cho, Trauma-gun, Saitama 350-12, Japan

Kurimoto Iron Works Ltd. (1980), "POLYCON fibreglass reinforced plastic pipes for underground power lines", **Technical Report**, Osaka, Japan

Mani, K and Neelamegam, M., (1984), "Assessment of Sunflower Oil Refinery Building damaged due corrosion and advice on remedial measures", **Internal Report**, Structural Engineering Research Centre, Madras

Neelamegam, M., Ohama, Y., et al, (1983), "Comparison of properties of polymer mortar composites", **Indian Concrete Journal**

Okada, K and Ohama, Y., (1987), "Recent research applications of concrete-polymer composites in Japan", **Proceedings, V International Congress on Polymers in Concrete**, Brighton, England

64 DEVELOPMENT OF FRACTURE ENERGY FOR FRC MATERIALS

E. S. LARSEN
Danish Building Research Institute, Hørsholm, Denmark

Abstract
When evaluating the service life of FRC-materials, the development of fracture energy, measured in 3-point bending, is a proper and simple method.
Five different fibre reinforced composites have been examined covering cement and concrete matrices reinforced with steel, polypropylene, AR-glass and cellulose fibres.
The materials have been exposed to different laboratory climates (i.e. a special weather simulator and stable climate) and natural weathering in Denmark.
The development of fracture energy is very much dependent on the ageing and degradation mechanisms in action.
Steel and polypropylene FRC show tendencies to an increase in fracture energy with age the first years of the service life of the materials.
AR-glass FRC shows decrease in fracture energy with age despite the fact that polymer modification of matrix has been used to prevent embrittlement of the composite.
Autoclaved cellulose fibre reinforced cement also show tendencies to decrease in fracture energy due to ageing but with different degradation rate when compared with AR-GFRC.
Keywords: Service life prediction, fracture energy, ageing, FRC-materials, steel fibre, glass fibre, polypropylene fibre, cellulose fibre.

1 Introduction

Service life of structures of cementitious composites depends on a great number of factors. It is not only weathering, especially freezing and thawing, chemical exposure, abrasion, corrosion and reactive components, but also the inherent properties of the composite which must be considered.

To get specific information about the service life of the fibre reinforced cementitious composites a large test program was set up at the Danish Building Research Institute.

The purpose of the test program was to illustrate the behavior of different fibres in the matrix. The following types of fibres were considered:

1) Fibres unaffected by the environment in the matrix (steel and polypropylene).
2) Fibres affected by the environment in the matrix (glass fibre).
3) Fibres affected but with reduced deterioration due to modification of the matrix (glass fibres in a polymer modificated matrix and autoclave of cellulose fibre reinforced cement).

The following matrices were used:

1) Concrete, maximum size of aggregate is 16 mm.
2) Cement mortar, with or without polymer modification.
3) Autoclaved cement.

One of the best and most direct ways to evaluate the service life of a brittle composite is to determine the development of fracture energy with age. The fracture energy can be determined either in uniaxial tension or in bending, with bending tests being the simplest method (Hillerborg, 1985).

All types of materials in the test program was tested in a three-point bending test in an Instron 6022 Series Materials Testing Machine operated with strain control. The three-point bending test was stopped when deflection reached 2.50 mm or when the load reached zero after fracture; the deflection speed was 0.25 mm per minute.

When carrying out the three-point bending test the fracture energy is directly obtained as the area under the load-deflection curve divided by the cross-sectional area of the test specimen.

Toughness indices, the I_5 and I_{10} indices, corresponding to progressively increasing amounts of deflection and cracking were determined according to ASTM C1018 (1985).

2 Description of materials

2.1 Fibres

Four different types of fibres were used in the experiments:

1) Fibrillated high tenacity, high bond polypropylene fibres 12 mm long and an average fibre cross-section of 0.030 mm x 0.140 mm (Danish KRENIT fibres, type special).

2) Enlarged end steel fibres 18 mm long and a fibre cross-section of 0.3 mm x 0.4 mm (Australian EE-fibres).
3) Alkali resistant glass fibres (Cem-FIL II), prechopped mix of fibres with lengths of 18-36 mm.
4) High strength paper pulp cellulose fibres with a maximum length of 1 - 2 mm.

2.2 Cement
One autoclaved cellulose fibre reinforced cement was tested, B series. The fibre content was approximately 10% by weight.
These cellulose FRC boards were produced in a Hatscek process and delivered in plates of 6 mm x 1200 mm x 2400 mm which were then cut into specimens of 6 mm x 480 mm x 480 mm.

2.3 Mortars
Two different glass fibre reinforced mortars were investigated; One with polymer additives FORTON type (P-AR-GFRC) and one without polymer additives (AR-GFRC). Water cement-ratio was 0.29 and glass fibre content was 5% by weight. Both mortars were sprayed up in specimens of 10 mm x 1000 mm 1000 mm. The specimens were air cured (65% RH, 23°C) for 28 days before they were cut into smaller specimens of 10 mm x 480 mm x 480 mm and tested in natural weathering and laboratory aging tests.

2.4 Concretes
Different types of concrete were used in the tests.

1) F series: Concrete reinforced with 2 vol% polypropylene fibres, an equivalent water-cement ratio of 0.47 and an equivalent cement content (OPC + fly ash + micro silica) of 590 kg/m^3. The aggregate content was 1200 kg/m^3, maximum aggregate size was 16 mm.

2) SF series: Concrete reinforced with 1 vol% polypropylene fibres and 1 vol% steel fibres, an equivalent water-cement ratio of 0.44 and an equivalent cement content (OPC + fly ash + micro silica) of 510 kg/m^3. The aggregate content was 1400 kg/m^3, maximum aggregate size was 16 mm.

3) C series: One ordinary concrete without fibre reinforcement used as reference in some of the experiments. The concrete was with an equivalent water-cement ratio of 0.43 and an equivalent cement content (OPC + fly ash + micro silica) of 410 kg/m^3. The aggregate content was 1800 kg/m^3, and maximum aggregate size was 16 mm.

The SF, F and C series were mixed in a factory concrete mixer in a batch of minimum one cubic meter.
Specimens of 40 mm x 480 mm x 480 mm were cast from a mix of SF, F and C series. The SF and F series were heavily

vibrated to make them "look like" pumped concrete. The C series were normally vibrated.
After casting, the specimens were covered with a plastic sheet for 24 hours at room temperature. They were then demoulded, placed in a water tank and kept at 20°C for 28 days. After curing the specimens were exposed to natural weathering and laboratory ageing tests.

Further, two fibre reinforced concretes were mixed in the laboratory (each with a batch of 50 litres) with the same recipe as F and SF series, but normally vibrated. These are named as 90 and 85, respectively.

The specimens were cast in a size of 60 mm x 40 mm x 600 mm and covered with a plastic sheet for 24 hours at room temperature and then demolded and cut into specimens of 60 mm x 40 mm x 300 mm. Then half the specimens were cured in water at 20°C,and the other half of the specimens were cured in air at 65% RH and 23°C.

3 Description of tests

The investigated materials were exposed to three different climates. These were a weather simulator, a stable climate at 65% RH and 23°C, and natural weathering in an outdoor exposure rack facing south at an inclination of 45°. Additionally the two fibre reinforced concretes were exposed to a heat-moisture test and the glass fibre reinforced mortars were exposed to a hot water test.

The different types of frc-materials examined are shown in Table I, where also the type and duration of test are summarized.

3.1 Weather simulator

Laboratory weathering tests were carried out in a specially designed apparatus for ageing of building materials and components (in everyday talk called "the four seasons") shown in figure 1. The apparatus consists of four test frames surrounding a central chamber. The tested materials are placed in the frames. The climatic stresses are created by three boxes outside the central chamber. The central chamber is usually at rest, but it is moved a quarter turn (90°) at predetermined intervals.

In this way, the test specimens are subjected to the following test cycle:

- Radiation from sunlight lamps (UV-light), intensity of 1900 +/- 50 W/m², and simultaneous heating to an elevated temperature of 75°C +/- 5°C (black body surface temperature) which should be reached within half an hour.

Table I: Type and duration of tests on different types of frc-materials in the program.

Serie	Matrix	Fibre	V_f	Tests	Duration
B	Cement (mortar)	Cellu-lose	≈10 (by weight%)	Weather simulator Stable climate Natural weathering Freeze-thaw test	2 years (960, 1000 cycles) 2 years 2 years 112 cycles
G	Mortar with polymer additives	AR-glass	5.0 (by weight%)	Weather simulator Stable climate Natural weather Hot water test (50°C)	365 days (730 cycles) 365 days 365 days 365 days
G	Mortar without polymer additives	AR-glass	5.0 (by weight%)	Weather simulator Stable climate Natural weather Hot water test (50°C)	365 days (730 cycles) 365 days 365 days 365 days
F 90	Concrete D_{max}=16 mm	PP	2.0 (vol%)	Weather simulator Stable climate Natural weathering Heat-moisture test Freeze-thaw test	1.5 year (1000 cycles) 1.5 year 1.5 year 225 days (900 cycles) 56 cycles
SF 85	Concrete D_{max}=16 mm	PP + ST	1.0 + 1.0 (vol%)	Weather simulator Stable climate Natural weathering Heat-moisture test Freeze-thaw test	1.5 year (1000 cycles) 1.5 year 1.5 year 225 days (900 cycles) 56 cycles
C	Concrete D_{max}=16 mm	----	----	Weather simulator Stable climate Freeze-thaw test	1.5 year (1000 cycles) 1.5 year 56 cycles

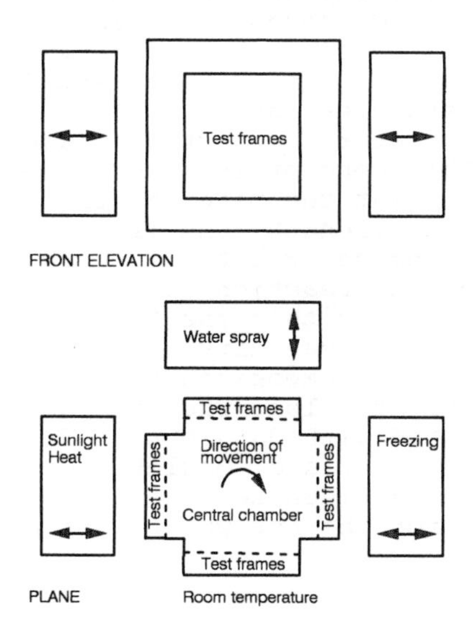

Figure 1: Weather simulator

- Wetting with a spray of demineralized water, 15 +/- 2 liters/m^2/hour.
- Cooling to an air temperature of -20°C +/- 5°C.
- Thawing at room temperature (23°C +/- 2°C, 40% +/- 10% RH) with possibilities of inspecting and changing the samples without stopping the test.

The fibre and unreinforced concretes, glass fibre reinforced mortar were tested in the weather simulator operating in a 12 hour cyclus. The cellulose fibre reinforced cement was tested in the simulator with one series operating in a 4 hour cyclus and another series operating in a 12 hour cyclus.

3.2 Heat-moisture test

For heat-moisture test (figure 2) of the fibre reinforced concretes a special apparatus was designed. The apparatus consists of a construction inclined at 25° to which the materials was attached. It was then alternately heated uniformly by radiant heat and wetted by water.
In this way the test samples was subjected to the following test cycle:

- Wetting with water, 10 liters/minut/m^2, at room air temperature for 2 hours and 55 minutes followed by an interval of 5 minutes rest.
- Heating to a temperature of 65°C +/- 5°C (measured on a black body surface) for 2 hours and 55 minutes followed by an interval of 5 minutes rest.

3.3 Hot water test

After the 28 days initial curing in stable climate the glass fibre reinforced composites were placed in a water bath at 50°C.
The concept behind this ageing test is that the strength loss at high temperatures can be used to predict strength loss at lower temperatures expected to occur over a long period of time. This concept is based on many assumptions, including the fact that the material is thermo-rheologically simple, i.e. the strength loss is directly related to the rate of some chemical reaction. Further it is assumed that the same chemical reaction for ageing mechanisms is the controlling factor within the temperature range generally 20°C to 80°C. The validity of these assumptions for glass fibre reinforced composites is at least indirectly established for strength loss when the data for accelerated ageing are correlated with data obtained from natural weathering sites (Litherland, Oakley and Proctor, 1981).
In Denmark with annual mean temperature of 8°C the acceleration factor for the hot water test at 50°C is approximately 100.

4 Results and discussion

4.1 Autoclaved cellulose fibre reinforced cement

Figure 3 shows examples of stress-deflection curves for the autoclaved cellulose fibre reinforced cement.
It is obvious that the material, when tested in three-point bending, is very much dependent on the condition of the material. An ovendry (50°C) specimen shows large strength and low fracture energy. On the contrary, a water saturated specimen shows low strength and high fracture energy. A material conditioned between ovendry and water saturated conditions will have a stress-deflection curve between the two extremes as exemplified in figure 3.

Examples of stress-deflection curves after 2 years of natural weathering in Denmark are also shown in figure 3. Both strength and toughness shows reductions due to ageing.

Comparing modulus of rupture from the 2 years natural weathering with 160 days (960 cycles in a 4 hours cyclus) and 500 days (1000 cycles in a 12 hours cyclus) in the weather simulator the reduction in strength is approximately the same (figure 4). It seems to be the dynamic exposure which influence this composite with a delamination of the plate material as a result.
The acceleration factor of the weather simulator for the autoclaved cellulose fibre reinforced cement can then be calculated as:

$$4\ \textit{hours cyclus:}\ \frac{730\ \textit{days (natural weathering)}}{160\ \textit{days (weather simulator)}} \approx 4.5$$

$$12\ \textit{hours cyclus:}\ \frac{730\ \textit{days (natural weathering)}}{500\ \textit{days (weather simulator)}} \approx 1.5$$

4.2 Glass fibre reinforced mortar

Examples of load-deflection curves of the alkali resistant glass fibre reinforced mortars are shown in figure 5.
Polymer modified alkali resistant glass fibre reinforced mortar (P-AR-GFRC) show higher strength and toughness at 28 days than the same product without polymer modification of matrix (AR-GFRC).
After 5 months of storage in hot water strength as well as toughness of the different products are significally reduced.

In figure 6 and figure 7 is shown modulus of rupture (MOR) and fracture energy of the two products as function of the time in Danish natural weathering and the hot water experiments where the corresponding age in Denmark is calculated assuming an acceleration factor of 100.

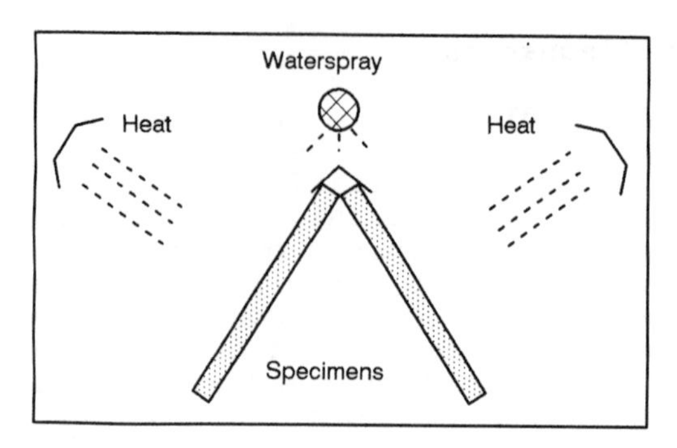

Figure 2: Heat-moisture test

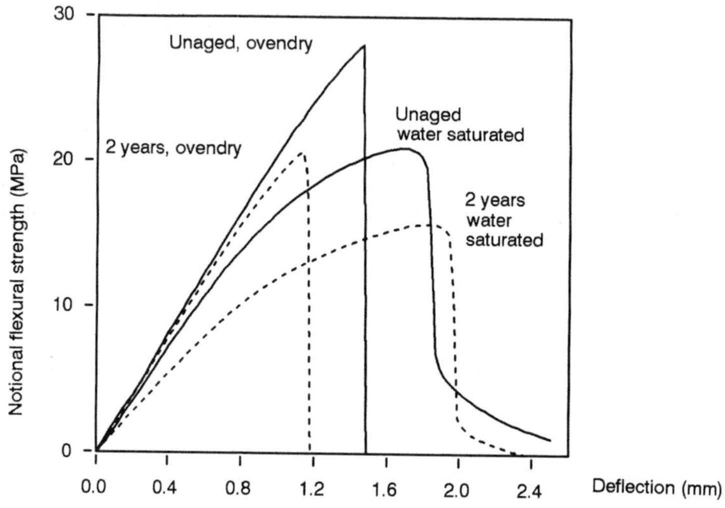

Figure 3: Examples on stress-deflection curves for autoclaved cellulose fibre reinforced cement.

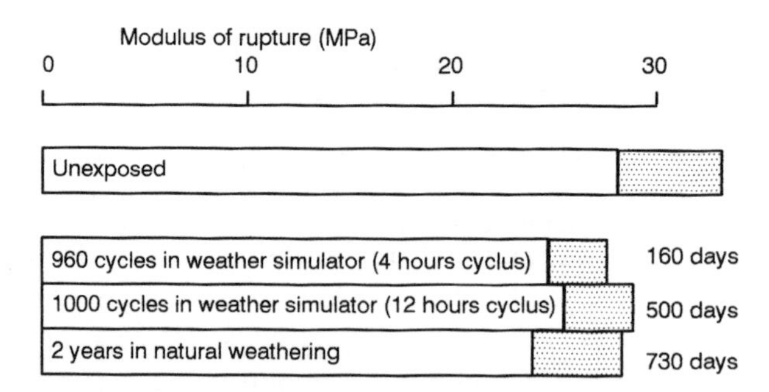

Figure 4: Comparison of modulus of rupture of autoclaved cellulose fibre reinforced cement aged naturally and in weather simulator.

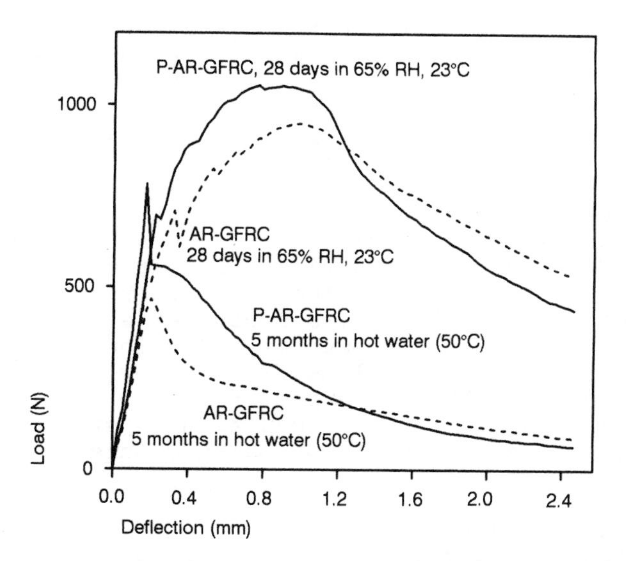

Figure 5: Examples on load-deflection curves for alkali resistant glass fibre reinforced mortar with and without polymer modification of matrix.

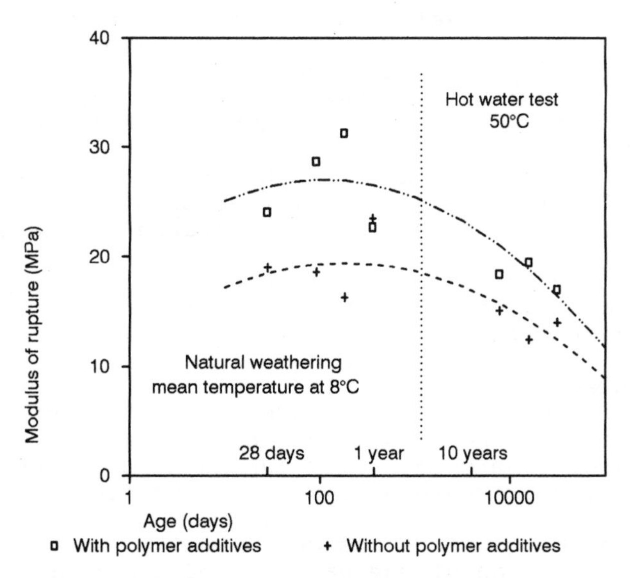

Figure 6: Development in MOR with age of alkali resistant glass fibre reinforced mortars with and without polymer modification of matrix.

The tendencies in both figures are that in short time the strength and toughnes increase with age. Later on large reductions occur in both properties.
The explanation could be that a continuing hydration process will increase the fibre bond with increase in strength and toughness as a result. Here the polymer modification show its advantage, since it prevents the growth of hydration products between the fibre filaments with a larger strength and toughness as a result. Several years of ageing result in alkali corrosion of the fibres and at this stage the polymer modification has no effect. In other words the ageing mechanisms are as follows (as Bentur et al, 1985):
1) In short time (5-30 years) growth of hydration products, if it is possible, between the glass filaments with reduction in strength and fracture as a result.
2) In prolonged ageing (>30 years) reduction in strength and fracture energy due to alkali corrosion of the glass fibres.

4.3 Fibre reinforced concrete

Fibre reinforcement of concrete results in load-deflection curves as exemplified in figure 8. Polypropylene fibres significantly increase fracture energy with little or none effect on MOR. Steel fibres increase both strength and fracture energy.

Figure 9 and 10 show the development in limit of proportionality (LOP) and fracture energy for the concrete reinforced with 2 vol% polypropylene fibres (F and 90 series). No matter the way of ageing, weather simulator, stable climate or natural weathering, the LOP and fracture energy increases.
Exactly the same picture is shown for the concrete reinforced with 1 vol% polypropylene and 1 vol% steel fibres. There seems not to be any significant difference in the way of ageing.
900 heat-moisture cycles have no influence in the development of strength and fracture energy. The tendencies are the same as for the different ways of weathering.

The mentioned tendencies are more clear when the development in fracture energy and strength are shown as function of the maturity of the concretes, i.e the increase in strength and fracture energy are only dependent of temperature and the age of the concrete (figure 11 and figure 12).

The development in the toghness indices I_5 and I_{10} are shown in figure 13 as a function of the maturity of the fibre reinforced concretes. Both concretes are very close to the linear-elastic-plastic material on the first part of the load-deflection curve (I_5 are close to 5) and I_{10} increases with age up against the value 10 which is the ideal value for the elastic-plastic material.

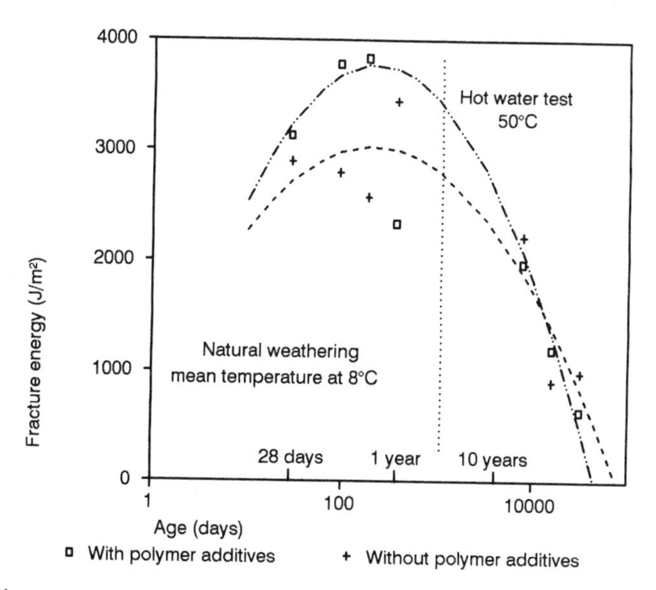

Figure 7: Development of fracture energy with age of alkali resistant glass fibre reinforced mortars with and without polymer modification of matrix.

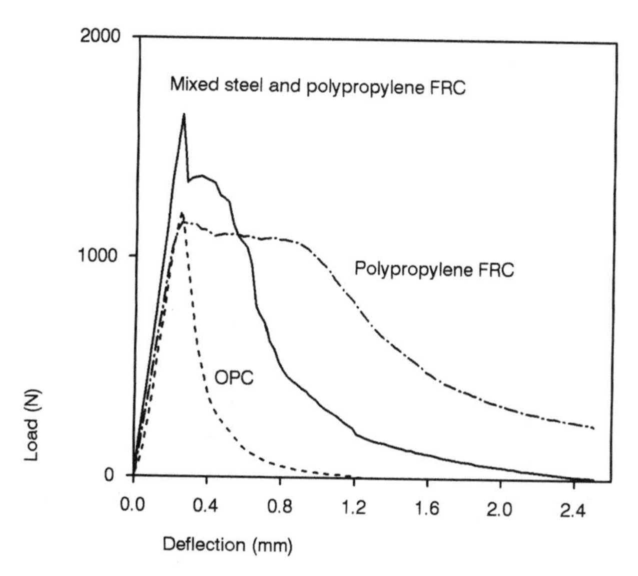

Figure 8: Examples on load-deflection curves for ordinary concrete, polypropylene fibre reinforced concrete, and steel fibre and polypropylene fibre reinforced concrete.

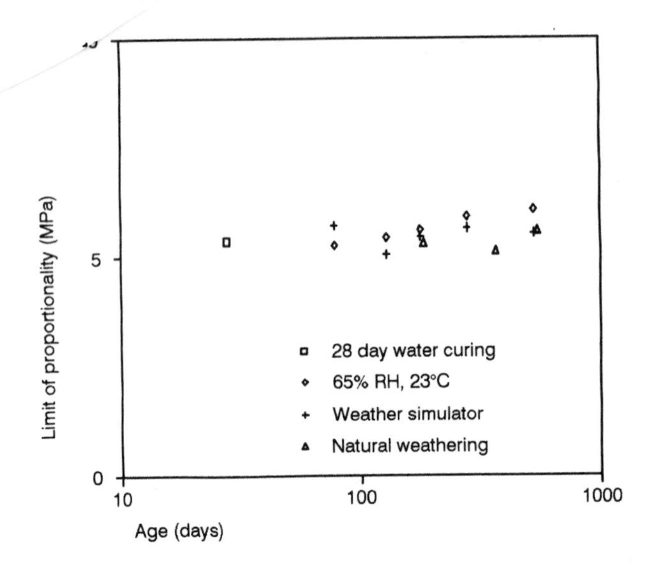

Figure 9: Development in limit of proportionality strength with age of polypropylene fibre reinforced concrete and with different ways of weathering as parameter.

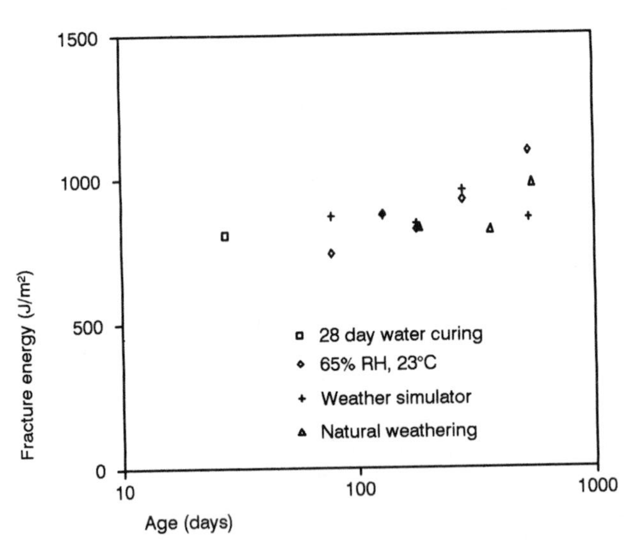

Figure 10: Development of fracture energy with age of polypropylene fibre reinforced concrete and with different ways of weathering as parameter.

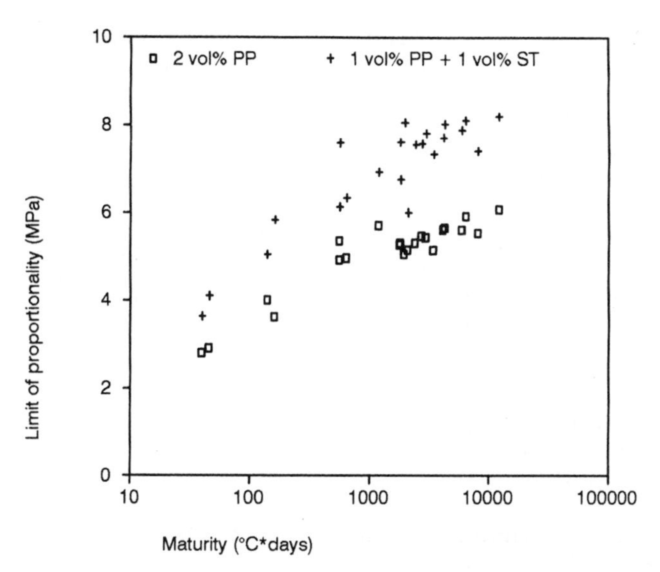

Figure 11: Development of limit of proportionality strength with maturity of polypropylene fibre reinforced concrete, and steel and polypropylene fibre reinforced concrete.

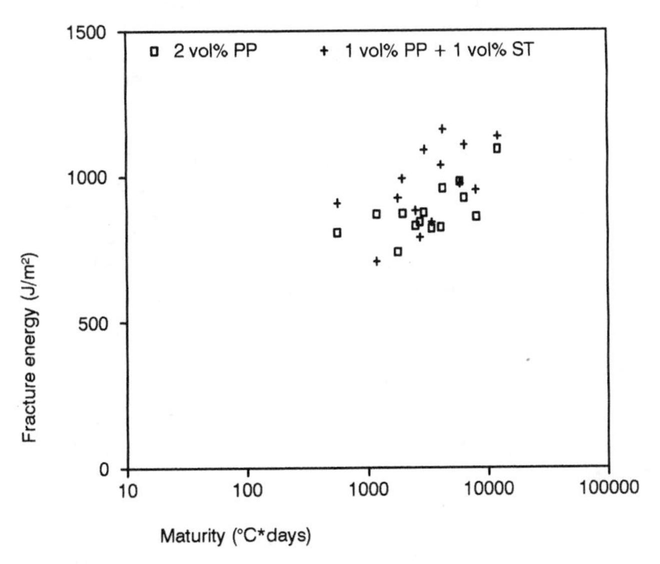

Figure 12: Development of fracture energy with maturity of polypropylene fibre reinforced concrete, and polypropylene and steel fibre reinforced concrete.

The conclusion is that continuing hydration of the two fibre reinforced concretes result in increased fibre bond of polypropylene and steel fibres with an increase in strength and fracture energy as a result.

5 Conclusions

If one normalize the development in fracture energy, so that the fracture energy at the age of 28 days is 1.0, it is possible to get a picture on how a fibre reinforced composite increases or decreases its fracture energy.

In figure 14 the decrease in fracture energy for the autoclaved cellulose fibre reinforced cement is shown. The results are from natural weathering in Denmark. As mentioned earlier there are large differences depending on the condition of the specimen when tested in three-point bending. In the figure the result from water saturated and ovendry specimens are shown. Assuming that the decrease in fracture energy will continue as exemplified, reduction of fracture energy to approximately 2/3 of its initial value after 40 years of natural weathering in Denmark can be predicted.

For the glass fibre reinforced mortar the fracture energy will be reduced to approximately 30-50% of its initial value after 35-45 years of natural weathering in Denmark. It is here assumed that the hot water test is valid, and it is the results from the hot water tests which are shown in figure 15.

The results from the two fibre reinforced concretes are presented in figure 16 together with ten years results from England on polypropylene fibre reinforced slates (Hannant, 1989). These products will continue with an increase in the fracture energy at least up to ten years after casting with approximately a 40% increase in fracture energy as result.

6 Acknowledgements

This study has been possible thanks to a grant from the Danish Research Academy, Århus, and the Danish Building Research Institute.

7 References

ASTM C1018-85: "Standard Test Method for Flexural Toughness and First-crack Strength of Fiber-Reinforced Concrete (using Beam with Third-Point Loading)". In: Annual Book of ASTM Standards 1986. Vol. 04.02 Concrete and Mineral Aggregates. Philadelphia, American Society For Testing And Materials. 1986. pp. 650-657.

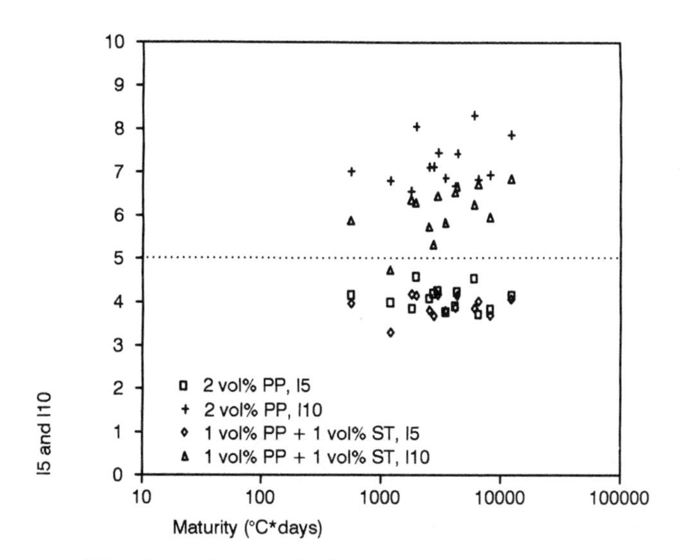

Figure 13: Development in toughness indices with maturity. The indices are determined according to ASTM 1018. The results are from SF, F, 85 and 90 series.

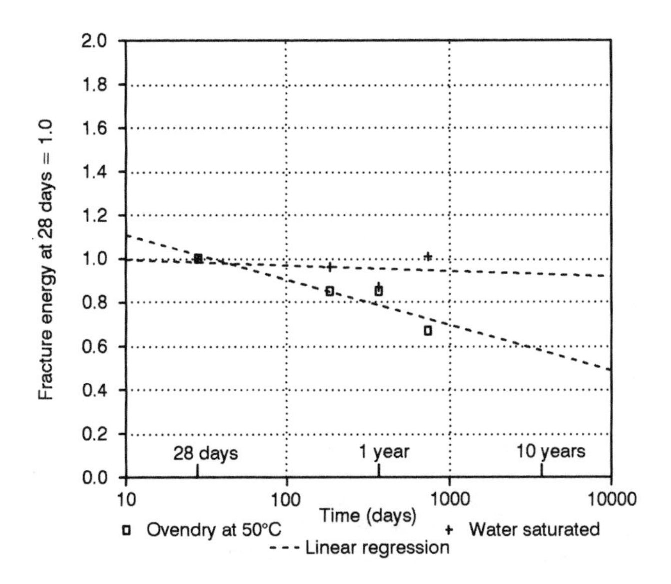

Figure 14: Development of normalized fracture energy with age of the autoclaved cellulose fibre reinforced cement.

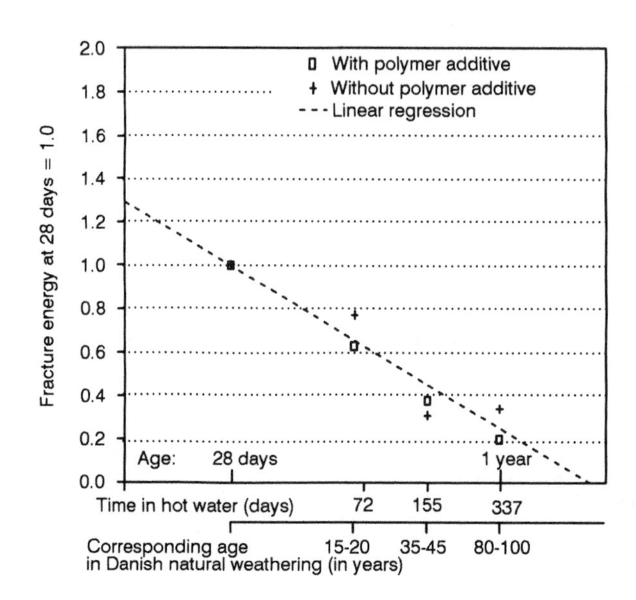

Figure 15: Development of normalized fracture energy with age of alkali resistant glass fibre reinforced mortars with and without polymer modification of matrix.

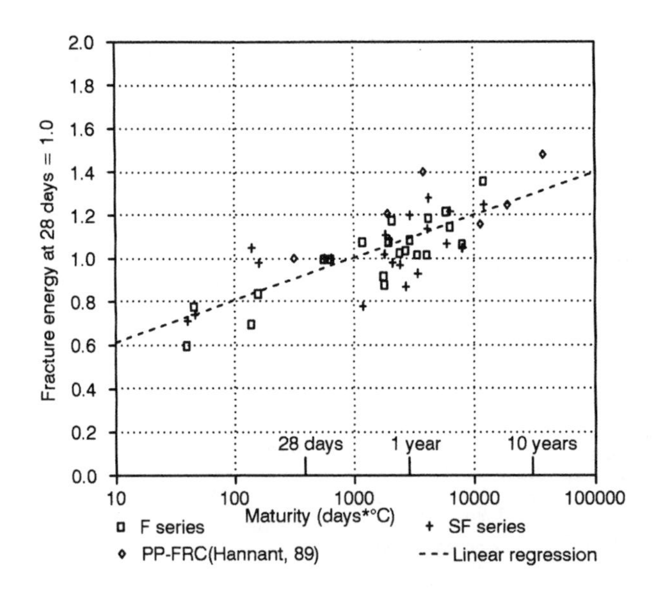

Figure 16: Development of normalized fracture energy with maturity of polypropylene fibre reinforced concrete and polypropylene and steel fibre reinforced concrete.

Bentur, A.; Ben Bassat, M. and Schneider, D.: "Durability of glass fiber reinforced cements with different alkali resistant fibers". J. Amer. Ceram. Soc. Vol. 68. 1985. pp. 203-208.

Hannant, D. J.: "Ten year flexural durability tests on cement sheets reinforced with fibrillated polypropylene networks". In: Fibre Reinforced Cement and Concretes: Recent Development (Ed. R. N. Swamy, B. Barr). Elsevier Applied Science, 1989. pp. 563-572.

Hillerborg, A.: "The theoretical basis of a method to determine the fracture energy, G_F, of concrete". Materials and Structures, Vol. 18, No. 106. 1986. pp. 291-296.

Litherland, K. L.; Oakley, D. R. and Proctor, B. A.: "The Use of Accelerated Ageing Procedures to Predict the Long Term Strength of GRC Composites". In: Cement and Concrete Research. Vol. 11, No. 3. 1981. pp. 455-466.

65 FREEZE-THAW RESISTANCE OF FRC MATERIALS

E. S. LARSEN
Danish Building Research Institute, Hørsholm, Denmark

Abstract
The freeze-thaw resistance of polypropylene fibre reinforced concrete was tested according to Swedish Standard SS 13 72 44: Concrete testing - Hardened concrete - Frost-resistance.
The freeze-thaw testing was carried out on concrete specimens with different air contents and fibre volumes. Additionally the freeze-thaw resistance of a steel plus polypropylene fibre reinforced concrete was tested.
The results show that the pore structure of the matrix is the most significant factor affecting the freeze-thaw resistance. The fibre content does not seem to influence in general the freeze-thaw resistance of fibre reinforced cementitious materials.
Keywords: Freeze-thaw resistance, FRC-materials, polypropylene fibre, steel fibre, microstructural analysis, SS 137244

1 Introduction

In natural weathering the effect of frost depends on the previous history of the composite, particularly on the combination of circumstances that determine its water content and the distribution of water among its components.

In the following a short review will be given of the freeze-thaw resistance of different fibre reinforced composites and next the results from a freeze-thaw test programme will be presented.

Gram, Fagerlund and Skarendahl (1978) used the critical degree of saturation method (Fagerlund, 1977) on steel fibre reinforced concrete. The results showed that the degree of saturation at which the initial frost damage took place is not significantly changed when fibres are added. But degradation due to freeze-thaw when exceeding the critical degree of saturation, takes place at a much slower rate in fibre reinforced composites than in unreinforced

Fibre Reinforced Cement and Concrete. Edited by R. N. Swamy. © 1992 RILEM.
Published by E & FN Spon, 2-6 Boundary Row, London SE1 8HN. ISBN 0 419 18130 X.

composites as a result of the crack arresting properties of the fibres.
On basis of data from 20 years of laboratory testing and natural weathering Hoff (1987) concluded that fibre reinforced concrete must be air entrained to obtain proper freeze-thaw resistance. The fibres may prolong the life of non-air entrained concrete subjected to continous wetting and freezing and thawing action. The non-air-entrained concrete will deteriorate significantly within the normal design life of the structure.
In the large program of Balaguru and Ramakrishnan (1986, 1988) the following could be concluded on the freeze-thaw resistance of steel fibre reinforced concrete:

- The air content is the most significant parameter for the freeze-thaw resistance.
- If air content is the same, the durability of fibre reinforced concrete and plain concrete is similar.
- With the same air entrainment, the frost resistance may be improved by increasing the cement content and reducing the water-cement ratio.
- With a water-cement ratio of more than 0.4 and a cement content less than 415 kg/m^3, a minimum of 6.0 percent air, preferably 8.0 percent, should be used to avoid deterioration under freeze-thaw cycling according to ASTM C666.

Proctor (1980) reported results from freeze-thaw tests on glass fibre reinforced composites. The conclusions were that freeze-thaw conditions do not pose a significant problem in the use of GRC.

The frost resistance of cellulose and natural fibre reinforced composites depends on the composition of the matrix according to Dinwoodie and Paxton (1989) and Sarja (1989). Harper (1982) tested cellulose fibre reinforced composites according to BS 690 and the results did not show any deterioration. At vacuum water saturation there were reductions in strength and stiffness after 100 cycles of freeze-thaw. The present author (1991) tested cellulose fibre reinforced composites immersed in water exposed to temperature cycling between -20°C and + 20°C and the results showed some reductions in strength, stiffness and fracture energy after 112 cycles. However, these materials normally show a proper freeze-thaw resistance in natural weathering.

2 Test program

2.1 Materials

The fibre type used was fibrillated high tenacity, high bond polypropylene fibres. The fibres were 12 mm long with an average fibre cross-section of 0.030 mm x 0.140 mm (Danish KRENIT fibres, type special).

The Australian EE steel fibres were 18 mm long and had a cross-section of 0.3 mm x 0.4 mm.

Fifteen series (66, 63, 40, 31, 33, 57) of concretes were mixed in a factory mixer (one cubic meter per series), reinforced with with 0, 0.5, 1.0, 1.5, 2.0 vol% polypropylene fibres respectively and three levels of air content. The concretes were cast in blocks in size of 300 mm x 300 mm x 200 mm. After casting the blocks were stored in water for 21 days. Immediately after 3 cylinders, 100 mm in diameter, were drilled from each block. The cylinders were sawn in test specimens of 100 mm in diameter and a height of approximately 50 mm.
From each series one specimen was used for structural analysis in optical microscope and five specimens were used for freeze-thaw test.

Additionally the following mixtures were cast:
- One concrete reinforced with 2 vol% polypropylene (F-series).
- One concrete reinforced with 1 vol% polypropylene + 1 vol% steel fibres (SF-series).
- One concrete reinforced with 1 vol% polypropylene fibres (not named)
- One concrete without fibre reinforcement (C-series).

The F, SF and C series were cast as specimens with dimensions of 40 mm x 480 mm x 480 mm. The SF and F series were heavily vibrated to make them "look like" pumped concrete. The C series was normally vibrated.
After casting, the specimens were covered with a plastic sheet for 24 hours at room temperature. They were then demolded placed in a water tank and kept at 20°C for 21 days.

In the laboratory were further mixed two fibre reinforced concretes according to the same recipe as SF and F series, but normally vibrated, named as 85 and 90, respectively. Additionally a series (not named) reinforced with one vol% polypropylene fibres was casted. The three series all amounted to 50 litres.

In all the concretes used for the test program the water-cement ratio was kept approximately at 0.45. Depending on the mix, the equivalent cement content (OPC + fly ash + micro silica) varied between 410 kg/m^3 - 590 kg/m^3 and the aggregate content between 1200 kg/m^3 - 1800 kg/m^3 with a maximum aggregate size of 16 mm.

2.2 Freeze-thaw test

Testing was performed according to the Swedish Standard SS 13 72 44: "Concrete testing - Hardened concrete - Frost resistance" (1988).
Every freeze-thaw test consists of five specimens with a freeze-thaw surface area of 7850 mm^2.

The 21 days old and water saturated specimens were covered with a moisture prof and thermal insolation material as shown in figure 1. The specimens were then stored with pure water on the freeze-thaw surface for another 7 days. Fifteen minutes before placing the specimens in the freeze-thaw apparatus the freeze-thaw surface was exposed to a 3 mm thick layer of 3 per cent NaCl-solution. This salt solution was protected from evaporation with a plastic foil covering the specimens.

The specimens were then exposed to freeze-thaw according to a specified temperature cycle shown in figure 2. The temperature was measured in the salt solution. The temperature was raised to above zero for at least 7 but not more than 9 hours.

After 7, 14, 28, 42 and 56 freeze-thaw cycles the following procedure was followed:

- Scaling from the freeze-thaw surface was collected with a brush and washed with water.
- New salt solution in a 3 mm thick layer was applied to the freeze-thaw surface and the specimen was returned to the freeze-thaw apparatus.
- The scaling material was dried at 105°C until constant weight, note that the salt should not be counted.
- For each specimen the following is calculated: M_n/A (kg/m^2) where M_n is the total weight of scaling in mg after n freeze-thaw cycles. A is the freezing area in mm.

According to SS 13 72 44 the freeze-thaw resistance is defined as:

Very good:
If no specimen has a larger scaling than 0.1 kg/m^2 after 56 cycles.

Good:
If the avarage value of scaling after 56 cycles is smaller than 0.5 kg/m^2 and M_{56}/M_{28} is less than 2.

Acceptable:
If the average value of scaling after 56 cycles is smaller than 1.0 kg/m^2 and M_{56}/M_{28} is less than 2.

Non acceptable:
If the demands to acceptable freeze-thaw resistance are not fulfilled, or if the requirements for being rated acceptable are not met.

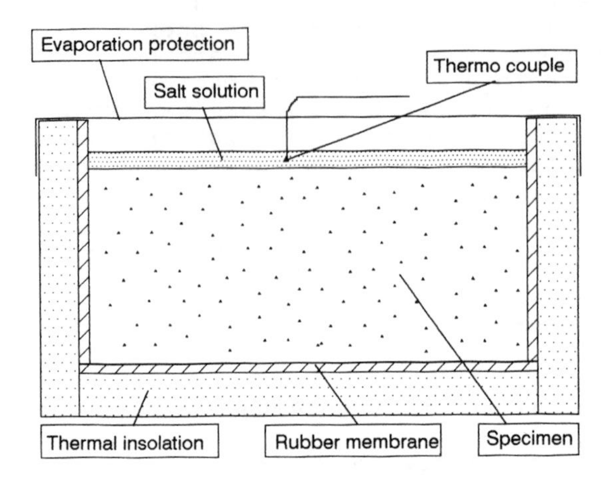

Figure 1: Test specimen for freeze-thaw test according to the Swedish standard SS 137244

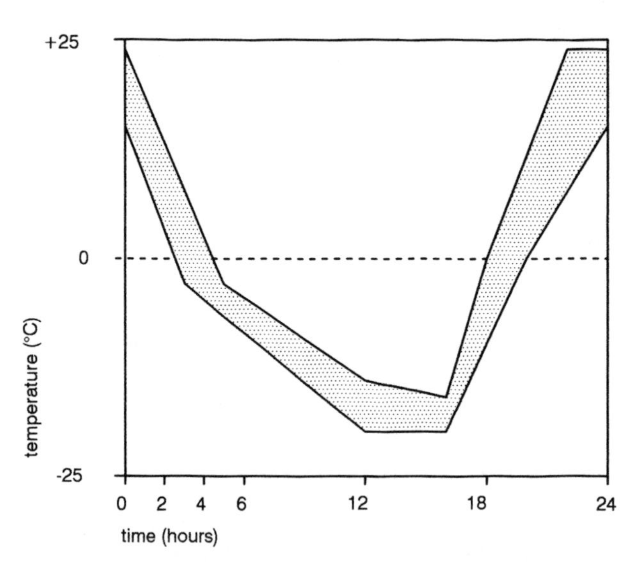

Figure 2: Temperature cycle used in the freeze-thaw tests according to SS 137244.

3 Results and discussion

Results from pressuremeter measurements on air contents and microstructural analysis on air-void characteristics are shown in table 1.
In table 2 the scaling from each series after 7, 14, 28, 42 and 56 cycles respectively are presented. The results in the table are mean values of five specimens with standard deviation in brackets.

Freeze-thaw test of fibre reinforced concretes' outer surface results in scaling which is larger than from a sawn surface. The creation of the disadvantageous concrete skin structure could be due to the heavy vibration of the concretes and it seems to increase with the fibre amount. There is no significant difference between scaling of the two vol% polypropylene fibre reinforced concrete (F series) and, the one vol% polypropylene + one vol% steel fibre reinforced concrete (SF series).

The two heavily vibrated fibre reinforced concretes (SF and F series) have lesser scaling compared to the normally vibrated and laboratory produced fibre reinforced concretes (85 and 90 series). Therefore the production method and handling of the concretes have large influences on the freeze-thaw resistance.

None of the tested concretes, 28 days old at the start of freeze-thaw, could be characterized as very good according to the Swedish standard. However, 500 cycles in a weather simulator on the unreinforced concrete seems to have a beneficial effect on the freeze-thaw resistance. This could be due to strengthening of the matrix or rather to the fact that the surface skin of the concrete could be scaled off due to the weathering in the simulator.

The results presented in figure 3 - 6 are freeze-thaw tests on sawn concrete surfaces.

Figure 3 shows the amount of scaling after 56 freeze-thaw cycles as function of the air content. The results show no significant fibre influence on the freeze-thaw scaling. The more air entrainment the less scaling is the most significant result.
More than 6% air in the fresh concrete results in good and acceptable freeze-thaw resistance of the concretes according to the Swedish Standard.

In figure 4 is shown the total scaling after 56 freeze-thaw cycles as a function of air content in the cement paste measured on the fresh and hardened concretes. The measurements show that there seems to be a tendency to a smaller air content in the hardened concrete compared to the fresh concrete.

Table 1: Results from measurement of air content in fresh concrete and air void characteristics in hardened concrete.

Series	V_f (vol%)	Fresh concrete	Hardened concrete			
		Air content (%)	Air content (%)	Air content bubbles > 2 mm (%)	Specific surface (mm^{-1})	Spacing factor (mm)
66-1	0	2.7	1.7	0	29	0.24
66-2	0	3.3	1.8	0	27	0.24
63-3	0	6.5	4.3	0.1	36	0.12
40-1	0.5 PP	1.6	2.3	0.5	19	0.30
40-2	0.5 PP	3.7	2.9	0.2	24	0.24
40-3	0.5 PP	6.0	6.4	0.9	25	0.15
31-1	1.0 PP	2.1	1.4	0	42	0.20
31-2	1.0 PP	4.0	2.5	0.2	23	0.25
31-3	1.0 PP	11.0	11.1	1.1	17	0.12
33-1	1.5 PP	1.6	2.3	0.4	25	0.25
33-2	1.5 PP	3.3	3.0	0	24	0.27
33-3	1.5 PP	8.2	6.8	0.5	21	0.20
57-1	2.0 PP	1.6	2.9	0.7	16	0.35
57-2	2.0 PP	7.0	4.7	0.2	22	0.26
57-3	2.0 PP	8.3	5.7	0.4	21	0.23
C	0	9.0	---	---	---	---
---	1 PP	12.0	---	---	---	---
90	2 PP	12.0	---	---	---	---
F	2 PP	6.0	---	---	---	---
85	1 PP + 1 ST	10.0	---	---	---	---
SF	1 PP + 1 ST	6.0	---	---	---	---

Table 2: Results from freeze-thaw tests according to SS 137244 on unreinforced and fibre reinforced concrete.

Series	Scaling (kg/m²) as function of the number of freeze-thaw cycles, Mean value and standard deviation.					Freezing surface
	7	14	28	42	56	
66-1	0.19 (.10)	0.43 (.16)	0.81 (.21)	1.10 (.27)	1.88 (.49)	Sawn
66-2	0.22 (.12)	0.39 (.18)	0.69 (.21)	0.95 (.21)	1.40 (.21)	Sawn
63-3	0.18 (.19)	0.27 (.23)	0.34 (.24)	0.40 (.27)	0.46 (.28)	Sawn
40-1	0.14 (.04)	0.43 (.07)	1.00 (.08)	1.79 (.25)	2.59 (.40)	Sawn
40-2	0.05 (.01)	0.14 (.06)	0.40 (.14)	0.59 (.21)	0.83 (.34)	Sawn
40-3	0.07 (.02)	0:19 (.04)	0.42 (.03)	0.51 (.02)	0.61 (.02)	Sawn
31-1	0.09 (.02)	0.47 (.05)	1.16 (.13)	1.75 (.21)	2.45 (.31)	Sawn
31-2	0.10 (.02)	0.43 (.05)	0.93 (.08)	1.48(.25)	2.13 (.30)	Sawn
31-3	0.06 (.02)	0.14 (.05)	0.26 (.07)	0.38 (.07)	0.47 (.08)	Sawn
33-1	0.25 (.05)	0.61 (.09)	1.18 (.12)	1.74 (.17)	2.18 (.13)	Sawn
33-2	0.19 (.03)	0.48 (.07)	0.98 (.09)	1.39 (.08)	1.76 (.14)	Sawn
33-3	0.12 (.06)	0.28 (.10)	0.52 (.13)	0.70 (.17)	0.94 (.34)	Sawn
57-1	0.21 (.09)	0.48 (.13)	0.93 (.23)	1.29 (.30)	1.59 (.35)	Sawn
57-2	0.12 (.06)	0.28 (.09)	0.58 (.08)	0.78 (.09)	0.96 (.09)	Sawn
57-3	0.18 (.13)	0.32 (.18)	0.58 (.27)	0.73 (.29)	0.89 (.28)	Sawn
C[1)]	0.03 (.01)	0.04 (.01)	0.05 (.01)	0.05 (.01)	0.05 (.01)	Free
C	0.07 (.05)	0.10 (.05)	0.17 (.09)	0.18 (.09)	0.21 (.14)	Free
---	0.10 (.03)	0.21 (.05)	0.42 (.01)	0.61 (.06)	0.76 (.08)	Free
90	0.13 (.03)	0.28 (.03)	0.72 (.17)	0.98 (.26)	1.21 (.31)	Free
F	0.01 (.01)	0.03 (.02)	0.08 (.05)	0.13 (.08)	0.15 (.10)	Free
---	0.02 (.01)	0.08 (.02)	0.20 (.03)	0.30 (.06)	0.40 (.08)	Sawn
90	0.03 (.01)	0.10 (.03)	0.23 (.06)	0.37 (.08)	0.48 (.13)	Sawn
SF	0.02 (.01)	0.07 (.03)	0.24 (.12)	0.57 (.34)	0.67 (.37)	Free
85	0.13 (.02)	0.25 (.04)	0.47 (.16)	0.65 (.23)	0.79 (.26)	Free
85	0.04 (.01)	0.15 (.03)	0.33 (.08)	0.43 (.10)	0.53 (.13)	Sawn

1) Aged 500 cycles in "weather simulator", i.e. 300 days old before freeze-thaw tested.

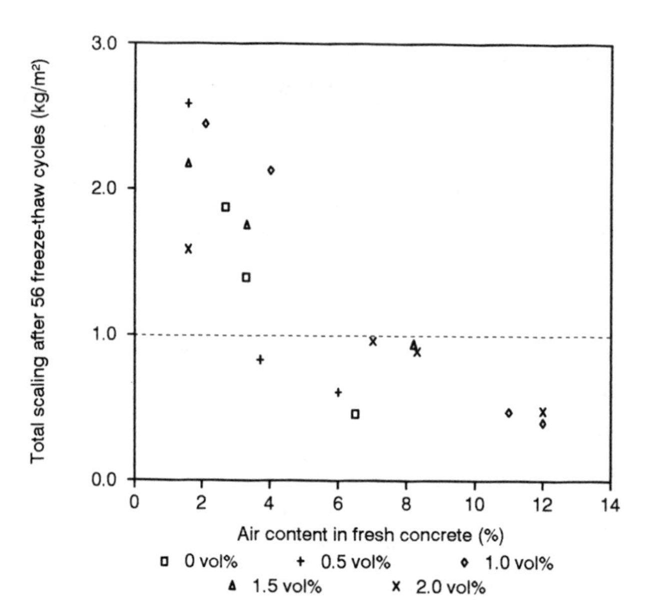

Figure 3: Total scaling after 56 freeze-thaw cycles as function of the air content in the fresh concrete.

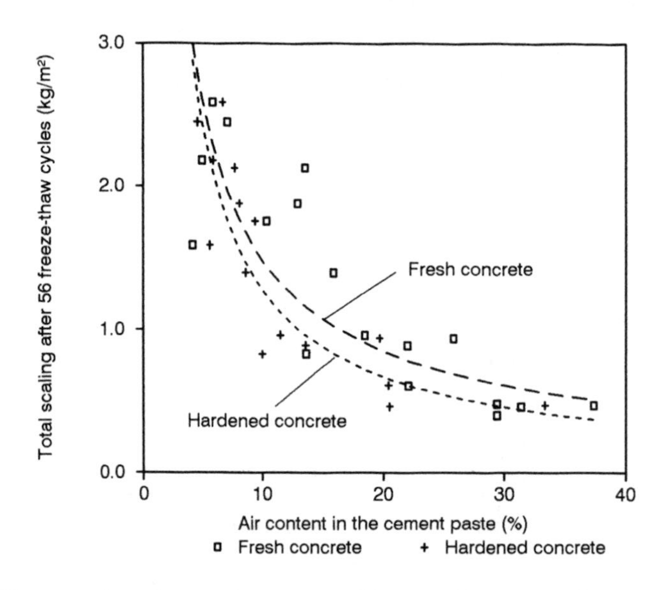

Figure 4: The totale scaling after 56 freeze-thaw cycles as function of the air content in the fresh and hardened cement paste.

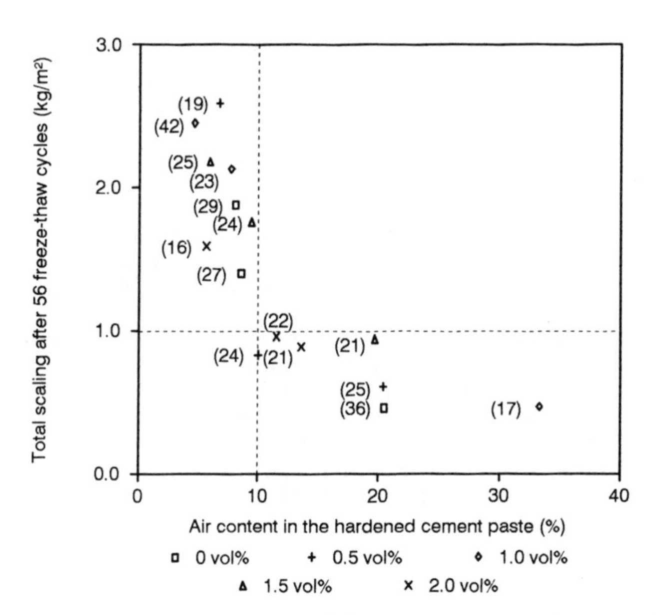

Figure 5: The total scaling after 56 freeze-thaw cycles as function of the air content in the hardened cement paste. The specific surface of the air bubbles are given in brackets.

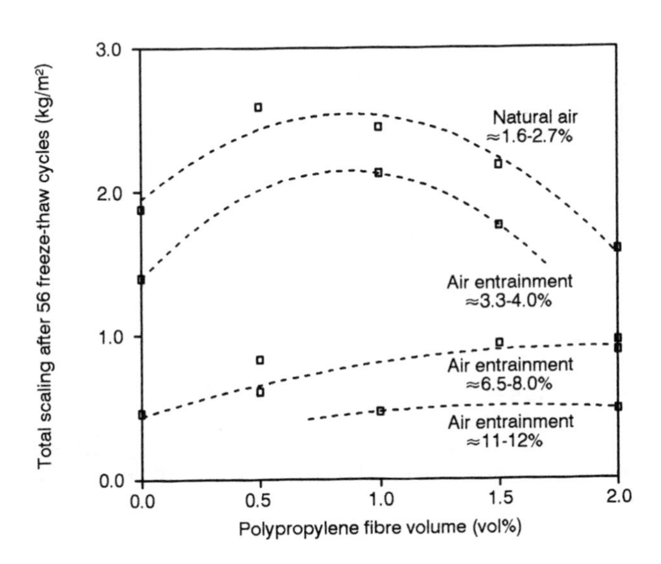

Figure 6: The total scaling after 56 freeze-thaw cycles as function of the fibre reinforcement volume (polypropylene fibres) and different air contents in the fresh concrete.

In figure 5 the total scaling after 56 freeze-thaw cycles as function of air content in the hardened cement paste is shown. Good-acceptable frost resistance is achieved if the cement paste air content is larger than 10% and the air-bubbles are well distributed. In brackets are given the specific surface of the air bubbles. It can be seen that the distribution of the air was very well in all mixes. It is not possible in this investigation to conclude anything about how the freeze-thaw resistance depends on the specific surface and the spacing factor (figures 7 and 8).

If air entrainment is not used this results in poor freeze-thaw resistance of the concretes. However it seems that the fibres in concretes with very low air content first result in poor freeze-thaw resistance at low fibre contents, but with higher fibre contents the scaling seems to decrease, so with a fibre content of 2 vol% polypropylene fibres the scaling is approximately the same as with no fibre reinforcement at all (figure 6).

4 Conclusion

It is the composition of the matrix which influences the freeze-thaw resistance in a fibre reinforced composite.

The overall conclusion is that the freeze-thaw resistance of fibre reinforced concrete does not differ significantly from an unreinforced concrete.

Test on quality concretes (W/C 0.4 - 0.5) indicates that a proper air entrainment and an air content of at least 6% results in good freeze-thaw resistance.

Test on polypropylene fibre reinforced quality concretes indicates that a proper and well distributed air entrainment of 10% in the hardened cement paste results in acceptable freeze-thaw resistance.

5 Acknowledgement

This study has been possible thanks to a grant from the Danish Research Academy, Århus and the Danish Buildning Research Institute. The project has also been supported financially by the Danish Materials Development Program on Cementbased Composites.

6 References

Balaguru, P. N. and Ramakrishnan, V.: Freeze-thaw durability of fibre reinforced concrete". In: ACI Journal. May-June 1986. pp. 374-382.

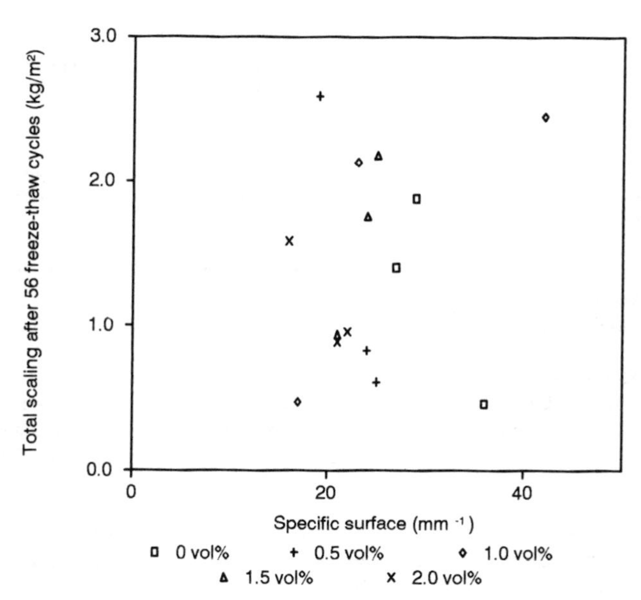

Figure 7: The total scaling after 56 freeze-thaw cycles as function of the specific surface of the air bubbles.

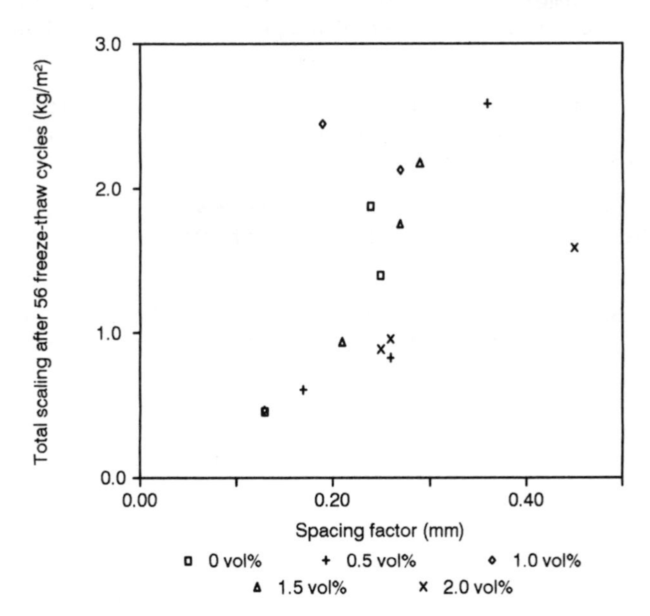

Figure 8: The total scaling after 56 freeze-thaw cycles as function of the spacing factor of the air bubbles.

Balaguru, P. and Ramakrisnan, V.: "Properties of Fibre Reinforced Concrete: Workability, Behaviour under Long-Term Loading, and Air-Void Characteristics". In: ACI Journal. May-June 1988. pp. 189-196.

Dinwoodie, J. M. and Paxton, B. H.: "A Technical assessment - cement bonded wood particle board". In: Construction and Building Materials. Vol 3, No. 1. 1989. pp. 14-21.

Fagerlund, G: "The critical degree of saturation method of assessing the freeze-thaw resistance of concrete".In: Materials and Structures. Vol. 10, No. 58, 1977. pp. 217-253.

Gram, H.-E.; Fagerlund, G. and Skarendahl, Å.: "Testing frost resistance of fibre concrete". In: Testing and Test Methods of Fibre Cement Composites (Ed. Swamy). RILEM Symposium, England, 1978. Lancaster, Construction Press, 1978. pp. 503-509.

Harper, S. "Developing asbestos-free calcium silicate building boards". In: Composites. Vol. 13. 1982. pp. 123-138.

Hoff, G.: "Durability of Fiber Reinforced Concrete in a Severe Marine Environment (SP 100-55)" In: ACI SP 100, Vol. 1 (Ed. Scanlon). Detroit, American Concrete Institute 1987. pp. 997-1041.

Proctor, B. A.: "Properties and performance of GRC". In: CI80 Fibrous Concrete, the Concrete Society. The Construction Press, Lancaster. p. 69-86.

Sarja, A.: "Structural wood fibre concrete technology". VTT Symposium 105. Helsinki, Technical Research Centre of Finland, 1989. pp. 49-59.

Swedish Standard SS 137244: "Betongprovning - Hårdnad betong - Frostresistens" (in Swedish). Byggstandardiseringen 1988.

66 CHLORIDE RESISTANCE OF SFRC

D. SINGHAL, R. AGRAWAL and B. D. NAUTIYAL
Department of Civil Engineering, Institute of Technology, Banaras Hindu University, Varanasi, India

ABSTRACT

Steel fibre reinforced concrete (SFRC) is being used as a construction material and extensive research work has been carried out in this area during the last three decades. As a result, SFRC is finding wider applications in cement concrete industry. Recently it has been used in marine structures, industrial structures, sewer pipes, highway and airfield pavements, chemical storage tanks etc. SFRC, when used in these structures, may be subjected to aggressive environment containing chloride and sulphate salts. Sulphate salts cause concrete corrosion and chloride salts are responsible for steel corrosion. Therefore, randomly distributed fibres in SFRC are more prone to corrosion by chloride salt.

This investigation has been undertaken to study the resistance of SFRC against chloride salts. For this purpose, two hundred forty cubes of M-20 grade concrete, each of 100mm size, were cast and subsequently saturated in sodium chloride solutions. The chloride concentrations used were 19,38,76 and 152 gm/litre which were varied in multiples of 19 gm/litre ie. approximately equivalent to the sea-water chloride concentration. Simultaneously, thirty reference cubes each of SFRC and conventional concrete having same size were cured in tap-water. The cubes were tested in compression at the age of 14,28,60,120,270 and 540 days. The chloride ion concentrations and pH values were obtained at three different depths of the tested cubes. The results indicate that steel fibres in SFRC are more effective when SFRC had been exposed to lower chloride ion concentration only.

Keywords: Steel fibre concrete; compressive strength; chloride diffusion; pH value

INTRODUCTION

Research carried out so far has established SFRC as a construction material of improved engineering properties (1,2). Even though several engineering properties of SFRC directly related to strength have been well exposed yet its durability aspect in terms of its resistance against aggressive exposure has not been fully

Fibre Reinforced Cement and Concrete. Edited by R. N. Swamy. © 1992 RILEM.
Published by E & FN Spon, 2-6 Boundary Row, London SE1 8HN. ISBN 0 419 18130 X.

investigated. A survey of the studies on chemical resistance of SFRC (3) shows that only limited data is available on this aspect (4-8) including chloride diffusion process of SFRC (9-11). Mangat and Gurusamy (9-11) have studied the chloride diffusion process in SFRC using different steel fibres at sea-water chloride concentration only. SFRC may face chloride exposure of varied concentration when used in offshore structure, highway pavement, chemical storage tank etc. The chloride concentration in the hardened cement concrete depends on the concentration of chloride salts present in the surrounding solution (12). Since the corrosion process is accelerated by the presence of chloride salts in concrete (13), it is quite relevant to study the chloride diffusion process in SFRC. Therefore, this investigation has been undertaken to study as to how the chloride salts penetrate into SFRC. The chloride salt concentration to which SFRC is exposed right from the fresh state has been varied upto eight times the sea-water chloride salt concentration.

To achieve this, three hundred cubes of 100mm size each, were exposed to increased chloride concentration and observations related to compressive strength, chloride ion concentration at different depths of cubes alongwith pH variations were made.

MATERIALS AND MIX PROPORTION

The concrete constituents used were ordinary Portland cement, fine aggregate, coarse aggregate and mild steel rectangular fibres. Ordinary Portland cement confirmed the requirements of IS:269-1976 (14). Fine and coarse aggregates confirmed the gradings of IS:383-1970 (15). Coarse aggregate had maximum size of 10mm. The fineness modulus of coarse and fine aggregates were 6.24 and 2.57 respectively. The rectangular steel fibres used were of the size 0.37mm x 0.50mmx 30mm with an aspect ratio of 61 which were easily available. Fibre content was 1.0% by volume or 3.10% by weight of concrete ie. 78.31 kg-mass/m^3. The concrete mix used was of M-20 grade having constituents proportion of 1:1.70:1.77 (by weight) and cement content of 462 kg-mass/m^3. The mix was designed as per SP:23-1982 (16). The percentage of fine aggregate was kept about 50% of the total aggregate to achieve better distribution of fibres.

EXPERIMENTAL DETAILS

In order to study the chloride resistance of SFRC subjected to increased chloride concentration right from the fresh state, four solutions of sodium chloride having chloride concentrations of 19,38,76 and 152 gm/litre were used. These solutions designated as NaCl(C), NaCl(2C), NaCl(4C) and NaCl(8C) respectively were prepared taking 19 gm/litre as the approximate chloride concentration of sea-water (17). The pH values of the solutions used ie. NaCl(C), NaCl(2C), NaCl(4C) and NaCl(8C) were 11.52, 11.16, 10.90 and 10.75 respectively.

Cubes of 100mm size each were cast in well-oiled steel moulds. Various ingredients of concrete were hand mixed in dry state. Fibres were sprinkled through a sieve to achieve uniform distribution of fibres and lastly water was added. Cubes were filled in three layers of concrete and compaction was done using table vibrator. Cubes were covered with wet gunny bags for 24-hours after casting. The cubes were then taken-out from moulds and kept in different solutions having above stated chloride concentration for saturation at room temperatre. The other thirty cubes each of SFRC and conventional concrete were saturated in tap-water at room temperature for reference.

The average of five samples was considered while evaluating the compressive strength at different ages of saturation. The cubes were saturated in different solutions for 14,28,60,120,270 and 540 days. Cubes removed from the solutions were kept in open air for about two hours before testing so as to attain surface dry condition.

From the cubes tested at 540 days, samples of concrete were taken out from the depth of 0, 25 and 50mm from the cube surface. The samples were ground and 10gm of the material passing through 150 micron sieve was collected. The material was mixed with 10ml and 20ml distilled water for determining pH value and chloride salt concentration respectively. Further, it was shaken for two hours on a mechanical rotor. The solution was further allowed to settle down for twentyfour hours and then filtered. The chloride content and pH value of the solutions were then measured with the help of UC-204 digital water checker (make - Central Kagaku Co. Ltd., Japan) and pH meter (make - Control Dynamics, Bangalore, India) respectively. The results have been plotted in Figs 1-6.

RESULTS AND DISCUSSION

Fig.1 shows the plot between the compressive strength and the age of saturation for conventional concrete cured in different chloride solutions and in tap-water. It is apparent from the figure that the compressive strength of the cubes saturated in water is maximum at 540 days. It may also be seen from Fig.1 that the presence of chloride has accelerated the compressive strength of the cubes saturated in NaCl(C) and NaCl(2C) upto about 90 days. The rate of increase in strength reduces after 270 days. The reason for this loss in strength may be due to the formation of Friedel's salt ie. $C_3A\ CaCl_2.10H_2O$* in chloride exposure (18). The formation of Friedel's salt takes place when chloride salts reacts with calcium hydroxide present in the concrete. These cubes as such have less strength when their strength is compared with the strength of cubes saturated in tap-water. The cubes saturated in tap-water have no Friedel's salt formation and as such the strength attained by these cubes is maximum.

* CaO = C Al_2O_3 = A

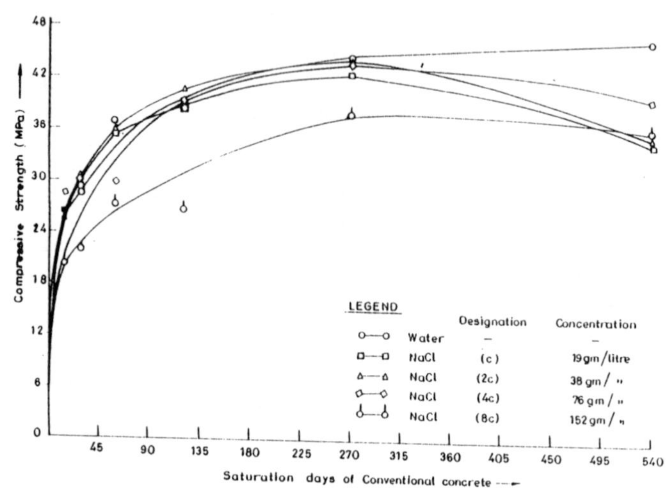

Fig. 1. Compressive strength vs saturation days of conventional concrete

The cubes cured in higher chloride concentrations of 76 and 152 gm/litre have followed the trend of retarded compressive strength as compared to the reference cubes right from the initial stage. The cubes have attained maximum strength at 270 days. The rate of loss of strength in case of these cubes beyond 270 days is not as sharp as in case of these cubes in NaCl(C) and NaCl(2C). This results in higher strength of these cubes at 540 days. The cubes saturated in NaCl(8C) exhibited the lowest strength throughout except for last few days of saturation. Therefore, it may be concluded that the cubes of conventional concrete when exposed to chloride environment right from the fresh state have finally lower compressive strength. The loss in compressive strength is more rapid in case of cubes having accelerated strength during initial saturation period.

Fig.2 shows the plot of compressive strength of SFRC against saturation period. From this figure, it is apparent that the SFRC cubes saturated in different solutions have greater 540 days strength than the corresponding cubes of conventional concrete. This is because of the reason that the SFRC cubes have slow loss of strength after 270 days as compared to conventional concrete. The increase in the strength of SFRC cubes saturated in water, NaCl(C), NaCl(2C), NaCl(4C) and NaCl(8C) after 540 days saturation period is found to be 3.70%, 18.10%, 34.75%, 13.53% and 2.50% respectively as compared to the strength corresponding conventional concrete cubes. The increased strength of SFRC indicates the effectiveness of fibres in all the chloride concentrations. However, an increase in strength of 34.75% is significant in the case chloride concentration of 38 gm/litre. Thus the effectiveness of fibres increases upto a concentration of 38 gm/litre and it reduces thereafter.

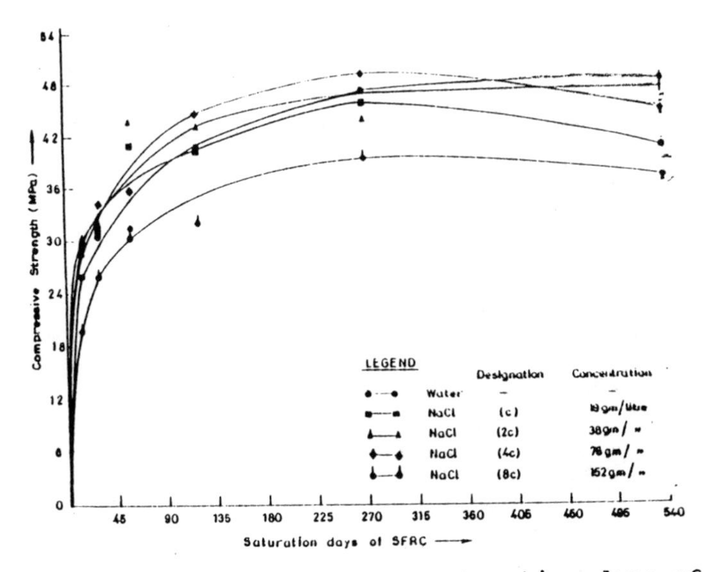

Fig. 2. Compressive strength vs saturation days of SFRC

Fig.3 shows the plot of chloride concentration in conventional concrete against depth of penetration. It is evident from this figure that the chloride concentration has increased all along the depth with the increase in the concentration of surrounding solution. However, the rate of decrease in concentration along the depth of cube is faster for higher surrounding concentrations and also in the zone near to the surface. The reduction in the chloride concentration observed at the centre of cubes corresponding to the depth of 50mm has been observed to be 14.70%,37.35%,42.71%,70.00% and 63.84% for water as well as for solutions NaCl(C), NaCl(2C),NaCl(4C) and NaCl(8C) respectively.

Fig.4 shows the plot of chloride concentration in SFRC at various depths of the cube. Here also, the chloride concentration in the cube has increased at all the depths with the increase in chloride concentration of surrounding solutions. However, the chloride concentration has decreased throughout the depth in comparison to the chloride concentration of the conventional concrete cubes except at 25mm depth of cubes saturated in NaCl(8C). Therefore, it can be inferred that the fibres are effective in arresting the entry of chloride salts even in the vicinity of surface of SFRC cubes. The effectiveness of fibres at the highest concentration is doubtful as observed from the chloride concentration curve of NaCl(8C). The reason of such high concentration of chloride salts at 25mm may be that concrete is severely damaged due to very high chloride concentration as the strength of the cubes cured in NaCl(8C) has remained lowest all through, as discussed earlier. However, the steeper slopes of the curves for NaCl(C) and NaCl(2C) for SFRC cubes show more effectiveness of fibres in arresting the

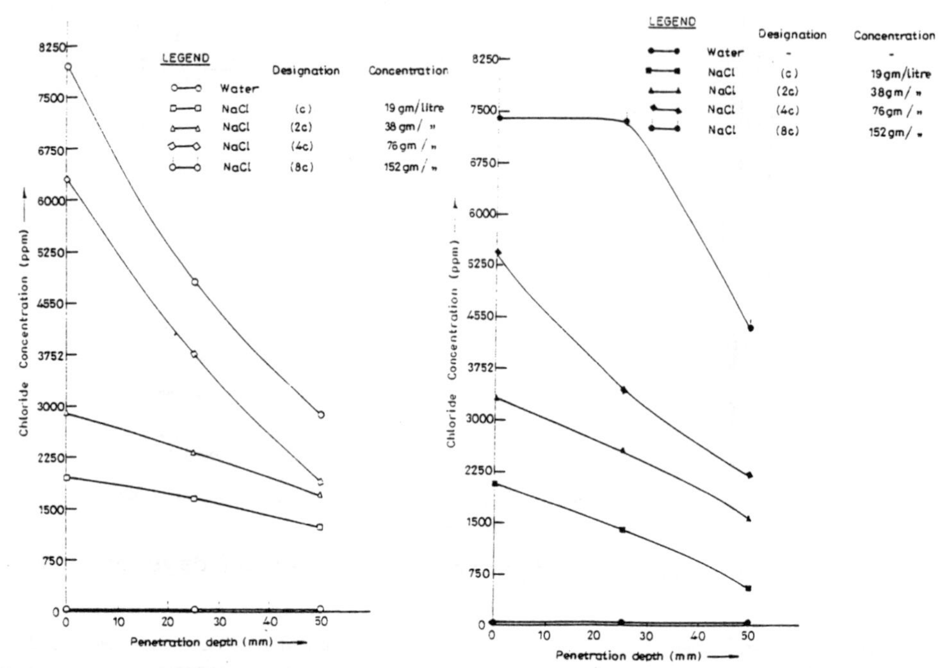

Fig. 3. Chloride concentration vs penetration depth of conventional concrete

Fig. 4. Chloride concentration vs penetration depth of SFRC

chloride concentration than that for the corresponding reference cubes. The reductions in the chloride concentration in SFRC cubes at its centre ie. corresponding to the depth of 50mm have been observed to be 14.7%, 72.0%, 52.7%, 59.2% and 50.0% for water as well as for the surrounding solutions of NaCl(C), NaCl(2C), NaCl(4C) and NaCl(8C) respectively. Therefore, it is evident that the percentage reduction in chloride concentration has decreased when the cubes are saturated in solutions of higher chloride concentration say 76 gm/litre and above.

Figs 5 and 6 show the plots of pH value at various depths of the cubes. It is apparent from the plots that pH values of SFRC cubes at various depths are higher than corresponding values for the conventional concrete upto the chloride concentration of 38 gm/litre ie. solution NaCl(2C). The pH values at various depths of the cubes saturated in NaCl(4C) are almost same for both SFRC and conventional concrete. However, the pH values are lower at all depths of SFRC cubes saturated in NaCl(8C) when compared to corresponding conventional concrete cubes. Therefore, it is concluded that the SFRC has more alkalinity than the conventional concrete upto the chloride concentration of 38 gm/litre and as such the fibres are more effective upto this concentration only.

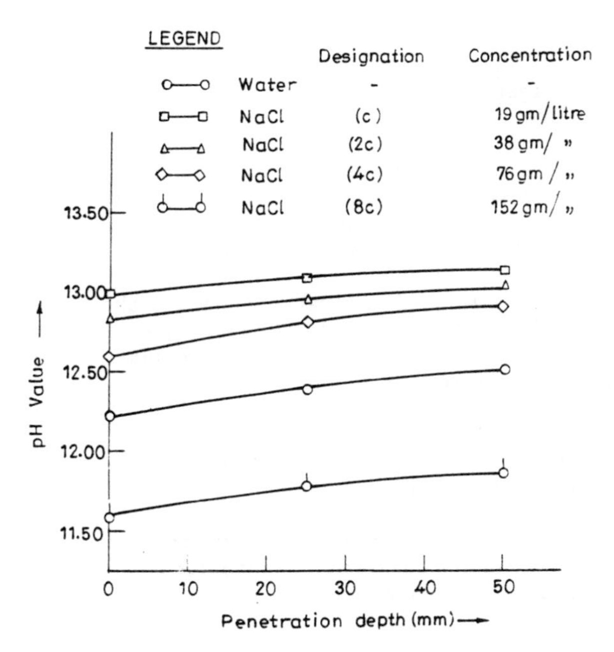

Fig.5. pH VALUE vs. PENETRATION DEPTH OF CONVENTIONAL CONCRETE.

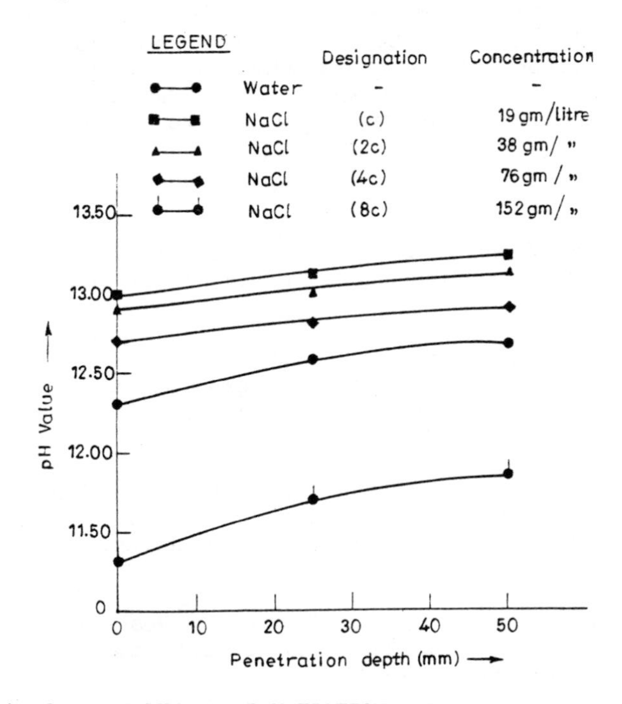

Fig.6. pH VALUE vs. PENETRATION DEPTH OF SFRC.

CONCLUSIONS

Based on the investigation carried out as above following conclusions are being drawn:

1. Conventional concrete when exposed to chloride environment in fresh state has lower compressive strength than referal concrete. This loss in strength is more rapid if this concrete has accelerated strength during initial saturation period.

2. Addition of fibres is effective at all the chloride concentrations and its presence retard the loss of strength. However, the fibres are significantly effective upto a chloride concentration of 38 gm/litre.

3. Fibres are more effective in arresting the chloride penetration upto the concentration of 38 gm/litre. SFRC has also more alkalinity upto this concentration.

4. It is observed that the conventional concrete and SFRC have maximum strength at 270 days of saturation. The strength however falls with increased days of saturation.

ACKNOWLEDGEMENT

The first author of the paper sincerely conveys his thanks to C.S.I.R. and U.G.C., New Delhi for the financial assistance provided. The authors of the paper sincerely convey their thanks to Departments of Civil Engineering, and Chemical Engineering, Institute of Technology, Banaras Hindu University to conduct this investigation.

REFERENCES

1. Agrawal R. and Singhal D., "Non-Linear Behaviour of Steel Fibre Reinforced Concrete," Proceedings of 14 OWICS, Singapore, 1989, pp 7-14.

2. Karthikeyan O.H., Kumar V., Singhal D. and Nautiyal B.D., "Fibres for FRC - Their Properties, Applications and Mixing: A Review Report," Indian Concrete Institute Bulletin, 34, 1991, pp 37-49.

3. Singhal D., Agrawal R. and Nautiyal B.D., "Chloride and Sulphate Resistance of SFRC - Research Needs," Proceedings of Fibre Reinforced Concrete and its Applications, Madras, 1991, pp 11-1 to 11-10.

4. Mangat P.S. and Gurusamy K., "Corrosion Resistance of Steel Fibres in Concrete under Marine Exposure," Cement and Concrete Research, V.18, 1988, pp 44-54.

5. Hoff G.C., "Durability of Fibre Reinforced Concrete in Severe Environment," Proceedings of International Symposium on Concrete Durability, ACI SP-100, V.1, American Concrete Institute, Detroit, 1987, pp 997-1041.

6. Mangat P.S. and Gurusamy K., "Steel Fibre Reinforced Concrete for Marine Applications," Proceedings of 4th International Conference on the Behaviour of Offshore Structures, Delft, July 1985, pp 867-879.

7. Hannant D.J., "Additional Data on Fibre Corrosion in Cracked Beams and Theoretical Treatment of the Effect of Fibre Corrosion on Beam Load Capacity," Proceedings of RILEM Symposium, 1976, pp 533-538.

8. Mangat P.S. and Gurusamy K., "Long-Term Properties of Steel Fibre Reinforced Marine Concrete," Materials and Structures, V.20, 1987, pp 273-282.

9. Mangat P.S. and Gurusamy K., "Chloride Diffusion in Steel Fibre Marine Concrete," Cement and Concrete Research, V.17, 1987, pp 385-396.

10. Mangat P.S. and Gurusamy K., "Chloride Diffusion in Steel Fibre Reinforced Concrete Containing PFA," Cement and Concrete Research, V.17, 1987, pp 640-650.

11. Mangat P.S. and Gurusamy K., "Pore Fluid Composition under Marine Exposure of Steel Fibre Reinforced Concrete," Cement and Concrete Research, V.17, 1987, pp 734-742.

12. Midgley H.G. and Illuston J.M., "The Penetration of Chlorides into Hardened Cement Pastes," Cement and Concrete Research, V.14, 1984, pp 546-558.

13. Nautiyal B.D. and Singhal D., "Durability of Cement Concrete against Sulphate and Chloride Exposure - A Review," Proceedings of "Durability of Concrete and Cement Products," All India Seminar, Nagpur, 1990, pp III-20 to III-24.

14. IS:269-1976, "Specifications for Ordinary and Low Heat Portland Cement," Bureau of Indian Standards, New Delhi.

15. IS:383-1970, "Specifications for Coarse and Fine Aggregates from Natural Sources for Concrete," Bureau of Indian Standards,

16. SP:23-1982, "Handbook of Concrete Mixes," Bureau of Indian Standards, New Delhi.

17. Liptak B.G., "Environmental Engineer's Handbook," V.1, 1974, "Water Pollution".

18. Jain N.K., "Some Experimental Studies on the Behaviour of Portland Cement and Concrete in Sea-Water," Ph.D. Thesis, Department of Civil Engineering, I.I.T., New Delhi, 1984.

67 RESISTANCE OF OVER 1-YEAR-OLD SFRC EXPOSED TO LONG-TERM EROSION-ABRASION LOADING

J. ŠUŠTERŠIČ
Institute for Testing in Materials and Structures, Ljubljana, Slovenia

Abstract
The results of investigations into the erosion-abrasion resistance according to CRD-C 63-80 test method of over 1-year-old SFRC specimens are discussed in the paper. Nine mix proportions were used. The w/c ratios were varied from 0.30 to 0.65. The volumetric percentage of hooked steel fibres were varied from 0.25 to 2.0 vol.% at the w/c of 0.30 and at the others the quantity of fibres was constant. In addition, mixes without fibres were made at each w/c.

The loss of mass was measured every 12 hrs up to 72 hrs. After a period of time the same specimen was exposed to the erosion-abrasion loading again up to the next 72 hrs. This cycle of 72 hrs was repeated four times.

Performance of long-term loaded concrete is described in more detail with a coefficient of erosion-abrasion loss according to time Lt_c.

The experimental results obtained show that the long-term erosion-abrasion resistance of SFRC is improved more by an increase in the volumetric percentage of low hardness steel fibres and by an increase in quality bonds between fibres and cement paste.
Keywords: Erosion-Abrasion Resistance, Steel Fibre Reinforced Concrete, Water-Cement Ratio.

1 Introduction

During the last five years, the research project on erosion-abrasion resistance of concrete and steel fibre reinforced concrete (SFRC) has been undertaken in our Institute, Šušteršič et al. (1991a) and (1991b), and it is being continued. The results of this research and investigations were used to prepare mix proportions for erosion-abrasion resistant concrete for spillways of the power-plants on the river Sava in Slovenia.

Erosion-abrasion action of waterborne particles in the stilling basins is duplicated quite well by the underwater test method CRD-C 63-80. This method was developed in the Structures Laboratory, U.S. Army Engineer Waterways Experi-

Fibre Reinforced Cement and Concrete. Edited by R. N. Swamy. © 1992 RILEM.
Published by E & FN Spon, 2-6 Boundary Row, London SE1 8HN. ISBN 0 419 18130 X.

ment Station and it is described in detail in References Liu (1980), (1981) and Liu and McDonald (1981).

Circulating water moves the steel grinding balls on the surface of a concrete specimen, producing the desired abrasion effects. The water velocity and the agitation effect are not sufficient to lift the steel balls off the surface of the concrete specimen to cause any significant impact action against the surface, Liu (1980), (1981), Liu and McDonald (1981).

Relative abrasion resistance of different concrete types (conventional concrete with different w/c ratio, Liu (1980), (1981), polymer concrete, Liu (1980), (1981), concrete placed underwater, Hester et al. (1989), fibre reinforced concrete (FRC), Liu (1980), Liu and McDonald (1981), Šušteršič et al. (1991a)), silica fume - mortars without and with fibres, Berra et al. (1989), concrete with different aggregate types, Liu (1980), (1981) and concrete with different types of surface treatment, Liu (1980), (1981) was evaluated by this underwater erosion-abrasion test method.

Two exactly opposite conclusions of the investigations into the erosion-abrasion resistance of SFRC are presented in:

(a) References Liu (1980) and Liu and McDonald (1981): a comparison of the performance of SFRC and concrete not containing fibres clearly showed that SFRC was less resistant to erosion-abrasion than concrete of the same aggregate type and water-cement ratio without fibres;
(b) Reference Berra et al. (1989): best results for erosion-abrasion resistance were obtained in a mortar containing Blast Furnace cement, silica fume and steel fibres; Reference Šušteršič et al. (1991a): the experimental results obtained show that steel fibres are adequate for applications requiring an erosion-abrasion resistance concrete because the erosion-abrasion resistance according to CRD-C 63-80 is improved by an increase of volumetric percentage of steel fibres at a constant w/c.

The factors which influenced these opposite conclu-sions are the mix proportions, the bond between fibres and cement paste and steel hardness of fibres.

Two factors which might have influenced the poor performance of the SFRC subjected to erosion-abrasion are discussed in Reference Liu and McDonald (1981). The second factor is: "When FRC is subjected to erosion-abrasion, the film of surface mortar resists the erosion-abrasion forces initially; but as the surface mortar is worn away, the fibres are exposed. The water flow and the movement of the abrasive charges in the test environment cause the exposed fibres to vibrate. As the fibre vibrates, it introduces large stresses in the concrete caused by stress concentration. These large stresses contribute to further deterioration of the concrete around the fibres. The behaviour was indicated by the deteriorated concrete around the circumference of the fibres on the surface of the test specimens."

This discussion shows, that the steel fibres used should have a high hardness (over 550 HB approximately), if we take into account too our investigations where the same effect of fibres was observed.

On the other hand, the steel fibres, which were used in investigations presented in Reference Šušteršič et al. (1991a) and in this paper, have a low hardness (250 - 270 HB approximately). So these fibres became flat by striking of balls, which were moved by the circulating water and they did not pull out the fibres from the concrete, because a quality bond between the fibres and the cement paste was achieved.

The erosion-abrasion resistance of the investigations presented in all the discussed References was measured on 28-day-old specimens, which were exposed to the erosion-abrasion loading up to 72 hrs (1 cycle).

The performance of over 1-year-old SFRC specimens exposed to the long-term loading (up to 288 hrs - 4 cycles of 72 hrs) was presented in this paper. The performace of two types of concrete surface was estimated:

(a) intact surface up to the date of testing and
(b) deformed surface of concrete specimens, which were loaded first after 28 days (up to 72 hrs) and then (over 1 year) they were exposed to the long-term loading.

2 Details of experimental program

The mix proportions used are presented in Table 1.

Table 1. The mix proportions used to make the SFRC specim

Mix designation	w/c ratio	Cement (kg/m)	Fibres (vol.%)
A1	0.30	550	0.00
A2	0.30	550	0.25
A3	0.30	560	0.50
A4	0.30	580	1.00
A5	0.30	600	2.00
A6	0.42	400	0.00
A7	0.42	400	0.50
A8	0.65	250	0.00
A9	0.65	260	0.50

These mix proportions are the same as they were used in References Šušteršič et al. (1991a) and (1991b) where the properties of fresh and hardened SFRC are presented too.

Hooked steel fibres were used. The length of the fibres was l=32 mm, and their diameter was d=0.50 mm. Thus, the aspect ratio of the fibres was l/d=64.

Portland cement with 15 % blast-furnace slag, gravel aggregate (mainly limestone) in gradings of 0/4 mm, 4/8 mm and 8/16 mm and superplasticizer were used for all the mixtures. The content of cement was varied so that the workability of fresh concrete was constant approximately.

The concretes were prepared in a 50 l laboratory mixer with a vertical shaft. Compaction was carried out by means of a vibrating table. The specimens were cured at 95 % relative humidity and at a temperature of 20° C. A specimen was placed in a tank of water before the test to saturate with water completely.

Two series of SFRC specimens (A1 - A9) were tested according to underwater test method CRD-C 63-80:

1st series: specimens designated A1a to A9a with intact surface until the beginning of the test were exposed to the erosion-abrasion loading up to 72 hrs at the specimen age of 390 days (1 year and 2 months approximately); then they were exposed again to the same loading cycle of 72 hrs every following two months (up to the joint number of cycles - 4);

2nd series: specimens designated A1b to A9b were tested at first at the age of 28 days; but the next test began not later then at the specimen age of 600 days (1 year and 8 months approximately); then they were tested again every following 12 days (up to the joint number of cycles - 4).

3 Evaluation of erosion-abrasion resistance

The underwater test method CRD-C 63-80 can only be used to determine the relative resistance of a material to the erosion-abrasion action of waterborne particles, Liu (1981).

Mass of the surface-dry specimen is weighed and erosion-abrasion loss is calculated every 12 hr and at the end of the test period by the following equation:

$$L = \frac{M_i - M_f}{M_i} \times 100 \qquad (1)$$

where:
L ... erosion-abrasion loss, (mass%)
M_i ... mass of the surface-dry specimen before tests
M_f ... mass of the surface-dry specimen after test.

When all results of erosion-abrasion loss of a concrete type are compared with the time at which these results were obtained, a good linear correlation between erosion-abrasion loss and time is found (Fig. 1). It can be expressed with the equation:

$$L = C + Lt_c \times t \tag{2}$$

where:
L ... erosion-abrasion loss
C ... constant
t ... time
Lt_c ... angle between loss/time line and parallel line to the absciss or coefficient of erosion-abrasion loss which gives a rate of wear and erosion-abrasion loss respectively of concrete surface, and it is expressed with equation:

$$Lt_c = \frac{L}{t} \tag{3}$$

A decrease in an incline into an erosion-abrasion/time line according to the absciss represents an improvement in erosion-abrasion resistance of tested concrete (Fig. 1).

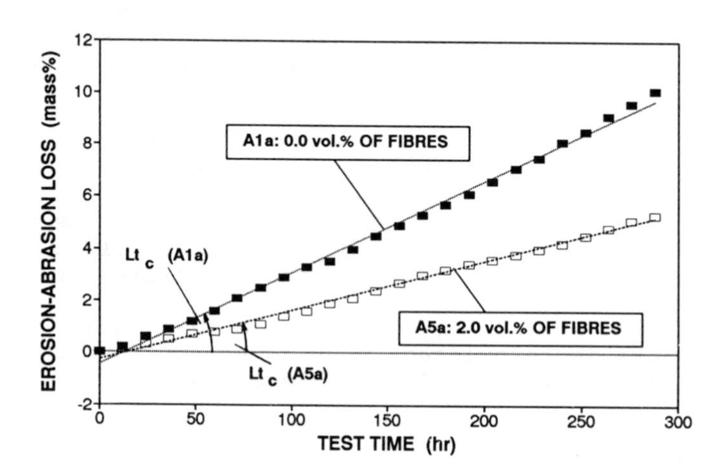

Fig. 1. Linear correlation between erosion-abrasion loss of concrete designated A1a and A5a and time.

4 Results and discussion

The obtained results (erosion-abrasion loss) of underwater test CRD-C 63-80 of SFRC specimens of the 1st series and 2nd series at beginnings and end of every cycles of 72 hrs are shown in tables 2 and 3 respectively.

Table 2. Results of underwater test CRD-C 63-80 on concrete specimens of 1st series

TEST TIME	AGE OF CONCR.	LOSS OF MASS (mass%) OF CON. SPEC. DESIGNATED								
(hrs)	(days)	A1a	A2a	A3a	A4a	A5a	A6a	A7a	A8a	A9a
0	390	0.0	0.0	0.0	0.0	0.0	0.0	0.0	0.0	0.0
72		2.1	2.1	1.3	1.6	0.9	2.7	2.6	5.2	4.3
84	450	2.5	2.4	1.6	2.1	1.1	3.2	2.8	6.3	5.2
144		4.5	4.3	3.6	3.7	2.4	5.6	5.2	11.8	9.8
156	510	4.9	4.8	4.1	4.2	2.7	5.8	5.8	12.2	11.1
216		7.1	6.8	6.2	6.1	3.8	8.2	8.4	17.4	17.1
228	570	7.5	7.1	6.7	6.4	4.0	8.6	8.8	18.7	17.6
288		10.1	8.7	8.8	8.3	5.3	10.9	10.8	23.0	20.5

Table 3. Results of underwater test CRD-C 63-80 on concrete specimens of 2nd series

TEST TIME	AGE OF CONCR.	LOSS OF MASS (mass%) OF CON. SPEC. DESIGNATED								
(hrs)	(days)	A1b	A2b	A3b	A4b	A5b	A6b	A7b	A8b	A9b
0	28	0.0	0.0	0.0	0.0	0.0	0.0	0.0	0.0	0.0
72		3.4	2.9	3.1	2.4	1.5	3.8	3.0	9.4	7.2
84	600		3.5			1.7	4.3	3.2		
144			5.5			2.8	5.6	5.0		
156	612		6.0			3.0	6.2	5.4		
216			8.5			4.3	8.6	7.2		
228	624		8.9			4.5	9.0	7.6		
288			11.2			5.6	11.2	8.6		

The calculated values of Lt_c (equation (3)) of the same specimens are shown in tables 4 and 5.

4.1 Effects of concrete age

When the concrete with and without fibres from the 2nd series is exposed after 28 days to the erosion-abrasion loading up to the 72 hrs the loss is greater than the loss of the same concrete from the 1st series after 390 days (Fig. 2).

Table 4. Lt_c of concrete specimens of 1st series

UP TO THE TIME (hrs)	Lt_c (mass%/hr) x 10^{-2} OF CON. SPEC. DESIGNATED								
	A1a	A2a	A3a	A4a	A5a	A6a	A7a	A8a	A9a
72	2.9	3.0	1.8	2.3	1.3	3.6	3.6	7.0	6.1
288	3.5	3.1	3.2	2.9	1.9	3.8	3.9	8.4	7.8

Table 5. Lt_c of concrete specimens of 2nd series

UP TO THE TIME (hrs)	Lt_c (mass%/hr) x 10^{-2} OF CON. SPEC. DESIGNATED								
	A1b	A2b	A3b	A4b	A5b	A6b	A7b	A8b	A9b
72	4.7	4.2	4.2	3.5	2.1	5.3	4.3	13.0	10.0
288	-	3.8	-	-	2.0	3.7	3.0	-	-

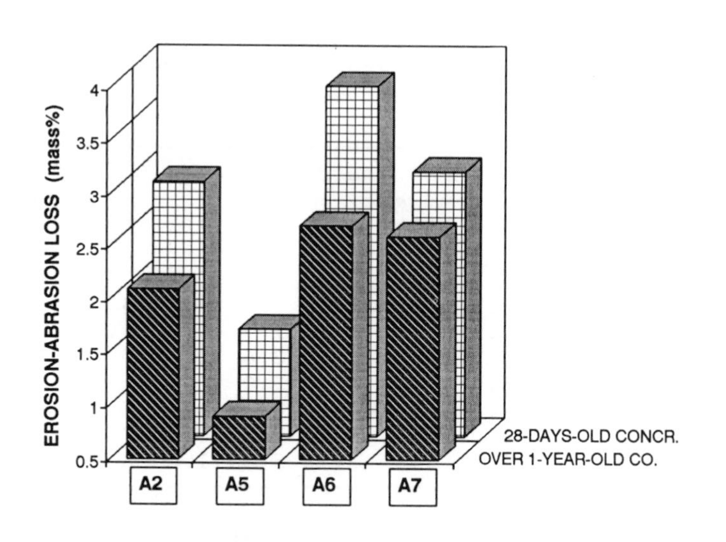

Fig. 2. Erosion-abrasion loss of concrete from 1st series: A2a, A5a, A6a and A7a and from 2nd series: A2b, A5b, A6b and A7b up to 72 hrs.

After these tests, the relative loss of concrete from the 2nd series is smaller than the loss of concrete from 1st series (Fig. 3).

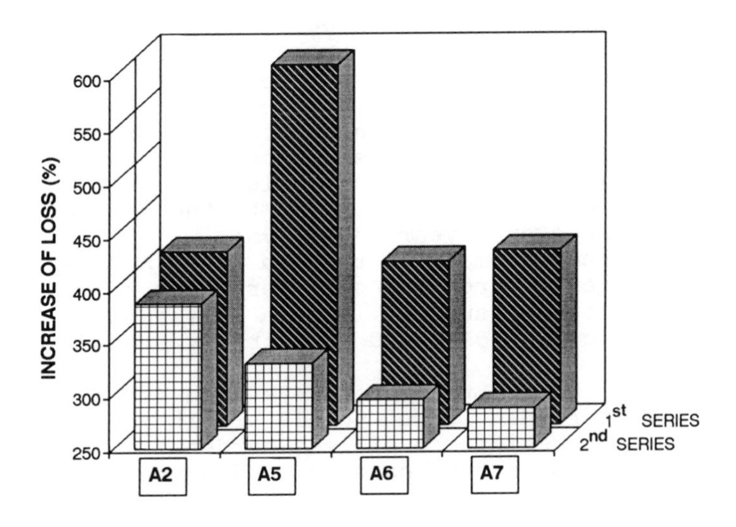

Fig. 3. Relative erosion-abrasion loss of concrete from 1st and 2nd series according to loss up to 72 hrs.

Performance of concrete with and without fibres from 1st and 2nd series according to the age of concrete can be clearly seen as in Fig. 4, where the diagrams of erosion-abrasion loss of concrete specimen A5 are shown for example, as in tables 4 and 5.

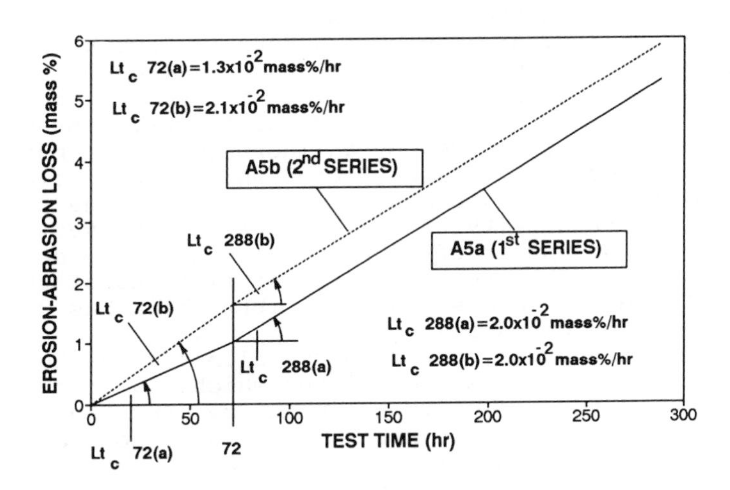

Fig. 4. Lines loss/time of concrete A5a and A5b.

The erosion-abrasion loss of concrete with intact surface after 28 days is greater than the loss of concrete with intact surface after 390 days. This conclusion is expected and it is true both for the concrete without fibres and for the SFRC, but at different levels of resistance. If the concrete which was exposed to the erosion-abrasion loading after 28 days is exposed again to this loading after over 1 year (600 days), its resistance is increased and it is approximately the same as the resistance of the concrete which was exposed for the first time to the erosion-abrasion loading after over 1 year (390 days).

Lt_c of concrete from 1st series up to 72 hrs is smaller than Lt_c of concrete from the same series up to 288 hrs. It means that the intact surface is more resistant to erosion-abrasion loading than deformed and already eroded surfaces respectively.

On the other hand, Lt_c of concrete from 2nd series up to 72 hrs is greater than Lt_c up to 288 hr. It means that the erosion-abrasion resistance of less hardened concrete is smaller than the same more hardened and already eroded concrete.

4.2 Effects of fibre content

The improvement in the erosion-abrasion resistance of SFRC with used steel fibres was already reported in Reference Šušteršič et al. (1991a) and it is mentioned in the introduction to this paper. The same effects are obtained from the investigations of these SFRC after over 1 year and at the repeated loading cycles of 72 hrs. The resistance of SFRC is improved by an increase in fibre content. It can be seen from test results of concrete specimens designated A1a to A5a and A1b to A5b respectively which are given in tables 2 to 5.

The decrease in erosion-abrasion loss according to the fibre content is indicated by diagrams in Fig. 5. Evident erosion-abrasion resistance of SFRC according to fibre content is shown in Fig. 6. Resistance of concrete is improved by the presence of low hardness steel fibres. Improvement of resistance is increased by an increase in the number of fibres on an eroded surface. Fibres which become flat by striking of balls create a steel coating over the concrete surface. Higher density of the coating is achieved by an increase in the number of fibres and by an increase in fibre flatness. The greatest measured flatness of the fibre used is 2.5 mm. It means that the fibre which has diameter of 0.5 mm is extended five times.

SFRC specimens were cured in the climatic chamber between the test cycles. The fibres which lay on the eroded surface began to corrode in the meantime. But the fibres were worn by the previous striking of balls so much that corosion had no noticeable influence on the resistance of fibres against the repetition of striking and on the erosion-abrasion resistance of SFRC respectively.

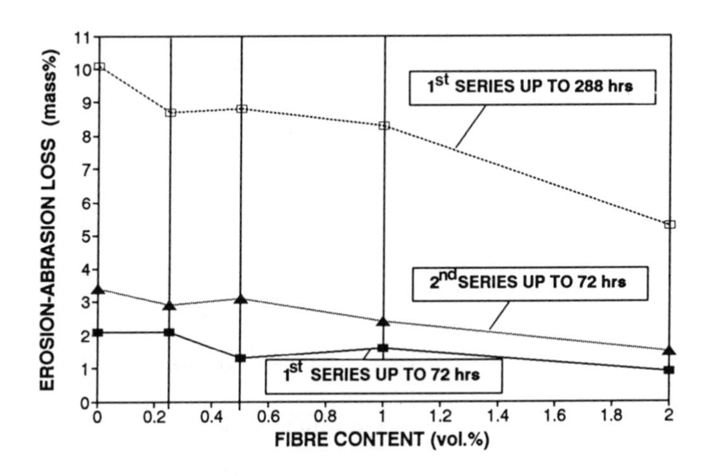

Fig. 5. Erosion-abrasion loss according to fibre content at the w/c = 0.30.

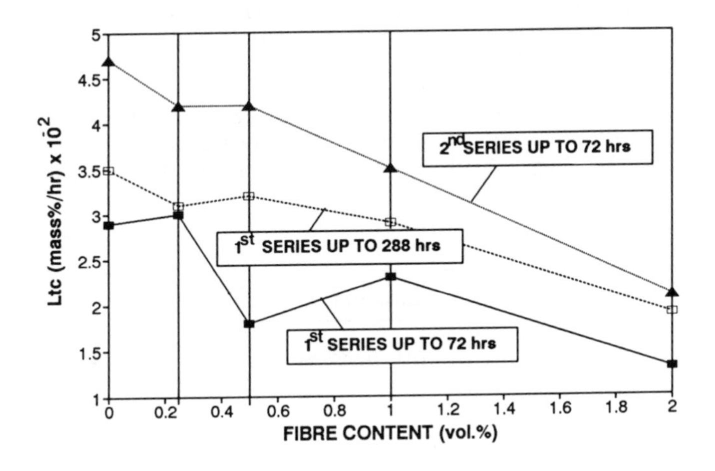

Fig. 6. Lt_c according to fibre content at the w/c = 0.30.

4.3 Effects of water-cement ratio

The quality of hardened cement paste and the bond between particles of aggregate and fibres with cement paste is improved by a decrease in w/c. Therefore, the erosion-abrasion resistance of concrete without fibres and SFRC is also improved by a decrease in w/c.

An increase of erosion-abrasion loss of concrete without fibres (designated A1a, A6a and A8a) and SFRC with 0.5 vol.% of fibres (designated A3a, A7a and A9a) according to increase of w/c is shown in Fig. 7.

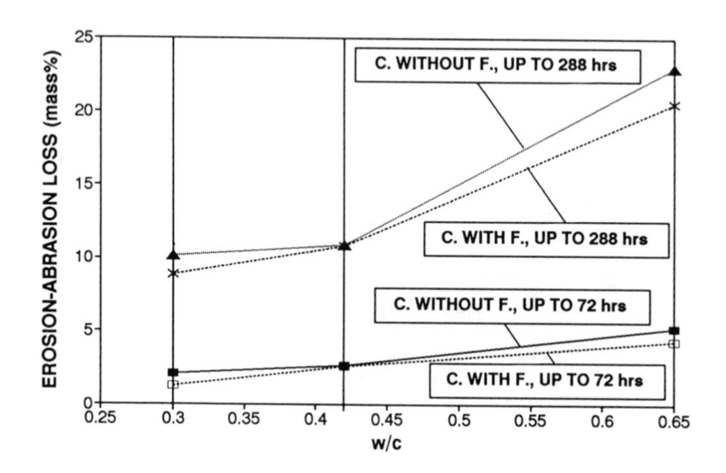

Fig. 7. Erosion-abrasion loss of concrete with and without fibres according to w/c.

A decrease of erosion-abrasion resistance of concrete with and without fibres according to increase of w/c is indicated too in Fig. 8 where the diagrams Lt_c - w/c are shown.

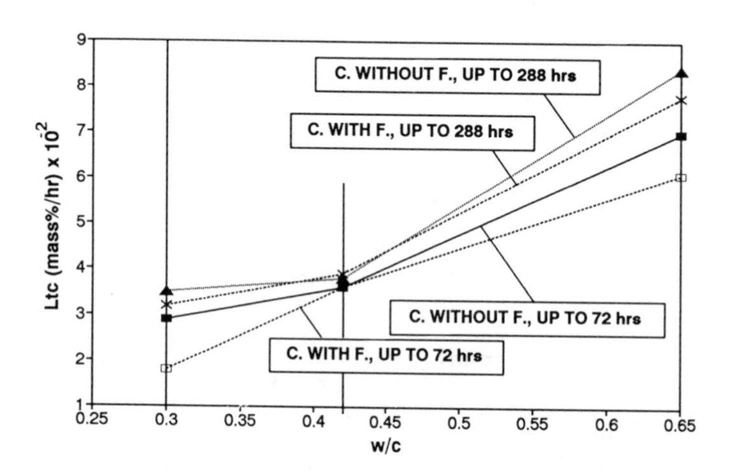

Fig. 8. Lt_c of concrete according to w/c.

5 Conclusions

Performance of concrete which is exposed to long-term loading is described more clearly by the coefficient of erosion-abrasion loss according to time Lt_c. Log-term erosion-abrasion resistance of concrete with or without fibres, which was exposed for first time to erosion-abrasion loading after over 1 year, is the same approximately as concrete which was exposed for the first time to erosion-abrasion loading at the age of 28 days and then exposed to the long-term loading after over 1 year.

An intact surface of over 1-year-old concrete is moro resistant to erosion-abrasion loading than deformed and already eroded surfaces respectively. But the erosion-abrasion resistance of an intact surface of less hardened concrete is smaller then the same more hardened and already eroded surface.

Long-term erosion-abrasion resistance of SFRC is improved by an increase of fibre content and by a decrease in w/c.

6 References

Liu, T.C. (1980) **Technical Report No. C-78-4**. U.S. Army Engineer Waterways Experiment Station, Vicksburg.

Liu, T.C. (1981) Abrasion Resistance of Concrete. **ACI Jour.** Sept.-Oct., 341 - 350.

Liu, T.C. and McDonald, J.E. (1981) Abrasion-Erosion Resistance of Fiber-Reinforced Concrete, **Cem. Con. and Aggr.**, 2 (3), 93 -100.

Hester, W.T.,Khayat, K.H. and Gerwick,Jr., B.C. (1989) Properties of Concretes for Thin Underwater Placements and Repairs, in **ACI SP 114** (ed V.M. Malhotra), V.1, pp. 713 - 731.

Šušteršič J., Mali E. and Urbančič S. (1991a) Erosion-Abrasion Resistance of Steel Fibre Reinforced Concrete, in **ACI SP 126** (ed V.M. Malhotra), V.2, pp. 729 - 743.

Berra M., Ferrara G. and Tavano S. (1989) Behaviour of High Erosion-Resistant Silica Fume-Mortars for Repair of Hydraulic Structures, in **ACI SP 114** (ed V.M. Malhotra), V.2, pp. 827 - 847.

Šušteršič J.,Rebić M. and Urbančič S. (1991b) Testing of SFRC by the Schmidt Rebound Hammer, in **MRS Symp. Proc. V.211** (eds S. Mindess and J. Skalny), pp. 33 - 38.

68 ASSESSMENT OF STEEL FIBRE CONCRETE EXPOSED FOR 14 YEARS

T. HARA, M. SHOYA and K. KIKUCHI
Nihon University, Koriyama, Fukushima-ken, Japan

Abstract
This paper presents the results of tests to determine the properties of the surface layer of concrete applied to steel fibre reinforced concrete after 14 years of exposure. Eight different mixtures were used in this study. They were designed to resolve the primary factors influencing tensile strength of steel fibre concrete which were found to be aspect ratios and fiber content.

It is found that the mix design variables with respect to the pull-off strength near the surface layer of specimens have been confirmed by the volume fractions of steel fibres as a threshold level. As an alternative, depth of carbonation and corrosion of steel fibers were smaller than that expected.
Keywords: Steel Fibre Reinforced Concrete, durability, surface layer, assessment, in-situ tests

1 Introduction

The addition of steel fibres in concrete is an effective technique for improving the toughness, fatigue resistance, impact resistance and flexural strength. However, an interaction of Portland Cement Concrete with the environment is usually significantly different from the response of other building and construction materials under similar conditions of exposure. Then, the corrosion of steel fibres in concrete would raise serious doubts as to the durability of steel fibres owing to their high surface area/volume ratio and low cover.

The main difference is due to fact, that an aging process in concrete is associated with the potential for a long term hydration of Portland Cement, resulting in the physical, chemical and mechanical changes of the structure and the properties of concrete. Therefore durability is not a bulk property like strength, but in the first place a surface property, determined by composition and properties of the surface layer, Kreijger (1984).

The authors have conducted both, laboratory and field

Fibre Reinforced Cement and Concrete. Edited by R. N. Swamy. © 1992 RILEM.
Published by E & FN Spon, 2-6 Boundary Row, London SE1 8HN. ISBN 0 419 18130 X.

investigations, on the causes of various deteriorations and have conducted tests on surface layers in order to assess the state of damage in concrete structures. The investigation was then extended to relate the information which was obtained from steel fibre concrete exposed for 14 years to surface layer characteristics for durability assessment. This paper reports the results obtained so far in assessing the non-destructive and partially-destructive test methods.

2 Experimental Methods

2.1 Exposure Conditions

In August 1977, test beams (100mm x 250mm section, 3.0m long) which have different fibre content and aspect ratios as the variables, were exposed at the north side of the laboratory building in the College of Eng., Nihon University. This college is located in the middle of the northern part of Japan and about 60 km inland from the Pacific Ocean. The condition of exposure provide mostly on an average of over 80 cycles of freezing and thawing per year, Hasegawa and Fujiwara (1988). After 14 years, the specimens were removed from the exposure site and placed in the controlled climate room for tests at an average temperature of 20°C and 80 percent R.H.

2.2 Properties of Steel Fibre Concrete before Exposure

Flat type of fibres of three different aspect ratios which were 50 (0.3x0.3x15mm), 60 (0.5x0.5x30mm) and 100 (0.3x0.3x30mm), were used in mixes having different fibre contents as shown in Table 1. Results shown in Table 2, include that the use of short fibres increased the strength.

Table 1. Mix Proportions and Properties of Fresh Concrete

Specimen V_f (vol%)	Unit Weight (kg/m³)		S.F (kg/m³)	Aspect Ratio	Slump (mm)	Air (%)
	S	G				
0	1133.9	518.2	0.00	–	188 ~ 200	3.7 ~ 4.5
0.5	1125.7	513.3	39.25	50	95 ~ 159	4.0 ~ 5.0
				60	82 ~ 166	4.4 ~ 4.6
				100	24 ~ 40	3.9 ~ 4.4
1.0	1116.7	509.2	78.50	50	15 ~ 55	4.8 ~ 5.8
				60	74 ~ 122	4.8 ~ 5.9
				100	0 ~ 27	3.8 ~ 5.0
1.5	1107.8	505.1	117.75	60	5 ~ 24	4.5 ~ 4.7

note : G_{max}=13 mm , W/C=60.0 % , s/a=70.0 %
W=240 kg/m³ , C=400 kg/m³

Table 2. Strengths of Specimens before Exposure

Vf (vol.%)	Aspect Ratio	f'c (MPa)	fdt (MPa)	fb (MPa)
0	-	22.09	1.97	3.16
0.5	50	24.37	1.85	3.39
	60	24.36	2.13	4.11
	100	24.29	2.47	3.75
1.0	50	26.82	2.48	3.77
	60	24.49	2.43	4.49
	100	27.49	2.54	4.14
1.5	60	29.24	2.63	4.90

note f'c : Compressive Strength
fdt : Direct Tensile Strength
fb : Flexural Strength

2.3 Testing of Exposed Specimens

The measured characteristics and test methods are as follows:

(1) Visual examination and rating system
(2) Pulse velocity measurements
(3) Rebound number measurements
(4) Pull-off tensile strength
(5) Recovering speed (Rapid air permeability)
(6) Depth of carbonation
(7) Depth of penetration of chloride ion
(8) Content of chloride ion

The test apparatus for the pull-off tensile test is shown in Fig. 1. A circular groove 5 mm in depth and 73 mm in diameter, was cut into the surface of the exposed specimen using a dry coring bit. A steel disk was bonded to the surface with epoxy adhesive. Forty eight hours later, the pull-off strength tester was applied to pull off the concrete.

The apparatus for the rapid air permeability test which can be alternatively applied as a water absorption test, is shown in Fig. 2. For this test, a hole of 35 mm depth and 10 mm diameter, was drilled in the surface of the specimens, and a vacuum pump was used to remove the air from the hole. Then, the time required for the difference between vacuum gage reading to increase from 8.0 kPa to 10.7 kPa was measured. The recovering speed was calculated by dividing the increase in the gage reading, 2.7 kPa, by the recorded time.

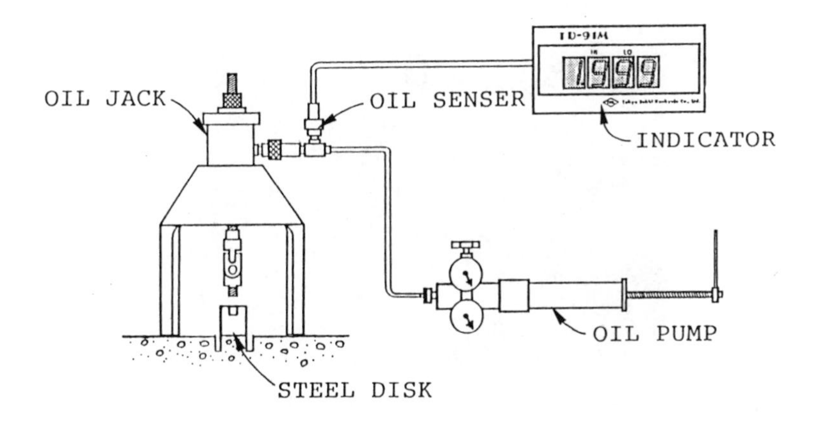

Fig.1. Outline of Testing Equipment in Pull-off Test

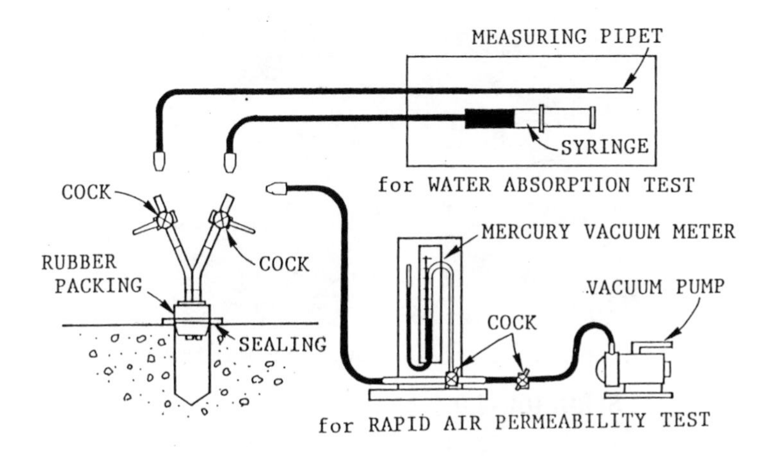

Fig.2. Outline of Testing Equipment of Water Absorption and Rapid Air Permeability Test

3 Results and Discussion

3.1 Visual Examination

Results for the specimens submitted for visual examination are shown in Figures 3 to 5. In accordance with ACI 201 (1984), scaling was rated visually on light, i.e. loss of surface mortar without exposure of coarse aggregate. On Figures 4 and 5, the corrosion of steel fibres, as a function of the number of fibres, is shown separately for the scaling rate.

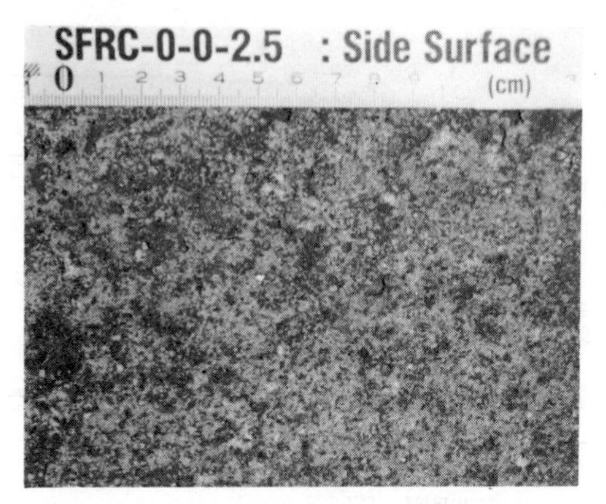

Fig.3. Scaling of Surface (Plain Concrete)

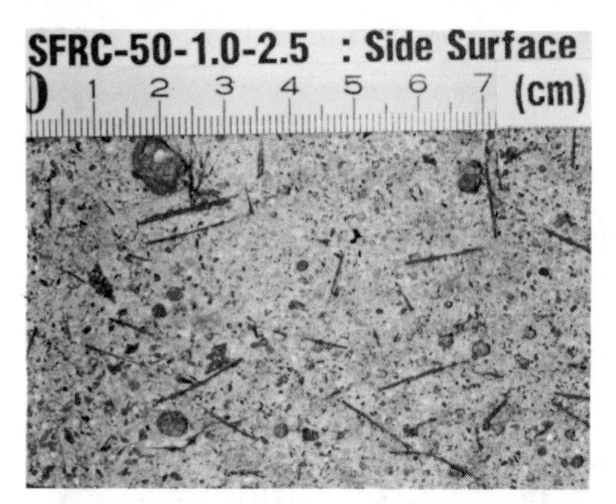

Fig.4. Scaling of Surface (Aspect Ratio=50, 1.0%)

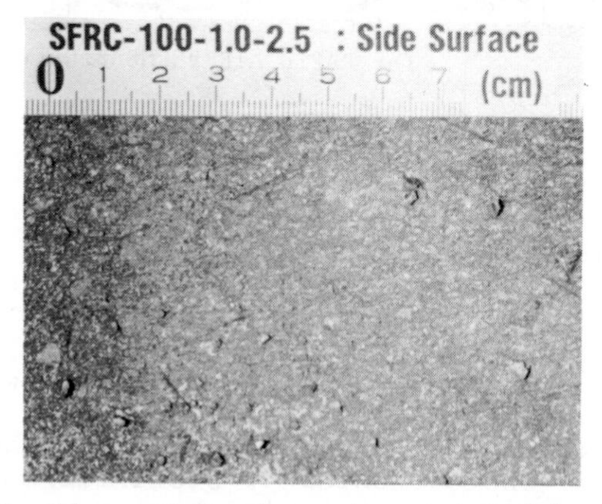

Fig.5. Scaling of Surface (Aspect Ratio=100, 1.0%)

3.2 Ultrasonic Pulse Velocity and Rebound Number

The procedures which were adopted for the ultrasonic pulse velocity test followed BS 4408: Part 5, (1974). The test specimens which have 73 mm in diameter and 100 mm length, were made by a dry coring bit. For each beam three sets of results were recorded. Results are summarized in Table 3. It can be seen that the difference between the pulse velocity values for the eight mixes is very small. This suggests that the ultrasonic pulse velocity which measured by the direct method, as used in the present study, was not sensitive to evaluate the qualities of the surface layer. This apparent insensitively may have been caused by the character of steel fibre concretes as the use of the direct method of transmission.

Table 3. Results of Ultrasonic Pulse Velocity

V_f (vol%)	Aspect Ratio	Ultrasonic Pulse Velocity V_p ($\times 10^3$ m/sec)
0	-	4.198 (C.V.=1.02%)
0.5	50	4.002 (C.V.=0.30%)
	60	4.090 (C.V.=0.71%)
	100	4.083 (C.V.=2.38%)
1.0	50	4.067 (C.V.=2.58%)
	60	4.017 (C.V.=2.69%)
	100	4.303 (C.V.=0.60%)
1.5	60	4.069 (C.V.=0.49%)

The adopted procedure for the rebound hammer test also followed the main requirements of the standard method of BS 4405 (1974), using both, NR and P types. A total of 20 readings was recorded. Four sets of readings were taken on each beam. The results in Fig. 6 show that the rebound index is almost the same value to the presence of lightly scaled surfaces as the use of the NR type of the Schmidt hammer. Similarly, the P type of hammer lead to the same tendency in the performance of the surface layer which excepted the aspect ratio of 60.

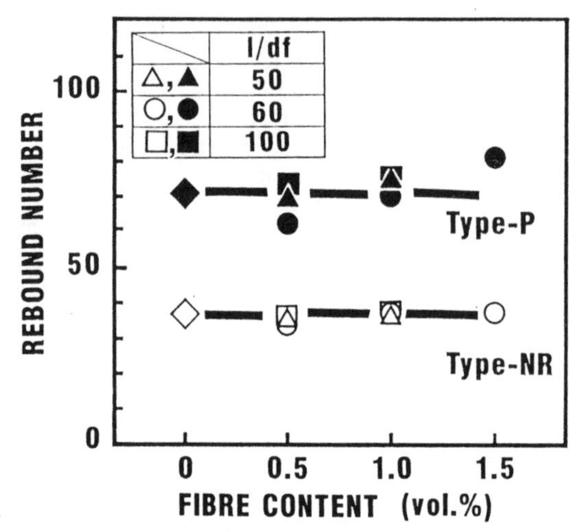

Fig.6. Average Rebound Number vs. Fibre Content

3.3 Pull-off Tensile Strength

Fig. 7 shows the average pull-off strength versus the volume fractions of the fibre constituents which can be expressed by the law of mixture. The surface layer strength by pull-off method will be divided into the two phases. The specimens, which have the volume fractions of the steel fibres less than the value of 0.5, decreased as increasing the fiber content as in Fig 7. However, there are clear indications that the surface layer strength of the specimens with the volume fractions of over 0.5 can be expected to an improvement in the presence of steel fibres. Similarly, the use of pull-off method which is sensitive to surface variation, lead to an improvement in the performance of an assessing technique.

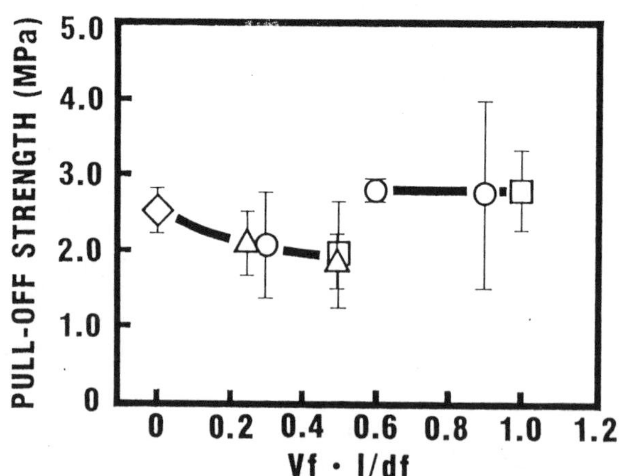

Fig.7. Average Pull-off Strength of Surface Layer

3.4 Recovering Speed in the Rapid Air Permeability Test

Fig. 8 shows the recovering speed in the rapid air permeability versus the fibre content and the aspect ratios. It can be seen that, generally, the recovering speed of steel fibre concretes is smaller than that of the specimens without steel fibers. The recovering speed of the specimens with steel fibre is roughly 70 percent of the speed which compared to the specimen without steel fibres. For the steel fibre concretes, there is a slight increase in speed as the amount of fibre increases.

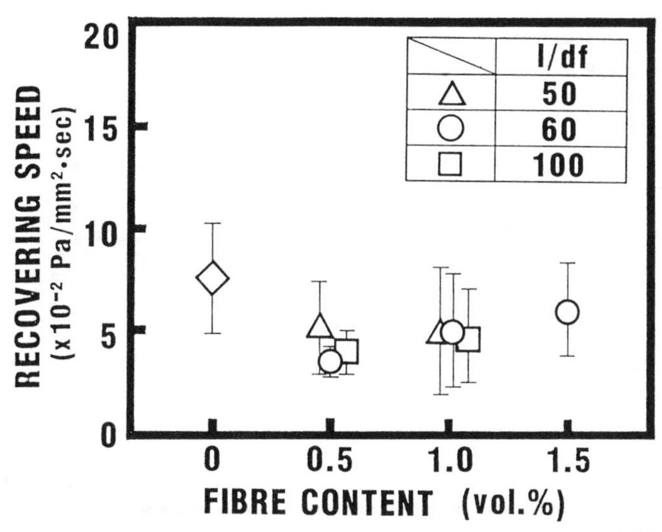

Fig.8. Recovering Speed in Rapid Air Permeability Test

3.5 Carbonation and Intrusion of Chloride Ion

The mean depths of carbonation, which used the drilled cylinder specimens with 73 mm in diameter and 100 mm in length, are given in Fig. 9. The carbonation rate of plain concrete was observed to be 5.8 mm which was advantageously affected by the addition of steel fibres. The depth of carbonation for plain concrete estimated using the empirical equation proposed by Kishitani (1971) is about 8.4 mm which is relatively close to the measured values. The mean depths of penetration of chloride ion showed same tendency which decreased as increasing the fiber content as is shown in Fig. 10.

Whilst the depth of carbonation and penetration of chloride ion are inversely related to the volume fractions of the fibre constituents, it is recognized that the differences between the dimension of fibre 0.3x0.3 mm and 0.5x0.5 mm. These results may have been affected by the surface area of the steel fibres, the difference being rather less in the larger amount of fibres.

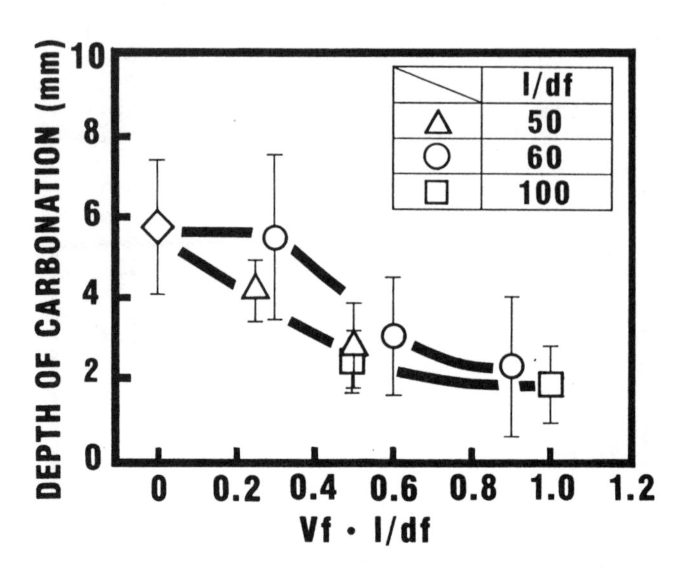

Fig.9. Average Depth of Carbonation

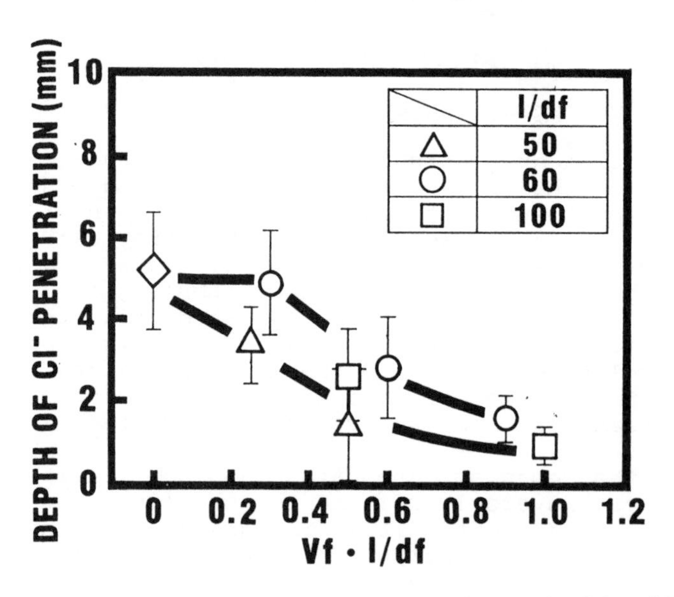

Fig.10. Average Depth of Penetration of Chloride Ion

Fig. 11 shows the average content of total chloride ion at a depth of 5 mm from the surface to 30 mm depth. Because of the location of exposure site, smaller amount of the content was observed compared to the results in marine splash zone, Kobayashi et al. (1990). For the plain concrete, the maximum content have appeared at the mean depth of carbonation. The peak value in the distribution of chloride ion may have been caused by the concentration of sulphate ions which induced by carbonation of concrete as is shown in the research works by Kobayashi et al. (1989).

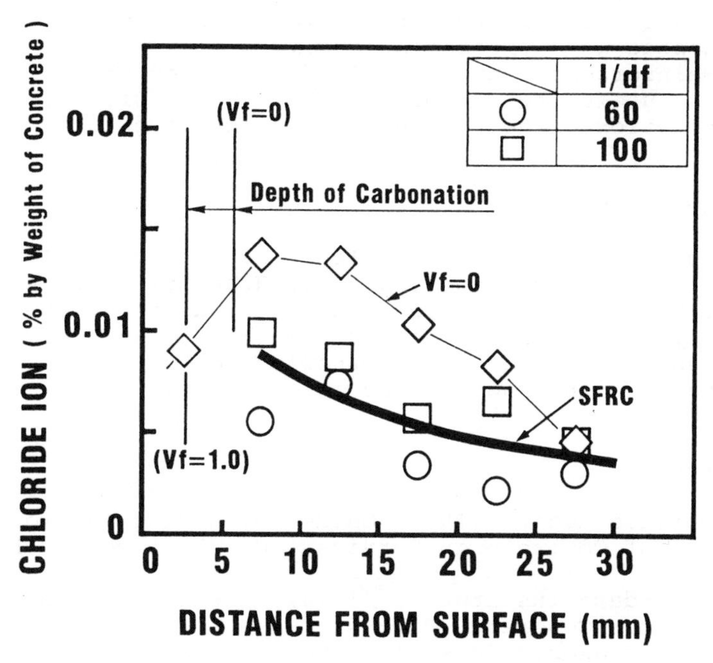

Fig.11. Distribution of Content of Chloride Ion

Figures 9, 10 and 11 approximately represent the state of steel fibres. From the results and observations in this investigation, it is apparent that breakdown of the passivity, resulting in pitting corrosion, was difficult to occur in the fibre concrete specimens. It has alternatively shown by Kobayashi et al. (1990) that the state of passivity of steel fibres has been explained by the electrochemical concept of the corrosion protection mechanism of steel fibre concrete. Owing to their nature, the maximum cathodic area for steel fibres is limited. Then, even though corrosion is initiated, it is probable that the subsequent rate of corrosion will be very small.

4 Conclusions

The following conclusions can be drawn from the steel fibre concretes exposed for 14 years and tested for surface properties.

(1) The scaling was rated visually on light, and the corrosion of steel fibres only limited near the surface of specimens.
(2) Ultrasonic pulse velocity was not sensitive as the use of the direct method of transmission.
(3) The results of the pull-off tests showed that in the surface zone of specimens, the limited volume fractions of the steel fibres may have been suggested as a threshold level.
(4) Reinforcement by randomly distributed steel fibres was useful to minimize the air permeability, carbonation, penetration and content of chloride ion.

5 Acknowledgement

The authors acknowledge Mr. K.Tsuchida and Mr. M.Tsukamoto who are research students, and Mr. H.Asami, Mr. K.Nakaue and Mr. T.Watanabe who are graduate students at the Nihon University for their enthusiastic work.

The authors are also acknowledged to Assoc. Prof. Y.Tsukinaga for his advice and information.

6 References

ACI Committee Report (1984) Guide for Making a Condition Survey of Concrete in Service, ACI 201.1 R-68 (Revised 1984).

British Standard Institution (1974) Non-destructive Methods of Test for Concrete, Part 5: Measurement of the Velocity of Ultrasonic Pulse in Concrete, BS 4408: Part 5.

ibid. (1974) BS 4408: Part 4.

Hasegawa,T. and Fujiwara,T. (1988) Frost Damage, Chapter 5 (in Japanese), Giho-Do Shuppan, Feb., 71-77.

Kishitani,K. (1971) Corrosion of Rebars in Concrete, Cement and Concrete, the Cement Association of Japan, No.289.

Kobayashi,K. et al. (1989) Migration and Concentration of Sulphate Ions induced by Carbonation of Concrete-Part 1, Seisan-Kenkyu, Vol.41, No.12, Dec., 52-55.

Kobayashi,K. et al. (1990) The Effect of Steel Fiber Reinforced Concrete on the Corrosion Behavior of Reinforcing Steel embedded in Concrete Members Exposed to Marine Environment, proc. of JSCE No.414/V-12, Feb.,195-203.

Kreijger,P.C. (1984) The Skin of Concrete: Composition and Properties, Materials and Structures, Vol.17, No.100, July-August, 275-283.

69 PERFORMANCE EVALUATION OF SFRC PAVEMENTS

S. K. KAUSHIK and R. M. VASAN
Civil Engineering Department, University of Roorkee, Roorkee, India

Abstract
It has recently been established that steel fibre reinforced concrete matrices (1,2) offer the possibility of reduced pavement thickness or higher load carrying capacity, reduced deflections and crack widths. The composite matrix, exhibits a substantially higher flexural strength. At design loads normally encountered in highway pavements, the stresses induced in the SFRC pavements are much lower than the flexural strength of the matrix. At higher loads, SFRC pavements are capable of bearing stresses much beyond the flexural strength of the matrix. Under the present investigation, the performance of SFRC pavements containing fibre volume between 0.5-2% has been evaluated under static plate load tests. SFRC pavements having a nominal thickness of 100 mm used in the investigation have shown a remarkable improvement in the load carrying capacity.
Keywords : Steel Fibres, Pavement, Stress, Flexural Strength, Fibre Volume.

1 Introduction

The introduction of closely spaced uniformly dispersed steel fibres in concrete provide a crack arrest mechanism. The strength of the composite matrix increases linearly with an increasing fibre volume fraction. The rate of gain in strength is a maximum upto an optimum fibre content of 1.25% (1,2) beyond which the rate of gain in strength is not substantial. This may be attributed to the non availability of sufficient mortar required to coat the surface of the fibres or nesting of fibres at a percentage higher than the optimum value. Half scale pavement sections consisting of fibre volume 0.5, 1.0, 1.25, 1.5 and 2% tested in this investigation exhibited excellent performance.

2 Experimental Investigation

Half scale SFRC pavement slabs 1.8m x 1.8m x 0.1m size were laid directly over well compacted subgrade. The SFRC mix was designed by ACI method (3). Hooked steel fibres having 0.456 mm dia. and aspect ratio 80 were used in the SFRC mixes. The strength charac-

Fibre Reinforced Cement and Concrete. Edited by R. N. Swamy. © 1992 RILEM.
Published by E & FN Spon, 2-6 Boundary Row, London SE1 8HN. ISBN 0 419 18130 X.

teristics of SFRC mixes used in the invesatigation are shown in Table 1. The properties of steel fibres are given in Table 2.

Table 1. Strength Characteristics of PCC and SFRC Mixes

Sl. No.	Mix Designation*	Fibre Volume (%)	Strength Characteristics: Compressive strength (N/mm^2)	Flexural strength (N/mm^2)
1	M0	0.00	28.700	6.16
2	M1	0.50	29.980	8.02
3	M2	1.00	31.794	9.85
4	M3	1.25	32.680	10.04
5	M4	1.50	32.800	9.95
6	M5	2.00	31.312	8.61

* Mix properties (1:1.95:1.95) by wt. of dry mix
w/c ratio = 0.60
fineness modulus of fine aggregate = 2.42
coarse aggregate = 6.00

Table 2. Properties of Steel Fibres

Sl. No.	Properties	Value
1	Type	M.S. black annealed wire
2	Shape	Hooked
3	Gauge	26 SWG
4	Average diameter, mm	0.46
5	Young's modulus of elasticity, N/mm^2	216×10^3
6	Yield strength, N/mm^2	313.5
7	Ultimate strength, N/mm^2	410.0
8.	Ultimate bond stress between fibres and concrete, N/mm^2	1.525

3 Performance of SFRC Pavement Slabs

The characteristics of PCC pavement SPC_1 and SFRC pavements SPC_3-SFC_7 and the underlying subgrade are given in Table 3. A comparison of the flexural stress strength ratios of these pavements at a design load of 41 kN under edge and corner loading conditions for experimental values and those obtained theoretically from Meyerhof and Westergaard analyses have been presented in Table 4.

Table 3. Characteristics of PCC, SFRC Pavements and Subgrade

Sl. No.	Pavement Type	Pavement Mod. of Elast. kN/mm^2	Poisson's ratio	Subgrade Mod. of Elast. N/mm^2	Subgrade Poisson's ratio	Mod. of Sub. R N/mm^3	Ep/Ks Ratio MM 10000
1.	PCC Pavement SPC1	27.31	0.1514	16.97	0.250	0.0384	71.12
2.	SFRC Pavement SFC3	29.15	0.1587	28.41	0.305	0.0642	45.40
3.	SFRC Pavement SFC 4	30.21	0.1668	25.46	0.305	0.0575	52.54
4.	SFRC Pavement SFC 5	30.68	0.1700	28.76	0.305	0.0650	47.20
5.	SFRC Pavement SFC 6	31.05	0.1709	20.96	0.305	0.0414	75.00
6.	SFRC Pavement SFC 7	28.54	0.1698	40.71	0.305	0.0920	31.02

From the test results of PCC and SFRC pavements, a considerable reduction in the ratio of stress to flexural strength is observed. The ratios are found to be 0.450 and 0.568 in case of SFRC pavement having an optimum fibre volume of 1.25% for the edge and corner loading conditions respectively. The load carrying capacity of SFRC pavements upto the level of flexural strength alongwith deflections, strains and stresses is given in Table 5.

The test results indicate that SFRC pavements (SFC_3-SFC_7) could take loads 70 kN, 78.81 kN, 100 kN, 120 kN and 90.94 kN respectively at the level of the flexural strength of the composite matrix. The deflections were found to be within the allowable limits for rigid highway pavements.

It was further observed that an increase in the fibre concentration in SFRC pavements results in a reduction in the induced strainn at the same load and a significant increase in the strain bearing capacity. At a fibre concentration of 1.25% by volume, SFRC pavements could bear strains of 2.44×10^{-4} and 2.92×10^{-4} at 100 kN and 69.84 kN loads under edge and corner loading conditions respectively without showing any failure. This shows the enormous potential of SFRC pave-

Table 4. Comparison of Flexural Stress Strength Ratios of PCC and SFRC Pavements (1.8mx1.8mx0.1m) For Design Wheel Load of 41 kN Under Edge & Corner Loading

Pavement type	Mod. of Elasticity	Flex. Strength	Meyerhof Theory*				Westergaard Theory*				Experimental*			
			Edge		Corner		Edge		Corner		Edge		Corner	
	(N/mm^2)	(N/mm^2)	Stress	Ratio	Stress	Ratio	Stress	Ratio	Stress	Ratio	Stress	Ratio	Stress	Ratio
SPC_1	27.31	6.162	2.751	0.446	42.927	0.697	5.792	0.940	47.920	0.777	5.998	0.974	8.608	1.397
SFC_3	29.15	8.020	2.669	0.333	38.750	0.483	5.343	0.974	42.960	0.536	5.572	0.695	8.412	1.048
SFC_4	30.21	9.850	2.698	0.274	39.493	0.400	5.494	0.672	44.640	0.453	5.180	0.526	7.909	0.800
SFC_5	30.68	10.040	2.677	0.268	38.946	0.390	5.387	0.539	43.440	0.435	4.491	0.450	5.670	0.568
SFC_6	31.05	9.950	2.739	0.275	40.507	0.407	5.719	0.575	47.120	0.474	4.905	0.493	7.343	0.738
SFC_7	28.54	8.610	2.591	0.300	36.800	0.427	4.961	0.576	38.440	0.446	5.844	0.678	7.362	0.815

* Ratio of Flexural Stress/Strength.

ments to carry heavy loads and bear substantially higher strains. The potential advantages of SFRC in pavement applications lies in it's reported (2) life expectancy as shown in Fig. 1.

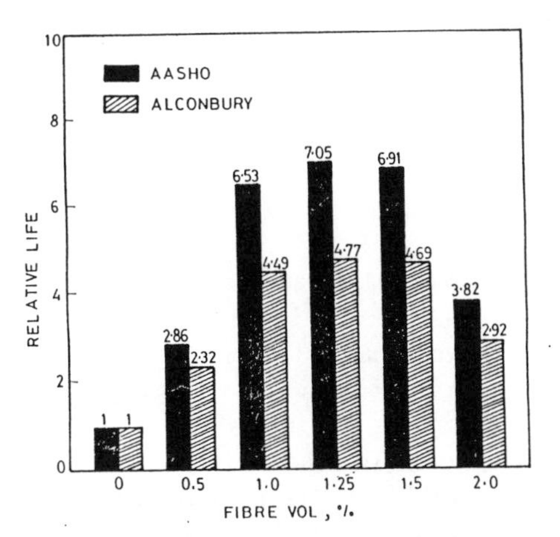

FIG. 1—RELATIVE LIFE FOR CONSTANT THICKNESS

4 Conclusions

Steel fibre reinforced concrete is a potentially advantageous composite material which can be employed in highway and airfield pavements by virtue of its higher load carrying capacity and the possibility of laying it in thin sections. SFRC pavements exhibit excellent performance under central, edge and corner loading conditions when laid directly over compacted subgrade. The significant improvement in the capacity to bear higher strains points out to the capacity of these pavements to bear higher load repetitions and therefore greater life expectancy of these pavements.

5 References

Kukreja, C.B. (1983) "Fibre Reinforced Concrete", **Ph.D. Thesis**, University of Roorkee, India.

Vasan, R.M. (1989) "Investigation of Steel Fibre Reinforced Concrete Pavements", **Ph.D. Thesis**, University of Roorkee, India.

Recommended Practice for Selecting Proportions for Normal and Heavy Weight Concrete (1977), **ACI 211-1-77**, American Concrete Institute, Detroit, Michigan.

Meyerhof, G.G. (1962). "Load Carrying Capacity of Concrete Pavements", **J. of Soil Mechanics and Foundation Division**, ASCE, SM3, Vol. 88.

Wsstergaard, H.M. (1926). "Stresses in Concrete Pavements Computed by Theoretical Analysis", **Public Roads**, Vol. 7, No. 2.

IRC (1988), Guidelines for the Design of Rigid Pavements for Highways, **The Indian Roads Congress**, New Delhi, India.

70 HIGH DURABILITY GFRC USING LOW-ALKALI, LOW SHRINKAGE (CGC) CEMENT

M. HAYASHI
Nippon Electric Glass Co. Ltd, Shiga, Japan
T. SUENAGA
Kajima Corporation, Tokyo, Japan
I. UCHIDA
Chichibu Cement Co. Ltd, Saitama. Japan
S. TAKAHASHI
Central Glass Co. Ltd, Tokyo, Japan

Abstract
Low-alkaline, low-shrinkage cement made with calcium silicates, calcium sulfoaluminate, anhydrite, and blast furnace slag (CGC cement) have recently been developed in Japan for use in glass fiber reinforced concrete (GFRC). This paper presents the results of comprehensive tests which evaluated the mechanical properties, durability, and dimensional behaviors of GFRC made with CGC (GFRC-CGC). The flexible anchor pull-out strength of the material is also discussed. Also, the durability of GFRC-CGC was ascertained by means of accelerated aging and natural weathering tests.
Key words: low-alkaline, low-shrinkage cement, CGC, AR glass fiber, GFRC-CGC, accelerated aging, natural weathering, durability, flexible anchor, dimensional stability, bending strength, tensile strength, accelerated carbonation, resistance to freezing/thawing cycle

1 Introduction

Seven years ago CGC was introduced to the Japanese market by Chichibu Cement Co., Ltd. as a product able to dramatically improve the durability of GFRC in terms of mechanical strength and dimensional stability. CGC is characterized by the absence of calcium hydroxide formation and a very low concentration of OH^- ions in the hardened cement paste (Tanaka 1985).

In order to confirm the suitability of GFRC-CGC as a construction material, various tests were conducted on the stability of its performance by Kajima Corporation, Nippon Electric Glass Co., Ltd, Chichibu Cement Co., Ltd, and Central Glass co., Ltd. The tests included those on bending strength (accelerated aging and natural weathering), tensile strength (accelerated aging and natural weathering), resistance to freezing/thawing cycles, pull-out strength of flexible anchors (accelerated carbonation and natural weathering), and dimensional behavior (wet/dry cycles and natural weathering).

Fibre Reinforced Cement and Concrete. Edited by R. N. Swamy. © 1992 RILEM.
Published by E & FN Spon, 2-6 Boundary Row, London SE1 8HN. ISBN 0 419 18130 X.

2 Experimental program

2.1 Materials and mix proportions

The properties of the low-shrinkage, low-alkaline cement (CGC) and the high-zirconia, alkali-resistant glass fiber (AR-glass fiber), made up of 20% zirconia by weight, used in the experiments are shown in Tables 1 and 2. The mix proportions of the GFRC-CGC are shown in Table 3. The AR-glass fiber content of the GFRC-CGC was about 5% except for the material used for the freezing/thawing test.

2.2 Casting and curing

2.2.1 Test specimen casting

Large plates were cast by the direct spray method, after which specimens were cut to the shapes and dimensions described below. Specimens were initially steam cured at 40°C for eight hours, then demolded at the age of one day and stored in a conditioning chamber at 20°C and 65% RH until testing.

2.2.2 Full-scale panel casting for dimensional behavior test

GFRC-CGC panels were cast by the direct spray method and steam cured at 40°C for eight hours. After demolding they were stored indoors until the start of exposure to natural weathering.

Table 1 Specific gravity, specific surface area, and chemical components of cements

Cement	Specific Gravity	Specific Surface Area (cm^2/g)	Chemical Components					
			SiO_2	Al_2O_3	Fe_2O_3	CaO	R_2O	SO_3
CGC	2.96	4500	23	11	1	48	0.5	9
OPC	3.15	3100	22	5	3	65	0.7	2

Table 2 Properties of AR glass fiber

Item	Property
ZrO_2 content	20% by weight
density	2.7×10^3 kg/m^3
fiber diameter	13.5 μm
strand tensile strength	1.4 GN/m^2
Young's modulus	74 GN/m^2
strain to failure	2%

Table 3 Mix proportions of GFRC-CGC

W/C	S/C	Water-Reducing Agent	ARGF Content (% by weight)	Setter*	Cement
0.325	0.66	C x 1% by weight	5.0±0.5 3.5±0.5	C x 0.3 – 0.5% by weight	CGC

*Setter: citric acid

2.3 Test methods

The methods employed in the experiments are summarized in Table 4.

2.3.1 Durability tests for bending strength and tensile strength

Accelerated aging tests were conducted on specimens after they were immersed in 50°C water from the age of four weeks for specified periods and then dried in a room for two to three days before testing. Specimens tested after natural weathering were exposed to natural weathering in Saitama, Japan from the age of four weeks.

2.3.2 Flexible anchor pull-out strength (after natural weathering or accelerated carbonation)

The shape and dimensions of specimens used for the flexible anchor pull-out test are shown in Figure 1. The specimens had been air-cured at 20°C, 60% RH for four weeks after they were steam cured at 40°C for eight hours and were stripped from the mold. They were then subjected to outdoor exposure in Saitama, Japan, or accelerated carbonation. The conditions of accelerated carbonation were: 10% CO_2 concentration, 40°C, and 40% RH.

2.3.3 Drying shrinkage

Base lengths of the GFRC-CGC specimens were measured with a contact

Table 4 Test program

Test Item		Specimen Dimensions (mm) length	width	thickness	Testing Method and Conditions
Bending Strength Durability	accelerated aging	250	50	10	three-point loading, immersion in 50°C water
	natural weathering				weathering in Saitama, Japan
Tensile Strength Durability	accelerated aging	380	40	10	direct tension, immersion in 50°C water
	exposure to three different environ-ments				direct tension, storage indoors, outdoors, and in 20°C water
Pull-Out Strength for Flexible Anchor	accelerated carbonation	anchor diameter: 13			10% CO2, 40°C, 40%RH for 13 weeks
	natural weathering	bonding pad: 140x300x15			weathering in Saitama, Japan
Length Change	drying shrinkage	250	50	10	20°C, 60%RH
	dry and wet cycles	400	100	10	20°C, 30%RH/ 20°C, 90%RH
Length Change of Full-Scale Panels		3700	2000	13, 22	weathering in Shiga, Japan
Distortion of Full-Scale Panels		3700	2000	13, 22	weathering in Shiga, Japan
Resistance to Freezing and Thawing		400	75	75	ASTM C 666

gauge at the age of one day. They were then stored in a room at 20°C, 60% RH, and their lengths were measured at specified intervals.

2.3.4 Change in length with dry/wet cycles

After seven days of curing at 20°C and 65% RH, the specimens detailed in Table 4 were immersed in water at 20°C for seven days. Their lengths were then measured by comparator at the age of fourteen days. Differences in length were measured at intervals a total of eight times during four repetitions of the following cycle in the conditioning room: fourteen days at 20°C, 30% RH, and then fourteen days at 20°C, 90% RH.

2.3.5 Change in length and distortion of full-scale panels

The panels tested were 3.7 m long and 2.0 m wide, and were designed for a building twenty meters high. Table 5 shows the types of panels tested and their finish materials. Measurements of panel length were conducted using a Humboldt-type strain gauge and gauge plugs attached to panel backs as shown in Figure 2. Base lengths were measured at the age of one day and changes in length were then measured at specified intervals.

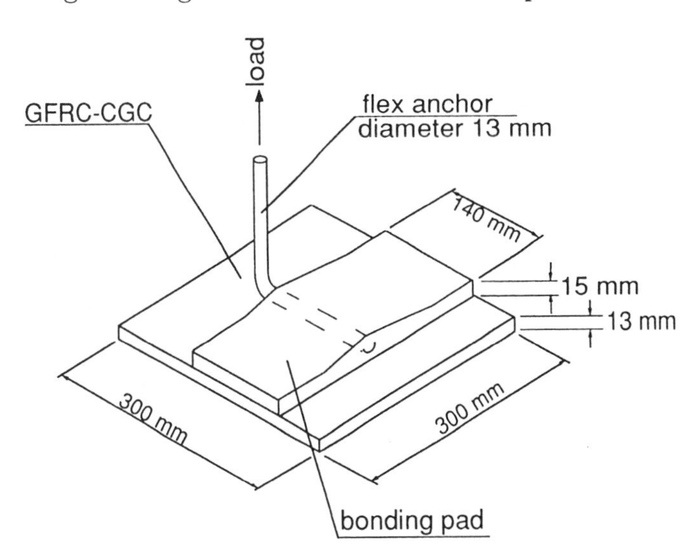

Figure 1 Flexible anchor

Table 5 Test panels: panel types and finish materials

Finishing Material	Panel Type	
	steel stud stiffened	rib-reinforced
ceramic tile	1	1
exposed aggregate	1	1
acrylic coating	1	1

Note: Panels were designed for a 20m-high building.
Panel dimensions: 2000mm wide x 3700mm high
GFRC-CGC thickness: 13mm (steel stud stiffened)
22mm (rib-reinforced)

Panel distortion was measured by transit at specified intervals, using calibrated gauges as shown in Figure 3. The gauges were attached to panel surfaces as shown in Figure 4. Before being exposed to outdoor weathering in Shiga, Japan, they were cured indoors until the age of four weeks.

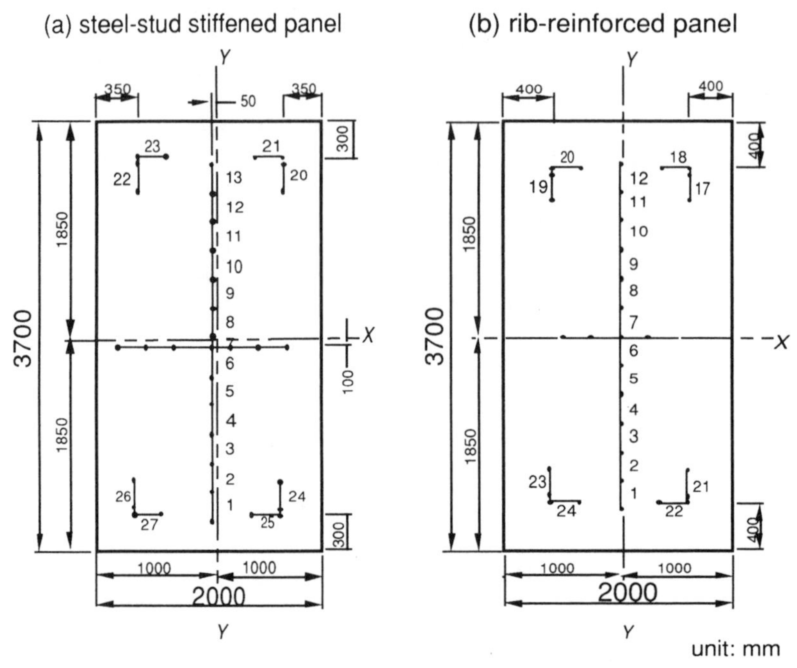

Fig. 2. Positions of gauge plugs attached to panel backs for length measurements

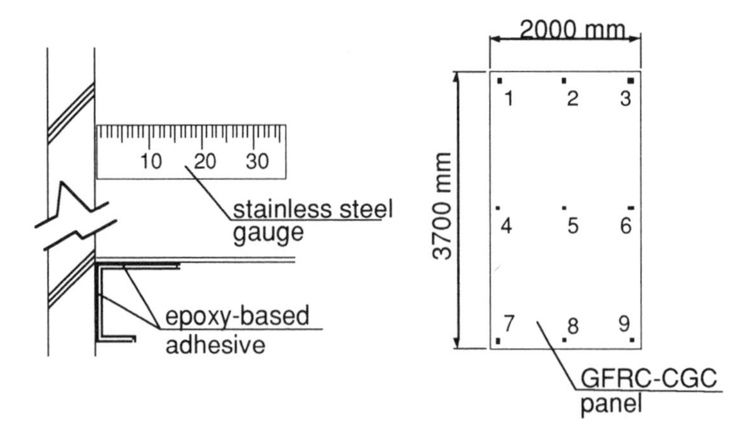

Fig. 3. Detail of calibrated measurement gauges

Fig. 4. Positions of distortion gauges

2.3.6 Resistance to freezing/thawing cycles

The test was performed according to ASTM C 666. Elongation, relative dynamic elastic modulus, and weight were measured. The specimens were two months old at the initiation of the test.

3. Test results

3.1 Bending strength

Figure 5 shows the results of bending tests on GFRC-CGC specimens after accelerated aging by immersion in 50°C water. The MOR (modulus of rupture) after one year of immersion (estimated to be equivalent to about 50 years of outdoor exposure in Japan (Proctor 1982)) was roughly 90% of the original value. This demonstrates the excellent MOR durability of GFRC-CGC, and was confirmed by the results of natural weathering tests over up to five years shown in Figure 6.

3.2 Tensile strength

The results of tensile strength tests on GFRC-CGC conducted after accelerated aging by immersion in 50°C water are shown in Figure 7. The UTS (ultimate tensile strength) after one year of immersion was about 100% of the original value, indicating the excellent tensile strength durability of GFRC-CGC.

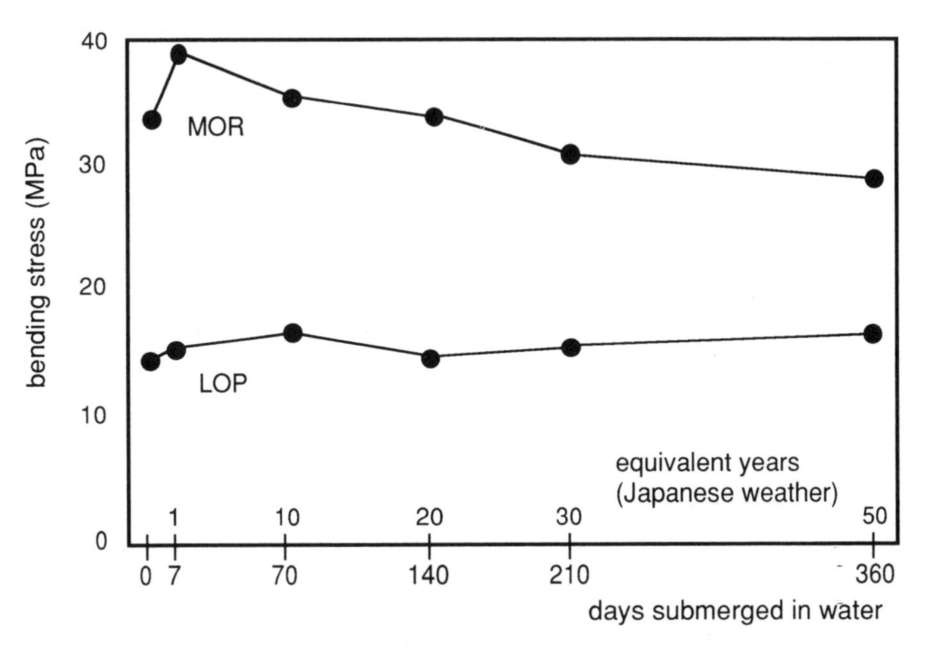

Fig. 5. Change in bending stress for GFRC-CGC stored in water at 50°C

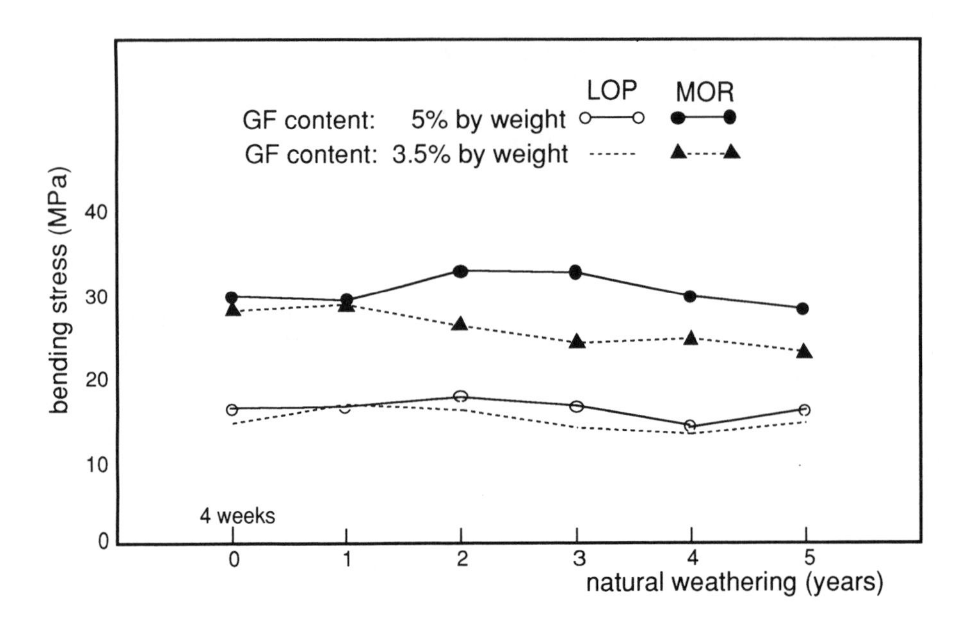

Fig. 6. Bending stress for GFRC-CGC after five years of exposure to natural weathering

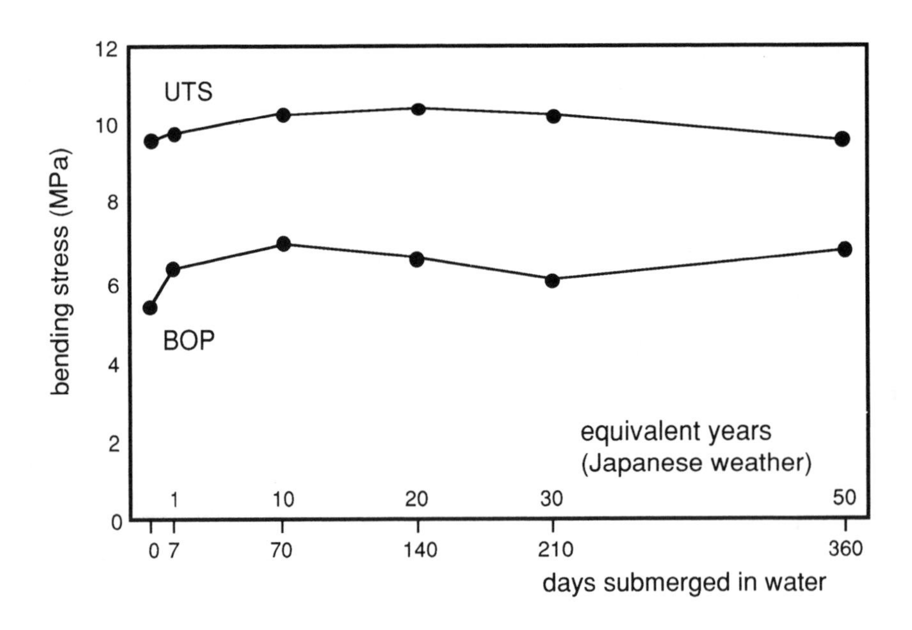

Fig. 7. Change in tensile stress for GFRC-CGC stored in water at 50°C

Specimens were exposed to three different storage conditions for long-term tensile strength tests: outdoors, under water, and indoors at 20°C and 60% to 70% RH. Figure 8 shows the results of tensile strength tests performed after up to three years of storage under these conditions. The

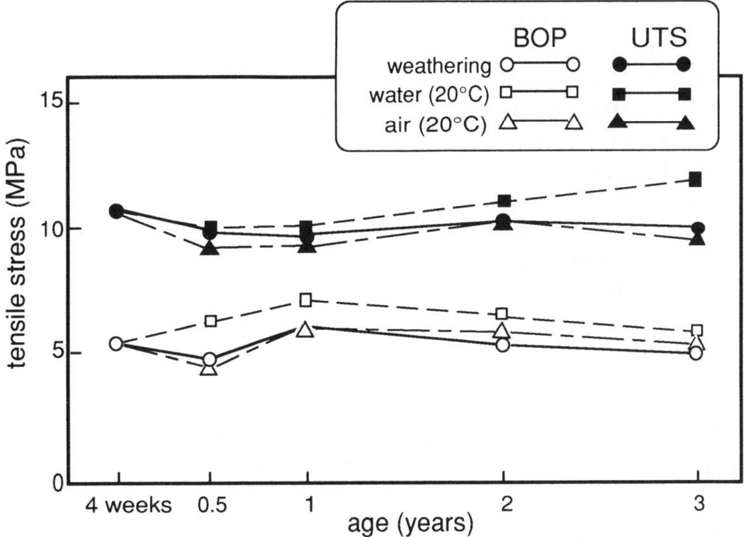

Fig. 8. Tensile stress for GFRC-CGC after three years of exposure to specific environments

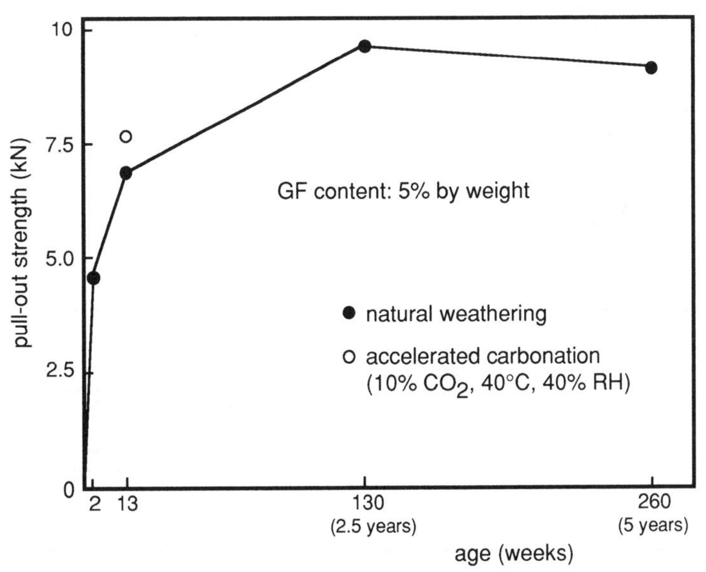

Fig. 9. Pull-out strength for flexible anchors after five years of exposure to natural weathering and accelerated carbonation

residual rate of UTS of specimens kept under water at 20°C for three years was about 110%, and that of specimens stored outdoors and indoors was about 95%.

3.3 Flexible anchor pull-out strength

Values for pull-out strength of flexible anchors are shown in Figure 9. While the pull-out strength of the specimens was 4.5 kN per anchor at the age of four weeks, values measured after 5 years of outdoor exposure and after 13 weeks of accelerated carbonation were 9.0 and 7.5 kN per anchor, respectively. The thirteen weeks of accelerated carbonation, under conditions of 10% CO_2 concentration, 40°C, and 40% RH, are estimated to be equivalent to 35 years of exposure to natural carbonation outdoors in Japan (Yoda 1977). The results indicate that GFRC-CGC has sufficient durability for flexible anchor pull-out strength, a property which is critically important for the structural stability of GFRC-CGC panel construction using steel studs.

3.4 Drying shrinkage and dimensional change associated with wet/dry cycles

Figure 10 shows the results of drying shrinkage measurements of GFRC-CGC specimens stored at 20°C and 60% RH. Shrinkage was about 3×10^{-4} after one year, about half that of conventional precast concrete. The difference is due to the larger amount of ettringite formed in GFRC-CGC at early ages.

The results of length-change measurements performed on GFRC-CGC specimens exposed to cycles of drying at 20°C, 30% RH and wetting at 20°C, 90% RH are given in Figure 11. Length change during the cycle was 2×10^{-4}, indicating that GFRC-CGC has good dimensional stability.

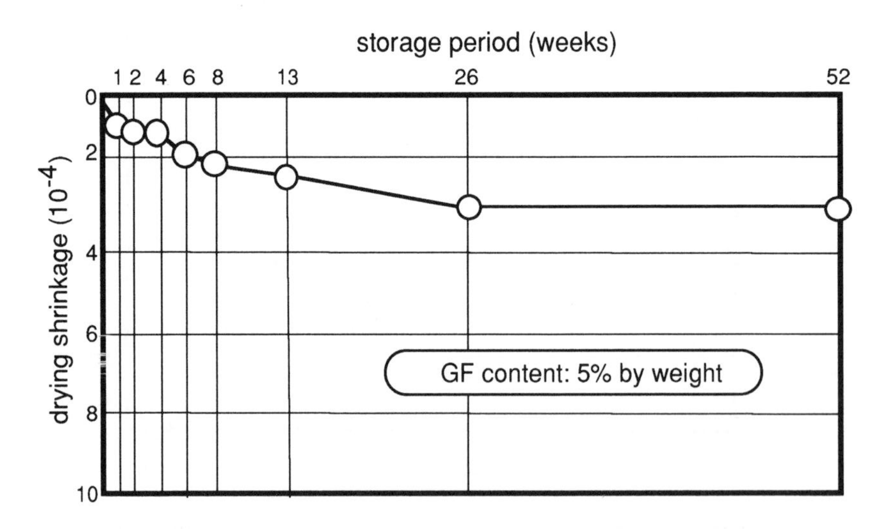

Fig. 10. Drying shrinkage of GFRC-CGC stored at 20°C, 60% RH

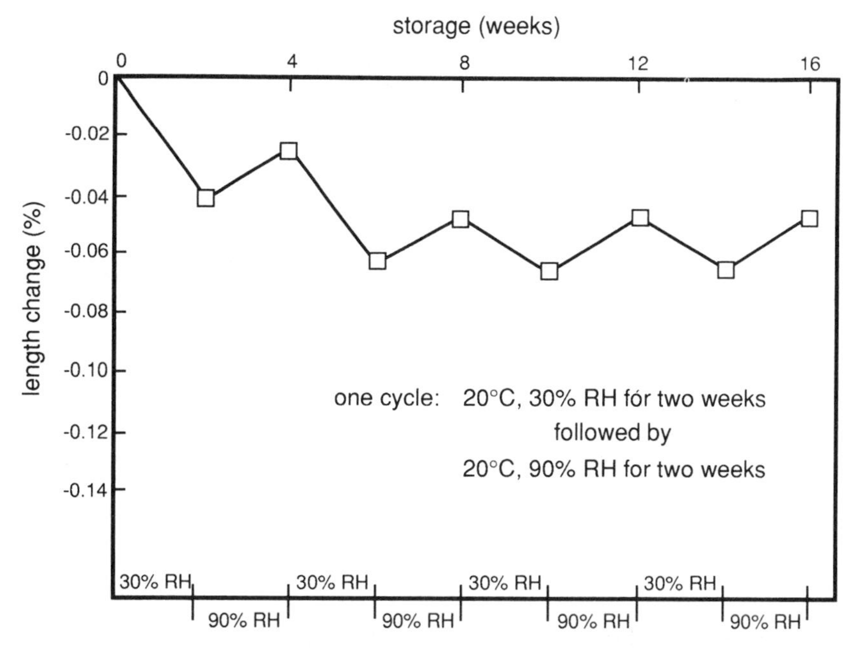

Fig. 11 Length change of GFRC-CGC during cycles of drying at 20°C, 30% RH and wetting at 20°C, 90% RH

3.5 Change in length of full-scale panels

Measurements were made over a period of four years along the center of the rear surfaces of panels (*y*-axis) to determine the ratio of length change of full-scale GFRC-CGC panels. The ratios obtained (change in length divided by total base length), which are shown in Figure 12, were –2.2 to -4.6×10^{-4} at the age of 26 weeks. These favorable results agree with those obtained in shrinkage tests performed on small specimens. After four years, the ratios ranged from 0 to -4.5×10^{-4}, indicating satisfactory dimensional stability.

The ratios obtained from measurements taken in winter (at the ages of 6 and 18 months) were greater than those obtained from measurements taken in summer (at the ages of one, two, and four years) by 2 to 5×10^{-4}. Since GFRC-CGC's coefficient of thermal expansion is 10 to 15×10^{-6}/°C, the length change ratios obtained from winter measurements should be 3 to 4.5×10^{-4} greater than those from summer measurements if the difference between summer and winter temperatures were 30°C. Thus the differences in the ratios obtained from summer and winter measurements are presumed to be due largely to the differences in prevailing temperatures at the time the measurements were taken.

The ratios were somewhat larger for steel stud panels than for rib-reinforced panels. Almost no differences in length change ratio were observed to be associated with the type of panel finish material.

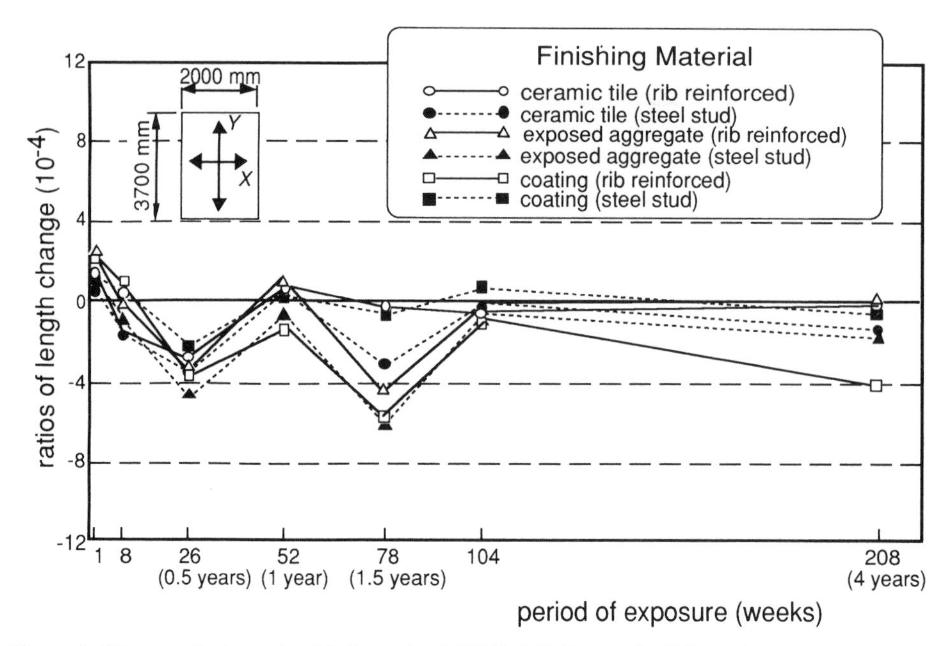

Fig. 12 Change in length of full-scale GFRC-CGC panels (*Y*-axis)

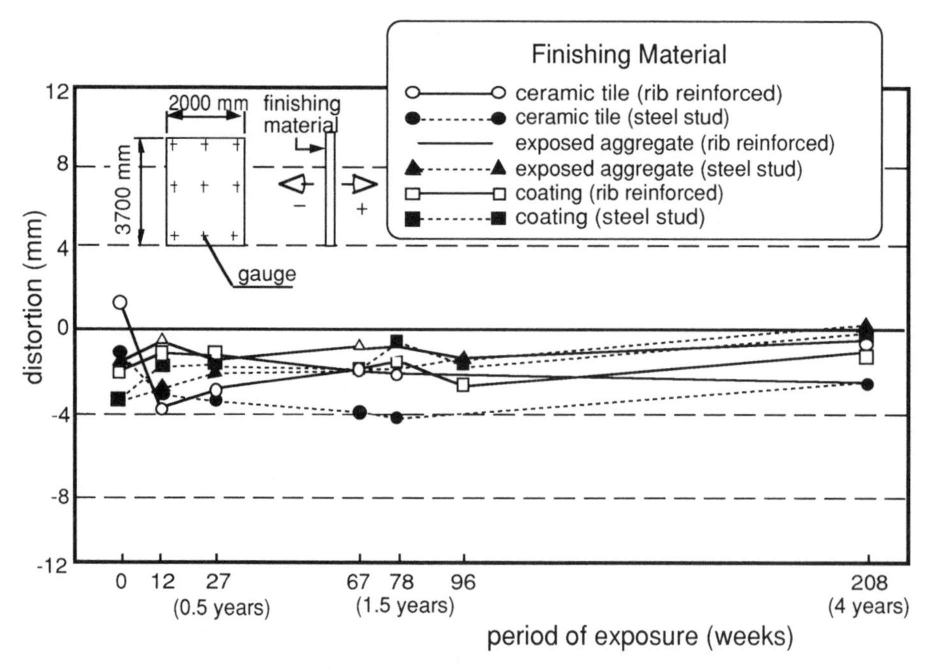

Fig. 13. Distortion of full-scale GFRC-CGC panels

3.6 Distortion of full-scale panels

Panel distortion, as measured along the center of the panels, is shown in Figure 13. Distortion was minimal: 0 to –3 mm at the age of four years (negative values indicate a convex finish surface).

Unlike the case for length change measurements, almost no differences were associated with the season in which distortion was measured. This is apparently due to the fact that temperature-caused dimensional change occurred throughout a panel's thickness and did not affect distortion.

Tile-finished panels originally showed slightly greater distortion than other panels. This is due to the difference in the drying shrinkage ratios of tile and GFRC-CGC being greater than those of the exposed aggregate or coating finishes and GFRC-CGC. However, the maximum distortion of the tile-finished panels during a four-year period was only 4 mm (1/925), and the use of tile-finished panels can be recommended.

No differences were associated with different panel types.

3.7 Resistance to freezing/thawing cycles

Figures 14 and 15 show the results of freezing/thawing cycle tests performed with freezing and thawing in water, and freezing in air and thawing

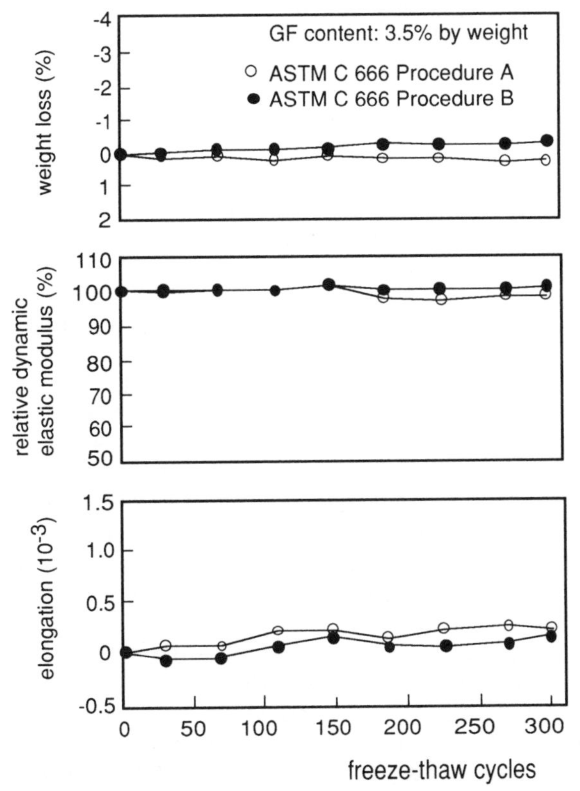

Fig. 14. Resistance to freezing and thawing of GFRC-CGC

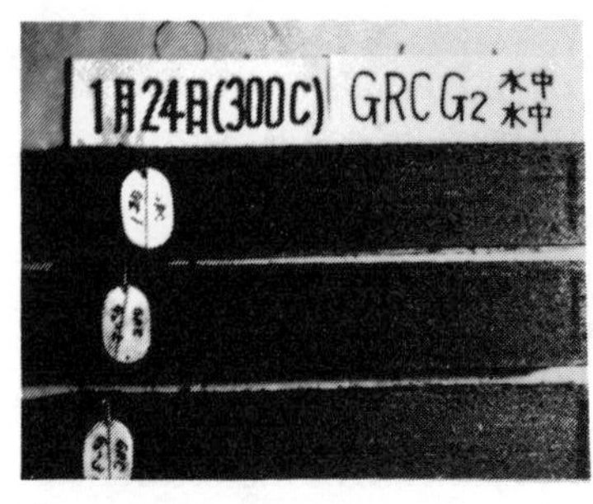

Procedure A
freezing and thawing in water

Procedure B
freezing in air, thawing in water

Fig. 15 Photographs of GFRC-CGC specimens after 300 cycles of test freezing and thawing

in water (ASTM C 666 procedures A and B). GFRC-CGC's relative dynamic elastic modulus after 300 cycles was close to 100%, and the material showed very little change in length and weight after 300 cycles. No scaling was observed after 300 cycles. The tests confirmed that GFRC-CGC possesses good frost resistance.

4 Conclusions

The results obtained from the comprehensive testing of the performance of GFRC made with low-alkaline, low-shrinkage cement and high-zirconia AR- glass fiber (GFRC-CGC) demonstrate the following points.

1. The bending strength of GFRC-CGC has excellent stability, as confirmed by both accelerated-aging (in 50°C water) and natural-weathering (in Saitama, Japan) tests
2. The tensile strength of GFRC-CGC also has excellent stability, as confirmed by accelerated-aging (in 50°C water) and three different exposure tests (natural weathering in Saitama, Japan; submerged in 20°C water; and indoors at 20°C and 60% to 70% RH).
3. The flexible anchor pull-out strength of GFRC-CGC after five years of outdoor exposure in Saitama, Japan was 9.0 kN per anchor, and was 7.5 kN per anchor after 13 weeks of accelerated aging at 10% CO_2, 40°C, and 40% RH. These values are 2.0 and 1.7 times those measured at the age of four weeks.
4. GFRC-CGC displayed the minimal drying shrinkage of 3×10^{-4} after one year of storage at 20°C and 60% RH. Length change during cycles of 20°C, 30% RH and 20°C, 90% RH was only 2×10^{-4}. The dimensional stability of GFRC-CGC was also confirmed by measurements of length change in full-scale GFRC-CGC panels exposed to outdoor weathering.

5. The distortion of full-scale GFRC-CGC panels as measured along the center line was 0 to –3 mm after four years of natural weathering in Shiga, Japan. The maximum distortion of tile-finished panels over four years was only 4 mm (1/925), and the use of tile-finished panels can be recommended.
6. GFRC-CGC retained its initial relative dynamic elastic modulus values, length, and weight after 300 freezing/thawing cycles conducted according to ASTM C 666.

5 References

Proctor, B. A. Oakley, D. R. and Litherland, K. L. (1982) Developments in the Assessment and Performance of GRC over 10 Years. **Composites** 13, pp. 173-179.

Tanaka, M. and Uchida, I. (1985) Durability of GFRC with Calcium Silicates—$C_4A_3\overline{S}$-$C\overline{S}$-Slag Type Low Alkaline Cement, in **Proc. PCI Symp. Durability of GFRC**, ed S. Diamond, Chicago, pp. 305-314.

Yoda, A. (1977) Neutralization of concrete: Erosion of steel and its countermeasure. **Concrete Engineering**, 15[9], pp. 34-36 (in Japanese).

71 DURABILITY OF MEDIUM-ALKALI GLASS FIBER REINFORCED SUPER LOW-ALKALI CEMENT

YANG QINGJI, HUANG DANENG and YANG GUOYING
China Building Materials Academy, Beijing, China

ABSTRACT
As we know, medium–alkali glass fiber reinforced portland cement is not durable, so, for a long time, the GRC researchers have been thinking that the reason is that there is only an interaction between $Ca(OH)_2$produced from the hydration of the cement and the silica skeleton of the glass fiber and this interaction may lead to the destruction of the glass fiber.

We have done some studies on the durability of medium–alkali glass fiber reinforced various cement. In doing such tests as X–ray diffraction analysis, SEM analysis, Laser Raman Spectrum analysis and a lot of tests of the macromechanics of the GRC, we have found that Al–O bond of the fiber breaks at first, then comes the destruction of the Si–O bond.According to this, we advance that the destruction of the medium–alkali glass fiber in the cement matrix may be divided into 3 stages: the hydrolysis of the alkali oxides in the fiber network; the breaking of the Al–O bond and the destruction of the Si–O bond. Thus, the tensile strength of the glass fiber decreases considerably.

Based on the above knowledge, we have provided the theoretical basis for using medium–alkali glass fiber as reinforcement of cement matrix. The GRC made with medium–alkali glass fiber and low–alkalinity cement developed by us turns out to be with good durability, which has been proved through ten years′natural weathering test and one and a half years′accelerating test.

A series of this kind of GRC products have been developed and have showed their good durability in the application.

KEY WORDS: super low–alkalinity cement, medium–alkali glass fiber, PH value, Durability

1 Preface

Glass fiber reinforced cement (GRC) is a relatively new building material developed at the beginning of the 70′s in the world. With the advantages of lightweight, high strength, multifunction, multiapplication, this kind of material is particularly suitable to be processed to complex shaped products with thin wall. GRC is highly praised as one of the pioneer building materials in the world.

Because of the failures by reinforcing portland cement with medium–alkali glass fiber in the 50′s, researchers in many countries have carried out their researches from many aspects, such as coated medium–al-

Fibre Reinforced Cement and Concrete. Edited by R. N. Swamy. ©1992 RILEM.
Published by E & FN Spon, 2-6 Boundary Row, London SE1 8HN. ISBN 0 419 18130 X.

kali glass fiber, adding active additions to cement for descreasing its alkalinity, etc. But they all transfer to use very expensive alksli–resistant glass fiiber because the reinforced strength decreases rapidly with the methods above.

Abroad, especially in the United States and the United Kingdom, GRC products are processed by alkali–resistant glass fiber reinforced portland cement. But they can't compietely solve the problems on the durability and brittleness of the glass fiber. The technical route of our country is reinforcing low–alkalinity cement with alkali–resistan glass fiber, so our GRC products are advanced in durability.

During the "sixth five–year plan", our academy had attempted to reinforce I–type low–alkalinity cement with medium–alkali glass fiber, but we have to give it up because of the problems on the durability and some other problems.

In the researching of using medium–alkali as reinforcing material for cement, the House Building Materials and Concrete Institute of our academy developed out super low–alkali glass fiber, this small glass fiber cement sheet will almost keep its strength no matter in 50℃ accelerated corrosion test or in long–term water immersing test. The glass fiber doesn't show any corrosions and brittle fracture when being observed under the electron macroscope. So the hope of using medium–alkali glass fiber as reinforcing material has been ignited again. The paper describes the long–term durability of medium–alkali glass fiber reinforced super low–alkalinity cement from both micro–structure and macro mechanical properties.

2 Experiment

2.1 Main Raw Materials

2.1.1 cement

525# super low–alkalinity cement and naturially–made ordinary portland cement, early strength cement and I–type low–alkalinity cement with the same grade.

2.1.2 glass fiber

Medium–alkali glass fiber are provided by;

Shanghai Yaohua Glass Factory

Qinhuangdao Yaohua Glass Factory

Tianjin General Glass Fiber Plant

alkali–resistant glass fiber is provided by;

The Research Institute of Glass Fiber, China Building Materials Academy

Followings are their chemical composition

type\%	SiO_2	CaO	Na_2O	K_2O	ZrO_2	TiO_2	Al_2O_3	MgO
alkali–resistant glass fiber	60.0	4.5	12.5	2.5	14.5	6.0	/	/
medium–alkali glass fiber	67.0	9.5	12.0	0.5	/	/	6.2	4.2

2.2 Specimen Preparation and Testing

2.2.1 accelerating test on durability for glass fiber reinforced cement

The specimens are made by glass fiber reinforced cement paste with water / cement ratio of 0.3 and specification of 10 × 30 × 120mm (see Figure 1).

Glass fiber is continuously oientated at thetensile field inside the specimen.

After moulded, specimens are coated with plastic sheet and set in the laboratory for 24 hours. Then, demoulded and are accelerately aged in a moist–heating aging box at the temperature of 50 ± 1℃, and with the moisture of more than 90% inside (the specimens should often be watered in order to keep moisture). In due time, fetch a group of specimens and test their bending strength by cement mortar bending machine. The span of specimen is 100 mm, load only at the middle point.

2.2.2 durability after long–term water immersion

Specimens are 1:1 mortar with the water / cement ratio of 0.35 and prepared with same method above. After demoulding, set them in the water tank in a standard curing room and keep water temperature between 20 ± 3℃, then fetch a group of specimens and test thear bending strength with the same method above in due time.

2.2.3 determinating the PH value of the cement paste

Mix cement and water according to the ratio of 1:10 or 1:4 to prepare cement paste. Shake to due time, then filtrate them and measure the PH values of filtrate with type PHS–3 PH meter.

2.2.4 determinating the strength of glass fiber

Glass fiber with certain length (30cm) are immersed in solutions with different PH values for various time. Then they are tested tensile strength by a fan–shaped tensile machine of type WPM–100.

2.2.5 analysis of hydrated products

X–ray diffractometer (D / max–ⅢA, Japan LIGAKU Company Ltd.)
SEM electron microscope (ASM–SX Japan SHIMADZU Company Ltd.)
thermogravimetric analyser (Dupont 1090 type combined automatic thermal analyser, U.S.A.)

2.2.6 pore structure of cement mortar

The specimens are 1:1 mortar with the water / cement ratio of 0.35. Their pore structure is tested by POROSIMETER series of 200 type pore meter made in Italy when standard curing to due time.

2.2.7 SEM electron microscope are used to observe the appearance of glass fiber (ASM–SX Japan SHIMADZU)

2.2.8 Ramanooru 1000 Type Laser Raman Spectrum Analyser of France is used to determine the structure of glass fiber.

3 Test Contents and Results

3.1 Hyrated Products and PH Values of Super Low–alkalinity Cement

3.1.1 hydrated products of super low–alkalinity cement

The hydrated products which begin to be obtained from 30 minutes after adding water to hydration to 1 year have the hydrated characteristic peaks of 9.662, 5.615, 4.951 and 4.690 by x–ray analysing, which means that the crystal phase is calcium sulphoaluminate hydrate of low–alkalinity. Furthermore, only ettringite crystal and unhydrated cement particles can be observed by electron microscope.

The combined thermal analyser are used to thermogravimeter the hydrated products of non–crystal phase after hydrating for 1 to 28 days. Because their weights are lost at 276.2, 259.9, 268.2, 278.5, and 278.4℃ which are the dehydration temperature of $Al(OH)_3$, we can conclude that the hydrate of non–crystal phase is gel of alumina hydrate.

In order to fully prove that the hydrates of super low–alkalinity cement are ettringite and gel of alumina hydrate only, we have analysed the white deposites accumulating on the surface of long age cement stone by x–ray diffractometer.The results have again confirmed that the hydrates are ettringite and $Al(OH)_3$.

3.1.2 PH value of super low–alkalinity cement paste

The PH value of cement paste depends on the constituents of hydrated mineral phase. Super low–alkalinity cement is the lowest, generally lower than 10.5, because its hydrates are $C_3A.3C\overline{S}.H_{32}$and $Al(OH)_3$and no $Ca(OH)_2$, please see Figure 2 for the PH values at different hydrating ages.

3.2 Pore Structure of Super Low–alkalinity Cement

The porosity and the distribution of various diameter pores are the important characteristics for cement stone,which can decide a series of other characteristics.

The porosity of super low–alkalinity cement nortar determined by 200 type pore meter is in Figure 3.

The mortar specimens with the water / cement ratio of 0.35 and cement / sand ratio of 1:1 have total porosity of $0.0458cm^2/g$ on the 28th day and pores whose diameters are less than 1000Å about 80%. Pores whose diameters are around 1000Å will be more than 90% after one year.

J.F.Young, who has sumbitted the concept of limit value of effective pore diameter, thinks that the impermeability and some other properties have close relation with $r\infty$. When pore diameter is $< r\infty$, the impermeability is good. $r\infty$ value can be determined through test but different researchers will obtain a littlea different results. $r\infty$ from P.K. Meta is 1320Å which means that the pore will be harmful if the diameter is larger than 1320Å. This view can explain why the super low–alkalinity cement has the strong capabilities of impermeability,frost–resistance and corrosion–resistance to resist from various media and why although in low–alkalinity protect steel from corrosion of weather moister in order to durably composite with glass fiber.

3.3 The Relation Between Chemical Attack Resistance and Chemical Composition,Micro Structure of Medium–alkali Glass Fiber

3.3.1. the alakli resistance of glass fiber

The chemical stability of glass fiber depends on the resistance to chemical attack of glass fiber and the type and the characteristic ofattacking media. The glass fiber including alkali–resistant and medium–alkali will suffer from different force of attack when the solutions with different PH values. Generally, two methods are used to indicate the alkali resistance of glass fiber, i.e.

(1) the decrease in the strength of glass fiber in different basic solutions and cement filtrates.

(2) the decrease in diameter of glass fiber in different basic solutions and cement filtrates.

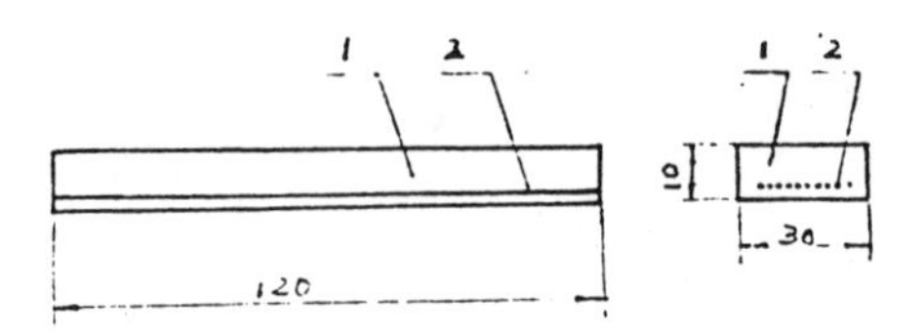

Figure 1. structure of glass fiber reinforced cement specimen
1—cement matrix
2—glass fiber

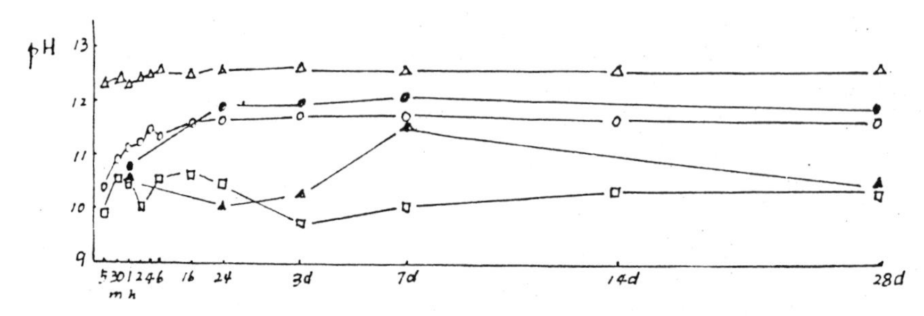

Figure 2. PH values at different hydrating ages(solid to liquid ratio is 1:10)

△— ordinary portland cement
○—high–alumina cement
●— salphoaluminate early strength cement
▲— I type low –alkalinity cement
□— super low –alkalinity cement

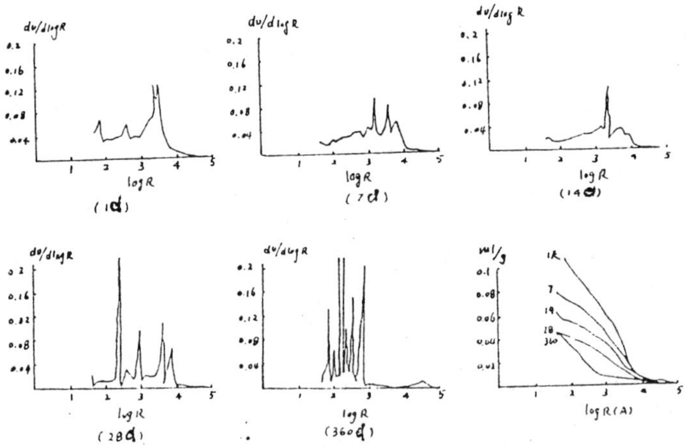

Figure 3. pore structure of mortar and total proposity of super low–alaklinity cement

We have used the first method. The test results are in Figure 4.

Many authors have investigated the resistance to chemical attack of glass fiber from various aspects, but no investigations which can systematically study the effect of different PH values on attack of glass fiber were published before. Our study can conclusively prove the following:

(1) The alkali–resistant glass fiber is good in alkali resistance, betterthan medium–alkali glass fiber in any conditions.

(2) You must take care that PH11 is the turning point indicating the alkali resistance of medium–alkali glass fiber when selecting the composited cement matrix. But this is only the necessary not the sufficient condition .(this will be discussed later) whether or not the durability of GRC is good can't be guranteed only with this condition.

3.3.2. chemical compositions and micro–structure of glass fiber

Glass is a special state forming between crystal and liquid Silicon–oxygen tetrahedron is the structural unit to form silicate glass. With the increase of O / Si ratio, Silicon–oxygen network. such as chain, layer and framework structre will be formed. And the oxidates of alkali metal and alkali earth metal spread in the gap of it in ion state of R^+or R^{2+}. In the silicate glass of sadium and calcium, network is a composite formation of layer and framework structure, mainly layer structure. Figure 5 is the structure.

In the silicate glass of sodium and calcium, Al_2O_3oxidate is different from R_2O, RO and SiO_2because Al^{3+}cannot only form aluminium–oxygen tetrahedron with 4 coordination to replace part of $[SiO_4]$ to enter network for repairing and consoliolating the network (see Figure 6) but also aluminium–oxygen octahedron with 6 coordination filling in the gap of network Generally, when $Na_2O / Al_2O_3$is > 1 or batching with feidspar, Al^{3+}has 4 coordinations and enter the framework in the unit of alumium–oxygen tetrahedron $[AlO_4]$.

The chemical compositions of medium–alkali glass fiber of our country can satisfy this requirement. Its structure obviously belongs to that showed in Fig.6. We are the first who use laser Raman spectrum to analyse GRC to verify this view. The analysing results indicate that about 5% Al^{3+}can form$[AlO_4]$ and enter the network and about 1% Al^{3+}form $[AlO_6]$ and fill in the gap of network.

3.4. Ourability of Medium–alkali Glass Fiber Reinforced Super Low–alkalinity Cement

3.4.1. The hydrates of super low– alkalinity cement can protect medium–alkali glass fiber

When hydrated, super low–alkalinity cement will produce a large amount of $Al(OH)_3$which can protect medium–alkali glass fiber or release the corrosion.

$Al(OH)_3$, a typical amphoteric oxidate, has the following ionization equation

$$H_2O+AlO_2^-+H^+ \rightleftharpoons Al(OH)_3 \rightleftharpoons Al^{3+}+3OH^-$$

acidic ionization basiz ionization

When the PH value in solution is > 7 , $Al(OH)_3$will be acidic ionization, which can produce $H^+$to further decrease the PH value of cement paste . No doubt , this can benefit glass fiber from not being corrosively attacked. Otherwise, ionization process can produce big ionic group of AlO_2^-which can't easily enter the gap of glass fiber network but will adhere the surface of glass fiber , just like a screening layer, to resist the exchange of ions.

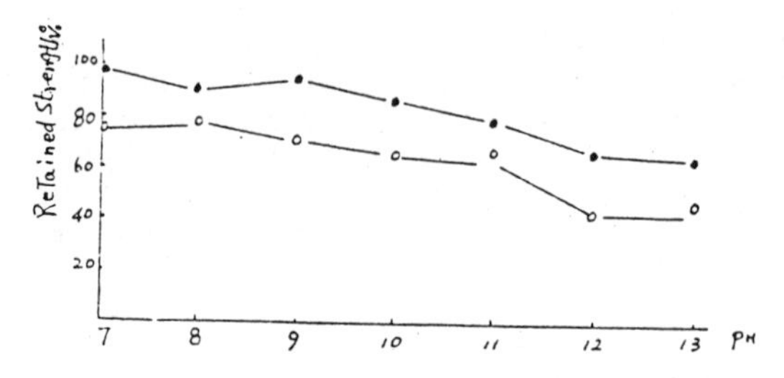

Figure 4. the retained strength after being attacked in the solutions with different PH values.
●—alkali–resistant glass fiber
○—medium–alkali glass fiber

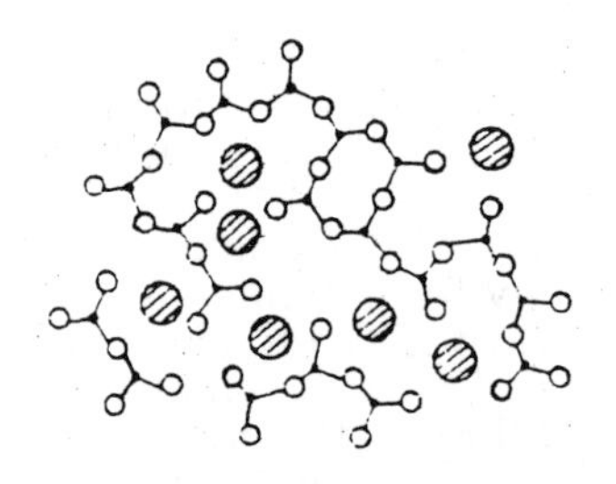

• Si^{4+} ○ O^{2-} ● R^{+},R^{2+}
Figure 5. glass network structure of R_2O–RO–SiO_2

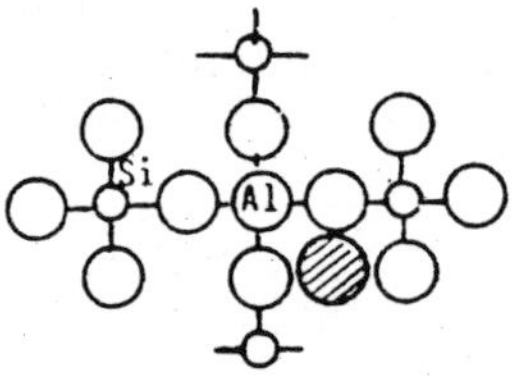

Figure 6. [AlO_4] insteady of [SiO_4] to enter network structure

This idea has been proved through macro and micro analysis .
(1) The retained strength of medium–alkali glass fiber in $Al(OH)_3$solution is the highest.

The medium–alkali glass fiber is laid in the sdutions of new $Al(OH)_3$, old $Al(OH)_3$, saturated solution of $Ca(OH)_2$and distilled water respectively for determining the retained strength (new $Al(OH)_3$refers to the one prepairing at laboractory in time and old $Al(OH)_3$is bought at the chemist's shop). The test results show that the best one is new $Al(OH)_3$see Figure 7.
(2) After one–year accelerated corrosive test at 50℃ the structure of glass fiber inside the small sheet is analysed by Raman spectrum (omit the picture). The results indicate, that, aluminium–oxyen octahedron [AlO_6] in glass fiber can reach about 8% and aluminium–oxygen tetrahedron [AlO_4] in network frame remain same amount, which has forcefully confirmed from micro–structure that the $Al(OH)_3$can protect the medium–alkali glass fiber and explain why with the increase of [AlO_6] and remedy of the existing surface defects the strength of glass fiber in the solution of $Al(OH)_3$can be higher than that of the original serength.

3.4.2. The appearance and interface of medium–alkali glass fiber inside the small sheet observed by electron microscope

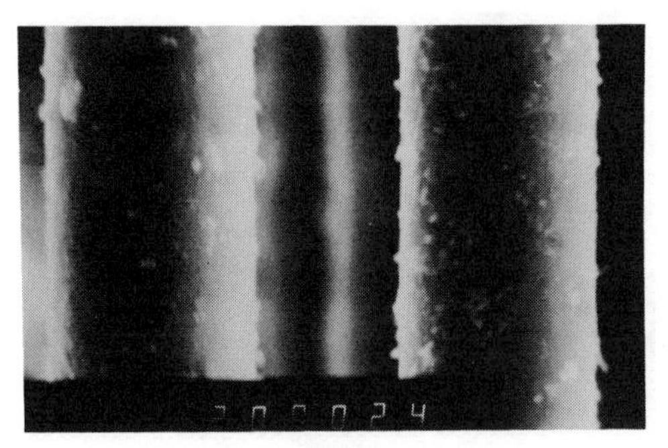

×2000
50℃ RH > 95%(one year)
picture 1

×2000
50℃ RH > 95% (one year)
picture 2

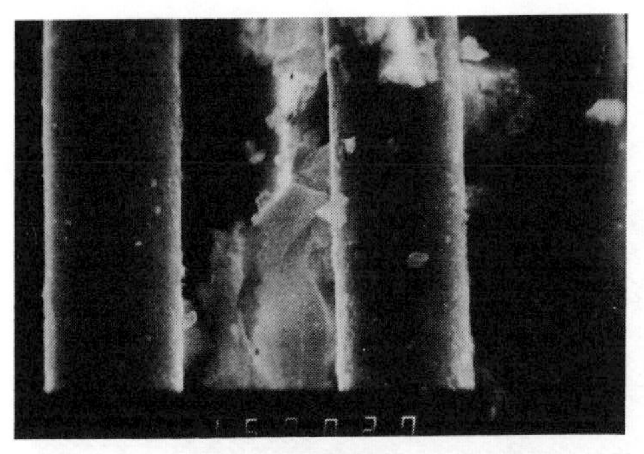

×2000
50℃ RH > 95%(16 months)
picture 3

×1500
50℃ RH > 95%(16 months)
picture 4

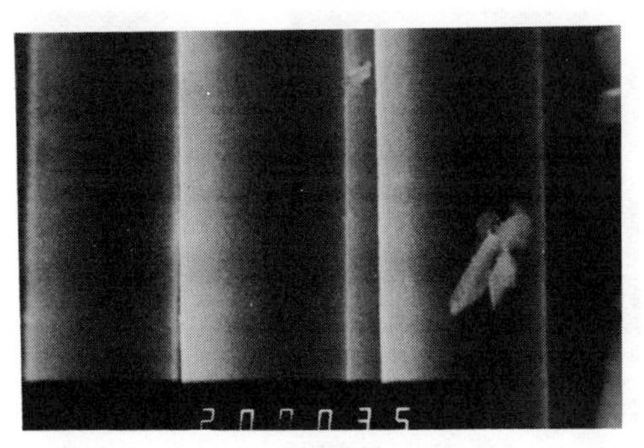

×2000
50℃ RH > 95%(18 months)
picture 5

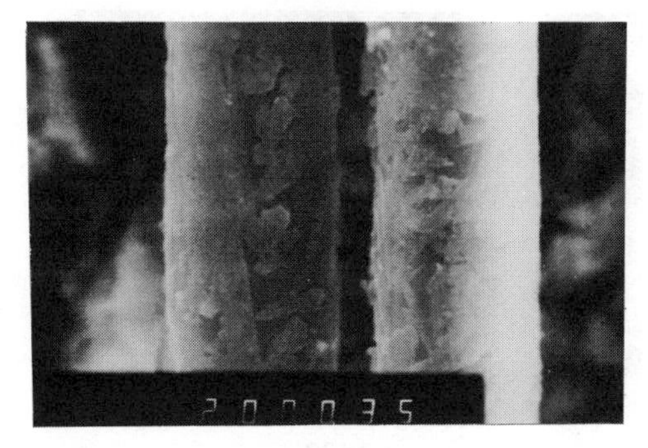

×2000
50℃ RH > 95%(18 months)
picture 6

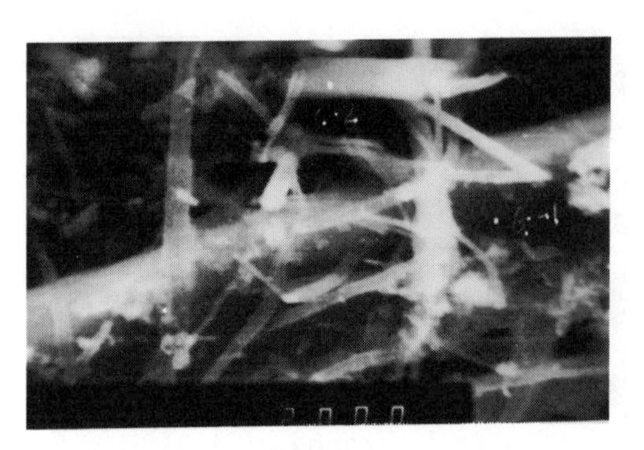

× 2000
900℃ (heating two times)
picture 7

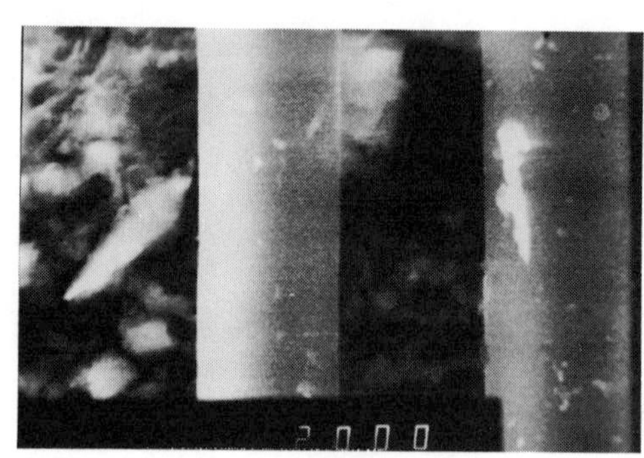

× 2000
100℃ RH100%
picture 8

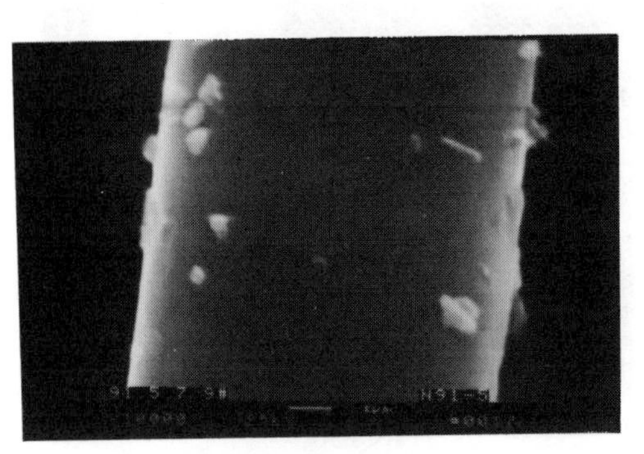

× 10000
in water 9 years
picture 9

× 2000
in water 9 years
picture 10

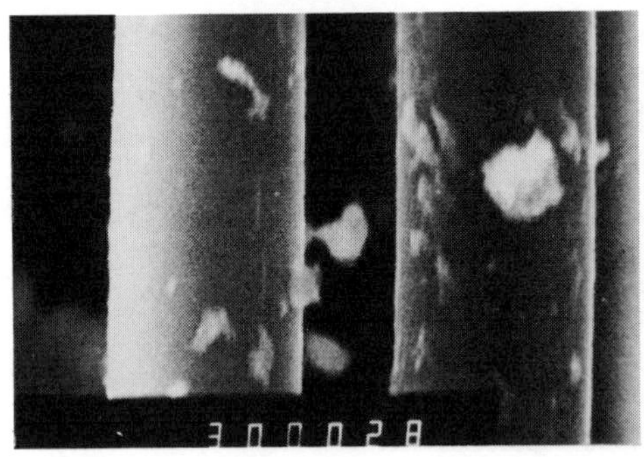

× 3000
in water 10 years
picture 11

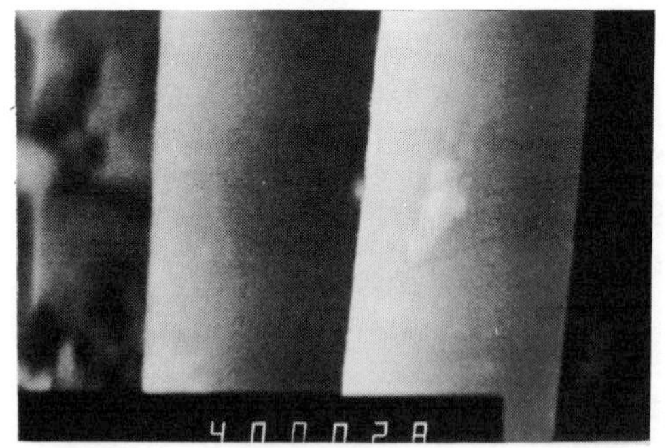

× 4000
in water 10 years
picture 12

The twelve pictures above are the micro structure of glass fiber suffering from different corrosions. The glass fiber has good appearance and no corrosive signs.

3.3.3. retained strength (%) of medium–alkali glass fiber in super low–alkalinity cement paste

(1) The retained strength (%) of medium–alkali glass fiber in super low–alkalinity cement paste under the temperature below 80℃ is in Figure 8 and 9 (△ marks are the super low–alkalinity cement paste)

At the same time, the strength changes of medium–alkali glass fiber and alkali–resistant glass fiber in I–type low–alkalinity cement paste, saturated $Ca(OH)_2$and portland cement paste have been measured, The results show that the retained strength of medium–alkali glass fiber in the super low–alkalinity cement paste is not only higher than that of in saturated $Ca(OH)_2$, portland cement paste and I–type low–alkalinity cement paste but also higher than alkali–resistant glass fiber′s.

An interesting phenomena observed at 80℃ immersion test is that either the medium–alkali or the alkali–resistant glass fiber will remain relevently higher strength in saturated $Ca(OH)_2$than in portland cement paste because of the saturated $Ca(OH)_2$. But further investigation should be carried out to vertify whether or not the phenomena above can be explained that the corrosion of glass fiber from portland cement is reduced to the $Ca(OH)_2$produced in hydration. We think that it is worthly impantant to pay attention to the general alkalinity of portland cement and the catalytic effect of K^+, Na^+.

(2) The retained strengths of glass fiber of long–term cement immersing at 50℃ are in Figure 10.

3.4.4. Retained Strength of Medium–alkali Glass Fiber Reinforced Super Low–alkalinity Cement Sheet Under the Hydrothermal of 50℃ (see Figure 11)

3.4.5. Retained strength of medium–alkali glass fiber reinforced super low–alkalinity cement mortar after long–term water immersing,

cement / sand 1:1, water / cement 0.35

The test results are in Figer 12.

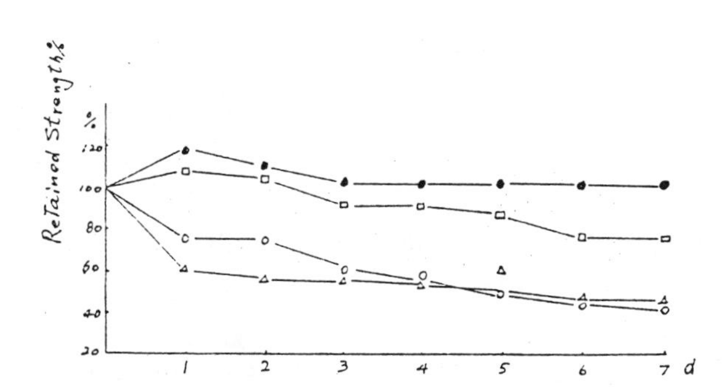

Figure 7. the retained strength of medium–alkali glass fiber in different solutions (80℃)

●—new $Al(OH)_3$ □—old $Al(OH)_3$

○—saturated $Ca(OH)_2$ △—distilled water

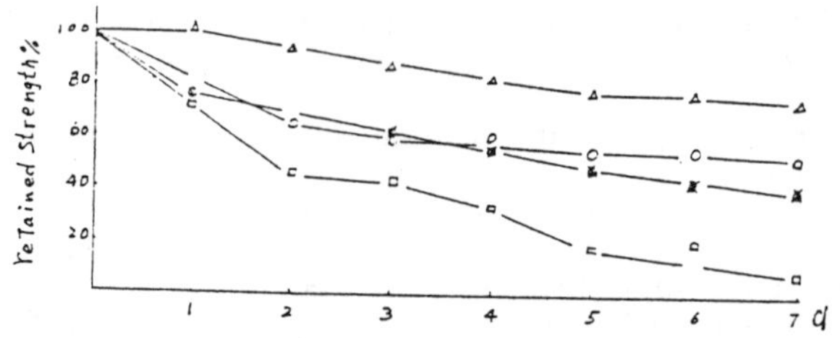

Figure 8. the retained strength of medium–alkali glass fiber in different cement pastes (80℃)

△—super low–alkalinity cement ■— saturated $Ca(OH)_2$solution

○— Ⅰ–type low–alkalinity cement □— portland cement

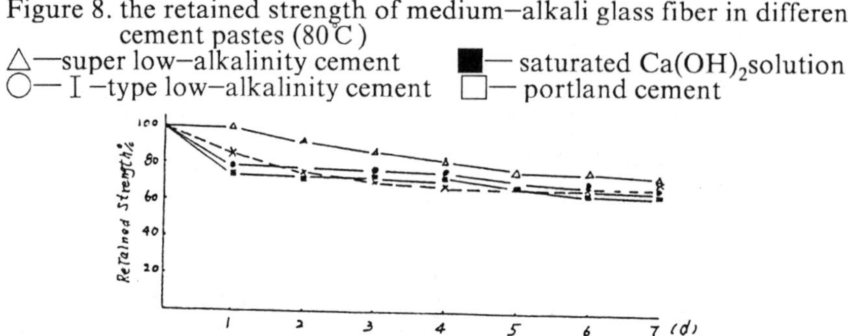

Figure 9. Comparation of retained strength between the medium–alkali glass fiber in super low–alkalinity cement and the alkali–resistant glass fiber in other cement postes (80℃)

△ super–medium–alkali glass fiber ■ portland–AR–glass fiber

● Ⅰ–type–AR–glass fiber × –$Ca(OH)_2$–AR–glass fiber

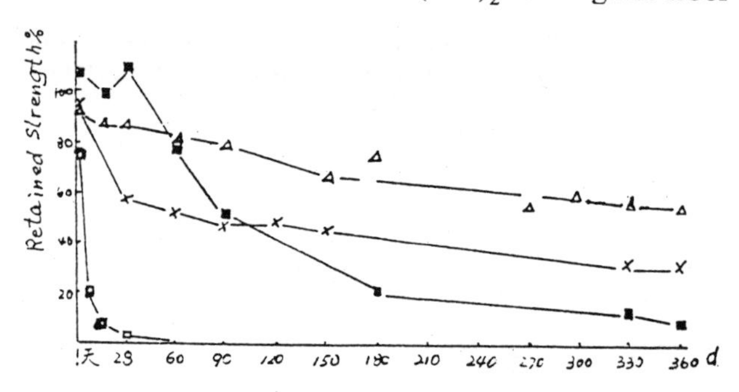

Figure 10. the retained strength of glass fiber immmersing in different cement pastes

(50℃ solid to liquid ratio 1:4)

□ OPC–medium alkali–glass fiber ■ OPC–AR–glass fiber

× Ⅰ–type–medium alkali–glass fiber △ super–medium alkali–glass fiber

notice:(1) The super low–alkalinity cement paste are prepared acccording to solid / liquid

(2) The solid to liquid ratio of Ⅰ type cement and portland cement paste are 1:10 All the data are extracted from Xue Junganetal "Effect of Low–alkalinity Cement on Medium–alkali Glass Fiber" (Journal of the China Silicates Society No.9,1981)

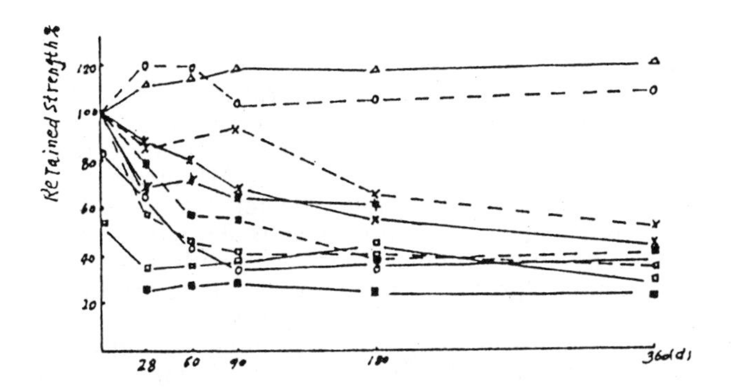

Figure 11. time–retained strength for different glass fiber sheet
△ super–medium alkali–glass fiber   × early strength–AR–glass fiber
···○ I –type–AR–glass fiber   —○ I –type–medium alkali–glass fiber
···■ modified OPC–AR–glass fiber   ···□ OPC–AR–glass fiber
—■ modified OPC–medium alkali–glass fiber   —□ OPC–medium alkali–glass fiber
—¤ early strength–medium alkali–glass fiber (coated two times)

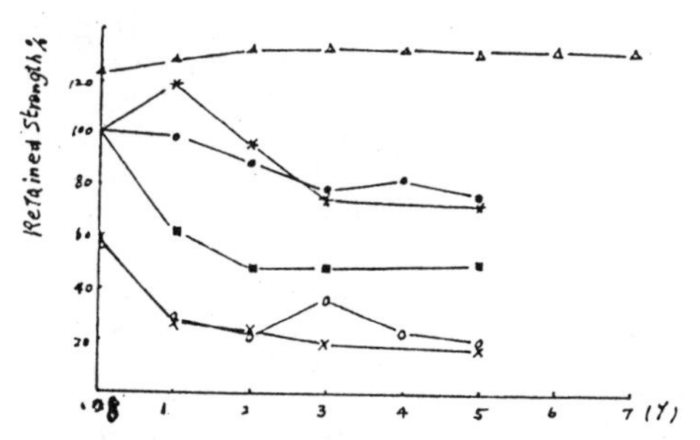

Figure 12. time–retained strength of different glass fiber sheet (long–time water immersing)
■ OPC–AR–glass fiber   △ super–medium alkali–glass fiber
● early strength–AR–glass fiber   ○ early strength–medium alkali–glass fiber
※ I–type–AR–glass fiber   × I–type–medium alkali–glass fiber

4 Discussion

4.1 The Corrosive Process of Glass Fiber in Cement Paste

The researchers all over the world hold a rather current idea to the corrosive process and principle of glass fiber in cement paste. Mr. Xue Junxuan, as their representive, has pointed out "The main reason that the glass fiber is corroded in cement is that the $Ca(OH)_2$producing during the cement hydration reacts with the silicon–oxygen frame in glass fiber and forms a new material–hydrated calcium silicate gel. This irreversible reaction can damage the structure of glass fiber, as a result lose their strength."

The laser Raman Spectrum has proved that aluminium–oxygen tetrahedron $[AlO_4]$ exists inside the frame of glass fiber, so that the damage principle of medium–alkali glass fiber in cement medium can't be simply explainted as "the $Ca(OH)_2$producing during the cement hydration reacts with the silicon–oxygen frame in glass fiber and damages the structure of glass fiber", otherwise it is also difficult to explain some phenomena of test.

We think that the corrosive process of medium–alkali glass fiber in cement can be generally divided into three stages:

the first stage:

Ions exchange between the alkali metal ion R^+on the surface of glass fiber and the H^+or H_3O^+in cement paste.

$$R^+\text{— glass}+H^+=H^+\text{—glass}+R^+\text{(solution)}$$
$$R^+\text{— glass}+H_3O^+=H_3O^+\text{—glass(solution)}$$

In fact, the reaction above if the hydration process of alkali metal oxides in glass, whose rate depends on the diffuse rate of ion R^+. The R^+dissolving out of the glass does not damage the structure of glass fiber but only make them lose part of strength. But if the dydration were continuons, the gap of network would increase and the structure would be loosed, which will finally result in the destruction of structure. In glass composting, R^{2+}or R^{3+}metal oxides and double–alkali or multi–alkali etc. are often used to produce pressing effect and decrease the diffusing rate of ion R^+. So, if Al^{3+}filled in the gaps as aluminium–oxygen octahedron $[AlO_6]$, the water resistance of glass fiber could be increased. Furthermore, the hydration of R^+can explain why in Figure 4 the retained strength of PH at 8 is higher than that of at 7.

While the R^+or R^{2+}at the surface of glass fiber are dissolving out, ion R^+in the cement paste are adhered to the surface of glass fiber and enter into their gaps. Researches on GRC before paid more attention to the corrosion from OH– but ignored that from varions cations. In fact, the corrosion of glass fiber is related to not only the density of OH– but also the cation types, i.e. various R ions can result in much different corrosion degree for the glass fiber at the same PH value.

The test results from W.M.Poley demonstrate that different cations have different corrosing degree in the basic solutions of the same PH value. Their order is as follows.

$$Ba^{2+} > Sr^{2+} > NH_4^+ > Rb^+ \approx Na^+ \approx Li^+ > N(CH_3)_4^+ > Ca^{2+}.$$

That $Ca^{2+}$is the smallest can be explained as:

(1) $SiO_2$is difficult to be dissolved in $Ca(OH)_2$

(2) The solubility of new–born $CaSiO_3$is low.

We think that it is proboble the temperature can influence the solubility of $Ca(OH)_2$a bnormally.

the second stage:

With the deepening of the corroded surface, the frame of glass fiber will be disrupted. Bond breaks at aluminium–oxygen tetrahedron[AlO_4] first, i.e. [AlO_4] changes to [AlO_6], then transfers from network to gap.

The view above is different from that before. But if considering the bond strength and the stability of $Al_2O_3$and SiO_2in the solutions with different PH values, we can understand why Al–O bond breaks earlier than Si–O bond.

First, the sinde bond energy of Si– O is 106 kcal / mol, and Al–O is 79–101 kcal / mol. The second, we can compare the stability of $Al_2O_3$and SiO_2in various solutions with different PH values. Figure 13 and 14 are the therodynamic results of $Al_2O_3$and SiO_2, respectively.

In the solutions, the stable values of $Al_2O_3$and $SiO_2$are about 10.8 and higher than 12, respectively (the paper only discuss the range of basic solutions), which is somewhat different from those, generally 10.6–13.0 for Al_2O_3, 12–13 for SiO_2, in relevent glass or chemical handbook. They depend on the ageing degree (crystallized degree) of born oxides, (new born oxides with lower PH values) which can explain why the retained strength of medium–alkali glass fiber sharply when PH value is higher than 11 in Figure 4.

the third stage:

With the co–effect from many aspects, the frame [SiO4] will be damaged finally. First, some gaps are formed in glass fiber when the corrosion proceeds to a certain degree, then H_2O will enter inside the structure of glass fiber.

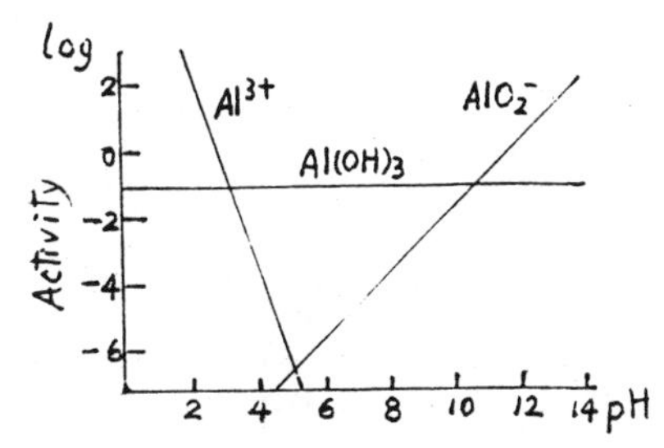

Figure 13. the stabilities of Al_2O_3in the solutions with different PH values different PH values.

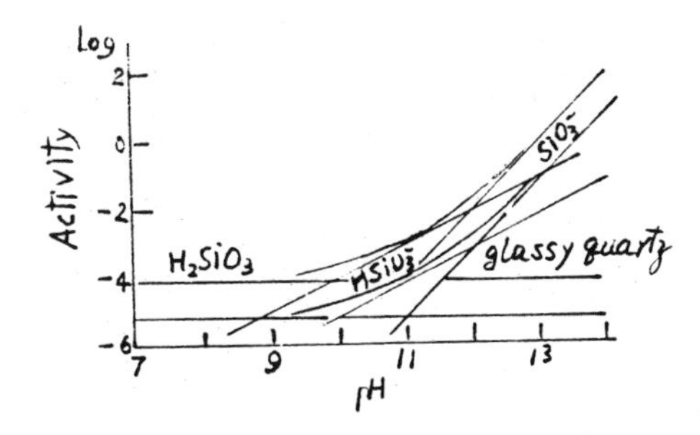

Figure 14. the stabilities of SiO_2in the solutions with different PH values different PH values.

$$-\overset{|}{\underset{|}{Si}}-O-\overset{|}{\underset{|}{Si}}-+H_2O=2(-\overset{|}{\underset{|}{Si}}-OH)$$

With the proceeding of the hydration, all the existing four oxygen bridges sounding the atom Si change to OH because H_2O can destroy the silicon–oxygen skeleton directly. The product $Si(OH)_4$is a polar molecules which can polarize the other molecules sounding and adsorb them to form $Si(OH)_4.nH_2O$ (or $SiO_2.XH_2O$). That is a highly diffused SiO_2-H_2O system usually called silica gel. Except parts of them are dissolved in water solution, most silica gel will adhere to the glass surface to form a loose layer of film where the diffusing speed of R^+or H_2O is much more rapidly than that in no corroded glass. So the corrosion will be proceeding until all the glass fiber is damaged unless the corrosive media of R^+or H_2O could not be provided continuonsly.

The destructive principle of cement paste to glass fiber is similar to that of strong alkali solution to glass fiber. The strong alkali solution destories the glass with the similar way to that of a small amount of water to glass for a long time. Both were imputed to water. The hydration and neutralization of H+ or OH– to the film of silica gel react stronglier in basic solution with high OH– density.

Long–term corrosion of small amount of water to glass is much more serious than that in fluiding water became of the corrosion medium of water.

$$-\overset{|}{\underset{|}{Si}}-ONa+H^++OH^-\xrightarrow{\text{hydration}}HO-\overset{OH}{\underset{OH}{Si}}-OH+NaOH$$

silica gel

$$Si(OH)_4+NaOH\xrightarrow{\text{neutralization}}[Si(OH)_3O]Na+H_2O$$

The three stages above can not be separated, completely for the first and the second stage, certainly became the alkalinity of corrosive of media.

As we all know, all the chemical reactions begin from the interfaces. The surface defects can increase the cohesive strength between glass fiber and cement but they also can increase the corroding speed. So it is best for GRC plants to choose new–making glass fiber.

4.2 Brittleness of GRC

Alkali–resistant glass fiber and medium–alkali glass fiber reinforced portland cement has another problem of brittleness after weather unexposing for 3–5 years, so have the medium–alkali or alkali–resistant reinforced early strength low–alkalinity cement (see pictures 13–19).

$\times 1000$
medium–alkali glass fiber–portland cement
50℃ RH>95% 7 days
picture 13

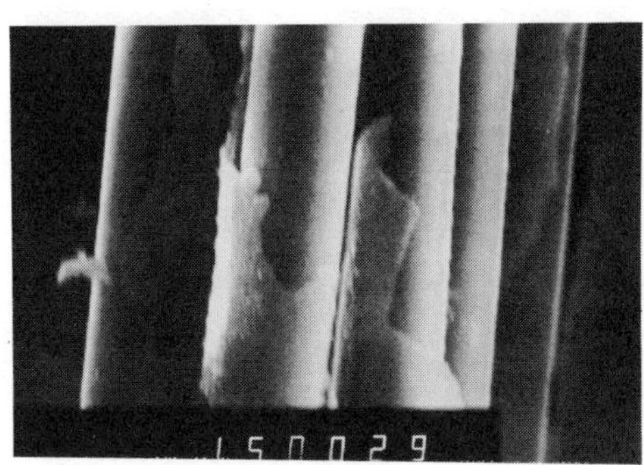

$\times 1500$
medium–alkali glass fiber–portland cement
50℃ RH>95% 7 days
picture 14

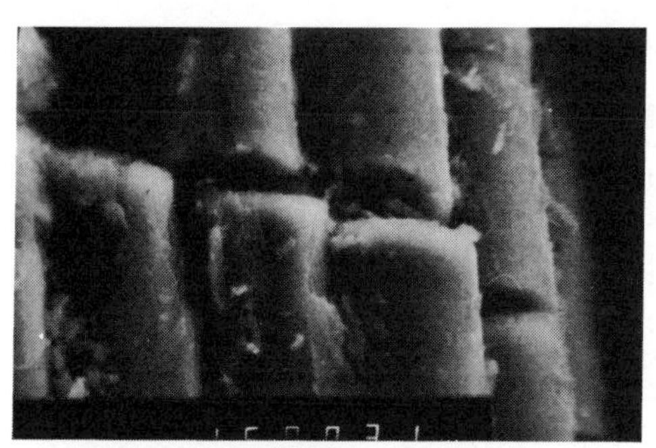

$\times 1500$
Ar–glass fiber–portland cement
50℃ RH>95% 9.5 months
picture 15

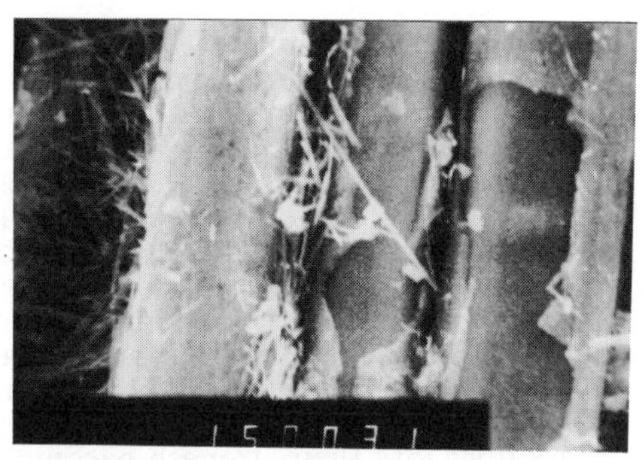

$\times 1500$
Ar–glass fiber–portland cement
50℃ RH>95% 9.5 months
picture 16

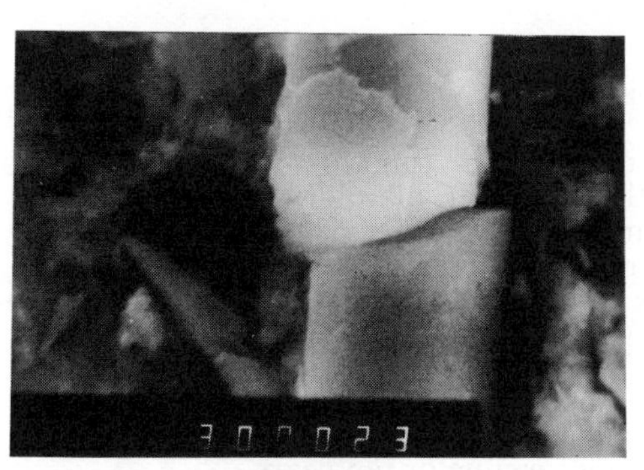

$\times 3000$
medium–alkali glass fiber–early strength low–alkalinity cement
50℃ RH>95% 5 months
picture 17

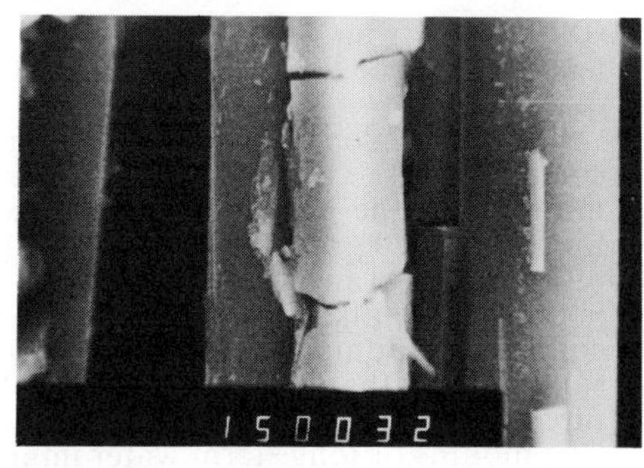

$\times 1500$
medium–alkali glass fiber–early strength low–alkalinity cement
50℃ RH>95% 9 months
picture 18

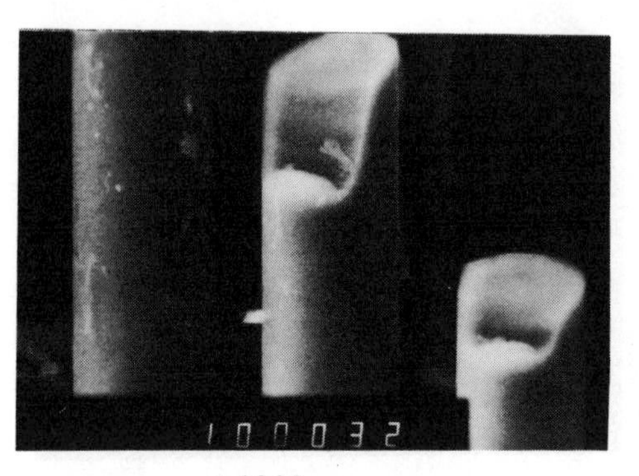

× 1000
medium–alkali glass fiber–early strength low–alkalinity cement
50℃ RH > 95% 9 months
picture 19

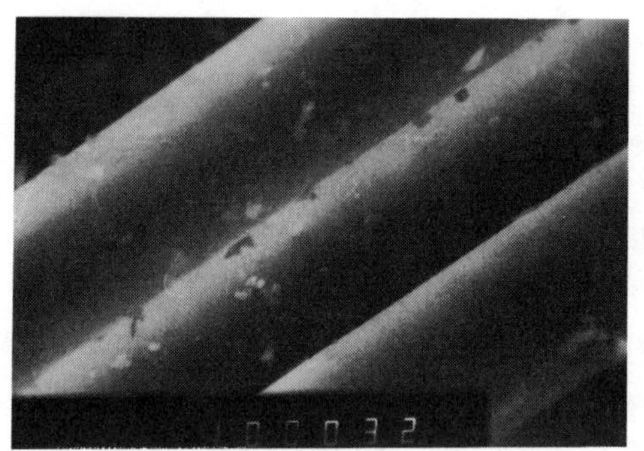

× 1000
Ar–glass fiber–early strength low–alkalinity cement
50℃ RH > 95% 9 months
picture 20

"For the principle of brittleness, there isn't a consensus opinion now. Some researchers think the main reason is that the alkali–resistant glass fiber is corroded by basic solution coming from the holes on cement matrix. Some others opine that large amount crystal of calcium hydroxide are deposited inside the holes of alkali resistant glass fiber strands and grow sounding the filament, so the glass fiber strand lose their deformability. Another opinion is that both corrosions from basic solution and deposites of calcium hydroxide stone exist together. Although their views are not identical, increasing the alkali resistance of glass fiber, removing the calcium hydroxide stone and largely decreasing the alkalinity of cement liquid phase are thought to be the key for inceasing the durability of GRC"①. After investigating, we feel that for GRC composite, at the meanwhile that glass fiber lose their strength became of the corroding from cement matrix, the brittleness will be relevantly increased. They are the two reactions which occur at the same time and are interdependent but with different reacting speeds.

Notice
① adapted from Mr. Sheng Rongxi "Current State and Development Potential for Fiber Reinforced Cement Composites" No.10, 1986

The brittleness of glass fiber itself is a function to chemical comprositions. Generally, the larger the diameter of added cations, the more brittle the glass fiber is. For example, the loading of brittle fracture for Li_2O is 170 gram, while CaO is 70 gram. This is became the ionic radius of Li^+ is 0.78Å and the radius of Ca^{2+} is 1.06Å. No $Ca(OH)_2$ are discovered in super low–alkalinity cement, but a great deal of aluminium hydroxide colloid. The ionic radius of Al^{3+} is 0.57Å and its brittle fracture work is surely higher than that of Ca^{++} (no value has been found), so no brittle fracture of the glass fiber occurs after specimen's accelerated corrosion for sixteen months or long–term water immersing for nine or ten years.

5 Conclusion

5.1 An obvious turning point exists during alkali solution corrodes medium–alkali glass fiber. When PH value is less than 11, corroded degree is low and when PH value is larger than 11, corrosion will considerably increase.

Suffering corrosion of medium–alkali glass fiber in portland cement can not simply impute to $Ca(OH)_2$. It is also significantly related to water and the surface condition of glass fiber (such as micro craker, defect etc.) when considering the corrosion from ion OH–, we should pay more attention to the cations combining with OH– at the same time became different cations will corrode glass fiber in very different degrees.

The corrosion of medium–alkali glass fiber can be divided into three stages. The first is ion R^+or R^{++}in the structure network of glass fiber interact with ion H^+or H_3O^+in solution to corrode the surface of glass fiber. The second is that Al–O bond are broken by various influences. The breaking of the Si–O bond indicates the final stage. During this third stage,the structure network of glass fiber are destroyed completely. Then the glass fiber will lose the strength and tonghness sharply and suffer from brittle damage.

5.2 The hydrates of super low–alkalinity cement are $C_3A.3CS.32H$ and $Al(OH)_3$collide and no $Ca(OH)_2$. During hydration, PH value keeps low and a great deal of $Al(OH)_3$collide adhere to the surface of glass fiber to decrease corrosion from alkali solution densify the hardened cement mortar and increase the impermeability so that water and other media are not able to go through easily.

5.3 The corrosion of super low–alkalinity cement to medium–alkali glass fiber is much lighter than that of portland cement and early strength low–alkalinity cement to alkali–resistant glass fiber. The GRC small sheet made by medium–alkali glass fiber reinforced super low–alkalinity cement can remain its strength for seven years after being accelerated aging test under 50℃ and long–term water immersing test. No corrosion and fracture are observed by electron microscope. The composite wall panel (4955 × 3500mm) made by this material, half–wave tile (0.95 × 1.71m, 0.95 × 2.85m) can keep their good durability after five or six years.

references

1.解决玻纤增强水泥混凝土耐久性的正确之路 曹永康 1983.9

2.玻纤增强水泥的耐久性及其使用范围问题 曹永康 1984.2 中国建材院院刊

3.低碱度水泥对玻纤侵蚀机理的研薛君干等 79 年全国新材料经验交流会汇编

4.低碱度水泥对中碱玻纤作用机理的研究 薛君干 81.9 中国硅酸盐学报

5.论改善玻璃纤维增强水泥的长期耐久性的技术途径 沈荣熹 87.5 中国建材院院刊

6.Maiumdar 博士讲学详细提纲 陆慧棠等译 1987.9

7.玻纤的试验研究 陆慧棠 79 年全国新材料经验交流会汇编

8.有关玻纤增强水泥耐久性的几个问题探讨 杨庆吉 1986 年中国建材院学术年会

9.玻纤在水泥石中破坏机理的探讨 杨庆吉 1983 年院学术年会

10.超低碱度水泥的研究与应用 杨庆吉 86.12

11.超低碱度水泥作基材的 GRC 耐久性的研究 杨庆吉 86.12

12.The Handfook of glass Manufacture Volumel books for Zndustang,Inc.1974

13.The physical properties of glass D. G. Holloway 1973
14.Glaschemic Werner Vogre 1979
15.Fiber Cement and Fiber Concretes D.J.Hannant 1978
16.玻璃形成学邱关明,黄良钊编著 1987.11
17.Glas Natur,Strukur and Eigenscbaften H.Scholje 1977
18.纤维混成复合材料之应用 台湾 张志纯 1990.6
19.ガラス表面の物理化学 1979
20.Intre duction to the physical chemistry of the Victreous state P.BALTA.E.BALTA 1976
21.玻璃工艺原理 浙江大学等 1981.12
22.玻璃工艺学 武汉工业大学等 1982.1
23.1981 北京国际玻璃讨论会论文集
24.1981 北京国际玻璃讨论会论文集
25.GRC 玻纤增强水泥出国考察科技论文集 1983.4
26.GRC 专集(三) 1990.11
27.试论我国 GRC 工业的特色与现状 沈荣熹 1990.11

72 GLASS FIBER REINFORCED LIGHTWEIGHT CONCRETE MODIFIED WITH POLYMER LATEX

LIN WEIWEI and WANG HENG
Department of Materials, Zhejiang University, Hangzhou, People's Republic of China

Abstract

This paper discusses the effect of polymer latex on mechanical, micro-structural and durable properties of glass fiber reinforced concrete (GRC). By SEM, XRD, TG, DTA and mechanical properties measurement, it is found that polymer latex can reduce the content of calcium hydroxide in the composite, retard the corrosion of glass fiber and improve the durability of lightweight GRC. At the same time, the strength of the composite is greatly improved due to the good adhesion, toughness and pore-filling ability of the polymer latex film.

Introduction

Glass fiber possesses the characteristics of high mechanical strength, non-combustibility and cheapness, so there has been interesting in the use of glass fiber to reinforce the cement and concrete. It is known that the alkali-resistance ability of glass fiber is quite poor and its mechanical strength will decrease after corrosion by alkalis in cement. At present, the practical ways to improve the durability of GRC are: changing glass fiber composition to improve its alkali-resistance or using low-alkali cement. In order to make lightweight GRC and improve its durability, our work has concentrated on selecting low-density active aggregates and addition of polymer latex to obtain good quality concrete.

Experimental

The active aggregates used were floating beads extracted from fly ash. The cement was 425 ordinary portland cement produced by Hangzhou Cement Factory. Non-alkali glass fiber cloths were from Hangzhou Glass Factory. Polymer latexes were polyvinylidene di-chloride co-polymer (PVDC) and chloroprene rubber (CR).

These materials in proper ratios were mixed, and 2-ply glass cloths were bedded, then made into concrete boards.

Fibre Reinforced Cement and Concrete. Edited by R. N. Swamy. © 1992 RILEM.
Published by E & FN Spon, 2-6 Boundary Row, London SE1 8HN. ISBN 0 419 18130 X.

The cured concrete boards were cut to obtain the rectangular standard samples for the flexual strength measurement, the sample size was 120 x 15 x 10mm (According to Chinese Standard for GRP, GB1449-83).

Scanning electron micrographs were taken to analyse the micro-structure of the samples by JSM-T20. DTA/TG were done on DT-30 (Shimadzu Co.) at the heat rate 10°C/min. X-ray diffraction plots were obtained using D/MAX-RA (Rigaku Co.) to analyse the reactions in the composite system.

Results and Disscusion

1. Improvement of composite mechanical strength by polymer latex.

More and more industries need lightweight, high mechanical strength concrete products. For this reason, floating beads were selected as aggregate. Floating bead is a hollow active sphere with high specific surface area, low density (0.2-0.4 g/cm^3), low thermal co-efficient, and the main constituent of beads are SiO_2 and Al_2O_3[1]. It is evident that there are alkali-aggregate reactions and CSH gel formed on floating bead surface (Fig.1). The products possess good thermal insulation, non-combustibility and lightweight characteristics.In order to get high mechanical strength, polymer latex was added into these systems. Table I shows the effect of polymer latex on mechanical strength. By virtue of the good adhesion, toughness and pore-filling ability of the polymer latex film, the flexual strength of the composites is greatly increased.

Table I Mechanical Strength of Glass Fiber Reinforced Lighweight Concrete Modified With Polymer Latex

Polymer Latex	Glass Fiber	Flexual Strength (Mpa)	Density (g/cm^3)
Without	Without	3.85	1.013
PVDC	Without	6.71	0.92
	With	8.30	0.95
CR	Without	5.22	0.81
	With	10.80	0.90

2. The effect of polymer latex on reaction, micro-structure and durability of lightweight GRC.

It is well known that during cement hydration, CSH gel is formed (Fig.2). At the same time calcium hydroxide about 6-12% of total cement weight, is liberated. Although

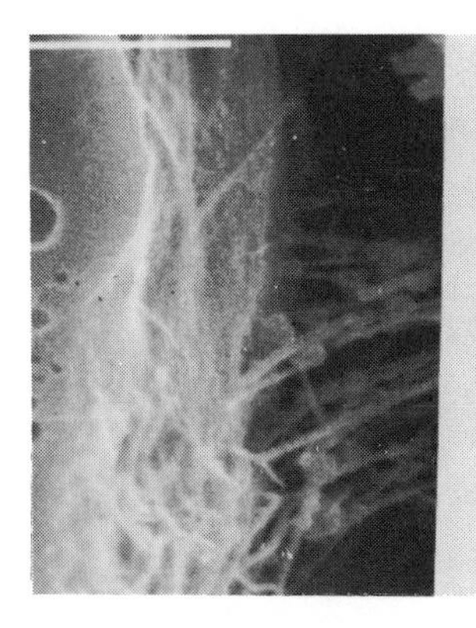

Fig. 1. Needle-like hydrate products (CSH) on floating bead surface, 7 days. SEM x8000.

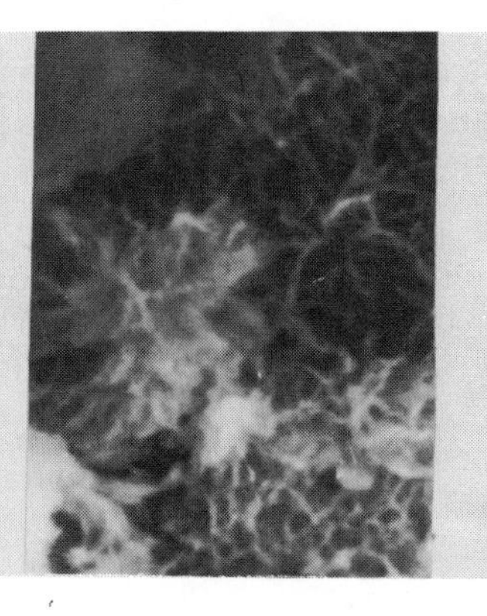

Fig. 2. Hydrated product of floating bead/cement system, 28 days. SEM x4000.

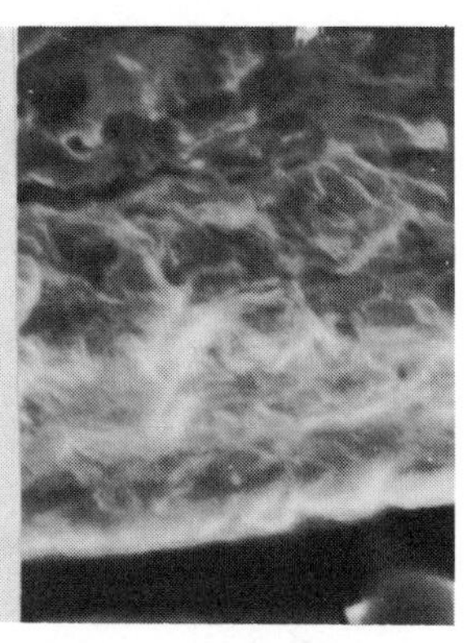

Fig. 3. Surface of glass fiber in composite after immersion in $Ca(OH)_2$ saturated solution, 30 days. SEM x4000.

Fig. 4. Hydrated products of floating bead/cement system modified with PVDC, 28 days. SEM x4000.

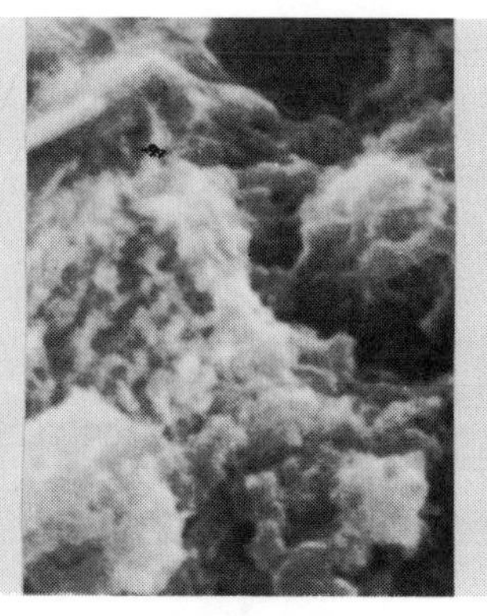

Fig. 5. Hydrated products of floating bead/cement system modified with CR, 28 days. SEM x4000.

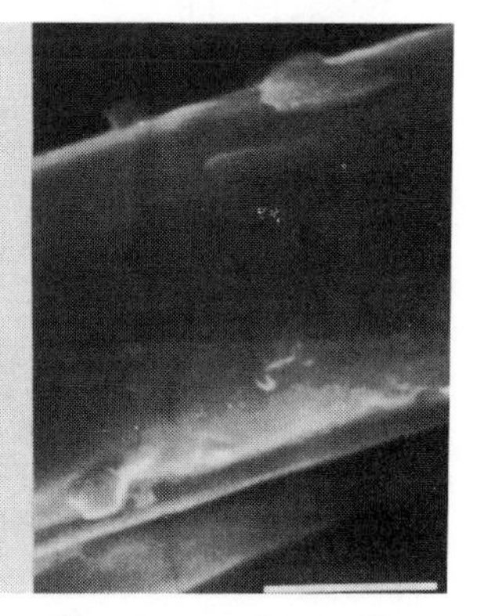

Fig. 6. Surface of glass fiber in composite modified with PVDC after immersion in $Ca(OH)_2$ saturated solution. SEM x4000.

alkaline materials such as $Ca(OH)_2$ can react with floating beads to form CSH gel, it can also corrode glass fiber to produce calcium silicate and sodium silicate (see Fig.3). The result is that the composite strengths decrease graduately.

When polymer latex was added, water in polymer latex could take part in the hydration reaction of cement, and polymer latex film was formed [2]. Polymer latex film can envelope glass fiber,floating bead and some unhydrated cement particles (Fig.4, 5), protect glass fiber surface and retard the corrosion of glass fiber. Fig. 6 shows that the polymer film envelopes glass fiber, impedes the contact between the fiber and the alkalis, so hinders the corrosion of glass fiber. Therefore glass fiber's alkali-resistance ability is increased and the durability of the composite is improved.

Due to the polymer latex film's effect of envelopment and pore-filling ability, the cement hydration is retarded

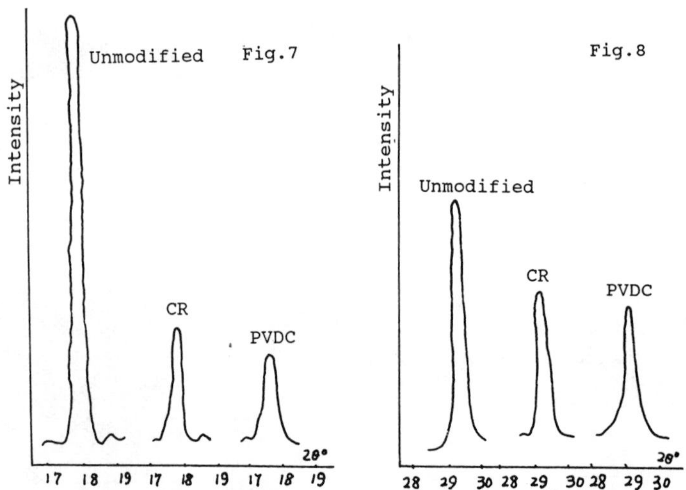

Fig.7 Typical plots of X-ray diffraction patterns for the charaterization of $Ca(OH)_2$ in unmodified and PVDC, CR modified floating bead/cement systems.

Fig.8 Typical plots of X-ray diffraction patterns for the charaterization of $CaCO_3$ in unmodified and PVDC, CR modified floating bead/cement systems.

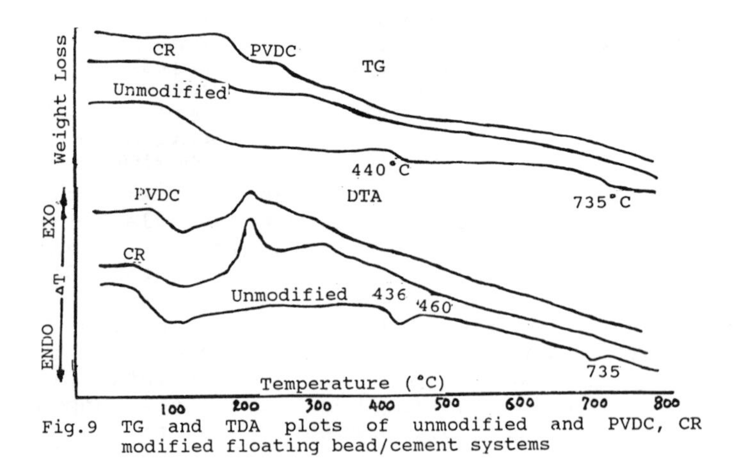

Fig.9 TG and TDA plots of unmodified and PVDC, CR modified floating bead/cement systems

and the amount of calcium hydroxide decreases, it can be evidenced by XRD and DTA/TG analyses. The characteristic X-ray diffraction peaks of $Ca(OH)_2$ and $CaCO_3$ are about $2\theta=18^\circ$ and 29°, respectively. From Fig.7, 8, it is found that the intensity of characteristic peaks of $Ca(OH)_2$ or $CaCO_3$ in the modified systems are lower than that in the unmodified systems. This is in agreement with published report[3].

DTA/TG analyses are shown in Fig.9. Glass fiber and floating bead are thermally stable and they do not decompose before 800°C. But CSH, $Ca(OH)_2$, $CaCO_3$, polymers and other hydrates are thermally unstable, they will decompose, oxidize, cross-linking, and lose water under elevating temperatures. It is known that the decomposition temperatures of $Ca(OH)_2$ and $CaCO_3$ are about 450°C and 700°C respectively. From Fig.9, it can be seen that without the polymer latex, there are endothermic peaks of $Ca(OH)_2$ and $CaCO_3$ decomposition at about 450°C with 5 wt% loss and 725°C with 2 wt% loss respectively.But after materials are modified with the polymer latex such as PVDC or CR, the endothermic peaks and weight-losses at above temperatures are not obvious, it shows that the content of $Ca(OH)_2$ is reduced. As a result, the corrosion of glass fiber is effectively retarded, and the durability of glass fiber reinforced lightweight concrete is improved.

Conclusions

From above analyses and explanation, it is very clear that polymer latex such as PVDC or CR can retard the cement hydration and decrease the content of calcium hydroxide. Polymer latex film possesses good adhesion, toughness and pore-filling ability, so that the strengh of the composite can be greatly improved and the corrosion of glass fiber can be impeded.

Reference

1. Lin Weiwei, Sun Guofang, Zheng Min, Proceeding of the Sixth International Congress on Polymer in Concrete, International Academic Publishes, Shanghai, P.R. China, Sept.24-27, 1990, pp.487-491.
2. Y.Ohama, ACI Materials Journal, pp.511-518, Nov. - Dec., 1987.
3. M.U.K.Afradi, Y.Ohama, M.Zafar.Iqabal, K. Kemum, The International Journal of Cement Composite and Lightweight Concrete, Vol.11, No.4 (1989), pp.235-244.

* The Project Supported by National Natural Science Foundation of China

73 EFFECTS OF QUALITY CONTROL PRACTICE ON GFRC DURABILITY – US EXPERIENCE

D. M. SCHULTZ, J. J. ROLLER, T. L. WEINMANN and R. G. OESTERLE
Construction Technology Laboratories, Inc. (CTL), Skokie, Illinois, USA

Synopsis
Thin walled, non-load bearing exterior building panels of glass fiber reinforced concrete (GFRC) are manufactured by the spray-up process. Controlled factory conditions with strict attention to quality control from original design to stripping and handling of panels are required to assure manufacture of a high quality product with long-term durability. Lessons learned in evaluation of GFRC facade failures, as well as observations made during quality assurance monitoring of new GFRC panel applications, have formed the basis for this first hand experience on the effects of quality control practice on GFRC durability.
Keywords: Durability, Glass-fiber Reinforced Concrete, Building Cladding Panels, Quality Control, Failure Analyses, Mix Design, Compaction, Thickness Control, Curing Practice, Carbonation

1 Introduction

GFRC has been employed as facade cladding in the United States of America for over a decade. U.S.A. manufacturers have pioneered a number of innovative manufacturing techniques for GFRC facades including procedures related to mix design, curing, use of facing materials, and most notably, the introduction of the steel stud framing system. These innovations and adaptations of the early European technology have allowed GFRC to find a proper niche in the American building market. Unfortunately, this growth has not been without its setbacks, not unlike many new construction materials.

The authors have had the opportunity to investigate a number of major U.S. failures of GFRC facade cladding systems. It is important to note that failure is defined by the authors as "a condition of the facade which is leading to or has led to a loss of adequate or intended function prior to the intended life span of the product." This paper focuses on the experiences gained in these investigations and the role that quality control practice in manufacture has played in the lack of durability exhibited in these failed facade cladding systems.

2 Facade investigations

A number of extensive facade investigations have been performed by Construction Technology Laboratories, Inc. (CTL) throughout the United States. In all instances the investigation was pursued as a result of premature or unexpected deterioration of the GFRC facade system. While it would be of some benefit to discuss in detail the exact

Fibre Reinforced Cement and Concrete. Edited by R. N. Swamy. © 1992 RILEM.
Published by E & FN Spon, 2-6 Boundary Row, London SE1 8HN. ISBN 0 419 18130 X.

locations of each of the specific investigations, it is not possible to do so since the majority of the investigations involve litigation. On the other hand, the various observations and conclusions can be discussed in more generic terms so as to broaden the knowledge and understanding of aging and durability of GFRC.

In nearly all instances, premature problems with durability of the glass fiber reinforced concrete have been associated with early cracking. However, the cracking was not necessarily noticeable with the unaided eye. In some instances, fine interconnected microcrack networks allowed water penetration through the facade panels to such an extent as to cause damage to interior elements. In other instances, the premature durability problems were observable through bowing or warping of the panels themselves. Often such distortion was a precursor to readily visible cracking as natural aging of the GFRC took place. None of the investigations performed to date involved problems caused by excessive interpanel joint movement.

Two additional categories of durability-impairing mechanisms have been observed in the United States over and above the general cracking phenomenon more commonly seen. The first is discoloration of the exterior surface of the facade panel which has more often than not been considered a simple aesthetic problem but has more recently been associated with excessive carbonation of the GFRC facade panel. The second is delaminations between the GFRC and facing material adhered to the GFRC.

Panel cracking, distortion, delamination, and discoloration often occur simultaneously, making the determination of the proximate cause of the deterioration and durability problem more difficult to ascertain. In some instances the reported problems began as early as the construction phase of the building. In other instances the reported problems began a number of years after the building was completed. In these latter instances, the panel defects were latent in nature in that they did not become apparent until the panel began to age.

3 Effects of quality control practice

Various aspects of the design and manufacture of the GFRC have been linked with durability problems noted in the field. While these are separated by category in this paper, it is recognized that they seldom occur as isolated phenomena. It is most often the case that errors and/or omissions in quality control practice occur in a number of categories conjunctively. For convenience, however, they will be discussed as isolated phenomena.

3.1 Mix design

While the mix designs for the GFRC are reasonably uniform throughout the U.S. industry, there are variations from manufacturer to manufacturer. These variations in the GFRC mix design can result in significant variations in the material properties. Significant variation in the material properties of facing mixes must also be anticipated. Use of "typical" values for certain material properties in panel design can result in gross inaccuracies in magnitude of calculated stresses causing severe problems in the future durability of the product. Material properties must be clearly defined prior to design and manufacture of the panels. An adjunct to this situation occurs when a cementitious facing material is used in combination with the GFRC. Compatibility of the various material properties between the two materials must be firmly established prior to the manufacture of the panels and must be included in the design calculations. Instances have occurred where improper or incomplete material investigations were performed resulting in basic incompatibilities between the facing material and the GFRC from the onset. This situation can lead to warping, cracking, and in some instances, even delamination of the facing material from the GFRC. Such defects will often remain latent until the material is aged.

3.2 Compaction

Lack of compaction of the GFRC has not been noted to be a widespread problem in the investigations performed by the authors. Rather, it occurs in isolated instances and generally causes either leaking of the facade at those locations and/or delaminations within the material as a result of poor compaction between the layers. Poor compaction can also result in increased depth of carbonation causing material compatibility problems. Compaction problems are often associated with panel skin discontinuities such as returns or reveals. Therefore, it is apparent that special attention must be given to these areas in manufacture.

3.3 Thickness control

Lack of proper thickness control has been observed in every facade investigation performed by the authors. However, in general, the lack of thickness control has not been in and of itself the primary cause for the observed deterioration. Failure of design engineers to adequately address the effects of panel skin movements resulting from thermal and moisture variations has resulted in excessive skin stresses even in areas where the design skin thickness is maintained or exceeded. Poor thickness control in conjunction with design deficiencies has often resulted in early panel cracking, occurring well before the GFRC has significantly aged.

While specified panel tolerances have included thickness requirements for the GFRC and the architectural facing mix, the quality control requirements intended to assure that the panel tolerances are being met have been found to be inadequate in a number of instances. The simple pin gauge which has been used to ensure proper thickness works well for a single skin panel with no facing mix. However, this may not be an adequate technique if a cementitious face mix is used on the GFRC panel. In order to ensure proper GFRC thickness, separate thickness measurements of the facing mix and GFRC + facing mix must be made. More specifically, the thickness of the facing mix must be known and controlled in order to ensure proper GFRC thickness. As indicated in Fig. 1, special attention must be given to panel skin returns or reveals as it is particularly difficult to control thickness in these areas.

3.4 Curing

The 1981 Recommended Practice for Glass Fiber Reinforced Concrete Panels required a seven day moist cure. Subsequent to the issuance of that document, work was performed at CTL which determined that an acrylic thermoplastic copolymer dispersion could be used as a curing agent, thus obviating the need for a moist curing facility. This latter technique has been predominant in U.S. practice since the mid-1980s. However, prior to the introduction of the acrylic thermoplastic copolymer as a curing agent, the required seven day moist cure was not always performed adequately by the manufacturers in the United States. Consequently, premature durability problems which have been investigated by the authors, have in some instances been related to lack of adequate moist curing when the panels were first made. Obviously, if incomplete or inadequate curing has taken place, effects on strength of the in-situ product can cause both short and long term durability problems to occur. Certain investigations have suggested that lack of curing or improper curing techniques utilized in some structures have led to extensive and serious deterioration in the form of excessively deep carbonation of the GFRC and cementitious face mix, as shown in Fig. 2. Considering the variation in material properties between carbonated GFRC and uncarbonated GFRC, it is recognized that an incompatibility within the material itself

Fig. 1. Improper thickness control at panel skin reveal.

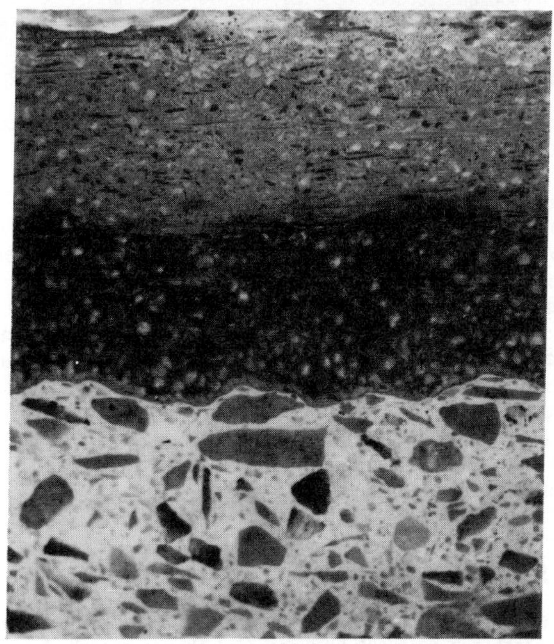

Fig. 2. Excessively deep carbonation of GFRC and facing mix.

can develop if the depth of carbonation becomes excessive and represents a major portion of the thickness of the GFRC.

Similarly, if the GFRC has been properly cured but the cementitious face mix has not been properly cured, then the face mix can carbonate excessively causing an incompatibility between the two materials to develop, regardless of whether or not the two materials were initially established to be compatible through proper mix design. As indicated in Fig. 3, when such phenomena are present, deterioration in the forms of severe cracking, bowing, and delamination of the constituent materials have occurred.

Fig. 3. Cracking and delamination of facing mix resulting from improper curing and other manufacturing deficiencies.

3.5 Stripping and erection stresses

In early GFRC applications, it was suggested that the high early strength of GFRC could be utilized to resist the high stresses associated with stripping and erection. If in fact this was allowed to occur, and the proportioned elastic limit (PEL) of the material was exceeded during stripping and erection, cracking of the cement matrix occurred. If conditions do not foster autogenous healing of these cracks, then the cracks will become latent defects, generally unnoticeable at the time of manufacture and erection, but will become evident as the GFRC naturally ages and loses strength. There is some evidence to suggest that autogenous healing can occur during very early stages of the curing process. However, CTL has performed investigations which have determined that cracking of the matrix which occurs more than a day after the panels are first manufactured will most probably not heal.

4 Retrospective analysis

In view of the above problems which have occurred as a result of ineffective or improper quality control, one must ask how can such things occur? An obvious culprit is inappropriate attention to industry-established quality control procedures. In fact, often this is the case; but a more important aspect which may not be fully understood by specifiers and manufacturers alike is that quite often quality control requirements and quality control technology in general often lag behind new developments in the application of the material itself. The introduction of facing materials onto single skin GFRC panels was an instance where quality control requirements and applicable technology lagged behind. Importance of issues such as thickness control and material compatibility were not thoroughly understood by specifiers and manufacturers. As a result, various problems related to ineffective or improper quality control practice occurred in various structures before proper quality control guidelines could be published. For the innovative manufacturer this is a risk assumed by him if in fact he proposes to use new techniques. On the other hand, the architect who specifies new or advanced techniques outside the common technology available to the industry is also assuming the risks in its application if the quality control related to the product cannot be established or properly ensured.

In retrospect, it has also been recognized by the authors and in the literature in general that accountability within the building professions, in particular the architectural and structural engineering practitioners, as opposed to the manufacturer, has not been clearly defined. In addition, relative to GFRC practice, technical communications between architects and structural engineers and the manufacturer and his engineer have been lacking. Even today, the architects and structural engineers often treat the design of GFRC cladding as if it requires little or no technical basis. As a consequence, a separation of the responsibilities and a separation of the tasks has occurred in the design analysis, construction, installation, and inspection and maintenance of GFRC in the United States. While the problems are particularly acute in this area for GFRC, they most certainly exist in the design, manufacture, and installation of nearly all cladding materials in this country. In general, structural engineers and architects with significant technical expertise only enter the picture when failures have occurred. As a consequence, structural engineers and architects need to recognize the importance of their participation in the design and detailing of GFRC, particularly where a failure of such an installation can lead to a safety hazard. Historically in the U.S. it has been assumed by the architect and structural engineer of record that the structural design of the GFRC remains in the hands of the manufacturer of the panels. While this is common practice and may be applicable for certain facade systems, it is not the case for GFRC facade systems.

5 Recommendations

While numerous detailed recommendations can be made relative to additional quality control requirements and quality control practice in general, the single most important aspect is a recognition of the importance of and commitment to proper quality control by the manufacturer himself. The manufacturer further needs to fully understand or appreciate the risks associated with departure from common practice. In the U.S. new application development in GFRC has been aggressive in the last decade. However, supporting research to establish the appropriateness of the new or innovative applications has been sorely lacking. While it can be stated that this is not unlike any new construction material, it has become apparent to the authors that the consequences of errors in this area are far more significant than other more forgiving facade materials.

6 Concluding remarks

While this paper has focused on negative aspects of quality control practice and problems related to durability of GFRC in the United States, it is acknowledged by the authors that numerous entities in the United States have sought to ensure the proper development and use of GFRC. In this vein, it should be mentioned that the Precast/Prestressed Concrete Institute (PCI) of the United States has recently issued a "Manual for Quality Control for Plants and Production of Glass Fiber Reinforced Concrete Products." This manual, which has been prepared as a guideline for the manufacture of glass fiber reinforced concrete, represents an extensive and concerted effort by this industry to ensure the proper use of GFRC and ensure the proper quality control necessary for its manufacture. The authors, having been involved in the development of this document, along with the principal subcommittee members working on the manual, feel that it represents a significant step in ensuring the proper use and manufacture of GFRC in the United States. The work of this subcommittee is hereby acknowledged by the authors.

7 References

Journal of Performance of Constructed Facilities, "Cladding Design: Whose Responsibility?," American Society of Civil Engineers, Vol. 5, No. 3, August 1991, pp. 208-217.

PCI Committee on Glass Fiber Reinforced Concrete Panels, "Manual for Quality Control for Plants and Production of Glass Fiber Reinforced Concrete Products," Prestressed Concrete Institute, 1991, Chicago, IL, 168 pp.

PCI Committee on Glass Fiber Reinforced Concrete Panels, "Recommended Practice for Glass Fiber Reinforced Concrete Panels," Prestressed Concrete Institute, 1981, Chicago, IL, 33 pp.

74 AGING AND CRACKING OF COMPOSITE GFRC WALL PANEL SKINS ON METAL STUD FRAMES IN THE UNITED STATES

D. W. PFEIFER, E. A. ROGALLA and W. J. NUGENT
Wiss, Janney, Elstner Associates, Northbrook, IL, USA

Abstract
Composite skins of GFRC combined with other materials such as ceramic tile, brick, terra cotta and conventional exposed aggregate concrete have developed cracks early in their life when utilized on metal stud frames. Extensive strain-relief tests and other jobsite and laboratory testing have determined that material incompatibility of these two-material skins has caused out-of-plane bowing potential and very large stresses as a result of differential shrinkage and other differential volume changes. Physical evidence and finite element analyses indicate that these stresses are far in excess of allowable factored tensile stresses allowed by the 1987 PCI Recommended Practice for Glass Fiber Reinforced Concrete Panels.
Keywords: GFRC, Walls, Cracks, Composite materials, Finite element analysis.

1 Introduction

Composite glass fiber reinforced concrete panels (GFRC) have been used in the United States to provide an architecturally pleasing facade that is lightweight and versatile. Typical fabrication involves first fabricating the outside face of the wall panel skins with either exposed aggregate face mix concrete, ceramic tile, terra cotta or brick. GFRC is then sprayed against these materials to form the second composite layer. The total design skin thickness is usually 19 to 25 mm (0.75 to 1.0 in.). A light gage metal stud framing system is attached to the GFRC layer for later attachment to a building frame.

During the previous four years, Wiss, Janney, Elstner Associates, Inc. (WJE) has studied many composite GFRC facades in the United States that have experienced serviceability problems as a result of cracking and sometimes excessive bowing. These efforts have combined

Fibre Reinforced Cement and Concrete. Edited by R. N. Swamy. © 1992 RILEM.
Published by E & FN Spon, 2-6 Boundary Row, London SE1 8HN. ISBN 0 419 18130 X.

extensive and costly field documentation and testing, laboratory testing, analysis, and literature review to understand their behavior. Our efforts have revealed that composite panels are often prone to high stresses and distress when attached to metal stud framing systems. These potential problems have not been adequately emphasized and addressed in previous design guidelines.

2 United States GFRC Panel Design Guidelines

The *Recommended Practice for Glass Fiber Reinforced Concrete Panels* [1], published in 1987 by the Prestressed Concrete Institute (PCI), is considered by designers and producers in the United States to be the state-of-the-art design reference. This publication shows the use of concrete face mixes, terra cotta, ceramic tile, and clay brick as components of composite GFRC panels. This manual states that the panel design must consider gravity, wind, earthquake, thermal, moisture, and shrinkage effects.

This document recommends that the framing system supporting the GFRC skins be flexible enough in the plane of the skin to permit essentially unrestrained movement of the skin for volumetric changes caused by thermal and moisture conditions. No additional guidance is given regarding calculations for thermal and moisture conditions. It is assumed by designers that if the framing system used to support the skin is sufficiently flexible to prevent planar restraint, and if sufficient space is left around the edges to permit expansion without interference, then significant stresses will not develop as a result of the strain from volume changes created by thermal, moisture and shrinkage effects.

Steel stud frames are designed for stripping, handling, and in-place loads. They provide vertical support of the panels, and provide support to resist out-of-plane loads such as wind pressures and earthquake-induced forces. These frames are typically attached to a structural frame with two vertical load-bearing connections and additional lateral connections.

To attach the GFRC composite skins to the steel stud frames and to provide support for out-of-plane forces such as wind, "flex anchors" are typically attached to the wall panel skin with GFRC bonding pads, and welded to the steel stud frame. These anchors are typically bent rod, oriented in the horizontal plane, with sufficient rigidity in the vertical and out-of-plane direction, to provide gravity and lateral support. These anchors provide little restraint against horizontal in-plane skin movement and minimize the forces that develop in the skin from planar movements of the skin.

In larger panels, heavier gravity anchors may be used at the bottom corners of the panel to support the dead weight of the panel. These supports typically are plates oriented vertically, to maximize stiffness in the vertical direction and minimize stiffness in the plane of the skin. Similarly, seismic anchors may be installed in the upper corners, often steel plate oriented in the horizontal plane. However, even in larger panels, flex anchors provide the primary support against wind loads.

The design method requires that factored combinations of dead, live, wind, earthquake, and moisture loads be considered. While some of the load combination equations include thermal and moisture effects, these particular equations are typically ignored by designers, because the thermal and moisture effects are thought to be negligible and adequately accounted for if the panel is not significantly restrained in the plane of the skin. It is interesting to note that the PCI equation for load combinations does not assume simultaneous thermal and moisture effects, even though they are often present simultaneously. Shrinkage stresses are not considered in the load combination equation.

3 Significant Observations from WJE Field and Laboratory Investigations

3.1 Coefficient of Thermal Expansion of GFRC and Other Materials

The thermal coefficient of expansion has been measured on a wide range of GFRC materials from across the United States and from Australia. These tests were made utilizing Whittemore mechanical strain gage buttons attached to opposite sides of the specimens. Most tests were made on materials at various moisture conditions but with a multiple layer of plastic wrapping to minimize drying shrinkage and weight changes during the temperature change periods which were commonly 8 to 16 hours.

The tested GFRC had cement/sand ratios ranging from 3:1 to 1.3:1. Some contained Forton and high-range water reducers, while others contained neither. Some were 2 to 3 years old, some were only several months old. The test results were similar in many respects, even though they ranged widely in composition and age. The main conclusions are as follows:

1. The coefficient of thermal expansion of GFRC can vary from 10 to 27 x 10^{-6}/°C (5.8 to 15.0 x 10^{-6}/°F).
2. The thermal coefficient for water-saturated GFRC was small at about 10.8 x 10^{-6}/°C (6 x 10^{-6}/°F).
3. Significant non-reversible thermal strains developed on normal moisture content GFRC, but were eventually

eliminated during long-term testing or storage. It was further observed that during such a 3-month test program, the non-reversible strain effect could be measured more than once. Lack of full strain recovery was never measured during similar tests on conventional face mix concrete, terra cotta, or ceramic tile.

4. When thermal testing permitted GFRC to undergo weight changes due to the environmental conditions of the laboratory test chamber, the test results are significantly lower. This was expected since drying shrinkage will occur simultaneously with thermal expansion during unprotected thermal heating.
5. The observed unusual behavior of GFRC can be compared to similar observations made by Powers [2] and Helmuth [3], as early as 1958 and 1961, respectively, for cement paste.
6. Tests on conventional face mix concrete, ceramic tile, and terra cotta determined coefficients of about 9, 7.2 and 6.8 x 10^{-6}/°C (5, 4 and 3.8 x 10^{-6}/°F), respectively. These materials never produced non-reversible strains during the thermal excursion tests.

These data show that the design guidance provided in the past by numerous authorities underestimated the potential for thermally-induced strains in GFRC. While most design guidance utilized a low thermal coefficient range of 7 to 12 x 10^{-6}/°C (3.9 to 6.7 x 10^{-6}/°F), the BRE increased the design value to 20 x 10^{-6}/°C (11.1 x 10^{-6}/°F) in 1984, and PCI increased the design range to 10.8 to 16.2 x 10^{-6}/°C (6 to 9 x 10^{-6}/°F) in 1987. It is noteworthy that BRE [4] observed in 1976 that values as high as 36 x 10^{-6}/°C (20 x 10^{-6}/°F) had been measured under "certain conditions of atmospheric humidity."

Therefore, when GFRC is composite with conventional face mix concrete, ceramic tile, clay brick or terra cotta, a large potential for out-of-plane bowing occurs due to differential thermal expansion properties. Under certain moisture conditions, the GFRC can develop up to 3 to 4 times the thermal expansion potential when compared to the other materials.

4 Drying Shrinkage of GFRC and Other Materials

Drying shrinkage tests were made in Florida on low w/c ratio GFRC with a cement/sand ratio of 1.3:1, Forton, 5 percent glass and a high-range water reducer. Specimen types included pure GFRC, pure face mix concrete, composites of GFRC and ceramic tile, and composites of GFRC and face mix concrete. The test specimens were 15 cm

(6 in.) wide and 35.5 cm (14 in.) long, and ranged from 9.4 to 32 mm (0.37 to 1.26 in.) thick. The pure GFRC specimens were about 15.2 mm (0.60 in.) thick, and the pure face mix concrete was 9.4 mm (0.37 in. thick. The ceramic tile was 10 mm (0.40 in.) thick with a 12.7 mm (0.50 in.) thick GFRC layer, for a total composite skin thickness of 22.9 mm (0.90 in.). The composite concrete face mix and GFRC specimens varied in total thickness from 20 to 32 mm (0.79 to 1.26 in.), with GFRC thicknesses from 12.7 to 17.3 mm (0.50 to 0.68 in.).

Photographs of the testing are shown in Fig. 1. After overnight wet burlap curing, Whittemore gage plugs were applied on each side of the specimens, to permit out-of-plane bowing effects to be measured. They were then exposed to the out-of-door environment and measurements were made over a 100-day period. Some of the specimens were stored in the Florida environment for about one month and then stored in the WJE laboratories at 23°C (73°F) and 50 percent RH. The major observations from this differential shrinkage strain study were as follows:

1. The pure face mix concrete reached a stable shrinkage of about 600 x 10^{-6} after about 100 days.
2. The pure GFRC material in Florida reached a stable shrinkage of about 900 x 10^{-6} after 100 days.
3. The GFRC mortar without glass fiber reached a stable shrinkage of about 900 x 10^{-6} after 100 days. The glass fiber had no effect on shrinkage potential.

Fig. 1 - Long-term differential shrinkage study in Florida on composite GFRC skins

4. When GFRC was composite with ceramic tile, large differential strain gradients developed. For the Florida and laboratory environments, the measured strain differentials after 100 days were about 1250 and 1550 x 10^{-6}, respectively. Such differential strains would create the following out-of-plane bowing potential for an unattached 23 mm (0.90 in.) thick composite ceramic tile skin in the Florida environment.

Unattached skin length (m) / (ft)	Out-of-plane bowing potential (cm) / (in.)
3.05 / 10	6.4 / 2.5
4.57 / 15	14.3 / 5.6
6.10 / 20	25.4 / 10.0
7.62 / 25	39.7 / 15.6

The data illustrate the huge potential for out-of-plane bowing when GFRC is attached to ceramic tile. For specimens stored at 23°C (73°F) and 50 percent RH, the bowing potential at 100 days was about 25 percent greater.

5. When GFRC was composite with conventional concrete face mix, strain gradients of 100 to 1000 x 10^{-6} were measured at 100 days under the Florida environment. Larger gradients of 300 to 1200 x 10^{-6} were obtained from laboratory tests at 50 percent RH. While the ratio of GFRC thickness to total skin thickness was relatively constant at about 0.60, the strain gradient increased as the total skin thickness increased. As such, even when two shrinkable portland cement based materials are composite, the potential for out-of-plane bowing for skins with total thicknesses of about 25.4 mm (1 in.) and with a GFRC thickness of 15.9 mm (0.63 in.) could be 600 to 900 x 10^{-6}. These strain gradients create large bowing potentials about 50 to 75 percent of those tabulated above for ceramic tile faced skins.

The PCI GFRC committee reported drying shrinkage data on GFRC made with a cement to sand ratio of 1.18:1, a water to cement ratio of 0.28; 5 percent Forton and a high-range water reducer. Silica sand with six different gradations were evaluated. The specimen size was 25.4 x 25.4 x 165 mm (1 x 1 x 6.5 in.). Shrinkage tests after 1 day curing in the molds was at 21°C (70°F) and 50 percent RH. All six formulations had essentially the same 35-day shrinkage of about 1950 x 10^{-6} with a coefficient of variation of only 3.5 percent. Other low w/c ratio mixes were made with one sand gradation but with cement to sand ratios of 1.25, 1.0, 0.75 and 0.50 to 1. These four mixtures produced 35-day shrinkage values of about 1700, 1600, 1500 and 1300 x 10^{-6}, respectively.

These cumulative shrinkage data from WJE and PCI activities illustrate that low w/c ratio GFRC formulations with Forton, and with high-range water reducers produce shrinkage potentials of 900 to 2000 x 10^{-6}, depending upon climate and mix proportions. Even with low cement to sand ratios of 1.3:1 or 1:1, shrinkage potentials at 50 percent RH are high at 1100 to 2000 x 10^{-6}.

5 Wetting and Drying of GFRC and Other Materials

Of particular concern to bowing potential of composite GFRC skins is the total lack of expansion and contraction of ceramic tile, terra cotta, and brick when subjected to wetting and drying processes. This volumetric stability during wet/dry cycles can be contrasted to the very high expansion and contraction of GFRC when subjected to the same wet/dry cycles. Laboratory tests of samples soaked in water for 24 hours indicated that the differential strain gradient across a 15.2 mm (0.60 in.) thick ceramic tile/GFRC composite skin (normally air dry) was about 600 to 800 x 10^{-6}. Similar water soaking tests on terra cotta/GFRC composite skins indicated differential strain gradients of about 300 to 400 x 10^{-6}. When GFRC is composite with face mix concrete, the potential for differential strain is much less than for the above mentioned clay-based products.

6 Aging of GFRC

Aging tests have been made on various age GFRC materials from different regions of the United States. We believe the best determination of the degree of aging (loss of ductility) is made from measurements of flexural strain capacity with ASTM C947 flexural strength testing, either air dry or water soaked prior to testing, as shown in Fig. 2.

Our review of strain gage data from Nippon Electric Glass Co. and our WJE 10-year collection of strain gage data shows that new, unaged GFRC has a flexural strain capacity of about 6000 to 12000 x 10^{-6}.

One aging test series on new GFRC utilized 82°C (180°F) water storage for 7 days, followed by 4 hours of water soaking just prior to test. The average artificially-aged flexural tensile strain capacity was 650 x 10^{-6} with a coefficient of variation (CV) of 28 percent. The unaged, new GFRC average strain capacity was 12500 x 10^{-6}, with a CV of 47 percent. The average aged strain capacity was 5 percent of the unaged, new GFRC strain capacity. The average aged MOR of 10.6 MPa (1536 psi), with a CV of 4.1 percent, was 58 percent of the new GFRC MOR of 18.4 MPa (2666 psi), with a CV of 13.8 percent.

Fig. 2 - Modulus of rupture tests on instrumented pure GFRC and composite ceramic tile/GFRC specimens

This test series showed that flexural strain was a much more sensitive indicator of aging than MOR test results.

Another series tested pure GFRC and composite GFRC/face mix concrete skins that were about 3 years old. These older specimens were then aged in 71°C (160°F) lime-saturated water for 14 days, and then tested in the air-dry condition. The artificially aged flexural strain capacity averaged 600 x 10^{-6}, with a CV of 23 percent. The averaged aged MOR was 7.9 MPa (1140 psi).

Another series compared relatively new GFRC with the same GFRC that had been aged in lime-saturated water at 71°C (160°F) and 82°C (180°F) for 30 days. All samples were soaked in water for 24 hours just prior to flexural testing. The MOR of the new and aged GFRC were as follows:

	MOR	CV
New	21.3 MPa (3092 psi)	8.7 percent
71°C (160°F)	11.8 MPa (1711 psi)	5.2 percent
82°C (180°F)	10.5 MPa (1519 psi)	5.5 percent

These data show that the 71°C (160°F) aging reduced the new GFRC MOR by 45 percent while the 82°C (180°F) aging reduced the new MOR by 51 percent. The strain capacity of the new GFRC was about 6400 x 10^{-6} and the aged strain capacities from both hot water conditions was about 500 x 10^{-6}.

The flexural strain capacity of naturally aged GFRC has also been measured. One series of tests from a 3-year old facade in Florida resulted in an average flexural strain capacity of about 1200 x 10^{-6} with a CV of 64 percent, ranging from 410 to 2770 x 10^{-6}. Assuming strain capacities of 12000 x 10^{-6} for new GFRC and 700 x 10^{-6} for fully aged GFRC, these tests indicate that the 3-year old GFRC had been extensively aged by the Florida environment. Since the full potential strain loss would be about 12000 - 700 = 11300 x 10^{-6}, the average as-is strain capacity of 1200 x 10^{-6} represented an aging loss of 96 percent of the full potential loss. Even the specimen with the highest strain capacity was approximately 82 per-cent aged. These naturally aged GFRC materials that maintained an average strain capacity of 1200 x 10^{-6} had a MOR of 11.9 MPa (1720 psi) with a CV of 23 percent. The naturally aged face mix concrete layer from the same 3-year old project that utilized a composite GFRC/concrete skin had an average flexural strain capacity of 335 x 10^{-6} with a CV of 24 percent, and an average MOR of 6.6 MPa (960 psi) with a CV of 15 percent.

Another test series on 6-year old naturally aged GFRC was undertaken. These tests did not utilize the water soaking prior to the flexural test. The naturally-aged GFRC MOR averaged 10.1 MPa (1470 psi) with a CV of 47 percent. The average flexural strain capacity of these naturally-aged GFRC specimens was 870 x 10^{-6} with a CV of 83 percent. These data indicated that the natural aging process has essentially fully aged this GFRC in 6 years.

Another test series on 5-year old GFRC was also undertaken. These specimens also were not water soaked prior to testing. The naturally aged GFRC MOR averaged 17.2 MPa (2500 psi) with a CV of 25 percent, and the proportional elastic limit (PEL) averaged 7.9 MPa (1150 psi) with a CV of 19 percent. The measured strain capacities for these naturally aged specimens ranged from 1400 to 9000 x 10^{-6}, while averaging 4900 x 10^{-6} with a CV of 61 percent. The percent aging based upon loss of strain capacity ranged from 94 percent to 25 percent, with an average loss of about 65 percent. The loss of strain capacity appeared to correlate to exposure, with the largest losses associated with sun exposed surfaces.

These numerous tests show that artificially aged GFRC produced MOR values that ranged from about 7.9 MPa

(1150 psi) to 11.7 MPa (1700 psi) while averaging about 10.3 Mpa (1500 psi). The aged MOR values from various investigated projects fall in line with the 9.0 to 13.8 Mpa (1300 to 2000 psi) values suggested in the 1987 PCI document [1].

The naturally aged MOR values from these same investigated projects ranged from about 10.3 Mpa (1500 psi) to 17.2 Mpa (2500 psi) while averaging 13.1 Mpa (1900 psi). The new GFRC strain capacities of 6000 to 12000 x 10^{-6} compared favorably with other published data.

7 Wall Panel Bowing Behavior

In-plant storage - A typical GFRC/face mix concrete composite skin wall panel was instrumented with taut string lines, as shown in Fig. 3, to measure the bowing behavior of the skin caused by normal daily temperature changes while in yard storage without restraint from the building connection details. The wall panel was 2.64 m (104 in.) tall and 4.03 m (159 in.) long. The vertical metal stud framing was spaced at about 0.50 m (20 in.) on centers. Six flex anchors were attached to each metal stud at about 0.46 m (18 in.) centers. The total skin thickness was about 25 mm (0.97 in.), with 15.2 mm (0.60 in.) of face mix concrete on the outer surface and 10.2 mm (0.40 in.) of GFRC on the interior surface.

Surface and air temperatures on both sides of the skin were measured at each measurement time with thermocouples. The bowing was measured at midlength of the tight string with a steel rule to the nearest 0.40 mm ($\frac{1}{64}$ in.).

Fig. 3 - Typical testing for bowing on wall panels in yard storage

Bowing measurements were made at the string midlengths between the following horizontal and vertical lines, as shown in Fig. 4.

Horizontal	Vertical
A to B	A to E
C to D	B to F
E to F	

These April 13 measurements were made on cool morning periods and then again on the same day in the late afternoon. Typical data are tabulated below.

Temperature conditions (°C)			Measured bowing of skin (mm)				
Air	Face mix	GFRC	A-B	C-D	E-F	A-E	B-F
9	7	7	0	0	0	0	0
22	26	26	-3.6	-4.8	-3.4	0	0

The above data show that the 25 mm (1 in.) thick skin heats up very uniformly even though the sun heated the back surface of the skin (GFRC) in the morning and the front face of the skin (FM) in the afternoon, with shadows on opposite sides during each period. For this full-day test, the temperature of the composite skin increased 19°C (34°F) from early morning to late afternoon. The skin bowed inward along horizontal lines from the early morning readings and produced a concave surface when viewed from the front of the panel. However, the skin did not bow along vertical lines that parallel the direction of the metal studs attached to the skin. This non-uniform inward bowing behavior shows that the metal stud framing system, even though not yet

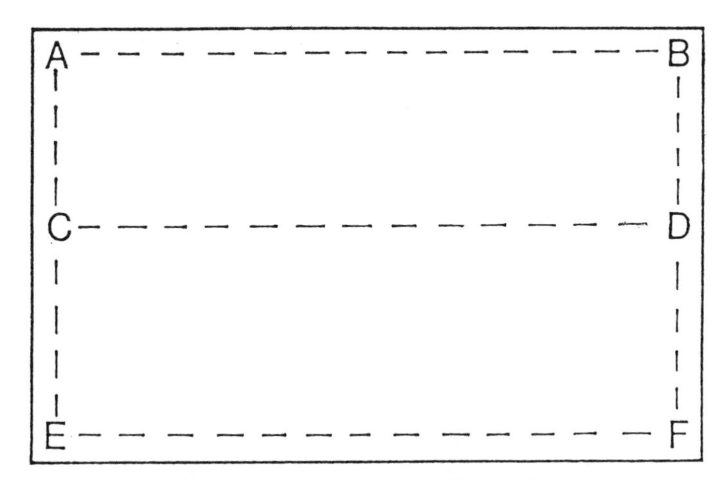

Fig. 4 - Location of string lines for wall panel bowing study at yard storage and jobsite

attached to the structure offers significant restraint to thermal bowing in the vertical direction, while allowing some degree of freedom and bowing in the horizontal direction.

The inward bowing of the composite skin and wall panel system is caused by the GFRC material on the back of the skin having a much higher thermal coefficient of expansion than the concrete face mix.

This measured bowing shows that other differential volume changes, such as differential drying shrinkage of two skin materials, will create unanticipated potential for bowing and stresses in the skin which will be highly dependent on the metal stud framing system as well as the length of time the wall panel is stored in the yard prior to erection and attachment to the building frame. This shows that a panel that would be erected on a very cool or hot day could lock in significant thermally-induced stresses into the skin from that particular day, as well as lock in long-term differential drying shrinkage effect stresses from yard storage.

Attached to structure - A number of wall panels, for the same building discussed above during yard storage, were measured for bowing and thermal behavior when attached to this multistory framed structure as shown in Figs. 5 and 6.

A west side ground level wall panel was measured for thermal bowing, as shown in Fig. 5. Measurements were made on 6 days using a single horizontal string line at midheight of the wall panel. This wall panel had essentially the same overall dimensions as the wall panel studied in the yard storage discussion. The following are typical bowing data.

Temperature conditions (°C)				
Date	Air	Face mix	GFRC	Measured bowing of skin (mm)
March 31	19	17	17	0
March 31	28	48	50	-1.2
April 5	20	19	19	0
April 5	33	54	54	-1.2
April 7	14	13	13	0
April 7	22	32	34	-0.40

On these three typical days when the composite skin increased uniformly in temperature by about 32, 35 and 20°C (58, 63 and 36°F), the inward bowing at midlength along the horizontal string line (about 4 m long) was only 1.2, 1.2 and 0.4 mm (0.047, 0.047 and 0.016 in.), respectively. A comparison of the building panel behavior on April 7 that experienced a 20°C (36°F) temperature increase with the April 13 yard storage behavior that experienced a 19°C (34°F) temperature increase shows that the panel installed in the building (of the same size) bowed inward along the horizontal

Fig. 5 - Typical bowing measurement tests on jobsite panels

Fig. 6 - Typical bowing measurements in concrete shear wall region

midlength location only 10 percent as much as the panel in yard storage; i.e., 0.40 mm versus 4.8 mm (0.016 in. versus 0.188 in.). Thus, the panel attachment system to the building frame restrained 90 percent of the horizontal bowing tendency that occurred in the in-plant yard storage condition for the same temperature change. This shows that additional unanticipated stresses develop in the skins after they are attached to the building.

A number of other wall panels that were attached to the structure were measured for horizontal and vertical bowing, as indicated in Fig. 6. These panels were attached to steel or concrete framed portions of the structure, and concrete shear walls. All panels were 2.74 mm (108 in.) tall and were 4 to 4.87 m (158 to 192 in.) long. The data are tabulated below.

Steel Frame – 4.4 m (174 in.) long

Temperature conditions (°C)			Measured bowing of skin (mm)				
Air	Face mix	GFRC	A-B	C-D	E-F	A-E	B-F
13	14	14	0	0	0	0	0
22	38	39	-0.51	-0.51	-0.76	+0.76	-0.41

Concrete Frame – 4.0 m (158 in.) long

Temperature conditions (°C)				Measured bowing of skin (mm)				
Room	Air	Face mix	GFRC	A-B	C-D	E-F	A-E	B-F
A	22	22	22	0	0	0	0	0
A	–	42	42	-0.25	-0.41	-1.02	-0.41	-0.25
B	22	22	22	0	0	0	0	0
B	–	42	42	+0.51	-0.25	-0.76	+0.25	+0.41

Concrete Shear Walls – 4.87 m (192 in.) long

Temperature conditions, °C				Measured bowing of skin (mm)				
Level	Air	Face mix	GFRC	A-B	C-D	E-F	A-E	B-F
6	19	19	–	0	0	0	0	0
6	27	32	–	-1.27	0	-0.51	0	-0.41
7	19	19	–	0	0	0	0	0
7	27	32	–	-0.41	0	-0.25	+0.25	-1.02

These cumulative data show that the wall panels when attached to the steel or concrete frames were rigidly restrained from thermal bowing in both the vertical or horizontal directions. The horizontal bowing was extremely small. As anticipated, the vertical bowing was even smaller. The measured bowing was about L/5000, significantly less than the L/360 limit suggested by PCI.

Bowing of Unattached Composite Skins – Jobsite bowing tests were also undertaken on small sections of composite skins that had been sawcut from several wall panels. These test specimens were instrumented with mechanical gage plugs on opposite faces to measure differential strain and bowing of a totally unattached skin. These specimens were hung on wall panels at the site, as shown in Fig. 5, so that wall panel behavior could be compared to the behavior of the removed section, under the same temperature and humidity conditions. The bowing measurements were typically determined as the difference in behavior between readings at 6:30 a.m. and 5:30 p.m.

The measured average data from five pieces of skin taken at random from this building and then tested over a 3-day period in April are tabulated below.

These cumulative data show that for a 19.4°C (35°F) temperature change, a 4.04 m (159 in.) unattached 23 mm (0.9 in.) thick composite skin would bow about 10 to 14 mm (0.39 to 0.57 in.). The same skin when attached to a metal frame in yard storage would bow about 4.6 mm

Spec No.	Total thickness (mm)	GFRC thickness (mm)	Face mix thickness (mm)	Temperature change (°C)	Measured strain gradient (10^{-6})	Calculated bow* (mm)
1	17.3	7.6	9.6	19.4	170	20.0
2	21.6	13.2	8.4	19.4	105	9.9
3	22.6	13.2	9.4	19.4	160	14.5
4	28.5	19.6	8.9	19.4	95	6.9
5	30.2	8.6	21.6	19.4	130	8.6

*For a length of skin of 4.04 m (159 in.) and a constant skin temperature change of 19.4°C (35°F) each day

(0.18 in.) in the horizontal midlength location, and 0 mm in the vertical orientation following the metal studs. Therefore, yard storage condition provided full restraint to vertical bowing and about 50 to 70 percent restraint to horizontal bowing. When a similar sized panel was attached to the structure, the same 19.4°C (35°F) temperature change produced horizontal bowing of about 0.64 mm (0.025 in.) and even smaller vertical bowing. Therefore, when attached to the structure, the total restraint to bowing was about 95 percent.

8 Strain Relief Tests

A series of strain relief tests were made to determine the magnitude of the relieved stresses in the skin of crack- free wall panels as a result of removing the steel frame fixity. These stresses are created by long-term differential drying shrinkage, creep, wetting and drying,

Fig. 7 - Typical wall panel after strain-relief tests

Fig. 8 - Mechanical and electrical strain gages attached to ceramic tile and GFRC surfaces for strain relief tests

and thermal effects of the two-material skin. These tests were made when the skin temperature was about 23°C (73°F) so that thermally induced daily stresses would not be included. The strains on the surface of the exterior concrete face mix surface or ceramic tile, as well as the strains on the surface of the interior GFRC material, were measured after sawcutting, as shown in Figs. 7 and 8. The strains were measured with a Whittemore strain gage and/or electrical gages.

The relieved stresses in these materials were calculated from the strain data by using the elastic modulus values measured in our laboratories for the GFRC, ceramic tile and face mix concrete. These strain relief tests on crack-free skins indicated that the relieved stresses were typically as follows.

Skin type	Horizontal direction	Vertical direction
Face mix concrete	1.4 to 6.6 MPa (200 to 950 psi—C)	1.4 to 5.5 MPa (200 to 800 psi—C)
GFRC	0 to 6.6 MPa (0 to 950 psi—T)	0.7 to 3.1 MPa (100 to 450 psi—T)
Ceramic tile	9.6 MPa (1400 psi—C)	8.3 MPA (1050 psi—C)
GFRC	4.5 MPa (650 psi—T)	4.8 to 5.5 MPa (700 to 800 psi—T)

C = compression T = tension

These tests show that significant tensile stresses are relieved in the GFRC layer, even after long-term periods where creep has occurred.

9 Design Considerations

9.1 Composite Interaction

Composite GFRC panels require special design considerations to account for behavior problems unique to composites. Unsuspecting to most GFRC panel designers, large stresses develop in the composite system shortly after fabrication and before any loads are externally applied to the panels. The earliest stresses are caused by incompatibilities between the different materials as a result of differential shrinkage. Additional induced stresses include incompatibilities of coefficients of thermal and moisture expansion, and by thermal and moisture gradients within a layer.

Any out-of-plane restraint provided by 1) the natural internal restraint of an unattached two-material skin, 2) the external steel stud frame system restraint, and 3) the external restraint from the connection details to the building structure, which all attempt to keep the skin flat, can result in high tensile and compressive stresses. As previously discussed, an unattached 6.1 m (20 ft) long ceramic tile/GFRC skin has the potential to bow out-of-plane 254 mm (10 in.) from differential shrinkage effects in 100 days. Temperature changes, wet/dry cycles, etc., are significant additive out-of-plane effects to the differential shrinkage stresses.

9.2 Elastic Composite Analysis

Elastic analysis equations for composite circular plates can be found in several engineering texts. These equations assume elastic homogeneous materials, and because circular shapes are assumed, do not provide means of calculating stress concentrations that often develop in the corners of non-circular sections. Finite element analyses performed by WJE has shown that stress concentrations in corners are often several fold the magnitude of interior stresses. The equations also assume uniform unrestrained strain (the strain that would develop without any restraint) in each layer and cannot account for gradients within one layer. The modulus of elasticity of one or more material is reduced to account for creep. These equations assume that there is no planar restraint, and the edges are either free or fixed against all rotation. Free edges approximate the installation of a composite panel without any external restraint against bowing, while fixed edges estimate the behavior of a panel restrained against bowing such as a panel supported with regularly spaced flex anchors.

It is important to note that the elastic equations and typical finite element analysis assumes linear behavior of materials. It is our opinion, supported by strain-relief testing of actual composite panels, that linear analysis does provide reasonable estimates of stresses that may develop. Inelastic cracking will only provide local relief of stresses.

For most combinations of composite GFRC skins analyzed by WJE, the highest stresses occur at the bond-line mating surfaces. This is caused by large shear forces that are transferred across the mating surface to keep the adjoin-ing surfaces together. Because these forces act at the interface instead of the centroid of each layer, this results in an equivalent membrane force and moment on each layer. The stresses on the opposite surfaces of each layer often varied linearly from tension to compression.

Example 1 – Composite tile and GFRC – For example, consider the following composite panel, unrestrained in the plane of the panel, constructed with tile and GFRC.

Material	Thickness mm (in.)	Elasticity MPa (ksi)	Poisson ratio	Unrestrained shrinkage strain	Coefficient of thermal expansion (/°C)	Temperature change °C (°F)
GFRC	14(9/16)	12000* (1740)	0.2	-0.001 500	0.000 021 6	-30 (-54)
Tile	14(9/16)	60000 (8700)	0.2	0	0.000 007 2	-30 (-54)

* Adjusted for creep

It is important to note that the following calculations do not include any moisture effects which are additive. The following stresses (MPa, tensile stresses denoted positive), linear across each layer, are calculated for a completely unrestrained panel region such as at unrestrained edges.

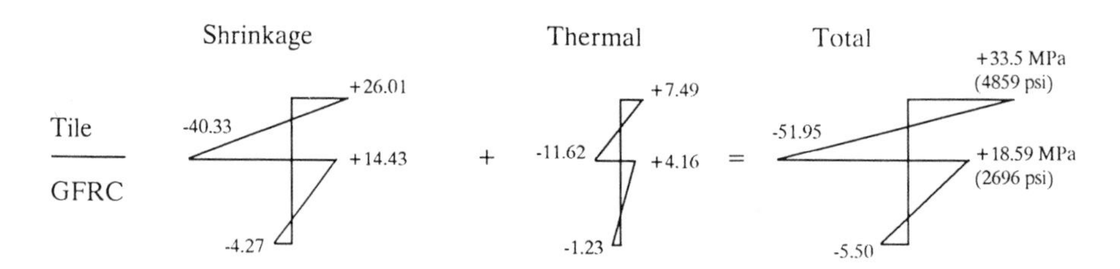

And the following stresses are calculated if the panel is restrained against rotation along the edges, or at interior locations with connections that keep the skin planar.

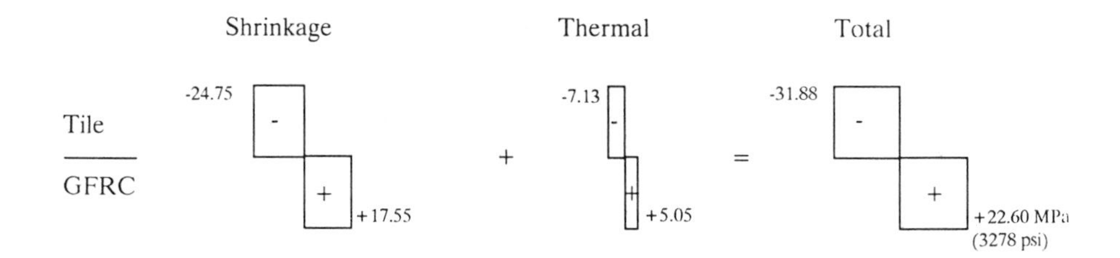

In an unrestrained skin 4 m (13.1 ft) long and 2 m (6.56 ft) high, elastic analysis indicates that the panel will bow approximately 223 mm (8.8 in.) and 64 mm (2.5 in.) for shrinkage and thermal loading, respectively. If both effects are combined, a total bowing of 287 mm (11.3 in.) is calculated for the unattached or unrestrained skin.

Example 2 - Composite Face Mix and GFRC - For another example, consider a composite panel constructed with GFRC and face mix concrete, unrestrained in the plane of the panel, with the following properties:

Mate-rial	Thickness mm (in.)	Elasticity MPa (ksi)	Poisson ratio	Unrestrained shrinkage strain	Coeffi-cient of thermal ex-pansion (/°C)	Temperature change °C (°F)
GFRC	14(9/16)	12000* (1740)	0.2	-0.001 500	0.000 021 6	-30 (-54)
Face mix	14(9/16)	15000* (2175)	0.2	-0.000 500	0.000 009 9	-30 (-54)

* Adjusted for creep

The following calculations also do not include any moisture effects which are additive. This example results in the following stresses for a completely unrestrained panel. These stresses are linear across each layer.

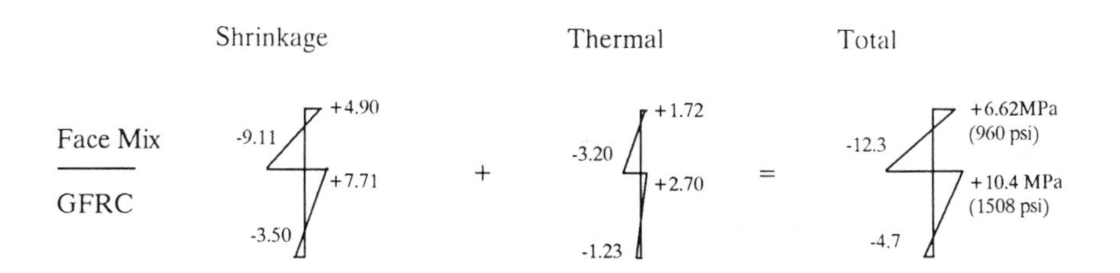

And the following stresses are calculated if the panel is restrained against rotation along the edges, or at interior locations with connections that keep the skin planar.

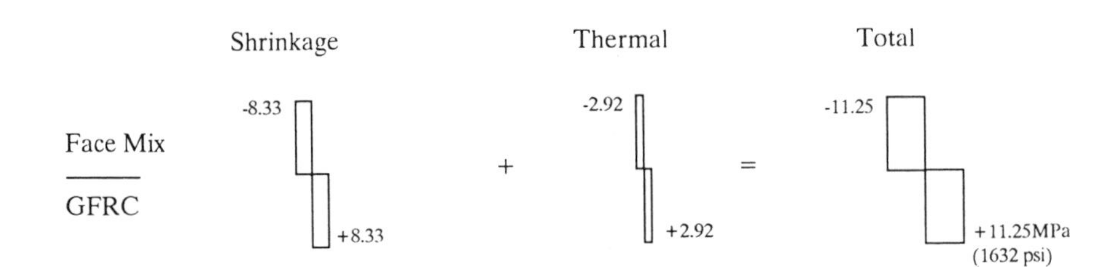

For a panel the same size as the GFRC/tile panel of Example 1, calculated bowing potentials for shrinkage and thermal loading are 134 mm (5.3 inches) and 47 mm (1.8 inches), respectively. If combined, a total bowing potential of 181 mm (7.1 inches) is calculated for the unattached skin.

While most design equations utilize flexural stresses, the above examples show that most of the skin surface is subjected to direct tension stresses, not flexural stresses.

The allowable stresses in the GFRC skin due to ACI 318 loads, thermal, and moisture effects would be as follows, when using the PCI suggested strength reduction factor ϕ = 0.67 and the shape factor S of 1.0.

Flexural or tension strength coefficient of variation, %		Allowable flexural or tension stresses due to factored loads and unfactored volume changes, MPa	
		f_{yr} = 6.2	f_{yr} = 10.3
10	Flexural	3.8	5.7
20		2.3	3.8
30		1.1	1.8
10	Direct tension	1.4	2.3
20		0.9	1.5
30		0.7	0.7

A comparison of the allowable stresses tabulated above with the calculated stresses from realistic finite element studies utilizing creep effects and our strain-relief tests shows that the in-service stresses from restrained out-of-plane bowing potentials are well above the naturally or artificially aged flexural or tension strengths. These comparisons explain the cracking and bowing problems observed at a number of sites where various composite material skins were utilized. While not discussed in this paper, a number of full-size wind load tests were made, as shown in Fig. 9, at one of the jobsites on cracked and uncracked wall panels in accordance with ASTM E330 to further study the implications of cracks on in-place structural behavior.

Fig. 9 - Suction load test on cracked and uncracked GFRC panels per ASTM E-330

10 References

[1] Prestressed Concrete Institute, *Recommended Practice for Glass Fiber Reinforced Panels*, PCI Committee on Glass Fiber Reinforced Concrete Panels, 1987, Chicago, Illinois.

[2] Powers, T.C., *The Physical Structure and Engineering Properties of Concrete*, Research and Development Laboratories of the PCA Research Department, Bulletin 90, July 1958, Chicago, Illinois.

[3] Helmuth, R.A., *Dimensional Changes of Hardened Portland Cement Paste Caused by Temperature Changes*, Research and Development Laboratories of the PCA Research Department, Bulletin 129, 1961, Chicago, Illinois.

[4] Building Research Establishment Paper CP 38/76, *A Study of the Properties of Cem-Fil/OPC Composites*, June 1976, Garston Watford, England.

75 EVALUATION OF PERFORMANCE DURABILITY OF PLASTIC FIBRE REINFORCED CONCRETE COMPOSITE MANHOLE COVER SYSTEM

O. P. RATRA
Polycrete - PFRC, Vasant Kunj, New Delhi, India

Abstract

Field evaluation of performance under conditions of use, of precast manhole cover system using PFRC composite has been studied over a period of ten years. The conditions of use included exposure of PFRC manhole covers/system of various shapes and dimensions over manhole chambers (to sewage gases from below) and impact load/vibrations of moving vehicular traffic on busy city roads in various parts of India, including coastal regions. The evaluation was against the conventional steel reinforced concrete for similar products. During evaluation, the PFRC composite manhole cover system demonstrated that: (i) it could continue to withstand impact load/vibrations of moving vehicular traffic; (ii) it could retain structural stability of the manhole cover system; and (iii) it did not show any signs of debonding of concrete from plastic fibre reinforcement, inspite of routine maintenance of the system, confirming thereby the satisfactory keying/bonding capacity of fibres with concrete.

Key Words: PFRC composite, twisted fibrillated polypropylene film fibre, **manhole cover system**, depth and seating, **precast slab with integral frame, wide-mesh mat**, impact load and vibrations, evaluation under conditions of use, **performance durability**.

1 Introduction

Evaluation of performance durability, in other words the useful service life of a product, is better assessed when the product is installed in a place or under conditions of use. The materials and the design of the product together contribute in its operational performance. The materials could be single entity like steel, timber, or plastics, or composites in nature such as cement concrete using various types and forms of reinforcements. Reinforced cement concrete composites using conventional steel bars or fibres, including organic-natural and synthetic and designed into precast products, have been studied and reported in literature. Technologically, the study and use of fibre reinforced concrete has engaged the attention of scientists and engineers all over the world for the last over three decades. Applications-wise, interesting developments have been reported and found in commercial use. Among the fibres as reinforcement, plastics come next to steel. Fibrillated polypropylene film fibres, in different forms/twists/orientation, used both as primary & secondary reinforcement in cement concrete have attained successful stage of adaptability, among the known plastics fibres, both in academic and commercial fields, and applications developed and evaluated[1-5]. Performance evaluation studies in the field of PFRC composite as reported to date have been of short-term nature and restricted mostly to laboratory- scale samples and conclusions arrived at accordingly[6-7]. Shah [8] has reported that use of continuous aligned fibres enhances the tensile load carrying capacity of cement based matrixes.

Fibre Reinforced Cement and Concrete. Edited by R. N. Swamy. © 1992 RILEM.
Published by E & FN Spon, 2-6 Boundary Row, London SE1 8HN. ISBN 0 419 18130 X.

The gradual corrosion of steel reinforcement is a known phenomenon whether as bars or fibres in cement concrete composite products, in particular when these are in use in coastal regions, or exposed to sewage gases, more so when these are subjected to impact load and vibrations and these develop cracks. Protective measures are advised in such circumstances using known coatings (Zinc or epoxy resin) to prolong the useful service life of such composites. There are several precast concrete products whose useful service life suffers due to gradual corrosion of steel reinforcement under conditions of use. This could be due to inadequate concrete cover, poorly compacted concrete (or shrinkage cracks), improper placements of reinforcement, and carbonation of concrete as and when the protection provided by concrete breaks down (due to cracking under impact load) the steel reinforcement corrodes which sets up internal tensile stresses within concrete. This in turn causes cracking, spalling and staining of concrete and thereby de-bonding of reinforcement and reduction in cross section of steel thus affecting strength of the structure. Examples are precast concrete pipes (inclding vent pipes), pavement slabs, manhole covers, fence posts, railings etc.
It is a recorded fact that plastics contribute in upgrading the performance of cement concrete composites. Plastics fibres being corrosion resistant, have been studied and found to be of practical value in place of steel reinforcement in a number of precast concrete products, and their performance under conditions of use found satisfactory. The choice-type and form of plastic fibres, their loading and orientation in cement concrete composites contribute significantly in achieving the ultimate strength properties in the desired pre-cast concrete product.

2 Significance of Precast PFRC Products: Manhole cover system:

The author initiated developmental studies during 1980-81, with fibrillated polypropylene (PP) film fibre, specially twisted to varied breaking strengths for use in cement concrete composites. The PP twisted fibre of the type (single twist) was used as reinforcement in the form of chopped fibres (50-75 mm length), together with double twisted string/strands made into wide-mesh mat. The cement used was Ordinary Portland and Slag type, with stone-dust as the fine aggregate, and stone chips (10- 12 mm size) as the course aggregate. The concrete mix design was M 20 and M 25, using tilting drum mixer. The PP fibre loading including wide-mesh mat was upto 0.5% by weight; the chopped fibres mixed in the concrete mixer carefully by sprinkling while the wide-mesh mat was placed at appropriate level during casting of the product. The PFRC mix so designed was used in the development and manufacture of precast manhole cover system (precast slab with integral frame of desired clear opening and matching manhole cover of required depth/seating), as an alternative to conventional steel reinforced concrete, and cast- iron manhole covers/frames *(Fig.1)*. The significance of the work involved centred around development of suitable manhole cover system using PFRC composite which would not attract 'pilferage' of the product in cities and towns (as reported in respect of cast-iron manhole covers/frame) but perform satisfactorily in the field, on busy city roads, under impact load and vibrations of vehicular traffic.

The process of manufacture of precast load-bearing manhole cover assembly using PFRC, is covered under Indian Patent No.164486, by the author [9]. Varied designs of precast PFRC manhole cover and the slab with integral frame-circular, rectangular and square, including matching manhole covers for existing cast-iron frames, were designed, produced and evaluated. Initially,doubts were raised by the engineers, in respect of the technical suitability of plastics fibres as reinforcement, and thus the ultimate performance value of manhole covers using PFRC, in tune with Indian standard/British Standard specifications (IS:1726, and BS:497) and the design parameters based on 5-tonne wheel load of static vehicle, and 12.5-tonne wheel load of moving vehicle. Though resorting to load testing as per the respective Indian Standard Specification IS:1726-1991 (Third Revision), was one route to assess the technical suitability of precast concrete manhole covers, but the field conditions-the installation on site, frequent maintenance of the system, and the vehicular impact and vibrations etc-could hardly be simulated in a laboratory for as diverse a product as PFRC composite, thus giving an opportunity of appreciating the contributions of each element involved. This necessitated systematic field trials and evaluation of precast PFRC manhole covers/and the system on busy city roads, in New Delhi, Bombay, Mithapur (Gujarat), Calcutta, and Jamshedpur-TISCO(Bihar), besides towns and cities in Punjab and Haryana. The earliest field sample trials were initiated for the benefit of New Delhi Municipal Committee, and Municipal Corporation of Delhi during March-December 1982. These were followed in Calcutta and other areas in West Bengal (1983-91); Mithapur (1985-90); Jamshedpur (1984-91); and Bombay 1986-90).

3 Field Trials and Evaluation:

As already indicated, that field trials and evaluation of finished products could hardly be simulated on laboratory scale. In this particular case, the use of precast PFRC manhole cover (with MS surround to protect the edges, and two MS rods as liftinghandles) when observed under conditions of use in a frame over manhole chamber, presents altogether varied environmental and impact load/vibration factors, both from below and above the manhole cover, not to speak of the operational conditions of the manhole chamber and its location on busy city roads.

The flexibility of the twisted PP fibres together with wide-mesh mat used in PFRC, their keying factor, and their corrosion resistance together contribute in obtaining ductility and resilience in the PFRC composite thus achieving performance in the field, which steel as reinforcement in cement concrete composite for similar products though at times over-designed has been found to fail in several cases under site conditions in less than six months *(Fig.2)*. The test load (breaking load) stipulations for PFRC composite manhole covers/system, were much different for the same depth and seating in the frames, but the overall field performance was equally satisfactory and even in certain locations better than conventional steel reinforced concrete, and cast-iron manhole covers/frames, apparently because of difference in composite materials nature and thus the ultimate results in precast products. The following **Table** illustrates the dimensions and test loads of the precast PFRC manhole covers/system which were tried and evaluated at different locations in the field:

Table

Sl. No.	Location		Dimensions / **Test Load**	In use since	Last visually examined on
1.	2.		3.	4.	5.
1	**New Delhi** (Municipal Committee)				
	i)	Motilal Nehru Marg	Manhole cover 650 (75) mm dia. / **10 tonnes**	29.12.1982	03.02.1992
	ii)	Dr.RMLHospital Road Junction	Precast Slab 1000x1000x150 mm with integral frame of 560 mm dia. clear opening & matching manhole cover 640 (80) mm dia. / **12 tonnes**	04.08.1984	04.02.1992
	iii)	Desh Bandhu Gupta Road (Municipal Corporation)	Manhole cover 640 (80) mm dia. / **12 tonnes**	Jan.1985	19.08.1990

1.	2.	3.	4.	5.
2.	**Jamshedpur** (TISCO Township)	Rectangular manhole cover 670x520 ------------x 80 mm 590x440 ------------------------ **7.5 tonnes**	08.06.1984	20.12.1990
3.	**Mithapur** (Gujarat) Tata Chemicals Township	Manhole cover 585(70) mm dia. ------------------------ **5 tonnes**	Early 1985	19.07.1990
4.	**Calcutta** (Municipal Corporation) i) Hazra-Garcha crossing	Precast slab 1000x1000x180 mm with integral frame of 580 mm dia clear opening and matching manhole cover 685/660(100) mm dia ------------------------ **20 tonnes**	08.01.1986	29.03.1991
	ii) Garihat Road	Lamphole Cover 460/430(130) mm dia ------------------------ **35 tonnes**	22.04.1990	29.03.1991
		Manhole Cover 645/605(140) mm dia ------------------------ **31 tonnes**	22.04.1990	29.03.1991
5.	**Bombay** (Municipal Corpoation of Greater Bombay) August Kranti Marg	Precast slab 1000 mm dia.x180 mm dia with integral frame of 600 mm dia clear opening and matching manhole cover 685/660 (100) mm dia. ------------------------ **20 tonnes**	24.01.1986	23.07.1990

The above field evaluations are illustrated in Figure No.3 - 11.

4 Observations and Conclusions

Over a period of time and under moving vehicular traffic impact load, the operative surface of the PFRC manhole covers/systems, was found abraded, as is normally the case with any city road surface. During frequent maintenance of the manhole covers i.e. their removal and placement in position with the tools available with the local authority departments, chipping of edges near the MS surround and the cup-hole for lifting handles, was observed. In certain cases, the top cement layer got used up so much that PP fibres were visible on the operative surface of the manhole cover. However, the top operative surface of the manhole cover was not found to have worn out more than 2 mm in depth at the busiest traffic junctions, as observed in Calcutta, New Delhi and Bombay during the period ranging between 6-10 years. Inspite of this routine operative surface deterioration under moving vehicular traffic, there has not been any case of collapse or structural failure of the precast PFRC manhole covers under conditions of use, as examined during February 1992, in New Delhi. In coastal areas, like Bombay (Maharashtra) and Mithapur (Gujarat) there was visible corrosion of the MS surround, the PFRC composite did show signs of cracks in Mithapur but the fibre bonding was intact. The toughness of the PFRC composite and its ductility to absorb impact load and vibrations of moving vehicular traffic, was visible at all the above locations - an added advantage over conventional steel reinforced concrete composite.

As against the 35 tonne stipulated test load for heavy-duty/extra heavy duty manhole cover (ISS/BSS) for cast iron manhole covers, the precast PFRC manhole covers of 10 to 20-tonne test loads have performed satisfactorily under severe moving vehicular traffic conditions as observed in New Delhi, Calcutta and Bombay for a period ranging between 6-10 years. The use of chopped twisted PP fibre together with wide-mesh mat of the double twist, as reinforcement established their satisfactory keying/bonding with concrete thereby ensuring toughness and ductility of the composite - its characteristic capacity of absorbing impact load and vibrations much desired in manhole covers applications and their use under conditions on busy city roads.

The field evaluation of performance durability of precast PFRC manhole cover system has established that:(i) it could continue to withstand impact load/vibrations of moving vehicular traffic; (ii) it could retain structural stability of the manhole cover system under tensile load probably due to the presence of continuous aligned fibers; and (iii) it did not show any signs of debonding of concrete from plastic fibre reinforcement used, inspite of routine maintenance of the system confirming thereby satisfactory keying/bonding capacity of fibers with cement concrete composite.

REFERENCES

1. **Hannant, D.J.,** *Fibre Cements and Fibre Concretes.* John Wiley, Chichester, 1978.
2. **Hannant, D.J., Zonsveld, J.J. & Hughes, D.C.,** Polypropylene film in cement based materials. *Composites*, 8(1978), 83-8.
3. **Naam, A.E.,Shah, S.P., & Thorne, J.L.,** Some Developments in Polypropylene Fibres for Concrete. *ACI SP-81, American Concrete Institute, Detroit, 1984*, pp.375-96.
4. **Kosa, K.,Naaman, A.E.,&Hansen,W.,** Durability of Fibre Reinforced Mortar, *ACI Materials Journal/May-June* 1991, pp.310-19.
5. **Nanni, A.,&Meramarian, N.,** Distribution and Opening of Fibrillated Polypropylene Fibres in Concrete. *Cement & Concrete Composites* 13(1991) 107-114.
6. **Ratra, O.P.,** Plastics Fibre Reinforced Concrete (PFRC) Composite Manhole Cover Technology. *Fibre Reinforced Cements and Concrete*: Recent Developments by R.N.Swamy and Barr; Elsevier Applied Science, 1989,pp.630-39.
7. **Ratra, O.P.,** Durability of Plastics Fibres As Reinforcement in Cement Concrete Composites. Polymers in Concrete; ICPIC 1990, Shanghai, International Academic Publishers, 1990, pp.434- 41.
8. **Shah, Surendra P.,** Do fibers increase the tensile strength of cement based matrixes ? *ACI Materials Journal, November-December 1991*, pp 595-602
9. **Ratra, O.P.,** *Indian Patent No. 164486.*

Fig. 1. Precast PFRC manhole cover/system.

Fig. 2. Relative performance durability of conventional steel reinforced concrete manhole cover system.

Fig. 3. Precast PFRC manhole cover in cast iron frame used in New Delhi since 1982.

Fig. 4. Precast PFRC manhole cover system installed in New Delhi in 1984.

Fig. 5. Precast PFRC manhole cover in cast iron frame used in New Delhi since 1985.

Fig. 6. Precast rectangular PFRC manhole covers in matching cast iron frames in Jamshedpur since 1984.

Fig. 7. Keying/bonding characteristics of twisted PP fiber as reinforcement in concrete composite manhole cover (manually damaged).

Fig. 8. Precast PFRC manhole cover in cast iron frame, showing corroded MS surround and cracks in PFRC composite under extreme coastal region environment, Mithapur, Gujarat.

Fig. 9. Precast PFRC manhole cover installed in Calcutta in 1986.

Fig. 10. Precast PFRC manhole cover system used in Bombay since 1986.

Fig. 11. Precast PFRC lamphole/manhole cover used as replacement cover in cast iron frames in Calcutta.

PART NINE

SYNTHETIC FIBRES AND REINFORCEMENT

76 MECHANICAL CHARACTERISTICS OF CHOPPED FIBER REINFORCED CEMENT COMPOSITES MAINLY USING CARBON FIBER

T. FUKUSHIMA
Building Research Institute, Ministry of Construction, Tsukuba, Japan
K. SHIRAYAMA
Kogakuin University, Tokyo, Japan
K. HITOTSUYA
Onoda Cement Co. Ltd, Tokyo, Japan
T. MARUYAMA
Nippon Concrete Industry Co. Ltd, Shimodate, Japan

Abstract
Mechanical characteristics of chopped fiber reinforced cement composites (FRC) reinforced with four types of new fibers, namely, PAN-based high tension type(PAN-HT) and pitch-based general purpose type(pitch-GP) carbon fibers, aramid fiber and alkali-resistant glass fiber were examined for various fiber contents and aspect ratios by bending, compressive and tensile tests. Their mechanical chracteristics are compared and evaluated under common experimental conditions. It was found that tensile and flexural characteristics of each FRC are greatly improved by the increase in fiber content and aspect ratio. Fiber effective coefficient which determines the effectiveness of fiber reinforcment in tensile strength of each FRC by composition law has the tendency of monotonous decrease with the increase in fiber content and aspect ratio. Under the same fiber content and aspect ratio conditions it shows the smallest values in PAN-HT carbon fiber and the greatest values in pitch-GP carbon fiber, which is the tendency in inverse proportion to the tensile strength of each fiber. Under the same matrix and fiber content conditions, PAN-HT carbon chopped fiber reinforced cement composite (PAN-CFRC) has the very high flexural strength and high elasticity, but have the very low toughness. Aramid chopped fiber reinforced cement composite (AFRC) shows the toughest characteristics, but has rather low flexural strength and elastic range. Pitch-GP carbon chopped fiber reinforced cement composite(Pitch-CFRC) and alkali-resistant glass chopped fiber reinforced cement composite(GFRC) have the flexural characteristics good for building materials of general performance with the normal strength and toughness.
Keywords : Chopped Fiber Reinforced Cement Composite, PAN-based High Tension Type Carbon Fiber, Pitch-based General Purpose Type Carbon Fiber, Aramid Fiber, Alkali-resistant Glass Fiber, Flexural Behaviour, Tensile Behaviour, Commpressive Behaviour,

1 Introduction

Chopped fiber reinforced cement composites(chopped fiber-FRC) reinforced with various types of new chopped fibers such as carbon,

Fibre Reinforced Cement and Concrete. Edited by R. N. Swamy. ©1992 RILEM.
Published by E & FN Spon, 2-6 Boundary Row, London SE1 8HN. ISBN 0 419 18130 X.

aramide, alkali-resitant glass and vinylon fibers have the possibility of their application for light, high-strength and highly-corrosion-resistant exterior and roofing building materials. They are especially applicable in severe environmental conditions such as in marine structures and high rise buildings as well as in ordinary ones. They have attracted many researchers from the viewpoint of interest both in their fundamental properties and in their application. The situation that asbestos fiber reinforced cement composites would be restricted or prohibited in use because of their cancer problems is one of the greatest reasons why there have been recently increasing research on chopped fiber-FRC using such new fibers as replacement of asbestos fiber.

There have been so far many reports on the mechanical characteristics of various kinds of chopped fiber-FRC(Ali et al.(1972), Akihama et al.(1984), Akihama et al.(1985a), Ohama et al.(1985b), Makitani et al. (1988)). However, experimental conditions have not yet been unified, so that it is impossible to compare these expeimenntal data, and understand mechanical characteristics of these FRC.

In this report for four types of chopped fiber-FRC reinforced with PAN based high tension type(PAN-HT) and pitch based general pupose type(pitch-GP) carbon fibers, aramid and alkali-resistant glass fibers(we would like to name these FRC as PAN-CFRC, pitch-CFRC, AFRC and GFRC hereafter) were evaluated using similar experimenntal conditions(Fukushima et al.(1990), Fukushima et al.(1991).)

2 Experimental methods

2 . 1 Materials

Table 1 shows the fundamental properties of four types of new chopped fibers used in this experiment, and in Table 2 is shown the mix proportion of chopped fiber-FRC. Ordinary portrand cement (specific gravity(S.G.); 3.17)

Table 1. Types and fundamental properties of new chopped fibers

Type of fiber	fiber length (mm)	specific gravity	fiber diameter (μ)	tensile strength (MPa)	Young's modulus (MPa)	elongation (%)
PAN-based carbon fiber	3, 6, 12	1.75	7	35.3	2,303	1.4
pitch-based carbon fiber	3, 6, 10	1.60	14	7.8	372	—
aramid fiber	3, 6, 12	1.45	12	30	696	4.4
alkali-resistant glass fiber	3, 6, 12	2.55	12	14.7~19.6	720	—

Table 2. Mix proportion of chopped fiber- FRC

water cement ratio W/C (%)	sand cement ratio S/C (%)	weight of materials (g/3ℓ)					fiber content (vℓ%)
		water (W)	cement (C)	sand (S)	high performance water-reducing agent (C×1.0%)	defoaming agent (C×0.05%)	
40	25	1,464	3,660	915	36.6	1.8	1, 2, 3

and Japanese standard sand (S.G. ; 2.63) were used. As admixtures high performance water-reducing agent(S.G. ; 1.20) and defoaming agent(S.G. ; 1.39) were also used.

2 . 2 Preparation of test specimens

Mixing was done by using an omni-mixer with a capacity of 50 liter. At first cement, defoaming agent and chopped fiber were premixed, thereafter high performance water-reducing agent and water were added. Sand was added in the end to obtain the mortar . Test specimens were steam cured for three hours at 65°C.

2 . 3 Testing methods

Flow test and unit volume weight test were done by applying the specifications of JIS R5201 and JIS A6201. By using test specimens of the shape of 5 x 40 x 3 cm, bending test was done by applying the bending test method of GRC specified by the Japanese GRC industy Association(deflection speed ; 2mm/min, span ; 30cm). Tensile test was done for test specimens of the shape of 4 x 38 x 1 cm also by applying the tensile test method of GRC specified by the Association(steel plate adhesion method, cross head speed ; 2mm/min). For test specimens of the shape of 4 x 4 x 4 cm, compressive test was carried out using the compressive test method of the Association.

3 Results and discussion

3 . 1 Results of flow test and unit volume weight test

Table 3 and Fig. 1 show the results of flow test and unit weight test. In any FRC consistency(flow) increses in proportion to fiber length and fiber content, but proper mixing still could be done.

3 . 2 Results of bending test

Table 4 and Figs. 2-7 show the results of bending test. Although the flexural strength of pitch-CFRC shows a gradual decrease with aspect ratio, that of PAN-CFRC, AFRC and GFRC becomes greater with aspect ratio and fiber content(Figs. 2,3). Figs. 4, 5 show the relationships of flexural toughness vs. aspect ratio and fiber content. The same tendency can be found. Fig. 6 show the flexural stress-deflection curves of four types of FRC with fiber content of 2% and fiber length of 6mm. PAN-CFRC has very high flexural strength, but it does not have so much tenacity , and their flexural stress at proportion limit and maximum flexural strength have almost the same value, so that flexural toughness of PAN-CFRC is very low. On the other hand, as shown in Fig. 7, AFRC shows at first rather low flexural strength and toughness for low fiber content and, but the flexural strength is high as compared to CFRC for higher fiber contents. The flexural toughness becomes remarkabley improved, which is considered to be due to the high adhesiveness to cement paste and high elongation of aramid fiber. Figs. 4, 5 shows the drastic increase in flexural toughness of AFRC with the increase in aspect ratio anf fiber content. The flexural strength of GFRC shows the maximum value for fiber content of 2% and fiber length of 6mm for the mix proportion of mortar used in this experiment.

Table 3. Results of flow test and unit weight test

fiber	fiber length (mm) – fiber content (v %)	flow (mm)	unit weight A (g/ℓ)	estimated unit weight B (g/ℓ) *	absolute volume A/B×100(%)
PAN	3－2	209	2026	2021	100.2
	6－1	223	2052	2024	101.1
	6－2	206	2040	2021	100.9
	6－3	192	2012	2018	99.7
	12－2	197	2029	2021	100.3
Pitch	3－2	220	2042	2018	101.2
	6－1	235	2044	2022	101.1
	6－2	187	2001	2018	99.2
	6－3	165	1982	2014	98.4
	10－2	186	2006	2018	99.4
Aramid	3－2	186	2013	2015	99.9
	6－1	211	2034	2021	100.6
	6－2	172	2022	2015	100.3
	6－3	145	2008	2010	99.9
	12－2	144	2006	2015	99.6
Glass	3－2	unmeasurable(over300)	2074	2037	101.8
	6－1	unmeasurable(over300)	2049	2032	100.8
	6－2	188	2047	2037	100.5
	6－3	158	2050	2042	100.4
	12－2	180	2052	2037	100.7
	plain	unmeasurable(over300)	2046	2037	100.9

* Unit weight of SN defoaming agent is not taken into account.

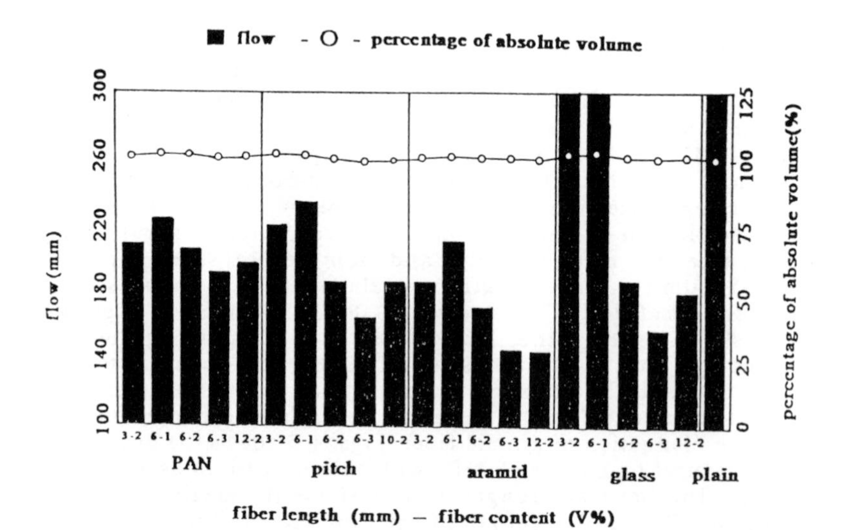

Fig. 1. Flow values and percentages of absolute volume of various chopped fiber- FRC

Table 4. Results of bending test

fiber	fiber length (mm) fiber - content (v/v %)	flexural toughness* (N・m)	stress at proportion limit (MPa)	maximum flexural strength (MPa)	reduced flexural strength (MPa)	flexural Young's modulus (GPa)
PAN	3−2	0. 180	7. 37	7. 48	2. 68	26. 8
	6−1	0. 161	7. 20	7. 21	2. 54	26. 9
	6−2	0. 180	7. 37	7. 41	2. 56	26. 2
	6−3	0. 281	9. 21	9. 25	3. 56	30. 9
	12−2	0. 272	8. 68	8. 68	3. 14	26. 1
Pitch	3−2	0. 573	5. 63	6. 78	3. 39	21. 9
	6−1	0. 195	4. 64	4. 80	2. 28	23. 2
	6−2	0. 460	4. 83	6. 27	3. 13	23. 0
	6−3	0. 350	5. 69	6. 08	2. 88	24. 6
	10−2	0. 142	5. 02	5. 89	2. 96	25. 5
Aramid	3−2	0. 305	5. 30	5. 56	2. 71	24. 3
	6−1	0. 485	5. 22	6. 09	3. 16	25. 9
	6−2	0. 759	6. 37	7. 28	3. 77	25. 6
	6−3	1. 475	7. 34	9. 65	5. 17	28. 4
	12−2	1. 789	6. 69	9. 10	4. 84	23. 8
Glass	3−2	0. 149	5. 43	5. 44	2. 11	24. 0
	6−1	0. 176	5. 85	5. 88	2. 36	25. 9
	6−2	0. 655	6. 86	9. 06	4. 29	28. 5
	6−3	0. 365	7. 30	8. 12	3. 60	29. 4
	12−2	0. 374	6. 19	7. 93	3. 48	27. 6
	plain	0. 085	6. 53	6. 53	1. 90	24. 5

* Values estimated from the areas under the stress-deflection curves until the 80% reduction of the maximum load occurs

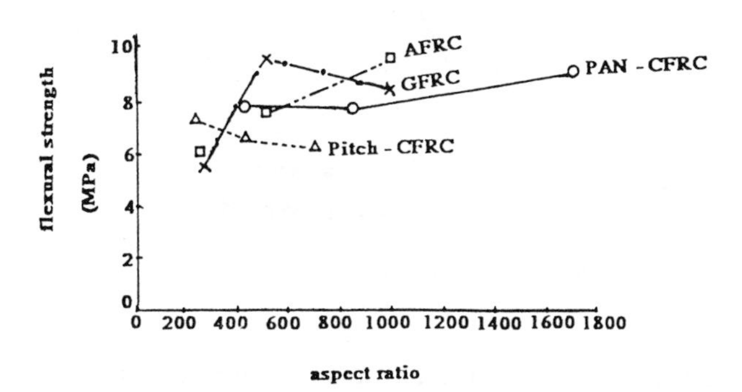

Fig. 2. Relationship of flexural strength vs. aspect ratio in various FRC

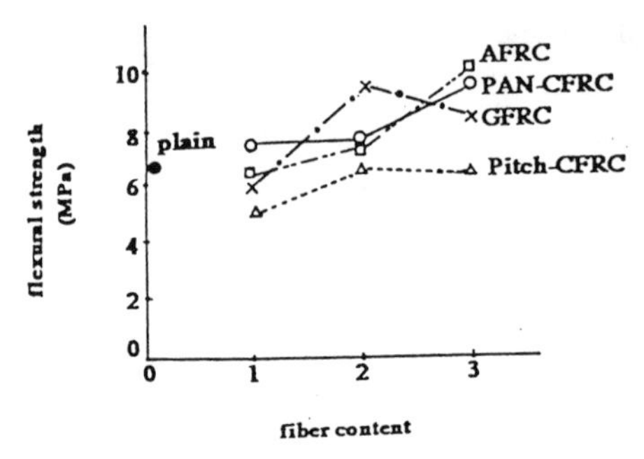

Fig. 3. Relationship of flexural strength vs. fiber content in various FRC

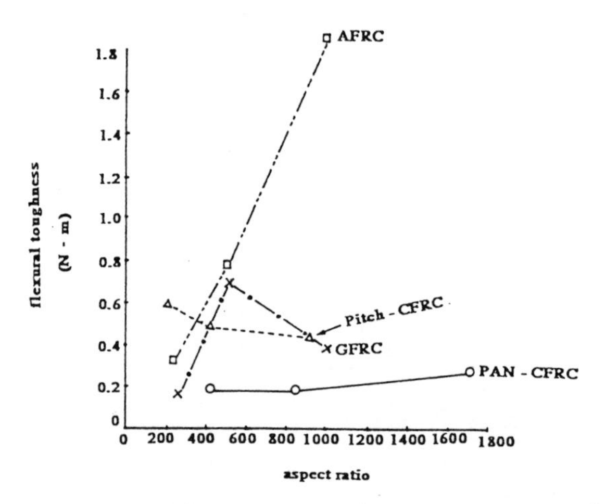

Fig. 4. Relationship of flexural toughness vs. aspect ratio in various FRC

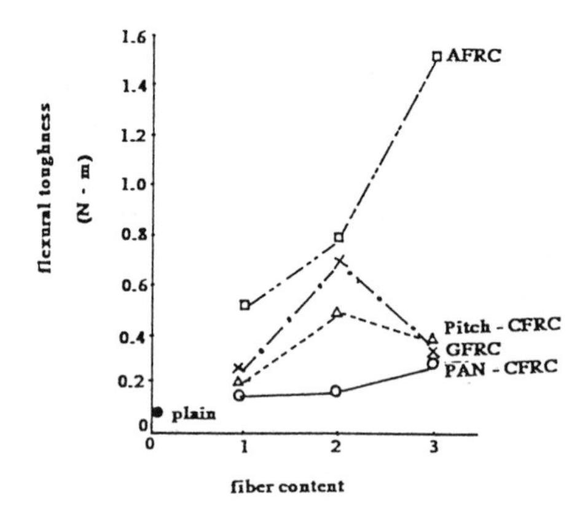

Fig. 5. Relationship of flexural toughness vs. fiber content in various FRC

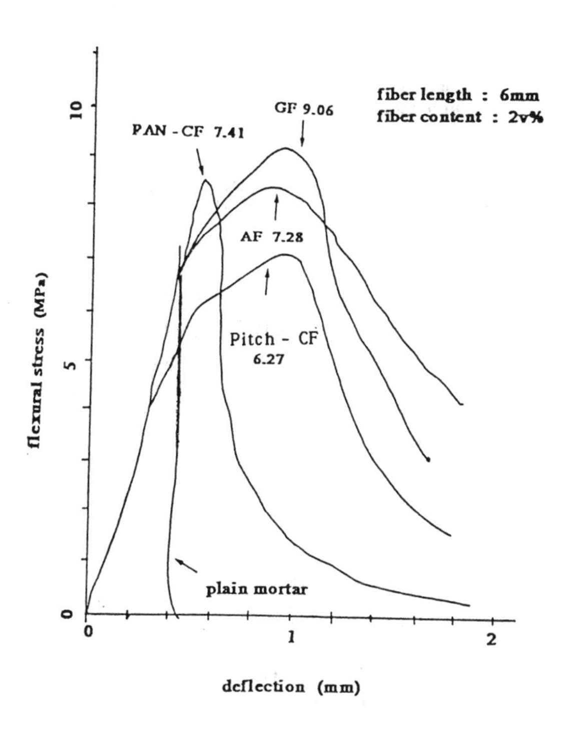

Fig. 6. Flexural stress - deflection curves of various types of FRC

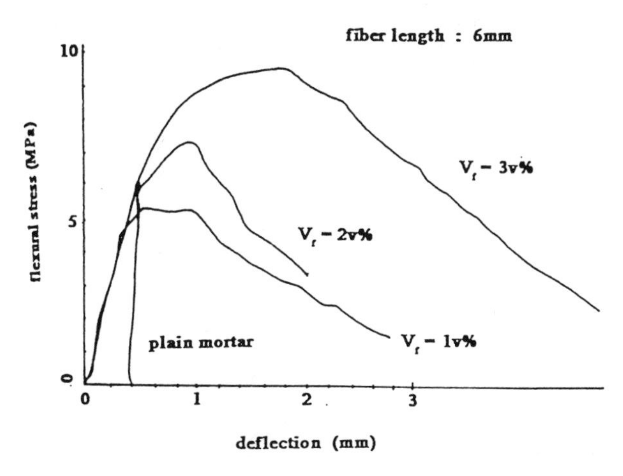

Fig.7. Change of Flexural stress-deflection curves with fiber content of AFRC

The flexural Young's moduli of all FRC are about the same as the flexural Young's modulus of matrix mortar , and that they do not depend on fiber content and aspect ratio.

3 . 3 Results of tensile test

Table 5 and Figs. 8-15 show the results of tensile test. Although the tensile strength of pitch-CFRC shows a gradual decrease with aspect ratio, that of PAN-CFRC, AFRC and GFRC becomes greater with aspect ratio and fiber content(Figs. 8, 9). Figs. 10, 11 show the relationships of tensile Young's modulus vs. aspect ratio and fiber content. Tensile Young's moduli were obtained from tensile stress-strain curves. Strains in this experiment were obtained using the diplacements of the cross head, so that values of tensile Young's modulus shown are smaller than the ordinarily reported values. Fig. 12 shown the tensile stress-strain curves of four types of FRC with fiber content of 2% and fiber length of 6mm. Each FRC show the tensile strength mostly corresponding to each fiber tensile strength, but in spite of the fact

Table 5. Results of tensile test

fiber	fiber length (mm) fiber content (v/v %)	tensile stress at proportion limit (MPa)	ultimate tensile strengh (MPa)	tensile Young's modulus (GPa)
PAN	3－2	4. 21	4. 21	3. 32
	6－1	2. 08	2. 09	2. 62
	6－2	2. 83	2. 83	2. 61
	6－3	3. 65	3. 66	3. 57
	12－2	4. 46	4. 46	2. 97
Pitch	3－2	2. 24	2. 45	4. 07
	6－1	1. 49	1. 50	2. 06
	6－2	2. 11	2. 22	2. 88
	6－3	2. 18	2. 37	3. 44
	10－2	1. 19	1. 55	1. 74
Aramid	3－2	1. 63	1. 67	2. 27
	6－1	2. 27	2. 56	3. 17
	6－2	2. 14	2. 48	2. 51
	6－3	3. 48	3. 65	3. 24
	12－2	2. 67	2. 76	3. 44
Glass	3－2	1. 39	1. 45	1. 99
	6－1	1. 74	2. 05	2. 39
	6－2	1. 45	1. 62	2. 13
	6－3	3. 07	3. 28	2. 62
	12－2	2. 71	3. 15	3. 15
	plain	1. 71	1. 71	1. 95

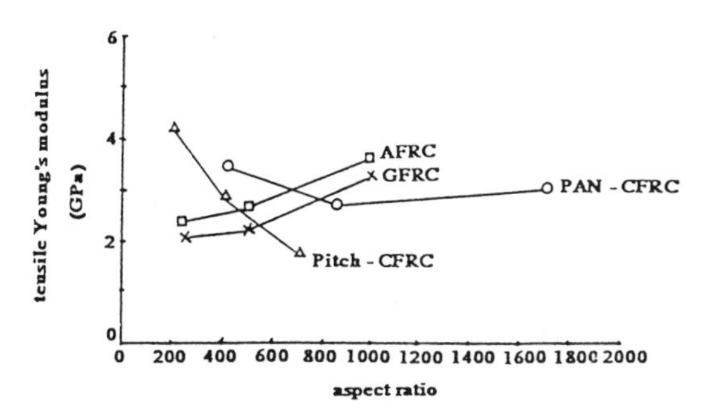

Fig. 8. Relationship oftensile strength vs. aspect ratio in various FRC

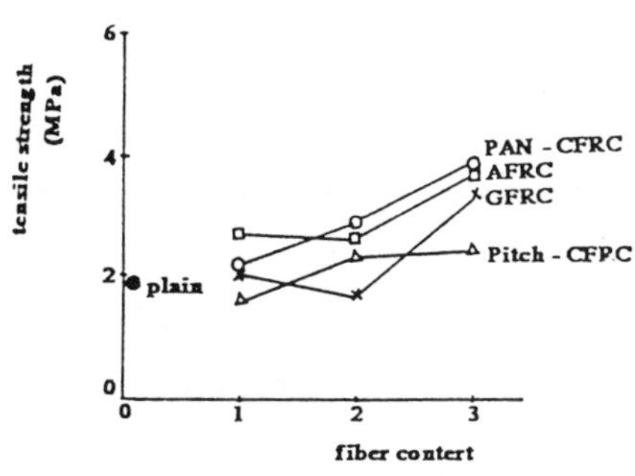

Fig. 9. Relationship of tensile strength vs. fiber content in various FRC

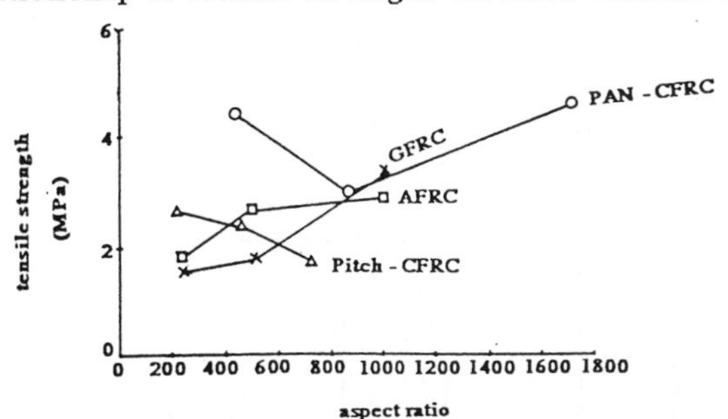

Fig. 10. Relationship of tensile Young's modulus vs. aspect ratio in various FRC

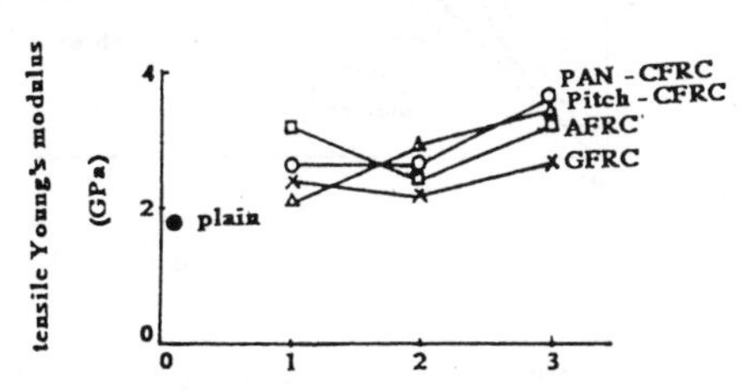

Fig. 11. Relationship of tensile Young's modulusvs. fiber content in various FRC

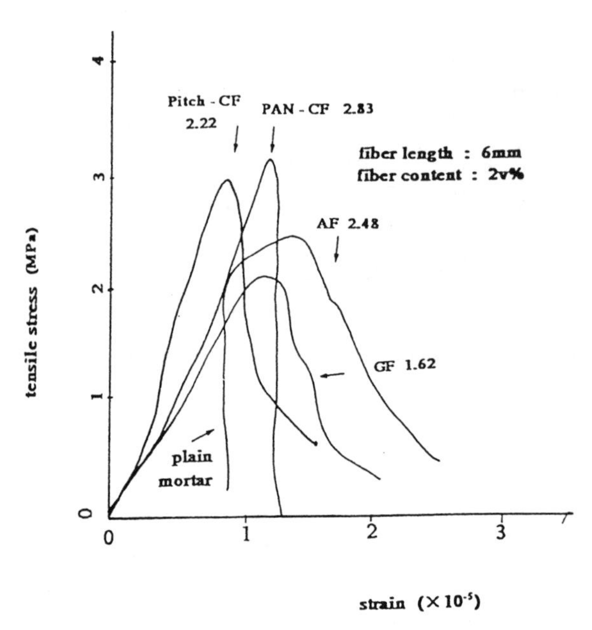

Fig. 12. Tensile stress-strain curves of various types of FRC

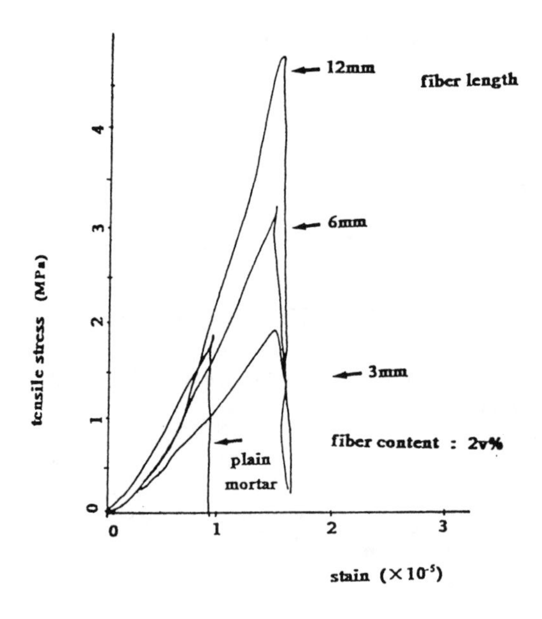

Fig. 13. Change of tensile stress-strain curves with fiber length of PAN - C FRC

that pitch-GP carbon fiber has more than 4 times lower tensile strength than that of PAN-HT carbon fiber, pitch-CFRC has the tensile strength comparable to that of PAN-CFRC. This could be due to the high adhesion to cement paste of pitch-GP carbon fiber compared to PAN-HT carbon fiber. As described later, this can be explained by the comparison of fiber effective coefficients both carbon fibers. Fig. 13 shows the change of tensile stress-strain curve of PAN-CFRC with fiber length. It could be seen that fiber length has a great infleuence on the tensile behaviour.

Generally, the tensile strength of each FRC can be predicted from each fiber tensile strength by composite law as follows :

$$F_t = \alpha (L / D) V_f \ \sigma_f \qquad (1)$$

, where F_t is the tensile strength of chopped fiber-FRC, α ; fiber effective coefficient, L ; fiber length, D ; fiber diameter, V_f ; fiber content, σ_f ; fiber tensile strength ,and L/D expresses the aspect ratio of fiber.

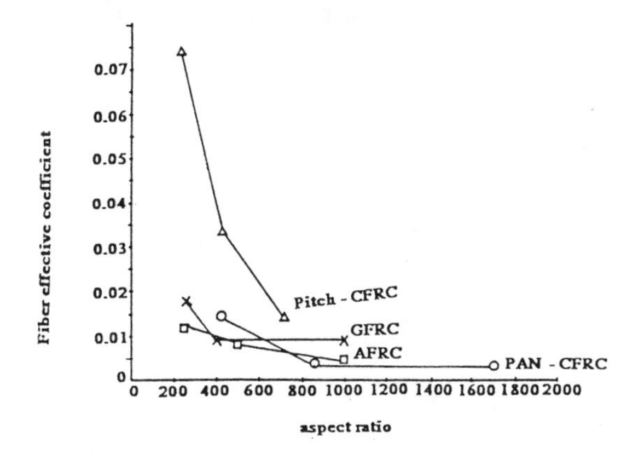

Fig. 14. Relationship of fiber effective coefficient vs. aspect ratio in various FRC

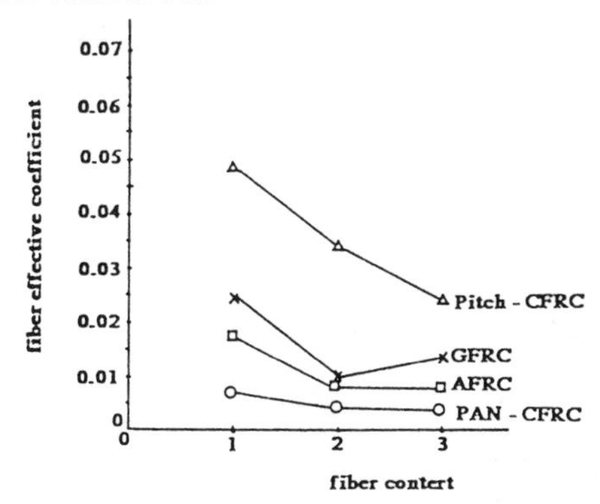

Fig. 15 Relationship of fiber effective coefficient vs. fiber content in various FRC

If this composition law can be perfectly applied for various chopped fiber-FRC, α becomes constant, but various factors have infleuences on the tensile strength of chopped fiber FRC, and in real situations α is not always a constant. Figs. 14, 15 show the relationships berween fiber effective coefficient and aspect ratio and fiber content. It is found that fiber effective coefficients of each FRC decrease monotonically with aspect ratio and fiber content, and that the fiber effective coefficients of pitch-GP carbon fiber have the greatest values, and those of PAN-HT carbon fiber have the smallest values. Aramid and glass fiber have the medium values. This variation is in inverse proportion to the tensile strength of each fiber, and is considered to be due to the difference of adhesion among fibers to cement paste, that is the high adhesion to cement paste.

3 . 4 Results of compressive test

Table 6 and Figs. 16-18 show the results of compressive test. Although the compressive strength of pitch-CFRC shows a gradual decrease with aspect ratio anf fiber content, that of PAN-CFRC, AFRC and GFRC remains almost constant(Figs. 16, 17). The compressive strength of pitch-CFRC is 26 % less as compared to that of plain mortar. Both PAN-CFRC and GFRC shows an average reduction in compressive strength of 7% as compared to that of plain mortar.

Table 6. Results of compressive test

fiber	fiber length (mm) fiber content (v/v %)	compressive strengh (MPa)	compressive Young's modulus (GPa)
PAN	3－2	48.0	16.9
	6－1	46.2	15.3
	6－2	48.3	16.4
	6－3	44.9	17.7
	12－2	46.3	13.8
Pitch	3－2	44.4	12.0
	6－1	43.6	14.0
	6－2	37.7	13.8
	6－3	31.8	19.2
	10－2	36.8	13.0
Aramid	3－2	37.6	12.7
	6－1	35.0	14.6
	6－2	38.1	13.3
	6－3	35.0	16.7
	12－2	38.1	12.4
Glass	3－2	35.0	13.8
	6－1	49.3	19.4
	6－2	42.7	12.8
	6－3	46.6	21.9
	12－2	49.0	13.6
	plain	50.2	17.7

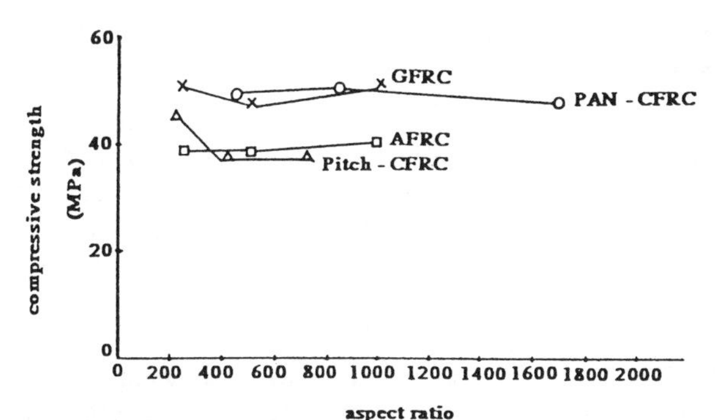

Fig. 16 Relationship of compressive strength vs. aspect ratio in various FRC

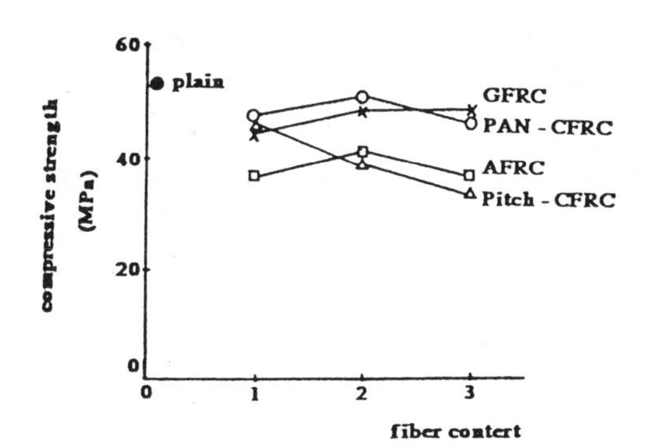

Fig. 17 Relationship of compressive strength vs. fiber content in various FRC

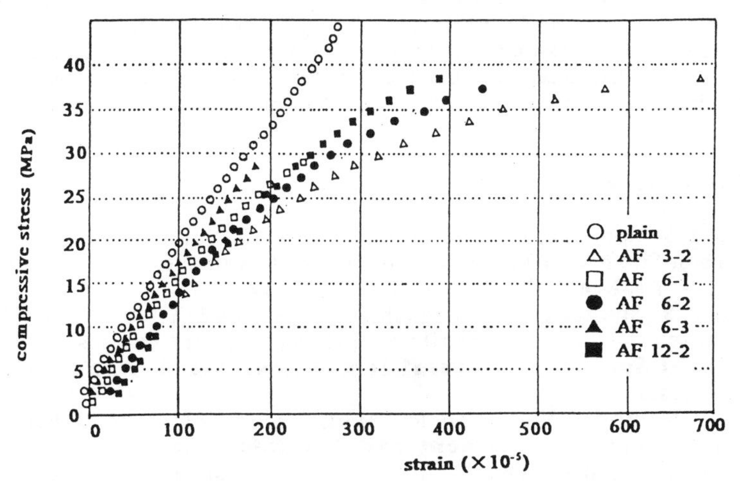

Fig. 18 C ompressive stress - strain curves of AFRC

Consequently, as far as the compressive strength is concerned, the fiber reinforcement has negative influence, and does not contribute to the increase in the compressive strength .

Fig.18 shows the compressive stress-strain curves of AFRC.

As for the compressive Young's moduli of each FRC, those of pitch-CFRC and AFRC are a little smaller than that of plain mortar, but those of PAN-CFRC and GFRC have almost the same value as that of plain mortar.

4 Conclusion

By unifying the mix proportion of mortar matrix as water cement ratio of 40% and sand cememt ratio of 25%, and by using test specimens of the same shapes and the same testing methods, mechanical characteristics of four types of chopped fiber reinforced cement composite reinforced with PAN-based high tension type and pitch-based general purpose type carbon fibers , aramid and alkali-resistant glass fiber(PAN-CFRC, pitch-CFRC. AFRC, and GFRC) were examined by bending, tensile, and compressive tests. The results of show that PAN-CFRC is mostsuitable for buildingapplications which need both high strength and high elastic limit, and AFRC is most suitable for building components which need both high strength and high tenacity. Pitch-CFRC and GFRC have the flexural characteristics good for building materials of general performance with normal strength and toughness.

5 Acknowlegement

Various cosiderations into planning of these experiments were done in the " Special Committee for Joint Research on Effective Application of Fiber Reinforced Cement and Concrete Composites as Building Materials and Components "(Chairman: Shirayama Kazuhisa, an honorary professor of Tsukuba University and a professor of Kogakuinn University) set in the Building Research Association in order to promote the Joint Research between Building Research Institute and the Association. The authors would like to heartily express their thanks to the members of Committee.

6 References

Akihama, S., Suenaga, T. and Banno, T. (1984) Mechanical Properties of Carbon Fiber Reinforced Cement Composite and the Application to Large Domes, **Kashima Corporation KICT Report**, 53 .

Akihama, S., Suenaga, T., Yamaguchi, M., Nakagawa, Y., and Suzuki, K. (1985a) Characteristics of aramid fiber reinforced cement composite(AFRC), **Proc. Annual Sym. Architectural Institute of Japan, A**, pp. 33-34(in Japanese).

Ali, A. J. Majundar, M. A. and Rayment, D. L. (1972) Carbon Fiber Reinforced Cement. **Cement and Concrete Research** , 2, pp. 201-212.

Fukushima, T. , Shirayama, K. , Hitotsuya, K. and Maruyama, T.(1990) Evaluation of Mechanical Characteristics of New Chopped Fiber Reinforced Concrete, **CAJ Proceedings of Cement/Concrete**, 44, pp.570-575 (in Japanese).

Fukushima, T., Shirayama,K., Hitotsuya, K. and Maruyama, T. (1991) A Study on Evaluation of Mechanical Characteristics of New Chopped Fiber Reinforced Concrete Mainly Using Carbon Fiber--Influences of Aspect Ratio and Fiber Content , **Proc. Annual Sym. Architectural Institute of Japan, A**, pp. 643-644(in Japanese).

Makitani, E. , Machida, E. , and Aburada, K. (1988) Fundamental Experiment about the Flexural and Drying-shrinkage Behaviours of New GRC", **Proc. Annual Sym. Architectural Institute of Japan, A**, pp. 65-66 (in Japanese).

Ohana, Y. , Amano, M. , and Endo, M. :(1985b) Properties of Carbon Fiber Reinfoeced Cement and Silica Fume, **Concrete International**, 7, pp. 58-62(1985).

77 EXPERIMENTAL STUDY OF CARBON FIBER REINFORCED CEMENT COMPOSITE USING SUPER LOW CONTRACTILE ADMIXTURE

K. YODA and T. SUENAGA
Kajima Technical Research Institute, Chofu-Shi, Tokyo, Japan
T. TAMAKI and J. MORIMOTO
Denki Kagaku Kogyo Co. Ltd, Chiyoda-Ku, Tokyo, Japan

Abstract
Carbon Fiber Reinforced Cement Composite (CFRC) has recently become a popular curtain wall material in Japan. CFRC must be heat cured when it is manufactured, but heat curing limits the varieties of finish and the dimensions of curtain wall panels, and makes quality control difficult. This study was conducted to develop a product called New CFRC which would not need heat curing and for which quality control would be easier than is the case for ordinary CFRC. The most notable characteristic of New CFRC is its binder, which is a mixture of normal portland cement and Super Low Contractile Admixture (SLCA). The hydration products of New CFRC are more consistent and thus quality control is easier. This report presents the results of tests on properties such as consistency, compressive strength, bending strength and drying shrinkage of this New CFRC. The results of a preliminary study on its ageing and durability when used as a curtain wall material are also described.
Key words: CFRC, admixture, heat curing, high early strength, drying shrinkage, durability

1 Introduction

Building construction methods have recently become modernized and efficient in Japan. Some building components are commonly being prefabricated, curtain wall panels among them. Curtain wall panels are made of precast concrete, metal, or CFRC, the latter having been developed by KAJIMA CORPORATION of Japan [1].

CFRC curtain wall panels are lighter and can be produced in more shapes than these made of precast concrete, and are less expensive than metal panels. CFRC curtain walls have gradually become more common recently. CFRC panels must be further improved and a better curing method is particularly necessary.

Currently, two curing methods are being used as follows:

Fibre Reinforced Cement and Concrete. Edited by R. N. Swamy. © 1992 RILEM.
Published by E & FN Spon, 2-6 Boundary Row, London SE1 8HN. ISBN 0 419 18130 X.

(1) steam curing and autoclave curing;
(2) steam curing only.

Method (1) restricts the varieties of finish and the dimensions of curtain wall panels. Natural building stone is damaged by autoclave curing, and autoclave size limits the size of panels. Method (2) demands strict control to maintain consistent CFRC quality. New CFRC should thus be easier to cure while maintaining consistent quality. The outstanding characteristic of New CFRC is its binding agent, which is composed of normal portland cement and Super Low Contractile Admixture (SLCA). SLCA consists of calcium aluminate and inorganic sulfate. The hydration of calcium aluminate occurs first and produces ettringite, which makes possible high early strength and stable durability [2, 3].

This report presents the results of studies on the basic properties and durability of New CFRC and compares them with those of ordinary CFRC. The results of a preliminary study into characteristics such as the ageing stability of New CFRC when used as a curtain wall material are also noted. New CFRC is intend to be used in high performance curtain walls.

2 Experiments

2.1 CFRC ingredients and their properties

The ingredients of each CFRC tested are shown in Table 1.

Cements: Normal portland cement and high early strength portland cement were used. Both were prepared to satisfy JIS R 5201 specifications (Physical Testing Methods for Cement).

Aggregates: micro-balloons and silica powder (type 1 and 11) were used. Aggregate components and properties are detailed in Table 2.

Admixture: The components and properties of the admixture are shown in Table 3. Its main components were calucium aluminate and inorganic sulfate.

Carbon fiber: The dimensions and properties of the carbon fiber are shown in Table 4.

2.2 Mix Proportions

The mix proportions of each CFRC are shown in Table 5.

2.3 Mixing

An Omni-mixer was used, and average mixing time was 5 minutes. Specimens for the bending tests were molded in accordance with JIS R 5201 specifications. Specimens for compressive tests were molded in accordance with JIS A 1108 specifications (Method of Test for Compressive Strength of Concrete). Specimens for the length change test were molded in accordance with JIS A 1129 specifications (Method of Test for Length Change of Mortar and Concrete).

Table 1. CFRC ingredients

Material	New CFRC	A1 (ordinary CFRC)	A2 (ordinary CFRC)
Cement	normal portland cement	high early strength portland cement	high early strength portland cement
Water	Plain water	Plain water	Plain water
Admixture	Super Low Contractile Admixture fly ash	- -	- -
Aggregate	silica powder 11 -	silica powder 1 -	silica powder 1 micro-balloons
Fiber	carbon	carbon	carbon
Chemical Admixture	methyl cellulose hydroxycarboxylic acid air entraining agent	methyl cellulose antifoaming agent	methyl cellulose antifoaming agent

Table 2. Aggregates

Aggregates	Chemical Components(%) SiO_2	Al_2O_3	Others	Specific gravity	Particle size (μm)
Silica powder 1	94.4	1.4	4.2	2.67	0 ~ 80
Silica powder 11	92.0	3.0	5.0	1.28	0~297
Micro-balloons	67.0	14.0	19.0	1.00	0~150

Table 3. Admixtures

Admixtures	Chemical Components(%) SiO_2	Al_2O_3	CaO	SO_3	Specific gravity	Specific surface (cm^2/g)
Super Low Contractile Admixture	-	18.9	39.2	33.6	2.90	6000
fly ash	57.4	23.4	5.7	0.9	2.11	3570

Table 4. Carbon fiber

	Diameter (μm)	Specific gravity	Tensile strength (MPa)	Modulus of elasticy (GPa)	Elongation (%)
Pitch-based carbon fiber	18.0	1.63	765	37	2.1

Table 5. Mix proportions of CFRC

CFRC	water cement ratio(%) [$W/(C+A)$]	aggregate cement ratio [$S/(C+A)$]	fiber volume content Vf(%)	fiber length lf(mm)	Unit Weight (kg/m^3)						
					cement	water	Super Low Contractile Admixture	fly ash	silica powder (1,11)	micro-balloons	chemical admixture
New	50.0	0.10	2.0	3.0	656	400	144	267	185(11)	-	11.5*
A1	81.2	0.84	2.0	3.0	650	528	-	-	545(1)	-	7.2**
A2	73.7	0.43	2.0	3.0	716	528	-	-	207(1)	99	7.6**

* methyl cellulose, hydroxycarboxylic acid, and air entraining agent
**methyl cellulose and antifoaming agent

2.4 Casting and curing
All CFRC specimens were cast in a climate-controlled room at 20 °C and 60% RH. The molds were stripped from all specimens 24 hours after pouring. New CFRC specimens were cured under the above conditions until they reached the age specified for the tests. Other CFRC specimens were cured with steam at 40 °C for 7 hours, then autoclaved at 180 °C and 10 atm for 5 hours, and then cured in air at 20 °C and 60% RH. Specimens for outdoor exposure tests were placed outdoors after 28 days.

2.5 Specimens shapes and dimensions, and testing methods
Flow test: Performed in accordance with JIS R 5201 (Physical Testing Methods for Cement).

Unit weight: Measured in accordance with JIS A 1116 (Method of Test for Unit Weight and Air Content of Fresh Concrete).

Compressive test: The cylinder used for the compressive strength test was 5 cm in diameter and 10 cm in length. The test was carried out with a universal testing machine that had a capacity of 100 tons, in accordance with JIS A 1108 (Method of Test for Compressive Strength of Concrete) specifications.

Bending test: The size of the specimen was 16 cm (L) X 4 cm (W), X 4 cm (T). Testing method was Center-point bending test with 10 cm long span. An Instrolon testing machine with a cross head speed of 0.5 mm per minute was used, and the load-deflection curve was recorded by the machine's internal recorder.

Length Change: The size of the specimen was same as Bending test. A comparator was used in accordance with JIS A 1129 (Method of Test for Length Change of Mortar and Concrete).

3 Results and discussions

3.1 Properties of Fresh CFRC

Flow (consistency): The relationship between flow and time are shown in Figure 1. All fresh CFRCs flowed 200±20 mm and change was negligable up to 60 minutes. Thus, the consistency of New CFRC can be maintained by using same casting method and handling time as used with ordinary CFRC.

Unit weight: Approximate unit weights are shown in Figure 1. The unit weight of New CFRC was roughly 1600kg/l, about the same as that of ordinary CFRC A2 and about 200kg/l lighter than that of ordinary CFRC A1.

3.2 Properties of hardened CFRC

Bending stress: Bending stress-deflection curves are shown in Figure 2. The bending stress of New CFRC was much larger than that of the ordinary CFRCs. Deflection and absorbed energy (area under the curve) were much larger for New CFRC than for the ordinary CFRCs. The Young's modulus of New CFRC was smaller than that of CFRC. These results arise from differences in the results of hydration among these cements. The density of hydrates in air-cured New CFRC is less than that of ordinary autoclaved-cured CFRCs, since autoclave curing produces higher hydrate density.

Compressive Strength: The compressive strengths of the CFRCs are shown in Figure 3. The compressive strength of

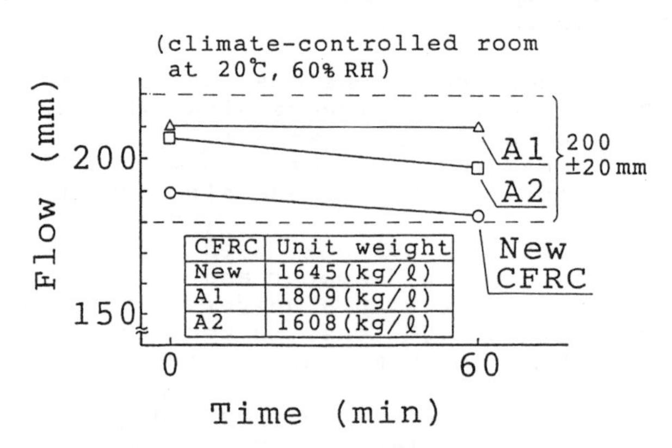

Fig.1. Relationship between flow and time

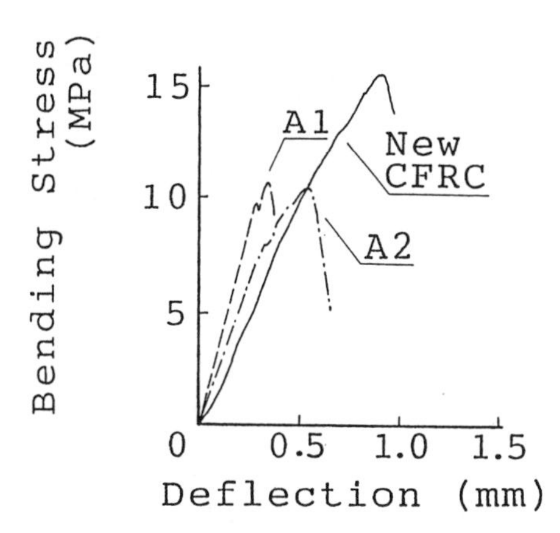

Fig.2. Bending stress-deflection curves

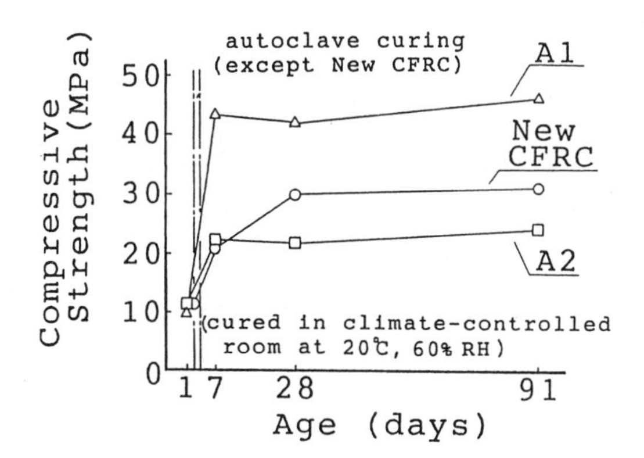

Fig.3. Compressive Strength

New CFRC cured for one day was 12.3 MPa, and thus molds can be stripped from this material one day after pouring without resorting to accelerated curing by heating. The compresive strength of New CFRC continued to increase, whereas the compressive strength of ordinary CFRCs remained constant after autoclave curing. This is, again, due to the results of hydration process differing between New CFRC and orinary CFRCs.

The relationship between compressive strength and time is also shown in Figure 4. The compressive strength of New CFRC gradually increased, especially when it was outdoors. Outdoor curing produces better results than indoor curing with regard to the compressive strength of New CFRC.

Length change: Length changes of CFRCs are shown in Figure 5. The length change of New CFRC was -4.32×10^{-4}

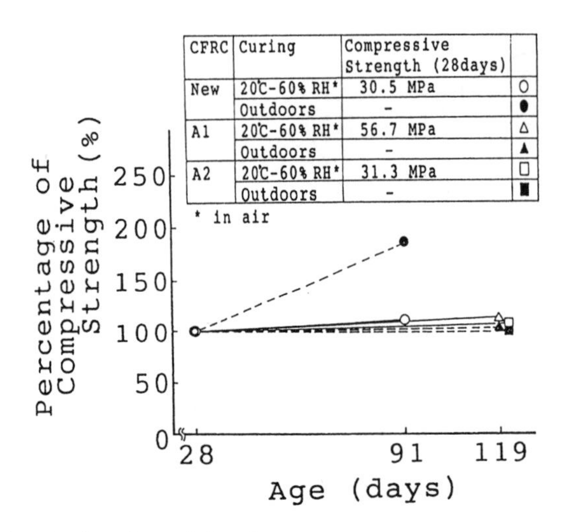

Fig.4. Relationship between compressive strength and curing condition

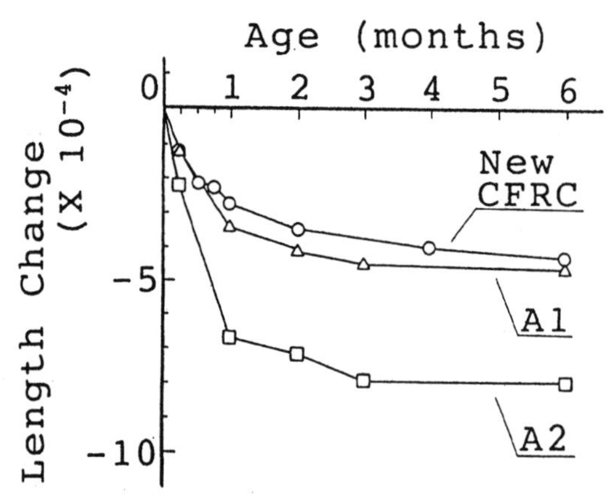

Fig.5. Drying shrinkage

at 6 months. Since length change is caused by drying shrinkage, the drying shrinkage of New CFRC must be less than that of ordinary CFRC.

3.3 Full scale model specimen subjected to outdoor exposure

As illustrated by Photo 1, full scale curtain wall panels made of New CFRC (2 m X 5 m) were manufactured and then exposed to natural weather for about 6 months. There was no warp age or flexure of the panels. Small cracks due to drying shrinkage were observed, but such cracks ordinarily occur in all concrete products. New CFRC thus appears to be a stable curtain wall material.

Photo 1. Curtain wall made of New CFRC panels

4 Conclusion

(1) The flow of all CFRCs tested changed little up to about 60 minutes. The consistency of New CFRC is the same as that of ordinary CFRC.
(2) The unit weight of New CFRC is roughly 1600kg/l. It is lighter than ordinary CFRC.
(3) The bending strength of New CFRC is higher than that of ordinary CFRC. The deflection and absorbed energy of New CFRC are much larger than those of ordinary CFRC.
(4) New CFRC develops adequate compressive strength in a single day. Molds can be stripped one day after pouring without accelerated curing by heating. The compressive strength of New CFRC increased gradually, especially when cured outdoors.
(5) The length change of New CFRC is -4.32×10^{-4} at 6 months. It has less drying shrinkage than ordinary CFRC.
(6) Full-scale New CFRC curtain wall panels (2 m X 5 m) were exposed to outdoor weathering for about 6 months. There was no warp or flexure. New CFRC is a stable curtain wall material.

5 References

[1] Akihama, S., et al. "Mechanical Properties of Carbon Fiber Reinforced Cement Composite and The Application to Large Domes", KICT Report, No. 53, pp. 31-56, July 1984.
[2] Yoda, K., et al. "Carbon Fiber Reinforced Concrete Using Newly Developed Low-Contractile Admixture", Proceedings of the Japan Concrete Institute Vol. 12, No. 1, pp. 1053-1056, June 1990. (in Japanese)
[3] Yoda, K., et al. "Durability of Carbon Fiber Reinforced Concrete Using Newly Developed Low-Contractile Admixture", Proceedings of the Japan Concrete Institute Vol. 13, No. 1, pp. 757-760, June 1991. (in Japanese)

78 DEVELOPMENT OF PANELS MADE WITH CONTINUOUS CARBON FIBRE REINFORCED CEMENT COMPOSITE

T. HATTORI, K. SUZIKI, T. NISHIGAKI and
T. MATSUHASHI
Technology Research Center, Taisei Corporation, Yokohama, Japan
K. SAITO and K. SHIRAKI
Toho Rayon Co. Ltd, Tokyo, Japan

Abstract
A new type of continuous carbon fibre reinforced cement composite (CFRC) with high strength, durability and high fire resistance was developed by using PAN-type carbon fibre and cement as matrix. This new material is fabricated by similar technologies used in resin-based composites such as prepregging lay-up, filament winding and pultrusion. By using cement as matrix, one of the characteristics of the CFRC is its high fire resistance, compared with ordinary resin-based composites. On the other hand, it was very difficult to increase the fibre content in the cement matrix, because the mean diameter of ordinary cement (20-30 μm) is too large compared with that of carbon fibre (7 μm). This problem was solved by using fine cement and high water reducing admixture.
This paper presents the characteristics of this new material, such as fire resistance, flexural strength, frost resistance and shape stability. The development of the cross-ply laminated panel made with CFRC and its application to the external wall of the actual building are also described.
Keywords: Continuous Carbon Fibre, Cement Matrix, Panel, Application.

1 Introduction

The continuous carbon fibre reinforced composites made with organic resin matrix (FRP) are widely used as aircraft wing materials and many types of sporting goods thanks to their high strength and light weight. However, it is well known that FRP has only low heat or fire resistance. In order to use the continuous carbon fibre reinforced composites for building materials which should have the high fire resistance, we selected cement paste as matrix.

In this new composite(CFRC), PAN-based high strength type carbon fibre (Besfight HTA-7-1200 manufactured by Toho Rayon) is used. The properties of this fibre are shown in Table 1. The diameter, tensile strength and elastic modulus are 7 μm, 3630 MPa and 235 GPa, respectively. In order to increase the fibre content in cement matrix and to improve the bond strength between carbon fibre and cement matrix, fine cement and silica fume are used. The high water reducing admixture is also used to obtain good workability. Water cementious ratio including silica fume is 24% by weight, and the compressive strength of cement matrix is more than 80 MPa.

Fibre Reinforced Cement and Concrete. Edited by R. N. Swamy. © 1992 RILEM.
Published by E & FN Spon, 2-6 Boundary Row, London SE1 8HN. ISBN 0 419 18130 X.

Fig. 1 shows the process for making unidirectional sheet (prepreg) of the CFRC. In this process, each individual fibre is uniformly distributed into the cement matrix without void formation, and the maximum fibre content of 20% in volume can be obtained.
Fig. 2 shows the cross-section of the typical CFRC by scanning electron microscope. Those round shapes show individual carbon fibres and that they are distributed uniformly in the cement matrix.

Table 1. Properties of PAN-based Carbon Fibre

Diameter	7 μm
Elongation	1.6%
Tensile Strength	3630 MPa
Elastic Modulus	235 GPa
Specific Gravity	1.77
Specific Heat	711 J/kgK
Thermal Conductivity	17.4 W/mK
Thermal Expansion Coeff.	$0.1 \times 10^{-6}/℃$

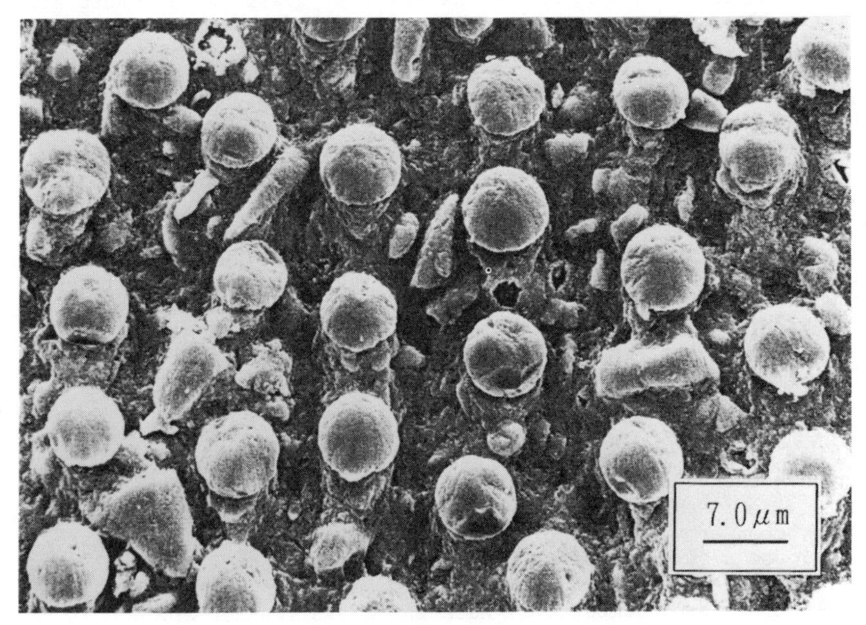

Fig.2. Cross-section of CFRC by S.E.M.

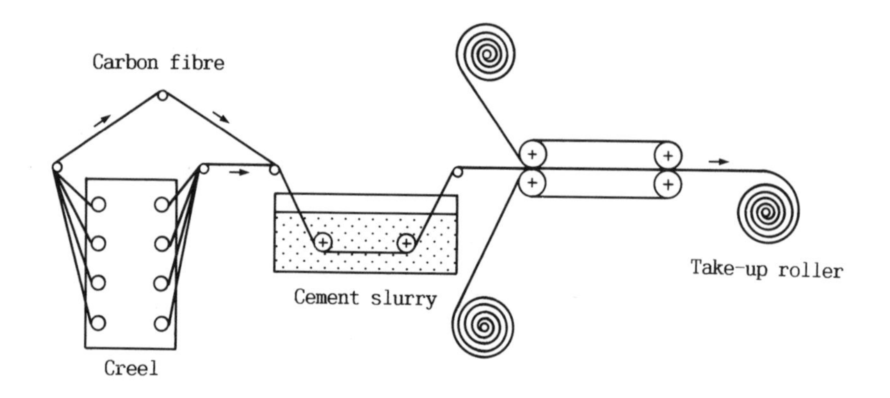

Fig.1. Production Process of CFRC

2 Development of CFRC panel

2.1 Manufacturing

Table 2 shows the mix proportions of cement matrix. In this case, SHIRASU balloon (hollow glass micro spheres from the SHIRASU that is one kind of acid volcanic glass) is added to increase the frost resistance. The unit weight and viscosity of fresh cement paste are 1.9 kg/l, 10 poise respectively.

The designed fibre content of prepreg is V_f=4.5% by volume. The reason why high fibre content is not required is that V_f=4.5% is sufficient for the curtain wall panel which must have flexural rigidity as well as flexural strength.

The prepregs, 300mm in width and 2.5mm in thickness, were produced and laminated to four ply sheet on the flat plate mold.

Both unidirectional[0,0,0,0 degree] and bidirectional[0,90,90,0 degree] four ply sheets were produced, where 0 degree or 90 degree means the direction of carbon fibre of each laminated prepreg. The laminated sheets were kept to set at room temperature 20-25℃ for 2 days while preventing vaporization of water in the laminates. After the setting, the products were cured in hot water (50℃) for 2 days. Then, they were kept in the room, where temperature and relative humidity were 25℃ and 80% respectively, more than four weeks before the following several tests were carried out.

2.2 Fire Resistance

As the components of CFRC are inorganic incombustible substances, CFRC is also incombustible material. The superior fire resistance of CFRC has been proved by simple heating test compared with mild steel. Tables 3 and 4 show the official data of the CFRC bidirectional panel tested by JIS A 1321. From these test data, CFRC panel has been recognized as the Incombustible Material Class I under dry condition.

Table 2. Mix Proportion of Cement Matrix

	Mix Proportion (by weight)
Water	26
Fine Portland Blast-furnace Slag Cement (mean diameter : 8 μm)	100
Silica Fume (mean diameter : 0.1 μm)	10
SHIRASU Balloon (mean diameter : 40 μm)	2.5
High Water Reducing Admixture	5

Table 3. Base Material Test by JIS A 1321 *)

Increase of Temperature	39℃
Reduction of Weight	14.4%
Judgment	passed

*):The specimen(40×40×50mm) was exposed in 750℃ hot air, then the increase of temperature after 20 minutes must be below 50℃.

Table 4. Surface Test by JIS A 1321 **)

Coefficient of Smoke (<30)	1.2
Reduction of Weight	3.2%
Time of Flaming after gas burner was stopped (<30 sec.)	0 sec.
Temperature of Back Side	104℃
Melting or Cracking	none
Judgment	passed

**):The specimen(220×220mm) was heated by gas burner and electric heater, then the smoking, flaming and deformation were checked.

2.3 Flexural strength and fatigue strength

The static flexural test was carried out under the following conditions.

simple supported 3-point flexural test : Span 200mm(l)
size of specimen : 300(length)×20(width:b)×10(thickness:t)mm
bidirectional[0,90,90,0] four ply panel
The direction of the fibre in the top and the bottom lamina is longitudinal.
number of specimens : 8
span depth ratio : 20

The test results are shown in Table 5, where flexural strength and flexural elastic modulus are calculated as following.

$\sigma_b = P_{max}\, l/4Z$

$E_b = Pl^3/48I\delta$

where, P_{max} : maximum load
P : half of maximum load ($P_{max}/2$)
δ : deflection of the mid-span when load is P
Z : modulus of section ($bt^2/6$)
I : moment of inertia of cross section ($bt^3/12$)

The fracture modes of all specimens were the break of carbon fibres in tensile lamina. The contribution ratios of fibres to the strength and modulus in Table 5 were obtained from comparing the test data with the theoretical values. The theoretical values were calculated with the assumption that the specimens were the sandwich panels made with upper and lower lamina of CFRC composite and core two laminas of only cement matrix. With this assumption, the Z and I of sandwich panel were reduced to 92.8% of above-mentioned Z($bt^2/6$) and I($bt^3/12$). The contribution ratios of flexural strength and flexural elastic modulus were 71% and 93% respectively, and it was recognized that the superior potential of carbon fibre worked out well.

The flexural fatigue test was carried out by using Instron-1331 testing machine. The size of specimen was the same as in above-mentioned flexural test, and the flexural repeated loading was continued with 0.3Hz. The number of specimens was 9, and the amplitudes of flexural repeated stress were in the range of 70-95% of the static flexural strength. Fig. 3 shows the test results in semilog S-N diagram. The marks with arrow in Fig. 3 show the specimens which did not break by the increase of deflection more than 20mm when the testing machine was controlled to stop automatically. Besides, those unfailed specimens were subjected to the static flexural test, and the reduction of flexural strength by their repeated loading history was scarcely recognized. From Fig. 3, it is estimated that the fatigue strength of CFRC at 10^7 cycles is approximately 70% of its static flexural strength.

Table 5. Flexural Test (number of specimen : 8)

Flexural strength			Flexural elastic modulus		
average value MPa	standard deviation MPa	contribution ratio of fibre %	average value GPa	standard deviation GPa	contribution ratio of fibre %
107	10.2	71	29.5	0.72	93

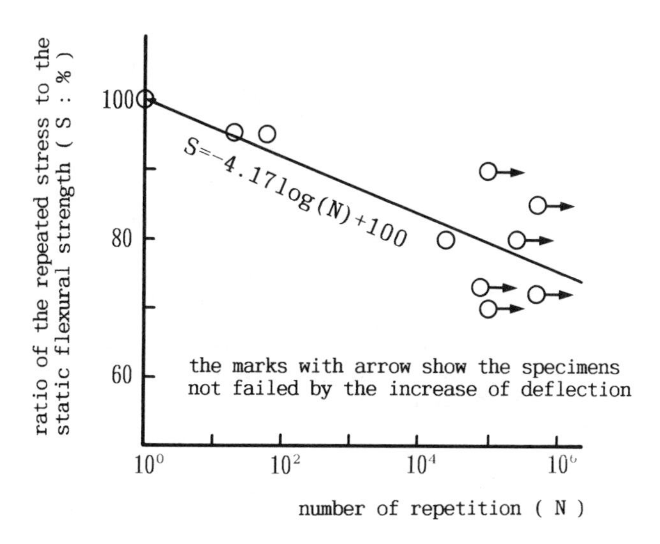

Fig.3. Flexural Fatigue Strength in Semilog S-N Diagram

2.4 Frost resistance

The frost resistance of CFRC was investigated by the freezing and thawing test according to ASTM C 666. The specimens were made from bidirectional panel, and their size was the same as the above-mentioned flexural specimens. The cycles of freezing and thawing were as follows, and after 50, 100, 150 cycles the static flexural tests were carried out.

→ -5℃ → 5℃ →
3hrs 3hrs 3hrs 3hrs

The test results are shown in Table 6. Both in flexural strength and flexural rigidity, the reductions affected by freezing and thawing were not entirely recognized. It was considered that the reason why the CFRC showed superior frost resistance without dosage of the air entrained admixture could be explained from the low water cementious ratio and the cushion effect of the SHIRASU balloon.

Table 6. Freezing and Thawing Test

Cycles of freezing and thawing	Flexural strength MPa	Flexural elastic modulus GPa
(static)	(107)	(29.5)
50	106	30.2
100	113	30.6
150	100	29.7

2.5 Shape stability

The matrix of the CFRC is almost neat cement paste without aggregate. So, the shrinkage of the matrix is considerably large in spite of its low water cementious ratio. To improve the shape stability of the CRFC, the organic cement-shrinkage-reducing agents (CSR) represented by lower alcohol alkylene oxide adducts was added to cement matrix and unidirectional and bidirectional four ply panels were made. The contents of the CSR were 1% and 2% of cement by weight in the mix proportions (Table 2). The size of specimen was 160mm(length)×40mm (width)×10mm(thickness), and the panels were cut in this size after 2 days curing. Then, after another 2 days curing in hot water (50℃), the specimens were cured in water (20 ℃) till the seventh day, when they were moved to the room (20℃, 60% R.H.) and the length change has been measured till 13 weeks by comparator method according to JIS A 1129. The specimens are shown in Table 7, where the fibre direction 0 or 90 coincides with the length direction or the width direction of the specimens, respectively. That is, the specimens named [0,0,0,0] or [90,90,90,90] were made by cutting from unidirectional panels, and those named [0,90,90,0] or [90,0,0,90] were from bidirectional panels.

The test results are shown in Fig. 4, 5, 6. In the case of [90,90,90, 90] in which the fibres make a right angle with the measuring direction, the shrinkage strain is largest (0.14% at 13 weeks, without CSR). This value is considered to be almost same as the shrinkage strain of the matrix, because the restraint to shrinkage by fibres is scarcely expected. On the other hand, this restraining effect is obvious in the case of [0,0,0,0] in which the shrinkage strain (0.068% at 13 weeks, without CSR) is half the value of [90,90,90,90].

The shrinkage strain of [0,90,90,0] or [90,0,0,90] without CSR are about 65% of that of [90,90,90,90]. By adding CSR with 1% or 2% of cement by weight, the shrinkage strains reduced remarkably to 60-80% of those without CSR. But, there was no obvious difference between 1% dose and 2% dose.

Table 7. Specimens of Shrinkage Test

content of CSR*) to cement by weight	direction of carbon fibre	number of each kind of specimen
0,1,2%	[0,0,0,0]	3
0,1,2%	[90,90,90,90]	3
0,1,2%	[0,90,90,0]	3
0,1,2%	[90,0,0,90]	3

*):the organic cement-shrinkage-reducing agents represented by lower alcohol alkylene oxide adducts

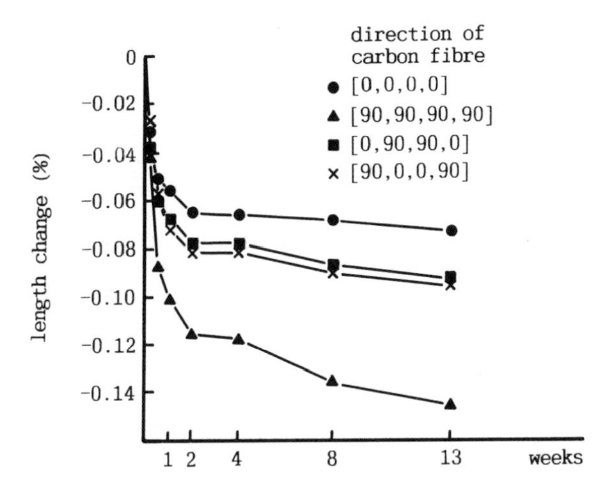

Fig.4. Shrinkage Test of CFRC (without CSR)

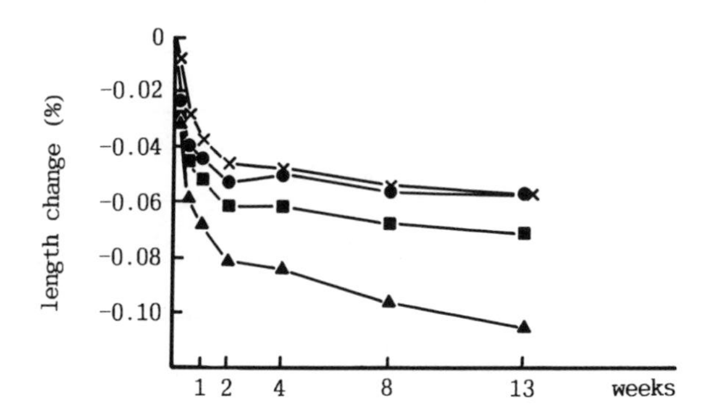

Fig.5. Shrinkage Test of CFRC (CSR : 1% of cement by weight)

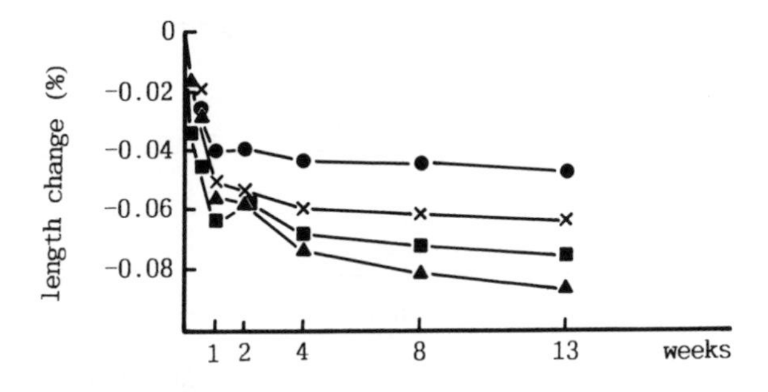

Fig.6. Shrinkage Test of CFRC (CSR : 2% of cement by weight)

3 Application to the external wall

The CFRC panels were applied to the external finishing wall of the laboratory building in Biotechnology Research Center, Taisei Corporation. This building was built in August 1990 and located near northern verge of Tokyo Bay, where the wind and salt attack by sea water are severe. The building is one-storey steel frame structure with the floor area of 277 m^2, and the CFRC panels were applied to all its external walls which area was 230 m^2 . The standard module of the panel was 890mm×890mm, and the characteristics of the panel are shown in Table 8. The content of carbon fibre is 4% by volume and the thickness is 12mm made by four cross ply laminas [0,90,90,0].
The design wind load is 2000 Pa, and the flexural stress in panel calculated by this load is 8.4 MPa with assumption that panel is simply supported at two opposite edges. This working stress by wind has enough safety as compared with the flexural strength of the panel (96 MPa in static, 67 MPa in fatigue). The maximum deflection of the panel by the wind load is estimated to be 4.0mm, which is smaller than half of the design allowable deflection 10mm. Thus, the CFRC panel has enough safety both in the flexural strength and rigidity.

Table 8. Characteristics of CFRC Panel

Fibre Content	4% by volume
Matrix	Table 2 with CSR (1% of cement by weight)
Size	890×890mm
Thickness (four plied)	12mm [0,90,90,0]
Specific Gravity	2.0
Flexural Strength	96 MPa
Flexural Elastic Modulus	28.5 GPa
Shrinkage Strain	0.07%
Water Absorption Ratio	5% by weight
Thermal Conductivity	1.16 W/mK
Thermal Expansion Coeff.	$4\times10^{-6}/℃$
Fire Resistance	Incombustible Material Class I
Others	High Frost Resistance Screw Available Electromagnetic Shielding effect

The critical point of the panel in the design is in its supporting detail as following.

The external surface of the panel was polished by fine abrasive paper in its manufacturing process, and the beautiful hairlines of carbon fibre appeared glittering in the sun, and then it was coated with the clear acrylic silicone paint to increase its durability. It looks rather like well polished natural stone. This unique beautiful texture was highly appreciated by the designer, and the fixing detail like stone works was required in which the fixing pieces had to be invisible from outside.

The two fixing details were devised as shown in Fig. 7, made with stainless steel. In Type A which is named cut end supporting system, the narrow circular arc ditch with thickness 2mm and depth 11mm is digged in the cut end of panel by diamond blade. In this system, the panel is supported by inserting the ditch directly into the supporting disc (1.0mm in thickness) with adjusting cap nut. In Type B named back supporting system, the receiving plates are screwed to the back side of panel and the panel is jointed to the supporting disc through this receiving plates. Thus, it is one of the characteristics of CFRC panel that the screw is available.

The strengths of the two supporting systems against the wind load were investigated by the static and repeated loading. The test results are shown in Table 9. Failure mode of every specimen was the delamination of the CFRC panel, and the slip out of screw was not recognized in all specimens of Type B.
The maximum loads of the static strength test of Type A and Type B were 583N and 1808N respectively, and the strength of the latter was 3.1 times higher than that of the former.

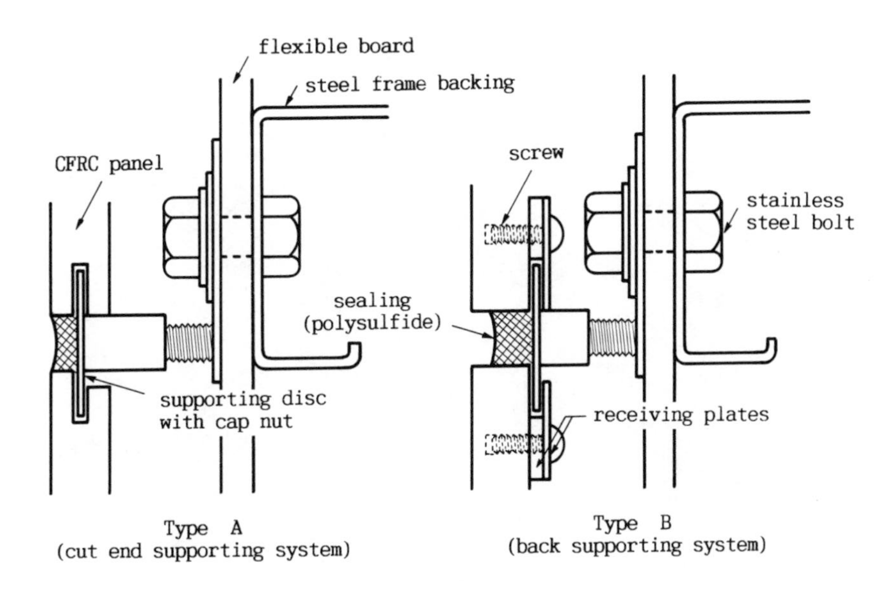

Fig.7. Fixing Details

Table 9. Bearing Test of Fixing System

Type of fixing detail	static loading			repeated loading	
	number of specimens	average strength	standard deviation	number of specimens	load ratio to static strength not failed
TYPE A	20	583 N	51.1 N	6	70%
TYPE B	10	1808 N	90.2 N	3	50%

Fig.8. General View of the Building

Fig.9. Close-up of the Corner

The fatigue test was carried out by the repeating of constant load corresponding to the range of 50, 60, 70, 80% of the static strength. Each loading range was repeated 1000 times, and the highest loading range when the specimen did not fail was defined as the fatigue strength as followings.

Type A : 70% of static strength (408 N)

Type B : 50% of static strength (904 N)

Each CFRC panel was supported with four fixing points ; two points at the bottom side and the other two points at the top side. As the reaction at one fixing point against the design wind load is 405 N, both Type A and Type B meet the design requirement. In the actual work Type B was used mainly for safety.

The panels were fixed to the building as the direction of surface carbon fibre to be vertical, and the every joint (10mm in width) was sealed with polysulfide. Fig. 8 and 9 show the photographs at the completion, and the round special pieces of panel are used at the corner. According to the follow-up observation after one year from the completion, the maximum warping is 0.7mm and the initial brilliance has been kept entirely.

4 Conclusion

A new type of continuous carbon fibre reinforced cement composite (CFRC) was developed by using PAN-type carbon fibre and cement as matrix. The superior characteristics of the CFRC, such as fire resistance, flexural properties, frost resistance and shape stability were experimentally clarified. The panels made with this new material were applied to the external wall of the actual building successfully.

5 References

Saito,H. Kawamura,N. and Kogo,Y. (1989) Development of carbon fibre reinforced Cement. 21st International SAMPE Technical Conference, Atlantic City,New Jersey.

Nishigaki,T. Suzuki,K. Matsuhashi,T. and Sasaki,H. (1991) High strength continuous carbon fibre reinforced cement composite(CFRC). The Third International Symposium of Brittle Matrix Composites, Warsaw.

Japanese Industrial Standard(JIS) A 1321. Testing Method for Incombustibility of Internal Finish Material and Procedure of Buildings.

Japanese Industrial Standard(JIS) A 1129. Methods of Test for Length Change of Mortar and Concrete.

79 LONG-TERM LOADING TESTS ON PPC BEAMS USING BRAIDED FRP RODS

T. OKAMOTO, S. MATSUBARA, M. TANIGAKI and
K. HASUO
R & D Division, Mitsui Construction Co., Tokyo, Japan

Abstract
Recently the application of resin-impregnated high-strength fiber rods(FRP) as concrete reinforcement or prestressing tendons are actively studied. The aims of these investigations are to improve durability, increase strength, or to improve the electro-magnetic characteristics of concrete structures. But usually FRP rods involve some problems such as low modulus and low bond strength to concrete. The authors have proposed braiding the fibers to improve the bond properties of FRP, and have found that partially prestressing was effective for the beams using FRP which had low sectional rigidities.

The authors conducted long term loading tests on partially prestressed concrete beams reinforced and prestressed with braided aramid fiber rods to investigate the long term reliability of these fiber rods.

The test results demonstrated that partially prestressing was effective in improving long term serviceability such as deflections and cracking. Computational method to estimate long term deflection and crack width for the beams using FRP are also discussed.
Keywords: Aramid Fiber, Braiding, Partially Prestressed Concrete, Long Term Loading, Deflection, Cracking

1 Introduction

High-strength synthetic (e.g. aramid and carbon) and glass fibers usually embedded in a resin matrix are used to produce fiber-reinforced-plastic (FRP) reinforcements and tendons for concrete members. These FRP are generally characterized by high-strength, low rigidity, low bond strength to concrete, corrosion resistance, and non-conductance etc. Recently, application of the FRP rods to improve durability[1][2], strength,[4] impact resistance[3], and electro-magnetic characteristics[4] of concrete structures is actively studied. But this new material does not have so much experience in practical application and its long

Fibre Reinforced Cement and Concrete. Edited by R. N. Swamy. © 1992 RILEM.
Published by E & FN Spon, 2-6 Boundary Row, London SE1 8HN. ISBN 0 419 18130 X.

term reliability as concrete reinforcement is still not established.

The authors have proposed braiding fibers to improve bond strength to concrete and tensile strength efficiency of the rods[5]. However, concrete members reinforced with these FRP rods show high bearing capacities, they have serviceability problems such as deflection or crack width because of low rigidity or low bond strength of the FRP. The authors have found that partially prestressing was effective to control deflections of such beams having low sectional rigidities[5][6].

The objectives of this project are to ensure the long term reliability of the FRP and to acquire basic data for the design of FRP reinforced beams through the long term loading tests on concrete beams reinforced and prestressed with braided FRP rods.

2 Braided Aramid FRP Rods

The braided FRP rods are composed of aramid fibre and epoxy resin. Fig.1 shows the surface appearance. The properties of these rods are listed in Table 1. These rods are elastic and their modulus is about one third of that of steel.

Fig.1. Braided aramid FRP rods

Table 1. Properties of braided aramid FRP rods

Type	Weight (g/m)	Ave. diameter (mm)	Sectional area (mm^2)	Fiber content (%)	Tensile strength (MPa)	Young's modulus (GPa)	Elongation (%)
K64	58	8	42	65~70	1638	64	2.0~2.3
K96	92	10	66		1530	62	
K128	125	12	90		1432	61	
K192	188	14	135		1411	60	
K256	250	16	180		1393	56	

Fig.2 shows the method and the results of bond test. The suffix s to the rod name indicates that sand is adhered on the surface of the rod. A sand adhered rod has almost the same bond capability as a deformed steel bar. Fig.3 shows the tensile creep characteristics of these rods. After 1000 hours, the ratio of the creep strains to the initial strains was about 12% regardless of tensile force intensity. This value is about three times greater than that of steel. It was observed from linear variation of creep ratio with time on logarithmic scale that most of creep strain occurs in first few weeks. The relaxation losses of the rods were similar to creep characteristics. Further details of the material properties have been described in Ref.[5].

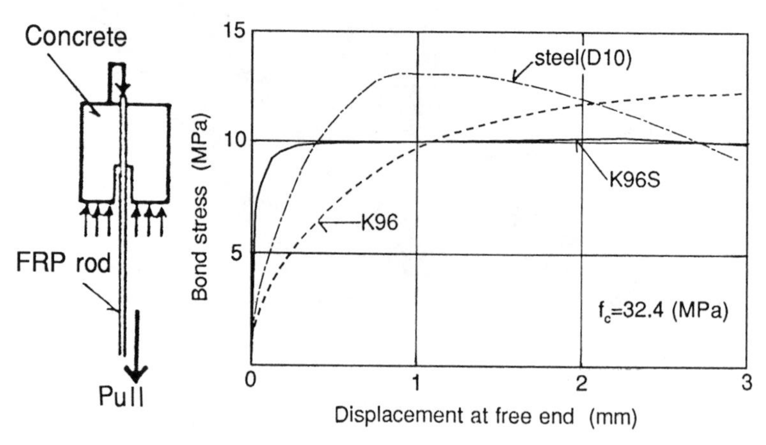

Fig.2. The bond characteristics of braided aramid FRP rods

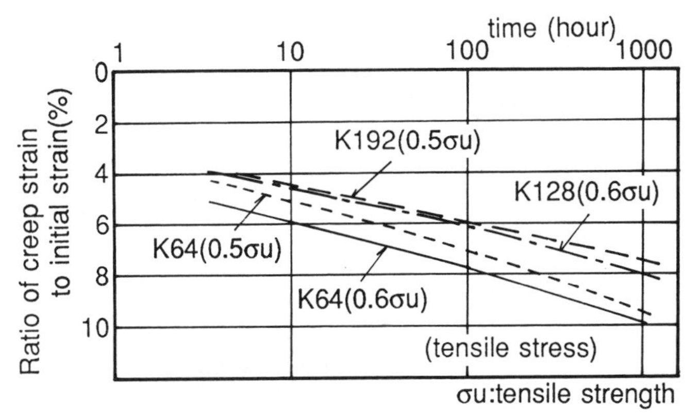

Fig.3. The creep characteristics of braided aramid FRP rods

3 Test Program

3.1 Specimens

Six specimens were fabricated as shown in Table 2. The specimens No.1 to 4 are model beams reinforced and prestressed with braided aramid FRP rods which are 360cm in length. As shown in Fig.4, they have rectangular cross sections with 22.5cm in width and 30cm in depth. The type K128S of braided aramid FRP is used as main reinforcement and the types K192 or K192S are used as prestressing tendons. The specimens No.5 and 6 are full scale beams reinforced and prestressed with braided aramid FRP rods which are 1000cm in length. As shown in Fig.4, the cross sections of these specimens are T-shaped having 35cm in width, 80cm in depth, 150cm in slab width, and 12cm in slab thickness.

Table 2. Specimen series

No	Main reinforcement	Prestressing tendons				Constant load
		Type	Effective Prestressing force (KN)	Bottom fiber effective prestress (MPa)	Tensioning	
1	3-K128S	2-K192S	49.4	1.47	Pretension	1.0 P_{cr} 28.6 KN
2			98.0	2.94		1.0 P_{cr} 37.4 KN
3						1.5 P_{cr} 56.2 KN
4		2-K192			Postension	1.0 P_{cr} 37.4 KN
5	4-K256S	5-K256	490.0	4.79		0.67 P_{cr} 13.9+9.8 KN/m
6			274.0	2.13		1.0 P_{cr} 13.9+9.8 KN/m

P_{cr}: Initial cracking load

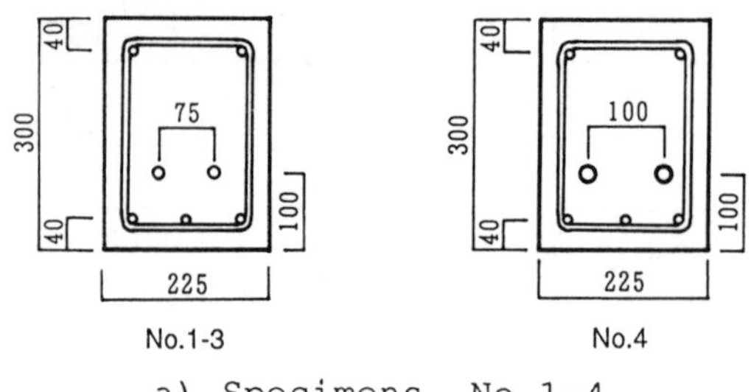

a) Specimens No.1-4

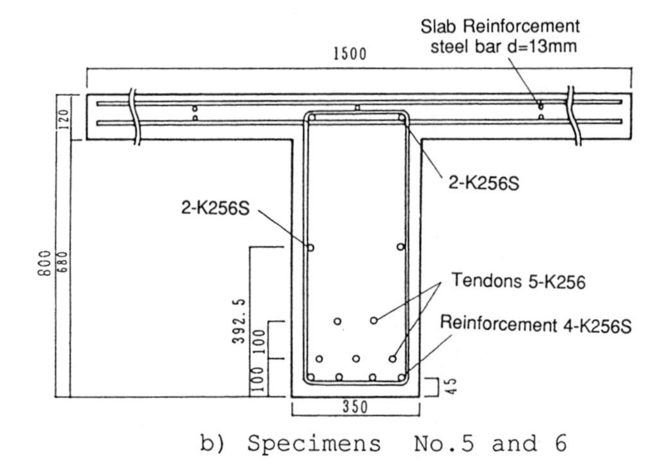

b) Specimens No.5 and 6

Fig.4. Cross sections of specimens

The type K256S of braided aramid FRP is used as main reinforcement and type K256 of braided aramid FRP is used as prestressing tendon. The shear reinforcement of No.1 to 4 consisted of steel stirrups and that of No.5 and 6 consisted of spiral shaped FRP of type K96.

3.2 Fabrication

The specimens No.1 to 3 are pretensioned beams and No.4 to 6 are postensioned beams being grouted with usual non-shrink cement grout. The pretensioning tendons are sand adhered type and the postensioning tendons have no sand on their surface. The transfer lengths of the pretensioned beams were less than 40 times of tendon diameters. The postensioned beams have anchorages of sleeve and wedge system, and CCL jack was used as tensioning device (see Fig.5). The specimen No.1 was induced 1.47MPa of bottom fiber concrete compressive prestress and the specimens No.2 to 4 were induced 2.94MPa at 14 days after casting of concrete. The specimen No.5 was designed considering the

Fig.5. Prestressing method for specimens No.5 and 6

application for beams of warehouse subjected to dead load and live load. The design conditions of the specimen No.5 are shown in Fig.6 and Table 3. The specimens No.5 and No.6 were induced 4.79MPa and 2.13MPa of bottom fiber stress respectively. Prestressing was done at 43 days after casting of concrete. The amount of reinforcement and induced prestress of the specimen No.5 was determined to satisfy the following conditions,

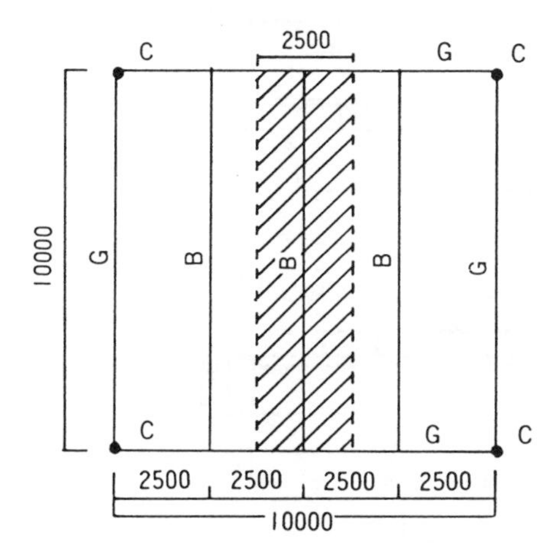

Fig.6. Designed portion for specimen No.5

Table 3. Design load for specimen No.5

Dead load D.L.	5.59 KN/m^2
Live load L.L.	4.90 KN/m^2
Serviceability limit state load	D.L.+0.8L.L.
Ultimate limit state load	D.L.+1.8L.L.

:Long term deflection is less than 0.4% of beam span.
:Flexural capacity is three times greater than the ultimate limit state design load at initial time (e.g. t=0).

The strength of concrete and the material compositions are listed in Table 4.

Table 4. Concrete

Specimen	Cement	Water cement ratio (%)	Compressive strength (MPa)		Tensile strength (MPa)	Young's modulus (MPa)
			at prestressing	at loading		
No.1~4	High early strength portland cement	38.5	37.6	40.7	3.19	29900
No.5,6	Normal portland cement	45.0	43.1	44.6	3.02	34300

3.3 Loadings and Measurements

The methods to apply constant static long term loadings are shown in Fig.7. The specimens No.1 to 4 were supported by two rollers subjected to two point loading at both sides by two springs connected to reaction steel beams (Fig.8). The specimens No.5 and 6 were subjected to uniformly distributed load by concrete elements as shown in Fig.9.

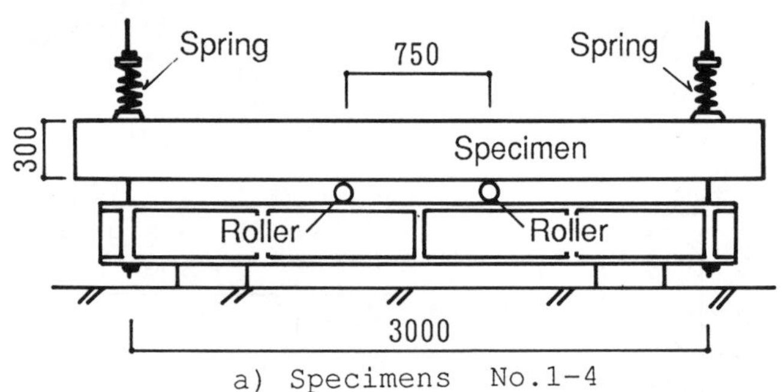

a) Specimens No.1-4

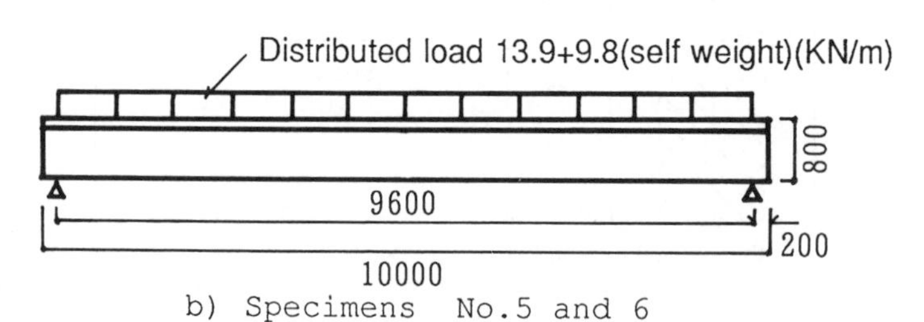

b) Specimens No.5 and 6

Fig.7. Long term load application

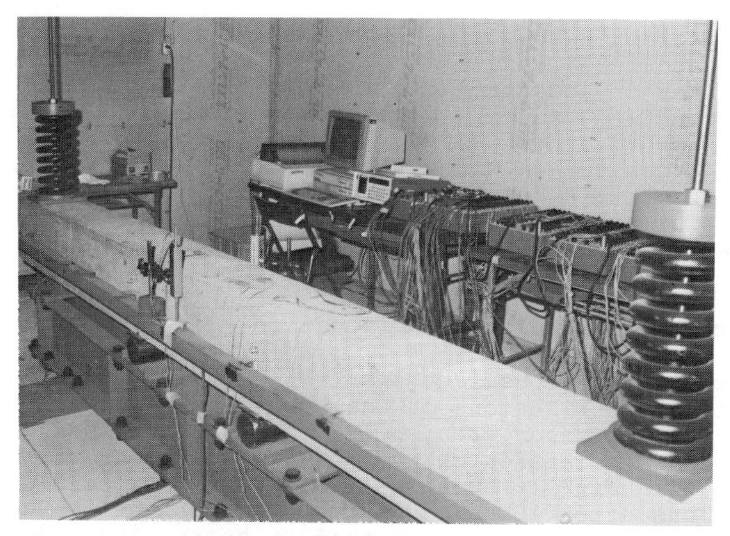

Fig.8. Testing setups of specimens No.1-4

Fig.9. Testing setups of specimens No.5 and 6

Load intensities for the specimens No.1, 2, and 4 are equal to their initial cracking load P_{cr} and for No.3 is $1.5P_{cr}$. The specimens No.5 and 6 are subjected to loads $0.67P_{cr}$ and $1.0P_{cr}$ respectively. P_{cr} was calculated from concrete tensile strength and compressive prestress assuming that the effective prestress was 0.8[6]. The loads were applied at 21 days after casting the specimens No.1 to 4 and at 51 days after casting the specimens No.5 and 6.

Every specimen was set in a room having constant temperature of about 25°C and constant humidity of about 70%. Deflections of the specimens were measured by linear displacement transducers having sensitivity of 0.01mm and the strains of concrete and FRP are measured by wire strain gauges. Crack width was measured by displacement meters which were 20cm in length and had 1000μ/mm sensitivity. The data acquisition was controlled by micro computer with constant measuring intervals.

4 Experimental Results and Discussion

4.1 Curvatures and Deflections

The strain distributions of the specimens No.1 to 3 measured at the midspan cross sections are shown in Fig.10. The midspan deflection changes of all specimens with time are shown in Fig.11. Specimens No.1 to 4 show stable deflections after 10000 hours elapse. The specimen No.5 has no crack at present(e.g. t=15000 hours). When comparing specimen No.1 with No.2, and specimen No.5 with No.6, the effects of prestress on controlling deflections are remarkable. The solid lines in Fig.10 and 11 are calculat-

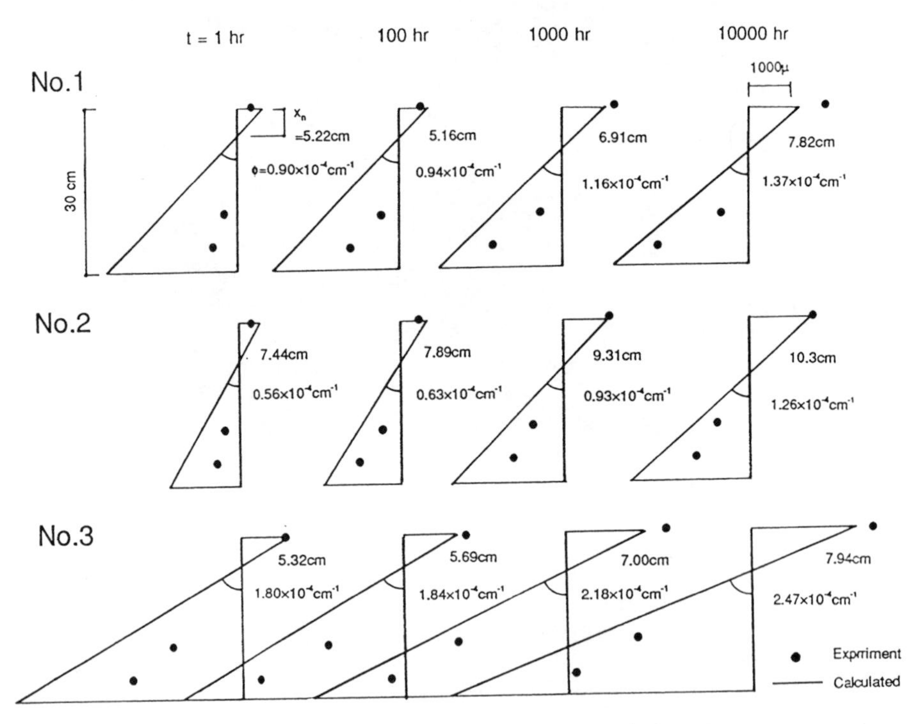

Fig.10. Change of curvatures with time

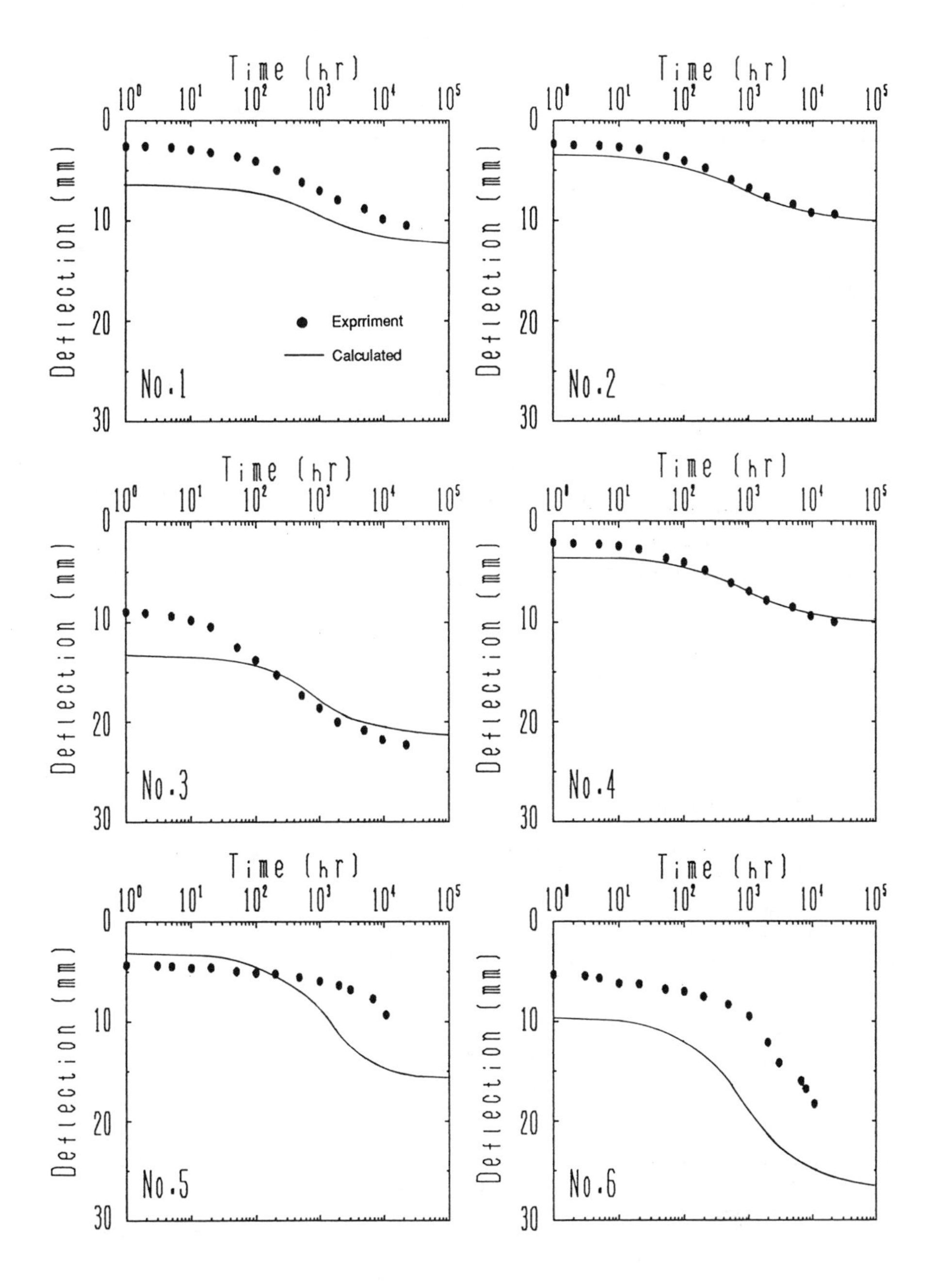

Fig.11. Change of deflections with time

ed curvatures ϕ_t and deflections δ_t. As expressed by Eq.1, the time dependent curvature ϕ_t is summation of the curvature due to creep strain of concrete $\phi_{cr,t}$ and the curvature due to drying shrinkage of concrete $\phi_{sh,t}$. $\phi_{cr,t}$ is calculated considering that the young's modulus of concrete can be modified according to its creep coefficient as shown in Eq.3, and $\phi_{sh,t}$ is calculated by Eq.4. The suffix t in Eq.1 to 5 indicates the time-dependence of these parameters.

$$\phi_t = \phi_{cr,t} + \phi_{sh,t} \tag{1}$$

$$\phi_{cr,t} = M_L / E_t I \tag{2}$$

$$E_t = E_c / (1+\psi_t) \tag{3}$$

$$\phi_{sh,t} = \varepsilon_{sh,t} b x_n^2 / 2I \tag{4}$$

$$\delta_t = \int\int_0^{L/2} \phi_t dx \tag{5}$$

The following assumptions were employed in writing these equations.

(1) The tensile resistance of concrete can be ignored.
(2) Concrete behaves elastically and its creep coefficient ψ_t and dry shrinkage $\varepsilon_{sh,t}$ can expressed as[7],

$$\psi_t = \alpha \times 0.75T / (1.5 + 0.25T) \tag{6}$$

where α = 1.0 for normal portland cement
= 0.7 for high early strength cement
T : time in week

$$\varepsilon_{sh,t} = 1.25 \times 10^{-4} \psi_t \tag{7}$$

As shown in Figs.10 and 11, the curvatures and deflections in early stage were overestimated in calculations, because the tension stiffening effects were not taken into account. But as time elapses, good agreement can be seen between computed and experimental results.

4.2 Crack Width and Crack Spacing

Fig.12 shows the crack formation of the specimens No.1, 2, 3, and 6. The specimens No.1 to 4 had a few cracks and No.6 had one crack at loading(e.g. t=0) and after then the number of cracks started increasing gradually. The specimen No.5 has no crack at present (e.g. t=15000hr). Figs.13 and 14 show the comparison between calculated and experimental values of the stabilized average crack spacing and the average crack width respectively. The average crack width was measured on the surface of concrete at the same height as main reinforcement. The average crack width is calculated by the Eq.8[8]

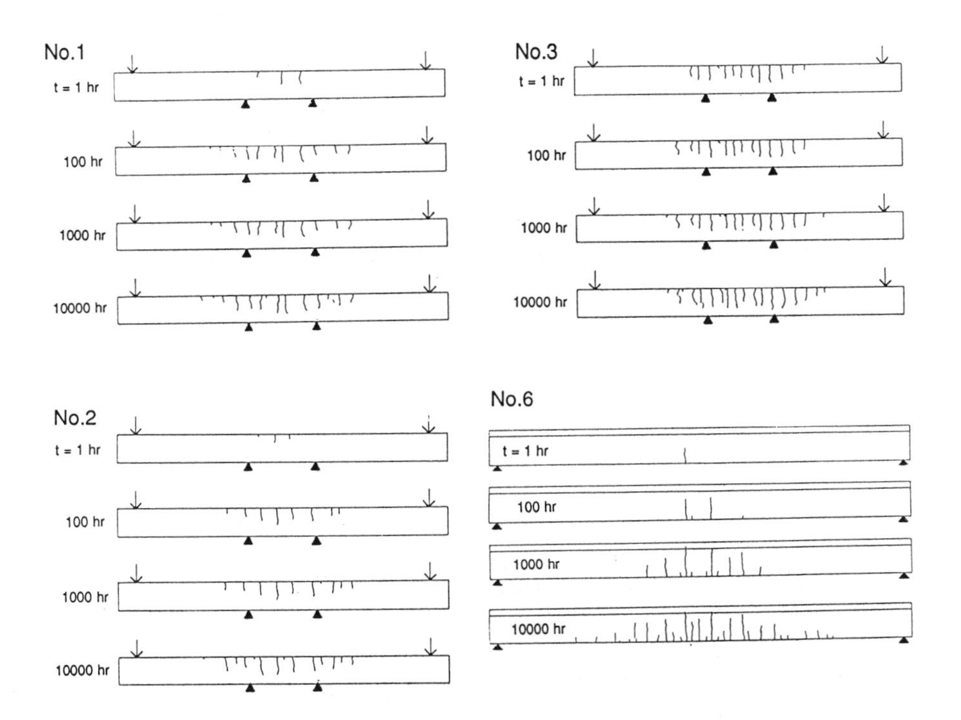

Fig.12. Crack formations with time

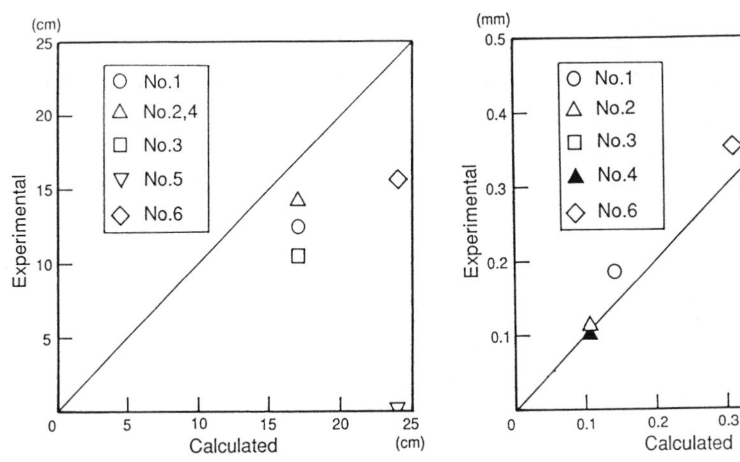

Fig.13. Stabilized average crack spacing (at t=10000 hr)

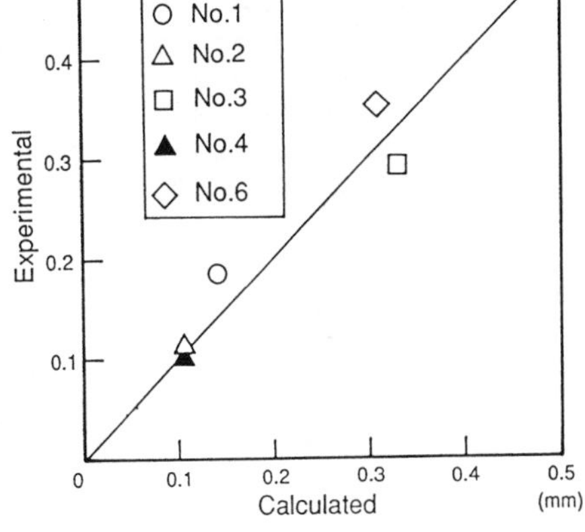

Fig.14. Average crack width (at t=10000 hr)

$$w_{av}=l_{av}(\varepsilon_{s2}-\beta\varepsilon_{sr2}-\varepsilon_{sh}) \quad (8)$$

where the average crack spacing was expressed as Eg.9[7].

$$l_{av}=2(c+0.1s)+0.1d/p_e \quad (9)$$

Although, the calculated crack spacings are longer than the observed values, the calculated crack widths are almost the same as the measured values. This is because the sand-adhered braided aramid FRP rods have similar bond properties to deformed steel bars (see Fig.2). Hence, the method of estimation of crack width, which was developed for deformed steel bars, can be used for the beams using sand-adhered braided FRP rods.

5 Conclusions

The following conclusions were drawn from the test results and discussions on the concrete beams reinforced and prestressed with braided aramid FRP rods subjected to long term constant static loadings:

1) The beams showed stable deflections after the elapse of about 1 year. Hence the FRP rods are reliable for long term application.
2) The modulus at cracked sections of the beams reinforced with FRP are rather low because of the low modulus of the FRP. However, partially prestressing was effective in controlling the long term service abilities of the beams.
3) The computational method which takes the creep and shrinkage of concrete into account and ignores the tensile resistance of concrete can be used to estimate long term deflection and curvature of the beams.
4) The method of estimation of long term crack width which was developed for conventional deformed steel bars can be used for the beams using sand-adhered braided aramid FRP because they have bond properties similar to deformed steel bars.

Acknowledgment

The authors wish to express their sincere thanks for assistance from Shinko Wire Co., Ltd. and Kobe Steel (KOBELCO) Co., Ltd. in the production of test specimens.

Notation

b : beam width (cm)
c : cover thickness (see Fig.15) (cm)
d : diameter of reinforcement (see Fig.15) (cm)

E_c : Young's modulus of concrete (GPa)
E_t : modified Young's modulus of concrete at time t (GPa)
f_c : compressive strength of concrete (MPa)
f_{ct} : tensile strength of concrete (MPa)
I : moment of inertia at cracked section (cm^4)
l_{av} : stabilized average crack spacing (cm)
M_{cr} : initial cracking moment (see Fig.16) (KN.m)
M_L : moment due to constant static loading (see Fig.16) (KN.m)
P_{cr} : initial cracking load (KN)
p_e : equivalent tensile reinforcement ratio (see Fig.15)
s : distance of reinforcements (see Fig.15) (cm)
t : time (t=0 at loading) (hour)
w_{av} : average crack width (mm)
x_n : neutral axis location (cm)
β : 0.38 for long term loading
δ_t : midspan deflection of beams at time t (mm)
$\varepsilon_{sh,t}$: concrete shrinkage at time t
ε_{s2} : strain of reinforcement at crack (see Fig.16)
ε_{sr2} : strain of reinforcement at crack when cracking forces reached to f_{ct} (see Fig.16)
$\phi_{cr,t}$: curvature due to concrete creep at time t (cm^{-1})
$\phi_{sh,t}$: curvature due to concrete shrinkage at time t (cm^{-1})
ϕ_t : curvature at time t (cm^{-1})
ψ_t : creep coefficient at time t

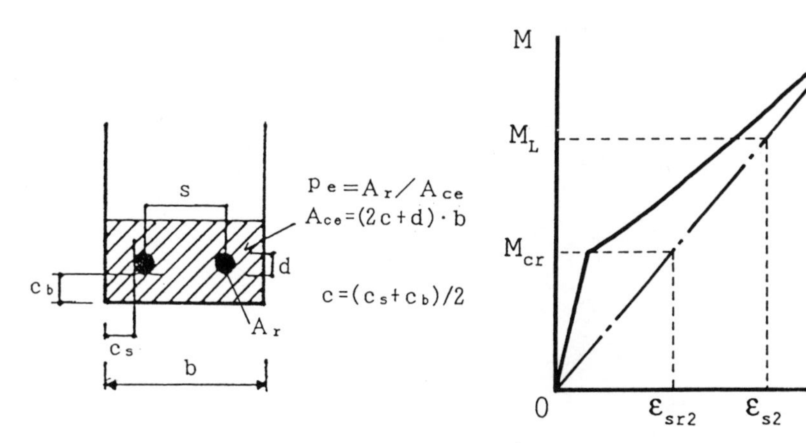

Fig.15. Notation

Fig.16. Moment-curvature relation

References

[1] Konig,G.,and Wolff,R., "Heavy Duty Composite Materials for Prestressing Concrete Structures," **Proc., IABSE Symposium**, Paris-Versailles, France, 1987, 6pp.

[2] Gerritse,a.,and Schurhoff,H.J., "Prestressing with Aramid Tendons," **Proc., 10th FIP Congress**, New Delhi, India, 1986, 7pp.

[3] Mikami,H.,Kato,M.,Tamura,T.,and Kishi,N., "Experimental Study on Dynamic Behavior of Concrete Slabs Reinforced with Braided AFRP Rods under Impact Loads," **Proc., International Symposium on Concrete Engineering**, 1991, 6pp.

[4] Miyata,S.,Wakui,H.,Tottori,S.,and Terada,T., "Shear Capacity of PC Beams with Spiral FRP Reinforcement," **Proc., 11th FIP Congress**, Hamburg, vol.1, 1990, pp.R22-25

[5] Tanigaki,M.,Okamoto,T.,Tamura,T.,Matsubara,S.,and Nomura,S., "Study of Braided Aramid Fibre Rods for Reinforcing Concrete," **Proc., 13th IABSE Congress**, Helsinki, 1988, pp.15-20

[6] Tanigaki,M.,Okamoto,T.,and Endo,K., "Flexural Behaviour of Concrete Beams Reinforced with Braided High Strength Fibre Rods," **Proc., 11th FIP Congress**, Hamburg, vol.2, 1990, pp.T78-T81

[7] **Architectural Institute of Japan**, "Recommendations for Design and Construction of Partially Prestressed Concrete," 1986

[8] **CEB-FIP**, "CEB-FIP Model Code 1990"

80 BENDING BEHAVIOUR OF UNBONDED PRESTRESSED CONCRETE BEAMS PRESTRESSED WITH CFRP RODS

H. HAMADA and T. FUKUTE
The Port and Harbour Research Institute, Ministry of Transport, Yokosuka, Japan
K. YAMAMOTO
The Third Port and Harbour Construction Bureau, Ministry of Transport, Sakaiminato, Japan

Abstract
Recently, some port and harbour structures in Japan have been found to be heavily damaged by salt attack. Especially, the RC (Reinforced Concrete) type slab and beam members of super-structures of wharves are damaged. This deterioration problem has become not only engineering problem but also one of the most serious social problems in Japan. As one of the solutions for this problem, the use of corrosion-free new materials in the place of steel bars have been discussed and some experimental works have been carried out. Among corrosion-free new materials, CFRP (Carbon Fiber Reinforced Plastic) is considered to be the most excellent as construction materials, because CFRP rod has higher strength and lower relaxation compared to the ordinary prestressing steel tendons. Recently, new type wharve structures shown in Fig.1 are discussed. This structure will be fabricated with PC beams which are prestressed with CFRP rods. CFRP rods are used not only as prestressing tendons of precast type pretensioning beams, but also as the lateral unbonded prestressing cables. One advantageous feature of this structure is as follows; when some of the beams are damaged it can be replaced by new one. Therefore it will contribute to the economical maintenance. In order to realize this new type wharve structures, the experimental work described herein is carried out. The main part of this work is the bending tests of beam specimens prestressed with CFRP rods. In this paper, the test procedure and results are described, and the comparison of the test results and linear analysis results are discussed.
Keywords: CFRP rods, Unbonded, Bending test, Effective prestress,

1 Introduction

Unbonded prestressed concrete beams prestressed with CFRP rods are tested to failure in bending as the basic studies of the application of CFRP rods into the new type wharve structures as shown in Fig.1. There are two different kinds of beam specimens, namely, one of them (Non-Block type) has no joint and the other (Block type) has four joints (composed of five units). The conception of the two different types of beam specimens is shown in Fig.2. The Block-type specimen is a model of a lateral element of new type wharve structures. The

Fibre Reinforced Cement and Concrete. Edited by R. N. Swamy. © 1992 RILEM.
Published by E & FN Spon, 2-6 Boundary Row, London SE1 8HN. ISBN 0 419 18130 X.

lateral element is shown in Fig.3. The main purposes of this study are as follows:

(1) To evaluate the bending behavior of the unbonded prestressed concrete beams prestressed with CFRP rods.

(2) To evaluate the stability of effective prestress of these beam specimens.

For these purposes, a linear analysis was carried out.

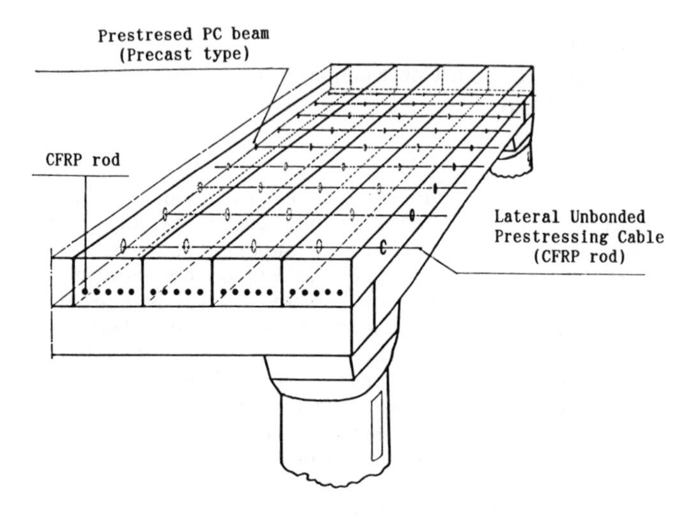

Fig.1 An image of new type wharf structure

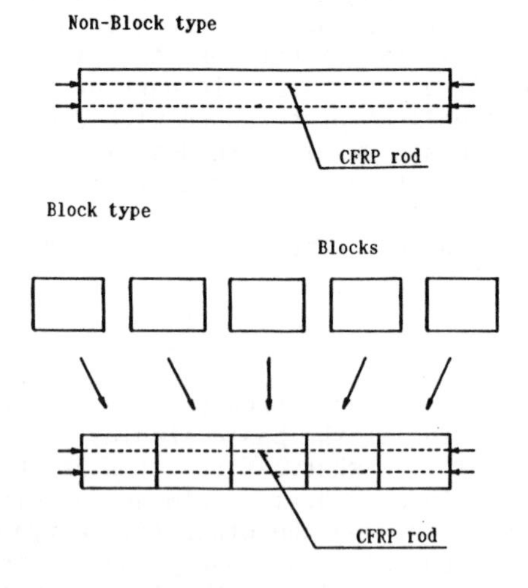

Fig.2 Block type specimen and Non-Block type specimen

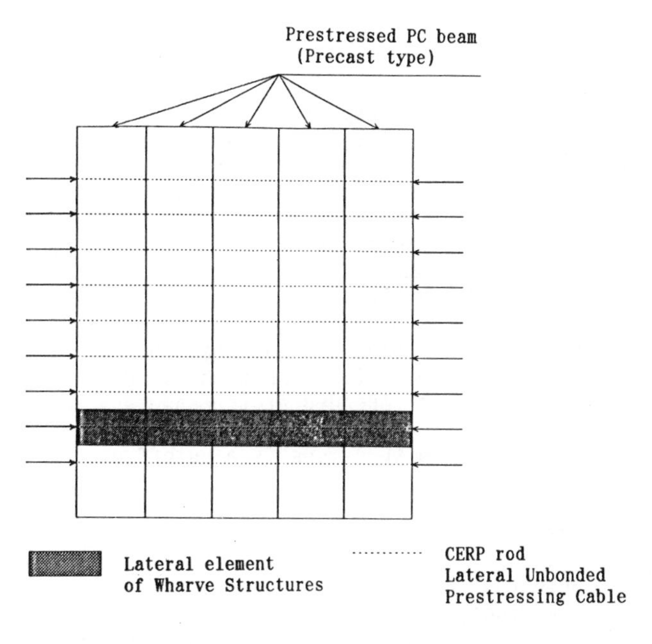

Fig.3 Plan of new type wharf structure

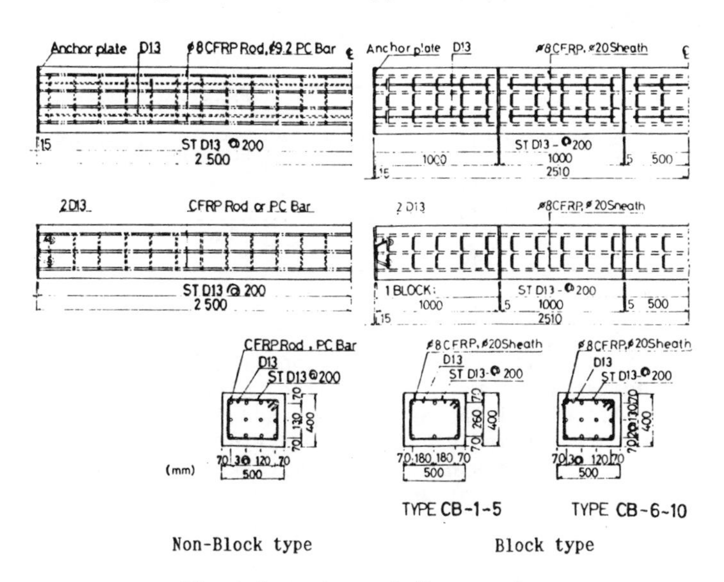

Fig.4 Geometry of the specimen

2 Test procedure

2.1 Specimen

All specimens tested in this study are listed in Table 1. In total, twelve specimens are tested. Two specimens are Non-Block type, and ten specimens are Block type; eight specimens were used for bending tests and four specimens are used for long-term tests to investigate time-dependent behaviour of the effective prestress. Ten Block type specimens are divided into two groups characterized by the number of CFRP rods used in each specimen. The geometry of the specimens is shown in Fig.4. As shown in Fig.4, the length of the specimen is 5000mm, the cross section is 500mm width and 400mm height. Block type specimen is composed of five blocks. The space between the blocks is filled with cement mortar 5mm thick. In both Non-Block type and Block type specimens, stirrups are embedded at the same spacing of 200mm. All specimens are prestressed with CFRP rods without grouting, and the design failure mode of these specimens is a cutoff of CFRP rod.

Table 1. Specimens

No.	Type	Number of CFRP rods	Effective Prestress (Mpa)	Note
CI-1	Non-Block	12	2.53	Bending test
CI-2			2.44	
CB-1	Block	6	1.23	
CB-2			1.23	
CB-3			1.23	
CB-4			1.21	Long-term test
CB-5			1.21	
CB-6		12	2.46	Bending test
CB-7			2.46	
CB-8			2.46	
CB-9			2.40	Long-term test
CB-10			2.40	

2.2 Concrete

For concrete mixing, high early strength portland cement, river sand, river gravel and tap water ware used. Also, admixtures (Water reducing agent, Air entraining agent) were added. Mix proportion of concrete is presented in Table 2. After casting, the concrete was moisture cured until 28 days old. Compresive strength and modulus elasticity of the concrete (at 28 days old) are presented in Table 3.

Table 2 Mix proportion of concrete

Gmax (mm)	Slump (mm)	Air (%)	W/C (%)	s/a (%)	Unit content (kg/m^3) W	C	S	G	Ad.	Note
20	80	4	34.7	39.0	160	460	671	1097	8.28	Non-Block type
20	80	4	39.0	42.0	164	420	753	1088	7.56	Block type

Table 3 Compressive strength and Modulus of elasticity of concrete

	Compresive strength (Mpa)	Modulus of elasticity (Mpa)
Non-Block type	60.27	23912
Block type	70.36	38220

2.3 CFRP rod

The CFRP rod used in this study consisted of pitch carbon fibres. The characteristics of CFRP rod are presented in Table 4 and Table 5. Fig.5 shows stress-strain relationship of CFRP rods. As shown in Fig.5, the relationship is linear until failure. The end anchorage used in these specimens is shown in Fig.6 and Fig.7.

Table 4 Characteristics of CFRP rod

Composition	64.7% (Carbon fibre) 35.3% (Epoxy resin)
Diameter (mm)	8.0
Specific gravity	1.57
Unit weight (g/m)	76.9
Tensile strength (N/mm^2)	1431
Modulus of elasticity (N/mm^2)	152880
Elongation at failure (%)	0.95

Table 5 Relaxation of CFRP rod

Permanent	Relaxation (%)		
Load	100 hours	1000 hours	30 years
0.67xPu	2.61	2.80	3.65
0.56xPu	2.23	2.50	3.35
0.44xPu	1.91	2.10	2.76

Pu : tensile strength
30 years : estimated value

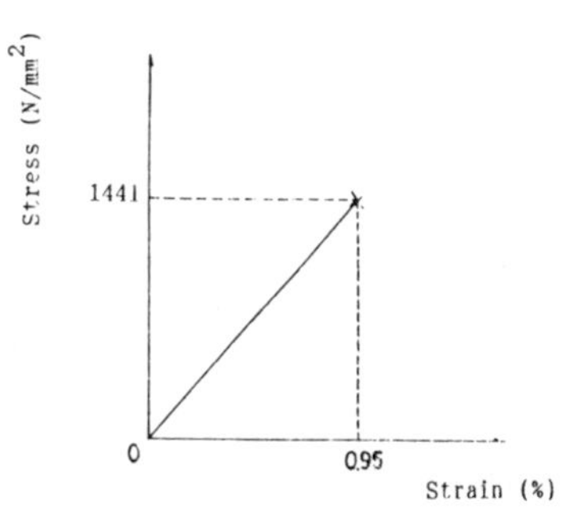

Fig.5 Stress-strain curve of CFRP rod

Fig.6 End anchorage

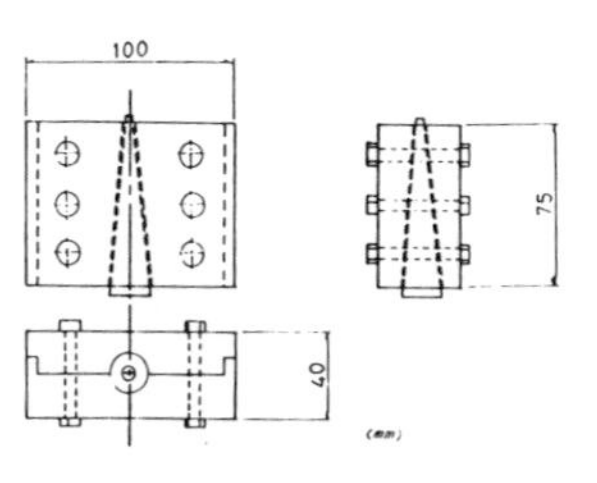

Fig.7 End anchorage

2.4 Bending test

The bending test setup is shown in Fig.8 and Fig.9. AS shown in Fig.8, bending test is carried out under static third-point loading with a 4200mm span. During loading, following measurements were made;

(1) cracking load,
(2) crack width,
(3) crac propagation,
(4) deflection,
(5) strain of concrete,
(6) strain of tendon,
(7) load at failure.

Eight beams (CI-1,2 CB-1,2,3, CB-6,7,8) were loaded until failure, and four beams (CB-4,5, CB-9,10) were loaded until cracking followed by long term evposure. The detailed description about long term test is given later.

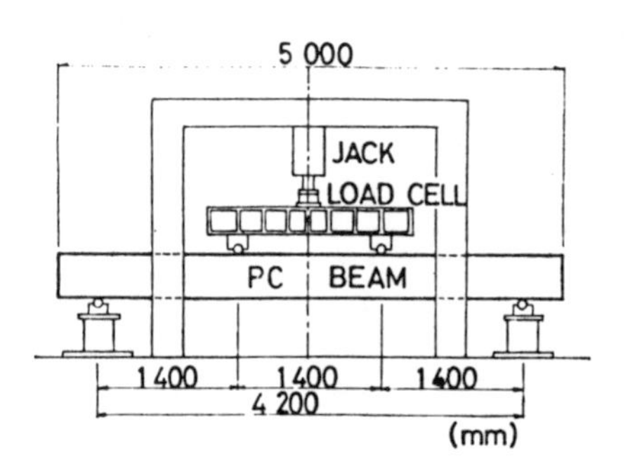

Fig.8 Bending test setup

Fig.9 Bending test setup

2.5 Linear analysis

The analysis of the cross sectional stress in unbonded prestressed beam specimen is complicated, because strictly speaking, the Bernoulli-Navier assumption is not valid for the unbond prestressed beam. In this study, the authors adopted different methods for two stages of loading procedure. The two stages of loading procedure are defined as follows, the first stage is before cracking and the second stage is after cracking. For the first stage, the ordinaly analysis method [1] is adopted, and for the second stage the modified method is adopted. The modified method is established on the following assumptions,

(1) The strain of concrete is proportional to the distance from neutral axis,

(2) The maximum compressive strain of concrete is proportional to the bending moment,

(3) The elongation of CFRP rod is equal to the integration of the concrete strains at the same depth as CFRP rod over the beam length,

(4) The friction between CFRP rod and concrete is neglected.

Analyses are carried out only for the Non-Block type beam specimens.

2.6 Long-term test (Time dependence of effective prestress)

Four specimens (CB-4,5, CB-9,10) were used for the long term test. At first, these beams were loaded by bending until cracking. Just after cracking, these beams were unloaded and followed by long term exposure. The long term exposure consisted of two stages, namely the first stage and the second stage. The first stage is an indoor exposure and the second stage is an outdoor exposure. During the exposure, compressive strain of concrete was measured by Calson type gauge instrument embedded in the specimens. The location of this gauge instrument is shown in Fig.10. After 1 year exposure, these four specimens are loaded by bending until failure. The bending setup is shown in Fig.8.

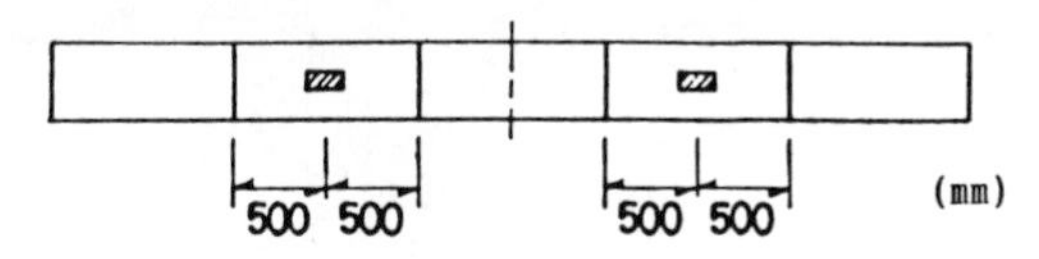

▨ : Calson type gauge instrument

Fig.10 Location of Calson type gauge instrument

3 Test results

3.1 Bending behavior

Cracking pattern

Fig.11 shows some examples of the typical cracking pattern of the specimens. As shown in this figure, the cracking pattern of Non-Block type specimen and Block type specimen are quite different. In all Block type specimens, the first cracks ocurred at the joints in the midspan and developed along the joints. On the other hand, in all Non-Block type specimens, the crack pattern does not show such a pattern.

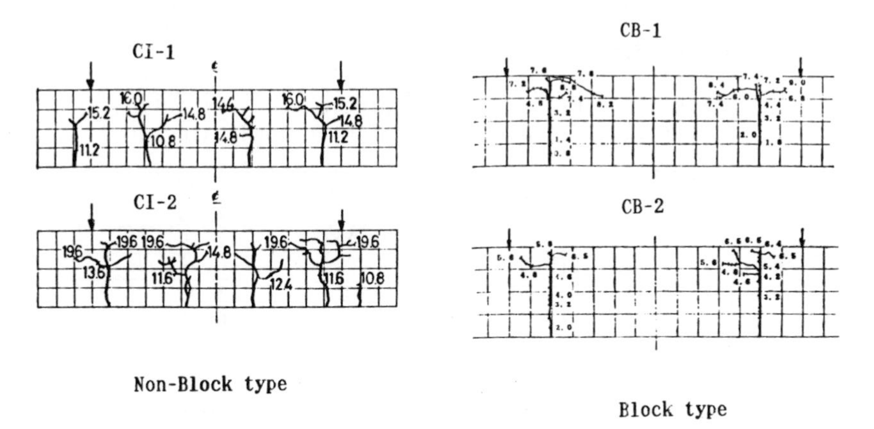

Fig.11 Cracking pattern

Failure mode
All specimens including Non-Block type and Block type showed the same failure mode. All specimens broke down when one of the CFRP rods fractured. The moment-deflection curves do not show plastic transformation untill failure, in other words, the failure mode of all specimens was brittle. An example of the failure state of the Block type specimens is shown in Fig.12.

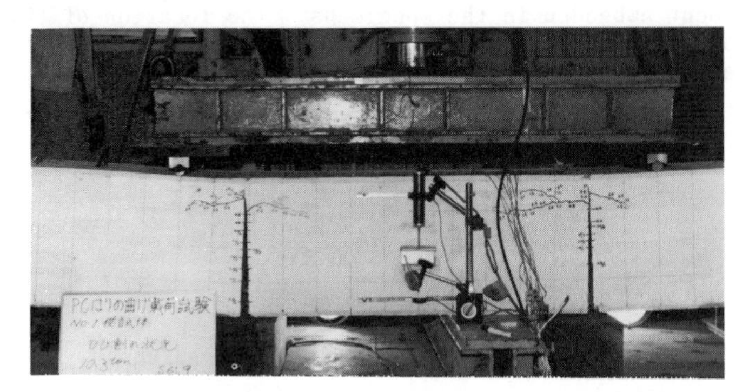

Fig.12 Failure state of Block type specimen

Ultimate bending load
Table 6 presents the ultimate bending load, both experimental results and calculated results. The calculation is carried out based on the assupmtion that the failure of the specimen is followed by the fracture of the CFRP rod. The average ultimate load of the Non-Block type specimens is a little larger than that of the Block type specimens, however the difference is quite small. The ratios of the calculated ones to the experimental ones range as follows:

(1) 0.92 to 1.09 in Non-Block type specimens,
(2) 0.87 to 1.10 in Block type specimens.

The calculated values are a little larger than experimental ones.

Table 6 Ultimate bending load

No.	Type	Number of CFRP rods	Experimental Ultimate Load (kN)	Calculated Ultimate Load (kN)	Experimental / Calculated
CI-1	Non-Block	12	311.17	339.12	0.92
CI-2			370.80		1.09
CB-1	Block	6	197.51	178.88	1.10
CB-2			175.15		0.98
CB-3			180.74		1.01
CB-4			177.01		0.99
CB-5			163.97		0.92
CB-6		12	311.17	357.75	0.87
CB-7			318.62		0.89
CB-8			337.26		0.94
CB-9			311.17		0.87
CB-10			348.44		0.97

Load-deflection relationship
Some examples of load-deflection relationship are shown in Fig.13. The calculation is carried out based on the assumption mentioned earlier. As shown in this figure, in the case of Non-Block type specimens calculated line coincides with experimental line very well until cracking, however after cracking calculated deflection is larger than experimental one. In the case of Block type specimens, the tendency is almost the same as the case of Non-Block type specimens, however the difference between calculated line and experimental one is much smaller than that of Non-Block type specimens. The difference in experimental curve between Non-Block type and Block type may be due to the joints of Block type specimens. The larger the load becomes, the wider the opening of joints becomes. These results show that both Block type and Non-Block type specimens behave elastically until cracking, however after cracking they do not show elastic behavior. It is also said that the calculation method adopted in this study gives a little larger values in deflection.

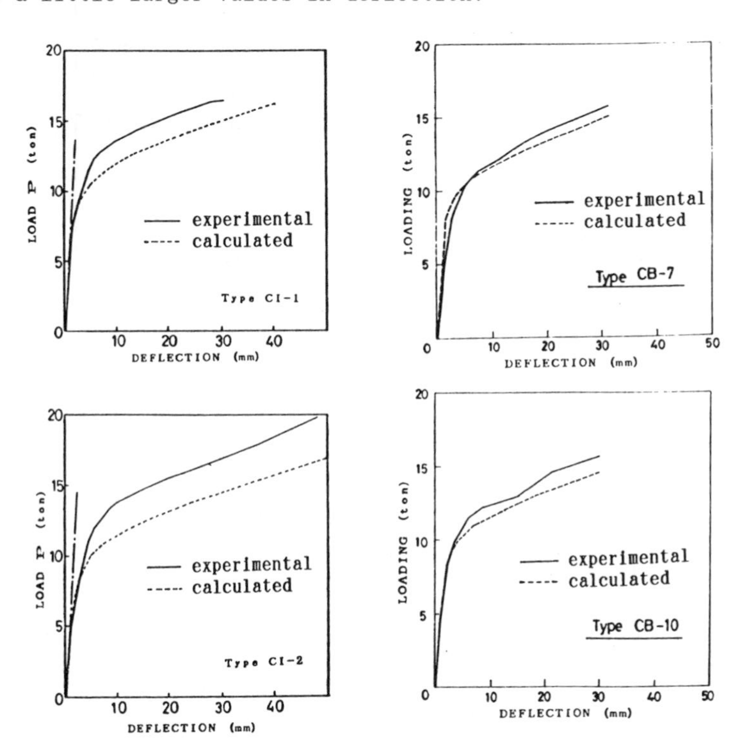

Fig.13 Load-deflection relationship

3.2 Time dependence of effective prestress
Fig.14 shows the relationship between the age from the introduction of prestressing force and the increase of compressive strain caused by the creep and drying shrinkage of concrete. The experimental values were measured with Calson type gauge instruments embedded in the

specimens. The calculated values are obtained by using the design formula of JSCE (Japan Society of Civil Engineering)[2]. As mentioned earlier, there are two exposure stages (the indoor exposure and the outdoor exposure). The calculated line is also different for the two stages. For indoor exposure the calculation is carried out based on the relative humidity of 40%R.H., and for outdoor exposure it is based on the relative humidity of 70%R.H. As shown in Fig.14, all experimental results do not agree with calculated values. Fig.15 shows the relationship between the age from the introduction of prestressing force and the ratio of the effective prestressing force to initial prestressing force (Pe/Pt). The calculated line in this figure is obtained by using the JSCE method[2], then as the relaxation of the CFRP rods "3.52%" is estimated from Table 5.

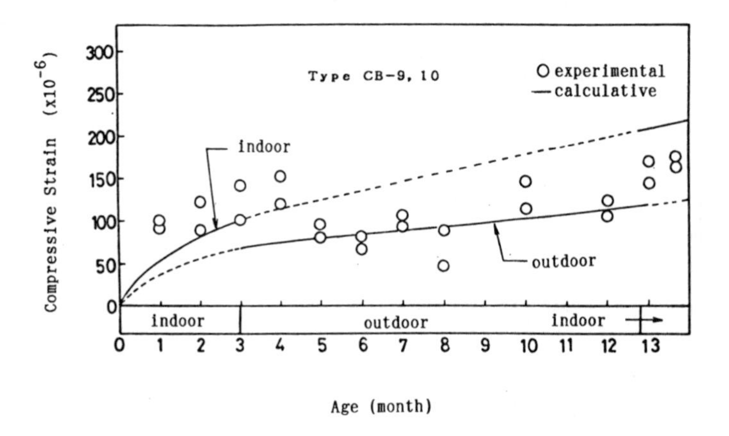

Fig.14 Time dependent of compressive strain of concrete

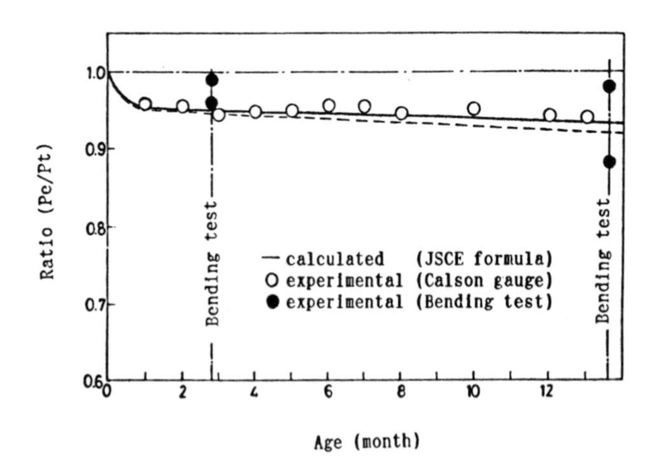

Fig.15 Time dependent of ratio of effective prestress to initial prestress

An experimetal data (symbol : o) is calculated by using the data of compressive strain of concrete. Another experimental data (symbol : •) is obtained by bending test. These calculated and experimental data agree with each other very well. From the long term exposure test results, it can be said that for 1 year exposure the effective prestress is stable and can be calculated by JSCE method.

4 Conclusions

From the test results, following conclusions are derived.

(1) The bending behavior of the beam specimens prestressed with CFRP rods is almost same as that of the beams prestressed with prestressing steel tendons.

(2) The failure of both the Block type and the Non-Block type specimens is caused by the fracture of the CFRP rod, and the failure mode is brittle.

(3) The average ultimate bending load of the Block type beam specimens is a little smaller than that of the Non-Block type beam specimens.

(4) The average deflection of the Block type specimens is nearly the same as that of the Non-Block type specimens at the maximum load.

(5) The unbonded prestressed beam specimens are analysed by the linear method which has some calculative assumptions, and the analysed results coincide well with the experimental results.

(6) The effective prestress of the Block type beam specimens remains stable at least for 1 year after the introduction of prestresing force.

(7) It is concluded that the beams prestressed with CFRP rods csn be designed for bending.

5 Acknowledgement

The authors greatly acknowledge Dr. Otsuki (Associate professor of Tokyo Institute of Technology) for technical advice and Mr. Ohashi (Penta-ocean Construction Co., Ltd,) for experimental aids. Without their help, this paper could have been not completed.

6 References

[1] Okada, Kamiyama; "The Design of Prestressed Concrete", Kokuminkagakusya Co., 1984. (in Japanese)

[2] Standard Specification for Design and Construction of Concrete Structures, Part 1 (Design), JSCE, 1986.

PART TEN

SPECIAL FIBRES, NATURAL FIBRES

81 STUDIES ON WHITE MICA FLAKES AS REINFORCEMENT FOR CEMENT PASTE

SHEN RONG-XI and WANG WUXIANG
China Building Materials Academy, Beijing, China

Abstract
White mica flakes with high aspect ratio were used as reinforcement for portland cement paste. Mica reinforced cement (MRC) specimens were produced by using 'vacuum–dewatering and pressing' method. It was found that the orientation of mica flakes has great influence on flexural strength and toughness of MRC. In order to fully utilize the reinforcing effect of mica flakes, they should be distributed in two dimensions in MRC. An optimum volume fraction of mica flake exists corresponding to the maximum flexural strength and toughness of MRC. There is an optimum compaction pressure in pressing the vacuum dewatered MRC. In order to improve the interfacial bond between mica flake and cement matrix, three different means were tested: addition of polymer emulsion, mica flake surface treatment with hydrofluoric acid and mica flake surface treatment with coupling agent. By comparison, the first means gave the best effect. Based on the above tests, the function of white mica flake, the mica flake–cement matrix interface and the failure mode of MRC have been discussed.
Keywords: White mica (Muscovite mica), Mica flakes,Cement matrix, Aspect ratio, Mica reinforced cement, Flexural strength,Toughness.

1 Introduction

Mica flakes posses a series of attractive properties, such as high Young's modulus, strong alkali–resistance, low water absorption, excellent thermal stability etc. Furthermore, mica flakes are rich in natural resources and low in costs. Therefore, mica flake maybe a potential alternative to asbestos, whose dusts are regarded as harmful to human's health, for producing fibre reinforced cement products.

J.J. Beaudoin has reported that addition of optimum amounts of suzorite mica (phlogopite type) flakes with high aspect ratio increases the flexural

Fibre Reinforced Cement and Concrete. Edited by R. N. Swamy. ©1992 RILEM.
Published by E & FN Spon, 2-6 Boundary Row, London SE1 8HN. ISBN 0 419 18130 X.

strength and toughness of hardened portland cement or high–alumina cement. The Composites R & D Department of Kurary Co. in Japan has reported that addition of suzorite mica flakes is useful to improve the dimension stability of pulp–cement or asbestos–pulp–cement sheets. However, most researchers have used phlogopite as reinforement for cement matrices.

This study aimed at exploring the main factors influencing the reinforcing effectivness of white mica (muscovite) flakes in cement composite, seeking the way to improve the interfacial bond between mica flakes and cement matrix as well as discussing the reinforcing mechanisms of mica flakes in cement composite.

2 Test program

2.1 Materials

The following materials were used:

(a) Cement — A grade 525, type R portland cement. Its chemical composition (%) was as follows: SiO_2–21.54, Al_2O_3–5.32, Fe_2O_3–3.63, CaO–63.33, MgO–1.08, Na_2O–0.16, K_2O–1.06, SO_3–2.18, TiO_2–0.28. Its Blaine fineness was 340 m^2 / kg.

(b) White mica flake — Two sizes of white mica flake, 16 mesh and 20 mesh were supplied by Lingshou Vermiculite Processing Plant. Thier average aspect ratio (diameter / thickness)was 102 and 90 respectively. Thier chemical composition (%) was as follows: SiO_2–47.35, Al_2O_3–34.30, K_2O–9,61, Na_2O–1.09, Fe_2O_3–1.72, TiO_2–0.08, CaO–0.16, MgO–0.74. An optical photomicrograph of mica flakes and SEM photomicrograph of a mica flake are shown in Fig.1 and Fig.2, respectively. The cleavage planes and the cracks on the flake show clearly in Fig.2.

Fig.1 Optical photomicrograph of 20 mesh white mica flakes (x44)

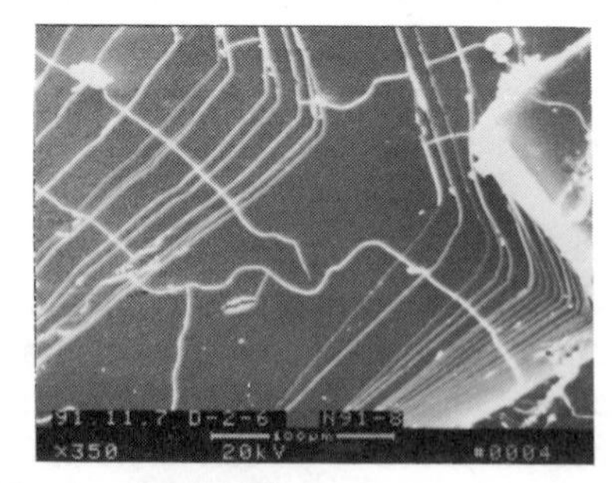

Fig.2 SEM photomicrograph of a white mica flake

(c) Polymer emulsion — A vinyl acetate–ethylene copolymer emulsion designated as BJ–707. Its main technical datum were as follows: solid content–54.5%, viscosity–750 CPS, pH–4.5, average particles size–0.2~2μ.

(d) Hydrofluoric acid — A reagent type hydrofluoric acid of concentration 40.

(e) Coupling agent — A methacryloxy–silane designated as KH–570. Its specific density was 1.069 (25℃).

2.2 Mica flake surface treatment

(a) HF treatment — 40% HF solution with treatment time of 60s, 300s and 450s and 20% HF solution with treatment time of 180s, 300s, 600s and 900s were used on mica flake surface treatment. After the treatment mica flakes were washed by water. The surface of mica flake treatd by HF solution was examined by optical microscope.

(b) Coupling agent — A certain amount of silane was hydrolyzed for 15 min in deionized water, whose pH had been adjusted to about 5 with acetic acid. Time of treatment was determined as the moment when a certain amount of mica flakes were added to the silane solution. The suspension was continuosly agitated. Thereafter, the treated mica flakes were washed with deionized water and then the water was partly removed by centrifugal apparatus. Finally, the flakes were dried approximately 105℃ for about 2 hours.

2.3 Specimen preparation

Cement was mixed with mica flakes according to certain mix ratio by hand. After adding the water, which was measured in the ratio of water:solid = 1.2:1, all components were mixed into a homogeneous slurry by a mixer. The slurry was then dewatered in a special steel mould with a porous bottom which was connected to a vacuum pump. Mica reinforced cement (MRC) specimen was made either by multilayer casting or by monolayer casting according to the requirement. The vacuum pressure did not usually exceeded–0.06 MPa. The total thickness of the specimen was 12 mm. For multilayer casting the thickness of each layer was about 1.2 mm. For monolayer casting the thickness of the layer was equal to the total thickness of the specimen. After dewatering the specimen was demoulded and transferred to a steel pressing mould. The specimen in the pressing mould was pressed under certain compaction pressure by using a hydraulic press. The size of the pressed specimen was 50 × 9 × 250 mm. After pressing the specimen was taken out from the mould and covered with plastic sheet for 24 ± 2 hr. Thereafter the specimen was put into a moist curing chamber till the age of 28 days. The specimen with polymer was cured in the moist chamber for 6 days, then put into a chamber, at a temperature of 20 ± 3℃ and relative humidity of 50 ± 2%, till the age of 28 days.

2.4 Mixes

Six series of MRC specimens were made for various purposes:

Series 1 — To study the effect of mica flake orientation on the mechanical properties of MRC, the monolayer specimens and multilayer specimens were produced either with mica flake of 16 mesh size or with that of 20 mesh size with mica flake content of 7 vol.–%, and compaction pressure of 6 MPa.

Series 2 — To study the effect of mica flake content on the mechanical properties of MRC, the specimens with different mica flake content were produced for each size of mica flake. The casting method was multilayer and compaction pressure was 6 MPa.

Series 3 — To study the effect of compaction pressure on the mechanical properties of MRC, the specimens pressed at different pressures were produced. The size of mica flake was 16 mesh, mica flake content was 7 vol.–% and casting method was multilayer.

Series 4 — To study the effect of polymer addition on the mechanical properties of MRC, the specimens with different polymer additions were produced. The size of mica flake was 16 mesh, mica flake content was 7 vol.–%, casting method was multilayer and compaction pressurce was 6 MPa.

Series 5 — To study the effect of mica surface treatment with HF solutions of different concentrations and different times on the mechanical properties of MRC, the specimens were produced with mica flake of 16 mesh size. The mica flake content was 7 vol.–%, casting method was multilayer and compaction pressure was 6 MPa.

Series 6 — To study the effect of mica surface treament with KH–570 of different concentrations and different times on the mechanical properties of MRC, the specimens were produced under the same conditions as series 5.

2.5 Testing

(a) Flexural test — All the specimens were cured for 28 day and their two surfaces were polished prior to test. Flexural test (mid–span loading) was carried out on a WD–5 type universal testing machine using a cross–head speed of 1.0 mm / min. The span of the specimen was 210 mm. Load–deflection curve was recorded and used to determine flexural strength and toughness of the specimen. The area under the load–deflection curve up to twice the deflection corresponding to the maximum load was integrated and designated as 'S'. The toughness was given by 'S' divided by the cross section area of the specimen. In all cases, at least five specimens were tested.

(b) SEM examination — The fracture surface of the representive samples cut from MRC specimens after flexural test was examined using ASM–SX+EDAX type scanning electronic microscope.

3 Test results

3.1 Mica flake orientation

The test results describing the effect of mica flake orientation on flexural strength and toughness of MRC are listed in Table 1. It can be clearly seen from the Table that the mechanical properties of multilayer MRC is much better than that of monolayer MRC. Fig.3 and Fig.4 show the distribution of mica flakes on the fracture surface of monolayer MRC and multilayer MRC respectively. There is a striking contrast between Fig.3 and Fig.4: in monolayer MRC the mica flakes are distributed randomly in three dimensions, while in multilayer MRC almost all the mica flakes are distributed in two dimensions. It is the main reason why the reinforcing effectiveness of the mica flakes embedded in multilayer MRC is higher than that of the same mica flakes embedded in monolayer MRC. It can also be seen from Table 1 that the reinforcing effectiveness of 20 mesh mica flakes is higher than that of 16 mesh mica flake, only for the multilayer MRC. The reinforcing effectiveness of these two sizes mica flake is almost the same for monolayer MRC.

Table 1. Effect of mica flake orientation on flexural strength and toughness of MRC

Size of Mica flake (mesh)	Aspect ratio of mica flake	Casting method	Flexural strength (MPa)	Toughness (KJ / m^2)
16	102	Monolayer	10.88 ± 0.67	0.34 ± 0.04
16	102	Multilayer	14.36 ± 1.04	0.44 ± 0.06
20	90	Monolayer	10.89 ± 1.10	0.30 ± 0.03
20	90	Multilayer	17.11 ± 0.75	0.51 ± 0.02

3.2 Mica flake content

The test results describing the effect of mica flake content on flexural strength and toughness of MRC are shown in Fig.5 and Fig.6, respectively.

From these Figures it can be seen that both, flexural strength and toughness, increase to a maximum value as the mica flake content is increased, and then decrease with the further increase of the mica flake content. It seems that a high mica flake content may give unfavourable influence on the bond between flakes and cement matrix inducing the decrease of flexural strength and toughness of MRC. From Fig.5 and Fig.6 it can also be seen that the mica flake content corresponding to the maximum flexural strength and toughness may decrease as the mesh size of the mica flake increases.

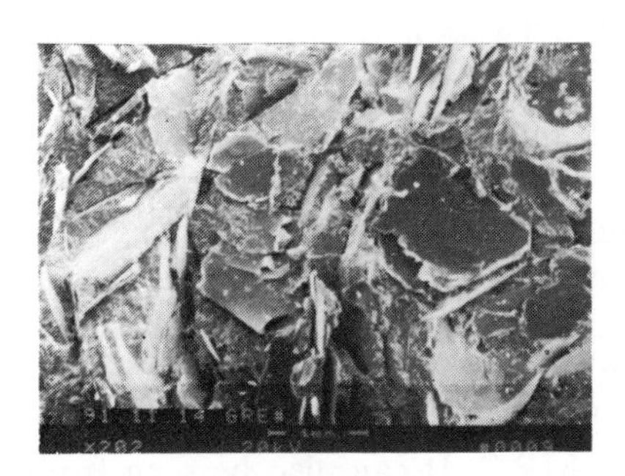

Fig.3 Distribution of mica flakes on the fractures surface of a monolayer MRC

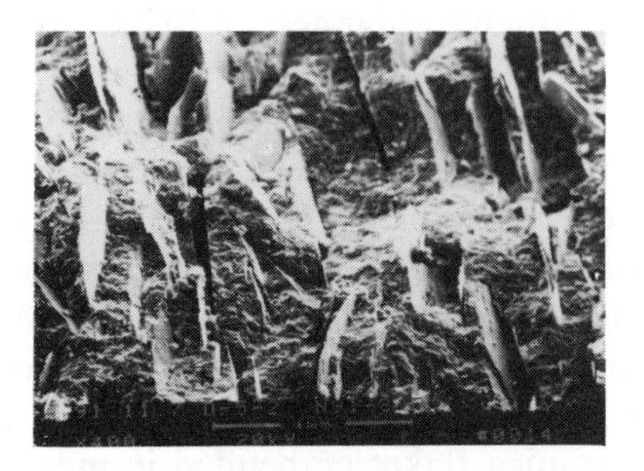

Fig.4 Distribution of mica flakes on the fractures surface of a multilayer MRC

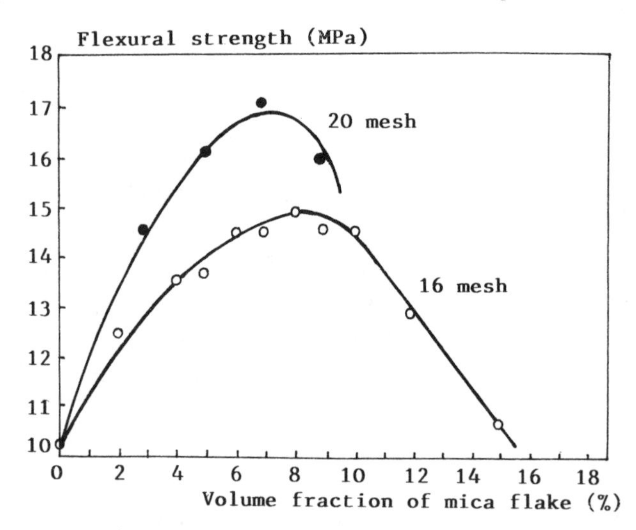

Fig.5 Flexural strength of MRC versus mica flake content

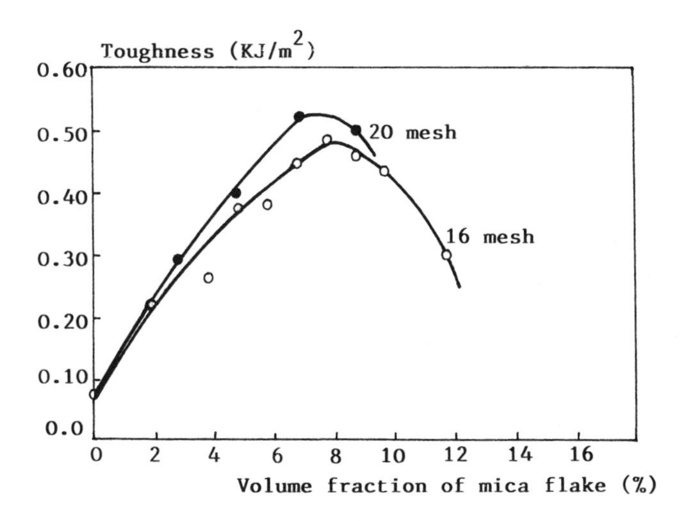

Fig.6 Toughness of MRC versus mica flake content

3.3 Compaction pressure

The test results describing the effect of compaction pressure on flexural strength and toughness of MRC are shown in Fig.7 and Fig.8, respectively. It seems that there is an optimal compaction pressure for the vacuum–dewatered MRC specimen. Excessive pressure may induce internal defects in the MRC specimens.

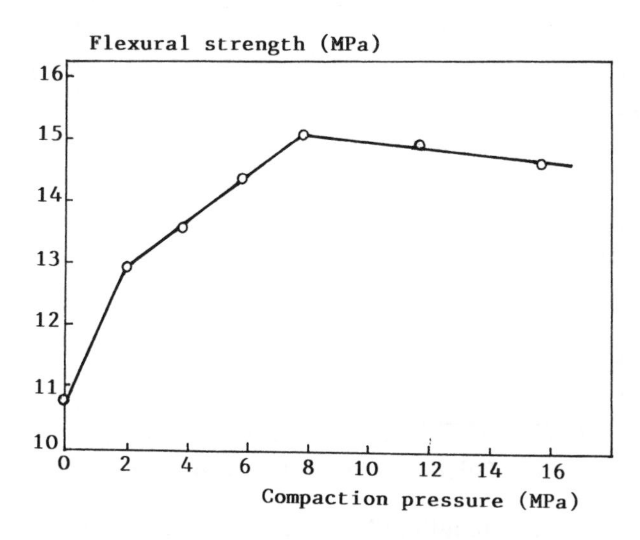

Fig.7 Flexural strength of MRC versus compaction pressure

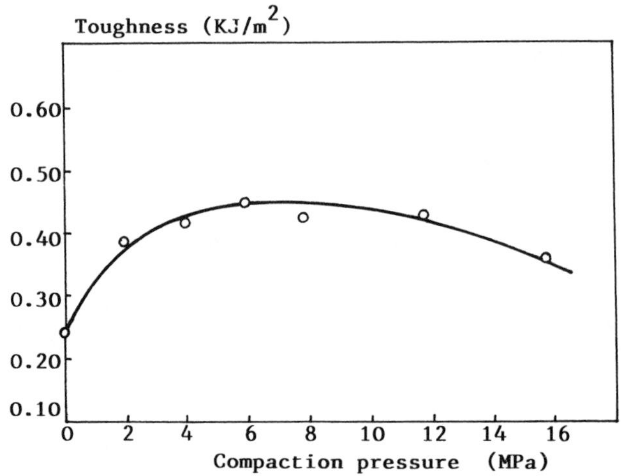

Fig.8 Toughness of MRC versus compaction pressure

3.4 Polymer addition

The test results describing the effect of BJ–707 content on flexural strength and toughness of MRC are shown in Fig.9 and Fig.10 respectively. It can be seen that there maybe an optimum polymer content under certain conditions. When excessive polymer is added, the flexural strength of MRC decreases and the toughness of MRC increases very slightly.

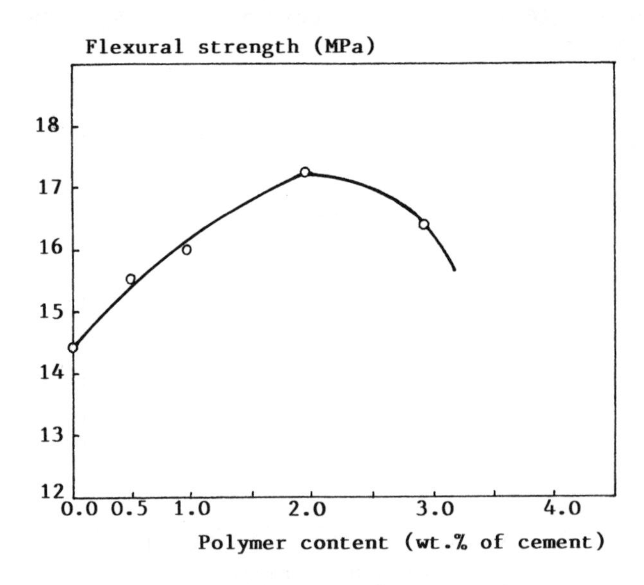

Fig.9 Flexural strength of MRC versus polymer content

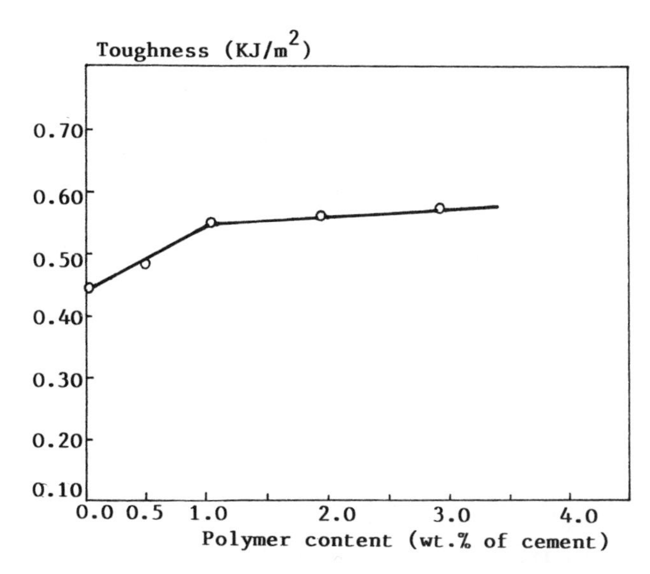

Fig.10 Toughness of MRC Versus polymer content

3.5 Mica flake surface treatment with HF solution

The test results describing the effect of mica flake surface treatment with HF solution on flexural strength and toughness of MRC are shown in Fig.11 and Fig.12, respectively. Fig.11 shows that mica flake treated with 20% HF solution for 300s or 40% HF solution for 60s maybe helpful in increaing flexural strength of MRC. However a long period of treatment with these solutions causes a decrease in strength. Fig.12 shows that mica flake treated with HF solution in either concentrations does not contribute to the increase in toughness of MRC, may even decrease in a case where mica flake was treated with 40% HF solution for more than 60s.

Fig.13 shows the surface of a white mica flake treated with 40% HF solution for 60s. It can be cleary seen that there are several etching pits on the surface. These etching pits may contribte to increase the bond between the flakes and the cement matrix and thus the flexural strength of MRC will obviously increase. Fig.14 shows the surface of a white mica flake treated with 40% HF solution for 450s. It is obvious that most of this flake was seriously attacked by HF. It is easy to understand that the weakened mica flakes may lead to a sharp decrease in flexural strength and toughness of MRC.

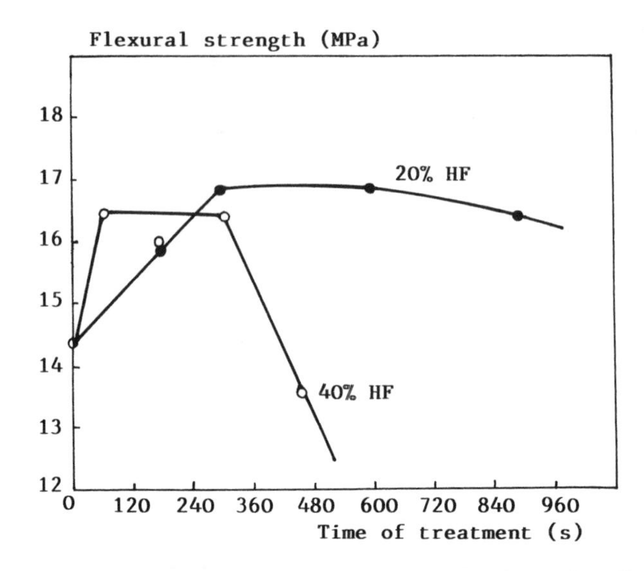

Fig.11 Flexural strength of MRC containing mica flakes treated with HF solution versus time of treatment

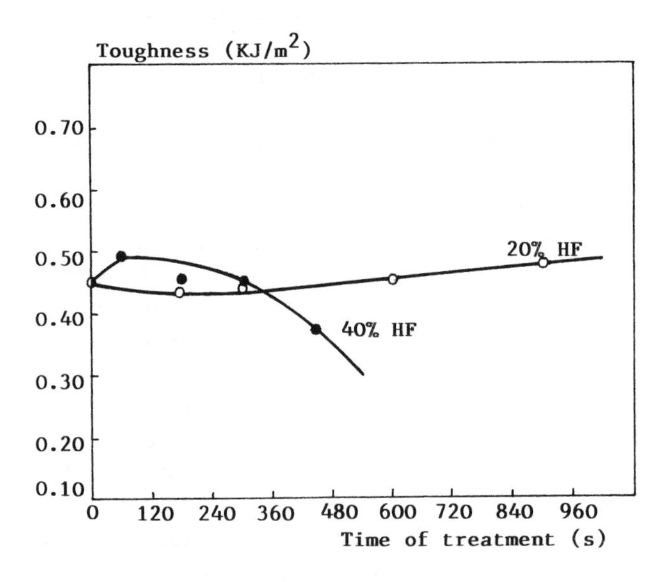

Fig.12 Toughness of MRC containing mica flake treated with HF solution versus time of treatment

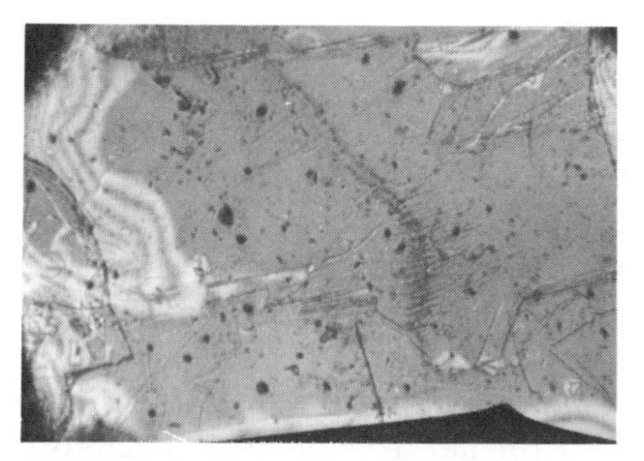

Fig.13 Surface of a white mica flake treated with 40% HF solution for 60s(x88)

Fig.14 Surface of a white mica flakes treated with 40% HF solution for 450s(x88)

3.6 Mica flake treatment with silane

The test results describing the effect of mica flake surface treatment with silane KH–560 on flexural strength and toughness of MRC are presented in Table 2.From the Table it can be seen that the effectiveness of mica surface treatment with KH–570 for flexural strength of MRC depends on both the concentration of the silane and the time of treatment. It seems that the most suitable concentration of KH–570 is 1.0% and the corresponding time of treatment is 1800s. The mica flake surface treatment with KH–570 has almost no effect on the toughness of MRC.

Table 2. Effect of mica flake surface treatment with KH–570 on flexural strength and toughness of MRC

Sample No.	Concentration* (%)	Time of treatment (s)	Flexural strength (MPa)	Toughness (KJ/m^2)
S0	—	—	14.36±1.0	0.44±0.06
S1	0.5	3600	14.87±1.41	0.43±0.05
S2	0.5	5400	14.82±0.52	0.41±0.03
S3	1.0	1800	16.76±0.30	0.47±0.02
S4	1.0	3600	16.27±0.59	0.46±0.02
S5	1.0	5400	16.02±0.05	0.44±0.03
S6	1.5	1800	15.73±1.07	0.46±0.04
S7	1.5	3600	15.38±0.50	0.42±0.04
S8	1.5	5400	15.58±0.62	0.42±0.04

* The unit of concentration is gram of KH–570 / 100 gram of mica flake.

4 Discussion

4.1 Function of white mica flakes in MRC

Though there is a great difference between the configuration of the mica flake and that of the fibre, the function of the former in cement composite is similar to that of the latter. When mica flakes with high aspect ratio (diameter / thickness) are distributed uniformly in the cement matrix, they may act as crack arrestors. From Fig.15 it can be seen clearly that the propagation of a microcrak existing in the cement matrix is arrested by the mica flake and it has to make a detour for further development. Therefore the mechanical properties of MRC maybe improved markedly in comparison with the unreinforced cement matrix.

Similar to fibres, the reinforcing effectiveness of mica flake is depended on the following main factors: flake orientation, flake content, average aspect ratio of the flakes, interfacial bond between flake and cement matrix etc. It seems that the influence of orientation of the mica flake is even greater than that of the fibre. If the surface of mica flake is parallel to the direction of the stress induced by external force, its reinforcing effectiveness may reach the maximum. On the contrary, if the surface of a mica flake is perpendicular to the direction of the stress, it does not give any reinforcing effect to the composite, or it even maybe detrimental, due to its delamination along the cleavage planes.

According to our test results the reinforcing effectiveness of 20 mesh mica flakes is greater than that of 16 mesh mica flakes. There maybe explained by the average flake spacing, which is smaller in the former than in the latter.

4.2 Mica flake–cement matrix interface

The mica flake–cement interface is similar to the fibre–cement interface, as there is $Ca(OH)_2$ enriched region at the interface between the mica flake and cement matrix (Fig.16). This is attributed to the water film surrounding the mica flake and later deposition of $Ca(OH)_2$ crystals within it. It is obvious that the $Ca(OH)_2$ enriched region is the weakness point in the cement composite. In order to increase the interfacial bond between the mica flake and cement matrix, the thickness of the water film surrounding the mica flake should be reduced. There are sevral ways to do it, such as suitable compaction pressure applied to the vacuum–dewatered MRC, adding certain fine dispersive and active additives (silica fume, zeolite etc); adding certain polymer emulsion etc.

Surface treatment of mica flakes with HF solution or coupling agent may contribute to the improvement of the mica flake–cement matrix interface.

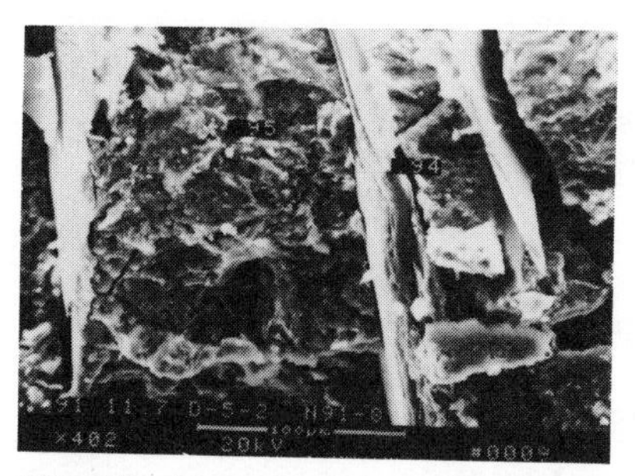

Fig.15 Propagation of a micro-crack in matrix is arrested by mica flake and makes a detour

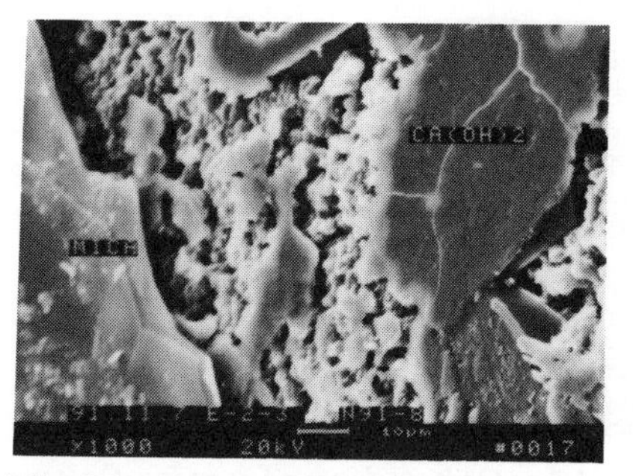

Fig.16 An enrichment of $Ca(OH)_2$ at the interface between mica flake and cement matrix

However, the effectiveness of those surface treatment means is not the same as that of polymer addition. It seems that the polymer in the cement matrix may also contribute to inhibiting the propagation of the microcrackes in the cement matrix.

4.3 Failure mode of MRC

Based on a large amounts of tests, the fundamental failure mode of MRC can be described as follows:

Under the action of an external force, the cracks in the cement matrix occur when the strain in MRC exceeds the ultimate strain of the cement matrix. The propagation of the cracks is arrested by the mica flakes embedded in the cement matrix and increase their tortousity for further advance. The force is transferred from the cement matrix to the mica flakes through the interfacial bond between the flake and the matrix. While the interfacial bond once reaches to the ultimate bond strength, the debonding of the mica flakes occures, and subsequently the flakes, which are bridging across the main crack in the matrix, will pulled–out from the matrix. Both the mica flake–cement matrix debonding and flake pull–out shall dissipate energy. As a consequence, the tensile strength and (also the flexural strength) and the toughness of MRC will increased in comparison with the hardened plain cement paste.

Fig.17 shows debonding of a mica flake from the surrounding cement matrix. Fig.18 shows the pull–out end of a mica flake from the broken cement matrix. Fig.19 shows delamination of a mica flake along its cleavage planes after pulled–out from the cement matrix.

The improvement of interfacial bond between the mica flakes and cement matrix, such as obtained by adding polymers, increasing surface roughness of the flakes with HF, adhering a silan moleculars monolayer to the flake surface etc, maybe helpful in increasing the energy dissipating

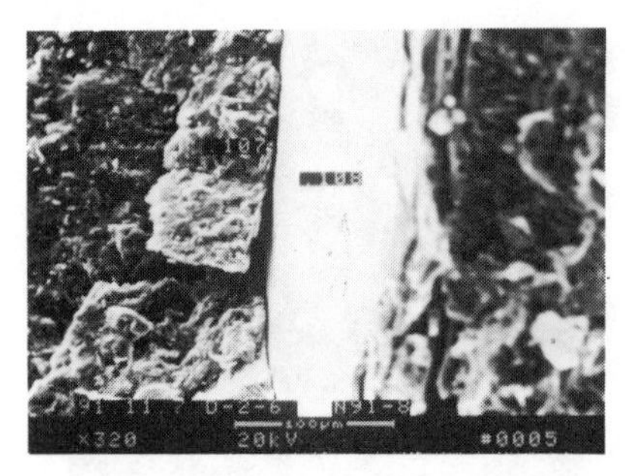

Fig.17 Debonding of a mica flake from the surrounding cement matrix

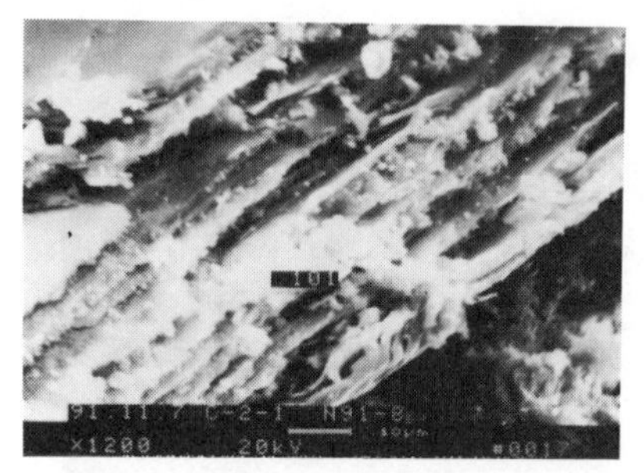

Fig.18 Pull–out end of a mica flake from the broken cement matrix

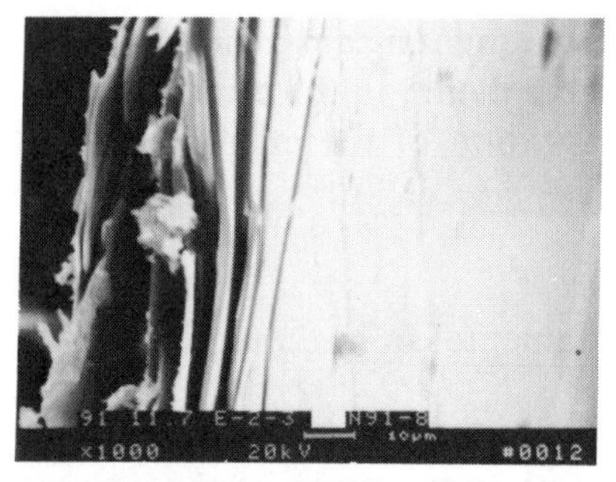

Fig.19 Delamination of a mica flake along the cleavage planes after pull out from cement matrix

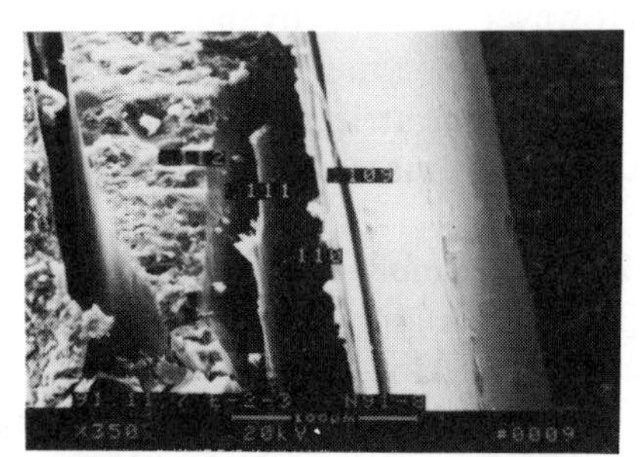

Fig.20 Part of a mica flake embedded in polymer modified cement matrix has pulled away, the remainder still bond with the matrix

both, by mica flake–cement matrix debonding and by flake pull–out. Fig.20 shows part of a mica flake embedded in the polymer modified cement matrix which has pulled away during failure in flexure of the composite. However the remainder of the flake is still bonded with the matrix. It proves again that polymer addition is a most effective means for enhancing the interfacial bond between mica flakes and the cement matrix.

5 Conclusions

1. The function of white mica flakes in a cement composite is the same as fibres, i.e. they act as crack arrestors increasing the tortousity of the advancing cracks.

2. Orientation of mica flakes has greater influence on the mechanical

properties of MRC. In order to utilize their reinforcing effect as much as possible, they should be distributed in two dimensions in the composite.

3. The average aspect ratio (diameter / thickness) and the average mesh size of the flakes have similar significance in enhancing the mechanical properties of the composite.

4. An optimum mica flake content exists for obtaining maximum flexural strength and toughness. This value is 7 vol.–% for 20 mesh mica flakes and 8 vol.–% for 16 mesh mica flakes.

5. There is an optimum compaction pressure in pressing the vacuum–dewatered MRC. Over this value flexural strength and toughness of MRC may decrease.

6. Adding a small amount of polymer may cause increase in flexural strength and toughness of MRC.

7. Mica flake surface treatment with HF solution or silane may increase flexural strength of MRC, but have no effect on its toughness.

8. The main failure mode of MRC is debonding and pull–out of mica flakes, due to the failure of bond between mica flakes and the cement matrix.

Ackowledgements

The authors gratefully acknowledge the fund supported by both, National Nature Science Foundation of the People's Republic of China and China State Administration of Building Materials Industry, for this study.

References

Beaudoin, J.J. (1982) Properties of high alumina cement paste reinforced with mica flakes. **Cement and Concrete Research,** 2, 157–166.

Beaudoin, J.J. (1982) Properties of portland cement paste reinforced with mica flakes. **Cement and Concrete Research,**2, 153–160.

Beaudoin, J.J. (1985) Effect of mica surface treatment on mechanical properties of mica–flake–reinforced cement composites. **Cement and Concrete Research,**4, 637–643.

Beaudoin, J.J. (1990) **Handbook of Fiber–Reinforced Concrete: Principles, Properties, Developments and Applications,** Noyes Publications, Park Ridge, New Jersey.

82 GLASS WOOL WASTE-REINFORCED GYPSUM: EFFICIENCY OF COMPACTION AND COMPOSITE MECHANICAL PROPERTIES

O. BRAGA Jr
IPT/EPUSP, Sao Paulo-SP, Brazil

Abstract
Parings from the production of glass wool mats can be incorporated into beta-hemihydrate gypsum to yield a new material with different mechanical properties. This paper reports the influence of casting pressure on composite flexural strength.
Increasing fibre contents, viz 3 % by wt, 6 % by wt and 7 % by wt, were added to plaster of Paris by the premixing method. Some test specimens were compacted on a vibrating table (50×50 cm) at 2850 rpm, and some were compacted on a hydraulic machine (Mohr & Fedderhaff 1960 kN) at 0.1 MN/m2, 0.2 MN/m2 and 0.3 MN/m2. The fibre composite was dried at (40±4) °C and conditioned at (45±2) % RH at a temperature of (20±3) °C.
The main conclusion was that better performance in terms of strength is achieved with a more efficient compaction technique.
Keywords: Glass-reinforced Gypsum, Wool Waste Fibre, Plaster of Paris, Compaction Technique, Flexural Strength.

1 Introduction

Glass wool waste-reinforced gypsum has a potential for use in the construction industry, especially in indoor applications. This multiphase material should have better cohesion during transport, be less affected by humid environments than plain gypsum and have improved fire and impact resistance, just as is reported(1,2) for E-glass- reinforced gypsum.

The fibre, glass wool waste, is supplied in the form of knot-like tangles, which are difficult to fiberise. The asbestos cement treatment, for instance, cannot be applied as the glass fibres might easily get broken during the process. Thus the density of the composite is an aspect to be investigated for an effective use of the reinforcing action of this fibre waste.

Possible compaction techniques essentially involve the extraction of excessive air and water from the fresh composite mixture and this may be done by manual compaction through vigorous shaking of the filled mould against a fixed support, by mechanical vibration on a suitable vibrating table, by vaccuum extraction of air and water, by hydraulic pressing and by an extrusion process.

Fibre Reinforced Cement and Concrete. Edited by R. N. Swamy. ©1992 RILEM.
Published by E & FN Spon, 2-6 Boundary Row, London SE1 8HN. ISBN 0 419 18130 X.

This paper reports the analysis of the influence of casting pressure on composite flexural strength. This research was carried out at the Instituto de Pesquisas Tecnológicas do Estado de São Paulo (IPT) and the Escola Politécnica da Universidade de São Paulo (EPUSP), both in Brazil.

2 Materials

Two brands of beta-hemihydrate gypsum were used as matrices. In this research they were designated as A (from Trindade-PE, Brazil) and B (from Araripe-PE, Brazil). Matrix properties are given in Table 1.

The reinforcement was glass fibres of borosilicate type in the form of wool waste. Fibre properties are given in Table 2.

Soluble salts were detected in the fibre waste as impurities. The effect of these salts upon the setting time of the hemihydrate matrix is given in Table 3.

Washing powder of sodium lauryl-sulphate type was used as set retarder.

3 Experimental programme

Glass/plaster ratio was varied as 0 % by wt, 3 % by wt, 6 % by wt and 7 % by wt. Composite flexural strength was assessed in an INSTRON 1115 testing machine under three-point loading over a span of 270 mm. Crosshead speed was 3 mm/min. Mix proportions are given in Table 4.

4 Composite production

The mixture process consisted of initially pouring the water and retarder into the bowl of a planetary bench mixer of 5.0 dm3 capacity, then adding the fibres and mixing for 2 minutes at slow speed. The hemihydrate material was added and mixed for 1 minute. Then the mixer was put off for approximately 30 seconds and its paddles were scraped clean. Now the mixer was put on again and a total mixing time, from the addition of water, was reached of 5 minutes.

Consistence was taken as (250±10) mm spread in the flow table, obtained for 15 jolts of the table in 15 seconds.

The fresh mixture of hemihydrate brand A plus glass wool waste was cast in moulds in the form of sheets and compacted on a vibrating table of 50×50 cm at 2850 rpm. The moulded material was stored for 24 hours under normal indoor conditions, then each sheet was demoulded and oven dried at (40±4) °C to constant weight. The sheet was eventually sawn into 6 specimens 15×60×300 mm per series of fibre content. Storage was at (45±2) % RH, (20±3) °C until the time of testing.

The fresh mixture of hemihydrate B was cast in a mould 150×150×300 mm over a steel tray, which was put over the lower plate of a Mohr & Fedderhaff 1960 kN hydraulic machine. Then 0.1 MN/m2 pressure was applied for each fibre series. Additional pressures of 0.2 MN/m2 and 0.3 MN/m2 were applied to specimens with 6 % by wt fibre content. The hardened material was cut and sanded down to dimensions of 15×60×300 mm

Table 1. Properties of beta-hemihydrate gypsum

State	Parameter		Brand A	Brand B
Powder	Density		2.60 g/cm3	2.64 g/cm3
	Unit weight		0.60 g/cm3	0.60 g/cm3
	Surface area (Blaine apparatus)		738 m2/kg	714 m2/kg
	Sieve analysis (% retained, individual)			
	0.840 mm sieve		0	0
	0.420 mm sieve		0	0
	0.210 mm sieve		5	8
	0.105 mm sieve		18	12
	< 0.105 mm sieve		77	80
	Chemical analysis (%)			
	free water		0.00	1.36
	combined water		4.70	5.60
	SiO2 + insolubles		0.70	1.60
	Fe2O3 + Al2O3		0.19	0.18
	CaO		38.70	37.12
	MgO		0.62	0.06
	SO3		53.00	55.00
	CO2		1.14	0.17
	Composition (%)			
	hemihydrate (CaSO4.1/2H2O)		71.1	90.3
	anhydrite (CaSO4)		23.4	5.37
	CaCO3		1.04	0.23
	MgCO3		1.30	0.13
	Fe2O3 + Al2O3		0.19	0.18
	insolubles		0.70	1.60
Paste	Density		1.69 g/cm3	1.67 g/cm3
	Standard consistency (ml water/100 g hemihydrate)		46.7	51.0
	Setting time			
	retarded	Beginning	12'25"	8'28"
		End	20'48"	13'10"
	unretarded	Beginning	6'15"	5'10"
		End	12'05"	9'20"
Solid	Average compressive strength		21.6 MN/m2	21.0 MN/m2
	Average flexural strength		8.2 MN/m2	7.4 MN/m2

Table 2. Properties of glass wool waste fibre

Formulation range (%)	
SiO_2	60-70
Na_2O	12-15
CaO	6-9
MgO	3-4
B_2O_3	3-5
Fe_2O_3	0.2-0.4
Al_2O_3	1.0-3.0
Density	2.56 g/cm^3
Soluble salts	CO_3, Na, K
Concentration of salts	0.07 %

Table 3. Effect of fibre soluble salts upon the setting time of beta-hemihydrate gypsum A*

Setting time of gypsum plaster		
	Water of fibre impurities	Plain water
Beginning	6'25"	6'15"
End	14'00"	12'05"

*There was a problem of material supply at the time of the experimental programme.

Table 4. Mix proportions

Mix (kg/kg) [hemihydrate:fibre:w/p]		Fibre volume (%)	
Brand A hemihydrate	Brand B hemihydrate	A	B
1:0.00:0.495	-	-	-
1:0.03:0.536	1:0.03:0.562	1.9	1.9
1:0.06:0.590	1:0.06:0.679	3.7	3.5
1:0.07:0.657	1:0.07:0.772	4.1	4.0

5 Discussion of data

The modulus of rupture (MOR) of the composite was calculated assuming elastic behaviour in flexure. Results are presented in Tables 5 and 6 with the standard error of the mean and the coefficient of variation (CV).

Reduction of around 35 % on mixture water was observed for 0.1 MN/m2 pressing, and MOR of this composite (Table 5) was virtually enhanced as compared to the MOR of the composite compacted on the vibrating table at 2850 rpm (Table 6).

Table 5. Mechanical properties of pressure-compacted composite

Centre-point loading, span 270 mm
4 specimens/series 15×60×300 mm, hemihydrate B

Mix (kg/kg) [hemihydrate:fibre:w/p]	Vf (%)	Pressure (MN/m2)	MOR (MN/m2)	CV (%)	Young's modulus (GN/m2)
1:0.03:0.562	1.9	0.1	10.1±0.2	4.2	3.7±0.4
1:0.06:0.679	3.5	0.1	8.8±0.6	13.5	2.5±0.1
		0.2	8.0±0.3	7.1	2.3±0.2
		0.3	9.8±0.2	4.5	3.1±0.2
1:0.07:0.772	4.0	0.1	7.1±0.2	5.8	1.7±0.3

Table 6. Mechanical properties of mechanically-compacted composite

Centre-point loading, span 270 mm
6 specimens/series 15×60×300 mm, hemihydrate A

Mix (kg/kg) [hemihydrate:fibre:w/p]	Vf (%)	MOR (MN/m2)	CV (%)	Young's modulus (GN/m2)
1:0.00:0.495	-	8.1±0.8	25.3	7.9±0.7
1:0.03:0.536	1.9	6.0±0.4	15.3	2.2±0.2
1:0.06:0.590	3.7	5.3±0.1	4.3	1.4±0.0
1:0.07:0.657	4.1	5.6±0.1	2.6	1.1±0.1

The decrease in composite MOR with the addition of fibres may be related to the critical fibre volume* not having being achieved as the fibres are parings of uneven length and diameter. Moreover, some balling up of these fibres does occur in the mixture process, which results in non-uniform coating fo the reinforcement with the plaster slurry.

As the two brands of hemihydrate gypsum have roughly equal mechanical strength (Table 1), in this study we are trying to compare the strength of composites made out of hemihydrate brand A with that of composites made out of hemihydrate brand B (cf footnote to Table 3).

From the linear part of the load-deflection plots obtained in the bending test the composite Young's modulus was calculated (Tables 5 and 6). It was then observed that the strain capacity of the unreinforced

*Defined(3) as the minimum volume of fibres which, after matrix cracking will carry the load the composite sustained before cracking.

matrix was improved with the addition of fibres, both i n the case of compaction on the vibrating table and on the hydraulic machine.

From the strength and stiffness data, it can be seen that the fibres did work as knot-like tangles, in this manner introducing a considerable amount of voids into the matrix. Yet, casting presssure permitted an increase in MOR and a decrease in Young's modulus of composite.

At 6 % fibre content by weight, upon increasing the pressure from 0.1 MN/m2 to 0.2 MN/m2 and then to 0.3 MN/m2 a continuous increase in composite MOR and Young's modulus would be expected, but at 0.2 MN/m2 pressure there is discrepancy of mechanical behaviour (cf Table 5). The uneven size distribution and concentration of the fibre waste in the matrix must have some relevance to this discrepancy.

6 Conclusions

Efficient compaction is very important for the strength of the fibre-reinforced material. Then according to the end use the production method has to be carefully chosen so that a composite with optimum mechanical performance may result.

Along with the efficiency of compaction, the production method of the fibre composite material should take into account a technique for efficiently dispersing and orienting the wool waste in the gypsum matrix as well as some suitable means of improving the fibre-matrix bond strength (through alkali attack on the fibre, for instance).

Hydraulic pressing permitted an improvement in composite strength, and because of this it is expected that thinner sections could be designed for a component that would thus be lighter and stronger.

7 Acknowledgements

This study was developed in the laboratories of IPT as part of a Dissertation for the degree of M Eng at EPUSP, São Paulo, Brazil. A scholarship was provided by a governmental institution called Coordenação dos Programas de Aperfeiçoamento de Pessoal de Ensino Superior (CAPES).

We are especially indebted to Dr F A S Dantas, Dr M A Cincotto and Dr V Agopyan from IPT/EPUSP for their helpful technical guidance in this research.

8 References

1. AGOPYAN, V. The Preparation of Glass-reinforced Gypsum by Premixing and its Properties under Humid Conditions. PhD Thesis, King's College, University of London (1983)
2. ALI, M A & GRIMER, F J. Mechanical Properties of Glass Fibre-reinforced Gypsum. Journal of Materials Science, vol 04:389-395 (1969)
3. HANNANT, D J. Fibre Cements and Fibre Concretes. Wiley, Chichester, UK (1978)

83 METAL-REINFORCED CEMENT MATERIALS

R. F. RUNOVA and S. Ye. MAKSUNOV
Kiev Civil Engineering Institute, Kiev, Ukraine

Abstract
In the present paper it has been shown that using various metals (as fibres as powders) and special technology procedure revealed new possibilities of cement composites. The main reason is development of physical interaction between metal and cement (or, largely, dispersed hydrated cement) particles under normal temperature during their simultaneous plastic deformation, including compacting (contact hardening). It has been shown that the main factors which define the degree of instantaneous interaction are state of crystalline structure of substance, plasticity of metal, and compacting pressure. The results of long standing observations for metal-reinforced materials are given. The main advantages of the materials are the possibility of exploitation by improvement of strength and water-resistance immediately after compaction, low energy and metal consumption
Keywords: Amorphous Silicates, Metal Powders, Metal-containing By-products, Compacting.

1 Introduction

The key point on fibre-reinforced materials is interaction between fibre and matrix [5]. As shown in literature [2, 3, 4], strain transfer from matrix to fibre is determined by dense bond in the contact zone between fibre and matrix. This bond, first of all, is the result of chemical processes in the cement matrix. Besides, surface state of fibre is important [3]. There is not full understanding of conditions determining the formation of fibre-matrix bonds despite intensive investigations of contact zone in the hydrated fibre-cement composites.

As reported from other studies, [1, 2], amorphous substances (e.g. hydrated cement) possess considerable quantity of free energy, which promotes formation of contact-condensation materials. They are based on physical interaction between particles resulting in the formation of water-resistant body immediately after compaction. As shown earlier, [1], in such materials, metal powders can be effectively used. In that case metal-reinforced materials assume special character and investigation of joint compacted metal-silicate mixtures is able to give new information about the rules of the interaction between metal and silicate particles in different conditions.

Fibre Reinforced Cement and Concrete. Edited by R. N. Swamy. © 1992 RILEM.
Published by E & FN Spon, 2-6 Boundary Row, London SE1 8HN. ISBN 0 419 18130 X.

The study of metal-matrix interaction in pressed metal-reinforced materials taking into consideration some special features of the silicate matrix is the main goal of the present paper.

2 Experimental Details

2.1 Raw materials

In order to prepare metal-reinforced materials portland cement grade M400, complying with the standard rules of the USSR, and calcium hydrosilicates of C-S-H(I) group were used as starting materials. Calcium hydrosilicates were produced by wet grinding of lime and sand (water/solids ratio (w/s) is 1) and hydrothermal treatment of suspension (w/s = 6) under the temperature 175°C during 8-50 hours. The treatment time changing results in production of hydrosilicates with different degree of crystallinity. The degree of crystallinity was estimated by density: amorphous substance is characterised by lower density [1, 2]. The density of calcium hydrosilicates is given in Table 1.

Table 1. Density of calcium hydrosilicates

Duration of hydrothermal treatment (hours)	Density (g/cm^3)
8	1.34
10	1.43
30	1.61
40	1.75
50	1.83

Commercially available technical grade copper and iron powders, as well as metal-containing slurry were used for reinforcement of silicate matrix. Metal-containing slurry is the by-product formed under polishing and sharpening of metal products. It consists of metal particles (60-100% by mass), abrasive grains (Al_2O_3, SiO_2) and lubricant-coolant. The metal particles are presented by twisted microfibre with length 200-500 mkm and diameter 30-60 mkm.

2.2 Preparation of samples

Starting materials (cement, calcium hydrosilicates, copper and iron powders) were passed through 200 mesh sieve. Metal-containing slurry after filtration was air dried at 150°C in the closed box.

Homogeneous mixtures of the powder composition were made by dry ball milling for 4 hours.

Test samples in the form of cylinder of 32 mm diameter and 32 mm length were cold pressed in a steel die at 40-1000 MPa pressure with dwell time of one minute.

Cement-based samples were treated in boiling water for 4 hours.

3 Results and Discussion

As can be seen from Fig. 1, strength-metal content curves for the composites based on calcium hydrosilicates are determined by the density of silicates and plasticity of metal. It is characteristic that using calcium hydrosilicates with density less than 1.61 g/cm^3 changes the behaviour of strength-metal content curves. Comparing with composites based on high density calcium hydrosilicates of composites based on low density hydrosilicates is increased by 2-3 times. It can be clearly observed on copper-based materials (copper is characterised by higher plasticity). As far as hydrosilicates are different in density, that is in the state of crystalline structure, strengthening of reinforced materials is defined by the structural state of the silicate.

The compacting pressure changes lead to changes in the behaviour of strength-metal content curves also. As shown in Fig. 2, strengthening can be observed under the critical pressure more than 300 MPa for the copper-based materials and 500 MPa for the iron-based materials.

The analysis of density-pressure curves for the copper-silicate mixtures (Fig. 3) reveals that their compacting reduces under pressure of 300 MPa.

The results given in Fig. 3 may be connected with increasing of the share of perfect interparticle contacts, their strengthening, that make worse compacting mixtures.

Strengthening of pressed metal-silicate composites is the result of formation of interphase contacts. The main factors which define the degree of interaction are state of crystalline structure of silicates, plasticity of metal, simultaneous plastic deformation of metal and silicate powders, and compacting pressure.

Using this developed technology procedure to produce cement-iron composites, increasing the compressive strength up to 350 MPa (Fig. 4).

The moulding of traditional fibre-reinforced materials by mechanical compacting decreases the effect of reinforcement because of steel fibre failure and cracking on their surface. Therefore, it is interesting to use twisted fibres, which get smoothed out during compacting providing reinforcement of the material. In this case, strain transfer from matrix to fibre is determined by the presence of metal-silicate contacts. Study of composites based on calcium hydrosilicates and metal-containing slurry reveal that their flexural strength has increased by two times in comparison with strength of matrix (Fig. 5).

Tests show that bond strength between fibre and matrix further develops during storage under normal conditions, resulting in strengthening of samples (Table 2)

4. Conclusion

State of the crystalline structure of silicate is the main factor determining the formation of interphase bonds during the simultaneous mechanical compaction of metal and silicate powders. Using the amorphous silicates leads to creation of metal-silicate bonds ensuring strain transfer from matrix to reinforcement elements and results in strengthening of composites. Compacting and further treatment of cement-based materials provide about 10 times increase of compressive strength.

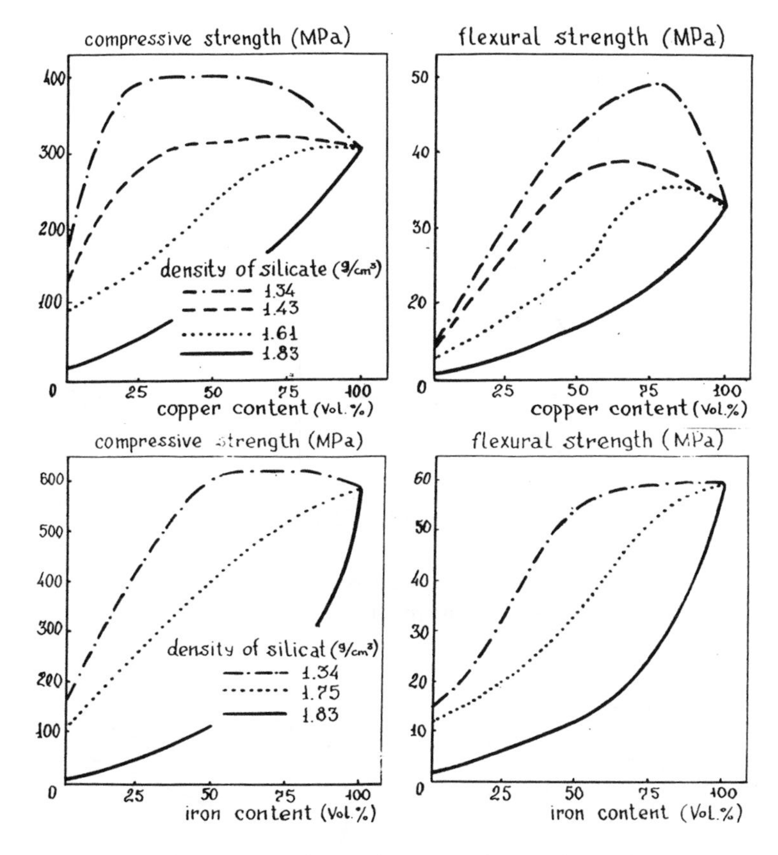

Fig. 1 Strength-metal content curves of materials based on calcium hydrosilicates

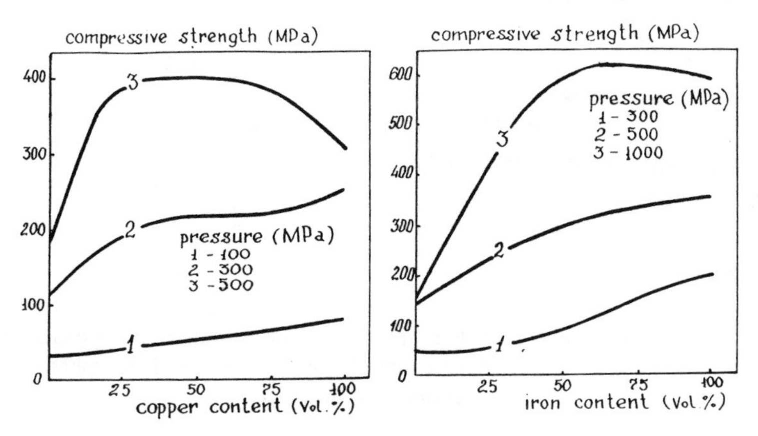

Fig. 2 The influence of compacting pressure on strength-metal content curves

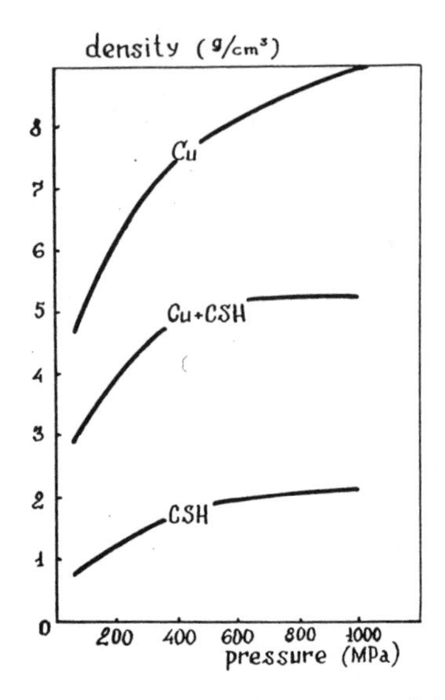

Fig. 3 Density-pressure curves of the copper-silicate mixtures

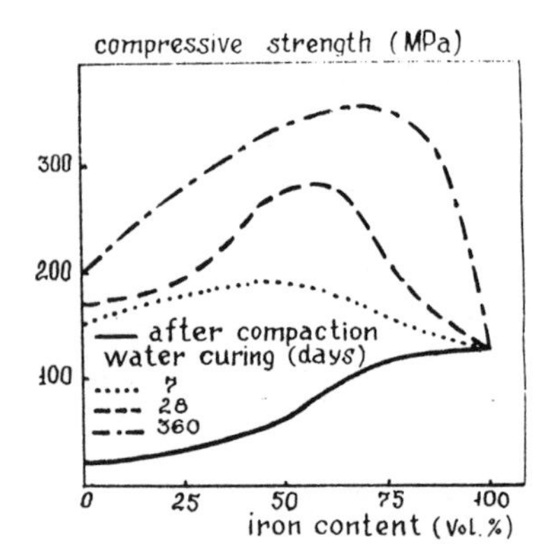

Fig. 4 The influence of iron content and curing conditions on compressive strength of cement-based composites

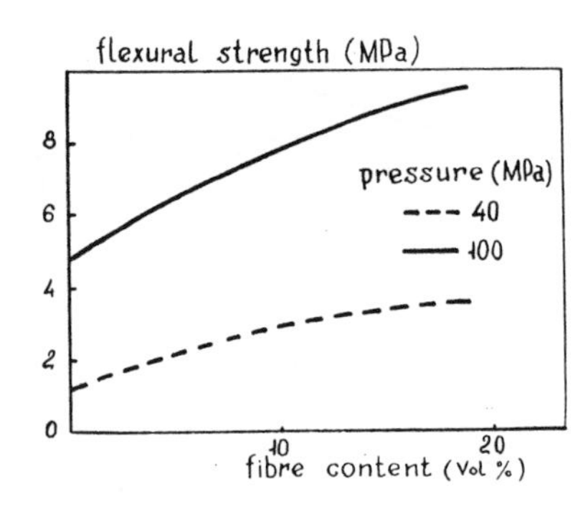

Fig. 5 Strength-fibre content curves of composites

The results further indicate that high quality reinforced materials for different applications could be made from metal-containing by-products and amorphous silicates by using dry compacting of component materials.

Table 2 Influence of storage time on strength of composites based on metal-containing slurry

Number	Pressure (MPa)	Fibre content (vol %)	Storage time (years)	Compressive strength (MPa)	Flexural strength (MPa)
1	40	10	-	10.3	2.8
2			0.5	13.1	3.0
3			1.0	15.0	3.1
4		18	-	17.3	3.8
5			0.5	19.1	4.3
6			1.0	21.3	5.0
7	100	10	-	21.7	7.8
8			0.5	23.7	8.1
9			1.0	26.2	9.4
10		18	-	42.0	9.5
11			0.5	43.1	10.0
12			1.0	51.1	12.1

5. References

1 Glukhovsky, V D, Runova, R F and Maksunov, S Ye (1991) Binders and Composite Materials of Contact Hardening. High School Publishing House, Kiev, Ukraine.

2 Felts, A (1986) Amorphous and Glassy Inorganic Solids. World Publishing House, Moscow, USSR.

3 Homeny, J, Wallace, W L and Ferber, M K (1990) Silicon Carbide Whisker/Alumina Matrix Composites: Effect of Whisker Surface Treatment on Fracture Toughness, J Am Ceram Soc, 73, 394-402.

4 Naaman, A A and Homrich, J R (1989) Tensile Stress-Strain Properties of SIFCON. ACI Mater J, 86, 244-251.

5 Ramachandran, V S, Feldman, R F and Beandoin, Y Y (1982) Concrete Science, Treatise on Current Research, Heyden

6 Shah, S P and Young, J F (1990) Current Research at the NSF Science and Technology Center for Advanced Cement-Based Materials, Ceram Bul, 69, 8, 1319-1331.

84 MECHANICAL PROPERTIES OF SISAL FIBER-MORTAR COMPOSITES CONTAINING RICE HUSK ASH

B. CHATVEERA
Department of Civil Engineering, Thammasat University, Pathumthani, Thailand
P. NIMITYONGSKUL
Division of Structural Engineering and Construction, AIT, Bangkok, Thailand

Abstract
The most important properties of materials used in construction are their strength and durability. Natural fibers appear to be prospective reinforcing materials in the matrix. This experimental investigation looks into the possibility of using organic natural fibers as reinforcement in thin concrete sheets because natural fibers are inexpensive and easily available in developing countries. The constituent materials are sisal fibers, sand, rice husk ash and cement. The main parameters are the fiber-cement ratio, unit weight and fiber length. The water-cement ratio and aggregate-cement ratio by weight are kept constant. The cementitious material contains 70% of Ordinary Portland Cement and 30% of rice husk ash by weight. Test results indicated that the optimum mix of the composite was a mix having 1 cm fiber length with fiber-cement ratio of 0.160 by weight and unit weight of 2150 kg/m^3. This optimum mix resulted in highest strength both in compression and flexure. The sisal fiber-mortar composites using rice husk ash as pozzolana showed a higher modulus of resilience as well as modulus of toughness as compared to mortar boards. It indicated that the combination of using sisal fiber reinforcement and rice husk ash can improve the ductility of the composites.
Keywords: Cement, Composite, Ductility, Flexure, Resilience, Rice husk ash, Sisal fibers, Toughness.

1 Introduction

The properties of a composite are dependent on the content and properties of the constituent materials, on the properties of the interfaces and on the homogeneity and isotropy in the distributions of the phases. A knowledge of the fiber content, the distribution and orientation of the fibers, amongst other parameters, is important for correlating the composition of a fiber composition with its properties and controlling the quality of finished products. Natural fibers are prospective materials and their use until

Fibre Reinforced Cement and Concrete. Edited by R. N. Swamy. © 1992 RILEM.
Published by E & FN Spon, 2-6 Boundary Row, London SE1 8HN. ISBN 0 419 18130 X.

now has been more traditional than technical. Cementitious materials in the form of mortar or concrete are attractive for use as constructional materials. Their deficiencies lie in their brittle characteristics, poor tensile and impact strength and in their susceptibility to moisture movements. Reinforcement by fibers can offer a convenient practical and economical method in overcoming these deficiencies. The investigations began on the possibility of manufacturing products of natural fiber reinforced concrete whose properties are comparable to those of asbestos-cement. The replacement of asbestos fiber is seen as an area of priority research, particularly in developing countries due to the associated health problem. Construction engineers and builders apply appropriate technology to utilize as effectively and economically.

In 1979 a three years project on material durability was initiated at the Swedish Cement and Concrete Research Institute. Sisal fiber concrete became brittle. The reason for this was that the fibers were decomposed in alkaline environment of concrete. The best results were obtained with a reduction of alkalinity by replacing a part of Ordinary Portland Cement with a highly active. During the last few years pozzolana based on rice husk ash, properties characterized by high reactivity and it can be used as cement substitute.

The present study is aimed at investigating the mechanical properties of sisal fiber-mortar composites containing rice husk ash hereinafter referred to as RHA. To achieve this, the physical and mechanical properties of the individual component have to be investigated. Therefore, the experimental program can be divided into four major parts as follows:

(a) Properties of sand and RHA
(b) Properties of sisal fibers
(c) Properties of cementitious material
(d) Properties of sisal fiber-mortar composite boards

Since no optimum mix proportion is known, the optimum fiber-cement ratio and unit weight of the composite will be determined based on the properties under compression, flexure and post-cracking behavior of each mix. The parameters adopted are the fiber-cement ratio by weight, unit weight and fiber length. The water-cement ratio and aggregate-cement ratio by weight are kept constant, equal to 0.7 and 2.0 respectively. The cementitious material contains 70% of ordinary Portland cement and 30% of RHA by weight.

The assumptions made in the theoretical considerations and the derivation of the expressions for E_{cc} and M_u are given in Appendix.

2 Experimental Programme

2.1 Part A : Preparation of Sisal Fiber-Mortar Composite Boards

2.1.1 Fabrication of Mould for Casting

Since the mixture of sisal fiber and mortar is a lightweight composite and highly compressible, the casting process should therefore be done under pressure. The formwork was made of wood and a 1.6 mm thick steel sheet. Details and dimension of the formwork are shown in Fig.1.

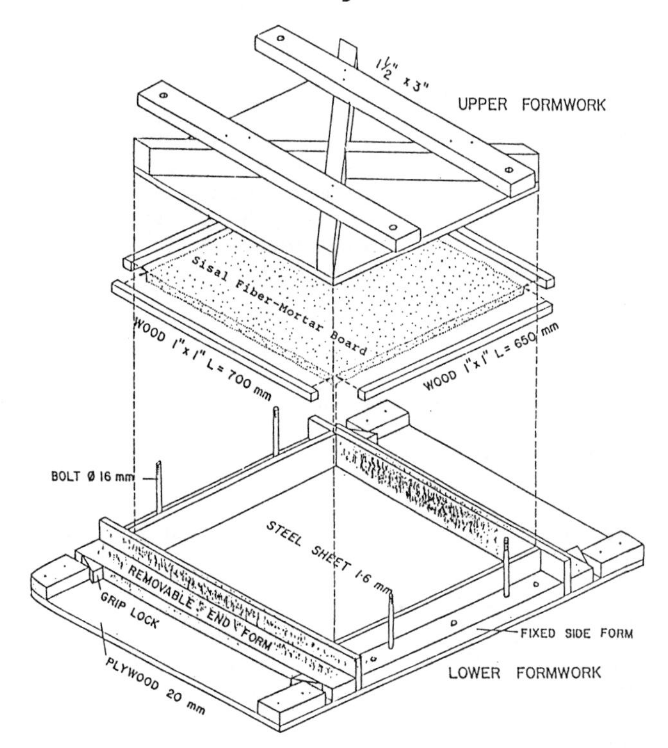

Fig.1. Three dimensional view of assembly of formwork

2.1.2 Materials

(a) Portland Cement : Ordinary Portland Cement Type I was used.

(b) Rice Husk Ash : The rice husk ash was obtained by burning rice husk in the ferrocement incinerator and ground by using the grinding machine developed by We (1981) and Islam (1981).

(c) Mixing Water : Ordinary tap water was used.

(d) Fine Aggregate : Natural river sand passing ASTM sieve number 8 was used in the mortar.

(e) Fiber : The length of sisal fiber is about 1.2-1.5 m. Its diameter ranges from 0.05-0.22 mm and the moisture content was approximately 25%.

2.1.3 Casting of Sisal Fiber-Mortar Composite Board

A factor k was introduced to compensate for evaporated water and material waste, the proportion of each component was calculated as follows :

$$\left[1.25\left(\frac{W_f}{W_c}\right)+3.7\right]W_c = 10.5625kU_n \tag{1}$$

where W_f, W_c and U_n are the weights of sun-dried sisal fiber, cementitious material and unit weight respectively.

The mortar was poured over the sisal fiber and then was manually mixed in such a way that all fibers were uniformly coated with sand and cement paste. The mixing was done in two operations. When the mixing procedure was finished, the mixture was then placed in the formwork and pressed by hands in order to check the uniformity. The casting process had to be done under pressure by means of a hydraulic cylinder. The loads required for casting boards having unit weights of 1950, 2050 and 2150 kg/m^3 were 1000, 3750 and 5000 kg respectively. The formwork was tightly fastened by four bolts to keep the specimen under pressure for 24 hours. Subsequently, the specimen was removed from the formwork and kept in the laboratory environment for 28 days. Two boards were required for each specimen and cut into small pieces depending on the type of testing.

2.2 Part B : Experimental Evaluation

2.2.1 Properties of Materials

Sieve analysis for the grading curve and fineness test were conducted as well as the determination of its moisture content and specific gravity.

2.2.2 Properties of Individual Sisal Fiber

(a) Specific Gravity : The apparent specific gravity is the ratio of the mass, based on oven-dried condition, of unit volume of a material at certain temperature to the mass of gas-free distilled water having the same volume and temperature.

(b) Moisture Content and Water Absorption : Fibers were weighed and oven-dried for 24 hours and weighed again. The ratio of the difference in weight to the oven-dried weight expressed in percent is call moisture content.

For the water absorption, fibers had to be immersed in water for 24 hours and laid on an absorbent cloth so that a saturated surface dried condition was achieved. All saturated fibers were weighed and then dried in the oven for 24 hours. The ratio of the difference in weight to the oven-dried weight is call water absorption.

(c) Tensile Strength and Elasticity : Since sisal fiber is small, special clamps were therefore designed for this purpose to fix the sisal at both ends during loading. The initial load was 300 gm.

2.2.3 Properties of Cementitious Material
(a) Bulk Density : After 28 days, all dimensions were measured and weights were recorded. The bulk density was expressed as weight of specimen divided by its volume.
(b) Axial Compressive Strength : The specimen cubes were loaded, the stress-strain curve and test data were obtained.
2.2.4 Properties of Sisal Fiber-Mortar Composite Boards
(a) Moisture Content : Those specimens having a size of 100x150x25 mm were weighed and then kept in the oven for 24 hours. Moisture content was determined from the ratio of the difference in weight to the oven-dried weight.
(b) Water Absorption : By submerging the specimens under water, after 2 hours, the specimens were allowed to drain for 10 minutes. The specimens were weighed. The increase in weight of specimens was obtained and water absorption calculated.
(c) Compressive Strength : Three specimens having dimension of 100x150x25 mm were tested for each varying parameter. The specimen was loaded until the specimen failed.
(d) Modulus of Elasticity : The modulus of elasticity was obtained from slope of each curve.
(e) Flexure Test : Three specimens having dimension of 150x580x25 mm were used. The sample was simply supported over a span of 500 mm and subjected to mid-point loading until a deflection of 5% of the span length was observed. The flexural strength was expressed in term of modulus of rupture. Other relevant properties determined were " resilience " which is defined as the energy absorbed within elastic range and " toughness " which is defined as the energy absorbed during application of load up to fracture.

3 Results and Discussion

3.1 Properties of Sisal Fiber

The physical and mechanical properties of sisal fibers were determined and shown in Table 1. As a natural organic fiber, it is rather difficult to obtain the exact value of the ultimate tensile strength or modulus of elasticity of sisal fiber, because of the variation in cross section, moisture content, age of sisal fiber and some defects existing in the fiber.

3.2 Properties of Cement Matrix

Ten cube specimens were tested and the results are tabulated in Table 2. The results of strength and modulus of elasticity were rather low compared to those of Ordinary Portland Cement on the same basis. This is due to the substitution of rice husk ash because of its advantage in reducing the alkalinity of cement paste which improves the durability of natural fiber.

Table 1. Physical and mechanical properties of sisal fibers

Description	Mean Value
Diameter of sisal fiber, mm.	0.05-0.22
Specific gravity of sisal fiber	1.0025
Moisture content of sun-dried fiber, %	25.00
Water absorption, %	56.25
Ultimate tensile strength of sisal fiber, MPa	275.43
Modulus of elasticity of sisal fiber, MPa	15000

Table 2. Physical and mechanical properties of cement matrix

Description	Mean Value
Bulk density of hardened cement paste, gm./cc.	1.895
Ultimate compressive strength, MPa	22.50
Modulus of elasticity, MPa	7000

3.3 Properties of Sisal Fiber-Mortar Composites

3.3.1 Moisture Content and Water Absorption : The specimens which have high fiber-cement ratios showed higher percentage of moisture content and water absorption than those of low fiber-cement ratios due to the high fiber content. By the fact that the composite having higher unit weight possesses lower percentage of air voids, the specimens with low unit weights would have higher percentage of moisture content and water absorption than those of high unit weights.

3.3.2 Axial Compression Test

(a) *Compressive Strength* The strength of specimens increased with the increase in unit weight, this can be explained by the fact that when the composites have higher unit weight, they contained more fibers as well as matrix and water in the same volume. This results in a denser composite and the amount of air void is decreased. For a particular unit weight, it was noted that the strength varied with fiber-cement ratio and the optimum strength was obtained at the same fiber-cement ratio of 0.160 by weight in case of fiber length equals to 1 cm. The reason why the composite with low fiber-cement ratios of 0.150 and 0.155, which contained more cement and less fiber, gave lower strength was that the consistency of the mix was rather high. When the cement paste was mixed with sisal fiber in the mixing process, balling occured and it became more difficult to the disperse the cement paste uniformly. In case of fiber length

Table 3. Average moisture content

Fiber Length, cm.	Fiber/cement ratio	Moisture content %, unit weight of, (kg/m³) 1950	2050	2150
1	0.150	3.185	4.375	3.883
	0.155	2.607	3.366	3.877
	0.160	3.303	5.041	4.833
2	0.150	4.836	4.672	4.165
	0.155	4.517	3.934	3.612
	0.160	4.576	2.590	2.512

Notes : Dimension of specimen = 100x150x25 mm.
For Fiber/cement ratio = 0 ;
Average moisture content = 6.508 %

Table 4. Average water absorption

Fiber Length, cm.	Fiber/cement ratio	Water absorption %, unit weight of, (kg/m³) 1950	2050	2150
1	0.150	18.364	18.195	17.199
	0.155	18.326	14.476	17.062
	0.160	20.486	14.960	17.181
2	0.150	17.848	18.029	17.403
	0.155	16.224	16.481	20.173
	0.160	17.460	19.211	18.572

Notes : Dimension of specimen = 300x300x25 mm.
For Fiber/cement ratio = 0 ;
Average water absorption = 9.441 %

equals to 2 cm, the optimum strength was obtained at the same fiber-cement ratio of 0.150 by weight. The reason why the composite with high fiber-cement ratios of 0.155 and 0.160, which contained less cement and more fiber, gave lower strength was that the mix was too dry and the amount of cement paste was insufficient to coat all the fibers. Therefore the optimum mix of the composite was a mix (1cm) which had fiber-cement ratio of 0.160 by weight and a unit weight of 2150 kg/m³. The highest compressive strength of specimens was 3.323 MPa compared to those of the mortar board of 10.595 MPa.

(b) *Modulus of Elasticity* The strength reduction factor K_m, modulus of elasticity from experiment and predicted modulus of elasticity from Eq.3 are shown in Table 5. By applying a factor ϕ', equals to 0.1 to the modulus of elasticity of composites. The modulus of elasticity from experimental results showed a similar trend as that of the compressive strength. A difference between experimental and theoretical results of 4.22% was found in the comparison due to the fact that less amount of air voids was presented in the specimens and more accurate results can be predicted.

3.3.3 Flexure Test : For a particular unit weight, it is observed that the highest strength was obtained at the fiber-cement ratio of 0.160 which was the same as the results obtained from the compression test in case of fiber length equals to 1 cm. In case of fiber length equals to 2 cm, the highest strength was obtained at the fiber-cement ratios of 0.155 and 0.160 and unit weights of 1950 and 2050 kg/m^3, respectively. Therefore, the optimum mix which gives the highest flexural strength is a mix (1cm) having fiber-cement ratio of 0.160 by weight and unit weight of 2150 kg/m^3, Similar results were also obtained in the compression test. The highest value of the modulus of rupture of sisal fiber-mortar composite was found to be 2.224 MPa compared to that of mortar board of 3.427 MPa. The flexural strength expressed in terms of modulus of rupture, the results from flexure test are shown in Table 7. By applying a factor ϕ equals to 0.5 to the ultimate moment of the section.

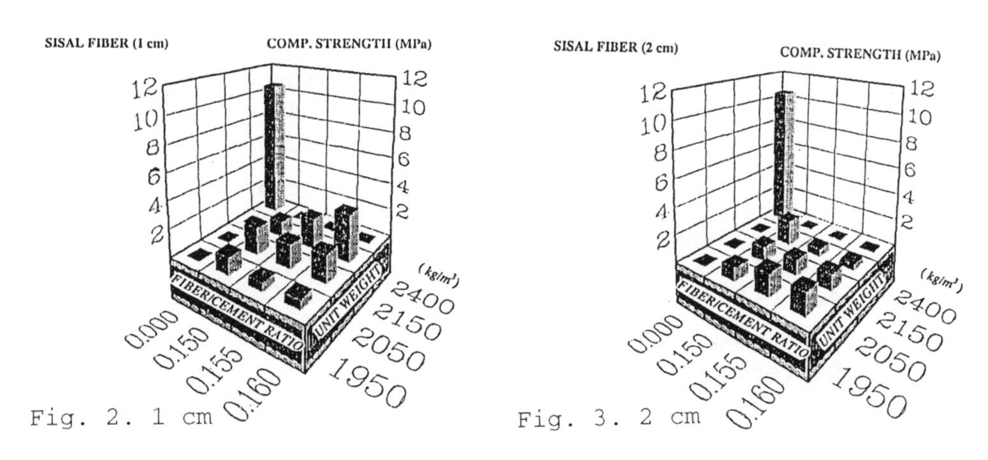

Fig. 2. 1 cm Fig. 3. 2 cm

Influence of fiber-cement ratio and unit weight on compressive strength of sisal fiber-mortar composite.

Table 5. Comparison of modulus of elasticity between theoretical and experimental results

Fiber length, cm.	Fiber/ cement ratio	Unit weight kg/m^3	K_m from Eq. (5)	$\phi' E_{cc}$ Eq. (3) MPa	E_{cc} from Exper. MPa	Difference %
1	0.150	1950	.059	68.98	149.3	116.38
		2050	.090	89.52	66.7	25.53
		2150	.053	65.05	125.0	92.16
	0.155	1950	.028	49.35	61.5	24.70
		2050	.079	83.04	137.2	65.24
		2150	.101	97.65	106.7	9.24
	0.160	1950	.031	52.20	50.0	4.22
		2050	.092	92.49	44.4	51.95
		2150	.157	135.4	454.5	235.63
2	0.150	1950	.044	59.03	103.9	76.24
		2050	.050	63.01	90.9	44.28
		2150	.084	85.60	67.0	21.73
	0.155	1950	.076	81.11	233.3	187.67
		2050	.054	66.50	83.3	25.31
		2150	.027	48.69	28.0	42.49
	0.160	1950	.072	79.28	96.4	21.58
		2050	.055	68.05	108.0	58.70
		2150	.021	45.60	40.0	12.27

Notes : ϕ' = Strength reduction factor for modulus of elasticity of the composite is assumed to be equal to 0.1

For Fiber/cement ratio = 0 ;

K_m from Eq.(5) = 0.471

E_{cc} from Eq.(3) = 3297 MPa

E_{cc} from Exper. = 754.72 MPa

Difference = 77.11 %

Table 6. Calculation of centroid of the compressive zone, X_c

Fiber length, cm.	Fiber/ cement ratio	Unit weight kg/m^3	ε_b (%)	ε_u (%)	X_c, mm. from Eq. (8)
1	0.150	1950	1.67	4.00	7.956
		2050	2.00	4.00	6.978
		2150	1.90	4.00	7.967
	0.155	1950	1.50	4.00	9.162
		2050	1.25	4.00	7.750
		2150	2.30	4.00	6.680
	0.160	1950	1.70	4.00	8.895
		2050	1.60	4.00	7.279
		2150	1.25	4.00	6.187
2	0.150	1950	1.54	4.00	8.521
		2050	2.00	4.00	7.979
		2150	2.20	4.00	7.013
	0.155	1950	1.45	4.00	7.701
		2050	1.65	4.00	8.168
		2150	1.60	4.00	9.101
	0.160	1950	2.00	4.00	7.506
		2050	1.50	4.00	8.307
		2150	2.00	4.00	8.951

Notes : For fiber/cement ratio = 0 ;
ε_b = 0.80 %
ε_u = 1.28 %
X_c = 12.50 mm.

Table 7. Comparison of modulus of rupture between theoretical and experimental results

Fiber length, cm.	Fiber/ cement ratio	Unit weight kg/m^3	M_u Eq. (9) N-m	ϕM_u N-m	$\sigma_r = \frac{\phi M_u}{z}$ MPa	σ_r from Exper. MPa	Diff. %
1	0.150	1950	44.8	22.38	1.43	0.24	83.22
		2050	60.7	30.33	1.94	0.31	84.02
		2150	40.4	20.19	1.29	0.79	38.76
	0.155	1950	24.0	11.98	0.77	0.73	5.20
		2050	57.8	28.92	1.85	0.73	60.54
		2150	65.0	32.50	2.08	0.71	65.87
	0.160	1950	25.9	12.97	0.83	0.89	7.23
		2050	64.1	32.03	2.05	0.58	71.71
		2150	93.7	46.83	3.00	2.22	26.00
2	0.150	1950	35.4	17.70	1.13	0.81	28.32
		2050	38.2	19.09	1.22	0.50	59.02
		2150	56.8	28.40	1.82	0.86	52.75
	0.155	1950	55.6	27.81	1.78	1.49	16.29
		2050	41.9	20.93	1.34	0.89	33.58
		2150	23.0	11.52	0.74	0.58	21.62
	0.160	1950	51.8	25.88	1.66	0.45	72.89
		2050	43.1	21.53	1.38	1.49	7.97
		2150	17.8	8.89	0.57	0.50	12.28

Notes : ϕ' = Strength reduction factor for flexure is assumed to be equal to 0.5

For Fiber/cement ratio = 0 ;

M_u = 351.56 N-m

ϕM_u = 175.78 N-m

$\sigma_r = \frac{\phi M_u}{z}$ = 11.25 MPa

σ_r from Exper. = 3.43 MPa

Difference = 69.51 %

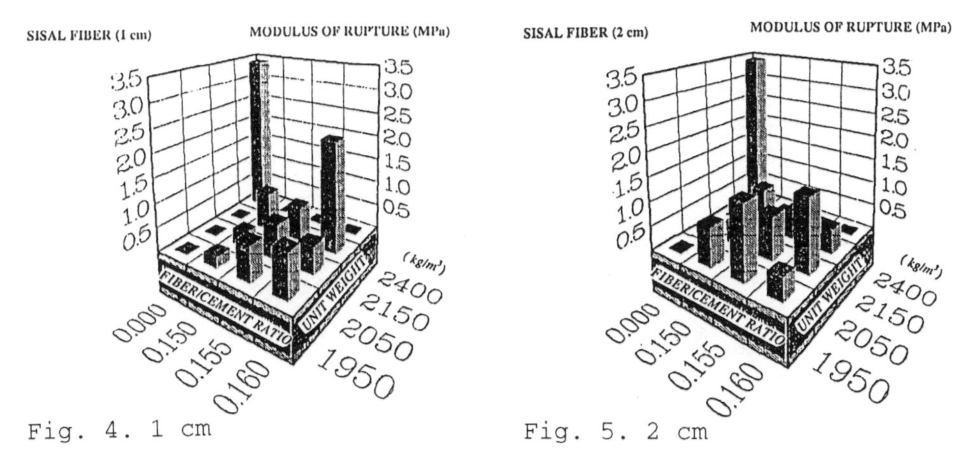

Fig. 4. 1 cm Fig. 5. 2 cm

Influence of fiber-cement ratio and unit weight on modulus of rupture of sisal fiber-mortar composite.

3.3.4 Ductility

(a) *Resilience* The modulus of resilience was determined from the area under the load-deflection curve within the elastic range represented by area *OAB* of Fig.6. Comparing the maximum modulus of the resilience of sisal fiber-mortar composite and that of mortar board, the maximum modulus of resilience of the mix was 0.248 N-m whereas that of the mortar board was 0.0054 N-m. Based on highest strength both in compression and bending properties, the optimum mix was a mix having 1 cm fiber length with fiber-cement ratio of 0.160 by weight and unit weight of 2150 kg/m^3. The modulus of resilience of the optimum mix was 0.180 N-m.

(b) *Toughness* The modulus of toughness is expressed in term of the work performed in deforming a material up to fracture and therefore is represented by the area *OACD* under the stress-strain diagram shown in Fig.6. Most of the mortar board failed before a maximum deflection of 25 mm while the sisal fiber-mortar composite board failed at a maximum deflection of 25 mm, the specimen could still sustain high

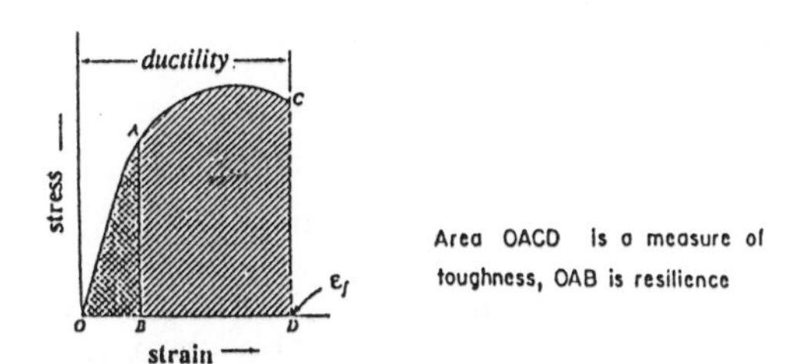

Fig.6. Diagrammatic representation of toughness and resilience

Table 8. Actual volume fraction of fiber, matrix and air void

Fiber length, cm.	Fiber/ cement ratio	Unit weight kg/m^3	Actual volume fraction %, of fiber (V_f)	matrix (V'_m)	air void (V_v)
1	0.150	1950	5.31	61.03	33.66
		2050	5.31	58.89	35.80
		2150	5.32	56.36	38.32
	0.155	1950	5.48	53.91	40.61
		2050	5.47	60.95	33.58
		2150	5.48	59.02	35.50
	0.160	1950	5.64	57.60	36.76
		2050	5.64	63.75	30.61
		2150	5.64	61.32	33.04
2	0.150	1950	5.31	59.87	34.82
		2050	5.31	56.65	38.04
		2150	5.32	59.91	34.77
	0.155	1950	5.48	65.08	29.44
		2050	5.47	60.21	34.32
		2150	5.48	56.18	38.34
	0.160	1950	5.64	57.60	36.76
		2050	5.64	63.75	30.61
		2150	5.64	59.55	34.81

Notes : For fiber/cement ratio = 0 ;
Actual volume fraction of,
Fiber (V_f) = 0 %
Matrix (V'_m) = 78.38 %
Air void (V_v) = 21.62 %

load at that deflection. The highest value of toughness for the mix was found to be 0.723 N-m, where that of the mortar board was only 0.0196 N-m. Based on highest strength both in compression and bending properties, the optimum mix was a mix (1cm) with fiber-cement ratio of 0.160 by weight and unit weight of 2150 kg/m^3. The toughness of the optimum mix was 0.300 N-m.

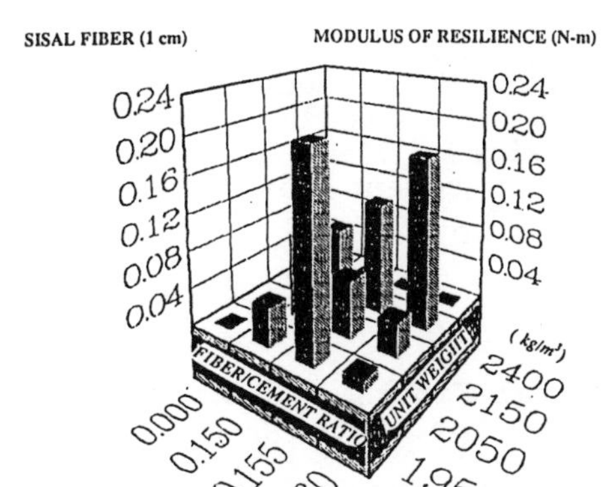

Fig. 7. Influence of fiber-cement ratio and unit weight on modulus of resilience of sisal fiber-cement composite. (1 cm)

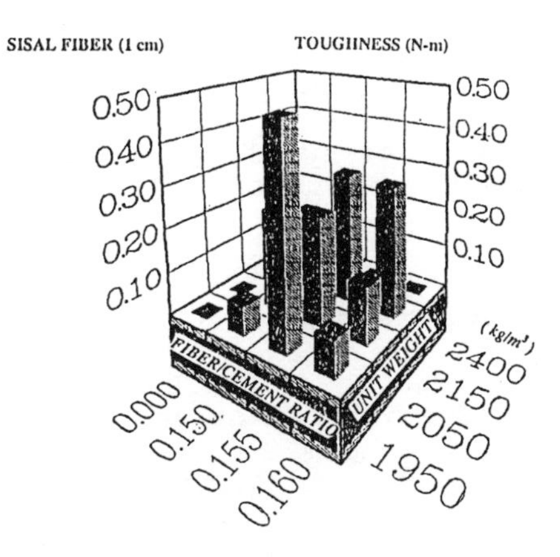

Fig. 8. Influence of fiber-cement ratio and unit weight on modulus of toughness of sisal fiber-mortar composite. (1 cm)

4 Conclusions

The following conclusions can be drawn regarding the behavior of sisal fiber-mortar composite boards.

(a) The modulus of elasticity of the sisal fiber-mortar composite which is a function of the compressive strength, can be predicted by using the Rule of Mixture with a discrepancy of 4.22%. However, it is valid only for the composite (1cm) having a unit weight of 1950 kg/m^3.
(b) The presence of air voids has remarkable influence on the strength of composite, by the introduction of a reduction factor ϕ which is found to be 0.5 for flexural strength and a reduction factor ϕ' which is found to be 0.1 for moduli of elasticity of composite.
(c) Based on highest strength both in compression and bending properties, the optimum sisal fiber-cement ratio of sisal fiber-mortar composite (1cm) is experimentally found to be 0.160 by weight, and unit weight is 2150 kg/m^3. Moisture content of composites was found to be 4.8% and water absorption was 17.2% by weight.
(d) Although the ultimate strength or modulus of rupture of a mortar board was higher than that of a sisal fiber board, it was brittle while sisal fiber board was more ductile.
(e) The modulus of resilience of sisal fiber composite is higher than that of mortar board especially at the optimum sisal fiber-cement ratio, unit weight and fiber length. Moreover, its toughness is considerably higher. Therefore, the sisal fiber-mortar composites are remarkably more ductile than mortar board.

(f) Sisal fiber-mortar composites can be satisfactorily used as ceilings, partition walls and formwork. However, they should not be directly subjected to moisture due to the effect of swelling of fibers in composites. Therefore, they should be coated with a thin layer of cement mortar when used.

5 References

Gram, H.E. (1983) Durability of Natural Fibers in Concrete. **Swedish Cement and Concrete Research Institute**, S-100 44, Stockholm.

Gram, H.E. and Nimityongskul, P. (1987) Durability of Natural Fiber in Cement-based Roofing Sheets. **Journal of Ferrocement**, Vol.17, No.4, October.

Islam, S. (1981) Grinding Methods and Their Effects on Reactivity of RHA. **M.Eng. Thesis no.ST-81-7**, AIT., Bangkok.

Lin, T.H.J. (1987) Some Properties of Fiber-Reinforced, Polyester-Based Composites, made with Coir, Jute and Sisal Fibers. **U.M.I. Dissertation Information Service.**

Nilsson, L. (1975) Reinforcement of Concrete with Sisal and Other Vegetable Fibers. **Swedish Council for Building Research**, Document D14.

Shafiq, N. (1987) Durability of Natural Fiber in Mortar containing RHA. **M.Eng. Thesis no.ST-87-20**, AIT., Bangkok.

We, A.B. (1981) Production of RHA and its Application in Mortar and Concrete. **M.Eng. Thesis no.ST-81-20**, AIT., Bangkok.

Wu, M.Z. (1989) Use of Sisal as Fiber Reinforcement in Rice Husk Ash Mortar. **M.Eng. Thesis no.ST-89-20**, AIT., Bangkok.

6 Appendix

In this study, the fiber is to be mixed with rice husk ash to produce a composite building board.

6.1 Assumptions

The following assumptions are made in the analysis:

(a) The sisal fibers are considered as discrete fibers which are randomly oriented and the mortar which contains rice husk ash acts as matrix.

(b) The sisal fiber has an equal probability of being oriented in any direction in one plane and parallel in the other plane.

(c) The sisal fiber is uniformly distributed in the mortar matrix.

(d) The sisal fibers are firmly bonded by the matrix so that no slippage occurs at the interface of the fiber and matrix.

6.2 The Rule of Mixtures

In a composite material consisting of a matrix reinforced with uniformly dispersed unidirectional continuous fibers, since no slippage occurs at the interface.

$$E_c = E_m V_m + E_f V_f \tag{2}$$

in which V_m and V_f denote the volume fractions of matrix and fibers respectively, and E_c, E_m and E_f represent the moduli of elasticity of composite, matrix and fibers respectively.

Considering the elongation of fiber in Fig.9, by small-strain theory.

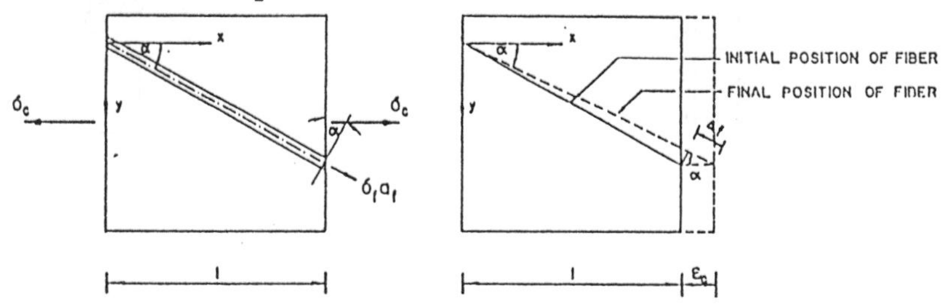

Fig.9. Material element under axial load

$$E_{cc} = K_m E_{mc} V_m + 0.375 E_f V_f \tag{3}$$

where E_{cc}, E_{mc} and E_f denote the moduli of elasticity of composite, matrix and fiber in compression respectively.

In tension, by introducing the reduction factor for the tensile strength of matrix due to the presence of air voids.

$$E_{ct} = K_m E_{mt} V_m + 0.375 E_f V_f \tag{4}$$

in which E_{ct} and E_{mt} are the moduli of elasticity of composite and matrix in tension respectively.

In compression, the ultimate strength of composite may be expressed by neglecting the contribution of fibers as

$$\sigma_{cuc} = K_m \sigma_{muc} V_m \tag{5}$$

in which σ_{cuc} is the ultimate compressive strength of composite and σ_{muc} is the mean value of the crushing strength of cement matrix.

In the elastic range, the relationship of modulus of rupture, σ_r, and ultimate moment, M_u, can be expressed as

$$\sigma_r = 6\frac{M_u}{bh^2} \tag{6}$$

where b and h denote the width and depth of specimen respectively.

At ultimate condition, the strain & stress distribution is as shown in Fig.10.

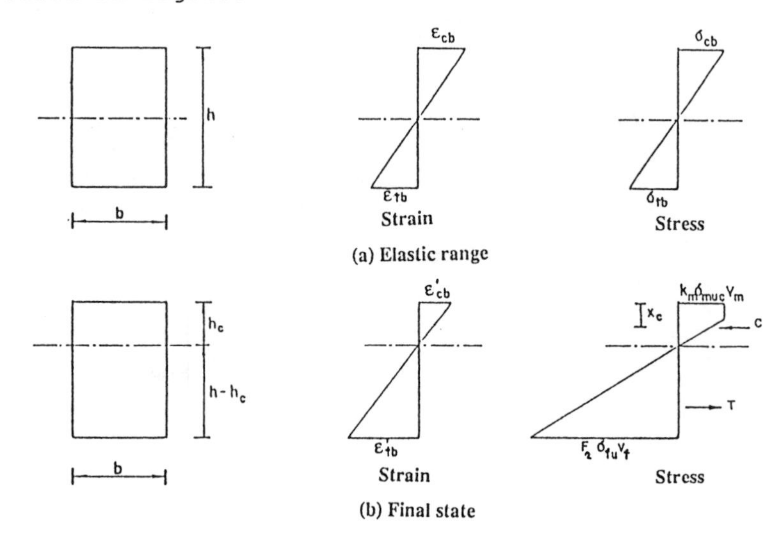

Fig.10. Stress-strain distribution at different stages of loading

The depth of compressive zone, h_c, can be expressed as

$$h_c = 0.25\sigma_{fu}V_f\frac{h}{\left[K_m\sigma_{muc}V_m\frac{(2\epsilon_u-\epsilon_b)}{[2\epsilon_u]}+0.25\sigma_{fu}V_f\right]} \tag{7}$$

where ε_b and ε_u are the compressive strains of composite at the end of elastic range and at the ultimate stage respectively.

The centroid of compressive zone is located at distance X_c from top surface and can be written as

$$X_c = \left[\frac{[3\epsilon_u^2-\epsilon_b(3\epsilon_u-\epsilon_b)]}{[3\epsilon_u(2\epsilon_u-\epsilon_b)]}\right]h_c \tag{8}$$

Finally, the ultimate bending moment, M_u, can be expressed as

$$M_u = 0.25\sigma_{fu}V_f b(h-h_c)\left[\frac{2}{3}h+\frac{1}{3}h_c-X_c\right] \tag{9}$$

where σ_{fu} is the ultimate tensile strength of fiber.

85 EFFECT OF SIMPLE TREATMENTS OF MALVA FIBRES FOR THE REINFORCEMENT OF PORTLAND CEMENT MORTAR

M. J. E. OLIVEIRA and V. AGOPYAN
Escola Politécnica, University of São Paulo, São Paulo, Brazil

Abstract
Vegetable fibres are presented as alternative reinforcement for low-cost construction materials therefore they have been studied in several countries, mainly from the Third World. This paper presents the study of malva (Urena *lobata* Linn.) fibre used as reinforcement of an alkaline matrix. It aims is to evaluate the composite behaviour with time. The behaviour of composites prepared with fibres submitted to treatment with simple washing (with ordinary washing-up liquid) and impregnated with linseed-oil is presented and their results are compared with composite obtained with untreated fibre. The final discussions are based on the results of tensile and impact strengths of the composites.
Keywords: Malva Fibres, Vegetable Fibres, Durability, Fibre Reinforced Mortar, Fibrecement.

1 Introduction

Vegetable fibres have been used for reinforcement in many types of mortars and concretes. For instance, in Brazil building components were developed with coir and sisal fibres, and also incipient research work was done with jute, sugar-cane bagasse and bamboo fibres (Agopyan, 1988). The malva fibres production in Brazil have been increased due to their good properties for sacking. Moreover, based on intensive research work, the malva plantation processes and fibre properties have been improved. Due to this prospective large availability of malva fibres, their use as brittle matrix reinforcement is proposed (Agopyan, 1988). As it happens with all vegetable fibres, durability is the major concern for the use of these fibres; consequently its improvement is proposed by conditioning the fibres in simple solutions.

The research work has been carried out with Ordinary Portland Cement mortar reinforced with malva fibres. The effect of the treatments is analyzed by the changes on the tensile and impact strengths of the composite up to 120 days.

Fibre Reinforced Cement and Concrete. Edited by R. N. Swamy. © 1992 RILEM.
Published by E & FN Spon, 2-6 Boundary Row, London SE1 8HN. ISBN 0 419 18130 X.

2 The malva fibre

Malva from the Malvaceae family is a common plant in tropical countries like Brazil, which has grown as a commercial plant in the North of the country (Amazon Region). The malva is a bushy plant, ranging from 1 to 3 meters tall, branching out enough when it grows apart. In Brazil, it is known as "malva", "vassourinha", "guanxuma", and "guanxima"(Medina, 1959). This plant demands for its quick and satisfactory growth a humid and hot climate with alternative periods of sun and rain. During its vegetable cycle that lasts from 150 to 180 days it requires rain precipitations from 125 to 250 mm, mostly at a temperature range of 18 to 32° C. The malva plant grows better in sandy or clay sandy humus soils with pH around 5.7 and with low content of potassium (K) and phosphorus (P). An acceptable production of malva is about 2000 kg by hectare which results about 120 kg of dry fibre.

The fibre is removed from the bark of the trees between the cambium layer and the outer layer bark. The characteristics of Brazilian fibres compared with those of other countries are presented in Table 1. The average tensile strength of the Brazilian fibre is 159 MPa.

Table 1. Characteristics of the malva fibres from several countries (Oliveira, 1989 and Samuel, 1948)

Characteristics	Brazil	Gambia	Zimbabwe	India
humidity (%)	9.3	10.6	10.8	9.9
hydrolysis (%)				
alfa	10.3	9.8	8.5	12.3
beta	18.2	16.3	11.8	16.3
acid purification (%)	1.9	1.3	1.0	3.0
cellulose contend (%)	76.0	74.6	80.5	73.5
length of fibril (mm)	4.0	-	-	3.7

3 Action of the alkaline environment on the vegetable fibre

The composites reinforced with vegetable fibres have usually the Ordinary Portland Cement as the main binder material that possesses high alkalinity, pH around 12.4, which reduces the durability of the vegetable fibre in the long run due to degradation of fibre compounds as follows:

a) cellulose: the final groups of molecular chain are separated from polymeric chain of the cellulose in alkaline medium. This fact depends on the exposure conditions such as temperature, pressure and pH. The oxidation of the cellulose can be possible by the presence of OH^- resulting in the isosaccarin acid (CH_2OH) (Gram, 1983). The polymerization degree of the cellulose is not quite damaged, but in higher temperatures, around 75°C, the reaction is remarkable intensified.

b) lignin: through the electronic microscopy it can be seen how the alkaline environment dissolves the lignin that acts as a binder for the cellulose. The lignin begins to soften at 70°C - 80°C. At 120°C it is partly liquid. The lignin has no resistance for the alkaline medium.

c) hemicellulose: it has similarities with cellulose, so it follows the same degradation mechanisms. However it has a lower polymerization degree than the cellulose, therefore it is more easily damaged by alkaline environment.

There are small amounts of other compounds such as pectin and cellubiose, which are soluble in ordinary water.

4 Experimental work

4.1 Fibre treatments

From preliminary studies, fibre treatments were selected by their low cost, easy way of processing and low effect on the fibre structure. The two adopted treatments consist of a simple washing of the fibres (with ordinary washing-up liquid) and impregnation of the fibres with linseed-oil.

4.2 Surface washing

The ordinary washing-up liquid was used in order to remove part of the soluble substances at fibre surface as well as the residues and dust. Such procedure makes fibres rough, so it also improves the fibre-matrix bond. The exposed parts of the fibres are more resistant to the attack of alkalinity.

A little reduction of the hemicellulose and the lignin on surface of the fibres can expose the fibrils giving better fibre-matrix bond and increasing the fibre life span since the cellulose is less attacked by the alkaline environment than the hemicellulose.

4.3 Linseed-oil impregnation

Another treatment used was the linseed-oil as superficial protection for malva fibre. The idea was to impregnate with linseed-oil at a temperature of (21 ± 2) °C. After finishing they were chopped to be employed in the manufacturing of the composite. The excess linseed-oil was removed from malva fibre surface with tissues before the chopping.

4.4 Control samples

Specimens with the matrix reinforced with untreated fibres and also unreinforced matrix were prepared for comparison.

4.5 Raw materials

The raw-materials used in the preparation of specimens were:

a) malva fibres with characteristics and properties as

mentioned in section 2;

b) commercially available washing-up liquid and linseed-oil;

c) Ordinary Portland Cement, following Brazilian standards, with compressive strength of 32 MPa (at the age of 28 days) and fineness of 260 m^2/kg (Blaine);

d) natural dry sand with density of 2590 kg/m^3, bulk density of 1410 kg/m^3 and fineness of 2.33.

4.6 Composites

They had a cement-mortar matrix with cement-sand ratio of 1:2 by weight. The water-cement ratio was 0.49 which gave a good workability (230 mm in flow-table). The amount of fibres was 2.6% of the weight of the cement. After treatment, the fibres were chopped resulting 40 mm long strands.

4.7 Preparation of the specimens

The materials are mechanically mixed in ordinary concrete pan-mixer with capacity for 120 litres. The materials were put in the following order: dry sand, fibres added little by little; about one third of the water, cement and the rest of the water then the materials were mixed for about four minutes.

The moulds of the 40 mm x 40 mm x 160 mm and 100 mm x 100 mm x 20 mm sizes were filled up into two layers with manual compaction. After this, the specimens were mechanically compacted on a vibrating table with a frequency of 1750 rpm for 120 seconds. After casting the specimens, they were placed in a wet chamber at a temperature of (21 $\pm$ 2) ^{o}C and relative humidity higher than 95%. 144 specimens were prepared for bending test (ASTM C 239-79) and 48 specimens for impact test (falling ball). The flow-index (FI) of the composites were determined by the Brazilian method which is similar to a former American one (ASTM C 124-66).

5 Results and discussion

5.1 Flow index (FI)

The results of flow index are shown in Table 2:

Table 2. Flow index

Series	Treatment	FI (%)
MO	impregnation with linseed-oil	175
MN	untreated fibre	121
ML	surfaced washing	127
MS	unreinforced matrix	230

5.2 Tensile strength

Specimens of 40 mm x 40 mm x 160 mm were tested at 7, 14, 28, 60, 90 and 120 days. The average strengths of 144 specimens submitted to bending test are presented in Table 3. The Figure 1 shows the development of strength up to 120 days. The composites MN and ML presented similar tensile strength development curves up to 28 days. After 60 days, mortar reinforced with fibres treated with the linseed-oil (MO) had its strength curve quite below the curves from other series.

Table 3. Average tensile strength (MPa)

Series	Age (days)					
	7	14	28	60	90	120
MO	2.8	9.1	4.1	4.4	4.2	4.2
MN	4.8	5.1	6.8	7.6	6.5	6.5
ML	5.0	5.9	7.0	8.2	7.7	7.4
MS	2.6	9.2	4.9	4.7	4.8	4.8

5.3 Impact test

Experiments were carried out on plates measuring 100 mm x 100 mm x 20 mm by the fall of a 0.98 kg steel ball from a height of 400 mm, increasing 50 mm each trial. The Table 4 presents the average results obtained for impact test. The Figure 2 shows the curves for MN and ML series in comparison to the curve MS for specimens without reinforcement. The curve for MO Series shows that the fibre treatment with linseed-oil was not efficient.

Table 4. Average applied impact energy (J)

Series	Age (days)	
	28	90
MO	10.1	10.7
MN	17.7	16.9
ML	24.2	23.6
MS	4.9	4.9

5.4 Final remarks

It can be concluded from this study that major fibre deterioration occurs by alkaline environment during the liquid phase of hydration of the Ordinary Portland Cement during the first days. The hydroxyl can break the polymeric chain of fibre cellulose by oxidation.

The results showed that composite reinforced with fibre washed with ordinary washing-up liquid behaves better than other composites.

The tensile strength of composites from MN and ML series show similar behaviour up to 60 days. Afterwards there is a higher decrease of strength of these composites in comparison to the specimens from the series MS and MO. The washed fibre sample develops a good matrix-fibre bond up to

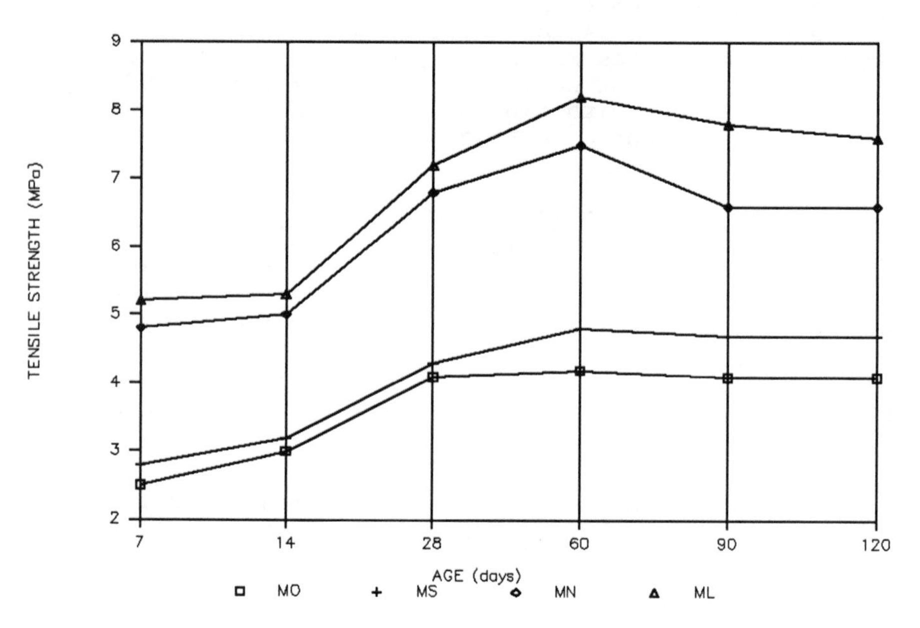

Fig.1. Tensile strength

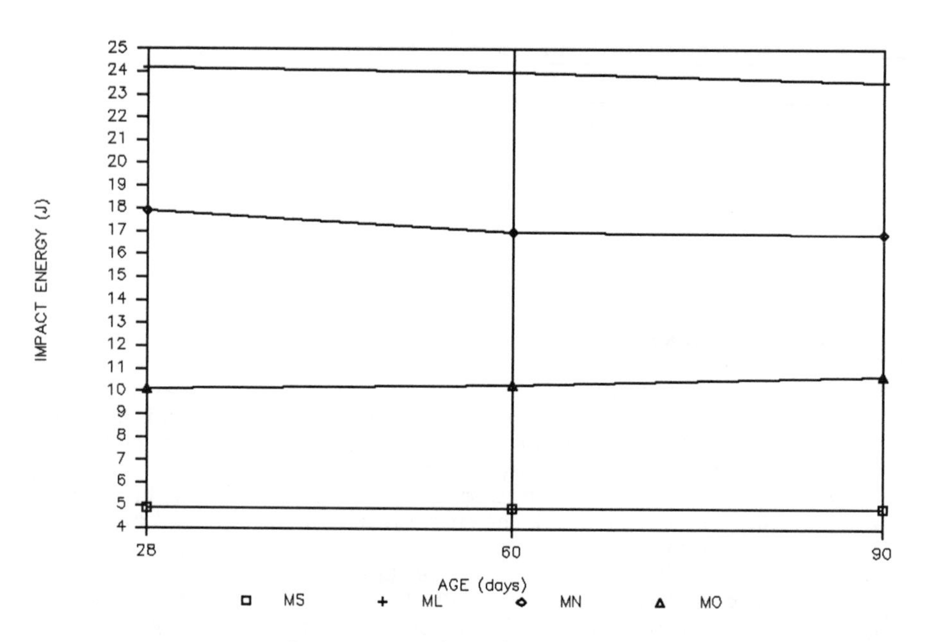

Fig.2. Applied impact energy

60 days, probably due to chemical reaction and by intumescence of the fibre.

It seems to be possible to improve the durability of the composites by simple treatment of malva fibre as this treatment must acts only during the early ages. Later, the cement hydration rate decreases and the OH^- concentration is also reduced, and consequently the fibre cellulose degradation OH^- is minored.

This simple fibre surface treatment seems to be another suitable and inexpensive solution for the increase the durability of vegetable fibre composites in addition to the previous suggestions for sealing the composites (Gram, 1983) or using BFS based binder (John et al., 1990).

6 References

Agopyan, V. (1988) Vegetable fibre reinforced building materials - developments in Brazil and other Latin-American countries, in **Natural Fibre-Reinforced Cement and Concrete** (Concrete Technology Design 5, ed. R. N. Swamy), Blackie, Glasgow. pp. 208-242.

Gram, H.-E. (1983) **Durability of natural fibres in concrete.** Swedish Cement and Concrete Research Institute, Stockholm. pp. 50-63.

John, V. M.; Agopyan, V. and Derolle, A. (1990) Durability of blast furnace slag based cement mortar reinforced with coir fibres, in **Vegetable Plants and their Fibres as Building Materials** (RILEM Proceedings 7, ed. H. S. Sobral), Chapman and Hall, London. pp. 87-97.

Medina, J. C. (1959) **Plantas fibrosas da flora mundial.** Instituto Agronômico, Campinas-SP.

Oliveira, M. J. E. (1989) **A Influência dos tratamentos simples nas fibras empregadas em reforço de argamassa de cimento Portland.** MEng. Dissertation, Civil Construction Engineering Department, Escola Politécnica, University of São Paulo, São Paulo.

Samuel, P. (1948) **La culture et le rouissage de l'Urena lobata.** (Buletin Agricole du Congo Belge no. XXXIX)

Acknowledgments

The results presented is this paper are part of the Dissertation of the first author. The authors would like to thanks the Faculty of Engineering in Guaratinguetá (UNESP) and the Instituto de Pesquisas Tecnológicas do Estado de São Paulo - (IPT) where part of the tests were performed.

86 COIR FIBRE REINFORCED CEMENT BASED COMPOSITE PART 1: MICROSTRUCTURE AND PROPERTIES OF FIBRE-MORTAR

G. SHIMIZU and P. JORILLO Jr
College of Science and Technology, Nihon University, Tokyo, Japan

Abstract
This paper describes the microstructure and the basic mechanical properties of both the reinforcing coir fibre and the fibre-cement composite. Results of SEM micrograph analysis revealed insights on the microstructure of fibre, its decomposition behaviour, and the interfacial characteristics between fibre and matrix. Results of mechanical properties showed significant improvement in post-cracking strength, toughness and ductility of the matrix due to the reinforcing capability of the coir fibre. Properties in compression and flexure were summarized in simple empirical equations in the form of composite mechanics approach.
Keywords: Coir fibre, Compression, Composite, Flexure, Microstructure, Modulus of elasticity, SEM, Stress-strain, Toughness

1 Introduction

The potential use of cement based products reinforced with natural fibres from a technical and economic point of view is very significant especially in the developing countries. Its relative cheapness and ready availability are some of its many distinct advantages, which when fully utilized can directly lead to the conservation of energy and scarce resources. Technical advantages of natural fibres in cement matrices like controlling of tensile cracking, improvement of post cracking strength, and toughness have been reported (Agopyan,1988; Swamy,1990a; Jorillo et.al., 1991a).

In spite of these advantages, researches and practical applications are still very limited. One of the drawbacks in the use of natural fibres is its inherent low modulus of elasticity and low-durability characteristics especially in an alkaline medium (Swamy,1990a). Also, scarcity of precise information on the structure of natural fibres, interaction with the matrix, and the properties of the whole composite system itself are the major factors that contribute to this slow development (Gram,1984). This paper describes these three major points on a particular natural fibre composite material, that is coir (coconut husk) fibre cement composite. The objective of this paper is to provide comprehensive information and microstructural data on the structure of fibre, the region of intimate contact between fibre and matrix, and also the structure of fibre exposed to different environmental conditions. Engineering properties of both the reinforcing fibre and the composite were likewise examined and discussed in this paper.

Fibre Reinforced Cement and Concrete. Edited by R. N. Swamy. © 1992 RILEM.
Published by E & FN Spon, 2-6 Boundary Row, London SE1 8HN. ISBN 0 419 18130 X.

2 Microstructure of Fibre and Composite

Investigations of the structure of the composite by scanning electron microscope (SEM) is still relatively recent (Coutts,1988), even in the case of the well advanced wood fibre reinforced concrete. Microstructural studies can provide both insights and understanding to the structure-property behaviour of the composite system up to the engineering level.

2.1 Structure of Coconut husk fibre

This type of seed fibre is extracted from the husk (mesocarp) of a coconut fruit (Cocos Nucifera Linn) through the combined process of retting, mechanical and manual extraction. The extracted fibre length ranges from 100 to 280 mm. The fibre diameter varies from 0.1 to 0.55 mm, and the specific gravity from 1.12 to 1.15.

A single strand of fibre may be thought of as a composite mate rial itself which is composed of elongated oriented reinforcement material (that is filamentous fibre cells which are principally cellulose), and a matrix (mostly lignin, hemicellulose and extractives) that holds these cells in place. From SEM micrographs in Figs. 1 to 4, the cross-section of a fibre shows that each strand contains 30 to more than 200 individual cells with diameters 15-25μm and lengths of 0.25 to 1.0 mm. The central cavity in each cell, the lumen has a rounded to an elliptical cross section of diameter varying from 5 to 10μm. The spiral angle of the fibrils seen in Figs. 3 to 4 ranges from 30 to 40°. Inherent greater extensibility of the coir fibre is attributed to this fibril formation characteristics. In Fig.5 the central lacuna and the membrane that separates the fibre cells from each other (middle lamella), can be clearly seen at 1000 x magnification. These materials are approximately 71% lignin (Satyanaryana et.al., 1981a), and its chemical decomposition due to the dissolving action of the alkaline pore solution is one of the prime cause in the changes in mechanical characteristics of the fibre.

The cell walls are thin to fairly thick of about 5 to 18μm, see Figs.2 to 6. The outer surface consists of epidermal cell layers of gum type material which gives the fibre the dense and relatively close outer covering. The inherent resistance of coir fibre against rotting and decomposition is attributed to this outer structure characteristics. Also, the presence of pores and globular protrusions (tyloses) of diameters of 8 to 15μm is visible (Fig.8). This is a special attribute to the bonding properties of coir fibre.

Slate (1976a) conjectured that the poor performance of the coir fibre he used in his experiment maybe due to some of the coconut pith material left in the fibres due to the crude form of extraction. This pith material shown in Fig.9 appears weak, and may indeed affect not only the hydration of cement but also may weaken the interfacial shearing stresses which can consequently lower the performance of the fibre in transmitting stresses.

2.2 Fiber-matrix interface characteristics

Another major factor which controls the overall performance of a composite system, i.e. mechanical properties and mode of failure is the bonding and the condition of the interface of fibre and matrix. An understanding of the microstructure of this region of intimate contact will provide not only insights on the mechanism of bonding but also the ability and efficiency of the fibres to delay and control tensile microcracking in a composite system.

2.2.1 Mechanical bonding

It was observed from the SEM micrographs in Figs. 8 to 12 that the mechanism of bonding of coir fibre and cement matrix is generally a combination of mechanical interlocking and physicochemical bonding. The outer surface of the fibre is relatively smooth, but the presence of tyloses throughout the surface can be seen. These pro trusions can offer extra anchoring points by which the fibre can accept stresses from the matrix. The imprints of the tyloses and the whole topography of fibre surface in the dense matrix (Figs. 7 and 10) confirms the mechanical interlock provided by the fibre. Also, the hollow sections of fibres can collapse to ribbons, develop a helical twist, and be excessively contorted during mixing and casting. These can consequently result to an additional frictional property which can further enhance the effectiveness of fibre in transmitting stresses.

2.2.2 Physicochemical bonding

In this SEM study, the microstructural features of the fractured surface of the composite came from a 56-day old (1:1:0.6, i.e.,C:S:W/C proportion) and one-year old coir fibre mortar (1:1:0.65) specimen.

Crystal growth deposit at the surface of fibre from a 1-year old specimen as seen in Figs. 11 and 12 shows the presence of hexagonal lamellar crystal of portlandite (Calcium Hydroxide or CH), and the formation of porous gel. A part of this gel is conjectured to be a product of the reaction of lignin (oxidation of end groups) or the remaining phenol or tannin in the fibre surface and the hydroxyl group of the cement matrix.

Figs. 13 and 14 show the dense matrix consist mostly of continuous CH crystals which replicated the topography of the reinforcing fibre. From these figures, discontinuities are evident in some portion, and if magnified to 5000 times the hydration product of calcium-silicate hydrate (CSH), large CH crystals and ettringite can be found. Near the layer of the interface (Fig.16) the amount of CH decreases with an increasing distance from the reinforcing fibre, and there is a formation of a relatively porous layer (CSH gel and ettringite) paralleling the fibre near the interface. The same observation was reported by Coutts(1988)in the case of wood fibre cement.

SEM micrograph in Figs. 10 to 18 provide insights on the ability of the fibre to arrest a propagating microcrack. It appears that a microcrack running across a fibre with a small angle of inclination seems to propagate parallel to the fibre length and later branches out. These crack formations can lead to psuedo-debonding.

2.3 Decomposition of coir fibre

The following are the conditions by which the coir fibres are exposed: (1) 14 days in cement saturated water, (2) 14 days in sea water, (3) 1-year water retting, (4) 1-year in 1% H_2SO_4 solution, (5) inside a 1-year old mortar (1:1:0.65) water cured, and (6) inside the same mortar but air cured.

Fiber conditioned in cement saturated water became dark brown possibly due to the reaction between the OH^- ions in cement and lignin in the fibre. An SEM micrograph (Fig. 19) revealed the presence of crystal deposits at the surface confirming this impression. Also, the alkali has started to leach-out the fatty acids that forms the waxy cuticle layer in the outer surface. On the other hand, the surface of fibre exposed to seawater at same 14-day period seem less intact compared to the one conditioned in alkaline medium (Fig.20). Fiber exposed to continuous water retting for 1 year became dark brown and black in color, and upon closer examination (Fig. 21) it shows excessive scaling of the outer surface which maybe due to bacteria break down of the components (probably hydrolysis of pectin by enzymes). Likewise, a corrosive medium, e.g. dilute acid solution had the same effect to the fibre under the same period.

Fiber inside a cement matrix for 1 year can provide understanding on the actual rate of decomposition of the fibre. Fibre inside a water cured mortar shows evidence of denser deposit of CH at the surface compared to the fibre inside an air dried sample. Although the outer surface of the fibre seems still to be intact, there is already some evidence of initial precipitation of cement hydration products in the lumen of the fibre to a few micron thickness. This deposit shown in Fig.22 can gradually thicken, fill these openings (lumen and pores) and ultimately cause fibre embrittlement.

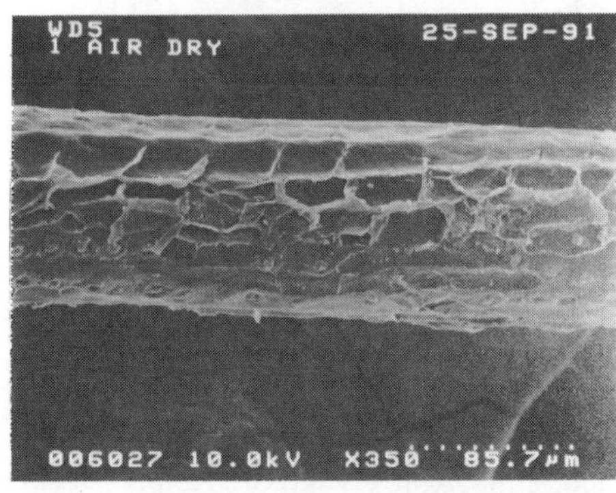

FIG 1 - SEM micrograph of a newly extracted air dried coir fiber

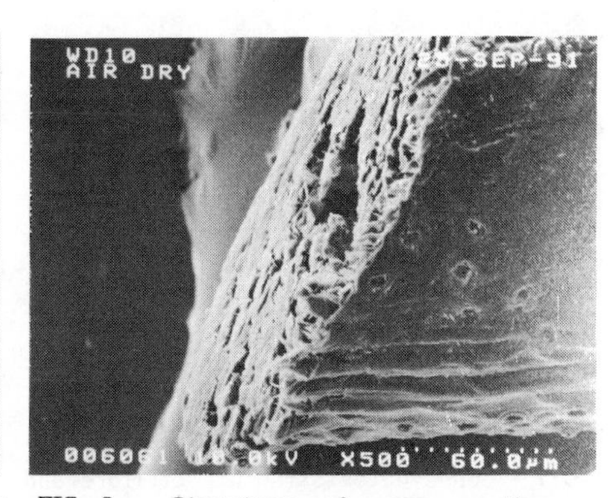

FIG 2 - Structure of cross section outer surface of fiber

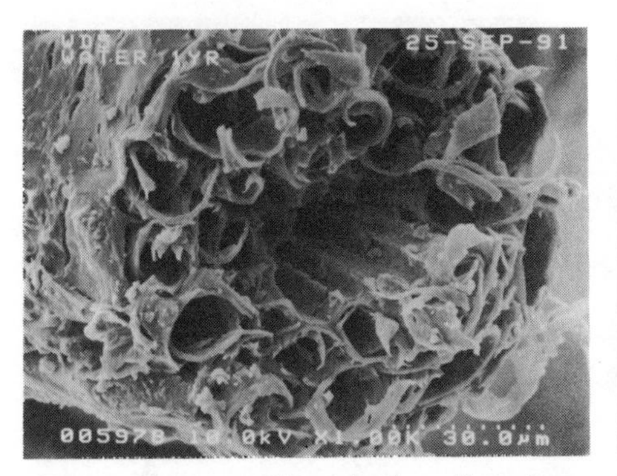

FIG 3 - Cross section showing the structure of central lacuna, fibre cells, middle lamella and lumen.

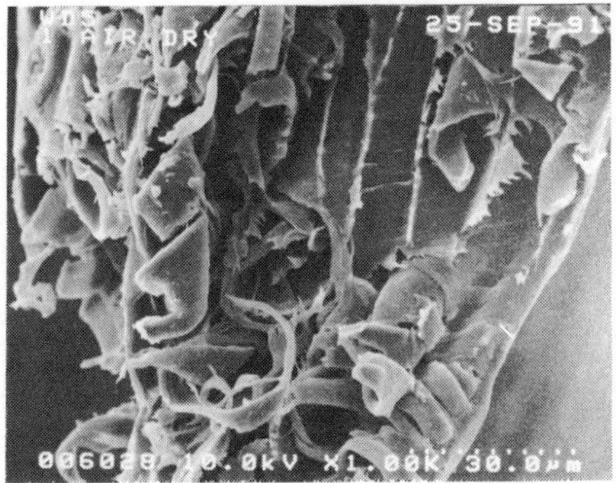

FIG 4 - Spiral structure of fibrils

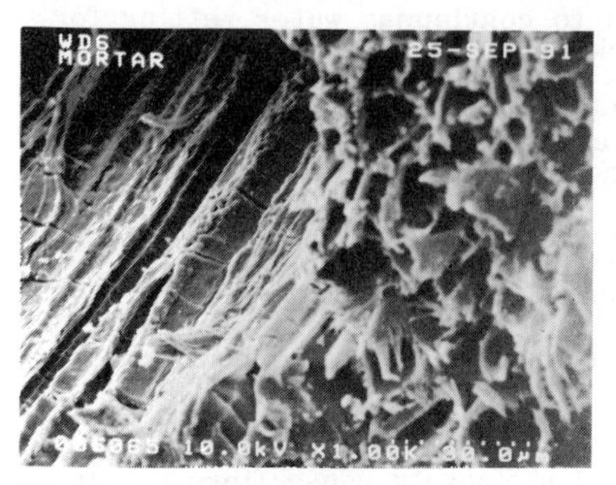

FIG 5 - Detailed structure of central lacuna and middle lamella

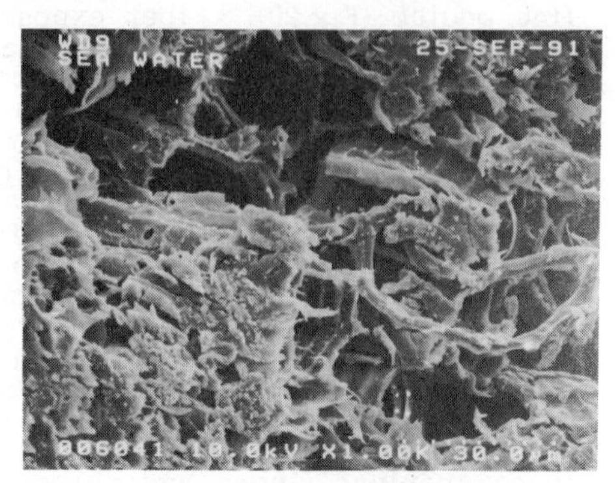

FIG 6 - Internal structure of fibre cells and middle lamella

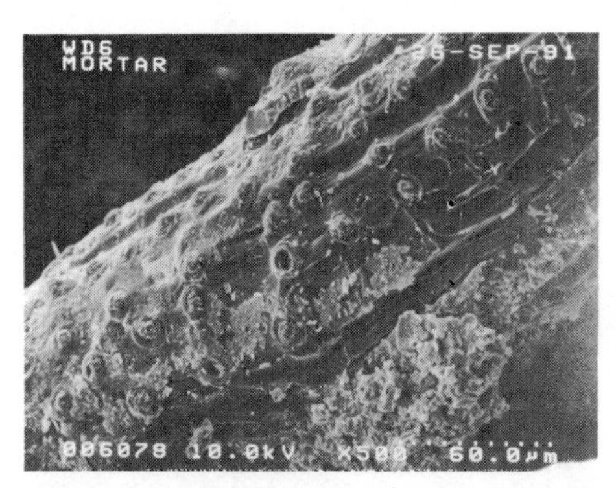

FIG 7 - Dense outer structure of fibre and distribution of pores and globular protrusions (tyloses)

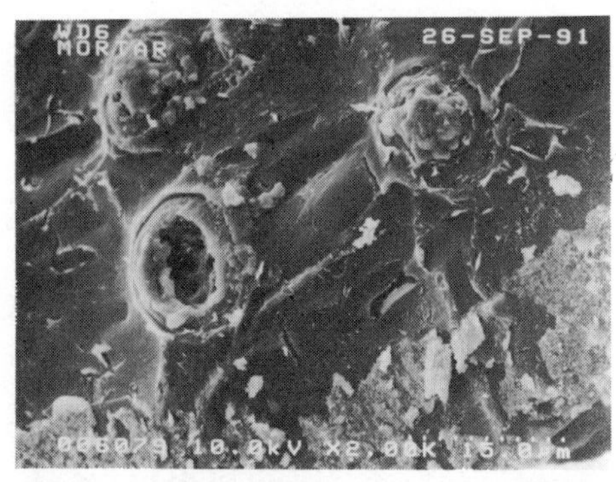

FIG 8 - Magnified structure of pores,tyloses, and waxy cuticle outer surface layer

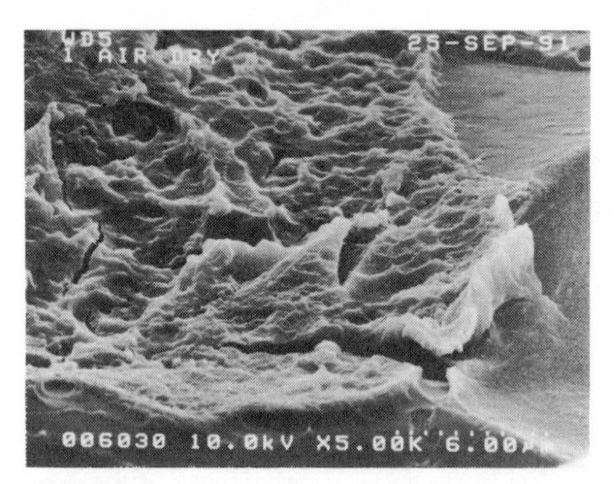

FIG 9 - Less dense and discontinuous structure of the layer between the coconut pith material and the fiber surface

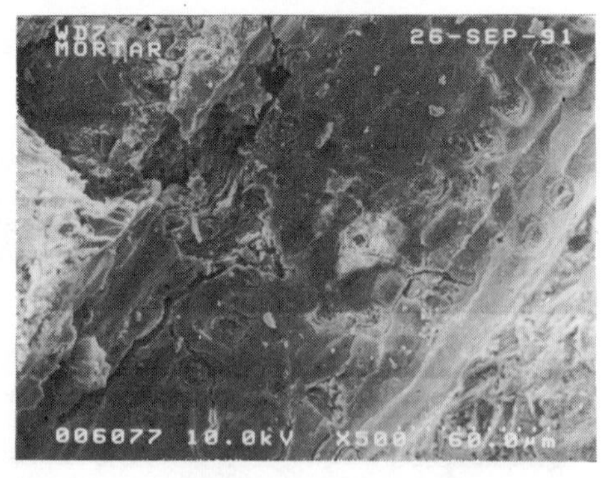

FIG 10 - Replicated topography of fiber surface in fig.7 to the dense matrix

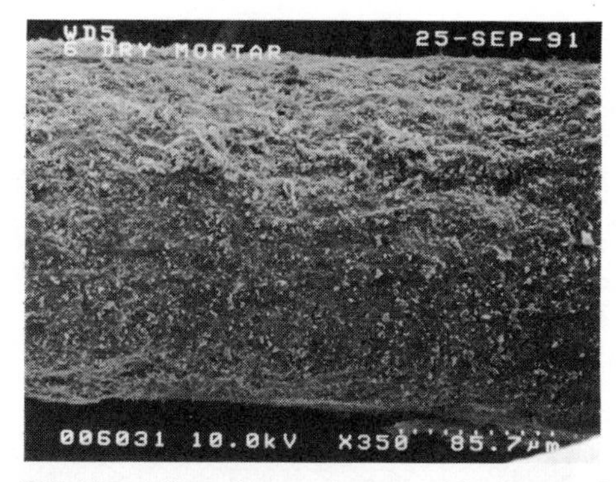

FIG 11 - Fiber from a 1-year old air cured mortar specimen showing the deposit of hydration products

FIG 12 - Hydration products magnified fro Fig.11 showing hexagonal CH plate an porous gel.

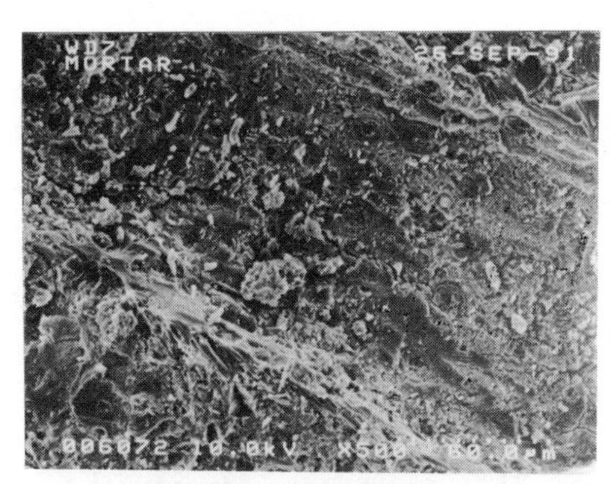

FIG 13 - Replicated topography of fiber in the matrix with some discontinuities

FIG 14 - Microstructure of hydration products taken from the discontinuities in fig 13

FIG 15 - Dense CH at fiber interface and porous layer of CSH gel and ettringite parallel the interface

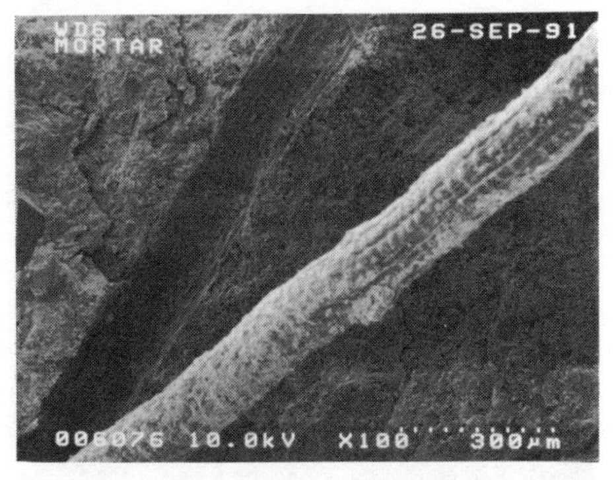

FIG 16 - Microcrack propagation at the interface of fiber

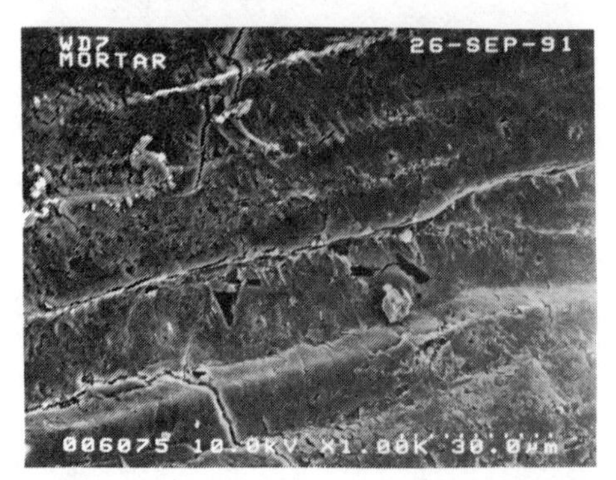

FIG 17 - Microcrack propagating parallel to the fibre length

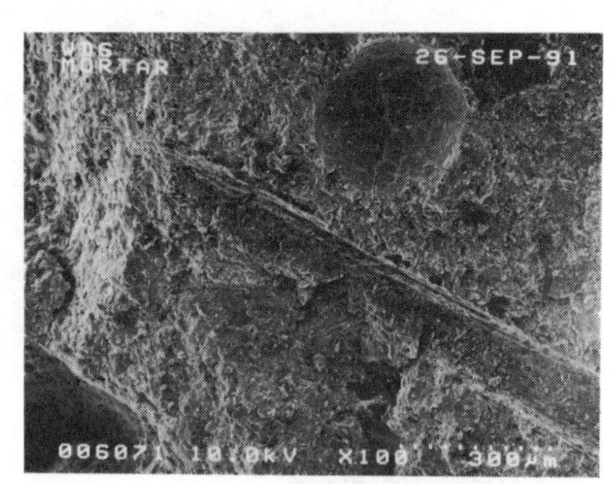

FIG 18 - SEM micrograph showing the failure of fiber to arrest a microcrack

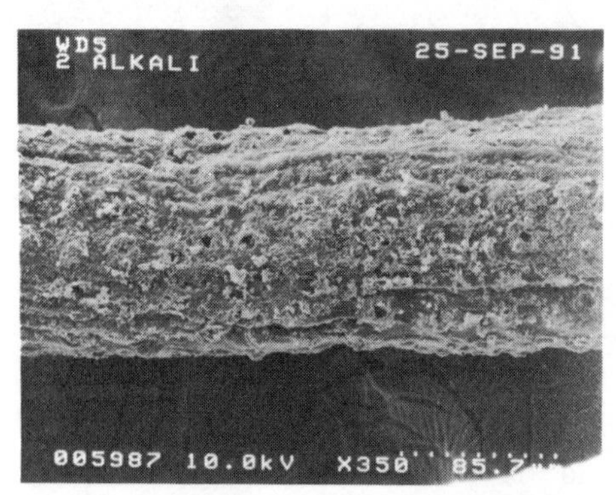

FIG 19 - Surface of fiber after 14 days in saturated cement water

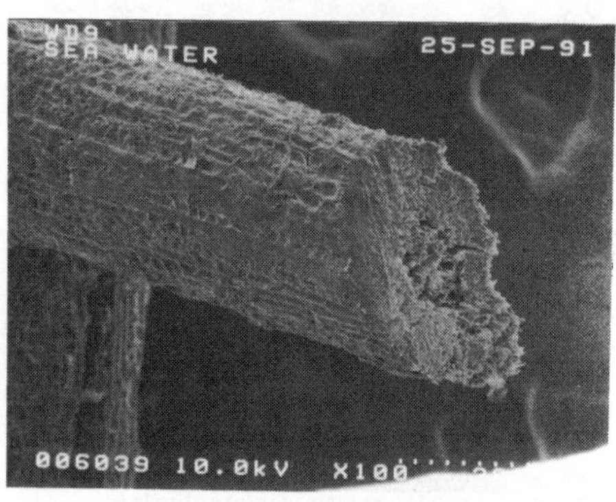

FIG 20 - Fiber after 14 days in seawater

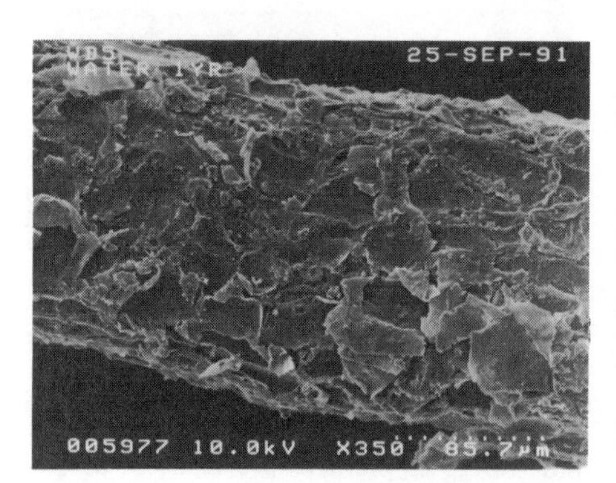

FIG 21 - Fibre after 1-year of water retting

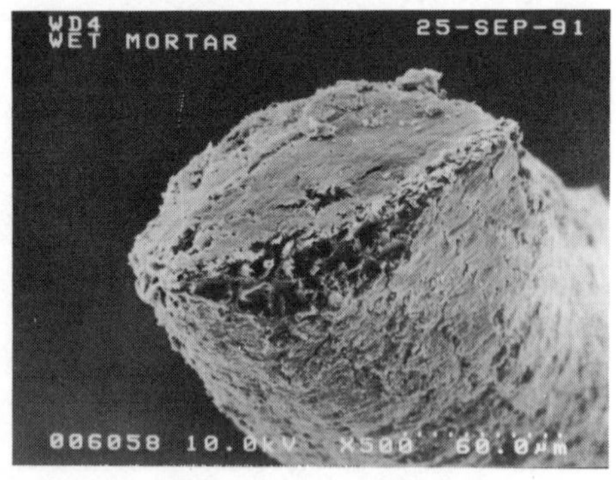

FIG 22 - Fibre in a 1-year old mortar showing initial precipitation of hydration products in the lumen and in the outer surface.

3 Mechanical Properties of Fibre and Fibre-cement Composite

3.1 Experimental programme

3.1.1 Materials

The coir fibre used in this study came from a matured air-dried coconut fruit (Typica Variety) harvested for copra in Laguna, Phil ippines. The husks were retted in tap water for about a month and afterwards the fibres were manually extracted. Saturated surface dry (SSD) condition was uniformly attained using the centrifuge method, i.e., dry spinning for 5 minutes. The SSD weight is approximately 2.15 times that of an air-dried. As for the mortar constituents, an ordinary portland cement with specific gravity of 3.15 was used as the binding material; an ordinary river sand from Fuji river with a maximum size of 2.5mm, specific gravity of 2.60 and absorption of 1.4% as fine aggregate, and sulphonated naphthalene formaldehyde condensates as superplasticiser.

3.1.2 Outline of Experiment

The program is divided into three subseries as follows:

SERIES 1: Investigation of the tensile properties (fracture strength, elongation and modulus of elasticity) of the reinforcing fibre. A fibre tensile test (ASTM D-3822,D-3379) was carried out at a 50 mm gage length and tested at a strain rate of 30 mm/min. Tensile test of fibres exposed to different conditions were likewise taken.

SERIES 2: Investigation of the effect of fibre reinforcing parameters i.e., fibre volume fraction (Vf) and fibre length (or aspect ratio, L/d) to the properties of the composite in compression (ASTM C-39) and flexure (ASTM C-78). A 100 ø x 200 cylinder and 50 x 50 x 300 mm beam specimens were used for compression and four-point flexure test respectively. Lengths of 3, 6, and 9 cm (approximate aspect ratio of 100, 200, and 300 respectively) were dispensed at different volume fraction of 0.8, 1.5, 2.5, 3.5, and 4.5% to the mortar mixture having a mix proportion of 1:1:0.4 and 1:1:0.7. Two method of specimen fabrication were adopted namely, random mixing for short fibres (3M40, 6M40, 9M40 and 3M70 series) and hand-laying for long and continuous fibres (LM40 series). The subseries are coded, for example a 3M40 series means a 3 cm length fibre in mortar with W/C=40%.

SERIES 3: Investigation of the effect of matrix quality to the compressive property of the composite. For this series a constant fibre length of 3 cm and two volume fractions of 1 and 3% were dispensed in a different kinds of matrix proportions, that is W/C ratio of 40, 50, 60, and 70% at a constant flow value of 180$\pm$10 mm.

3.2 Discussion of Results

3.2.1 Properties of the reinforcing fibre

About 40 samples were tested to obtain the fracture strength, initial modulus of elasticity and elongation data for each fibre diameter category. Each sample was mounted on a cardboard holder and tested at a constant strain rate. The scatter in strength, elasticity and elongation are shown in histogram form in Figs. 23 thru 25. The averaged properties of fibres are summarized in Table 1. The measured fracture strength and initial modulus of elasticity calculated in this study are slightly low compared to other previous works [1,6]. It is because the fibres used in this study came from a matured seed harvested for copra and not from the seed with the kind and age specifically used for coir production. The wide scatter of strengths is due to many factors, a major portion of which is due to the inherent flaws in the fibre itself.

For fibres conditioned in (a) cement saturated water for 56 days, (b) seawater for 56 days, and (3) 1-year tap water retting, showed no evidence of marked strength reduction. This implies that the method of retting of coconut husks does not have any major effect to the resulting tensile properties of the extracted fibres. This is in agreement with the results obtained by Satyanaryana et.al(1981a). Surface scaling observed from SEM micrographs (Figs.19-22) at a much earlier period only shows superficial changes in the structure of fibre. However, it is conjectured that such superficial changes may accelerate fibre embrittlement through the precipitation of hydration product in fibre structure openings.

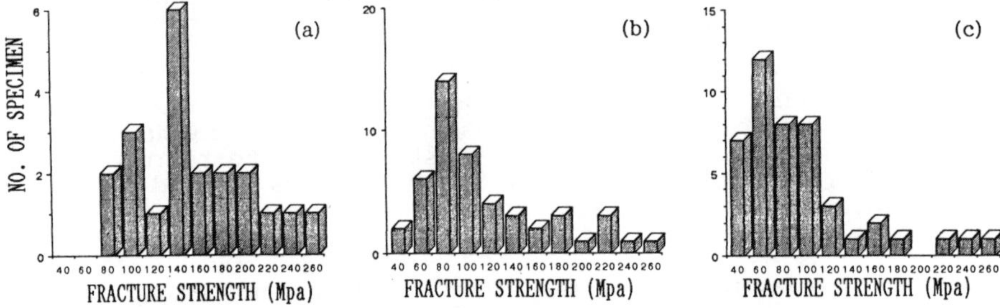

FIG 23- Histogram of fibre tensile strength (a) for d=0.15-0.2 (b) d=0.2-0.3, (c) d=0.4

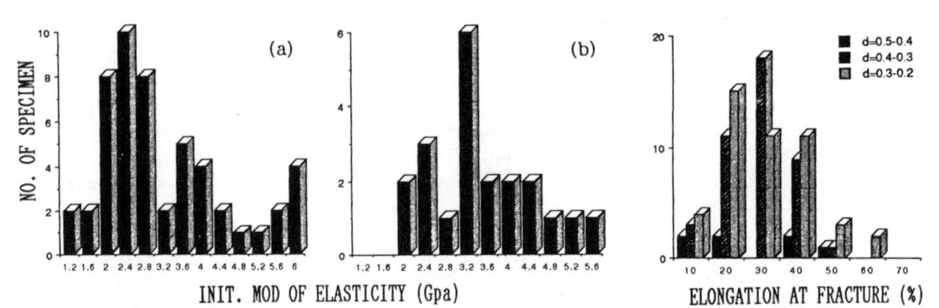

FIG 24 - Histogram of fibre initial modulus of elasticity (a) d=0.2-0.1 (b) d=0.2-0.3

FIG 25 - Histogram of elongation at fracture

TABLE 1 Tensile Properties of Coir Fibre at different diameter

DIAMETER (mm)	MEAN STRENGTH (Mpa)	STANDARD DEVIATION (Mpa)	COEFF.OF VARIATION (%)	MEAN ELASTICITY (Gpa)	STANDARD DEVIATION (Gpa)	COEFF.OF VARIATION (%)	MEAN ELONGATION (%)	STANDARD DEVIATION (%)	COEFF VAR. (%)
0.45-0.4	55.9	21.6	38.7	1.555	0.792	50.9	26.5	14.5	54.7
0.35-0.3	71.7	34.4	20.8	2.040	1.231	60.3	23.5	7.7	32.8
0.25-0.2	108.1	59.4	54.9	2.855	1.143	40.0	24.6	11.2	45.4
0.15-0.1	155.9	73.2	46.9	4.245	1.933	45.5	25.4	25.3	50.0

3.2.2 Properties Fibre-cement Composite under Compression

The entire stress-strain curve generated during the compression test sums up all the important information of an engineering material. Hence, not only the basic characteristic parameters were evaluated but also the entire stress-strain curve.

It was observed during test that the failure of specimen was gradual, and in spite of the occurrence of excessive vertical cracks the specimen still did not break into pieces. The following points were observed from the experimental results of the compressive stress-strain curves: (1) slight changes in the slope of the elastic portion, (2) peak strain at ultimate point, (3) shallow slope of the post-cracking portion and (4) increased area under the stress-strain diagram. These experimentally observed properties were correlated with the fibre reinforcing parameters and matrix properties in order to derive a simple prediction equation which can be used in calculating the basic compressive properties, and in the analytical generation of the entire stress-strain diagram.

Characteristic parameters

In the analysis of data, several combinations of the independent variable (e.g., Vf, Vf(L/d), Vf/σ c, etc.) were examined to improve the correlations of the observed data and the prediction equation.

Modulus of Elasticity

A slight reduction in the secant modulus of elasticity (at 45% of the ultimate compressive strength) can be observed in Fig. 26 due to the addition of fibres. This may be due to the following factors: (1) increased porosity brought about by the intertwined fibre which inhibited the effective compaction of the mixture, (2) increase in local stress and strain concentration due to elasticity mismatch, i.e. relative weaknesses between cement paste, aggregate and fibres, and (3) inherent low-modulus of elasticity and high Poisson ratio of the fibres. The effect of matrix quality and fibre volume fraction is most evident from the figure. Hence a regression analysis using a linear fit line equation was carried out to correlate the secant modulus (Ec) with these independent parameters.

$$Ec = 560.70\ \sigma c_m + 3593.7\ Vf(L/d) \qquad (R=0.895) \qquad (1)$$

where σc_m, Vf, L/d represent the matrix compressive strength, fibre volume and approximate aspect ratio respectively.

Ultimate Compressive strength and corresponding strain

Fig.26 shows that at a fibre volume of not greater than 2.5% the effect to the compressive strength is minimal, within the range of $\pm$5% only. However, in excess of this volume a reduction in strength of about 10% for every 0.5% fibre volume increase occurred. It is clear from the figure that the variations in strength of the series 3M40, 6M40 and 9M40 never differs by more than 10% regardless of fibre volume, and thus can be considered insignificant. A strength (σc_c) prediction equation in the form of composite mechanics approach is given,

$$\sigma c_c = 0.8665\ \sigma c_m(1-Vf) - 2.61\ Vf(L/d) \qquad (R=0.98) \qquad (2)$$

Increase in matrix strength and fibre volume resulted to an increase in peak strain ranging from 10 to 30% (Fig.28). The presence of fibre with good bond has obviously contributed to this increase. Although the exact mechanism of failure of fibre and matrix is not clearly known, it is generally believed that the fibres may have hold the failed cement crystals or some of the fibres in perpendicular orientations may have controlled the lateral expansion, thus allowing the system to carry additional strain. This higher peak strain property entails higher degree of toughness due to the ability to absorb additional energy prior to failure. The correlation of peak strain (εp_c) with key parameters is given as,

$$\varepsilon p_c = 106.9 \, Vf + 4.484 \, \sigma c \quad (R=0.769) \quad (3)$$

All other characteristic parameters, e.g., modulus of elasticity (Ep) taken at ultimate point, stress and strain at inflection point (ε_i, σ_i), and at an arbitrary last point (ε_l=15000, σ_l) are all correlated with the key parameters and summarized in Table 2.

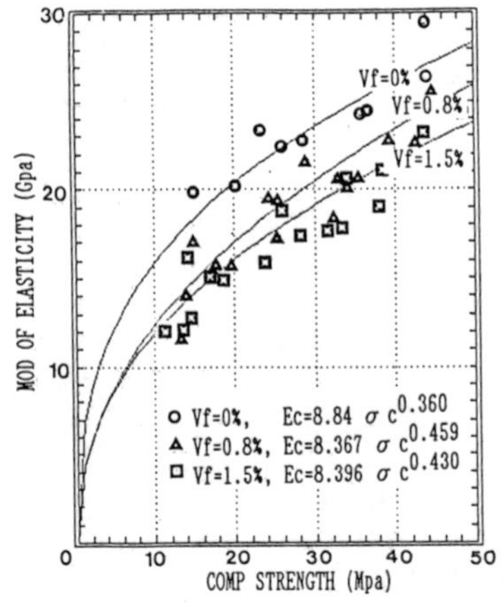

FIG 26 - Elasticity-Strength relationship

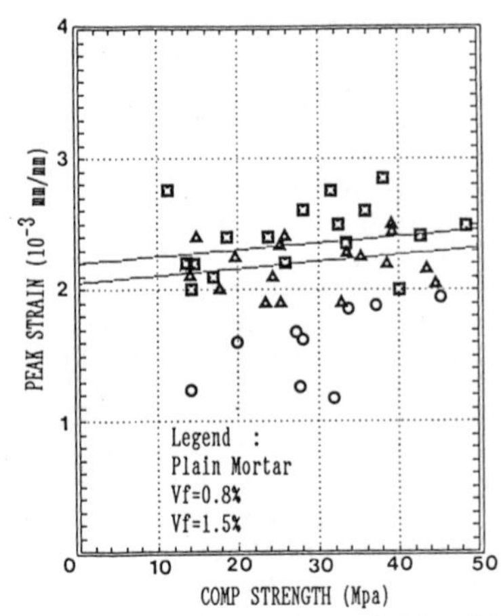

FIG 27 - Peak strain relationship

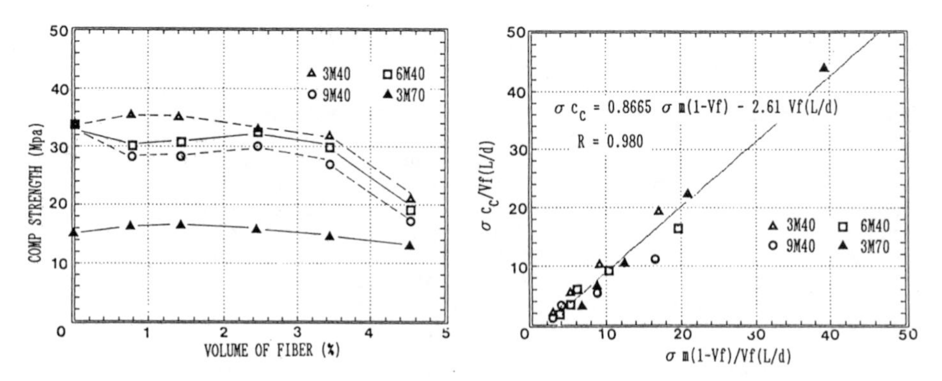

FIG 28 - Ultimate compressive strength relationship

TABLE 2 Summary of Prediction Eqautions for Compressive Properties

LINEAR EQUATION PARAMETERS y = C1 + C2 x					Correlation Coefficient (R)
Ec/(Vf(L/d)) =	3593.7	+	560.7	σ m/(Vf(L/d))	R=0.935
σ c/(VF(L/d)) =	-2.61	+	0.867	σ m(1-Vf)/(VF(L/d))	R=0.980
ε p/(VfL/d) =	106.87	+	4.484	σ c/(Vf(L/d))	R=0.766
Ep/(Vf(L/d)) =	256.1	+	428.6	σ c/(Vf(L/d))	R=0.979
σ i/(Vf(L/d)) =	19.44	+	0.0704	σ c/(Vf(L/d))	R=0.974
ε i/Vf =	41.7	+	47819	Vf/σ c	R=0.782
σ l/(Vf(L/d)) =	2.349	+	0.087	σ c/(Vf(L/d))	R=0.634

Stress-strain curve

In order to characterize the stress-strain curve of coir fibre concrete, an analytical model developed by Sargin (1971), and later applied to steel fibre concrete by Fanella et.al.(1985a) was used in this study. It is expressed in normalize form as,

$$Y = \frac{Ax + Bx^2}{1 + Cx + Dx^2} \qquad (4)$$

where $Y = \sigma / \sigma p$, and $x = \varepsilon / \varepsilon p$ refer to the normalize with peak stress and strain used in y and x coordinates respectively. A, B, C and D are constants that can be derived from the boundary conditions based on ascending and descending properties. Fig. 29 shows this relationship schematically.

In spite of the limited number of variables chosen, a comparison of the observed and the analytically generated curve in Fig.30 shows good correlations. However, caution should be taken in extrapolating the data beyond the limited chosen variables (age=7-28, Vf=0.5-4%, L/d=100-300, and $\sigma c \leq 50$Mpa) because it may show unrealistic curves. Further evaluation of other variables are still being undertaken. This analytical model can be used as a first hand approximation of the compressive properties of a coir-fibre cement mortar composite.

BOUNDARY CONDITIONS

ASCENDING

1. dY/dx = Ec/Ep at x=0, Y=0
2. Y=0.45, x=0.45, curve passes 0.45
3. Y=1, x=1, curve passes peak
4. dY/dx=0 at x=1, Y=1

DESCENDING

1. Y=1, x=1, dY/dx=0
2. Y=σ i/σ c, x=ei/ep,
3. $d^2Y/dx^2 = 0$ at infletion
4. curve passes final pt

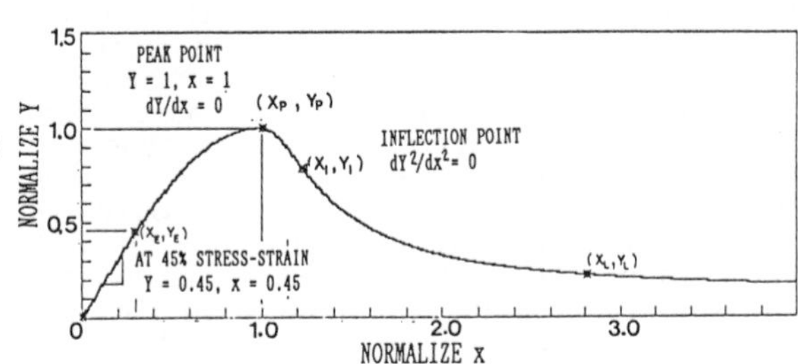

FIG 29 - Schematic diagram of non-dimensionalize stress-strain curve

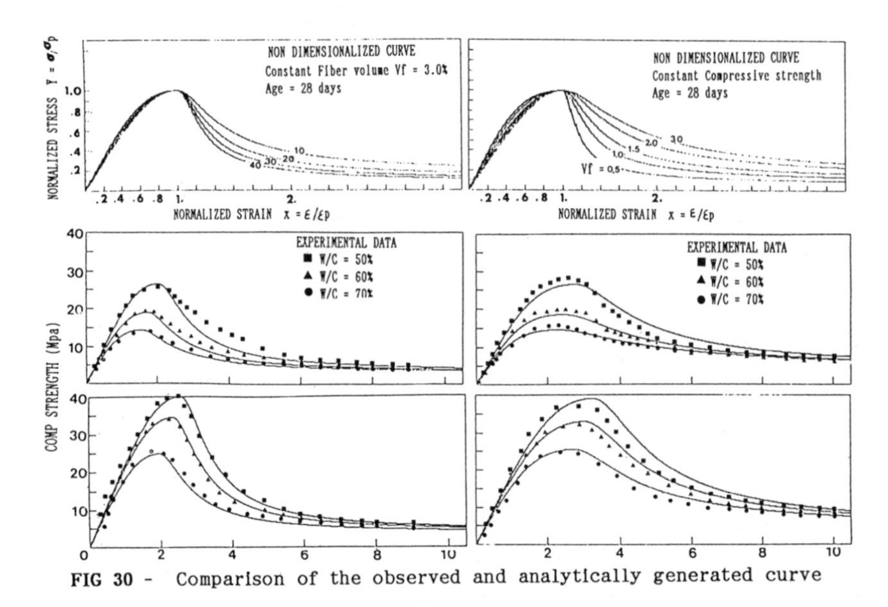

FIG 30 - Comparison of the observed and analytically generated curve

3.2.3 Properties of fibre-cement composite Flexure

In the case of plain concrete beam an increasing strain in the tension part causes cracks to be formed followed by a sudden failure, but in a fibre-cement composite, sudden failure did not occur due to the presence of fibres spanning the cracks. These fibres maintain the equilibrium of the system, as well as the structural integrity of the member. Fig. 31 (a) and (b) show typical load-deflection diagrams of a short-randomly-oriented fibre reinforced composite (3M40 series) and a continuous hand-laid fibre reinforced composite (LM40 series). Such behaviour can also be seen in the case of fibrillated polypropylene and low volume steel fibre composite (Hannant, 1978).

Fig. 32 shows the trend of the ultimate flexural load with increasing fibre volume. A nominal improvement in strength of about 6-12% for short randomly-oriented fibre, and a significant increase of 25% in the case of continuous fibre. This can be attributed to the effectiveness and efficiency of a continuous fibre to transmit stresses to the matrix. It is also observed that at random orientation the effect of aspect ratio (L/d) seems minimal, and variation in strength differs by around 10% only. This may be due to the curling and bundling of fibres during mixing which limit the effectiveness of the entire length in transmitting stresses. A regression analysis correlating the composite flexural strength (σb_c) and the fibre reinforcing parameter (Vf, L/d) and matrix strength (σb_m) was carried-out, and shown in Fig 35. The resulting fit line equation with a correlation coefficient of 98.7% is given,

$$\sigma b_c = 0.9542\ \sigma b_m(1-Vf) + 0.22\ Vf(L/d) \qquad (R=0.987) \qquad (5)$$

A useful insight on the bond strength properties of the reinforcing fibre can be calculated by comparing Eq. 5 to the simplified equation (Eq.6) derived by Swamy et.al(1974a),

$$\sigma b_c = \sigma b_m(1-Vf) + 0.82\tau \quad Vf(L/d) \qquad (6)$$

$$0.82 \; \tau \quad Vf(L/d) = 0.22 \; Vf(L/d), \quad \text{and} \quad \tau = 0.28 \text{ Mpa}$$

where τ is the averaged interfacial shearing stress. This value is quite similar to the experimental result obtained by Cook et.al.(1978a) in a partial embedded fibre bond test which is 0.21 Mpa. This value of τ confirms the notion that the stress level at the initiation of pullout is well below the stress level required to cause the yielding of fibre which is about 49 Mpa. Hence the effectiveness of the natural fibre to impart ductility and integrity to a structural member during failure through gradual fibre pullout does not diminish in spite of its inherent low tensile strength.

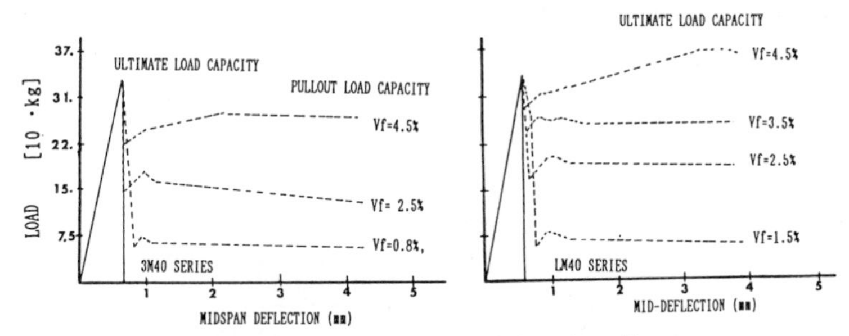

FIG 31 - Typical Load deflection diagram

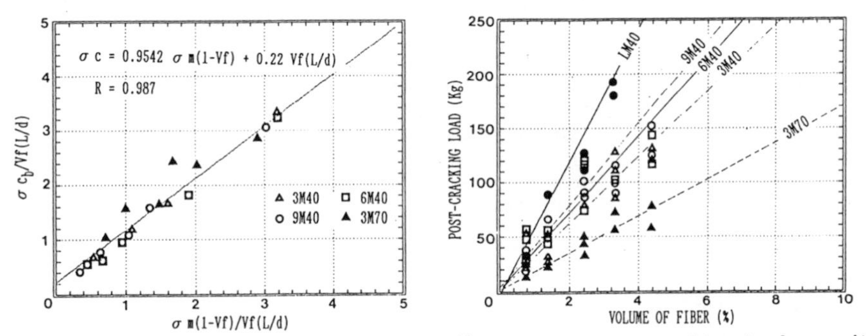

FIG 32 - Trend of Ultimate flexural strength

FIG 33 - Post-cracking load capacity

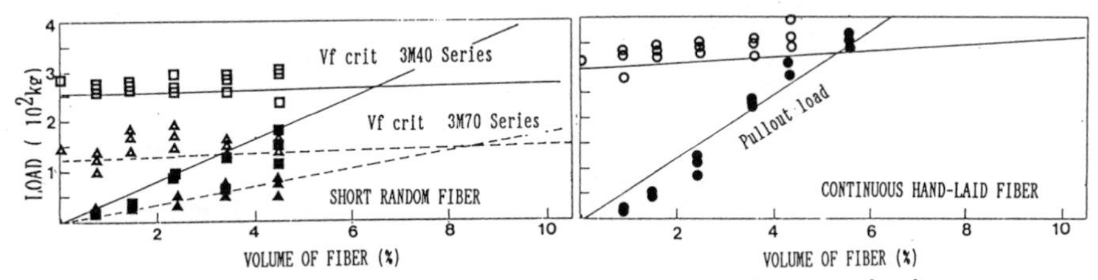

FIG 34 - Location of Vf critical for flexural strengthening

Post-cracking strength
When the first crack strength is reached, either a continuous increase in load carrying capacity or a sudden drop of load with a maintained residual load will occur depending on the properties of fibre and matrix, as well as the specimen preparation. For short randomly oriented coir fibre, a drop of load with a maintained residual load proportional to the fibre content occurred. Fig.33 shows that the effect of fibre length at random orientation (3M40, 6M40, and 9M40) to the post-cracking strength is also very minimal which may be again due to the curling of fibre. As for the effect of matrix quality, it is very significant from the point of view of post cracking strength as seen from the load carrying capacity of the 3M70 series. Needless to say, a good quality matrix ensures stronger bond and ultimately a relatively higher post cracking strength. In the case of continuous fibre, post-cracking strength exceeds the first crack strength at Vf greater than 3.5%.

Critical Fiber Volume
A clear-cut method of Vf critical determination suggested by Hannant (1978) was adopted in this study. It can be seen from the figure that the post-cracking strength greatly influenced the intersection point by which the critical fibre volume for flexural strengthening occurs. Fig.34 shows the location of Vf critical for different mix series. For short random fibres in a matrix with a w/c ratio of 40% the value of Vf critical is roughly 6.0% regardless of aspect ratio. However, for a composite with a continuous fibre (LM40 series) having a higher post-cracking strength gives a lower Vf critical of 4.5%. The significance of matrix strength to the value of Vf critical is also reflected in this graph, wherein due to the relatively shallow slope of the 3M70 series the point of intersection is much farther bringing the Vf critical to about 8.0%.

Toughness Index
The first crack toughness in bending was computed in accordance with ASTM C-1018. The toughness index (T.I.) was measured up to 5.5 times the deflection at the first crack. It is evident that the use of fibre has increased the value of toughness by more than twofold (100%), and transition from an elastic brittle material to an elastic pseudo-plastic material can be achieved by an increasing fibre volume fraction.

4 Conclusion

This paper indicates that the utilization of coconut husk fibres derived from the residues of copra industry can provide solutions to two significant problems, viz., elimination of solid wastes, and the provision of valuable construction material. Based from this study the following conclusions on the properties of coir fibre cement composite were:

1. The microstructure studies of fibre, its structure characteristics, interfacial properties with the cement matrix revealed insights on its ability as a reinforcing material. SEM micrographs have shown the decomposition characteristics of fibre exposed to simulated and natural conditions, the state of bonding between fibres and matrix, and efficiency of fibres to delay crack propagation. All of these can provide understanding to the behaviour of a fibre-cement composite system up to the engineering level.

2. Observed properties in compression like peak strains, toughness and ductility showed significant improvement due to the addition of coir fibres. Equations summarized in Table 2 can be used as a first-hand approximation of the basic compressive properties of a coir fibre cement mortar composite. These equations can also be satisfactorily used together with Eq.(4) in the analytical generation of the stress-strain diagram. Improvement in the toughness and ductility can be seen from the shallow slope of the post-cracking portion and the entire area under the stress-strain diagram.

3. Observed properties in flexure have likewise shown enhancement in toughness, and ductility of the whole system due to addition of fibres. In spite of the slight improvement in flexure strength, the increase in post-cracking strength proportional to the fibre content maintained the structural integrity of the whole system. Comparison of the properties between the short random fibres and continuous hand-laid fibres have shown the merits of each method in terms of the first crack strength, post-cracking strength and value of Vf critical.

Acknowledgement

The authors are grateful to the staff of KAO Corporation Wakayama Research Laboratory, Japan, for their assistance in obtaining the SEM micrographs used in this study.

5 References:

Agopyan, V.,(1988) **Vegetable fibre reinforced building materials-developments in Brazil and other Latin American countries,** Concrete Technology & Design Vol.5 (ed. RN Swamy), London, pp.208-242.

Cook, D.J.; Pama, R.P.; Weerasingle, H.L.D.,(1978) **Coir fibre reinforced cement as a low cost roofing material,** Building and Environment, Vol.13, Great Britain, pp.193-198.

Coutts,RSP., (1988)**Wood fibre reinforced cement composites,** Concrete Technology & Design, Vol.5, (ed. RN Swamy) London, pp.1-58.

Fanella,D.; and Naaman,A.,(1985) **Stress-strain properties of fibre rein forced mortar in compression,** in ACI July, pp.475-483.

Gram, H.E.,(1984) **Durability of natural fibres,** SAREC report R2, Sweden, pp.65-100.

Hannant, D.J., (1978)**Fibre cement and fibre concretes,** London, 219.

Jorillo, P.Jr., and Shimizu G., (1991)**Fresh and mechanical properties of short discrete coir fibre in cement based matrix,** Int. Confer ence on Concrete Technology, (ed.Razak), Malaysia 1991,pp.3-13

Sargin, M., (1971)**Stress-strain relationship for concrete and analysis of structural concrete sections,** Study No.4 solid mechanics Div., Univ of Waterloo.

Satyanaryana, K.G.; Kulkarani, AG.; and Rohtagi PK.(1981) **Natural fibre as valuable resource for materials in the future of Kerala,** Material Science and Technology in the Future of Kerala, (ed.Satyanaryana, et.al.), Trivandrum, pp. 127-128.

Slate, F.O., (1976)**Coconut fibres in concrete,** Engineering Journal of Singapore Vol.3 No.1, Singapore 1976, pp.51-54.

Swamy, R.N.,(1990) **Vegetable fibre reinforced cement composites-a false dream or potential reality?** 2nd Int RILEM Symposium on Vegetable Plants and their fibres as building materials, (ed. HS Sobral), Brazil, pp.3-8.

Swamy, R.N., and Mangat, S.P.,(1974)**A theory for flexural strength of steel fibre reinforced concrete,** in Cement and Concrete Research Vol.4, pp.315-325.

87 COIR FIBRE REINFORCED CEMENT BASED COMPOSITE PART 2: FRESH AND MECHANICAL PROPERTIES OF FIBER CONCRETE

P. JORILLO and G. SHIMIZU
College of Science and Technology, Nihon University, Tokyo, Japan

Abstract
Comprehensive test data on the fresh and mechanical properties of concrete reinforced with short randomly oriented coir fibres are presented. Three types of cement matrices were used, and the combined effect of fibre parameters on workability, compressive properties, flexural and tensile stress-strain were examined and discussed in this paper. Findings showed that in spite of a nominal increase in flexural, tensile and compressive strengths, a significant improvement in toughness, ductility and post-crack strength of the composite occurred. Properties were summarized in simple empirical expression in the form of composite mechanics approach.
Keywords: Coir fibre, Compression, Ductility, Durability, First-crack, Flexure, Modulus of elasticity, Peak strain, Tension, Workability

1 Introduction

The major advantage of fibre reinforcement is to impart additional energy absorbing capability so as to transform a brittle material into a pseudoductile material. Fibres in concrete serve as crack arrestor which can create a stage of slow crack propagation and gradual failure (Swamy et.al,1974a). At post-crack stage, the presence of fibres spanning the cracks usually inhibits the sudden collapse of a member giving ample warning of imminent failure. Previous researches has also reported such advantages imparted by the natural fibres to the concrete material (Aziz et.al,1988; Agopyan,1988). Appropriate short-range applications like pre-cast formworks or shutterings for embankment foundations, and even in long-range applications like pipes and cladding panels can adopt such kinds of fibres as a support or main reinforcement.

If, such are the intended applications, then it is necessary to study exhaustively the properties of such composite material. This paper describes the results of the comprehensive experimental evaluation of the fresh and mechanical properties of coir-fibre composite with gravel-concrete as the matrix phase. The objective of this paper is to provide information on the effect of this kind of low-modulus natural fibre to the properties of the composite. Effects of the major fibre reinforcing parameters in combination with different matrix qualities to the compressive properties, flex ural load-deflection diagram, and tensile stress-strain behaviour are discussed, and summarized in simple empirical expressions.

Fibre Reinforced Cement and Concrete. Edited by R. N. Swamy. © 1992 RILEM.
Published by E & FN Spon, 2-6 Boundary Row, London SE1 8HN. ISBN 0 419 18130 X.

2 Experimental Programme

2.1 Materials

The properties of cement, fine and coarse aggregates used in this study are shown in Table 1. The structure and properties of coir fibre used in this experiment are described in detail in Part 1 (Shimizu and Jorillo,1992a). A lignosulphonic acid-polyol compound was used as a water reducing and an air-entraining admixture, and was added into the mixture at a dosage rate of 2.5-3.0 cc/kg to attain an air content of about 4%.

2.2 Mixing and moulding of specimens

The constituent materials were mixed in a forced-mixing type mixer in the following order: (1) half of the dry materials and three-fourths water, (2) gradual introduction of fibres (half-volume), and (3) addition of remaining water and fibres. The total mixing time was about 7 minutes on average. The mixture are designed for a workability of 10-14 seconds inverted slump time (ASTM C-995). The results of mix proportioning are shown in Table 3. All specimens were cured in $20^{o}C$ water until the testing age of 7 and 28 days.

2.3 Outline of Experiment

A systematic experimental program was carried out to determine the influence of fibre reinforcing parameters i.e. fibre volume (Vf), and aspect ratio (L/d) in combination with different matrices to the mechanical properties of coir fibre concrete. The test detail is shown in Table 2. Each series are coded, for example a 3C50 series refers to a 3 cm fibre in a concrete mixture with the W/C = 50%.

TABLE 1 Material Properties

MATERIAL	SPECIFIC GRAVITY	DESCRIPTION
CEMENT	3.15	FINENESS:3260 cm/g CONSISTENCY:28.1%
SAND	2.60	FUJI RIVER SAND, ABS:1.60% MAX SIZE:2.5mm,FM:2.41
GRAVEL	2.62	KINU RIVER GRAVEL,ABS:1.40% MAX SIZE:20mm,FM:6.26

TABLE 2 Test Details and coding

MIX CODE SERIES	LENGTH (mm)	VOLUME FRACTION (%)
3C40	30	0.5, 1.0, -- 2.5
3C50	30	0.5, 1.0, 1.5, 2.5
3C60	30	0.5, 1.0, -- 2.5
1C50	15	0.5, 1.0, -- 2.5
6C50	60	0.5, 1.0, -- 2.5

TABLE 3 - Mix proportion and Fresh Properties

MIX SERIES	VF (%)	QUANTITIES (per cu.m.) CEMENT (kg)	WATER (kg)	SAND (kg)	GRAVEL (kg)	FIBER (kg)	FRESH PROPERTIES SLUMP (cm)	INV.SLUMP (sec.)	AIR (%)	DENSITY (kg/m3)
3C40	0.5	627	251	1063	709	10	17.5	12.3	3.8	2660
	1.0	613	245	1134	643	20	11.5	13.8	3.9	2656
	2.5	715	286	1092	470	37	11.4	16.5	3.9	2603
3C50	0.5	456	558	1122	828	8	12.0	11.0	3.9	2642
	1.0	572	286	1095	617	28	9.5	17.8	4.1	2597
	2.5	625	313	938	625	40	16.0	19.0	3.0	2540
3C60	0.5	372	224	1191	840	11	9.0	10.2	3.1	2637
	1.0	395	237	1164	760	17	11.0	10.0	4.2	2574
	2.5	485	291	1161	632	32	12.5	25.5	2.8	2550
1C50	0.5	408	218	1020	791	8	17.0	8.0	3.5	2262
	1.0	430	215	930	650	15	16.5	7.8	3.1	2241
	2.5	561	281	841	484	38	17.5	11.0	2.5	2205
6C50	0.5	408	204	956	741	8	10.0	14.0	2.7	2318
	1.0	442	221	955	667	15	9.5	15.8	2.1	2300
	2.5	567	283	850	487	31	5.0	30.0	2.0	2218

Fresh properties of fibre concrete mixture like slump, inverted slump time, air content, and unit weight were measured in accordance with ASTM C-143, C-995, C-231, and C-138, respectively.

Properties in compression was obtained from a 100 ϕ x 200 mm specimen tested at a stress rate of 3Mpa/min in a stiff hydraulic testing machine. Strain was measured from the 100 mm middle-third height by a linear variable differential transformer (LVDT) compressometer, and was plotted with the load in the x-y recorder.

Flexural properties were measured from a four-point flexure test of a 100x100x400 mm size beam specimen at a constant load rate of 1.5 Mpa/min. The load monitored with a load cell, was directly plotted as a function of mid-span deflection on an x-y recorder. To measure the onset of first crack, 60 mm electrical resistant type strain gages were placed at the compression and tension sides of the constant moment span of the beam. Two types of bending tests were carried out. Test series 1 examined the effect of fibre parameters like fibre volume and aspect ratio with different grades of matrices to the flexural properties. Test series 2 examined the effect of the use of partial fibre concrete, i.e. depths of d/4, d/2 and 3/4 d, to the bending properties.

Indirect tensile test were carried out in accordance with ASTM C-496. Direct tensile properties was measured from a bone-shaped specimen with tensile forces exerted through a pair of inserts at both ends. Details of specimen and test set-up is described elsewhere (Shimizu,1980). Tensile test were carried out at a load rate of 0.4 Mpa/min. To evaluate the stress-strain behaviour as well as the onset of first crack, strains were measured by strain gages.

A coir-fibre concrete (3C50 series) with 3 cm length fibre at a volume fraction of 1.0% were tested for durability against freezing and thawing in accordance with ASTM C-666.

3 Discussion of Results

3.1 Effect to Fresh Properties

It was observed from the trial mixes that the inclusion of coir fibres in fresh concrete significantly altered the rheological behaviour in terms of its stability and mobility. The coir fibre concrete stiffened the mixture resulting in an increased stability or cohesion, but also to an apparent reduced workability. In this study, stability and mobility characteristics were measured by the 'static' slump test and the 'dynamic' inverted slump cone test respectively. Effect of fibre addition to the mix proportioning and fresh properties are summarized in Table 3.

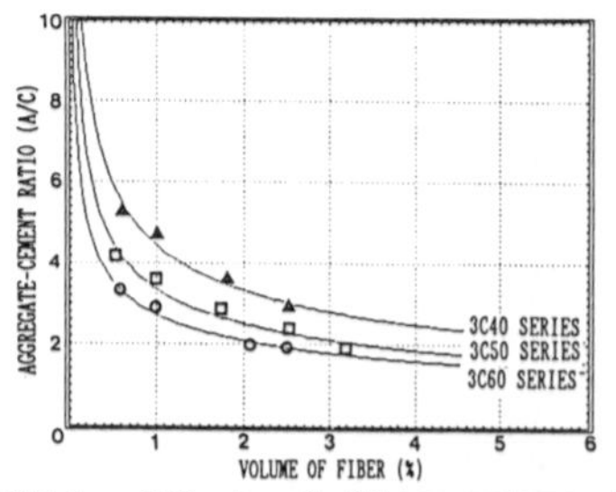

FIG.1 Effect of fiber volume to aggregate reduction

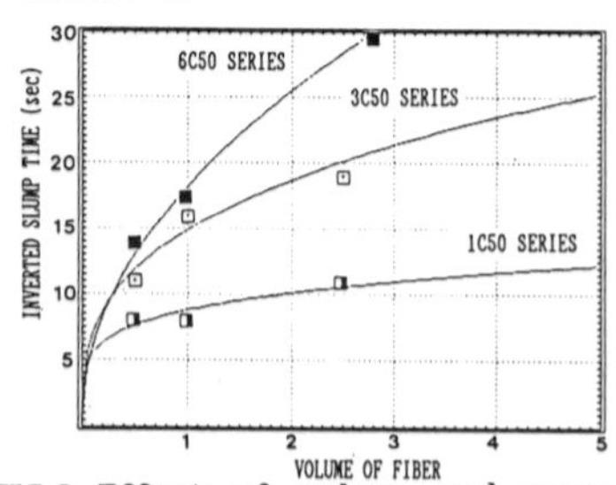

FIG.2 Effect of volume and aspect ratio to inverted slump time

If the workability is kept constant at an increasing fibre volume, it is necessary either to increase the dosage of the water-reducing admixture or reduce the proportion of the aggregates in order to meet such requirement. In this experiment, it was opted to reduce the proportion of the aggregate rather than increase the dosage of admixture. Fig.1 shows the reduction of the aggregate-cement (A/C) ratio that is necessary in order to meet the required workability of 10-14 seconds inverted slump time. The effect of fibre content at an amount of less than 30 kg/m^3 (or Vf$\leq$1.0%) to the reduction of A/C ratio appears minimal. However in excess of this fibre content, the mixture necessitates further reduction in the A/C ratio because of its harsh and unworkable characteristics. The phenomenon of fibre balling for this particular method of mixing containing a typical fibre length of 30mm was observed at a Vf $\geq$ 1.5%. Fig.2 clearly shows marked reduction in the workability, as seen from the increase in the inverted cone time due to an increasing fibre length and volume fraction. This confirms the well-known trend of loss of workability with increase in fibre content and aspect ratio. This loss are due to the curling of fibres, especially with this kind of flexible fibres. Other factors like the proportions of the constituents, and particle interference between fibre and coarse aggregates also affect the overall rheological behaviour of the fibre mixture.

3.2 Mechanical properties of coir fibre concrete

Combined effects of the key parameters to each of the properties in compression, flexure and tension like their modulus of elasticity (Ec), stress and strain at onset of first crack (σ_{1st}, ε_{1st}), at ultimate point (σ_c, ε p), post-cracking strength (σ_{cu}), ductility and toughness (T.I.) were statistically correlated and summarized.

3.2.1 Effect to Compressive properties

Structural integrity of the brittle concrete material was greatly enhanced due to the reinforcement of coir fibres. Gradual failure after the peak stress was observed. The occurrence of shallow slope in the descending portion of the stress-strain curve (Fig.3) confirms this behaviour. A typical trend for common fibre concrete (e.g., polypropylene and low-modulus carbon) can also be seen in coir-fibre composite, that is the effect of fibre inclusion to compressive strength at a low volume fraction is minimal. It can be observed in Fig.4(a) that variation in strengths for fibre volume of not greater than 1.5% is at the range of 11% only. In excess of this volume a marked decrease in strength of about 5% for every 0.5% volume increase can be expected. Such reduction in strength may be due to increased porosity brought about by insufficient compaction of the high fibre volume mixture. Also at a high fibre volume content, it was necessary to reduce further the proportion of the aggregates to meet the workability requirement. This may have consequently resulted to a weakening of the fibre-matrix bond due to the lack of sufficient matrix necessary for efficient fibre embedment.

Slight changes in strengths can be seen for series 1C50, 3C50, and 6C50 in Fig 4(a). Among the three series only the 1C50 showed no reduction in strength even at high fibre volume. This can be attributed to the low aspect ratio of fibres (approximate L/d=50) which enabled it to be uniformly dispensed into the mixture without curling and fibre balling. The 6C50 series on the other hand showed the greatest strength reduction among the three, and may be caused by fibre bundling and ineffective consolidation at high fibre content. Such behaviour shows the indirect effect of aspect ratio of coir fibres to the compressive strength of the composite.

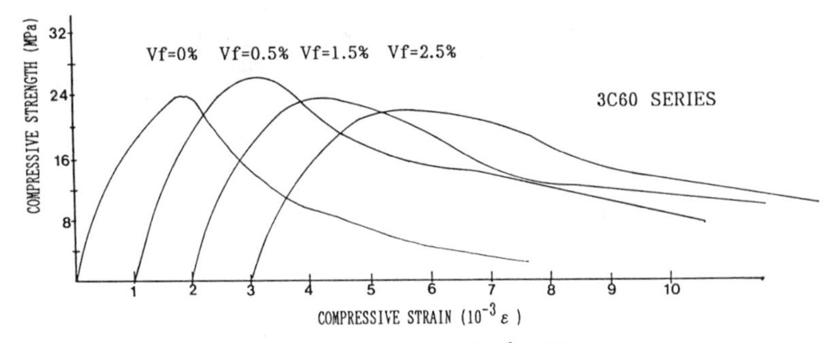

FIG.3 Typical stress-strain curve

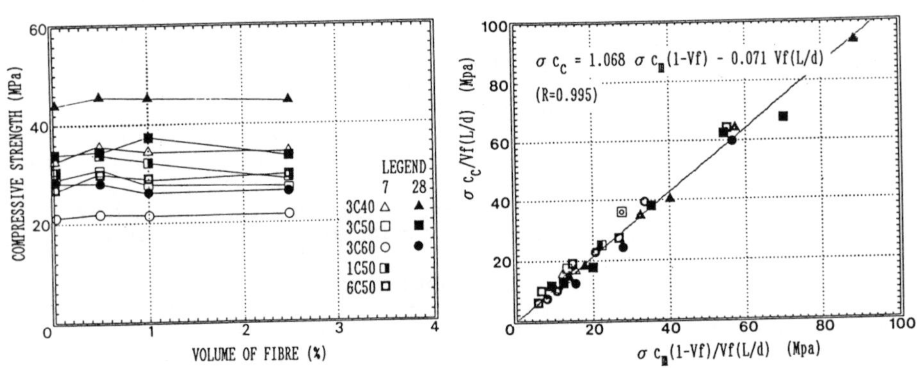

FIG. 4(a) Trend of compressive strength

FIG. 4(b) Correlation with fibre and matrix parameters

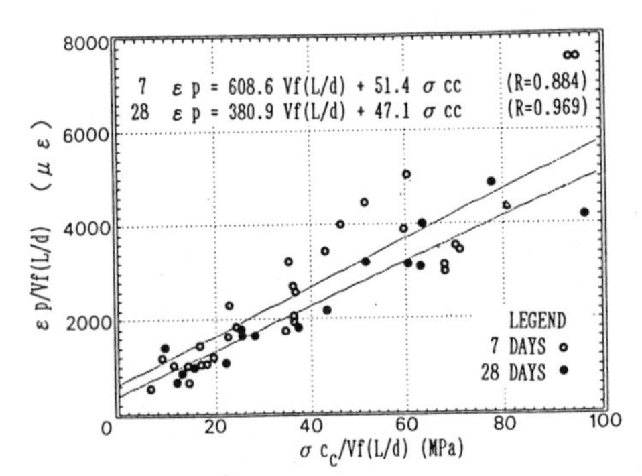

FIG. 5 Relationship of peak strain and reinforcing parameters

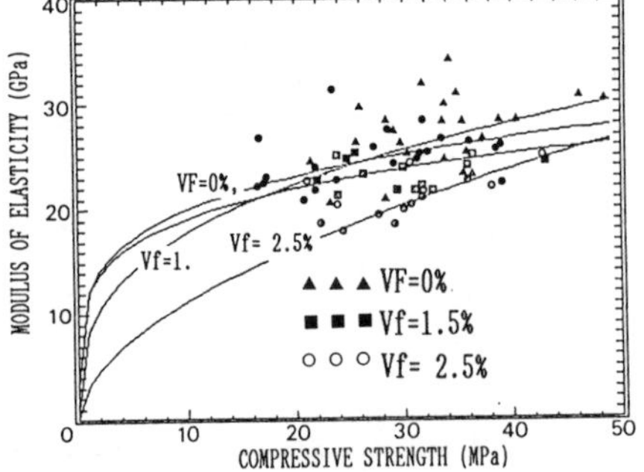

FIG. 6 Relationship of Modulus of Elasticity and major parameters

The ultimate compressive strength (σc_c) was correlated with the fibre reinforcing parameters (VF, L/d) and matrix strength (σc_m) in the form of composite mechanics approach. The correlation coefficient (R) of the best fit-line shown in Fig.4(b) is 99.5%.

$$\sigma c_c = 1.068\ \sigma c_m(1-Vf) - 0.0712\ Vf(L/d) \qquad (R=0.995) \qquad (1)$$

where the subscripts c and m refer to a particular property of the composite and matrix respectively. The notation σc_c represents a particular stress (σ_c) of the composite material.

Strain at ultimate compressive strength

Despite of the negligible improvement observed in the compressive strength of coir fibre concrete, modest enhancement to the peak strain occurred due to the inclusion of fibres. This property en tails higher degree of toughness due to the increased area under the stress-strain diagram. Fig.5 shows the trend of peak strain with increasing fibre reinforcing index and matrix strength. Such improvement in peak strain may be brought about by the presence of fibres which stitch and limit the microcracks, thus delaying its concentration in the direction of the principal strain (Rossi et.al., 1987a). Also from the figure, a decrease in peak strain at 28 days age can be seen, which may be brought about by an increasing brittleness on the part of the matrix.

Modulus of Elasticity

The secant modulus of elasticity (Ec_c) was obtained at 45% of the ultimate compressive strength. Fig.6 shows the elasticity and strength relationship for different fibre volume mix series. It can be observed from the figure that the effect to the modulus of elasticity is negligible for fibre content of up to 1.0%. However, a marked reduction can be seen at a fibre volume of 2.5%. If the values of the secant modulus are plotted against the fibre reinforcing index, an increasing index (VfL/d) will show a decreasing trend in the compressive modulus of elasticity. Although there are many factors which causes such behaviour, it is believed that the porosity of the material due to ineffective consolidation of the mixture brought about by the use of excessive fibre length and volume fraction is the major factor which causes such reduction. A simple prediction expression given in eq.2 can be used as a first-hand approximation of the modulus of elasticity of the composite.

$$Ec = 751.4\ \sigma c_c + 2105\ Vf(L/d) \qquad (R=0.961) \qquad (2)$$

Degree of Toughness in compression

The area under the stress-strain diagram represent the energy absorbed by the material during loading. Toughness index (T.I.comp) was calculated as the ratio of the area under the stress-strain diagram of fibre concrete and the base concrete. In this study, the reference strain was taken at peak (εp_c) and at 10,000 $\mu\varepsilon$. Fig.8 shows the effect of the reinforcing index to the T.I. in compression. A significant toughness enhancement of about 100% and 25% occurred for T.I. measured at 10000 $\mu\varepsilon$ and at peak strain, respectively. This significant improvement is brought about by the increased peak strain and in the occurrence of a shallow slope in the descending portion of the stress-strain curve.

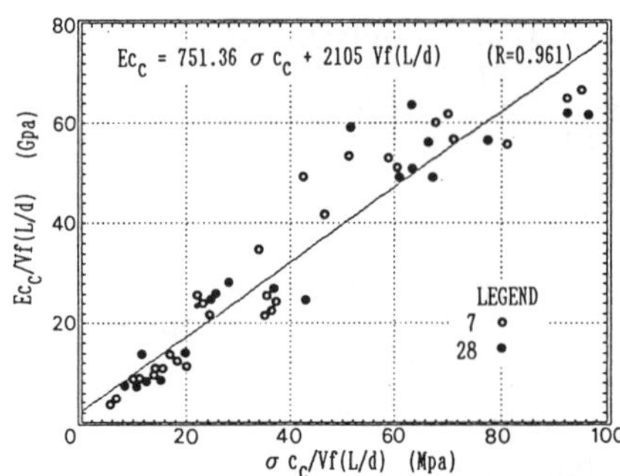

FIG. 7 Correlation of elasticity with fibre and matrix parameters

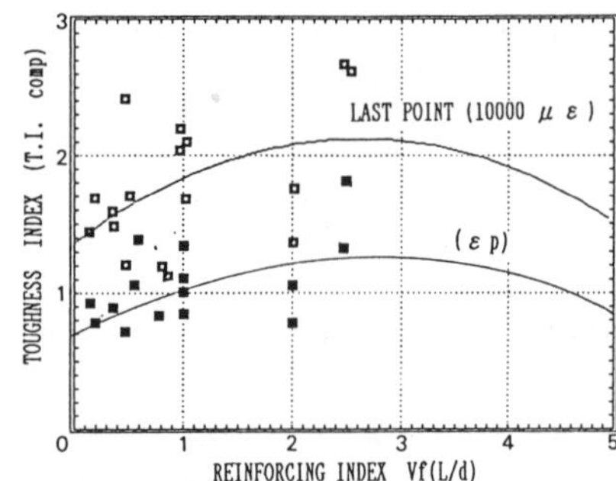

FIG. 8 Toughness index in compressive stress system

Despite of the wide scatter of data, appearance of an optimum amount at a range of Vf(L/d)=3.0 can be seen. It is believed that an excessive fibre length or volume will ultimately reduce the compressive strength properties. It was also observed that a relatively higher compression toughness occurred for a ductile matrix series (W/C=60%) series, which may be due to the combined ductility effect of the matrix and fibre, see Fig.3.

3.2.2 Effect to Flexural properties of Coir fibre concrete (Test Series 1)

Typical flexural load-deflection curves of coir fibre concrete are shown in Fig. 9 for different fibre contents. Higher fibre dosage improves the post-cracking strength without significantly affecting the first-crack strength. The coir fibres in the matrix were very effective in controlling the growth of major cracks, thus preventing the sudden failure and collapse of the specimen. In Fig.10(a), a nominal increase in flexural strength of about 8-12% occurred up to a fibre volume of 2%. Such trend is comparable to the case of polypropylene fibre composites (Hannant,1978) where contribution to flexural strength is minimal. As for the effect of length, variation in strength at different aspect ratio is only 8% on average. The flexural strength is related to the plain matrix bending strength (σb_m) and fibres reinforcing parameters, and given in Eq.3. The correlation coefficient (R) of the fit line (Fig.10(b)) is 0.9945.

$$\sigma b_c = 1.102\ \sigma m_c + 0.0898\ Vf(L/d) \qquad (R=0.994) \qquad (3)$$

This equation gives not only the prediction for the flexural strength of fibre concrete but also insights on the bonding characteristics between fibre and matrix. From a simple derivation, carried out by Swamy et.al.(1974), the average interfacial shearing stress (τ) at ultimate can be calculated as,

$$0.82\ \tau\ Vf(L/d) = 0.0898\ Vf(L/d), \qquad \tau = 0.11\ \text{Mpa}$$

This value is about half of that obtained from the results calculated from a coir fibre-mortar composite in Part 1 (Shimizu,Jorillo, 1992a). This finding is in agreement with the results reported by Swamy et.al (1974a) in the case of steel fibre concrete, where they noted the progressive reduction in the interfacial bond strength as the matrix changed from a mortar to a gravel-concrete matrix.

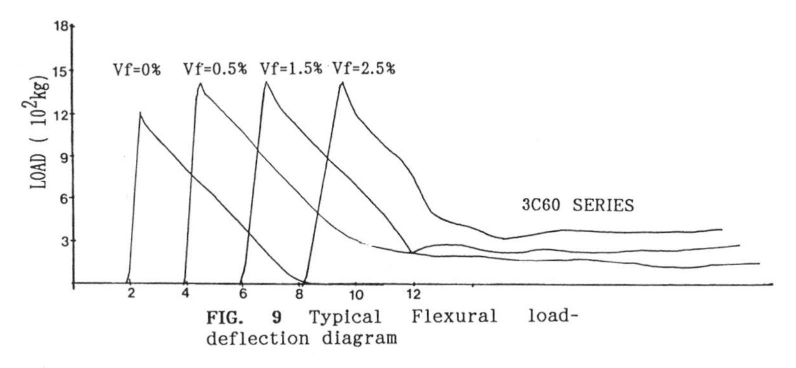

FIG. 9 Typical Flexural load-deflection diagram

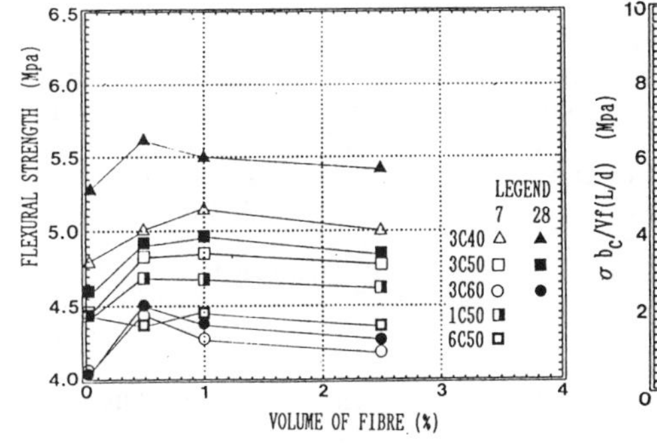

FIG.10(a) Trend of Flexural strength

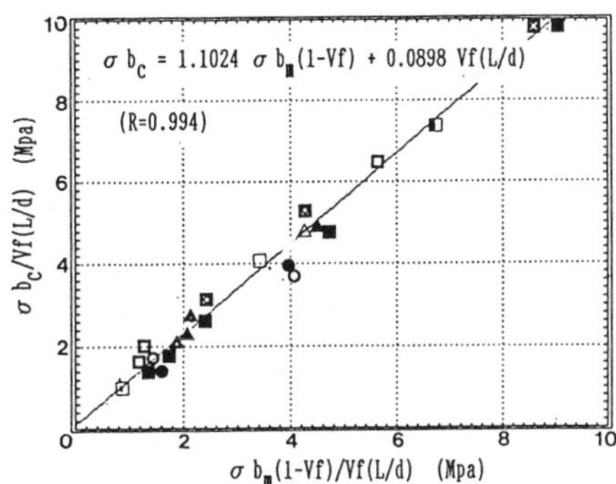

FIG. 10(b) Correlation of flexural strength with major parameters

Onset of First-crack

Onset of the first crack was measured from a point in a flexural stress-strain diagram (both compression and tension portion of the beam) where the curve started to deviate from linearity(Swamy et.al.,1975a). Although the exact mechanism of the occurrence of the first-crack is quite complex, it is generally believed that the flaws and microcracks start to propagate and concentrate in the direction of the principal strain, which gradually reduces the stiffness of the material, thereby causing the non-linear behaviour in the stress-strain[6]. It was observed that the effect of the reinforcing index to the value of the first crack strength is negligible. This implies the minimal or no contribution of coir fibres prior to the cracking stage. Actually such findings are common for low-modulus fibres like polypropylene fibres(Hannant,1978). Even in the case of high modulus steel fibre, Johnson(1974) reported that improvement in the first crack strength is limited to less than 10%.

Strain at ultimate and first-crack strength

Enhancement of the peak strain due to the inclusion of fibres into the matrix was also observed under the flexural stress system. Also, improvement in strain was most visible in the tensile portion of the beam, and this can be attributed to the more efficient contribution of fibres in a tensile stress system compared to the compression system. The measured maximum strain in the tensile zone ranged between 200-400$\mu\varepsilon$ compared to the value of 150 $\mu\varepsilon$ of the plain concrete, see Fig.11. This ability to sustain additional strain without collapse is a unique property of the fibre concrete [9], and this can also be seen in the case of low-modulus coir fibre composite. The maximum tensile strain (εf_c) is summarized in the following equation,

$$\varepsilon f_c = 39.7\ \sigma b_c + 24.31\ Vf(L/d), \qquad (R=0.946) \qquad (4)$$

Strains measured at the tensile and compressive zone at onset of the first crack showed no significant changes even at an increasing reinforcing index. It is conjectured that since the fibre is of low-modulus in nature, significant contribution of fibre to mechanical properties prior to cracking may not be visible.

Flexural modulus of Elasticity
The flexural modulus of elasticity was measured from the linear portion of the stress-strain curve up to the first crack, and were summarized in Fig.13(a) and (b). The secant modulus varied between 25 to 35 Gpa, and was very similar to the range of results obtained from a plain concrete. The law of mixture predicts an increase in the modulus of elasticity of the composite (Ec) with addition of the fibres into the matrix, especially in the case of high modulus fibres[8]. As shown in the following equation,

$$Ec/Em = 1 + Vf\ \{(Ef/Em)-1\} \qquad (5)$$

where Ec, Em and Ef represents the modulus of elasticity of the composite, matrix and fibre respectively. Due to the low-modulus character of coir fibre, and from the results obtained from compression test, it was presumed that a reduction in the flexural modulus of elasticity would occur. On the contrary, no reduction in the elasticity was seen even at a fibre volume content of 3.0%, and instead a slight increasing trend proportional to the reinforcing index occurred. A prediction equation of the flexural modulus having a correlation coefficient of 99.02% is given as,

$$Ec = 6205.7\ \sigma b_c - 633.8\ Vf(L/d) \qquad (R=0.990) \qquad (6)$$

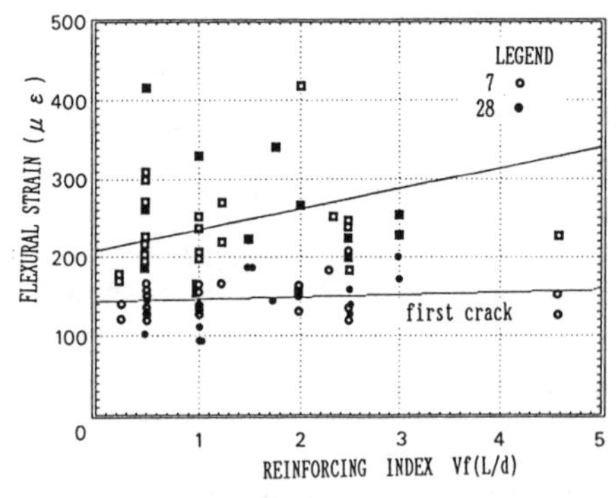

FIG. 11 Relationship of flexural strain with fibre index

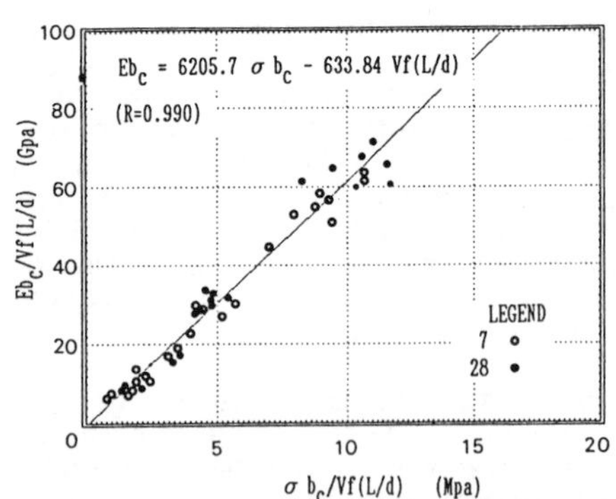

FIG.12 Correlation of flexural mod. of elasticity with major parameters

Post-cracking strength and location of Vf critical

Sudden failure was prevented by the presence of fibres spanning the cracks in the beam specimens, and a gradual increasing deflection due to fibre de-bonding and pullout occurred, as shown in the SEM micrographs in Fig. 17. This behaviour is reflected from the presence of a maintained residual load in the load-deflection diagram of fibre concrete. In Fig. 13, the effect of fibre volume, aspect ratio, and matrix quality can be clearly seen. Effect of matrix quality seems negligible from the point of post cracking strength and reinforcing index relationship, implying the relative similarities in the interfacial shear stress characteristics for a range of W/C ratio of 40 to 60%. In the case of effect of length, remarkable improvement in post cracking strength can be observed at a lower aspect ratio (1C50 series). This can be attributed to the larger surface area of fibre and their higher incidence of passing a vertical crack. As expected, the 6C50 series showed a relatively less post-cracking strength compared to other series. This confirms the overall action of the surface area of fibres in a state of de-bonding and pullout to the post-cracking strength. Hence, a larger fibre surface area in a state of pullout will ultimately result to a higher post-cracking load capacity.

In Fig.14, the intersection of the lines of the ultimate flexural load and the post-cracking load at an increasing fibre volume gives the location of the critical fibre volume (Vf_{crit}) required for flexural strengthening ($\sigma b_c \geq \sigma_{cu}$). From this simple method(Hannant,1978), actual Vf_{crit} in fibre concrete was found to vary from 7 to 8%. Increase in the value of the Vf critical can be attributed to the decrease in the interfacial shear stress and in the decrease of the efficiency of the post-cracking strength.

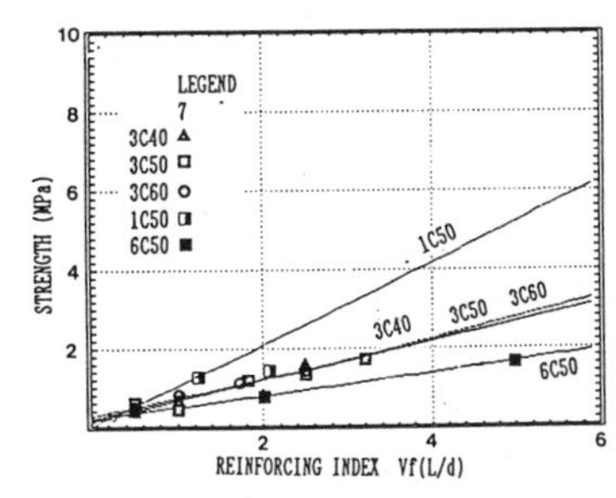

FIG.13 Effect of Vf and L/d to the post crack capacity of composite

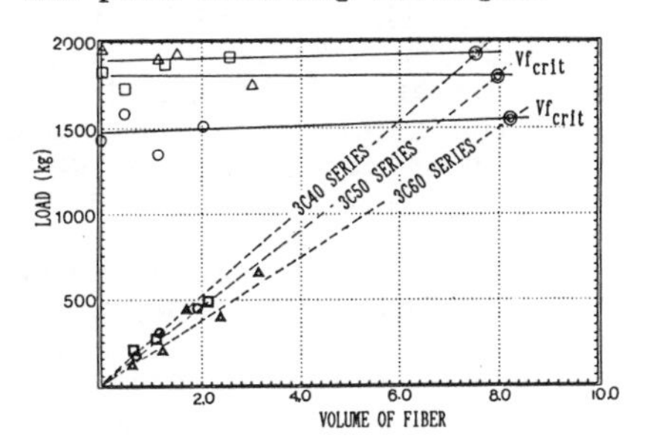

FIG.14 Location of Vf critical for flexural strengthening

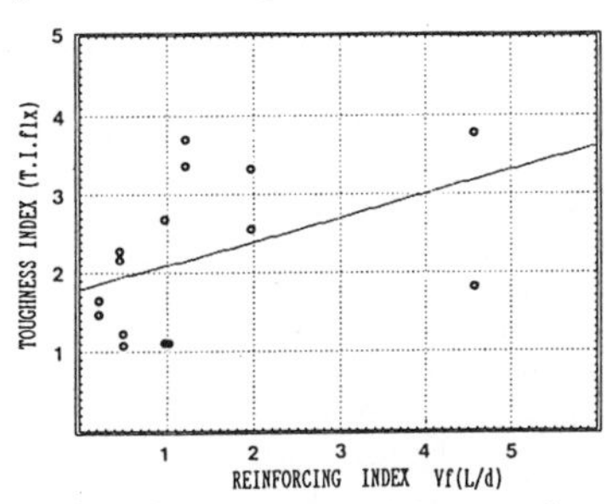

FIG.15 Toughness index under a flexural stress system

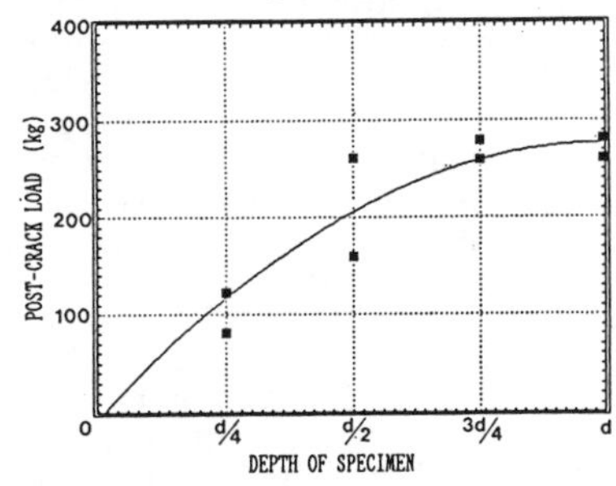

FIG.16 Effect of depth of tensile skin layer to post crack load

(a) failure by fracture

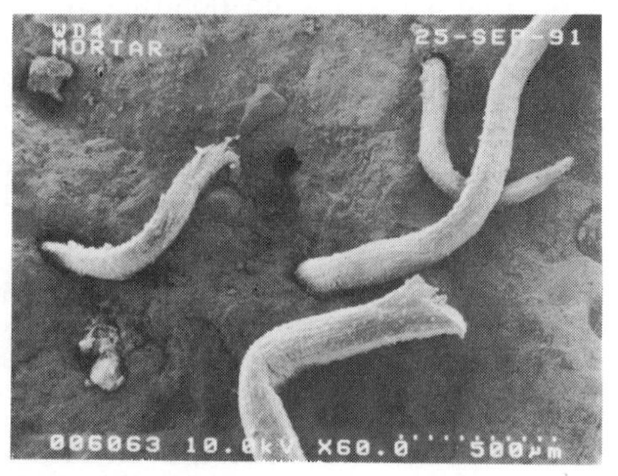

(b) combined pullout and fracture

FIG.17 SEM micrograph of coir fiber

Energy absorbing capability of FRC in Flexure

The first crack toughness in flexure was computed based on the definition of the ASTM C-1018 on fibre reinforced concrete. The toughness index ($T.I._{flx}$) was measured up to 5.5 times the deflection at the first crack. This index has a minimum value of 1 for an elastic-brittle material and a maximum value of 10 for an elastic-perfectly plastic material. It is evident from Fig. 15 that the addition of fibres increased the toughness by more than 100% for a fibre volume of 1.0% only, and it increases in proportion to the value of the reinforcing index. Such toughness not only implies ductility and fracture toughness enhancements but also assurance of safety and integrity of a structural element prior to failure.

3.2.3 Effect to Flexural properties (Test Series 2: Partial fibre concrete)

As seen from the results of Test series 1, increase in flexural strengths and modulus of elasticity is nominal, but the resulting crack control, strain capacity enhancements, and ductility can be used in appropriate structural design to improve the serviceability and integrity of a conventional structural element(Swamy et.al., 1975). One such application is the provision of a fibre tensile skin-layer in a conventional structural member. Hence, the use of partial fibrous concrete, i.e., utilization of fibrous material in the tensile zone only, were investigated and compared with the behaviour of a plain and full fibrous concrete material.

The effect to the ultimate strengths (or equivalent elastic flexural stress= $6M/bd^2$) is also nominal. However, the most significant contribution of a fibre concrete tensile skin-layer is the retention of the residual load, even at a thin depth of d/4, see Fig.16. Presence of post-cracking strength assures of greater toughness, ductility, and structural integrity to a conventional structural element. A higher post-cracking load (residual load) carrying capacity can be easily attained by the use of a higher fibre volume content in the tensile skin layer.

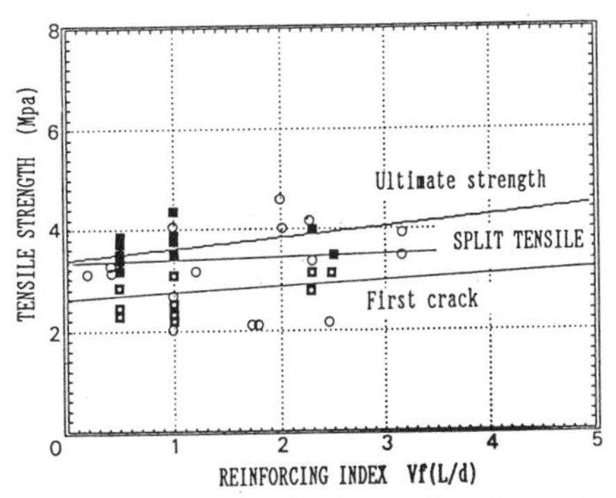

FIG. 18 Trend of tensile strength

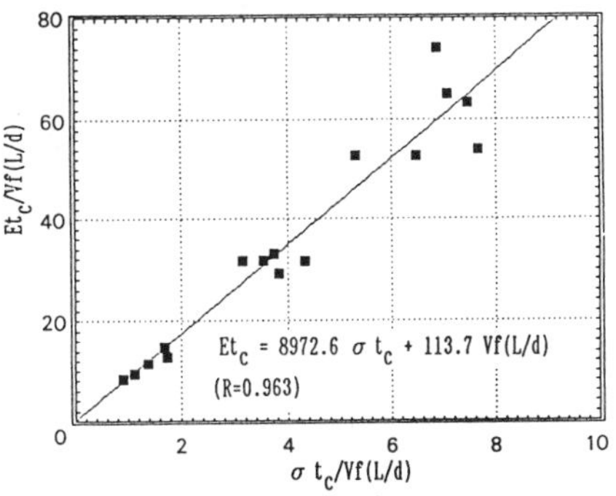

FIG. 19 Correlation of Tensile mod of elasticity with major parameters

3.2.4 Effect to the Tensile properties of fibre concrete

Indirect tensile properties

Results of the first crack strength in split tension with an increasing fibre reinforcing index are shown in Fig.18. The first crack in split tension was found to be about 76% of the corresponding flexural value. This is very similar to the 80% in plain concrete reported by Mindness and Young(1981). From the test results, it is very difficult to differentiate the effect of aspect ratio and fibre content to the first crack strength. This is due to the inherent low-modulus character of fibres which showed no significant contribution to mechanical properties prior to cracking. Also, according to Nanni (1991), confinement of fibres bridging the primary cracks by the matrix in a high compressive stress field is one of the many factors that makes it difficult to differentiate the effect of different fibre parameters.

Direct tensile properties

Typical relationship between fibre reinforcing index and tensile strength are shown in Fig 18. Although the correlation coefficient is quite low, the general tendency shows enhancement in strength at 28 days age. The linear regression line indicates a nominal increase of about 18% in the ultimate tensile strength. Likewise, a slight increase of about 10% was observed for the first crack.

The modulus of elasticity calculated from stress-strain diagram in direct tension shows similarities with the results obtained from the stress-strain in flexure tests, that is, no reduction in the elasticity at an increasing reinforcing index. In the case of direct tension, a slight improvement in proportion to the index Vf(l/d) occurred, see Fig.19. The secant modulus of elasticity in direct tension is likewise correlated with the matrix property and fibre parameters, and given as,

$$Et_c = 8972.6\ \sigma t_c + 113.7\ Vf(L/d) \qquad (R=0.963) \qquad (7)$$

It is interesting to note that unlike in compression and flexure, there was no observable increase in peak strain found with an increasing fibre index in the direct tension test results. It is possible that the fibre volume range adopted may be so low to have any significant effect to the peak strain in this kind of test.

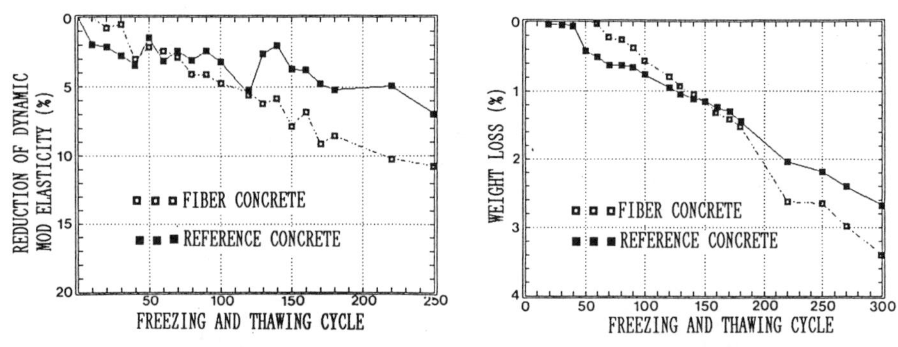

FIG.20 Freezing and thawing regime (a) Loss in weight (b) reduction in the dynamic modulus of elasticity

3.2.5 Durability against freezing and thawing

Failure of the material exposed to severe environmental conditions may take in the form of loss in strength, disintegration, or a combination of two. When a concrete material is subjected to freezing, water in the capillary pores will start to expand and cause increase in hydraulic pressure in its vicinity. Fibers can offer reinforcement so as to limit the cause of local cracking due to this hydraulic pressure expansion. In this study, the capability of the coir fibre to offer any reinforcement against freezing and thawing was investigated.

Results of the freezing and thawing test are shown in Figs. 20 to 21. The results is an average obtained from two specimens. Comparison of the weight loss and the reduction of the dynamic modulus of elasticity between fibre concrete and the reference air-entrained concrete showed practically the same behavioural characteristics under the freezing and thawing regime. Weight loss was accompanied by surface scaling, the degree of which is the same as to that of the reference concrete. Variations in the reduction of the dynamic modulus of elasticity is about 10% only, and can be considered negligible.

4 Conclusion

The following conclusion can be drawn from the experimental study related to the fresh and mechanical behaviour of coir fibre composite with gravel-concrete matrix.

1. Addition of fibres to the concrete resulted in a greater stability and cohesiveness, as well as an apparent reduced workability. Workability in terms of the mixture stability and mobility was greatly affected by increasing fibre volume and aspect ratio.
2. Mechanical properties in compression, flexure and tension were correlated with major parameters, and presented in simple empirical expressions in the form of composite mechanics approach. Those equations can be use as a the first-hand approximation of the basic properties for design.
3. Nominal improvement ranging from 8 to 12% in flexural and tensile strengths occurred. However, in the case of ultimate com pressive strength, no observable significant improvement was seen.
4. The reinforcing index Vf(L/d) has the most significant effect in flexural and tensile test results, as manifested from an improve-

ment in peak strain in proportion to the fiber reinforcing index. Enhancement in peak strain ranging from 10 to 20% occurred in flexure, tension and compression.

5. Significance of aspect ratio was seen not only in the resulting workability of the mixture but also to the properties in com pression. The effect of fibre length to modulus of elasticity and ultimate compressive strength was noted. In the case of flexure, the effect of overall surface area of fibres spanning the crack at the post-peak stages governs the resulting mode of failure and post-cracking load capacity.

6. Toughness in compression and flexure showed significant im provement from the ultimate stage up to the serviceability requirement e.g., 10000 $\mu\varepsilon$ in compression and 5.5 times the first crack in flexure. Improvement ranging from 100 to 200% occurred with an increasing reinforcing index.

7. Durability characteristics of fibre concrete during freezing and thawing showed that the coir-fibre cement composite possessed enough resistance in a severe environmental condition. Behavioural pattern is very similar to that of the reference air-entrained concrete.

Acknowledgement

The authors express their gratitude to the Kisitani Trust for the International Research of Building Materials and Fire Protection Technology for their support and assistance.

5. Reference

Agopyan, V.,(1988) **Vegetable fibre reinforced building materials-Developments in Brazil and other latin American countries,** Concrete Technology & Design Vol.5 (ed.Swamy) London, pp. 208-242

Aziz,M.A.; Paramasivam,P. and Lee, S.L.,(1988) **Concrete Reinforced with natural fibres,** in concrete Technology & Design Vol.2 (ed. R.N. Swamy), London, pp. 106-135

Hannànt,D.J.,(1978) **Fibre cement and fibre concretes,**London,219 pp.

Johnston,C.D.,(1974) **Steel Fibre reinforced mortar and concrete: A review of mechanical properties,** Proc of Fiber reinforced con crete, SP-44, USA, pp.127-142

Mindness S and Young, J.F, **Concrete** USA 1981

Nanni A. **Pseudo-ductility of fibre reinforced concrete**in Journal of Materials in CE Vol 3. USA 1991, pp 78-91

Rossi,P.; Acker,P., and Malier,Y,(1987), **Effect of steel fibres at two different stages: the material and the structure,** in Materials and Structure vol.20 France, pp.436-439

Shimizu,G and Jorillo P.,(1992) **Coir fibre reinforced cement based composite: part 1 - Microstructure and properties of fibre-mortar,** 4th RILEM Int. Symposium on Fibre Concrete, Sheffield July.

Shimizu G., (1986),**Experimental study on the tensile behaviour of concrete (in Japanese),** in Proc of Architectural Institute of Japan, Japan, pp.155-158.

Swamy R.N., and Mangat S.P., (1974),**A theory for flexural strength of steel fibre reinforced concrete,**in Cement and concrete Research Vol.4, pp. 315-325

Swamy, R.N., and Mangat P.S.,(1975),**The onset of cracking and duc tility of fibre concrete,** in Cement and concrete Research Vol 5, USA, pp.37-53

88 TRANSITION ZONE OF HARDENED CEMENT PASTE AND VEGETABLE FIBRES

H. SAVASTANO Jr and V. AGOPYAN
Escola Politécnica, University of São Paulo, São Paulo, Brazil

Abstract
Transition zones of composites with portland cement without any addition and random short monofilament vegetable fibres are studied. The effects of the type of fibre (coir, sisal or malva), water-cement ratio (from 0.30 to 0.46) and the age of the composite (up to 180 days) on this zone characteristics are analyzed and also compared with the mechanical properties of the composites.
Keywords: Vegetable Fibres, Composites, Fibrecement, Fibre-Matrix Bond, Transition Zone, Coir Fibres, Malva Fibres, Sisal Fibres.

1 Introduction

The transition zone of the hardened cement paste and vegetable fibres is in fact a portion of the paste very close to the fibre with a thickness less than 100 micrometers and properties not similar to those of the rest of the matrix.

This study aims to characterize this zone through a microstructural analysis of the porosity and of the products resulting from the cement hydration close to the fibres. It is also intended to show the variation of the results of these characteristics due to effect of the type of fibre, water-cement ratio and the age of the composite.

2 Transition zone

Farran (1956) in his paper, which was one the first publications on the interface of cementitious composites, informs that the calcium hydroxide crystals may present a preferential direction in the paste-aggregate interface. This was justified by Maso (mentioned by Monteiro, 1985) as a result of the differential diffusion of the ions liberated due to the cement hydration in this zone.

Fibre Reinforced Cement and Concrete. Edited by R. N. Swamy. © 1992 RILEM.
Published by E & FN Spon, 2-6 Boundary Row, London SE1 8HN. ISBN 0 419 18130 X.

The studies of this interface have been carried on with various materials in contact with hydrated cement paste. Regarding the fibrous materials, Bentur et al. (1985) admit that there is a double layer film close to the fibre (steel fibre in their case) with a thickness of about 1 micrometer. The first layer consists of a continuous film on the fibre surface of portlandite crystals with their axis perpendicular to this surface. The second layer consists of a film of hydrated calcium silicate, but this film is not continuous and is very difficult to recognize with a microscope. On the double layer film they found a portlandite zone with a thickness of 10 micrometers. Covering this zone there is a porous transition zone with a thickness from 10 to 20 micrometers, which becomes more compacted as far as from the fibre. All the authors conclusions were done for fracture surfaces observed with a SEM with X-ray dispersive energy analyzer.

The large amount of calcium hydroxide and ettringite in the transition zone may be due to the acceleration of the through solution process of the calcium, sulfates and aluminates ions in this zone as quoted by Monteiro et al. (1985) for the aggregate and cement mortar interface.

The use of SEM with backscattering electrons image analyzer combined with image analyzer for the study of microstructure of the transition zone is presented by Scrivener et al. (1988) for the characterization of the transition zone of high strength concretes (with or without silica fume) and they obtained the amount and distribution of anhydrous cement, pores and calcium hydroxide.

3 Vegetable fibres

Vegetable fibres are more affected by weathering and the alkaline medium than man-made fibres. Sisal and coir fibres have their tensile strength reduced to half when immersed in calcium hydroxide solution during 28 days (Agopyan, 1988). However this degradation can be reduced with low alkaline binders, for instance, coir fibres in blast furnace slag based cement mortar presented, after one year of natural weathering, similar mechanical strength to those of non exposed fibres (John et al., 1990).

Nevertheless, even when the fibres have their properties unchanged, the composites have their strengths slightly reduced with time. The transition zone must be the reason of the strength alteration, as it changes its porosity and thickness with the age of the composite, due to hydration of the cement.

4 Experimental work

The transition zone has been studied through its microstructure characterization and mechanical tests of the composites.

4.1 Raw materials

The portland cement used in this research work has no addition except gypsum, the vegetable fibres are coir (Cocos *nucifera* Linn.), malva (Urena *lobata* Linn.), and sisal (Agave *sisalana* Perrine), without any chemical treatment.

The main properties of the fibres are presented in Table 1. The values are average of large number of samples with high coefficients of variation. Due to the chopping process, the length of the fibres varies a lot, but been always from 15 to 30 mm long.

It is necessary to point out the importance of water absorption of these fibres on the transition zone. About 60% of the total absorption occurs during the first 15 minutes of immersion for all of the fibres. Therefore the absorption of the fibres will induce a flow of water in the matrix, moreover as the adsorption of water on fibre surface must also occur, the water-cement ratio should be considerable changed in the vicinity of the fibres.

Although the fibres have a large water absorption they are used oven dried (60°C) in this work, because it has been previously demonstrated that the control of added water by wetting is not possible besides this fact do not change the final composite mechanical behaviour (Savastano Jr., 1987).

Table 1. Main characteristics of the fibres

Property	coir	sisal	malva
Permeable pore volume (%)	56.6	60.9	74.2
Water absorption (%)	93.8	110.0	182.2
Elongation at rupture (%)	37.7	5.2	5.2
Tensile strength (MPa)	107	363	160

4.2 Composites

Eighteen series of specimens of vegetable fibre reinforced cement paste and six series of the unreinforced matrix were prepared to be tested at the ages of 7, 28, 90 and 180 days. The water-cement ratio was 0.38 but for the age of 28 days specimens with 0.30 and 0.46 were made.

The selected fibre volume is 4% which is the largest amount of fibre that allows a proper mixing without bundling of the fibres. The composite is mixed in an ordinary mortar pan mixer, compacted on a vibrating table and cured in a wet chamber until the date of the test.

4.3 Bending test

The adopted method is the four point bend test, following the RILEM (1984) recommendation for this type of material. The strength obtained from this test must be considered as a estimated one because for its calculation the moving of the neutral axis is not took into account. For this reason the energies spent during this test is also mentioned. The absorbed energy from a specimen is equivalent to the surface area under the load - displacement curve while the specific energy is the absorbed energy divided by the cross section.
The research work includes other mechanical tests such as direct tensile, indirect tensile and pull-out tests which are not mentioned in this paper.

4.4 Microstructural analysis

The transition zone characterization is done using an electron microscope and X-ray energy dispersion. For these tests, the hydration of the composites is stopped with acetone and the use of a vacuum pump at specific ages. The dried specimens are kept in a desiccator with silica gel and soda lime which absorb the water and carbon dioxide, respectively.

4.4.1 Microscopy

For SEM observation the samples are impregnated with a low viscosity epoxy resin and then polished with a diamond paste. The purpose of this impregnation is to avoid the removal of low hardness particles of the cement paste and also to keep the original structure of the pores. The samples are recovered with graphite as it does not affect the X-ray analysis.

4.4.2 X-ray analysis

X-ray analysis is a quick method to qualify the composition of a microregion. For quantitative determination it is necessary to compare the results with those of standard samples.

These analysis in connection with SEM can be done by dispersive energy in three ways: elements distribution from X-ray images, analysis of one line by mechanical scanning and the analysis of one single point.

5 Results

5.1 Bending tests

The relation of the mechanical properties with the water-cement ratio is better showed in the Figures 1 and 2. As can be seen the strength is reduced with the increase of the water-cement ratio but there is not much differences between the composites and the unreinforced matrix behaviour. However when the specific energy variation is taken into account the mechanical performance of the matrix is quite different from the composites but the energy consumed to break the specimens seems not to have changed with the water-cement variation of the matrix.

Figures 3 and 4 present the effect of the age on the specimens mechanical performance with water-cement ratio of 0.38. As was expected only the unreinforced matrix presents an increase in strength, the composites suffer a reduction in this value due to fibre decay, mainly because of the

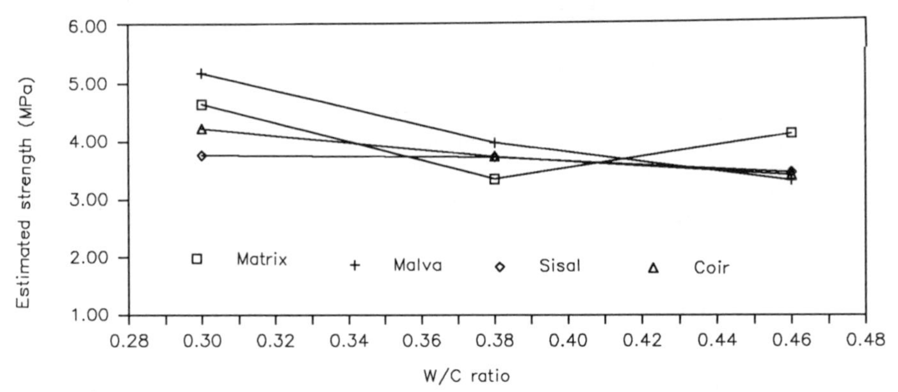

Fig.1. Bending test: effect of water-cement ratio on estimated strength.

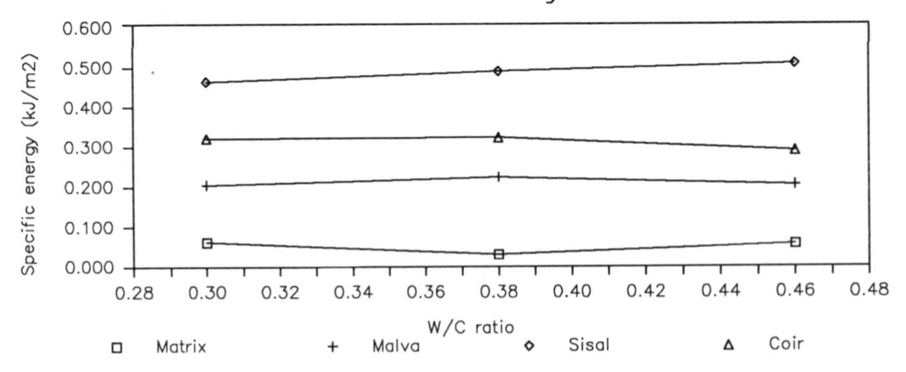

Fig.2. Bending test: effect of water-cement ratio on specific energy.

severe humid conditions of the curing of the specimens. It is clearly noticed in these figures that the composites not only have their strength reduced but also their ductility has a sharp decrease of the specific energy happened. According to the previous studies, where the mechanical behaviour of the composites is analyzed as well as the microstructural deterioration of the fibres is observed, the coir fibres seem to be the more durable ones (Agopyan, 1988; John et al., 1990 and Savastano Jr., 1987).

5.2 Microstructural analysis
Some micrographs of backscattering electron image are presented in this paper for discussion.

For the malva composite, at the age of 7 days and with a water-cement ratio of 0.38, the transition zone is highly porous and has portlandite macrocrystals at the thickness of about 50 micrometers (Figure 5). This zone is clearly fissured probably due to lower strength than the matrix itself. Some microcracks cross even the portlandite crystals and can also be noticed calcium hydroxide plates in a perpendicular direction from the analyzed surface.

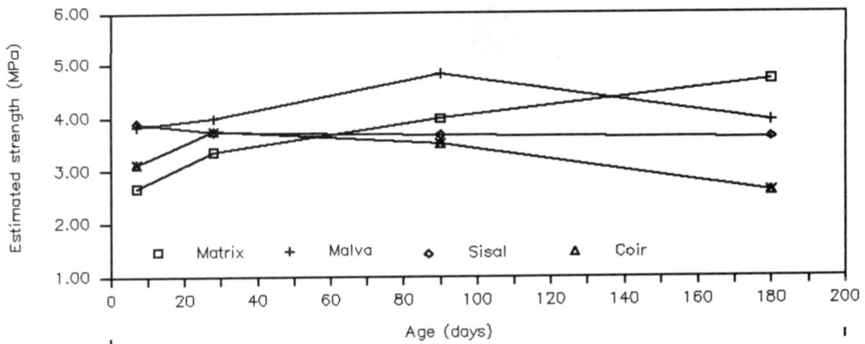

Fig.3. Bending test: effect of the age on estimated strength.

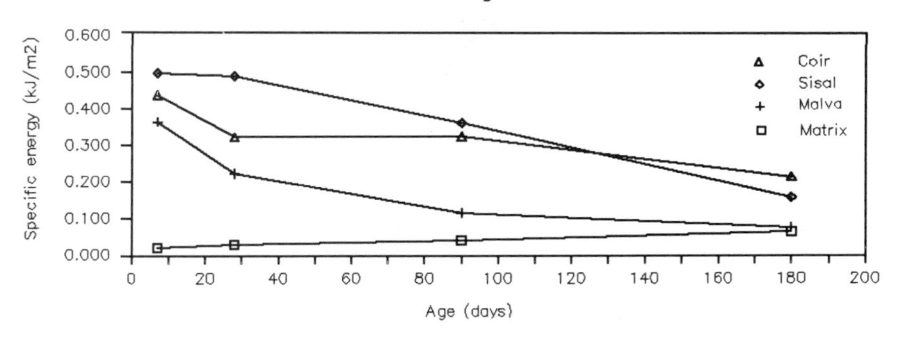

Fig.4. Bending test: effect of the age on specific energy.

The sisal fibre is also debonded from the matrix with water-cement ratio of 0.38 at the age of 7 days. The same behaviour is observed for coir fibres (Figure 6) where the porous transition zone is still attached to the fibre. The debonding of the vegetable fibres is expected because of great volume variation of the fibres due to water absorption and loss.

The transition zone for the three types of the vegetable fibres apparently becomes more dense with the increase of the age of the composite, as can be seen in Figures 7 and 8 for malva and coir fibres, respectively, for the age of 180 days. In these figures the debonding of the fibres is more visible than taht from the previous figures.

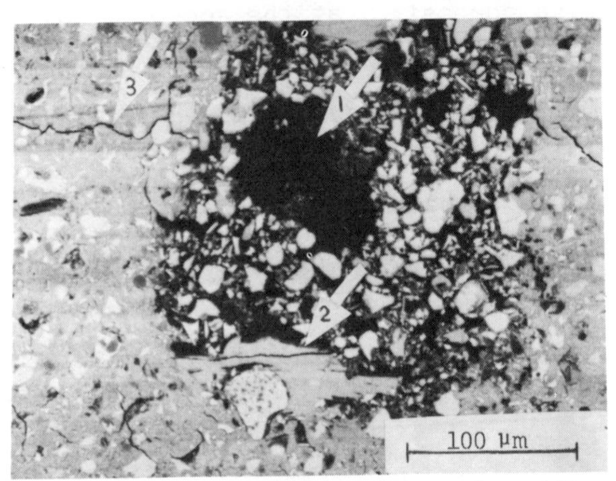

Fig.5. BSE image of composite with malva fibre, water-cement ratio of 0.38, age of 7 days (259x). Arrow 1: malva fibre; arrow 2: portlandite crystal; arrow 3:crack.

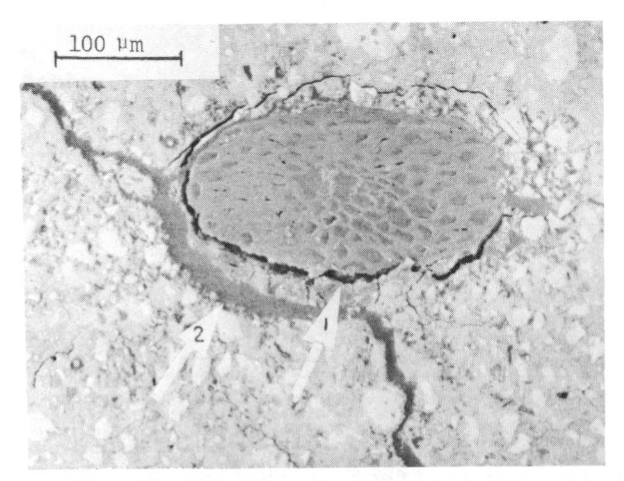

Fig.6. BSE image of composite with coir fibre, water-cement ratio of 0,38, age of 7 days (208x). Arrow 1: fibre-matriz debonding; arrow 2:crack

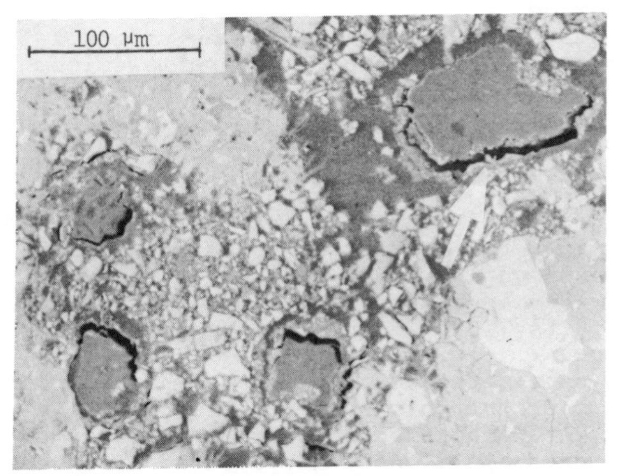

Fig.7. BSE image of composite with malva fibre, water-cement ratio of 0,38, age of 180 days (287x). The arrow indicates fibre-matrix debonding.

Fig.8. BSE image of composite with coir fibre, water-cement ratio of 0,38, age of 180 days (354x). The arrow indicates fibre-matrix debonding

6 Final remarks

The results of the bending tests are adherent to the theory and previously known results (Agopyan, 1988) although they do not reflect the actual behaviour of the composites because of the effect of the curing which has accelerated the fibre damage.

Due to high water absorption of the vegetable fibres at the first minutes a significant water flow close to the fibres may occur therefore the transition zone thickness is higher than those obtained with non absorbed fibres like the steel ones, where the fibres actually induce a wall effect in the matrix. The increase of water-cement ratio in the vicinity of the vegetable fibres may help to increase the shrinkage of the matrix in this region and therefore induce the microcraking of this porous region. Nevertheless this phenomenon can be fully altered if the casting process is modified. For instance, for composites casted under pressure, Coutts (1987) found a dense matrix close to the wood fibres.

If it is considered porosity as an indicator of the transition zone, the thickness of this region is about 50 micrometers. It is also found in this zone large portlandite crystals without any defined arrangement and not accumulated on the surface of the fibres. The amount of ettringite in the transition zone is similar to the matrix.

The porosity of the transition zone is reduced with the increase of the age of the samples however the fibre decay becomes more evident therefore the fibre itself is probably the weakest part of the composite and the reason of the reduction of fibrecement ductility.

As was mentioned the fibre debonding from the matrix must be due to its great volume variation. The main effect of this phenomenon is the reduction of energy absorption during mechanical test so another factor to the reduction of the ductility of the composite.

The research work is still in progress aiming to correlate the transition zone characteristics with the mechanical performance of the composites.

7 References

Agopyan, V. (1988) Vegetable fibre reinforced building materials - developments in Brazil and other Latin-American countries, in **Natural Fibre-Reinforced Cement and Concrete** (Concrete Technology Design 5, ed. R. N. Swamy), Blackie, Glasgow, pp. 208-242.

Bentur, A.; Diamond, S. and Mindess, S. (1985) The microstructure of the steel fibre-cement interface. **J.Mater. Sci.**, 20, 3610-3620.

Coutts, R. S. P. (1987) Fibre-matrix interface in air-cured wood-pulp fibre-cement composites. **J. Mater. Sci. Letters**, 6, 140-142.

Farran, J. (1956) Contribution minéralogique à l'étude de l'adherence entre le constituants hydratés des ciments et

le matériaux erodés. **Revue de Matériaux de Construction**, 490/491/492, 155-172.
John, V. M.; Agopyan, V. and Derolle, A. (1990) Durability of blast furnace slag based cement mortar reinforced with coir fibres, in **Vegetable Plants and their Fibres as Building Materials** (RILEM Proceedings 7, ed. H. S. Sobral), Chapman and Hall, London, pp. 87-97.
Monteiro, P. J. M. (1985) **Microstructure of concrete and its influence on the mechanical properties.** PhD Thesis, Department of Civil Engineering, University of California, Berkeley.
Monteiro, P. J. M.; Maso, J. C. and Ollivier, J. P. (1985) The aggregate-mortar interface. **Cement and Concrete Research**, 15, 953-958.
RILEM Technical Committee 49 TFR (1984) Testing methods for fibre reinforced cement based composites. **Matériaux et Constructions**, 17, 441-456.
Savastano Jr., H. (1987) **Fibras de coco em argamassas de cimento portland para produção de componentes de construção civil.** MEng. Dissertation, Civil Construction Engineering Department, Escola Politécnica, University of São Paulo, São Paulo.
Scrivener, K. L.; Bentur, A. and Pratt, P. L. (1988) Quantitative characterization of the transition zone in high strength concretes. **Advances in Cement Research**, 1.

Acknowledgments

The authors would like to thank Professor Paulo S. C. Pereira da Silva of the Metallurgical and Materials Engineering Department of EPUSP for his help with the microstructural analysis. They are also in debt to the IPT - Instituto de Pesquisas Tecnológicas do Estado de São Paulo where the main research work has been carried on with the help of Dr. Francisco A. S. Dantas, Dr. Maria A. Cincotto and Mr. Pedro C. Bilesky.

89 DURABILITY STUDIES ON COIR FIBRE REINFORCED CEMENT BOARDS

L. K. AGGARWAL
Central Building Research Institute, Roorkee, India

Abstract
Coir fibre reinforced cement board, a new material which could be used as an alternative to wood and wood based products, was subjected to accelerated durability tests along with resin bonded particleboards and plywood (commercial products) to study the durability of the newly developed material as well as to generate comparative durability data. The physico-mechanical properties like thickness swelling, water absorption, bending strength and internal bond strength of the exposed samples were determined before exposure and at varying intervals upto the completion of weathering cycles. The results obtained show that the thickness and internal bond strength of coir fibre cement boards remain unchanged and there is slight change in water absorption and bending strength while the resin bonded particleboards showed appreciable increase in thickness and water absorption and decrease in bending strength and internal bond strength under similar exposure conditions. In boiling water test, plywood samples showed delamination and disintegration and there is no such effect on coir fibre cement boards. Therefore, it can be inferred on the basis of above studies that the developed boards will be more durable in service.
Keywords: Coir Fibre, Ordinary Portland Cement, Durability, Accelerated Weathering, Plywood, Particleboards, Properties, Water Absorption, Thickness, Bending Strength, Internal Bond Strength, Boards.

1 Introduction

The annual consumption of industrial wood in India is estimated (9) at about 27.58 million cubic meters (m.cu.m.), out of which 25-30% is used by the building industry in the form of wood (5.0 m.cu.m.) and wood based products like plywood, particleboards and fibre boards (2.33 m.cu.m.). Need to develop alternative materials was necessary because of scarcity and rising cost of wood and wood based products. Coir fibre reinforced cement board, a newly developed

Fibre Reinforced Cement and Concrete. Edited by R. N. Swamy. © 1992 RILEM.
Published by E & FN Spon, 2-6 Boundary Row, London SE1 8HN. ISBN 0 419 18130 X.

material, could be an alternative to the above mentioned products for various applications in buildings. The physico-mechanical properties of the developed boards have already been reported (1), but the effect of weathering on the properties of boards are unknown. Studies to measure the changes in properties of boards on weathering are important, since the cellulosic fibres are liable to degradation (3,5) on exposure to natural weathering. The investigations on the durability of cellulosic fibre reinforced cement composites have shown that degradation of cellulosic fibre by alkali from the cement matrix depends upon the pretreatment (4,8) of fibres prior to inclusion in the composites.

Coir fibre reinforced cement boards along with commercially available particleboards and plywood were subjected to accelerated weathering cycles and changes in their properties, were determined at various intervals of exposure. The results obtained are discussed in the paper.

2 Experimental

Coir fibre reinforced cement boards and commercially produced resin bonded particleboards and plywood (BWP) were used in the present studies. Coir fibre cement boards were produced using coir fibre and ordinary portland cement following the method described elsewhere (1). The density of the boards was in the range of 1375 ± 25 Kg/m^3.

The accelerated durability programme included three types of accelerated cycles, designated as A, B and C. The accelerated cycles A and B were carried out following the procedures laid down in IS:2380-1963 (6) and cycle C as per IS:1734-1983 (7). In brief, these tests comprises of the following steps;

A) One cycle - 4 hours soaking in water + 20 hours drying in oven at $50 \pm 1T^oC$ - 30 Cycles.
B) One cycle consisting of 1 hour in water at 49 ± 1^oC + 3 hours in steam and water vapour + 20 hours at 20 ± 1^oC + 3 hours heating in dry air at 99 ± 2^oC + 3 hours in steam and water vapour + 18 hours heating in dry air at 99 ± 2^oC - 20 cycles.
C) Boiling in water - 200 hours.

Coir fibre cement boards were subjected to all the three accelerated cycles while resin bonded particleboards to cycles A and B and plywood to cycle C only.

2.1 Sample Preparation and Test Methods

About 60 samples of the following sizes were cut from the coir fibre cement boards and resin bonded particleboards after discarding 25 mm wide border from each side. The samples were mixed and then randomly chosen for the accelerated cycles A and B. Density of all the samples was determined before exposure. Six samples were withdrawn at various intervals of exposure for each test.

Bending strength = 250x75 mm
Internal bond strength = 50x50 mm
Water absorption and thickness = 100x100 mm

In boiling water test 150x150 mm size samples of plywood and coir fibre cement boards were used.

Thickness swelling, water absorption, bending strength and internal bond strength (tensile strength perpendicular to the plane of the board) were determined following the methods described in IS:2380-1963 (6). Water absorption was measured by immersing the samples in water for 24 hours. Bending strength was determined at a span of 200 mm and the crosshead speed of testing machine (Zwick testing machine - model 1474) was 5 mm/min. Internal bond strength was determined at the crosshead speed of 1 mm/min.

3 Results and Discussion

The properties of the coir fibre cement boards before and at various intervals of exposure to cycles A and B are given in Tables 1 and 3 respectively and the results obtained for resin bonded particleboards in Table 2. The condition of the coir fibre cement boards and particleboards before and after the completion of 30 and 15 accelerated cycles designated as A as well as 10 and 5 cycles designated as B is shown in Figs. 1 and 2 respectively.

The coir fibre cement boards showed 12-13% increase in water absorption and 9-10% increase in bending strength but no change in thickness and internal bond strength after 30 cycles exposure to cycle-A (Table-1)

Table 1 Properties of Coir Fibre Cement Boards Exposed to Cycle-A

Number of Cycles	Property			
	Thickness mm	Water Absorption (24 Hrs),%	Bending Strength N/mm^2	Bond Strength N/mm^2
0	12.30±0.20	15.00±0.80	9.85±0.55	0.37±0.02
5	12.25±0.15	15.50±0.90	10.05±0.45	0.37±0.20
10	12.20±0.20	16.30±0.60	10.60±0.40	0.39±0.03
20	12.20±0.20	16.65±0.45	10.65±0.45	0.38±0.02
30	12.20±0.20	16.90±0.30	10.80±0.30	0.37±0.03

The values for resin bonded particleboards after 15 cycles exposure are 20-22% increase in thickness, 52-53% increase in water absorption and 40-50% reduction in bending strength and internal bond strength (Table 2).

Table 2 Properties of Resin Bonded Particleboards Exposed to Cycles A and B

Number of Cycles	Property			
	Thickness mm	Water Absorption (24 Hrs),%	Bending Strength N/mm^2	Bond Strength N/mm^2
CYCLE A				
0	12.10±0.05	19.00±0.60	20.10±0.50	0.48±0.02
5	13.00±0.10	23.50±0.70	17.70±1.20	0.42±0.03
15	14.60±0.15	33.50±0.60	10.60±1.30	0.27±0.04
CYCLE B				
0	12.10±0.05	19.00±0.60	20.10±0.50	0.48±0.02
5	14.80±0.20	36.50±1.30	10.20±1.80	0.25±0.03

The increase in the bending strength of coir fibre cement boards with the increase in exposure period may be due to further hydration of cement in the coir fibre cement boards. But, in case of resin bonded particleboards the reduction in bending strength and internal bond strength is due to the adhesion failure between the wood particles and resin matrix and/or breakdown of wood particles, which had also resulted in the disintegration and thickness swelling of the boards, Fig. 1.2.

The effect on the properties of coir fibre cement boards when subjected to accelerated durability cycle designated as cycle B is 14-15% increase in water absorption and 9-10% increase in bending strength but no change in thickness and internal bond strength of the boards after 10 cycles of exposure (Table 3).

Further exposure, 20 cycles, showed reduction in bending strength and internal bond strength along with increase in thickness and water absorption of boards. The increase in bending strength of the boards upto 10 cycles of exposure may be due to further hydration of cement in boards. The reduction in bending strength as well as internal bond strength after 20 cycles of exposure indicates the possibility of breakdown of the fibres/or disruption of fibre cement bond (2) due to alternate swelling and shrinkage of the fibres. The disruption of fibre cement bond has also resulted in the increase in thickness of boards

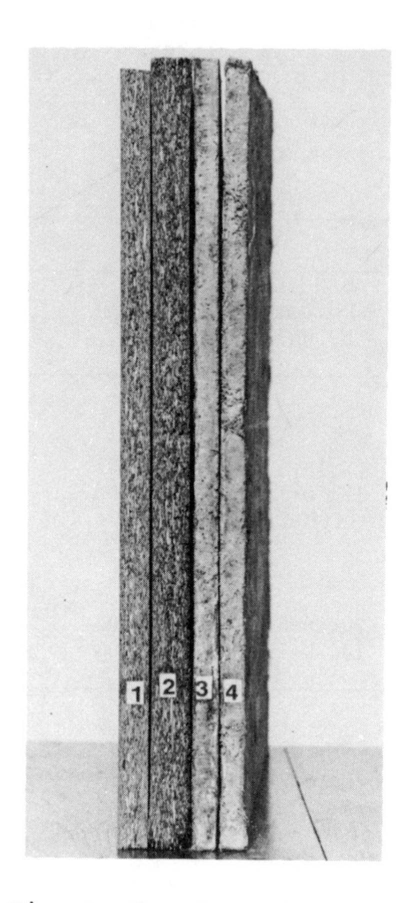

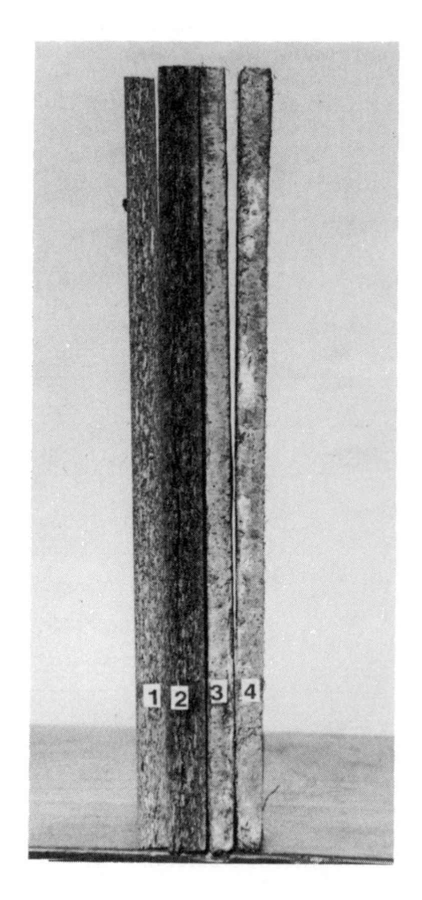

Fig. 1. Boards exposed to accelerated durability cycle A.
1.1 Particle board - unexposed
1.2 Particle board after 15 cycles exposure
1.3 Coir fibre cement board - unexposed
1.4 Coir fibre cement board after 30 cycles exposure

Fig. 2. Boards exposed to accelerated durability cycle B.
2.1 Particle board - unexposed
2.2 Particle board after 5 cycles exposure
2.3 Coir fibre cement board - unexposed
2.4 Coir fibre cement board after 10 cycles exposure

from 12.3 mm to 12.4 mm after 20 cycles of exposure (Table 3).

The resin bonded particleboards when exposed to accelerated durability cycle designated as cycle B showed 22-25% increase in thickness, 64-66% increase in water

Table 3 Properties of Coir Fibre Cement Boards Exposed to Cycle-B

Number of Cycles	Property			
	Thickness mm	Water Absorption (24 Hrs),%	Bending Strength N/mm^2	Bond Strength N/mm^2
0	12.30±0.20	14.85±0.75	10.00±0.60	0.38±0.02
5	12.20±0.20	15.75±0.55	10.55±0.35	0.40±0.02
10	12.20±0.20	17.00±0.60	10.90±0.40	0.38±0.03
20	12.40±0.25	18.10±0.80	9.20±0.80	0.30±0.03

absorption, 45-50% reduction in bending strength and internal bond strength after only 5 cycles of exposure (Table 2). The changes in the properties of the particleboards are due to the adhesion failure between the wood particles and matrix and/or breakdown of wood particles as mentioned earlier. The disintegration of particles and swelling of the board due to the adhesion failure is shown in Fig.2.4.

The durability of coir fibre cement boards and plywood was compared by carrying out the boiling water test, generally recommended for exterior grade plywood. Plywood samples showed delamination and disintegration of plies after 80 hours of boiling in water while coir fibre cement boards showed no such effect even after 200 hours boiling in water.

It can be inferred from the above studies that coir fibre reinforced cement boards showed more resistance to various alternate drying and wetting cycles and hence would be more durable than resin bonded particleboards and plywood in service. The developed boards can, therefore, be recommended for both internal and external applications in buildings.

4 Applications

The developed coir fibre cement boards can be used for various applications in buildings such as panelling of door and window shutters, partitioning, false ceiling, cladding etc. These boards can be sawn and drilled using normal wood working tools and can be painted like wood and wood based products. The use of boards as panelling material for door shutters is shown in Figs. 3-4. The door shutter before and after painting is shown in Fig. 3 and Fig. 4 respectively. These door shutters have been installed in a recently constructed building in the Institute to study

Unpainted (Fig. 3., left) and painted (Fig. 4., right) door shutter with coir fibre cement boards.

their performance in unexposed and exposed conditions.The weight of the door shutter, made by using sheesham wood for frames and coir fibre cement boards as panels, is about 19-20 kg/m^2. These shutters are 15-25% cheaper in comparison to good quality wood shutters.

5 Acknowledgement

The work reported in this paper forms a part of normal research programme at the Central Building Research Institute, Roorkee and is published with kind permission of the Director.

6 References

1. Aggarwal, L.K. Studies on Cement Bonded Coir Fibre Boards. **Cement & Concrete Composites** (Under publication).
2. Beech, J.C., Hudson, R.W., Laidlaw, R.A. and Pinion, L.C. (1974) Studies of the Performance of Particleboard in Exterior Situations and the Development of Laboratory Predictive Tests. **Building Research Establishment Current Paper** CP77/74, B.R.E., Garston, U.K.
3. Cook, D. J.(1980) Natural Fibre Reinforced Cement & Concrete - Recent Developments, Advances in Cement Matrix Composites in **Proceedings of the Symposium on L.Materials Research Society,** pp.251-258.
4. Gram, H. E. (1983) Durability of Natural Fibres in Concrete in **Swedish Cement Concrete Institute,** 1.83, pp 255.
5. Mansur, M.A. and Aziz, M.A. (1982) A Study of Jute Fibre Reinforced Cement Composites. **International Journal Cement Composites and Lightweight Concrete,** 4, 75-82.
6. Methods of Test for Wood Particleboards and Boards from other Lignocellulosic Materials -IS:2380(1963) **Indian Standards Institution,** New Delhi.
7. Methods of Test for Plywood -IS:1734(1983) **Indian Standards Institution,** New Delhi.
8. Sharman, W.R. and Vautier, B. P. (1986) Accelerated Durability Testing of Autoclaved Wood Fibre Reinforced Cement Sheet Composites. **Durability of Building Materials** 3, 225-275.
9. Van Stithi par Report (1987) published by **Forest Research Institute,Dehradun.**

90 CARBOHYDRATE CONTENT IN OIL PALM TRUNK AND ITS INFLUENCE ON SOME CHARACTERISTICS OF CEMENT-BONDED PARTICLE BOARD

S. RAHIM
Forest Research Institute Malaysia, Kuala Lumpur, Malaysia
M. A. ZAKARIA
Science University of Malaysia, Penang, Malaysia

Abstract

During replanting of the oil palm which is a major economic crop of Malaysia, a huge amount of waste is generated. This study was carried out to determine the feasibility of using this waste material in the form of felled trunks for cement-bonded particleboard manufacture.

A major setback in using the oil palm trunk (OPT) lies in its high carbohydrates content. Freshly felled OPT were found to have more sucrose than glucose and fructose. The total amount of sugars in OPT of age 35 years and above was in the range of 1.8 to 13.3 % (based on dry weight of wood) at various height and zones of the tree. The starch content was found to be in the range of 0.4 to 25.5 %. In general, the amount of sugars in OPT was more concentrated at the central zone of the palm while the starch content was more concentrated at the higher level of the tree.

Among the various pre-treatments to reduce the carbohydrates content, soaking of the OPT chips in water for 3 days was adequate to produce cement-bonded particleboard with consistent and acceptable physical and mechanical properties.

Keywords : Oil palm trunk, Cement-bonded particleboard, Carbohydrates, Sugars, Starch, Pre-treatments, Physical and Mechanical properties.

Introduction

The oil palm (Elaeis guinensis Jacq.) is the most important crop in Malaysia. By the end of 1990, oil palm plantations covered the biggest plantation area in this country of about 2.07 million ha, equivalent to 38.7 % of total agricultural area (Anonymous, 1991). Currently, Malaysia is the largest palm oil producer and contributes

Fibre Reinforced Cement and Concrete. Edited by R. N. Swamy. © 1992 RILEM.
Published by E & FN Spon, 2-6 Boundary Row, London SE1 8HN. ISBN 0 419 18130 X.

about 57.6 % of the total supply of palm oil in the world. Intensive plantation establishment was started in the 1960's and since the average economic life of the oil palm is estimated at around 25 years, a large amount of the trunks is expected to be available during replanting by 1985 onwards. Mohamad Husin *et al.* (1986) estimated that the availability of the oil palm trunk due to replanting was about 1.3 million tonnes (dry matter) in 1990 and would increase steadily to over 7.0 million tonnes toward the year 2000. Currently, the felled oil palm trunks are just left to decay or burnt in the plantation without any commercial application. This practice is unhealthy as it creates environmental pollution and encourages fungal and insect infestation to the young trees.

Much research has been conducted to determine the feasibility of using this abundantly available ligno-cellulosic material for value-added products. Khozirah *et al.* (1991) reported on the several studies including the anatomy, processing and preservation, pulp & paper, animal feed and wood-based panel products carried out since 6 years ago. However, this oil palm trunk has yet to be used commercially due to several problems such as difficulty in processing, high moisture content and high susceptibility to fungal and insects.

This paper discusses the carbohydrates content of this material and its influence on the hydration of cement. Various types of pre-treatment were carried out on the oil palm trunk (OPT) in order to reduce its carbohydrates content prior to cement-bonded particleboard (CBP) manufacture. The mechanical and physical properties of CBP produced from pre-soaked and piling stored OPT flakes are also discussed.

2 Materials and methods

2.1 Oil palm trunk

Several felled oil palm trees (OPT) (35 years of age and above) were used for this study. In the first trial, three trees of age 38 years were acquired from the Palm Oil Research Institute of Malaysia (PORIM) plantation in Serdang, Selangor. The trees, belonging to the Tenera variety, were felled by chain-saw. Each tree was divided into six height levels namely A, B, C, D, E & F for sugars and starch content analyses (refer Figure 1). The average diameter at breast height and total height of the tree up to the first frond were 59.2 cm and 12.8 m respectively.

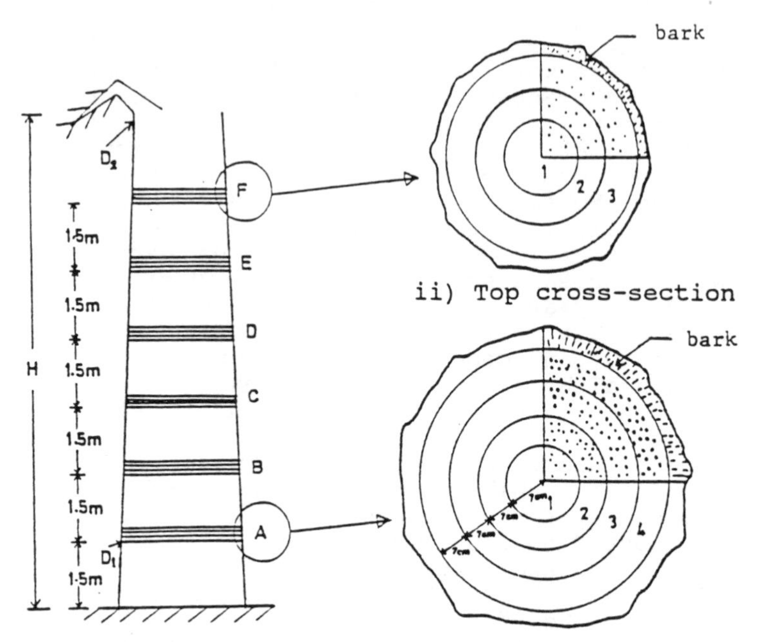

i) Oil palm tree iii) Bottom cross-section

Fig. 1. Sampling of oil palm trunk in Trial 1

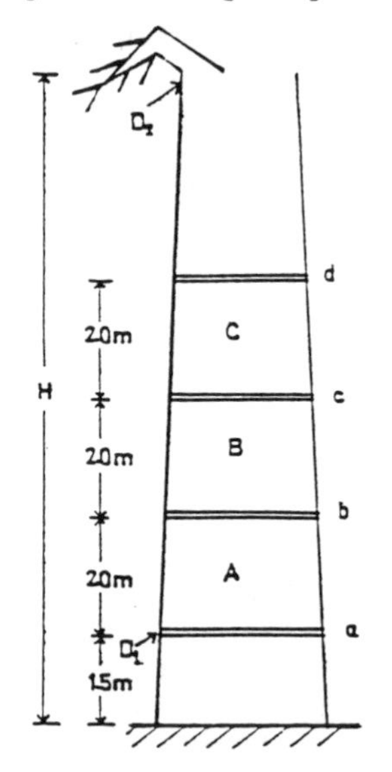

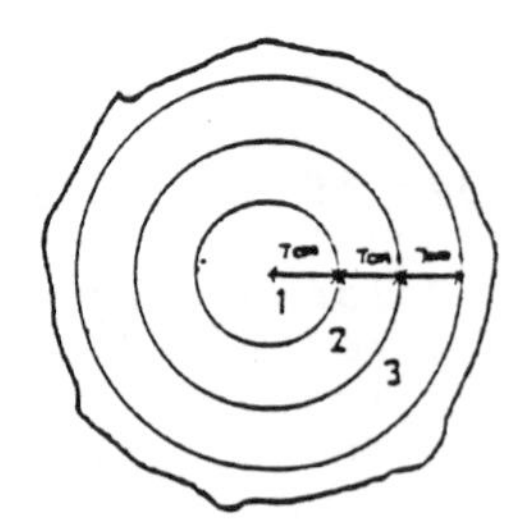

i) Oil palm tree ii) Cross-section of the trunk.

Fig. 2. Sampling of oil palm trunk in Trial 2

In the second trial, four trees of age 35 years were obtained from the Minyak Estate, Selangor. The trees belonging to the Dura variety were push-felled by an excavator and later divided according to Figure 2. The average diameter at breast height and total height of the tree were 44.5 cm and 11.06 m respectively.

2.2 Sugars and starch analyses

Fresh OPT was ground into 200 mesh size and dried in a desiccator over concentrate sulphuric acid for at least 3 days prior to sugars and starch content determination. Sugars content in OPT was determined by HPLC as recommended by Simatupang (1987). A combination of Aminex HPX-87P column with double-distilled water as eluant and a refractive index detector was used to determine monosaccharide or disaccharide. The determination of starch was carried out by the spectrophotometer procedure of Humphrey and Kelly (1961). Two replications were carried out on each sample.

2.3 Hydration test

A Shinko Thermorecorder (Model TR 206) was used to determine the hydration properties of Portland cement on the addition of OPT particles, chemical additives, and pre-treated OPT particles. The mixing ratio of cement : OPT particles : water was at 200g : 5g : 83.5ml as specified by the American Standard (ASTM C186-62, 1982) and Simatupang (1979).

Three replications were made at each condition and six pairs of thermocouple wire Type J and Dewar flask were used to measure the heat evolved from the reaction vessel which was read off from the recorder.

2.4 Processing and pre-treatments of OPT

The OPT billets obtained from trial 2 were split-cut by a Wehrhahn gang-saw into about 50 mm planks before being fed into a Pallmann drum-chipper. The OPT chips were then flaked by a Scotz-Ritz disintegrator. The OPT flakes were divided into 2 portions for pre-soaking and storage in piles for 0 to 14 days. A certain amount of OPT flakes was taken every alternate day to be used as wood aggregate for CBP manufacture.

2.5 Cement-bonded particleboard (CBP) manufacture

A total of 13 series of CBP were manufactured from OPT flakes pre-soaked and stored in piles. CBP produced from fresh OPT flakes was used as control. The OPT flakes taken from pile storage was tested for its moisture content and directly mixed with Portland cement. The OPT flakes pre-soaked in water were slighty dried before being mixed with Portland cement. Five pieces of CBP were manufactured from each series with a dimension of 45 cm x 45 cm x 10 mm and target density 1300 kg/m^3. The wood/cement ratio was at 1 : 2.5 and 2 % (based on cement weight) of aluminium sulphate and sodium silicate were added as an cement accelerator. The boards were pressed up to 25 kg/m^2 specific pressure and clamped together in a hardening chamber set at the temperature of $65 \pm 5^{o}C$. After 24 hours, the boards were declamped and stacked at room temperature until about 28 days prior to testing.

2.6 Board testing

All boards were tested according to the Malaysian Standard specification for Wood Cement Board (MS 934 , 1986). The tests were for bending strength (MOR), internal bond, water absorption and thickness swelling.

3 Results and discussion

3.1 Carbohydrates analysis

The HPLC chromatogram showed that sucrose, glucose and fructose were the main sugars available in the OPT. The distribution of sugars and starch within the OPT taken in Trial 1 is given in Table 1.

In the fresh OPT, sucrose was the major sugar available followed by glucose and fructose (Figure 3). The total amount of sugars determined in this batch of OPT was in the range of 1.80-9.80 % (based on dry weight of wood). Most of these sugars were concentrated at the central zone of the stem. The reason for this is possibly due to the fact that more parenchymatous tissues are found in the central zone, decreasing toward the outer zone. The distribution of sugars along the height of the stem did not show a clear trend.

Starch content analysis showed that the amount of starch remained practically constant up to the height of 6.0 meter and suddenly rose at 7.5 meter height and above. In general, more starch is available at the central zone compared to the outer zone.

Carbohydrates analysis on OPT in Trial 2 is given in Table 2. The results show that the amounts of total sugars and starch in this batch of OPT were in the range of 2.73-13.26 % and 0.44-9.55 % respectively, which were comparable to the amounts of carbohydrates in the OPT of the earlier trial. The total amount of sugars was higher in the central zone of the stem at all height levels of the tree (Figure 4) which is almost identical to the earlier determination. This study indicates that OPT consists of a high amount of carbohydrates which is expected to impede cement hydration.

Table 1. Carbohydrates content analysis in oil palm trunk taken from Serdang, Selangor (Trial 1)

Height from ground (m)	Zone	Average Sugars content (%)			Total Sugars (%)	Average Starch (%)
		Sucrose	Glucose	Fructose		
1.5	A 1	5.18	2.89	0.58	8.65	0.76
	A 2	3.22	2.14	0.52	5.88	0.69
	A 3	1.62	1.07	0.30	2.97	0.69
	A 4	2.63	2.61	0.88	6.12	0.44
3.0	B 1	3.10	2.80	0.69	6.59	0.81
	B 2	2.96	1.79	0.64	5.39	0.64
	B 3	1.47	0.67	0.43	2.57	0.67
	B 4	2.60	1.30	4.41	8.31	2.30
4.5	C 1	6.10	2.13	0.69	8.92	2.30
	C 2	3.54	1.29	0.84	5.67	1.16
	C 3	2.60	0.51	0.51	2.29	1.06
6.0	D 1	3.92	3.47	1.03	8.42	1.52
	D 2	2.62	1.09	0.83	4.54	1.45
	D 3	0.85	0.49	0.46	1.80	1.42
7.5	E 1	6.27	1.86	1.18	9.31	10.67
	E 2	2.02	1.05	1.00	4.05	5.24
	E 3	4.22	0.57	0.55	5.34	1.70
9.0	F 1	6.89	1.54	1.37	9.80	25.51
	F 2	5.02	1.52	1.63	8.17	8.85
	F 3	3.48	0.87	0.93	5.28	5.30

Note : Sugars and starch content were based on dry weight of OPT particles and the average of 3 trees.

Table 2. Carbohydrates content analysis in oil palm trunk taken from Minyak Estate, Selangor (Trial 2)

Height from ground (m)	Zone No.	Average Sugars content (%)			Total Sugars (%)	Average Starch (%)
		Sucrose	Glucose	Fructose		
1.5	A 1	3.91	2.97	1.77	8.65	0.49
	A 2	3.16	1.51	1.42	6.09	0.51
	A 3	1.45	0.67	0.78	2.90	1.76

3.5	B 1	5.21	4.77	1.82	11.80	0.49
	B 2	3.37	2.11	1.54	7.02	0.44
	B 3	1.54	0.69	0.95	2.73	1.87
5.5	C 1	5.02	4.24	1.71	10.97	1.05
	C 2	3.00	2.00	1.67	6.67	1.53
	C 3	2.83	0.70	1.04	4.57	2.14
7.5	D 1	6.85	4.30	2.11	13.26	4.60
	D 2	3.60	1.59	1.87	7.06	9.55
	D 3	2.85	0.90	0.98	4.73	9.07

Note : Sugars and starch content were based on dry weight of OPT particles and the average of 4 trees.

3.2 Effect of OPT flakes on cement hydration

The hydration of cement was completely delayed when 5 g of fresh OPT particles (dry weight) was incorporated in cement-water system. There was hardly any hydration peak even after 45 hours of reaction. This result indicates that OPT has a very strong retardation effect on cement setting. This is probably due to high amount of carbohydrates available in OPT.

Several methods such as pre-soaking OPT particles in cold & hot water and incorporating chemical accelerators in the OPT particles-cement-water system were tried out. All these pre-treatments improved the hydration temperature and reduced the hydration time of Portland cement to certain extents. Soaking the OPT particles in water was most effective to get rid of most of the sugars and other water soluble components and would thus improved the bonding properties between cement and the OPT particles (Figure 5).

3.3 Effect of sugars on hydration of cement

The addition of 0.1-3.0 % (based on cement weight) of three major sugars such as sucrose, glucose and fructose in the cement-water system caused the reduction of hydration temperature and the elongation of hydration time of Portland cement (Figure 6).

The retardation effect of sucrose and glucose was stronger than that of fructose since the addition of 0.2 % of the former sugar completely eliminated the hydration temperature even after 50 hours. This confirms the earlier results of Thomass and Birchall (1983) who discovered that sucrose and raffinose were excellent retarders for cement and glucose, maltose, lactose and cellobiose were moderate retarders.

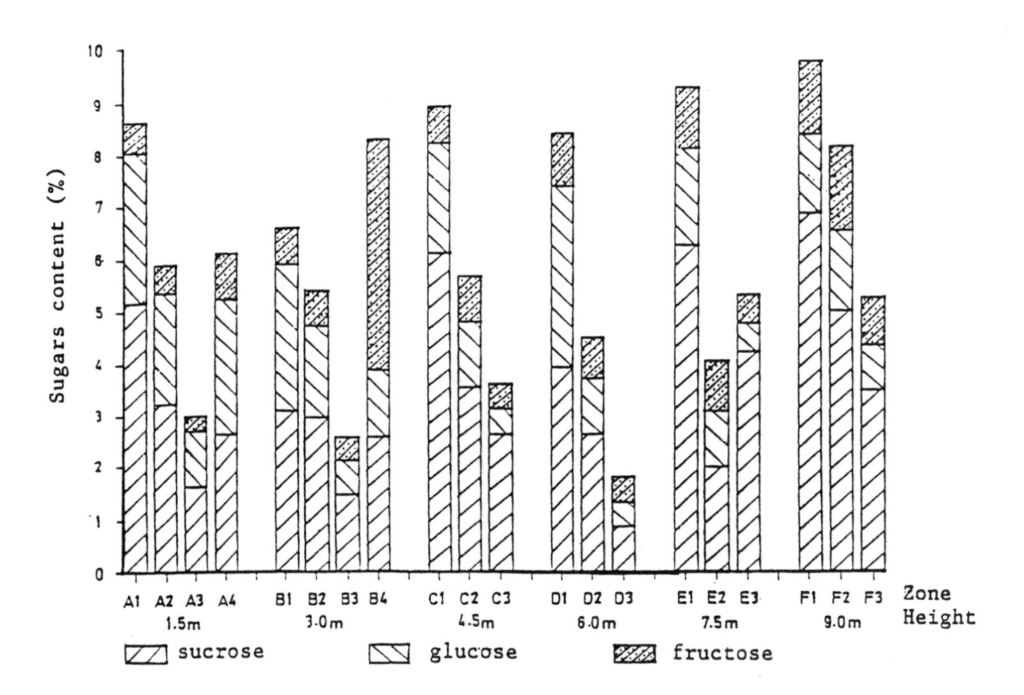

Fig. 3. Sugars content distribution in oil palm trunk in Trial 1.

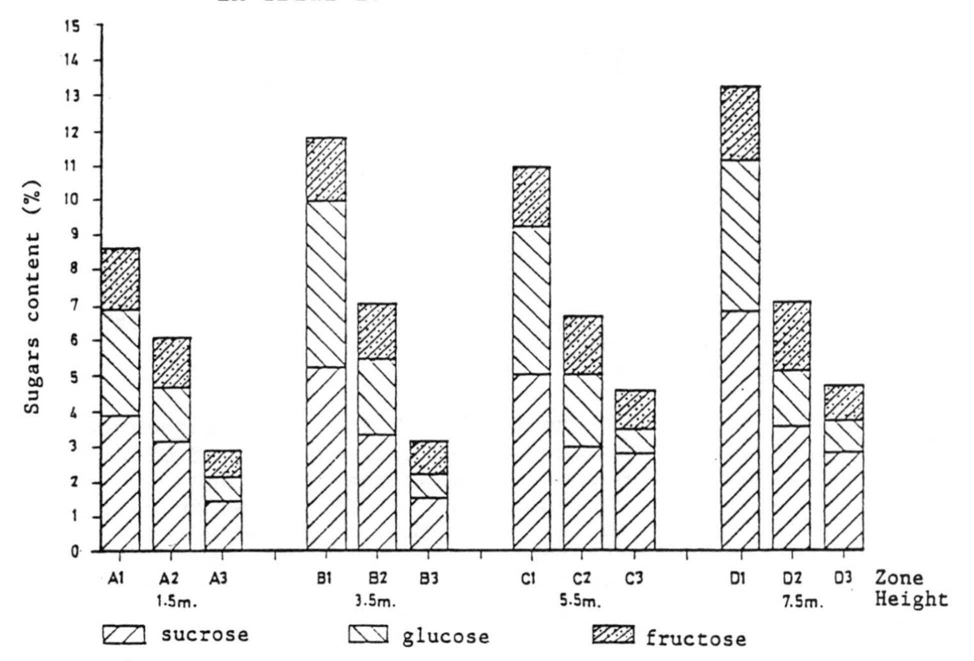

Fig. 4. Sugars content distribution in oil palm trunk in Trial 2

Since fresh OPT contains high amounts of sucrose and glucose, it is therefore impossible to develope a good bonding between fresh OPT flakes and cement in CBP manufacture. In other words, the OPT flakes must undergo several pre-treatments in order to reduce the sugars content before this material can be used as wood aggregate for CBP manufacture.

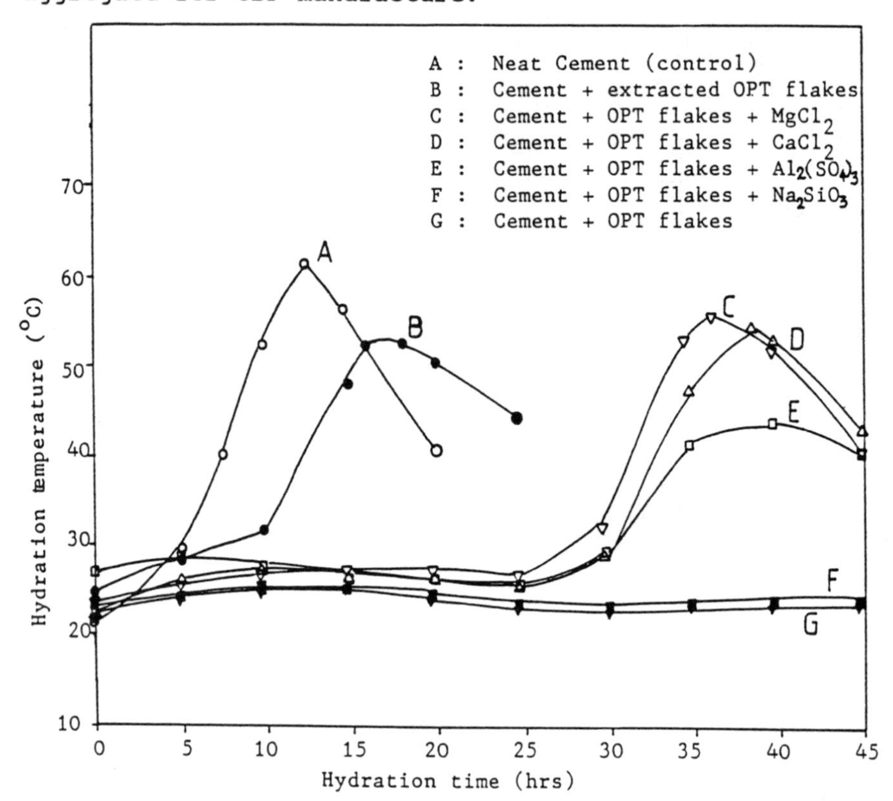

Fig. 5. Hydration properties of Portland cement with the addition of OPT flakes and chemical accelerators.

3.4 Effect of pre-treatments of OPT for CBP manufacture

Several pre-treatments such as natural storage of OPT logs and plank in the open, soaking the OPT planks in cold and hot water, pre-soaking the OPT flakes in cold water and pile storing the OPT flakes were conducted. Storage of roundwood in the open is the most common and economical method practised in the CBP manufacturer. In some cases, the wood is converted to wood chips or flakes and stored in pile form in order to reduce the

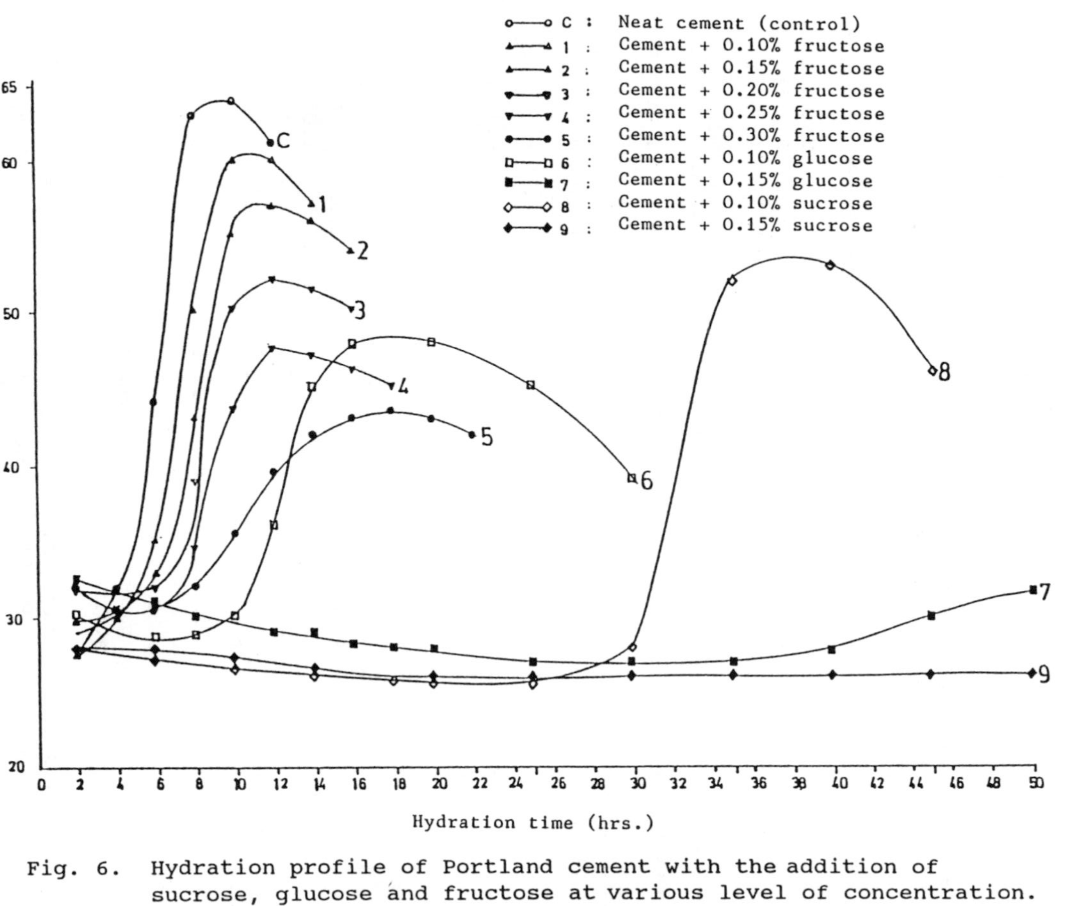

Fig. 6. Hydration profile of Portland cement with the addition of sucrose, glucose and fructose at various level of concentration.

carbohydrates content at a faster rate. However, this study showed that natural storage is unsuitable for OPT because this material contains very high amounts of carbohydrates and moisture and is therefore very susceptible to fungi and decay agents. Decayed OPT cannot contribute to the strength properties of CBP.

Soaking of OPT flakes in water was found more suitable because most of the sugars dissolved in water could be removed and the OPT could be stored longer. The strength properties of CBP manufactured from OPT flakes pe-soaked in water and from pile-stored OPT are given in Table 3.

Table 3. Effect of pre-soaked and pile storage of OPT flakes on the mechanical and physical properties of CBP

Type of treatment	Duration of treat-ment (day)	Bending strength (MPa) n=16	Tensile strength (MPa) n=32	Water Absorption (% w/w) n=16	Thickness Swelling (% w/w) n=16
CONTROL	0	3.29(0.60)	0.18(0.06)	27.63(2.27)	1.04(0.56)
PRE-SOAKED IN WATER	3	8.95(1.00)	0.69(0.20)	10.83(1.23)	0.74(0.42)
	5	8.20(0.98)	0.73(0.18)	12.92(1.65)	0.75(0.37)
	7	8.89(1.16)	0.97(0.24)	7.82(1.31)	0.42(0.38)
	10	7.98(1.20)	0.76(0.21)	9.57(2.29)	0.64(0.34)
	12	7.72(0.74)	0.68(0.21)	9.48(1.83)	0.29(0.17)
	14	9.79(1.08)	0.72(0.17)	7.88(1.67)	0.91(0.37)
STORED IN PILING FORM	2	5.81(0.59)	0.35(0.09)	17.21(3.66)	0.62(0.81)
	5	5.27(0.82)	0.28(0.07)	20.92(4.18)	0.44(0.29)
	7	4.89(0.43)	0.24(0.06)	22.27(4.46)	0.55(0.28)
	9	4.72(0.68)	0.24(0.05)	21.71(3.69)	0.58(0.25)
	11	5.13(0.58)	0.23(0.06)	24.59(3.50)	1.04(0.42)
	14	3.87(0.64)	0.18(0.04)	26.06(1.88)	1.48(0.70)
Malaysian Standard		9.00	0.50	NA	2.00

Notes : n = no. of test specimen
( ) figure in parentheses is standarad deviation
NA data not available

A rapid improvement of the strength properties of CBP manufactured from OPT flakes soaked in water for 3 days was observed compared to those of the control. With longer pre-soaking of the OPT flakes in water the mechanical properties of CBP remained practically constant and almost satisfied the Malaysian Standard specifications for Wood Cement Board (MS 934, 1986).

The strength properties of CBP manufactured from OPT flakes stored in pile form showed a slight improvement at the beginning before dropping toward the 14 days of storage. Within this period, the colour of the OPT flakes had turned black indicating that black stain had already attacked this material. This study revealed that pre-soaked OPT flakes in water is necessary for CBP manufacture.

4 Conclusion

This study revealed that OPT consist of a considerable amount of sugars and starch which inhibit cement setting. However, soaking of the OPT flakes in water improved the strength properties of CBP significantly compared to other methods of pre-treatment.

5 References

ASTM (1982) American Standard Specification for Portland Cement, **ASTM C186-82**

Anonymous, (1991) **Statistics on Commodities.** Ministry of Primary Industries Malaysia, 12-52.

Humphrey, F.R. and Kelly, J. (1961) A method for the determination of starch in wood, **Anal. Chim. Acta.**, 24, 66-70.

Khozirah, S., Khoo, K.C. and Abd Razak, M.A. (1991) Oil palm stem utilisation. **FRIM Research Pamphlet.**, 107.

Mohamad, H., Abd Halim, H. and Tarmizi, A.M. (1986) Availability and potential utilisation of oil palm trunks and fronds up to the year 2000, **PORIM Occasional Paper.**, 20.

MS 934 (1986) Malaysian Standard Specification for Wood Cement Board. **Standard & Industrial Research Institute of Malaysia**, MS 934 : 1986.

Simatupang, M.H., (1987) **Consultant Report on Wood Cement Board** submitted to Forest Research Institute Malaysia (FRIM)

Thomas, N.L. and Birchall, J.D., (1983) The retarding action of sugars on cement hydration, **Cement and Concrete Research**, 13, 830-842.

91 DEVELOPMENT OF SISAL CEMENT COMPOSITES AS SUBSTITUTE FOR ASBESTOS CEMENT COMPONENTS FOR ROOFING

M. SAXENA, R. K. MORCHHALE, A. N. MESHRAM and A. C. KHAZANCHI
Regional Research Laboratory (CSIR), Bhopal, India

Abstract
The present paper explores the possibilities of utilisation of renewable vegetative fibre (sisal) in place of non-renewable and carcinogenic mineral fibre (asbestos) in cement matrix for roofing in housing. This investigation deals with the development of sisal cement composites by optimisation of various parameters such as water cement ratio, fibre length, casting pressure, curing time, shape and gauge of steel mesh by conducting the flexural strength test. The comparative studies of sisal fibre cement sheet and asbestos sheet are carried out by measuring their water absorption, breaking load, weathering (indoor and outdoor), water impermeability, acid resistance and frost cracking. Finally, it may be concluded that the breaking load of the sheet at a span of one meter comes to 70-80% of AC sheet. But due to its repairability, less water absorption, impervious nature, easy processing, safety factor of human life and economy, sisal fibre cement sheet has a great scope for substitution of health hazardous and carcinogenic asbestos cement sheets.
Keywords: Sisal Cement Composites, Vegetable fibres, Roofing , Asbestos, Housing,Weathering.

1 Introduction

The acute shortage of building materials demand, the Research & Development work for development of new building materials using locally available resources. India requires 8 million tons per annum fibre cement product for housing (Karade SR et al,1990). Presently asbestos cement sheet contain asbestos fibre, which apart from being carcinogenic is also an imported materials [(Anon,1987),(Harding J,1987),(Anon),(Anon,1981),(Anon,1982),(Zyrer PS,1985)]. The deficiency of high quality asbestos and due to its health hazardous nature , the search for asbestos replacement has been going on for many years. The main surge of activity dates from about the mid 1970s

Fibre Reinforced Cement and Concrete. Edited by R. N. Swamy. © 1992 RILEM.
Published by E & FN Spon, 2-6 Boundary Row, London SE1 8HN. ISBN 0 419 18130 X.

(Crabtree,1986). A lot of research work has been conducted into the possible replacement of asbestos with various synthetic fibres such as glass,kevlar,carbon, nylon and polypropylene [(Alford NM,1985),(Hanrant DG,1978),(Schupack M,1986),(Swamy RN et al,1984)]. However a little published material is available in the scientific literature concerning the use of natural fibres as reinforcement (Gram HE,1988).An extensive survey of the literature regarding natural fibre in cement based products has been made by Gram [(Gram HE,1983),(Gram HE et al,1984)].Keeping in view, the abundant availability of renewable resources, natural fibre is taken into consideration for the investigation (Khazanchi AC et al,1989).The main drawback in the use of natural fibre is the durability in composite (Agopyan V et al,1989).

Out of the various natural fibres sisal fibre was chosen as a substitute of asbestos fibre, due to its ultimate tensile strength, yield, reasonable price and easy cultivation in any climate (Navin Chand et al).Main aim, in the present study, is to develop a "sisal fibre-cement composite" using renewable local resource for roofing element, to substitute health hazardous asbestos fibre.

2 Materials

2.1 Sisal fibre

In the present study mechanically extracted (by Raspador Machine) sisal fibre was procured from Khadi & Village Industries Commission, Bilaspur.The composition of sisal leaf & fibre and their availability in different countries, is shown in Tables 1 & 2 respectively.

TABLE 1

COMPOSITION OF A SISAL LEAF & FIBRE

S.NO.	CONSTITUENT DETAILS	% CONTENT
	A. LEAF	
1.	Dry fibre	4.00
2.	Sisal pulp & waste	11.00
3.	Moisture	85.00
	B. FIBRE	
1.	Cellulose	78.00
2.	Lignin	8.00
3.	Wax	2.00
4.	Carbohydrates & pectins	10.00
5.	Ash	2.00

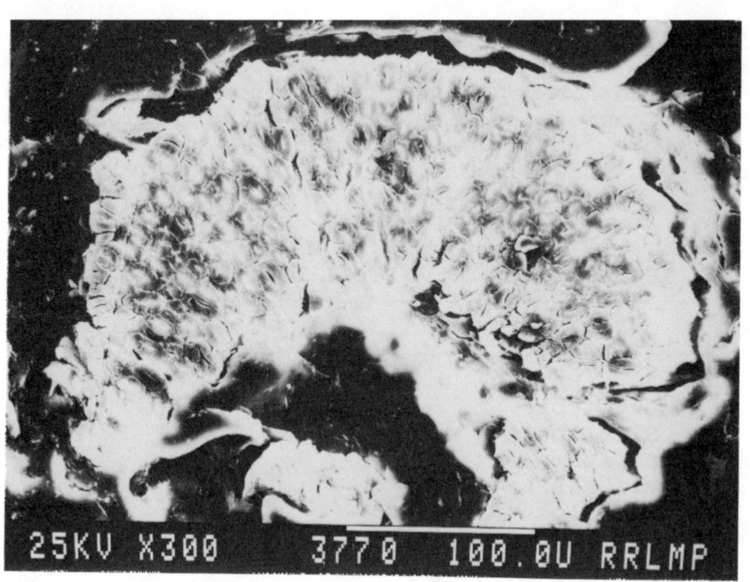

FIG.1 SISAL PLANT & CROSS SECTION OF SISAL FIBRE.

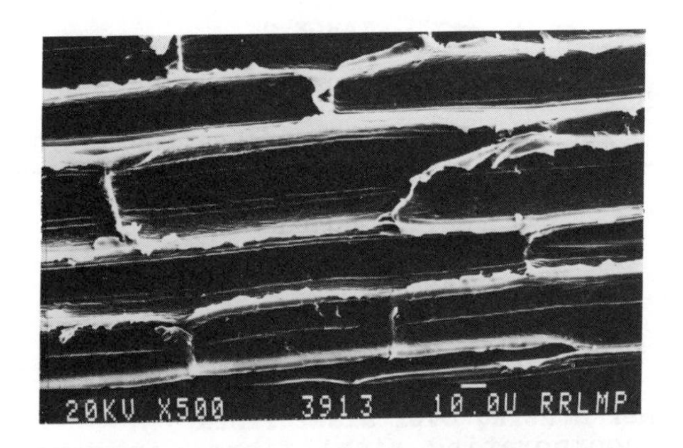

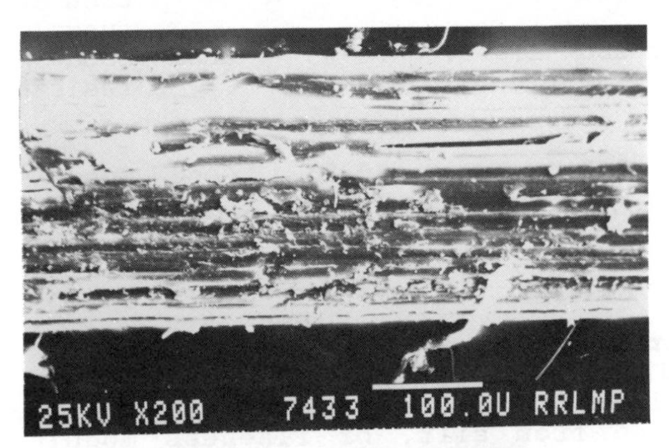

FIG. 2 SISAL FIBRE.
A. UNTREATED B. 5% ACRYLIC COATED
C. 5% POLYSTERENE COATED

TABLE 2

PRODUCTION OF SISAL FIBRE

S.NO.	COUNTRY	PRODUCTION IN 1000 TONS
1.	Tanzania	197
2.	Kenya & Uganda	46
3.	Mozamique & Angola	81
4.	Brazil	100
5.	Indonesia	27
6.	India	21

2.2 Polymer coating over sisal fibre

Before incorporating sisal fibre in cement matrix, fibre was characterized for mechanical strength and bonding with cement matrix. To make the fibre alkali resistant and also to improve its adhesion with cement, the thermoset and thermoplastic coatings were tried keeping in view the workability of mixture.

TABLE 3

PROPERTIES OF POLYMER COATED AND UNCOATED SISAL FIBRE

Sisal Fibre	UTS MPa	Strain %	Moisture absorption %	Bond with cement	Weathering	Alkali effect
Uncoated	250-300	3.5-4.5	55-60	Poor	Deterioration and fungal	Deterioration
Acrylic 5% coated	250-350	2.5-3.0	30-35	Good	Good-resistance	Good-resistance
Polystyrene 5% coated	250-325	2.7-3.0	30-35	Good	Good-resistance	Good-resistance

2.3 Other Ingradients

The "Ordinary Portland Cement" confirming to IS: 269 - 1976,(Vikram brand) and Locally available "Narmada Sand" free from silt, of fineness modulus 2.45 and particles passing through 1 mm aperture of IS sieve was used for making composite. The mild steel mesh of hexagonal (20 & 28 gauge) and square (20 & 25 gauge) opening and liquid water proofing cum plasticizer under the trade name of "Conplast Prolopin 421 IC" confirming to IS specification 2645 were used. (0.3% by weight of cement)

3 Method

To get the optimum desirable engineering properties and workability of composites, the flat samples of 15 cm x 15 cm size were cast in hydraulic press of 100 tons capacity, after considering the following variables : Water cement ratio, Fibre length, Casting pressure, Curing period, Shape and gauge of steel mesh, by measuring the flexural strength of the samples using Instron testing machine (INSTRON 1185, England).

3.1 Water cement ratio

The water cement ratio was varied from 0.4 - 0.5 to obtain the optimum flexural strength.

3.2 Fibre length

The samples were prepared using sisal fibre of varying length in the range of 5 - 15 mm.

3.3 Casting pressure

The sample casting was done at pressure varying from 0 - 220 Kg/cm^2

3.4 Curing period

The effect of curing on the cast samples was observed by immersing them in water for a period of 7,14 and 28 days.

3.5 Shape and gauge of steel mesh

To study the effect of shape of opening and gauge of steel mesh, the samples were prepared using hexagonal and square mesh.

After finalising the said variable and to correlate the engineering properties of asbestos-cement and sisal fibre-cement sheet the following tests were conducted as per IS-5913,1970 on cast sheet of size 2.00 x 1.05 mt.

(a) Flexural strength
(b) Water absorption
(c) Weathering
(d) Water impermeability
(e) Chemical resistance
(f) Frost cracking

4 Process of casting sisal fiber cement sheet

The process of production of fibre cement sheets did not require sophisticated machinery,simple hydraulic press was used for making sheets.In the process,sisal fibres were cut into small length and treated with polymeric solution for use in cement matrix.These treated fibres were weighed and mixed with cement, sand,water proofing cum plasticizer and water thoroughly.The mixed was spread over a flat sheet and then shifted over to a corrugated mould and pressed in a hydraulic press.Then the sheet was demoulded and kept for a period of 28 days water curing.

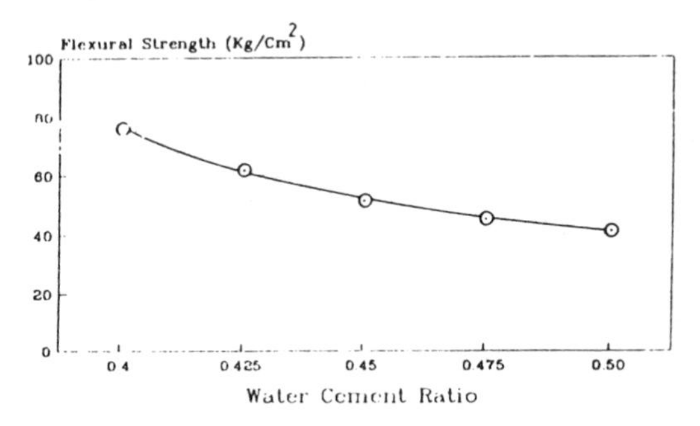

FIG. 3 EFFECT OF WATER CEMENT RATIO ON FLEXURAL STRENTH.

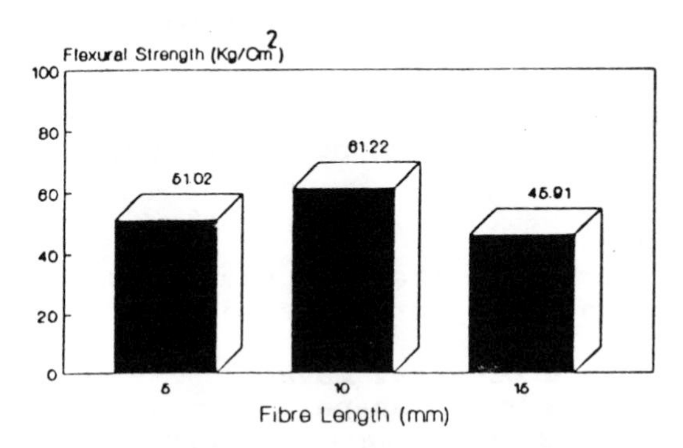

FIG. 4 EFFECT OF FIBRE LENGTH ON FLEXURAL STRENGTH.

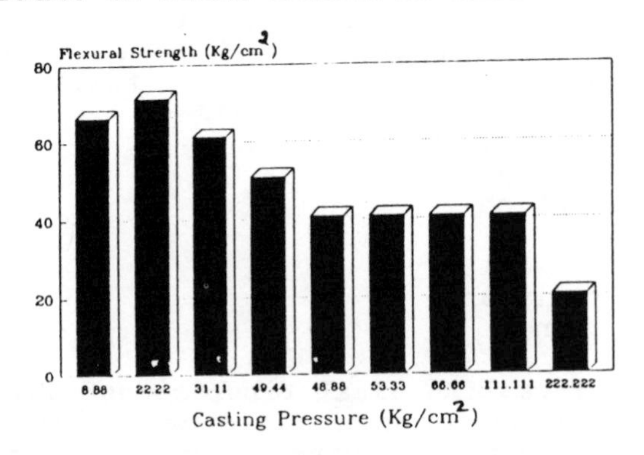

FIG. 5 EFFECT OF CASTING PRESSURE ON FLEXURAL STRENGTH.

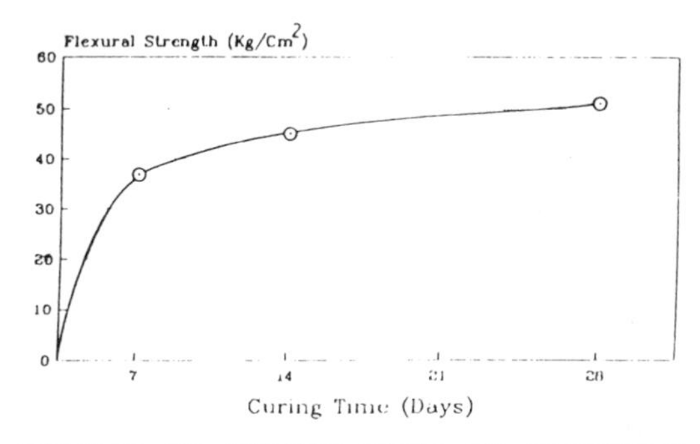

FIG. 6 EFFECT OF CURING TIME ON FLEXURAL STRENGTH.

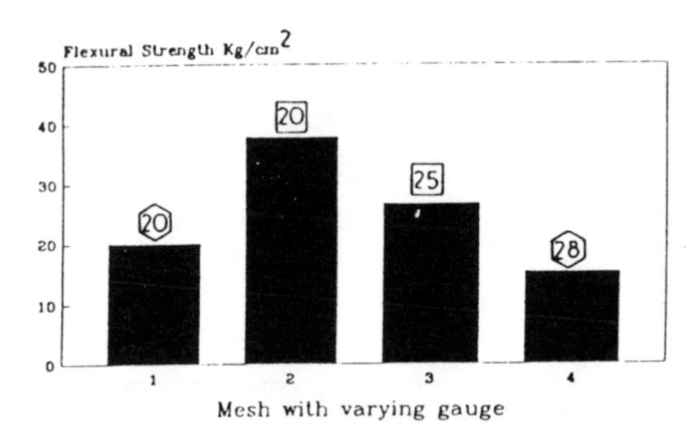

FIG. 7 EFFECT OF WIRE MESH DESIGN & GAUGE ON FLEXURAL STRENGTH.

5 Result & Discussion

5.1 Polymer coating

To improve the adhesion of fibre with matrix, earlier (Sanadi AR et al) limited work was done using 5% alkali solution for dewaxing of fibres. But the use of alkali deteriorates the fibres,(Gram H,1988) at the same time lump formation takes place during mortar processing. Therefore in the present study the dewaxing of sisal fibre was done using organic solvent to avoid the above mentioned problems. The results of polymer coatings show that, the thermosets were not suitable to coat the fibre due to its sticky nature and its post curing, resulting in the loss of flexibility. Among the thermoplastics, acrylic and polystyrene could be used due to their easy dissolution in organic solvent. In the present study 3 - 8% solid content coatings were applied, but only 5% solid content coating has given desirable results. The higher percentage of coating was ruled out due to lump formation during coatings, as a result coated fibre was not suitable for processing of cement mortar. Application of polymeric coating has resulted in an improvement of ultimate tensile strength, moisture absorption, alkali resistance and bonding with cement mortar as given in Table 3. The scanning electron microstructure of fibre uncoated, coated and immersed in boiled extract of cement solution (pH - 12.3, 90 days duration) are shown in Fig. 2. It has been observed that even after a 90 days exposure in alkali there was no change in the coating.

5.2 Water cement ratio

As shown in Fig. 3, the maxmium strength is observed at 0.4 water cement ratio and it gradually decreases, with increase in water cement ratio. The 0.4 water cement ratio was optimized for the development of composite, which is also fulfilling the water requirement of cement for consistancy, complete hydration of cement and workability,becouse from the mix design concepts,the 26% water is required for consistency in case of Ordinary Portland Cement and 14% water is required for workability of mortar during processing time.It was the main idea for keeping water cement ratio minimum 0.4 for the investigation.

5.3 Fibre length

As observed from Fig. 4 the maximum strength was at 10 mm fibre length & at the same time the obtained mix was workable. The workability of mortar increases at 5 mm fibre length, but the possibility of debonding of fibre also increases, as a result it will not serve its purpose to arrest the microcracks of cement during shrinkage. In case of 15 mm fibre length, the mix was not workable at 0.4 water cement ratio. Finally 10 mm fibre length was used for the present investigation.The main objective of feeding fibre in the matrix to arrest the micro cracks during change of temperature.

5.4 Casting pressure

Fig. 5 shows that the maximum flexural strength is at 22 Kg/cm^2 pressure. If the pressure was changed on either sides, the decrease in strength was noticed. The following mechanisms act during pressure application :

i. As the pressure was increased successively between 0 - 22 Kg/cm^2 an increase in strength was observed. It was due to the compaction of particles taking place without squeezing out the water from the green matrix, resulting in the availablity of the required water for complete hydration of cement.

ii. The further increase in pressure from 22 - 220 Kg/cm^2 has resulted in decrease in strength. This is,due to the compressive stresses developed on the particle in addition to the squeezing out of water from the green composite. These compressive stresses when released after demoulding attribute to the volume change, swelling and porosity development, due to the disturbance generated by stress after completing the final setting time.Therefore, the casting in the present study was done at 22 Kg/cm^2 to get the optimum flexural strength.

5.5 Curing period

The observed result in Fig. 6 shows that there was approximately 30% increase in flexural strength of 28 days cured samples as compared to 7 days cured samples. The present study follows the same pattern of curing as in concrete technology.

5.6 Shape and gauge of steel mesh

From the bar diagram in Fig. 7, it is observed that the square mesh of 20 gauge shows higher strength then the 20 gauge hexagonal mesh. But due to workability problem during casting of corrugated sheets with square mesh in the present study 20 gauge hexagonal mesh was selected. The incorporation of steel mesh serve the purpose of reinforcement.

The indoor weathering test shows that there was no effect on the sample after 1500 hours exposure visually. Also the outdoor test of the sheet was not showing any change physically after one and half year exposure. The acid resistance and frost cracking properties were also well within the acceptable limit. Although the breaking load of the sisal fibre cement product is around 60-80% of A.C. product, but due to its repairability, less water absorption, impervious nature, easy processing, safety factor of human life and economy,sisal fibre cement sheet has a great scope for the substitution of health hazardous and carcinogenic asbestos cement sheet.

6 Conclusions

6.1 Technology based on utilization of a local renewable resource (sisal) as a substitute of carcinogenic and imported asbestos fibre.

6.2 Hexagonal mesh is used keeping in view, the possibility of repair after the impact, which is not possible in the case of A.C.Sheets.

6.3 The safety factor of human life is higher from the production as well as utility point.

6.4 Product is 25% cheaper than the A.C. Sheet.

6.5 Initial investment for establishment of production plant is less as compared to that of A.C. Sheets for the same capacity of production.

6.6 Due to simplicity of present technology for casting, it may act as a source of income generation to rural mass.

7 Acknowledgement

Authors are gratefull to Prof. T.C.Rao,Director RRL Bhopal for incouragement and guidance. Thanks are also due to Ministry of Urban Developement,New Delhi for financial support.

7 References

Agopyan V, Cincotto MA, Derolle A (1989)
Durability of vegetative fibre reinforced materials, Proc. 11th Triennial Congress CIB 89, Paris

Alford NM (1985)
Production of fibre reinforced cementitious composition,US Patent,4, 528, 238.

Anon (1987)
Asbestos use steady despite bans in West,chem.Ind. 16th Feb, 101.

Anon
Aborigines and Asbestos, New Scientist, 29 March,1020.

Anon (1981)
Double standards, Asbestos in India,New scientist, 26 Feb, 522 - 523

Anon (1982)
Turkish cancer deaths - It was asbestos after all, New Scientist, 6 May, 353.

Crabtree J.D.(1986)
Past, present and future developments in industrial fibre cement technology, Keynote speech, 2nd edn. Rilem Symp. Sheffield, 15 July.

Gram HE (1988)
"Durability of natural fibre in concrete" Swamy RN "Natural fibre reinforced cement and concrete" Vol 5,p.143.

Gram HE (1983)
Durability of natural fibres in concrete CBI Research for 1.83, Swidish Cement & concrete researc Institute, Stockholm, pp 255.

Gram HE (1984)
Persson H & Skarendahl A, Natural fibre concrete,Sarecc Report R2, Stockhome, pp 139.

Gram H (1988)
Durability of natural fibres in concrete in Natural fibre reinforce cement & concrete Vol.5, Ed, Swamy RN, Blackie.

Hanrant DJ (1978)
Fibre cement and fibre concrete, John Wiley & Sons, New York.

Harding J (1987)
Superior pollution - Asbestos in the Great lakes,New Ecolgist 2,Mar/April,p 47-50.

Karade S.R.et.al (1990)
"Developement of New Building Material using waste materials for construction Industry" Procs. of 5th National Convention of Civil Engineers,Jan Calcutta.

Khazanchi AC, Saxena M, Morchhale RK (1989)
Surfaces modification of Plant fibres for development of fibre/cement/clay/polymer composite for housing, Int.COnf. for Housing, 1989, UK.

Navin Chand, Verma KK, Saxena M, Khazanchi AC,
Material science of Plant based material of MP for housing, Material science & Technology in the future of MP.

Sanadi AR, Prassad SV, & Rohatgi PK
"Natural fibre and agro waste as fillers and reinforcement in polymer composite".Material Science & Technology in the future of M.P.

Schupack M.(1986)
Three Dimensionally reinforced fabric concrete, US Patent 4, 617, 219.

Swamy RN et al. (1984)
New reinforced concrete, Survey Divn.press.

92 FIBRE–MATRIX INTERACTIONS IN AUTOCLAVED CELLULOSE CEMENT COMPOSITES

B. de LHONEUX
Redco n.v., Kapelle-op-den-Bos, Belgium
T. AVELLA
Unité des Eaux et Forêts, Université Catholique de Louvain, Louvain-la-Neuve, Belgium

Abstract
Two aspects of cellulose-cement interactions are described in this study.
On one hand, the mechanical properties of the composites are observed to be dependent on fibre properties such as tensile strength and degree of polymerisation. As the fibres are partially degraded by the severe curing conditions, the stability of the fibre during autoclaving is of great importance.
On the other hand, some fibres which present a high solubility in alkaline solutions influence the structure of the surrounding matrix, as reflected by the XRD analysis.
Keywords: Cellulose, Cement, Autoclaving, Fibre-Cement, Zero-Span Tensile Strength, Degree of Polymerisation, Porosity, Bending Strength, X-Ray Diffractometry, Free Lime, Kraft Pulp, Softwood, Hardwood, Cotton, Abaca.

1 Introduction

Several new fibre-cement and fibre-concrete building products have been developed in recent years, based on organic fibres (cellulose, polypropylene, polyvinyl-alcohol, polyacrylonitrile, ...) or mineral fibres (glass, carbon, steel, ...). The fibre-cement industry has already produced significant quantities of several of these materials in different parts of the world, following extensive research and development work (Harper 1982, Studinka 1983, 1989). The use of cellulose in this field has been privileged due to its price being lower than that of synthetic fibres and to its relative stability in autoclaving conditions.

The cellulose fibres are obtained from a large variety of plant materials (wood, annual plants), using different

Fibre Reinforced Cement and Concrete. Edited by R. N. Swamy. © 1992 RILEM.
Published by E & FN Spon, 2-6 Boundary Row, London SE1 8HN. ISBN 0 419 18130 X.

production techniques (chemical, mechanical, thermo-mechanical, chemi(thermo)mechanical pulping). The resulting fibres have a wide spectrum of chemical, physical and mechanical properties.

For fibre-cement applications, the use of different fibres has been described:

- softwood and hardwood chemical (kraft) pulps (Andonian et al. 1979, Coutts and Ridikas 1982, Rockwool 1979, Ametex et al. 1988, Cape Universal Claddings 1983, Procter and Gamble 1989).
- softwood mechanical, thermomechanical (TMP) and chemithermomechanical (CTMP) pulps (Coutts and Michell 1983, Coutts 1986).
- wastepapers (Coutts 1989)
- Manilla hemp (abaca) (Coutts and Warden 1987)
- Cotton (Cape Universal Claddings 1981, GAF Corporation 1977)
- New Zealand flax (Coutts 1983)

Many more fibres exist which are still to be evaluated in this technology.

Most of these fibres are produced by paper pulp manufacturers. While these fibres are well characterized for this application, little is known about the relations existing between the properties of the fibres and those of the fibre-cement composites.

The aim of the present study is to analyse these relations, in the case of autoclaved fibre-cement composites. Autoclaving is a steam curing technique allowing a reaction between a finely ground quartz and cement. It occurs at steam temperatures up to 200°C during several hours. Industrial products manufactured using this technique comprize flat and corrugated sheets for roofing and cladding, fire insulating boards, ...

In a first step, emphasis is placed on fibre tensile strength, as the prime function of the fibres is to reinforce the matrix. In a second step, one analyses how the resistance of the fibres to autoclaving conditions influences their reinforcing potential. Finally, the influence of certain fibres on matrix synthesis is briefly described.

2 Materials and Methods

2.1 Fibres

- Softwood kraft pulps: Pinus pinaster (Portugal), Pinus patula (Swaziland), Pinus sylvestris (Scandinavia), Pinus radiata (New Zealand).
- Hardwood kraft pulps: Betula spp (Finland),

Eucalyptus globulus (Portugal, Spain), Eucalyptus grandis (Brazil).
- Manilla hemp (abaca) pulps: Musa textilis (Philippines).
- Cotton linters pulps: Gossypium spp (UK).
- Softwood (Pinus spp) and hardwood (Eucalyptus globulus) chemimechanical pulps, from France and Spain respectively.
- Flax (Linum usitatissimum): raw (unpulped) fibre (Belgium).
- Oxidized pulps: a hardwood unbleached kraft pulp is oxidized using potassium dichromate ("K" pulps) or a sequence of potassium dichromate and acid chlorite ("KH" pulps). Different levels of oxidation are obtained by varying the concentration of dichromate. In a decreasing order of intensity, the following pulps were produced: K1, K2, K3, K4 and K1H, K2H, K3H and K4H. The experimental details and the fibres properties are described elsewhere (de Lhoneux et al. 1991a). In the first series, mostly carbonyl functions are generated, while they are reoxidized in carboxyl functions in the second case.

2.2 Fibre test methods

- Zero span tensile strength (ZST): five laboratory paper sheets of approximately 60 g/m^2 are manufactured using a "Rapid Köthen" laboratory sheet machine. Eight 90 x 20 mm paper strips are cut from each sheet and conditioned at 65 % RH and 20°C for 24 hrs. They are tested using a Pulmac zero-span tensile tester. Results are expressed in breaking length (km), i.e., the length of fibre leading to rupture under its own weight.
- Degree of polymerisation: viscosimetric determination using an Ubbelohde viscosimeter and cadoxene or sodium ferritartrate as solvents. In the first case, intrinsic viscosity is calculated according to Martin's formula and DP according to Brown's formula (Brown 1967). In the second case, the intrinsic viscosity is calculated using Schultz-Blaschke equation (Jayme, El-Kodsi 1968).
- Wet classification: Bauer-Mc-Nett classifier fitted with Tyler mesh screens (4M, 14M, 35M, 100M and 200M).
- Solubility in NaOH 1 % solution: Tappi 212 m - 44.
- Stability in autoclaving conditions: fibres are autoclaved in lime saturated water following a method described elsewhere (de Lhoneux et al. 1991a). Zero-span tensile strength or degree of polymerisation is analysed before and after autoclaving.

2.3 Composites manufacturing and testing

Laboratory composites are manufactured using a filter press and the following composition: cellulose fibres (6 %), matrix (94 % - lime/silica/cement 2/3/1). The cellulose fibres are dispersed using a laboratory desintegrator and are not refined.
The composites are cured for 48 hr in 95 % RH at ambient temperature and autoclaved at 7 bars for 20 hours afterwards.

The following tests are performed:

- 3 points bending test: span = 146 mm; speed = 2 mm/min; water saturated composites.
- Density, porosity: gravimetry.
- Free lime: method of Schläpfer and Bukowski (Seidel 1964).
- X-ray diffractometry: Cu Kα, λ = 1.5418 Å.

3 Results

3.1 Influence of fibre properties on composite bending strength

In the present study, the fibre tensile strength is measured using the zero-span tensile test. Although some contribution of interfibre bonding within the sheets to the measured strength cannot be avoided in the dry state, it gives a fair approximation of the fibre intrinsic tensile strength. It is indeed extremely difficult to determine the tensile strength of individual wood fibres due to their small dimensions (1-4 mm length, 10-40 µm diameter). Additionally the heterogeneous nature of the wood necessitates the testing of numerous fibres.

Fibre strength is influenced by several factors: the angle of the microfibrils with the fibre axis (Page et al. 1972), the presence of defects or weak points along the fibre (Furukuwa et al. 1974), the pulping type and yield (Leopold and Thorpe 1968), the degree of polymerisation (Conrad et al. 1951), refining (Mc Intosh and Uhrig 1968), etc....

In the present series, a wide range of fibre tensile strengths and of composite bending strengths is observed. By plotting one against the other (Fig. 1), a roughly linear relation is observed. At the lower end of the curve, one finds a chemimechanical pulp, a cotton pulp and some softwood kraft pulps. In the middle range are softwood and hardwood kraft pulps, while highest strengths are reached using some kraft and Manilla hemp pulps.

The lower strength of the cotton linters pulp most likely results from a high angle of the microfibrils axis with the main fibre axis. In the case of the chemimechanical pulps, the fibres have a rather low cellulose content (about 50 %) and, thus, a rather low strength. In the case of the kraft pulps, the differences of strength are to be sought in the differences of fibre structure (type of wood, growth conditions, ...) and of pulping conditions. Manilla hemp fibres have a low angle between microfibrils and fibre axes, and thus, a high strength.

This observed relation indicates that the mode of failure of the composites is, at least partially, fibre rupture. This, in turn, implies a strong fibre-matrix bond allowing these short fibres (1-6 mm) to be brought to rupture. This has indeed been observed by electron microscopy (Coutts and Kightly 1982; Bentur and Akers 1989; de Lhoneux et al. 1991b).

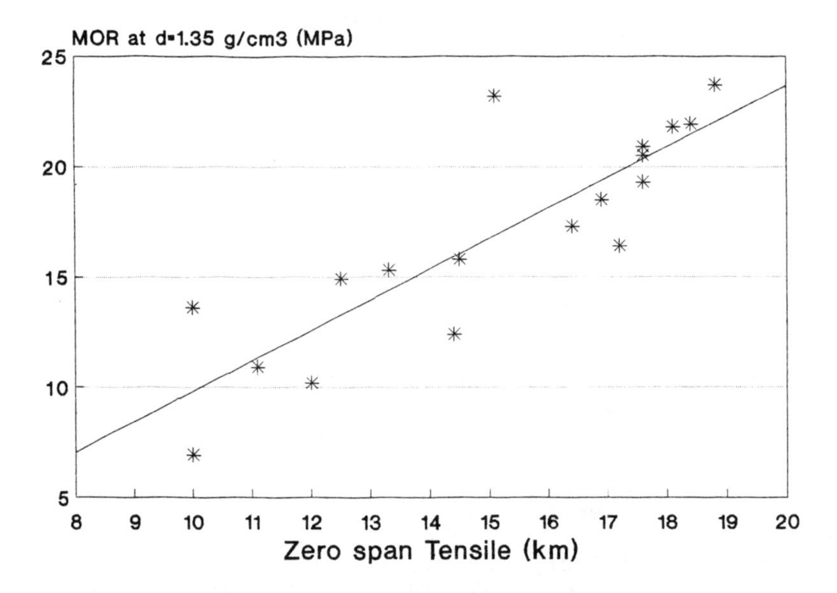

Fig 1. Relation between the bending strength of autoclaved laboratory composites and the zero-span tensile strength of different pulps.

The importance of fibre strength is further evidenced by a similar relation observed using three length fractions of a hardwood kraft pulp fractionated using a Bauer Mc Nett classifier (Fig. 2). The longest fibres have the highest tensile strength and provide the highest reinforcement.

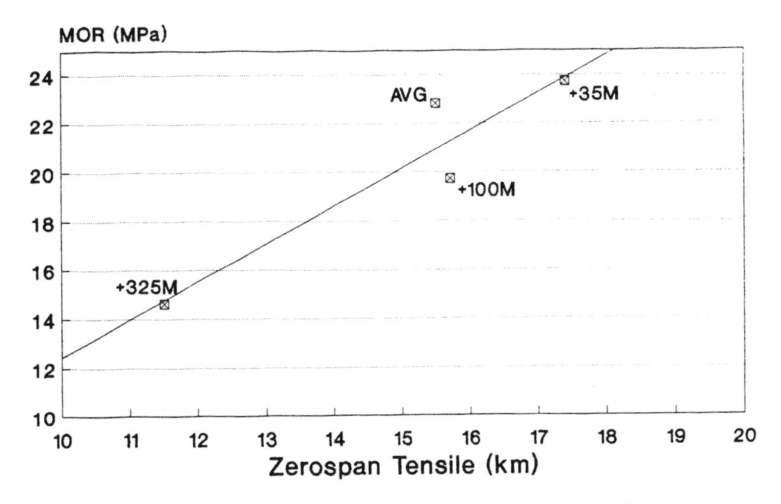

Fig 2. Relation between the modulus of rupture of the composites and the zero-span tensile strength of three length fractions of a hardwood kraft pulp. Avg = unfractionated pulp.

However, fibre tensile strength is not the only parameter influencing the modulus of rupture of the composites. In the following example, two pulps of the same origin having a similar zero-span tensile strength but a different degree of polymerisation do have a much different reinforcing potential:

	Degree of polymerisation (DPw)	1 % NaOH solubility (%)	ZST (km)	MOR (MPa)
pulp A	3190	1.9	15.8	20.1
pulp B	1370	5.2	15.4	15.2

This indicates that the chemical structure of the fibres largely influences their ability to reinforce composites. In particular, chemically degraded pulps loose their reinforcing potential.

3.2 Fibre stability during autoclaving and reinforcing potential

In an another part of the study (de Lhoneux et al. 1991a), the influence of autoclaving on fibre properties was investigated.
Decreases of strength and of degree of polymerisation were observed when fibres were autoclaved in alkaline

solutions, the importance of which being dependent on fibre type and chemical structure. In the present work, composites were manufactured using the same fibres which were described in that study.

In a first series of four pulps, the relations observed between the modulus of rupture of the composites and the zero-span tensile strength of the fibres before and after autoclaving are different (Fig. 3). A better fit is observed in the second case than in the first, indicating that the stability of the fibres during autoclaving influences their reinforcing potential.

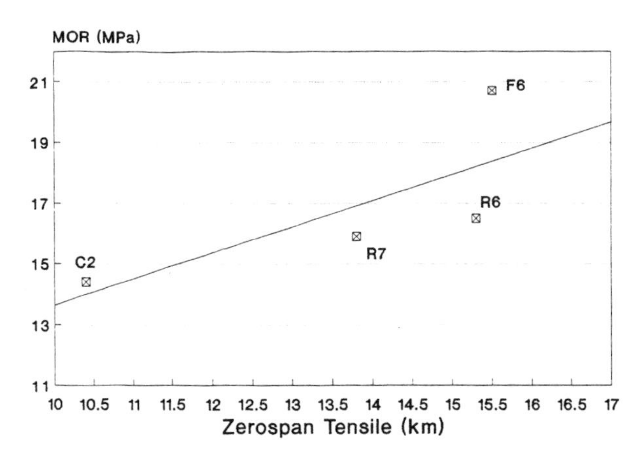

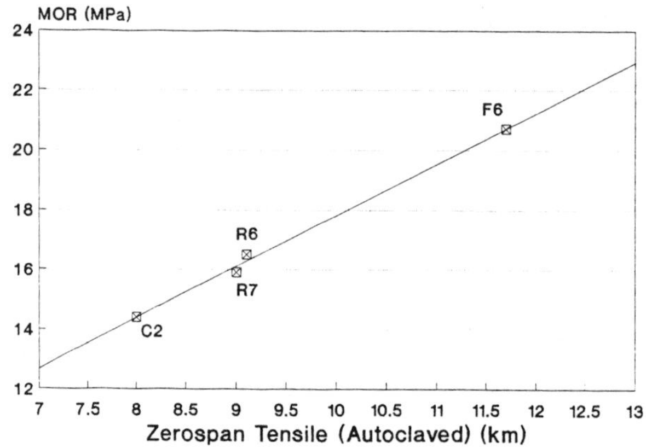

Fig. 3. Relation between the modulus of rupture of autoclaved laboratory composites and the zero-span tensile strength of the fibres before (a) and after (b) autoclaving. F6 = hardwood kraft pulp; R6, R7 = softwood kraft pulps; C2 = cotton pulp

Similarly, the degradation of differently oxidized fibres was studied, by measuring their degree of polymerisation before and after autoclaving. In the present series, the strength of the composites manufactured using the oxidized fibres shows a direct relation with their residual degree of polymerisation (after autoclaving) (Fig. 4).

These relations indicate that the chemical structure of the pulps is of major significance for their reinforcing potential. In order to effectively reinforce, the fibres should not only have a high tensile strength but should keep it during autoclaving. A high degree of polymerisation is important for this purpose in order that, after autoclaving, enough strength is maintained for effective reinforcement. The positive relation between these two parameters is non linear, i.e., the tensile strength is less sensitive to variation of the D.P. in the highest range of this parameter (Wakeham 1955).

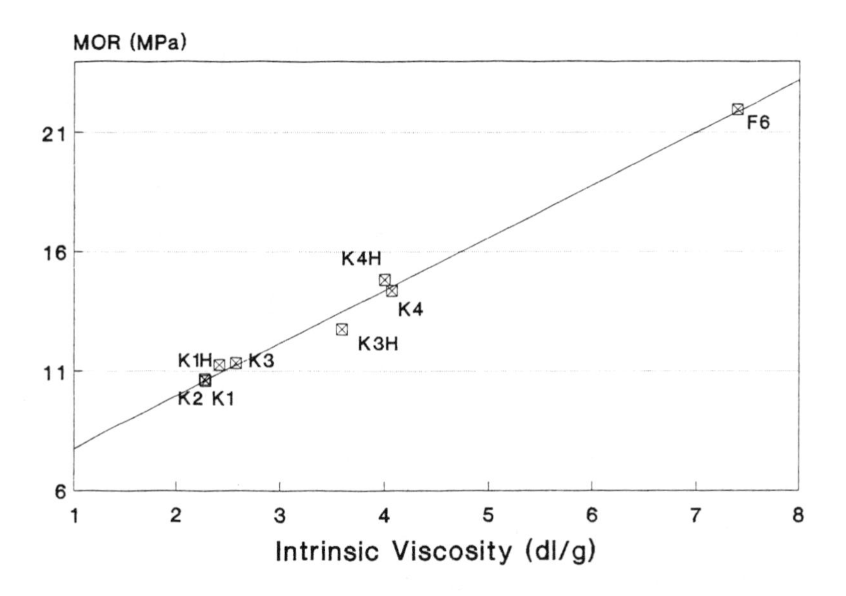

Fig. 4. Relation between the modulus of rupture of the composites and the residual intrinsic viscosity (after autoclaving) of differently oxidized fibres. (For definition of symbols of fibres, see experimentals).

3.3 Influence of fibres on matrix structure

The fibres which were oxidized present a high solubility in alkaline solutions. The question arizes whether this can influence the curing of the matrix in the composites. Such interferences are known in the case of cement cured at ambiant temperature. In particular, soluble sugars are known to retard cement hydration.

From the evolution of the strength properties and of the free lime content of the composites in function of the time of autoclaving (Table 1), interferences with the matrix curing can indeed be observed.

Table 1. Mechanical properties (bending test), density and free lime content of composites reinforced by oxidized and reference cellulose fibres at different autoclaving times. LOP = limit of proportionality, MOR = Modulus of rupture, Ipl = work of fracture, E = Young's modulus, CaO_l = free lime. F6 = unoxidized (reference) pulp; Kl and KlH: see experimentals.

Sample	Autoclaving time (h)	LOP (MPa)	MOR (MPa)	Ipl (Jm-2)	E (GPa)	Density d (g/cm^3)	CaO_l %
	2.5	5.99	10.60	1579	5.8	1.36	4.90
F6	5	7.15	12.47	1653	6.2	1.33	1.75
	10	9.68	16.92	1347	8.2	1.33	0.35
	20	13.50	21.96	899	8.7	1.33	0.30
	2.5	3.23	4.02	31	3.8	1.33	6.50
Kl	5	4.54	5.75	48	4.4	1.29	4.35
	10	6.68	7.76	70	6.6	1.28	0.35
	20	9.26	10.65	85	8.5	1.36	0.35
	2.5	3.72	4.35	33	4.7	1.34	8.00
KlH	5	5.53	6.37	38	6.0	1.34	5.90
	10	6.41	7.35	47	6.6	1.34	0.70
	20	9.81	11.24	81	9.2	1.35	0.35

The decrease of the free lime content is slower in presence of the highly soluble fibres. Simultaneously, the increase of the Young's modulus and of the LOP is slower. Obviously, the MOR and the work of fracture of the composites are much lower than for the reference (unoxidized) fibre, due to the lower performances and the degradation of the fibres in the autoclave. After 20 hr of autoclaving, the free lime content of the composites is, however, at a similar low level.

From the X-ray diffractograms, the slower consumption of lime can also be seen (Fig. 5a). Additionally, after 20 h, the 2.97, 5.42 and 11.3 Å peaks of Tobermorite/CSH-I are less important in comparison with the 4.26 Å peak of quartz (Fig. 5b). The content of well cristallized Tobermorite thus appears to be less important in this case.
The retardation of the lime-silica reaction is probably to be related to the complexing action of sugars on lime preventing its reaction with silica.

In another study, Petitpas (1987) has shown that more unreacted quartz remained in the composites when low quality celluloses were used. In the present case, this parameter was not specifically analysed.

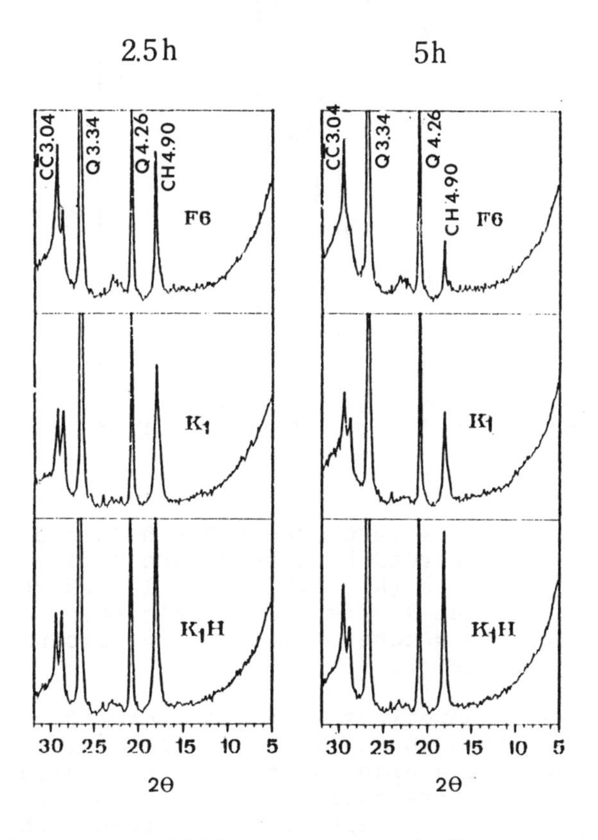

Fig. 5a. X-ray diffractograms (2θ = 5 - 32°) of composites reinforced by a reference (F6) and by oxidized (K1, K1H) fibres after 2.5 and 5 h of autoclaving, at 7 bars. C$\bar{C}$ = calcite, CH = lime, Q = Quartz.

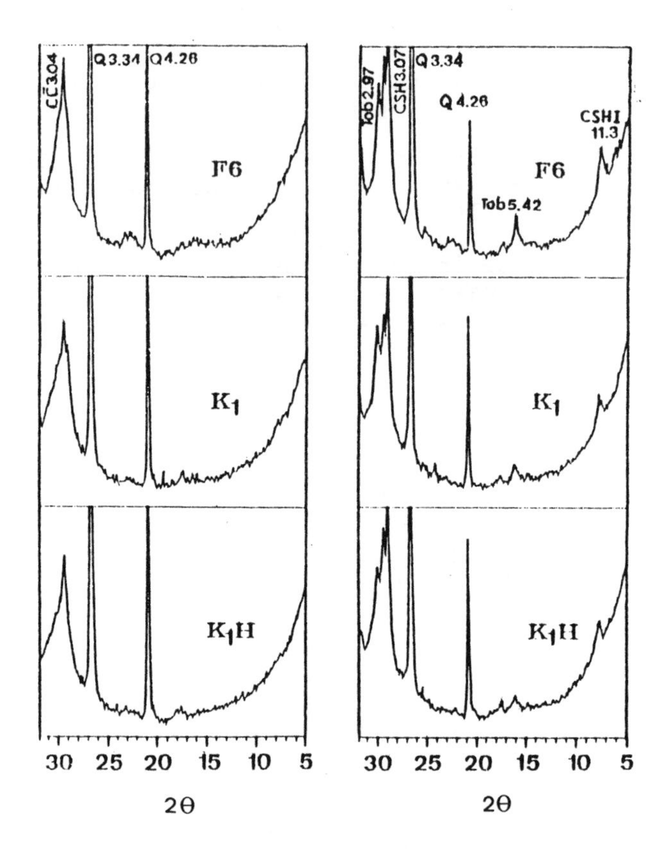

Fig. 5b. X-ray diffractogram (2θ = 5-32°) of composites reinforced by a reference (F6) and by oxidized fibres (Kl, KlH) after 10 and 20 hr of autoclaving at 7 bars. CC̄ = calcite, Q = quartz, Tob = tobermorite, CSH = calcium silicate hydrate.

In a similar way, composites were manufactured using a mechanical softwood pulp and raw (unpulped) flax fibres. These fibres present a high solubility in alkaline solutions. X-ray diffractograms of the composites (Fig. 6) show a reduced 11.3 Å peak, in comparison with a composite made using a kraft pulp.
Although further work is necessary in order to determine the exact mechanisms, these results indicate that interferences are to be expected in the case of highly soluble fibres.

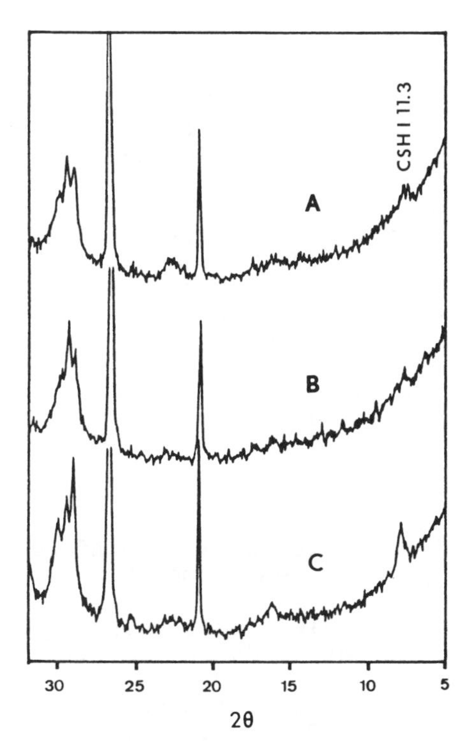

Fig. 6. X-Ray diffractograms (2 θ = 5 - 32°) of composites reinforced by an unpulped flax fibre (A), a mechanical softwood pulp (B) and a hardwood kraft pulp (C).

4 Conclusions

The properties of autoclaved cellulose-cement composites are largely influenced by fibre properties. A relation has been found between fibre strength and composite modulus of rupture. Further data indicate that the stability of the fibres in the autoclave is of prime importance for the strength of the composites. Finally, interferences with the matrix synthesis are observed in the case of the highly soluble fibres, i.e. degraded kraft pulps or non chemical pulps.

5 Aknowledgement

The financial support of the Belgian Institute for the Encouragement of Research in Industry and Agriculture (IRSIA-IWONL) is gratefully aknowledged.

6 References

Ametex A.G., Financière Eternit S.A., Eternit Werke Ludwig Hatschek and Redco N.V. (1988) Faserbewehrter Formkörper und Verfahren zu seiner Herstellung. Eur. Pat. 0287962.

Andonian R., May Y.W., and Cotterell B. 1979. Strength and fracture properties of cellulose fibre reinforced cement composites. Int. J. Cem. Comp. 1, 3, 151-158.

Bentur A. and Akers S.A.S. (1989) The microstructure and ageing of cellulose fibre reinforced autoclaved cement composites. Int. J. Cem. Comp. Lightw. Concr. 11, 2, 111-115.

Brown W. (1967) The Cellulose Solvent Cadoxen. Svensk Papperstion. 70, 15, 458-461.

Cape Universal Claddings Ltd. (1981) Building boards and sheets. Eur. Pat. 0033796 A2.

Cape Universal Claddings Ltd. (1983) Boards and sheets. Eur. Pat. 0068741.

Coutts R.S.P. (1983) Flax fibres as a reinforcement in cement mortars. Int. J. Cem. Comp. Lightw. Concr. 5, 4, 257-262.

Coutts R.S.P. (1986) High yield wood pulps as reinforcement for cement products. Appita 39, 1, 31-35.

Coutts R.S.P. (1989) Waste paper fibres in cement products. Int. J. Cem. Comp. Lightw. Concr. 11, 3, 143-147.

Coutts R.S.P. and Kightly P. (1982) Microstructure of autoclaved refined wood fibre cement mortars. J. Mater. Sci. 17, 1801-1806.

Coutts R.S.P. and Michell A.J. (1983) Wood pulp fiber-cement composites. J. Appl. Polym. Sci. Appl. Polym. Symp. 37, 829-844.

Coutts R.S.P. and Ridikas V. (1982) Refined wood fibre-cement products. Appita 35,5, 395-400.

Coutts R.S.P. and Warden P.G. (1987) Air-cured abaca reinforced cement composites. Int. J. Cem. Comp. Lightw. Concr. 9, 2, 69-73.

Conrad C.M., Tripp V.W., Mares T. (1951) Thermal degradation in tire cords. Part 1. Effects on strength, elongation and degree of polymerisation. Text. Res. J. 21, 726-739.

de Lhoneux B., Avella T., and Garves K. (1991a) Influence of autoclaving on chemical pulp fibre properties for fibre-cement applications. Holzforschung 45, 1, 55-60.

de Lhoneux B., Baes E. and Avella T. (1991b) Ultrastructural aspects of fibre-matrix bond in cellulose cement composites. In **Interfaces in New Materials** (Eds P. Grange and B. Delmon) Elsevier applied science, London, New York, pp. 129-138.

Furukawa I., Saiki H., and Harada H. (1974) A micro tensile-testing method for single wood fiber in a scanning electron microscope. J. Electr. Microsc. 23, 2, 89-97.
Gaf Corp. (1977) Cotton-cement articles. US Pat. 4,040,851.
Harper S. (1982) Developing asbestos-free calcium silicate building boards. Composites 13, 2, 123-128.
Jayme G. and El-Kodsi G. (1968) Uber ein neues Verfahren zur Herstellung praktisch Sauerunempfindlicher Cellulose-EWNN (MOD) - Lösungen für Viscositätsmessungen und damit erhaltene Ergebnisse. Papier 22, 3, 120-124.
Leopold B. and Thorpe J.L. (1968) Effect of pulping on strength properties of dry and wet pulp fibers from Norway Spruce. Tappi 51, 7, 304-308.
McIntosh D.C. and Uhrig L.O. (1968) Effect of refining on load-elongation characteristics of loblolly pine holocellulose and unbleached kraft fibers. Tappi 51, 6, 268-273.
Page D.H., El-Hosseiny F., Winkler K. and Bain A. (1972) The mechanical properties of single wood - pulp fibres. 1. A new approach. Pulp Pap. Mag. Can. 73, 8, 72-77.
Petitpas J.-M. (1987) Contribution à l'étude des materiaux silico-calcaires renforcés avec des fibres de cellulose. Mémoire C.N.A.M., Paris
Procter and Gamble Cy (1989) Cellulose fiber-reinforced structure. Eur. Pat. 0297857.
Rockwool International A/S (1979) Produit cimentaire renforcé par des fibres et son procédé de préparation. Belg. Pat. 879.275.
Seidel K. (1964) **Handbuch für das Zement Labor.** Bauverlag Wiesbaden, Berlin.
Studinka J. (1983) Faserzement ohne Asbest. Neue Zürcher Zeitung 286, 61 (7.12.1983).
Studinka J. (1989) Asbestos substitution in the fibre-cement industry. Int. J. Cem. Comp. Lightw. Concr. 11, 2, 73-78.
Wakeham H. (1955) Mechanical properties of cellulose and its derivatives, in **Cellulose and Cellulose Derivatives** (Eds E. Otts and H.M. Spurlin), Intersci. Pub. New York.

93 LONG-TERM DURABILITY AND MOISTURE SENSITIVITY OF CELLULOSE FIBER REINFORCED CEMENT COMPOSITES

P. SOROUSHIAN and S. MARIKUNTE
Michigan State University, East Lansing, MI, USA

Abstract
Cellulose fiber reinforced cement composites provide the highest performance-to-cost ratio among fibrous cement composites considered for the replacement of asbestos. Cellulose fibers being fairly strong and stiff are particularly suited for the reinforcement of thin-sheet cement products. The research reported herein is concerned with the effects of repeated wetting-drying, freezing-thawing, and also variable moisture conditions on the performance characteristics of cellulose fiber reinforced cement composites. The effectiveness of pozzolans in improving the moisture sensitivity of cellulose fiber reinforced cement composites were also assessed.

An experimental study was undertaken in order to investigate the performance of cellulose fiber reinforced cement composites containing 1% and 2% mass fractions of kraft pulp. The results generated in this investigation were indicative of the adverse effects of wetting-drying cycles on toughness characteristics of composites. Microstructural studies confirmed that the precipitation of cement hydration products within cellulose fiber cores (petrification) and at interface zones is the key deterioration mechanism in composites. The flexural strength of composites, however, was not adversely influenced by the repeated cycles of wetting and drying. Cellulose fiber reinforced cement composites performed desirably under repeated freeze-thaw cycles.

High moisture contents were observed to reduce the flexural strength and increase the flexural toughness of cellulose fiber reinforced cement composites. Microstructural investigations showed that adverse effects of moisture on fiber-to-matrix bond are the main factors contributing to the moisture-sensitivity of composites. Pozzolanic admixtures were found effective in reducing the moisture-sensitivity of composites.
Key Words: Cement, Composite Materials, Deflection, Durability, Flexure, Pulp, Moisture-Sensitivity, Reinforcement, Strength, Toughness, Cellulose Fibers.

Fibre Reinforced Cement and Concrete. Edited by R. N. Swamy. © 1992 RILEM.
Published by E & FN Spon, 2-6 Boundary Row, London SE1 8HN. ISBN 0 419 18130 X.

1 Introduction

Cellulose fibers have been used for many years as an additive in the conventional asbestos cement industry. Some asbestos cement replacement products also utilize cellulose fibers. In these cases, cellulose fibers contribute mainly processing benefits rather than reinforcement. Recent studies have shown that the reinforcement ability of cellulose fibers is quite good relative to other potential asbestos replacements (Vinson et. al. 1990, see Figure 1). Cellulose fibers are effective in increasing the fracture energy of cementitious matrices, and also in enhancing the tensile and flexural strength, toughness, and impact resistance of the material (Coutts 1988, CSIRO 1981, Coutts 1987, Soroushian et. al. 1990). It is important to ensure that the improvements in material properties of cement achieved through cellulose fiber reinforcement would be retained over long time periods in actual exposure conditions. There are, however, concerns regarding the long-term durability and moisture resistance of cellulose fiber-cement composite (Coutts 1984, Sharman et. al. 1986, Akers et. al. 1989). Much of the available test data on durability of cellulose fiber reinforced cement deal with the use of natural fibers (e.g., sisal and coir) in cement-based materials (Gram 1983). The main thrust of this research is to assess the long-term durability and moisture sensitivity of cement composites reinforced with chemically processed cellulose fibers (kraft pulps).

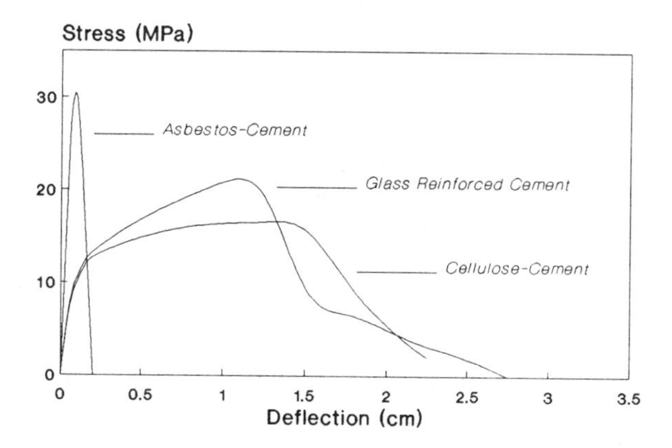

Figure 1. Flexural Performance of Saturated Cellulose Fiber-Cement composite Compared with Glass Fiber Reinforced Cement and Asbestos Cement.

Cellulose fiber reinforced cement composites have found commercial applications for the production of thin-sheet products, including sidings, soffits, curtain walls, and backer boards.

2 Background

A key concern with the use of cellulose fiber reinforced cement composites relates to the long-term durability and moisture sensitivity of fibrous cement products, particularly under severe exposure conditions. Much of the available test data on durability of cellulose fiber-cement composites deal with the use of natural fibers (e.g., sisal and coir). Accelerated wetting-drying (rain-heat) test results on natural fiber reinforced concrete cause considerable drop in flexural strength and toughness with ageing (Gram 1983). This can be attributed mainly to the attack on the lignin in natural fibers by the alkaline pore water of concrete. Pore water reacts with the lignin and hemicellulose existing in the middle lamellae of cellulose fibers, thus weakening the link between the individual fiber cells in the natural fibers (Gram 1983, Figure 2).

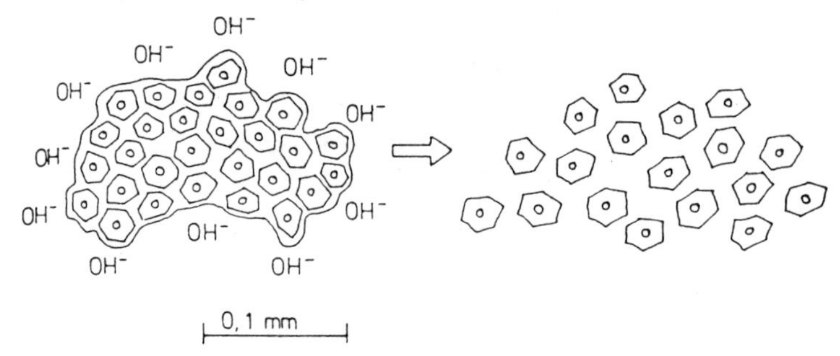

Figure 2. Schematic Sketch of the Decomposition of Sisal Fibers in the Alkaline Pore water of Concrete.

Cellulose fibers obtained by chemical pulping processes are the dominant types used for the production of thin-sheet cement products in developed countries. These cellulose fibers contain negligible amounts of lignin and are thus expected to better withstand alkaline attack in cement. There are, however, other ageing mechanisms which may lead to the embrittlement of cellulose fiber reinforced cement composites. In particular, in the presence of moisture, the gradual filling of the fiber cell cores with hydration products and the densification of matrix in the vicinity of fibers may encourage brittle breakage of fibers under stress, leading to the embrittlement of cellulose fiber reinforced cement composites (Bentur et. al. 1989). This phenomenon, in the case of chemical pulps with low lignin contents, can lead to increased flexural strength and modulus of elasticity of cellulose fiber-cement composites, while reducing their toughness.

Wetting of concrete may lead to losses in compressive strength as a result of the dilation of cement gel by adsorbed water and also breaking of Si-O-Si bonds which consequently reduce cohesion of the solid particles. Conversely, when upon drying the wedge-action of water ceases, an apparent increase in strength of the specimen is recorded. Resoaking of oven-dried specimens in water reduces their strength to the value of continuously wet-cured specimens, provided that they have hydrated to the same degree. The variation in strength due to drying is thus a reversible phenomenon (Neville 1963, Mindess 1981).

Cellulose fiber reinforced cement composites when compared with plain concrete, are highly sensitive to moisture variations. Considerable differences in flexural strength and flexural toughness values are observed when the specimens are tested at different moisture contents (Vinson et. al. 1990, Coutts 1984). There is a general tendency for flexural strength to decrease and flexural toughness to increase with increasing moisture content in cellulose fiber reinforced cement composites. This has been attributed to moisture effects on the fiber-to-matrix bond strength and also on the properties of cellulose fibers.

3 Experimental Program

3.1 Mix proportions and sample preparation

The cellulose fibers used in this investigation were Southern Softwood Kraft pulp (Procter and Gamble Cellulose) with an average length of 3.0 mm. The Canadian Standard Freeness (CSF), a measure of the level of beating of cellulose fibers (Tappi Standard), was 700 for these fibers.

The fiber mass fractions and matrix mixture proportions used in this study are introduced in Table 1 (the mixtures incorporating pozzolans were used only in moisture-sensitivity tests). The water and superplasticizer contents were varied in order to achieve reasonable fresh mix workability characteristics represented by a flow (ASTM C-230) of 60±5% at 1 minute after mixing.

Cellulose fiber reinforced composites were manufactured in this study using a regular mortar mixer. The mixing procedure adopted was as follows: (1) Add cement, sand

Table 1. Mixture Proportions and Fresh Mix Flow

Mix	Binder	Water/Binder	Superplasticizer/ Binder	Set Accelerator/ Binder	Flow (%)
Plain	Cement I	0.30	-	0.02	58±1
1% Kraft Pulp	Cement I	0.35	0.01	0.02	58±2
2% Kraft Pulp	Cement I	0.43	0.02	0.02	48±1
2% Kraft Pulp	Cement : Silica Fume 85 : 15	0.43	0.03	0.02	55±2
2% Kraft Pulp	Cement : Fly Ash 70 : 30	0.35	0.02	0.02	55±2

and 70% of water and mix at low speed (140 RPM) for about 1 minute or until a uniform mixture is achieved; (2) Turn the mixer to medium speed and gradually add the fibers and the remainder of water and superplasticizer into the mixture as the mixer is running at medium speed (285 RPM) over a period of 2-5 minutes depending on the fiber content, taking care that no fiber balls are formed; (3) Add set accelerator and mix for 1 minute at medium speed; and (4) Stop the mixer and wait for 1 minute, and then finalize the process by mixing at high speed (450 RPM) for 2 minutes. It should be noted that commercial cellulose fiber reinforced cement composites incorporate relatively high fiber contents and are produced by the slurry-dewatering techniques. Composites with relatively low fiber contents were used in this investigation for fundamental studies on the mechanisms of ageing and moisture effects in cellulose fiber reinforced cement composites.

3.2 Test methods

The fresh fibrous cement mixtures were tested for: (1) flow (ASTM C-230); and (2) air content (ASTM C-185).

From each mix, molded specimens were manufactured for wetting-drying, freeze-thaw and flexure tests in the hardened state. Panel specimens (280 mm x 400 mm x 10 mm) were manufactured for wetting-drying tests. The flexural specimens were prisms with 38.1 mm square cross sections and total length of 152.4 mm and the freeze-thaw specimens were prisms with 76.2 mm x 101.6 mm cross section and length 406.4 mm. The broken flexural specimens were used later for the specific gravity, water absorption and hardened material air content tests. All the specimens were compacted by external vibration and were kept inside their molds underneath wet burlap covered with a plastic sheet for 24 hours. They were then demolded and moist cured for 5 days before being air cured in a regular laboratory environment of about 40 to 60% RH and 19 to 25 deg. C temperature until the testing age of 28 days.

An accelerated wetting-drying test procedure was developed in order to study the ageing behavior of cellulose fibers under the environmental effects stimulating the alkaline pore water attack on cellulose fibers (e.g., conditions involving repeated exposure to rain and sun shine). For this purpose an accelerated wetting-drying test chamber (climate box) was developed. The 10 mm thick panel specimens are subjected in the climate box to moisture by spraying for 1/2 hour until the capillary pores are filled with pore water; the panels are then heated to reach 82 deg. C (180 deg. F) and the temperature is maintained at this level for a sufficiently long period (5 1/2 hours) to dry out the capillary pore system. Under these conditions the fibers come in contact with the alkaline pore water of cement during the moistening phase, the decomposition products which are formed as a result of the reaction between the fiber components and the pore water are transported away from the fiber during the drying phase.

Six panel specimens were subjected to this accelerated ageing test. After 0, 12, 24, 30, 60 and 120 cycles a panel specimen was taken out and minimum of ten flexural specimens (38.1 mm x 152.4 mm x 10 mm) were cut out using a diamond saw. Flexural tests were then performed according to the Japanese specification JCI-SF (Japanese Concrete Institute 1984).

In order to assess the resistance of cellulose fiber reinforced cement composites to freezing and thawing the specimens were subjected to repeated cycles of freezing and thawing in water (ASTM C-666). An average of two specimens for each mix were subjected to a maximum of 300 cycles and tested for fundamental transverse frequency over a span of 304.8 mm at a regular interval of 20 cycles. The nominal freezing and thawing cycle consists of alternately lowering the temperature of the specimens from 4.4 deg. C to -17.8 deg. C in 3 3/4 hour and raising it from -17.8 deg. C to 4.4 deg. C in 1 1/4 hour. The total cycle consisted of 5 hours in which 25% of the time was used for thawing.

A series of standard conditions for testing were established in order to investigate the effects of moisture content. Samples were conditioned in the following environments: (a) in the laboratory with relative humidity of 40 to 60% and 19 to 25 deg. C; (b) in an oven at 116 deg. C for 24 hours and then cooled in the laboratory; and (c) in water for 48 hours prior to testing, and the excess water removed with a cloth.

An average of 10 specimens with a specific mix design were manufactured for tests at each moisture condition. The specimens were selected equally from two different batches in order to accurately measure variability of the results. Five test specimens for each moisture condition was obtained from each batch. A batch represented a block in the experimental design. Thus the between batch variability could be estimated in the analysis of variance of the test data.

The void content, specific gravity and water absorption of the hardened materials were also assessed using the broken flexural specimens (ASTM C-642).

3.3 Flexural test setup

The flexural tests were performed according to the Japanese specification JCI-SF. The Japanese method of measuring flexural deflections (Japanese Concrete Institute 1984, Figure 3) is particularly effective in reducing errors associated with the rigid body move-

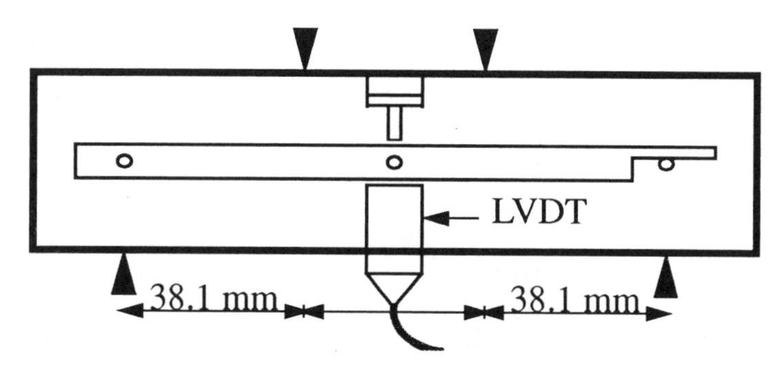

Figure 3. Japanese Standard Flexural Test Set Up.

ments of specimens and local deformations at the supports and loading points.

An important consideration in flexural tests performed on fiber reinforced cement composites is the measurement of energy absorption capacity, defined as the area underneath the load-deflection curve. The Japanese fixtures which monitor flexural deflections during the test give accurate results for energy absorption calculations. According to the Japanese specification JCI-SF 4, Flexural toughness is defined as the area underneath the flexural load-deflection curve up to a deflection equal to the span length divided by 150 which is 0.76 mm in the flexural tests of this investigation.

4 Test Results And Discussion

This section presents the generated test data in two parts. First the results regarding the effects of cellulose fibers on the water requirement, setting time, unit weight, air content, permeable void content, specific gravity and water absorption capacity of cement based materials are briefly reviewed. Then the observed effects of wetting-drying, freezing and thawing and moisture condition on the flexural behavior of composites are discussed and the results of statistical analyses of the test data are presented.

Table 2 presents the test data on fresh mix air content and hardened material physical properties of plain and cellulose fiber reinforced materials. Cellulose fibers are observed in Table 2 to reduce the specific gravity of fresh and hardened cement-based materials. The reduction in specific gravity was 12% when 2% mass fraction of kraft pulp was added. This may be attributed in part to the increased air content in the presence of fibers. The increase in fresh mix and hardened material air contents were 7% and 6%, respec-

Table 2. Fresh Mix Air Content and Hardened Material Physical Properties.

Mix	Air Content (%)		Specific Gravity (%)		Water Absorption (%)
	Fresh	Hardened	Fresh	Hardened	
Plain	21.07±0.25	22.75±0.30	2.01±0.01	2.00±0.01	11.07±0.12
1% Kraft Pulp	21.38±0.13	23.28±0.25	1.94±0.03	1.91±0.01	11.95±0.12
2% Kraft Pulp	22.50±0.26	24.18±0.07	1.77±0.01	1.75±0.04	13.66±0.06
2% Kraft Pulp (Silica Fume)	19.27±0.18	20.07±0.14	1.71±0.02	1.70±0.01	12.78±0.07
2% Kraft Pulp (Fly Ash)	20.21±0.16	21.06±0.19	1.62±0.03	1.61±0.02	12.81±0.14

tively, with the addition of kraft pulp at 2% mass fraction. The increase in air content could be attributed to the difficulty in compacting cement composites incorporating fibers, which leads to increased entrapped air content.

The water absorption capacity of the cement based materials is observed in Table 2 to increase with the addition of fibers. The increase in water absorption over that of plain cementitious matrix was 23% at 2% mass fraction. This could be attributed to the higher void content of matrix and also the relatively high absorption of water by cellulose fibers.

In order to see if there is any relationship between the fresh mix air content, and the hardened material void content, specific gravity and water absorption, correlation coefficients were calculated and analyzed statistically. The results showed that the correlation is significant at a confidence level of 95% for all the paired combinations of properties. The above results of correlation studies indicate that the effects of cellulose fibers on water absorption and specific gravity at least partly result from the corresponding effects on the void content of fresh and hardened materials.

4.1 Ageing under wet-dry cycles

The effects of cellulose fiber reinforcement on the flexural load-deflection behavior of unaged and aged cement-based materials are shown in Figures 4 (a) and (b). The effects of cellulose fiber reinforcement on flexural strength and toughness (defined as the area under the flexural load deflection curve) for plain matrix and cellulose fiber reinforced composite at different wetting-drying cycles are presented in Figures 5 (a) and (b), respectively.

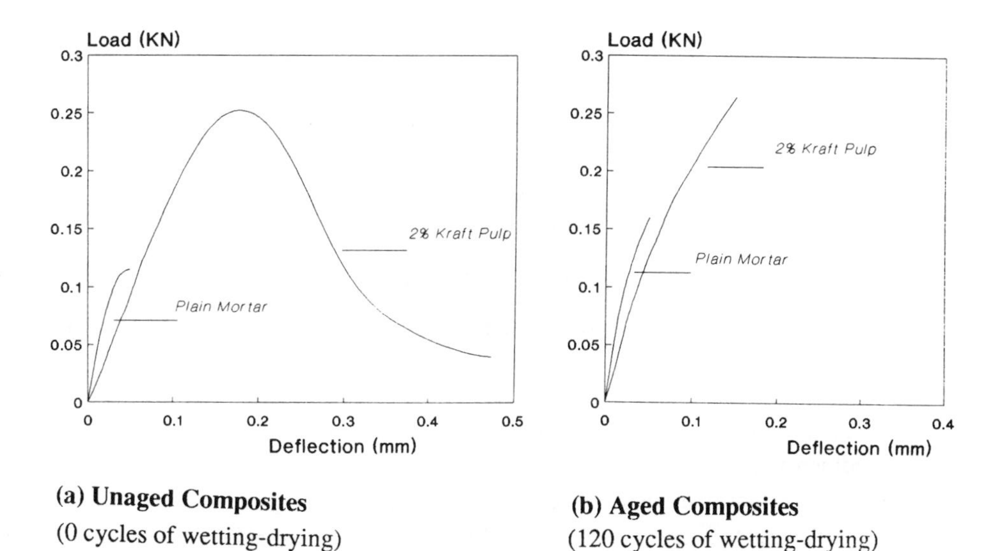

(a) Unaged Composites
(0 cycles of wetting-drying)

(b) Aged Composites
(120 cycles of wetting-drying)

Figure 4. Typical Flexural Load-Deflection Behavior.

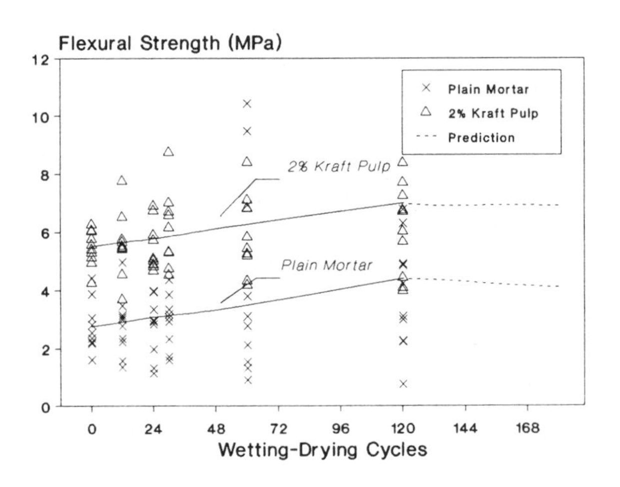

(a) Flexural Strength

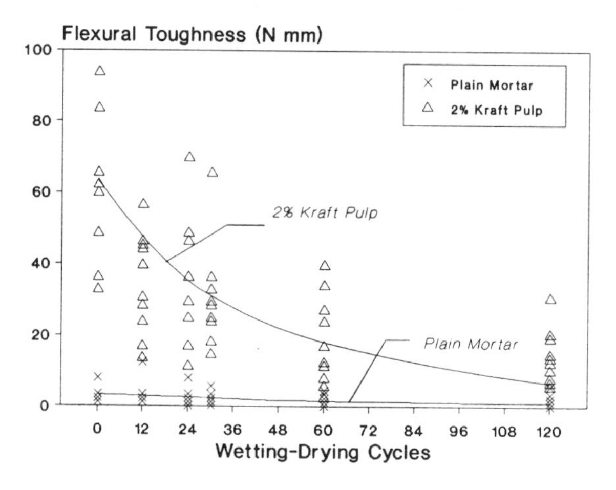

(b) Flexural Toughness

Figure 5. Effects of Accelerated Weathering on Flexural Performance (Regression Analysis).

One may derive the following conclusions from the test data regarding the effects of fiber addition and ageing cycles on flexural strength and toughness.

(1) The flexural strength of unaged fiber reinforced cement composites increased by 100% and flexural toughness by more than 20 times at 2% fiber mass fraction over plain specimens.

(2) In aged specimens (after 120 cycles of wetting-drying), the specimens with 2% cellulose fiber had average flexural strength and toughness values, respectively, 58% and

8 times higher than plain specimens.

(3) Wetting-drying cycles tend to increase the flexural strength and stiffness of both plain and fiber reinforced cement composite. The increase in flexural strength after 120 cycles of wetting and drying was 39% for plain specimens and 12% for cellulose fiber reinforced cement composites. Analysis of variance of test results at 95% level of confidence, however, indicated that, considering the random experimental errors, changes in flexural strength of plain and fibrous cement composites with wetting-drying cycles were not significant. Statistical time series analysis with regression forecasting predicted a slight drop in flexural strength for both plain and fibrous composites with continued ageing cycles wetting and drying.

(4) Accelerated wetting-drying cycles caused a drop in flexural toughness for both plain and fibrous specimens. On the average, drops of 44% and 77% were observed after 120 cycles. Analysis of variance of test data confirmed the significance of drop in flexural toughness only for fibrous composites at 95% level of confidence. In this case, it was also noted that the initial 12 cycles of wetting-drying caused significant drops in flexural toughness, and consequent cycles caused only gradual drops in toughness; this was also confirmed at 95% level of confidence.

4.2 Ageing Under Freeze-Thaw Cycles:

The resistance of plain and fibrous specimens to rapidly repeated cycles of freezing and thawing are presented in Figure 6. This figure presents the relative dynamic modulus of elasticity of the material as a function of the number of repeated freeze-thaw cycles.

The following conclusions could be derived from the test data regarding the effects of fiber addition on resistance to freezing and thawing. It should be noted that the speci-

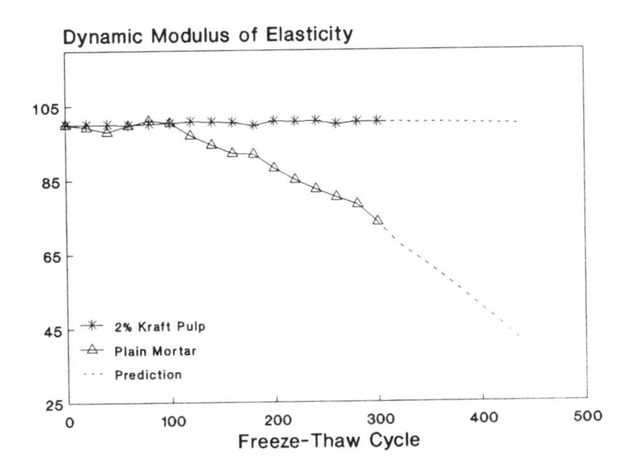

Figure 6. Effects of Freezing and Thawing on Relative Dynamic Modulus of Elasticity.

mens tested here were not air-entrained.

(1) The relative dynamic modulus of elasticity of plain cementitious matrices decreased under repeated cycles of freezing and thawing. The average drop in relative dynamic modulus was 25% after 300 cycles of freezing and thawing, and it was found to be statistically significant at 95% level of confidence.

(2) The relative dynamic modulus of fibrous composites marginally increased under repeated freeze-thaw cycles. Much of the increase was found in earlier cycles, while later cycles showed gradual decrease in relative dynamic modulus of elasticity. However, the analysis of variance test results at 95% level of confidence showed no significant change in relative dynamic modulus of cellulose fiber reinforced cement composites at 2% fiber mass fraction.

(3) Statistical time series analysis with regression forecasting predicted a slight drop in relative dynamic modulus of elasticity of fibrous composites, and a severe drop in that of plain matrices, with continued cycles of freezing and thawing.

4.3 Moisture-Sensitivity

The effects of cellulose fiber reinforcement on the flexural load-deflection behavior of cement-based materials are shown in Figures 7 (a) and (b) and (c) for different moisture conditions. The effects of fiber mass fraction on flexural strength and toughness for different moisture conditions are presented in Figures 8 (a) and (b), respectively.

The following conclusions could be derived from the test data regarding the effects of moisture condition and fiber mass fraction on flexural behavior, noting that the specimens used in moisture -sensitivity test were thicker than those used in durability studies:

(1) Flexural strength in air-dried condition increased with the addition of 1% fiber

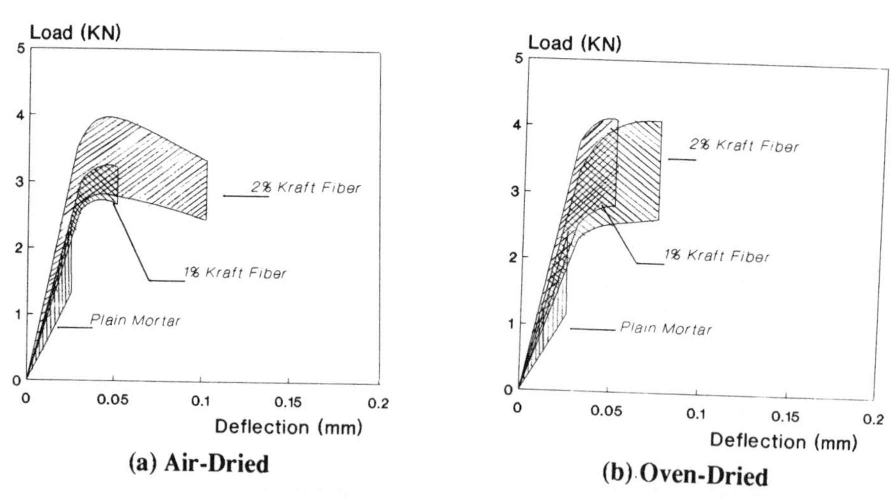

Figure 7. Typical Flexural Load-Deflection Behavior (range for 10 curves).

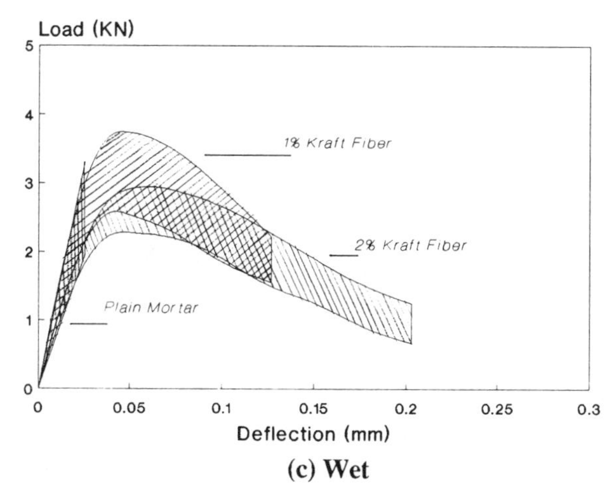

(c) Wet

Figure 7 (cont'd). Typical Flexural Load-Deflection Behavior (range for 10 curves).

mass fraction and then decreased when fiber content was increased to 2%; fracture toughness, however, increased continuously with increasing fiber content. The composite with 1% fiber mass fraction had, on the average, 99% increase in flexural strength and 126% increase in flexural toughness over the plain matrix. The increase in flexural toughness at 2% fiber mass fraction was even more significant, but the average increase in flexural strength at 2% fiber mass fraction was 71% over that of plain specimens. Analyses of variance of test results at 99% confidence level showed that the effect of fiber content was significant on both flexural strength and toughness.

(2) Oven drying of the composite caused only a slight increase in flexural strength in cellulose fiber composites; plain oven dried mortar specimens showed approximately 8% increase in flexural strength over the air dried ones. Wetting, on the average, reduced the flexural strength by 45% for kraft pulp at 2% fiber mass content when compared with air dried ones. Analysis of variance test results at 99% confidence level showed that only wetting of fibrous composites incorporating 2% fiber content was significant on flexural strength (where strength was reduced).

(3) Oven drying of the specimens reduced flexural toughness of the composites by about 50% for kraft pulp at 2% fiber mass content, while wetting of specimens produced a considerable increase in fracture toughness values (by as much as 212% above that for air-dried specimens in the case of kraft pulp at 2% fiber mass content). Analysis of variance test results at 99% confidence level showed that the effect of moisture condition on flexural toughness was highly significant at all fiber contents (where toughness was increased with wetting and reduced with oven drying).

4.4 Matrix Modification

The effects of pozzolans on flexural strength and toughness of cellulose fiber reinforced

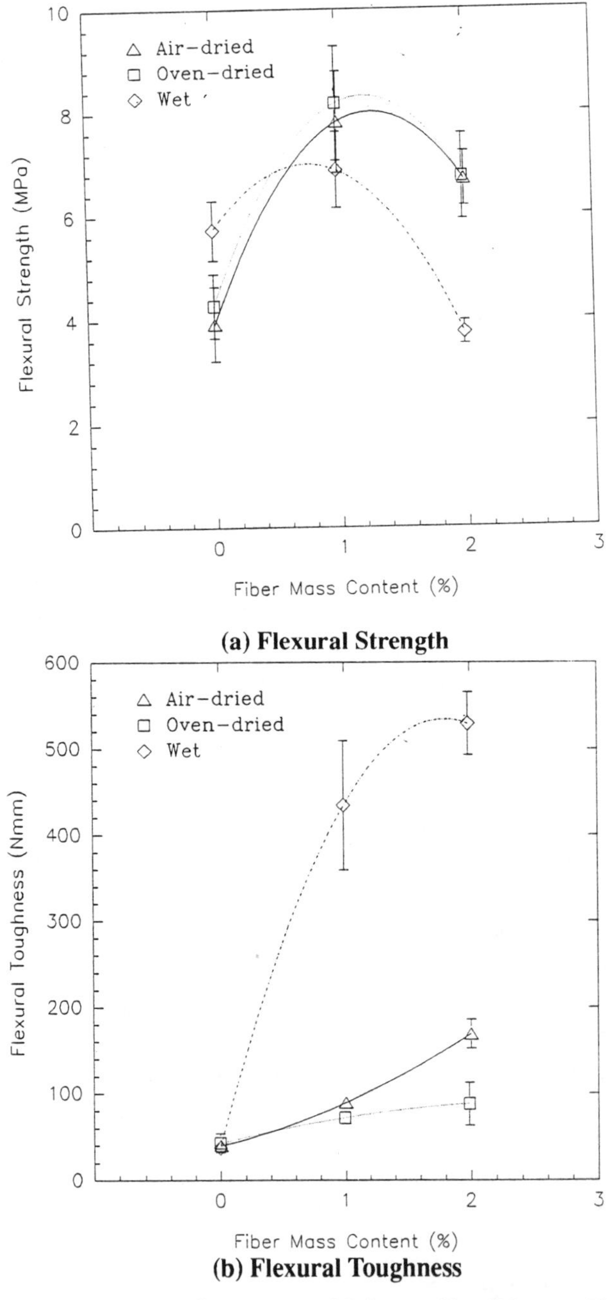

(a) Flexural Strength

(b) Flexural Toughness

Figure 8. Effects of Fiber Mass Content and Moisture Condition on Flexural Performance.

cement composites with 2% fiber mass fraction are presented in Figures 9 (a) and (b), respectively. Pozzolans (30% Silica fume and 15% Fly ash) are observed to increase the flexural strength of the composite while slightly reducing toughness. These observations were confirmed through statistical analyses of results at 95% level of confidence; these effects are more pronounced with silica fume. The highest increases in flexural strength in the presence of pozzolans were obtained in the wet condition; silica fume and fly ash increased flexural strength on the average by 20% and 10%, respectively, in the air-dried and oven-dried conditions, and by 40% and 30% in the saturated conditions. The drop in toughness in the presence of pozzolans was almost constant in different moisture conditions and with different pozzolans; the average drop in toughness was about 15%.

In the presence of pozzolans no statistically significant difference was observed between the flexural behavior of air-dried and oven-dried composites; wetting still had adverse effects on flexural strength and positive effects on flexural toughness at 99.9%

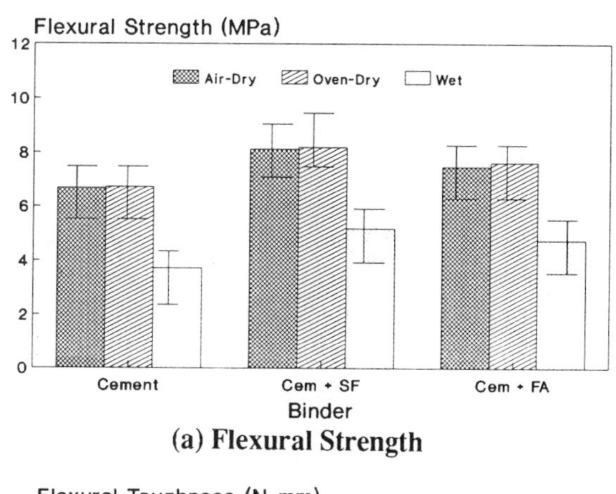

(a) Flexural Strength

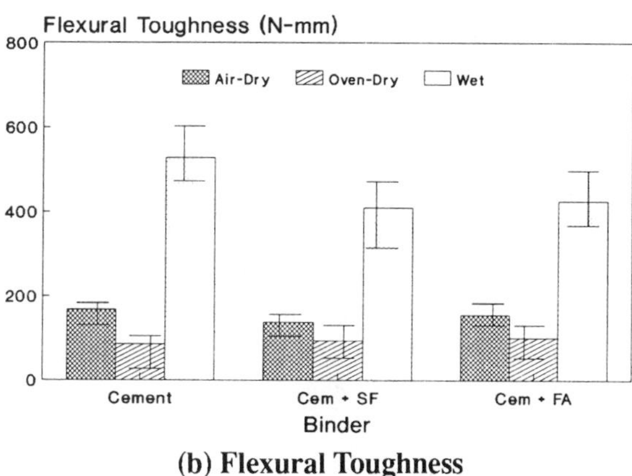

(b) Flexural Toughness

Figure 9. Effects of Pozzolan and Moisture Condition on Flexural Performance.

level of confidence. There was, however, an interaction between binder composition (i.e., the presence of pozzolans) and wetting effects. Pozzolans reduced the moisture-sensitivity of cellulose fiber reinforced cement composites; while without pozzolans wetting reduced flexural strength by an average of 45% below that in air-dried composites, in the presence of silica fume or fly ash there was a drop of 40% in flexural strength upon wetting. The average increase in flexural toughness with wetting was 3.2 times without pozzolans and 2.8 times with fly ash or silica fume. The effects of pozzolans on reducing moisture-sensitivity were confirmed at 95% level of confidence through factorial analysis of variance of test results.

5 Microstructural Studies

The fiber-to-matrix interfacial bond in fiber-cement composites can be mechanical or chemical, or a combination of both. In the case of cellulose fiber reinforced cement composites both mechanical bonds and chemical bonds play an important role.

Scanning Electron Microscopic studies of the fracture surfaces of the broken cellulose fiber reinforced cement composites were conducted with emphasis placed on the interface zone behavior. The results showed that failure occurs by the dual mechanisms of fiber fracture and fiber pull-out in air-dried composites (Figure 10 a). Figure 10 (b) shows that fiber fracture dominates the failure mechanism of oven-dried samples while, as shown in Figure 10 (c), fiber pull-out dominates failure in wet samples. Hence, oven-dried samples are expected to be relatively strong but brittle because energy dissipation associated with fiber pull-out is lost by oven-drying; wet specimens, on the other hand, are relatively weak but tough with considerable energy absorption capacity associated with fiber pull-out.

At the pH of the cement (>12.5) C-OH, Si-OH, Ca-OH bonds are present in the composite which contribute to hydrogen bonds taking place between cellulose fibers at the matrix or between cellulose fibers themselves. Wet and dry cellulose fibers have about the same tensile strength, but the stiffness is ten times greater in dry condition. Thus, an oven-dried composite has stiff, contorted fibers locked into the cement matrix and held together by a number of hydrogen bonds. This system when stressed can load the fibers until failure occurs. A wet sample on the other hand has swollen conformable fibers. The hydrogen bonds between fibers and matrix or between the fibers themselves are broken by insertion of water molecules. Under stress this system allows the fibers to pull out of the matrix. However, due to the swelling, considerable frictional forces may be present and thus fiber pull-out would be associated with substantial frictional energy dissipation (Coutts et. al. 1982).

In order to study the behavior of cellulose fiber reinforced cement composites under ageing, scanning electron microscopic studies were performed on the fracture surfaces of

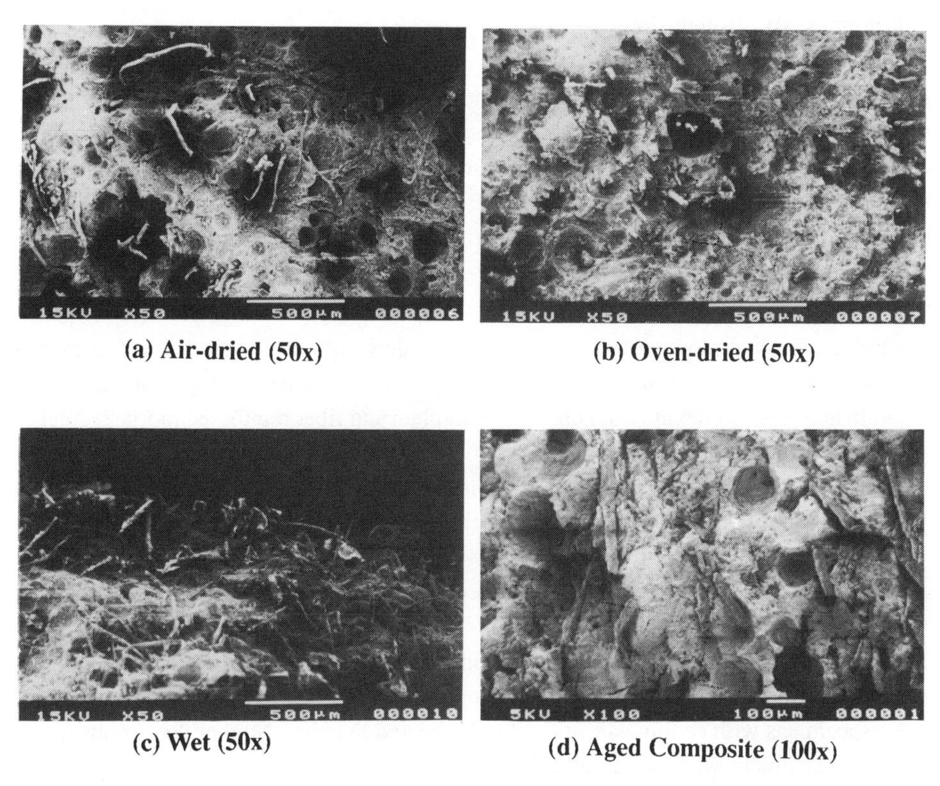

(a) Air-dried (50x) (b) Oven-dried (50x)

(c) Wet (50x) (d) Aged Composite (100x)

Figure 10. Fracture Surface of Wood Fiber Reinforced Cement composites.

aged and unaged specimens. The dominant mode of fracture in aged composite was fiber fracture (Figure 10 d), while unaged specimens showed a combination of fiber pull-out and fiber fracture (Figure 10 a), and the matrix around fibers in unaged specimens was relatively porous. After ageing, the hollow fibers were often found to be filled with dense hydration products and showed little or no debonding from the matrix. This microstructure is often referred to as brittle petrified, and this carbonation conditions have been suggested to the accelerated formation of this microstructure (Bentur et. al. 1989). Petrification makes fibers stronger and stiffer but more brittle; these changes together with the densening of interface zones and excessive fiber-to-matrix bonding can illustrate the increased strength and embrittlement of the composite with ageing. Petrified fibers with excess bonding tend to rupture rather than pull-out at fracture surfaces, thus eliminating the desirable toughness characteristics associated with frictional energy dissipation during fiber pull-out.

6 Summary And Conclusions

An experimental study was conducted in order to assess the effects of weathering and moisture condition on the flexural strength and toughness of cellulose fiber reinforced cement composites incorporating 1% and 2% mass fraction of kraft pulp. Comprehensive sets of replicated flexural test data were generated and were evaluated statistically using the analysis of variance and multiple comparison techniques. Also in durability tests efforts were made to predict the future trends beyond test periods, using statistical time series analyses. The materials used in this investigation were manufactured using a regular mortar mixer. The following conclusions could be derived from the generated test data.

(1) Repeated wetting-drying cycles caused an increase in the average flexural strength and reduced flexural toughness of both plain and fiber reinforced mortars. Analysis of variance of test results at 95% level of confidence, however, indicated that the increase in flexural strength with ageing was not significant, while the reduction in flexural strength was found to be significant only in fibrous composites.

(2) Repeated cycles of freezing and thawing caused a reduction in the relative dynamic modulus of plain specimens (confirmed statistically at 95% level of confidence), while fibrous composites showed marginal (statistically insignificant) improvements. Statistical time series analysis with regression forecasting predicted a slight drop in the relative dynamic modulus of elasticity of fibrous composites and a severe drop in that of plain specimens with continued cycles of freezing and thawing. Neither the fibrous nor the plain specimens were air-entrained.

(3) Wetting had highly significant effects on reducing flexural strengths at 2% fiber mass fraction, but this effect was not significant at 1% fiber content. Air-dried composites with 2% fiber mass fraction lost, on the average, 45% of their flexural strength upon wetting. The effects of oven drying on flexural strength were statistically not significant at 95% level of confidence. Wetting and oven drying had highly significant effects on, respectively, increasing and decreasing the flexural toughness of fibrous composites. The flexural toughness of wet composites with 2% fiber mass fraction were more than 3 times that of air-dried materials.

(4) Addition of pozzolans (silica fume or fly ash) led to increased flexural strength of composites with 2% fiber mass fraction, while slightly reducing flexural toughness at all moisture conditions. These effects were confirmed statistically, at 95% level of confidence; they were more pronounced with silica fume and at saturated condition.

(5) In the presence of pozzolans no statistically significant difference was observed in the flexural behavior of air-dried and oven-dried composites; wetting caused adverse effects on flexural strength and positive effects on flexural toughness values, at 99.9% level of confidence. The presence of pozzolans led to reduced moisture-sensitivity of cellulose fiber reinforced cement composites.

(6) Microstructural studies indicated that wetting leads to increased tendency toward

fiber pull-out rather than rupture of fibers at fracture surfaces, while oven drying promotes fiber rupture. This may illustrate why increased moisture content tends to produce higher flexural toughness values and lower flexural strengths in cellulose fiber reinforced cement composites. In the case of aged composites, petrification of fibers and densening of interface zones caused tendencies toward increased strength and embrittlement with ageing.

7 Acknowledgment

Financial support for the performance of this research project was provided by the U. S. Department of Agriculture (Eastern Hardwood Utilization Program) and the Research Excellence Fund of the State of Michigan. The fibers used in this project were provided by The Procter and Gamble Cellulose Company. These contributions are gratefully acknowledged. The authors are also thankful to Dr. Otto Suchsland from the Forestry Department of Michigan State University for his continuous support of research activities in the area of cellulose fiber-cement composites. The technical support provided by the Composite Materials and Structure Center of Michigan State University is also gratefully acknowledged.

8 References

Akers, S. A. S. and Studinka, J. B. (1989) Ageing Behavior of Cellulose Fiber Cement Composites in Natural Weathering and Accelerated Tests. **International Journal of Cement and Light Weight Concrete**, 11(2), 93-97.

Bentur. A. and Akers, S. S. S. (1989) The Microstructure and Ageing of Cellulose Fiber Reinforced Cement Composites Cured in a Normal Environment. **International Journal of Cement Composites and Light Weight Concrete**, 11(2), 99-109.

Coutts, R. S. P. (1988) Sticks and Stones. **Forest Products Newsletter**, (CSIRO Division of Chemical and Wood Technology (Australia), 2(1), 1-4.'

Coutts, R. S. P. (1987) Air-Cured Wood Pulp, Fiber/Cement Mortars. **Composites**, 18(4), 325-328.

Coutts, R. S. P. (1984) Autoclaved Beaten Wood Fiber Reinforced Cement Composites. **Composites**, 15(2), 139-143.

Coutts, R.S.P. and Kightly, P. (1982) Microstructure of Autoclaved Refined Wood Fiber Cement Mortars. **Journal of Materials Science**, 17, 1801-1806.

CSIRO. (1981) New-A Wood Fiber Cement Building Board. **CSIRO Industrial Research News (Australia)**, 146.

Gram, H. E. (1983) Durability of Natural Fibers in Concrete. **Swedish Cement and Concrete Research Institute**. Stockholm, 255.

Japanese Concrete Institute. (1984) **JCI Standards for Test Methods of Fiber Reinforced Concrete** Report No. JCI-SF, 68.

Mindess, S. and Francis, J. Y. (1981) **Concrete**, Prentice-Hall, 422-425.

Neville, A. M. (1963) **Properties of Concrete**, John Wiley & Sons, Inc., 409-415.

Procter and Gamble Cellulose Paper Grade. **Wood Pulp HP-11**. Technical Bulletin, Memphis, Tennessee.

Sharman, W.R. and Vautier, B.P. (1986) Accelerated Durability Testing of Autoclaved Wood Fiber Reinforced Cement-Sheet Composites. **Durability of Building Materials**, 3, 255-275.

Soroushian, P. and Marikunte, S. (1990) Reinforcement of Cement-Based Materials with Cellulose Fibers. **Thin Section Fiber Reinforced Concrete and Ferrocement**, Publication SP-124, American Concrete Institute, 99-124.

Tappi Standard. **T803m-50**, Technical Association, Pulp and Paper Industry, New York.

Vinson, K. D. and Daniel, J. I. (1990) Specialty Cellulose Fibers for Cement Reinforcement. **Thin Section Fiber Reinforced Concrete and Ferrocement**, American Concrete Institute Publication SP-124, 99-124.

94 PERFORMANCE PROPERTIES OF SISAL-FIBER-REINFORCED ROOFING TILES IN THE IVORY COAST

O. SANDE
ENSTP, Yamoussoukro, Ivory Coast
O. DUTT and W. LEI
IRC/NRC, Ottawa, Canada
D. TRA BI YRIE
ENSTP, Yamoussoukro, Ivory Coast

Abstract
The manufacture of sisal-fiber-reinforced cement tiles introduced in the Ivory Coast with the help of the International Labour Organization (ILO) was investigated. Their quality was found to be generally poor because of inadequate quality control. Various properties of tiles obtained from five different plants were tested. Their composition and manufacture and various physical and mechanical properties were compared. Recommendations are made for improving future products.
Key Words: Cement tile, Damp-proofing, Fiber-reinforced tile, Impact load, Nib strength, Roofing tiles, Sisal fiber.

1 Introduction

The common phrase of "having a roof over one's head" communicates sense of security and comfort derived from the protected environment. While a roof is a symbol of shelter, technically, it is one of the most significant parts of a house. Its prime function, to cover the living dwelling and working areas from inclement weather, is achieved by a myriad of shapes, construction systems, and materials that frequently vary with societies, cultures and climatic conditions. Also their structural support systems are numerous which may consist of beams, girders, slabs, trusses, corrugated sheets, arches, domes, shells etc. selected from functional and architectural consideration. Similarly, materials may be brick, concrete, steel, clay tile, slate, bamboo, thatch, reed, stabilized soil, etc. Functionally, all roofs are either shingled or sloped for shedding water or flat and covered with waterproof systems or membranes.

Most developing countries have an acute shortage of suitable housing. For example, the Ivory Coast is a West African country, mostly agrarian, that needs some

Fibre Reinforced Cement and Concrete. Edited by R. N. Swamy. © 1992 RILEM.
Published by E & FN Spon, 2-6 Boundary Row, London SE1 8HN. ISBN 0 419 18130 X.

25,000 dwellings every year. In their efforts to overcome this situation, maximum use is made of local resources and available materials in the construction of durable and low cost roofs. A few craft industries for producing fiber-reinforced cement tiles were established with the help of International Labour Organization (ILO). They are now making tiles in accordance with the recommendations of ILO.

A preliminary study done on tiles manufactured in one of the factories showed performance inadequacies. Unfortunately, the factory's quality control of the product was only carried out during the first few months of the installation of the plant. Accordingly, there was a need to determine the properties of tiles now being manufactured and sold on the market. It was also necessary to develop a procedure establishing a quality control program.

The study involved four plants located in Abidjan, the capital of the Ivory Coast, and one from the inland part of the country. To determine product uniformity sample tiles made at different periods were taken from two of the five selected plants.

Fifty tiles of 6 mm and 8 mm thickness from each plant were obtained. All tiles had been aged for at least 45 days which is above the minimum of 15 days curing recommended by ILO; i.e., 7 days in water or under a canvas cover plus 8 days under a shelter. The samples were tested for a number of physical properties considered to be essential for the performance of tiles.

2 Materials and Mixing

The manufacturing method and materials are practically the same at all the sites. The following materials are used for making tiles:

- sand, from locally available deposits;
- imported sisal fiber, chopped into suitable lengths;
- portland cement from local grinding of imported clinker (there is only one grade of cement);
- pigments (imported);
- water.

The ingredients are first mixed dry with a shovel before adding water. An electric vibrating table is used for moulding to obtain the required compactness, thickness and density of the mortar. This table is designed such that after compaction on the flat surface the material is gently let drop on the corrugated base board which gives the tile its required shape.

The quantities of the constituents used in four different plants and with those recommended by ILO are shown in Table 1. The proportions of sand and cement expressed as sand/cement (s/c) ratios are higher than the recommended range of 8 to 21%, but the quantity of fiber conforms to ILO recommendations.
Only one case, generally is more sand and less fiber used per unit of mortar.
With the exception of one plant, the amount of added water is left to the judgement of the worker.

3 Physical Characteristics

3.1 Weight

All the tiles in each series were weighed.
Table 2 shows that in the case of the 6 mm tiles, there is greater batch to batch variability in mean weight in production in the same plant than between lots made in different plants. With the 8 mm tiles there is still some variation between batches from the

Table 1. Composition of tiles based on one sack of cement (50 kg).

	Quantity (kg)					Proportions		
Plant *	Cement	Sand	Fiber	Pigment	Water	Sand/ Cement	Fiber/ Sand + Cement %	Pigment/ Cement %
A	50	125	1.25	1.25	?	2.5	0.71	2.5
B	50	125	1.1	1	?	2.5	0.67	2
C	50	135	1.1	1.5	?	2.7	0.59	3
D	---	---	---	---	---	---	----	---
E	50	151	0.85	1	34	3.02	0.42	2
ILO**	50	125	1.1	0.9-1.35	32.5	2.5	0.67	2 to 3

* Plant Designation ** ILO Recommendations

Table 2. Mean weight of tiles and tiled roofs.

	Thickness (mm)	Mean weight of tile (g)	Std. Dev. (g)	Mean weight of tiles (kg/m²)	Mean weight of roof (kg/m2)
A1 (1)	6	1490	42.7	18.6	29.6
A2 (1)	6	2042	45	25.5	36.5
B	6	1793		22.4	33.5
C1 (1)	8	2861	104.3	35.8	46.8
C2 (1)	8	2936	129.5	36.7	47.7
D	8	2865	1146.6	35.8	46.8
E	8	2242	103	28	39
Average	6	1775		22.2	33.2
Average	8	2726		34.1	45.1

(1) Tiles from the same plant but made at different periods.

same plant but tiles from plant E (Table 2) differ notably from C and D in mean weight, although its range is smaller on the absolute basis.

The mean weight of the tile 6-mm thick is 1775 g with a mean range of 300 g. For 8-mm tiles, the mean weight is 2725 g, with a mean range of 520 g. The mean range as percent of mean weights are 14% and 19% for tiles 6 mm and 8 mm respectively.

3.2 Weight of tiles and roofing

12 1/2 tiles make a square meter of roofing. Studies carried out on various buildings covered with fiber-reinforced roofing tiles have shown that a "useful frame" weighing 11 kg/m^2 needs 700 kg/m^3 of timber.

As shown in Table 2, mean values of 22.2 kg/m^2 and 34.1 m^2 m correspond to tiles of 6 and 8 mm respectively, i.e. 5 to 7 times the weight of a square meter of galvanized iron and steel roofing sheets.

For roofs covered with tiles of 6 and 8 mm (tiles + frame) the values are 33.2 kg/m^2 and 45 kg/m^2 respectively, i.e. 4 to 6 times the weight of square meter of an iron and steel roofing.

3.3 Water absorption

Water absorption is determined by drying the specimens at 70°C until constant weight P1 is obtained. The specimen is immersed in water at ambient temperature until constant weight P2 is obtained. In order to determine P2, the test specimen is taken out of the water, quickly surface dried and weighed. The porosity is obtained by multiplying the weight difference (P2 - P1)/P1 by 100.

Table 3 shows that the mean porosity for all the series is 10.5%.

The less porous tiles are from plant B (6 mm) and plant D (8 mm) with 6.8% and 9.3% of porosity.

If groups B and D are excluded, the average value for the other groups is 11.5%.

Table 3. Water absorption of tiles.

Plant Designation	Mean %	Variation %
A1	11.8	0.6
A2	11.6	0.6
B	6.8	2.7
C1	11.6	1
C2	11.3	1.9
D	9.3	0.8
E	11.1	1
Mean fof all tiles	10.5	

3.4 Damp-proofing

The test is performed on full tile laid horizontally with the ends of the corrugations blocked by mortar or plywood cut to shape and the edges sealed with silicone (Figure 1). The enclosed space is filled with water so that the water is full to the corrugation. The tile is considered dampproof if there is no sign of dampness on the underside within 24 hours. The test duration was extended to 72 hours in some cases to verify damp-proofing.

Testing of dampproof qualities show that only tiles B (6 mm) and tiles D (8 mm), did not have any sign of dampness even after 72 hours. All other tiles tested were at least 80% damp after only 2 hours of contact with water.

It is possible that for better damp-proofing characteristics of tiles B and D is due to the application of a coating during the manufacturing.

4 Mechanical Characteristics

4.1 Impact Test

The test is performed with a 200 g metallic ball dropped from different heights on a tile laid flat on a table (Figure 2). Although ILO had recommeded 20 cm as the height of the dropping of weight, it was increased progressively to 100 cm in order to find the height at which the micro-cracks or the rupture of the tile will occur. This methodology has given the results shown in table 4.

The height from which the dropped weight causes damage is at least equal to 50 cm, i.e. more than twice the ILO recommended value.

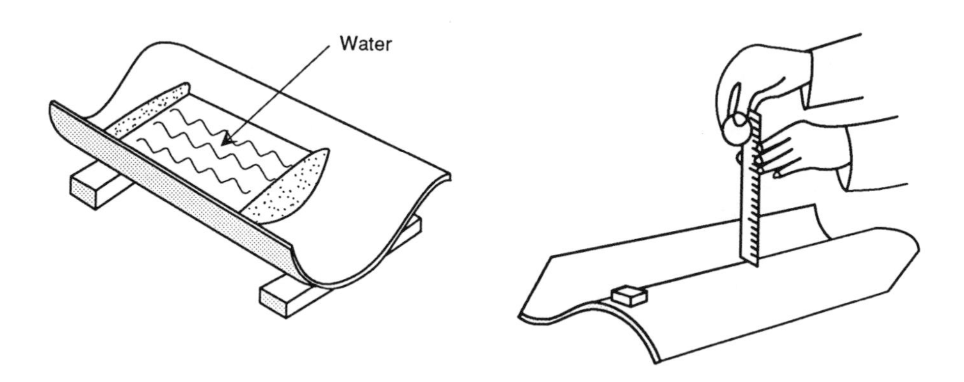

Figure 1: Damp proofing test

Figure 2 : Impact test

Table 4. Impact test (%) tiles destroyed or unacceptable according to the height of fall.

Plant *	Thichness (mm)	% of tiles broken due impact of load drop (cm)							% of acceptable samples as per ILO
		20	40	50	60	70	80	100	
A1	6	100							0
A2	6		33	67					100
B	6		33	50			17		100
C1	8			17	17		33		100
C2	8				33			33	100
D	8							17	100
E	8			33	67				100

* Plant Designation

4.2 Nib Load Test

The Nib load test is determined with a simple arrangement that keeps the tile in horizontal position on a table with an overhang of 50 mm as shown in Figure 3. The fixture system includes a steel hook and a wire loop with which a load of 20 kg is suspended to the heel.

If the tile remains intact, which means that neither the heel breaks, nor the wire pulls out of the heel, after at least 20 seconds of loading, the products are considered to have passed the test.

Table 5 shows that all 6 mm tile failed with the 20 kg load. The failure rate of the 8 mm tiles was 0 percent for D and E, and 20 and 40 percent for C2 and C1 respectively. Six mm tiles were tested also with 10 kg and 15 kg weight. It was noticed that all of them passed with a 10 kg load while with 15 kg loads the failure rate was 80 percent for tiles A1 and 20 percent for A2 and B.

Table 5. Nib load test and bending test.

Plant *	Tile Thickness (mm)	Nib load test # % of tiles broken due to each load %			Bending test #		
		10 kg	15 kg	20 kg	Average rupture load (kg)	Variation (kg)	ILO+ (%)
A1	6	0	80	100	30	0	0
A2	6	0	20	100	64	30	100
B	6	0	20	100	84	15	100
C1	8	-	-	40	81	37	83
C2	8	-	-	20	72	30	100
D	8	-	-	0	87	-	100
E	8	-	-	0	55	15	50

* Plant designation
# Each test result represents mean of six replicates tested
+ Acceptable according to ILO recommendations

4.3 Bending Test

The resistance to bending is determined by a device fabricated locally which enables to vary the flexural load as desired
(Figure 4). The load is applied perpendicular to the length of the tile for at least 10 seconds. ILO recommends 30 kg and 50 kg loads for 6 mm and 8 mm tiles respectively.
In order to study the overloading conditions, the tiles were loaded to rupture or to a maximum of 90 kg. The results in table 5 show that according to ILO recommendations, tiles A1 (6 mm) failed the bending test, whereas other tiles produced acceptable results.
Tiles B (6 mm), C1 (8 mm) and D (8 mm) gave excellent strength; they either supported or went above the average rupture load of 80 kg.

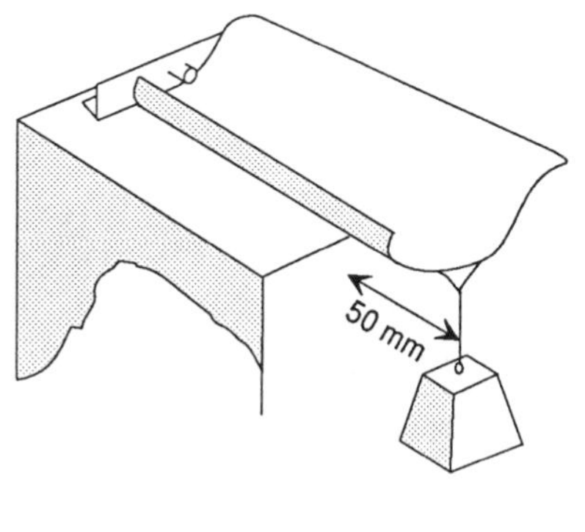

Figure 3 : Nib load test

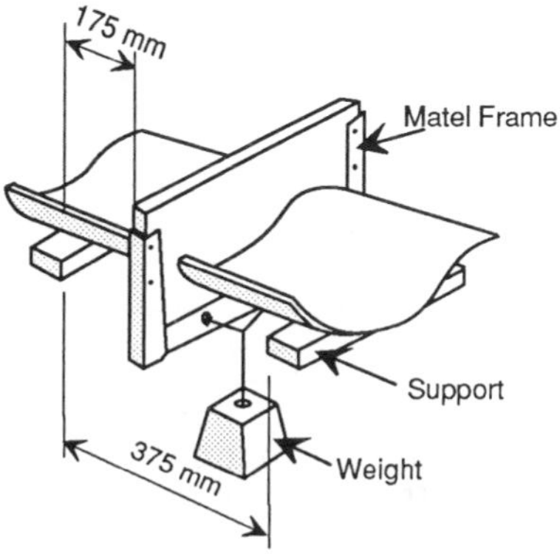

Figure 4 : Bending test

5 Discussion and Conclusion

The physical and mechanical properties of fiber-reinforced roofing tiles manufactured in Ivory Coast vary from one plant to another and in some cases from batch to batch from the same plant.
This study shows that the majority of plants do follow ILO recommendations. The quantity of mixing water used in the mortar is generally more than prescribed.
The wide range between the weights of tiles shows a lack of uniformity in their thickness. The vibrating time has not been clearly defined, which causes mortar segregations.
Porosity in the tiles is a major problem. Tiles B and D show the least amount of absorption. That improves

their flexural property as seen in table 5. According to ILO recommendations, the mechanical properties for 8 mm tiles are generally satisfactory. However, it may be worth noting that during an earlier study carried out on the tiles manufactured under similar conditions in one of the plants, the failure of tiles in the bending was 50% (2), whereas for the same plant their rose to 83% and 100% at a different time, and from one batch to another. This shows the lack of proper quality control. The plants should conform to the recommended compositions and quality control program.

6 Aknowledgement

The authors greatfully aknowledge the assistance of Cecilia Monastario and Craig Campbell of IRC in performing laboratory analyses. This research on vegetable Tiles in Ivory Coast, is funded by the International Development Research Center, Ottawa, Canada.

7 References

1. Brys, G., "Création d'une unité de production de tuiles en fibro-mortier", Report of assignment in Côte d'Ivoire, Geneva, ILO, 1987.
2. Sande, O., Dutt, O., Dembelé, S., "Matériaux pour toitures en Côte d'Ivoire", 12e World Building Congress, May 1992, Montréal, Canada, 4 pages.

95 COCONUT AND SISAL FIBRE REINFORCED CEMENT AND GYPSUM MATRICES

C. W. de A. PIRES SOBRINHO
CNPq/Pernambuco State Technology Institute, Recife, Brazil

Abstract
The simple incorporation of vegetable fibres of coconut and sisal in brittle matrices modifies their rupture behaviour. This enables the development of reserve of resistance of the composites after cracking and increases their usefulness. This paper presents comparative results of prismatic specimens of composites subjected to flexural and impact test by varying matrix type, fibre length, fibre/matrix relative volume and mixture processing. The results indicate that those elements interact in different manner in the behaviour of the composites and that their influences can be assessed adequately by flexural tests as an alternative form to the direct tensile test.
Keywords: Fibre Reinforced Matrix, Flexural Strength, Tensile Cracking Strength

1 Introduction

In the field of composite materials, mainly those used as panels in building, special attention should be given to the possibility of modifying the behaviour of the brittle matrices by incorporation of vegetable fibres Laws (1971), Hannant (1978), CEPED (1982). The introduction of fibre in gypsum or cement matrices enables the development of reserve of resistance of these composites after cracking. Many factors interact in the composites properties: matrix type, fibre type and characteristic, fibre/matrix relative volume, and mixture process, moulding and storage conditions of the composites. This paper presents comparative results of prismatic specimens of composites subjected to flexural and impact test, varying the matrix type, fibre length, fibre/matrix relative volume and mixture processing. The results indicate that these elements interact in different manner in the behaviour of the composites and that their influence can be assessed adequately by flexural tests, as an alternative form to the direct tensile test.

2 Materials and methods

Two types of matrices were defined: the gypsum plaster matrix "G", with a water/gypsum ratio fixed at 0.5 and 0.5% of citric acid added (as retarder of the

Fibre Reinforced Cement and Concrete. Edited by R. N. Swamy. © 1992 RILEM.
Published by E & FN Spon, 2-6 Boundary Row, London SE1 8HN. ISBN 0 419 18130 X.

setting time), and the comparative matrix "C" of the cement plaster with water-cement ratio fixed at 0.5.

Two vegetable fibre types were used: the base fibre "S" of sisal fibre (diameter about 0.22 mm and tensile strength about 449.0 MPa); and the comparative fibre "C" of the coconut (diameter about 0.34 mm and tensile strength about 56.1 MPa).

This paper reports the influences of the following parameters: range of the fibre/matrix relative volume by 1%, 2% (base), and 3%; range of the fibre length by 2 cm, 4 cm (base), 6 cm and 8 cm.

Concerning the mixture processes, two methods were followed: base process "M" of the mass when the fibre was introduced in the mixer after matrix components; the comparative process "A" of the fibre that was aligned at the time of placing of the layers in the mould to form a sandwich (plaster, fibre, plaster, fibre, plaster).

The specimen materials had internal dimensions of 25x25x280 mm and were subjected to mechanical vibration with a frequency of 10,000 hz for 1 minute and followed by 28 days in an environment of adequate ventilation and temperature range from 28 to 30°C.

The notation adopted in this paper concerns 5 elements that relate directly to: type of matrix, relative volume of fibre/matrix, type of fibre and its length, and mixture process (adopted). Example:

G2S4M = Gypsum plaster + 2% volume of sisal fibre with 4 cm length incorporated into the matrix.

3 Test Details

Tests were performed on the tensile strength of sisal and coconut fibre according to ASTM D-3379/74. A tension machine type WOLPERT was used to control deformation and sensitivity was 0.05 N. Tests of simple flexural strength were performed on prismatic specimens, giving "loading" vs "deformation" diagrams. A HOUNSFIELD TENSOMETER with deformation control was used, with an advance of 1.25 mm/min and accuracy of 5 N.

Impact tests were performed on prismatic specimens, using a Charpy pendulum type VEB WERKTOPFPRUFMASCHINEN with a capacity of 0.30 kN.m (absorbed energy) and accuracy of 2.5 N.m.

4 General Remarks

The majority of published papers on fibre-matrix, refer to characteristic elements defined by tests of simple axial tension, Laws (1971), Hannant (1978), Savastano (1978), Gram (1983). The direct application of pure tensile test is very difficult because generally there is sliding due to grips system and/or secondary tension appear in adjacent zones.

Besides, tension appearing on specimens built of these materials is primarily due to flexural loading. According to that, this paper discusses the influence of composite elements of fibre-matrix using simple flexural tests and impact test.

Many parameters can be evaluated by a simple flexural test: loading, bending moment and modulus of rupture, during elastic and plastic zones.

Comparative evaluation of loading and bending moment is independent of adopted assumption for the distribution of internal tension of specimen cross section. On the other hand, associated evaluation of modulus of rupture in the tension or compressive regions, is based on hypothesis about stress distribution over the cross section. The maximum theoretical stress attained by fibres on the most strengthened cross-section is named traction modulus. The following Fig. 1 shows the scheme of load application, the diagrams of applied loads and the hypothetical stress distribution of tension over the cross section, on the elastic and plastic zones (before and post cracking).

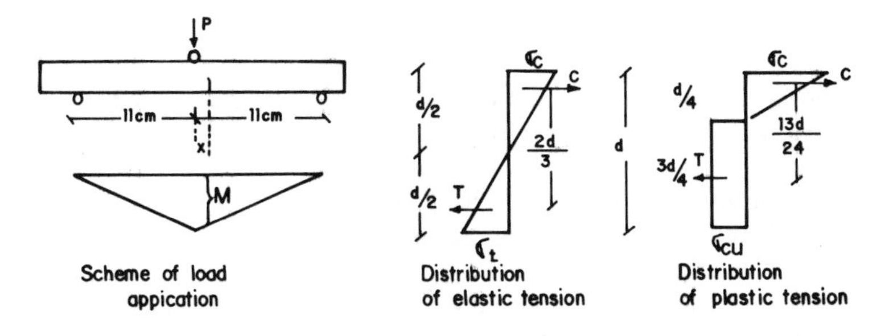

Fig. 1 Hypothetical distribution of tension

In each, the moment applied "M" can be determined by the expression:

$$M = \frac{P(11-|X|)}{2} \qquad \text{(I)}$$

In the elastic zone, before the first crack, the tensile cracking strength can be determined by the expression:

$$M_e = \sigma_t \frac{d.b}{4} x \frac{2.d}{3} \qquad \sigma_t = \frac{6M}{b.d^2} \qquad \text{(II)}$$

In the yield zone, characterized by maintenance of load level, the modulus of rupture can be determined by the expression III:

$$M_p = \sigma_{cu} \frac{3.d.b}{4} \frac{13d}{24} \qquad \sigma_{cu} = \frac{32.M}{13.b.d^2} \qquad \text{(III)}$$

By comparison these moduli do not reveal the internal resistance of the cross section, due to different hypotheses about tension distribution of actual region adopted in each case. In this respect a comparative norm has been defined among these modulii by taking into account the maintenance of the resistance level, after and post cracking, Fig. 2.

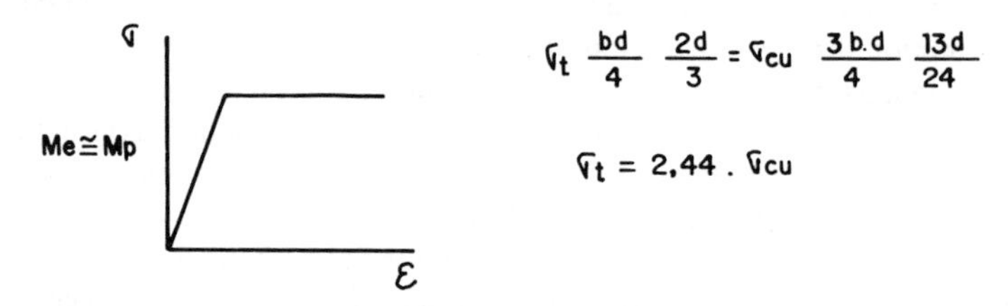

Fig. 2 Hypothetical distribution of tension

5 Discussion and analysis of results

Figure 3 shows the average behaviour of prismatic specimens subjected to flexural loads, where each series was formed by 6 tests on average.

As can be seen, in all there is a change in the behaviour of composites, with changes in the length of fibre, showing that, the greater the length, greater is the yielding plateau and the greater is the resistance gain to bending strength (in the loading of the first crack and/or the yield zone). This behaviour is due to the fact that the greater the fibre length the greater will be the bond to the matrix, by increase of confinement.

On the other hand, gypsum composites with coconut fibre present a greater yielding plateau level than the sisal fibre ones but at lower resistance levels. Cement composites with coconut fibres show greater resistance levels than sisal.

The fibre-matrix mixing process is more efficient in the gypsum-sisal composites, maybe due to the greater fibre-matrix contact, as the sisal fibre is more absorbent than coconut fibre.

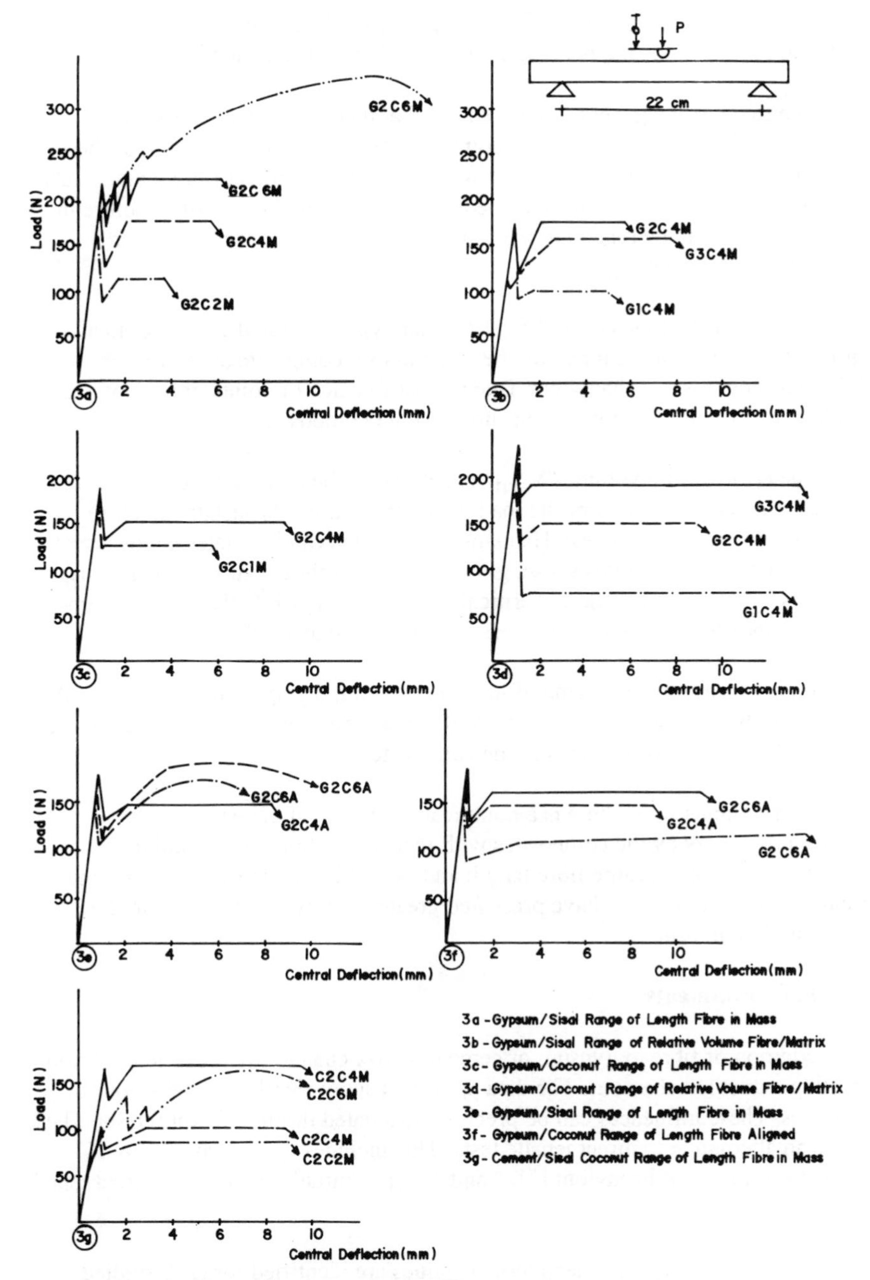

Fig. 3 Flexural test results for fibre-matrix composites

Figures 4 and 5 show the influence of the fibre length and fibre-matrix volume ratio on the resisting bending moment and modulus of rupture.

The trend of the general graphs shows that there are points on singular regions where there is a change in the composite behaviour. The meaning of these characteristic elements can be related to what is known as critical fibre length "1c", defined as twice the length of fibre embedment which would cause fibre failure in a pull out test. This can be related to the fibre length that composites sustain in a flexural test before cracking.

In this sense, graphs 4a and 5a show that gypsum-sisal at 2% the critical length is in the range of 5 cm and on the cement-sisal composite at 2% (graphs 4c and 5e) this length is less than 4 cm. Due to insufficiency of points, it has not been possible to determine "1c" for the gypsum-coconut composites.

The critical fibre volume "V_{crit}" is defined as the relative volume of fibre which, after cracking, will carry the load which the composite sustained before cracking in the direct tensile test, Hannant (1978). This can be related in this same sense, in a flexural test. In this sense, graphs 4b and 5b show that for gypsum-sisal composites, with fibre of 4 cm, the critical volume is 2%, while that for gypsum-coconut composites, the same characteristic is in the range of 3%.

Figure 6 shows the scheme of impact test and the graphs show a trend in the value of absorbed energy in the specimen of gypsum matrix of varying lengths, type of fibre and their relative volumes in the composites.

These graphs show there is a nonlinear proportional increase, in the absorption of energy by the composites with increase of fibre length and their relative volume. For the same fibre length and same fibre-matrix relative volume, the gypsum-coconut composites have presented greater energy absorption than the gypsum-sisal composites.

6 Final comments

Simple addition of fibre to gypsum or cement matrix changes their brittle behaviour permitting a ductile behaviour. The data presented show clearly this behaviour of composites. These influences can be precisely evaluated through flexural tests. These tests are an alternative to direct tensile tests. This includes the determination of characteristic values of behaviour ("1c" and "V_{crit}), initially defined for direct tensile tests.

The trend shows that characteristic values are identified for each studied composite. It is possible to have benefits and increase the utility potential of these composites.

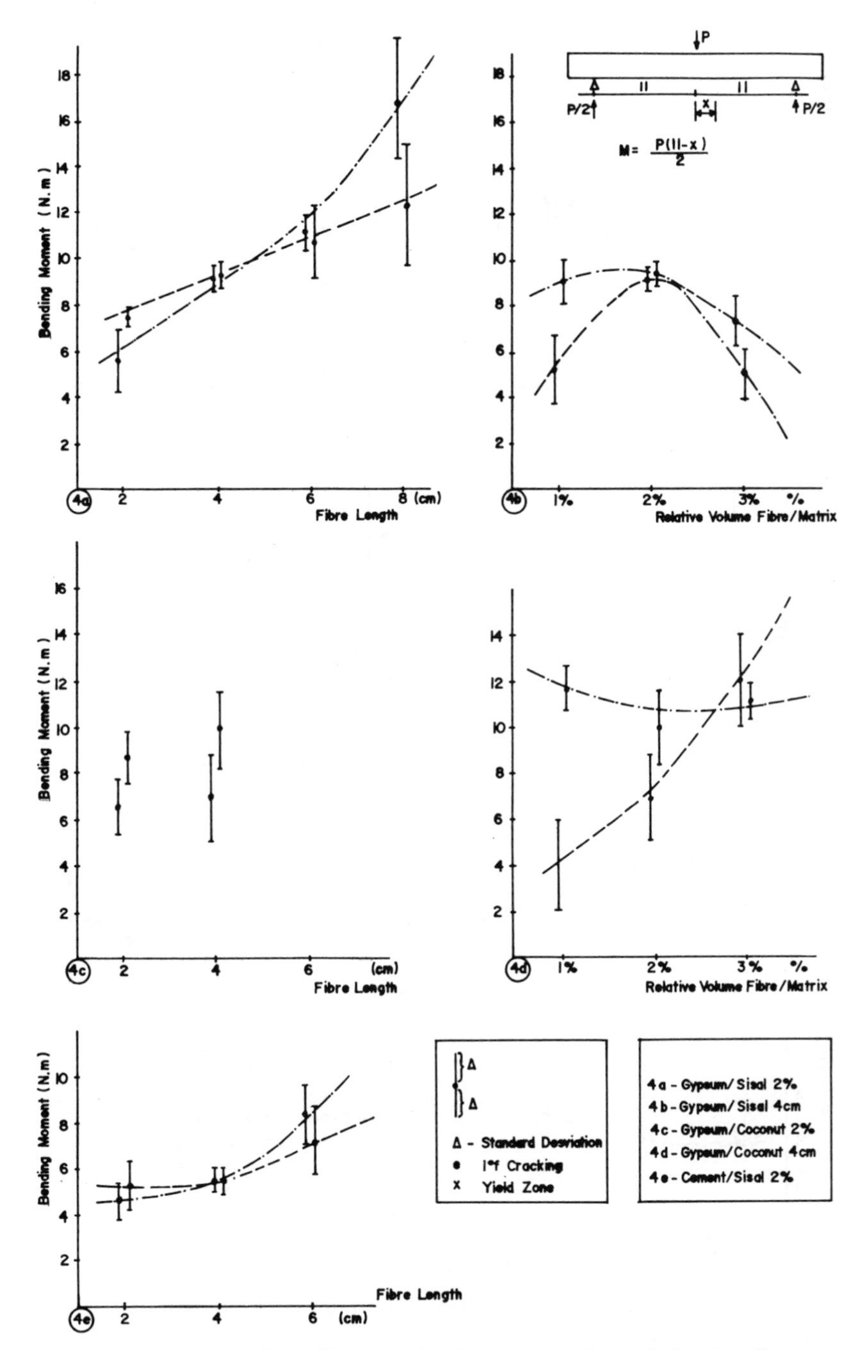

Fig. 4 Influences of the fibre-matrix elements on the resisting bending moment

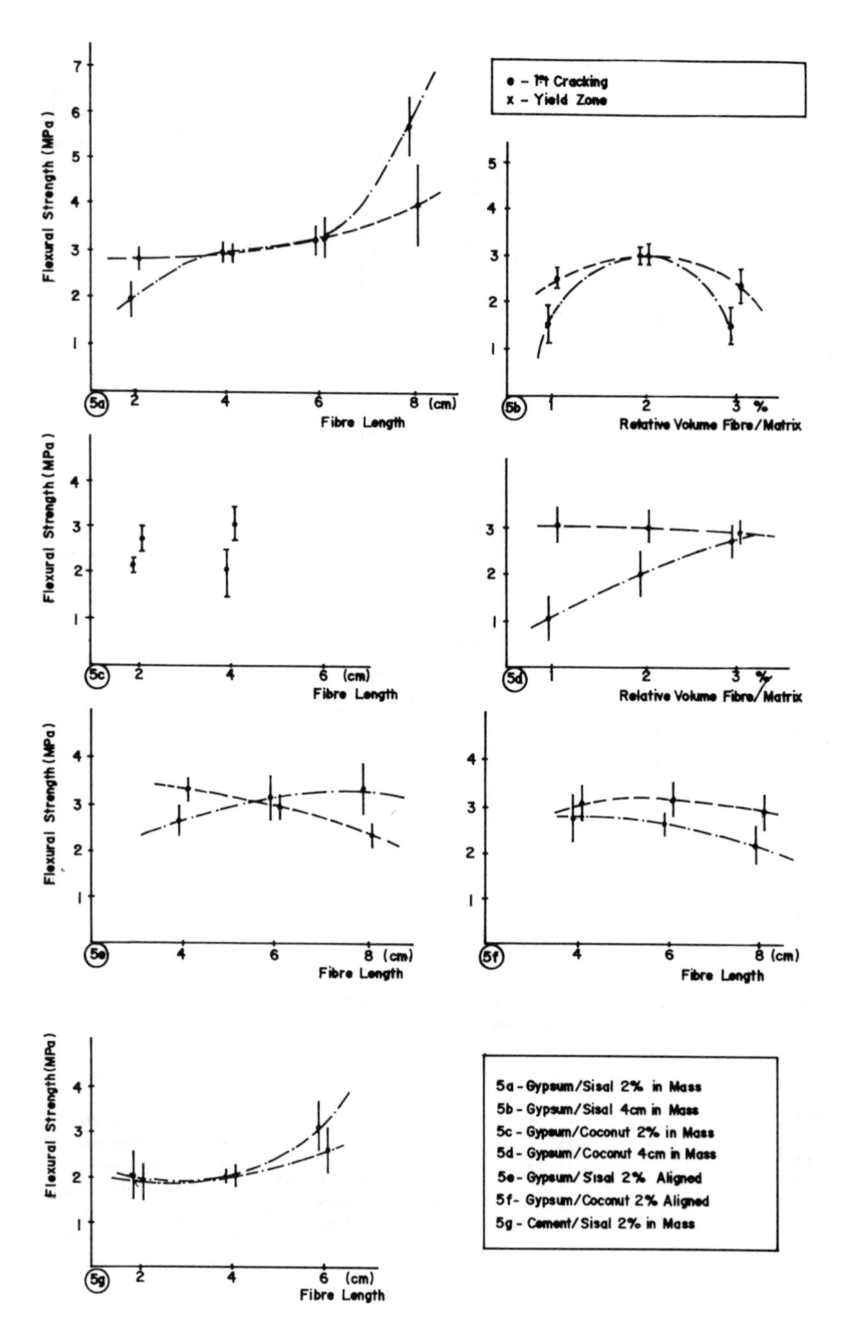

Fig. 5 Influences of the fibre length and fibre-matrix ratio on the modulus of rupture

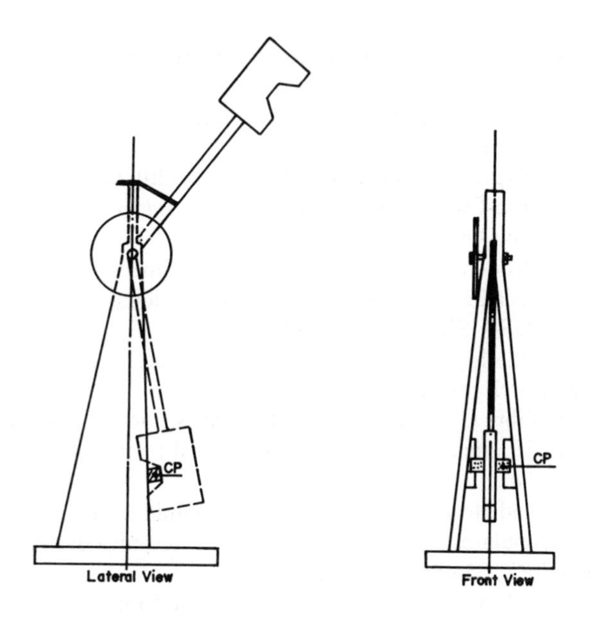

(a) Scheme of Charpy pendulum

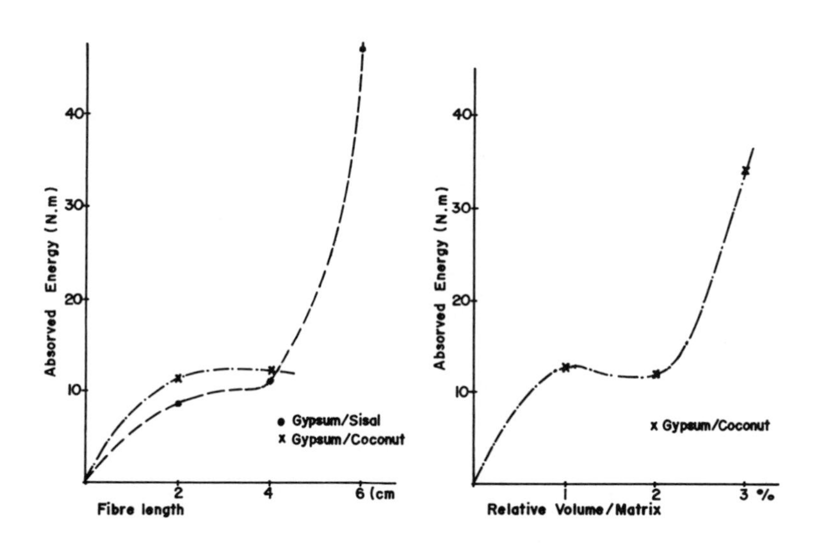

(b) Absorbed energy

Fig. 6 Impact test

7 References

Agopyan, V and Swamy, R N (eds) (1978) Natural Fibre Reinforced Cement and Concrete. Blackie, Glasgow 208-242.

Carvalho Filho, A C (1988) Argamassas Reforcadas com Fibras de Sisal - Comportamento Mecãnico à Flexão

Centro de Pesquisas e Desenvolvimento CEPED (1982) Utilização de Fibras Vegetais no Fibro-Cimento e no Concreto Fibra, Banco Nacional de Habitação, Rio de Janeiro 120 p.

Dantas, F A S (1978) Concretos de Baixo Consumo de Cimento Reforçados com Fibras - Propriedades e Análise da Fissu raçâo devido a Retraçâo. Tese de Doutorado EPUSP, São Paulo SP, 214p.

Gram, H-E (1983) Durability of Natural Fibre in Concrete. Swedish Cement and Concrete Research Institute, Stockholm, 230 p.

Hannant, D J (1978) Fibre Cements and Fibre Concretes, John Wiley, 219 p.

Laws, V (1971) The Efficiency of Fibrous Reinforced of Brittle Matrics, J Phys D, vol 4, 1773-1746.

Pires Sobrinho, C W and Horowitz, B (1987) Pontencialida des do Gesso para Habotações Populares - Caracteriza ção Fisico-Mecânica das Pastas, Anais I Habitec CIB/IPT, São Paulo SP, 121-138 p.

Savastano Jr, H (1978) Fibras de Côco em Arganassas de Cimento Portland para Produção de Componentes de Construção Civl, Disteração de Mestrado EPUSP, São Paulo, 201 p

96 PHYSICAL-MECHANICAL PROPERTIES OF FIBRE CEMENT ELEMENTS MADE OF RICE STRAW, SUGAR CANE BAGASSE, BANANA RACQUIS AND COCONUT HUSK FIBRES

A. RAMIREZ-CORETTI
Costa Rica Institute of Technology, Cartago, Costa Rico

Abstract
Fiber-cement specimens were tested for their basic physical and mechanical properties. Four organic fibers (rice straw, sugar-cane bagasse, banana raquis, and coconut husk), previously treated by chemical and mechanical methods, were combined each separately in a relatively dry cement mix. Three different fiber percentages were selected. Paired specimens were stored under indoor ambient conditions and tested in the same manner after six years to study the effect of time on the fibers and properties of the composite. No significant differences in strength relative to the un-reinforced specimens were observed, yet fiber-cement specimens were more ductile. Physical properties did vary according to the fiber percentage.
Keywords: Cement, Fiber-cement, Flexural Strength, Morphology, Organic Fibers, Physical Properties, Reinforcing Materials, Stiffness.

1 Introduction

Fiber reinforced cement can be considered as a material where relatively short discontinuous fibers are randomly distributed throughout the matrix in order to overcome the problems brought about by the low tensile strength and strain capacity of a plain cement mix. It has been assumed that the chief function of the fibers is to inhibit the propagation of cracks through the brittle matrix and to enable stress to be transfered across cracked sections allowing the composite to retain some post crack strength and to withstand deformations much greater than can be sustained by the matrix alone. The type of fiber is considered to determine the properties of the fiber-reinforced cement **(17)**.

Local organic fibers represent an available resource for fiber reinforced materials in developing countries. Non organic fibers are relatively expensive, and make it

Fibre Reinforced Cement and Concrete. Edited by R. N. Swamy. © 1992 RILEM.
Published by E & FN Spon, 2-6 Boundary Row, London SE1 8HN. ISBN 0 419 18130 X.

difficult to provide adequate solutions to the housing needs of such areas of the world.

Different fibers and production methods have been studied worldwide for those purposes, yet, results obtained respond to particular situations such as: fabrication process, testing methods, specimen size, w/c ratios, fiber types, etc. Difficulties of this kind indicate that research should be strongly related to the characteristics of the elements intended to be produced by each country. Thus, production and testing procedures should be very close to the expected performance, local technology, and final use of the fiber-cement elements.

Fabrication procedures range from simple casting of the fiber-cement mix in the molds, to vibration or pressure techniques applied to the specimens while the mix is still fresh. In order to consider the simplest of all the manufacturing procedures, plain cement paste was combined with four different fibers abundant in this region: rice straw, banana raquis, sugar-cane bagasse, and coconut husks. These last two have been considered by other authors, but following different procedures. The purpose of this study was to establish the benefits of these fibers relative to each other, and to plain un-reinforced cement mixes.

2 Some factors that determine fiber-cement properties

Previous studies conducted by da Silva **(14)** with five organic fibers (sisal, coir, bamboo, piassava, and sugar-cane bagasse) and mortar mixes, concluded that an increase in the ratio of aggregates in the matrix produced greater mixing difficulties. Also a great decrease of the fiber capacity to increase the tensile or flexural strengths of the composite was observed. Pressure was applied as it was considered to make a more efficient use of fibers, when compared to vibration and compaction methods, in increasing the flexural strength. Other studies **(2,5,6)** also determined that pressing the mix in the molds gives stronger specimens.

Adding fibers above a certain amount, even for the pressed elements, does not increase significantly the flexural strength. Lower w/c ratios give higher flexural strength, but as the fiber content increases, the Modulus of Rupture tends to decrease. The use of fibers as reinforcement provides increased safety due to the post-cracking ductility, assuring slow rupture, without a separation of the damaged parts. Yet, this too is subject to the effective fiber-matrix bond strength.

The volume of fibers is strongly related to the properties of the composite **(4)**. The ideal volume in sisal-cement specimens was between 3 an 7 % by volume **(2)**. Flexural strength increased when longer fibers of sisal were used. However, above 30 mm, placement became difficult, causing decrease in strength and scatter of the test results. Low w/c ratios and long fibers make mechanical mixing difficult, and hand mixing must be used. Thus, for randomly placed fibers, there must be a limit to the length of the fibers.

Ryder (referred to in **8**) confirmed that the length of fibers to be used in fiber-cement ranges from 1.0 mm to 40.0 mm. Longer fibers may be used, but they should be placed in a defined orientation. Works with agave fibers 50 to 75 mm long **(4)**, randomly allocated in the matrix, reported good results.

Organic fibers have low Young's Modulus. This property has been found to increase the impact strength of mortar mixes **(3,7)**. Water absorption increases with fiber volume, differing with the type of fiber used. Also, specific gravity of the matrix decreases, but this decrease is not significant over the range studied. Coconut fibers have been found suited for low strength, lightweight, high thermal insulation in low cost construction materials in the manufacture of non load bearing walls, ceiling and roofing in residential, industrial and comercial buildings **(12,15)**. Also, it is resistant to both fungal and bacterial decomposition **(15)**.

Tezuka **(17)** worked with piassava, coconut, and jute in cement concrete mixes. Fiber length was approximately 50 mm, and fiber percentages ranged from 1 to 6 by volume. Fibers were added in dry ambient condition, and a high w/c ratio was used (1,274). Problems in workability of the mixture were evident, and attributed to surface area, size and shape of the fibers. The inclusion of fibers in the matrix improved the flexural strength of the specimens, reaching maximum values at 2 and 3% by volume content of fibers. Fiber content above those values produced a drop in strength due to unworkable and segregated mixes. For values less than 2% by volume, the flexural strength of the fiber concrete were close to that of plain concrete.

Krenchel **(10,11)** suggested that organic fibers may be used in cement composites, but volumes of 15 to 20 percent by volume are required to give the composite suitable strength properties. However, due to the hygroscopic nature of the fibers, they vary with moisture content and may rot if kept moist for long periods. Also, they can not tolerate heating beyond 100 to 120 ^{O}C. He suggested that it was not likely that organic fibers would make possible a direct replacement of asbestos fibers, but that they may be included with asbestos fibers in asbestos cement production. This however has proved to be incorrect, since other organic fibers are been used to make fiber-cement boards and roof elements in this country with good results.

Besides the absence of a detailed description of the way the fiber volume was determined, the studies consulted do not mention an statistical analysis of the results, thus, there is no indication of whether the differences in strength between plain and fiber reinforced specimens were statistically significant or not.

3 Experimental procedure

Portland Cement (Type I) was used as the binding material. Cement-fiber mixes were prepared to exclude other variables such as aggregate size, fiber distribution around aggregate (difficulties produced by the presence of aggregates), and to use procedures followed with elements made with other fibers. Fiber bundles (true fibers are much smaller as shown by the morphological studies) were 10 to 20 mm average. They were cut using various procedures. The length was established with previous tests that provided the range within which fibers mixed well with the cement paste. The only fiber bundles that had shorter lengths were bagasse. That was due to a previous extraction process at the sugar mill, giving shorter length fiber bundles.

Cut fiber bundles were treated in a light solution (2%) of sodium hydroxide for 48 hours. That treatment helped to remove part of the lignin, rendering fibers less stiff and easier to mix. The material was beaten to separate fiber bundles producing a fibrous mass, washed to remove chemicals and fines, and then dried.

A first series of tests was conducted to determine the most workable and strongest possible mix. Four water to cement ratios (by weight) were selected : 0.20, 0.30, 0.40, and 0.50; two fiber conditions: natural and water saturated surface dry; and five fiber content percentages: 1, 2, 3, 4, and 5.

Adding natural unsaturated fiber to the mix complicated matters due to water requirements. Natural fiber absorbs water necessary for the cement reaction. Parallel tests were conducted to determine fiber water absorption. Rice straw, bagasse, banana raquis, and coconut fibers absorbed 179, 226, 246, and 258 percent of water (respectively) after 24 hours saturation in water.

Due to difficulties of unsaturated fibers, a second group of tests used saturated surface dry fibers, so that water to cement ratios could be kept constant, avoiding loss of reaction water by absorption into the fibers. Complementary tests determined that 4 and 5 percent fiber content made the cement paste and fiber mix difficult to place by the casting process selected. Thus, optimum fiber content (by weight of cement) was between 1 and 3 percent.

The equivalent volume relations, that resulted from those combinations are presented in Table 1.

The water to cement ratios studied gave workable mixes that ranged between 0.30 and 0.40. An additional set of tests determined an optimum ratio of 0.35.

Table 1.

Equivalent volume relations of fiber by weight of cement

Fiber percent (by weight of cement)		1	2	3	4	5
Volume relation	$\left(\frac{\text{fiber volume}}{\text{paste volume}}\right)$	9.33	18.66	27.99	37.32	46.65

Once variables were set test specimens were made by simple placement in the molds. No pressure was applied to densify or compact them. Size (242 x 76 x 8 mm) and testing procedures were determined using ASTM D-1037-78 **(1)** as a general guide. Load application speed was set to 0.5 mm/min.

Based on the results of the test program previously described, the following parameters and levels for the variables were set:

Water to cement ratio	=	0.35
Fiber condition	=	water saturated, surface dry condition
Fiber percentage	=	0, 1, 2, and 3 percent (by weight)
Fiber types	=	banana raquis, coconut husk, sugar cane bagasse, rice straw
Dependent variables	=	water absorption, volume variation due to water absorption, unit weight, Modulus of Rupture, and Modulus of Elasticity.

Strength properties were conducted on samples cured for 28 days and on paired samples after a six year period of exposure to natural indoor ambient conditions. Morphological properties of the source fibers were conducted following microscopy procedures **(16)**. Samples for the morphological properties were prepared following the Jeffrey Method. A random three block, 4 x 4 factorial experimental design was selected. An Analisis of Variance was conducted to determine whether the test results were significantly different. Factors considered were fiber type and fiber percentage. Moisture content during test was used as covariable. Materials for each cement-fiber combination were randomly selected, as was the fabrication and testing order.

4 Discussion of results

4.1 Morphological Properties

Banana raquis is characterized by having very long average fibers (Table 2). Sugar-cane bagasse is characterized by medium length fibers. Rice straw and coconut husks have short average fiber lengths. These fibers, as measured, have not been submitted to any mechanical treatment. When treated mechanically their length will be shortened. In banana raquis, medium to very long fibers predominate. In bagasse, medium size and long fibers are more abundant. Rice straw and coconut have a large percentage of medium and short fibers. According to the flexibility factor of Peteri, they show very similar characteristics. A higher Factor indicates a better possibility of fibers to intercross each other making a more consistant fiber mix. Runkell Factor shows that these fibers can be considered from good to excellent for paper production. A smaller Factor indicates better characteristics **(13)**.

Table 2.

Morphological properties for the four fiber types.

Fiber type	Fiber Length Distribution (%)				Mean length (mm)	Diameter (microns)	Wall thick. (microns)	CF	RF
	Short	Medium	Long	Very Long					
Banana raquis	2	12	25	61	1,89	65,00	7,10	29	0,22
Bagasse	18	37	26	19	1,23	53,30	8,35	23	0,46
Coconut husk	68	17	13	2	0,88	40,40	11,60	22	0,57
Rice straw	57	30	11	2	0,92	24,40	4,55	38	0,59

Short - $0,1 \leq L < 0,9$ mm
Medium - $0,9 \leq L < 1,2$ mm
Long - $1,2 \leq L < 1,6$ mm
Very Long - $1,6 \leq L$ mm

CF - Coef. Flexibility Peteri
RF - Runkel Factor

4.2 Physical Properties

4.2.1 Unit Weight

Fiber percentage and type influence this property. Unit weights differ for each fiber percentage, been lighter as the fiber content increases (as expected). Of the fiber types used, those made with coconut, rice straw, and bagasse weigh alike, but are heavier than those made with banana raquis for a 3 percent fiber content (Table 3, Figure 1).

4.2.2 Volumetric Variation

Fiber percentage has an effect on the volume change of the specimens, increasing with the amount of fiber. However, there are no significant differences among fiber types (Table 3, Figure 2).

4.2.3 Water Absorption

Fiber percentage and type influence this property. Water absorption is higher for those elements made with a 3% fiber content. No significant differences were observed between those containing 0, 1, and 2 % fiber content. Those specimens made with banana raquis absorbed more water than those made with coconut, rice straw or bagasse (Table 3, Figure 3). No significant differences were observed among this last three.

4.3 Mechanical Properties

4.3.1 Specimens cured for 28 days

Modulus of Rupture

Differences were observed among the results obtained, percentage and fiber type have a significant effect (Table 3, Figures 4 and 6). Fiber percentages of 1 and 2 percent have no beneficial effect on the strength of the specimens with respect to the control specimens (0 % fiber content). The type of fiber does have an effect also. Although bagasse, coconut, banana raquis and the control elements do not present considerable differences in strength, they are significantly stronger than rice straw. Ductile failure was observed, holding together until the last fibers failed or pulled-out (unlike the control specimens). Failure was more ductile for coconut fiber specimens, and very similar for the other three.

Modulus of Elasticity

Fiber percentage affects the flexibility of the specimens, but no differences were observed regarding fiber type. Specimens with no fiber, and those with fiber contents of 1 and 2 percent are stiffer than those containing 3 percent fiber (Table 3, Figure 7).

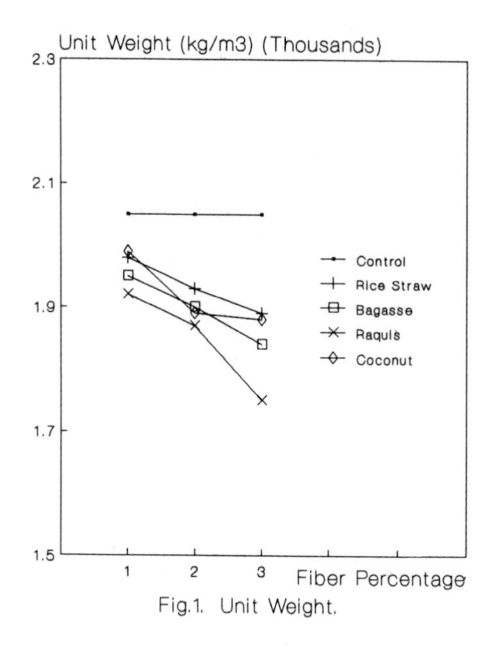

Fig.1. Unit Weight.

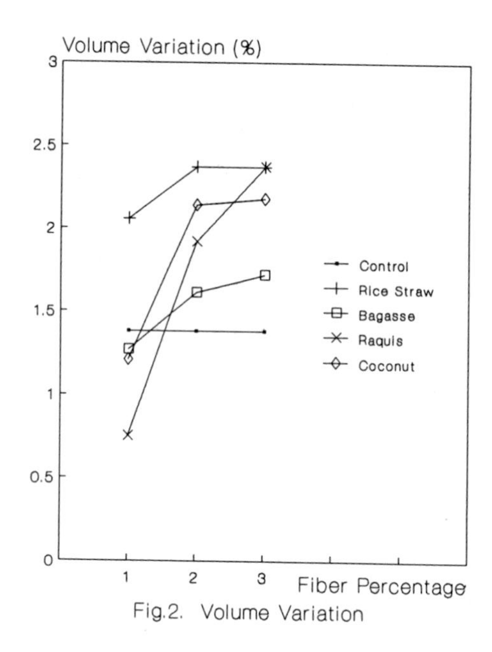

Fig.2. Volume Variation

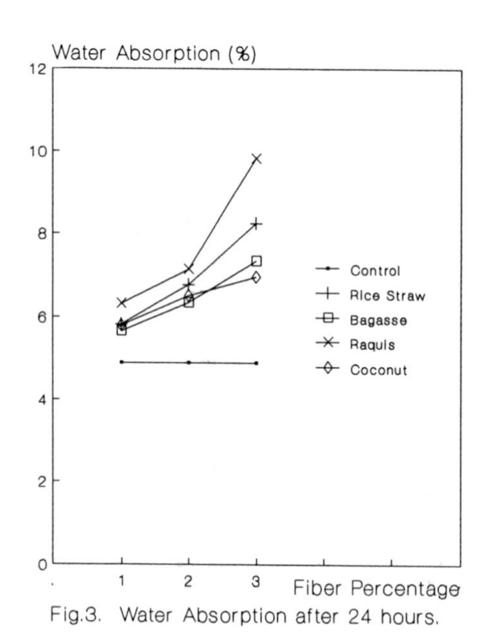

Fig.3. Water Absorption after 24 hours.

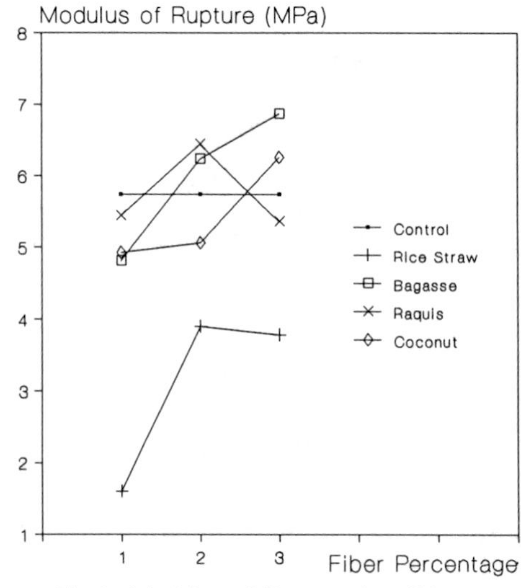

Fig.4. Modulus of Rupture for different fibers and percentages. After a 28 day curing period.

Table 3.

Physical and Mechanical properties for the different fibers

Fiber Type (%)	Fiber content (%)	Modulus of Rupture 28 days (MPa) (*)	Modulus of Rupture 6 years (MPa) (*)	Modulus of Elasticity 28 days (GPa) (*)	Modulus of Elasticity 6 years (GPa) (*)	Physical Properties UW (kg/m3) (*)	Physical Properties VV (%)	Physical Properties WA (%)
No fiber	0	5,74	8,24	9,16	16,03	2050	1,38	4,89
Banana raquis	1	5,44	8,77	7,79	16,35	1920	0,75	6,32
	2	6,44	8,90	7,24	12,41	1870	1,92	7,14
	3	5,36	7,89	6,11	13,11	1750	2,37	9,82
Bagasse	1	4,81	7,92	7,78	16,81	1950	1,27	5,66
	2	6,23	8,03	8,20	12,63	1900	1,61	6,34
	3	6,86	8,94	7,38	14,77	1840	1,72	7,34
Coconut husk	1	4,93	6,80	9,40	15,79	1990	1,21	5,80
	2	5,06	8,04	7,45	13,77	1890	2,14	6,51
	3	6,26	7,89	7,95	13,62	1880	2,18	6,97
Rice straw	1	1,60	8,93	10,19	12,29	1980	2,06	5,82
	2	3,90	8,93	8,18	14,27	1930	2,37	6,78
	3	3,78	9,20	5,64	11,38	1890	2,37	8,25

* Average Moisture content during test of 11,46%

UW - Unit weight

WA - Water absorption after 24 hours

VV - Volumetric Variation after 24 hours saturation in water

4.3.2 Specimens cured for 28 days and tested after 6 years

Modulus of Rupture

Small differences were observed for fiber contents or 2 and 3 percent (Table 3, Figures 5 and 6). Strength properties in general were higher than those obtained with the specimens tested after 28 days, but this can be attributed to the increase in strength of the cement mix with time. Although strength values differed slightly, those specimens with fiber did show ductile failure, unlike the brittle and

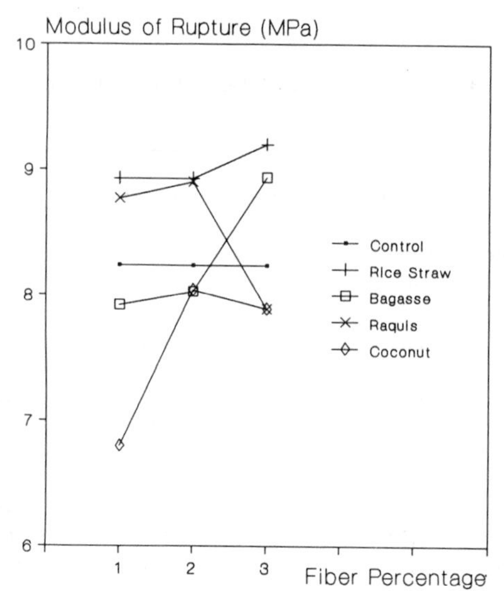

Fig.5. Modulus of Rupture for different fibers and percentages. After 6 years.

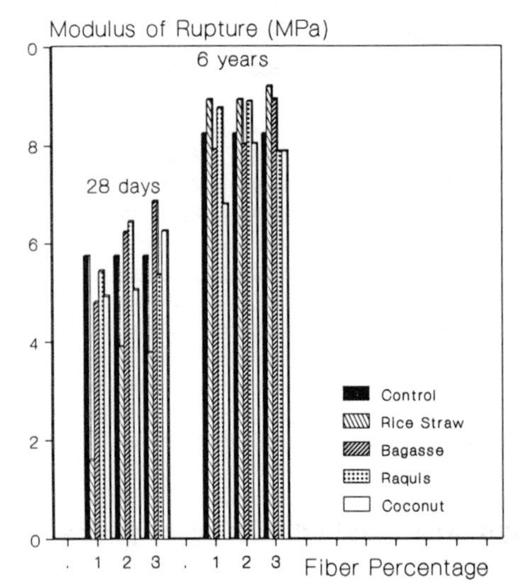

Fig.6. Modulus of Rupture for different fibers and percentages. After 28 days and 6 years.

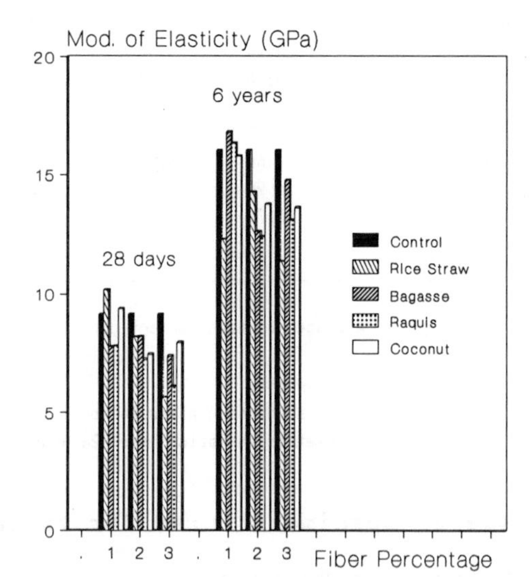

Fig.7. Modulus of Elasticity for different fibers and percentages. After 28 days and 6 years.

sudden mode observed in specimens with no fiber content. The specimens failed until the last fibers broke or pulled-out. Those reinforced with coconut husk fiber were relatively more ductile than the other three types.

Modulus of Elasticity
Fiber type and content affects the behavior of the specimens. Those made with coconut and rice fibers were less stiff than the other types for a fiber content of 1 percent. For the other fiber contents, no appreciable differences were observed between fiber types (Table 3, Figure 7).

5 Conclusions

- Fibers and fiber-cement specimens considered show differences among them in morphological and physical properties.

- Fibers of rice straw, banana raquis, sugar-cane bagasse, and coconut husks had little effect in improving the composite's strength. They behaved very similar to un-reinforced specimens after long time exposure to natural ambient indoor conditions. Yet, ductile behavior was observed, and was better at fiber contents above 2%, providing added safety for impact loads, and less stiff specimens.

- Failure of the fiber-cement specimens was progressive. Rice straw and banana raquis specimens failed basically by breakage of the fibers. Bagasse and coconut husk specimens failed by both pullout and breakage of the fibers.

- Water saturated fibers should be used to avoid problems of workability and to maintain an specified water-to-cement ratio.

- Fiber percentages should be over 2% by weight of cement. Percentages above 3% showed difficulty in mixing and placement. For higher fiber contents the use of pressure is recommended. Approximate volume percentages (volume of fiber to volume of cement mix) for the above ranged between 19 and 28 %.

6 References

(1) ASTM. **Evaluating the properties of wood-base fiber and particle panel materials.** ASTM D 1037 -78.

(2) Centro Interamericano de Vivienda (1963) **Suelo-cemento, su aplicacion en la edificación.** Bogotá.

(3) Centro de Pesquisas e Desenvolvimiento (u.d.) **Castilha de Construcao con solo-cimento.** Projeto Tecnologias do Habitat Camacari. Camagari-Br, 27 p.

(4) ________________ (1978) **Manual de Construcao con solocimento.** Camacari, Brazil, 132 p.

(5) ________________ (1980) **Silo de ferro-cimento.** Camacari-Brazil, 9 p.

(6) ________________ (1980) **Utilizacao de fibras vegetais no fibro-cimento e no concreto-fibra.** Banco Nacional da Habitacao. Departamento de Pesquisas. Rio de Janeiro, 151 p.

(7) ________________ (1981) **Cobertura e Componentes para habitacao popular.** Banco Nacional de Habitacao. Departamento de Pesquisas. Rio de Janeriro, 226 p.

(8) ________________ (u.d) **Ensaio de formas para solocimento.** Camacari, Brazil.

(9) Ghavani, Khosrow; Van Hombeeck, Ricardo (1980) Application of coconut husk as a low cost construction material, in **Low-Cost and Energy Saving Construction Materials.** Vol. 1. Rio de Janeiro, Brazil. July 9-12, pp 53-60.

(10) Hannant, D.J. (1978) **Fiber Cements and Fiber Concretes.** Wiley. New York, 217 p.

(11) Krenchel, H. (1974) Fiber-reinforced brittle matrix material, in **Fiber-reinforced Concrete.** ACI Publication SP-44, pp. 45-78.

(12) Mohan, D.; Singh, S. M. (1976) Low-Cost Construction Materials from Coconut husk and Rice husk, in **New Horizons in Construction Materials** (ed. H. Y. Fang), ENVO Pub, Penn, pp. 477-487.

(13) Runkel, R. O. (1952) Pulp from Tropical Wood, in **TAPPI 34**, pp. 174-178.

(14) Silva Guimaraes, Suely da (1984) Experimental mixing and moulding with vegetable fiber reinforced cement composites, in **Low-Cost and Energy Saving Construction Materials.** Vol. 1. Rio de Janeiro, Brazil. July 9-12, pp. 37-51

(15) Surgh, S. M. (1978) Cocunut Husk - A versatile Building Material. **J. Ind. Acad. Wood Science.** V.9. No.2, pp. 80-87.

(16) TAPPI. **Fiber-length of pulp by projection.** T 232 cm-85.

(17) Tezuka, Yasuko; et al. (1984) Behavior of natural fibers reinforced concrete, in **Low Cost and Energy saving construction materials.** Vol. 1. Rio de Janeiro, Brazil. July 9-12, pp. 37-51

PART ELEVEN

FERROCEMENT

97 PREDICTION OF POST-CRACKING BEHAVIOUR OF FERROCEMENT BOX GIRDER ELEMENTS BY FEM

N. M. BHANDARI
Civil Engineering Department, University of Roorkee, India
V. K. SEHGAL
Regional Engineering College, Kurukshetra, India
S. K. KAUSHIK and D. N. TRIKHA
Civil Engineering Department, University of Roorkee, India

Abstract
In spite of widely greater use of ferrocement construction there is no rational method of analysis presently available to predict the complete response of the ferrocement thin plated structures in the pre and post cracking phases. The present paper describes a FEM formulation employing modified 'EI' approach for the non-linear analysis of ferrocement structures. The proposed technique has been rational and computationally economical. Four ferrocement box girders have been cast and tested destructively upto failure under monotonically increasing u.d.l. for the simply supported conditions. The girders were tested under symmetrically and assymetrically placed distributed load over the top flange. The comparison of the predicted and the test results establish the accuracy and the economic viability of the proposed method of analysis.
Keywords : Ferrocement Box Girder, Post Cracking Behaviour, Finite Element Method, Material Models, Failure Criteria, Crack Pattern.

1 Introduction

Thin ferrocement elements in different spatial form or shapes are being increasingly used for flooring/roofing elements in buildings on the basis of experimental investigation of their behaviour. However, an analytical procedure for prediction of the complete response of such complex structures through pre and post cracking ranges until failure to assess the true factor of safety against collapse and serviceability requirements is yet to emerge. Such a procedure should consider composite nature of ferrocement, non-linear material characteristics, anisotropy arising from unequal Young's modulii in the two principle directions arising from different stress levels, topological changes in the member caused by cracking/crushing of mortar or yielding of steel and the deformability of cross-section under applied loads. The finite element method is probably the only technique capable of such an analysis.

Analysis of ferrocement elements in the past have been attempted using either the conventional R.C. theory (1-3) or by treating ferrocement as a composite material (4-5). These methods are incapable of handling complete structures or predicting cracking patterns.

In the present paper, results of an analytical cum experimental

Fibre Reinforced Cement and Concrete. Edited by R. N. Swamy. © 1992 RILEM.
Published by E & FN Spon, 2-6 Boundary Row, London SE1 8HN. ISBN 0 419 18130 X.

study to investigate the complete behaviour upto failure of a ferrocement box girder are presented. Based on finite element method, a continuously deteriorating 'whole' element approach (6), which is a variation of the modified 'EI' approach has been proposed. A suitable four noded rectangular flat shell element capable of representing membrane and bending actions has been chosen for idealising the structure. The effect of material non-linearity and tension stiffening of mortar has been included.

For establishing the validity of the proposed formulation and feasibility of box beams for large spans, four simply supported ferrocement single cell box girders of span 4.58 m have been cast and tested under symmetrically and asymmetrically placed loads on the span.

2 Finite Element Formulation

In the past, the two approaches used to analyse concrete structures by finite element method are the 'modified EI' approach (6,7) in which cracked elements are assigned rigidities according to M-χ relations based on cracked beam concept and the layered element approach (8,9) in which every element is assumed to consist of a number of parallel layers having uniform material properties. The computational accuracy and the economic viability of the modified 'EI' approach has been demonstrated for R.C. plated structure (10). In the present study, the same approach has been further modified to get a more efficient algorithm.

A flat four noded shell rectangular element with six degrees of freedom (dof) has been chosen (11) for the non-linear analysis, Fig. 2. The element is assumed to consist of a single mortar layer in the uncracked stage with smeared layers for wire mesh/skeletal steel reinforcement. After cracking the cracked and the uncracked depths of the mortar layer are considered separately. The cracking criterion requires determination of strains across the element depth. In accordance with the Kirchoff's assumptions, the strains $\{\epsilon\}$ at distance z from a reference plane are given by

$$\{\epsilon\} = \{\epsilon_0\} + z\{\chi_0\} \tag{1}$$

where $\{\epsilon_0\}$ and $\{\chi_0\}$ are strains and curvatures at the corresponding point on the reference plane.

The stiffness matrix $[k]^e$ for a flat shell element is of the form :

$$[k_e] = \begin{bmatrix} [k_m] & [k_{mb}] \\ [k_{bm}] & [k_b] \end{bmatrix} \tag{2}$$

where m, b and mb or bm stand for membrane, bending and membrane-bending interaction. The interaction terms exist due to unsymmetrical material properties about the reference plane resulting from unsymmetrically reinforced section or cracking/crushing of mortar, yielding of steel etc.

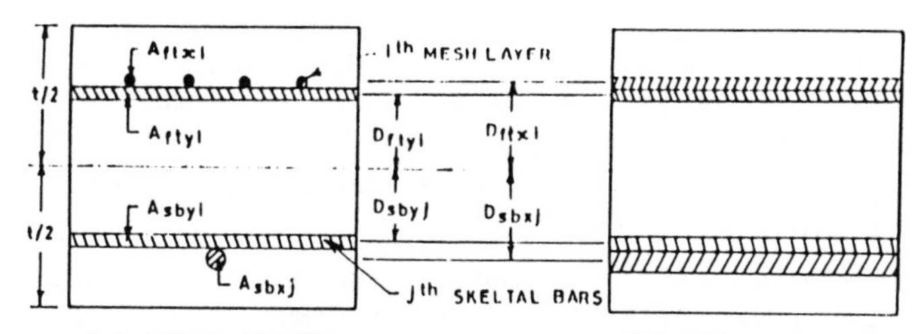

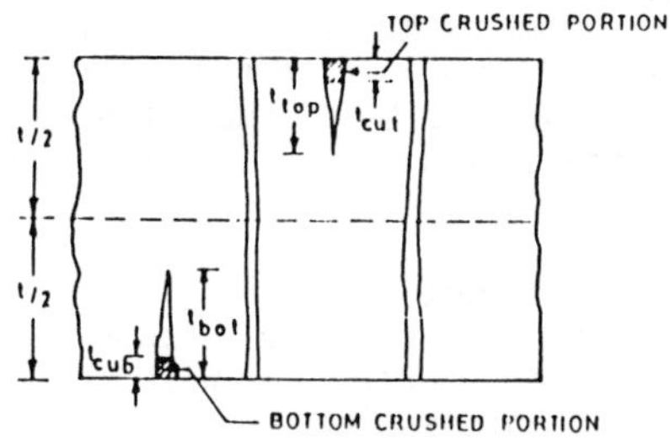

FIG.1 (a) TYPICAL FERROCEMENT SECTION
(b) FERROCEMENT IDEALIZED SECTION WITH EQUIVALENT DISTRIBUTED WIRE MESH AND SKELETAL STEEL REINFORCEMENT
(c) TYPICAL CRACKED SECTION

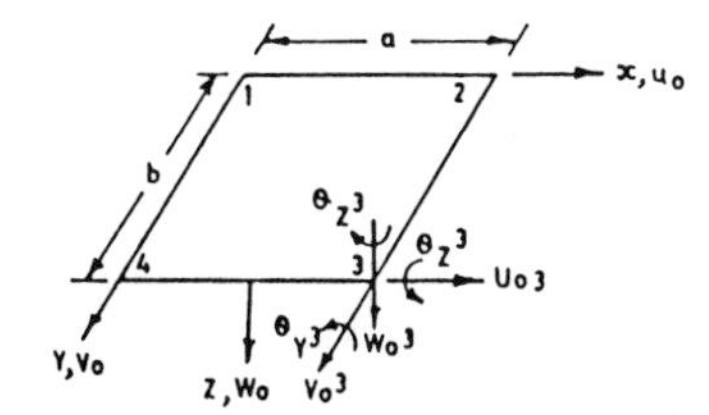

FIG. 2-RECTANGULAR FLAT SHELL ELEMENT (24 d.o.f)

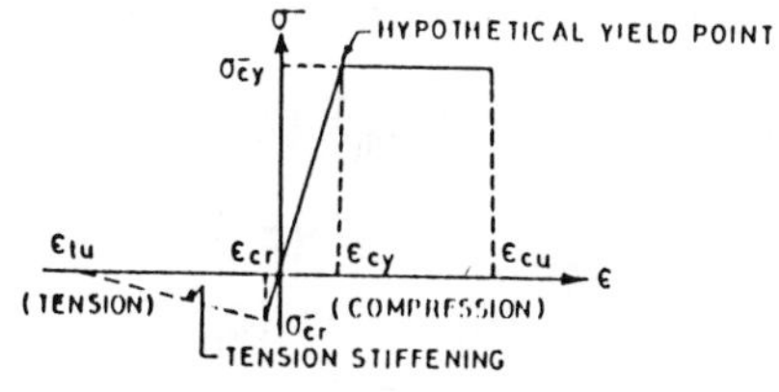

FIG. 3-IDEALIZED UNIAXIAL STRESS STRAIN CURVE FOR MORTAR (PRESENT STUDY)

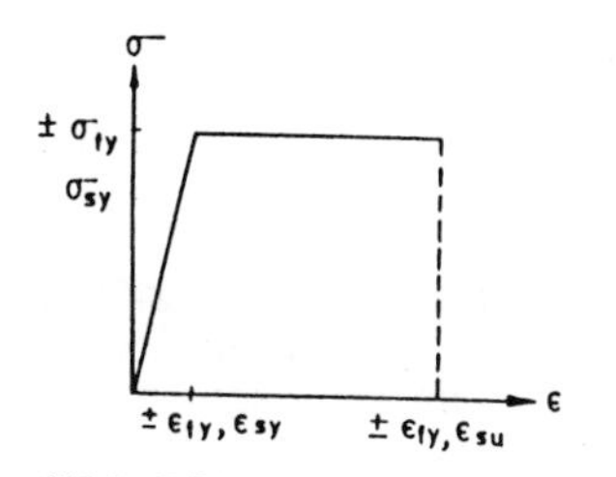

FIG. 4-IDEALIZED STRESS-STRAIN CURVE FOR WIRE MESH & SKELTAL STEEL

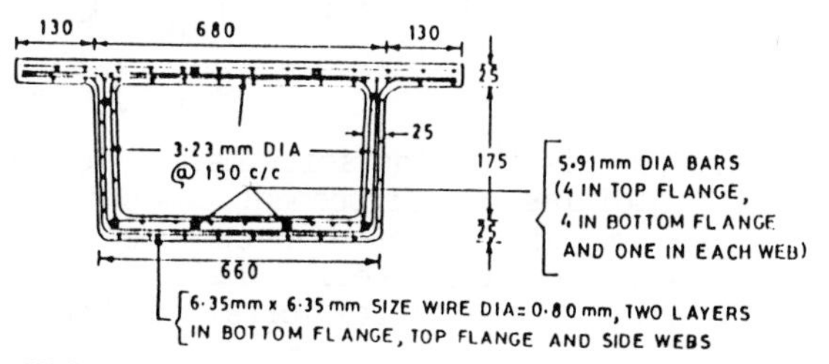

FIG. 5-SECTIONAL & REINFORCEMENT DETAILS OF BOX GIRDER

3 Material Models and Failure Criteria

a) Mortar

The uniaxial stress-strain relationship for mortar has been idealised as linearly elastic, perfectly plastic in compression and linearly elastic in tension as shown in Fig. 3. The mortar is assumed to crack/yield when the maximum principal tensile/compressive strains reach ultimate cracking strain, ε_{cr}/yield strain ε_{cy}. The mortar is assumed to be crushed when the principal compressive strain reaches the value of crushing strain (ε_{cu}).

Tension stiffening effect in mortar has been incorporated thus gradually relieving the stress normal to the crack as shown in Fig. 3.

At the yielding/cracking of mortar, the elastic modulus and Poisson's ratio have been reduced to zero along the direction of the principal compressive strain/normal to the crack and the shear modulus reduced to 0.4 G. Further, the crushed mortar is assumed to lose all its stiffness and material matrix [D'] is reduced to null matrix.

b) Wire mesh and skeletal steel

A linearly elastic-perfectly plastic stress strain relationship, as shown in Fig. 4, has been assumed for wire mesh as well as skeletal steel both in tension and compression.

4 Rigidities of Ferrocement Composite

In a plane stress problem membrane forces $\{F\}$ and moments $\{M\}$ at any point are given by relation :

$$\{F\} = [D_{mm}]\{\varepsilon_o\} + [D_{mb}]\{\chi_o\}, \quad \text{and}$$

$$\{M\} = [D_{mb}]\{\varepsilon_o\} + [D_{bb}]\{\chi_o\} \qquad (3)$$

where $[D_{mm}]$, $[D_{mb}]$ and $[D_{bb}]$ are tangent rigidities due to membrane action, membrane-bending interaction and bending action respectively. The three rigidities are given by following relations :

$$D_{mm} = \int [D]dz \; ; \qquad D_{mb} = D_{bm} = \int [D]z\,dz \quad \text{and}$$

$$D_{bb} = \int [D]z^2\,dz \qquad (4)$$

where [D] is the elasticity matrix for the material.

Rigidities of ferrocement composite have been obtained by summing up the rigidities of the component materials namely mortar, wire mesh and skeletal steel. Their evaluation in elastic stage poses no problem as the entire mortar thickness is uncracked.

Cracked/crushed stage : Based on the strain criterion, the affected depths due to crushing or cracking at top and bottom, t_{top} and t_{bot} (See Fig. 1), are calculated. If mortar cracks or yields/crushes in both the principle directions, then the larger of the two cracked/yielded depths is adopted for determining the rigidities. The rigidity of such a section has been determined by adding the rigidities of its contributing components. For example the membrane rigidity of such a section $[D_{mm}]'$ is given by ;

$$[D_{mm}]' = [D_{mm}]_{uc} + [D_{mm}]'_{tc} + [D_{mm}]'_{bc} + [D_{mm}]'_{s} \qquad (5)$$

where $[D_{mm}]'_{uc}$, $[D_{mm}]'_{tc}$, $[D_{mm}]'_{bc}$ and $[D_{mm}]'_{s}$ are the rigidity contributions of the uncracked mortar thickness, top cracked/yield mortar thickness, bottom cracked/yielded mortar thickness and the reinforcing steels. $[D_{mm}]$ is defined explicity below :

$$[D_{mm}]_{uc} = \frac{E_m (t_{bt} + t_{tp})}{1 - \nu^2} [D] \qquad (6)$$

$$[D_{mm}]'_{tc} = (t_{top} - t_{cut}) [D'] \qquad (7)$$

$$[D_{mm}]'_{bc} = (t_{cub} - t_{bot}) [D''] \qquad (8)$$

$$[D'_{mm}]_{s} = [D_s] \begin{bmatrix} (A_{tx} + A_{bx}) & 0 & 0 \\ 0 & A_{ty} + A_{by} & 0 \\ 0 & 0 & 0 \end{bmatrix} + [D_f] \begin{bmatrix} (A_{ftx}+A_{fbx}) & 0 & 0 \\ 0 & (A_{fty}+A_{fby}) & 0 \\ 0 & 0 & 0 \end{bmatrix} \qquad .....(9)$$

where $t_{tp} = \frac{t}{2} - t_{top}$ and $t_{bt} = \frac{t}{2} - t_{bot}$, $[D']$ and $[D'']$ are the modified material matrices for the mortar state. Other terms are defined in Fig. 1.

The membrane-bending interaction and the bending rigidities of the cracked/yielded mortar sections have similarly been determined (12), but not reproduced here.

5 Nonlinear Analysis Approach

An incremental iterative procedure has been used for the nonlinear analysis. The total load on the structure is applied in a suitable number of load increments and within each load increment, iterations are carried out till equilibrium is satisfied between the applied load and the forces induced within the elements. Thus during ith iteration, the unbalanced forces are determined using the updated tangent stiffness matrix $[K_T]_{i-1}$ based on stress/strain/cracked/yielded/crushed state of the materials and the total displacements at the end of the (i-1)th iteration. Convergence and divergence (for failure indication) checks have been applied on both unbalanced forces and the energy norms and the iterations are performed till the desired convergence is obtained. However, if convergence is not achieved after a specified number of iterations, the unbalanced forces are added to the next load increment. The failure is indicated by the divergence of the solution or the stiffness matrix becoming non-positive definite.

6 Experimental Programme

Four identical single cell ferrocement box girders G_1 to G_4 of cross section shown in Fig. 5., 4.88m long, have been cast to provide experimental evidence of viability of ferrocement box beams for long spans and to validate the theoretical approach proposed for analysis. Ordinary Portland Cement and coarse sand with a fineness modulus of 2.8 has been used in the ratio 1:2.5 by weight for the mortar, with water/cement ratio by weight of 0.45. A locally available superplasticizer equal to one percenet by weight of cement has been added. Bottom flange and side webs were first cast on specially prepared masonry mould plastered from inside and cured for 4-5 days. Top flange was then cast by providing 10-12 cm of overlap of mesh and skeletal steel reinforcement projecting from the web top. Mortar was compacted with the help of surface vibrator. The girders were moist cured for 28 days. Precast R.C.C. diaphragms of 50 mm thickness were provided at 150 mm from each end on completion of curing.

Reinforcement in the form of two layers of galvanized woven square wire mesh of dia. 0.8 mm and opening size 6.35 mm, longitudinal skeletal steel bars of 5.91 mm dia and transverse skeletal steel of 3.23 mm dia, was provided (Fig. 5).

Average values of compressive strength, direct tensile strength, modulus of elasticity and Poisson's ratio of the mortar were 21.7 N/mm^2, 2.3 N/mm^2, 14.0 kN/mm^2 and 0.2 respectively. For the wire mesh, 5.91 mm dia and 3.23 mm dia skeletal steel bars the average values of the yield stress, the ultimate strength and the Young's modulus of elasticity were determined as 440 N/mm^2, 520 N/mm^2, 146.2 kN/mm^2, 440 N/mm^2, 605 N/mm^2, 200.0 kN/mm^2 and 380 N/mm^2, 490 N/mm^2, 210 kN/mm^2 respectively.

These girders were tested under simply supported end conditions over a span of 4.58 m. Monotonically applied uniformly distributed load (udl) was applied in the form of brick layers. Two girders (G1, G2) were tested under udl over the entire top flange and the other two (G3, G4) under udl over half top flange width spread over full span.

7 Test Results

a) Girders G1, G2 subjected to udl over entire top flange

In both the girders, the first visible cracks appeared in the bottom flange and side webs near mid span at a load of 3.09 kN/mm^2 (3 brick layers). The maximum crack width observed in both the girders at this load was 0.1 mm. At a load of 4.12 kN/mm^2 (4 brick layers), a large number of cracks appeared and those formed earlier widened and penetrated further. Many cracks in the bottom flange transversed its full width. The maximum crack width in G1 and G2 was noted to be 0.15 mm 0.20 mm respectively. At the last and seventh brick layer (7.22 kN/m^2), the deflections increased appreciably and the maximum crack width in the two girders reached 0.6 mm and 0.8 mm respectively.

The experimental load versus deflection and the load versus longitudinal tensile strain at 25 mm above the soffit for the mid span sections are shown in the Fig. 6 and Fig. 7 respectively. The non-linearity in the behaviour is noticed at loads of 2.67 kN/m^2 and 2.38 kN/m^2 for the girders G1 and G2 respectively. The average test first crack load is thus about 2.52 kN/m^2.

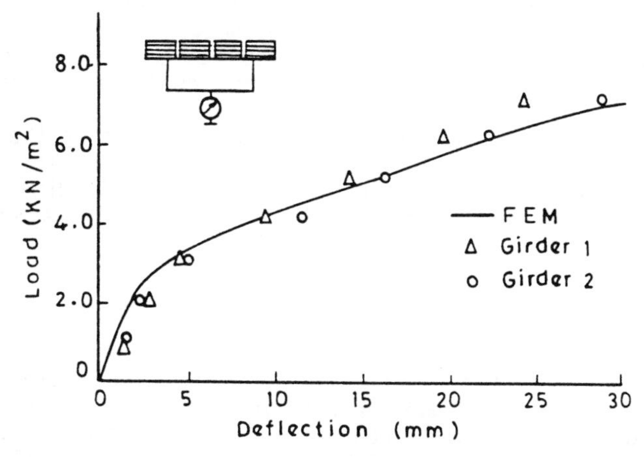

Fig. 6. Load vs mid span deflection for Girders 1, 2

b) Girders G3, G4 subjected to udl over half top flange width on the full span

In girder G3, the first visible cracks appeared at mid-span in the bottom flange below the loaded portion after the application of six brick layers (6.19 kN/m^2). The maximum crack width was 0.05 mm. The first visible cracks in girder G4 appeared at five brick layers

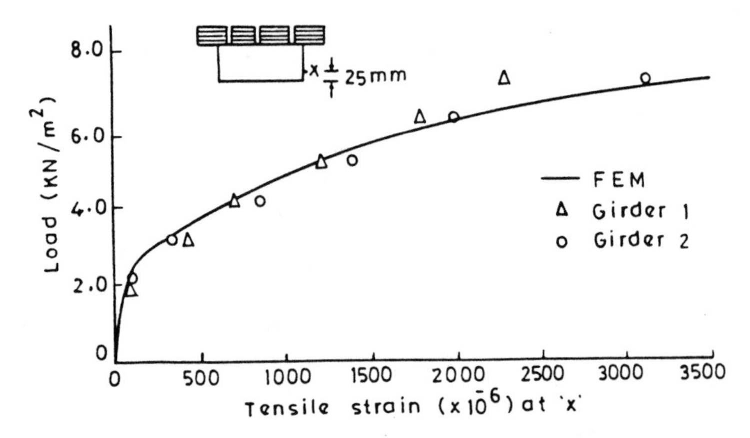

Fig. 7. Load vs mid span long. tensile strain for Girders 1, 2

(5.15 kN/m^2) and the maximum crack width was 0.05 mm. With the application of subsequent brick layers, new cracks appeared in the bottom flange and side webs and those formed earlier widened and penetrated further. A total of nine brick layers (9.28 kN/m^2) were applied on both the girders and the further loading stopped because of large deformations. The maximum crack widths in the two girders at this load level were 0.20 mm and 0.22 mm respectively. The average spacing of cracks in the central part of the bottom flange were about 50 mm and 80 mm respectively. No shear cracks or longitudinal cracks at the flange-web junctions were observed.

The experimental load versus maximum deflection under the loaded webs at the mid span section is shown in Fig. 8 for both the girders G3 and G4. The variation of the load versus the longitudinal strain at a point 25 mm above the bottom of the loaded web is shown in the Fig. 9.

8 Theoretical Analysis

For the validation of the proposed theoretical formulation for predicting the post cracking behaviour of ferrocement plates structures, the four ferrocement box girders G1-G4, tested as described above, have been analysed. The finite element idealization of girders G1/G2, taking advantage of the symmetry, is shown in Fig. 10 for only quarter girder. There are 56 nodes and 44 elements. The non-linear analysis has been carried out for the self weight and the monotonically applied superimposed load in six steps each step comprising of one brick layer except the first load step of two brick layers. The theoretically predicted load versus deflection and the load versus tensile strain at the mid span section are shown in Fig. 6 and Fig. 7 respectively along with the test results. The first crack load was predicted by extrapolation at a load of 2.19 kN/m^2.

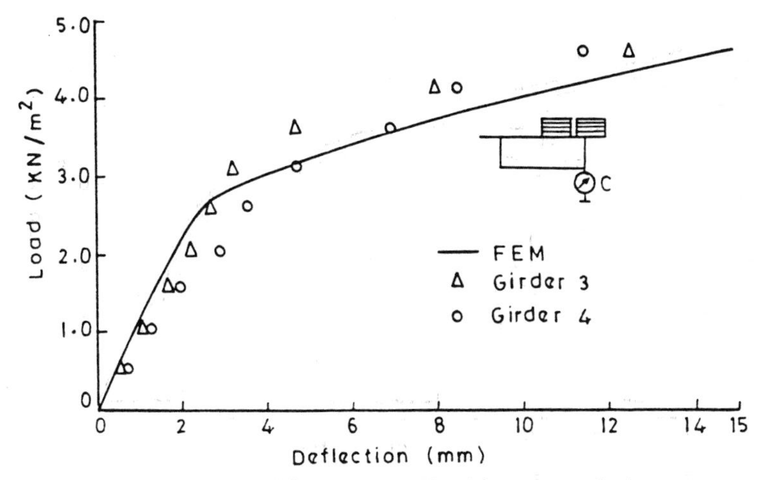

Fig. 8. Load vs mid span deflection for Girders 3, 4

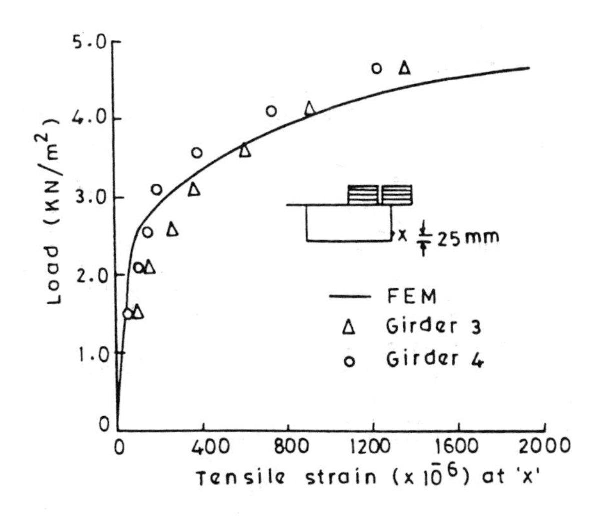

Fig. 9. Load vs mid span long. tensile strain for Girders 3, 4

The failure was indicated in the theoretical analysis during the seventh brick layer by the divergence of the solution. The yielding of the skeletal steel was indicated in the sixth brick layer.

Due to asymmetrically applied load on the half flange width of the girders G3 and G4, the one half of the girder has been discretized, as shown in Fig. 11, having 80 nodes and 72 elements. The load has been applied in six steps each of one brick layer except the first load step which was equal to 4 brick layers. Theoretically first crack load is predicted at 4.15 kN/m^2 and the failure is indicated by the divergence of the solution in the last step i.e. the ninth brick layer.

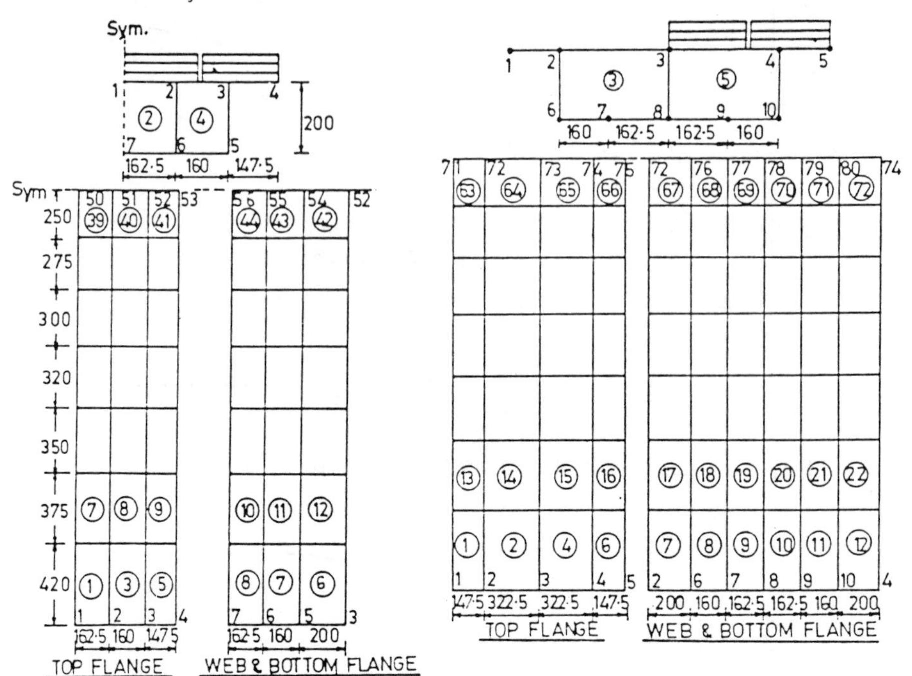

Fig. 10. Discretization for Girders 1 & 2

Fig. 11. Descretization for Girders 3 & 4

The load versus vertical deflection of bottom of the loaded web and the load versus the tensile strain at 25 mm above the bottom of loaded web at the mid span section is shown in the Fig. 8 and 9 respectively along with the test results of the two girders.

The present mathematical modelling cannot predict crack widths.

9 Comparison of Theoretical and Experimental Results

On the basis of the results presented, the following observations are made :

(a) Girders G1 and G2 : The predicted deflections upto four brick layers (4.12 kN/m^2) are about 78 percent of the average experimental deflections of the two girders whereas at five and six brick layers these are about 87 and 98 percent respectively. At seven brick layers, the predicted deflections become large thus indicating the failure of the girder.

The predicted first crack load of 2.19kN/m^2 and the experimental first crack load of 2.52 kN/m^2 show a good agreement.

At mid span the predicted and the experimental strains, Fig. 7, are small in the uncracked range and are in good agreement. After cracking of the mortar, the strains increase at a much faster rate. The predicted and the experimental strains are in agreement upto five brick layers. The yielding of the skeletal steel is predicted at six brick layers. The predicted strain at six brick layers is more than the experimental strain by about 43 percent due to sudden loss of rigidity due to the yielding of skeletal steel as the mortar had already cracked throughout the depth. At seven brick layers, the yielding of wire mesh is also predicted.

At the recommended crack width of 0.1 mm for ferrocement structures (13), the load carried by both G1 and G2 girders is 3.09 kN/m^2. At this load level, the average span/deflection ratio is about 960 which is about four times the limiting ratio of 250 from serviceability point of view. At a span/deflection ratio of 250, the average load taken by the girders is about 5.70 kN/m^2 which is close to the load at which yielding of the skeletal steel at the mid span is predicted. The crack width serviceability criterion is thus more critical than the deflection criterion.

The theoretical analysis predicts location, direction as well as depth of cracks. The predicted crack-patterns at various load levels compare well with the experimental crack-pattern, as seen at maximum applied load of seven brick layers in Fig. 12. The theoretical analysis also predicts longitudinal cracks on the inside surface of the top flange at mid span which could not be observed in tests. The analysis also predicts longitudinal tensile cracks on the top surface at the junction with the side webs in the mid span region. Experimentally, the cracks on the top surface, if any, could not be seen due to the brick loading and these cracks closed down after unloading.

(b) Girders G3 and G4 : From Fig. 8, it is seen that the predicted deflections show a good comparison with the experimental deflections in the elastic and cracked stages, but at the failure load of 9.28 kN/m^2, these are 33 percent higher.

As seen in Fig. 9, the predicted strains agree reasonably well with the experimental strains in the elastic and the cracked stages, except towards the end after the yielding of the skeletal steel when the predicted strains are somewhat higher.

At the recommended crack width of 0.1 mm, the loads taken by the two girders G3 and G4 are 7.22 kN/m^2 and 6.19 kN/m^2 respectively, and the average maximum deflection is about 5.0 mm which corresponds to a span/deflection ratio of about 920. At the maximum applied load of nine brick layers, the maximum crack widths in the two girders were 0.20 mm and 0.22 mm respectively, and the span/average deflection ratio was about 380 which is still greater than 250.

The predicted crack-pattern of the bottom flange and side webs shows a good agreement with the experimental crack-pattern, as seen in Fig. 13. The theoretical analysis also predicts the longitudinal cracks at mid-span on the inside surface of the loaded portion of the top flange. This could not be confirmed experimentally.

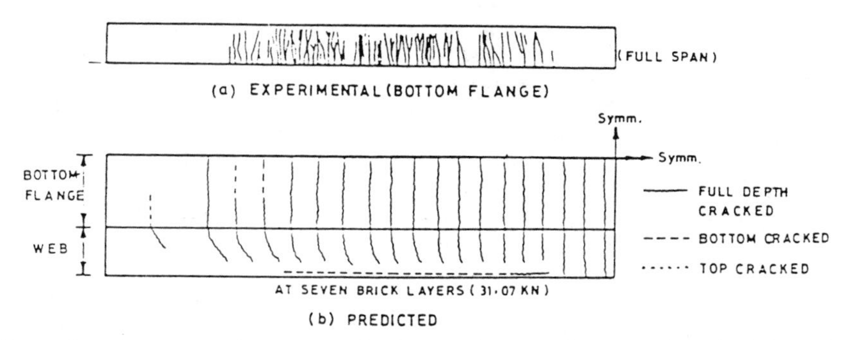

Fig. 12. Crack patterns of single cell box girder subjected to udl over entire top flange

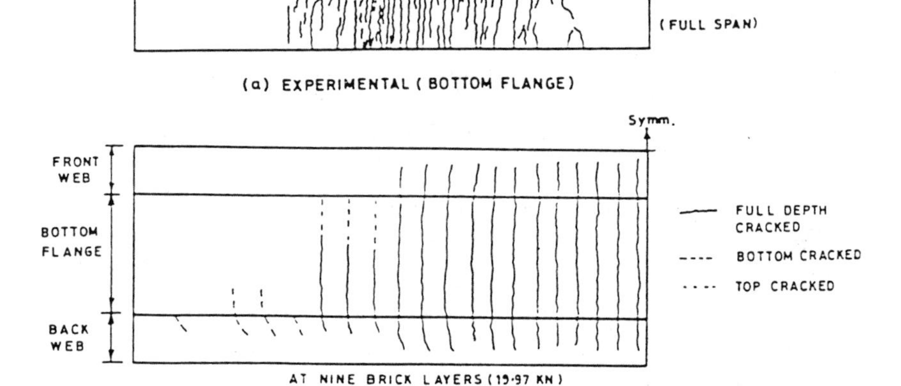

Fig. 13. Crack patterns of single cell box girder subjected to udl over half width and full span

10. Conclusions

Based on the results of the theoretical and the experimental study reported herein, the following conclusions may be drawn :

i) The proposed formulation which employs simple material models is capable of satisfactorily predicting the overall behaviour of ferro-cement structures in the post cracking phase including features like cracking, crack penetration, and yielding of steel.
ii) The proposed simplified EI approach is more economical than the layered approach, although some approximations have been made in the formulations.
iii) The predicted deflections and strains agree reasonably well with the experimental results till yielding of the reinforcement is indicated. Beyond this, the predicted values are somewhat larger.

iv) A comparison of the predicted first crack load with the first visible crack in the tests indicates that the former is lower by about 12 to 24%, when the loading ranges from a case of pure bending to bending combined with maximum torsion.
v) Serviceability in ferrocement box beams is governed by limiting crack width and not deflections. At the recommended crack width of 0.1 mm, the box beams have taken nearly same load placed symmetrically or at extreme eccentricity and shown similar deflections. This establishes the high torsional rigidity of ferrocement box beams.

vi) The experimental study clearly establishes feasibility of covering large spans in ferrocement by using box beams from view points of load carrying capacity and crack free construction.

11 References

Logan, D. and Shah, S.P. (1973) Moment capacity and cracking behaviour of ferrocement in flexure, **ACI J. Proceedings**, Vol. 70, No. 12, pp. 799-804.

Balaguru, P.N., Naaman, A.E. and Shah, S.P. (1977) Analysis and behaviour of ferrocement in flexure, **Proceedings ASCE**, Vol. 102, ST 10, pp. 1937-1951.

Guide for the Design Construction and Repair of Ferrocement, reported by ACI Committee 549, **ACI Structural Journal**, (1988), pp. 325-351.

Huq, S. and Pama, R.P. (1978) Ferrocement in flexure : Analysis and design, **J. of Ferrocement**, Vol. 8, No. 3, pp. 143-167.

Kaushik, S.K., Trikha, D.N., Kotdawala, R.R. and Sharma, P.C. (1984) Prefabricated ferrocement ribbed elements for low-cost housing, **J. of Ferrocement**, Vol. 4, No. 4, pp. 347-364.

Agarwal, S.H. (1979) Post cracking behaviour of R.C. plated structures, **Ph.D. Thesis**, University of Roorkee, Roorkee.

Bell, J.C. (1970) A complete analysis for R.C. slabs and shells, **Ph.D. Thesis**, Canterburry University, New Zealand.

Hand, F.R., Pecknold, D.A. and Schnorbrich, W.C. (1972) A layered finite elements non-linear analysis of R.C. plates and shells, **Str. Research Series No. 389**, Deptt. of Civil Engg., University of Illinios.

Lin, C.S. (1973) Nonlinear analysis of R.C. slabs and shells, **Structural and Material Research Report No. UCSESM 73-7**, Deptt. of Civil Engg., University of California, Berkley.

Jain, O.P., Trikha, D.N.and Agarwal, S.H. (1980) A finite element solution of post cracking behaviour of R.C. slabs, Advances in concrete slab Technology, Pergamon Press.

Gibson, J.E. and Mitwally, M.H. (1976) An experimental and theoretical investigation of model box beams in perspex and microconcrete, **The Structural Engineer,** Vol. 54, No. 4, pp. 147-151.

Sehgal, V.K. (1989) Behaviour of ferrocement box girder elements, **Ph.D. Thesis**, University of Roorkee, Roorkee.

ACI Comittee 549 (1982) State-of-the-art Report on ferrocement, **Concrete International** : Design and Construction, Vol. 4, No. 8, pp. 13-38.

98 FATIGUE CHARACTERISTICS OF WIREMESH REINFORCED MORTAR IN A CORROSIVE ENVIRONMENT

G. SINGH and M. FONG L. IP
Department of Civil Engineering, University of Leeds, UK

SYNOPSIS

A study on the fatigue behaviour of wiremesh reinforced ferrocement in a corrosive environment has been carried out through experimental work. Composites were loaded to the serviceability limit of 207-N/mm^2 stress in the outermost reinforcement after 28 days of normal curing. With the load sustained, they were stored in the environment chambers for various durations before being tested for static and fatigue flexure.

P-S-N relationships are established for the wire and the ferrocement composites. At 5% probability of failure, the fatigue resistance of the composites which had been subjected to sustained loading in the environment chamber of up to three months duration (1350 cycles of exposure) was found to be lower than that of the 28 days old control specimens. Whereas, composites with duration of seven months (3150 cycles of exposure) showed higher fatigue endurance than the control specimens for maximum stress level of 68% or lower. In contrast to the results given in the Report of the ACI Committee 549, the fatigue life of wire in air was found to be lower than that in ferrocement composites. Besides, an empirical formula is proposed to depict the relationships for the composites investigated, between crack width, cycle ratio and fatigue life.

KEYWORDS Accelerated, Corrosion, Cracks, Fatigue, Ferrocement, Flexural, Marine, Mortar, Probability, Wiremesh

INTRODUCTION

Wiremesh reinforced mortar is one of a variety of composites popularly known as ferrocement. As implied in the name it combines a ferrous product with cement in such a way that a multiphase composite is formed whose geometrical, physico-chemical and mechanical characteristics are quite different from those of the ordinary reinforced concrete [1,2]. This highly versatile material is formed into thin plates (including those with folds and double curvature) from cement rich mortar in which reinforcing elements are finely dispersed in the form of wire meshes/nets or expanded metal etc, with cover ranging from 3 to 7 mm (approx.). High passivity of the cement

Fibre Reinforced Cement and Concrete. Edited by R. N. Swamy. © 1992 RILEM.
Published by E & FN Spon, 2-6 Boundary Row, London SE1 8HN. ISBN 0 419 18130 X.

rich mortar (made with approximately 2mm limit on the maximum size of sand particles) combined with the low water-cement ratio (< 0.4) produce a matrix that is highly resistant to corrosion and abrasion. The ratio of cover thickness to diameter of the wires is very favourable in ferrocement.

Comparison of ferrocement with the more familiar and conventional materials reveals some of its distinctive characteristics. Figure 1 compares, schematically, the flexural behaviour of ferrocement with that of an equivalent gunite, the brittleness of which is contrasted by the very pronounced "ductility" of the former; vastly greater energy is required to cause failure. This leads to structures of high reliability and safety [3]. Figure 2 shows typical relationships between crack width and reinforcement stress in

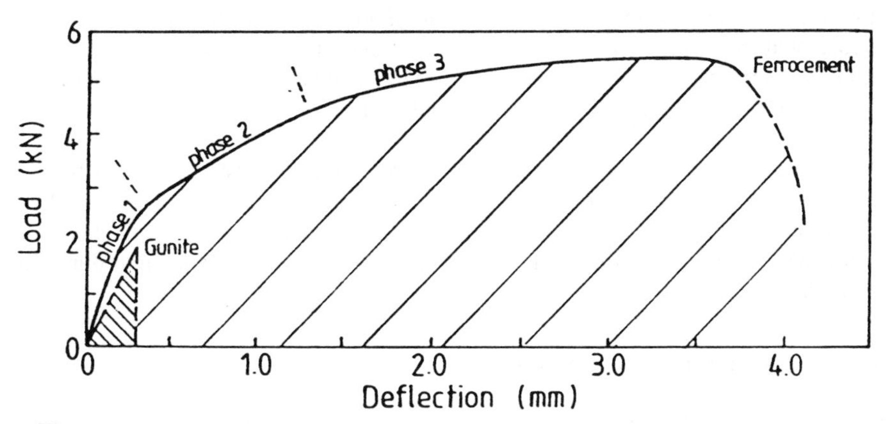

Figure 1 Load-deflection behaviour of ferrocement compared with that of Gunite

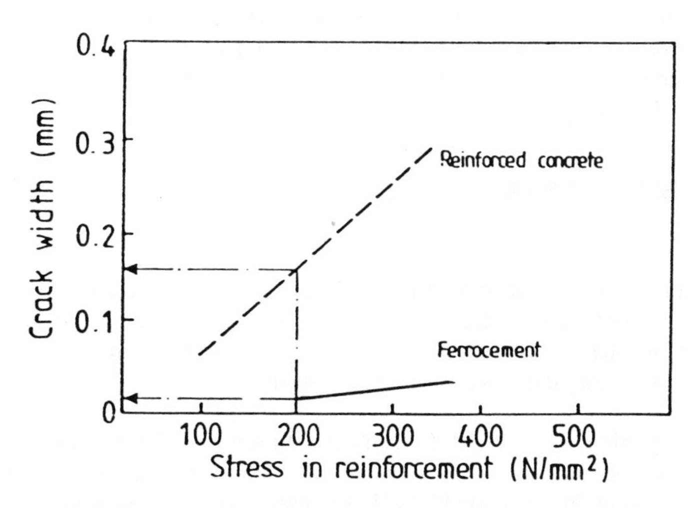

Figure 2 Flexural crack width-steel stress behaviour of ferrocement compared with that of reinforced concrete

the extreme fibres for reinforced concrete and ferrocement subjected to flexure. This figure is based on the American Concrete Institute (ACI) report 549 [2] and the authors' observations. It is important to note that at a steel stress of 200 N/mm^2 the crack widths in ferrocement are of the order of 1/8 of those in conventional reinforced concrete.

The concept of this composite, as we know it now, was put forward in 1847 by Lambot and was revived a hundred years later by Nervi who christened it by its present name. The present wave of interest in the UK can be dated to the sixties. The catalogue of examples of successful applications of this material all over the world are impressive. The first author of this paper conducted his first ferrocement fatigue tests in 1970 [4] and was directly involved in the De Menil Museum project in Houston [5] which is, in his view, the most innovative and elegant application of ferrocement [6] by the famous architect Renzo Piano. He used over 300 units of 12m x 2m size (approx.) which were manufactured by the then Ferrocement Laminates Ltd in Leeds (UK) and shipped to the USA in containers.

During the past few decades, ferrocement has attracted worldwide attention because of its proven suitability for marine structures and its potential as a constructional and rehabilitation material. While its mechanical properties are reasonably well established, relatively little work has been done on the fatigue and durability properties.

SCOPE OF THE INVESTIGATION

As part of a continuing programme of research into the endurance characteristics of ferrocement in normal and corrosive environment the authors have studied the effect of a corrosive environment on the fatigue life a ferrocement composite with a view to develop an improved understanding of this material and to help in the design of structures. Besides, an empirical formula is proposed to depict the relationship between crack width and cycle ratio.

MATERIALS, SPECIMEN PREPARATION AND TESTING

Materials

Only one type of reinforcement was used for all the specimens. It was galvanised drawn square weldmesh of 6.35mm x 6.35mm x 0.71mm diameter. The ultimate strength of reinforcement was 403 N/mm^2. The yield strength and modulus of elasticity were 345 N/mm^2 and 140 kN/mm^2 respectively.

The mixture proportion was 1:2.5:0.5 (cement:sand:water). The grading of the quartzite sand complied with Zone 2 limits of BS 882 (part 2) : 1973. Ordinary Portland cement, complying with BS12:1971 was used. The average cube strength was 60 N/mm^2 after 28 days of fog room curing.

Specimen Preparation

The specimen size was 350mm x 125mm x 30mm thick. Six layers of weldmesh were used as reinforcement, thus giving a percentage of reinforcement of about 1.14. No skeletal steel was used and 5mm spacers were fastened onto the first layer of reinforcement to obtain a cover of 5mm.

The specimens were cast horizontally in groups of five in a steel mould. For each group of specimens cast, control specimens of mortar were cast at the same time. They consisted of six 100mm cubes and six 500mm x 100mm x 100mm beams. Immediately after casting, the ferrocement and control specimens were covered with hessian and polythene sheeting for 24 hours at a temperature of about 19°C. The specimens were then demoulded and transferred to the curing room for further curing of 27 days at 20± 1°C and relative humidity of 98%.

Corrosive Environment and Preloading

This system was similar to that used by Ravindrarajah and Paramasivam [7]. It provided an accelerated marine weathering condition. Loaded specimens were stored in this system for various durations. Splash zone is the most severe marine condition. It is just above the high tide level where there are build-ups of salt-spray, wetting-drying and freeze-thaw cycles. The main features of this sytem were the cyclic temperature and moisture environments. The duration of each cycle was 90 minutes, which included a 60 minutes dry phase of warm air (70°C) and a 30 minutes phase of full immersion (40°C). 3.5% NaCl solution was used to simulate sea water and act as a wetting medium in the system.

To ensure that the specimens were subjected to severe corrosive conditions, they were preloaded before placing into the storage tanks. The specimens were loaded to the maximum allowable service limit. The State-of-the-art Report on Ferrocement [2] gives some guidleines for performance criteria. In this study, the concerning allowable limits were tensile stress in steel reinforcement of 207 MPa and crack width of 0.05mm.

Testing Rig

The testing rig consisted of a four point arrangement with a span of 300mm and a constant bending moment zone of 120mm. An upward jacking force was applied by a 20kN hydraulic jack through a diaphragm. The whole set up was enclosed in a perspex box. A pump drew the salt solution from the tray at the bottom and supplied it on the top in the form of a spray through a perforated copper pipe.

Strain and Deflection Measurement: The strain measurements were done with demountable transducers made with spring steel and strain gauges, and suitably water-proofed.

Deflection was measured by using an ordinary stainless steel linear variable differential transformer, fitted with brass coverings to avoid water ingress.

Data Logging System: It consisted mainly of a micro computer, an interface and a bank of amplifiers. Eight channels which comprised of six from strain transducers, one from the LVDT and one from the load cell were used. An interactive computer program was developed to capture, retrieve, amplify and store the results.

TESTING PROGRAMME AND PROCEDURES

A total of 60 specimens were tested for fatigue in flexure. They included four groups of fifteen specimens, one for each of the following curing conditions:

1) 28 days of normal curing: no preloading : control specimens
2) 28 days of normal curing plus one month in the marine environment (450 cycles of exposure)
3) 28 days of normal curing plus three months in the marine environment (1350 cycles of exposure)
4) 28 days of normal curing plus 7 months in marine environment (3150 cycles of exposure)

Failure was defined as fracture of the outermost reinforcement. The specimens which did not fail after two million cycles were tested to failure under static loading.

The designations of the various groups of specimens for static and fatigue testing are given in Table 1, a and b.

Flexural fatigue tests were carried out at a frequency of 5Hz under constant cyclic load. The maximum cyclic load corresponded to the stress induced in the outermost reinforcement expressed as a percentage of its ultimate strength. For group CONT and COR1M, five specimens were tested for each maximum stress level (nominal) of 75%, 65% and 55%. The minimum stress level (nominal) of 12.5% was used for all tests. About one-third of the specimens were runouts in each of the two groups. Therefore, higher maximum stress levels of 80%, 70% and 60% were used for COR3M and COR7M. The minimum calculated stress level remained nominally the same.

A continuous spray of 3.5% NaCl solution was applied onto the specimens during the flexural fatigue tests. The permanent and cyclic deflection, the load and cyclic strains were measured by the data logging system automatically. Tests ran uninterrupted except when crack and permanent strain measurements were taken. In general, each interruption was about five minutes. Specimens that did not fail after 2 million cycles (run-outs) were tested to static flexural failure.

After the specimens were broken either due to fatigue or static failure (of run-outs), the cover to outermost reinforcement and the specimen thickness were measured.

Table 1 Test Programme and specimen designations

1a: Static

Age of normal curing	Duration in environment chamber	Number of Specimens	Nomenclature
28 days	NIL	5	CONSTA
28 days	1 month	5	COR1MSTA
28 days	3 months	5	COR3MSTA
28 days	7 months	5	COR7MSTA

1b: Fatigue

Age of normal curing	Duration in environment chamber	Number of specimens	Nominal maximum stress level (% of fsu)	Nomenclature
28 days	nil	5	55	CONT
		5	65	
		5	75	
28 days + 1 month	nil	5	70	CONT1M
28 days	1 month	5	55	COR1M
		5	65	
		5	75	
28 days + 3 months	nil	5	70	CONT3M
28 days	3 months	5	60	COR3M
		5	70	
		5	80	
28 days + 7 months	nil	5	70	CONT7M
28 days	7 months	5	60	COR7M
		5	70	
		5	80	

The reinforcements exposed at the fracture surface were examined for evidence of rust.

RESULTS AND DISCUSSION

Static

Figure 3 gives a comparison of the average load-deflection curves obtained for the control and the preloaded specimens after various degrees of exposure. It can be seen that the load at which the initial linear portion of the curves end increases with an increase in duration of preloading period. Moreover, the ultimate strength increases with increasing duration of preloading. It is generally accepted that application of

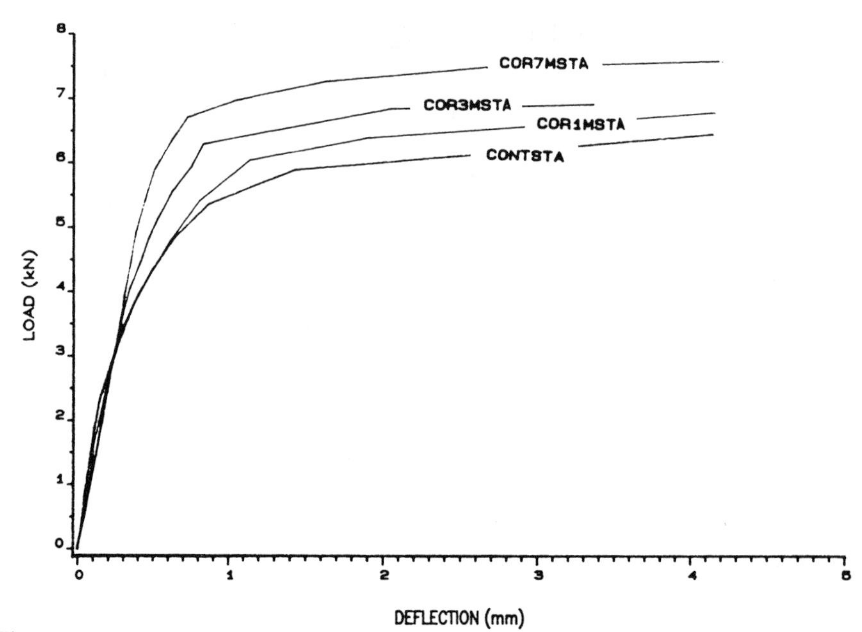

Figure 3 Load-deflection curves for control and preloaded specimens: not subjected to repeated loading prior to static test

sustained or dynamic loading to reinforced concrete prior to further static or dynamic test results in improved strength [8]. Such improvement is observed only if the sustained load level is below 75% of the static strength. The trend seen in Figure 3 may partly be explained by the increases in the mortar and the bond strength with age under sustained loading. This is supported by the fact that the final crack spacing increased (accompanied by fewer cracks) with increasing strength of specimens.

Comparison of the static load-deflection curves of the run-out specimens with those of the specimens of the same curing history but which were not previously subjected to fatigue loading showed the following:-

For control specimens tested after 28 days of normal curing the ultimate static strength and initial stiffness of cyclically preloaded specimens was distinctly higher than those of specimens which were not. This is in accord with the general experience.

For specimens which were previously subjected to sustained load under aggressive environment the ultimate static strength and initial stiffness of cyclically preloaded specimens was significantly but marginally higher than those of specimens which were not. This may be due to the fact that the beneficial effects of stress relief and stress conditioning provided by the repeated preloading are diminished by the stress corrosion effects in the aggressive environment.

Fatigue

Fatigue failure is defined herein as the fracture of a specimen. The stress in the outermost layer of the reinforcement is expressed as a percentage of its ultimate strength. To improve the estimate of this stress specimen size and reinforcement location were established after failure. The calculations were based on the transformed uncracked and cracked section theory for reinforced concrete; after making allowance for the number and locations of the various layers of mesh.

Table 2 shows the life of all the specimens together with their nominal and adjusted (calculated) maximum stress levels. The very significant differences between the nominal and the adjusted stress values point to the importance of adjustment based on actual post-failure measurements on specimens and use of Modified Goodman diagram for making allowance for minimum (12.5%) stress level in the loading cycle.

Table 2 Fatigue results for ferrocement specimens

SPECIMENS IN ENVIRONMENT CHAMBER FOR VARIOUS DURATIONS				
Stress range	Specimen	Maximum stress level (% of fsu) nominal	adjusted	Failure cycle
low	COR1M62S1	55	61.9	2000000
	COR1M62S2	55	61.9	2000000
	COR1M62S3	55	62.1	2000000
	COR1M65S1	55	65.1	2000000
	COR1M67S1	65	66.6	2000000
	COR1M67S2	65	66.6	75320
middle	COR1M69S	75	69.0	59260
	COR1M70S	65	70.0	2000000
	COR1M71S1	65	70.8	548310
	COR1M71S2	55	71.2	344770
	COR1M72S	65	72.3	50300
high	COR1M78S	75	77.9	53470
	COR1M82S1	75	81.6	110180
	COR1M82S2	75	82.1	33150
	COR1M85S	75	84.5	45680
low	COR3M61S	60	61.0	2000000
	COR3M62S	60	61.8	164830
	COR3M64S	60	63.6	2000000
	COR3M65S	60	65.4	2000000
	COR3M68S	60	68.5	411860
middle	COR3M71S1	70	70.8	548310
	COR3M71S2	70	71.2	344770
	COR3M76S	70	76.3	2000000
	COR3M80S	80	80.4	10630
high	COR3M83S	70	83.4	3700
	COR3M84S1	80	84.0	16640
	COR3M84S2	70	84.2	41020
	COR3M85S	80	85.0	15800
	COR3M86S	80	85.6	27630
	COR3M90S	80	90.4	13500
low	COR7M68S	60	67.6	2000000
	COR7M70S1	70	69.9	2000000
	COR7M70S2	60	70.4	2000000
	COR7M71S	70	71.3	181210
	COR7M72S1	70	71.9	2000000
middle	COR7M72S2	60	72.2	74210
	COR7M73S	70	72.9	124450
	COR7M76S	70	76.0	17600
	COR7M78S	60	78.1	2000000
	COR7M81S1	80	80.9	34270
high	COR7M81S2	80	81.3	16900
	COR7M83S1	80	72.5	2530
	COR7M83S2	80	73.1	23830
	COR7M85S	60	84.9	29030

CONTROL SPECIMENS				
Stress range	Specimen	Maximum stress level (% of fsu) nominal	adjusted	Failure cycle
low	CONT60S	55	60.1	2000000
	CONT63S1	55	63.1	2000000
	CONT63S2	55	63.2	2000000
	CONT63S3	65	63.4	779660
	CONT65S	55	64.7	2000000
middle	CONT66S	55	66.5	2000000
	CONT68S	65	67.7	2000000
	CONT71S	65	71.0	581650
	CONT74S	65	74.1	147980
	CONT75S	75	75.5	70410
high	CONT79S	75	78.7	37540
	CONT80S1	65	79.9	101630
	CONT80S2	75	80.0	121320
	CONT87S	75	87.2	13370
	CONT92S	75	92.2	8250
middle	CONT1M71S	70	70.7	384350
	CONT1M73S	70	73.3	89080
	CONT1M74S	70	73.6	142190
	CONT1M77S1	70	76.8	52790
	CONT1M77S2	70	77.0	182060
middle	CONT3M64S1	70	64.1	494030
	CONT3M64S2	70	64.4	189090
	CONT3M65S	70	65.0	2000000
	CONT3M69S	70	69.0	496780
	CONT3M75S	70	74.7	27630
middle	CONT7M71S1	70	70.8	2000000
	CONT7M71S2	70	70.8	2000000
	CONT7M73S	70	72.8	471840
	CONT7M74S	70	74.1	143500
	CONT7M76S	70	76.1	152690

Fatigue is a highly stochastic process. The scatter of results is much wider than that of the static ones. Therefore, appropriate statistical and probability analyses are essential in order to draw meaningful conclusions. Singh et al [9], Weibull [10] and Wirsching and Yao [11] have emphasised that it is important to include the probability aspect into the S-N curves, thus leading to P-S-N curves. The significance of the probability aspect could be demonstrated by the study conducted by Hawkins and Heaton [12]. They investigated the fatigue behaviour of welded wire fabric by testing 59 specimens of weld-section and compared their results with those of ordinary reinforcing bars. At 50% and 95% probability of failure, wire fabric had higher fatigue endurance than reinforcing bars, therefore, wire fabric was preferred. However, at 5% probability of failure, reinforcing bars were preferred. In view of the fact that 5% probability of failure is more relevant for design purposes, the reversal of preference indicates the significance of the probability aspect.

P-S-N relationships were obtained by first performing a least square fit to all the data for any group, with stress as the dependent variable. Weibull's [10] method was then used to obtain various probabilities of failure for different stress levels through having first obtained the P-N regression plots.

The 5% probability of failure lines for each group are presented in Figures 4a and b. These figures show the effect of exclusion and inclusion of run-out test results. For fatigue of ferrocement, Karasudhi et al [13] and Singh et al [9] included runouts in the linear regression for the S-N relationship, whereas McKinnon and Simpson (14) excluded runouts in their analysis. Cornelissen (15) and Singh and Ip (16) prefer the former approach because excluding runouts may result in an non-conservative regression analysis.

Also shown in these figures are the fatigue characteristics of wire tested in the air. It is obvious that in all cases the wire performance is significantly inferior. This confirms the suggestion of Bennett et al [17,18], Singh et al [9] and Singh and Ip [16] that it may be because only a limited number of wire sections, located at the cracks, are subjected to the maximum stress level.

In addition, the mortar cover protects even the preloaded steel reinforcement from the corrodent. This is confirmed, dramatically, by the results shown in Figure 5. All wire specimens (without preloading) which had been subjected to ten days of corrosion failed in the first fatigue loading cycle.

This conclusion regarding the relative performance of the wire in the air is not in agreement with that drawn by the ACI Committee 549 on ferrocement [2] which based its conclusion on a combination of results; from various researchers, of single tests on several types of reinforcing steel and ferrocement.

Preloaded ferrocement specimens subjected to aggressive environment for upto three months showed shorter life at all stress levels (although at the lower stress levels the difference is insignificant) but at lower stress levels (< 68%) the specimens exposed

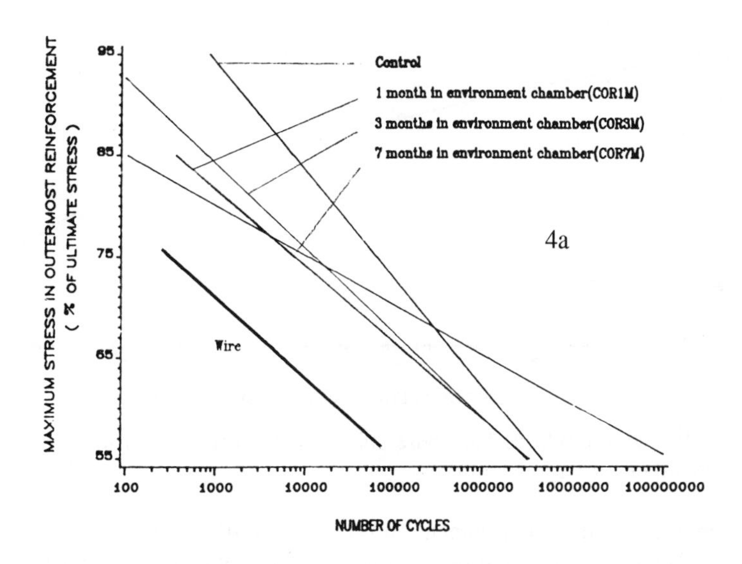

5% probability of failure --- with runouts

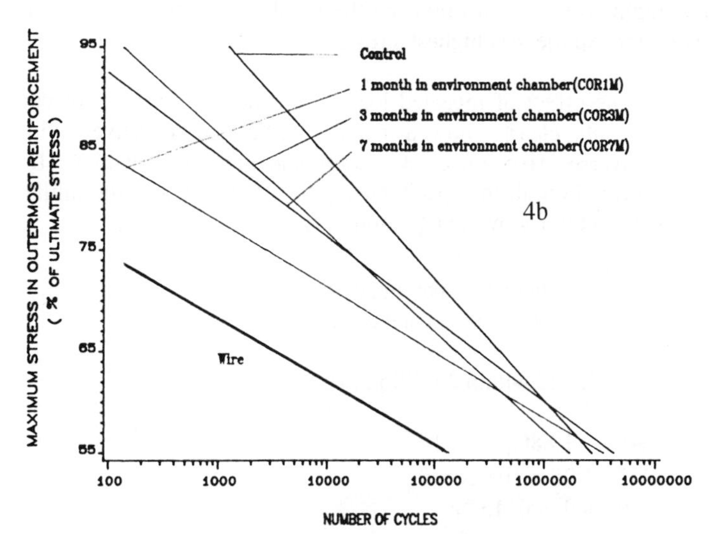

Figure 4 Stress-life (S-N) relationships for ferrocement and wire in air at 5% probability of failure: with and without runouts

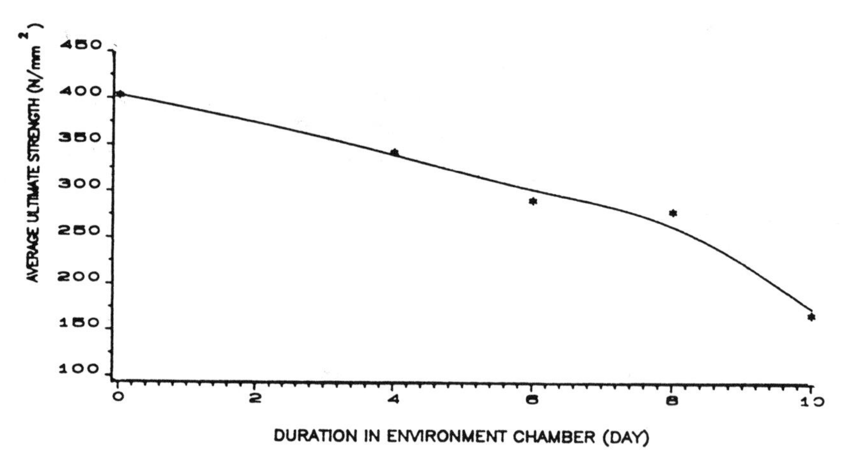

Figure 5 Ultimate strength of bare wire subjected to corrosion in the environment chamber for various durations

to seven months of this environment showed longer life when compared with the control specimens. The shallower slopes of the preloaded specimens may be attributable partly to the stress relieving and conditioning due to preloading and the higher average temperature in the environmental chambers leading to greater mortar and bond strength which in turn reduced the numbers of cracks or sections at which the reinforcement experiences highest stress.

Figure 6 shows the effect of repeated loading on the development of crack widths. Comparison of results clearly showed that the accelerated corrosive environment did not have any adverse effect on crack widths and their propogation. Least square regression performed on all the specimens, ignoring the data in the initial portion of the tests, resulted in the following equation [19]:

$Y = Ae^{Br^C}$ where Y = average crack width
A,B,C are constants

For specimens that failed within 2 million cycles

$$A = 0.0486 + 1.88C_1 - 1 \times 10^{-7}N$$
$$B = 1.772 - 1.28 \times 10^{-6} N > 0$$
$$C = 0.0199 + 0.000115 N$$

where C_1 = average crack width of the end of the first loading cycle.

The empirical equation proposed by Balagura et al [20] is also plotted on this figure. It gives a gross underestimation of the crack widths.

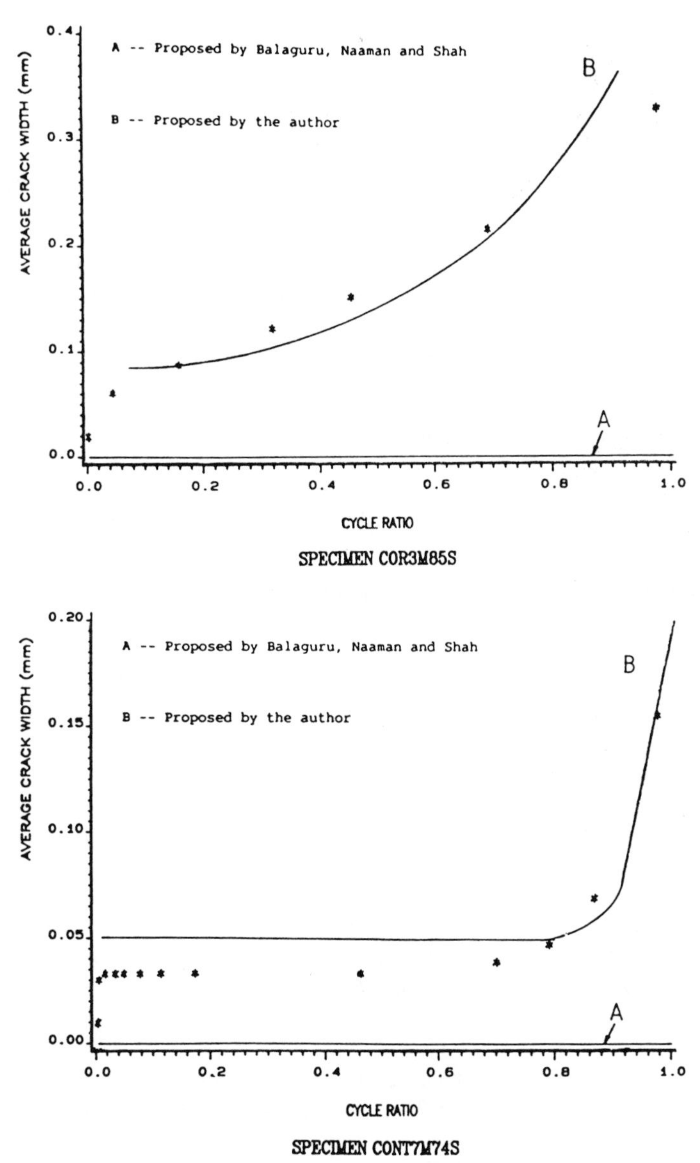

Figure 6 Typical empirically derived relationships between crack width and cycle ratio compared with experimental plots

CONCLUSIONS

At 5% probability of failure the static and fatigue performance of specimens subjected upto 7 months of accelerated aggressive environment did not appear to have been adversely affected in the working stress range.

The fatigue life of reinforcement as part of the composite was very significantly higher than that in the air.

Where knowledge of crack widths is deemed to be essential the model proposed in this paper should give satisfactory predictions for the composite tested.

REFERENCES

1. Huq, S. and Pama, R.P., 'Ferrocement in tension: Analysis and design,' Journal of Ferrocement 8 (3), 1978.

2. ACI, 'State-of-the-Art Report on ferrocement,' Report No. 549R-82, 1982.

3. Singh, G., 'Risk and Reliability Appraisal on microcomputers,' Chartwell-Bratt, UK 1987.

4. Singh, G. 'Flexural response of a ferrocement to constant deflection repeated loading,' Inter. Symp. on Ferrocement, RILEM/ISMES, Bergamo, 1981.

5. Menil Museum, The Architectural Review, March 1987.

6. Singh, G. Full scale test report on the behaviour of "Leaf-truss" to the Menil Foundation, 1983.

7. Ravindrarajah, R.S., and Paramasivam, P. 'Influence of weathering on ferrocement Properties,' In *Proceedings of the Second International Symposium on Ferrocement, Bangkok: International Ferrocement Information Center.,* 1988, pp 1237-1247.

8. ACI, 'Considerations for Design of Concrete Structures Subjected to Fatigue Loading,' ACI Committee 215, ACI Journal, Title no.71-10, March 1974 pp 97-121.

9. Singh, G., Bennett, B.W., and Fakhri, A. 'Influence of Reinforcement on Fatigue of Ferrocement,' The International Journal of Cement Composites and Lightweight Concrete, Vol. 8, No. 3, Aug. 1986, pp 151-164.

10. Weibull, W., 'Fatigue Testing and Analysis of Results,' Pergamon Press, 1961.

11. Wirsching, P.H. and Yao, J.T.P., 'Statistical Methods in Structural Fatigue,' Journal of the Structural Division, Proceedings of the American Society of Civil Engineers, ST6, June 1970.

12. Hawkins, N.M., and Heaton, L.W., 'Fatigue Characteristics of Welded Wire Fabric,' Abeles Symposium, Fatigue of Concrete, SP41-9, ACI, 1974, pp 183-202.

13. Karasudhi, P., Mathew, A.G., and Nimityongskul, P., 'Fatigue of Ferrocement in Flexure,' Journal of Ferrocement, Vol. 7, No. 2, Oct. 1977, pp 80-95.

14. McKinnon, E.A., and Simpson, M.G., 'Fatigue of Ferrocement,' Journal of Testing and Evaluation, JTEVA, Vol. 3, No. 5, Sept. 1975, pp 359-363.

15. Cornelissen, H.A.W., 'Fatigue Failure of Concrete in Tension,' Heron, Vol. 29, No. 4, 1984.

16. Singh, G. and Ip, L., 'Influence of a Marine Environment on Flexural Fatigue Properties of a Ferrocement,' International Conference on Performance of Concrete in Marine Environment, Canada 1988, pp 525-542.

17. Bennett, E.W., Fakhri, N.A., and Singh, G., 'Fatigue Characteristics of Ferrocement in Flexure,' ACI Journal, title no. 82-10, March-April 1985, pp 129-134.

18. Bennett, E.W., Fakhri, N.A., and Singh, G., 'Discussion of ACI Committee 549 Report, State-of-the-Art Report on Ferrocement, ACI 549R-82,' Concrete International, Dec. 1983.

19. Singh, G., and Ip, L., 'Effect of Repeated Loading on Crack Width of Ferrocement,' Journal of Ferrocement, Vol. 21, No. 2, 1991, pp 119-128.

20. Balaguru, P.N., Naaman, A.E., and Shah, S.P., 'Fatigue behaviour and design of ferrocement beams.' *Journal of the Structural Division* 105(ST7):1979 1333-1345.

99 CORROSION PERFORMANCE OF FERROCEMENT STRUCTURES 1972-1990

V. K. GUPTA
Civil Engineering Department, University of Roorkee, India
V. K. TIWARI
Metallurgical Engineering Department, University of Roorkee
P. C. SHARMA
SERC, Ghaziabad, India
S. K. KAUSHIK and D. N. TRIKHA
Civil Engineering Department, University of Roorkee, India

Abstract
Several ferrocement structures were cast during the period 1972-1978 using hand casting and semi-mechanised fabrication processes. Galvanised and Ungalvanised steel meshes were used in their manufacture. These structures have been exposed to several non-industrial environmental conditions of varying degrees. A few of the structures were cracked and exposed to similar environmental conditions for comparative performance.

Samples of the wire meshes were drawn from the structures in 1983, 1984, 1987 and October 1990 to study the extent of corrosion. The extent of corrosion was found to have doubled between 1984 and 1987. The data for the period 1987 to October 1990 shows a decrease in the rate of corrosion indicating development of a passive corrosion product layer and stabilisation of the corrosion process over a period of 15-18 years. The rate of metal penetration is also found to be assuring in all cases.
Keywords : Corrosion, Ferrocement, Wire Mesh, pH-Value, Metal Penetration.

1 Introduction

Corrosion in wire mesh in ferrocement structures fabricated between 1972 and 1978 subjected to moderate to severe environmental conditions have been studied by Trikha et al. (1,2,3) at ages of 3000 to 4500 days. Ferrocement strip specimens have also been studied by Trikha et al. (3) under simulated marine environment at a temperature of 25-30°C and salt concentration of 1N to 6N. All the above studies made use of ordinary portland cement. The corrosion behaviour of galvanised and ungalvanised wire mesh in ferrocement using ordinary portland cement (with or without rice husk ash) and sulphate resistant portland cement have been studied by Lukita et al. (4). This study determined the extent of corrosion and ultimate tensile strength using 12 mm and 18 mm galvanised hexagonal wire mesh under simulated marine environment and sulphuric acid attack. The ungalvanised mesh was obtained by immersing the galvanised mesh in 1:1 hydrochloric acid solution for approximately two minutes to remove the zinc

Fibre Reinforced Cement and Concrete. Edited by R. N. Swamy. © 1992 RILEM.
Published by E & FN Spon, 2-6 Boundary Row, London SE1 8HN. ISBN 0 419 18130 X.

coating. The weight loss of the wire mesh stabilised after immersion for two minutes. In this study, the mortar cover ranged from 0 to 4 mm. The ratio of cement to sand was 1:2 by weight, W/(C+RHA) ratio varied from 0.53 to 0.63 for different type of mortar with RHA replacement of 10% to 30% by weight. All the specimen were subjected to 0, 10 and 20 immersion cycles.

The nature of the corrosion phenomena has been discussed in detail by Trikha et al. (1) and Chalisgoankar (5). Total corrosion (Ce), the rate of corrosion per day (Cr) and percentage loss of strength per year (plspy) have been defined by Trikha et al. (1,3). The corrosion resistance scale based on the metal penetration per year has been given by Hhlig (6).

2 Samples From Prototype Structures

As in the past, mesh wire samples of 100 mm length were drawn from nine existing prototype structures described in Table 1. These samples were drawn from the same locations as for the previous samples.

The nine prototype structures can be classified into several groups based on a common feature. For example, structures 1,2 and 9 have been manufactured using a semi-mechanised process while structures 6,7 and 8 are manually cast. In structures 3 and 9 no fine sand has been used while in structure 4 a mixture of fine and coarse sand in equal proportions has been used. Considering the importance of mechanical vibrations in obtaining a dense mortar matrix, structures 3 and 4 have been cast without and with mechanical vibrations respectively. It is a well known fact that painting of external surfaces of structures provides better protection to the reinforcement against corrosion effect as compared to unpainted surfaces. Keeping this in view structures 1,2 and 6 were painted on the external surface while all the other structures were unpainted.

The weight losses of the various specimens due to corrosion are given in Table 2 for the periods ending November 1983, July 1984, August 1987 and October 1990. Table 3 gives the extents of corrosion (Ce) and Table 4 the rates of corrosion (Cr). The pH values of the mortar matrix as in October 1990 are also listed in Table 4. The percentage losses in strength per year are given in Table 5. The procedure for determining these quantities, described in detail by Trikha et al. (1), is summarised in Appendix I.

3 Discussion of Results

A study of the results of the first group of structures i.e. 1,2,9 and 6,7,8 shows that the extent of corrosion (Ce) as well as the rate of corrosion (Cr) are in general comparable for structures manufactured by mechanised process as well manual technique. The same trend is observed in the percentage loss in strength per year (plspy).

In the structures manufactured by the semi-mechanised process i.e. 1,2 and 9, the first two were painted on the external surface,

Table 1. Details of prototype ferrocement structures used for corrosion studies

Struct. Number	Ferrocement Structure	Date of Fabrication and Exposure Condition	Surface Treatment	Reinforcement and Visual Condition
1	2	3	4	5
1	A Gas Holder cast with semi-mechanised process *D = 1.2 m *H = 1.06 m *t = 1.2 cm	March 1974, used in a plant upto 1977. then taken out and kept inverted in the open	Bitumen painting on external surface only, inside unpainted	Two layers of galvanised iron square mesh. General corrosion of zinc coating, white corrosion products formed at the square mesh, base mild steel also corroded.
2	A Gas holder cast with semi-mechanised process. D = 1.2 m H = 1.06 m t = 1.2 cm	January 1974, used in a plant upto 1980, then taken out and kept inverted in the open.	Green shade popurothene paint on external surface, inside unpainted.	One layer of fly mesh and two layers of galvanised chicken mesh. Both fly and hexagonal meshes corroded, white corrosion products formed, mild steel mildly corroded.
3	A Floating Tray hand cast without mechanical vibration D = 1.25 m H = 100 mm t = 15 mm	Sept. 1974, kept floating in a large tank as demonstration object till 1980, then taken out and exposed to weathering condition. Cement/Coarse sand = 1:2 No fine sand.	Original Mortar surface, unpainted.	6 mm dia bars & 2 layers of galvanised iron mesh. Mild steel corroded with the production of brown & brownish black products.
4	A plate used for flexural test series at SERC Roorkee (cracked), Vibrated during casting. (Mortar mix : cement/fine sand/ coarse sand = 1:1:1)	April 1973, tested in 1973 itself, broken into pieces in Jan. 74 for observing the deterioration. Samples taken from the intact portion covered with mortar, sample taken on 8 Nov. 1983.	None	Galvanised iron mesh. Only zinc coating corroded with the production of white corrosion products, Mild steel uncorroded.
5	A plate used for flexure test series at SERC (Cracked)	April 1973, tested in 1973 itself, broken into pieces in Jan. 1974 for observing deterioration. Test samples collected from the end of the mesh left exposed to environment since Jan. 1974.	None	Galvanised iron mesh. Quite appreciable corrosion even chips of corroded material not visible at several places.
6	A bin, capacity 1/2 ton hand cast.	June 1972, kept exposed to weathering conditions since then.	Painted outside with bital paint	Black mild steel wire reinforcement. Medium corrosion of the square mesh. Particularly more at the junction of the square mesh.
7	A roofing unit (Folded plate) hand cast without vibration.	Jan. 1976 , exposed to weathering conditions, since then.	None	Two layers hexagonal mesh with G.I. wire reinforcement. Zinc coating corroded, white corrosion products, Mild steel slightly corroded at very few junction points.
8	A rectangular water tank, hand cast, capacity = 400 L size = 800mmx800mmx 810mm wall thickness = 15 mm	October 1973, tank surface cracked and exposed to weather conditions since 1973.	None	Two layers of hexagonal mesh and G.I. wire mesh. Hexagonal mesh corroded at the zinc coated surface with white corrosion products. Mild steel only slightly corroded at very few places.
9	A cylindrical water tank, cast using SERC semi-mechanised process. capacity = 2500 L D = 1.2 m H = 2.0 m t = 10 mm	October 1973, tested repeatedly upto 1975 for strain measurements by filling and emptying the tank several times. (Mortar-cement:coarse sand : 1:2, No fine sand used)	None	Black mild steel square mesh in 3 layers. Mild steel mesh corroded particularly at the junctions of the square mesh.

* D = Diameter ; H = Height ; t = Shell thickness.

Table 2. Weight loss

Sample Number	Original Weight (gms)	Loss in weight (gms) till			
		Nov. 1983	July 1984	August 1987	October 1990
1	0.84139	0.04474	0.03989	0.09556	0.09341
2a*	0.15890	0.00676	0.00931	0.01141	0.01052
2b**	0.15890	0.00958	0.00971	0.01677	0.01823
3	0.49800	0.01888	0.02010	0.06698	0.05865
4	0.49800	0.01862	0.01895	0.07440	0.06306
5	0.49800	0.02929	0.03431	0.07620	0.01220
6	0.49800	0.06244	-	-	0.08219
7	0.15890	0.01148	0.00981	0.04525	-
8	0.16901	0.00688	0.00783	0.02055	0.01736
9	0.49800	0.90415	-	-	0.05193

* flyash ** Hexagonal mesh

Table 3. Extent of corrosion

Sample Number	Extent of corrosion, Ce (md) till							
	Nov. 1983		July 1984		August 1987		October 1990	
	Period*	Ce	Period	Ce	Period	Ce	Period	Ce
1	3525	1139	3765	1032	4920	2473	6055	2416
2a	3585	412	3825	509	4980	698	6155	642
2b	3585	588	3825	593	4980	1026	6155	1113
3	3285	660	3525	705	4680	2343	5815	2047
4	3830	651	4070	663	5225	2602	6360	2207
5	3830	1026	4070	1201	5225	2665	6360	426
6	4195	1511	-	-	-	-	-	-
7	2825	703	3065	601	4220	2770	6725	2872
8	3650	405	3890	463	5045	1211	6180	-
9	3650	928	-	-	-	-	6180	-

* Period in days.

used galvanised mesh reinforcement and well graded sand while the last structure was unpainted, used ungalvanised mesh reinforcement and only coarse-sand. It is seen that the extent and rate of corrosion in structure 9 are almost twice that of structure 2. This is also clearly brought out by the smaller pH value of the mortar for structure 9 as compared to structures 1 and 2.

In the group of hand cast structures, structure 6 was painted externally and used ungalvanised mesh reinforcement with structures 7 and 8 were unpainted and used galvanised mesh reinforcement. A study of the data for this group shows that use of ungalvanised reinforcement produces, 1.5 to 2 times heavier rate of corrosion

Table 4 . Rate of corrosion and pH values

Sample Number	Rate of corrosion, Cr (mdd) till								pH Values in Oct. 1990
	Nov. 1983		JUly 1984		August 1987		October 1990		
	Period*	Cr	Period	Cr	Period	Cr	Period	Cr	
1	3525	0.323	3765	0.274	4920	0.503	6055	0.399 (0.686)	11.31
2a	3585	0.115	3825	0.133	4980	0.140	6155	0.182	-
2b	3585	0.164	3825	0.155	4980	0.206	6155	0.105	11.36
3	3285	0.201	3525	0.200	4680	0.501	5815	0.352 (0.353)	11.04
4	3830	0.170	4070	0.163	5225	0.498	6360	0.347	8.9
5	3830	0.268	4070	0.295	5225	0.510	6360	0.067	-
6	4195	0.535	-	-	-	-	-	-	8.02
7	2825	0.249	3065	0.196	4220	0.656	6725	0.427 (0.930)	-
8	3650	0.111	3890	0.119	5045	0.240	6180	-	10.3
9	3650	0.456	-	-	-	-	6180	-	9.07

* Period in days. Values indicated in parenthesis refer to mesh exposed to atmosphere.

Table 5. Percentage loss in strength per year

Sample NUmber	Percentage loss in strength per year, plspy till							
	Nov. 1983		July 1984		August 1987		October 1990	
	Period*	pispy	Period	plspy	Period	plspy	Period	plspy
1	3525	0.55	3765	0.46	4920	0.84	6055	0.67
2a	3585	0.43	3825	0.60	4980	0.69	6155	0.40
2b	3585	0.61	3825	0.58	4980	0.77	6155	0.69
3	3285	0.42	3525	0.42	4680	1.05	5815	0.74
4	3830	0.36	4070	0.34	5225	1.04	6360	0.73
5	3830	0.56	4070	0.62	5225	1.07	6360	0.14
6	4195	-	-	-	-	-	-	-
7	2825	0.93	3065	1.36	4220	2.46	6275	-
8	3650	0.41	3890	0.43	5045	0.88	6180	0.61
9	3650	-	-	-	-	-	6180	-

* Period in days.

even though protected by a coat of paint as compared to unpainted structures using galvanised wire meshes.

A comparison of the extent and rate of corrosion for the second group of structures i.e. 3,9, and 4, clearly brings out the importance of the use of well graded sand in the mortar matrix. The corrosion instructure 4 with well graded sand is found to be one-half to one-third of that for structures 3 and 9 which used only coarse sand. The pH values for all the three structures are more or less the same.

Considering the effects of mechanical vibrations for compaction of the mortar matrix, the data for structures 3 and 4 indicate that the extent and rate of corrosion are in general lower by 10-20% when mechanical vibrations are used. The drop in pH value for structure 4 is higher than for structure 3, despite the use of mechanical vibrations. This is probably because structure 4 was cracked and the atmospheric pollutants had access to the reinforcement.

The beneficial effects of painting external/internal surfaces is evident by the lower extent and rate of corrosion of structures 2 as compared to the other unpainted structures. Structure 6, though painted, was reinforced with ungalvanised steel and therefore had a higher rate of corrosion.

The pH values in Table 4 for the nine prototype structures as obtained in October 1990 are probably not much different from the values existing in November 1983 or earlier, because the environment in and around Roorkee is primarily unpolluted due to few industries and low traffic density. The values range between 8 to 11.36 as compared to a standard pH value of 13 for a fresh cement based alkaline medium.

The depth of metal penetration computed for various samples drawn give values which are in the range 6.28×10^{-5} to 9.57×10^{-4}. These are far less than the value of 1.3×10^{-2} as specified for good corrosion resistance by Hhilg (6).

In overview, the rate of corrosion in August 1987 was earlier found to be 1.5 to 2.5 times the corrosion level in November 1983 (3). The values obtained in October 1990 for all the samples drawn are invariably less by about 20-50% as compared to the values obtained in August 1987. This indicates that the extent and rate of corrosion in these ferrocement structures are stabilising after elapse of almost 14 to 17 years of their fabrication.

4 Conclusions

On the basis of the data monitored for 14-17 years old structures over a period of seven years and subjected to varying environmental conditions of temperature, humidity, pollution and biochemical attacks, the following conclusions may be drawn :

(1) The corrosion effects in various structures with thicknesses varying from 10-15 mm over a period of 14-17 years, appear to be stabilizing, indicating that the corrosion effects in ferrocement structures are no more dangerous that those in reinforced concrete structures. This is true even in manually cast structures without external painting, provided they are well cast.

(2) The ferrocement structures should preferably be designed on no crack basis as far as possible to enhance their life. However, small crack width of the order of 0.05 mm which can close by auto-

geneous healing may be permitted as recommended by ACI Committee 549 (7).

(3) Production of strong and sound ferrocement structures require the use of mechanical casting processes, good compaction, galvanised mesh reinforcement, properly graded sand for the mortar, a minimum cover of 4-5 mm and water proofing coatings.

5 References

Trikha, D.N. Sharma, S.P. Kaushik, S.K. Sharma, P.C. snd Tiwari, V.K. (1984) Studies in ferrocement structures, **J. Ferrocement**, 19, 221-233.

Trikha, D.N. Kaushik, S.K. Gupta, V.K . Tiwari, V.K. and Sharma, P.C. (1985) Studies on corrosion behaviour of ferrocement structures, **Proc. Second International Symposium on Ferrocement**, Bangkok, Thailand, 621-632.

Kaushik, S.K. Gupta, V.K. Tiwari, V.K. and Sharma, P.C. (1988) Corrosion performance of ferrocement structures 1972-1987, **Proc. Third International Symposium on Ferrocement**, New Delhi, India, 142-152.

Lukita, M. Austriaco, L.R. and Nimityoungskul, P. (1987) Corrosion behaviour of wiremesh in ferrocement, **Proc. International Correspondence Symposium on Ferrocement Corrosion**, Bangkok, Thailand, 3-19.

Chalisgoonkar, R. Corrosion of steel in concrete and ferrocement, ibid, 21-29.

Hhlig, H.H. (1971) Corrosion and corrosion control, John Wiley and Sons Inc.

ACI Committee 549 (1988) Guide for design, construction and repair of ferrocement, **ACI Structural Journal**, 325-351.

6 Appendix I

Total corrosion, $Ce = \frac{(W_o - W_f)}{\pi dL}$

Rate of corrosion/day, $C_r = C_e/t$

Loss of weight (mg)/sq. Decimeter surface Area/day

$$C_r = \frac{(W_o - W_f) \times 1000}{\pi dLt} \text{ (mdd)}$$

Percentage loss of strength/year

$$\text{plspy} = \left(\frac{W_o - W_f}{W_o}\right) \times \frac{100 \times 365}{t}$$

where,

W_o	=	Original weight of sample, gms
W_f	=	Final weight of sample after removal of corrosion products, gms
d	=	Diameter of wire sample, dm
L	=	Length of wire sample, dm
t	=	Age of the structure, days

CORROSION RESISTANCE SCALE (6)

Corrosion resistance	cpy
Good	0.013
Satisfactory	0.013 - 0.13
Unsatisfactory	0.13

where,

cpy = Metal penetration in cms of penetration per year

100 PERFORMANCE OF TWIN CELL FERROCEMENT BOX GIRDER ROOF/FLOOR STRUCTURES

V. K. SEHGAL
Regional Engineering College, Kurukshetra, India
N. M. BHANDARI and S. K. KAUSHIK
Civil Engineering Department, University of Roorkee, India

Abstract
The use of box shapes is eminently suited for flexural elements of large spans due to their high rigidities. They also have the advantage of being able to accommodate services in them, as well as maintaining a uniform temperature. This paper reports the behaviour and performance of a twin cell ferrocement box girder structure of width 2.36 m and effective span 4.58 m in the uncracked and cracked stages. In the uncracked stage, the girder was loaded and unloaded under various combinations of symmetric and unsymmetric loads. The girder was later subjected to monotonically increasing sustained loads of short durations upto near ultimate load. The girder has been analysed by finite element method under dead load and monotonically increasing live loads. The experimental results in the elastic stage show good comparison with the theoretical results. The effect of sustained loads of short durations is maximum in the initial portion of the cracked range. However, the ultimate load of the girder is unaffected by the sustained loading.
Keywords: Ferrocement Box Girder, Finite Element Method, Loading, Strains, Deflections, Crack Width, Creep.

1 Introduction

The ferrocement box girders can be used as roofing/flooring elements for large spans due to their high rigidities. The box girders can also be used in situations where flat top surface is the requirement. They have the advantage of being able to accommodate services in them, as well as maintaining a uniform temperature.

The investigation into the behaviour of single cell ferrocement box girders (1,2) has shown that they have large load distribution capacity. The experimental results showed that irrespective of the mode of load application (whether uniformly distributed load applied over the entire top flange or over half width of top flange and full span length), the first crack load was about the same. Also, the maximum deflection at mid span at first crack load for both the loading cases was nearly the same.

To cover a roof/floor, one may have to join two or more single cell ferrocement box girders. In the present work, the twin or double cell box girder was obtained by joining two single cell ferrocement

Fibre Reinforced Cement and Concrete. Edited by R. N. Swamy. © 1992 RILEM.
Published by E & FN Spon, 2-6 Boundary Row, London SE1 8HN. ISBN 0 419 18130 X.

box girders at the level of top compression flange. The behaviour and performance of the twin cell box girder has been investigated in the uncracked and cracked stages. In the uncracked stage, the girder was loaded and unloaded under various combinations of symmetric and unsymmetric loads. The girder was later subjected to monotonically increasing sustained loads of short durations upto near ultimate loads.

The analysis of the twin cell ferrocement box girder has been done under dead loads and monotonically increasing live loads (3,4). For ferrocement composite, the element is assumed to be consisting of single mortar layer in the uncracked stage, uncracked and cracked mortar in the cracked stage and smeared layers of wire mesh and skeletal steel. A rectangular flat shell element capable of representing membrane action, bending action and the interaction between membrane and bending actions is adopted. The degrees of freedom per node are u,v and $\partial v/\partial x$ (i.e., θ_z) for the membrane action and w, $\partial w/\partial y$ (i.e., θ_x) and $-\partial w/\partial x$ (i.e., θ_y) for the bending action respectively. The stress-strain relation for mortar is assumed to be linearly elastic-perfectly plastic in compression and linearly elastic in tension upto the cracking strength. The stress-strain relation for wire mesh and skeletal steel is assumed to be linearly elastic-perfectly plastic with the same yield stress and modulus of elasticity in tension and compression. Only material non-linearity due to cracking of mortar, tension stiffening effect of mortar between the cracks and the nonlinear stress-strain relationships for the mortar, wire mesh and skeletal steel is considered. Geometrical nonlinearity, bond slip between the reinforcement and mortar, time dependent effects and thermal effects are not considered.

2 Experimental programme

The twin cell box girder having reinforcement and sectional details as shown in Fig. 1 was obtained by first casting two single cell box girders and then joining the two at the level of the top flange. The two single cell box girders were cast (1,2) in such a manner that

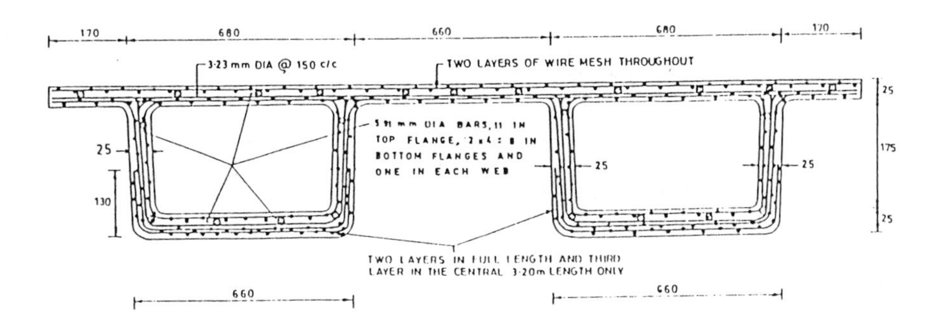

Fig. 1. Reinforcement and sectional details of double cell box girder

both the wire mesh layers and transverse skeletal steel bars from one edge of the top flange were projecting beyond the cast portion. The girders were lifted from the mould and placed on the supports in such a way that the distance between the inner webs of the two girders was kept equal to the width of the bottom flange of each cell. After placing the girders in the required position, the mortar of the cast edges was chipped and the loose mortar particles removed. The wire mesh layers and transverse skeletal steel bars of the girders were lapped. One longitudinal skeletal steel bar was provided in the lapped portion. The overlaps of the top mesh layer, bottom mesh layer and transverse skeletal steel were staggered. The average width of the top joint was about 240 mm and the average overlap length of the top and bottom wire mesh layers was about 110 mm. For casting the joint, two wooden planks with adequate support from the bottom were placed below the joint. The gaps between the cast edges and the planks were closed with waste cotton fibres. The joint of the double cell box girder prior to casting is shown in Plate 1. The mortar was placed over the joint portion, pressed through the mesh reinforcement and finally compacted using a portable surface vibrator. The wooden planks were removed from the bottom next day and the soffit of the joint was given smooth finish. The precast reinforced concrete diaphragms were provided at the ends between the two cells. The joint portion was cured for 28 days.

The girder was tested under uniformly distributed load applied over the top flange. The uniformly distributed load was applied in the form of brick layers. The girder was subjected to loading and unloading under various combinations of symmetric and unsymmetric loads to study its behaviour in the elastic stage.

The girder was later subjected to monotonically increasing sustained loads of short durations. The sustained loading was in the form of brick layers applied over the entire flange. The girder was loaded upto six brick layers (66.7 kN or 6.16 kN/m^2).

The strength properties of the materials used for casting the double cell box girder are shown in Table 1.

Table 1. Material strength properties

Cement-Sand Mortar	
Cement : Sand : Water (by weight)	1 : 2.5 : 0.45
28 days average crushing strength of 100 mm cubes	21.70 Mpa
28 days direct tensile strength of briquettes having minimum cross-section of 100 mm x 100 mm	2.30 Mpa
Modulus of elasticity	14 kN/mm^2
Poisson's ratio	0.20

Contd.......

Table 1. (Contd..........)

Reinforcement	Yield strength (Mpa)	Ultimate strength (Mpa)	Modulus of elasticity (kN/mm^2)
0.8 mm dia. GI square wire mesh of size 6.35 mm x 6.35 mm	440	520	146.2
5.91 mm dia. longitudinal skeletal steel	440	605	200.0
3.23 mm dia. transverse skeletal steel	380	490	210.0

3 Test results and discussion

3.1 Loading in the uncracked stage

The double cell box girder under various combinations of symmetric and unsymmetric load cases behaved as one single unit by undergoing downward deflections along the entire length and width. The loading and unloading paths under various loading conditions were quite close. In the present paper, the test results of one symmetric load case and one unsymmetric load case are shown below:

Symmetric Load Case : Girder subjected to four brick layers (4.51 kN/m^2) on the two cells (Fig. 2).

Due to the above loading, the deflections and strain variation at the mid span cross-section are shown in Figs. 2(a) and 2(b). The experimental values of the deflections and strains at symmetrical points were close. Therefore, average values of the deflections and strains of symmetrical points are plotted on one half of the cross-section. Considering the deflections at mid span (Fig. 2(a)), the predicted deflections show a good comparison with the experimental deflections. The maximum variation is at the cantilever end where the predicted deflection is less than the average experimental deflection by about 14 per cent.

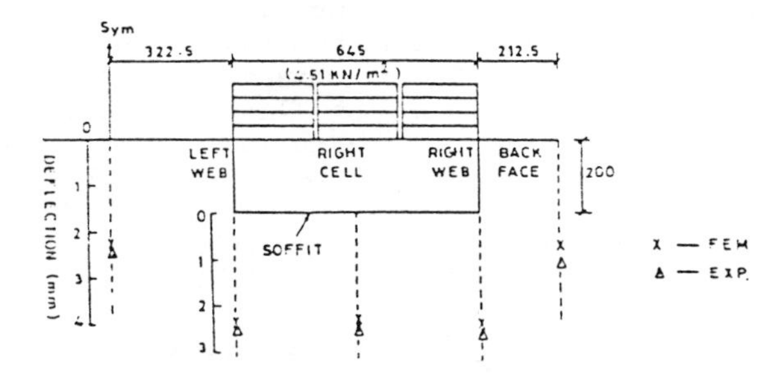

Fig. 2(a). Deflections across the cross-section at mid span.

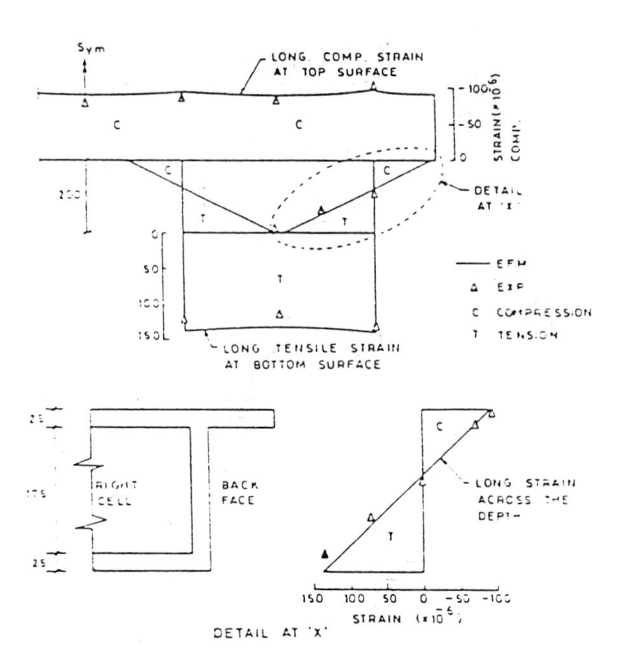

Fig. 2(b). Strain variation across the cross-section at mid span

The variation of longitudinal strains at mid span cross-section is shown in Fig. 2(b). Considering the longitudinal compressive strains at the top surface, the predicted strains are comparable with the experimental strains and the maximum variation is at the centre of the top flange where predicted strain is more by about 13 per cent. Considering longitudinal tensile strains on the bottom surface of the cell, the predicted strains are more than the experimental strains and the maximum variation is at the centre of the bottom flange where predicted strain is more by about 16 per cent. Considering the variation of the longitudinal strains across the girder depth, the predicted strains are comparable with the experimental strains across the entire girder depth except at 25 mm above the soffit where predicted strain is less by about 17 per cent.

Unsymmetric Load Case : Girder subjected to three brick layer (3.08 kN/m^2) over half width of top flange and full span length.

Due to the above loading, the deflections and strain variation at the mid span section are shown in Figs. 3(a) and 3(b). Considering mid span deflections (Fig. 3(a)), the predicted deflections are less than the experimental deflections below the loaded portion. The maximum difference is below the cantilever end of the loaded portion where predicted deflection is about 22 per cent less than the experimental deflection. This difference continues to decrease towards the unloaded side and changes sign below the unloaded portion. However, the magnitude of the deflections in this portion are small.

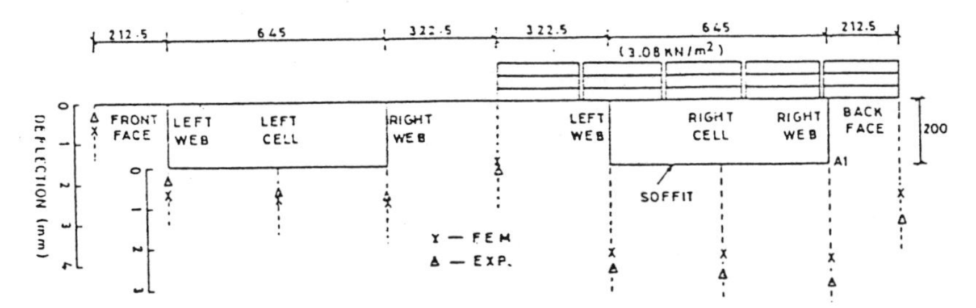

Fig. 3(a). Deflections across the cross-section at mid span.

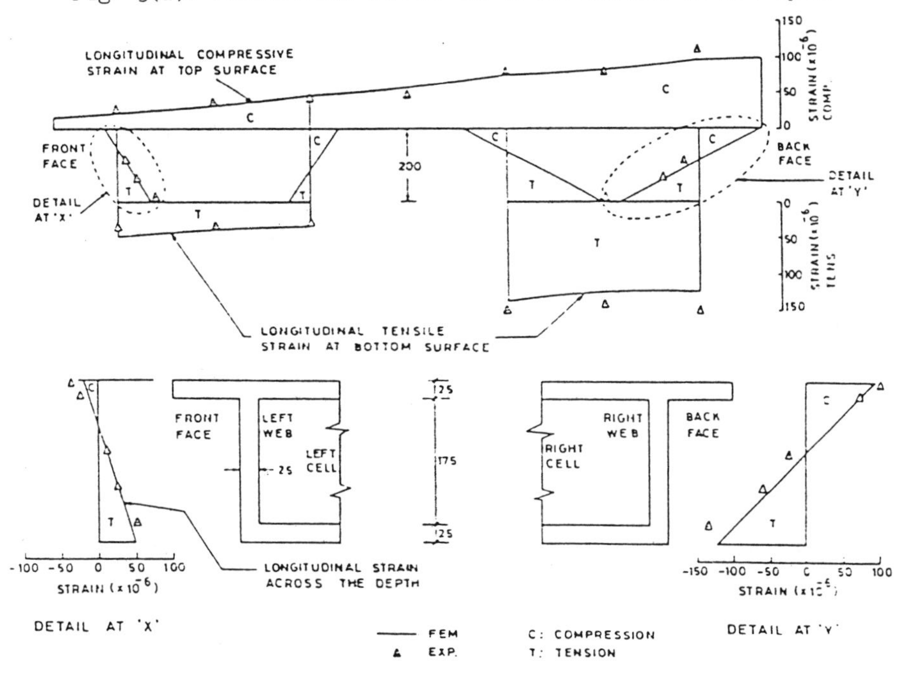

Fig. 3(b). Strain variation across the cross-section at mid span.

The variation of longitudinal strains at mid span cross-section is shown in Fig. 3(b). Considering the longitudinal compressive strains at the top surface, the predicted and the experimental strains show good agreement across the entire width of the top flange. The maximum variation is at the centre of the top flange where predicted strain is more by about 22 per cent. The predicted longitudinal tensile strains on the soffit of the bottom flanges are less than the experimental strains in the loaded cell while these are more in the unloaded cell portion. The maximum variation is below the left web of the unloaded cell where predicted strain is more by about 40 per cent. However, the magnitude of the strains in the unloaded cell is only about one-third of that in the loaded cell.

Considering the variation of longitudinal strains across the girder depth, the predicted strains show good agreement with the experimental strains. The maximum variation is at 25 mm above the soffit on back face (loaded web) where predicted strain is less than the experimental strain by about 26 per cent.

3.2 Girder subjected to monotonically increasing sustained loading

After a rest of about one month, the double cell box girder was subjected to monotonically increasing sustained loads of short durations. Each brick layer was allowed to remain on the girder for a few days till the deflections stabilized, i.e., the increase in deflections in the previous 24 hours being less than 0.05 mm. The girder was loaded upto six brick layers in this manner in 75 days time. The mid span deflections stabilized for one, two, three, four, five and six brick layers in 3, 11, 23, 10, 13 and 15 days respectively.

The validity of the nonlinear finite element method as described elsewhere (3,4) has been checked with the published experimental/analytical results of various researchers as well as experimental results of the ongoing investigation on box girder elements. Therefore, it has been considered fit to generate the experimental load-deflection/strain results for double cell box girder under monotonically increasing loads by the computer program developed. The predicted behaviour can be assumed to represent within ± 20 per cent variation, the experimental load-deflection/strain response under monotonically increasing loads. The predicted results under monotonically increasing loads have been compared with the experimental results under monotonically increasing sustained loads of short durations.

A comparison of the experimental mid span deflections under sustained loads with the predicted deflections under monotonically increasing loads is shown in Fig. 4. The experimental deflections under sustained loads are on the flexible side of the deflections under monotonically increasing loads upto the predicted yielding of skeletal steel occurring at about four and half brick layers (4.64 kN/m^2). The predicted deflections at six brick layers are large as compared to the deflection under sustained loads due to the predicted yielding of the skeletal steel and wire mesh layers of the bottom flanges.

Due to sustained loading, the deflections stabilized rapidly in the uncracked range. At three brick layers a large number of cracks appeared and it took maximum number of days (twenty three) for deflections to stabilize. At four, five and six brick layers, it took lesser number of days for deflections to stabilize. Defining the creep deflection coefficient in this case as the ratio of time dependent deflections to the deflection after the application of load (including the sustained deflections at lower load levels), the maximum creep deflection coefficient was 0.39 in the initial portion of the cracked range (i.e., at three brick layers) while it was 0.11, 0.09 and 0.08 at four, five and six brick layers respectively.

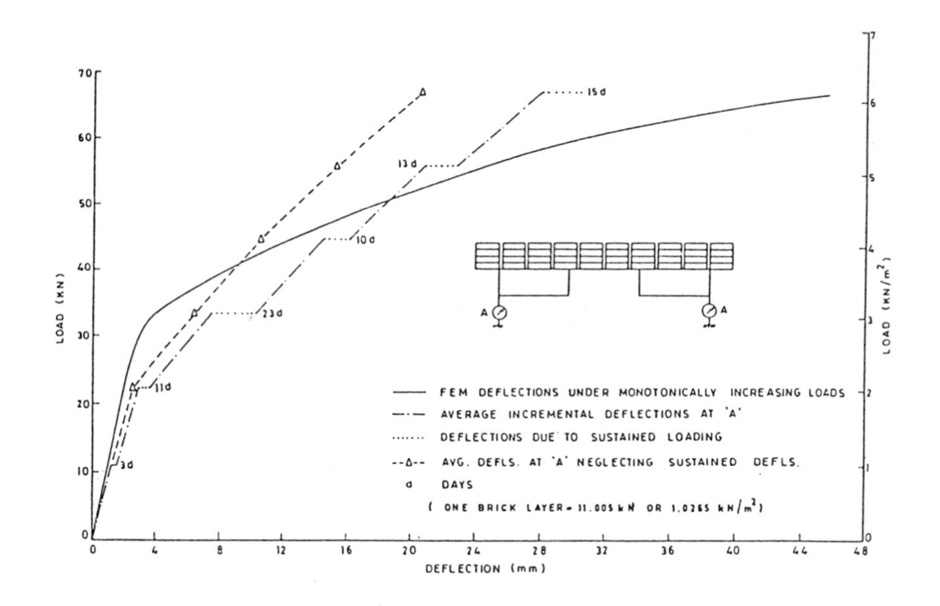

Fig. 4. Load vs. mid span deflection under outer webs 'A'.

If sustained deflections at all the load levels are neglected, then the experimental deflections in the cracked range are highly on the stiffer side of the predicted deflections under monotonically increasing loads. Thus, the instantaneous deflection due to the application of the load increment (having sustained loading at lower load levels) appears to be less than the instantaneous deflection which would have occurred if the load had been applied in monotonically increasing manner. The increased stiffness of the girder may be due to compaction of the mortar matrix under sustained loading. Similar behaviour under sustained loading at lower load levels was observed by Ravindrarajah and Tom (5) and Raisinghani and Sai (6).

A comparison of the longitudinal tensile strains at 25 mm above the soffit at mid span due to sustained loads and due to values predicted under monotonically increasing loads is shown in Fig. 5. The strains due to sustained loads are on the higher side of the strains due to monotonically increasing loads upto the predicted yielding of skeletal steel occurring at about four and half brick layers of load. Defining creep strain coefficient as the ratio of time dependent strain to the strain after the application of load (including the strain due to sustained loading at lower load levels), the tensile creep strain coefficient is maximum (0.62) at three brick layers. The creep strain coefficient reduces at higher load levels,

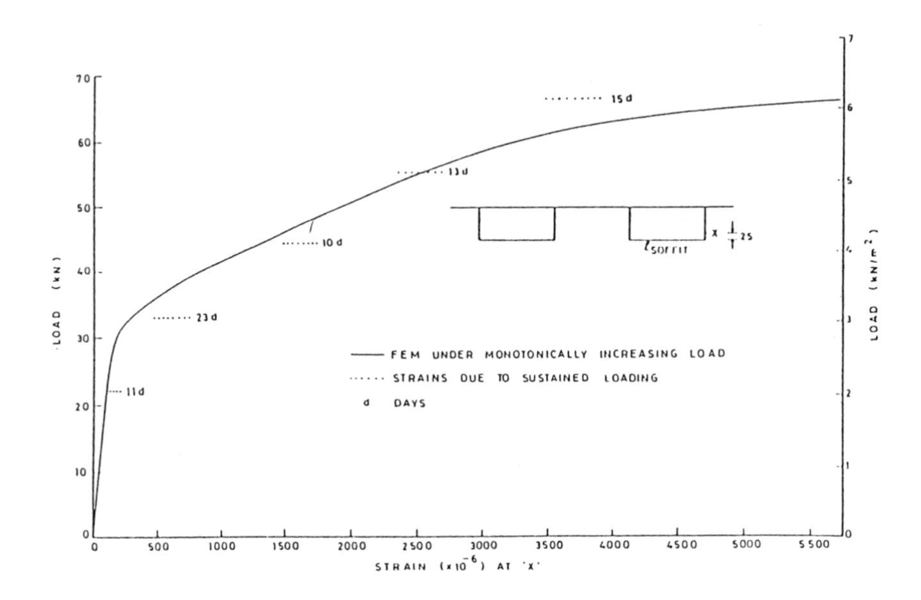

Fig. 5. Load vs. longitudinal tensile strain 25 mm above the soffit

i.e., 0.18, 0.14 and 0.12 at four, five and six brick layers.

A comparison of the longitudinal compressive strains at the top surface of the girder due to sustained loads and due to monotonically increasing loads is shown in Fig. 6. The compressive strains due to sustained loads are on the higher side of the compressive strains due to monotonically increasing loads upto five brick layers. At six brick layers, the predicted compressive strains due to monotonically increaing loads are on the higher side of the compressive strains due to sustained loading due to predicted yielding of skeletal steel and wire mesh layers of the bottom flanges. In this case also, the compressive creep strain coefficient is maximum at three brick layers (0.41) and less at four, five and six brick layers (i.e., 0.27, 0.15 and 0.27).

4 Crack spacing and crack width

In the cracked stage, the effect of sustained loading leads to an increase in the width of cracks, formation of new cracks and extension of cracks formed earlier. The average crack spacing after the sustained loading of six brick layers was about 50 mm in the central

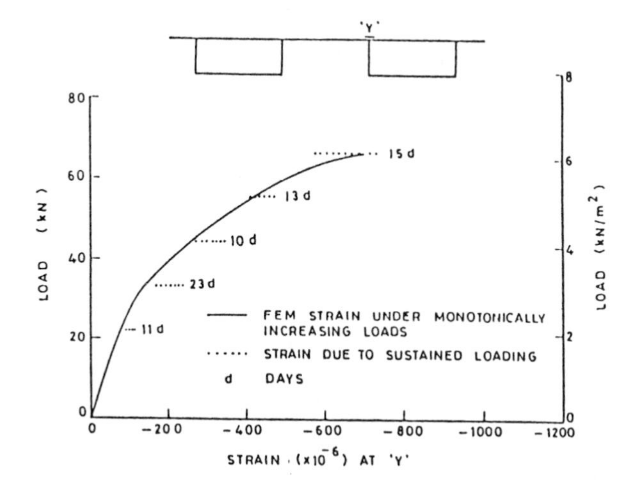

Fig. 6. Load vs. longitudinal compressive strain at top surface.

part of the bottom flanges. The crack-pattern of the bottom flanges and side webs after the sustained loading of six brick layers is shown in Plate 2.

The load versus maximum crack width curve of the girder is shown in Fig. 7. Defining the creep crack width coefficient as the ratio of the increase in crack width due to sustained loading

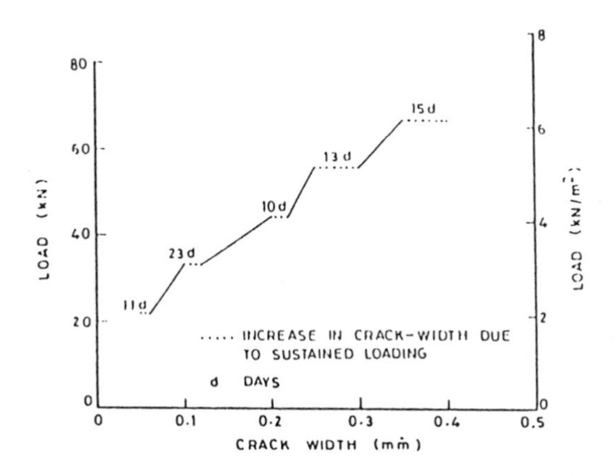

Fig. 7. Load vs. maximum crack width curve.

to the crack width after the application of load (including the increase in crack width due to sustained loading at lower load levels), the creep crack width coefficients at three, four, five and six brick layers are 0.20, 0.10, 0.20 and 0.14 respectively.

Including the time dependent effect on the crack width, the load taken by the double cell box girder at the recommended crack width of 0.1 mm (7) (Fig. 7) is about 29 kN (or 2.68 kN/m^2). At this load level, the span/deflection ratio is about 600. At span/deflection ratio of 250, the load taken by the girder is about 48 kN (or 4.44 kN/m^2) which is close to the load at which yielding of skeletal steel is predicted. At this load level, the maximum crack width in the girder is about 0.25 mm.

5 Conclusions

(i) The double cell box girder under the action of symmetric and unsymmetrical loads in the uncracked stage has behaved as one single unit by undergoing downward deflections along the entire length and width.

(ii) The effect of a sustained loading of short duration on deflections and strains is maximum in the initial portion of the cracked range.

(iii) The instantaneous deflection of the girder is reduced due to the sustained loading at lower load levels as compared to the instantaneous deflection that would have occurred under monotonically increasing loads.

(iv) The limit state of serviceability is governed by the maximum crack width and not the deflection.

(v) The ultimate load of the girder is not affected by the mode of loading, i.e., monotonically applied instantaneous or sustained loads of short durations.

(vi) The sustained loading increases the width of the cracks and the region of crack formation.

6 Acknowledgement

The authors are grateful to the Department of Science and Technology, New Delhi, for financial support to this project.

7 References

Kaushik, S.K., Gupta, V.K. and Sehgal, V.K. (1988) Performance evaluation of ferrocement box girder elements for roofs and floors. **J. of Ferrocement,** Vol. 18, No. 4, 413-420.

Sehgal, V.K., Bhandari, N.M., and Kaushik, S.K. (1988) Ferrocement box girder elements for rocks and floors. Proc. **3rd Int. Conf. Ferrocement,** New Delhi, 551-560.

Sehgal, V.K. (1989) Behaviour of ferrocement box girder elements, **Ph.D. Thesis**, Deptt. of Civil Engineering, University of Roorkee, Roorkee.

Bhandari, N.M., Sehgal, V.K., Kaushik, S.K. and Trikha, D.N. (1991) Post cracking behaviour of ferrocement box girder elements by FEM, (Paper communicated to **J. of Ferrocement**).

Ravindrarajah, R. and Tom, T.C. (1983) Dimensional stability of ferrocement. **J. of Ferrocement,** Vol. 13, No. 1, 1-12.

Raisinghani, M. and Sai, A.S.R. (1984) Creep and fatigue characteristics of ferrocement slabs, **J. of Ferrocement,** Vol. 14, No. 4, 309-322.

ACI Committee 549 (1982) State-of-the-Art report on ferrocement, Concrete International : Design and Construction, Vol. 4, No. 8, 13-28.

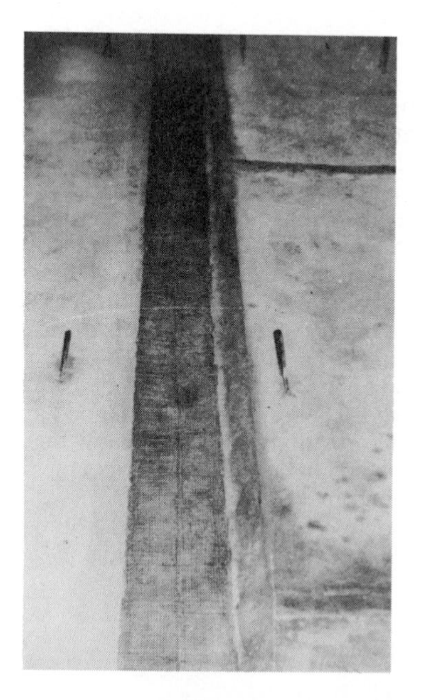

PLATE 1

PLATE 2

101 BEHAVIOUR OF REINFORCED CONCRETE-FERROCEMENT COMPOSITES

M. HOSSAIN, R. P. PAMA and P. NIMITYONGSKUL
Asian Institute of Technology, Bangkok, Thailand

Abstract
The flexural behavior of six composite beams made of low strength concrete and encased in U-shaped ferrocement forms made of high strength mortar reinforced with square welded wire mesh, were compared with two reference beams by conducting flexural tests under two symmetrical point loads, in simply supported conditions.

This paper deals with an experimental and analytical investigation of the ultimate moment capacity, first cracking moment, ultimate shear load capacity and deformational behaviors of these beams. The only variable parameter of this study was the number of layers of wire mesh used in ferrocement form. The scope was to determine the effect of wire mesh and the skeletal steel bars as flexural reinforcement and shear force reinforcement.

In computing the ultimate moment capacities of the composite beams, conventional reinforced concrete theory was used. The test results are in good agreement with the calculated moments, shears and deformations. They also show that the wire mesh is fully effective and a full composite action between the ferrocement form and the core concrete is attained up to failure load.
Keywords: Beams (supports), Composite materials, Crack spacing, Crack width, Deflection, Ferrocement, Ultimate moment, Volume fraction.

List of symbols

a	= shear span
a_1	= equivalent stress block depth
Ac, Am	= compressive area of the core concrete and mortar of ferrocement form
A_s, A'_s	= area of main steels and compression steels respectively
A_{sk}, A_{st}, A_w	= area of transverse skeletal steel, stirrup reinforcement and transverse wire respectively
A_{ski}, A_{wi}	= area of longitudinal skeletal steels and longitudinal wires in the ith layer, located in the tension zone
A'_{ski}, A'_{wi}	= area of longitudinal skeletal steels and longitudinal wires in the ith layer, located in the compression zone
b	= width of the beam
b_1	= width of the compression face of the uncracked and cracked transformed section of the composite beam, as defined in Fig.6
b_2	= width of the tension face of the uncracked transformed section of the composite beam, as defined in Fig.6
C	= total compressive force in the compression zone
d	= effective depth

Fibre Reinforced Cement and Concrete. Edited by R. N. Swamy. © 1992 RILEM.
Published by E & FN Spon, 2-6 Boundary Row, London SE1 8HN. ISBN 0 419 18130 X.

ds, $d's$ = distance of the main and compression reinforcement from the top of the section respectively

d_{ski}, d_{wi} = distance of the ith layer skeletal steels and wires from the top of the section respectively

E = modulus of elasticity

E_c, E_m, E_s, E'_s, E_{sk}, E_w = moduli of elasticity of concrete, mortar, main steel, compression steel, skeletal steel, and the wire mesh respectively

E_{comp} = composite modulus of elasticity

E_{cr} = cracked modulus of elasticity

f_c = equivalent concrete stress

f'_c, f'_m = companion cylinder strength of concrete and mortar respectively

f_{cm} = equivalent ultimate compressive strength of the concrete area

f_{comp} = modulus of rupture of the composite

f_s = stress in the main reinforcement

f_{sy}, f_{sky}, f_{sty}, f_{wy} = yield strength of the main steel, skeletal steel, stirrup and wire mesh respectively

F_s = force carried by the steel

h = total depth of the beam

h'_c = depth of the compression zone at failure

h'_t = depth of the tension zone at failure

h_{c1}, h_{c2} = depth of the compression zone of the uncracked and cracked transformed section respectively

I = moment of inertia

I_{eff} = effective moment of inertia

I_g = moment of inertia of the gross section

I_{cr}, I_{tr} = moment of inertia of the cracked and uncracked transformed section

M = moment

M_{cr} = first cracking moment

M_{max} = maximum moment at any stage of loading after first cracking

M_u = ultimate moment capacity of the beam

n_m, n_s, n'_s, n_{sk}, n_w = modular ratios of mortar, main steel, compression steel, skeletal steel and wire mesh respectively

t_{sk}, t_{st}, t_w = spacing of transverse skeletal steels, stirrups and transverse wires respectively

T = total tensile force in the tension zone

T_s, T_{sk}, T_w = tensile force resisted by main steels, longitudinal skeletal steels, and longitudinal wires of the mesh respectively

V_c, V_m, V_s, V'_s, V_{sk}, V_w = volume fractions of the concrete, mortar, main steel, compression steel, longitudinal skeletal steel and longitudinal wires of the mesh respectively

V_{cu} = shear load capacity by direct transfer in the uncracked compression zone

V_{sk}, V_{st}, V_w = shear force resisted by skeletal steel, stirrup and wire mesh respectively

w_0 = percentage of tensile reinforcement

Y_{cr}, Y_{tr} = depth of the tension zone of the cracked and uncracked transformed section respectively

z = lever arm

ε_c, ε_s = strain in concrete and in steel

σ_{c1}, σ_{c2} = extreme compression fiber stress in the uncracked and cracked

	range respectively
σ_{t1}, σ_{t2}	= extreme tension fiber stress in the uncracked and cracked range respectively
$\Delta_{mid}, \Delta_{quar}$	= deflections at mid span and quarter span respectively

1 Introduction

The main world-wide applications of ferrocement construction to date have been for silos, water tanks, roofing elements, wall panels, water pipes and mostly boats. However, the universal availability of the basic ingredients of ferrocement, steel mesh and concrete created interest in the potential application of this material for other structural members. Since no formwork is required as in conventional reinforced concrete construction and can be fabricated into almost any desired shape, ferrocement is sometimes used as formwork for complicated structures where other conventional formworks are very difficult to fabricate. It becomes more advantageous when the ferrocement formwork is made integral with the main reinforced concrete member as it will give additional strength to the original structure. Besides serving first as casting forms, they also serve as concrete cover for the main reinforcement.

In order to promote the use of ferrocement as formworks for various reinforced concrete structures, considerable research must be conducted. But unfortunately, very few experimental and analytical investigations have been carried out so far, to study the mechanical properties of the composite i,e, ferrocement form integrated with the reinforced concrete member. But, individually, a wide range of experimental data, mathematical models and equations are available on their mechanical properties [Rajagopalan and Parameshwaran (1975), Suryakumar and Sharma (1975), Logan and Shah (1975), Lee et al. (1972) and Pama et al. (1974)].

In this paper, the various mechanical properties of composite beams made of low strength concrete encased in ferrocement forms made of high strength mortar are investigated and are compared with those of the reference beams which are just like other reinforced concrtete beams. A systematic series of experiments were carried out to determine the flexural, shear, deformational and crack formation properties of both the composite and the reference beams. The only variable of this study was the reinforcing parameter which was taken as the number of layers of mesh used in the ferrocement forms. The number of layers were varied at 2, 4 and 6 layers of wire mesh. Other parameters, mainly shear span to depth ratio (a/h), compressive strength of both the mortar and the concrete, main reinforcement ratio (ρ), compression reinforcement, and web reinforcement were kept constant. The shear span to depth ratio (a/h) was fixed at 3 to ensure flexure failure mechanism.

An analytical approach for calculating the ultimate flexural tensile strength, cracking moments, shear load capacity and deformational characteristics of the composite beams as well as the reference beams is also presented in this paper. The predicted values based on the proposed equations were also compared to the actual behaviors (flexural, shear and deformational) obtained experimentally.

2 Experimental program

2.1 Details of specimen

Six composite beams (beams B1 through B6) and two reference beams (beams B7 and B8) were tested. All the test specimens were rectangular beams of 0.14 m X 0.3 m X 2.9 m with a clear span of 2.7 m which were loaded by two symmetrical concentrated forces. For all the beams the shear span to depth ratio were kept constant to 3 to obtain flexure-failure mechanism. All the six composite beams made of low strength concrete encased in U shaped 25 mm ferrocement forms, made of high strength mortar were compared with two reference beams having the same compressive strength as the core of the composite beams.

2.2 Materials

Steel reinforcement	Two 20 mm defomed bars (as main reinforcement) and two 12 mm deformed bars (as hanging bar) were used for all the beams. Round bars of 6 mm were used both as web reinforcement and skeletal steel.
Wire mesh	Galvanized welded wire mesh with square openings of 12.7 mm (0.8 mm average diameter of wires) were used for reinforcing the ferrocement forms of the composite beams. In beams (B1,B2), (B3,B4) and (B5,B6) one, two and three layers of specified wire mesh were used respectively.
Coarse aggregate	Crushed stones with a maximum size of 10 mm were used.
Cement	Ordinary portland cement (ASTM TYPE I) were used.
Sand	Natural river sand passing no.7 sieve were used.
Mix proportions	For the mortar mix, cement to sand ratio of 1:2 by weight was used. For the concrete mix, cement to sand to coarse aggregate ratio of 1:2:3 by weight was used. Water cement ratio for both the mortar and the concrete was 0.5.

2.3 Fabrication

For the fabrication of reference beams, conventional wooden form-work was used. Although for the fabrication of usual ferrocement constructions, no form work is required, but in this case, for fabricating the U-shaped ferrocement forms, a special type of formwork were needed. This formwork was essential for the accurate dimension, thickness and shape of the ferrocement form. The main problem associated with the fabrication of this shape of ferrocement form was to demould it after being hardened. For this reason, the form work was made in such a way that it can be dismantled part by part to facilitate the demoulding operation. The cross section of the formwork used for this purpose is shown in Fig.1. The surface of the fabricated ferrocement forms were

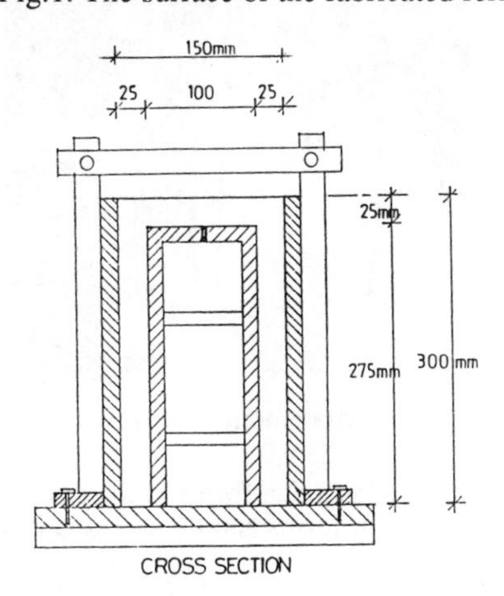

Fig.1. Formwork used for preparation of composite beams.

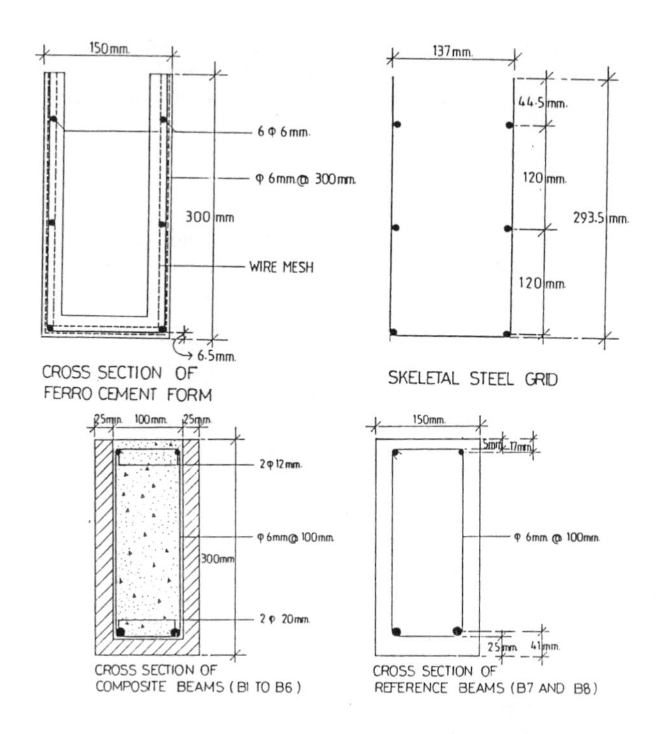

Fig.2. Cross section of ferrocement form, reference and composite beams.

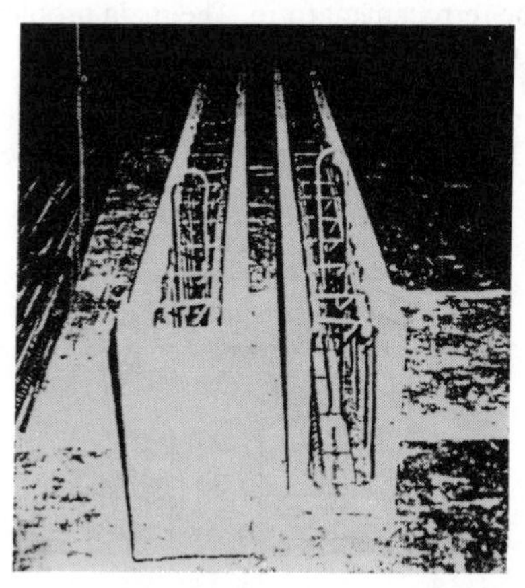

Fig.3. Fabricated ferrocement forms with reinforcement.

relatively smooth due to the use of wooden forms in the casting operation. Besides, no connectors between the ferrocement forms and the encased beam were used to get the worst testing conditions.

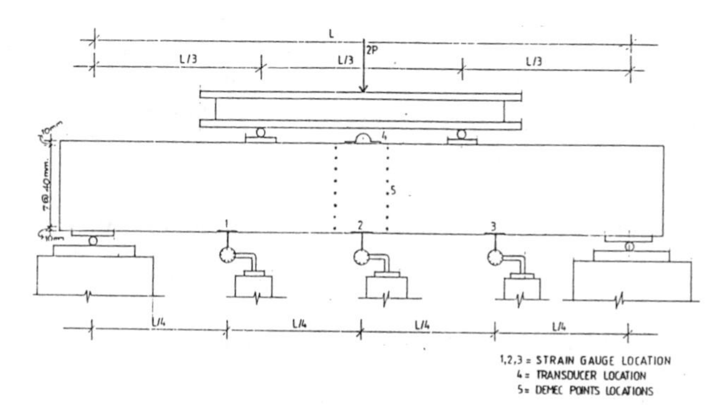

Fig.4. Test arrangement for beam specimens.

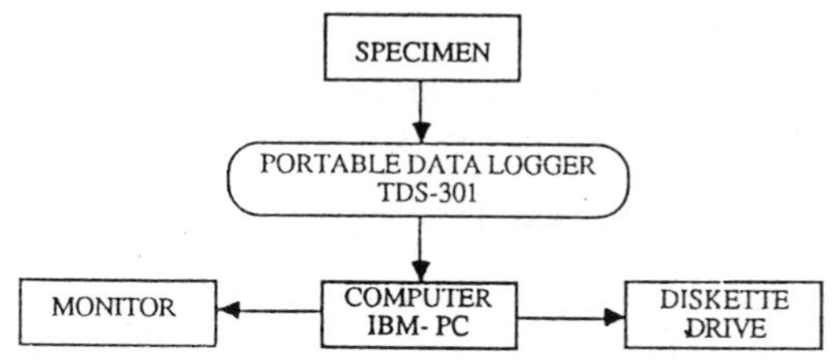

Fig.5. Arrangement for data collection, storing and monitoring.

2.4 Testing procedure

The beams were simply supported over a span of 2.7 m. The load was applied to the specimens equally at the third points, in order to give an area of pure bending over the central 0.9 m portion of the specimen. For this purpose, an I shape steel beam (serving as a spreader beam) was used to apply the two symmetrical point loads on the beams which were provided with simple supports (Fig.4). To measure the mid span and quarter span bottom fiber strains, 120 mm electrical resistance strain gages were attached to the appropriate concrete surface. To measure the mid span top fiber strain a TML strain gage type transducer was used. Vertical deflection at mid span and the quarter span of the beams were measured by using electronic dial gages having 20 mm travel and least count of 0.01 mm. Schematic diagram for the test set-up is shown in Fig.4.

The tests for the beams were carried out by subjecting the beams to monotonically increasing loads using a hydraulic jack and a load cell having a capacity of 500 KN. Deflections and strain readings were recorded accordingly. The loading rate was kept slow enough so that dynamic effects could be avoided, but not too slow to avoid creep effects. Aside from the beam properties, the data observed were the load increments and the corresponding deflections, strains and cracking behaviors. The data collection, storing and monitoring was arranged as illustrated in Fig.5. After failure, careful visual inspection of the beam was done.

3 Theoretical considerations

3.1 Ultimate moment capacity

The theoretical ultimate moment capacity of the composite beam sections are calculated by means of the conventional reinforced concrete theory. The ultimate moment capacity of a section may be defined as the maximum moment that the section can resist immediately before the attainment of either yield strain by the tension steel (for under reinforced section) or the ultimate strain of the extreme compression fibre (for over reinforced section) [Fergusion (1958)]. In the present analysis the following assumptions are made:

(a) As all the beams are under reinforced, the contribution from all sort of reinforcement in the compression zone can be neglected at the ultimate stage.
(b) The distribution of strains across the depth of the section is linear at all stages of loading.
(c) Equivalent rectangular stress block depth in compression is equal to 0.85 times the neutral axis depth.
(d)Equivalent ultimate compressive strength of the concrete area in the compression zone can be defined, by assuming that the strain in the concrete and the mortar at their interface is equal, as,

$$f'_{cm} = (f'_c A_c + f'_m A_m) / (A_c + A_m) \quad (1)$$

where f'_c and f'_m are the companion cylinder strength of concrete and mortar respectively.A_c and A_{cm} are the compressive area of the core and ferrocement encasing respectively.
(e)Failure stress in the stress block of the composite beam is assumed to be $0.85f'_{cm}$.

The stress block approach is used as it is much easier for analysis. The theoretical values were computed on the basis of the steel bars and the wire mesh positions measured from the tested composite beams. Without knowing the position of the neutral axis, it is not possible to tell how many longitudinal wires are located in the tension zone, along the depth of the beam. So, for the first trial only those longitudinal wires, which are located at the level of the lowermost longitudinal skeletal steel will be considered.

Now the compressive force acting on the section is,

$$C = 0.85 f'_{cm} a_1 b \quad (2)$$

Again the total tensile force in the tension zone is,

$$T = T_s + T_{sk} + T_w$$
$$= A_s f_{sy} + \Sigma A_{ski} f_{sky} + \Sigma A_{wi} f_{wy} \quad (3)$$

in which the suffices s, sk, w, y and i refer to the main reinforcement, skeletal steel, wire mesh, yield strength and layer identification number respectively while T, f, A, a_1 and b represent respectively the tensile force, yield strength, area, equivalent stress block depth and width of the beams.

From equilibrium considerations,

$$a_1 = (A_s f_{sy} + \Sigma A_{ski} f_{sky} + \Sigma A_{wi} f_{wy}) / (0.85 f'_{cm} b) \quad (4)$$

According to assumption (c),

$$h'_c = a_1 / 0.85 \quad (5)$$

$$h'_t = h - h'_c \quad (6)$$

where h'_c, h, h'_t represent the neutral axis depth at failure from the extreme top fiber, total depth of the section and the depth of the tension zone at failure respectively.

Now the actual value of the total tensile force in the tension zone and the corresponding equivalent stress block depth can be calculated by including the wire areas situated in the tension zone with the previous wire area A_{wi} and using this new value in Eqn.3 and 4. It will be seen that the new value of the equivalent stress block depth is very much closer to the previous value. So no more trial will be needed.

Now, let d be the distance from the extreme top fibre to the resultant of all the tensile forces, then,

$$d = (A_s f_{sy} d_s + \Sigma A_{ski} f_{sky} d_{sky} + \Sigma A_{wi} f_{wy} d_{wy}) / (A_s f_{sy} + \Sigma A_{ski} f_{sky} + \Sigma A_{wi} f_{wy}) \quad (7)$$

where d_s, d_{ski} and d_{wi} represent respectively the distance of the main reinforcement, distance of the ith layer skeletal steel and the distance of the ith layer wires, all from the top of the section.

The ultimate moment capacity of the composite beams can be calculated as,

$$M_u = T(d - a_1/2) \quad (8)$$

The same equations can also be used for determining the ultimate moment capacity of the reference beams by dropping the terms of the skeletal steels and the wires.

3.2 Flexural analysis of composite beams by the principle of transformed section

Both the composite and the reference beams can be treated as homogeneous if all sort of reinforcements and the mortar are considered to be replaced by an equivalent area of concrete, so placed, and of such an amount as to produce the same effect as those steels and mortar in resisting the bending moment. According to this transformed area concept, the equivalent concrete area is,

$$A_c = (E_s / E_c) A_s = n_s A_s \quad (9)$$

where E_s, E_c and n_s are the modulus of elasticity of steel, modulus of elasticity of concrete and the modular ratio for the main steel respectively.

In a composite section, for the mortar of the ferrocement form (A_m), compression steels (A'_s), skeletal steels (A_{ski}) and the longitudinal wires of the meshes(A_{wi}), the corresponding equivalent concrete areas to be replaced by $n_m A_m$, $n'_s A'_s$, $n_{sk} A_{ski}$ and $n_w A_{wi}$ respectively, all in the same horizontal planes as their respective positions. Where n_m, n'_s, n_{sk} and n_w are the modular ratios for the mortar, compression steels, skeletal steel and the wire mesh respectively andwill be equal to E_m / E_c, E'_s / E_c, E_{sk} / E_c and E_w / E_c respectively, while E_m, E'_s, E_{sk} and E_w are their corresponding modulii of elasticity. It is to be noted that any sort of steel that lies in the compression zone of the section will result in a transformed area of $(n_s - 1)A_s$ since the area A_s originally occupied by the steel has also been replaced by the concrete.

As the analysis is to be carried out in the uncracked and cracked range , both the cracked and uncracked transformed sections are to be figured out. The resulting uncracked and cracked transformed sections for the composite beams are shown in Fig. 6. The transformed area for the wire mesh have not been shown in the figure, as the figure will then be too clumsy.

3.2.1 Uncracked transformed section

Let h_{cl} be the depth of the neutral axis from the top of the uncracked transformed section (Fig.6). The value of h_{cl} can be calculated by taking moments of the areas about the neutral axis,

$$\begin{aligned}(b_1 h_{cl}^2)/2 &= (b_1/2)(h - h_{cl} - t)^2 + b_2 t(h - h_{cl} - t/2) + (n_s - 1)A_s(d_s - h_{cl}) \\ &+ (n'_s - 1)A'_s(d'_s - h_{cl}) + \Sigma(n_{sk} - 1)A_{ski}(d_{ski} - h_{cl}) \\ &+ \Sigma(n_w - 1)A_{wi}(d_{wi} - h_{cl}) \end{aligned} \quad (10)$$

After locating the neutral axis position, the moment of inertia of the uncracked transformed section I_{tr} can be calculated from the following equation.

$$\begin{aligned}I_{tr} &= (b_1 h_{cl}^3)/3 + (b_1/3)(h - h_{cl} - t)^3 + (b_2/12)t^3 + b_2 t(h - h_{cl} - t/2)^2 \\ &+ (n_s - 1)A_s(d_s - h_{cl})^2 + (n'_s - 1)A'_s(d'_s - h_{cl})^2 + \Sigma(n_{sk} - 1)A_{ski}(d_{ski} - h_{cl})^2\end{aligned}$$

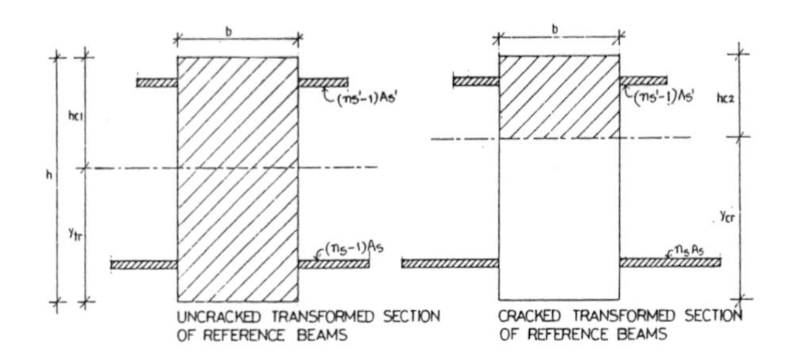

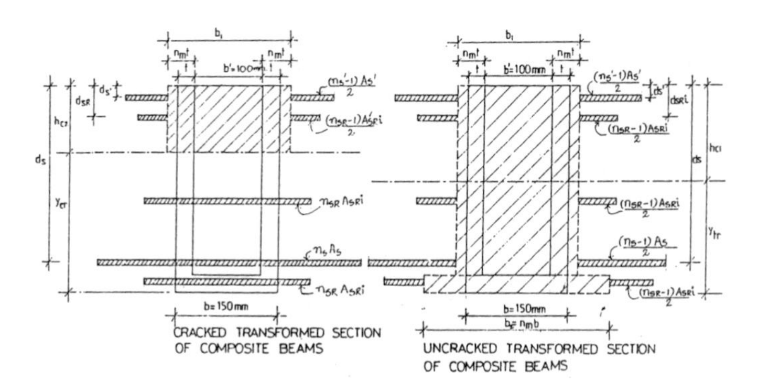

Fig.6. Uncracked and cracked transformed sections of the reference and composite beams.

$$+ \Sigma(n_w - 1)A_{wi}\,(d_{wi} - h_{c1})^2 \tag{11}$$

3.2.2 Cracked transformed section

Let h_{c2} be the depth of the neutral axis from the top of the cracked transformed section. In this case, for the longitudinal wires which will be located in the compression zone, the transformed area to be used will be $(n_w - 1)A_{wi}$ and for the wires in the tension zone, the transformed area to be used will be $n_w A_{wi}$. So, for the first trial only those longitudinal wires, which are located at the same level of the lower most longitudinal skeletal steel, will be considered. And as these wires are located in the tension zone, the transformed area to be used will be $n_w A_{wi}$. The value of the trial h_{c2} can be calculated by taking the moment of the areas about the neutral axis.

$$(b_1 h_{c2}^{2})/2 = n_s A_s\,(d_s - h_{c2}) + (n'_s - 1)A'_s\,(d'_s - h_{c2}) + \Sigma(n_{sk} - 1)A'_{ski}\,(d_{ski} - h_{c1})$$
$$+ \Sigma n_{sk} A'_{ski}\,(d_{ski} - h_{c2}) + \Sigma n_w A_{wi}\,(d_{wi} - h_{c2}) \tag{12}$$

From this trial valueof h_{c2}, it is possible to determine how many longitudinal wires are in the compression zone and how many are in the tension zone. Now for the second trial, Eqn.16 can again be used by just adding one term $\Sigma(n_w - 1)A'_{wi}\,(d_{wi} - h_{c2})$ in the right hand side of the equation. Again, solving for h_{c2} will show that this value is very close to the previous one. So no more trial will be needed.

The moment of inertia for the cracked transformed section I_{cr} can be calculated as,

$$I_{cr} = (b_2 h_{c2}^{3})/3 + (n'_s - 1)A'_s\,(d'_s - h_{c2})^2 + n_s A_s\,(d_s - h_{c2})^2$$

$$+ \Sigma(n_{sk} - 1)A'_{ski} (d_{ski} - h_{c2})^2 + \Sigma n_{sk} A_{ski} (d_{ski} - h_{c2})^2$$
$$+ \Sigma(n_w - 1)A'_{wi} (d_{wi} - h_{c2}) + \Sigma n_w A_{wi} (d_{wi} - h_{c2})^2 \quad (13)$$

3.3 Composite modulus of elasticity

The composite modulus of elasticity, E_{comp} of the composite beams can be computed using the Law of Mixtures as,

$$E_{comp} = E_c V_c + E_m V_m + E_s V_s + E'_s V'_s + E_{sk} V_{sk} + E_w V_w \quad (14)$$

where V_c, V_m, V_s, V'_s, V_{sk}, V_w are the volume fractions of the concrete, mortar, main tension steels, hanging bars, longitudinal skeletal steels and longitudinal wires respectively. The value of this modulus of elasticity applies only when the section is completely uncracked.

Flexural specimens, will have a compression zone in which it is uncracked right up to failure and a tension zone which will be cracked right after the first crack. So, in the cracked range, the modulus of elasticity of the compression zone is still given by Eqn.14. For the tension zone, the cracked modulus of elasticity, E_{cr} is obtained by dropping the terms $E_c V_c$ and $E_m V_m$ from Eqn.14. Thus,

$$E_{cr} = E_s V_s + E'_s V'_s + E_{sk} V_{sk} + E_w V_w \quad (15)$$

3.4 First cracking moment

For theoretical calculation of the first cracking moments of the beams, it will be considered that as the applied moment is increased, the tensile stress of the extreme fiber will reach the ultimate tensile strength of the composite, f_{comp} and tensile cracking will occur. The value of f_{comp} is calculated based on ACI code formulation [ACI code 318-71 (1971)] as,

$$f_{comp} = 7.5 \, (E_{comp} / E_m) (f'_m)^{0.5} \quad (16)$$

where f'_m is the ultimate compressive strength of the mortar, in psi.

Thus, the cracking moment is,

$$M_{cr} = f_{comp} \, (I_{tr} / Y_{tr}) \quad (17)$$

where, Y_{tr} is the depth from the neutral axis to the extreme tension fiber for the uncracked transformed section.

3.5 **Calculation of deflections and stresses**

The load deflection curve of the composite beams subjected to monotonically increasing bending moment can be approximated as trilinear as shown in Fig.7. The stress-strain diagrams for the composite beams are also idealized as elastic-perfectly plastic in compression and trilinear in tension as shown in Fig.8.

3.5.1 Uncracked Range

From elastic consideration, the values of the deflections at mid point and at the quarter point can be obtained as,

$$\Delta_{mid} = (23PL^3) / (648EI) \quad (18)$$

$$\Delta_{quar} = (29PL^3) / (1152EI) \quad (19)$$

where Δ_{mid} and Δ_{quar} are the mid point and quarter point deflections. In the uncracked range, the deflections can be calculated by replacing E and I by E_{comp} and I_{tr} respectively in Eqns.18 and 19. Compressive stress, σ_{cl} and the tensile stress, σ_{tl} at extreme fibers in the uncracked range can be calculated from classical beam theory as,

$$\sigma_{cl} = Mh_{cl} / I_{tr} \quad (20)$$

$$\sigma_{tl} = MY_{tr} / I_{tr} \quad (21)$$

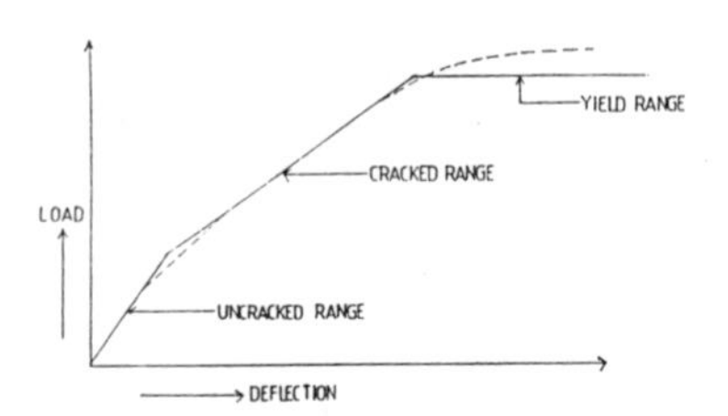

Fig.7. Assumed load deflection curve for the beam in flexure.

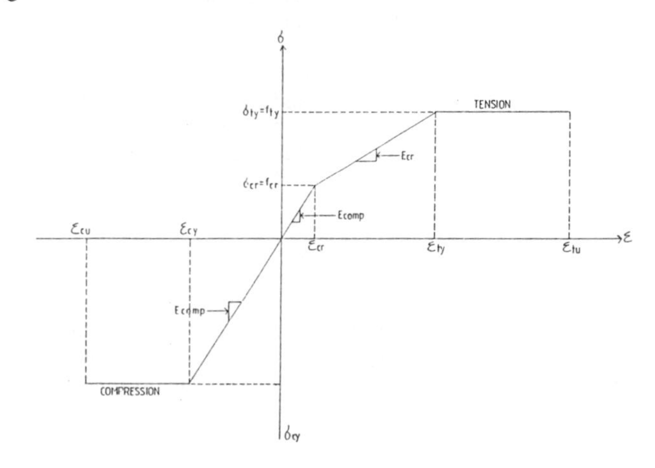

Fig.8. Assumed stress-strain distribution for the beams in flexure.

3.5.2 Cracked Range

In this range, the deflections can be calculated from elastic consideration using and equivalent flexural rigidity, as suggested by Branson [Branson (1963)]. The effective moment of inertia for the section can be determined from the expression,

$$I_{eff} = I_{cr} + (M_{cr} / M_{max})(I_{tr} - I_{cr}) < I_g \tag{22}$$

Here, M_{cr} is the cracking moment, M_{max} is the maximum moment at any stage of loading after first cracking. Eqns. (18) and (19) can also be used for deflection calculations in the cracked range by just replacing E and I by E_{comp} and I_{eff} respectively. Compressive stress, σ_{c2} and the tensile stress, σ_{t2} at extreme fibers in this range can be calcuted as,

$$\sigma_{c2} = Mh_{c2} / I_{cr} \tag{23}$$

$$\sigma_{t2} = (MY_{cr} / I_{cr})(E_{cr} / E_{comp}) \tag{24}$$

3.6 Shear load capacity

The ultimate shear load is determined by the direct transfer in the uncracked compression zone, direct tension in the web steel (stirrups), direct tension in the skeletal steel and direct tension in the wire mesh. Aggregate interlock and dowel action of the main flexural reinforcement are neglected.

For the first case, the shear load capacity by the direct transfer in the uncracked compression zone is determined according to the empirical formula of Rafla

[Rafla (1971)] after being modified by Pruyssers [Pruyssers (1986)] . The final formula, only valid for a/d >2.5, will be

$$V_{cu} = 0.62(a/d)^{-0.62} f'_{cm} (w_0)^{0.3} d^{-0.2} bd \tag{25}$$

where a, d, w_0 and b represent respectively the shear span, effective beam depth, percentage of tensile reinforcement and the width of the beam.

For the second, third and fourth cases, the shear load capacity can be calculated accordingly to the ultimate shear design method of web reinforcement.

$$V_{st} = (A_{st} f_{sy}\, z) \,/\, t_{st} \tag{26}$$

$$V_{sk} = (A_{sk} f_{sky}\, z) \,/\, t_{sk} \tag{27}$$

$$V_{w} = (A_{w} f_{wy}\, z) \,/\, t_{w} \tag{28}$$

where t_{st} , t_{sk} , t_w are the spacing of thestirrups, skeletal steel and wire mesh respectively and z is the lever arm.

Hence, the total shear load capacity can be calculated as,

$$V_u = V_{cu} + V_{st} + V_{sk} + V_w \tag{29}$$

4 Results and discussions

4.1 Crack pattern

Fig.9 shows the crack patterns of the beam specimens at failure. It can be seen that, in the composite beams only the flexural cracks were developed. But in reinforced concrete beams, the flexural cracks, that formed betwween the load point and the support, turned into diagonal cracks prior to failure. These diagonal cracks were not formed in the composite beams due to the crack arrest mechanism contributed by the

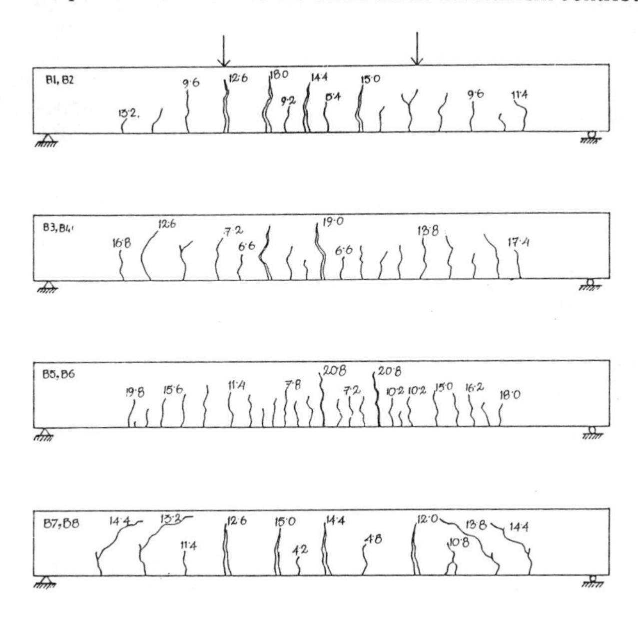

Fig.9. Crack patterns of reference and composite beams at failure (load stage shown in tf).

wire mesh in the ferrocement forms. Cracks are more numerous and crack width is less in the composite beams as the reinforcement in the ferrocement forms are more finely distributed. These cracks close considerably after the load is removed.

4.2 Comments on composite action

Before the occurrence of first crack, full composite action was observed between the ferrocement form and the core concrete. After the first crack a partial seperation between them was observed.

4.3 Load-deflection curve

Fig. 10 shows the experimental load vs mid span deflections of all the beams. It can be seen that the flexural rigidity, which is the slope of the load deflection curves, are higher for the composite beams as compared to the reference beams and this flexural rigidity increases with the increase in number of layers of wire mesh used in the ferrocement form. From Fig. 10, one more interesting thing can be observed. The rate of increase in the flexural rigidity decreases with the increase in number of layers of wire mesh. So, it can be concluded that 2 layers of wire mesh is enough for reinforcing the ferrocement forms.

4.4 Comparison of first cracking moments

Comparison of the various cracking moments show a satisfactory agreement between the calculated and measured first cracking moments (Table 3 and Fig.11). The average ratio obtained M_{cr} (me)/M_{cr} (th) for the reference beams (B7,B8) is 1.24 and that for the composite beams (B1,B2), (B3,B4) and (B5,B6) are 1.21, 1.19 and 1.33 respectively. Observed cracking moments of the composite beams (B1,B2), (B3,B4) and (B5,B6) are 6.7%, 6.7% and 20.63% higher than that of the reference beams, respectively.

4.5 Comparison of ultimate moment capacity

Theoretical ultimate moments of the beams, computed according to the conventional reinforced concrete theory, are compared to the observed ultimate moments and is shown in Table 3 and Fig. 12. It can be seen that the theoretical values are on the conservative side. Hence, the conventional reinforced concrete theory can be safely used for the computation of the ultimate moment capacities of the composite beams. The observed ultimate moments of the composite beams (B1,B2), (B3,B4) and (B5,B6)

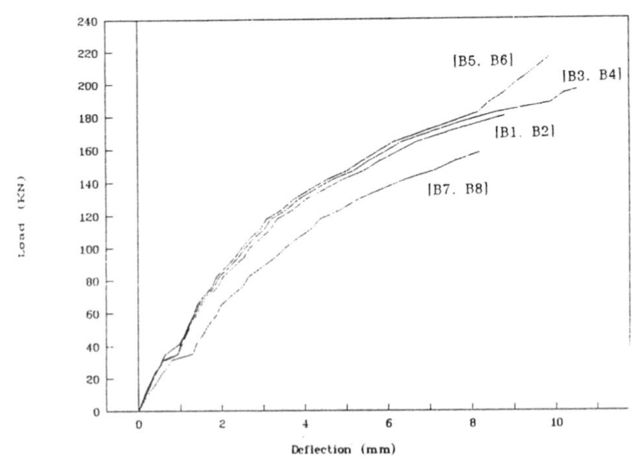

Fig.10. Load vs mid span deflection curves for all the beams.

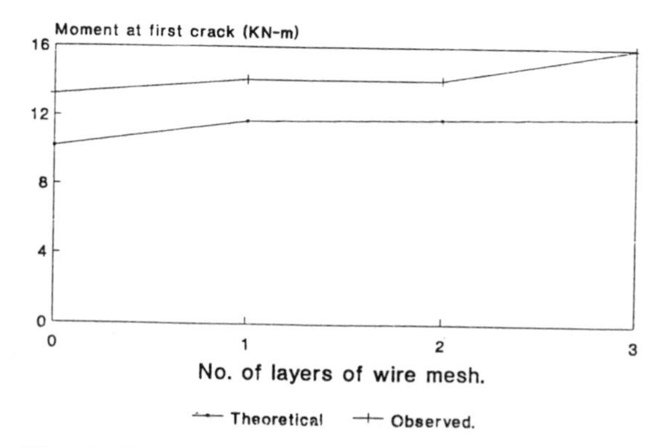

Fig.11. Comparison between theoretical and observed first crack moments for different beams.

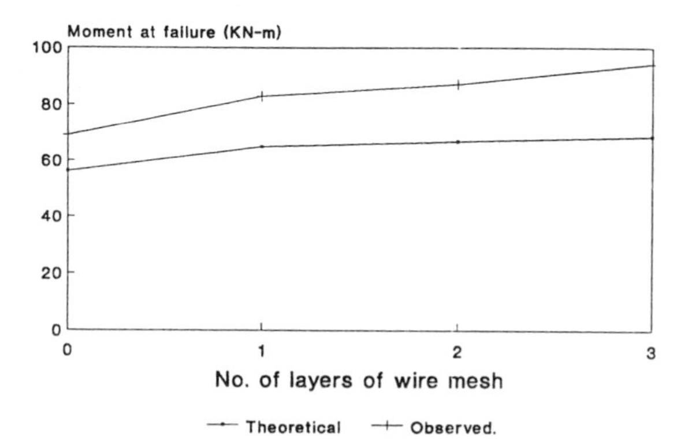

Fig.12. Comparison between theoretical and observed ultimate moments for different beams.

are 20%, 26.2% and 36.6% higher than that of the reference beams, respectively. Thus it can be concluded that the ultimate moment capacities of the composite beams encased in ferrocement forms show very superior performance, as compared to the ordinary reinforced concrete beams. While, in case of first cracking moments, they don't show that much superior performance.

4.6 Comparison of ultimate shear load capacity

Theoretical ultimate shear load capacities of the beams are compared to the observed shear load capacities and is shown in Table 4 and Fig. 13. It can be seen that the observed shear load capacities are smaller than the calculated shear load capacities for all the beams. This can be justified by showing that all the beams failed in flexural failure mode before attaining the ultimate shear load. The shear load capacities the composite beams (B1,B2), (B3,B4), (B5,B6) are 20%, 26.2%, 36.6% higher than that of the reference beams, respectively.

Table 1. Details and designation of beam specimens

Specimen designation	Number of specimen	Wire diameter (mm)	Mesh opening	Number of layers
B1, B2	2	0.8	0.5 in x 0.5 in	2
B3, B4	2	0.8	0.5 in x 0.5 in	4
B5, B6	2	0.8	0.5 in x 0.5 in	6
B7, B8	2	0.8	0.5 in x 0.5 in	--

Table 2. Volume fractions, moduli of elasticity and moments of inertia of the beams

Beam	Volume fractions						Moduli of elasticity		Moments of inertia	
	V_c	V_m	V_s	V_s'	V_{sk}	V_w	E_{comp}	E_{cr}	I_{tr}	I_{cr}
B1 B2	0.592	0.384	0.014	0.005	.00377	.00123	36666.67	4850.61	468.0×10^6	196.53
B3 B4	0.592	0.383	0.014	0.005	.00377	.00246	36880.49	5102.57	470.9×10^6	202.10
B5 B6	0.592	0.381	0.014	0.005	.00377	.00369	37056.16	5354.53	473.7×10^6	207.58
B7 B8	0.592	--	0.014	0.005	--	--	32249.00	3800.00	405.24×10^6	169.23

Table 3. Comparison between theoretical and measured ultimate and first cracking moments of the beam specimens

Beam	Skin reinforcement	Ultimate capacity (KN-m)		First cracking moments (KN-m)			
		Theoretical M_u(th)	Measured M_u(me)	Theoretical M_{cr}(th)	Measured M_{cr}(me)	$\frac{M_u(th)}{M_u(th)}$	$\frac{M_u(th)}{M_u(th)}$
B1	2 layer	64.96	81.14	11.7	14.11	1.25	1.21
B2	2 layer	64.96	84.67	11.7	14.11	1.30	1.21
B3	4 layer	66.79	88.2	11.84	14.11	1.32	1.19
B4	4 layer	66.79	86.1	11.84	14.11	1.30	1.19
B5	6 layer	68.49	91.64	11.98	15.96	1.34	1.33
B6	6 layer	68.49	97.02	11.98	15.96	1.42	1.33
B7	0	56.25	67.03	10.21	12.44	1.19	1.22
B8	0	56.25	71.09	10.21	14.02	1.36	1.26

Table 4 Comparison between theoretical and measured shear load capacities of the beam specimens

Beam	Theoretical shear capacity (KN)				Total $V_u(th)$ (KN)	Shear at failure $V_u(me)$ (KN)	$\frac{V_u(th)}{V_u(th)}$
	V_{cu}	V_{st}	V_{sk}	V_w			
B1	41.52	42.03	14.01	10.17	107.73	90.16	0.84
B2	41.52	42.03	14.01	10.17	107.73	94.08	0.87
B3	42.07	41.4	13.80	20.03	117.3	98.00	0.84
B4	42.07	41.4	13.80	20.03	117.3	95.65	0.82
B5	42.52	40.79	13.60	29.61	126.52	101.82	0.81
B6	42.52	40.79	13.60	29.61	126.52	107.80	0.85
B7	37.77	43.92	--	--	81.69	74.48	0.91
B8		43.92	--	--	81.69	78.99	0.97

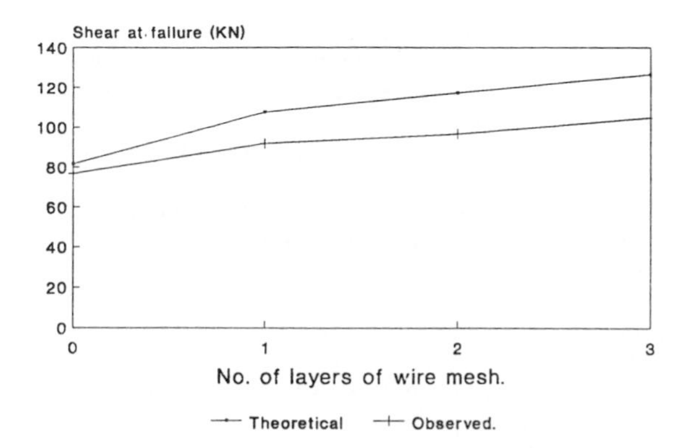

Fig.13. Comparison between theoretical and observed ultimate shear loads for different beams.

5 Conclusions

On the basis of the experimental results and by comparing with the theoretical values, the following conclusions may be drawn:

(a) Conventional reinforced concrete theory can be adopted for predicting the ultimate moment capacities of the composite beams.
(b) The theoretical values of the ultimate moment capacities and first cracking moments of all the beams are on the conservative side as they are smaller than the observed values.
(c) The composite beams (B1,B2), (B3,B4) and (B5,B6) show 20%, 26.2% and 36.6% higher ultimate moment capacities than the reference beams (B7,B8), respectively. But the difference in cracking moments were minor between the reference and the composite beams.

(d) The observed shear load capacities are smaller than the calculated values for all the beams, as they all failed in flexural failure mode before attaining the ultimate shear load. The shear load capacities of the composite beams are higher than the reference beams.
(e) The rate of increase in the flexural rigidity, ultimate moment capacity, shear load capacity for the composite beams decreases with the increase in number of layers of wire mesh, used in the ferrocement forms. So, from economical point of view and from the consideration of easiness of fabrication, it is better to use 2 layers of wire mesh, for reinforcing the ferrocement forms, instead of using more number of layers.
(f) Composite action between the core and the ferrocement form was fully obtained until crack appearance. Beyond that stage and up to failure, a partial separation was observed. From the cracking behaviors, it was found that the presence of wire mesh in the composite beams decreases the crack width and the crack spacing. With the increase in number of layers of wire mesh, both the crack width and the crack spacing decreases.

6 References

____. (1971) **ACI code 318-71**, American Concrete Institute

Branson, D.E. (1963) Instantaneous and time-dependent deflections of simple and continuous reinforced concrete beams, Report No.7, **Alabama Highway Research Report**, Barcan of Public Roads.

Fergusion Phil, M. (1958) **Reinforced Concrete Fundamentals.** John Wiley & Sons Inc., pp 48-52.

Lee, S.L. Raisinghani, M. and Pama, R.P. (1972) Mechanical properties of ferrocement, **FAO Seminar on the Design and Construction of Ferrocement Fishing Vessels**, Weilington, New Zealand, pp 20.

Logan, D. and Shah, S.P. (1975) Moment capacity and cracking behavior of ferrocement in flexure. **Proceedings of the ACI Journal**, 70(12), 155-164.

Pama, R.P. Sutharatna Chaiyaporn, C. and Lee, S.L. (1974) Rigidities and strength of ferrocement, in **Proceedings of the First Australian Conference on Engineering Materials**, Sydney, Australia, pp 287-308.

Pruyssers, A.F. (1986) Shear resistance of beams based on the effective shear depth, **University of Technology Report No.5-86-1**, Delft.

Rafla, K. (1971) **Empirical Formula for the Design on Shear Force Resistance of Reinforced Concrete Beams.**

Rajagopalan, K. and Parameshwaran, V.S. (1975) Analysis of ferrocement beams. **Journal of Structural Engineering**, India, 2(4), 155-164.

Suryakumar, G.V. and Sharma, P.C. (1975) An investigation into the flexural behavior of ferrocement. **Journal of Structural Engineering**, India, 2(4), 137-144.

102 DESIGN OF FERROCEMENT COMPOSITE COLUMNS

K. K. SINGH, S. K. KAUSHIK and A. PRAKASH
Department of Civil Engineering, University of Roorkee, India

Abstract
A ferrocement concrete composite (FCC) column is defined as one having a plain concrete or a reinforced concrete core and an outer casing of ferrocement. When subjected to an axial load it compresses in the vertical direction and tends to expand in the lateral direction due to Poisson's ratio effect. However, the mesh reinforcement, opposes the lateral expansion and imposes compressive stresses on the core. Due to this confining effect the structural behaviour of this column is different-both strength and ductility increase.
Keywords: Ferrocement, Composite Columns, Confined Concrete, Failure Load, Buckling.

1 Introduction

The behaviour of concrete in direct compression is modified if it is subjected to compressive pressures in the transverse direction as well. These are confining pressures. Experimental and theoretical investigations of behaviour of confined concrete are reported extensively in research literature. Some of the significant findings are covered here.

Confinement may be classified as active or passive. In active confinement, pressures are applied by external loading devices or hydrostatically. The pressures can be applied and varied independently of the direct compression. Hence active pressure is more appropriate for experimental study of confined concrete.

Passive confining pressures are those which develop due to the presence of reinforcement in the form of hoops, ties, spirals etc. Concrete is subjected to compressive strains in the direction of direct compression. As a result of Poisson's ratio effect, tensile strains are introduced in the transverse directions without the introduction of corresponding stresses. The presence of ties etc. restricts the transverse strains. The reinforcement is itself subjected to tensile strains and stresses and exerts equal and opposite stresses i.e. comparessive, on the concrete. Thus, the passive case corresponds to the actual case in R.C. columns.

Experimental investigations have used both active and passive cases to study the behaviour of confined concrete. The earliest investigations were in the 1920's. Pfister (1964), Somes (1976),

Fibre Reinforced Cement and Concrete. Edited by R. N. Swamy. © 1992 RILEM.
Published by E & FN Spon, 2-6 Boundary Row, London SE1 8HN. ISBN 0 419 18130 X.

Burdette and Hilder (1975), Ahmad and Shah (1988b), Mander et al. (1988b) have made significant studies on passive confinement. Mills and Zimmerman (1970), Palaniswamy and Shah (1974) and others have investigated active confinement. Furlong (1967), Bertero and Moustafa (1976), Knowles (1970) etc. have dealt with confinement of concrete in hollow steel tubes. Considering the reported experimental results Palaniswamy and Shah (1974), Kotsovos and Newman (1978), Ahmad and Shah (1982a) and Mander et al. (1988a) have proposed theoretical relations for stress-strain behaviour of confined concrete. The results of these studies may be summarised as follows:

Both compressive strength and ductility improve with confinement, the improvement being directly proportional to the confining pressures.
There are two possible failure modes - the first where the usual tensile spliting occurs. This is when ties or hoops spacing exceeds cross-sectional dimensions of the column making confinement ineffective or the core concrete in the tube shrinks excessively and confinement does not occur.
The second is due to the the crushing of the concrete matrix. This occurs when confining stresses are large.
When area of core is small as compared to the total sectional area, the expected strength increase does not materialize.

From the above it is clear that ferrocement can also be used for confinement of concrete in columns. The possible applications being for the purposes of prefabrication, repair and strengthening of existing structures and for insitu construction. The first reported study on confinement with ferrocement is by Sandowicz and Grabowski (1981). They observed improvement in ductile behaviour but no significant increase in strength. The first successful applications of ferrocement for confinement are by Singh et al. (1988) and Balaguru (1988). The following section is based on these references and subsequent work of the authors (1989 a & b).

2 Test results

The tests conducted by the authors cover the following :

Effect of quantity of mesh
Circular cross-section
Square cross-section
Effect of grade of concrete in the core
Effect of grade of mortar
Eccentricity of loading
Strengthening of tested specimens
Inelastic buckling.

Some of the above tests are exhaustive and conclusive while for some the tests conducted so far are limited and further tests are in progress. Many of the above test results have already been reported elsewhere by the authors and so are reported below very briefly :

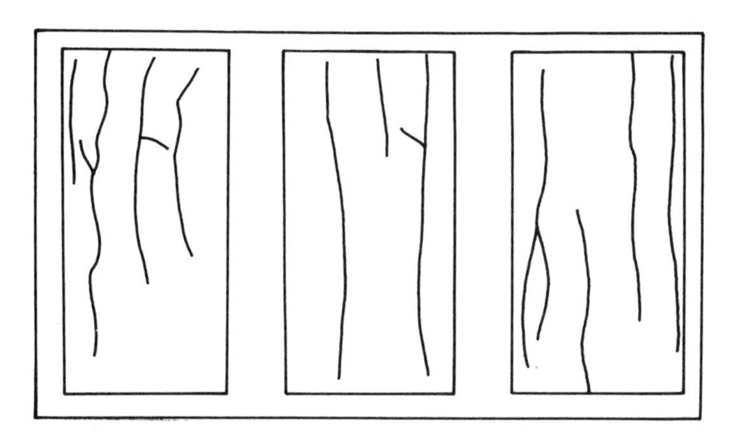

Fig. 1. Vertical cracks on compressive face.

The strength and ductility both increase linearly with quantity of mesh. The failure mode is the typical vertical splitting failure. The first visible cracks develop at about 95% of maximum load. The steepness of the descending portion of the curve reduces with increase in quantity of mesh. In this region the horizontal strands of the mesh snap and cover concrete spalls.

Confinement is uniform and very effective in a circular section. In a square column the corresponding strength increase is less being approximately 0.75 to 0.8 of the strength increase in the circular specimen. A similar effect is observed in case of ductility. In square columns visible cracks occur at about 90% of maximum load.

Moderate variations in strength of core concrete do not effect confinement behaviour significantly. However, in case of low grade concrete the strength increase is higher.

Grade of mortar in the ferrocement shell does not significant effect strength of columns, but for low grade mortar the visible cracks develop earlier.

Even for eccentric loading the ferrocement encased specimens gave higher strength and ductility. There is a linear increase with mesh quantity. The compressive and tensile forces develop vertical and horizontal cracks respectively (Fig. 1 and 2). Ductile behaviour is brought out by the fact that the specimens develop significant curvature and sustains the load at a high level beyond the maximum load.

Some plain concrete specimens were tested to maximum load taking care to avoid disintegration. These were repaired by wrapping mesh reinforcement and applying mortar. On retesting, strengths obtained even with one and two mesh layers, were higher than original values.

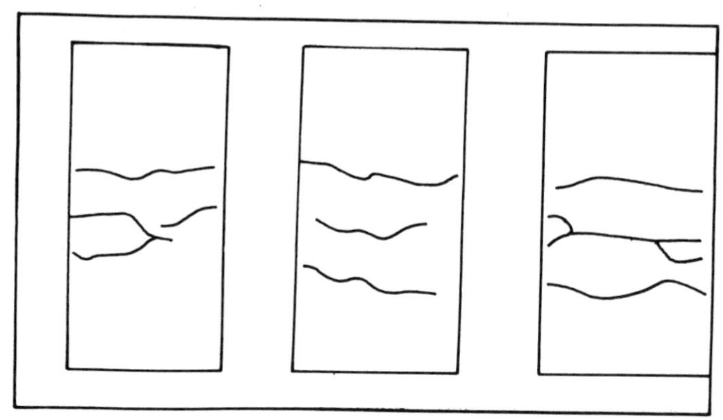

Fig. 2. Horizontal cracks on tensile face.

Production of FCC columns poses some difficulties depending on the sequence of construction. When core is cast first and the mesh is wrapped around the core and plastered, mortar penetration is difficult, if the mesh spacing is fine and number of layers is large. In such cases the expected strength increase could not be obtained. Hence the mesh used must have thicker wires with a larger spacing so that less number of layers are needed. Instead of manual plastering either mechanical devices or shortcreting may be tried.

For insitu columns separate plastering is not necessary. The concrete should contain a higher proportion of sand, cement and superplasticiser so that when poured into the core and vibrated the mortar component penetrates the mesh from the inside and encapsulates it fully.

3 Axially loaded short columns

If a concrete cylinder of radius 'b' is subjected to a uniform radial compressive stress 'p' then its compressive failure stress is known to increase by 4.2 p. Let this cylinder be provided which several layers of circular mesh reinforcement of mean radius 'a' as shown in Fig. 3 and let area of horizontal strands of mesh per unit height of cylinder be 'A_m' on each face and 'σ_y' the yield stress of steel. At its yield point the mesh exerts a uniform radial pressure on the core $p = A_m \sigma_y / a$.

The failure load (P_u) of this cylinder is

$$P_u = \pi(b^2 \cdot \sigma_o + 4a \cdot A_m \sigma_y)$$

where σ_o is the compressive strength of concrete and the second term in the parenthesis corresponds to strength increase due to confinement.

The shape of stress-strain curves of plain and FCC cylinders are shown in Fig. 4. The initial slope of the curves is same. Also, they do not differ significantly upto about $0.8\ \sigma_o$. The top most curve for FCC is for a higher mesh content.

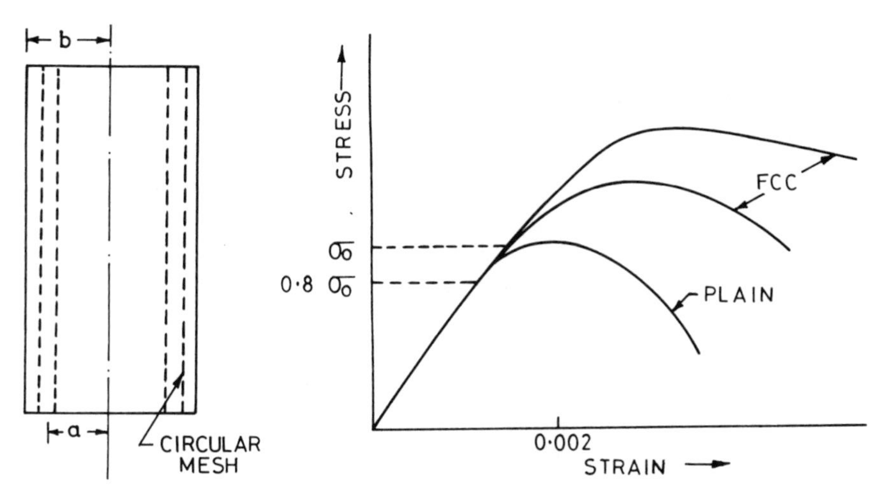

Fig. 3. Cylinder with mesh.   Fig. 4. Stress-strain curve.

Tests on square specimens had given somewhat lower values of strength increase. This is because confinement of core is not uniform. If this time 'b' and 'a' represent the side of the specimen and the core respectively then failure load is

$$P_u : b^2 \sigma_o + 3.2 a \cdot A_m \sigma_y \,.$$

where the second term again represents the strength increase due to confinement.

4 Column buckling

According to Indian Standards a short column is one in which the aspect ratio (effective length to lateral dimension) is less than 12. A short FCC column can drive the full benefit of strength and ductility increase due to confinement. However, for ratio above 12 the column is slender and buckling governs.

There are two types of buckling possible i.e. elastic or inelastic. The elastic buckling load can be directly determined by the Euler's formula. For two columns with the same aspect ratio the buckling load depends upon the Young's modulus i.e. the initial slope of the stress-strain curve of concrete. From Fig. 4, it is obvious that confinement does not change the initial slope of the curve. Hence, confinement does not modify the elastic buckling load. This type of buckling is possible only for very slender columns (aspect ratio above 35 for concrete of strength 200 kg/cm^2). Such columns are rare.

Inelastic buckling governs in the range of aspect ratio 12 to 35. For real columns the tangent modulus formula $P_{cr} = \pi^2 E_t I/L^2$ gives a fairly good prediction of the buckling load. Here E_t is tangent modulus, L is effective length and I is moment of inertia of the

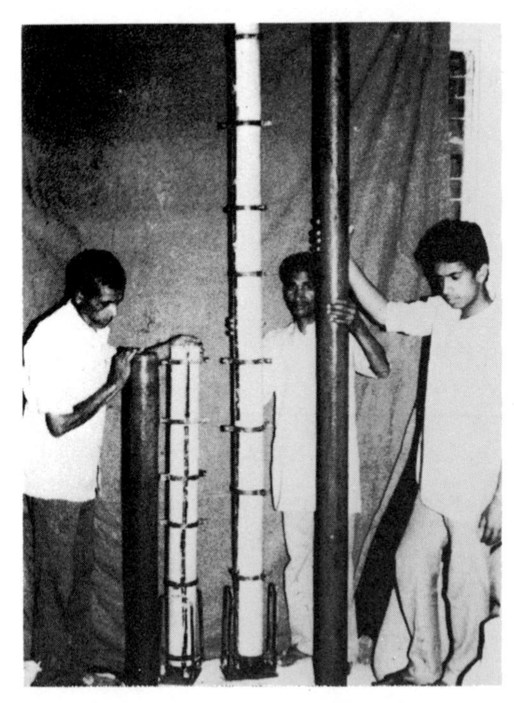

Fig. 5. Slender FCC columns with moulds.

section. As E_t is a variable the buckling load is determined iteratively. The confined and unconfined curve are almost identical till $0.8\,\sigma_o$. Upto this value the inelastic buckling load will be same for both columns. By representing the stress-strain curve of concrete by Hognestad's parabolic equation the buckling load may be determined without recourse to iteration from the graph. The value of aspect ratio corresponding to $0.8\,\sigma_o$ is 24. This means that inelastic buckling of an FCC column is higher than that of an RC column in the slenderness range 12 - 24. Some slender FCC columns (shown in Fig. 5 with moulds) were tested for buckling with both ends hinged (Fig. 6). The failure was sudden and accompanied by an explosive noise. Snapping of horizontal wires occurred at the critical section.

5 Design of FCC columns

To establish the design guidelines of FCC columns a comparision has to be made with reinforced concrete columns. The popular basis of design is the limit state concept using partial safety factors and load factors. It is assumed that strain in concrete corresponding to maximum and failure stress are 0.002 and 0.0035 respectively. For confined concrete the stresses as well as corresponding strains are higher. Hence design would have to account for the higher strains otherwise the benefit of higher strength cannot be utilized.

There is a difference in confinement by closely spaced ties, hoops, spirals etc. or by wire mesh. In the former case the concrete cover is thick while in the later it is thin about 3 mm or so.

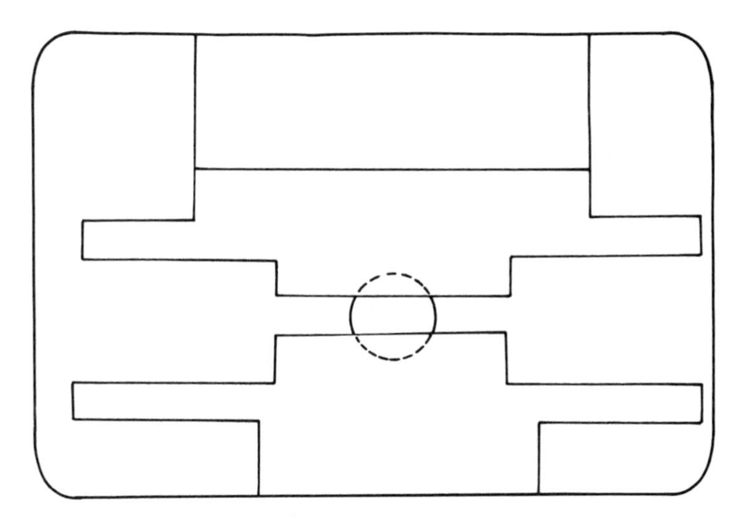

Fig. 6. Ball and sockets for hinged end.

The thick cover cracks and spalls before development of full confinement. In FCC columns the cover does not develop visible cracks before attainment of 90% of the maximum load and spalls only after the maximum load. Some FCC columns were prepared with thick cover i.e. 1.5 cm or so. Here cover cracked and spalled before development of full confinement. Hence, there is a significant advantage of FCC columns with proper amount of cover. The design principles for such columns are proposed as below for the consideration of the profession :

Appropriate value of partial factor of safety should be assumed for the mesh.
The core may be assumed to develop 90% of maximum confined stress and the corresponding strain. This limitation is placed to avoid cracking outside the core. The outside area is assumed to be unconfined.
For eccentric load vertical strands of mesh on the tension side may be considered to carry tensile forces along with the steel reinforcement.
Steel reinforcement may be treated as in RC columns. Similarly in other respects the design of FCC columns would be similar to that of RC columns.

It is planned to produce design charts on the basis of above assumptions.

6 Conclusions

The test results presented here and elsewhere clearly show that confinement by ferrocement leads to improved behaviour of columns. Thus, such columns may be used with advantage particularly in prefabrication and for repair/strengthening of columns.

The buckling load of FCC columns is higher than the corresponding RC columns for a considerable range of slenderness ratio.

It is hoped that the proposed design principles pass the scrutiny of the profession and are adopted in practice.

7 References

Ahmed, S.H. and Shah, S.P. (1982) Complete triaxial stress-strain curves for concrete. **J. of Str. Engg.**, ASCE, No. ST4, 728-742.

Ahmed, S.H. and Shah, S.P. (1982) Stress-strain curves of concrete confined by spiral reinforcement. **ACI J.**, 484-490.

Balaguru, P. (1988) Use of ferrocement for confinement of concrete. **Third Int. Symp. on Ferrocement**, New Delhi, India, 52-58.

Bertrero, V.V. and Moustafa, S.E. (1970) Steel encased expansive cement concrete columns. **J. of Str. Engg.**, ASCE, 2267-2282.

Burdette, E.G. and Hilsdorf, H.K. (1971) Behaviour of laterally reinforced concrete columns. **J. of Str. Engg.**, ASCE, Vol.97, No. ST2, 587-602.

Furlong, R.W. (1967) Strength of steel-encased concrete beam columns. **J. of Str. Engg.**, ASCE, 113-124.

Knowles, R.B. and Park, R. (1970) Axial load design for concrete filled steel tubes. **J. of Str. Engg.**, ASCE, 2125-2153.

Kotsovos, M.D. and Newman, J.B. (1978) Generalized stress-strain relations for concrete. **J. of Engg. Mech.**, ASCE, EM4, 845-855.

Mander, J.B., Priestly, M.J.N. and Park, R. (1988) Theoretical stress-strain model for confined concrete. **J. of Str. Engg.**, ASCE, Vol.114, No. ST8, 1804-1825.

Mander, J.B., Priestly, M.J.N. and Park, R. (1988) Observed stress-strain behaviour of confined concrete. **J. of Str. Engg.**, ASCE, Vol.114, No. ST8, 1827-1849.

Mills, L.L. and Zimmerman, R.M. (1970) Compressive strength of plain concrete under multiaxial loading conditions. **ACI J.**, 802-807.

Palaniswamy, R. and Shah, S.P. (1974) Fracture and stress-strain relationship of concrete under triaxial compression. **J. of Str. Engg.**, ASCE, Vol.11, No. ST5, 901-916.

Sandowicz, M. and Grabowski, J. (1981) The properties of composite columns made of ferrocement pipes filled with concrete. Tested in Axial and Eccentric Compression. **Int. Symp. on Ferrocement,** Bergamo, Italy, L/93 - 99.

Singh, K.K., Kaushik, S.K. and Prakash, A. (1988) Ferrocement composite columns. **Third Int. Symp. on Ferrocement,** New Delhi, India, 296-305.

Singh, K.K. (1989) Concrete confinement with ferrocement. **National Seminar on Special Concrete**, Mussorie, India.

Singh, K.K, Kaushik, S.K. and Prakash, A. (1989) Ferrocement encased columns. **Institution of Engineers (U.P.).**, India, 69th Annual Number, 4045, Lucknow, India.

Somes, N.F. (1976) Compression tests on hoop reinforced concrete. **J. of Str. Engg.**, ASCE, 1903-1916.

Pfiser, J.F. (1964) Influence of ties on the behaviour of reinforced concrete columns. **ACI J.**, 521-535.

103 FIBROUS FERROCEMENT: PERFORMANCE OF CRIMPED STEEL FIBER FERROCEMENT PLATES UNDER BENDING

L. F. de SILVA
University of Sao Paulo, Sao Carlos, Brazil

Abstract
An investigation has been made on the flexural tests of crimped steel fiber ferrocement plates in comparison with traditional ferrocement. The results of first cracking and ultimate loadings of flexural tests are discussed.

Observations on the cracking development were made and the load-deflection diagrams were also determinated.

The steel fiber ferrocement plates have shown a better behaviour than conventional ferrocement, with higher ultimate and cracking flexural strengths, better cracking distribution and load-deflection diagram configuration, showing a higher toughness of fiber reinforced composites.
Keywords: Fiber Reinforced Concrete, **Fibrous Ferrocement**, Crimped Steel Fibers, Flexural tests, Reinforced Mortar.

1 Introduction

The idea to reinforce brittle concrete and mortar matrices, in order to obtain more homogeneous and ductile materials is relatively old. From current development of technology and science, man has obtained his objective with the development of new materials and new construction techniques. Two technologies, fiber reinforced concrete and ferrocement, are studied in this work. The utilization of fibers in ferrocement plates is already employed in many countries around the world, where the researches about steel fiber ferrocement combined with polymers, admixtures and additions are commom.

The main goal of this work is to investigate the flexural behaviour of ferrocement plates with welded steel meshes, with crimped steel fibers and with both meshes and fibers simultaneously, aiming to use them in thin structures in civil engineering as components for covering and sealing in general.

Fibre Reinforced Cement and Concrete. Edited by R. N. Swamy. © 1992 RILEM.
Published by E & FN Spon, 2-6 Boundary Row, London SE1 8HN. ISBN 0 419 18130 X.

2 Experimental Program

2.1 Materials

The materials used in this work are those related in the specification of the Brasilian Association of Technical Standards (ABNT). The cement was a "ARI" (High Initial Strength) cement type. The sand had a fineness modulus of 2.17, maximum size 2.4mm, bulk density, 1.48 kg/dm3 and specific gravity of 2.60 kg/dm3.

The fiber reiforcement consisted of carbon steel formed by sheets of retangular section with crimped section with dimensions of 0.2mm thickness, 2.3mm width, and three lengths of 25.4mm (1"), 38mm (1 1/2") and 51mm (2"), with rough surfaces and splines (Figure 1), a by product of steel sponge manufacturing used for cleaning. The fibers had an equivalent diameter of 0.76 mm so that the fibers aspect ratios (L/D) were respectively 33, 50 and 67 for the 25.4mm, 38mm and 51mm lay fibers.

Welded meshes were used as main ferrocement reinforcement, with mesh opening of 25mm X 50mm (wire 2mm diameter) or mesh opening of 50mm X 50mm (wire 2.5mm).

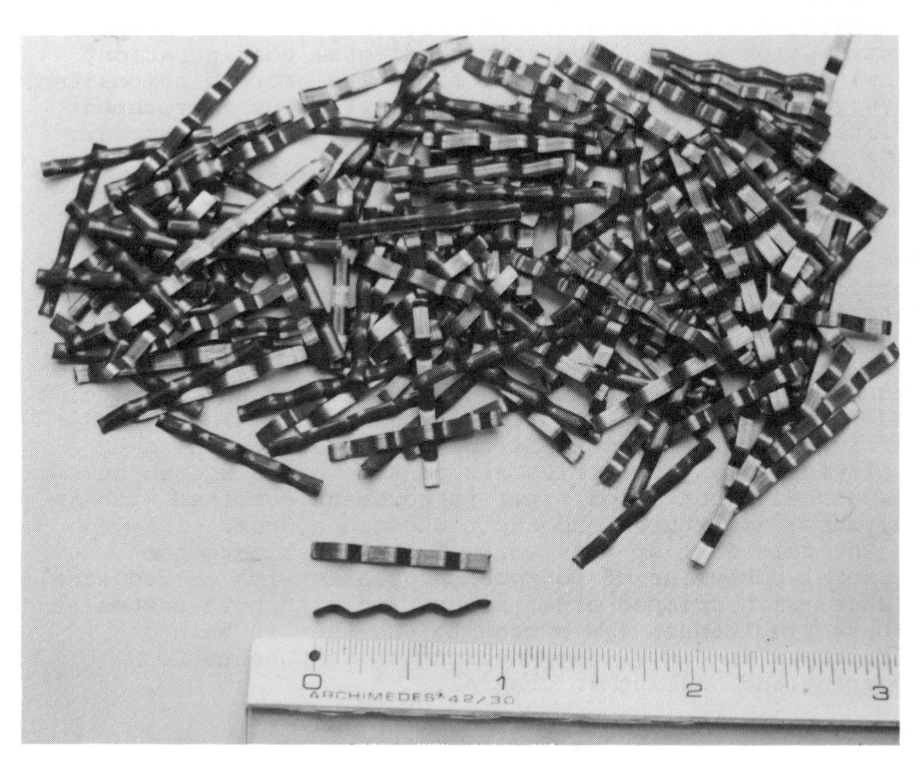

Figure 1 - Crimped Steel Fibers - 25.4mm (1")

2.2 Mortar mix design

Based on preliminary characterization studies, the mortars were chosen with the following mass mix proportions cement : sand: 1.0 : 2.0 with w/c=0.40 and 3% volume (234 kg/m3 of concrete) of fibers.

2.3 Dimensions and production of plates

The plate dimensions used in the flexural tests were 1.00 m length X 210mm width X 15mm or 25mm thickness.

The mixtures were made into a simple concrete mixer with inclined axis, cast in metalic moulds with the specified dimensions, and compacted on vibrating table, obtaining in this way a good superficial and internal uniformity for all the mixtures. For the mixtures without fibers, cylinder specimens of 50mm diameter X 100mm height were made for compression strength control tests, at 1, 3, 7 and 28 days, the resulting strengths were 24.8 , 36.4 , 40.3 and 47.4 MPa. After casting, the slabs were covered with plastic and at the following day, removed from their molds and immersed in water for curing. The tests were made after 7 days, and some hours before the test, the plates were removed from water to dry in open air, to mark the support and load application points, and the cracking observation area,and to provide a suitable means of observing the cracks.

Twelve series of tests were carried out, each series formed by four plates, obtained by variation of the fiber type, slab thickness and mesh type. A total of fourty-eight plates, as shows in Table 1, were thus tested. The first five series had 15mm thickness with fibers 25.4mm (1") and 38mm (1 1/2") lengths , with and without mesh and one series without fibers and with 25mm X 50mm mesh (wire 2mm). The same patern was adopted for the seven remaining series, using fibers of 1 1/2" and 2" length and 25mm thickness. In this case, two more groups of plates with double reinforcement of 50mm X 50mm opening mesh were made.

Table 1. Series of plates used in flexural tests

n.	Thick(mm)	Series	Description
1	15	3/1-NM	3% vol.fiber 1" - without mesh
2	15	3/1.5-NM	3% " 1 1/2"- without mesh
3	15	3/1-1M	3% " 1" - with 1 mesh
4	15	0/0-1M	Without fiber - with 1 mesh
5	15	3/1.5-1M	3% vol.fiber 1 1/2"-with 1 mesh
6	25	3/1.5-NM	3% vol.fiber 1 1/2"-without mesh
7	25	3/2-NM	3% " 2" - without mesh
8	25	3/1.5-1M	3% " 1 1/2" -with 1 mesh
9	25	3/2-1M	3% " 2" - with 1 mesh
10	25	0/0-1M	Without fiber - with 1 mesh
11	25	0/0-2M	Without fiber - with 2 meshes
12	25	3/1.5-2M	3% vol.fiber 1 1/2"- 2 meshes

2.4 Flexural Test

Figure 2 shows the sketch of four points plates flexural tests with free span of 900mm, and the location of mesh reinforcement in the plate. In order to facilitate the observation of cracking, the tests were carried out with tension face uppermost.

Five mechanical deflectometers were placed, two in the basis, one in the half span and two in the third parts for the reading of the displacements. An inductive deflectometer placed in the middle of the span allowed the plotting of the load-deflection diagram, during the test, through one XY register (Hewlett Packard - model 7004-B), as well as, the determination of the first cracking load at the same time it was occuring. The displacements readings were done by loading stages, in number and load value each stage separated in function of the final expected value for each series, containing the amount of ten stages. Reaching the second stage, the sample was unloaded back to the first stage, continuing the test from this point.

From the first cracking observation, by the graphic deflection in the register, or using magnifying glasses, in many other following stages, the cracking and their evolution in the same sample were observed and registered by the number of the correponding stage.

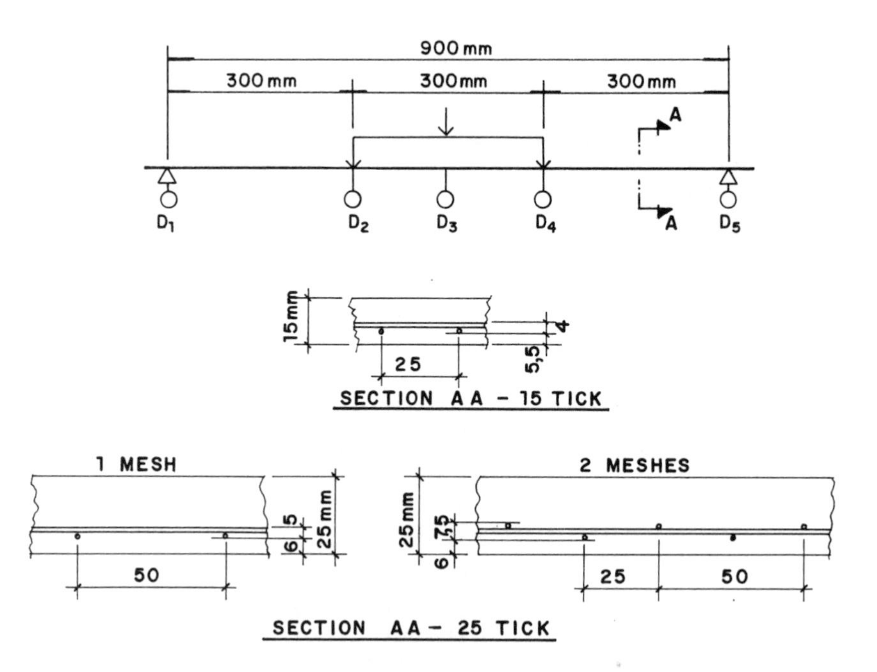

Figure 2 - Loading sketch and location of meshes

In many of these stages, the opening of the respectives cracks, were measured with a graduated magnifying glass, mainly in the central third of the plate. Due to the great pieces displacement, with exception of those without mesh reinforcement, the deflectometers reached their maximum course point, leaving to be removed before fracture. In the case of the inductive deflectometer, it was possible to maintain it until just before the end of the test.

2.5 Results and Discussion

The results of the flexural tests of the plates, containing the first cracking and ultimate loads and moments, are shown in the Tables 2 and 3, and in Figures 3 and 4. Figures 5 and 6 show typical load-deflection diagrams of the slabs with and without fibers.

Table 2. Flexural tests of plates - 15mm tick

Series	Load (kN) Pr *	Pu	Flerural Moment (kN.cm) Mr	Mu
0/0-1M	0.24	0.69	3.65	10.35
3/1-NM	0.28	0.30	4.20	4.43
3/1.5-NM	0.32	0.35	4.73	5.18
3/1-1M	0.47	1.05	7.00	15.80
3/1.5-1M	0.40	1.06	6.00	15.83

Table 3. Flexural tests of plates - 25mm tick

Series	Load (kN) Pr	Pu	Flexural Moment (kN.cm) Mr	Mu
0/0-1M	0.80	1.93	12.00	28.95
0/0-2M	0.63	2.66	9.38	39.86
3/1.5-NM	0.81	0.86	12.19	12.94
3/2-NM	0.88	0.95	13.13	14.29
3/1.5-1M	1.23	2.43	18.38	36.45
3/2-1M	1.10	2.24	16.50	33.56
3/1.5-2M	1.15	2.70	17.25	40.50

Legend

0/0-1M : Ferrocement without fibers with 1 mesh
0/0-2M : Ferrocement without fibers with 2 meshes
3/1-NM : Mortar - 3% volume fiber 1" - without mesh
3/1.5-2M: Ferrocement 3% vol. fiber 1 1/2" with 2 meshes
*Pr : Cracking Load , Pu : Ultimate Load
Mr : Cracking Flexural Moment, Mu : Ultimate Flex. Moment

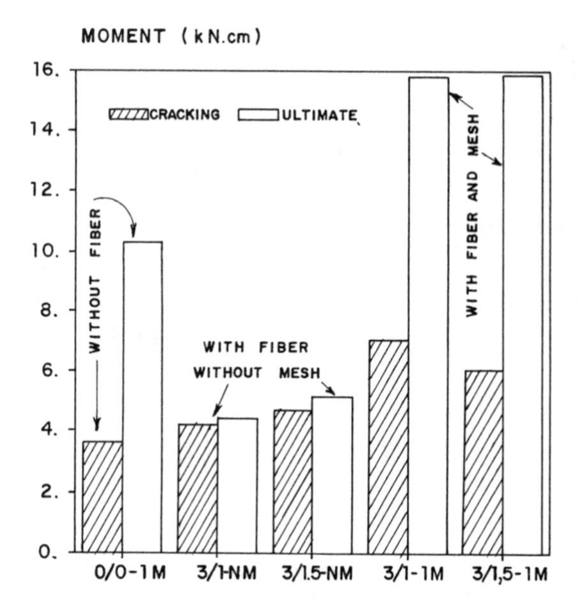

Figure 3 - Cracking and ultimate moments - 15mm tick

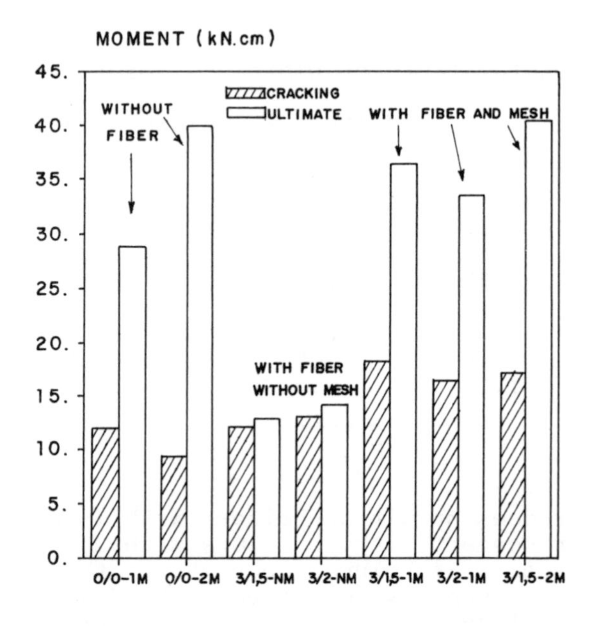

Figure 4 - Cracking and ultimate moments - 25mm tick

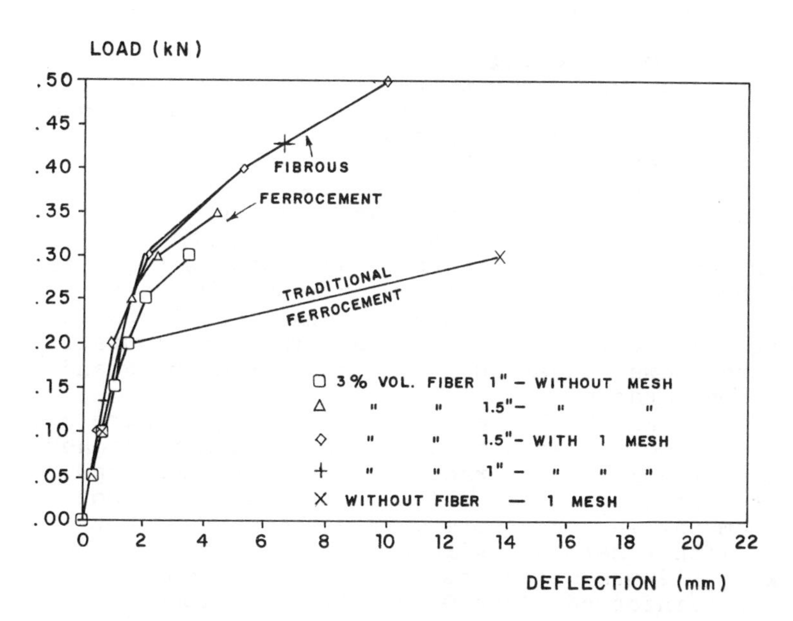

Figure 5 - Comparative Diagram Load-Deflection - 15mm thick

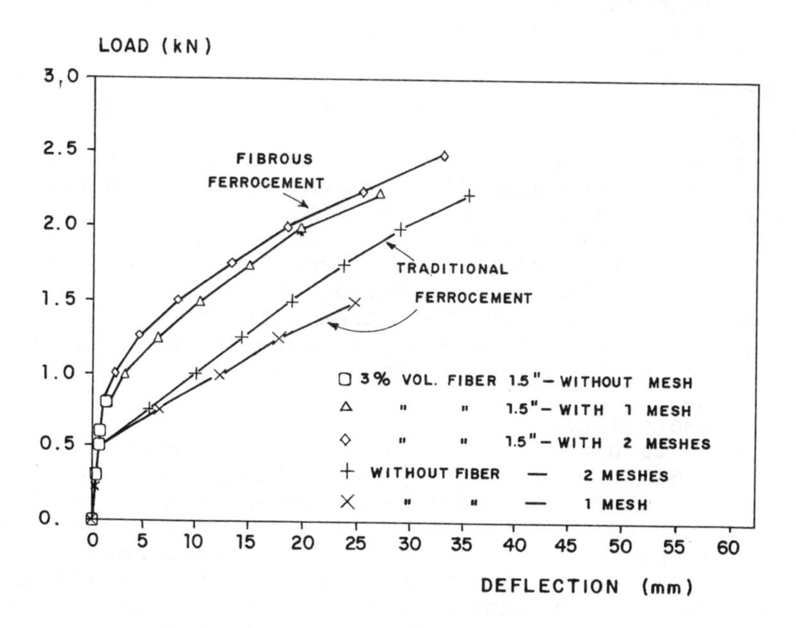

Figure 6 - Comparative Diagram Load-Deflection - 25mm Tick

The analysis of these figures and tables, leads to the following discussion:

2.5.1 State of cracking formation

In the plates with 15mm thickness, the first cracking flexural forces and moments are on average 25% higher for the slabs with fibers than those without fibers.

It was observed that the samples with fibers and without mesh showed only one crack where the rupture occurred and whose cracking flexural loads and moments are very near to the ultimate loading (close to 90% of these). The comparison of the results from the plates without fiber, with the results of the ones with mesh and fiber reinforcement, show cracking flexural forces and moments higher for the fibrous composites. The configuration of cracking shows higher quantity, smaller opening and smaller gaps among cracks.

In the 25mm thick plates, the cracking flexure strengths and moments of those with only fibers are 5% higher than those of traditional ferrocement with one mesh and 34% higher , on average, in relation to the plates mounted with double mesh. In this sense, it was observed that the plates with fibers and with double mesh reinforcement, have cracking flexure strengths and moments 7% smaller than the similar reinforced slabs with only one welded steel mesh.

These observations come to corroborate previous researchs that samples with higher conventional reinforcement have smaller cracking flexure forces and moments. Still with reference to 25mm thick plates, those containing fibers and one mesh, show first cracking flexure strengths 45% higher than those traditional ferrocement with one mesh, and 85% higher, on average, than those with double mesh.

In respect to the configuration of cracking, the plates of 25mm with fibers and without mesh, showed only one or two cracks, while in the samples with mesh, with or without fibers, the quantity of cracks is much bigger.

In general, the traditional ferrocement showed, on the tension face of the plate, parallel cracks in the transverse direction. In the plates with fibers and mesh, the number of cracks increased, in parallel directions and inclined in relation to the transverse axis.

2.5.2 Ultimate limit state

In relation to ultimate flexure forces and moments, the plates of 15mm just with fibers have their values 47% smaller than those of traditional ferrocement while the ones with fibers and mesh show ultimate flexure forces and moments 53% higher in relation to the ferrocement.

For the plates of 25mm thickness, with fibers and double reinforcement, the values of the ultimate flexure forces and moments are 11% higher in relation to those with only one mesh.

The ferrocement samples without fibers have, on average, the values of the ultimate flexure strengths 40% higher than the fibrous ones without mesh and 20% smaller than the fibrous ones with one mesh and with the same size order for double mesh reinforcement.

2.5.3 Flexural behaviour in relation fiber length
In both types of thickness plates, the shorter fibers showed better behaviour in relation to cracking and ultimate flexure forces and moments. In the plates of 15mm thickness, the fibers of 25.4mm (1") length led to values 17% higher than those containing 38mm (1 1/2") length fibers and craking strength with the same order of magnitude.

In the plates of 25mm thickness, the fibers of 38mm length led to cracking flexure forces and moments 12% higher and ultimate values 8% higher than those reinforced with fibers of 51mm length.

2.5.4 Load-Deflection Diagrams
It was observed in the load-deflection diagrams, that there are clear differences of behaviour among plates with fibers (without mesh), traditional ferrocement and fibrous ferrocement.

In the ferrocement plates (without fibers) there is an accentuated deflexion, when the first crack occurs, which doesn't happen with the fibre reinforced plates, where the deflection is much smoother. These ones show a bigger area under the load-deflection curve implying a bigger toughness compared to the unreinforced ones with fibers. Taking as an example, Figures 5 and 6, in which it is observed a better behaviour of the reinforced fibers plates, in relation to these ones without fibers.

It is observed that fibers (38mm) reinforced samples with one mesh provide a higher toughness (higher area under load-displacement curve) in relation to the ones without fibers and double welded mesh reinforcement.

3 Conclusion

It was demonstrated that the Fiber Reinforced Composites show a better general behaviour compared with the similar traditional Ferrocement, showing the best results from the following aspects:

a) higher cracking and ultimate flexure strengths of the fiber and mesh reinforced composites were observed in relation to the ferrocement;

b) there is a "delay" in the occurance of the cracking, with better distribuition and smaller cracking openings, giving a better protection of the main reinforcement of fiber reinforced structures;

c) higher toughness (higher area under the load-deflection curve) of fiber ferrocement. It is observed that fibers (38mm) reinforced plates (25mm tick), with one mesh reinforcement, provide a higher toughness in relation to the ferrocement without fibers with double welded mesh reinforcement.

d) It is necessary to observe that there is absolute necessity to conjugate the traditional steel mesh reinforcement with the fiber material, in order to obtain better performance, being these formers considered as an additional reinforcement. In other words, the slabs reinforced only with fibers have, in general, a behaviour bellow the traditional ferrocement.

4 References

American Concrete Institute (1982) State-of-the-art report on ferrocement. **Concrete International.** Detroit, pp. 13-38.

American Concrete Institute (1982) State-of-the-art report on fiber reinforced concrete. in: _______. **ACI Manual of Concrete.** Detroit, pp.1-22. (ACI Committee 544.1R-82).

Associacao Brasileira de Normas Tecnicas, ABNT (1989), **Projeto e execucao de argamassa armada.**(NBR 11173), Rio de Janeiro, 19 p.

Hanai, J. B. (1989) **Argamassa armada:** fundamentos tecnologicos para projeto e execucao. Sao Carlos, Escola de Engenharia de Sao Carlos-Universidade de Sao Paulo - EESC-USP, 261 p. (Tese de Livre Docencia).

Hanai, J. B. (1981) **Construcoes de argamassa armada:** situacao, perspectivas e pesquisas. Sao Carlos, EESC-USP, 359p. (Tese de Doutorado).

Naaman, A.E. (1985) Fiber reinforced for concrete, in **Concrete International**, Detroit, 7(3): pp. 21-25.

Ohama,Y. & Shirai,A. (1985) Prediction of flexural strength of polymer ferrocement with steel fibers, in **International Symposium on Ferrocement**, 2, Bangkok - Thailand, pp.14-16.

Silva, L.F. (1990) **Comportamento a Flexao de Placas de Argamassa Armada com Fibras de Aco Onduladas.** Sao Paulo, Escola Politecnica da Universidade de Sao Paulo-USP, 145 p. (Tese de Doutorado).

Swamy, R.N. & Spanos, A. (1985) Deflection and Cracking Behavior of Ferrocement with Grouped Reinforcement and Fiber Reinforced Matrix, **ACI Journal**, 82(1), pp. 79-91.

Swamy, R.N. & Hussin M.W. (1990) Flexural Behaviour of Thin Fiber Reinforced and Ferrocement Sheets, in **ACI SP-124 - Thin Section Fiber Reinforced Concrete and Ferrocement**, pp. 323-356.

104 ENHANCEMENT OF FERROCEMENT PROPERTIES USING STEEL FIBRE ADDITIONS TO MORTAR

D. ALEXANDER
New Zealand

SYNOPSIS
The addition of steel fibres to ferrocement mortar reduces the fine subdivision requirements of the primary steel reinforcement. The mechanism that leads to this reduction occurs at the onset of cracking, and is principally due to the sharing of load between the primary reinforcement and the wire fibres bridging the crack, which reduces the load on the primary reinforcement and hence the bond length required to transfer this load back into the concrete on either side of the crack.
Keywords: Ferrocement, Fibres, Flexural Strength, Composite materials, Cement, Structural Design.

EVOLUTION OF HIGH TENSILE WIRE REINFORCED FIBROUS FERROCEMENT

Ferrocement is typically a steel reinforced cement mortar composite used in thin plate and shell applications, and which is characterised by its ability to provide corrosion resistance to the substrate steel, even though thin covers are employed, while maintaining levels of stress in the steel substantially higher than can be used in reinforced concrete.
Its corrosion performance is obtained through the use of:

- dense low permeability mortars
- high cement ratios providing a high PH environment for the steel reinforcement
- protective coating of the steel reinforcement by galvanising
- restricting design loads to within a seviceability range that ensures that only fine cracks with widths less than 0.05mm occur.

In orthodox ferrocement this last criterion is met by using multi-layers of small diameter weld mesh reinforcement with the area and size of reinforcement complying with minimum volume fractions and specific surface area requirements primarily aimed at producing a close-spaced crack regime.

Fibre Reinforced Cement and Concrete. Edited by R. N. Swamy. © 1992 RILEM.
Published by E & FN Spon, 2-6 Boundary Row, London SE1 8HN. ISBN 0 419 18130 X.

From a theoretical perspective the minimum volume fraction ensures that when a crack forms in flexure there is sufficient area of steel bridging the crack to prevent yield occurring when the tensile force carried by the concrete before cracking is transferred to the steel. The specific surface area requirement works on the assumption that the rate of transfer of bond force between the steel and the concrete is a function of the boundary surface area, and that the tensile force applied to the bridging steel at the crack interface should be transferred back into the concrete in as short a distance as possible to minimise the length of steel being differentially strained and thereby the crack width. A shorter bond length also promotes more closely spaced crack regimes.

During the 1977-1978 period the author's practice experimented with the addition of steel fibres to ferrocement in conjunction with high tensile wire reinforcing.

The motive for the experiment was the observed flexural strength limitations of ferrocement when using conventional fine wire mild steel meshes. This limitation arises from the difficulty experienced in introducing sufficient area of steel into the moment effective outer layer of mesh, due to the fine diameters of wire used and to the comparatively low yield stress of mild steel, which together with the packing constraints required for mortar penetration results in the substantially under-reinforced nature of this form of ferrocement.

The need to improve the strength characteristics arose from the application of ferrocement to increasingly larger vessels, barges and floating structures where the provision of a substantial elastic overload capacity to accommodate collision impact was concluded to be more important than ductility; for although the energy of impact could be absorbed through non-elastic ductile deformation, the watertight integrity of the hull was destroyed at low levels of impact.

The difficulty in replacing mild steel wires with high tensile steel wires (UTS 1550 MPa and greater) was that it is neither easy nor cheap to form high tensile steel meshes that have a satisfactory performance in ferrocement. For example in ordinary woven steel meshes, the wires tend to straighten at an early stage of the loading resulting in premature crack opening. Therefore the placement of discrete high tensile wires was adopted but it was obvious that the number of wires had to be reduced from that used in the mild steel meshes if the use of HT wire was to be economical, that is larger diameter wires have to be used which also introduces the potential to overcome the deficiency of area in the outer layer of reinforcement.

To compensate for the increased crack spacing, which is observed to occur with an increase in wire diameter, and the accompanying reduction in the permissible ferrocement flexural design moment, fine 14.55mm long enlarged end steel wire fibres were added to the mortar at the rate of 5% by weight. These were expected to bridge the cracks and carry part of the load during crack propagation which would otherwise have been carried by the principal reinforcement and should theoretically result in a reduced bond length, smaller crack spacings and reduced crack width.

OBSERVATIONS

The results of this research were published in the ACI Journal of Materials and Applications in 1979 (Reference 1) The test plates were subjected to strain controlled bending with crack openings being monitored using a gratule microscope. It was observed that the crack spacings and crack patterns for plates using 2mm and 2.5mm diameter high tensile steel wires with wire fibres added to the mortar were very similar to those that developed in traditional multi layer mild steel mesh reinforced ferrocement plates.
Using a 0.05mm surface crack width as the limit to the serviceability range, the high tensile steel and wire fibre reinforced plates exhibited improved service moments because more steel could be concentrated in the surface layer and consequently at a better lever arm, and because the outer layer of tension reinforcement could be taken to in excess of 400 MPa for both the 2mm and 2.5mm diameter high tensile wires before the crack serviceability criterion was exceeded. (This was not available for the mild steel reinforcement because yield occurred).
Also by using a high tensile steel balanced section design it was possible to obtain ultimate moments on the plates of between 2½ and 3 times the service moments thus achieving the objective of a substantial near-elastic overstrength capacity.
These experimental observations were confirmed in a previously unpublished testing report carried out by Australian Wire Industries of Sydney Australia (now Aquilla Steel Co Ltd), the suppliers of the enlarged end steel fibres, and which is reproduced as an appendix with their kind permission. It is interesting because it is illustrative of the improvement of crack performance that arises from adding wire fibre.
In a paper published in the Journal of Ferrocement 1990 (Reference 2) this same phenomenon was reported for the effect of steel fibre additions to mortar used with coarse wire mild steel mesh reinforced ferrocement. In the associated experimental work 2mm diameter wire meshes with 25 and 50mm apertures in a wire fibre mortar showed crack spacings and patterns similar to those occurring when 1mm

diameter 12mm x 12mm aperture mesh reinforcing was used in plain mortar.

DISCUSSION

When these results were first published in 1979 the emphasis of the programme was on the development of a stronger form of ferrocement, rather than examining the theoretical basis of its performance. Even though there has been substantial progress in the understanding of the behaviour of ferrocement in the intervening decade a quantitative understanding of the mechanism is still difficult.

The relationship between maximum crack width and the reinforcement characteristics may be outlined as follows:

Max crack width = f(ave. crack spacing, tension face strain)
Ave. crack spacing = f(bond length)
Bond length = f(wire diameter), force transfer rate/unit surface area, additional force in the wire to be transferred back to the concrete)

although the ave. crack spacing is often dominated by the presence of crack initiators such as cross wires in mesh.

The most obvious effect of the addition of fibres is to introduce a new class of reinforcement which acts in conjunction with the primary wire reinforcement to bridge the cracks. Theoretically the relative load carrying contribution of each type of reinforcement can be determined by using the moment at the point of cracking in an elastic cracked section analysis. However there are difficulties in doing so first because the effective area of the wire fibre must take into account fibre orientation and its load carrying efficiency when the crack intercepts some fibres close to their ends. The second difficulty arises because compatibility between each type of reinforcement and the mortar at the cracked section is related by the variation of crack width which then requires that the mean bond length for each type of reinforcement, and an element length for the mortar must be estimated before strains can be derived.

All these factors introduce uncertainty into the estimation of the load carried by each type of reinforcement so that only an indicative result can be expected. By using this process it is indicated that the addition of wire fibres to a typical high tensile reinforcement configuration could reduce the force carried by the primary reinforcement at the point of crack initiation by as much as 50%.

This effect could in itself explain why the addition of wire fibres results in a reduced average crack spacing. However the addition of wire fibres to the mortar may also improve the force transfer rate since it is recognised that other mechanical properties of the mortar are modified by the addition of fibres. This would also help reduce the average crack spacing.

Appendix
Testing of Fibrous Ferrocement and Comparison of 18mm and 14.5mm Long EE Fibresteel

On Thursday, February 1, 1979, six panels were tested in flexure to determine the strength, deflection and cracking characteristics of the different composities. Three ferrocement panels primarily reinforced with high tensile wire reinforcement were tested and a comparison made to determine the effect of additional fibre reinforcement. Control panels of plain mortar and fibre reinforced mortar were also tested.
The panels were 1020mm long x 200 mm wide x 18 mm thick and they were cast in steel moulds.
The mix design was as follows:

	kg/m^3
Cement (Type A)	725
Sand (Fine Dune Sand)	1160
Water	305
Admixture (WRA - 250 mls/100 kg cement)	
Fibre (5% by weight)	115 (where applicable)

Two fibre sizes were used. These were 18mm x 0.4mm x 0.3mm EE fibre and 14.5mm x 0.4mm x 0.3 EE fibre. The high tensile wire was 2mm diameter spring wire while the transverse spacer wires were 4mm low carbon wire. The load/extension characteristic of the HT wire is shown in Diagram 1. The wire reinforcement in the particular panel is shown in Table 1.
The following panels were produced and tested:

(a) Plain mortar
(b) Fibre Reinforced mortar : 14.5mm x 0.4mm x 0.3mm EE Fibresteel (5%).
(c) Fibre Reinforced mortar : 18mm x 0.4mm x 0.3mm EE Fibresteel (5%).
(d) Mortar with HT Reinforcement.
(e) Mortar with HT Reinforcement plus 5% 14.5mm EE Fibresteel.
(f) Mortar with HT Reinforcement plus 5% 18mm EE Fibresteel.

The panels were cast on December 28 and 29, 1978, and wet cured for 5 weeks. The panels were tested approximately 6 hours after removal from the curing tank.
The panels were simply supported over a span of 950mm and loaded on two points at 200mm spacing as shown in Plate 1.

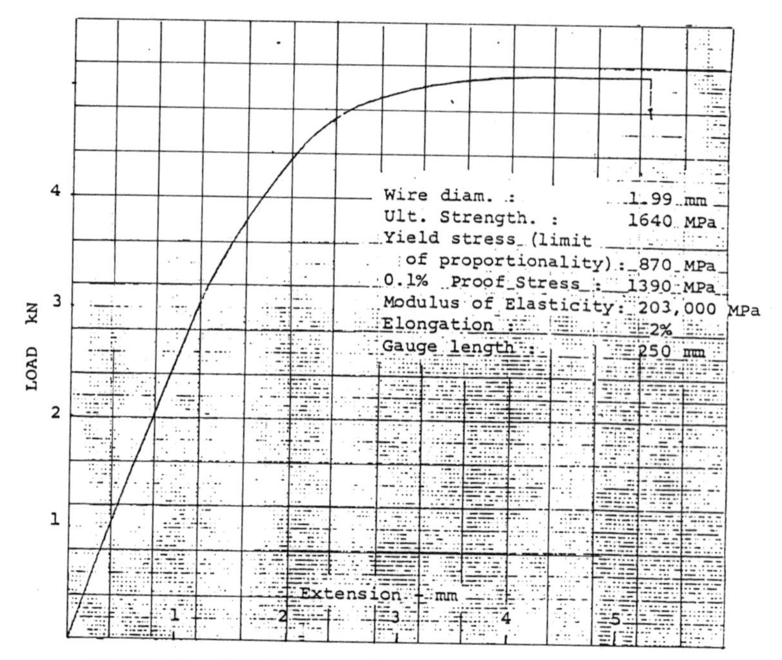

DIAGRAM 1 : HT WIRE LOAD/EXTENSION CHARACTERISTICS

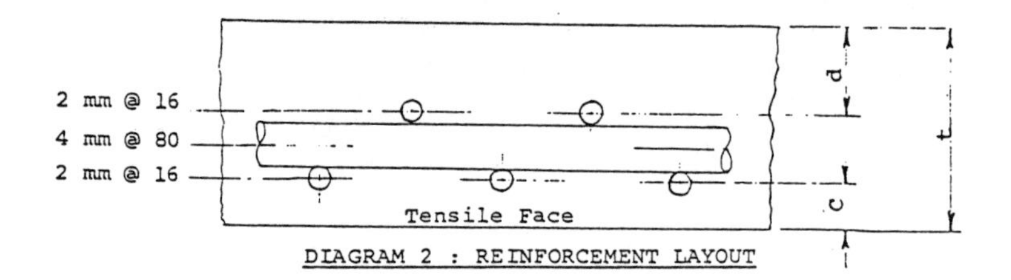

DIAGRAM 2 : REINFORCEMENT LAYOUT

TABLE 1 : ACTUAL WIRE POSITION

Panel	Panel Thickness (mm)	Cover c (mm)	Cover d (mm)
d) Plain Mortar HT Wire	18.18	4.5	8.0
e) 14.5 mm EE Fibre HT Wire	18.80	3.5	9.0
f) 18 mm EE Fibre HT Wire	19.00	5.0	8.0

The panels were closely monitored for cracking under load and the occurrence of initial visible cracking was noted. Due to the internal free moisture and surface dryness of the panels it was possible to observe the onset of surface microcracks by the presence of moisture lines on the surface and this was also noted (see Plate 2).
The following results were recorded:

Panel	Load at first appearance of moisture lines kN	Load at first visible crack kN	Ultimate load kN	Average Panel Thickness mm	Section Modulus mm^3
a) Plain Mortar	-	-	0.240	18.56	11,482
b) Fibre Reinforced 14.5 mm EE	*	0.565	0.565	19.15	12,230
c) Fibre Reinforced 18 mm EE	0.500	0.780	0.780	19.00	12,033
d) Plain Mortar HT Reinforced	0.370	0.950	3.270	18.18	11,017
e) 14.5 mm EE Fibre HT Reinforced	0.550	2.350	3.950	18.80	11,781
f) 18 mm EE Fibre HT Reinforced	0.750	2.100	3.050	19.00	12,033

* Observed but the onset of moisture lines not recorded.
The properties of the composite were calculated and were based on the panel cross-section and load/deflection characteristics. The steel and concrete stresses were also estimated.

PLATE 1 - LOADING ARRANGEMENT

PLANT 2 - SURFACE MICROCRACKING - MOISTURE LINES

Panel	Flexural Stress at first appearance of moisture lines MPa	Flexural Stress at first visible crack MPa	Ultimate Flexural Strength MPa	Steel Stress at first visible crack MPa	Steel Stress at Ultimate MPa	Concrete compressive Stress at Ultimate MPa
a) Plain Mortar	-	-	3.92	-	-	-
b) Fibre Reinforc. 14.5 mm EE	-	8.66	8.66	-	-	-
c) Fibre Reinforc. 18 mm EE	7.80	12.15	12.15	-	-	-
d) Plain Mortar HT Wire	6.30	16.17	55.65	360	1251	102.1
e) 14.5 mm EE HT Wire	8.75	37.40	62.87	790	1343	102.6
f) 18 mm EE HT Wire	11.69	32.72	47.53	780	1138	91.9

* All panels with HT wire reinforcement failed in compression.

The modulus of elasticity (secant modulus)of the three HT Wire panels up to the point of first visible crack was determined:

Plain Mortar/HT Wire	-	16.87 x 10^3 MPa
14.5mm EE Fibre/HT Wire	-	27.08 x 10^3 MPa
18mm EE Fibre/HT Wire	-	21.22 x 10^3 MPa

Observations and Conclusions

(a) The inclusion of fibres in high strength mortar significantly increased its flexural strength. 18mm EE fibre gave higher strength and "ductibility" than 14.5mm EE.
The same relationship was observed in the HT wire-reinforced panels with the onset of surface micro-cracks which were indicated by moisture lines on the surface. This point of microcrack propagation was directly related to the flexural strength of the fibre/mortar or mortar.

(b) In the two HT Wire panels with fibres, visible cracking occurred just prior to the wire reaching its yield (proportional limit). 5% of fibre reinforcement approximately doubled the load to first visible crack

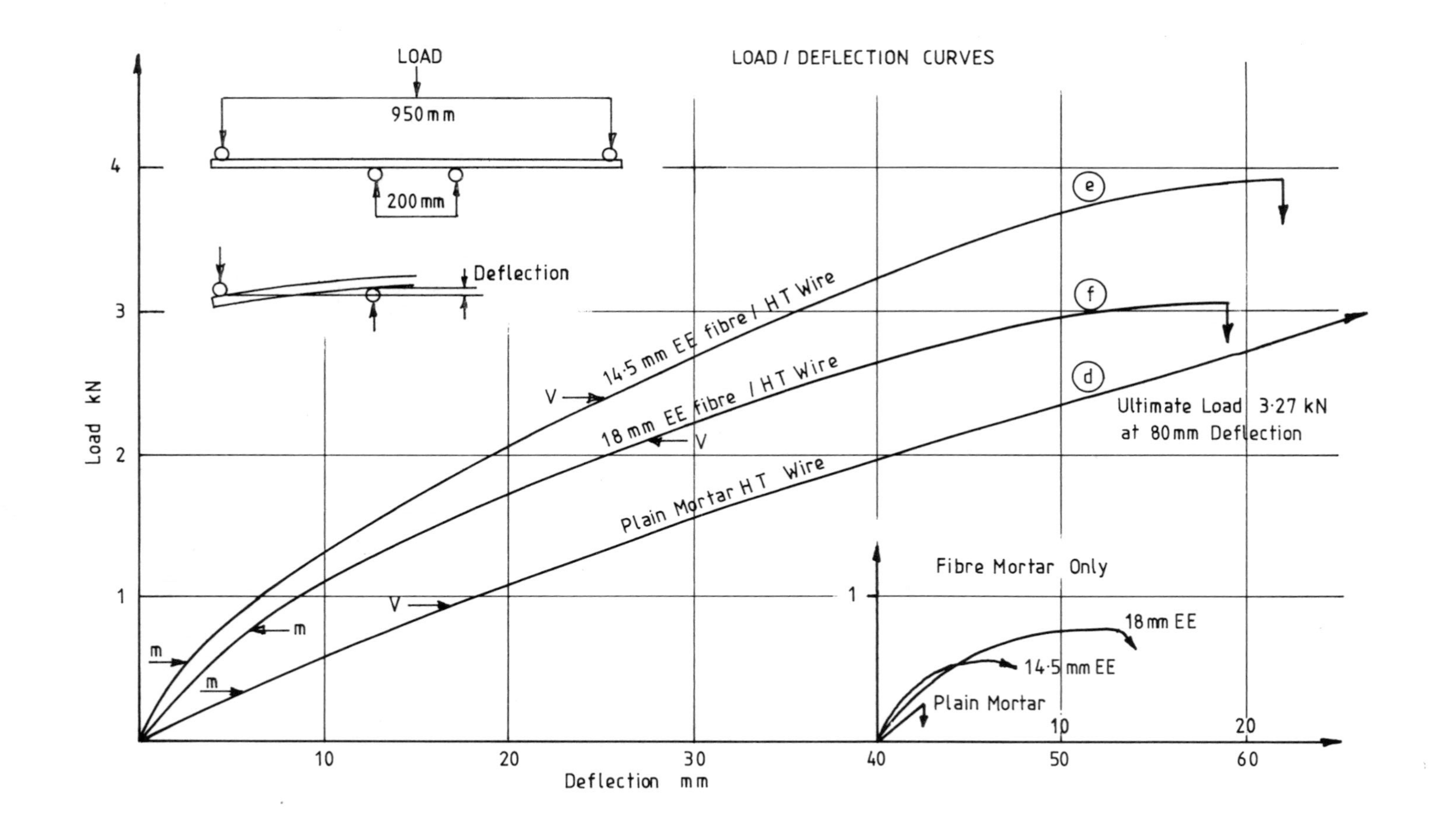
LOAD / DEFLECTION CURVES
LOAD
950 mm
200 mm
Deflection
Load kN
4
3
2
1
14·5 mm EE fibre / HT Wire
18 mm EE fibre / HT Wire
Plain Mortar HT Wire
V
m
e
f
d
Ultimate Load 3·27 kN
at 80 mm Deflection
Fibre Mortar Only
18 mm EE
14·5 mm EE
Plain Mortar
10
20
30
40
50
60
Deflection mm

in the HT panels.

(c) The two fibre reinforced (HT wire) panels were somewhat stiffer than the plain mortar panel, particularly in the initial loading stages. The difference in the two fibre reinforced panels is probably due to HT Wire location in the particular panel.

(d) The major transverse cracks at ultimate load were directly over the 4mm cross wires. There is probably a critical relationship between cross wire diameter and concrete cover in regard to major transverse cracking.

(e) The three panels with HT wire reinforcement failed in compression. By modifying the HT wire reinforcement density and position in the cross section based on this data, a panel could be designed which would be more "balanced". The steel stress at ultimate panel capacity was at about 70 to 80% of the HT Wire strength.

(f) The difference in ultimate capacity of the three HT Wire panels is mainly due to the actual HT Wire position and compressive strength of the mortar matrix.

REFERENCES

1. Atcheson Maurice and Alexander Douglas - Development of Fibrous Ferrocement - Ferrocement - Materials and Applications SP 61. American Concrete Institute Detroit 1979 pp 81-101

2. Hussin M W 1991 Defection on Cracking Performance of Fibrous Ferrocement Thin sheets - Journal of Ferrocement 21 (1) 31-41.

105 FERROCEMENT APPLICATIONS IN ISRAEL

E. Z. TATSA
Faculty of Architecture and Town Planning, Technion, Israel Institute of Technology, Haifa, Israel

SYNOPSIS
The paper summarizes research and development and mainly the use of ferrocement in Israel during the last decade and especially in the last two years. The first research poject was carried out in 1963, concentrating on the production of a folded beam structure. Afterwards, in the mid seventies extensive research and development work was directed towards industriacization on one hand and do it yourself methods on the other. In the eighties construction of spatial structures gathered momentum and a number of buildings were built. The paper describes some of these as well as the problems of introducting a new product to the construction industry.

Keywords: Spatial, Planar, Industrialized, DIY

INTRODUCTION
1974 was a turning point in the efforts to make use of ferrocement as a structural material. The oil crisis and its impact on the cost of cement led researchers at the Technion - Israel Institute of Technology to try and develop less cement consuming products in the building industry. Attempts were carried out in various directions- industrialized, semi-industrialized and do it yourself systems were tried. These included the folowing:

a) About thirty one family houses, each of an area between 50 and150 square meters in one or two levels.
b) 32 x 56 x 7.5 meters industrial facility.
c) Four 650 square meters shell buildings in an amusement park.
d) Columns for a five storey building.
e) Acoustic panels, floor tiles, decorative elelments.
f) Curtain walls for high rise buildings.
g) A 50 x 80 meters convention centre (now being designed).

These projects represent, by their variety, the yet unexploited potential in the use of ferrocement.

Fibre Reinforced Cement and Concrete. Edited by R. N. Swamy. © 1992 RILEM.
Published by E & FN Spon, 2-6 Boundary Row, London SE1 8HN. ISBN 0 419 18130 X.

FERROCEMENT IN SPATIAL CONSTRUCTION

All spatial structures built are based on a sandwich type section. The sandwich consists of two skins made of ferrocement (15 to 20mm) and an internal non structural core made of an insulating material (20 to 40 mm). The two skins are connected by ribs which are arranged in two directions and are located at distances up to one meter apart. Two methods have been used. The no form method in which the shape of the shell is achieved by flexible skeletal bars, some of which povide also the rib reinforcement. Figures 1a and 1b show a two level family house. Figures 2a, 2b and 2c show an agricultural storage facility. It should be noted that a light steel truss was used to create a large opening. For larger structures, especially those which are based on a repetitive geometrical shape an industrialized form was found to provide the best means of construction. A structure of this nature was built in the Superland amusement park near Tel-Aviv. The whole project consisted of four buildings. Each building had an octagonal floor plan (each side 11.0 meters long). Altogether 32 identical ridge elements and 32 valley elements were built using one form made of timber. The ridge elements were cast first and later the valleys were completed. Some openings of large sizes were introduced according to architectural requirements. Decorations were added by an artist later. These are also ferrocement elements. Figures 3a to 3j show the finished structure and during construction. Other spatial structures are a six meter prefabricated dome (figures 4a to 4d) and an acoustic panel (figure 5a, 5b).

PANEL CONSTRUCTION

Construction with planar ferrocement components presents many more advantages than using curved spatial elements because of the possibility to fully industrialize the system. The massive housing problems the world is facing now and in the near and far future calls for a system which is economical in material and workmanship and still provides good quality products. Also, it has to be universal in the sense that small building companies and individuals will be able to apply it. With these goals in mind a basic unit was developed. It is very simple to manufacture and together with cast in situ ribs creates a panel of any required reasonable size. The unit in figures 6a (flat form), 6b (cast flat), 6c (folded) was used to construct the slab shown in figure 6d. Note that the form already includes the unsulating material. Trying to evaluate the economic advantages of a panel system for housing for a potential market a study was made for a typical housing project in China. The results of this study with regard to material quantities are shown in Tables 1 and 2.

These results demonstrate the huge savings in cement enabled by the use of ferrocement. On a national economic basis this means cost savings of a very large scale.

SUMMARY

Instead of summarizing let us ask few questions.
- Is ferrocement a new material? No.
- Do we know enough in order to build dwellings in masses? Yes.
- Then why don't we do it?
Based on the Israeli and other experiences it can be said that when small quantities are involved it is much less problematic to introduce ferrocement. Industry in general is, however, reluctant to do so. There is the fear of the unknown or lack of information with regaard to construction itself. One can say that the potential and know how are there, but like in many other fields translating academic knowledge to industrial products is weak. Strengthening this link is the task of many.

ACKNOWLEDGMENT

The author wishes to thank all who helped summarize a decade of effort in trying to make good use of ferrocement. Special thanks to Dr. Y. Sarne, Eng. S. Maik, Arch. S. Yomtov and Arch A. Tibbi.

Table 1 Material quantities - basic data- multi apartment project.

a) Walls (per m^2 of walls)

Ttpe	Concrete m^3	Cement kg
Single layer F.C	0.038	14.0
Double layer F.C	0.058	20.0
Conventional concrete (full)	0.200	50.0
Hollow blocks	0.090	20.0

b) Floors (per m^2 of floor)

Type	Concret m^3	Cement kg
Single layer F.C	0.048	21.6
Double layer F.C	0.055	24.8
Conventional concret (full)	0.140	42.0
Prestresed hollow	0.120	42.0

c) Stairs prefabricated (per m^2 hor.)

Type	Concrete m^3	Cement kg
F.C	0.05	22.5
concrete	0.18	54.0

d) Parapets (per m^2 par.)

Type	Concrete m^3	Cement kg
F.C	0.035	15.8
Concrete	0.120	36.0

Table 2

MATERIAL QUANTITIES PER SQUARE METER OF FLOOR AREA
multi apartment project.

No	SYSTEM DESCRIPTION				CONCRETE m3	CEMENT kg
	WALLS	FLOORS	SRAIRS	PARAPET		
1	FC SLP	FC SLP	FC	FC	0.113	46.1
2	FC SLP	FC DLP	FC	FC	0.120	49.3
3	FC DLP	FC SLP	FC	FC	0.147	55.5
4	FC DLP	FC DLP	FC	FC	0.152	58.7
5	FC SLP	FC SLP	C	FC	0.125	49.0
6	C HB	FC SLP	FC	FC	0.194	53.2
7	C HB	FC DLP	FC	FC	0.202	56.4
8	C	FC SLP	FC	FC	0.367	74.2

9	C	FC DLP	FC	FC	0.374	77.4
10	C HB	C	C	C	0.302	79.5
11	C	C	C	C	0.474	126.5
12	C HB	C PRH	C	C	0.279	79.5
13	BRICKS*	FC SLP	FC	FC	0.054	24.2
14	BRICKS*	FC DLP	FC	FC	0.061	27.4
15	BRICKS*	FC SLP	C	C	0.065	27.8
16	BRICKS*	FC DLP	C	C	0.072	31.0
17	BRICKS*	C (full)	C	C	0.157	48.2
18	BRICKS*	C (pres)	C	C	0.137	48.2

* Bricks not icluded

FC - FERROCEMENT
SLP - SINGLE LAYER PANEL
DLP - DOBLE LAYER PANEL
HB - HOLLOW BLOCKS
C - FULL CONCRETE
PRH - PRESTRESSED HOLLOW

1a

1b

1. Construction of a one family house

2a

2b

2. Agricultural storage facility

2c

a. General view

3. Amusement park

3b

3c

3d

b. c. d. Finished buildings

e. Detail at top (opened shell)

f. g. Form

h. Valleys not yet cast

i. Casting completed

j. Internal detail

4a 4. Prefabricated dome

4b

4c

4d

5a

5b

5. Acoustic panel

6a

6b

6. Manufacturing of a plane panel

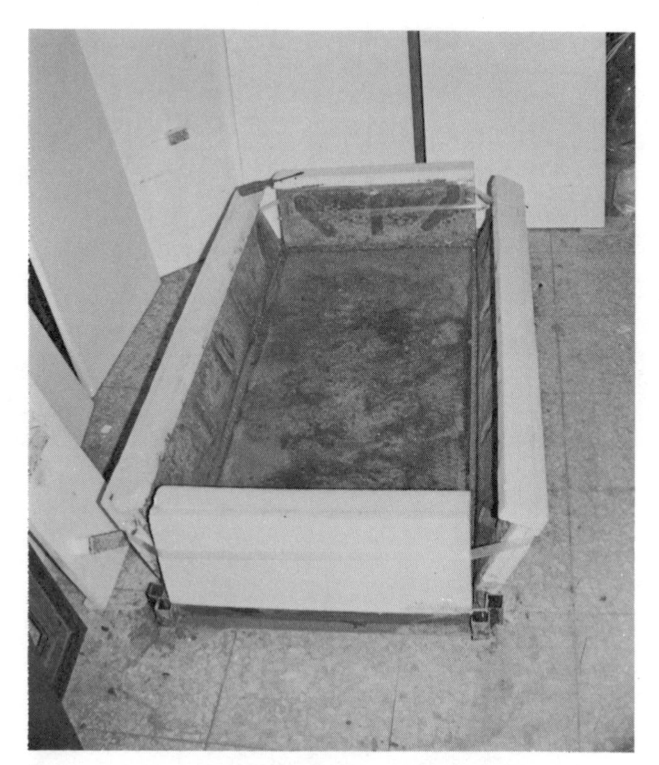

6c

106 PERFORMANCE APPRAISAL OF FERROCEMENT NORTHLIGHT FOLDED PLATES

S. K. KAUSHIK and V. K. GUPTA
Civil Engineering Department, University of Roorkee, India

Abstract
The use of prefabricated ferrocement elements for the construction of industrial buildings can lead to a better finish and strength as well as the efficient utilisation of materials. Substantial savings in time and money can also be affected. Precast ferrocement northlight folded plate roofs offer a distinct possibility for construction of industrial roofs. This paper presents an investigation on two specimens having a total effective span of 10 meters with a support in the middle of the span. The effective width of the element was 1.3 m and thickness 20 mm. The plates were subjected to an increasing uniformly distributed load upto a 4 kN/m^2. The experimental results were validated using Simpson's theory and the results correlate well. Based on the study, it is recommended that a load factor of 1.5 may be used for the design of ferrocement northlight folded plate roofs.
Keywords : Ferrocement, Northlight Folded Plate, Simpson's Theory, Crack Width, Strains, Deflection, Load Factor.

1 Introduction

The major problem in developing countries is to achieve cheaper units for houses and industries. The shortage of housing and industrial facilities can be overcome to a large extent by prefabrication of major roofing components. The prefabricated components have a better finish, strength and are economical. Ferrocement folded plate roofing elements offer the distinct possibility of being utilised for the above purpose.

Experimental studies on ferrocement folded plates have been reported by Narayanaswami (1), Basavarajaih (2), and Karasudhi (3). Several studies have recommended the use of trapezoidal cross section for the folded plate elements due to its structural efficiency (2,4). Basavarajaih (2) reported that V-shaped ferrocement folded plate units were suitable for precast roofs. The behaviour of long span triple V type precast ferrocement folded plate was tested to failure under a uniformly distributed load. It was concluded that the square meshes used were effective in controlling the cracks. The deflections at cracking load were 40% of the maximum deflection

Fibre Reinforced Cement and Concrete. Edited by R. N. Swamy. © 1992 RILEM.
Published by E & FN Spon, 2-6 Boundary Row, London SE1 8HN. ISBN 0 419 18130 X.

and the cracking load was 88% of the ultimate load.

Narayanaswami, Surya Kumar and Sharma (1) have suggested that the trough type elements can be analysed as a beam in which trough section is replaced by an inverted T-section. The element tested by them had a simply supported clear span of 3 m. These plates were 16mm thick and designed for a dead and live load of 3.2 kN/m^2 and 0.73 kN/m^2 of the plan area. The cost of the roofing per square meter of the plan area was worked out to Rs. 80/-.

Kaushik, Gupta and Singhal (4) have presented an investigation on the behaviour of the trapezoidal ferrocement folded plate elements of 20 mm thickness and span 3 and 5 meters respectively. The experimental results were validated using the Finite Strip Method (6). The plates were subjected to uniformly distributed load through layers of bricks, each layer representing a load of 1 kN/m^2. The maximum compressive and tensile strains recorded at the mid span were of the order of 500 and 1600 microns respectively. These results correspond to a live load of 8 kN/m^2 for the specimen of 5 m span. Based on the composite modulus of elasticity of 18.1 kN/m^2, the average experimental compressive stresses were 11.9 and 13.5 N/mm^2 for 3 and 5 m spans respectively. The theoretical stresses based on the finite strip method were 7.19 and 7.01 N/mm^2 respectively. At the limit state of crack width the average deflections were span/465 and span/335 for 3 and 5 m spans respectively. The overall governing design criterion was found to be limit state of crack width. The specimens carried the normal live load of 2 kN/m^2 for 5 m span and 4 kN/m^2 for 3 m span safely.

2 Object and Scope of the Investigation

The main object of the investigation (8) was to present a systematic study on the strength and behaviour of the northlight ferrocement continuous folded plates. The study consisted of casting and testing of two northlight continuous folded plates of 10 m span with a support at the middle. The plate thickness of both the specimen was 20 mm (Fig. 1). The specimens were made of 1:2 cement sand mortar of average compressive cube strength 23 Mpa with a w/c ratio 0.45 by weight. The plates were reinforced with two layers of 20 gauge G.I. Square Woven mesh with 6 mm mesh opening having a yield strength of 46 Mpa and breaking strength of 50.8 Mpa. The skeletal steel consisted of 6 mm mild steel bars placed at 300 mm centre to centre in both directions. The 0.2% proof stress of the bar was 59 Mpa (See Fig. 2 and 3). The plates were tested on brick masonry supports and subjected to uniformly distributed load applied through brick layer (Refer Fig. 4). The observations recorded were the general behaviour and porformance, deflection, crack widths and crack spacing. The first crack load and the loads at serviceability limit states of 0.1 mm crack width and deflection equal

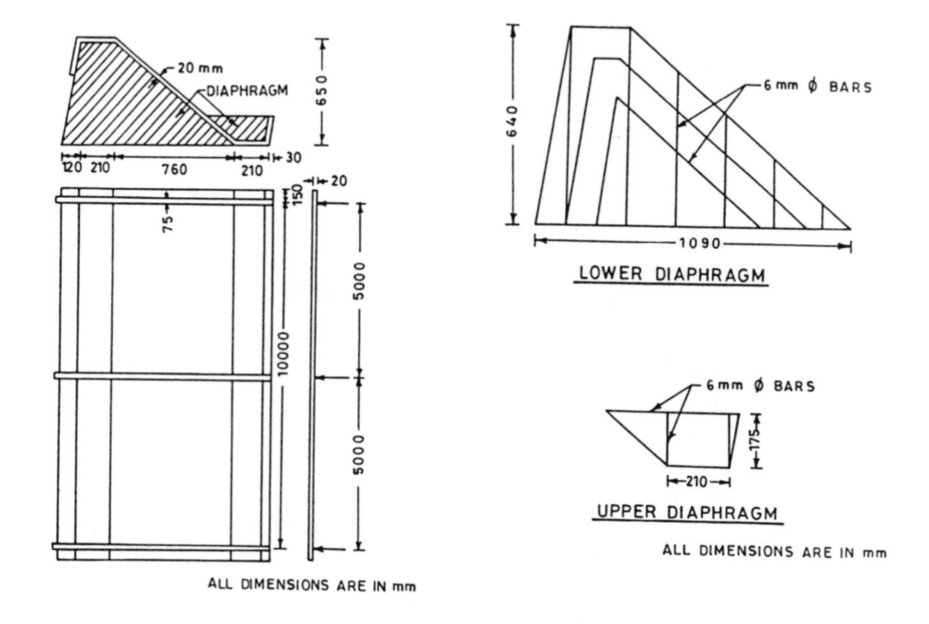

Fig. 1(a). Plan and section of folded plate.

Fig. 1(b). Reinforcement details - diaphragms.

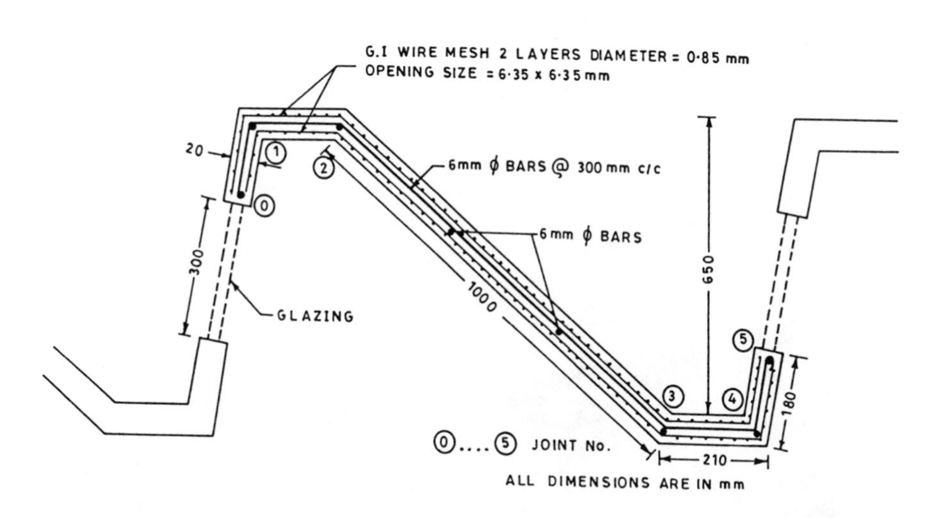

Fig. 2. Reinforcement details of northlight folded plate.

Fig. 3. Reinforcement laid on masonry mould.

Fig. 4. Northlight folded plate subjected to four layers brick load (4 kN/m^2).

to span/250 were monitored. The specimen and the control mortar cubes were wet cured for 28 days under gunny bags. The specimens were allowed to dry at a temperature of 12-18°C prior to testing. The theoretical validation was done using Simpson's method (6,7).

3 Presentation and Discussion of Results

The maximum crack width recorded for the two specimens at various levels of the live load are given in Table 1. As can be clearly seen both the specimens satisfied the crack width serviceability limit of 0.1 mm under a superimposed live load of 4 kN/m^2 (See Fig. 5 and 6).

Table 1. Maximum Experimental Crack Width at Different Load Levels

Live Load	Crack Width (mm)	
(kN/m^2)	Specimen 1	Specimen 2
1.0	No crack	No crack
2.0	No crack	No crack
3.0	0.055	0.040
4.0	0.105	0.095

The variation of the average compressive strains in the top and bottom flange of both the specimens varied linearly upto the average first crack load which was observed to be 2.75 kN/m^2. Beyond this load the strains deviated from linearity and achieved a maximum strain value of 550 microns in compression and 720 to 740 microns in tension at a load of 4 kN/m^2 (Table 2). The ultimate direct tensile strain for ferrocement is normally 16000 microns. The specimens were not loaded beyond 740 microns as the limit state of serviceability was reached. It is worth mentioning that specimens were not tested to complete destruction.

A study of the load deflection relation showed that the specimens behaved linearly upto the appearance of the first crack (Table 3). At the limit state of crack width (0.1 mm), the average deflection to span ratio was about 1/450, which is much lower than the value 1/250 specified by Indian code of practice (9).

The experimental stresses were calculated using the measured strains and the value of modulus of elasticity calculated from composite material approach (Table 4). The experimental and theoretical longitudinal stresses along the folded plate section at mid span and at the middle support were found to be in good agreement at a superimposed load of 0.75 kN/m^2 (Table 5). The experimental value were 90 to 98% of the theoretical values.

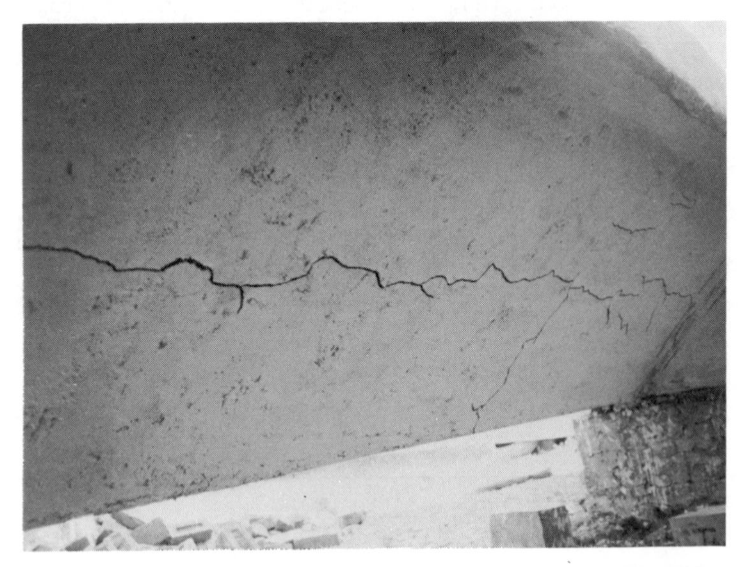

Fig. 5. Longitudinal crack at midspan on bottom of inclined plate.

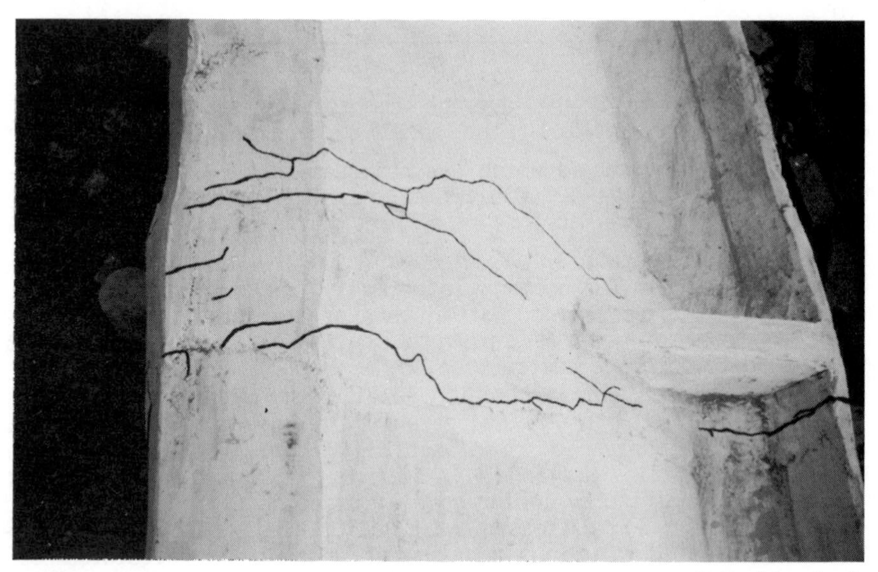

Fig. 6. Transverse cracks at middle support on top face of inclined plate.

Table 2. Variation of Average Experimental Strains at Mid Span

Live Load (kN/m^2)	Average Experimental Strains (Microns)			
	Top Flange		Bottom Flange	
	Specimen 1	Specimen 2	Specimen 1	Specimen 2
1.0	117.70	62.70	133.50	84.40
2.0	197.60	127.95	208.00	176.80
3.0	394.40	284.50	376.40	326.60
4.0	551.90	458.20	720.90	673.80

Table 3. Average Mid Span Deflection with Loads

Live Load (kN/m^2)	Average Mid Span Deflection (mm)					
	Specimen 1			Specimen 2		
	TF	BF	MIP	TF	BF	MIP
1.0	1.41	1.57	5.23	1.13	1.26	2.11
2.0	3.92	3.19	9.12	2.77	2.72	4.50
3.0	7.38	6.00	11.21	4.79	4.20	6.95
4.0	10.20	9.57	12.64	7.02	5.89	9.64

TF = Top Flange; BF = Bottom Flange; MIP = Middle of Inclined Plate

Table 4. Maximum Stress at Mid Span and Mid Support (Mpa)

Live Load (kN/m^2)	Mid Span				Mid Support			
	Strain (Microns)		*Stress (Mpa)		Strain (Microns)		*Stress (Mpa)	
	Spec 1	Spec 2	Spec 1	Spec 2	Spec 1	Spec 2	Spec 1	Spec 2
1.0	179.00	134.50	3.24	2.43	404.20	365.00	6.60	5.56
2.0	281.50	218.50	5.10	3.95	534.80	465.50	9.68	8.43
3.0	499.50	375.20	9.04	6.79	689.50	574.00	12.84	10.40
4.0	736.20	613.00	13.30	11.10	779.00	697.50	14.10	12.62

* $E = 1.81 \times 10^4$ Mpa

A study of the crack width and the deflection at a live load of 4 kN/m^2 shows that the deflections were well within the allowable codal limits and the crack widths were almost equal to the value prescribed by ACI Task committee 549 for design of ferrocement folded plates. Hence the limit state of serviceability of crack width is expected to be the governing design criteria.

Table 5. Longitudinal Stress Along Folded Plate Section at Mid Span and Mid Support Section

Joint	Longitudinal Stress (Mpa)					
	Mid Span			Mid Support		
	Spec 1	Spec 2	Theor.*	Spec 1	Spec 2	Theor.*
0	-	-	+3.33	-	-	+2.86
1	0.24	0.19	+0.21	-7.80	-7.35	-8.86
2	-1.85	-1.76	-1.89	7.68	6.95	+7.28
3	1.88	1.72	+1.90	-7.85	-7.25	-7.30
4	-0.39	-0.35	-0.36	8.24	7.95	+9.20
5	-	-	-3.12	-	-	-3.34

* Values corresponding to Service Live Load of 0.75 kN/m^2 as per IS:875-1987.

Defining the load factor as the ratio of load at the critical limit state of serviceability to the load causing the first crack, a factor of more than 1.5 is seen to be possible. Therefore the partial safety load factor prescribed by IS:456-1978 (10) are met with successfully. The cracked specimens were subjected to a water leakage test. Despite the fact that the specimens were cracked no leakage was observed.

4 Conclusions

Based on the limited experimental investigations carried out the following conclusions may be drawn.

1. The Simpson's method can be used to analyse ferrocement continuous folded plates. The computed results are found to compare fairly well with the test results.

2. Ferrocement continuous northlight folded plate of 5 m span can sustain a superimposed load upto 4 kN/m^2 at the limiting crack width of 0.1 mm.

3. The deflection to span ratio at the limiting crack width of 0.1 mm is 1/450 which is well below the limit of 1/250 prescribed by IS:456-1978.

4. A partial load factor of 1.5 may be used for superimposed load for the design of ferrocement northlight folded plate roofs.

5. The ferrocement northlight folded plate roof was found to be water tight during the water leakage test even though the specimens were cracked.

5 References

Narayanswamy, V.P., Suryakumar, G.V. and Sharma, P.C. (1981) Precast ferrocement trough element for low cost roofing, **J. Ferrocement**, 11, 71-84.

Basavarajaih, B.S., Vankatakrishna, H.V. and Rangnath, U.R. (1975) Experimental study of applicability of ferrocement for precast folded plate elements, **J. Structural Engg.**, 2.

Karasudhi, P., Pama, R.P., Nimityongskul, P. and Narayanaswamy, V.P. (1979) A training programme on ferrocement technology for Indonesia, **J. Ferrocement**, 9.

Kaushik, S.K., Gupta, V.K.and Singhal, A.K. (1988) Behaviour of ferrocement folded plate roofs, **Proc. of the Third International Symposium on Ferrocement**, New Delhi, 335-343.

Cheung, Y.K. (1976) Finite strip method on structural analysis, Structures and Solid Body Mechanics Series, Pergamon International Library.

Simpson, H. (1958) Design of folded plate roofs, **J. Structural Div.**, ASCE, 84, 1508-1/1508-21.

Ramaswamy, G.S. (1974) Design and construction of concrete shell roofs, Tata McGraw Publication Company Ltd., New Delhi, 264-302.

Sushant Kumar (1992) Behaviour of continuous ferrocement north light folded plates, **Master Dissertation**, Civil Engineering Department, University of Roorkee, Roorkee.

IS:875-1987, Loading standard for buildings, **Bureau of Indian Standards**, New Delhi.

IS:456-1978, Code of practice for plain and reinforced concrete, **Bureau of Indian Standards**, New Delhi.

AUTHOR INDEX

SUBJECT INDEX

This index has been compiled from the keywords provided by the authors of the papers, with some editing and revision. The numbers refer to the first page number of the relevant paper.

RILEM - The International Union of Testing and Research Laboratories for Materials and Structures - is an international, non-governmental technical organisation, founded in 1947, with a membership of over 900 in some 80 countries. It forms an institutional framework for cooperation by experts to:

* optimise and harmonise test methods for measuring properties and performance of building and civil engineering materials and structures under laboratory and service environments;

* prepare technical recommendations for testing methods;

* prepare state-of-the-art reports to identify further research needs.

RILEM members include the leading building research and testing laboratories from around the world, industrial research, manufacturing and contracting interests as well as a significant number of individual members, from industry and universities. RILEM's focus is on construction materials and their use in buildings and civil engineering structures, covering all phases of the building process from manufacture to use and recycling of materials.

RILEM meets these objectives though the work of its technical committees. Symposia, workshops and seminars are organised to facilitate the exchange of information and dissemination of knowledge. RILEM's primary output are the technical recommendations. RILEM also publishes the journal *Materials and Structures* which provides a further avenue for reporting the work of its committees. Many other publications, in the form of reports, monographs, symposia and workshop proceedings, are produced.

Details of RILEM membership may be obtained from RILEM, École Normale Supérieure, Pavillon du Crous, 61, avenue du Pdt Wilson, 94235 Cachan Cedex, France.

Details of the journal and the publications available from E & F N Spon/Chapman & Hall are given below. Full details of the Reports and Proceedings can be obtained from E & F N Spon, 2-6 Boundary Row, London SE1 8HN, Tel: (0)71-865 0066, Fax: (0)71-522 9623.

Materials and Structures

RILEM's journal, *Materials and Structures*, is published by E & F N Spon on behalf of RILEM. The journal was founded in 1968, and is a leading journal of record for current research in the properties and performance of building materials and structures, standardization of test methods, and the application of research results to the structural use of materials in building and civil engineering applications.

The papers are selected by an international Editorial Committee to conform with the highest research standards. As well as submitted papers from research and industry, the Journal publishes Reports and Recommendations prepared buy RILEM Technical Committees, together with news of other RILEM activities.

Materials and Structures is published ten times a year (ISSN) 0025-5432) and sample copy requests and subscription enquiries should be sent to: E & F N Spon, 2-6 Boundary Row, London SE1 8HN, Tel: (0)71-865 0066, Fax: (0)71-522 9623; or Journals Promotion Department, Chapman & Hall, 29 West 35th Street, New York, NY 10001-2291, USA, Tel: (212) 244 3336, Fax: (212) 563 2269.

RILEM Reports

1 **Soiling and Cleaning of Building Facades**
Report of Technical Committee 62-SCF. *Edited by L. G. W. Verhoef*
2 **Corrosion of Steel in Concrete**
Report of Technical Committee 60-CSC. *Edited by P. Schiessl*
3 **Fracture Mechanics of Concrete Structures - From Theory to Applications**
Report of Technical Committee 90-FMA. *Edited by L. Elfgren*
4 **Geomembranes - Identification and Performance Testing**
Report of Technical Committee 103-MGH. *Edited by A. Rollin and J. M. Rigo*
5 **Fracture Mechanics Test Methods for Concrete**
Report of Technical Committee 89-FMT. *Edited by S. P. Shah and A. Carpinteri*
6 **Recycling of Demolished Concrete and Masonry**
Report of Technical Committee 37-DRC. *Edited by T. C. Hansen*
7 **Fly Ash in Concrete - Properties and Performance**
Report of Technical Committee 67-FAB. *Edited by K. Wesche*

RILEM Proceedings

1 **Adhesion between Polymers and Concrete. ISAP 86**
Aix-en-Provence, France, 1986. *Edited by H. R. Sasse*
2 **From Materials Science to Construction Materials Engineering**
Versailles, France, 1987. *Edited by J. C. Maso*
3 **Durability of Geotextiles**
St Rémy-lès-Chevreuses, France, 1986
4 **Demolition and Reuse of Concrete and Masonry**
Tokyo, Japan, 1988. *Edited by Y. Kasai*
5 **Admixtures for Concrete - Improvement of Properties**
Barcelona, Spain, 1990. *Edited by E. Vazquez*
6 **Analysis of Concrete Structures by Fracture Mechanics**
Abisko, Sweden, 1989. *Edited by L. Elfgren and S. P. Shah*
7 **Vegetable Plants and their Fibres as Building Materials**
Salvador, Bahia, Brazil, 1990. *Edited by H. S. Sobral*
8 **Mechanical Tests for Bituminous Mixes**
Budapest, Hungary, 1990. *Edited by H. W. Fritz and E. Eustacchio*
9 **Test Quality for Construction, Materials and Structures**
St Rémy-lès-Chevreuses, France, 1990. *Edited by M. Fickelson*
10 **Properties of Fresh Concrete**
Hanover, Germany, 1990. *Edited by H.-J. Wierig*
11 **Testing during Concrete Construction**
Mainz, Germany, 1990. *Edited by H. W. Reinhardt*
12 **Testing of Metals for Structures**
Naples, Italy, 1990. *Edited by F. M. Mazzolani*
13 **Fracture Processes in Concrete, Rock and Ceramics**
Noordwijk, Netherlands, 1991. *Edited by J. G. M. van Mier, J. G. Rots and A. Bakker*
14 **Quality Control of Concrete Structures**
Ghent, Belgium, 1991. *Edited by L. Taerwe and H. Lambotte*
15 **High Performance Fiber Reinforced Cement Composites**
Mainz, Germany, 1991. *Edited by H. W. Reinhardt and A. E. Naaman*
16 **Hydration and Setting of Cements**
Dijon, France, 1991. *Edited by A. Nonat and J. C. Mutin*
17 **Fibre Reinforced Cement and Concrete**
Sheffield, UK, 1992. *Edited by R. N. Swamy*
18 **Interfaces in Cementitious Composites**
Toulouse, France, 1992. *Edited by J. C. Maso*